D0687980

Handbook of Biosurveillance

Handbook of Biosurveillance

Edited by

Michael M. Wagner, M.D., Ph.D.
Real-time Outbreak and Disease
Surveillance (RODS) Laboratory,
Center for Biomedical Informatics
University of Pittsburgh
Pittsburgh, Pennsylvania

Andrew W. Moore, Ph.D.
Robotics and Computer Science
School of Computer Science
Carnegie Mellon University
Pittsburgh, Pennsylvania

Ron M. Aryel, M.D., MBA.
Del Rey Technologies
Los Angeles, California

AMSTERDAM • BOSTON • HEIDELBERG • LONDON
NEW YORK • OXFORD • PARIS • SAN DIEGO
SAN FRANCISCO • SINGAPORE • SYDNEY • TOKYO
Academic Press is an imprint of Elsevier

Elsevier Academic Press
30 Corporate Drive, Suite 400, Burlington, MA 01803, USA
525 B Street, Suite 1900, San Diego, California 92101-4495, USA
84 Theobald's Road, London WC1X 8RR, UK

This book is printed on acid-free paper. ∞

Library of Congress Cataloging-in-Publication Data

Handbook of biosurveillance/edited by Michael M. Wagner, Andrew W. Moore, Ron M. Aryel. -- 1st ed.
p. cm.
Includes bibliographical references and index.
ISBN-13: 978-0-12-369378-5 (alk. paper)
ISBN-10: 0-12-369378-0 (alk. paper)
1. Public health surveillance. I. Wagner, Michael M. II. Moore, Andrew W., Ph.D. III. Aryel, Ron M.
[DNLM: 1. Population Surveillance--methods. 2. Disease Outbreaks--prevention & control.
3. Data Interpretation, Statistical. 4. Models, Statistical. WA 105 H236 2006]
RA652.2.P82H36 2006
362.1--dc22

2005035436

British Library Cataloguing-in-Publication Data
A catalogue record for this book is available from the British Library.

ISBN 13: 978-0-12-369378-5
ISBN 10: 0-12-369378-0

For information on all Academic Press publications
visit our Web site at www.books.elsevier.com

Printed in the United States of America
06 07 08 09 10 1 9 8 7 6 5 4 3 2 1

Contents

Preface

Over the past five years, the very real threat posed by emerging infections and bioterrorism has challenged and revolutionized the practice of disease surveillance. Many cities and countries are constructing new disease surveillance systems. Even the basic concept of surveillance data has changed, expanding dramatically beyond notifiable disease reporting to include sales of medications, emergency department registration records, and absenteeism. New methods have been introduced into routine practice, including time series analysis, spatial scanning, natural language processing, and real-time data collection. These fundamental changes in practice may account for the appearance of a new word—*Biosurveillance*—that is being used to refer to both this expanded notion of disease surveillance as well as the scientific field itself. The extent of the changes in the science and practice of biosurveillance is reflected in the contents of this book, which bear little resemblance to the contents of recent texts in public health surveillance, epidemiology, or hospital infection control.

The *Handbook of Biosurveillance* offers a unified, multidisciplinary examination of the field of biosurveillance as both a practice and a scientific field of research. The *Handbook* reviews the state-of-the-art and covers all facets of the emerging discipline, offering practical advice for organizations developing or acquiring biosurveillance systems as well as an in-depth discussion of the theoretical underpinnings of biosurveillance for researchers. We believe that this book is the first attempt to present a comprehensive unified exposition and definition of the field of biosurveillance and its many components.

We wrote the *Handbook* to assist individuals who work in biosurveillance to cope with the rapid changes in the field. The book is directed at this audience, which includes epidemiologists, public health physicians, infection control practitioners, researchers, and the information technology staffs of organizations that conduct biosurveillance. The book is also directed to a broader audience that includes policy makers, teachers, microbiologists, and clinicians. The audience that we have in mind also includes fire and police chiefs—who in this modern age find themselves participating in preparedness efforts, developing "all-threat" capabilities, making purchase and acquisition decisions, and writing funding proposals that include technical approaches to biosurveillance—as well as researchers. We wanted to make this field accessible to the interested lay person who wishes to better understand how the flow of people across international borders should be handled to make us less vulnerable to the potential ravages of infectious disease outbreaks as well as acts of bioterrorism. The book assumes little prior knowledge and each chapter begins with basic material and builds to more advanced material.

The future progress of the field of biosurveillance will depend on interdisciplinary teams of researchers and practitioners. Physicians, statisticians, emergency responders, policy makers, IT experts, and computer scientists need to speak the same language and share the broad vision of the field. We intend for this book to help act as a starting point for such interdisciplinarity.

We present biosurveillance from a global perspective and draw upon examples of biosurveillance systems being developed in Australia, Asia, Europe, and North America. For a discussion of the myriad organizations involved in biosurveillance (Part II), however, we made a pragmatic decision to discuss how biosurveillance is organized in the U.S. because of our first-hand experience. We also made a pragmatic decision to limit the scope of the *Handbook* to biosurveillance for threats to human and animal health, omitting discussion of plant diseases and agricultural terrorism, at least in this edition.

For instructors who consider the *Handbook* for course use, we note that the material is inherently interdisciplinary and is best taught by two or more instructors. To cover the entire content of the book, the instructors should be familiar with the disciplines of public health, epidemiology, statistics, artificial intelligence, economics, decision theory, hospital infection control, and medical informatics. However, it is eminently possible to cover large portions of the book with a smaller instructional staff. It is possible to cover the material in an intensive one-semester course by assigning selected chapters for self-study; a less intensive course might span two semesters or omit selected chapters.

The Handbook could also be used as the basis of a focused or advanced course in public policy, artificial intelligence, biomedical engineering, or computer science. Depending on the focus, parts of the book could serve as a general background for more intensive exploration of the other materials. For example, a course for engineers or computer scientists might use Parts I, II, and IV as general background for a more intensive exploration of the materials in Parts III, V and VI.

We believe that the emerging field of biosurveillance is of vital importance to the welfare of society. From a research perspective, we believe it is an intellectually deep and exciting area, with many opportunities for contributions by creative individuals and forward-thinking organizations from many disciplines. We sincerely hope and intend that *The Handbook of Biosurveillance* will play a part in accelerating the progress and interdisciplinarity of the field. Please send suggestion and comments to handbook.biosurveillance@gmail.com.

Michael M. Wagner
Pittsburgh, PA
Andrew W. Moore
Pittsburgh, PA
Ron M. Aryel
Kansas City, MO

Acknowledgments

This book represents the work and support of many people in addition to the editors and the chapter authors. We are particularly grateful to Daphne Henry who assembled and copyedited each chapter, shepherded the project through to completion, and contributed her unique literary talents in several chapters. We were fortunate to work closely with people at Elsevier throughout the project, including Luna Han, who identified the need for a book on this topic, and our production editor Pat Gonzalez.

The idea for this book grew from two reports commissioned by the Agency for Healthcare Quality and Research in 2000 under contract 290-00-0009 (with special thanks to Sally Phillips for her guidance and support). The following research grants also supported our research during the period 1999–2005, when we developed many of the ideas in the book: Alfred P. Sloan Foundation 2003-6-19; Centers for Disease Control and Prevention U90/CCU318753; Commonwealth of Pennsylvania ME-01-737 and SAP4000007765; Defense Advanced Research Projects Agency, Air Force Research Laboratory F30602-01-2-0550 (with specific thanks to Pete Estacio, Jim Reilly, Ted Senator, and David Siegrist for initiating the projects); Environmental Protection Agency (with special thanks to Dave Apanian and Kathy Clayton); National Library of Medicine 5 R29 LM06233 (with thanks to Milton Corn for having the vision to allow a redirection of the project), 1 R21 LM008278-01, 5 T15 LM/DE07059; National Science Foundation IIS-0325581; and U.S. Department of Agriculture GS06K97BND0710 (with special thanks for championing of the project by Cmdr. Kimberly Elenberg).

Contributors

Numbers in parentheses indicate the first page on which an author's contribution begins.

Brigham Anderson (235) Robotics Institute, Carnegie Mellon University, Pittsburgh, Pennsylvania

T.V. (Sunil) Anklekar (161) Clariant Corporation, Elgin, North Carolina

Ron M. Aryel (89) Del Rey Technologies, Los Angeles, California

Charles Brokopp (129) Coordinating Office of Terrorism Preparedness and Emergency Response, Centers for Disease Control and Prevention, Atlanta, Georgia

Murray Campbell (393) IBM T.J. Watson Research Center, Yorktown Heights, New York

Wendy W. Chapman (255) Center for Biomedical Informatics, University of Pittsburgh, Pittsburgh, Pennsylvania

Nick M. Cirino (143) Biodefense Laboratory, Wadsworth Center, New York State Department of Health, Albany, New York

Gregory F. Cooper (273) RODS Laboratory, Center for Biomedical Informatics, University of Pittsburgh, Pittsburgh, Pennsylvania

Kenneth L. Cox (183) US Air Force, Force Health Readiness, Tricare Management Activity/Deployment Health Support Directorate, Washington, DC

Kaustav Das (235) Center for Automated Learning & Discovery, Carnegie Mellon University, Pittsburgh, Pennsylvania

Denver H. Dash (273) Intel Research, Santa Clara, California

Sherry Daswani (465) RODS Laboratory, Center for Biomedical Informatics, University of Pittsburgh, Pittsburgh, Pennsylvania

Virginia Dato (13) Center for Public Health Practice, University of Pittsburgh, Graduate School of Public Health, Pittsburgh, Pennsylvania

Steve DeFrancesco (481) RODS Laboratory, Center for Biomedical Informatics, University of Pittsburgh, Pittsburgh, Pennsylvania

Christina Egan (143) Biodefense Laboratory, Wadsworth Center, New York State Department of Health, Albany, New York

Jeremy Espino (199) RODS Laboratory, Center for Biomedical Informatics, University of Pittsburgh, Pittsburgh, Pennsylvania

Per H. Gesteland (333) Inpatient Medicine Division, Department of Pediatrics, University of Utah School of Medicine, Salt Lake City, Utah

Louise S. Gresham (27) San Diego County, Health and Human Services Agency and Graduate School of Public Health, Epidemiology and Biostatistics, San Diego State University, San Diego, California

Kenneth W. Goodman (465) University of Miami, Miami, Florida

William R. Hogan (89) RODS Laboratory, Center for Biomedical Informatics, University of Pittsburgh, Pittsburgh, Pennsylvania

Vahan Grigoryan (423) RODS Laboratory, Center for Biomedical Informatics, University of Pittsburgh, Pittsburgh, Pennsylvania

Harvey Holmes (129) Bioterrorism Preparedness and Response Program, National Center for Infectious Diseases, Centers for Disease Control and Prevention, Atlanta, Georgia

Kevin Hutchison (481) RODS Laboratory, Center for Biomedical Informatics, University of Pittsburgh, Pittsburgh, Pennsylvania

Judith Hutman (465) RODS Laboratory, Center for Biomedical Informatics, University of Pittsburgh, Pittsburgh, Pennsylvania

Neil Jacobson (493) BBH Solutions, Inc., New York, New York

Heather A. Johnson (375) Center for Biomedical Informatics, University of Pittsburgh, Pittsburgh, Pennsylvania

Jeffrey Johnson (361) Health & Human Services Agency, Country of San Diego, San Diego, California

David Kirsh (361) University of California, San Diego, San Diego, California

Bruce Y. Lee (423) Section of Decision Sciences and Clinical Systems Modeling, Division of General Internal Medicine, University of Pittsburgh, Pittsburgh, Pennsylvania

Leslie Lenert (361) University of California, San Diego, San Diego, California

John D. Levander (273) RODS Laboratory, Center for Biomedical Informatics, University of Pittsburgh, Pittsburgh, Pennsylvania

Andrew W. Moore (217) School of Computer Science, Carnegie Mellon University, Pittsburgh, Pennsylvania

Daniel B. Neill (243) School of Computer Science, Carnegie Mellon University, Pittsburgh, Pennsylvania

Angieszka Onisko (405) RODS Laboratory, Center for Biomedical Informatics, University of Pittsburgh, Pittsburgh, Pennsylvania

Julie Pavlin (183) US Army, Uniformed Services, University of the Health Sciences, Washington, DC

Eric Resultan (129) Bioterrorism Preparedness and Response Program, National Center for Infectious Diseases, Centers for Disease Control and Prevention, Atlanta, Georgia

Loren Shaffer (51) Bureau of Health Surveillance, Department of Prevention, Ohio Department of Health, Columbus, Ohio

Jason R.E. Shepard (143) Biodefense Laboratory, Wadsworth Center, New York State Department of Health, Albany, New York

Richard Shephard (51) Australian Biosecurity Cooperative Research Centre for Emerging Infectious Disease, Brisbane, Australia

M. Cleat Szczepaniak (465) RODS Laboratory, Center for Biomedical Informatics, University of Pittsburgh, Pittsburgh, Pennsylvania

Fu-Chiang Tsui (199) RODS Laboratory, Center for Biomedical Informatics, University of Pittsburgh, Pittsburgh, Pennsylvania

Rita Velikina (67) Department of Epidemiology, School of Public Health, University of California Los Angeles, Los Angeles, California

Michael M. Wagner (3) RODS Laboratory, Center for Biomedical Informatics, University of Pittsburgh, Pittsburgh, Pennsylvania

Garrick Wallstrom (301) RODS Laboratory, Center for Biomedical Informatics, University of Pittsburgh, Pittsburgh, Pennsylvania

Weng-Keen Wong (217) School of Electrical Engineering and Computer Science, Oregon State University, Corvallis, Oregon

I
PART
The Challenge of Biosurveillance

CHAPTER

Introduction

Michael M. Wagner
RODS Laboratory, Center for Biomedical Informatics, University of Pittsburgh, Pittsburgh, Pennsylvania

1. INTRODUCTION

Biosurveillance is a process that detects disease in people, plants, or animals. It detects and characterizes outbreaks of disease. It monitors the environment for bacteria, viruses, and other biological agents that cause disease. The biosurveillance process *systematically* collects and analyzes data for the purpose of detecting cases of disease, outbreaks of disease, and environmental conditions that predispose to disease.

Detection of disease *outbreaks* is of particular importance. Unlike explosions, disease outbreaks are silent. Disease outbreaks sicken or kill individuals before they are detected. Disease outbreaks can inflict this damage quickly, and they can also spread quickly. The window of opportunity to limit this damage can be as brief as a few days in the worst case (Wagner et al., 2001).

The United States spends billions of dollars per year on various forms of biosurveillance. The major expenditures are for hospital infection control, public health surveillance, surveillance of the air and water, training, improvement of the information technology infrastructure for public health, and research.[1]

Biosurveillance is also a rapidly growing scientific field at the intersection of epidemiology, artificial intelligence, microbiology, computer science, statistics, system engineering, medicine, and veterinary medicine.

2. THE BIOSURVEILLANCE PROCESS

The biosurveillance process is a continuous one (Figure 1.1). An organization conducting biosurveillance collects and analyzes *surveillance data* continuously. The organization also faces decisions continuously about whether to act based on the results of these analyses.

The biosurveillance process involves a positive feedback loop: when the continuous collection and analysis of surveillance data identifies an anomalous number of sick individuals (or a single case of a dangerous disease), investigators collect additional information that feeds back into the analytic process, resulting in better characterization of the event. The improved understanding of the event may lead to more questions, which drive further collection of data and additional analyses. Concurrent with these cycles of data collection and analysis, the organization may initiate response actions such as vaccinations and quarantine to control the outbreak. The net effect of this process, when viewed over time, is a series of actions that lead to characterization and control of the outbreak.

3. THE SCOPE OF BIOSURVEILLANCE

The word *biosurveillance* is of recent origin.[2] *Biosurveillance* overlaps with two existing terms: *disease surveillance* and public health surveillance. These terms are defined as systematic methods for the collection and analysis of data for the purpose of detecting disease (Thacker and Berkelman, 1988; Halperin and Baker, 1992; Teutsch and Churchill, 2000).

As with any new word, we could speculate whether its invention and growing usage signals the appearance of a new field or simply reflects an inadequacy of existing terminology.

1 Although it is difficult to identify components in federal and other government budgets that correspond to biosurveillance, as these organizations have broader missions, the overall 2005 budget of the Centers for Disease Control and Prevention (CDC) was $8 billion and the median budget for a state health department in 2004 was $2.9 billion (Hearne et al., 2004). The total U.S. government civilian biodefense funding for 2005 was estimated at $7.6 billion (Schuler, 2004), with $452 million allocated for agricultural laboratories, monitoring, and research and $129 million allocated for air monitoring (*http://www.whitehouse.gov/omb/pdf/Homeland-06.pdf*). A 1975 estimate of the cost of hospital infection control programs in the United States (updated to 2005 dollars) was $261 million (Haley, 1977).

2 In fact, the current edition of the *Oxford English Dictionary* (OED) does not define the word *biosurveillance*. although it is in widespread usage, as evinced by Google search results (13,000 hits on May 8, 2005) as well as its routine use by government agencies, politicians, journalists, and academics. There is no doubt that *biosurveillance* has been inducted into the common vernacular. Even those without technical expertise or training in the field understand the term intuitively, just as they understand the meaning of *bioterrorism,* another word currently left undefined in the OED. The absence of a standard definition reflects the need to synthesize the multidisciplinary work being done in the field. Indeed, this book is our effort to present a unified approach to and understanding of biosurveillance.

Handbook of Biosurveillance
ISBN 0-12-369378-0

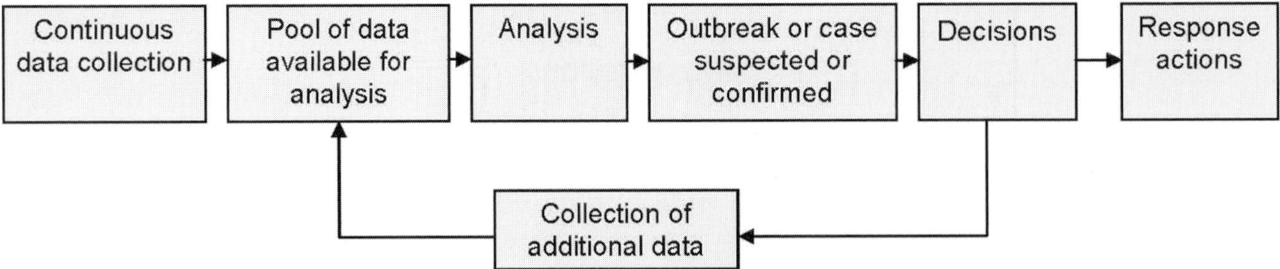

FIGURE 1.1 The biosurveillance process. When the continuous collection and analysis of surveillance data raises suspicion of an outbreak or a single case of a dangerous disease, biosurveillance personnel must decide whether to react to the information. They may decide to collect additional data that feed back into the analytic process, resulting in better characterization of the event. They may decide to take actions such as the issuance of a boil-water advisory (in the case of suspected water contamination), closure of a restaurant, or treatment of individuals with antibiotics or vaccines. Whenever the staff (or an automatic system) decides to collect additional data, the biosurveillance process exhibits a feedback loop. If the event is confirmed, staff will make many decisions over time about additional data to collect, directed by the analysis of data accumulated to that point (a positive feedback loop).

Instead, we simply state why, after considerable deliberation, we selected the term *biosurveillance* for this book. The terms *disease surveillance* and *public health surveillance* connote disease surveillance practiced by governmental public health. *Biosurveillance* allows us to broaden the scope of our discussion to include many other organizations that monitor for disease, such as hospitals, agribusinesses, and zoos. These organizations share the same basic goals of identifying individuals (people or animals) with disease, understanding disease transmission patterns in a population, elucidating the root causes of disease outbreaks, and monitoring the microbiological status of the environment. They collect similar types of data (clinical, microbiological, and environmental), use similar techniques to analyze the data, and they all face difficult decisions regarding how to react to the data. They often interact with each other to achieve the goals of disease detection and characterization. The similarities in goals and techniques suggested that we should unify them conceptually.

We also decided against *disease surveillance* and *public health surveillance* because these terms, to an epidemiologist, connote surveillance for noninfectious disease, child mortality, injury, cigarette smoking, and dental diseases such as enamel fluorosis (CDC, 2005). To keep what was already a very large topic manageable, we decided against discussing surveillance for these conditions. The principles and techniques that we discuss, nevertheless, apply to surveillance for any disease or condition.

Importantly, we decided against *disease surveillance* and *public health surveillance* because we consider outbreak characterization (i.e., determining the organism, source, route of transmission, spatial distribution, and number of affected individuals) a key process in biosurveillance. Epidemiologists may not consider outbreak characterization as falling under disease surveillance or public health surveillance (e.g., Buehler, 1998: Chapter 22; Teutsch and Churchill, 2000). To an epidemiologist, the process of public health surveillance detects an outbreak, and then an investigation characterizes it. *Biosurveillance,* as we use the term, encompasses both detection and characterization. As we will discuss later, processes that detect outbreaks also partially characterize them. Future advances in biosurveillance techniques will facilitate even better characterization of an outbreak at the time it is first detected. The blurring of the boundary between detection and characterization suggested that we should unify these processes conceptually.

4. FUNDAMENTAL PROPERTIES OF THE BIOSURVEILLANCE PROCESS

In addition to its continuous, cyclic nature, the biosurveillance process has several other properties, which are frequent themes in this book. These properties have implications for the design of a biosurveillance system, the types of computer systems that are required to support biosurveillance, and the training of individuals working in this field:

- Multidisciplinary
- Multiorganizational
- Time critical
- Probabilistic
- Decision oriented
- Data intensive
- Dependent on information technology
- Knowledge intensive
- Complex

4.1. Multidisciplinary

To conduct biosurveillance (i.e., to collect and analyze data to detect cases and outbreaks and characterize outbreaks), an organization must draw on the expertise of individuals with diverse professional backgrounds: epidemiologists, physicians, nurses, veterinarians, computer scientists, statisticians, water quality specialists, biologists, and microbiologists. Table 1.1 summarizes the expertise and training in different biosurveillance tasks (e.g., diagnosis of individuals and "diagnosis" of outbreaks) of many of the professionals that participate in biosurveillance. These individuals have different backgrounds

TABLE 1.1 Professional Training Related to Biosurveillance Tasks

Professional	Diagnosis* (Human)	Diagnosis* (Animal)	Epidemiological Analysis	Interpretation of Laboratory Tests	Environmental Monitoring	Information Technology‡	Other Training	Typical Degrees
Epidemiologist	Varies†	Varies	Extensive	Moderate	Moderate	Limited	Limited forensics	MPH, PhD, DPH
Physician§	Extensive	—	Limited	Extensive	—	—	Disease reporting	MD, DO
Nurse practitioner	Extensive	—	—	Extensive	—	—	Disease reporting	RN, BSN
Registered nurse	Moderate	—	—	Limited	—	—	Disease reporting	MSN
Radiologist	Extensive	—	—	Extensive		—	Disease reporting	MD, DO
Pathologist	Extensive	—	—	Extensive		—	Forensics, disease reporting	MD, DO
Medical technologist	Limited	—	—	Conducts of tests		Limited	Disease reporting	BS
Hospital ICP¶	Limited	—	Of hospitals	Extensive	Extensive	—	Disease reporting	BS
Veterinarian	Cross-over diseases	Extensive	Limited	Extensive		Minimal	Disease reporting	DVM,
Farm worker	—	Informal	—	—	—	—	—	Ag science
Microbiologist	—	—	—	Extensive	Varies	—	—	PhD
Medical examiner	Extensive	—	—	Extensive		None	Forensics	MD
Water inspector	—	—	—	—	Extensive	–	Water systems	BS
Food inspector	—	—	—	—	Extensive	–	Food systems	BS
Systems analyst	—	—	—	—	—	Extensive	Functional requirement definition	BS or MS
Database administrator	—	—	—	—	—	Extensive	—	BS or MS
Computer scientist	—	—	—	—	—	Extensive	—	PhD
Computer programmer	—	—	—	—	—	Extensive	—	BS or MS
Decision analyst	—	—	—	—	—	Extensive	Decision analysis	MS/PhD
Economist	—	—	—	—	—	—	Economics	PhD
Medical informaticist	Varies	Varies	Varies	Extensive	—	Extensive	Clinical information systems	PhD/PhD
Lawyer	—	—	—	—	—	—	Law	JD
Ethicist	—	—	—	—	—	—	Ethics	Varies

Extensive means that the skill is a primary objective of the professional training for this individual typically involving many courses and apprenticeship experience.

**Diagnosis* means that the individual is trained to establish a diagnosis from symptom, sign, and laboratory data.

†*Information technology* refers to training in programming, system architecture, and algorithms.

‡*Varies* Epidemiologists and medical informaticians often have dual degrees, including MD or DVM (Doctor of Veterinary Medicine) so the ability related to diagnosis may vary widely.

§*Physicians* refers to physicians that are practicing medicine. There are thousands of MDs who have expertise in other areas such as environmental testing (e.g., physicians who supervise environmental testing by industrial hygienists, Association of Occupational and Environmental Medicine [ACOEM]).

¶*Hospital ICP*: hospital infection control practitioner.

and may have difficulty communicating with each other unless their training includes exposure to the roles, methods, and concepts used by other individuals.

Many of these individuals have different "day jobs." Most physicians, for example, work in medical care, not biosurveillance. They, nevertheless, must be competent in the skills related to biosurveillance, such as diagnosing rare diseases and reporting the existence of a person with a communicable disease to the appropriate authorities. The professional and continuing education of physicians, veterinarians, and medical technologists must ensure that they have the necessary training in the diseases and procedures related to their biosurveillance roles—skills that will allow them to operate as components in a larger system whose goal is disease and outbreak detection and characterization. Conversely, a biosurveillance system must ensure that these individuals have the information that they need when they need it.

4.2. Multiorganizational

Biosurveillance of just a single city requires the cooperation of many organizations, including hospitals, infection control units within hospitals, laboratories within hospitals, medical practices, commercial laboratories, water suppliers, and health departments (Table 1.2).

In a region the size of the United States, there are more than 7,500 hospitals (U.S. Census Bureau, 2005), 40,000 long-term care facilities (National Center for Health Statistics, 1986), 160,000 water departments, and 185,000 clinical laboratories (CLMA, 2005). There are 3,000 local public health agencies and 60 state tribal, or territorial health departments (Hearne et al., 2004). In addition, 1.3 million farms carry livestock (Kellog, 2002), and there exist large numbers of manufacturers and distributors of food and drugs.

These organizations work collaboratively in the service of biosurveillance by communicating, exchanging data, and acting in concert during outbreaks. As we will see in Chapter 2, disease does not respect national boundaries. Outbreaks of diseases, such as severe acute respiratory syndrome (SARS) and influenza, spread quickly around the world. The set of organizations that may have to communicate and exchange information about an international outbreak may be 100-fold larger than the set that would have to collaborate for a nationwide outbreak, and the difficulty of coordination among these organizations is exacerbated by differences in languages, customs, and laws.

An engineer or a computer scientist reading the previous paragraphs would immediately characterize biosurveillance as a highly distributed process. The engineer would realize that each person and organization must perform specific functions for such an arrangement to work, and that were she to attempt to improve the process, significant attention would need to be paid to communication and coordination among the components.

TABLE 1.2 Responsibilities of Organizations for Monitoring of Environment, Detection of Cases, Detection of Outbreaks, and Outbreak Characterization

Organization	Monitoring Water, Food, Drug, Air	Case Detection	Outbreak Detection	Outbreak Characterization
Hospital	Primary in hospital	Primary	Primary in hospital, otherwise supportive	Primary in hospital, otherwise supportive
Laboratory	Supportive	Primary	Supportive	Supportive
Coroner	—	Primary	Supportive	Supportive
State/local health department	Primary	Primary (public health clinics/contact tracing)	Primary	Primary
CDC	Supportive	Supportive	Primary for multistate rare isolates	Primary for bioterrorism
WHO	Supportive	Supportive	Supportive	Supportive
FDA	Primary for drugs	Primary	Supportive	Primary for food or drug trace-back
Water Supplier	Primary (water)	—	Supportive	Supportive
EPA	Supportive, water	—	—	Supportive
Drug Manufacturer	Primary for drugs	—	—	Supportive
Food Manufacturer	Primary for food	—	—	Supportive
USDA	Primary for food, meat	—	—	Primary for agriculture
State department of agriculture	—	Through state laboratories	Supportive	Primary for agriculture
Large farm	Primary for farm	Primary for farm	Primary for farm	Primary for farm
Animal hospital	Primary in hospital	Primary	Primary in hospital, otherwise supportive	Primary in hospital, otherwise supportive
Zoo	Primary for zoo	Primary for zoo	Primary for zoo	Primary for zoo
DHS	Primary for air (bioterrorism)	—	—	Primary for bioterrorism
U.S. Postal Service	Primary for facilities	—	—	Supportive
DOD	Primary for DOD	Primary for DOD	Primary for DOD	Primary for DOD
Transit system	Primary for system	—	—	Supportive

CDC indicates Centers for Disease Control and Prevention; WHO, World Health Organization; FDA, Food and Drug Administration; EPA, Environmental Protection Agency; USDA, U.S. Department of Agriculture; DHS, Department of Homeland Security; DOD, Department of Defense; and Transit, airlines and mass transit systems.

4.3. Time Critical

Early detection of outbreaks is perhaps the most important requirement for a biosurveillance system. Morbidity and economic loss accumulate rapidly, beginning with the first sick individual. In a worst-case scenario, such as a surreptitious aerosol release of the organism *Bacillus anthracis* by a terrorist on a large city, hundreds of thousands of individuals would be exposed nearly simultaneously to a biological organism that is lethal and fast acting (Figure 1.2).

The implications of time criticality are profound for the design of biosurveillance systems. Reducing the time delay between the start of an outbreak and its detection is a key goal of research and development in biosurveillance. This requirement for early detection pervades the design of new biosurveillance systems, which are designed to collect and analyze new types of surveillance data in real time.

4.4. Probabilistic

Because early detection is important, biosurveillance increasingly involves analysis of novel types of data, such as sales of diarrhea remedies and numbers of visits to emergency departments for respiratory complaints. These data are more difficult to interpret than are definitive diagnoses (e.g., a patient has anthrax) because the former are not diagnostically precise. Detection of an outbreak depends on noticing an increase in the numbers of sales or visits relative to usual levels.

The challenge of early detection is that most outbreaks present weaker signals (increases) in the data streams earlier in the outbreak than they do in the middle of the event or after the event. This means that earlier detection requires the detection of smaller signals.

Early (and reliable) detection is necessarily probabilistic because early detection requires detection when signals are small and when few signal sources may yet be active. The goal is to detect a case or an outbreak before the signals are large enough and present in enough data streams for detection to be 100% certain.

The assessment of the probability that an anomalous event is occurring places demands on a biosurveillance system and its algorithms. Analytic techniques must handle multiple, independent, yet correlated data streams. The need for probabilistic detection from multiple data streams strongly suggests the need for a detection system based on Bayesian inference. A well-organized Bayesian approach allows for rational combination of many small indicators into a big picture. We discuss Bayesian methods in detail in this book.

4.5. Decision Oriented

Biosurveillance does not exist in a vacuum. Its purpose is to collect and analyze information that people use to guide decision making and action. Biosurveillance personnel make decisions under time pressure. They make decisions based on incomplete and uncertain information. Early in the course of an outbreak,

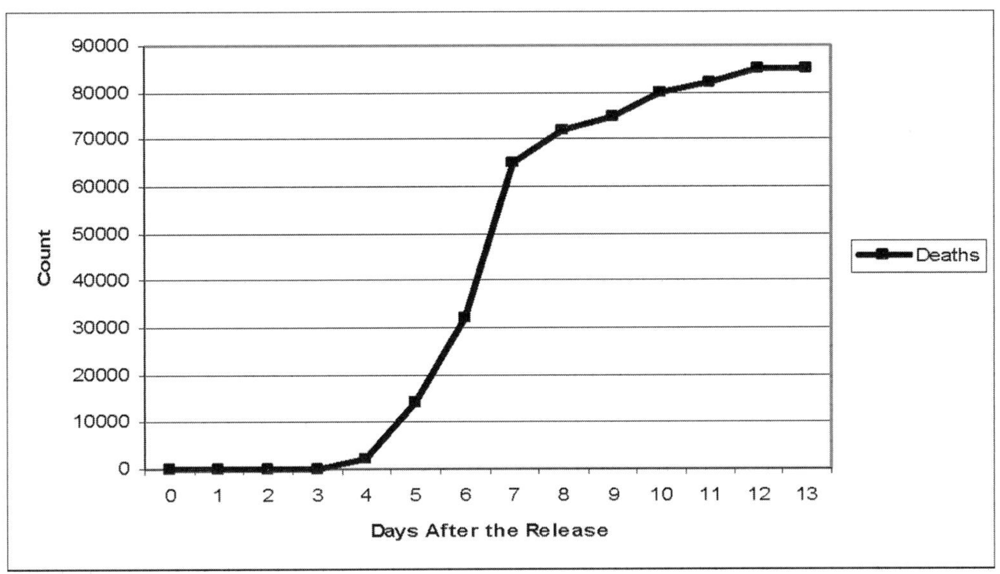

FIGURE 1.2 Hypothetical cumulative mortality from a surreptitious aerosol release of *Bacillus anthracis* by a terrorist on a major city. Such a release could expose hundreds of thousands of individuals nearly simultaneously to a biological organism that is lethal and fast acting. The window of opportunity to detect this event and administer antibiotics to those exposed is brief. (We estimated the shape of this curve from published data on the incubation period and mortality observed in the 1979 release of *B. anthracis* [*Kirov* strain] from Soviet Biological Weapons Compound 19 described in Chapter 2 and in the 2001 U.S. postal attacks).

they may not know the cause of the illness in patients, the number of affected individuals in the community, or the source of the infections. Nevertheless, they must form conjectures and hypotheses based on the available information and make decisions about how to direct resources to investigate, treat, and even quarantine individuals.

Psychological research has shown that human decision makers perform most poorly under conditions of uncertainty and time pressure. The effect of uncertainty on decision making can be profound as demonstrated by tabletop exercises (see Inglesby et al., 2001; O'Toole et al., 2002). Fortunately, the sciences of decision making and of economics provide methods to improve decision making under uncertainty. These methods elucidate the tradeoff between the risk of waiting and the cost of taking the wrong action. Biosurveillance organizations can use these methods to develop guidelines for such decision situations, or they can build these methods into computer systems that provide decision support to frontline personnel facing specific decisions. We discuss the science of decision making and economic studies in detail in Part V of this book.

4.6. Data Intensive

It is perhaps obvious, but worth stating, that biosurveillance is not a vaccine or drug that can save lives directly. Biosurveillance is a process that collects and analyzes data to guide the application of vaccines, drugs, quarantine, and other disease control strategies that can save lives.

The role of biosurveillance in disease control is to gather and process data—to collect, communicate, and analyze data. A chain of data-processing steps links raw surveillance data to "actionable information," as illustrated by Figure 1.3. The link between biosurveillance and response occurs at the point that biosurveillance personnel make decisions to act.

Each step in the chain may involve information systems and people–all of which must function effectively if an outbreak is to be quickly detected, characterized, and controlled. Any breakdown or delay in the chain can reduce the efficacy of the biosurveillance system and its ability to contribute, ultimately, to the prevention of mortality and morbidity.

4.7. Dependent on Information Technology

Societies, especially cities, have conducted biosurveillance in some form for centuries, so it is self-evident that organizations can conduct biosurveillance without the assistance of information technology. However, information technology is of increasing importance in biosurveillance because it can address the problem of time criticality. Information technology has the potential to speed up and improve the accuracy of almost every aspect of the biosurveillance process. Information technology can assist or fully automate data collection, transmission, storage, and communication. It can assist or partially automated even the most cognitively challenging steps–patient diagnosis, outbreak detection, outbreak characterization, and decision making.

4.8. Knowledge Intensive

Biosurveillance is a knowledge-intensive process. To diagnose a patient with an infectious disease, a physician must be familiar with the symptoms, signs, radiological characteristics, and laboratory tests for hundreds of diseases. This information can fill several large textbooks and requires years of study to master. A veterinarian must master an even larger body of knowledge as veterinary medicine concerns large numbers of animal species. An epidemiologist must similarly master a large body of knowledge, including a subset of human and animal diseases, as well as the subject of epidemiology, which concerns patterns of disease transmission. This knowledge also fills large textbooks, as does the knowledge required for the conduct of infection control in hospitals.

The human ability to master and apply large bodies of knowledge varies but, in general, is imperfect. Fortunately, there are technologies such as *diagnostic expert systems* and *knowledge-based* systems that professionals in many fields use to extend the range of their competencies. We discuss these technologies in Chapter 13.

4.9. Complex

The biosurveillance process is complex. There is complexity inherent in a system that distributes its functions over a large number of individuals and organizations. There is cognitive

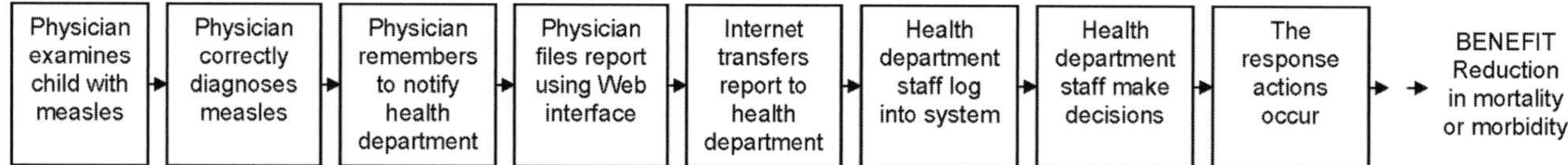

FIGURE 1.3 The indirect connection between information and benefit. A physician evaluates a child with measles. The physician must correctly diagnose the patient and remember to notify the local department of health. If the physician uses a Web-based disease reporting system, the local computer, the Internet, and the health department information systems must be functioning. Staff must review the report and make correct decisions about collection of additional information and appropriate control measures to institute. At some time later, the benefit of the information is realized by control of the outbreak and reduction in the level of morbidity and possibly mortality.

complexity inherent in reasoning and taking decisions from partial and uncertain information. There is complexity due to the number of biological agents that can cause disease and the myriad ways that they can present as outbreaks (e.g., air-borne pattern, food-contamination pattern, subway system—contamination pattern, mail system–contamination pattern, and building-contamination pattern).

This complexity makes it difficult to design a biosurveillance system. In the past, organizations and people have managed the complexity of biosurveillance by specialization and prioritization. *Specialization* is a divide-and-conquer technique in which people or organizations manage complexity by, for example, creating separate biosurveillance capabilities for communicable diseases and for water-borne diseases. Specialization is not without its drawbacks, as demonstrated by the existence of many specialized information systems that cannot interoperate. *Prioritization* refers to paying more attention to certain diseases, which is a polite way of saying that, to some extent, people manage complexity by sometimes ignoring it.

One of the key benefits of information technology in professional domains, such as engineering, medicine, and biosurveillance, is that it can help to manage complexity for the professional working in that field. By managing both data and knowledge, information technology can make previously impossible or Herculean tasks possible. Information technology is a way of managing the ever-increasing complexity of biosurveillance without relying as heavily on specialization and prioritization. We discuss information systems that manage data throughout this book, and we examine systems that assist biosurveillance personnel with analytic and cognitive tasks in Parts III and V.

5. BIOSURVEILLANCE SYSTEMS

The definitions of *biosurveillance, disease surveillance,* and *public health surveillance* all include the word *systematic.* A system is any organized way of doing something. Because of the numbers of individuals, organizations, and steps in the biosurveillance process, a basic property of biosurveillance is that it is systematic. A biosurveillance system may be manual, automated, or, more commonly, a mixture of manual and automated processes. Biosurveillance systems of all types exist. The systematic, process-oriented nature of biosurveillance can be represented diagrammatically, as illustrated in Figure 1.4, which represents a highly automated system. The developers of manual biosurveillance systems often represent the organization and flow of information in a system diagrammatically as well. The diagrammatic representation of biosurveillance systems finds its fullest expression in the concept of an *architecture* for a biosurveillance system. We discuss architecture in detail in Chapter 33.

6. SCIENTIFIC FOUNDATIONS OF BIOSURVEILLANCE

The scientific foundations of biosurveillance have evolved over the centuries, parallel to advances in medicine, microbiology, veterinary science, laboratory science, epidemiology, mathematics, and many other fields.

Over the past 5 years, the scientific foundations of biosurveillance have changed rapidly. Bioterrorism and the threat posed by emerging infectious diseases triggered this change by creating a new requirement—very early detection of disease outbreaks (Wagner et al., 2001). New techniques are being introduced rapidly from diverse scientific fields and

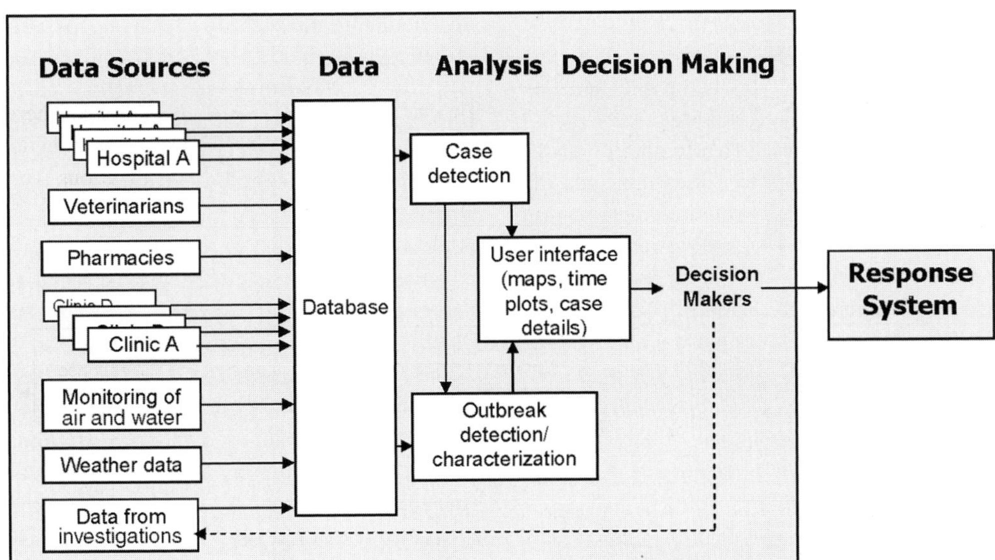

FIGURE 1.4 A generic biosurveillance system. The key elements of a biosurveillance system are data sources, a database, analysis, and decision making.

include mathematical models of the process of medical diagnosis and of decision making, as well as mathematical models of the process of "epidemic" diagnosis.

Perhaps the most important new techniques can be traced to Ledley and Lusted (1959), who first introduced the idea that medical diagnosis and decision making could be modeled mathematically. This idea spawned a large body of research about how physicians use diagnostic information, how a computer could represent medical knowledge, and how to construct computer programs that perform medical (and veterinary) diagnosis. Approximately 5 years ago, it became apparent that these same techniques could be applied to *epidemiological* diagnosis and decision making (Wagner et al., 2001). These techniques represent new core subject matter for the professional training of researchers and practitioners. We discuss these new approaches in Parts III and V.

This recent expansion of the scientific foundations of biosurveillance has been abrupt and large. The philosopher of science Kuhn termed such changes *paradigm shifts* (Kuhn, 1962). A paradigm shift is associated with changes in the curriculums of professional schools, the structure and functions of organizations, the appearance of new journals, the workforce, and the tables of contents in standard textbooks. There is evidence of such changes in biosurveillance (Logan-Henfrey, 2000; Yasnoff et al., 2001; Wagner, 2002).

7. OPEN RESEARCH PROBLEMS IN BIOSURVEILLANCE

Biosurveillance has been an active area of research since the seminal work of John Snow (1855). However, the recent requirement for very early detection caused a change in the direction (and intensity) of research. Researchers now focus on improving the timeliness and accuracy of case detection, outbreak detection, and outbreak characterization. Researchers are developing more rapid and accurate diagnostic tests, methods for sensing microbes or their effects in the environment, new detection algorithms that can extract maximum signal from early but noisy data, and research to identify types of surveillance data that provide an earlier indication of an outbreak.

Examples of the questions that research attempts to answer related to detection of individual cases of disease include the following: *What are the optimal data to collect to detect a* <u>*case*</u> *of disease* X? *What is the optimal analytic method to detect a case of disease X? What are the sensitivity, specificity, and timeliness of the current best methods for detecting a case of disease X?*

Research pursues the same set of questions for *outbreaks* of disease but also pursues additional questions, such as the following: *When can we expect to detect an outbreak of disease X that affects 1% of the population by analysis of some class of surveillance data? What is the smallest outbreak of X that we can detect?*

8. THE ROLE OF BIOSURVEILLANCE IN BIODEFENSE

Biodefense is a set of activities that together function to provide security against disease due to biological agents. Biosurveillance is one of these activities—along with sanitation, vaccination, quarantine, intelligence, interdiction (of terrorists and materiel), forensic science, and control of technologies used to create biological weapons.

Many organizations, in addition to biosurveillance organizations, play a role in biodefense, including governmental public health (in its response role), intelligence agencies, the police, the military, and pan-national organizations, such as the World Health Organization.

9. ORGANIZATION OF THE BOOK

We have organized this book into six parts.

Part I: The Problem of Biosurveillance comprises this introductory chapter and Chapters 2 through 4. Chapter 2 ("Outbreaks and Investigations") provides examples of outbreaks that have been investigated by governmental public health, hospital infection control, and the animal healthcare system. Chapter 3 ("Case Detection, Outbreak Detection, and Outbreak Characterization") provides an overview of the basic tasks of biosurveillance, explaining in detail the methods used to detect and characterize the outbreaks described in Chapter 2. Chapter 4 ("Functional Requirements for Biosurveillance") discusses biosurveillance from the perspective of a system analyst or engineer.

Part II: Organizations that Conduct Biosurveillance and the Data They Collect (Chapters 5–12) discusses governmental public health, the human healthcare system, the animal healthcare system, laboratories, water departments, the food and drug industries, and other organizations that conduct biosurveillance. The chapters discuss the types of professionals that work in these organizations, the organizations themselves, and the information systems used by these organizations.

Part III: Data Analysis (Chapters 13–20) discusses methods for detection of individual cases, methods for detecting anomalous numbers of cases in a population, and methods for elucidating characteristics of outbreaks. The first two chapters discuss algorithms for detection of individual cases ("Case-Detection Algorithms") and the simplest algorithms for detecting outbreaks ("Classical Time-Series Methods for Biosurveillance"). The last chapter (Chapter 20) discusses methods for evaluating both case-detection and outbreak-detection algorithms. The remaining chapters cover more advanced topics, including spatial scanning, multivariate analysis, atmospheric dispersion modeling, natural language processing, and Bayesian biosurveillance.

Part IV: Newer Types of Biosurveillance Data (Chapters 21–28) discusses what research has found about the value of newer types of biosurveillance data, such as school absenteeism, sales of over-the-counter medications, and data from sensors (including physiological sensors and remote sensing from space-based satellites). Because many of these types of data are still the subject of active research, we devote the first chapter in Part IV to research methods for evaluating surveillance data.

Part V: Decision Making (Chapters 29–31) discusses the types of decisions faced by biosurveillance personnel, the types of errors in judgment to which decision makers are prone, and formal methods for modeling decisions. Part V uses an extended example of a common decision problem that is currently at the forefront in the biosurveillance community: Whether and how to react to anomalies in newer types of surveillance data.

Part VI: Building and Field Testing Biosurveillance Systems (Chapters 32–37) covers implementation issues. To avoid hearing "Professor, you left out a whole bunch of stuff,"[3] the final part of this book covers pragmatic issues related to building biosurveillance systems. Although data and analysis are the foundations of biosurveillance, organizations that wish to build biosurveillance systems must attend to proper architectural design, use of standards, legal issues, and project management. Chapter 37 discusses methods for field testing of operational biosurveillance systems.

10. SUMMARY

We can perhaps best summarize this introductory chapter about the biosurveillance process and its role in biodefense with an analogy.

An individual or an organization that is operating a biosurveillance system is like a military commander who cannot see directly every threat he faces. The commander relies on "surveillance systems" that include reports from frontline units, reconnaissance, and sensing systems located in aircraft or in orbit to make tactical decisions. Both the military commander and a biosurveillance organization are decision makers who make high-stakes decisions under time pressure by using incomplete and uncertain information. Both the military commander and the biosurveillance organization have the ability to increase the information available (by sending out additional reconnaissance or by initiating an outbreak investigation), but that new information comes at a price, which involves both the cost of the investigation as well as the "cost" of waiting for additional information before acting. This latter cost can be quite high should either disease or a military opponent gain the upper hand during the delay. Interestingly, the technologies that the military and biosurveillance organizations use–signal processing, risk-benefit analyses, methods for decision support—are either already similar or converging.

Like most analogies, this one breaks down the further one goes into detail. The level of training required for the surveillance task in the military is less than that required for the biosurveillance task, and the data required by the commander and the biosurveillance organization are completely different.

Nevertheless, like a military commander, an organization that conducts biosurveillance is primarily action oriented and conducts biosurveillance as a means, not as an end. Biosurveillance is a means to the end of protecting health. It is an indispensable means to that end, which is absolutely dependent on the quality and timeliness of biosurveillance.

ADDITIONAL RESOURCES

Teutsch, S. and Churchill, R. (2000). *Principles and Practice of Public Health Surveillance.* Oxford: Oxford University Press. Provides an interesting account of the history of public health surveillance, as well as a description of the practice of this surveillance as of the time of writing.

Bennett, J. and Brachman, P., eds. (1998). *Hospital Infections.* Philadelphia: Lippincott-Raven. A textbook for hospital infection control practitioners.

O'Carroll, P., et al., eds. (2003). *Public Health Informatics and Information Systems.* New York, Springer. Discusses informatics principles as they relate to public health practice, as well as the current and future role of information technology in public health practice

REFERENCES

Buehler, J. (1998). Surveillance. In *Modern Epidemiology,* 2nd ed. (K. Rothman and S. Greenland, eds.). Philadelphia: Lippencott-Raven.

Centers for Disease Control and Prevention [CDC]. (2005). Surveillance for Dental Carries, Dental Sealants, Tooth Retention, Edentulism, and Enamel Fluorosis–United States, 1988–1994 and 1999–2002. *The Morbidity and Morality Weekly Report* vol. 54:1–48.

Halperin, W. and Baker, E. L. J., eds. (1992). *Public Health Surveillance.* Van New York: Nostrand Reinhold.

Hearne, S., Segal, L., Earls, M., and Unruh, P. (2004). *Ready or Not? Protecting the Public's Health in the Age of Bioterrorism.* Washington, DC: Trust for America's Health.

Inglesby, T. V., Grossman, R., and O'Toole, T. (2001). A Plague on Your City: Observations from TOPOFF. *Clinical Infectious Diseases* vol. 32: 436–445.

Kellog, R. (2002). Profile of Farms with Livestock in the United States: A Statistical Summary. Washington, DC: Natural Resources Conservation Service, U.S. Department of Agriculture. *http://www.nrcs.usda.gov/technical/land/pubs/livestockfarm.html.*

Kuhn, T. (1962). *The Structure of Scientific Revolutions.* Chicago: University of Chicago Press.

Ledley, R. S. and Lusted, L. B. (1959). Reasoning foundations of medical diagnosis. *Science* vol. 130:9–21.

Logan-Henfrey, L. (2000). Mitigation of bioterrorist threats in the 21st century. *Annals of the New York Academy of Sciences* vol. 916:121-33.

National Center for Health Statistics. (1986). Inventory of Long-Term Care Places/National Master Facility Inventory (NMFI) Public-Use Data Files. Hyattsville, MD: National Center for Health Statistics, Centers for Disease Control and Prevention.

3 Rodney Dangerfield playing Thornton Mellon in the film *Back to School.*

O'Toole, T., Mair, M., and Inglesby, T. V. (2002). Shining Light on "Dark Winter." *Clinical Infectious Diseases* vol. 34:972–983.

Snow, J. (1855). *On the Mode of Communication of Cholera.* London: John Churchill.

Teutsch, S. and Churchill, R. (2000). *Principles and Practice of Public Health Surveillance.* Oxford: Oxford University Press.

Thacker, S. and Berkelman, R. (1988). Public health surveillance in the United States. *Epidemiology Review* vol. 10:164–190.

U.S. Census Bureau. (2005). Facts for Features. CB05-FFSE.02-2. Bethesda, MD: U.S. Census Bureau. *http://www.census.gov/Press-Release/www/releases/archives/facts_for_features_special_editions/004491.html.*

Wagner, M. M. (2002). The space race and biodefense: Lessons from NASA about big science and the role of medical informatics. *Journal of the American Medical Information Association* vol. 9:120–122. *http://www.jamia.org/cgi/content/full/9/2/120.*

Wagner, M. M., et al. (2001). The emerging science of very early detection of disease outbreaks. *Journal of Public Health Management Practice* vol. 7:51–59. *http://www.ncbi.nlm.nih.gov/entrez/query.fcgi?cmd=Retrieve&db=PubMed&dopt=Citation&list_uids=11710168.*

Yasnoff, W. A., et al. (2001). A national agenda for public health informatics. *Journal of Public Health Management Practice* vol. 7:1–21. *http://www.ncbi.nlm.nih.gov/entrez/query.fcgi?cmd=Retrieve&db=PubMed&dopt=Citation&list_uids=11713752.*

2

CHAPTER

Outbreaks and Investigations

Virginia Dato
Center for Public Health Practice, University of Pittsburgh, Graduate School of Public Health, Pittsburgh, Pennsylvania

Richard Shephard
Australian Biosecurity Cooperative Research Centre for Emerging Infectious Disease, Brisbane, Australia

Michael M. Wagner
RODS Laboratory, Center for Biomedical Informatics, University of Pittsburgh, Pittsburgh, Pennsylvania

1. INTRODUCTION

In Ernest Hemingway's novel *The Sun Also Rises,* when Mike Campbell, a Scottish expatriate and World War I veteran, is asked how he went bankrupt, he replies, "gradually, then suddenly." Often, this is how outbreaks of disease appear. Outbreaks can wreak much havoc before they are noticed, and they can grow at an exponential rate before they are brought under control. Some outbreaks spread so far we call them *epidemics,* or *pandemics* if they encompass the entire planet. Some outbreaks have never been brought under control. The only certainty is that an outbreak becomes more difficult to stop through human intervention the longer it goes unnoticed.

Chapter 2 presents many examples of outbreaks that differ in their origins and in how they were detected, characterized, contained, or continued to develop. Some of these outbreaks have impacted not only health but history itself. These examples illustrate the basic goals, tasks, and activities of biosurveillance and provide a sense of how the methods of biosurveillance have evolved over time.

2. HISTORICAL OUTBREAKS

The lives of humans, animals, and microbial organisms have been irrevocably intertwined throughout evolution. Humans rely on these organisms to perform some of the basic functions of life; for instance, the organisms that reside in the gastrointestinal tract of humans participate in the digestion of food. Mitochondria, which provide the energy that fuels our cell processes (Penniston, 1997), evolved from bacteria that were incorporated into primitive cells during the early stages of evolution. Of course, not all interactions between microbial organisms and humans are favorable. Human (and animal and plant) populations have been battling these unseen living organisms throughout the course of history and, in many instances, losing. William H. McNeill (1989), in his book *Plagues and Peoples,* summarizes the scientific evidence of disease outbreaks that predate recorded history.

Hippocrates (460–377 BC) wrote some of the oldest surviving descriptions of disease outbreaks. He described an outbreak of mumps on the island of Thasos, now present-day Greece. Hippocrates also described what appear to be outbreaks of malaria, diphtheria, tuberculosis, and influenza (McNeill, 1989). Thucydides, a contemporary of Hippocrates, described an outbreak that decimated the Athenian army and many civilians in 430 to 429 BC (McNeill, 1989). When the outbreak ended, it left a weakened Athenian empire that soon fell to Sparta.

The Black Plague killed 25% to 50% of Europe's population between 1348 and 1351 (EyeWitness to History.com, 2001). Smallpox wiped out a large portion of the native population of Hispaniola (now Haiti and the Dominican Republic) and spread from there to Mesoamerica (present-day Mexico), contributing to the demise of the Aztecs (McNeill, 1989).

Until the 20th century, the world's urban population was so routinely devastated by infectious diseases that only constant in-migration from the countryside could maintain the population of growing cities (McNeill, 1989).

The 20th century brought technological and medical advancements—improved sanitation, vaccines, and antibiotics—that quickly reduced mortality due to infectious disease. Figure 2.1 demonstrates the steady decline in the death rate in the United States as infectious diseases were gradually controlled with these tools. The death rate reaches its lowest point in 1980, just after natural smallpox was eradicated and just before the appearance of AIDS.

3. THE 1918 PANDEMIC OF INFLUENZA

By the spring of 1918, when American forces joined the conflagration overseas, Europe had already endured four grinding years of trench warfare. When the tide turned in favor of the Allies in the fall of 1918, it appeared that the unimaginable carnage and devastation consuming the world's Great Powers would finally cease.

Then a new scourge appeared. Influenza caused a sudden increase in mortality in the United States (see Figure 2.1) and other countries (Reid et al., 2004). This outbreak was termed a *pandemic* because of its worldwide scope. It killed at least

ISBN 0-12-369378-0

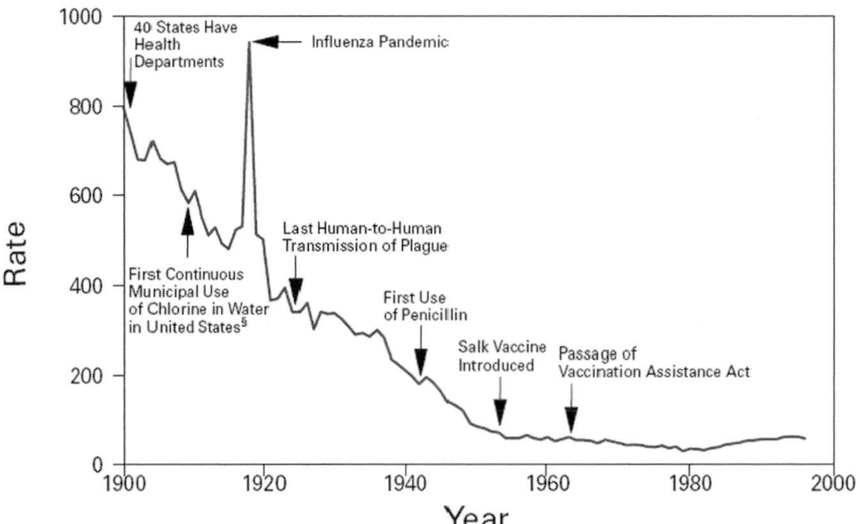

*Per 100,000 population per year.
†Adapted from Armstrong GL, Conn LA, Pinner RW. Trends in infectious disease mortality in the United States during the 20th century. JAMA 1999:281;61–6.
§American Water Works Association. Water chlorination principles and practices: AWWA manual M20. Denver, Colorado: American Water Works Association, 1973.

FIGURE 2.1 Crude death rate for infectious diseases—United States, 1900–1996. (Graph from CDC, 1999.)

21.5 million individuals, a number that dwarfs the combat casualties of World War I.

A U.S. military facility was an early victim of this outbreak and played an important role in the spread of disease. In Camp Funston (now Fort Riley,) Kansas, on March 11, 1918, 100 soldiers reported sick to the infirmary before noon (Barry, 2004). Figure 2.2 is an undated picture of the emergency hospital in Camp Funston.

Although ravaged by this outbreak, Camp Funston continued to train troops who then embarked to domestic and

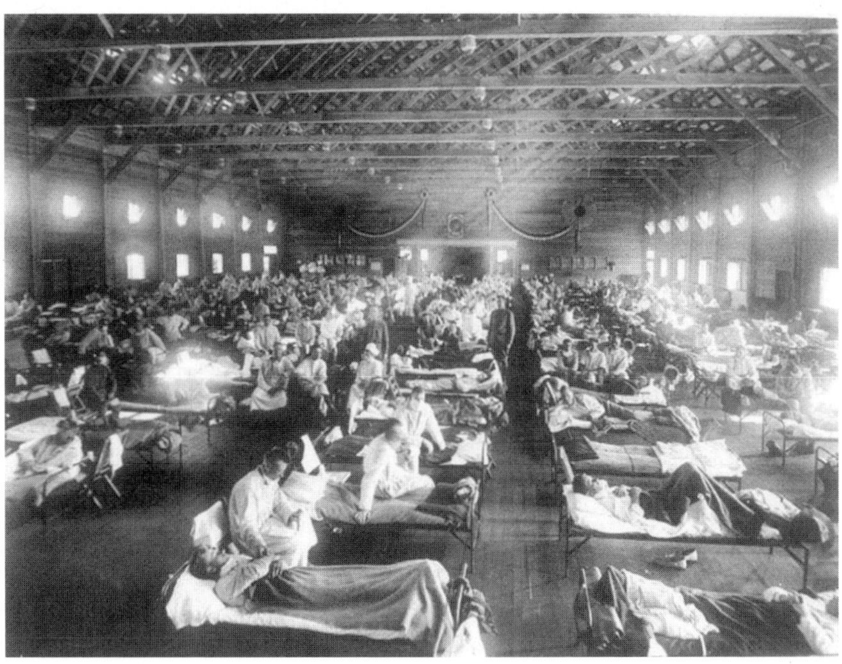

FIGURE 2.2 Emergency hospital during influenza epidemic, Camp Funston, Kansas. (Photo from the National Museum of Health and Medicine, Armed Forces Institute of Pathology, Washington, DC.)

overseas assignments, carrying with them a virus more deadly than their arms. Virus spread to the civilian populations of Europe and North America (Public Broadcasting Service, n.d.). The disease became known as "Le Grippe" or "Spanish flu" because Spain was not at war, and its domestic press reported on the outbreak.

Figure 2.3 shows the annualized weekly death rates in London, Paris, New York, and Berlin for October and November of 1918. All four cities experienced high mortality rates nearly simultaneously. There was another wave of mortality in early 1919 before this largest of all known epidemics finally burned out on its own, presumably having induced immunity in sufficient numbers of those who had survived.

Where did the outbreak in Camp Funston begin? After seven years of study, Barry (2004) concluded that it began in Haskell County, Kansas, located 300 miles to the west, where an outbreak of severe influenza was underway. Military recruits from Haskell County were inducted into the military at Camp Funston.

What made the 1918 influenza outbreak so deadly? To understand, it is necessary to know a little about influenza. Influenza is actually a family of viruses with three different types (A, B, and C) and many strains. Scientists create names for influenza strains in the A type based on two proteins (called *antigens* because they induce immunity in victims) that appear on the surface of the influenza virus and can be measured. Each antigen can take different forms, which are referred to as *subtypes:* Hemagglutinin (H) antigen with subtypes 1 to 15 and neuraminidase antigen (N) with subtypes 1 to 9. The 1918 virus was named H1N1 because it had on its surface subtype 1 of the hemagglutinin antigen and subtype 1 of the neuraminidase antigen.

All influenza A strains infect birds (and, for this reason, influenza A is also known as avian influenza), but only some of these strains are able to infect humans and other mammals and subsequently be passed directly from person to person (or animal to animal). There are two ways that a strain of influenza A can develop the ability to spread from person to person: (1) mutation in one or more genes, and (2) reassortment of genes. Reassortment is simply the exchange of genes between or among two or more different viruses to give birth to a hybrid virus. A prerequisite for reassortment is that the same cell in a bird or mammal must be simultaneously infected with two or more strains of influenza A. If a "parent" strain was capable of being transmitted from person to person, the descendant strain may inherit that ability. Reassortment is a common occurrence in influenza, and it is the key reason that influenza remains a major world health concern to this day.

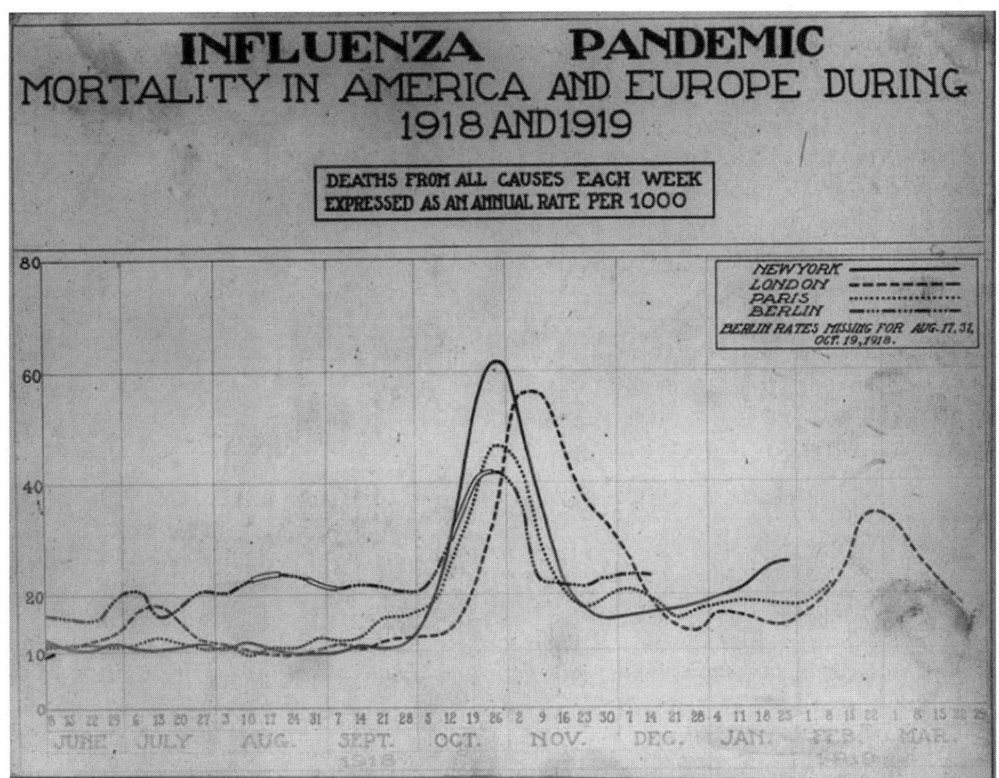

FIGURE 2.3 Influenza pandemic mortality in American and Europe during 1918 and 1919. (From the National Museum of Health and Medicine, Armed Forces Institute of Pathology, Washington, DC.)

To elucidate whether the 1918 strain was a product of reassortment or mutation, over the past decade, researchers have searched from the frozen tundra of Alaska to the archives of the Smithsonian Institution to find preserved tissues of birds and humans who died in 1917 and 1918. By using modern techniques of molecular biology, they isolated snippets of genetic material from influenza A viruses isolated from these specimens and compared them with each other. They found that the virus taken from individuals known to have died from the pandemic strain was not a direct descendant from the viruses that they found in birds from that period. The best theory is that the 1918 strain was the result of a mutation, a reassortment of an influenza A virus or viruses circulating in mammals such as horses or swine, or both (Reid and Taubenberger, 2003). Haskell County, Kansas, where animals outnumbered humans by a large factor, was just the type of location where influenza A strains from different mammalian species could converge to develop into the deadliest strain of influenza yet known to humankind (Barry, 2004).

Pandemics of influenza A occurred in 1957 and 1968. The virus strains causing these outbreaks were reassortments of genetic material between human and avian influenza virus strains circulating at the time (Reid and Taubenberger, 2003). An H2N2 strain caused the 1957 pandemic. An H3N2 strain caused the 1968 pandemic and is still circulating today (with gradual mutations over the years). Recently, the 1957 H2N2 strain was mistakenly included in a set of test strains sent to thousands of laboratories worldwide that participate in influenza monitoring for purposes of calibration. This error was potentially catastrophic because human beings born since the late 1960s do not have immunity to this strain (CDC, 2005).

The threat of future pandemics of influenza A is ever present because of the propensity of influenza to reassort or mutate into new forms for which humans do not have immunity. The influenza virus (as well as its control and treatment methods, such as vaccination, quarantine strategies, and antiviral treatments) remains an important area of research and development. Our knowledge is accelerating so rapidly that no textbook can remain current for long. Most of what is known about the 1918 influenza virus has been discovered in the past 5 years. Interested readers should consult the resources listed at the end of the chapter.

4. RECENT OUTBREAKS

By the 1950s, scientists were optimistic that advances in disease control would conquer the problem of communicable diseases (Lasker, 1997). Subsequent events, however, awakened a realization that infectious diseases will never be fully controlled (Lederberg J., 1992; CDC, 1998), as the outbreaks we describe next illustrate dramatically.

4.1. Lyme Disease (1975)

In the town of Lyme, Connecticut, in 1975, two children were diagnosed with the same rare disease—juvenile rheumatoid arthritis. Their mothers became aware of other similarly affected children in their community (American Museum of Natural History, 1998) and notified the local health department about this unusual circumstance. The local health department, suspecting the emergence of a new infectious disease, asked Dr. Allen Steere to investigate. His study of the children of Lyme produced several clues: the disease did not appear to spread from one person to another, it occurred most often in the summer when insect-borne disease was more common, and a rash often appeared before children developed arthritic symptoms, suggesting a tick-borne disease. In 1981, entomologist Willy Burgdorfer found the cause by looking at the digestive tracts of *Ixodes* ticks under a microscope. The bacterium he found, *Borrelia burgdorferi,* was named in his honor.

We now know that *B. burgdorferi,* and other closely related organisms, exist on many continents and likely have been causing disease in humans for hundreds of years. One reason for the unusual number of cases in Lyme was the fashionable practice of building new homes in wooded locations. This practice brought mice (a carrier of *B. burgdorferi*) and deer, which are necessary for tick survival, and humans (tick food) into proximity and created conditions in which the bacteria could infect humans in large numbers.

Lyme disease in humans is an example of a vector-borne disease. Malaria, the most prevalent of the vector-borne diseases, causes 1.5 to 2.7 million deaths annually, mostly in third world countries, and remains a major world health problem (Southwest Foundation for Biomedical Research, n.d.). Biosurveillance for vector-borne diseases is complex and involves monitoring the interactions of humans, animals, and insects.

4.2. Legionnaire's Disease (1976)

One Sunday in 1976, an American Legion official contacted Dr. Lewis D. Polk, director of the Philadelphia Health Department to report the deaths of eight veterans who had participated in a recent American Legion convention in the city (Lewis D. Polk, personal communication). The initial investigation found neither a causative organism nor a common source of exposure, and attendees of the convention continued to die (CDC, 1976). Philadelphia was celebrating the nation's bicentennial, leading to speculation that a chemical or biologic attack had occurred.

The strongest clue available was the Bellevue Stratford Hotel, one of the convention hotels. Many victims visited the hotel during the course of the convention or had walked within a single block of the hotel. However, it was puzzling that individuals who worked at the hotel did not get sick. Investigators compared individuals, both sick and healthy, who had some contact with the hotel. This analysis revealed only that the attack rate was higher in smokers and individuals who spent more time in the hotel lobby.

The cause of this outbreak was not found until 6 months later, when, in January 1977, researchers at the Centers for

Disease Control and Prevention (CDC) discovered a new class of bacteria by examining microscopic slides of guinea pigs injected with lung tissue from deceased patients. They named the bacteria *Legionella pneumophila.* Investigators believe that the water cooling tower on the roof of the hotel was the source of bacteria for the Philadelphia outbreak. Water from cooling towers sprays into the air as microdroplets, which may enter the body by inhalation. In the end, this outbreak caused an estimated 180 cases and 29 deaths (CDC, 1997).

Subsequent research has demonstrated that *L. pneumophila* had been causing outbreaks of pneumonia throughout the United States for years. Researchers thawed and tested serum samples saved from earlier outbreaks of pneumonic illness for which a cause had never been found. They found that *L. pneumophila* caused the outbreak of "Pontiac fever" in Pontiac, Michigan, in 1968 (CDC, 1997). A second large outbreak occurred in 1966 in a psychiatric hospital, causing 94 cases and 15 deaths (CDC, 1997).

Today, we know that *L. pneumophila* exists in nature as a common colonizer of water systems but rarely infects people. The American Legion convention brought together a large number of people whose lung defenses were weakened by age and the effects of cigarette smoking.

Effective measures to prevent, detect, and treat Legionnaire's disease now exist, so an outbreak from this organism with this degree of mortality is unlikely to occur again. However, outbreaks and sporadic cases secondary to aspiration of drinking water or inhalation of contaminated aerosols continue to occur (Pedro-Botet et al., 2002; Yu, 2002).

4.3. Anthrax Outbreak, Sverdlovsk, Soviet Union (1979)

In April 1979, people, and many animals, were dying in the Soviet city of Sverdlovsk from an unknown illness (Guillemin, 1999). The first clue that the organism causing the illness was *Bacillus anthracis* came from a pathologist, who found that the brain of a victim showed a cardinal's cap—bleeding at the top of the brain. A cardinal's cap is a pathognomic finding (a finding that allows a physician to conclude a diagnosis with certainty) for the disease anthrax (Mangold and Goldberg, 2000).

Soviet authorities maintained that the cause was consumption of contaminated meat. Although the Soviet Union had signed the Convention on the Prohibition of the Development, Production, and Stockpiling of Bacteriological (Biological) and Toxin Weapons and on their Destruction—effective March 26, 1975 (Goldblat, 1997)—experts in the West remained suspicious of the official explanation.

More than a decade later, a team lead by Mathew Meselson (Meselson et al., 1994) visited Sverdlovsk (then renamed Yekaterinburg), to interview survivors, witnesses, and pathologists. They reviewed tissue samples of lung and intestinal tract taken from victims of this outbreak. The tissue samples were more consistent with an inhalational form of anthrax rather than a gastrointestinal form, which would be expected if the infection were acquired by eating contaminated meat. They, therefore, postulated that the cause was airborne release of anthrax. To test this hypothesis, they analyzed the geographical distribution of sick individuals and the known wind conditions, concluding that the outbreak was most likely due to an airborne release of anthrax from a single location.

In 1992, Russian authorities admitted that there had been an accidental release of *B. anthracis* (*Kirov* strain) from Soviet Biological Weapons Compound 19. During a shift change, a laboratory technician had removed an air filter from a testing chamber, which the subsequent shift failed to replace (Israelyan, 2002). As a result, this outbreak was caused by an accidental release of material being tested for use in biological weaponry.

This outbreak highlights the difficulties of controlling disease in a world where bureaucratic organizations reward secrecy and restrict information. It is an interesting but unanswered question as to what extent their attempts to cover up the truth affected the morbidity and mortality of this outbreak. Because anthrax is not contagious, secrecy did not result in dissemination of disease outside of Sverdlovsk. However, in the words of Victor Israelyan, former Soviet ambassador, "Such a lack of transparency can have dire consequences in this new era of terrorism" (Israelyan, 2002).

4.4. AIDS (1981)

In 1981, doctors in Los Angeles, San Francisco, and New York City began to encounter homosexual patients with rare infections and diseases, including cytomegalovirus, *Pneumocystis carinii,* and Kaposi's sarcoma (CDC, 1981a,b; Drew et al., 1981). These rare diseases tend to occur in immunocompromised patients, a clue that pointed investigators to look for an infection or chemical exposure that damaged or interfered with cellular immunity (CDC, 2001).

A CDC team formed in June 1981. They named the disease acquired immunodeficiency syndrome (AIDS) and developed a working case definition. They used the working case definition to find infected individuals to interview and examine. They compared the behaviors and histories of these individuals with those of non-infected individuals, and concluded that AIDS could be contracted from blood transfusion, intravenous drug use, or sexual intercourse or could be passed from mother to child in utero. This information was the basis for the March 1983 recommendations on how to prevent infection, disseminated well before the actual cause of the disease—the human immunodeficiency virus (HIV)—was discovered in 1984 (Srikameswaran, 1999).

Subsequent research has elucidated that HIV existed long before AIDS was first recognized in 1981. The evidence indicates that HIV developed through mutation or recombination in monkeys before crossing over into humans. The earliest known human case, as determined by testing stored serum and tissue, was an adult living in the Democratic Republic of Congo in 1959 (Kanabus and Allen, 2005). In the absence of international travel and high-risk behaviors (multiple sex partners and sharing of needles), which produced clusters of cases in San Francisco and New York City, HIV could have remained undetected for many years, with the resultant deaths blending in with many infectious-disease deaths in areas with little health care.

The AIDS pandemic is an example of an outbreak that has never been brought under control. The Joint United Nations Program on HIV/AIDS estimates that 20 million people have died from AIDS, and an additional 38 million people are living with HIV. During 2003—19 years after the virus was isolated—an estimated five million people were newly infected with HIV (UNAIDS, 2004).

Two highly prevalent human infectious diseases—HIV and hepatitis C (Koop, 1998)—pose difficult biosurveillance problems because they have a long period of infectivity before the onset of symptoms. It is hoped that future advances in biotechnology will yield better methods for detecting infections during their asymptomatic periods.

4.5. Mad Cow Disease and Variant Creutzfeldt-Jakob Disease (1986)

In 1986, veterinary pathologists at the Central Veterinary Laboratory in Weybridge, United Kingdom, became suspicious that a new disease was killing cattle (Donnelly et al., 1999; Matravers et al., 2000). Their suspicions were aroused by the microscopic appearance of the brain from a cow that died after exhibiting progressive abnormalities of behavior and movement that resembled those of sheep affected with the disease *scrapie*. Because of the unnerving signs of this disease, it came to be known as mad cow disease.

When pathologists saw similar microscopic findings in two more cattle, they feared that a scrapie-type disease had emerged in cattle. The head veterinary epidemiologist for the United Kingdom, John Wilesmith, commissioned studies to find the origin of this new disease. As a starting point, investigators assumed the new disease was caused by an organism similar to the one that causes scrapie in sheep. Moreover, they hypothesized that an unusual transmission pathway—the eating of brains from infected animals—caused the spread of the disease. In the cattle industry, carcasses of livestock that are not fit for human consumption were rendered (cooked) into meat-and-bone meal, which was fed to animals as a protein supplement.

Their methods involved a series of observational studies and computer simulations. These studies suggested that a mass exposure of the cattle population to a new organism was occurring—most likely beginning in winter 1981/1982. They found that the risk of exposure was 30 times greater for one-month-old dairy calves than for adult cattle. This discovery, coupled with knowledge that dairy farmers remove calves from their mothers at one day of age and feed them powdered milk and high-protein supplements based on meat-and-bone meal, confirmed their conjecture about the route of exposure. The incidence of affected dairy herds also increased as herd size increased. Larger herds require more supplement feeds than do smaller herds, thereby increasing the risk of exposure.

Although cattle had been resistant previously to scrapie, it is now believed that a cow became infected with a mutant strain of the scrapie agent some time during the 1970s. The sick cow was recycled into meat-and-bone meal, resulting in the exposure of a large number of animals to the newly adapted scrapie agent.

Mad cow disease (more properly referred to as *bovine spongiform encephalopathy*) remained undetected for many years after the initial mutation in the 1970s. The delay in detection allowed the disease to establish within the cattle population. The conditions for disease transmission were broken in 1988, when regulators banned the use of animal protein sources for animal feeds, and the epidemic in cattle gradually abated.

For many years, however, cattle that were infected, but not yet symptomatic, entered the human food supply, resulting in a mass exposure of the human population to this new agent. This exposure set up conditions that allowed the disease to cross from cattle into humans. The public were aware of possible exposure to the agent through consumption of contaminated beef; therefore, when a new variant of Creutzfeldt-Jakob disease (vCJD) first appeared in humans in the United Kingdom during the 1990s, people saw the similarity to mad cow disease in cattle (Nathanson et al., 1997).

The total number of confirmed cases of human vCJD, from first identification to 2005, now approaches 150. The exact number of humans who will develop the disease is unknown. Additional routes of disease transmission (e.g., person to person through blood transfusion) are contributing to uncertainty in disease projections. Figure 2.4 shows a bovine spongiform encephalopathy epidemic curve.

The story of mad cow disease and vCJD demonstrates the need for better methods of biosurveillance of animals for the protection of humans. Although the investigation of mad cow disease occurred at breathtaking speed, the initiation of this

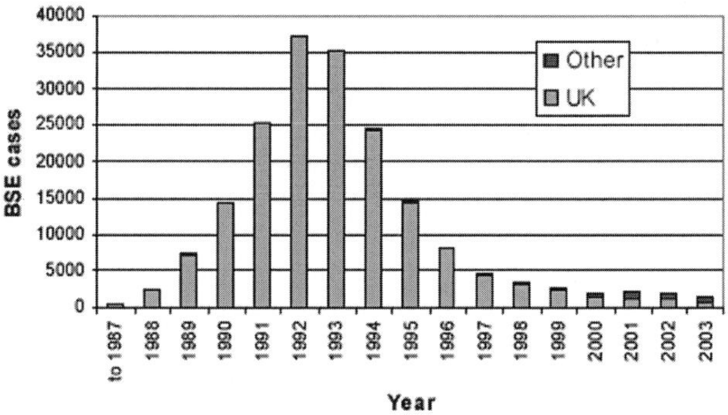

FIGURE 2.4 Bovine spongiform encephalopathy (BSE) epidemic curve for the United Kingdom (UK) and other countries. (Modified from European Commission, 2004.)

investigation only occurred in 1986, after the first discovery of the disease within the cattle population. It is very likely the disease existed undetected within the cattle population for at least a decade prior to this discovery. This prolonged period of cover allowed the outbreak to proceed unchecked, resulted in significant exposure of the human population, and contributed to movement of the disease into humans in the form of vCJD. Earlier discovery of the outbreak in cattle would have limited its economic impact and prevented it from spilling over into the human population.

4.6. Cryptosporidiosis Outbreak (1993)

The 1993 outbreak of cryptosporidiosis that occurred in Milwaukee is a striking example of how difficult it is for current biosurveillance systems to detect outbreaks in a timely manner. This outbreak, caused by a breakdown in the water filtration process at a water supplier, sickened an estimated 403,000 individuals (Mac Kenzie et al., 1994). Many individuals were sick by the time that public health authorities initiated an investigation, based on reports of widespread absenteeism among hospital employees, students, and schoolteachers owing to gastrointestinal illness. On April 7, laboratory tests confirmed the cause to be the parasite *Cryptosporidium parvum* (Mac Kenzie et al., 1994), and a boil-water advisory was issued.

Retrospectively, earlier indicators were an increase in sales of diarrhea remedies, noticed by a pharmacist on April 1 (also apparent in sales figures for sales of such products), and an increase in diarrhea-related calls to area nurse hotlines on April 2 (Rodman et al., 1997, 1998).

Attention to water treatment surveillance could have prevented this outbreak. The investigation discovered an improperly installed streaming current monitor and turbidity readings that were clearly elevated during the period when contaminated water was being supplied to the population (Mac Kenzie et al., 1994). Corso et al. (2003) estimated the total economic impact of this outbreak to be $96 million.

This largest of all U.S. waterborne outbreaks (Mac Kenzie et al., 1994) points to the need to augment current biosurveillance methods with methods that can detect large numbers of sick individuals who may not seek medical care. Cryptosporidiosis is a self-limited disease in immune-competent hosts. Most people do not get sick enough to go to a physician, the traditional discovery point for cases of diseases in biosurveillance.

4.7. Nipah Virus (1998)

In 1998, a province in Malaysia began reporting to Malaysian health authorities what were then felt to be cases of Japanese encephalitis. When control measures for Japanese encephalitis were less effective than expected, authorities became suspicious that the disease might not be Japanese encephalitis: The mortality from this disease was higher than expected (approximately 30%), cases occurred predominantly in adults instead of children, and there was seemingly no protection provided from vaccination for Japanese encephalitis or from mosquito control. Then, two additional clusters of disease developed—one in another region of Malaysia and one in abattoir workers in Singapore. At the same time, swine within Malaysia were also becoming ill. Veterinary authorities suspected the disease in swine to be classical swine fever (Chua, 2003).

Medical investigators observed an apparent association between the human disease and the swine disease. Many people affected by the disease worked on pig farms, and the infected abattoir workers in Singapore had recently processed pigs from one of the affected regions of Malaysia. These clues prompted investigators to undertake viral isolation studies from human victims. The emergence of a new disease was confirmed when a novel paramyxovirus was isolated from the brain of a human patient in 1999. By this time, the disease had caused illness in 300 people and more than 100 fatalities.

Further epidemiological studies indicated that the sick humans contracted their disease directly from swine. There was no evidence of person-to-person spread (Heymann, 2004). The commercial movement of infected pigs allowed the disease to establish within the second region of Malaysia and transfer into humans in Singapore.

Authorities believe this outbreak arose from the spillover of virus from its normal victims—the Pteropid fruit bat—into swine, and then from swine into man. No fundamental change to virus pathogenicity is believed to have preceded swine and human infection, just a change in circumstance that provided opportunity for swine to become exposed. The opportunity for infection to cross two species arose because a reduction in native forest caused by drought and slash-and-burn deforestation practices caused the migration of large numbers of fruit bats into orchards. At the same time, there was an increase in the number of orchards that were also used for swine farming. Swine were housed under the trees and fed fruit not suitable for marketing. This colocation of fruit bats and swine allowed this previously undescribed paramyxovirus virus to infect swine and then the human population (Chua, 2003).

4.8. Foot-and-Mouth Disease (2000)

Foot-and-mouth disease (FMD) is a highly contagious viral disease of cloven-hoofed animals such as cattle, deer, sheep, and swine. It is endemic in many parts of Asia, South America, Africa, and the Middle East. The virus can persist in contaminated material for prolonged periods. Outbreaks of FMD in regions previously free of the virus occur regularly around the world, resulting in significant threats to susceptible livestock production systems and requiring expensive, disruptive, and complex eradication programs. Fourteen countries from South America, the Middle East, Africa, Asia, and Europe have had or are experiencing a FMD outbreak.

One outbreak of FMD occurred in Korea in 2000. In early April, animals on a farm in North Korea grew lame rapidly, developed vesicles of the feet and mouth, and stopped eating. Veterinarians were concerned that FMD may have re-entered South Korea after an absence of nearly 70 years, and testing was undertaken. The positive FMD laboratory results led to movement restrictions for all livestock within a 20-km radius of an affected farm, the closure of all animal sale yards and abattoirs within the country, and the veterinary inspection of all farms within the affected region for signs of disease. Then slaughter of all animals from affected farms began. In addition, authorities took the controversial step of vaccinating animals from neighboring farms without disease in order to eliminate the infection and reduce its ability to spread from the region. Despite these interventions, over the next two months, the disease spread to two beef cattle farms situated 100 km away and to a dairy farm 150 km away from the initial outbreak. These events intensified fears that disease would soon enter the swine population and, with the presence of vaccinated animals, evade detection, thereby allowing FMD to become endemic within South Korea.

Control activities intensified. The army was deployed to operate checkpoints and to control animal movements; veterinary experts from around the world were employed and vaccination programs expanded. The slaughter of more than 2,000 cattle and pigs, vaccination of more than 1.5 million cattle and pigs, and the employment of 600 veterinarians restricted the outbreak. The inspection in 2000–2001 of 650,000 animals outside the vaccination zone did not find disease.

The virus causing FMD does not currently infect humans. The impact on human health is felt through its economic impact on human livelihood and trade. A recent outbreak of FMD in the United Kingdom in 2001 received enormous press attention and had severe economic impact on many industries, including agriculture and tourism (Scott et al., 2004). The organism causing FMD can be carried by the wind for distances up to 100 km, as discussed in Chapter 19.

4.9. Severe Acute Respiratory Syndrome (2003)

When Johnny Chen[1]—a Chinese-American businessman based in Shanghai—arrived in Vietnam on February 24, 2003, there was no indication this routine business trip would be different from any other. Chen worked for Gilwood Company, a small New York garment firm, and went to Hanoi to inspect the work on the blue jeans being manufactured by a local contractor (Cohen et al., 2003). To all appearances, Chen was the picture of health.

By February 26, the 49-year-old Chen was so gravely ill his colleagues rushed him to the Hanoi French Hospital. At the same time, Liu Jianlun, a 64-year-old Chinese medical professor, was already dying in a hospital in Hong Kong. The two unrelated and unacquainted men were suffering from a virulent respiratory ailment that resembled pneumonia and was characterized by high fever, cough, and body aches.

Jianlun had traveled to Hong Kong on February 20 to attend a wedding. While there, he stayed on the ninth floor of the Metropole Hotel. So had Johnny Chen.

Jianlun died in a hospital isolation ward on March 4, 2003. On March 5, after a week of ineffective treatment and at his family's request, Chen was moved to a facility in Hong Kong. He died eight days later.

What their doctors and the world health authorities did not know then was that Chen and Jianlun represented index cases in the outbreak of a new and lethal disease, later known as severe acute respiratory syndrome (SARS).

Figure 2.5 shows the chain of transmission linking these index cases with outbreaks of this disease that occurred throughout the world. Perhaps the most important chain is the Hanoi chain. On February 28th, the Hanoi French Hospital contacted Dr. Carlo Urbani, an infectious disease specialist with the World Health Organization (WHO), because physicians suspected Chen was infected with avian influenza (Reilley et al., 2003). Dr. Urbani and the staff at the hospital worked swiftly, instituting isolation measures and collecting respiratory and blood samples. Nevertheless, by March 10, 22 hospital workers in Hanoi had contracted the illness (Heymann, 2003).

On March 12, the WHO issued an unprecedented global alert regarding cases of atypical pneumonia in Vietnam, Hong Kong, and China. As a result of the alert, Toronto hospital emergency and infectious disease physicians (Varia et al., 2003) recognized that they had a case of SARS in their hospital, the son of case F (Figure 2.5), an older women who died at home on March 5.

The son had been admitted to the hospital on March 7 and placed under isolation with contact and droplet precautions on March 10. He died shortly after the alert was issued. Family members who visited him in the hospital became ill and were admitted to three Toronto hospitals on March 13. These relatives were placed under airborne, droplet, and contact precautions; these isolation measures were effective and no further disease transmission from these individuals occurred.

What doctors did not realize was that the son of case F had received nebulization treatment in the emergency room on March 7 that had facilitated the infection of a number of other patients. No one realized that these patients had SARS when they began developing fever and cough. Infection control experts

1 As a general practice, public health professionals do not reveal the names of victims or businesses involved in outbreaks unless necessary to prevent morbidity and mortality. In most cases confidentiality is protected by law and in others it is a professional commitment. An exception was made in this chapter for those names that are independently a part of the historical public record. No information from confidential investigations is included in this chapter.

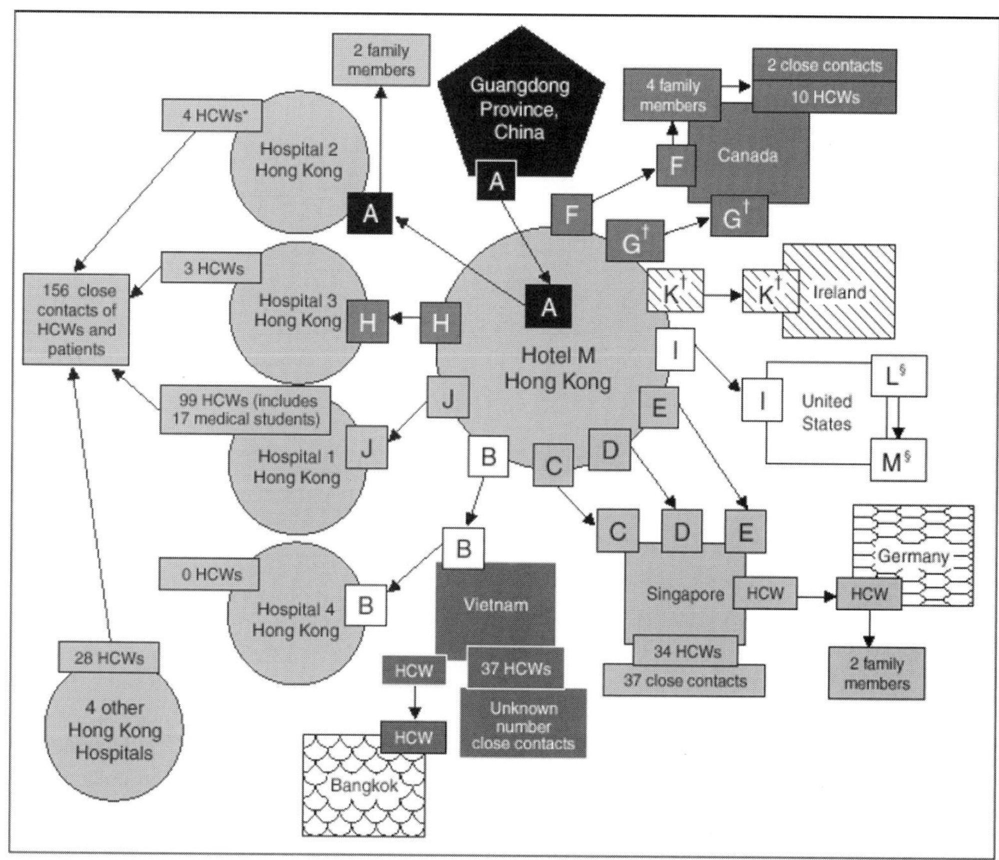

* Health-care workers.
† All guests except G and K stayed on the 9th floor of the hotel. Guest G stayed on the 14th floor, and Guest K stayed on the 11th floor.
§ Guests L and M (spouses) were not at Hotel M during the same time as index Guest A but were at the hotel during the same times as Guests G, H, and I, who were ill during this period.

FIGURE 2.5 Chain of transmission among guests at Hotel M, Hong Kong, 2003. (From CDC, 2003c.)

only gradually realized the extent of the problem when these patients infected additional patients, visitors, and hospital staff.

On March 22, the hospital infection control practitioners (ICPs) implemented contact and droplet precautions for all patients in the intensive care unit (ICU); the next day the ICU and emergency department were closed. By March 24, ICPs realized that SARS was potentially anywhere in the hospital and closed the hospital to new admissions, discharged patients, and quarantined off-duty staff at home.

ICPs considered every person in the hospital as potentially infectious. Staff wore equipment such as special masks at all times, even when interacting with other healthy-appearing staff. ICPs monitored the health of the hospital patients and staff to identify who was infected, how they became infected, and at what point in the course of the illness they were most infectious. This important information was quickly shared with the rest of the world. The Toronto investigation ultimately identified a total of 128 persons as probable or suspected cases of SARS, 17 of whom died.

Singapore (Gamage et al., 2005), Hong Kong (Lee and Sung, 2003; Joynt and Yap, 2004), and Taiwan (Hsieh et al., 2004) all experienced high rates of transmission of infection in healthcare facilities. Figure 2.6 is a picture of a Taiwanese emergency room taken several months after the end of the SARS outbreak, showing the level of precaution that existed.

By April 17, 2003, a pan-national research effort involving health care, academia, and governments identified a previously unknown coronavirus as the cause of SARS (Heymann, 2003) Isolation of the organism was the first step in developing a diagnostic test for SARS.

By the time this pandemic was brought under control (July 2003) there had been 8096 cases of SARS with 774 deaths (WHO, 2003b). Dr. Urbani, one of 1076 healthcare workers infected with SARS, died on March 29, 2003 (Reilley et al., 2003).

This outbreak highlights several important aspects of biosurveillance. First, hospitals may be agents for the spread of infectious diseases because they welcome all patients, including those who are infectious, immune compromised, or both.

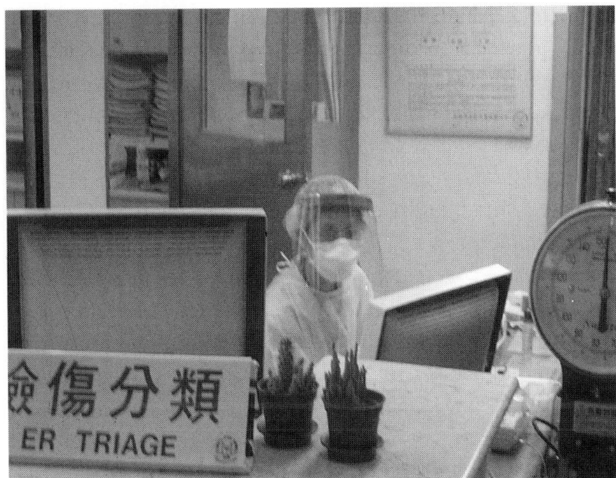

FIGURE 2.6 Photo of a Taiwanese emergency room taken several months after the end of the SARS outbreak.

By concentrating individuals with communicable diseases and individuals who are especially susceptible to communicable diseases, hospitals provide an ideal environment for the spread of these diseases. Effective infection control procedures are the first step in preventing transmission of infection. However, infection control measures sometimes break down or may simply be ineffective for novel agents. Surveillance for infections transmitted in hospitals (nosocomial) is an important component of biosurveillance.

Second, to its credit, the WHO had issued, on February 11, 2003, a routine communicable disease surveillance report about an outbreak of acute respiratory syndrome with 300 cases and five deaths in Guangdong Province (WHO, 2003a) This report was widely disseminated because of the capabilities of world telecommunication, and information about the SARS outbreak was shared quickly among scientists, public health officials, hospital infection control personnel, and clinicians. Communication technology played an important role in disseminating biosurveillance information worldwide.

Finally, the downside of technology is that every corner of the world is accessible to an infectious agent. Airplanes are capable of unintentionally carrying infectious agents half way around the world in less than a day. And because one day is less than the incubation period of most infectious diseases, it is exceedingly difficult, with presently available techniques, to prevent individuals infected in one city from infecting any other city on the planet. The implications of this fact for the organization of biosurveillance activities is profound.

4.10. The Largest U.S. Food-borne Hepatitis A Outbreak (2003)

During the week of October 26, 2003, Dr. Marcus Eubanks, an emergency room physician in Beaver County, Pennsylvania, treated six patients who had symptoms of hepatitis. The wife of the sixth patient told him that she was aware of three other individuals with similar symptoms with whom she had dined at a local restaurant. Dr. Eubanks notified the Pennsylvania Department of Health of this unusual event (Snowbeck, 2003).

Figure 2.7 shows the epidemic curve for this outbreak. A case-control study found that most of the infected restaurant patrons ate food items containing green onions (CDC, 2003a), and most were just becoming ill at the time the outbreak was detected. By the time the investigation was completed, 660 cases and three deaths were associated with the outbreak. The Pennsylvania Department of Health provided hepatitis A immune globulin to more than 9400 restaurant patrons and close contacts of infectious individuals in order to prevent additional cases of hepatitis A (Hersh, 2004).

The Food and Drug Administration conducted a formal trace-back investigation (Figure 2.8), which led to farms in Mexico (Food and Drug Administration, 2001). The investigation team found multiple sanitation problems on these farms (Food and Drug Administration, 2003).

This outbreak illustrates the need for monitoring that extends internationally to encompass the global food chain and the difficulty of ensuring high standards thousands of miles from the final consumers of food.

4.11. Severe Acute Respiratory Syndrome (2004)

The SARS story did not end in 2003. On April 5, 2004, a 20-year-old Beijing nurse developed a cold, a fever, and a cough (China, 2004). By April 7, she was sick enough to be admitted to a hospital. She did not improve, and on April 14, she was transferred to a second hospital, where she was placed into the ICU, and a SARS test was ordered. On April 22, an initial test returned positive. Hospital and public health staff did not wait for a confirmatory test to act. This nurse had been in two hospitals, and she could have infected many individuals in those hospitals. ICPs traced every person who had contact with the nurse to determine the source of her infection and to identify individuals who might be incubating the disease. One hundred seventy-one people were placed under observation (WHO, 2004).

One of the nurse's patients had been a medical student, who traveled to another province—Anhui—after discharge. Investigators found that the mother of this medical student had died in Anhui on April 19, 2004, after caring for the sick medical student. The medical student had recently worked with what was ostensibly inactivated SARS virus at the national research laboratory (ProMED-mail, 2004). Three other workers in her laboratory were tested and found to be infected.

This outbreak, which originated in a laboratory, is not unique. The last two cases of smallpox were the result of a laboratory exposure (Hogan et al., 2005), and the Boston Public Health Commission recently investigated a November 2004 outbreak of tularemia that originated in a laboratory (Barry, 2004). These and other outbreaks point to the need for enhanced surveillance of all individuals who work in laboratories where highly infectious agents are used or stored.

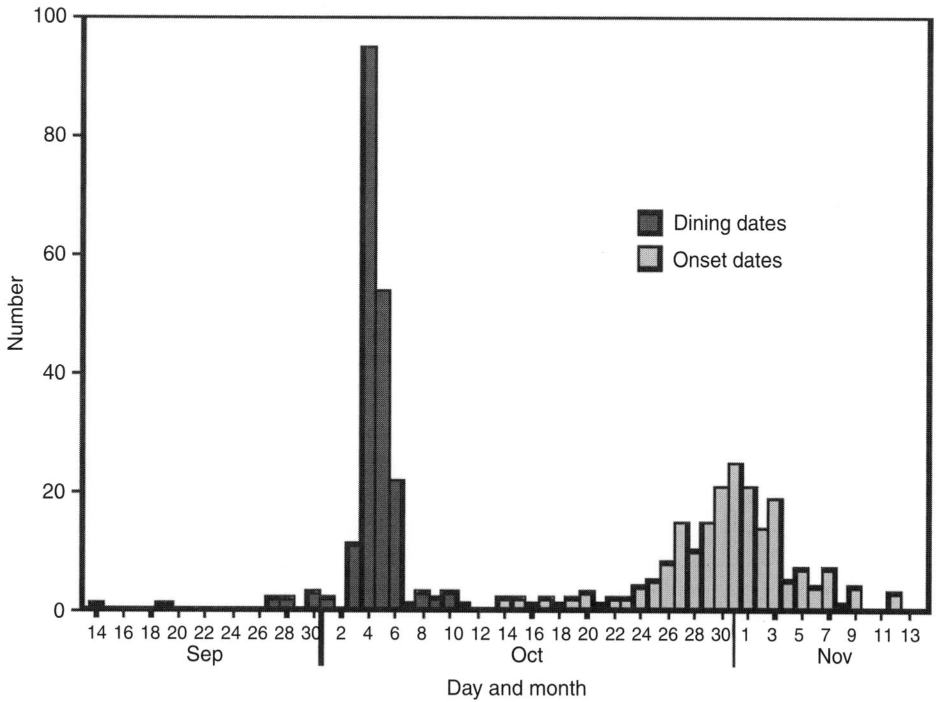

FIGURE 2.7 Number of hepatitis A cases by date of eating at restaurant A and illness onset, Monaca, Pennsylvania, 2003. (From CDC, 2003a.) * N = 206. Excludes one case-patient whose illness onset date was not available. Dining dates for three persons who ate at Restaurant A on October 15 (n=one) and October 17 (n=two) are not shown.

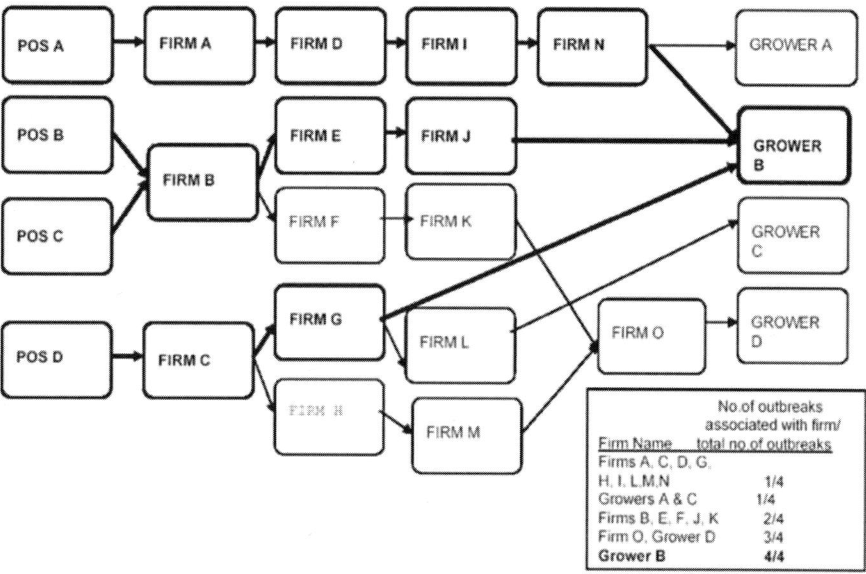

FIGURE 2.8 Sample trace-back investigation. (From Food and Drug Administration, 2001.)

5. DEFINITIONS OF "OUTBREAK" AND "EPIDEMIC"

So far in this book, we have used the terms *outbreak* and *epidemic* without definitions because they are commonly understood terms and a technical definition was not necessary. However, for completeness, here we provide more technical definitions.

Turnock (2001) defines an epidemic as "The occurrence of a disease or condition at higher than normal levels in a population." *Last's Dictionary of Epidemiology* defines an outbreak as "An epidemic limited to localized increase in the incidence of a disease, for example, in a village, town, or closed institution; upsurge is sometimes used as euphemism for outbreak" (International Epidemiological Association, 1995). Teaching material from the CDC considers the terms *outbreak* and *epidemic* synonymous but notes that public health practitioners might consider using the term *outbreak* in communications with the public to avoid causing panic (CDC, 2003b). In this book, we use the term *outbreak.*

We note that all definitions of *outbreak* (or *epidemic*) that we have encountered in published sources include a subjective element—"higher than normal levels in a population." In public health practice, it seems that an investigator or a health department will classify an exceedance as an outbreak when the expected impact on the health of a human or animal population—if the exceedance is not investigated—is greater than the cost and effort of investigation (ideally) or is greater than the priority of using those resources in some other way. This determination seems to depend not only on the magnitude of the exceedance but on the nature of the illness and other information that is available about the set of affected individuals.

Thus, the current published definitions of outbreak and epidemic are necessary (i.e., there must be an exceedance) but not sufficient (some exceedances do not qualify) conditions for labeling an exceedance an outbreak. Scientific readers may feel somewhat uncomfortable to realize that a key concept in this area of scientific study is not defined in unambiguous terms. In our experience, this ambiguity does not undermine the foundations of research. It is simply something to be aware of.

6. SUMMARY

Microbial organisms know no artificial boundaries of human society, and neither do many outbreaks. Novel and/or previously unknown organisms can come from a wide range of reservoirs (e.g., animal, insect, water, or soil). Organisms can mutate and jump from one animal species to another and result in outbreaks that can quickly spread around the world.

The outbreaks that we described identify many of the challenges of biosurveillance. HIV and mad cow disease illustrate the difficulty in detecting outbreaks of disease for which there is a long period of asymptomatic transmission of infectious agents. The hepatitis A and Nipah virus outbreaks demonstrate how infectious agents can be transmitted across national borders via the complex global food distribution system. SARS illustrates the problems faced by hospitals and other facilities housing sick and vulnerable individuals. The cryptosporidiosis outbreak shows the importance of water surveillance and how there can be delays in detection even for an extremely large outbreak when many individuals are not sick enough to go to hospitals. Finally, the anthrax outbreak demonstrates what can go wrong when humans intentionally manipulate organisms and, in addition, suppress information needed to control the outbreak.

Detecting and characterizing outbreaks requires enormous amounts of resources, skill, and knowledge. Mitigation of the dire consequences of outbreaks depends on rapid detection and characterization. The initial detection or key clue to understanding how to respond to the outbreak can come from many sources—astute citizens, physicians, veterinarians, laboratory investigations, autopsies, and pharmacists. In many cases, detection occurs late, with substantial cost in human suffering. The need for global cooperation and support for individual countries cannot be overstated because some countries simply do not have the resources to participate in biosurveillance.

The mortality curve in Figure 2.1 and the success in controlling recent outbreaks, such as SARS, tell a story about the developed world's success in combating microbes. There is little doubt that if the many individuals involved in biosurveillance worldwide were to abruptly cease in their efforts, a spike in mortality at least as large as that experienced in 1918 would be sure to follow. We need only look at developing countries to see the carnage that would result. Better methods are both necessary and possible.

ADDITIONAL RESOURCES

For information about current or recent outbreaks, consult the following resources.

- ProMEDmail (*http://www.promedmail.org*) is the best source for up-to-the-minute information from around the world. Sign up for e-mail alerts.
- The Clinician Registry for Terrorism and Emergency Response Update (*http://www.bt.cdc.gov/clinregistry/index.asp*) delivers timely e-mails on emergent issues, including infectious diseases, from a CDC service.
- *The Morbidity and Morality Weekly Report (http://www.cdc.gov/mmwr/)* provides epidemiologic information on on-going and recently completed investigations.
- *The World Health Organization Disease Outbreak News* (*http://www.who.int/csr/don/en/*) provides official information from around the globe, including infectious diseases.
- *Control of Communicable Diseases Manual* by David L. Heyman, ed. (Washington DC: American Public Health Association, 2004).

The following books examine how infectious diseases shaped the course of human history and the development of civilizations.

- *Guns, Germs, and Steel: The Fates of Human Societies* by Jared Diamond (New York: Norton, W. W. & Company, 1997).
- *Plagues and Peoples* by William McNeill (New York: Doubleday Books, 1977).

These two volumes offer detailed and accessible glimpses into the world of infectious disease and outbreak detection.

- *Control of Communicable Diseases Manual* by David L. Heyman, ed. (Washington, DC: American Public Health Association, 2004).
- *Eleven Blue Men and Other Narratives of Medical Detection* by Berton Roueche (Boston: Little Brown & Co, 1953).

ACKNOWLEDGMENTS

We would like to thank Daphne Henry for her patience, editing, and enhancing the readability of this chapter by providing the quote from Ernest Hemingway, as well as historical information for the SARS section. We would also like to thank William Hogan and Marian Pokrywka ICP for assistance with outbreak descriptions.

REFERENCES

American Museum of Natural History. (1998). Lyme on the Louse: What's Inside this Tick. *BioBulletin http://sciencebulletins.amnh.org/biobulletin/biobulletin/story986.html.*

Barry, J. M. (2004). The Site of Origin of the 1918 Influenza Pandemic and Its Public Health Implications. *Journal of Translational Medicine* vol. 2:3.

Centers for Disease Control and Prevention [CDC]. (1976). Respiratory Infection–Pennsylvania. *Morbidity and Mortality Weekly Report* pp. 244.

CDC. (1981a). Kaposi's Sarcoma and *Pneumocystis* Pneumonia among Homosexual Men–New York City and California. *Morbidity and Mortality Weekly Report* vol. 30:305–308.

CDC. (1981b). *Pneumocystis* Pneumonia–Los Angeles. *Morbidity and Mortality Weekly Report* vol. 30:250–252.

CDC. (1997). Respiratory Infection–Pennsylvania. 1977. *Morbidity and Mortality Weekly Report* vol. 46:49–50.

CDC. (1998). *Preventing Emerging Infectious Diseases : A strategy for the 21st Century.* Atlanta, GA: CDC.

CDC. (1999). Control of Infectious Diseases. *Morbidity and Mortality Weekly Report* vol. 48:621.

CDC. (2001). First Report of AIDS. *Morbidity and Mortality Weekly Report* vol. 50:429.

CDC. (2003a). Hepatitis A Outbreak Associated with Green Onions at a Restaurant–Monaca, Pennsylvania, 2003. *Morbidity and Mortality Weekly Report* vol. 52:1155–1157.

CDC (2003b), p. 24.

CDC. (2003). Update: Outbreak of Severe Acute Respiratory Syndrome–Worldwide, 2003. *Morbidity and Mortality Weekly Report* vol. 52:243.

CDC. (2005). Telebriefing Transcript: Update on Distribution of H2N2 Influenza Strain. *http://www.cdc.gov/od/oc/media/transcripts/t050413.htm.*

China, B. (2004). Suspected SARS Case in Beijing. *http://www.cnn.com/2004/WORLD/asiapcf/04/22/china.sars/index.html?headline?Suspected~SARS~case~in~Beijing.*

Chua, K. B. (2003). Nipah virus outbreak in Malaysia. *Journal of Clinical Virology* vol. 26:265–275.

Cohen, M., Fritsch, P., and Pottinger, M. (2003). Routine Trip Ends with Deadly Illness. *The Wall Street Journal*, March 19.

Corso, P. S., Kramer, M. H., Blair, K. A., Addiss, D. G., Davis, J. P., and Haddix, A. C. (2003). Cost of Illness in the 1993 Waterborne *Cryptosporidium* outbreak, Milwaukee, Wisconsin. *Emerging Infectious Diseases* vol. 9:426–431.

Donnelly, C. A., S., M. and Anderson, R. M. (1999). A Review of the BSE Epidemic in British Cattle. *Ecosystem Health* vol. 5:164–171.

Drew, W. L., Mintz, L., Miner, R. C., Sands, M., and Ketterer, B. (1981). Prevalence of cytomegalovirus infection in homosexual men. *Journal of Infectious Diseases* vol. 143:188–192.

European Commission. (2004). Report on the Monitoring and Testing of Ruminants for the Presence of Transmissible Spongiform Encephalopathy (TSE) in the EU in 2003, Including the Results of the Survey of Prion Protein Genotypes in Sheep Breeds. Brussels: Food and Feed Safety, European Commission.

EyeWitness to History.com. (2001). The Black Death, 1348. *http://www.eyewitnesstohistory.com/plague.htm.*

Food and Drug Administration. (2001). Guide to Traceback of Fresh Fruits and Vegetables Implicated in Epidemiological Investigations. *http://www.fda.gov/ora/inspect_ref/igs/epigde/epigde.html.*

Food and Drug Administration. (2003). FDA Update on Recent Hepatitis A Outbreaks Associated with Green Onions from Mexico. *http://www.fda.gov/bbs/topics/NEWS/2003/NEW00993.html.*

Gamage, B., Moore, D., Copes, R., Yassi, A., and Bryce, E. (2005). Protecting Health Care Workers from SARS and Other Respiratory Pathogens: A Review of the Infection Control Literature. *American Journal of Infection Control* vol. 33:114–121.

Goldblat, J. (1997). The Biological Weapons Convention–An Overview. *International Review of the Red Cross* vol. 318:251–265.

Guillemin, J. (1999). *Anthrax: The Investigation of a Deadly Outbreak.* Berkley, CA: University of California Press.

Hersh, J. (2004). Anatomy of an Outbreak. *http://www.cphp.pitt.edu/upcphp/ppt/Anatomy%20of%20an%20Outbreak-Hersh.ppt.*

Heymann, D. L. (2003). Update 27–One Month into the Global SARS Outbreak: Status of the Outbreak and Lessons for the Immediate Future. Geneva: World Health Organization.

Heymann, D. L., ed. (2004). *Control of Communicable Diseases Manual.* Washington, DC: American Public Health Association.

Hogan, C. J., Harchelroad, F., and McGovern, T. W. (2005). CBRNE–Smallpox. *http://www.emedicine.com/emerg/topic885.htm.*

Hsieh, Y. H., Chen, C. W., and Hsu, S. B. (2004). SARS outbreak, Taiwan, 2003. *Emerging Infectious Diseases* vol. 10:201–206.

International Epidemiological Association. (1995). *A Dictionary of Epidemiology.* Oxford: Oxford University Press.

Israelyan V. (2002). Fighting Anthrax: A Cold Warrior's Confession. *The Washington Quarterly* vol. 25:17–29.

The Joint United Nations Program on HIV/AIDS [UNAIDS]. (2004). Report on the Global AIDS Epidemic–Executive Summary. Geneva: UNAIDS.

Joynt, G. M. and Yap, H. Y. (2004). SARS in the Intensive Care Unit. *Current Infectious Disease Reports* vol. 6:228–233.

Kanabus, A. and Allen., S. (2005). The Origins of HIV and AIDSand the First Cases of AIDS. *http://www.avert.org/origins.htm.*

Koop, C. E. (1998). Hepatitis C–An Epidemic for Anyone. *http://www.epidemic.org/.*

Lasker, R. D. (1997). *Medicine and Public Health: The Power of Collaboration.* New York: The New York Academy of Medicine.

Lederberg J., Shope, R. E, and Oaks, S.C., eds. (1992). *Emerging Infections: Microbial Threats to Health in the United States.*, Washington, DC: National Academy Press.

Lee, N. and Sung, J. J. (2003). Nosocomial Transmission of SARS. *Current Infectious Disease Reports* vol. 5:473–476.

Mac Kenzie, et al. (1994). A Massive Outbreak in Milwaukee of *Cryptosporidium* Infection Transmitted through the Public Water Supply. *New England Journal of Medicine* vol. 331:161–167.

Mangold, T. and Goldberg, J. (2000). *Plague Wars: The Terrifying Reality of Biological Warfare.* New York: St. Martin's Press.

Matravers, L. P., Bridgeman, J., and Ferguson-Smith, M. (2000). The BSE Inquiry Report. *http://www.bseinquiry.gov.uk/index.htm.*

McNeill, W. H. (1989). *Plagues and Peoples.* New York: Doubleday.

Meselson, M., Guillemin, J., Hugh-Jones, M., Langmuir, A., Popova, I., Shelokov, A., and Yampolskaya, O. (1994). The Sverdlovsk Anthrax Outbreak of 1979. *Science* vol. 266:1202–1208.

Nathanson, N., Wilesmith, J., and Griot, G. (1997). Bovine Spongiform Encephalopathy (BSE): Causes and Consequences of a Common Source Epidemic. *American Journal of Epidemiology* vol. 145:959–969.

Pedro-Botet, M. L., Stout, J. E., and Yu, V. L. (2002). Legionnaires' Disease Contracted from Patient Homes: The Coming of the Third Plague? *European Journal of Clinical Microbiology and Infectious Diseases* vol. 21:699–705.

Penniston, K. (1997). Origin of Mitochondria in Eukaryotic Cells. *http://cas.bellarmine.edu/tietjen/images/origin_of_mitochondria_in_eukary.htm.*

ProMED-mail. (2004). SARS–Worldwide (18): China, Cases. *http://www.promedmail.org.*

Public Broadcasting Service. (n.d.). Influenza 1918: The First Wave. The American Experience. *http://www.pbs.org/wgbh/amex/influenza/peopleevents/pandeAMEX86.html.*

Reid, A. H., Fanning, T. G., Janczewski, T. A., Lourens, R. M., and Taubenberger, J. K. (2004). Novel Origin of the 1918 Pandemic Influenza Virus Nucleoprotein Gene. *Journal of Virology* vol. 78:12462–12470.

Reid, A. H. and Taubenberger, J. K. (2003). The Origin of the 1918 Pandemic Influenza Virus: A Continuing Enigma. *Journal of General Virology* vol. 84:2285–2292.

Reilley, B., Van Herp, M., Sermand, D., and Dentico, N. (2003). SARS and Carlo Urbani. *New England Journal of Medicine* vol. 348:1951–1952.

Rodman, J. S., Frost, F., -Burchat, L. D., Faser, D., Langer, J., and Jakubowski, W. (1997). Pharmaceutical Sales- A Method of Disease Surveillance?, *Environmental Health* pp. 8–14.

Rodman, J. S., Frost, F., and Jakubowski, W. (1998). Using Nurse Hot Line Calls for Disease Surveillance. *Emerging Infectious Diseases* vol. 4:329–332.

Scott, A., Christie, M., and Midmore, P. (2004). Impact of the 2001 Foot-And-Mouth Disease Outbreak in Britain: Implications for Rural Studies. *Journal of Rural Studies* vol. 20:1–14.

Snowbeck, C. (2003). How Sixth Hepatitis Case Put Focus on Chi-Chi's. *Pittsburgh Post Gazette.*

Southwest Foundation for Biomedical Research. SFBR Scientists Part of Team that Offers New Explanation for the Intercontinental Spread of Drug-Resistant Malaria. *http://www.sfbr.org/pages/news_release_detail.php?id?32.*

Srikameswaran.A. (1999). Timeline: Moments in HIV/AIDS History. *http://www.post-gazette.com/magazine/19990523aidstime9.asp.*

Turnock, B. J. (2001). *Public Health: What It Is and How It Works.* Gaithersburg, MD: Aspen Publishers.

Varia, M., Wilson, S., Srwal, S., McGeer, A., Gournis, E., Galanis, E., and Henry, B. (2003). Investigation of a Nosocomial Outbreak of Severe Acute Respiratory Syndrome (SARS) in Toronto, Canada. *CMAJ: Canadian Medical Association Journal* vol. 169:285–292.

World Health Organization [WHO]. (2003a). Acute Respiratory Syndrome in China. *Communicable Disease Surveillance & Response (CSR). http://www.who.int/csr/don/2003_02_11/en/.*

WHO. (2003b). Summary of Probable SARS Cases with Onset of Illness from 1 November 2002 to 31 July 2003. *Communicable Disease Surveillance & Response (CSR). http://www.who.int/csr/sars/country/table2004_04_21/en/.*

WHO. (2004). SARS: One Suspected Case Reported in China. *Communicable Disease Surveillance & Response (CSR). http://www.who.int/csr/don/2004_04_22/en/.*

Yu, V. L. (2002). Legionella Surveillance: Political and Social Implications–A Little Knowledge Is a Dangerous Thing. *Journal of Infectious Diseases* vol. 185:259–261.

3
CHAPTER

Case Detection, Outbreak Detection, and Outbreak Characterization

Michael M. Wagner
RODS Laboratory, Center for Biomedical Informatics, University of Pittsburgh, Pittsburgh, Pennsylvania

Louise S. Gresham
San Diego County, Health and Human Services Agency and Graduate School of Public Health, Epidemiology and Biostatistics, San Diego State University, San Diego, California

Virginia Dato
Center for Public Health Practice, University of Pittsburgh, Graduate School of Public Health, Pittsburgh, Pennsylvania

1. INTRODUCTION

In Chapter 2, we saw how some well-known outbreaks first came to the attention of investigators and how the investigators proceeded to elucidate the causative biological agent, source, route of transmission, and other characteristics of the outbreaks. In this chapter, we examine in detail how biosurveillance systems detect and characterize outbreaks. We introduce an important distinction between case detection and outbreak detection. We discuss *outbreak characterization,* which is the process by which investigators elucidate characteristics of an outbreak that are important for disease control (e.g., causative biological agent, source, and route of transmission). We break down and analyze biosurveillance in this manner so that it can be modeled more formally, which is a prerequisite to providing computer support. Table 3.1 summarizes the topics of this chapter and links them to examples in Chapter 2 and additional examples introduced in this chapter (marked by bold type). The table serves as both an outline and a summary of the chapter.

2. CASE DETECTION

The objective of *case detection* is to notice the existence of a single individual with a disease. We say that this individual is a *case* of the disease. The importance of case detection is that detection of an outbreak typically depends on detection of individual cases (Figure 3.1).

Many entities are involved in case detection. People (e.g., physicians, veterinarians, nurse practitioners, infection control practitioners, medical examiners, and pathologists) and laboratories detect cases. Biosurveillance organizations detect cases through surveillance systems and screening programs. Increasingly, computers detect cases.

Some methods of case detection use *case definitions,* a written statement of findings both necessary and sufficient to classify a sick individual as having a disease or syndrome (Figure 3.2). More commonly, however, the determination of whether an individual has a disease (or syndrome) is left to the expert judgment of a clinician.

2.1. Case Detection by Clinicians

Case detection by clinicians (physician, veterinarian, nurse practitioner, pathologist) is a by-product of routine medical and veterinary care. It works as follows: A sick individual seeks medical attention or is brought to a clinician, who establishes a diagnosis. If the diagnosis is considered a notifiable disease[1], the clinician reports it (for more information on notifiable diseases, see Chapter 5). If the sick individual is a person, the clinician reports the case to a state or local health department. If the sick individual is an animal, the clinician reports the case to a state department of agriculture (see Chapter 7). This mechanism of case detection played a role in the detection of every outbreak described in Chapter 2, with the exception of cryptosporidiosis.

A clinician establishes a diagnosis by collecting and interpreting diagnostic data, including symptoms, physical observations (e.g., rash or temperature), risk factors for disease (e.g., travel to a foreign country), pre-existing diseases in the individual (e.g., diabetes), results of microbiological tests, radiographic examinations, and autopsy findings. The interpretation of diagnostic data is a complex cognitive activity. The clinician first generates a *differential diagnosis,* which is a list of diseases that the patient could have given the information the clinician has thus far. The clinician then resolves the differential

1 A notifiable disease is a diagnosed condition for which a health statute requires reporting by physicians and laboratories. Health departments decide to make diseases notifiable based on their potential as threats to a community. We discuss notifiable diseases in Chapter 5.

ISBN 0-12-369378-0

TABLE 3.1 Methods for Detection of Cases, Clusters, Outbreaks, and Methods for Characterization of Outbreaks

Process	Method	Examples
Case detection	Diagnosis by clinician*	SARS, mad cow disease, anthrax, **meningitis, measles**
	Laboratory-based detection	Hepatitis A, **foodborne illness,** influenza
	Sentinel clinician	Influenza, **ILI**
	Drop-in surveillance	**Syndromes**
	Screening	SARS, **meningococcal meningitis**
	Computer detects case	**Hospital acquired illness, syndromes, pneumonia, tuberculosis, foodborne illness, laboratory notifiable diseases**
Outbreak detection	From individual case of unusual disease	Foot and mouth disease, **measles**
	Astute observer notices cluster†	Lyme disease, hepatitis A, AIDS, foot and mouth disease, cryptosporidiosis, SARS, Legionnaire's disease, **foodborne illness**
	Biosurveillance staff notices cluster	**Hospital-acquired illness**
	Computer notices cluster	**Influenza, ILI and other syndromes, listeriosis**
Outbreak characterization		
Biologic agent	Analysis of case data to narrow differential diagnosis, microbiological testing	Legionnaire's disease, AIDS, mad cow disease, Lyme disease and Nipah virus, **foodborne illness**
Disease	Clinical profile, case definition, incubation and infectious periods, attack rate, case mortality	Legionnaire's disease, SARS, Lyme disease, AIDS, Nipah virus, **anthrax 2001**
Source and route of transmission	Spatial analysis	Anthrax,‡ **cholera**
	Cohort or case-control study	Hepatitis A
	Trial of control method (e.g., vector control)	Nipah virus
	Food chain investigation	Hepatitis A
	Vector investigation	Lyme, **West Nile virus, encephalitis, monkeypox**
	Environmental investigation	Legionnaire's disease, **foodborne illness, anthrax 2001**
Number ill, number at risk, spatial distribution	Screening, contact tracing	All
	Spatial and temporal analysis	All
	Mathematical modeling	None

The examples are outbreaks discussed in Chapter 2 and additional examples that we discuss in this chapter (in bold type). SARS indicates severe acute respiratory syndrome; ILI, influenza-like illness.
*Includes physicians, nurse practitioners, veterinarians, and pathologists.
†Astute observer includes clinicians and lay observers.
‡The spatial analysis of the Sverdlovsk outbreak included a meteorological analysis.

diagnosis by ruling in (confirming) or ruling out (excluding) these diseases by further questioning, observation, and testing. The clinician draws on a large set of facts (medical knowledge) about the effects of disease on people or animals to generate and resolve a differential diagnosis. Clinicians acquire this knowledge during professional training and from textbooks of medicine (human or veterinary) and medical journals. The clinician also draws on available information about local disease prevalence, which may come to her attention through health alerts, morbidity and mortality reports, informal collegial consultations, and hospital surveillance information. In Chapter 13, we discuss how researchers have modeled clinical diagnosis mathematically and implemented these models in diagnostic expert systems.

Coroners, medical examiners, hospital pathologists, and veterinarians use a similar reasoning process when performing postmortem examinations to establish the cause of death. We further discuss the role of coroners and medical examiners in biosurveillance in Chapter 11 and the role of veterinarians in biosurveillance in Chapter 7. Postmortem examinations played a role in the detection of mad cow disease and anthrax (1979).

The strength of case detection by clinicians is that sick individuals seek medical care. Furthermore, clinicians are expert at diagnosing illness, which is fundamental to case detection.

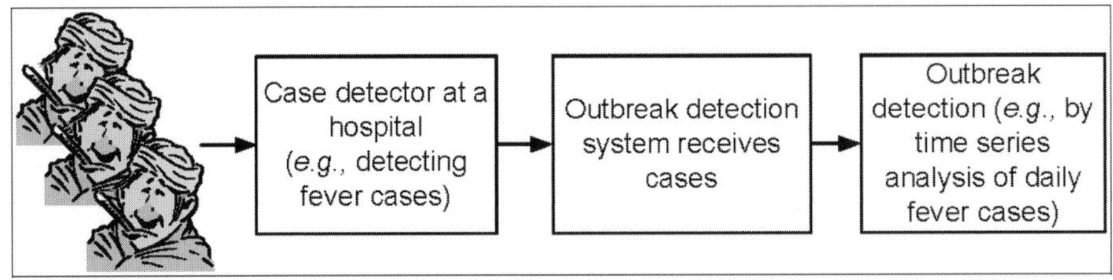

FIGURE 3.1 The relationship between case detection and outbreak detection. After the severe acute respiratory syndrome (SARS) 2003 outbreak, the Beijing center for disease prevention developed a biosurveillance system that comprised 61 "fever clinics" to which city residents were instructed to report if they developed fever (or were referred by emergency departments and physicians). Clinic staff entered the data from each clinic daily into a Web-based interface for central analysis. (Figure courtesy of Fu-Chiang Tsui.)

SARS-CoV disease classification

Probable case of SARS-CoV disease: in a person who meets the clinical criteria for severe respiratory illness and the epidemiologic criteria for likely exposure to SARS-CoV

Confirmed case of SARS-CoV disease: in a person who has a clinically compatible illness (i.e., early, mild-to-moderate, or severe) that is laboratory confirmed

FIGURE 3.2 Rules for classifying a patient as probable or confirmed severe acute respiratory syndrome (SARS). For expositional clarity, we split the Centers for Disease Control and Prevention (CDC) SARS case definition into two figures. Figure 3.3 contains definitions of the clinical, epidemiological, and laboratory criteria. The complete CDC SARS case definition is in Appendix C, which includes four additional rules for classifying patients as *SARS Reports under Investigation.* SARS-CoV refers to the coronavirus that causes the disease SARS. (From *http://www.cdc.gov/ncidod/sars/guidance/b/app1.htm.*)

Some limitations are that not every sick individual sees a clinician, clinicians may not correctly diagnose every individual they see, and clinicians may forget to report cases or fail to report cases in the time frame required by law (Ewert et al., 1994, 1995). Even when a clinician reports a case, the reporting occurs relatively late in the disease process. With some exceptions (e.g., suspected meningococcal meningitis, suspected measles, suspected anthrax) clinicians report cases only after they are certain about the diagnosis.

2.2. Case Detection by Laboratories

Case detection by laboratories is also a by-product of routine laboratory operation. Laboratories perform diagnostic tests. From the results of these tests, they often become aware of cases of notifiable diseases either before or at the same time as the clinician that ordered the test.

Laboratories play an increasing role in the detection of cases (including hepatitis A, cases of foodborne illness and influenza) as rapid, reliable, and specific laboratory tests are increasingly available. (Laboratories did not play a central role in the detection of outbreaks described in Chapter 2 because most of these outbreaks were caused by novel agents with no existing laboratory test.) The strength of laboratories as case detectors is that they are process oriented; therefore, they may report cases more reliably than can busy clinicians. A weakness is that there is not a definitive diagnostic test for every disease. A laboratory cannot detect a case unless a sick individual sees a clinician, who must suspect the disease and order a definitive test. Furthermore, not every individual for whom a test is ordered will comply and have the test done, and for some individuals with an illness, the test can be negative. Lag times for the completion of laboratory work can be substantial.

We further discuss the case detection role (and other roles) of laboratories in biosurveillance in Chapter 8.

2.3. Case Definitions

We next discuss several approaches to case detection that use case definitions. A case definition is a Boolean (logical) combination of patient findings, such as *the patient must have fever AND either cough OR pneumonia.* A case definition contains findings that are both necessary and sufficient for an investigator (or a clinician or research epidemiologist) to conclude that a patient has a disease.

Readers with backgrounds in public health should be quite familiar with case definitions.

Figure 3.2, for example, is the Centers for Disease Control and Prevention (CDC) case definition for both confirmed and probable cases of severe acute respiratory syndrome (SARS). Figure 3.3 contains definitions of the clinical, epidemiological, and laboratory criteria used in the case definitions. You can find additional examples of case definitions at Web sites operated by World Health Organization (WHO), CDC, and state and local departments of health (WHO, 2003; CDC, 2005).

2.4. Case Detection by Sentinel Clinicians

Health departments worldwide organize networks of sentinel clinicians to assist in monitoring influenza (Snacken et al., 1995, 1998; Fleming and Cohen, 1996; Zambon, 1998; Aymard et al., 1999; Manuguerra and Mosnier, 2000; Schoub et al., 2002). A sentinel clinician reports the number of individuals she sees per week who match a case definition for influenza-like illness (ILI). The California Department of Health Services provides school nurses with the following case definition for ILI: *fever (greater than 100.0°F) AND cough and/or sore throat.* The Global Influenza Surveillance Program of the Department of Defense uses a slightly different case definition: *fever (greater than 100.5°F) and either cough or sore throat or clinical radiographic evidence of acute nonbacterial pneumonia.* In some jurisdictions, the health department supplies the sentinel clinicians with rapid diagnostic tests for influenza, the results of which the clinicians also report.

The motivation for the sentinel clinician ILI system is earlier and more complete case finding of influenza to enable health departments to better control this common yet dangerous epidemic disease. Sentinel systems are not limited to influenza. The International Society of Travel Medicine and the CDC track diseases in patients who present to travel clinics (GeoSentinel, 2005).

2.5. Case Detection by Drop-In Surveillance

In the 1990s, the threat of bioterrorism led to the development of sentinel clinician–like capability for other diseases. This need

A. Clinical Criteria
Early illness: Presence of two or more of the following features: fever (might be subjective), chills, rigors, myalgia, headache, diarrhea, sore throat, rhinorrhea
Mild-to-moderate respiratory illness
Temperature of >100.4° F (>38° C) *and*
One or more clinical findings of lower respiratory illness (e.g., cough, shortness of breath, difficulty breathing)
Severe respiratory illness
Meets clinical criteria of mild-to-moderate respiratory illness, *and*
One or more of the following findings:
Radiographic evidence of pneumonia, *or*
Acute respiratory distress syndrome, *or*
Autopsy findings consistent with pneumonia or acute respiratory distress syndrome without an identifiable cause

B. Epidemiologic Criteria
Possible exposure to SARS-associated coronavirus (SARS-CoV)
One or more of the following exposures in the 10 days before onset of symptoms:
Travel to a foreign or domestic location with documented or suspected recent transmission of SARS-CoV *or*
Close contact with a person with mild-to-moderate or severe respiratory illness and with history of travel in the 10 days before onset of symptoms to a foreign or domestic location with documented or suspected recent transmission of SARS-CoV
Likely exposure to SARS-CoV
One or more of the following exposures in the 10 days before onset of symptoms:
Close contact with a confirmed case of SARS-CoV disease *or*
Close contact with a person with mild-moderate or severe respiratory illness for whom a chain of transmission can be linked to a confirmed case of SARS-CoV disease in the 10 days before onset of symptoms

C. Laboratory Criteria
Tests to detect SARS-CoV are being refined, and their performance characteristics assessed; therefore, criteria for laboratory diagnosis of SARS-CoV are changing. The following are the general criteria for laboratory confirmation of SARS-CoV:
Detection of serum antibody to SARS-CoV by a test validated by CDC (e.g., enzyme immunoassay [EIA]), *or*
Isolation in cell culture of SARS-CoV from a clinical specimen, *or*
Detection of SARS-CoV RNA by a reverse-transcription-polymerase chain reaction (RT-PCR) test validated by CDC and with subsequent confirmation in a reference laboratory (e.g., CDC)
Information regarding the current criteria for laboratory diagnosis of SARS-CoV is available at www.cdc.gov/ncidod/sars/labdiagnosis.htm.

FIGURE 3.3 Definitions of criteria used in severe acute respiratory syndrome (SARS) classification rules in Figure 3.2. (From *http://www.cdc.gov/ncidod/sars/guidance/b/app1.htm.*)

was most acute in cities that were hosting prominent events, such as the Olympic Games.

Drop-in surveillance refers to the practice of asking physicians in emergency departments to complete a form for each patient seen during the two- to six-week period surrounding a special event (County of Los Angeles, 2000; Arizona Department of Health Services, 2002; CDC, 2002b; Das et al., 2003; Moran and Talan, 2003). The clinicians record whether the patient meets the case definition for one or more syndromes of interest (Figure 3.4). Hospital staff, an epidemiologist, or an assistant then transcribes the data from the form into a database.

Sydney, Salt Lake City, and Athens used a surveillance variant during their Olympic Games that eliminated the need for physicians to fill out a form for each patient (Meehan et al., 1998; Dafni et al., 2004; Mundorff et al., 2004). Health department personnel visited each emergency department on a daily or more frequent basis and reviewed patient logs (and in some cases charts) to extract the required information.

The strength of drop-in surveillance (and sentinel clinician surveillance) is that it detects sick individuals on the day they first present for medical care; however, it is labor intensive.

Drop-in surveillance is a method of both case detection and outbreak detection. The drop-in surveillance team for the special event analyzes information from many emergency departments in the city on a daily basis for increased numbers of patients suggestive of an outbreak. Drop-in surveillance is initiated at least a week in advance of an event, if possible, so that a baseline rate of patients presenting with respiratory or diarrheal illness can be established.

2.6. Case Detection by Screening

Screening involves interviewing and testing people during a known outbreak to identify additional cases (or carriers of the disease). Screening is most often used for contagious diseases, for which it is important to find infected individuals to prevent further infections. A biosurveillance organization may use

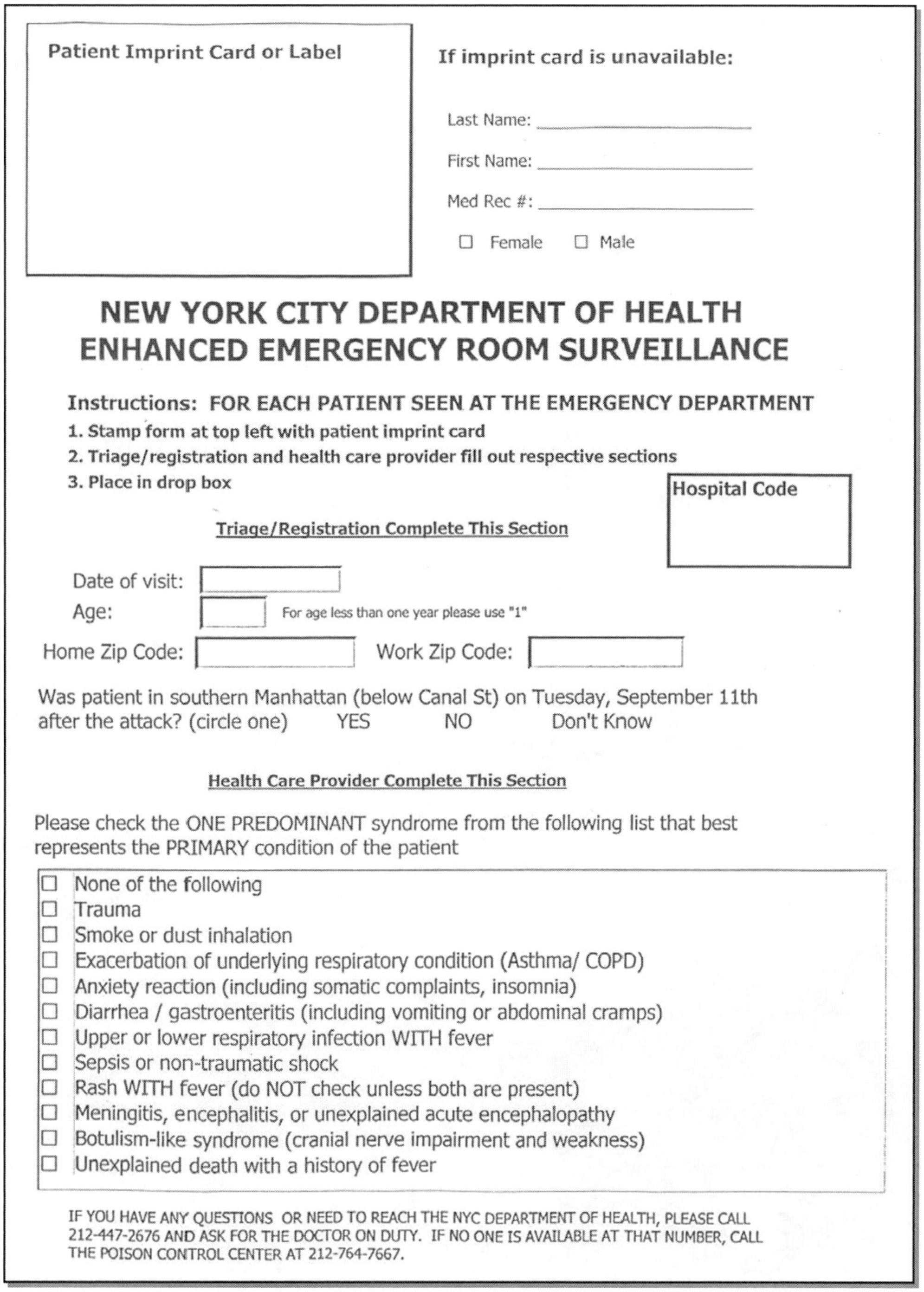

Patient Imprint Card or Label

If imprint card is unavailable:

Last Name: ____________

First Name: ____________

Med Rec #: ____________

☐ Female ☐ Male

NEW YORK CITY DEPARTMENT OF HEALTH
ENHANCED EMERGENCY ROOM SURVEILLANCE

Instructions: FOR EACH PATIENT SEEN AT THE EMERGENCY DEPARTMENT

1. Stamp form at top left with patient imprint card
2. Triage/registration and health care provider fill out respective sections
3. Place in drop box

Hospital Code

Triage/Registration Complete This Section

Date of visit: ☐

Age: ☐ For age less than one year please use "1"

Home Zip Code: ☐ Work Zip Code: ☐

Was patient in southern Manhattan (below Canal St) on Tuesday, September 11th after the attack? (circle one) YES NO Don't Know

Health Care Provider Complete This Section

Please check the ONE PREDOMINANT syndrome from the following list that best represents the PRIMARY condition of the patient

☐ None of the following
☐ Trauma
☐ Smoke or dust inhalation
☐ Exacerbation of underlying respiratory condition (Asthma/ COPD)
☐ Anxiety reaction (including somatic complaints, insomnia)
☐ Diarrhea / gastroenteritis (including vomiting or abdominal cramps)
☐ Upper or lower respiratory infection WITH fever
☐ Sepsis or non-traumatic shock
☐ Rash WITH fever (do NOT check unless both are present)
☐ Meningitis, encephalitis, or unexplained acute encephalopathy
☐ Botulism-like syndrome (cranial nerve impairment and weakness)
☐ Unexplained death with a history of fever

IF YOU HAVE ANY QUESTIONS OR NEED TO REACH THE NYC DEPARTMENT OF HEALTH, PLEASE CALL 212-447-2676 AND ASK FOR THE DOCTOR ON DUTY. IF NO ONE IS AVAILABLE AT THAT NUMBER, CALL THE POISON CONTROL CENTER AT 212-764-7667.

FIGURE 3.4 Drop-in surveillance form used in September and October 2001 in New York City. (From the New York City Department of Health and Mental Hygiene, New York, NY.)

screening in a focused manner (e.g., screening of all staff in a hospital), or it may deploy screening on a wide-scale basis. The scope of the screening effort depends on the nature of the outbreak. An outbreak of meningococcal disease in a hospital wing may require screening of only a few staff to find the person harboring the bacteria in their throat or nose. A disease such as SARS may warrant screening of tens of thousands of people.

During the SARS outbreak of 2003, many countries screened arriving and departing air travelers by using infrared thermal imaging devices in airports tuned to detect people with fevers (Figure 3.5). Similarly, hospitals and healthcare

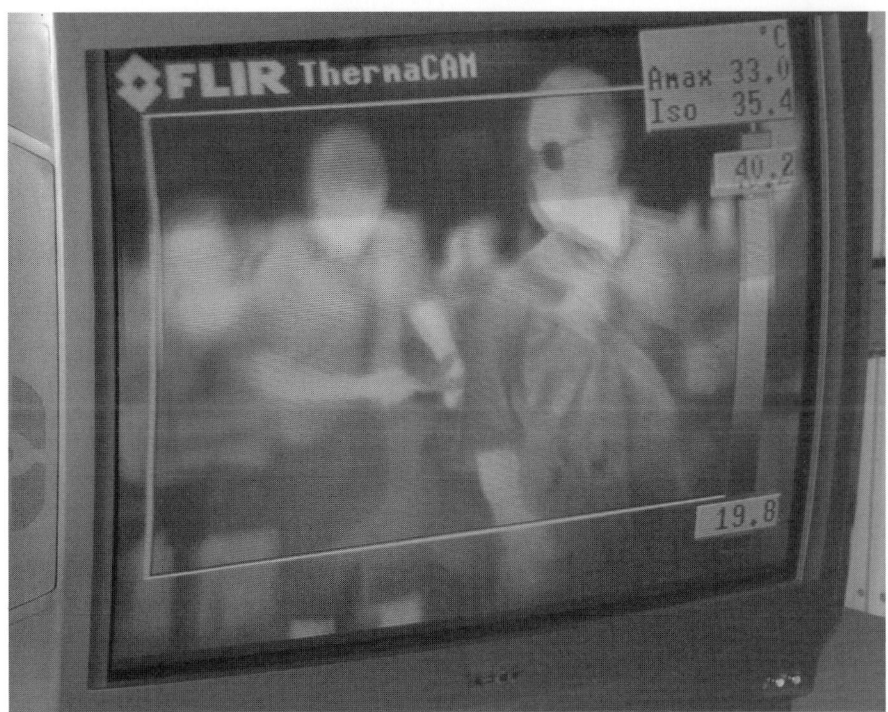

FIGURE 3.5 Thermal screening device at the Chiang Kai-shek Airport, Taipei. Lighter shades of grey correspond to higher temperatures (From Wagner.)

facilities used thermometers and questionnaires to screen people who wished to enter the facility, referring individuals with fever and respiratory illness to an isolation facility for more detailed screening. In Singapore, the government issued an electronic thermometer to every schoolchild; the child then measured his or her own temperature at school in the morning and afternoon.

2.7. Case Detection by Computers

As a result of the ever-expanding use of computers to collect and store clinical information, it has become possible for computers to detect cases by analyzing these data. Evans and colleagues (Evans, 1991; Evans et al., 1985, 1986, 1992, 1998) used computers to detect patients with infectious diseases in hospitals. Khan et al. (1993, 1995) demonstrated methods for automatic case finding for hospital infection control, and Jain et al. (1996) developed a tuberculosis case detector. Many organizations are creating electronic laboratory reporting systems, which automate case detection by laboratories (Effler et al., 1999; Overhage et al., 2001; Panackal et al., 2001; Hoffman et al., 2003).

Computerized case detection is most widely used at present, however, for detecting syndromes (DoD-GEIS, 2000; Lazarus et al., 2001; Lewis et al., 2002; Gesteland et al., 2003; Lombardo et al., 2003; Platt et al., 2003; Tsui et al., 2003, Espino et al., 2004; Heffernan et al., 2004; Nordin et al., 2004; Wagner et al., 2004; Yih et al., 2004; Chapman et al., 2005). A syndrome is an early presentation of illness. Almost all infectious diseases present initially as one of a small number of syndromes. Current computer-based case detection systems monitor for diarrhea, respiratory, influenza-like, rash, hemorrhagic, and paralytic syndromes.

In part, computers are widely used for detecting syndromes because of technical feasibility. Virtually all hospitals elicit a chief complaint from patients at the time that they register for service. The hospitals collect this information electronically and can provide it to a biosurveillance organization in a relatively uniform format.

Many of the above systems use techniques developed by the field of artificial intelligence to detect cases of disease (Cooper, 1989). We discuss these techniques in detail in Chapter 13.

2.8. Diagnostic Precision of Case Detection

We use the term *diagnostic precision* to refer to the nosological specificity of a diagnosis. For example, a physician may formulate (correctly) a relatively imprecise diagnosis of "pneumonia" after initial evaluation of a patient; subsequently, the physician may establish a more precise diagnosis of tuberculosis based on the results of laboratory testing. Diagnostic precision is not to be confused with diagnostic accuracy, which speaking loosely refers to whether the doctor was "right."

The range of diagnostic precision in medical practice (and biosurveillance) ranges from the very imprecise ("patient or animal is sick or dead") through intermediate levels of precision ("patient has respiratory illness with fever"), through organism-level diagnostic precision ("patient has *Mycobacterium tuberculosis*"), to the ultimate level, which is quite precise ("patient has *M. tuberculosis,* Beijing genotype, strain W"). As we all know from personal experience with the healthcare system, there may be considerable imprecision early in the course of a diagnostic workup about the diagnosis (and even at its conclusion). In general, the level of diagnostic precision improves over time as results of diagnostic tests become available.

For many decisions about the treatment of individual patients (e.g., surgery), the precision of diagnosis must be relatively high. In biosurveillance, however, the diagnostic precision of case detection can be lower—even as low as "sick" or "dead." As with medical care, the more diagnostic precision the better, although increased precision comes not only at the cost of further testing but also at a time cost due to the delay involved in waiting for results of the testing.[2]

The value of extremely precise case detection is that it can support detection of small or geographically diffuse outbreaks. Pulse-field gel electrophoresis (PFGE) of common pathogens now routinely matches outbreak victims separated by time and place. An outbreak that was not detected by any other method was a 2000 listeriosis outbreak: eight perinatal (three miscarriages/stillbirths) and 21 nonperinatal (median age 65) cases distributed over 10 states and seven months were only linked because of identical PFGE (PulseNet pattern numbers GX6A16.0014 by *Asc*1 and GX6A12.0017 by *Apa*1) and ribotyping (DUP-1053). A case-control study of 17 of the cases evaluating food eaten in the 30 days before illness found an association with consumption of a specific brand of deli turkey (CDC, 2000c).

The CDC National Food Borne Pathogen System serotypes every enteric isolate received to achieve the ultimate in diagnostic precision and the ability to detect very diffuse outbreaks in a nation with a population of 350 million (discussed in Chapters 5, 8).

We discuss the relationship between diagnostic precision and *detectability*—the smallest outbreak that a biosurveillance system can detect—later in this chapter and again in Chapter 20.

2.9. Investigations of Cases of Notifiable Diseases

Health departments investigate individual cases of notifiable diseases for four reasons: (1) to confirm that the case meets the case definition, (2) to determine whether there are environmental or other causes of the illness that can be remediated, (3) to identify other people who may have been exposed for antibiotic prophylaxis or vaccination, and (4) to educate or isolate communicable individuals so that their infection is not transmitted to others. When resources do not allow a case investigation on every notifiable disease, a health department must decide which reported diseases to investigate. Some investigations are so important (sexually transmitted diseases, tuberculosis) that the federal government provides substantial resources to health departments to ensure that sufficient resources are available for investigation.

The investigator may use CDC disease-specific reporting forms, department-generated interview forms, or computer-generated dynamic questionnaires to collect additional disease specific information from clinicians, infection control nurses, and/or patients. The questions explore the more common sources and exposures for the disease. If appropriate, the investigator contacts exposed individuals to provide information, screening, medication, and/or vaccination as appropriate to the disease and circumstances of exposure. Case investigations of notifiable diseases are an example of the feedback loop in Figure 1.1. If an outbreak is identified as a result of the case (or the analysis of subsequent cases), the case data already collected provide investigators a base of information for characterizing the outbreak.

3. OUTBREAK DETECTION

We use the term *outbreak detection* to refer to biosurveillance methods that detect the existence of an outbreak. A clinician may detect an outbreak by diagnosing a highly communicable disease such as measles or a rare disease such as anthrax. (A biosurveillance organization treats a single case of such a disease as evidence of an outbreak until proven otherwise.) An astute clinician may notice a cluster of cases, as happened in the 2003 hepatitis A outbreak, or a biosurveillance organization may detect an outbreak from analysis of surveillance data. Biosurveillance organizations are automating the collection and analysis of surveillance data, so a computer may detect an outbreak.

3.1. Outbreak Detection from an Individual Case of Highly Contagious or Unusual Disease

The 2004 SARS outbreak and the foot-and-mouth disease (FMD) outbreak in the United Kingdom, described in Chapter 2, illustrate a common means by which outbreaks are detected: that is a clinician, veterinarian, or pathologist encounters an individual with a rare disease. Outbreaks of measles, botulism, and tuberculosis often come to attention in this manner.

---

2 We note that clinicians and especially veterinarians working in agribusiness do not always work-up a case to the highest level of diagnostic precision due to cost-benefit considerations. For example, medical practice guidelines suggest that a clinician treating a woman with uncomplicated urinary tract infection may treat this relatively imprecise diagnosis without obtaining a urine culture (to establish a more precise bacteriological diagnosis) because the probability of curing the condition with a broad-spectrum antibiotic is high.

3.2. Outbreak Detection by an Astute Observer

The outbreaks of Lyme disease, hepatitis A, AIDS, cryptosporidium, SARS (2003), and Legionnaire's disease were detected by an astute observer who noticed a cluster of illness and reported its existence to a health department. Outbreaks caused by contamination of food are often discovered when affected individuals who have dined together phone each other upon waking up sick the next day, and one of them calls the health department.

3.3. Outbreak Detection by Biosurveillance Personnel

There are many examples of outbreaks that biosurveillance organizations detect through analysis of surveillance data. Some notifiable diseases, especially the enteric organisms that cause diarrhea, occur sporadically, and a single case report therefore does not constitute prima facie evidence of an outbreak. The New York State Department of Health detected an outbreak at a county fair after receiving reports of 10 children hospitalized with *Escherichia coli* 0157:H7 in counties near Albany, New York (CDC, 1999e). The Volusia County Health Department detected an outbreak when they received three reports of children with *Shigella sonnei* sharing a common exposure to a water fountain at a beach-side park (CDC, 2000d).

Hospital infection control units conduct similar surveillance of organisms of epidemiologic significance in the healthcare setting, such as antibiotic-resistant organisms, *Clostridium difficile*, and *Legionella pneumophila*. In addition, surveillance is done for hospital-acquired infections followed by trend analysis to access clustering of specific infection types (e.g., central-line associated blood-stream infections) and specific pathogens (e.g, *Klebsiella pneumoniae*).

3.4. Outbreak Detection by Computers

Increasingly, biosurveillance organizations use computers to analyze data to identify clusters of cases. These data may be cases reported by clinicians, veterinarians, or laboratories; aggregate data about the health of the population, such as sales of thermometers or diarrhea remedies; or a clinical data repository set up by a hospital for surveillance of nosocomial infections and levels of antibiotic-resistant organisms.

It is useful to note in the literature describing these approaches that the diagnostic precision of the data that are being analyzed by the detection algorithms can vary from notifiable diseases at the high end of diagnostic precision to "numbers of individuals absent from work" or "unit sales of diarrhea remedies" at the other end of the spectrum.

3.4.1. Automatic Cluster Detection from Notifiable Disease Data

Epidemiologists have long used the Serfling method to identify outbreaks of influenza retrospectively from pneumonia and influenza morbidity and mortality data (Serfling, 1963). But the use of computers to detect clusters in notifiable disease data is uncommon, perhaps because the necessary infrastructure is still being put into place in many jurisdictions. In current practice, epidemiologists use computers primarily to display and manipulate these data. The literature on automatic detection of clusters from notifiable disease data is, perhaps, as a result, relatively sparse at present (Hutwagner et al., 1997; Stern and Lightfoot, 1999; Hashimoto et al., 2000). A noteworthy exception is the use of clustering algorithms to analyze molecular fingerprints of enteric isolates (discussed above and in Chapter 8).

3.4.2. Automatic Cluster Detection from "Syndromic" Data

In contrast, there is a growing literature on the use of algorithms to detect clusters of cases or outbreaks with less diagnostically precise data, such as billing diagnoses.

Quenel and colleagues were the first to study detection of outbreaks from such data. They studied the sensitivity, timeliness, and specificity for detection of influenza outbreaks from 11 types of data (emergency home visits, sick leave reported to national health service, sick leave reported to general practitioners [GPs], sick leave reported by companies, sentinel GP visits, sentinel GP visits due to ILI, sentinel pediatrician visits, hospital fatality, influenza-related drug consumption, sentinel GP overall activity, and sentinel pediatrician overall activity).

Detecting outbreaks through analysis of cases of illness at an early stage is a relatively new approach for governmental public health and has been termed *syndromic surveillance*. Note that some investigators restrict the use of the term *syndromic surveillance* to methods for automatically detecting clusters of illness from case data, whereas other investigators use the term to also refer to monitoring of data aggregated from populations, such as total daily sales of thermometers, which are not case data (Buehler, 2004). Kelly Henning (2004) tabulated the various terms that have been used to refer to biosurveillance systems that provide early warning of disease outbreaks. Of these terms, we prefer "early warning systems" as it is the most descriptive of their functions

The rationale for early warning surveillance is as follows: Although the diagnostic precision of case detection is low, a highly unusual number of individuals with early symptoms consistent with a disease (e.g., 100 individuals from a single zip code presenting in 24 hours with fever and cough) may provide an early warning of an outbreak. The diagnostic precision can then be improved quickly by testing affected individuals to achieve a more precise diagnosis.

Sentinel ILI clinicians and drop-in surveillance are simple forms of early warning surveillance. In the past five years, there has been a marked trend to automate these surveillance activities to reduce the cost and possibly improve the performance. Although organizations still conduct drop-in surveillance during special events, the appropriate role for drop-in surveillance is limited to special events in cities that have not created equivalent automated capability or in areas where the surveillance requires additional data to improve diagnostic precision. Even in those settings, the current trend is to install an automated system in advance of the event and to supplement it with manual data collection from hospitals that cannot participate in the automated process, or to augment the data collected automatically with additional data collected manually to improve diagnostic precision.

3.5. How and How Well Are Outbreaks Detected?

Two studies have analyzed how existing biosurveillance systems have detected outbreaks. Dato et al. (2001, 2004) reviewed 43 well-known outbreaks, finding that 53% of the outbreaks were detected by health department staff through review of case reports from clinicians and laboratories, and 28% were detected by an astute clinician or person with knowledge of an outbreak in a school or work setting. An additional eight outbreaks (19%) were detected by laboratory networks using advanced testing and fingerprinting of specimens (three), by public sexually transmitted disease clinics (two), and by the military, another government, and a university (one each).

Ashford et al. (2003) reviewed 1,099 outbreak investigations conducted in the United States and abroad by the CDC's Epidemic Intelligence Service from 1988 to 1999. Of the 1,099 outbreaks, 399 (36%) were first recognized by healthcare providers or infection control practitioners. Health departments were the first to recognize 31% of the outbreaks. Other entities that recognized outbreaks were surveillance systems (5%), ministries of health (2.7%), nongovernmental organizations (2%), the WHO (1.5%), and the Indian Health Service (1.1%). Forty-nine (4.5%) of outbreaks were reported by other sources such as private clinics, laboratories, or private citizens. The study records were inadequate to establish the recognizing entity for the remaining 17% of outbreaks. The time delay from first case to recognition of the existence of an outbreak ranged from zero to 26 days. This study is also interesting because it analyzed 44 outbreaks caused by biological agents with high potential for use by bioterrorists.

Evidence indicates that some outbreaks are never detected, suggesting that there is room for improvement in current methods of outbreak detection. For example, the study by Dato et al. found multiple reports of outbreaks that involved contamination of nationally distributed products. However, the health departments of only one or two states detected these outbreaks, suggesting that outbreaks occurring in other states went undetected. The multistate outbreaks that were detected by only a few states involved commercially processed deli meat (CDC, 2000c), burritos (CDC, 1999c), orange juice (CDC, 1999b), parsley (CDC, 1999d), and dip (CDC, 2000e).

3.6. Diagnostic Precision and Outbreak Detection

The ability of a human or a computer to notice an anomalous number of cases above the background number of cases depends on the diagnostic precision of the surveillance data. For example, if a biosurveillance organization only collects information about the numbers of "sick" cattle in a feedlot (low diagnostic precision) and there are typically 500 sick cattle on the feedlot, an outbreak of FMD affecting 10 cattle will not stand out against the background level of sick cattle. If, however, the case data are diagnostically precise (e.g., the cases are confirmed diagnoses of FMD), one such animal in a data stream will stand out against the background level of zero. Figure 3.6 illustrates this concept for SARS surveillance,

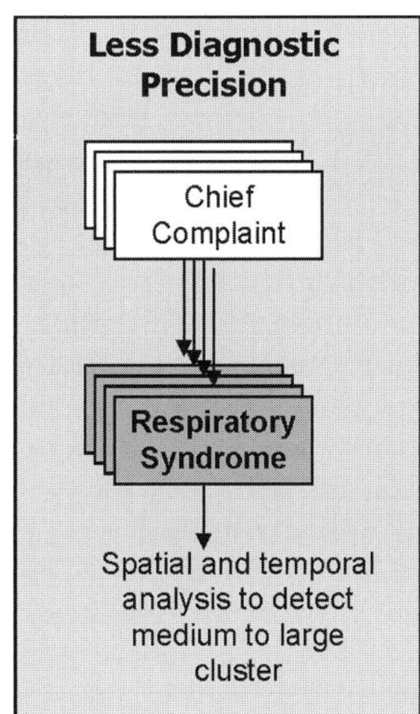

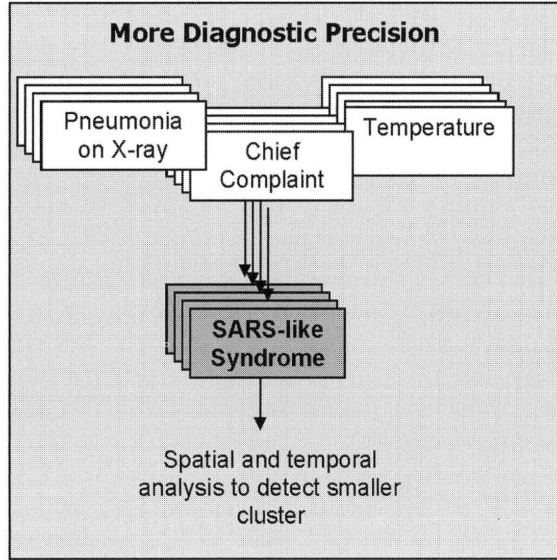

FIGURE 3.6 Diagnostic precision and minimum size of outbreak that can be detected. In this hypothetical example, the multiple boxes represent many cases in a population being detected automatically by computers from data available electronically. If the data available electronically support more diagnostically precise case detection, the size of a cluster that can be noticed above background levels will be smaller.

showing that if the diagnostic data available support a more diagnostically precise case detection (i.e., SARS-like syndrome rather than respiratory syndrome), then subsequent analysis of the case data is expected to detect smaller clusters of disease against the background levels of individuals presenting with respiratory illness.

3.7. Timeliness of Outbreak Detection

We close this section on methods for outbreak detection with a comment on the importance of timely detection of outbreaks. A biosurveillance system must detect an outbreak as quickly as possible to enable treatment of those already sick and to prevent further illness. The required timeliness varies by biological agent and route of transmission. Early detection is usually expensive, so the exact relationship between morbidity and mortality and time of detection for each type of outbreak is important. An outbreak of anthrax due to aerosol release, for example, must be detected within days of release or, ideally, at the moment of release because many people will sicken and die within days of the release. Therefore, significant resources should be expended to accomplish detection as close to day zero as possible. In contrast, detection of some diseases, even those as virulent as smallpox, as late as weeks from the onset of symptoms in the first case is still within the window of opportunity to reduce considerably mortality and morbidity (Meltzer et al., 2001).

4. OUTBREAK CHARACTERIZATION

We use the term *outbreak characterization* to refer to processes that elucidate the causative biological agent, source, route of transmission and other characteristics of an outbreak. These characteristics guide the treatment of victims and the application of control measures to prevent additional cases (e.g., by removing or isolating the source). Table 3.1 includes the complete list of outbreak characteristics and methods for their elucidation that we discuss.

As mentioned in Chapter 1, some outbreak characteristics may already be known at the time that an outbreak is detected. For example, if a biosurveillance organization detects an outbreak from analysis of notifiable disease data (which is largely organism-based reporting), it will already know the causative biological agent. If a participant in a church picnic reports an outbreak to a health department, that person may also report the source as macaroni salad, having "interviewed" most of the picnickers by phone before calling the health department. We expect the number of outbreak characteristics that are known at the time of outbreak detection to increase as biosurveillance systems collect increasing amounts of surveillance data on a continuous basis. The distinction between outbreak detection and characterization will continue to blur.

Nevertheless, a relatively crisp demarcation between the processes of outbreak detection and characterization exists. Health departments conduct disease surveillance to detect outbreaks, and they conduct investigations using different methods to characterize them. At present, the feedback loop in Figure 1.1 (Chapter 1) becomes quite active only after an investigation commences.

4.1. Outbreak Investigations

Outbreak investigations range in size from a small inquiry conducted by a single investigator to a major multinational investigation. A seasoned investigator may need only a 10-minute phone call to determine that a suspected outbreak is small, self-limited, and not worthy of additional investigation. An outbreak that is spreading rapidly and killing many individuals (such as the SARS outbreak in 2003) may warrant deployment of thousands of investigators and researchers.

When a health department suspects an outbreak based on any of the methods described in the previous sections, its staff typically initiates a preliminary inquiry to verify the available information and estimate the severity and scope of the event. The staff reviews notifiable disease records and available medical records and/or conducts open-ended interviews of a small number of individuals, asking questions and listening, quickly obtaining important information on signs/symptoms, source, and those who might have contracted the disease through contact with known cases. At this point, the staff decides if the problem is severe enough to launch a field investigation, a decision that is based on "The severity of the illness, the potential for spread, political considerations, public relations, available resources, and other factors" (CDC, 2002a). The staff also must decide whether to inform superiors and/ or request extra help, resources, or consultation. Extra investigators can divide and complete individual case investigations much more quickly than can one person.

The investigation team (or single investigator) then begins the process of interviewing all available patients and contacts. The investigators review other sources of information such as emergency department logs, pathology specimens, medical examiner records, entomological (insect) data, and animal health data (if they suspect the cause to be exposure to a sick animal). They might issue a health alert to physicians or the public requesting that similar cases be reported by healthcare providers or institutions. The investigators obtain blood, stool, urine or other specimens from affected individuals; collect materials that they suspect may be contaminated (e.g., food, water); and send samples to laboratories to be tested for organisms that may be involved based on the epidemiological information collected to that point.

The initial round of interviews and tests may yield a fairly complete characterization of the outbreak. The investigators may know the causative organism from tests done on the first infected individual, the source of the outbreak from commonalities identified among the cases identified to date, and even the complete set of affected individuals when the outbreak is geographically localized. If they do not, the outbreak investigation will continue to use many, if not all, of the analytical techniques that we will be discussing.

Throughout this process, investigators continuously formulate and refine hypotheses about outbreak characteristics that are not yet known (e.g., biological agent, source, and route of transmission), and seek to resolve *differential diagnoses* for the unknown characteristics by collecting additional information. As physicians do in clinical diagnosis, the investigators apply their knowledge of epidemiology to generate hypotheses and decide what additional information to collect. They may apply control measures suggested by the most likely and/or the most serious of the possible causes of the outbreak.

Outbreak investigations are labor intensive. Outbreak investigations are sometimes referred to as *shoe leather* epidemiology because investigators must visit numerous hospitals, homes, stores, and morgues during the course of an investigation. There are many opportunities to use information technology to improve the speed of this process and to extend the life of investigators' shoes. Significant portions of the case data that investigators assemble by hand are available electronically in clinical information systems (see Chapter 6). Opportunities also exist to provide cognitive support to investigators with their process of generating and efficiently resolving differential diagnoses of the biological agent as well as other outbreak characteristics.

4.2. General Analytic Techniques

We here provide a brief overview of general analytic techniques that investigators use to analyze case data collected during an investigation. Investigators use these techniques (e.g., spatial analysis) to elucidate outbreak characteristics.

4.2.1. Spatial Distribution of Cases

Investigators examine the spatial (geographic) distribution of cases as soon as possible. The spatial distribution of cases often provides a strong clue about the source of an outbreak. Because of the importance of spatial analysis, one of the first stories told to epidemiologists in training is the John Snow cholera story (Snow, 1855). Dr. John Snow, a London anesthesiologist and pioneer of the science of epidemiology, decided to test his hypothesis that cholera outbreaks were a result of contamination of the water supply, a view contrary to the medical beliefs of the time. He plotted the home address of people who died of cholera on a map of London; he also marked the location of neighborhood water pumps, which were the source of drinking water at the time. The striking cluster he found of cholera deaths centered on water pumps has become legendary. The concentration of cholera deaths around the Broad Street pump was twice the number of deaths in the rest of the city of London, with approximately 500 deaths in the neighborhood in 10 days. Figure 3.7 shows Snow's map with bars denoting people who died from cholera in buildings in the immediate vicinity of the Broad Street pump.

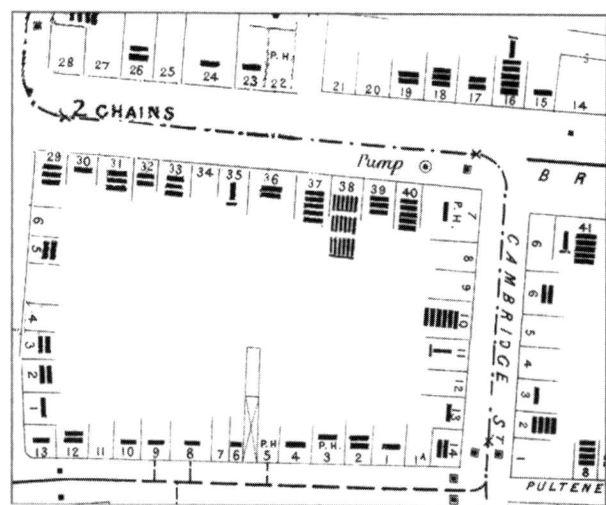

FIGURE 3.7 John Snow's cholera map showing 115 cholera deaths in the immediate vicinity of a pump on the corner of Broad Street and Cambridge Street. According to the legend, Snow advised unbelieving officials simply to remove the pump handle. (From *http://www.ph.ucla.edu/epi/snow.html.*)

At the beginning of an investigation, the investigators may only have the home address of each reported case. Therefore, the map they plot first is typically the home address of each case. During the course of an investigation, they may create many maps as they test hypotheses that the exposures may have occurred at work, school, a restaurant, fruit stand, or events such as conventions, picnics, and sporting events. Investigators may also map the location of individuals not affected. Snow did this and demonstrated that there were no cholera fatalities among brewery workers on Broad Street; these men had an allowance of free beer every day, which they apparently preferred to the water from the Broad Street pump.

Geographic information systems are modern descendants of Snow's painstakingly developed map. These systems partly automate spatial analysis. Spatial scans (see Chapter 16) are computer algorithms that more fully automate spatial analysis; these scans construct and search maps automatically for clusters of disease like that around the Broad Street pump. They can ask and answer questions such as the following: *If I were to map the people in a community who developed pneumonia in the past week by using their work addresses, would the cases cluster in particular hospitals?* This type of analysis would be very useful in SARS surveillance as SARS caused many hospital-based outbreaks in 2003. This type of analysis can be done routinely (e.g., daily or more frequently) even in the absence of a known outbreak as a method of outbreak detection. Spatial scans are an example of how the distinction between outbreak detection and characterization is blurring. When used for outbreak detection, spatial scans both find and spatially characterize outbreaks in one step.

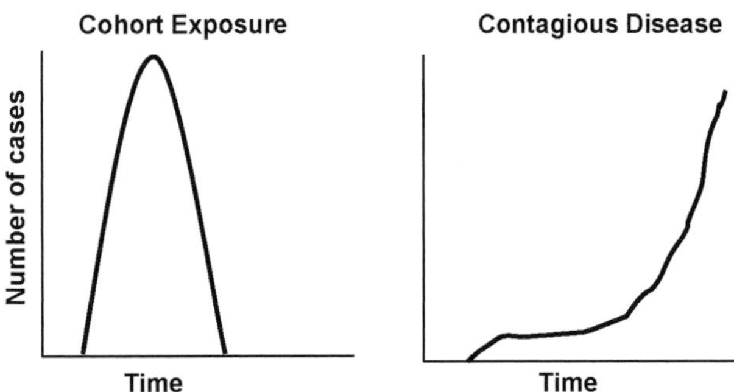

FIGURE 3.8 Hypothetical epidemic curves that would suggest a cohort exposure and a contagious disease.

4.2.2. Temporal Distribution of Cases

Investigators also examine the temporal distribution of cases as soon as possible by plotting an *epidemic curve,* which is a graph of the number of cases by date of onset of illness (Figure 3.8). The epidemic curve can provide a clue to the biological agent, source, and route of transmission. If the epidemic curve, for example, shows a sudden increase in cases, the investigator might suspect that the cause of the outbreak is contamination of food, air, or water and that the causative biological agent is more likely to be an agent with a propensity or ability to be transmitted in one of these ways (Figure 3.8, left) because such contaminations can infect a large cohort of individuals in a short period, producing a steep epidemic curve. If the epidemic curve rises more gradually, an investigator would suspect a communicable disease such as measles, in which the number of cases increases in an exponential fashion owing to successive generations of infection (Figure 3.8, right). A more level epidemic curve would suggest a continuous source of exposure, such as a persistently contaminated swimming pool.

4.2.3. Disease Incidence, Mortality Rates, and Attack Rates

Disease incidence is one measure of the magnitude of an outbreak (as are maps and epidemic curves). Disease incidence is the number of new cases in a population during a defined period such as a week. If disease incidence for every day or week during an outbreak is plotted, the result is an epidemic curve.

For lethal diseases, investigators gauge the severity (virulence) of the disease by the *case fatality rate,* which is the probability of death among diagnosed cases. Recall that investigators observed a 30% case fatality rate for the outbreak that they initially thought was Japanese encephalitis; however, the case fatality rate was highly atypical of Japanese encephalitis and led them to suspect a different disease. Investigators also compute other mortality rates. *Age-specific mortality rates,* for example, can help characterize an outbreak that is poorly understood by revealing that the disease affects the elderly or young with greater frequency or severity.

If investigators suspect an environmental exposure, they will calculate the *attack rate,* which is the fraction of people or animals exposed to a specific factor (e.g., macaroni salad or another infected individual) who subsequently contract the disease. If the attack rate in a population that is exposed to a specific factor is higher than a comparison group that is not exposed to the factor, it suggests a possible link between the factor and illness. If the analysis includes a comparison with a carefully matched control population of individuals known not to have the disease, the analysis is called a case-control study (described below). An investigator would conduct a case-control study if the less formal measurement of attack rate did not produce a definitive answer to the outbreak characteristic in question (e.g., if it did not point to macaroni salad, then the more formal case-control study likely would have).

4.2.4. Cohort and Case-Control Studies

An investigator conducts a *cohort* or a *case-control study* to test one or more hypotheses about some characteristic (oftentimes the source) of the outbreak. A cohort study compares the rate of illness of those exposed to specific factors (e.g., macaroni salad) to the rate of illness among those not exposed. A case-control study compares the frequency of specific factors in affected individuals relative to their frequency in unaffected individuals, controlling for age and other potentially confounding factors that may be correlated with the disease in question but do not cause it.

A cohort study is technically easier to conduct than is a case-control study. It is typically used for those outbreaks in which there is a small well-defined population available for interview. Examples of suitable cohorts are everyone who attended a wedding, a school, a camp, or a business conference or who ate at a specific restaurant on a given day. An investigator designs a form with three main types of questions: (1) contact and demographic information, (2) presence and onset of illness, and (3) specific exposures. For example, for a wedding at which gastrointestinal illness occurred, the questionnaire would include questions about sex, age, vomiting and diarrhea, every food item served or available (identified from menu's and the party coordinator), drinks, ice, and edible party favors. If the biological agent norovirus was suspected, exposure to public vomiting, other events such as the rehearsal dinner, and/or individuals known to be ill might also be asked. (An example of a form from a cohort study of a business conference is included in Appendix D.)

To conduct a case-control study, an investigator develops a questionnaire that covers all of the suspected sources and routes of transmission. Epidemiologists know from experience and knowledge of epidemiological patterns when to include items (e.g., intravenous drug abuse, food and water consumption, places visited, sexual practices, exposure to sick or dead animals or people, and travel history). Case-control studies invariably include age and sex, markers for socioeconomic status, race/ethnicity, occupation, disease history, and prior immunizations, in addition to questions on the exposures of interest in a specific investigation. The investigator then assembles a set of individuals with disease (cases) and a set without disease (controls). In the design of a case-control study, attention is given to matching controls to cases on known confounding variables such as socioeconomic status, age, and sex to remove these influences from the analysis.

The investigator then administers the questionnaire to each of the cases and controls. Required data may be collected or verified from medical records. The odds ratio (OR) for each factor is calculated, which is the ratio of the incidence rate in exposed individuals relative to that among unexposed individuals (Rothman and Greenland, 1998). If the OR is equal to one, it suggests that the factor is not causing the illness.

The investigation of the hepatitis A outbreak described in Chapter 2 involved a case-control study of food items served in the restaurant. Investigators interviewed individuals with hepatitis A and controls without hepatitis A who either had dined with case patients at Restaurant R or were identified through credit card receipts as having dined at Restaurant R during October 3 through 6. They found an OR of 24.2 for consumption of mild salsa with green onions, and an OR of 5.2 for consumption of chili con queso with green onions (CDC, 2003a). An OR of 24.2 indicated that people who dined at restaurant R and subsequently developed hepatitis A were 24.2 times more likely to have eaten mild salsa with green onions than were people who dined at the same restaurant but did not develop the disease.

Case-control studies depend on the ability of people to remember key historical details accurately such as what they ate. For the outbreak of hepatitis A, the investigators obtained a food history for a period of two to six weeks before onset of symptoms because the incubation period of hepatitis A is long. For this reason, investigators need to move quickly to develop and administer outbreak questionnaires. Investigators follow standard methods of interviewing to minimize bias in how they ask questions and to minimize recall and prevarication bias on the part of the interviewee (Kalter, 1992).

4.3. Outbreak Characteristics

This section discusses how investigators elucidate the following outbreak characteristics: biological agent; source; route of transmission; size; and, when the disease is new or unusual, the disease process itself.

4.3.1. Biological Agent

The causative biological agent is perhaps the single most important characteristic of an outbreak. It has immediate implications for treating the sick and focuses the search for the source and route of transmission as each biological agent has propensities and limitations in the environments in which it can reside and the mechanisms by which it can be transmitted. The investigators of Legionnaire's disease, AIDS, mad cow disease, Lyme disease, and Nipah virus did not know the causative biological agent and had great difficulty finding the sources and routes of transmission.

Although the biological agent is often known at the time that an outbreak is detected, for diseases that have recently crossed species or for rare diseases that clinicians do not routinely test for, it may not be known. Importantly, the recent trend toward monitoring surveillance data of lower diagnostic precision (e.g., sales of diarrhea remedies or numbers of individuals with flulike symptoms) has increased the number of situations in which the biological agent is not known when an outbreak is detected. In these situations, the differential diagnosis may be large (Table 3.2). When the biological agent is not known, investigators use the clinical symptoms of affected individuals to select laboratory tests to narrow down and ultimately identify the biological agent.

A significant amount of laboratory work may be required to identify the biological agent. As in the case of Legionnaire's disease, Lyme disease, and Nipah virus in which the organism was previously unknown, it may take considerable time to isolate the organism. Identification of a difficult-to-identify organism is largely a process of elimination. Laboratories use cultures, serological tests, immunohistochemistry, and nucleic acid probes to search for known organisms that are most

TABLE 3.2 Biological Agents and Toxins of Concern for a Large-Scale Aerosol Release

Biological Agent	Treatable?	Early Clinical Presentation
Bacteria		
Bacillus anthracis	Yes	Flulike
Brucella sp.	Yes	Flulike
Coxiella burnetti (Q fever)	Yes	Flulike
Francisella tularensis (Tularemia)	Yes	Flulike
Burkholderia mallei (Glanders)	Yes	Flulike
Histoplasmosis/coccidiomycosis	Yes	Flulike
Pseudomonas mallei	Yes	Flulike
Yersinia pestis	Yes	Flulike
Viruses		
Smallpox (aerosol release)	Yes (early vaccination)	Flulike, rash
Venezuelan equine encephalitis	No	Flulike, headache
Biological toxins		
Staph enterotoxin B	No	Flulike
Clostridium perfringens toxin	No	Respiratory distress/failure
Botulinum toxin	Yes	Double vision and paralysis
Ricin toxin	No	Cough, difficulty breathing

The differential diagnosis of a sudden, large increase in flulike illness includes the first 11 agents.

likely epidemiologically to be causing the illness. We discuss the full range of laboratory tests in Chapter 8.

The causative biological agent for some outbreaks is never found. Causative agents were not identifed for 16 outbreaks associated with burritos that affected approximately 1,700 individuals (CDC, 1999c). These outbreaks were eventually (epidemiologically) traced to two companies, resulting in the recall of two million pounds of burritos. In the study by Ashford et al. (2003), the causative biological agent was not found in 41 of the 1,099 (3.7%) investigations studied.

4.3.2. Characterizing the Disease

If the biological agent is unknown or if the disease itself is unusual in its presentation or severity, then characterizing the disease process becomes a priority for investigators. They will develop a working case definition, as was done by CDC and WHO for both AIDS and SARS, to enable additional case finding and to use in case-control studies. They will measure the incubation and infectious periods of the disease to bracket the period in which to search for causative factors and contacts.

Incubation Period. The *incubation period* is the time from infection of an individual to onset of clinical illness (Figure 3.9). The incubation period may vary from individual to individual based on health status and the dose of the biological agent to which the individual was exposed; therefore, investigators measure the average and range of the incubation period.

If the source or route of transmission of the infection is known, investigators measure the incubation period as the time from exposure to onset of symptoms. For contagious diseases, the time of exposure is the date at which an individual was exposed to an index case. For infections caused by contaminated materials, the time of exposure is the date that the contaminated material was ingested or otherwise entered the body of the victim.

The incubation period may provide a weak clue to the identity of the organism. Some classes of organisms such as HIV have long incubation periods.

Infectious Period. The *infectious period* is the time during the course of an individual's illness when he or she can transmit the disease to another individual. It usually does not provide a clue to the biological agent. Its importance is as a basis for developing guidelines for isolating infected individuals to prevent further infections. The beginning of the infectious period usually coincides with onset of symptoms because many diseases are transmitted by coughing, sneezing, diarrhea, or weeping skin lesions. There are exceptions, however, and the infectious period may begin before or after the onset of symptoms.

Investigators establish the beginning of the infectious period by analysis of dates of contact between infected individuals. In particular, investigators compare the dates of contact between an index case and the secondary cases that likely resulted from

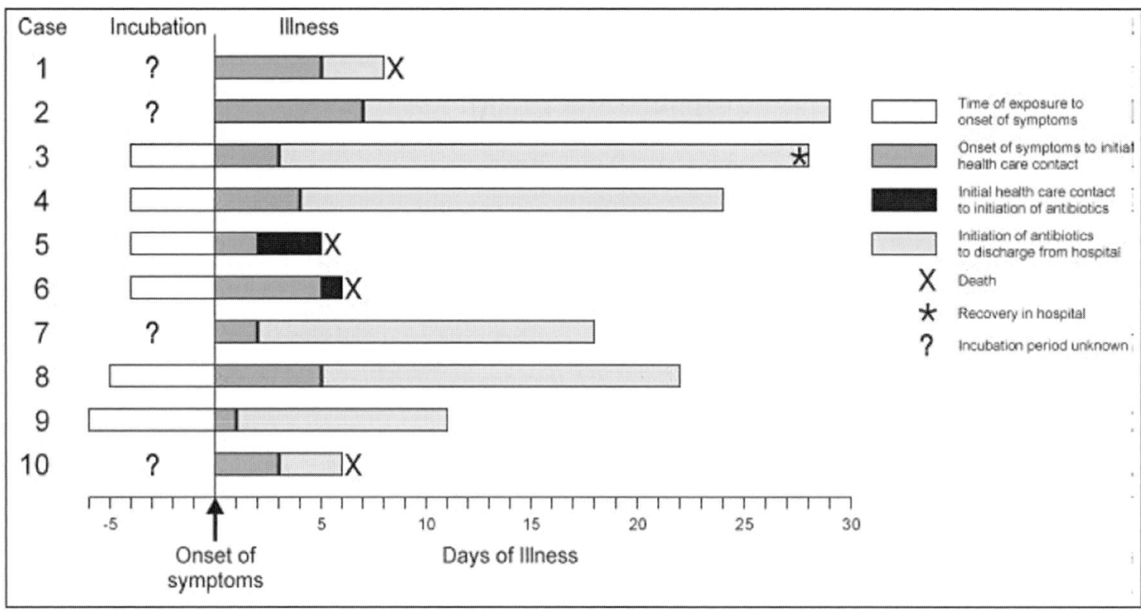

FIGURE 3.9 Incubation and symptomatic periods for 10 cases of inhalational anthrax. The *incubation period* is the time from infection or exposure to onset of symptoms (white bars to the left of the vertical line denoting day of onset of symptoms). The *symptomatic period* is the time from onset of symptoms to recovery or death (bars to the right of the vertical line). The infectious period may begin before of after the onset of symptoms. The infectious period for anthrax ends when the host develops sufficient antibodies to clear the infection, when the infection is eradicated through treatment, or when the body is cremated (Jernigan et al., 2001).

contact with that index case. The beginning and end of the infectious period for an index case are roughly the dates when that individual starts and stops infecting other individuals, respectively. As with the incubation period, investigators compute the average and range of the infectious period over many cases.

Another method of establishing the end of the infectious period is laboratory testing for agent-specific antibodies. When antibodies appear, the person is usually no longer infectious.

4.3.3. Source

The term *source* refers to the starting point in the path via which a biological agent is eventually conveyed into the body of a victim (Table 3.2). The source is critically important because removal or isolation of a source prevents further infections. In Hong Kong in 1997, an outbreak of avian influenza due to the H5N1 strain led to a small number of human cases with a high fatality rate. Fearing that the avian influenza would lead to a human pandemic, authorities sacrificed millions of chickens harboring the H5N1 strain (Sims et al., 2003).

There are many possible sources for outbreaks. The most common sources are food, water, other people, and animals.[3] We note that the source is simply the starting point in a path of transmission. Investigations seek to understand the entire path because it may contain many points at which they can apply disease control measures. We also note that the source attributed to an outbreak may not be the ultimate source. For example, the nominal source of the hepatitis A outbreak described in Chapter 2 was green onions from farms in Mexico. The source of contamination of the green onions is unknown. Operationally, the search for a source ends when a common early point in the path of transmission of the disease is found at which control measures can be applied to halt the outbreak.

Sometimes a source is never identified. The source was never identified for three outbreaks involving group A rotavirus (CDC, 2000b), *E. coli* O111:H8 (CDC, 2000a), and Norwalk-like virus (CDC, 2000f). Table 3.3 provides examples of potential sources and routes of transmission for microbes.

TABLE 3.3 Examples of Sources and Routes of Transmission

Source	Route of Transmission (Path)	Hosts
→ Terrorist	→ envelop → mail system → air	Postal workers, recipients of mail, people in buildings in which envelops were opened
→ Green onions	→ food → system → restaurant	Restaurant patrons
→ Contaminated source water (reservoir)	→ water *treatment plant* → water distribution system	Consumers of tap water
Poultry	→ air → persons → air	Close contacts of infected individuals

An arrow before the source means that there may be more proximate sources that remain unknown.

4.3.4. Route of Transmission

The term *route of transmission* refers to the path that connects a source of biological agent to sick individuals.[4] The route of transmission for the U.S. 2001 postal anthrax attack, for example, started in an unknown facility or facilities that manufactured the anthrax powder (the source) (Jernigan et al., 2002). Unknown individuals then transferred the powder into envelops and deposited them in mailboxes in or around Trenton, New Jersey. Mail sorting machines in postal processing and distribution centers compressed the envelops, thus expressing spores, which were sufficiently light to float in the air. The air carried the spores into the lungs of individuals working in the processing centers. After an incubation period, some of these individuals developed the disease inhalational anthrax. Postal workers delivered the envelopes to the addressees, who opened them, allowing the spores to float into the air in buildings. The addressees and other occupants of the buildings inhaled spores, and some developed inhalational anthrax. The spores infected the skin of other individuals, who developed the disease cutaneous anthrax. Investigators believe that cross-contamination of bulk mail resulted in a case of inhalational anthrax in a woman in rural Connecticut (Griffith et al., 2003). Figure 3.10 shows the route of transmission identified by investigators for the 2001 postal attack.

The final step in a route of transmission is always the point of entry into the host, which is biologically possible through only a finite number of entry points. Biological agents can enter humans (and other animals) through the respiratory tract (breathing or sniffing), the gastrointestinal tract (eating or per rectum), the skin, the eyes, sexual contact, medical procedures (surgical incision, transfusion, intubation), and nonmedical intravenous injections.

In contrast, the paths by which a biological agent can arrive at an entry point of an individual are virtually infinite. They include the air, any water (bottled water, city water supply, temporary water supplies at public events, swimming pools, hot tubs, dental office water), any food, the mail, manufactured products,

3 The term *reservoir* is somewhat synonymous with source, although it only refers to sources in which an organism lives and multiplies. The human body is a reservoir for many viruses and bacteria as are animals. Bats are a reservoir for rabies, and sheep a reservoir for anthrax. Human diseases that have an animal reservoir are called *zoonotic* diseases.

4 Epidemiologists use the term 'route of transmission' or 'mode of transmission' to refer to generic transmission patterns such as airborne, sexual, person-to-person, and food borne. When characterizing an outbreak, however, the goal is to elucidate in detail the particular path by which biological agents 'travel' from a source to a host (host is the term for an individual who is sick).

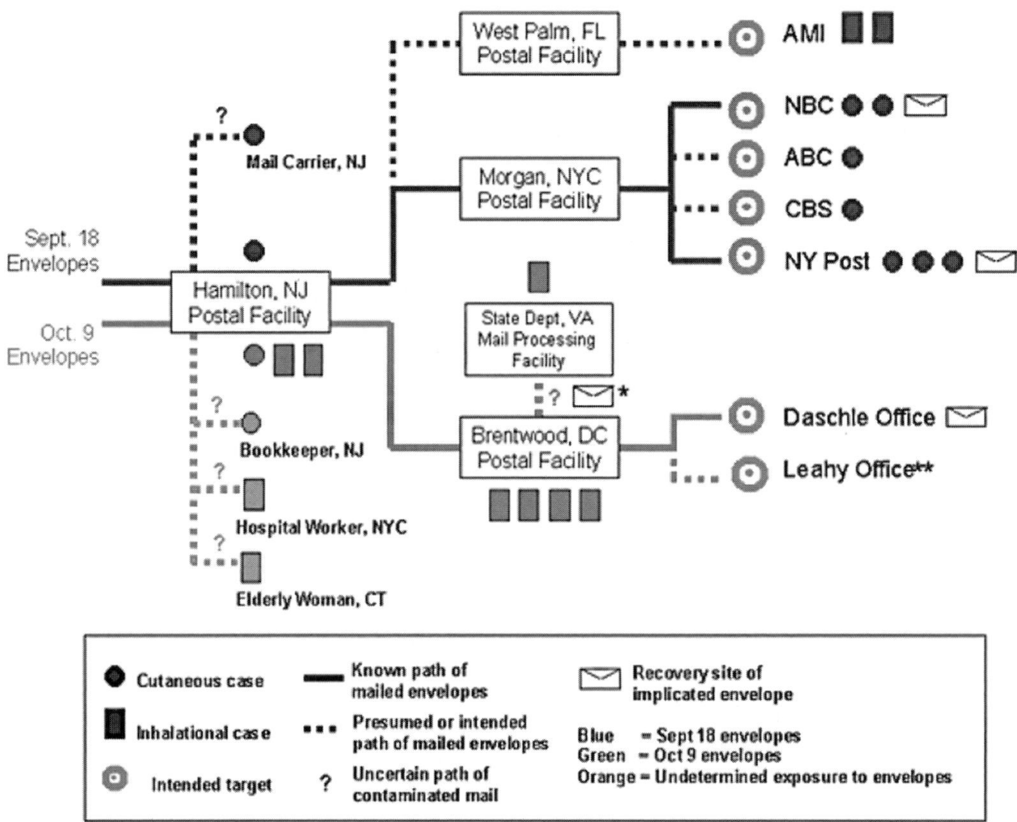

FIGURE 3.10 Cases of anthrax associated with mailed paths of implicated envelopes and intended target sites. NY indicates New York; NBC, National Broadcasting Company; AMI, American Media; USPS, United States Postal Service; and CBS, Columbia Broadcasting System. *Envelope addressed to Senator Leahy, found unopened on November 16, 2001, in a barrel of unopened mail sent to Capitol Hill. **Dotted line indicates intended path of envelope addressed to Senator Leahy. (From Jernigan et al., 2002.)

legal drugs, illegal drugs, medical instruments (surgery, endoscopic examinations, intravenous lines), blood products, another person, or an animal (including insects, snakes, and fish). With the advent of bioterrorism, the path is limited only by the ingeniousness of man, as evidenced by the murder of the expatriate Bulgarian writer and broadcaster Georgi Ivanov Markov by the Bulgarian secret police, who used an umbrella tip to inject a tiny platinum ball filled with the toxin ricin (http://en.wikipedia.org/wiki/Georgi_Markov).

4.3.5. *Methods to Elucidate Source and Route of Transmission*

Elucidating the source and route of transmission may be labor and time intensive. For example, the investigation that elucidated the source of the listeriosis outbreak described earlier involved a case-control study of 17 cases conducted by five states, two local health departments, and the CDC to identify potential common sources. The root source—a supplier of processed deli meat—was identified by visiting 13 stores to identify the supplier that they had in common (CDC, 2000c).

Environmental Investigations. Much of the dramatic decrease in U.S. crude death rate in the early part of the 20th century can be attributed to sanitary improvements in water, food, and sewage management (CDC, 1999a). Outbreaks may result when these practices break down or are not adhered to.

An environmental investigation may examine water and food sanitation, underground water, surface water, agriculture, and domestic or wild animals. When there is reasonable possibility that a facility may be involved in an outbreak, investigators request that sanitarians conduct an inspection or review of a facility. Sanitarians (also known as environmental health specialists), using a body of science developed through the past century, routinely inspect and advise food service facilities and recreational and potable water facilities to ensure that environmental safeguards are in place to prevent outbreaks and a return to 19th-century rates of infectious diseases. The sanitarian can quickly determine whether the facility is operating with no violations or practices that would cause an outbreak. If stronger evidence becomes available that a facility or specific environment is involved, the investigators may initiate a more extensive environmental investigation (Massachusetts Department of Public Health, 2005a,b).

More generally, an environmental investigation, depending on the problem at hand, explores the environments that provide reservoirs where agents can reside and multiply. When the causative biological agent and the source are unknown, as was the case during the 1976 Legionnaire's outbreak, an environmental investigation can be far-ranging.

Food service inspection and investigation methods are well developed as the result of accumulated experience with thousands of foodborne outbreaks. Hazard analysis critical control point (HACCP) is a "science based method to identify or prevent hazards which contribute to foodborne disease" (Massachusetts Department of Public Health, 2005a). *Critical control points* include the appropriate heating and cooling of food. The value of the HACCP method is that it can identify likely points in a path of transmission well in advance of full characterization of an outbreak. A malfunctioning refrigerator, for example, is both a clue to the potential source of an outbreak (staphylococcus can elaborate a toxin, which is heat stable and therefore not neutralized by subsequent cooking) as well as a point for immediate correction to prevent future problems. If a specific food is implicated by survey methods or microbiological analysis, the sanitarian will look very carefully at food preparation steps.

The investigators of the 2001 anthrax outbreak conducted environmental investigations in postal processing and distribution centers, offices, and homes to determine the presence of *Bacillus anthracis* and the paths by which it spread. For the environmental investigation related to the most unusual case–the 94-year-old woman in Connecticut discussed earlier–specialists assessed the patient's activities in her home and searched for letters she received in the prior two months, in addition to conducting sampling in and on the periphery of her home by using swabs on surfaces and high-efficiency particulate air vacuums (Griffith et al., 2003). Molecular subtyping identified the isolate from the 94-year-old woman as matching the isolates from the other anthrax patients infected through mail. The investigators did not find matching isolates in the woman's home or in any of the places she regularly visited. They did learn by going through her garbage that she regularly tore her bulk mail in half before discarding. And they found that bulk mail that was processed and delivered by her local mail distribution center had been processed in another post office in the 24 hours after heavily contaminated letters. Evidence that at least some of that bulk mail was cross-contaminated came when matching isolates were found on her local bulk mail processing machines. Investigators believe this woman's advanced age, medical condition, and habit of ripping junk mail in half before discarding it contributed to infection from a very low level of contamination of the mail she received. This explanation was the simplest and most biologically plausible.

The anthrax environmental investigations led to routine use of biohazard detection systems (BDSs) (Military Postal Service Agency, 2004) to identify mail contamination before mail distribution to the public.

Food Chain Investigation (Trace-Back and Trace-Forward). When investigators suspect or find a contaminated food item (based on microbiological analysis of a sample of the food or as the result of a case-control study), they trace backward through the food supply to identify the root source of the contamination. The trace-back begins when a sanitarian collects information about a product or food item from the restaurant, consumer, or retail seller. The necessary information includes brand name, product name, code/lot number, expiration/sell by/use by date, size/weight, package type, date of purchase, manufacturer and address, distributor name and address, and retail food establishment where purchased or consumed (Massachusetts Department of Public Health, 2005).

The U.S. Food and Drug Administration (FDA) conducted a trace-back study that led to green onions grown on farms in Mexico as the source of the hepatitis A outbreak in Pennsylvania (CDC, 2003a). The FDA then conducted an environmental investigation at the farms: "The investigation team identified issues of concern from interviews and observations at all four firms visited including items such as poor sanitation, inadequate hand washing facilities, questions about worker health and hygiene, the quality of water used in the fields, packing sheds, and the making of ice, any of which can have a role in the spread of infectious diseases such as hepatitis A" (FDA, 2003).

The complexity of a trace-back is evident from Figure 3.11. The Public Health Security and Bioterrorism Preparedness and Response Act of 2002 (FDA, 2004) requires that food producers, retailers, and restaurants maintain records to facilitate trace-back investigations. We discuss these regulations in more detail in Chapter 10.

Trace-forward investigations similarly track a product through the supply chain, but they do so in the forward direction; that is, from a starting point that may have been discovered by the trace-back process through the distribution system to the consumer. U.S. Department of Agriculture (USDA) and other entities conduct trace-forward investigations to find and remove contaminated products before they are distributed to consumers. Trace-forward can also identify people who have already been exposed, who are sick, or who may already have recovered or died from the illness.

Readers interested in more details about food-chain investigation should consult *The Guide to Trace Back of Fresh Fruits and Vegetables Implicated in Epidemiological Investigations* at *http://www.fda.gov/ora/inspect_ref/igs/epigde/epigde.html.* An example of a trace-forward protocol used by the USDA (for a plant disease) is at *http://www.aphis.usda.gov/ppq/ispm/pramorum/pdf_files/traceforwardprotocol.pdf.*

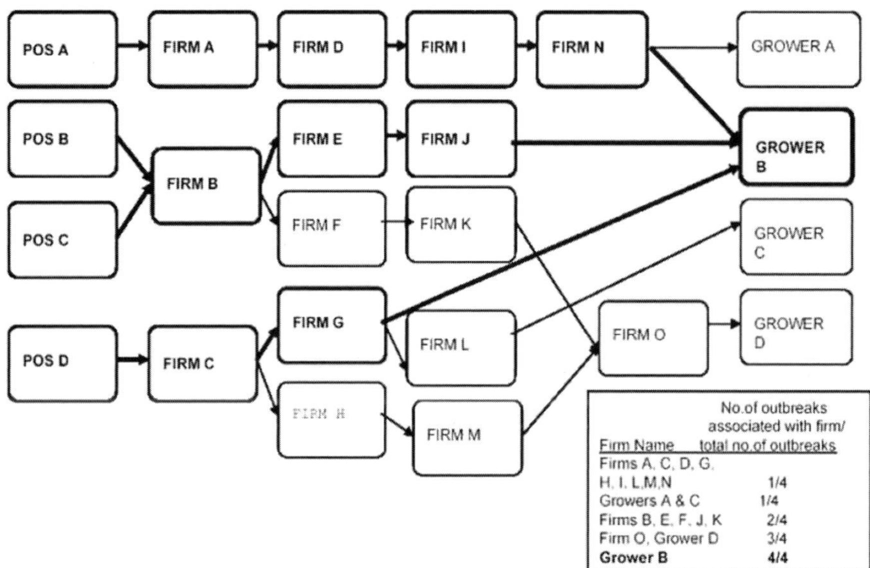

FIGURE 3.11 A hypothetical trace-back investigation involving four different points of service (POSs). POSs are restaurants or other retail stores (e.g., produce store) that sell or serve food or products believed to have caused an outbreak. In this example grower B was the ultimate source of produce for all four points of services. (From *The Guide to Trace Back of Fresh Fruits and Vegetables Implicated in Epidemiological Investigations http://www.fda.gov/ora/inspect_ref/igs/epigde/epigde.html.*)

Vector Investigation. A vector is an animal that can transmit a disease to humans. Many vectors are insects that depend on specific ecological conditions for survival. If the biological agent causing an outbreak is known and if it is known to be associated with vector-based transmission, an investigator will interview the patients and ask questions related to exposure to vectors, insect bites, animal bites, use of prophylaxis such as antimalarial drugs, and travel history before onset of illness. If travel is involved, the investigator may consult CDC travel advisory documents for current information on levels of vector-borne disease around the world.

If investigators suspect an exposure to a vector as responsible for disease, they will consult with environmental health specialists to discuss methods to identify the vector habitat and control the vector, especially for prevalent vector-borne diseases such as malaria that have no vaccine. A full investigation of a vector-borne disease normally requires rapid exchange of data and information among a multidisciplinary team of environmental health specialists, veterinarians, and epidemiologists. Environmental health expertise is needed to understand the complex transmission cycles involving a number of vectors (and usually reservoir hosts) and complex environmental controls (World Resources Institute, 1998).

In the spring of 2002, an outbreak of monkeypox in humans produced 71 cases (Ashford et al., 2003). The investigation was initiated based on the report of a three-year-old girl with a history of a prairie dog bite. Investigators used sales invoices to link all cases to a shipment of 38 prairie dogs sold at pet stores or at a swap meet (it is often difficult for investigators to obtain invoices transacted at venues such as a swap meet). A trace-back investigation (Figure 3.12) elucidated the path by which monkeypox was introduced into the United States. A shipment of exotic rodents from Africa made its way via an importer in Texas to its final destination in the midwest. The rodents were colocated temporarily with a colony of prairie dogs. Once the original shipment was identified, trace-forward investigations identified additional animal vendors and owners who purchased prairie dogs during the time frame of the suspect shipment.

4.3.6. Number of People Ill and Number of Persons at Risk

Early during an investigation, the investigators have considerable uncertainty about the number of sick individuals in the population and the number that are infected but not yet symptomatic. The investigators may have very worrisome questions about whether they have enough investigators, vaccine, or antibiotics on hand to manage the outbreak, and they may worry whether their control measures are sufficiently aggressive.

They must estimate the true spatial distribution and true epidemic curve, based on the information available (current set of cases identified, contacts, and known outbreak characteristics). To do this, they must understand the limitations of the biosurveillance systems in place (e.g., the notifiable

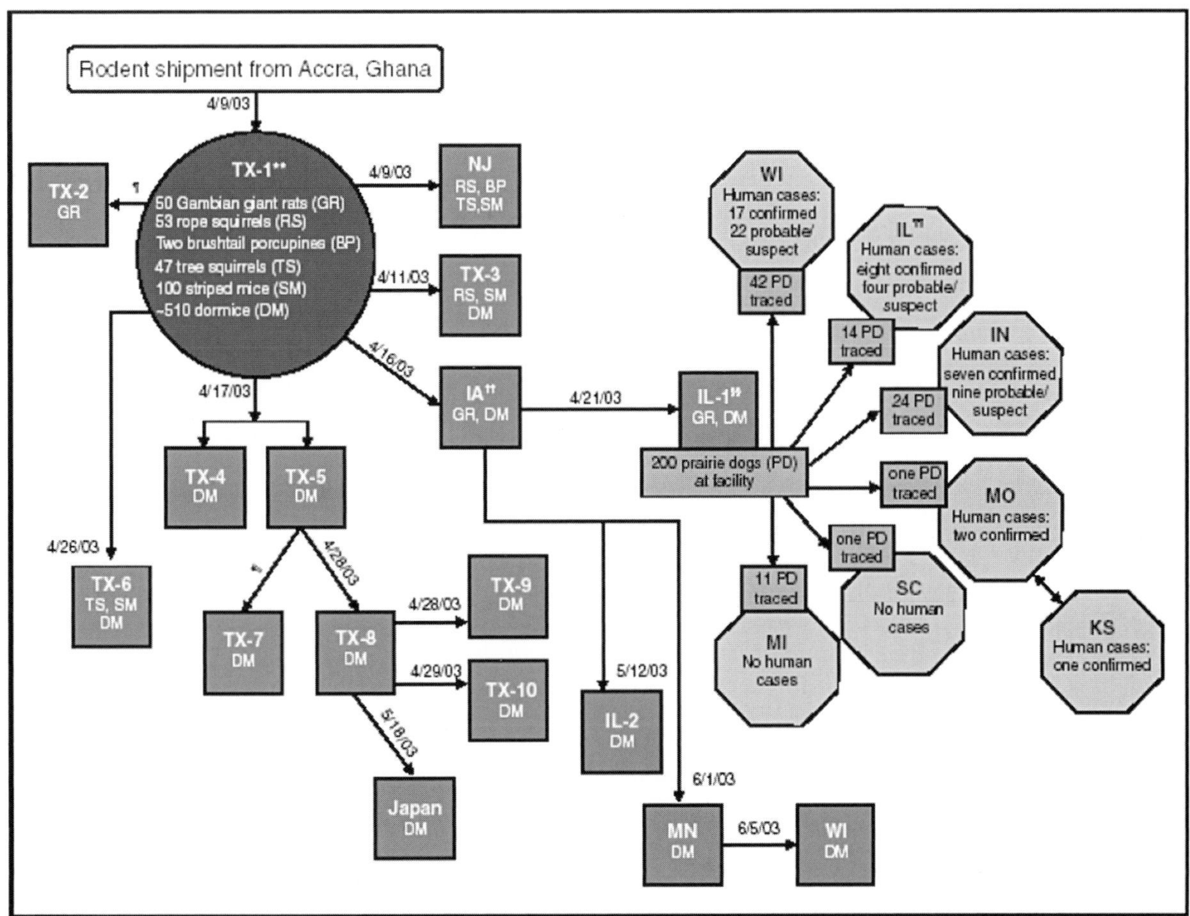

* Illinois (IL), Indiana (IN), Iowa (IA), Kansas (KS), Michigan (MI), Minnesota (MN), Missouri (MO), New Jersey (NJ), South Carolina (SC), Texas (TX), and Wisconsin (WI). Japan is included among sites having received shipment of rodents implicated in this outbreak.
† As of July 8, 2003.
§ Does not include one probable human case from Ohio; investigation is ongoing.
¶ Date of shipment unknown.
** Identified as distributor C in *MMWR* 2003;52:561–4.
†† Identified as distributor D in *MMWR* 2003;52:561–4.
§§ Identified as distributor B in *MMWR* 2003;52:561–4.
¶¶ Includes two persons who were employees at IL-1.

FIGURE 3.12 Result of a trace-back and trace-forward investigation of the 2003 Monkeypox outbreak. (From CDC, 2003b.)

disease system, any electronic laboratory reporting systems, and their screening procedures) and of their investigation. In particular, they must understand what fraction of cases their methods detect and what time delay may be present from date of infection. Their decision making related to logistics and control measures depends on an accurate assessment of both the state of the outbreak at the moment as well as projections of the future number of cases and their geographic distribution.

At present, the state of the art in real-time estimation of the magnitude and geographic scope of an outbreak is primitive. In current practice, investigators simply do their best to intensify surveillance to identify all cases so that the observed number of cases is as close to the real number of cases as possible. Any delays in case detection in the biosurveillance system compound the estimation problem. Mathematical models that can estimate the true parameters from observed parameters and knowledge of the delays and sampling efficiency of surveillance methods would likely be very useful, but this topic is an open area of research.

5. LEGAL, ETHICAL, AND PUBLIC RELATIONS ISSUES

Although the immediate purpose of a field investigation is to characterize and control an outbreak, investigators are cognizant that outbreaks often generate legal proceedings (Gregg, 2002).[5] Outbreak investigators and police investigators often talk to the same individuals and visit the same locations. They may even come into conflict over who gets to speak to an individual first or who has authority over a contaminated building. Police investigators depend on outbreak investigators for many of the clues that they need to identify and successfully prosecute the culprit. Biosurveillance systems must track chain of custody of evidence, especially in the laboratory. For these reasons, the public health workforce receives training in forensic epidemiology to handle legal issues that arise in the setting of joint investigations.

Investigators also must disclose information to the public about outbreaks. Ethical conduct of investigations includes the protection of individual information and confidentiality against disclosure of information (Coughlin and Beauchamp, 1996). Investigators understand that they depend on the public's trust to obtain cooperation and truthful answers to sensitive medical and behavioral questions in future investigations.

The public, politicians, and lawyers can influence the conduct of an investigation, especially investigations of outbreaks of a large number of people or in connection with sudden, mysterious illness with high mortality. The early HIV epidemic highlighted the complex bureaucracy and social agendas that come into play when an epidemic is investigated in an atmosphere charged with fear and prejudice (Shilts and Greider, 1987).

6. SUMMARY

In this chapter, we examined in detail how biosurveillance systems detect and characterize outbreaks. We described the overall process as comprising three distinct subprocesses—case detection, outbreak detection, and *outbreak characterization.* Although we described them as separate steps, one triggering the other, we expect that these processes will become more tightly integrated in the future and that the distinctions between these processes will blur.

Case detection is a front-line activity in biosurveillance, which is accomplished by diverse methods, including detection by clinicians, laboratories, screening programs, and, increasingly, computers. Outbreak detection and characterization depend on case detection. Outbreak detection is based on continuous analysis of human and animal data by people working in health departments, the animal healthcare system, and hospital infection control, as well as by astute citizens. *Outbreak characterization* is an intermittent process that is triggered by outbreak detection. It is the process by which investigators elucidate characteristics of an outbreak that are important for disease control (e.g., causative biological agent, source, and route of transmission). Characterization is based on intensive collection of additional data when an outbreak is suspected or confirmed. Outbreak characterization is the least automated process in biosurveillance at present, but the future role of automation is already recognized (e.g., a recent report from the National Defense University identifies *integrated automated event characterization system based on epidemiological, biological, and chemical models and artificial intelligence* as a key element in an advanced biosurveillance system (Thompson et al., 2005).

Each of these processes involves many individuals with different skills and many organizations with diverse and sometimes overlapping responsibilities. This situation is unlikely to change anytime in the near future, which is why we consider *multiorganizational* and *multidisciplinary* to be fundamental properties of biosurveillance that we must respect when designing biosurveillance systems.

Each of these processes is also data and knowledge intensive (also fundamental properties of biosurveillance). The processes depend not only on substantial data collection but also on mechanisms for the storage, distribution, and presentation of these data. When analyzing these data, people and, increasingly, computers, must bring enormous amounts of knowledge to bear. As in Chapter 1, the analytic processes can perhaps best be summarized by an analogy. The very best outbreak investigators have minds like the great Sherlock Holmes. They are capable of great leaps of insight that appear "elementary" only in retrospect. They arrive at the scene of an outbreak, assimilate the information available from patients and other observers, collect clues, and generate hypotheses about the culprit (the biological agent) and his accomplices (food, water, mail). Their secret: exhaustive knowledge about the modus operandi of hundreds of biological agents produced by scientific studies of past outbreaks. The very best investigators use this knowledge in a way that the clues that they receive and the evidence they collect (e.g., symptoms, test results, and epidemiological patterns) ultimately lead them to

5 The word forensics is derived from the Latin *forensis* meaning legal affairs. *Forensic attribution* assigns responsibility to an individual for an act (but not necessarily to a level required by a specific court as connoted by the term *forensic science*). Forensic attribution can be difficult as demonstrated by the two bioterrorist events that occurred in the U.S. Investigators did not identify the party responsible for the Salmonella outbreak in The Dulles, Oregon—the Bhagwan Shree Rajneesh cult. Although they considered intentional contamination as the source of the outbreak during their investigation, they rejected this idea because there were other plausible theories and no claims of responsibility, no motive, and no observed unusual behavior. Ultimately, an unrelated criminal investigation of the Bhagwan Shree Rajneesh cult uncovered the fact that the cult was responsible for the Salmonella outbreak. The party or parties responsible for the 2001 anthrax letters were never identified.

the biological agent and its source. The deductive techniques they use are numerous, and the selection is dictated by the problem at hand. The potential for formalizing and encoding this knowledge in computer-supported biosurveillance systems is significant.

ADDITIONAL RESOURCES

Centers for Disease Control and Prevention. (2005). Annotated Bibliography for Syndromic Surveillance. Atlanta: Epidemiology Program Office, CDC. *http://www.cdc.gov/epo/dphsi/syndromic/.*

Rothman, K. and Greenland, S., eds. (1998). *Modern Epidemiology.* Philadelphia: Lippincott-Raven Publishers. This work describes types of epidemiological studies such as case-control studies and field methods.

U.S. Department of Health and Human Services and Centers for Disease Control. Principles of Epidemiology: An Introduction to Applied Epidemiology and Biostatistics, 2nd ed. Atlanta: Epidemiology Program Office, CDC. *http://www.phppo.cdc.gov/PHTN/catalog/pdf-file/Epi_Course.pdf.* Readers with no formal training in epidemiology will find this an excellent basic introduction. Chapter 6 covers outbreak investigations.

Working Group on Foodborne Illness Control, *Foodborne Illness Investigation and Control Reference Manual.* Boston: Massachusetts Department of Public Health. *http://www.mass.gov/dph/pdf.* This is a comprehensive discussion of food-borne outbreaks, with chapter 7 examining environmental investigations in detail.

ACKNOWLEDGMENTS

We thank William Hogan for reading and commenting on this chapter.

REFERENCES

Arizona Department of Health Services. (2002). Syndromic Disease Surveillance in the Wake of Anthrax Threats and High Profile Public Events. *Prevention Bulletin* January/February 2002.

Ashford, D.A., et al. (2003). Planning against Biological Terrorism: Lessons from Outbreak Investigations. *Emerging Infectious Diseases* vol. 9:515–519. *http://www.ncbi.nlm.nih.gov/entrez/query.fcgi?cmd=Retrieve&db=PubMed&dopt=Citation&list_uids=12737732.*

Aymard, M., Valette, M., Lina, B., and Thouvenot, D., and the members of Groupe Régional d'Observation de la Grippe and European Influenza Surveillance Scheme. (1999). Surveillance and Impact of influenza in Europe. *Vaccine* vol. 17:S30–S41.

Buehler, J. (2004). Review of the 2003 National Syndromic Surveillance Conference: Lessons Learned and Questions To Be Answered. *Morbidity and Mortality Weekly Report* vol. 53:18–22.

Campbell, M., et al. (2004). An Evaluation of Over-the-Counter Medication Sales for Syndromic Surveillance.

Centers for Disease Control and Prevention [CDC]. (1999a). Control of Infectious Diseases. *Morbidity and Mortality Weekly Report* vol. 48:621–629. *http://www.ncbi.nlm.nih.gov/entrez/query.fcgi?cmd=Retrieve&db=PubMed&dopt=Citation&list_uids=10458535.*

CDC. (1999b). Outbreak of *Salmonella* Serotype Muenchen Infections Associated with Unpasteurized Orange Juice, United States and Canada, June 1999. *Morbidity and Mortality Weekly Report* vol. 48:582–585.

CDC. (1999c). Outbreaks of Gastrointestinal Illness of Unknown Etiology Associated with Eating Burritos, United States, October 1997–October 1998. *Morbidity and Mortality Weekly Report* vol. 48:210–213.

CDC. (1999d). Outbreaks of *Shigella sonnei* Infection Associated with Eating Fresh Parsley, United States and Canada, July–August 1998. *Morbidity and Mortality Weekly Report* vol. 48: 285–289.

CDC. (1999e). Public Health Dispatch: Outbreak of *Escherichia coli* O157:H7 and Campylobacter among Attendees of the Washington County Fair, New York, 1999. *Morbidity and Mortality Weekly Report* vol. 48:803. *http://www.cdc.gov/mmwr/preview/mmwrhtml/mm4836a4.htm.*

CDC. (2000a). *Escherichia coli* O111:H8 Outbreak among Teenage Campers, Texas, 1999. *Morbidity and Mortality Weekly Report* vol. 49:321–324.

CDC. (2000b). Foodborne Outbreak of Group A Rotavirus Gastroenteritis among College Students, District of Columbia, March–April 2000. *Morbidity and Mortality Weekly Report* vol. 49:1131–1133.

CDC. (2000c). Multistate Outbreak of Listeriosis, United States, 2000. *Morbidity and Mortality Weekly Report* vol. 49: 1129–1130.

CDC. (2000d). Outbreak of Gastroenteritis Associated with an Interactive Water Fountain at a Beachside Park, Florida, 1999. *Morbidity and Mortality Weekly Report* vol. 49:565–568.

CDC. (2000e). Outbreak of *Shigella sonnei* Infections Associated with Eating a Nationally Distributed Dip, California, Oregon, and Washington, January 2000. *Morbidity and Mortality Weekly Report* vol. 49:60–61.

CDC. (2000f). Outbreaks of Norwalk-like Viral Gastroenteritis, Alaska and Wisconsin, 1999. *Morbidity and Mortality Weekly Report* vol. 49: 207–211.

CDC. (2002a). Principles of Epidemiology: A CDC Self-Study Course. *http://www.phppo.cdc.gov/phtn/catalog/3030g.asp.*

CDC. (2002b). Syndromic Surveillance for Bioterrorism Following the Attacks on the World Trade Center, New York City, 2001. *Morbidity and Mortality Weekly Report* vol. 51:13–15. *http://www.ncbi.nlm.nih.gov/entrez/query.fcgi?cmd=Retrieve&db=PubMed&dopt=Citation&list_uids=12238536.*

CDC. (2003a). Hepatitus A Outbreak Associated with Green Onions at a Restaurant, Monaca, Pennsylvania, 2003. *Morbidity and Mortality Weekly Report* vol. 52:1155–1157.

CDC. (2003b). Update: Multistate Outbreak of Monkeypox, Illinois, Indiana, Kansas, Missouri, Ohio, and Wisconsin. *Morbidity and Mortality Weekly Report* 52:642–646. *http://www.cdc.gov/mmwr/preview/mmwrhtml/mm5227a5.htm.*

CDC. (2005). Case Definitions. *http://www.bt.cdc.gov/episurv/#casedef.*

Chapman, W.W., Dowling, J.N., and Wagner, M.M. (2005). Classification of Emergency Department Chief Complaints into Seven Syndromes: A Retrospective Analysis of 527,228 patients. *Annals of Emergency Medicine* vol. 46:445–455.

Cooper, G.F. (1989). Current Research Directions in the Development of Expert Systems Based on Belief Networks. *Applied Stochastic Models and Data Analysis* vol. 5:39–52.

Coughlin, S.S. and Beauchamp, T.L., eds. (1996). *Ethics and Epidemiology.* Oxford: Oxford University Press.

County of Los Angeles. (2000). Democratic National Convention: Bioterrorism Syndromic Surveillance. *http://www.lapublichealth.org/acd/reports/spclrpts/spcrpt00/DemoNatConvtn00.pdf.*

Dafni, U., et al. (2004). Algorithm for Statistical Detection of Peaks: Syndromic Surveillance System for the Athens 2004 Olympics Games. *Morbidity and Mortality Weekly Report* vol. 53:86–94.

Das, D., Weiss, D., Mostashari, F., Treadwell, T., McQuiston, J., Hutwagner, L., et al. (2003). Enhanced Drop-In Syndromic Surveillance in New York City following September 11, 2001. *Journal of Urban Health* vol. 80(Suppl 1):i76–i88. *http://www.ncbi.nlm.nih.gov/entrez/query.fcgi?cmd=Retrieve&db=PubMed&dopt=Citation&list_uids=12791782.*

Dato, V., Wagner, M., Allswede, M., Aryel, R., and Fapohunda, A. (2001). The Nation's Current Capacity for the Early Detection of Public Health Threats including Bioterrorism. Washington: Agency for Healthcare Research and Quality.

Dato, V., Wagner, M.M., and Fapohunda, A. (2004). How Outbreaks of Infectious Disease Are Detected: A Review of Surveillance Systems and Outbreaks. *Public Health Report* vol. 119:464–471. *http://www.ncbi.nlm.nih.gov/entrez/query.fcgi?cmd=Retrieve&db=PubMed&dopt=Citation&list_uids=15313109.*

DoD-GEIS. (2000). Electronic Surveillance System for Early Notification of Community-Based Epidemics (ESSENCE). *http://www.geis.ha.osd.mil/getpage.asp?page=SyndromicSurveillance.htm&action=7&click=KeyPrograms.*

Effler, P., Ching-Lee, M., Bogard, A., Man-Cheng, L., Nekomoto, T., and Jernigan, D. (1999). Statewide System of Electronic Notifiable Disease Reporting from Clinical Laboratories: Comparing Automated Reporting with Conventional Methods. *JAMA* vol. 282:1845–1850.

Espino, J.U., et al. (2004). Removing a Barrier to Computer-Based Outbreak and Disease Surveillance: The RODS Open Source Project. *Morbidity and Mortality Weekly Report* vol. 53:32–39.

Evans, R.S. (1991). The HELP System: A Review of Clinical Applications in Infectious Diseases and Antibiotic Use. *M.D. computing: computers in medical practice* vol. 8:282–288, 315.

Evans, R.S., et al. (1992). Computerized Identification of Patients at High Risk for Hospital-Acquired Infection. *American Journal of Infection Control* vol. 20:4–10.

Evans, R.S., et al. (1985). Development of a Computerized Infectious Disease Monitor (CIDM). *Computers and Biomedical Research* vol. 18:103–113.

Evans, R.S., et al. (1986). Computer Surveillance of Hospital-Acquired Infections and Antibiotic Use. *JAMA* vol. 256:1007–1011.

Evans, R.S., et al. (1998). A Computer-Assisted Management Program for Antibiotics and Other Antiinfective Agents. *New England Journal of Medicine* vol. 338:232–238. *http://www.ncbi.nlm.nih.gov/entrez/query.fcgi?cmd=Retrieve&db=PubMed&dopt=Citation&list_uids=9435330.*

Ewert, D.P., Frederick, P.D., Run, G.H., and Mascola, L. (1994). The Reporting Efficiency of Measles by Hospitals in Los Angeles County, 1986 and 1989. *American Journal of Public Health* vol. 84:868–899. *http://www.ncbi.nlm.nih.gov/entrez/query.fcgi?cmd=Retrieve&db=PubMed&dopt=Citation&list_uids=8192820.*

Ewert, D.P., Westman, S., Frederick, P.D., and Waterman, S. H. (1995). Measles Reporting Completeness during a Community-wide Epidemic in Inner-City Los Angeles. *Public Health Report* vol. 110:161–165. *http://www.ncbi.nlm.nih.gov/entrez/query.fcgi?cmd=Retrieve&db=PubMed&dopt=Citation&list_uids=7630992.*

Fleming, D. and Cohen, J. (1996). Experience of European Collaboration in Influenza Surveillance in the Winter 1993–1994. *Journal of Public Health Medicine* vol. 18:133–142.

GeoSentinel. (2005). The Global Surveillance Network of the ISTM and CDC. *http://www.istm.org/geosentinel/main.html.*

Gesteland, P.H., et al. (2003). Automated Syndromic Surveillance for the 2002 Winter Olympics. *Journal of the American Medical Informatics Association: JAMIA* vol. 10:547–554. *http://www.ncbi.nlm.nih.gov/entrez/query.fcgi?cmd=Retrieve&db=PubMed&dopt=Citation&list_uids=12925547.*

Gregg, M.B., ed. (2002). *Field Epidemiology.* Oxford: Oxford University Press.

Griffith, K., Mead, P., Armstrong, G., Painter, J., Kelley, K., and Hoffmaster, A. (2003). Bioterrorism-related Inhalational Anthrax in an Elderly Woman, Connecticut, 2001. *Emerging Infectious Diseases* vol. 9:681–688. *http://www.cdc.gov/ncidod/EID/vol9no6/02-0728.htm.*

Hashimoto, S., Murakami, Y., Taniguchi, K., and Nagai, M. (2000). Detection of Epidemics in their Early Stage through Infectious Disease Surveillance. *International Journal of Epidemiology* vol. 29:905–910.

Heffernan, R., Mostashari, F., Das, D., Karpati, A., Kuldorff, M., and Weiss, D. (2004). Syndromic Surveillance in Public Health Practice, New York City. *Emerging Infectious Diseases* vol. 10:858–864. *http://www.ncbi.nlm.nih.gov/entrez/query.fcgi?cmd=Retrieve&db=PubMed&dopt=Citation&list_uids=15200820.*

Henning, K.J. (2004). What is syndromic surveillance?, *Morbidity and Mortality Weekly Report* vol. 53(Suppl):5–11 *http://www.ncbi.nlm.nih.gov/entrez/query.fcgi?cmd=Retrieve&db=PubMed&dopt=Citation&list_uids=15714620.*

Hoffman, M., et al. (2003). Multijurisdictional Approach to Biosurveillance, Kansas City. *Emerging Infectious Diseases* vol. 9:1281–1286. *http://www.cdc.gov/ncidod/EID/vol9no10/03-0060.htm.*

Hogan, W.R., et al. (2003). Detection of Pediatric Respiratory and Diarrheal Outbreaks from Sales of Over-the-Counter Electrolyte Products. *Journal of the American Medical Informatics Association: JAMIA* vol. 10:555–562. *http://www.ncbi.nlm.nih.gov/entrez/query.fcgi?cmd=Retrieve&db=PubMed&dopt=Citation&list_uids=12925542.*

Hutwagner, L.C., Maloney, E.K., Bean, N.H., Slutsker, L., and Martin, S.M. (1997). Using Laboratory-Based Surveillance Data for Prevention: An Algorithm for Detecting *Salmonella* Outbreaks. *Emerging Infectious Diseases* vol. 3:395–400.

Ivanov, O., Gesteland, P.H., Hogan, W., Mundorff, M.B., and Wagner, M.M. (2003). Detection of Pediatric Respiratory and Gastrointestinal Outbreaks from Free-Text Chief Complaints. *Proceedings of American Medical Informatics Symposium* 318–322. *http://www.ncbi.nlm.nih.gov/entrez/query.fcgi?cmd=Retrieve&db=PubMed&dopt=Citation&list_uids=14728186.*

Jain, N.L., Knirsch, C.A., Friedman, C., and Hripcsak, G. (1996). Identification of Suspected Tuberculosis Patients Based on Natural Language Processing of Chest Radiograph Reports. *Proceedings of the Fall Symposium of the American Medical Informatics Association* 542–546.

Jernigan, D., et al. (2002). Investigation of Bioterrorism-Related Anthrax, United States, 2001: Epidemiologic Findings. *Emerging Infectious Diseases* vol. 8:1019–1028. *http://www.cdc.gov/ncidod/EID/vol8no10/02-0353.htm.*

Jernigan, J.A. et al. (2001). Bioterrorism-related Inhalational Anthrax: The First 10 Cases Reported in the United States. Bethesda, MD: Centers for Disease Control and Prevention.

Kahn, M.G., Steib, S.A., Fraser, V.J., and Dunagan, W.C. (1993). An Expert System for Culture-based Infection Control Surveillance. *Proceedings of the Annual Symposium on Computer Applications in Medical Care* 171–175.

Kahn, M.G., Steib, S.A., Spitznagel, E.L., Claiborne, D.W., and Fraser, V.J. (1995). Improvement in User Performance following Development and Routine Use of an Expert System. *Medinfo* vol. 8:1064–1067.

Kalter, H. (1992). The Validation of Interviews for Estimating Morbidity. *Health Policy and Planning* vol. 7:30–39. *http://www.ncbi.nlm.nih.gov/entrez/query.fcgi?cmd=Retrieve&db=PubMed&dopt=Citation&list_uids=10117988.*

Lazarus, R., Kleinman, K. P., Dashevsky, I., DeMaria, A., and Platt, R. (2001). Using Automated Medical Records for Rapid Identification of Illness Syndromes (Syndromic Surveillance): The Example of Lower Respiratory Infection. *BMC Public Health*, vol. 1:9. *http://www.ncbi.nlm.nih.gov/entrez/query.fcgi?cmd=Retrieve&db=PubMed&dopt=Citation&list_uids=11722798.*

Lewis, M., et al. (2002). Disease Outbreak Detection System using Syndromic Data in the Greater Washington DC Area. *American Journal of Preventive Medicine* vol. 23:180. *http://www.ncbi.nlm.nih.gov/entrez/query.fcgi?cmd=Retrieve&db=PubMed&dopt=Citation&list_uids=12350450.*

Lombardo, J., et al. (2003). A Systems Overview of the Electronic Surveillance System for the Early Notification of Community-based Epidemics (ESSENCE II). *Journal of Urban Health* vol. 80(Suppl 1):i32–i42. *http://www.ncbi.nlm.nih.gov/entrez/query.fcgi?cmd=Retrieve&db=PubMed&dopt=Citation&list_uids=12791777.*

Manuguerra, J. and Mosnier, A., on behalf of EISS (European Influenza Surveillance Scheme). (2000). Surveillance of Influenza in Europe from October 1999 to February 2000. *Eurosurveillance* vol. 5:63–68.

Massachusetts Department of Public Health. (2005a). Conducting an Environmental Investigation. *http://www.mass.gov/dph/pdf/ch7.pdf.*

Massachusetts Department of Public Health. (2005b). Food Protection Program. *http://www.mass.gov/dph/fpp/refman.htm.*

Meehan, P., Toomey, K.E., Drinnon, J., Cunningham, S., Anderson, N., and Baker, E. (1998). Public Health Response for the 1996 Olympic Games. *JAMA* vol. 279:1469–1473. *http://www.ncbi.nlm.nih.gov/entrez/query.fcgi?cmd=Retrieve&db=PubMed&dopt=Citation&list_uids=9600482.*

Meltzer, M.L., Damon, I., LeDuc, J.W., and Millar, J.D. (2001). Modeling Potential Responses to Smallpox as a Bioterrorist Weapon. *EID* vol. 7:959–969.

Military Postal Service Agency. (2004). Biohazard Detection System (BDS) Briefing Sheet. *http://hqdainet.army.mil/mpsa/nov_conf/ppt/pdf/BDS.pdf.*

Moran, G.J. and Talan, D.A. (2003). Syndromic Surveillance for Bioterrorism Following the Attacks on the World Trade Center, New York City, 2001. *Annals of Emergency Medicine* vol. 41:414–418. *http://www.ncbi.nlm.nih.gov/entrez/query.fcgi=cmd=Retrieve&db=PubMed&dopt=Citation&list_uids=12605212.*

Mundorff, M.B., Gesteland, P., Haddad, M., and Rolfs, R.T. (2004). Syndromic Surveillance Using Chief Complaints from Urgent-Care Facilities during the Salt Lake 2002 Olympic Winter Games. *Morbidity and Mortality Weekly Report* vol. 53:254.

Nordin, J.D., Harpaz, R., Harper, P., and Rush, W. (2004). Syndromic Surveillance for Measleslike Illnesses in a Managed Care Setting. *Journal of Infectious Diseases* vol. 189(Suppl 1): S222–S226. *http://www.ncbi.nlm.nih.gov/entrez/query.fcgi=cmd=Retrieve&db=PubMed&dopt=Citation&list_uids=15106115.*

Overhage, J.M., Suico, J., and McDonald, C. J. (2001). Electronic Laboratory Reporting: Barriers, Solutions and Findings. *Journal of Public Health Management and Practice* vol. 7:60–66. *http://www.ncbi.nlm.nih.gov/entrez/query.fcgi?cmd=Retrieve&db=PubMed&dopt=Citation&list_uids=11713754.*

Panackal, A.A., et al. (2001). Automatic Electronic Laboratory-Based Reporting of Notifiable Infectious Diseases. *Emerging Infectious Diseases* vol. 8:685–691.

Platt, R., et al. (2003). Syndromic Surveillance Using Minimum Transfer of Identifiable Data: The Example of the National Bioterrorism Syndromic Surveillance Demonstration Program. *Journal of Urban Health* vol. 80(Suppl 1):i25–i31. *http://www.ncbi.nlm.nih.gov/entrez/query.fcgi?cmd=Retrieve&db=PubMed&dopt=Citation&list_uids=12791776.*

Rothman, J. and Greenland, S. (1998). *Modern Epidemiology,* 2nd ed. Philadelphia: Lippincott Williams and Williams.

Schoub, B., McAnerney, J., and Besselaar, T. (2002). Regional Perspectives on Influenza Surveillance in Africa. *Vaccine* vol. 20(Suppl 2):S45–46.

Serfling, R. (1963). Methods for Current Statistical Analysis of Excess Pneumonia-Influenza Deaths. *Public Health Reports* vol. 78:494–506.

Shilts, R. and Greider, W. (1987). *And the Band Played On: Politics, People, and the AIDS Epidemic.* New York: St. Martins Press.

Sims, L., et al. (2003). Avian Influenza in Hong Kong 1997–2002. *Avian Disease* vol. 47:832–838.

Snacken, R., Bensadon, M., and Strauss, A. (1995). The CARE Telematics Network for the Surveillance of Influenza in Europe. *Methods of Information in Medicine* vol. 34:518–522.

Snacken, R., Manuguerra, J.C., and Taylor, P. (1998). European Influenza Surveillance Scheme on the Internet. *Methods of Information in Medicine* vol. 37:266–270. *http://www.ncbi.nlm.nih.gov/entrez/query.fcgi?cmd=Retrieve&db=PubMed&dopt=Citation&list_uids=9787627.*

Snow, J. (1855). *On the Mode of Communication of Cholera.* London: John Churchill.

Stern, L. and Lightfoot, D. (1999). Automated Outbreak Detection: A Quantitative Retrospective Analysis. *Epidemiology and Infection* vol. 122:103–110.

Thompson, K.M., Armstrong, R.E., and Thompson, D.F. for the National Defense University Center for Technology and National Security Policy. (2005). Bayes, Bugs, and Bioterroristis: Lessons Learned from the Anthrax Attacks. Washington, DC: Fort Lesley J. McNair.

Tsui, F., Wagner, M., Dato, V., and Chang, C. (2001). Value of ICD-9-Coded Chief Complaints for Detection of Epidemics. *Proceedings of the Fall Symposium of the American Medical Informatics Association* 711–715.

Tsui, F.-C., Espino, J.U., Dato, V.M., Gesteland, P.H., Hutman, J., and Wagner, M.M. (2003). Technical Description of RODS: A Real-Time Public Health Surveillance System. *Journal of the American Medical Informatics Association* vol. 10:399–408. *http://www.jamia.org/preprints.shtml.*

U.S. Food and Drug Administration [FDA]. (2003). FDA Update on Recent Hepatitis A Outbreaks Associated with Green Onions From Mexico. *http://www.fda.gov/bbs/topics/NEWS/2003/NEW00993.html.*

FDA. (2004). Protecting the Food Supply. *http://www.cfsan.fda.gov/~dms/fsbtac23.html.*

Wagner, M., et al. (2004). Syndrome and Outbreak Detection from Chief Complaints: The Experience of the Real-Time Outbreak and Disease Surveillance Project. *Morbidity and Mortality Weekly Report* vol. 53:28–31.

World Health Organization [WHO]. (2003). Case Definitions for Surveillance of Severe Acute Respiratory Syndrome (SARS). *http://www.who.int/csr/sars/casedefinition/en/.*

World Resources Institute. (1998). REPORT SERIES: World Resources 1998–99: Environmental Change and Human Health. *http://population.wri.org/pubs_description.cfm=PubID=2889.*

Yih, W.K., et al. (2004). National Bioterrorism Syndromic Surveillance Demonstration Program. *Morbidity and Mortality Weekly Report* vol. 53:43–49. *http://www.ncbi.nlm.nih.gov/entrez/query.fcgi?cmd=Retrieve&db=PubMed&dopt=Citation&list_uids=15714626.*

Zambon, M. (1998). Sentinel Surveillance of influenza in Europe, 1997/1998. *Eurosurveillance* vol. 3:29–31.

4

CHAPTER

Functional Requirements for Biosurveillance

Michael M. Wagner
RODS Laboratory, Center for Biomedical Informatics, University of Pittsburgh, Pittsburgh, Pennsylvania

Loren Shaffer
Bureau of Health Surveillance, Department of Prevention, Ohio Department of Health, Columbus, Ohio

Richard Shephard
Australian Biosecurity Cooperative Research Centre for Emerging Infectious Disease, Brisbane, Australia

1. INTRODUCTION

Chapters 2 and 3 described biosurveillance as the world has practiced it for the latter half of the 20th century. During that time, the basic methods for detecting cases, detecting outbreaks, and characterizing outbreaks changed little. The methods used to detect and characterize the 1975 Lyme disease outbreak and the 2003 severe acute respiratory syndrome (SARS) pandemic differed primarily in microbiological techniques (e.g., the increasing use of genetic analysis) and the speed at which outbreaks were investigated.

Around the beginning of the 21st century, however, researchers began to investigate new types of surveillance data and to automate methods for the collection and analysis of these data. A new requirement motivated these researchers—that of very early detection of disease outbreaks (Wagner et al., 2001a). The new methods were met with skepticism (Broome et al., 2002; Buehler et al., 2003; Reingold, 2003; Stoto et al., 2004). Their rate of adoption was slow until the fall of 2001, when the anthrax mail attacks in the United States effectively ushered in a new era of biosurveillance (Wagner, 2002).

We discuss these newer methods in detail in Parts III through V of this book. In this chapter, we discuss the redesign of biosurveillance systems from the perspective of an engineer or a system analyst. To satisfy the new requirement for very early detection, a designer must pay more attention to the *systematic* aspect of biosurveillance. The designer must examine how quickly outbreaks must be detected and characterized and design a system that can meet these requirements by collecting data that are available earlier than are case reports, collecting those data in real time from existing computer systems in hospitals and other organizations, and analyzing the data in real time.

2. FUNCTIONAL REQUIREMENTS AND SYSTEM SPECIFICATIONS

An engineer or a system analyst approaches design problems in a manner that is fundamentally different from how a doctor or an epidemiologist approaches diagnostic problems. Engineers and systems analysts first develop *functional requirements*, which for biosurveillance are specifications of the diseases that must be detected, the smallest size outbreak that must be detected, and the time frame within which detection must occur. From the functional requirements, the designer then develops *system specifications* and finally builds a system.

When designing and building a commercial information system, the elucidation of functional requirements is a first step in the process, and it is a prerequisite for subsequent steps. If the information system is an early warning system for missile attack, for example, the functional requirements might prominently feature the detection of the attack within several minutes of launch. In the case of a biosurveillance system for an aerosol release of anthrax, the functional requirements might similarly emphasize detection as quickly as possible, but no later than within days of release.

Although a biosurveillance system is fundamentally an information system, the process of functional-requirement specification is often less rigorous than in the commercial world. Even when organizations develop functional requirements (e.g., for an electronic disease reporting system), the requirements do not specify how quickly an outbreak of disease must be detected, rather they are formulated in terms of the data that should be collected, the properties of the user interface, and system security. The functional requirements of the Public Health Information Network (PHIN) of the Centers for Disease Control and Prevention (CDC) are an example of current functional requirement specifications (www.cdc.gov/phin). To our knowledge, no organization has published functional requirements derived from explicit consideration of timeliness requirements for specific diseases. System designers have been let off the hook, so to speak, by their customers for this—arguably, the most difficult—requirement. The two published analyses that considered timeliness requirements were partial analyses: one analyzed gaps in current biosurveillance systems (Dato et al., 2001), and the second analyzed the data requirements for earlier detection (Wagner et al., 2001b).

ISBN 0-12-369378-0

3. EXAMPLE: FUNCTIONAL REQUIREMENTS AND SPECIFICATIONS FOR ANTHRAX BIOSURVEILLANCE

To illustrate the approach that an engineer or systems analyst (henceforth referred to as *designers*) would take when designing a biosurveillance system, let us consider a special purpose system for detection of outbreaks caused by the organism *Bacillus anthracis,* a bacterium that can form a spore that can survive for extended periods in nature. The spores can infect humans or animals through the skin, through ingestion, and through inhalation. *B. anthracis* causes the disease anthrax, which was once a common disease among wool handlers but now is of concern as a terrorist threat (Henderson, 1999).

A designer tasked with the design of an anthrax biosurveillance system has available two outbreaks of anthrax from which to derive timeliness (and other) functional requirements. He has available the 1979 Sverdlovsk release of *B. anthracis* (*Kirov* strain) from Soviet Biological Weapons Compound 19, described in Chapter 2, and the 2001 U.S. postal attacks (Jernigan et al., 2001; Greene et al., 2002). The Sverdlovsk outbreak was a result of accidental release of anthrax spores into the air. Unknown individuals used envelopes containing anthrax spores to carry out the U.S. postal attacks.

3.1. Large Aerosol Release

Kaufmann et al. (1997) analyzed the available information from the Sverdlovsk outbreak and information about the disease anthrax. They demonstrated that the requirement for timeliness of detection of an aerosol release of anthrax is ideally the moment of release, but no later than 5 days after release.

Wagner et al. (2003b) used a taxonomy of surveillance data depicted in Table 4.1 to identify the types of surveillance data that might be available in this time window. They concluded that the system specifications for a biosurveillance system capable of meeting the time requirement would include components to obtain and process data from biosensors, *preclinical* data sources (e.g., sales of cough syrup), and early clinical data (e.g., symptoms and radiological reports). A system would have to collect and analyze these data in near real time with attention to corroborating and discriminating data from other sources, such as wind patterns and physical location of individuals in the days preceding onset of illness. They concluded that conventionally trained physicians could not be relied on to detect an outbreak of this type. In the Sverdlovsk outbreak, the earliest suspicion of anthrax came from an autopsy finding of a cardinal's cap (hemorrhagic meningitis, a pathognomonic finding for the disease anthrax) on the eighth day after the release (Abramova et al., 1993), by which time 14 individuals had died (Guillemin, 2001). At least six victims had their early symptoms dismissed by physicians as not serious, and 21 individuals had died by the time that the laboratory confirmation of anthrax was broadcast to area hospitals on the 10th day after the accidental release (Guillemin, 2001).

TABLE 4.1 Taxonomy of Surveillance Data

- *Preoutbreak data:* This category refers to data obtained during the period before the onset of the outbreak. Examples of data that might contribute to detection include intelligence that heightens suspicion or host factors such as vaccinations that determine susceptibility.
- *Attack, release, or exposure data* refers to data obtained at or very near the time of an accidental of intentional contamination. Data from this time might come from biosensor arrays, unauthorized airplane flights, or other activities.
- *Presymptomatic data (incubation period data)* refers to data obtained between the time that a person or animal becomes infected until the recognition of first symptoms. Examples of presymptomatic data are serology or cultures from presymptomatic individuals that are obtained serendipitously, through routine screening, or through enhanced screening because environmental conditions are favorable for an outbreak (e.g., there has been a natural disaster such as a flood).
- *Prediagnostic data* refers to data from the period between the onset of symptoms in an individual and when the illness becomes more fully developed and distinguishable from other illnesses. Examples of prediagnostic data include diarrheal symptoms or upper respiratory symptoms. Examples of data sources of potential value for the detection of individuals experiencing early symptoms include sales of over-the-counter cold medicines, vital signs, physical findings, and absenteeism.
- *Specific syndrome data* are data that either singly or in combination strongly suggest a specific agent. Examples include selected symptoms, histories of exposures, vital signs, physical findings, laboratory results, radiology results, and preliminary results from microbiology laboratories (e.g., Gram stains).
- *Diagnostic data* are data that are sufficient on their own to conclude that a patient has a disease. Examples include microbiology cultures or autopsy reports. Diagnostic data usually can be obtained during the specific syndrome period but can also be found during other periods through screening, routine testing, or testing of the environment.

Three additional categories of data in the table do not fit neatly on a timeline. They are:

- *Epidemiologic data,* which refers to the whereabouts of individuals before onset of disease, food and water consumption, contacts with affected individuals, and location information (work, home addresses; feedlot number). These data are often only available for analysis after the onset of the outbreak because they are not routinely collected, but some characteristics may be routinely recorded electronically in various databases.
- *Zoonotic data,* which refer to data from veterinary and public health sources. Examples include small animal deaths, positive mosquito pools for malaria, animal vaccination status, and sentinel chicken serology.
- *Environmental data* refer to data about the environment. Examples include weather, refrigeration temperature, ventilation plans, and water supply areas.

From Dato et al. (2001).

3.2. Postal Attack (2001)

After the fall 2001 attack, the U.S. Postal Service developed the biohazard detection system (BDS), an air monitoring system based on a DNA polymerase chain reaction (PCR) test for *B. anthracis* (U.S. Postal Service, 2004; Military Postal Service Agency, 2004). BDS attaches to mail-sorting machines. There is little doubt that the U.S. Postal Service designed this system with the functional requirement of early detection of an anthrax postal attack based on a careful postincident analysis of the 2001 postal attack. A mail-sorting machine is both a single point through which all mail passes (except locally routed rural mail and larger packages) and a nearly ideal device for expressing spores from all but the most tightly sealed envelops (because the sorting process compresses envelops), enabling the detection of spores by BDS.

A comprehensive system for detection of a postal attack, however, would require additional components to monitor human health to detect individuals that become infected via envelops and packages that were either well sealed or not sorted by a monitored mail sorting machine. The additional components would have to detect (1) an individual case of anthrax in a recipient of a single envelop, (2) a cluster of cases in a home or office in which the envelop was opened, or (3) a pattern of individuals or building clusters that might indicate an attack that used multiple letters or packages. The system would also have to monitor for suspicious cases in local postal facilities not defended by locally installed sensors either directly or through sensing upstream of their facility in the mail processing network.

3.3. Building/Vessel Contamination

An analysis by Wagner et al. (2003b) identified two additional anthrax attack scenarios: building/vessel contaminations and premonitory release. Specifically, a building or vessel contamination refers to the distribution of anthrax (or other agent) via the mechanical components in a building or ship. This is a serious threat because of our modern reliance on heating, ventilation, and air-conditioning systems, which can effectively disseminate spores throughout a building. The functional requirement for time of detection is similar to that of an outdoor release because many individuals would be exposed simultaneously. Recognition of a release contained within a structure, however, produces somewhat different requirements. Specifically, the requirement is the detection of a cluster of illnesses common to a relatively small number of individuals sharing a domicile, a place of employment, or a social facility. This detection requires recognition and analysis of these relationships. A detection system ideally would have access to data about heating, ventilation, and air-conditioning systems, identity of occupants, and hours of occupation. The building contamination with *B. anthracis* of a postal facility in New Jersey during the 2001 postal attack is an interesting example for study (Greene et al., 2002). Tracing the illnesses to the specific building required the knowledge of work times, responsibilities, and routines for a large number of postal workers.

3.4. Small Premonitory Release or Contamination

The term *premonitory release* refers to intentional or accidental infection of one or a limited number of individuals with an unusual organism such as *B. anthracis.* The functional requirement here is one of sensitivity for single cases and small outbreaks, not extreme timeliness. To detect a single case, a biosurveillance system must have extremely high case detection sensitivity, specificity, and diagnostic precision (or the prior probability must be extremely high, e.g., owing to intelligence information). A biosurveillance system would have to rely either on case detection by the healthcare system or on computer-based case detection. Computer-based case detection would have to be capable of diagnostic precision at least at the level of finding individuals with Gram-positive rods in the blood or cerebrospinal fluid and pneumonia on chest radiograph (which would be highly suggestive of anthrax). Examples of potential computer-based components include clinical information systems with decision support at the point of care, systems to monitor laboratory reporting of microbiology cultures; and free-text processing algorithms that scrutinize autopsy reports, newspaper stories, and obituaries for unusual deaths of animals or humans. If there are multiple cases, the demographics of the victims or the discovery of a geographic clustering of victims could help to identify a common cause with case detection at lower levels of diagnostic precision. In the absence of astute clinical diagnosis, it is likely that a single case of disease caused by a weaponized organism will progress to fatality. The requirements, therefore, include biosurveillance components (manual or automatic) that analyze unexplained deaths.

The problem of detecting a single case is identical to the problem of accurate diagnosis in medicine, and there is great deal of literature on clinical decision support describing relevant techniques, which is summarized in Miller (1994).

4. THE COMPLEXITY OF BIOSURVEILLANCE SYSTEM DESIGN

A designer of a comprehensive biosurveillance system for a city or a country would have to elucidate functional requirements for detection of hundreds of biological agents that can cause disease in animals and humans (and, through mutation, the number continues to expand). As in the case of anthrax, many of these agents can infect humans or animals through diverse pathways (amplified by the ingeniousness of terrorists), resulting in an almost infinite variety of outbreaks that a biosurveillance system must be capable of recognizing in a timely manner. This section provides but a sample of these biological agents, those identified by international and national organizations as being of the greatest concern. Our purpose in this section is to indicate the magnitude and complexity of the design problem.

4.1. Biological Agents that Threaten Human Populations

Table 4.2 is a list of biological agents that we created by consolidating lists developed by internationally recognized organizations and experts. Five of the primary sources listed bioterrorism threats; the sixth, nationally notifiable diseases (Wagner et al., 2003b). We sorted the list by the number of lists each threat appeared on to bring the consensus threats to the top. Note that several of the viral entries in the table represent classes of viruses that contain many individual viruses (e.g., the entry alphaviruses includes Venezuelan, eastern, western, and equine encephalomyelitis)

TABLE 4.2 Human Disease Threats

Threat	DTRA Bioterrorism List	CDC	NATO	Russian Experts Top 11 Threats	USAMRIID	Reportable List
Anthrax, inhalational	X	A list	X	Top 4	X	X
Botulism	X	A list	X	Top 4	X	X
Plague (pneumonic)	X	A list	X	Top 4	X	X
Smallpox	X	A list	X	Top 4	X	X
Tularemia (inhaled)	X	A list	X	X	X	X
Hemorrhagic fever viruses (e.g., Omsk, Korean, Ebola, Crimean-Congo, Marburg, Junin)	X	A list	X	X	X	
Brucellosis	X	B list	X		X	X
Glanders (Melioidosis)	X	B list	X	X	X	
Q fever (*Coxiella burnetti*)	X	B list	X	X	X	
Cholera	X	B list	X		X	X
Clostridium perfringens (epsilon toxin)	X	B list	X		X	
Ricin toxin	X	B list	X		X	
Lassa fever		A list	X		X	
Yellow fever		C list	X		X	X
Shigellosis	X	B list				X
Staphylococcal enterotoxins	X	B list	X			
Encephalitis (e.g., Russian spring summer, eastern equine, Saint Louis, West Nile, Venezuelan)	X		X		X	X
Alphaviruses (Venezuelan, eastern, western, equine encephalomyelitis)		B list			X	
Cryptosporidiosis		B list				X
Escherichia coli 0157:H7		B list				X
Salmonellosis		B list				X
Mycotoxins (trichothecene)	X		X		X	
Rickettsial diseases			X		X	X
Typhoid fever	X		X			X
Venezuelan equine encephalitis			X	X	X	
Hantaviral diseases		C list			X	
Tickborne hemorrhagic fever viruses		C list			X	
Tickborne encephalitis viruses		C list			X	
Tuberculosis		C list				X
Chikun gunya fever			X		X	
Diphtheria					X	X
Encephalomyelitis viruses	X				X	
Histoplasmosis			X			X
Influenza				X		X
Marine toxins	X				X	
Palytoxin			X		X	
Psittacosis			X			X
Rocky mountain spotted fever			X			X
Saxitoxin			X		X	
Tetrodotoxin			X		X	
Typhus (epidemic rickettsial)			X	X		
Typhus (scrub)			X	X		
Viral infections					X	X
Western equine			X		X	
Yersinia					X	X

This table is a merger of lists of disease threats developed by the Defense Threats Reduction Agency (DTRA), Centers for Disease Control and Prevention (CDC), North Atlantic Treaty Organization (NATO), interviews with Russian experts, United States Army Medical Research Institute for Infectious Diseases (USAMRIID) and nationally notifiable diseases.

Diseases appearing on only one list (not included in the table): Nipah virus (CDC C list); coccidiomycosis and dengue (NATO); Machupo (USAMRIID); acquired immunodeficiency syndrome (AIDS), amebiasis, Campylobacter, carbon monoxide poisoning, *Chlamydia trachomatis,* congenital rubella syndrome, food poisoning, giardiasis, *Haemophilus influenza* type B (HIB), hepatitis A, hepatitis B, hepatitis C, Kawasaki syndrome, Legionnaires' disease, leptospirosis, Lyme disease, lymphogranuloma venereum, malaria, measles, meningitis, mumps, neisseria gonorrhea, neisseria meningitis in blood or cerebrospinal fluid, pertussis, poliomyelitis, rabies, Reye syndrome, rheumatic fever, rubella, syphilis, tetanus, toxic shock syndrome, toxoplasmosis, and trichinosis (Reportable List).

4.2. Animal Diseases

The Office International des Epizooties (OIE) is the world organization for animal health (see Chapter 7) The OIE sets guidelines and provides recommendations to minimize the risk of the spread of animal diseases and pests while facilitating trade between nations. The OIE develops policy, standards, and techniques that member countries can apply to help protect themselves from animal diseases by establishing valid import barriers for the trade of certain animals or animal products. The OIE documents these standards in the International Terrestrial Animal Health Code and the International Aquatic Animal Health Code (Office International des Epizooties, 2004a,b). These standards include lists of diseases, organized by priority.

Table 4.3 includes a selection of OIE diseases that produce high morbidity/mortality, spread rapidly, and can easily cross national boundaries. These diseases represent the greatest potential to disrupt international trade. These diseases (formerly called *List A diseases*) can produce significant socioeconomic upheaval or present a major public health risk; therefore, OIE establishes more stringent requirements for demonstrating freedom from the disease agent.

The OIE also lists diseases that cause significant socioeconomic disturbance or health risk within an affected country but generally do not cross national boundaries (Table 4.4). These diseases (formerly called *List B diseases*) are associated with requirements for reporting and demonstrating freedom from the disease or agent that are less severe than those for the pandemic agents described above; nonetheless, they are still quite restrictive.

Commercial farming and trade of aquatic animals is increasing dramatically. The aquatic environment provides unique transmission modes for disease. Therefore, the OIE has developed an Aquatic Animal Code (Table 4.5). OIE includes diseases based on the potential for international spread and the potential for transmission to humans.

5. REDUCING COMPLEXITY: THREAT PATTERNS

Although to our knowledge no organization has attempted to develop functional requirements for an "all-biological threats" system, it makes for an interesting gedanken experiment (thought experiment), which may reveal whether such a system can even be specified given the large number of organisms and variability in their presentation, or how such a specification might best be accomplished.

To design an all-threats system, a designer would have to speak with many disease experts to understand the requirements for each disease. In particular, the designer would have to understand the required timeliness of detection for each disease and the smallest size outbreak that should be detected. Just understanding these functional requirements for all biological agents would be an enormous undertaking and would only represent a first step in the development of a system specification, which would require additional analysis for each threat of the types of surveillance data and analytic methods that could satisfy the requirements at lowest cost and effort.

There is another way, however, to manage this design complexity, which is to identify a smaller number of general "patterns" that a biosurveillance system must be capable of recognizing. This idea is not terribly new; the field has appreciated the *disease-independent* value of looking for patterns ever since John Snow used a spatial pattern of cases to elucidate the cause of a cholera outbreak in London in 1854 (Snow, 1855). What is novel here is taking that idea to its logical conclusion, which entails systematically examining all threats (biological agents and their various presentations as outbreaks) to identify a set of patterns (more multidimensional than spatial patterns) that most or ideally represent all of the patterns that a biosurveillance system must be capable of recognizing.

Table 4.6 is the result of such an analysis conducted by Wagner et al. (2003b) for the diseases listed in Table 4.2. Note that the following discussion relates only the threats to human health. A similar analysis would be required for diseases that threaten only animals.

The analysis resulted in a final set of nine patterns, each of which represents a fundamentally different pattern recognition problem for a biosurveillance system. For example, large-scale aerosol releases, in general, have the same requirements for a large-scale aerosol release of *B. anthracis* described above. Building contaminations in general have the same requirements as described for a building contamination with *B. anthracis*. They also searched the literature for actual outbreaks of diseases in each category for which detailed descriptions were available for study by system designers.

The previous section on the functional requirements and specifications for anthrax biosurveillance already discussed the first three of the nine patterns in Table 4.5. We discuss the other six patterns in the following sections.

5.1. Continuous or Intermittent Release of an Agent

The fourth pattern is that produced by a continuous or intermittent release of an infectious agent or biologic toxin over time. Under such conditions, individuals entering a limited area develop disease over an extended period. The Philadelphia Legionnaires outbreak was of this type, as would have been the attempt by the Aum Shinrikyo cult to sicken the population of Tokyo by disseminating continuously liquid anthrax slurry over a period of 4 days from the rooftop of their building, had it succeeded (Mangold and Goldberg, 2000). Pets and birds were sickened, but no human cases were reported.

Early detection of this type of outbreak would require analysis of surveillance data both spatially and temporally, that is, searching for clusters of diseases or syndromes in both

TABLE 4.3 Selected Office International des Epizooties (OIE) Former List A Diseases

Disease	Species	Distribution (see OIE HandiStatus at *http://www.oie.int/hs2/report.asp*)	Source/Transmission	Clinical Signs	Diagnosis
African horse sickness	Horses, mules, donkeys	Sub-Saharan Africa, occasional outbreaks in northern Africa, Middle East, and Europe	Arthropod vector. Disease is not transmitted from animal to animal.	Supraorbital swelling	Clinical signs can be confused with other diseases. Laboratory diagnosis essential. Serology. Culture of blood, spleen, lung, or lymph nodes.
African swine fever	Pigs	Sub-Saharan Africa, Europe, South America, Caribbean	Direct transmission from infected animals. Virus can shed from animal for several months after recovery and contaminate environment.	Peracute disease generally produces death in absence of signs following 1- to 2-week incubation period. Acute, subacute, and chronic forms produce signs of varying severity that include abortion, fever, and inappetence. Incoordination, cyanosis, dyspnea, vomiting, and bloody diarrhea may develop in more severe cases.	Serology. Clinical signs and gross pathology (e.g., hemorrhages in organs and lesions seen in reticuloendothelial system) are suggestive.
Bluetongue	Sheep, goats, cattle, deer, bighorn sheep	Africa, Middle East, China, United States, Mexico, southeast Asia, Australia, northern South America	Insect vector. Disease has seasonal occurrence based on conditions that promote vector survival.	Fever, facial edema, congestion, ulceration of mucous membranes. Tongue may become swollen and protrude from mouth. This hyperemia may extend to groin, axilla, and perineum. Signs in cattle are much milder.	Serology. Culture of infected tissue.
Classical swine fever (hog cholera)	Pigs	Europe, Asia, Central and South America	Direct contact and fecal–oral route. Maintained in endemic areas by carrier animals. Transplacental infection can lead to chronically infected piglets that shed significant amount of virus over their lifetime.	Sudden fever, depression, and inappetence in acute cases. Vomiting and diarrhea develop along with skin blotching and ocular discharge. Mild cases may only present with abortion and failure to thrive.	Serology. Hemorrhages and necrosis within the gut are common. Lesions are seen in reticuloendothelial system.
Contagious bovine pleuropneumonia	Cattle, buffaloes, sheep, goats	Africa, Asia, parts of Europe	Direct contact between animals. Agent does not persist in the environment.	Anorexia, fever, dyspnea, cough, and nasal discharge. Clinical signs not always present.	Serology. Culture from nasal discharge, pleural fluid or bronchial washings, lung tissue.
Foot and mouth disease (hoof and mouth disease)	Cattle, buffalo, signs highly pigs, sheep, goats, deer	Europe, Africa, Middle East, Asia, South America	Respiratory, mechanical transmission; virus can survive for weeks in bedding. Recovered animals may serve as carriers.	Fever, depression, inappetence, and vesicle emergence followed by painful ulcers.	Serology, clinical signs highly suggestive of disease.
Highly pathogenic avian influenza (fowl plague)	Domestic poultry, waterfowl, wild game birds, pigs, humans	All continents	Direct contact with infected birds or material (e.g., feces, meat).	Sudden increases in death rates within flocks before signs appear. Clinically affected birds are depressed and have ruffled feathers, inappetence, fever, and generalized weakness. Birds will also develop profuse diarrhea, their combs will become swollen, and they will have respiratory signs.	Necrotic foci in the spleen, liver, kidney, and lungs. Exudate may be present in air sacs. Virus can be isolated from tracheal and cloacal cultures.
Lumpy skin disease	Cattle	Africa, one confirmed outbreak in Israel in 1989	Transmission thought to be via insect vector.	Can be subclinical. In acute animals, persistence fever with nasal and ocular discharge. Lactating cattle will have a marked reduction in milk production. Nodules develop on head, neck, udder, and perineum. Swollen lymph nodes and animal may be reluctant to move.	Serology. Identification of capripox virions in biopsy material.

Newcastle disease	Domestic poultry, waterfowl, wild game birds, infrequently in humans	Worldwide	Contact with carrier bird or contaminated items (e.g., equipment, feed, personnel).	Marked depression, inappetence, decreased production, recumbency, and death. Edema of the comb and diarrhea are common. A nervous form may present with limb paralysis, head tremors, dyspnea, and coughing.	Necrosis and hemorrhage of gastrointestinal tract. Microscopic changes to brain. Virus isolated from cloacal culture of live birds.
Peste des petit ruminants	Small ruminants, especially goats; white-tailed deer have been infected experimentally	Africa, Arabian Peninsula, Middle East, southwest Asia	Transmitted from animal to animal via aerosol transmission.	Resembles rinderpest in cattle. Fever, ocular and nasal discharge, diarrhea, pneumonia, stomatitis. Oral lesions develop with excessive salivation. Does not produce clinical disease in cattle.	Serology. Tentative diagnosis based upon clinical signs. Linear hemorrhages (zebra stripes) present in large intestine. Necrotic enteritis, enlarged lymph nodes, necrotic lesions on spleen, and apical pneumonia.
Rift valley fever	Ruminants, dogs, cats, monkeys, humans	Africa, Middle East. Potential spread to other parts of world given wide host and vector range.	Disease in animals usually by mosquito vector. Human infection most commonly arises from contact with contaminated tissue and discharge.	Sever fever, vomiting, nasal discharge, bloody diarrhea beginning after 1 day after exposure, with 95% mortality in young animals. Adult animals often present with less severe signs. Severe influenza-like signs common in humans.	Liver of ruminants is enlarged, focally necrotic, and friable at necropsy. Serology.
Rinderpest	Cattle, buffalo, sheep, goats, pigs	North Africa, Middle East, India, parts of Asia	Respiratory, feces, and urine. Introduction of infected animal to naïve population generally results in explosive outbreak with high morbidity.	Nasal and ocular discharge followed by high fever, depression, restlessness, inappetence, and decreased production. Progresses with oral, respiratory, and gastrointestinal ulcers. Death typically 6 to 12 days after onset of signs.	Clinical signs are highly suggestive. Confirmed by serology in early stage of disease.
Sheep pox and goat pox	Sheep, goats	Africa, Middle East, Asia	Direct contact with infected animal saliva, nasal secretions, feces, or lesions. Inhalation of aerosols. Indirect via contaminated objects (e.g., vehicles, litter). Virus can survive for years in dried scabs.	Cases may be subclinical. Fever, depression, conjunctivitis, nasal discharge, and swelling of eyelids. Lesions evolving into papules form, beginning on hair/wool-free parts of body. Cough develops as papules in lungs cause pneumonia.	Serology. Identification of agent from full skin biopsy or lung lesions needed to differentiate from lumpy skin disease.
Swine vesicular disease	Pigs	Asian and European countries	Contact with infected animal, ingestion of raw waste and feed containing infected products. Transport in contaminated vehicle.	Sudden lameness in close animal groups. Slight fever. Development of vesicles on the snout, coronary band, and interdigital spaces. Young pigs may lose the horny hoof following vesicle rupture.	Serology. Culture of vesicular fluid, blood, or feces. Clinical signs easily confused with foot and mouth disease.
Vesicular stomatitis (Indiana fever)	Cattle, horses, pigs, sheep, wild animals, and rarely goats	Western Hemisphere	Ecology of agents not well understood. Arthropod-involved transmission cycle may be likely.	Very similar to foot and mouth disease. Brief febrile period and formation of papules and vesicles in mouth, udder, and coronary band. Profuse salivation.	Serology. Rapid laboratory diagnosis of animal disease important to distinguish from foot and mouth disease.

Transmissible diseases that have the potential for very serious and rapid spread, irrespective of national borders, that are of serious socioeconomic or public health consequences, and that are of major importance in the international trade of animals and animal products.

TABLE 4.4 Selected Office International des Epizooties (OIE) former List B Diseases

Disease	Species	Distribution (See OIE HandiStatus at *http://www.oie.int/hs2/report.asp*)	Source/Transmission	Clinical Signs	Diagnosis
Anthrax (woolsorter's disease, malignant pustule, malignant carbuncle)	Cattle, sheep, goats, deer, horses, fogs, pigs, humans	Worldwide	Spores from soil. Eating undercooked meat from infected animals, exposure to hide/wool of infected animals.	Fever, behavioral changes, seizures, hematuria, gastroenteritis, pharyngeal edema, In horses; colic, enteritis, followed by edema, hemorrhage and death.	Identification of etiologic agent in stained smears or culture of blood or aspirate from pustules.
Echinococcosis	Dog, cattle, sheep, pigs, goats, camels, horses, humans	Worldwide	Infection via fecal oral route. Dog is definitive host, shedding *Echinococcus* cestodes. Intermediate host infected via ingestion of contaminated water, food, or soil.	After ingestion, cestodes hatch and larvae encyst primarily in lungs or liver. Location, size, and number of cysts will dictate signs. Symptoms usually those associated with slow-growing tumor.	Identification of larval cyst in affected organ of intermediate host via sonogram or necropsy/autopsy. Identification of cestodes in dog feces.
Heartwater (cowdriosis, malkopsiekte)	All domestic and wild ruminants	Sub-Saharan Africa, Madagascar, Caribbean	Transmitted by *Amblyomma* ticks. Wild animals could play role as reservoir.	Sudden high fever (drops shortly before death), inappetence, diarrhea, lung edema; nervous signs develop gradually.	Culture of capillary endothelial cells of brain. Brain smears for dead animals.
Johne's disease (paratuberculosis)	All domestic and wild ruminants, horses, pigs, deer, alpaca, rabbits, foxs, weasels	Worldwide	Under natural conditions, ingestion of agent from contaminated environment. Infection can be spread to unborn fetus. Calves contract infection through milk of infected cow.	Slowly progressive wasting. Diarrhea that progressively becomes more severe. Diarrhea is less common in smaller ruminants.	Fecal smears. Fecal/tissue culture. Serology. Lesions in small intestine. Lesions progress to occur in cecum, colon, and mesenteric lymph nodes as disease progresses.
Leishmaniasis (Chiclero ulcer, buba, oriental sore, Aleppo boil, Baghdad sore, espundia)	Mammals	Southern Mexico to northern Argentina, Dominican Republic, South America, Middle East, Asia, northern China, northwest India, Africa	Vector-borne parasite. Rodents are reservoir. Once inside vertebrate host, organism invades cutaneous macrophages.	Begins as itchy lesion then forms papules and painless ulcers. Lesions may persist for a year or more.	Identification of parasite through lesion scraping or aspiration.
Leptospirosis (Weil's disease, swineherd's disease, cane-cutter's fever, mud fever, Stuttgart disease)	Cattle, pigs, horses, sheep, goats, dogs, many wild animals, including rodents	Worldwide	Environment is contaminated through infected animal's urine. Direct or indirect transmission through abraded skin or mucosa. Ingestion of contaminated water, soil, or foods.	May remain subclinical. Sudden onset of fever, anorexia, birth of weak/stillborn animals. Infertility. Some may present with jaundice. Chronic phase of disease may last months after clinical recovery.	Culture of agent from blood (early) or urine. Serology requires repeated samples.
New and Old World screwworm (traumatic myiasis)	Mammals, birds (rarely)	Old World: Africa, Gulf countries, Indian subcontinent, and southeast Asia to Papua New Guinea New World: southern United States to northern Argentina (area contracts in winter and expands in summer)	Female screwworm flies lay eggs on edges of wounds on live animals. Eggs laid on mucous membranes can result in infections of natural openings. Larvae burrow into flesh upon emerging. Larvae leave wound after maturing and purpurate in the soil where they further develop into an adult fly.	Extensive tissue destruction from burrowing larvae. Severe infections can result in death.	Identification of larvae collected from the deepest part of the wound

Pseudorabies (Aujesky's disease, mad itch)	All mammals, especially pigs, except humans and tailless apes	Worldwide except Canada and Oceania	Pigs are natural host and will remain latently infected after clinical recovery. Raccoons in the United States and wild boar in Europe may be healthy carriers. Airborne transmission possible.	Abortions and stillbirths. Nervous signs (young animals), respiratory disease (older animals). Intense itching.	Culture of agent from nasal fluid or tonsil biopsy from live animals, brain tissue from dead. Affected animals, other than pigs, do not survive long enough for marked seroconversion.
Q-fever (Balkan influenza, coxiellosis, abattoir fever, nine-mile fever)	Sheep, cattle, goats, cats, rabbits, dogs, humans (has been found in almost all domestic and wild animal species)	Worldwide	Domestic cycle of disease mainly exists in cattle, sheep, and goats. Infection through inhalation of aerosols from contaminated placenta, amniotic fluid, and excreta. In natural setting, agent is circulated between wild animals and ticks.	Spontaneous abortion and fever. Otherwise clinically inapparent. In man, sudden onset of fever, chills, profuse sweating, myalgia, and malaise following a 2- to 5.5-week incubation. Acute hepatitis also possible. Chronic disease mainly affects the cardiovascular system.	Serology since few laboratoriess have adequate installations and equipment to safely isolate the agent (Coxiella burnetii).
Rabies (hydrophobia, lyssa)	All mammals	All continents (except Oceania) with many rabies-free countries, including Japan, Ireland, the Netherlands, Portugal, Spain, and the United Kingdom	Inoculation (e.g., bite) or Infection can be obtained through inhalation of agent. Exposure to infected material.	Incubation period may be affected by amount of virus introduced. Reported as long as years. Anxious feeling, sensory alteration, headache, slight fever followed by extreme sensitivity to light, increased salivation, and pupil dilation. Liquids are violently rejected by muscular contractions as disease progresses. May prevail until death or develop generalized paralysis until death.	Diagnosis can only be made in laboratories utilizing brain tissue to identify Negri bodies microscopically or virus nucleocapsid antigen via ELISA techniques.
Trichinellosis (trichiniasis, trichinosis)	Pigs, rats, bears, other flesh-eating mammals, including humans	Worldwide, not confirmed in Australia or several tropical countries in Africa, Latin America, and Asia	Ingestion of cyst in contaminated meat. Larvae can live for months in badly decayed flesh.	No clinical disease in wild animals. Anorexia, nausea, vomiting, diarrhea. Followed by fever, muscle pain, swollen eyelids, headache and chills. Muscle pain may last several months.	Muscle biopsy. Identification of larvae. Clinical diagnosis difficult owing to nonspecific signs.

Transmissible diseases that are considered to be of socioeconomic and/or public health importance within countries and that are significant in the international trade of animals and animal products. ELISA indicates enzyme-linked immunosorbent assay.

TABLE 4.5 OIE Aquatic Animal List Diseases

FISH
Epizootic hematopoietic necrosis
Infectious hematopoietic necrosis
Oncorhynchus masou virus disease
Spring viremia of carp
Viral haemorrhagica septicemia
Channel catfish virus disease
Viral encephalopathy and retinopathy
Infectious pancreatic necrosis
Infectious salmon anaemia
Epizootic ulcerative syndrome
Bacterial kidney disease (*Renibacterium salmoninarum*)
Enteric septicemia of catfish (*Edwardsiella ictaluri*)
Piscirickettsiosis (*Piscirickettsia salmonis*)
Gyrodactylosis (*Gyrodactylus salaris*)
Red sea bream iridoviral disease
White sturgeon iridoviral disease
MOLLUSCS
Infection with *Bonamia ostreae*
Infection with *Bonamia exitiosus*
Infection with *Mikrocytos roughleyi*
Infection with *Haplosporidium nelsoni*
Infection with *Marteilia refringens*
Infection with *Marteilia sydneyi*
Infection with *Mikrocytos mackini*
Infection with *Perkinsus marinus*
Infection with *Perkinsus olseni/atlanticus*
Infection with *Haplosporidium costale*
Infection with *Candidatus* Xenohaliotis californiensis
CRUSTACEANS
Taura syndrome
White spot disease
Yellowhead disease
Tetrahedral baculovirosis (*Baculovirus penaei*)
Spherical baculovirosis (*Penaeus monodon*–type baculovirus)
Infectious hypodermal and hematopoietic necrosis
Crayfish plague (*Aphanomyces astaci*)
Spawner-isolated mortality virus disease

space and over time. This pattern requires a different kind of manipulation of epidemiological data, such as the use of detection algorithms based on cumulative sums or cut point statistics. It also requires detection algorithms capable of searching a large space of possible time intervals and spatial locations. Several chapters in Part III of this book discuss such algorithms.

5.2. Contagious Person-to-Person

The fifth pattern results from person-to-person transmission of a contagious disease, such as influenza, SARS, measles, or rubella (German measles). The timeliness requirement for this type of outbreak is not paramount because people become infected in waves. The recognition of this pattern involves biosurveillance components capable of collecting and analyzing social networks and contact information.

5.3. Commercially Distributed Products

The sixth pattern is contamination of commercially distributed products, especially food. Food contamination may be as simple as contamination at the site of preparation, which is what the Rajneesh cult attempted in The Dalles, Oregon, or as involved as tampering with distribution or production facilities (Torok et al., 1997). The timeliness requirement may be *anthrax-like,* as many individuals can be infected nearly simultaneously.

Improving the timeliness of detection of threats in this category requires biosurveillance components that can monitor the food supply directly for contamination. It also requires components that can correlate patterns of disease in a population with knowledge of food production and distribution systems.

TABLE 4.6 Nine Patterns

Pattern	Representative Outbreak	Other Threats
Large aerosol release	1979 Sverdlovsk release of *B. anthracis* (*Kirov* strain) from Soviet Biological Weapons Compound 19 (Mangold and Goldberg, 2000)	Weaponized anthrax, weaponized staph enterotoxin B, weaponized tularemia, weaponized botulism, weaponized *Coxiella burnetti,* Q Fever, weaponized *Pseudomonas mallei*: Glanders, weaponized *Clostridia perfringens* toxin, weaponized *Brucellae* sp.,weaponized, ricin aerosol, T2 mycotoxin aerosol, histo-coccidiomycosis
Building/vessel contamination	Epidemiologic investigations of bioterrorism-related anthrax, New Jersey, 2001 (Greene CM et al., 2002)	Any bioaerosol, marine toxin: saxitoxin, ciguatoxin, tetrodotoxin, palytoxin
Small premonitory release or contamination	Laboratory-acquired human Glanders, Maryland, May 2000 (CDC, 2000)	Weaponized anthrax, weaponized staph enterotoxin B, weaponized tularemia, weaponized botulism, weaponized *Coxiella burnetti,* Q Fever, weaponized *Pseudomonas mallei*: Glanders, weaponized clostridia perfringens toxin
Continuous or intermittent release of an agent	*Legionella,* Pennsylvania (CDC, 1976a,b)	*Legionella pneumophila,* histo-coccidiomycosis, *Chlamydia psittici*: psitticosis
Contagious person-to-person	Outbreak of Influenza A Infection Among Travelers–Alaska and the Yukon Territory, May–June 1999 (CDC, 1999a)	Influenza, variola, rubeola, rubella, mycobacterium tuberculosis, mumps, smallpox, diphtheria, *Haemophilus* influenza, mycobacterium leprae, *Neisseria meningititus,* group A strep: rheumatic fever, toxic shock, necrotizing fascitis

Continued

TABLE 4.6 Nine Patterns—*Cont'd*

Pattern	Representative Outbreak	Other Threats
Commercially distributed products	A large community outbreak of salmonellosis caused by intentional contamination of restaurant salad bars (Torok TJ et al., 1997)	Salmonellosis, *Shigella* sp., *Escherichia coli* 0157, *Brucella* sp., Staph enterortoxin *B. vibrio* cholera, *B anthracis,* toxic alimentary aleukia: T2 Mycotoxin, *Clostridium* botulinum: botulism, hepatitis A and C. Perfringins E toxin, ricin toxin, heavy metals: iron, mercury, arsenic, Nipah virus, marine toxin: saxitoxin, ciguatoxin, tetrodotoxin, palytoxin, trichinella: trichinosis, norwalk, cyclosporiasis
Water-borne	Large community outbreak of cryptosporidiosis due to contamination of a filtered public water supply (Hayes EB et al., 1989)	cryptosporidiosis, *Shigella* sp, camphylobacter, giardiasis, staph enterotoxin B, *Escherichia coli* 0157, botulism, bioterroristic, ricin toxin, entamoeba histolytica, cyclosporiasis
Vector/Host-borne	Outbreak of West Nile–like viral encephalitis, New York (CDC, 1999b,c)	Malaria, West Nile, yellow fever, dengue, *Yersinia pestis*–Tularemia, ebola, marburg, hantaviruses, Q fever, Glanders, melioidosis, Lassa, Machupo, Junin, Rift Valley, Crimean-Congo hemorrhagic fever, Hantaan, Alphaviridae: Venezuelan equine encephalitis, eastern equine encephalitis, western equine encephalitis, Chikungunya, Flaviviridea Encephalitis (e.g., Russian spring summer, eastern equine, St Louis, West Nile,Venezuelan), Nipah virus, rabies, Lyme, *Rickettsia* sp.: Rocky Mountain spotted fever, typhus
Sexual or parenteral transmission	Cluster of HIV-positive young women, New York, 1997-1998 (Anonymous, 1999)	HIV, *Neisseria gonorrhea, Hemophillus ducrei, Treponema pallidum, Chlamydia trachomatis,* hepatitis B and C

Reprinted from Journal of Biomedical Informatics, Vol. 36, Michael Wagner, Virginia Dato, John N. Dowling, and Michael Allswede, Representative Threats for Research in Public Health, pp. 177–188, Copyright 2003, with permission from Elsevier.

These analyses must be longitudinal to detect intermittent or ongoing contaminations and must consider that many foods can be stored for relatively long periods before consumption.

5.4. Water-Borne

The seventh pattern is contamination of the water supply (well or surface water), as illustrated by the example of cryptosporidium in Chapter 2. The timeliness requirement may be "anthrax-like," as a cohort of individuals can be exposed nearly simultaneously. To achieve the required timeliness of detection, a surveillance system would require in-line monitoring of the water system for contamination, as well as components that correlate the spatial distribution of disease data with the branching anatomy and vulnerabilities of the water supply system, to allow a subtle increase in cases to be noticed as early as possible.

5.5. Vector/Host-Borne

The eighth pattern results from disease transmission through an intermediate, nonhuman vector, such as mosquitoes or even contaminated blood products. Similar to diseases that are contagious and passed from human to human, outbreaks of this type do not have *anthrax-like* timeliness requirements. Detection of outbreaks in this category, however, requires that a biosurveillance system monitor for the presence of organisms in vectors (e.g., mosquitoes and blood products). It must also be capable of collecting and analyzing data about sick individuals and their exposures to vectors to find clusters of cases that otherwise would not be apparent above the background level of disease.

5.6. Sexual or Parenteral Transmission

The ninth and final pattern is that caused by a disease transmitted through sexual or other intimate contact, such as the sharing of needles. This pattern resembles the contagious person-to-person pattern, but the analysis separated this pattern because the data collection enabling the analysis of sexual contact patterns is difficult, requires sensitivity, and may infringe on legal rights. A key detection problem raised by this category is identifying a carrier who is infecting other individuals (either intentionally or unintentionally). The timeliness requirement is not severe, as illustrated by AIDS, but the difficulty in detecting cases and identifying contacts is high.

5.7. Strengths and Limitations of Using Patterns to Reduce Design Complexity

The goal of the above analysis was to identify a set of patterns that would dramatically reduce the complexity of designing biosurveillance systems. By using these patterns, a system designer avoids the paralysis induced by the complexity of developing specifications for hundreds of organisms.

Each of these nine patterns represents an important and different problem in detection. This set of patterns may be more useful to system designers than are priority lists, as in Table 4.2. Table 4.2 does not identify explicitly requirements to detect premonitory cases, building contaminations, or continuous releases; hence, a designer may overlook these requirements if Table 4.2 is used.

Wagner and colleagues noted that five of the nine patterns happen to correspond to organizational divisions within governmental public health (sexually transmitted, communicable, vector-borne, water-borne, foodborne disease). They conjectured that the specialization that has occurred in governmental public health (in response to complexity) over the years might have been shaped by similarities among diseases, especially with respect to the types of data and analyses required to detect and characterize outbreaks of these diseases.

It is worth noting that four of the nine patterns do not correspond to organizational divisions in health departments (*large-scale aerosol release, small premonitory release, building contamination, and continuous or intermittent release*). This observation suggests that these patterns may require special attention. We note that the Department of Homeland Security is addressing the large-scale aerosol release, and the U.S. Postal Service is addressing the problem of building contamination—at least for the tiny subset of buildings in the United States that they own.

It is also worth noting that designers can benefit from focusing their attention on an even smaller subset of patterns. Two patterns—the communicable person-to-person and the large-scale release pattern—cover many of the design issues raised by the larger set of patterns (Wagner et al., 2003a).

6. SPECIFYING BIOSURVEILLANCE DATA

As we discuss in Parts II and IV, a system designer has a large selection of surveillance data from which to construct a biosurveillance system. This abundance is a result of the increasing amounts of data collected electronically about the health of individuals and their purchasing, travel, attendance, and other behaviors (Sweeney, 2001).

For a designer, the task of data selection may be simple or complex. It is simple if the organization planning a system requests that the designer automate the collection of data that the organization is already receiving. It may be quite challenging if the organization specifies an engineering goal of earlier detection and leaves it to the designer to select data. In part, the designer's difficulty stems from the current lack of full understanding of the value of many of these types of data. The designer must weigh whether data can meet functional requirements for timeliness and accuracy of detection, the availability of the data, and the cost and efforts to acquire the data. Cost and effort may be dominant factors for data that either require effort on the part of many individuals or organizations to obtain the data, or data that are distributed among many computer systems, especially if the biosurveillance system will monitor at the national or international level. The chapters in Parts II and IV of this book provide detail about data availability and the cost and effort to obtain different types of data.

The complexity of the design process is further increased by the amount of data that must be specified, especially for a system that supports outbreak characterization. The data may include data about human health, animal health, environmental data, locations, and relationships of people. Just the data potentially needed to diagnosis a single sick human are staggeringly large–there are more than 4,000 symptoms and laboratory tests that bear on the diagnosis of disease in humans.

Appendix A provides tables that may be useful to system designers. This appendix contains tables of surveillance data that have been pivotal in detecting and characterizing outbreaks in the past, data that are required to satisfy published case definitions, and data currently being collected routinely for biosurveillance by governmental public health. The tables also include types of data that may be very early indicators of diseases, identified by a review of the literature on health psychology, especially the subliterature relevant to behaviors of ill individuals between the onset of symptoms and presentation (if ever) for medical care. The tables also identify data systems that routinely collect the data either for the purposes of biosurveillance or, more typically, for other purposes.

7. SUMMARY

Because of a new requirement for biosurveillance (very early detection of disease outbreaks), biosurveillance systems are undergoing a process of re-engineering and de novo construction. Designers of biosurveillance systems have an opportunity to adopt a more formal engineering approach that begins with specification of functional requirements. At this time, the understanding of such requirements for most diseases, however, is incomplete. There is a general recognition that systems should detect outbreaks "as early as possible"; however, this level of specification is less than ideal because designers must consider many tradeoffs between earliness and cost when designing systems. A designer may simplify the process of functional requirement specification by recognizing that a relatively small number of patterns can guide this process. The state of the art in functional requirement definition for biosurveillance is "perhaps the end of the beginning," as Winston Churchill so aptly put it.

This chapter is also the end of the beginning. We continue our discussion of the design of biosurveillance systems in Part II by reviewing the organizations that participate in biosurveillance and their missions. Part III discusses newer data analytic methods, Part IV considers new types of surveillance data that designers may consider, and Part V addresses the question of how biosurveillance systems can be designed to support decision making. We finally return explicitly to the topic of building biosurveillance systems in Part VI.

За мной, читатель! [Follow me, reader!] (Bulgakov, 2001).

ADDITIONAL RESOURCES

Five papers about high priority disease threats:

Arnon, S.S., et al. (2001). Botulinum Toxin as a Biological Weapon: Medical and Public Health Management. *JAMA* 285:1059–1070.

Dennis, D.T., et al. (2001). Tularemia as a Biological Weapon: Medical and Public Health Management. *JAMA* 285:2763–2773.

Henderson, D.A., et al. (1999). Smallpox as a Biological Weapon: Medical and Public Health Management: Working Group on Civilian Biodefense. *JAMA* vol. 281:2127–2137.

Inglesby, T.V., et al. (2000). Plague as a Biological Weapon: Medical and Public Health Management: Working Group on Civilian Biodefense. *JAMA* 283:2281–22-90.

Inglesby, T.V., et al. (2002). Anthrax as a Biological Weapon, 2002: Updated Recommendations for Management. *JAMA* 287:2236–2252. Review covering six papers.

ACKNOWLEDGMENTS

The material in this chapter evolved from studies commissioned by the Agency for Healthcare Research and Quality under its Bioterrorism Initiative; Contract No. 290-00-0009 Task Order No. 2 *Using Information Technology to Improve Preparedness for Bioterrorism.* We thank Dr. William Hogan for providing details about the 1979 Sverdlovsk outbreak and the *Journal of Biomedical Informatics* for permission to use material previously published.

REFERENCES

Abramova, F., Grinberg, L., Yampolskaya, O., and Walker, D. (1993). Pathology of Inhalational Anthrax in 42 Cases from the Sverdlovsk Outbreak of 1979. *Proceedings of the National Academy of Sciences, United States of America* vol. 90:2291–2294.

[Anonymous]. (1999). Cluster of HIV-positive young women, New York, 1997-1998. *Archives of Dermatology* vol. 135:1141–1142.

Broome, C., Pinner, R., Sosin, D., and Treadwell, T. (2002). On the Threshold. *American Journal of Preventive Medicine* vol. 23:229.

Buehler, J.W., Berkelman, R.L., Hartley, D.M., and Peters, C.J. (2003). Syndromic surveillance and bioterrorism-related epidemics. *Emerging Infectious Diseases* vol. 9:1197–1204.

Bulgakov, M. (2001). *Master and Margarita.* Saint Petersburg, Russia: Vita Nova.

Centers for Disease Control and Prevention [CDC]. (1976a). Follow-up on respiratory disease, Pennsylvania. *Morbidity and Mortality Weekly Report* 267.

CDC. (1976b). Respiratory Infection, Pennsylvania. *Morbidity and Mortality Weekly Report* 244.

CDC. (1999a). Outbreak of Influenza A Infection among Travelers, Alaska and the Yukon Territory, May–June 1999. *Morbidity and Mortality Weekly Report* vol. 48:545–46, 555.

CDC. (1999b). Outbreak of West Nile-like Viral Encephalitis, New York, 1999. *Morbidity and Mortality Weekly Report* vol. 48:845–849.

CDC. (1999c). Update: West Nile-like Viral Encephalitis, New York, 1999. *Morbidity and Mortality Weekly Report* vol. 48:890–892.

CDC. (2000). Laboratory-acquired Human Glanders, Maryland, May 2000. *Morbidity and Mortality Weekly Report* vol. 49:532–535.

Dato, V., Wagner, M., Allswede, M., Aryel, R., and Fapohunda, A. (2001). The Nation's Current Capacity for the Early Detection of Public Health Threats including Bioterrorism. Washington, DC: Agency for Healthcare Research and Quality.

Greene, C.M., et al. (2002). Epidemiologic Investigations of Bioterrorism-related Anthrax, New Jersey, 2001. *Emerging Infectious Diseases* vol. 8:1048–1055.

Guillemin, J. (2001). Detecting Anthrax: What We Learned from the 1979 Sverdlovsk Outbreak. In *Scientific and Technical Means of Distinguishing between Natural and Other Outbreaks of Disease* (M.R. Dando, G.S. Pearson, and B. Kriz, eds.) Boston: Kluwer Academic Publishers.

Hayes, E.B., et al. (1989). Large Community Outbreak of Cryptosporidiosis Due To Contamination of a Filtered Public Water Supply. *New England Journal of Medicine* vol. 320:1372–1376.

Henderson, D.A. (1999). The looming threat of bioterrorism. *Science* vol. 283:1279–1282.

Jernigan, J.A., et al. (2001). Bioterrorism-related Inhalational Anthrax: The First 10 Cases Reported in the United States. *Emerging Infectious Diseases* vol. 7:933–944.

Kaufmann, A., Meltzer, M., and Schmid, G. (1997). The Economic Impact of a Bioterrorist Attack: Are Prevention and Postattack Intervention Programs Justifiable? *Emerging Infectious Diseases* vol. 3:83–94.

Mangold, T. and Goldberg, J. (2000). *Plague Wars: The Terrifying Reality of Biological Warfare.* New York: St. Martin's Press.

Military Postal Service Agency. (2004). Biohazard Detection System (BDS) Briefing Sheet. *http://hqdainet.army.mil/mpsa/nov_conf/ppt/pdf/BDS.pdf.*

Miller, R.A. (1994). Medical Diagnostic Decision Support Systems: Past, Present, and Future: A Threaded Bibliography and Brief Commentary. *Journal of the American Medical Informatics Association* vol. 1:8–27.

Office International des Epizooties. (2004a). International Aquatic Animal Health Code. *http://www.oie.int/eng/normes/fcode/A_summry.htm.*

Office International des Epizooties. (2004b). International Terrestrial Animal Health Code. *http://www.oie.int/eng/normes/mcode/A_summry.htm.*

Reingold, A. (2003). If Syndromic Surveillance Is the Answer, What Is the Question? *Biosecurity and Bioterrorism: Biodefense Strategy, Practice, and Science* vol. 1:1–5.

Snow, J. (1855). *On the Mode of Communication of Cholera.* London: John Churchill.

Stoto, M., Schonlau, M., and Mariano, L. (2004). Syndromic surveillance: Is it Worth the Effort? *Chance* vol. 17:19–24.

Sweeney, L. (2001). Information Explosion. In *Confidentiality, Disclosure, and Data Access: Theory and Practical Applications for Statistical Agencies* (L. Zayatz, P. Doyle, J. Theeuwes, and J. Lane, es.) Washington, DC: Urban Institute.

Torok, T.J., et al. (1997). A Large Community Outbreak of Salmonellosis Caused by Intentional Contamination of Restaurant Salad Bars. *JAMA* vol. 278:389–395.

U.S. Postal Service. (2004). Biohazard Detection System Deployment Resumes: Statement of Azeezaly S. Jaffer, Vice President, Public Affairs and Communications, U.S. Postal Service. Release No. 38.

Wagner, M., Espino, J., Tsui, F.-C., and Aryel, R. (2003a). The Role of Clinical Information Systems in Public Health Surveillance. In *Healthcare Information Management Systems,* 3rd ed. (M. Ball, C. Weaver, and J. Kiel, eds.) New York: Springer-Verlag

Wagner, M., et al. (2001a). The Emerging Science of Very Early Detection of Disease Outbreaks. *Journal of Public Health Management and Practice* vol. 7:51–59.

Wagner, M. M. (2002). The space race and biodefense: Lessons from NASA about big science and the role of medical informatics. *Journal of the American Medical Informatics Association* vol. 9:120–122.

Wagner, M. M., Aryel, R., and Dato, V. (2001b). Availability and Comparative Value of Data Elements Required for an Effective Bioterrorism Detection System. Washington, DC: Agency for Healthcare Research and Quality.

Wagner, M.M., Dato, V., Dowling, J.N., and Allswede, M. (2003b). Representative Threats for Research in Public Health Surveillance. *Journal of Biomedical Information* vol. 36:177–188.

II
PART
Organizations That Conduct Biosurveillance and the Data They Collect

5
CHAPTER

Governmental Public Health

Rita Velikina
University of California Los Angeles, School of Public Health, Department of Epidemiology, Los Angeles, California

Virginia Dato
Division of Infectious Disease Epidemiology, Pennsylvania Department of Health, Harrisburg, Pennsylvania

Michael M. Wagner
RODS Laboratory, Center for Biomedical Informatics, University of Pittsburgh, Pittsburgh, Pennsylvania

1. INTRODUCTION

In the report *The Future of Public Health,* the Institute of Medicine defines public health as "what we, as a society, do collectively to assure the conditions for people to be healthy" (Committee for the Study of the Future of Public Health, 1988). Government at the federal, state, and local level plays an important role in keeping people healthy (Gostin, 2000). More than 3,000 governmental organizations at all three levels collect information about the health of populations for the purposes of *public health surveillance,* which, as discussed in Chapter 1, is both broader (includes chronic diseases) and narrower than is biosurveillance as we define the term in this book. In this chapter, we explore the organization of governmental public health, the types of data collected, and the information systems used to manage these data.

2. HISTORY OF PUBLIC HEALTH SURVEILLANCE

On May 22, 1869, a committee of the Massachusetts legislature asserted that "all governments since the time of Moses [Leviticus chapters 11–16] are established to protect the life and health of their people"; thus, the first state board of health and vital statistics in the United States was born (Rosenkrantz, 1972; Kaniecki and Asrti, 2004). But even before the United States of America formally existed, the 13 colonies fulfilled the recognized obligations of a sovereign state to take the necessary actions for the promotion and protection of the health and well-being of its inhabitants (Richards and Rathbun, 1993).

2.1. Notifiable Disease Reporting

Notifiable disease reporting (henceforth referred to simply as *disease reporting*) refers to a form of biosurveillance in which clinicians and laboratories report designated diseases (known as notifiable diseases) to governmental public health (henceforth referred to as *health departments*) at the time of diagnosis.[1]

Disease reporting in the United States dates to colonial America, when in 1741 the colony of Rhode Island required tavern keepers to report patrons with smallpox, yellow fever, and cholera to local authorities (Birkhead and Maylahn, 2000). In 1874, the Massachusetts Board of Health asked 168 physicians to notify them each week by postcard of the prevalence of 14 infectious diseases in their vicinity. Michigan instituted a similar voluntary system in 1876 and, in 1883, passed a law requiring immediate notification to the board of health of "smallpox, cholera, diphtheria, scarlet fever, or any other disease dangerous to the public health by householders, hotel keepers, tenants and physicians" (Birkhead and Maylahn, 2000). Massachusetts passed a similar law one year later and imposed a fine on those who failed to report.

From 1949 to 1970, disease reporting evolved into its modern form partly owing to the charismatic influence of Alexander Langmuir, chief of the Bureau of Epidemiology at the CDC[2] (Fumento, 2001). At the 1950 Association of State and Territorial Health Officials meeting, Langmuir was the driving force behind the revised specifications of the diseases to be

1 Although 'health department' is less precise than 'governmental public health organization,' it is more readable and understandable. We use the term 'health department' to refer to state and local health departments as well as to the CDC. We realize that the CDC is not considered a health department, but many of the points we make apply to all of these organizations. When information applies only to a local or state health department, we make the distinction clear in the text.

2 In 1946, the Communicable Disease Center was created from the Office of Malaria Control in War Areas, an agency that had been established in 1942 to limit the impact of malaria and other mosquito-borne disease on U.S. military personnel. In 1970, the CDC was renamed the Center for Disease Control to reflect responsibilities for noncommunicable disease problems. In 1992, CDC's name was changed to the Centers for Disease Control and Prevention. This change was enacted by Congress, as part of the Preventive Health Amendments of 1992, to recognize the CDC's leadership role in the prevention of disease, injury, and disability (*MMWR Morb Mortal Wkly Rep,* 1992).

Handbook of Biosurveillance
ISBN 0-12-369378-0

reported and the frequency of reporting (Committee on Health Promotion and Disease Prevention, 2003). In 1961, responsibility for the collection and publication of data on nationally notifiable diseases was transferred from the National Office of Vital Statistics to the CDC, which began publishing notifiable disease data in the *Morbidity and Mortality Weekly Report* (Committee on Health Promotion and Disease Prevention, 2003).

Since its early beginnings, disease reporting has expanded in the numbers of diseases reported, the entities responsible for reporting, and the methods by which health departments analyze and disseminate the results of the analyses.

2.2. Other Surveillance Systems

Although systematic disease reporting remains a mainstay of governmental public health, health departments have developed many systems to address specific needs.

The history of surveillance can be traced back to William Farr, the superintendent of the statistical department of the Registrar's General Office of England and Wales from 1839 to 1879. Farr focused on collecting vital statistics, on assembling and evaluating those data, and on reporting both to responsible health authorities and to the general public (Thacker, 2000).

In 1850, Lemuel Shattuck issued a report that linked infant and maternal mortality to communicable disease. He recommended a decennial (every 10 years) census, standardization of disease and death terminology, and collection of health data by demographics. He also applied his recommendations to program activities, such as immunization, school health, and smoking and alcohol abuse (Thacker, 2000).

In 1968, Alexander Langmuir presented a working paper at the 21st World Health Assembly on the concepts and practices of national and global surveillance of communicable disease. The paper was endorsed at the assembly by the 100 delegates present and identified three principal tenets of surveillance: "(1) the systematic collection of pertinent data, (2) the orderly consolidation and evaluation of these data, and (3) the prompt dissemination of results to those who need to know" (Thacker, 2000).

The 1968 World Health Assembly discussions reflect the broadened concept of surveillance and addressed the application to public health issues other than communicable disease. Since that time, health departments have expanded their focus, placing pediatric lead poisoning, congenital malformations, and injuries under surveillance (Thacker, 2000).

3. LEGAL BASIS FOR PUBLIC HEALTH SURVEILLANCE

The legal basis for public heath surveillance in the United States ultimately derives from the U.S. Constitution, which reserves for the states the authority and primary responsibility for protecting the health of the public. States pass legislation and develop health regulations to carry out this responsibility, but states vary in the extent to which they delegate this responsibility to local health departments.

3.1. U.S. Constitution

The U.S. Constitution (10th amendment) reserves for the states all powers not delegated to the United States or prohibited by it to the states (U.S. Constitution Online, 1995). Public health powers are clearly among those powers reserved for the states. As a result, states have primary authority and responsibility to protect the health of the public (Committee on Health Promotion and Disease Prevention, 2003). States can delegate these powers to the local government only if it is allowed by the state's constitution.

For most of its history, the U.S. Supreme Court has also interpreted the U.S. Constitution in favor of granting the federal government powers to protect the public's health and safety. In particular—under the federal authority to "regulate commerce…among several states" (U.S. Constitution, 1787) and other constitutional delineations of federal authority—the U.S. Supreme Court has understood the federal government to have roles in environmental protection, occupational health and safety, and food and drug purity (Gostin, 2000).

In practice, the federal government contributes to the public's health in six ways: "(1) setting health goals, policies, and standards; (2) financing; (3) protecting the public's health; (4) collecting and disseminating information about U.S. health and healthcare delivery systems; (5) building capacity for public health; and (6) managing services" (Committee on Health Promotion and Disease Prevention, 2003). In contrast to state and local public health agencies, the federal government has a limited role in the direct delivery of health and public health services, with the exception of military and Veterans Administration (VA) health care.

3.2. Legislative, Judicial, and Executive Branches of Government

The U.S. Constitution divides responsibility and authority among the legislative, judicial, and executive branches of the federal government. States, counties, cities, and towns have adopted similar schemes of governance.

In health matters, the legislative branches at all levels of government enact laws, develop health policy, and allocate monies for programs and infrastructure. The U.S. Congress, for example, has committees that oversee the activities of federal agencies, review the authorization of programs, and shape the appropriation of funds. Multiple committees in both the House of Representatives and the Senate have jurisdiction over programs and health-related activities (Committee on Health Promotion and Disease Prevention, 2003).

The executive branches at all levels of government establish health regulations and enforce health policy. State health departments, for example, develop codes that determine which diseases are notifiable, sanitation standards for restaurants, and standards for hospitals and doctor's offices. We discuss the administrative processes of government and the effect they have on development of biosurveillance systems at the end of this chapter.

The judiciary's task is to interpret laws and to adjudicate disputes (Gostin, 2000). The judiciary's function of interpreting the law to resolve legal disputes makes the courts' role in public health deceptively broad. Courts exert substantial control over public health policy by determining the boundaries of legislative and executive government power. The courts may decide whether a public health statute is constitutional, whether agency action is authorized by legislation, whether agency officials have gathered sufficient evidence to support their actions, and whether government officials and private parties have acted negligently. The judicial branch adjudicates constitutional claims regarding, for example, individual rights or federalism (Gostin and Hodge, 2002).

3.3. Health Codes

State and local governments have the authority to engage in a broad array of regulatory activities. They have the right to require businesses to meet standards for safety and sanitation through the institution of regulations, inspections, licenses, and nuisance abatements; to prevent individuals from engaging in unduly risky behavior that poses a danger to others; and to regulate the quality of health care provided in the public and private sectors (Committee on Health Promotion and Disease Prevention, 2003).

Each state and city enacts health laws and regulations (which we refer to collectively as *health codes*). These health codes, having evolved somewhat independently, may vary among jurisdictions in structure, substance, and the processes they specify for detecting, controlling, and preventing injury and disease. In fact, health codes across America differ so significantly in definitions, methods, and scope that they defy orderly categorization (Committee on Health Promotion and Disease Prevention, 2003).

Of particular importance, variations in health codes across jurisdictions create some variability in systems for public health surveillance. For example, HIV is a disease for which there is a significant difference in state reporting procedures. When laboratory tests for HIV first became available, many states began to require the reporting of HIV, the causal agent for AIDS, as it is standard public health practice to revise reporting practices when the causal agent for a disease is discovered. In several states, however, there was political opposition to HIV reporting because of the fear that public health officials would not protect the confidentiality of the reports and would use the information for improper purposes (Richards, 2005). Thus, although Connecticut (Connecticut Department of Public Health, 2002) requires physicians to report HIV cases, Hawaii only requires laboratories to report HIV infections by using a unique identifier (not by patient name) (Hawaii State Department of Health, 2005). As a result, in the event of an HIV outbreak, it would be virtually impossible to locate individuals in Hawaii in order to curtail the outbreak; however, in Connecticut, an outbreak may be more easily contained because the people involved may be more easily contacted.

In general, variability in methods of public health surveillance among jurisdictions makes control or understanding of outbreaks that cross jurisdictional boundaries difficult. Although there are often justifications for differences in health codes across states owing to different needs or circumstances, a certain amount of consistency is necessary to enable biosurveillance and coordinated responses to health threats by states (Committee on Health Promotion and Disease Prevention, 2003).

4. ORGANIZATIONAL STRUCTURE OF GOVERNMENTAL PUBLIC HEALTH

In the United States, there are more than 3,000 health departments, 60 state (or tribal or territorial) health departments, and the CDC at the federal level (Trust for America's Health, 2004). These health departments engage in a variety of activities to protect the public's health. The activities include monitoring the burden of injury and disease in the population through disease surveillance; identifying individuals and groups that have conditions of public health importance through testing, reporting, and partner notification; providing a broad array of prevention services, such as counseling and education; and helping ensure access to healthcare services for poor and vulnerable populations.

The structure of a health department is similar whether the organization is positioned at the federal, state, or local level. A secretary or director leads the organization and oversees numerous programs in addition to biosurveillance, which include such functions as communications, environmental health, laboratory, communicable diseases, and chronic diseases. Figure 5.1 is the structure of a representative health department.

The division of responsibility and authority between state and local health departments varies substantially by state. Some states have a centralized model (e.g., Arkansas, Florida, Georgia, and Missouri), meaning that the state health department has direct control and authority for supervision of local health departments. In other states (e.g., California, Illinois, and Ohio), local health departments function more independently; they are run by counties (rather than the state) and report directly to local boards of health or health commissioners. Still other states (e.g., Iowa and North Dakota) have no local health department, and the state health department provides all public health services directly (Committee on Health Promotion and Disease Prevention, 2003). In the following sections, we describe the typical division of responsibility between state and local health departments.

For biosurveillance, local and state health departments have primary responsibility for data collection and analysis. Although the CDC publishes goals and standards, local and state health departments implement biosurveillance systems (such as syndromic surveillance, discussed below) and develop analytical tools (Heffernan et al., 2004).

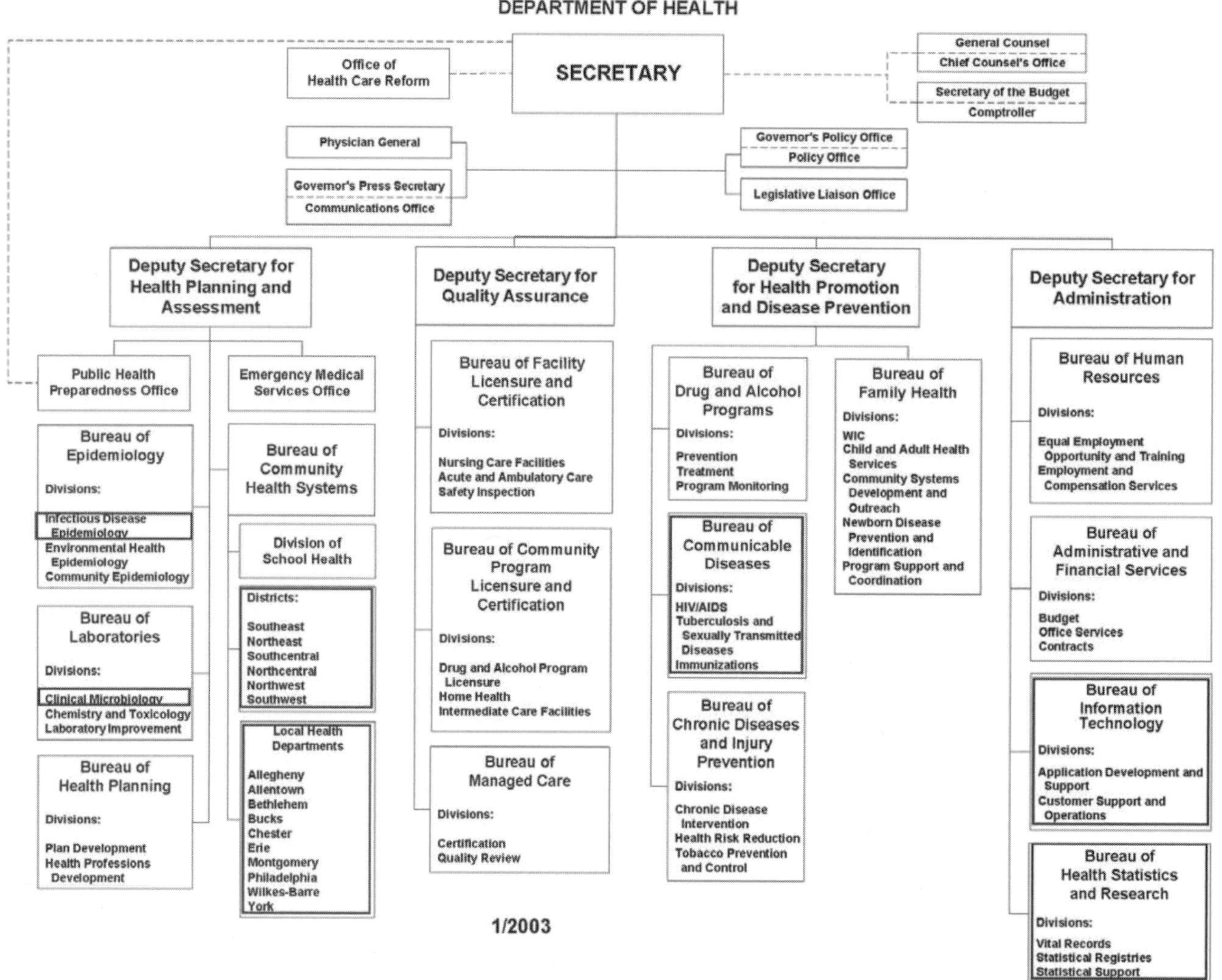

FIGURE 5.1 Illustration of the typical organization of a local, state, or federal health department. The subdivisions involved in biosurveillance directly are highlighted with bold outlines. (From the Pennsylvania Department of Health.)

4.1. State Health Departments

Every state has a health department with responsibility for public health activities. That health department may be a stand-alone department or a component of a department with broader responsibilities, such as a health and human services program.

In addition to operating infectious disease reporting systems, state health departments typically have divisions or units with responsibility for disease control, immunization, health education, and health statistics services. State health departments also license and regulate the healthcare system, for both institutions and individual providers that deliver healthcare services. States vary, however, in whether the state health department or some other governmental department has responsibility for programs such as mental health and substance abuse, environmental health, and Medicaid.

4.2. Local Health Departments

Local health departments are also responsible for a variety of activities. In addition to operating infectious disease reporting systems, responsibilities of local health departments include communicable disease control, community assessment, community outreach and education, environmental health services, epidemiology, food safety, health education, restaurant inspections, and tuberculosis testing. A smaller percentage of local health departments offer treatment for chronic disease, behavioral and mental health services, programs for the homeless, substance abuse services, and veterinary public health (National Association of County and City Health Officials, 2001).

4.3. Centers for Disease Control and Prevention

At the federal level, the lead agency related to biosurveillance is the CDC, a nonregulatory agency that is part of the

Department of Health and Human Services (DHHS). The CDC is responsible for collecting, analyzing, and disseminating national disease occurrence and mortality data to the public and to state and local health departments. Any request for CDC assistance must be accompanied by an invitation by a state health department.

The CDC promulgates goals and standards for biosurveillance systems and encourages their adoption by making them requirements for CDC-controlled federal funding to state and local health departments. The CDC also develops software for its own use and for health departments. Table 5.1 lists some agencies with biosurveillance responsibilities within the DHHS (Committee on Health Promotion and Disease Prevention, 2003).

5. SURVEILLANCE DATA

Health departments collect a variety of biosurveillance data. We begin this section by discussing vital statistics, which serve many functions such as monitoring increases in unexplained deaths. We also discuss notifiable disease reports. Both vital statistics and disease reports are among the earliest types of data collected systematically by governments for the purpose of biosurveillance. We also discuss sentinel surveillance.

5.1. Vital Statistics

A vital statistic is a record of a birth, death, or changes in civil status, such as marriage, divorce, and adoption (University of Washington, 2005). The basis for vital registration is rooted in state and federal law (Dean, 2000), which require local registrars, hospitals, and attending physicians to register births, deaths, and civil status changes.

Although each state is free to determine the required information and its preferred format (Birkhead and Maylahn, 2000), workgroups sponsored by the Division of Vital Statistics of the DHHS develop specifications that encourage uniform reporting so national comparisons are possible. Although the use of these specifications is voluntary, they have resulted in a fair degree of uniformity among the states.

Generally, health departments use birth and death data for biosurveillance, not civil status; therefore, we limit our discussion to these two vital statistics.

TABLE 5.1 Selected agencies which comprise the Department of Health and Human Services

Agency for Healthcare Research and Quality (AHRQ)
Centers for Disease Control and Prevention (CDC)
Agency for Toxic Substances and Disease Registry (ATSDR)
Food and Drug Administration (FDA)
Health Resources and Services Administration (HRSA)
Indian Health Service (IHS)
Nationals Institute of Health (NIH)
Substance Abuse and Mental Health Services Administration (SAMHSA)

5.1.1. *Death Certificates*

Physicians and medical examiners issue death certificates to certify death for legal purposes. A death certificate is required before a funeral home will bury an individual, a court will probate a will, any property of the deceased can be sold, and a variety of other civil functions. The death certificate includes the cause of death and the deceased age and place of death, which may be helpful in characterizing an outbreak and may give clues about the identity of the causative agent. Figure 5.2 is the 2003 national death certificate developed by the Division of Vital Statistics workgroup.

As discussed in Chapter 3, a health department may analyze disease-specific or overall death rates to detect and track outbreaks. For example, the CDC uses total deaths from pneumonia and influenza reported by selected cities weekly as a means of tracking outbreaks of influenza. Death certificates are the source of data for such analyses.

5.1.2. *Birth Certificates*

Various entities (e.g., clinicians, hospitals, midwives) record and certify births. Health departments issue birth certificates based on their certifications. Birth certificates have potential utility for detecting outbreaks due to infectious agents that may cause premature delivery (e.g., *Neisseria gonorrhoeae, Treponema pallidum,* herpes simplex, group b streptococcus) or birth defects (e.g., rubella, cytomegalovirus, *Toxoplasma gondii*). Birth certificates are only issued for live births; thus, birth certificates do not have potential usefulness in detecting increases in fetal deaths or miscarriages owing to infectious diseases such as HIV. There are databases, such as the Infant Mortality Database, that are formed by linking birth records and death records. These databases allow the detection of increases in deaths among infants, and the linkage provides additional information that may identify the cause.

Figure 5.3 is an example of a standard U.S. birth certificate.

5.2. Notifiable Disease Reports

As discussed in Chapter 3, the healthcare system (laboratories, hospitals, and physicians) within a health department's jurisdiction reports cases of notifiable diseases directly to the health department. Health departments decide to make diseases notifiable based on their potential as threats to a community. The list of diseases that are considered notifiable varies slightly by state (Koo and Wetterhall, 1996). To encourage uniformity across states, the Council of State and Territorial Epidemiologists (CSTE) and the CDC collaborate in establishing the *Nationally Notifiable Disease List.* These groups make recommendations annually for additions to and deletions from the list of nationally notifiable diseases. Diseases may be added to the list as new pathogens emerge and deleted as their incidence declines.

Most commonly, the healthcare system files reports of notifiable diseases manually through phone calls, mail, or faxes.

U.S. STANDARD CERTIFICATE OF DEATH

LOCAL FILE NO.

STATE FILE NO.

NAME OF DECEDENT ____________
For use by physician or institution

To Be Completed/Verified By: FUNERAL DIRECTOR

1. DECEDENT'S LEGAL NAME (Include AKA's if any) (First, Middle, Last)
2. SEX
3. SOCIAL SECURITY NUMBER
4a. AGE-Last Birthday (Years)
4b. UNDER 1 YEAR — Months | Days
4c. UNDER 1 DAY — Hours | Minutes
5. DATE OF BIRTH (Mo/Day/Yr)
6. BIRTHPLACE (City and State or Foreign Country)
7a. RESIDENCE-STATE
7b. COUNTY
7c. CITY OR TOWN
7d. STREET AND NUMBER
7e. APT. NO.
7f. ZIP CODE
7g. INSIDE CITY LIMITS? • Yes • No
8. EVER IN US ARMED FORCES? • Yes • No
9. MARITAL STATUS AT TIME OF DEATH • Married • Married, but separated • Widowed • Divorced • Never Married • Unknown
10. SURVIVING SPOUSE'S NAME (If wife, give name prior to first marriage)
11. FATHER'S NAME (First, Middle, Last)
12. MOTHER'S NAME PRIOR TO FIRST MARRIAGE (First, Middle, Last)
13a. INFORMANT'S NAME
13b. RELATIONSHIP TO DECEDENT
13c. MAILING ADDRESS (Street and Number, City, State, Zip Code)
14. PLACE OF DEATH (Check only one: see instructions)
IF DEATH OCCURRED IN A HOSPITAL: • Inpatient • Emergency Room/Outpatient • Dead on Arrival
IF DEATH OCCURRED SOMEWHERE OTHER THAN A HOSPITAL: • Hospice facility • Nursing home/Long term care facility • Decedent's home • Other (Specify)
15. FACILITY NAME (If not institution, give street & number)
16. CITY OR TOWN, STATE, AND ZIP CODE
17. COUNTY OF DEATH
18. METHOD OF DISPOSITION: • Burial • Cremation • Donation • Entombment • Removal from State • Other (Specify)
19. PLACE OF DISPOSITION (Name of cemetery, crematory, other place)
20. LOCATION-CITY, TOWN, AND STATE
21. NAME AND COMPLETE ADDRESS OF FUNERAL FACILITY
22. SIGNATURE OF FUNERAL SERVICE LICENSEE OR OTHER AGENT
23. LICENSE NUMBER (Of Licensee)

To Be Completed By: MEDICAL CERTIFIER

ITEMS 24-28 MUST BE COMPLETED BY PERSON WHO PRONOUNCES OR CERTIFIES DEATH

24. DATE PRONOUNCED DEAD (Mo/Day/Yr)
25. TIME PRONOUNCED DEAD
26. SIGNATURE OF PERSON PRONOUNCING DEATH (Only when applicable)
27. LICENSE NUMBER
28. DATE SIGNED (Mo/Day/Yr)
29. ACTUAL OR PRESUMED DATE OF DEATH (Mo/Day/Yr) (Spell Month)
30. ACTUAL OR PRESUMED TIME OF DEATH
31. WAS MEDICAL EXAMINER OR CORONER CONTACTED? • Yes • No

CAUSE OF DEATH (See instructions and examples)

32. **PART I.** Enter the chain of events—diseases, injuries, or complications—that directly caused the death. DO NOT enter terminal events such as cardiac arrest, respiratory arrest, or ventricular fibrillation without showing the etiology. DO NOT ABBREVIATE. Enter only one cause on a line. Add additional lines if necessary.

Approximate interval: Onset to death

IMMEDIATE CAUSE (Final disease or condition ——> resulting in death) a. ____________
Due to (or as a consequence of):
Sequentially list conditions, if any, leading to the cause listed on line a. Enter the **UNDERLYING CAUSE** (disease or injury that initiated the events resulting in death) **LAST** b. ____________
Due to (or as a consequence of):
c. ____________
Due to (or as a consequence of):
d. ____________

PART II. Enter other significant conditions contributing to death but not resulting in the underlying cause given in PART I

33. WAS AN AUTOPSY PERFORMED? • Yes • No
34. WERE AUTOPSY FINDINGS AVAILABLE TO COMPLETE THE CAUSE OF DEATH? • Yes • No
35. DID TOBACCO USE CONTRIBUTE TO DEATH? • Yes • Probably • No • Unknown
36. IF FEMALE: • Not pregnant within past year • Pregnant at time of death • Not pregnant, but pregnant within 42 days of death • Not pregnant, but pregnant 43 days to 1 year before death • Unknown if pregnant within the past year
37. MANNER OF DEATH • Natural • Homicide • Accident • Pending Investigation • Suicide • Could not be determined
38. DATE OF INJURY (Mo/Day/Yr) (Spell Month)
39. TIME OF INJURY
40. PLACE OF INJURY (e.g., Decedent's home; construction site; restaurant; wooded area)
41. INJURY AT WORK? • Yes • No
42. LOCATION OF INJURY: State: City or Town: Street & Number: Apartment No.: Zip Code:
43. DESCRIBE HOW INJURY OCCURRED:
44. IF TRANSPORTATION INJURY, SPECIFY: • Driver/Operator • Passenger • Pedestrian • Other (Specify)
45. CERTIFIER (Check only one):
• Certifying physician-To the best of my knowledge, death occurred due to the cause(s) and manner stated.
• Pronouncing & Certifying physician-To the best of my knowledge, death occurred at the time, date, and place, and due to the cause(s) and manner stated.
• Medical Examiner/Coroner-On the basis of examination, and/or investigation, in my opinion, death occurred at the time, date, and place, and due to the cause(s) and manner stated.
Signature of certifier: ____________
46. NAME, ADDRESS, AND ZIP CODE OF PERSON COMPLETING CAUSE OF DEATH (Item 32)
47. TITLE OF CERTIFIER
48. LICENSE NUMBER
49. DATE CERTIFIED (Mo/Day/Yr)
50. **FOR REGISTRAR ONLY**- DATE FILED (Mo/Day/Yr)

To Be Completed By: FUNERAL DIRECTOR

51. DECEDENT'S EDUCATION-Check the box that best describes the highest degree or level of school completed at the time of death.
• 8th grade or less
• 9th - 12th grade; no diploma
• High school graduate or GED completed
• Some college credit, but no degree
• Associate degree (e.g., AA, AS)
• Bachelor's degree (e.g., BA, AB, BS)
• Master's degree (e.g., MA, MS, MEng, MEd, MSW, MBA)
• Doctorate (e.g., PhD, EdD) or Professional degree (e.g., MD, DDS, DVM, LLB, JD)

52. DECEDENT OF HISPANIC ORIGIN? Check the box that best describes whether the decedent is Spanish/Hispanic/Latino. Check the "No" box if decedent is not Spanish/Hispanic/Latino.
• No, not Spanish/Hispanic/Latino
• Yes, Mexican, Mexican American, Chicano
• Yes, Puerto Rican
• Yes, Cuban
• Yes, other Spanish/Hispanic/Latino (Specify) ____________

53. DECEDENT'S RACE (Check one or more races to indicate what the decedent considered himself or herself to be)
• White
• Black or African American
• American Indian or Alaska Native (Name of the enrolled or principal tribe) ____________
• Asian Indian
• Chinese
• Filipino
• Japanese
• Korean
• Vietnamese
• Other Asian (Specify) ____________
• Native Hawaiian
• Guamanian or Chamorro
• Samoan
• Other Pacific Islander (Specify) ____________
• Other (Specify) ____________

54. DECEDENT'S USUAL OCCUPATION (Indicate type of work done during most of working life. DO NOT USE RETIRED).
55. KIND OF BUSINESS/INDUSTRY

REV. 11/2003

FIGURE 5.2 Main page of the 2003 revisions of the U.S. Standard Certificates of Death (The National Center for Health Certificates, 2003).

U.S. STANDARD CERTIFICATE OF LIVE BIRTH

LOCAL FILE NO. BIRTH NUMBER:

CHILD

1. CHILD'S NAME (First, Middle, Last, Suffix) | 2. TIME OF BIRTH (24hr) | 3. SEX | 4. DATE OF BIRTH (Mo/Day/Yr)

5. FACILITY NAME (If not institution, give street and number) | 6. CITY, TOWN, OR LOCATION OF BIRTH | 7. COUNTY OF BIRTH

MOTHER

8a. MOTHER'S CURRENT LEGAL NAME (First, Middle, Last, Suffix) | 8b. DATE OF BIRTH (Mo/Day/Yr)

8c. MOTHER'S NAME PRIOR TO FIRST MARRIAGE (First, Middle, Last, Suffix) | 8d. BIRTHPLACE (State, Territory, or Foreign Country)

9a. RESIDENCE OF MOTHER-STATE | 9b. COUNTY | 9c. CITY, TOWN, OR LOCATION

9d. STREET AND NUMBER | 9e. APT. NO. | 9f. ZIP CODE | 9g. INSIDE CITY LIMITS? • Yes • No

FATHER

10a. FATHER'S CURRENT LEGAL NAME (First, Middle, Last, Suffix) | 10b. DATE OF BIRTH (Mo/Day/Yr) | 10c. BIRTHPLACE (State, Territory, or Foreign Country)

CERTIFIER

11. CERTIFIER'S NAME: ____
TITLE • MD • DO • HOSPITAL ADMIN. • CNM/CM • OTHER MIDWIFE
• OTHER (Specify) ____

12. DATE CERTIFIED MM / DD / YYYY | 13. DATE FILED BY REGISTRAR MM / DD / YYYY

INFORMATION FOR ADMINISTRATIVE USE

MOTHER

14. MOTHER'S MAILING ADDRESS: • Same as residence, or State: City, Town, or Location:
Street & Number: Apartment No.: Zip Code:

15. MOTHER MARRIED? (At birth, conception, or any time between) • Yes • No
IF NO, HAS PATERNITY ACKNOWLEDGMENT BEEN SIGNED IN THE HOSPITAL? • Yes • No

16. SOCIAL SECURITY NUMBER REQUESTED FOR CHILD? • Yes • No | 17. FACILITY ID. (NPI)

18. MOTHER'S SOCIAL SECURITY NUMBER: | 19. FATHER'S SOCIAL SECURITY NUMBER:

INFORMATION FOR MEDICAL AND HEALTH PURPOSES ONLY

MOTHER

20. MOTHER'S EDUCATION (Check the box that best describes the highest degree or level of school completed at the time of delivery)
- 8th grade or less
- 9th - 12th grade, no diploma
- High school graduate or GED completed
- Some college credit but no degree
- Associate degree (e.g., AA, AS)
- Bachelor's degree (e.g., BA, AB, BS)
- Master's degree (e.g., MA, MS, MEng, MEd, MSW, MBA)
- Doctorate (e.g., PhD, EdD) or Professional degree (e.g., MD, DDS, DVM, LLB, JD)

21. MOTHER OF HISPANIC ORIGIN? (Check the box that best describes whether the mother is Spanish/Hispanic/Latina. Check the "No" box if mother is not Spanish/Hispanic/Latina)
- No, not Spanish/Hispanic/Latina
- Yes, Mexican, Mexican American, Chicana
- Yes, Puerto Rican
- Yes, Cuban
- Yes, other Spanish/Hispanic/Latina (Specify) ____

22. MOTHER'S RACE (Check one or more races to indicate what the mother considers herself to be)
- White
- Black or African American
- American Indian or Alaska Native (Name of the enrolled or principal tribe) ____
- Asian Indian
- Chinese
- Filipino
- Japanese
- Korean
- Vietnamese
- Other Asian (Specify) ____
- Native Hawaiian
- Guamanian or Chamorro
- Samoan
- Other Pacific Islander (Specify) ____
- Other (Specify) ____

FATHER

23. FATHER'S EDUCATION (Check the box that best describes the highest degree or level of school completed at the time of delivery)
- 8th grade or less
- 9th - 12th grade, no diploma
- High school graduate or GED completed
- Some college credit but no degree
- Associate degree (e.g., AA, AS)
- Bachelor's degree (e.g., BA, AB, BS)
- Master's degree (e.g., MA, MS, MEng, MEd, MSW, MBA)
- Doctorate (e.g., PhD, EdD) or Professional degree (e.g., MD, DDS, DVM, LLB, JD)

24. FATHER OF HISPANIC ORIGIN? (Check the box that best describes whether the father is Spanish/Hispanic/Latino. Check the "No" box if father is not Spanish/Hispanic/Latino)
- No, not Spanish/Hispanic/Latino
- Yes, Mexican, Mexican American, Chicano
- Yes, Puerto Rican
- Yes, Cuban
- Yes, other Spanish/Hispanic/Latino (Specify) ____

25. FATHER'S RACE (Check one or more races to indicate what the father considers himself to be)
- White
- Black or African American
- American Indian or Alaska Native (Name of the enrolled or principal tribe) ____
- Asian Indian
- Chinese
- Filipino
- Japanese
- Korean
- Vietnamese
- Other Asian (Specify) ____
- Native Hawaiian
- Guamanian or Chamorro
- Samoan
- Other Pacific Islander (Specify) ____
- Other (Specify) ____

Mother's Name ____ Mother's Medical Record No. ____

26. PLACE WHERE BIRTH OCCURRED (Check one)
- Hospital
- Freestanding birthing center
- Home Birth: Planned to deliver at home? • Yes • No
- Clinic/Doctor's office
- Other (Specify) ____

27. ATTENDANT'S NAME, TITLE, AND NPI
NAME: ____ NPI: ____
TITLE: • MD • DO • CNM/CM • OTHER MIDWIFE
• OTHER (Specify) ____

28. MOTHER TRANSFERRED FOR MATERNAL MEDICAL OR FETAL INDICATIONS FOR DELIVERY? • Yes • No
IF YES, ENTER NAME OF FACILITY MOTHER TRANSFERRED FROM: ____

REV. 11/2003

FIGURE 5.3 a, First page of the 2003 Revisions of the U.S. Standard Certificates of Live Birth.

Continued

MOTHER

29a. DATE OF FIRST PRENATAL CARE VISIT ___/___/___ MM DD YYYY • No Prenatal Care

29b. DATE OF LAST PRENATAL CARE VISIT ___/___/___ MM DD YYYY

30. TOTAL NUMBER OF PRENATAL VISITS FOR THIS PREGNANCY ________ (If none, enter "0".)

31. MOTHER'S HEIGHT ______ (feet/inches)

32. MOTHER'S PREPREGNANCY WEIGHT ______ (pounds)

33. MOTHER'S WEIGHT AT DELIVERY ______ (pounds)

34. DID MOTHER GET WIC FOOD FOR HERSELF DURING THIS PREGNANCY? • Yes • No

35. NUMBER OF PREVIOUS LIVE BIRTHS (Do not include this child)

35a. Now Living Number ____ • None

35b. Now Dead Number ____ • None

36. NUMBER OF OTHER PREGNANCY OUTCOMES (spontaneous or induced losses or ectopic pregnancies)

36a. Other Outcomes Number ______ • None

37. CIGARETTE SMOKING BEFORE AND DURING PREGNANCY
For each time period, enter either the number of cigarettes or the number of packs of cigarettes smoked. IF NONE, ENTER "0".
Average number of cigarettes or packs of cigarettes smoked per day.

	# of cigarettes		# of packs
Three Months Before Pregnancy	______	OR	______
First Three Months of Pregnancy	______	OR	______
Second Three Months of Pregnancy	______	OR	______
Third Trimester of Pregnancy	______	OR	______

38. PRINCIPAL SOURCE OF PAYMENT FOR THIS DELIVERY
• Private insurance
• Medicaid
• Self-pay
• Other (Specify) ________

35c. DATE OF LAST LIVE BIRTH ___/___ MM YYYY

36b. DATE OF LAST OTHER PREGNANCY OUTCOME ___/___ MM YYYY

39. DATE LAST NORMAL MENSES BEGAN ___/___/___ MM DD YYYY

40. MOTHER'S MEDICAL RECORD NUMBER

MEDICAL AND HEALTH INFORMATION

41. RISK FACTORS IN THIS PREGNANCY (Check all that apply)
Diabetes
• Prepregnancy (Diagnosis prior to this pregnancy)
• Gestational (Diagnosis in this pregnancy)
Hypertension
• Prepregnancy (Chronic)
• Gestational (PIH, preeclampsia)
• Eclampsia
• Previous preterm birth
• Other previous poor pregnancy outcome (Includes perinatal death, small-for-gestational age/intrauterine growth restricted birth)
• Pregnancy resulted from infertility treatment-If yes, check all that apply:
• Fertility-enhancing drugs, Artificial insemination or Intrauterine insemination
• Assisted reproductive technology (e.g., in vitro fertilization (IVF), gamete intrafallopian transfer (GIFT))
• Mother had a previous cesarean delivery If yes, how many ______
• None of the above

42. INFECTIONS PRESENT AND/OR TREATED DURING THIS PREGNANCY (Check all that apply)
• Gonorrhea
• Syphilis
• Chlamydia
• Hepatitis B
• Hepatitis C
• None of the above

43. OBSTETRIC PROCEDURES (Check all that apply)
• Cervical cerclage
• Tocolysis
External cephalic version
• Successful
• Failed
• None of the above

44. ONSET OF LABOR (Check all that apply)
• Premature Rupture of the Membranes (prolonged, ≥12 hrs.)
• Precipitous Labor (<3 hrs.)
• Prolonged Labor (≥20 hrs.)
• None of the above

45. CHARACTERISTICS OF LABOR AND DELIVERY (Check all that apply)
• Induction of labor
• Augmentation of labor
• Non-vertex presentation
• Steroids (glucocorticoids) for fetal lung maturation received by the mother prior to delivery
• Antibiotics received by the mother during labor
• Clinical chorioamnionitis diagnosed during labor or maternal temperature ≥38°C (100.4°F)
• Moderate/heavy meconium staining of the amniotic fluid
• Fetal intolerance of labor such that one or more of the following actions was taken: in-utero resuscitative measures, further fetal assessment, or operative delivery
• Epidural or spinal anesthesia during labor
• None of the above

46. METHOD OF DELIVERY
A. Was delivery with forceps attempted but unsuccessful?
• Yes • No
B. Was delivery with vacuum extraction attempted but unsuccessful?
• Yes • No
C. Fetal presentation at birth
• Cephalic
• Breech
• Other
D. Final route and method of delivery (Check one)
• Vaginal/Spontaneous
• Vaginal/Forceps
• Vaginal/Vacuum
• Cesarean
If cesarean, was a trial of labor attempted?
• Yes
• No

47. MATERNAL MORBIDITY (Check all that apply) (Complications associated with labor and delivery)
• Maternal transfusion
• Third or fourth degree perineal laceration
• Ruptured uterus
• Unplanned hysterectomy
• Admission to intensive care unit
• Unplanned operating room procedure following delivery
• None of the above

NEWBORN INFORMATION

NEWBORN

48. NEWBORN MEDICAL RECORD NUMBER

49. BIRTHWEIGHT (grams preferred, specify unit) ________ • grams • lb/oz

50. OBSTETRIC ESTIMATE OF GESTATION: ________ (completed weeks)

51. APGAR SCORE:
Score at 5 minutes: ________
If 5 minute score is less than 6,
Score at 10 minutes: ________

52. PLURALITY - Single, Twin, Triplet, etc. (Specify) ________

53. IF NOT SINGLE BIRTH - Born First, Second, Third, etc. (Specify) ________

54. ABNORMAL CONDITIONS OF THE NEWBORN (Check all that apply)
• Assisted ventilation required immediately following delivery
• Assisted ventilation required for more than six hours
• NICU admission
• Newborn given surfactant replacement therapy
• Antibiotics received by the newborn for suspected neonatal sepsis
• Seizure or serious neurologic dysfunction
• Significant birth injury (skeletal fracture(s), peripheral nerve injury, and/or soft tissue/solid organ hemorrhage which requires intervention)
• None of the above

55. CONGENITAL ANOMALIES OF THE NEWBORN (Check all that apply)
• Anencephaly
• Meningomyelocele/Spina bifida
• Cyanotic congenital heart disease
• Congenital diaphragmatic hernia
• Omphalocele
• Gastroschisis
• Limb reduction defect (excluding congenital amputation and dwarfing syndromes)
• Cleft Lip with or without Cleft Palate
• Cleft Palate alone
• Down Syndrome
• Karyotype confirmed
• Karyotype pending
• Suspected chromosomal disorder
• Karyotype confirmed
• Karyotype pending
• Hypospadias
• None of the anomalies listed above

56. WAS INFANT TRANSFERRED WITHIN 24 HOURS OF DELIVERY? • Yes • No
IF YES, NAME OF FACILITY INFANT TRANSFERRED TO: ________

57. IS INFANT LIVING AT TIME OF REPORT? • Yes • No • Infant transferred, status unknown

58. IS THE INFANT BEING BREASTFED AT DISCHARGE? • Yes • No

Mother's Name ________

Mother's Medical Record No. ________

REV. 11/2003

NOTE: This recommended standard birth certificate is the result of an extensive evaluation process. Information on the process and resulting recommendations as well as plans for future activities is available on the Internet at: http://www.cdc.gov/nchs/vital_certs_rev.htm.

FIGURE 5.3 *Cont'd* b, Second page of the 2003 Revisions of the U.S. Standard Certificates of Live Birth (The National Center for Health Certificates, 2003).

There is also a growing trend for automatic reporting via computer-to-computer interfaces. Health departments review, triage, aggregate, and investigate these reports.[3]

Figure 5.4 provides a sample disease reporting form.

5.3. Sentinel Surveillance Data

Health departments also receive data from sentinel physicians, as described in Chapter 3. An example of sentinel surveillance data is that collected for influenza. Health departments collect

3 Health departments also receive information from individual citizens, who may notice, for example, that many individuals attending a church picnic became ill the following day (Ruiz, 2004).

GEORGIA NOTIFIABLE DISEASE/CONDITION REPORT FORM
REPORT CASES BY MAIL, FAX OR PHONE TO DISTRICT HEALTH OFFICE
OR TO SENDSS (http://sendss.state.ga.us)

Disease/Condition ____ Medical Record Number ____

PATIENT DEMOGRAPHICS

Patient's Name

Last Name First Name MI

Patient's Address

Street

City State Zip+4 County

( ) Patient's Home Phone ( ) Patient's Work Phone ( ) Patient's Other Phone

Date of Birth / / Age Age Type: ☐ Yrs ☐ Mos ☐ Weeks ☐ Days ☐ Unk

Ethnicity: ☐ Hispanic ☐ Non-Hispanic ☐ Unknown

Sex: ☐ Male ☐ Female ☐ Unknown

Race: ☐ Asian ☐ Black/African-American ☐ Native American or Alaska Native ☐ Multiracial ☐ Native Hawaiian or Pacific Islander ☐ Other ☐ Unknown ☐ White

CLINICAL INFORMATION

Illness Onset Date / /

	Y	N	UNK
Hospitalized	☐	☐	☐
Emergency Rm	☐	☐	☐
Outpatient	☐	☐	☐

Died? ☐ N ☐ Y ☐ UNK Date of Death: / /

If hospitalized, complete:

Hospital Name	Admit Date	Discharge Date

LABORATORY INFORMATION *Report Hepatitis information in Viral Hepatitis box below

Specimen Collection Date	Test Name (ex. Culture, IFA, IGM, EIA)	Specimen Type (ex. Stool, Blood, CSF)	Result (ex. +/-, titer, Presumptive)	Species / Serotype	Lab Name

ADDITIONAL INFORMATION

	Yes	No	UNK
Pregnant	☐	☐	☐
Nursing Home or other Chronic Care Facility	☐	☐	☐
Child In Daycare	☐	☐	☐
Daycare Worker	☐	☐	☐
Prisoner/Detainee	☐	☐	☐
Food Handler	☐	☐	☐
Health Care Worker	☐	☐	☐
Outbreak Related	☐	☐	☐
Travel in Last 4 Weeks	☐	☐	☐

***VIRAL HEPATITIS Test Results** Date of test(s) ____

		Pos	Neg	UNK
Hepatitis A	Total anti-HAV	☐	☐	☐
	IgM anti-HAV	☐	☐	☐
Hepatitis B	HBsAg	☐	☐	☐
	Total anti-HBc	☐	☐	☐
	IgM anti-HBc	☐	☐	☐
Hepatitis C	anti-HCV (EIA)	☐	☐	☐
	anti-HCV signal to cut-off ratio ____			
	RIBA	☐	☐	☐
	HCV RNA (PCR, bDNA)	☐	☐	☐
All	ALT(SGPT) ____ AST (SGOT) ____			

REPORTER INFORMATION

Report Date / /

Reporter Name

Reporter Phone ( )

Reporter Institution

Physician Name

Physician Phone ( )

☐ Need More 3095 Forms

☐ Entered into SENDSS

Comments/Symptoms/Treatment:

Local Use Only

☐ Additional form completed Name:

State Use Only

Form 3095 (5-04)

FIGURE 5.4 A sample disease reporting form from Georgia (Georgia Department of Human Resources, 2004).

and analyze these data to determine the level of influenza activity and the strains of the virus that are circulating in the population. (Birkhead and Maylahn, 2000).

5.4. Data Collected During Investigations

As described in Chapter 3, health departments collect a wide range of information during an investigation. They collect patient demographic information such as name, age, and address; possible sources of the outbreak; potential contacts of the patient; exposure period; and results of laboratory testing.

Jurisdictions, including those in Pennsylvania, New York, and California, are currently working to streamline the data collection process and to make all data available in an integrated database.

6. GOVERNMENTAL PUBLIC HEALTH PROFESSIONALS

A health department is a biosurveillance organization and, as such, requires the expertise of many of the professions listed in Table 1.1 (Chapter 1). Although physicians working in health departments play a large role, epidemiologists, as well as specialists in laboratory medicine, play a prominent and expanding role.

6.1. Physicians

Until the 1970s, virtually all epidemiologists were physicians. Physicians working in health departments generally have both a medical and public health degree, and they may specialize in particular fields, such as sexually transmitted disease or vector-borne disease. From their medical training, they have expertise in the diagnosis and treatment of disease, and their public health education trains them in the methods of outbreak detection, investigation, and characterization. Many hold administrative positions that require them to concentrate on business-related issues rather than disease investigations.

6.2. Epidemiologists

Epidemiologists have expertise in the recognition of disease outbreaks in populations (epidemics). Epidemiologists receive training that is distinct from medically trained physicians, who are trained to recognize disease in an individual rather than a population. The training of an epidemiologist may include courses on questionnaire design, infectious disease epidemiology, and epidemiological methods. Their training also includes substantial coursework in statistical analysis, because recognition of outbreaks is accomplished largely by examination and interpretation of data, rather than of individual patients (although substantial knowledge of communicable and infectious diseases is also required).

6.3. Laboratorians

Laboratorians receive different levels of training. A microbiologist often has a doctoral degree. Medical technologists, who conduct laboratory tests, typically hold a bachelor's degree with a major in medical technology or in one of the life sciences, or they have a combination of formal training and work experience (American Medical Technologists, n.d.).

Though nearly 300,000 medical technologists currently work in the United States today (American Medical Technologists, n.d.), only a small and insufficient percentage work in public health, as evidenced by the introduction of the Medical Laboratory Personnel Shortage Act of 2005 (Shimkus, 2005) The technologists provide support for epidemiological studies and are a critical component of the disease surveillance resources of the public health infrastructure. Their work is important for disease surveillance, reporting, and recognition of infectious or toxic agents in populations and their environments. We discuss laboratories and the personnel who work in them in detail in Chapter 8.

7. INFORMATION SYSTEMS

Health departments receive vast quantities of data from laboratories, healthcare systems, and private citizens, as well as from their own investigations. Health departments and commercial information technology companies have been developing specialized information systems to manage these data since the 1970s. Many of the systems developed before 2000 were developed for specific organizational units within health departments, such as the Sexually Transmitted Disease Management Information System (STD-MIS) (Koo and Gibson Parrish II, 2000). As a result of differences in system design, terminologies used, and other formatting details, these systems could not automatically exchange data, even with systems within the same health department. In the late 1990s, these limitations led to a rethinking of how to develop and organize information systems used by health departments.

7.1. Systems Developed before 2000

As a result of the siloing of information-system functionality before 2000, health department staff often had to type the name, address, birth date, and other data about the same person into each system that needed the data. To file a weekly report of disease activity to the CDC, for example, a staff member of a state health department might have to print out (or read from a screen) data from a local system and type it into another system.

In 2003, the CDC conducted an internal inventory of its own information systems, identifying 120 surveillance systems in use at CDC (Figure 5.5). These systems collected data from various sources, including health departments, healthcare providers, laboratories, individuals, or medical records and birth and death certificates (Koo et al., 2003). We describe a representative sample of those systems to illustrate the lack of integration and consistency of biosurveillance systems in use before 2000.

7.1.1. National Electronic Telecommunications System for Surveillance

The National Electronic Telecommunications System for Surveillance (NETSS) was developed by CDC and the CSTE. NETSS supports the collection, transmission, analysis, and publication of weekly summaries of notifiable disease and injury activity in the 50 states, New York City, the District of Columbia, Puerto Rico, the Virgin Islands, Guam, American Samoa, and the Commonwealth of the Northern Mariana Islands.

NETSS depends on agreements and cooperation among a large number of organizations. CDC/CSTE agreements on notifiable conditions, protocols for formatting and transmitting data, and standard case definitions enable the required data to be collected in a standardized format. Designated staff members in each participating agency prepare and submit the data. Up until 1984, the staff transmitted only the total number of cases in its jurisdiction for each of 37 notifiable diseases (23 common diseases and 14 less frequent conditions). They reported this

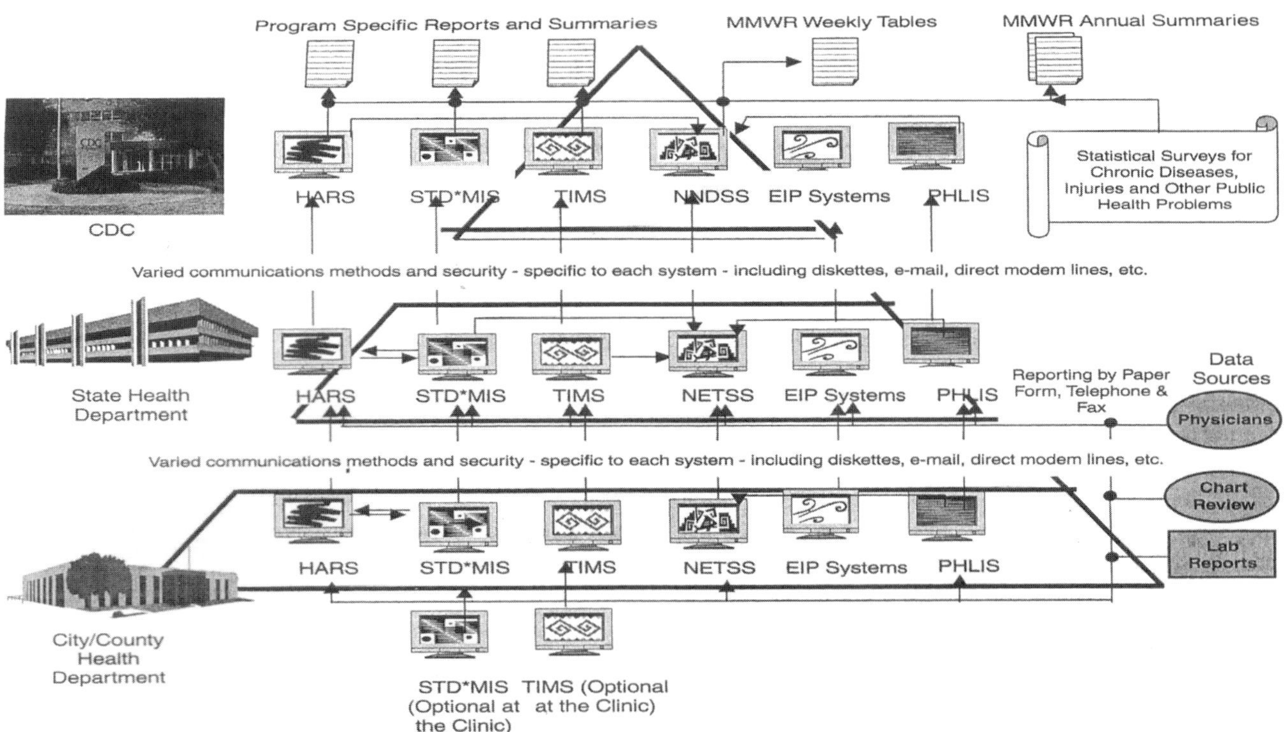

FIGURE 5.5 Existing surveillance information systems, data sources, and information flows. (From Koo et al., 383: figure 19.2. Copyright Springer-Verlag New York, 2003. Reprinted with kind permission of Springer Science and Business Media.)

information by telephone. In 1984, six state health departments began transmitting individual case records to CDC electronically via commercial telecommunications systems. Since 1989, all 50 states and some territories and cities have transmitted individual case records in this manner (CDC, 1991).

In January 1991, the CDC developed a new format for NETSS records. The new format provides for transmission of both individual and summary (aggregate) records. All states completed implementation of the new format in 1993 (CDC, 1991). NETSS is still operational.

7.1.2. Public Health Laboratory Information System

The Public Health Laboratory Information System (PHLIS) is a surveillance system that collects results of microbiology cultures. The National Center for Infectious Diseases and the Association of State and Territorial Public Health Laboratory Directors began developing PHLIS in 1988. By the summer of 1989, nine states were reporting *Salmonella* sp. isolates electronically to CDC and to state epidemiologists. States also report *Campylobacter* sp., *Mycobacteria* sp., and *Shigella* sp. isolates. As of December 1992, PHLIS was in use in more than 40 states (CDC, 2003a).

PHLIS reporting is limited to illnesses that are confirmed by culture and verified at a public health laboratory. After verification by the laboratory, the health department reports the information about the infection to the CDC (National Technical Information Service, 2005).

7.1.3. PulseNet

PulseNet is a nationwide surveillance system, operated by the CDC, that monitors food-borne diseases. PulseNet was developed after a large outbreak of foodborne illness caused by the bacterium *Escherichia coli* O157:H7 occurred in the western United States in 1993. In this outbreak, scientists at CDC performed DNA "fingerprinting" by pulsed-field gel electrophoresis (PFGE) and determined that the strain of *E. coli* O157:H7 found in patients had the same PFGE pattern as that of the strain found in hamburger patties served at a large chain of regional fast food restaurants. However, it took three weeks to trace the source of the *E. coli* contamination of hamburger meat to a single producer (CDC, 2003b).

Prompted by this event, the CDC, in collaboration with the Association of Public Health Laboratories (APHL), introduced PulseNet in 1998. PulseNet obtains samples of bacteria that have been isolated from patients with suspected foodborne illness. Laboratories located either in the healthcare system or in health departments submit samples to PulseNet (CDC, 2003b). After technicians take the DNA PFGE pattern, which determines the genotype (or other specific information on foodborne bacteria) of the bacteria, the result is stored in a centralized database, allowing scientists at public health laboratories throughout the country to compare the PFGE patterns of bacteria isolated from ill persons and determine whether the patterns are similar to those of bacteria isolated from other individuals or samples of food (CDC, 2003b).

7.1.4. Other pre-2000 CDC Systems

There were many other special purpose, "single-disease" systems identified in the 2003 survey. These single-disease systems included the National Malaria Surveillance Systems (Kachur et al., 1997), the 121-cities mortality reporting systems (CDC, n.d.) and the U.S. Influenza Sentinel Physicians Surveillance Network (CDC, 2005a).

7.2. Post-2000 Systems

The large (and proliferating) number of specialized surveillance systems that existed pre-2000 provided an impetus to re-engineer the information infrastructure of health departments. These systems could not interoperate and required health departments to enter the same data multiple times. The designers of newer information systems (which we refer to as *post-2000 systems*, although a few went into operation earlier than 2000) share the following design objectives: to reduce redundancy in data collection, to connect the divisions within a health department at a functional level, and to connect different health departments, as well as the CDC, for purposes of data sharing and better functional integration for outbreaks that cross jurisdictional boundaries.

The re-engineering of systems has been ongoing for approximately five years. At present, most health departments are focusing on the development of four types of systems that together will replace most of the functions of the existing base of diverse systems: National Electronic Disease Surveillance System (NEDSS), syndromic surveillance, electronic laboratory reporting (ELR), and a laboratory reporting network. As we

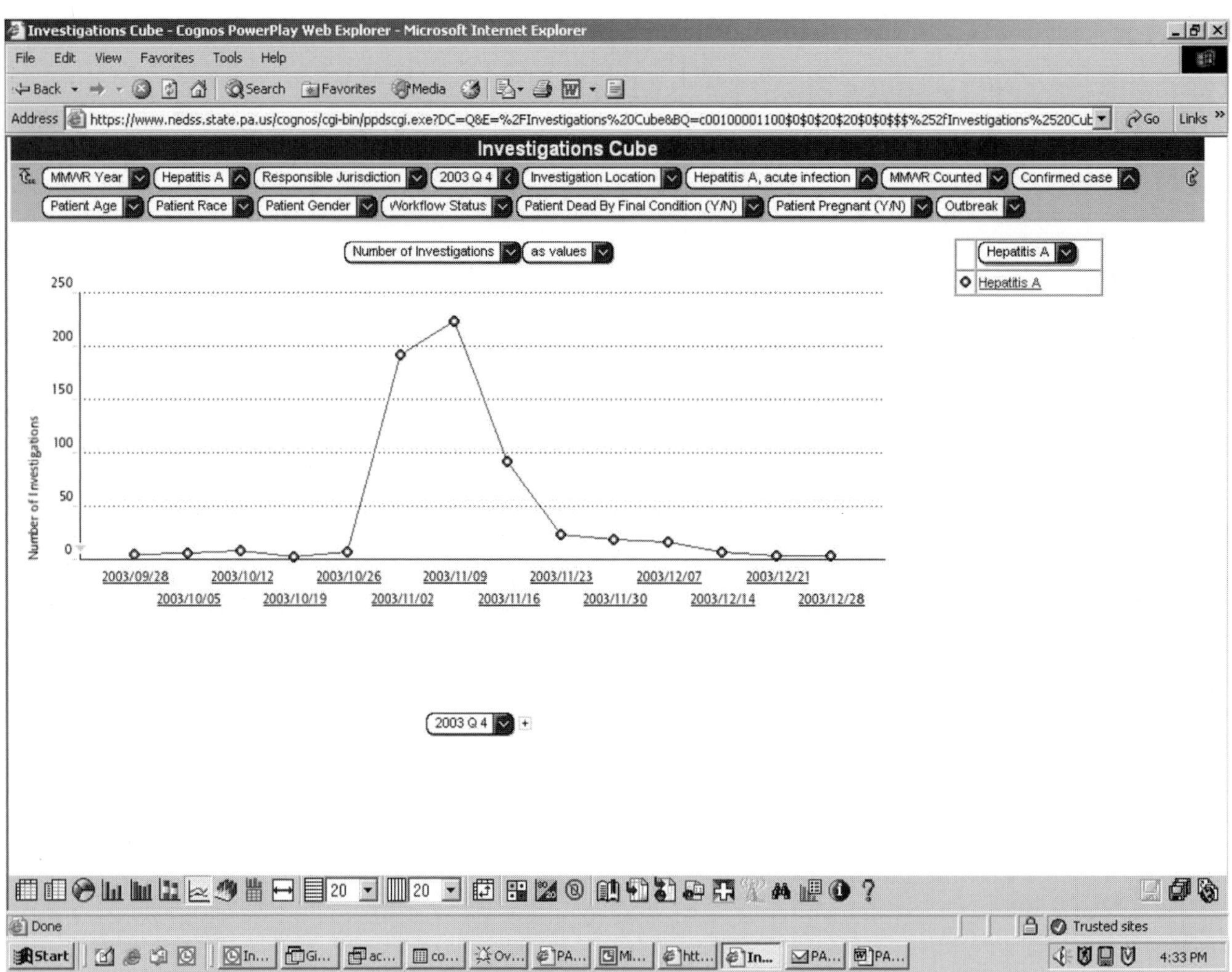

FIGURE 5.6 Pennsylvania NEDSS analysis and reporting. a, confirmed cases investigated fourth quarter 2003 by week investigation initiated. The large increase in November 2003 is owing to the hepatitis A outbreak described in Chapter 1.2.

Continued

will discuss, there is a pan-health department effort to standardize terminology and architecture that was originally referred to as NEDSS but is now called the Public Health Information Network (PHIN). Both of these initiatives involve standardization, system specification, and software development. The inclusion of this range of activities under names that suggest at first only a "system" or "network" may be somewhat confusing at first to readers.

7.2.1. National Electronic Disease Surveillance System

The recommendations of a 1995 report led the CDC to develop the concept of the NEDSS (Committee on Health Promotion and Disease Prevention, 2003).

NEDSS is a somewhat confusing name because it refers to three things. It is the name of software that the CDC develops (i.e., NEDSS base system), a set of specifications that a health department can follow to construct a system with similar functionality, and a "system of systems" (i.e., the network of interoperating systems in each state and city that has installed either the CDC-provided NEDSS system or a version that they have developed or acquired from another source) that will replace NETSS and many other pre-2000 systems.

To avoid confusion, we refer collectively to the CDC software and the software developed by states according to the blueprint as *NEDSS-component systems.* Figure 5.6 is a screen from a NEDSS-component system built by the Commonwealth of Pennsylvania. NEDSS-component systems include Web-based disease reporting, integration of laboratory data from the health department's laboratories, and collection and management of case data from investigations. It is important to note that NEDSS is designed to automate disease reporting first and foremost, so it is a replacement for NETSS and several disease-specific surveillance systems (e.g., the HIV/AIDS reporting system, the vaccine-preventable diseases, and systems for tuberculosis). It is intended to support disease reporting by the

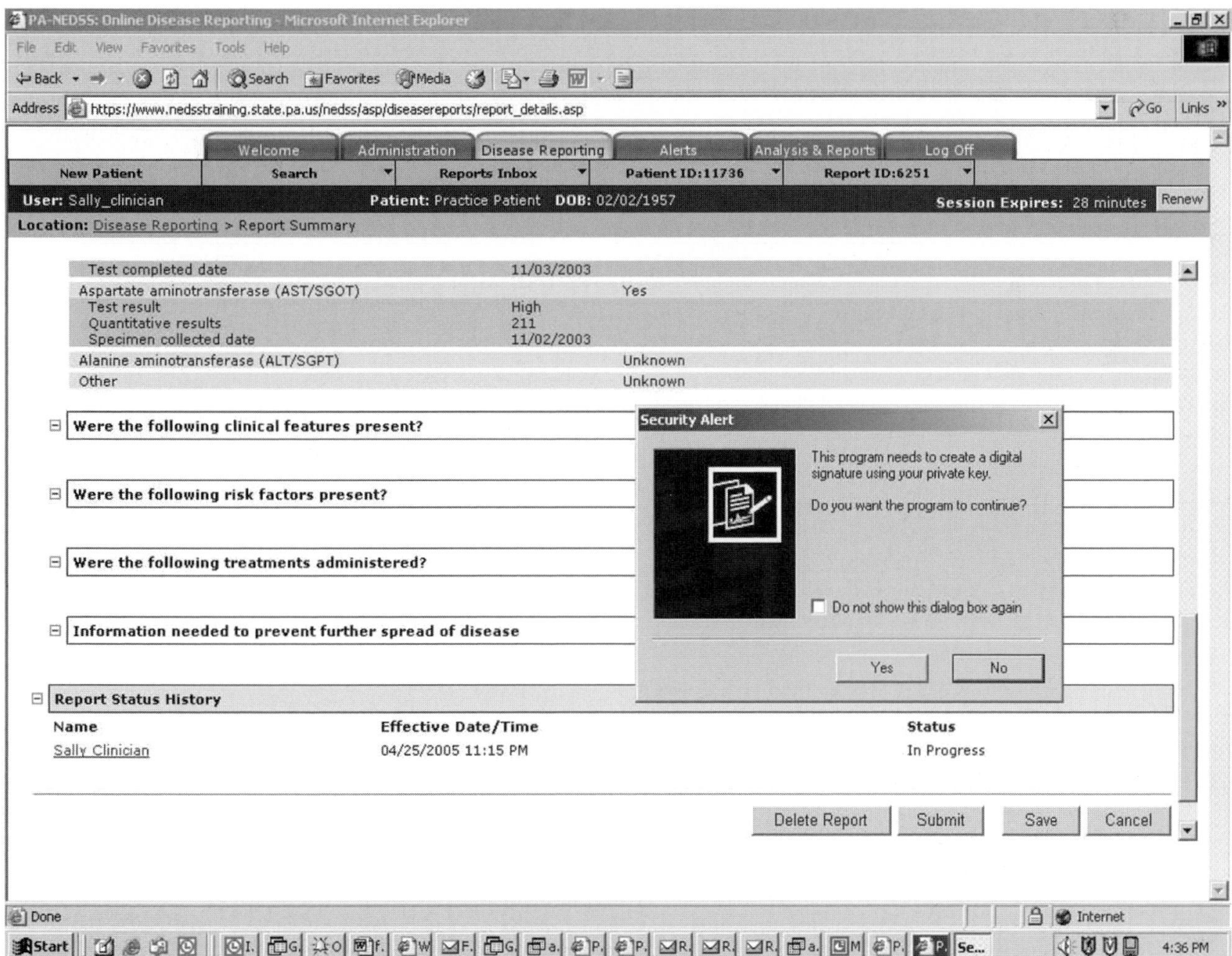

FIGURE 5.6 *Cont'd* b, Screenshot from a demonstration of how a clinician report of hepatitis A would have been signed by a clinician during the hepatitis A outbreak described in Chapter 1 (Pennsylvania Department of Health, 2005).

healthcare system to local and state health departments, investigations of reported cases, and reporting of cases to the CDC.

The NEDSS specifications include both architectural specifications (discussed in Chapter 33) and functional requirements. The architectural specifications cover eight elements: (1) Web browser–based data entry and management; (2) electronic HL-7 message processing; (3) integrated data repository, defining a single point of entry for receipt of data reported to the health department; (4) active data translation and exchange; (5) contemporary application programming practices, which permit, for example, the implementation of case definition logic at appropriate points; (6) data reporting and visualization, including the use of commercial off-the-shelf packages for analytical work and reporting capacity appropriate to different user groups; (7) shareable directory of public health personnel; and (8) consistent security standards to maintain the public health track record in protecting sensitive data (CDC, 2005d).

7.2.2. Early Warning Surveillance Systems

As discussed in Chapter 3, a variety of terms, including early warning and syndromic surveillance, have been used to refer to biosurveillance systems that attempt to detect outbreaks based on prediagnostic data (Mandl et al., 2004).

In 2003, approximately 100 sites throughout the country had implemented some form of syndromic surveillance (Buehler et al., 2003). The number of sites has increased since then.

Currently available syndromic surveillance systems, such as RODS (Wagner et al., 2004) and ESSENCE (Lombardo et al., 2004) use emergency department free-text chief complaints to group patient symptoms into syndromes. The systems aggregate counts of syndromes by day and monitor over time for statistical increases, which may provide early indication of an outbreak or potential bioterrorist event. RODS and ESSENCE are used by many health departments (Tsui et al., 2003; Henry et al., 2004). ESSENCE is used by the U.S. military globally. The RODS software is available as open source software under the GNU General Public License (downloadable from *openrods.sourceforge.net*). New York City (Das et al., 2003; Moran and Taln, 2003), Connecticut (Dembek et al., 2004) and Oregon (Townes et al., 2004) have implemented their own syndromic surveillance systems.

The Early Aberration Reporting System (EARS) (Hutwagner et al., 2003) is software that runs on a personal computer and allows epidemiologists to analyze syndromic data (unlike RODS and ESSENCE, EARS does not automate data collection from hospitals and other organizations). EARS is used by many local health departments.

BioSense is a syndromic surveillance system operated by the CDC (Loonsk, 2004). Biosense is one of three similarly named federal initiatives (the other two are Biowatch and Bioshield) to advance national bioterrorism preparedness. At present, Biosense collects ICD-9 coded registration data from the VA and the Department of Defense, and laboratory test orders (not results) from two large commercial laboratory companies.

7.2.3. Electronic Laboratory Reporting

ELR is the automated messaging of notifiable diseases from laboratory information management systems directly to a health department's reporting system (Wurtz and Cameron, 2005). A small number of health departments have developed ELR. States, such as Florida (Florida Department of Health, 2002), Hawaii (Effler et al., 1999), Oregon (Oregon Department of Human Services, n.d.), Pennsylvania (Pennsylvania Department of Health, 2005), and Texas (Texas Department of State Health Services, 2004) have established or have began testing ELR systems.

7.2.4. Laboratory Response Network

The Laboratory Response Network (LRN) was established by the CDC and became operational in 1999. The LRN maintains a laboratory network of health department, military, and international laboratories that can respond to bioterrorism, chemical terrorism, and other public health emergencies (CDC, 2005e). At present, 96 state and local public health laboratories constitute the 146 laboratories that are members of the LRN. Of these, 62 state, territorial, and metropolitan public health laboratories are part of the LRN's chemical testing network.

The CDC maintains a Web site that provides secure access for more than 1,000 LRN laboratory workers (CDC, 2005e). We further discuss the LRN and provide a map of LRN locations in Chapter 8.

7.3. Public Health Information Network

Although PHIN is not one of the four types of information systems just listed, we discuss it because it is the current incarnation of standards and specification work that began in NEDSS. Note that the word *network* in PHIN connotes not just an electronic network, such as the Internet, but software and computer systems needed to achieve a vision of more than 3,000 health departments in the United States functioning as a single virtual system. PHIN is a CDC-led initiative. PHIN creates specifications for data elements and promulgates or develops standards for exchange of data related to disease, investigations, and response. The PHIN specifications cover five functional areas: (1) *detection and monitoring,* which includes specifications for disease and risk surveillance and national health status indicators; (2) *data analysis,* which includes specifications to facilitate real-time evaluation of data; (3) *knowledge management,* which are specifications intended to facilitate access to reference materials and integrated distance learning materials; (4) *alerting and communications,* which are specifications to enable emergency alerting, routine professional discussions, and collaborative activities; and (5) *response,* which includes specifications that support the distribution of recommendations, and the administration of antibiotic prophylaxis and vaccinations (CDC, 2005b).

In conjunction with the PHIN standards, the CDC is beginning to develop software, such as the PHIN Messaging System (MS), which is a system used for secure message transport compatible with PHIN standards (CDC, 2005c).

7.4. Other Systems and Initiatives

7.4.1. Electronic Vital Record Systems

Health departments have relied on vital records as summative measures of the health of a population for two centuries. The current vital records system is labor intensive, requiring manual data entry, and is vulnerable to human error. The current system also entails time lags, which are not a limitation for the traditional uses of the data (estimating life expectancy and the other uses mentioned) but reduce their value for biosurveillance. Electronic registration of vital records may also reduce errors.

Electronic filing of death certificates is uncommon. In 1998, the City of New York commissioned the development and implementation of the New York Department of Health and Mental Hygiene's Electronic Death Registration System (EDRS). Although it was estimated that the system would be ready for full implementation by December 1999, it was not functioning as of 2003. In 2005, the new system entered a pilot phase (City of New York, 2003).

7.4.2. Health Alert Network

The Health Alert Network (HAN) is a CDC project that funds state and local health departments to improve their network infrastructure. The functional goal of the project is to create the capacity to broadcast and receive health alerts. A health alert is a notice that is issued to inform medical personnel about a certain urgent medical matter that requires their immediate awareness. The capability to send health alerts to front-line clinicians and laboratories is also part of the scope of HAN (Committee on Health Promotion and Disease Prevention, 2003).

Figure 5.7 provides two examples of a HAN alert, one issued by the CDC and one issued by the Pennsylvania Department of Health.

7.4.3. Secure Data Network

Under the Secure Data Network (SDN) project, the CDC promulgates standards and specifications for maintaining the confidentiality of data and providing a secure method for encrypting and transferring files from state health departments to a CDC program application via the Internet (CDC, 2002). The CDC Biosense system (Loonsk, 2004) uses such a network.

8. INTEROPERATING WITH OTHER ORGANIZATIONS

As we have discussed, disease surveillance by health departments involves daily interactions between health departments and many other organizations. During outbreak investigations, health departments interact with an even larger number of organizations, potentially every type of organization listed in Table 1.2 (see Chapter 1). In this section, we discuss interactions not discussed elsewhere in this book, including relationships among adjacent health departments and with law enforcement.

8.1. Other Health Departments

We have already discussed the "vertical" relationships between health departments operating at the local, state, and federal levels. There are also "lateral" interactions among state health departments, especially those that share geographic borders. There are similar interactions among local health departments. These organizations interoperate for many reasons; paramount is that communicable disease does not respect jurisdictional boundaries. Some specific examples of situations in which health departments interoperate include when an individual with a communicable disease, such as meningococcemia, is identified in one jurisdiction but exposed individuals live in another state, when an individual requiring directly observed treatment for tuberculosis moves to a different state, and when an individual becomes ill after returning from a vacation or trip in another state.

Importantly, many large population centers in the United States span jurisdictions. For example, the population centered on the city of Philadelphia extends into the states of Delaware, New Jersey, and Maryland. The population centered on New York City extends into New Jersey and Connecticut, with many individuals commuting to the city from eastern Pennsylvania.

The need to provide computer support for lateral communications and data exchanges among health departments is well understood. There are already basic capabilities in place. For example, the Epidemic Information Exchange (Epi-X), which is provided by the CDC, allows federal, state, and local epidemiologists; laboratories; and other public health colleagues to post health alerts to a Web site. Epi-X also supports secure, Web-based communication (e-mail) and allows for instantaneous notification, through e-mail, of urgent public health events. It also enables users to search the Epi-X database for information on outbreaks and unusual health events (Committee on Health Promotion and Disease Prevention, 2003).

The Metropolitan Medical Response System and the CDC Cities Readiness Project focuses on communication at the metropolitan level. Their goals are similar: to develop or enhance existing emergency communication systems to better handle a public health crisis, such as an act of bioterrorism. In particular, they focus on improving the communication and coordination among local law enforcement, fire, hazardous materials (hazmat), emergency medical services (EMS), hospital, public health, and other "first responders" (Metropolitan Medical Response System, n.d.). We include

DEPARTMENT OF HEALTH

Edward G. Rendell, GOVERNOR

Site Map | Contact Us | Subscribe

SEARCH Go!

You & Your Family's Health | Health Professional & Provider | About Your Health Department

Calvin B. Johnson, M.D., M.P.H., SECRETARY

YOU & YOUR FAMILY'S HEALTH

- Home
- Birth & Death Certificates
- Health Topics A-Z
- Healthcare Access Guide
- Family Health
- Men's Health
- Women's Health
- Infant & Children's Health
- Health Statistics
- Calendar Of Events
- Local Health Centers
- Pamphlets & Publications
- Quit Smoking Today
- Emergency Preparedness Guide
- Site Tools
- Contact Us

PROFESSIONALS & PROVIDERS

- Health Alert Network
- Health Research & CURE
- Healthy People 2010
- Recent Health Reports & Plans
- Nurse Aide Registry
- Primary Care Practice Opportunities
- Epidemiology
- Laboratories
- State Health Improvement Plan
- Community Health

Health Alert Network 2003

DEPARTMENT OF HEALTH

PENNSYLVANIA DEPARTMENT OF HEALTH
HEALTH ALERT #61

To: Health Alert Network
From: Calvin B. Johnson, M.D., M.P.H.
Secretary of Health
Date: November 3, 2003
Subject: **Large Outbreak of Restaurant-Associated Hepatitis A**

This transmission is a "Health Alert," conveys the highest level of importance; warrants immediate action or attention.

HOSPITALS: PLEASE SHARE THIS WITH ALL MEDICAL, INFECTION CONTROL, NURSING, LABORATORY, RADIOLOGY & PHARMACY STAFF IN YOUR HOSPITAL

LOCAL HEALTH JURISDICTIONS: PLEASE DISTRIBUTE AS APPROPRIATE

PROFESSIONAL HEALTH ORGANIZATIONS: PLEASE DISTRIBUTE TO YOUR MEMBERSHIP AS APPROPRIATE

EMS COUNCILS: PLEASE DISTRIBUTE AS APPROPRIATE

Date of Outbreak/Emergency: October 2003

Description of Outbreak/Emergency:
The Pennsylvania Department of Health (PADOH) is investigating an outbreak of

FIGURE 5.7 Health Alert Network. a, Screenshot of an official Pennsylvania Health Advisory related to the hepatitis A outbreak described in Chapter 2. This was distributed only to Pennsylvania clinicians and alerted clinicians about an immune globulin clinic and reminded them of their need to report hepatitis A cases (CDC, 2005f).

Continued

Web sites at the end of the chapter that discuss other groups that focus on facilitating communication and meetings between states and jurisdictions.

8.2. Law Enforcement

In the case of a bioterrorist or criminal act using a biological agent, criminal and epidemiological investigations occur concurrently. The steps necessary to identify a potential covert bioterrorism attack include a close coordination between those who collect and analyze medical and syndromic surveillance information and law-enforcement intelligence and case-related information. The best method for timely detection of a covert bioterrorist attack is early communication between the two communities and recognition of the extent and origin of the threat. For the Federal Bureau of Investigation (FBI), this recognition requires conducting a threat/credibility assessment, a process coordinated by the Weapons of Mass Destruction Operations Unit and FBI Headquarters, in conjunction with CDC and other federal agency experts (Butler et al., 2002).

The recent focus on biosurveillance requires health departments, national security communities, law enforcement, and EMS units to work closely together. This new requirement has motivated these organizations to develop new linkages and skills, to refine existing ones, and to foster new relationships. (Harper, 2000).

9. LIMITATIONS ON AND OF GOVERNMENTAL PUBLIC HEALTH

In the words of historian Lord Aston, "Power corrupts, and absolute power corrupts absolutely" (Powell, 2001). Therefore, the founding fathers of the United States divided power among three branches of government to create a system of checks and balances. However, these checks, balances, and other institutional mechanisms may slow or distort the development, implementation, and use of biosurveillance systems.

Funding decisions perhaps have the greatest effect. Biosurveillance funding competes for taxpayer financed funding with myriad other real and perceived needs. When the economy is poor, tax receipts go down and human service expenses,

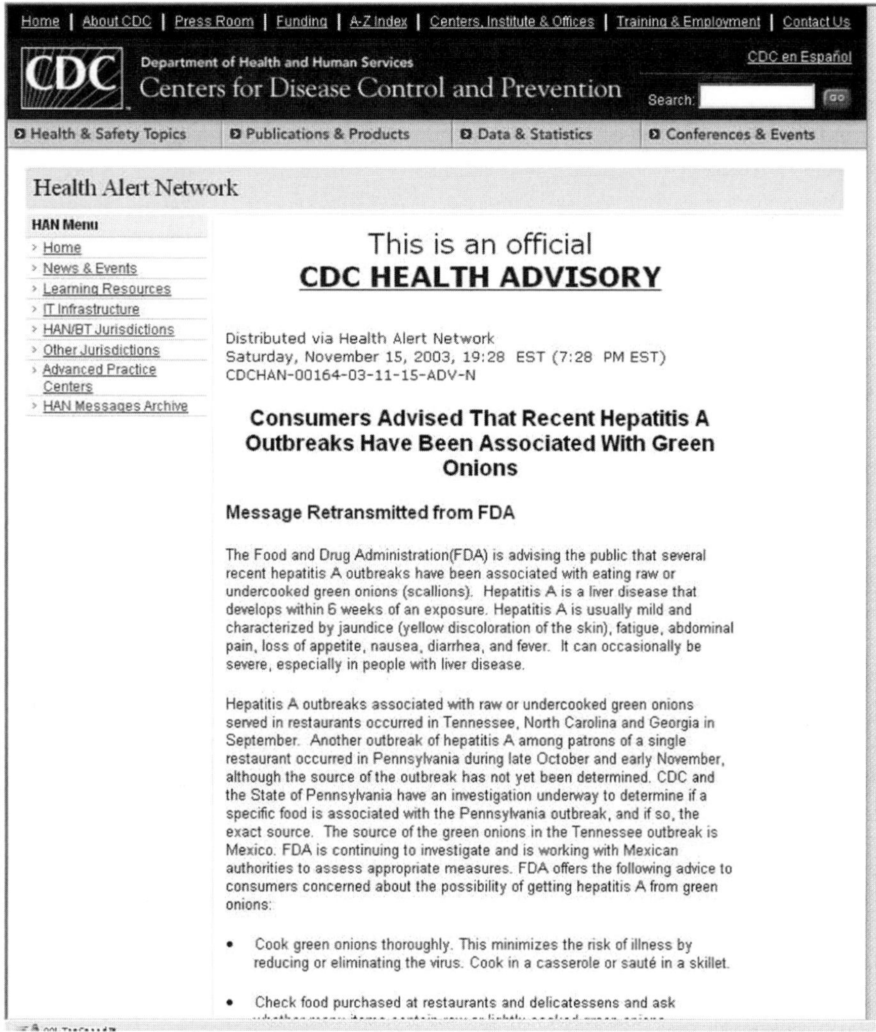
Home | About CDC | Press Room | Funding | A-Z Index | Centers, Institute & Offices | Training & Employment | Contact Us

CDC Department of Health and Human Services
Centers for Disease Control and Prevention

CDC en Español
Search: GO

Health & Safety Topics | Publications & Products | Data & Statistics | Conferences & Events

Health Alert Network

HAN Menu
- Home
- News & Events
- Learning Resources
- IT Infrastructure
- HAN/BT Jurisdictions
- Other Jurisdictions
- Advanced Practice Centers
- HAN Messages Archive

This is an official
CDC HEALTH ADVISORY

Distributed via Health Alert Network
Saturday, November 15, 2003, 19:28 EST (7:28 PM EST)
CDCHAN-00164-03-11-15-ADV-N

Consumers Advised That Recent Hepatitis A Outbreaks Have Been Associated With Green Onions

Message Retransmitted from FDA

The Food and Drug Administration(FDA) is advising the public that several recent hepatitis A outbreaks have been associated with eating raw or undercooked green onions (scallions). Hepatitis A is a liver disease that develops within 6 weeks of an exposure. Hepatitis A is usually mild and characterized by jaundice (yellow discoloration of the skin), fatigue, abdominal pain, loss of appetite, nausea, diarrhea, and fever. It can occasionally be severe, especially in people with liver disease.

Hepatitis A outbreaks associated with raw or undercooked green onions served in restaurants occurred in Tennessee, North Carolina and Georgia in September. Another outbreak of hepatitis A among patrons of a single restaurant occurred in Pennsylvania during late October and early November, although the source of the outbreak has not yet been determined. CDC and the State of Pennsylvania have an investigation underway to determine if a specific food is associated with the Pennsylvania outbreak, and if so, the exact source. The source of the green onions in the Tennessee outbreak is Mexico. FDA is continuing to investigate and is working with Mexican authorities to assess appropriate measures. FDA offers the following advice to consumers concerned about the possibility of getting hepatitis A from green onions:

- Cook green onions thoroughly. This minimizes the risk of illness by reducing or eliminating the virus. Cook in a casserole or sauté in a skillet.
- Check food purchased at restaurants and delicatessens and ask

FIGURE 5.7 *Cont'd* b, Screenshot of an official CDC Health Advisory related to the hepatitis A outbreak described in Chapter 2. This was distributed to clinicians nationwide via individual states (CDC, 2005g).

such as Medicaid and Temporary Assistance for Needy Families, increase. Even if a legislature appropriates money, there will sometimes be a freeze placed on spending mid-year because of revenue shortfalls and a prohibition many states have against deficit spending.

Lobbying by special interests, including businesses, universities, and nonprofit groups, also influences the allocation of funding. Members of legislatures are cognizant that they will face re-election, which depends on political contributions and votes. As a result, programs that attract both votes and contributions get funded; programs that don't may not. If the public does not perceive a need, it is difficult for elected officials to justify the allocation of public funds when other real and perceived needs are not met. Sometimes an elected official who was a champion of a program will not get re-elected. The newly elected official will have other priorities, and the program's funding will be in danger.

Sometimes legislatures allocate funding for activities that the executive branch did not ask for, or money is allocated but specified for a specific disease only. Developing systems to satisfy legislative requirements may take staff away from tasks that they consider more important. In addition, dedicated funding can not be used for anything else. Such earmarking of funds encourages the development of disease-specific surveillance systems at the expense of broader multipurpose systems.

Once funds are allocated, there is still the challenge of spending the money wisely within the stipulated time frame. Unspent money often reverts back to the general fund. This time frame is usually one year but can be less if a budget was passed late. One year is usually too short to plan, let alone develop a system. There are some mechanisms available to extend money or to guarantee continuing funds; however, administrators must be knowledgeable and careful.

Strict purchasing rules exist to eliminate political favoritism. They require administrators to develop specifications, submit them for review, and, finally, open them to bid. All decisions must be reviewed by other administrators (political appointees and civil service). Writing specifications is very difficult if you need to purchase something that does not yet exist or if the administrator is not knowledgeable about the specifications. Moreover, flexibility is limited, making it difficult to change specifications in response to newly identified needs.

These same problems exist when it comes to hiring. There are basically two types of positions: serve at will (or political appointees) and civil service. Serve-at-will positions are higher-level positions and easier to fill. However, the employees are likely to be released if the executive leader (president, governor, mayor) is not re-elected. Also, there is a temptation to appoint individuals who are politically connected even if they are not qualified.

Civil service developed to divorce hiring from politics and ensure that individuals are hired and promoted based on competency (Columbia Encyclopedia, 2005). Biosurveillance requires new skills that may not exist in current civil service classifications. Administrators must write new job descriptions, and the criteria for assessment must be developed. A time-consuming search for a qualified person adds to further hiring delays. Finally a uniquely qualified individual may need accommodations that are routine in the private sector (flexible hours, telecommuting) but are difficult for governments to offer.

Federalism is the term that defines the pattern of relationships between national, state, and local governments in the United States. The Constitution of the United States reserves certain powers for the federal government and delegates the remainder to the states. Although the Constitution is silent on the issue of the role of local governments, it has created a situation in the United States in which *the federal government has the money, state government has the legal authority, and local government has the problem.* This concept can be traced all the way back to Thomas Jefferson and the *Federalist Papers* (Nitzkin, 2001). What this essentially means is that money appropriated by the federal government may have to pass through three different bureaucracies, each with slightly different rules, until the money makes it to the local area where it is needed or must be spent.

If all of this sounds daunting, it can be. The best administrators usually find legal ways to make the system work including private–public collaborations and outsourcing to nonprofit organizations. At times they must make a choice between returning money (and not accomplishing the mission) or spending it less than optimally. However, for all of the difficulties that these checks and balances bring, most agree that it is worth the effort to preserve our democracy. Winston Churchill said it best: "Many forms of Government have been tried, and will be tried in this world of sin and woe. No one pretends that democracy is perfect or all-wise. Indeed, it has been said that democracy is the worst form of Government except all those others that have been tried from time to time" (The Quote Garden, 2005). Nevertheless, the slow rate of progress in advancing surveillance capabilities in the four years since the October 2001 anthrax attacks suggests that it may be worth rethinking the role of government in biosurveillance or finding a way to create a governmental organization, much like early National Aeronautics and Space Administration (NASA), with fewer constraints and a mission focused solely on improving technical biosurveillance capabilities.

10. SUMMARY

Governmental public health has primary responsibility for protecting the public's health. It plays a major role in biosurveillance along with water companies, hospitals, veterinarians, and other entities described in the following chapters. It has primary responsibility for investigation of outbreaks.

States and their local jurisdictions are responsible for the health of their populations, whereas the federal government is tasked with establishing goals and standards. The federal government has also been involved in setting goals and standards for information systems used by different state and local health departments.

Information system development focuses on four types of systems, which are being created in accordance with specifications that will ultimately allow them to interoperate or merge.

Despite the development of new biosurveillance systems based on the federal government's recommendations, public health still lacks a fully automated and integrated biosurveillance system. At present, the effort is focused on four types of systems: NEDSS, syndromic surveillance systems, ELR, and a laboratory network. These systems do not cover all of the functionality required, and they are also not well integrated. Each system requires its own password, database, data collection tool, user interface, and support system, making it cumbersome for the user and the information technologists in the health department who must maintain multiple systems.

The traditional role of health departments in biosurveillance has expanded beyond analysis and response to disease reporting since 2000. The use of information technology in health departments is also evolving rapidly. Recent reports and recommendations have reiterated the need for information to flow rapidly and seamlessly across sectors of the public health system (National Committee on Vital and Health Statistics, 2002). Moreover, a trend toward more effective use of information technology that began before the 2001 anthrax attack has greatly accelerated as a result of federal funding, resulting in re-engineering of biosurveillance systems. An integrated system, which is easily accessible and can be used and maintained by health departments at the federal, state, and local levels, would significantly enhance biosurveillance capability. This system should include established standards for consistent data collection and transmission practices and assurance of privacy protections. In addition, the system should afford the capacity for transmission of urgent health alerts across all levels of the public health system and implementation of data systems that facilitate reporting, analysis, and dissemination.

In addition to information systems, resources for a wide range of individuals, as in Table 1.1, are also required to develop the required functionality. Although a health department often has sufficient resources to develop needed biosurveillance capacities (and there are roughly 60 entities at the level, including territorial and tribal divisions), some of the 3,000 local health departments do not have the capacity. This observation suggests the need for regional cooperative efforts, perhaps centered around large population centers or, in less densely populated areas, at the state or even regional levels.

Such an integrated system, which includes the proper information systems and workforce infrastructure, may aid in the identification of a bioterrorist event by having the capacity to identify patterns and trends rapidly. In addition, rapid communication provided by the network will aid in information sharing and educating individuals and the community at large about critical issues.

ADDITIONAL RESOURCES

For additional information about data collected and information systems used by governmental public health, consult the following resources:

- The National Syndromic Surveillance Conference (*www.syndromic.org*). Provides selected information about syndromic surveillance.
- *Morbidity and Mortality Weekly Report (http://www.cdc.gov/mmwr/)*. Provides weekly summaries of notifiable diseases by disease and state.
- Epi-X *(http://www.cdc.gov/epix/)*. A CDC Web site with purpose of providing rapid access to health information. Sign up to receive daily notifications.
- Association Public Health Laboratories *(http://www.aphl.org/about_aphl/)*. Provides information about recent development, publications, and trainings related to public health laboratories.
- The Association of State and Territorial Health Officers *(http://www.astho.org/)*. An organization that formulates and influences sound public health policy, as well as assists state health departments in the development and implementation of programs and policies to promote health and prevent disease.
- The Council for State and Territorial Epidemiologists *(http://www.cste.org/)*. A group responsible for deciding which diseases should be reported nationally. It focuses on epidemiology and surveillance.
- The National Association for Public Health Statistics and Information Systems *(http://www.naphsis.org)*. Provides a forum for the study, discussion, and solution of problems related public health information systems that integrate vital records registration, public health statistics, and other health information.
- The Metropolitan Medical Response System *(https://www.mmrs.fema.gov/Main/About.aspx)*. Provides information about the development and enhancement of existing emergency preparedness systems. It is targeted to first responders.
- The CDC Cities Readiness Project *(http://www.bt.cdc.gov/cri/facts.asp)*. Addresses the delivery of medications and medical supplies during a public health emergency.
- The University of Miami Ethics Program *(http://privacy.med.miami.edu)*. Offers information about privacy and security issues for health data .

The following resources examine disease surveillance and information systems:

- Guegan, Y.N., Martin, E., Ward, E., Yasnoff, W., and Ripp L., eds. (2002). *Public Health Informatics and Information Systems.* New York: Springer-Verlag.
- Teutsch, S.M. and Churchill, R.E., ed. (2000). *Principles and Practice of Public Health Surveillance.* New York: Oxford University Press.

ACKNOWLEDGMENTS

We would like to thank Kirsten Waller for providing information about NEDSS.

REFERENCES

American Medical Technologists. (n.d). Career as a Medical Technologist. *http://www.amtl.com/site/epage/15307_315.htm.*

Birkhead, G.S. and Maylahn, C.M. (2000). State and Local Public Health Surveillance. In *Principles and Practice of Public Health Surveillance.* (S.M. Teutsch and R.E. Churchill, eds.). New York: Oxford University Press.

Buehler, J.W., Berkelman, R.L., Hartley, D.M., and Peters, C.J. (2003). Syndromic Surveillance and Bioterrorism-related Epidemics. *Emerging Infectious Diseases* vol. 9:1197–204.

Butler, J.C., Cohen, M.L., Friedman, C.R., Scripp, R.M., and Watz, C.G. (2002). Collaboration between Public Health and Law Enforcement: New Paradigms and Partnerships for Bioterrorism Planning and Response. *Emerging Infectious Diseases* vol. 8:1152–1156.

Centers for Disease Control and Prevention [CDC]. (1991). National Electronic Telecommunications System for Surveillance, United States, 1990–1991. *Morbidity and Mortality Weekly Report* vol. 40:502–503.

CDC. (1992). CDC: The Nation's Prevention Agency. *Morbidity and Mortality Weekly Report* vol. 41:833.

CDC. (2002). Secure Network Data Overview. *http://www.cdc.gov/irmo/ea/sdn.htm.*

CDC. (2003a). Public Health Laboratory Information System. *http://wonder.cdc.gov/wonder/sci_data/misc/type_txt/phlis.asp.*

CDC. (2003b). The National Molecular Subtyping Network for Foodborne Disease Surveillance. *http://www.cdc.gov/pulsenet/.*

CDC. (2005a). Weekly Update: Influenza Summary Report. *http://www.cdc.gov/flu/weekly/.*

CDC. (2005b). Public Health Information Network. *http://www.cdc.gov/phin.*

CDC. (2005c). Public Health Information Network. *http://www.cdc.gov/phin/messaging/systems/2003_04_23_An%20Overview%20of%20the%20PHINMS.pdf.*

CDC. (2005d). Public Health Infrastructure: A Status Report. *http://www.phppo.cdc.gov/owpp/docs/library/2000/PH%20Infrastructure.pdf.*

CDC. (2005e). Laboratory Response Network. *http://www.bt.cdc.gov/lrn/.*

CDC. (2005f). Pennsylvania Department of Health

Health Alert 61. *http://www.dsf.health.state.pa.us/health/cwp/view.asp?A=171&Q=239214.*

CDC. (2005g). *http://phppo.cdc.gov/HAN/ArchiveSys/ViewMsgV.asp?AlertNum=00164*

CDC. (n.d.) *http://www.cdc.gov/programs/resrch01.htm.*

City of New York. (2003). *Audit Report on the Development and Implementation of the Electronic Death Registration System By the Department of Health and Mental Hygiene.* New York: Office of the Comptroller.

Columbia Encyclopedia. (2005). Civil Service. *http://www.encyclopedia.com/html/section/civilser_History.asp.*

Committee for the Study of the Future of Public Health. (1988). *The Future of Public Health.* Washington, DC: National Academy Press.

Committee on Health Promotion and Disease Prevention, Institute of Medicine. (2003). *The Future of the Public's Health in the 21st Century:* Washington, DC: National Academy Press.

Connecticut Department of Public Health. (2002). Frequently Asked Questions about HIV Reporting. *http://www.dph.state.ct.us/BCH/AIDS/hiv_reporting.htm.*

Council of State and Territorial Epidemiologists. Division of Surveillance and Epidemiology, Epidemiology Program Office, Center for Disease Control and Prevention. (1991). Current Trends National Electronic Telecommunications System for Surveillance, United States, 1990-1991. *Morbidity and Mortality Weekly Report* vol. 40:502–503.

Das, D., et al. (2003). Enhanced Drop-In Syndromic Surveillance in New York City following September 11, 2001. *Journal of Urban Health* vol. 80:i76–i88.

Dean, A.G. (2000). State and Local Public Health Surveillance. In *Principles and Practice of Public Health Surveillance.* (S.M. Teutsch and R.E. Churchill, eds.) New York: Oxford University Press.

Dembek, Z.F., Carley, K., Siniscalchi, A., and Hadler, J. (2003). Hospital Admissions Syndromic Surveillance, Connecticut, September 2001–November 2003. *Morbidity and Mortality Weekly Report* vol. 24:50–52.

Effler, P., Ching-Lee, M., Bogard, A., Ieong, M.C., Nekomoto, T., and Jernigan, D. (1999). Statewide System of Electronic Notifiable Disease Reporting From Clinical Laboratories. *Journal of the American Medical Association* vol. 282:1845–1850.

Florida Department of Health. (2002). Florida Department of Health Epi Update. *http://www.doh.state.fl.us/disease_ctrl/epi/Epi_Updates/2002/eu101502.htm.*

Fumento, M. (2001). The My Hero Project: Dr. Alexander Langmuir. *http://myhero.com/myhero/hero.asp?hero=a_langmuir.*

Georgia Department of Human Resources. (2004). Notifiable Disease/Condition Report Form. *http://health.state.ga.us/pdfs/epi/notifiable/reportingform.05.pdf.*

Gostin, L.O. (2000). *Public Health Law: Power, Duty, Restraint.* Berkely: University of California Press.

Gostin, L.O. and Hodge Jr., J.G. (2002). *State Public Health Law: Assessment Report.* Washington, DC: Georgetown University Law Center.

Harper, J.L. (2000). Law Enforcement Issues in a Bio-event. *http://www.rand.org/nsrd/bioterr/harper.htm.*

Hawaii State Department of Health. (2005). Integrated Epidemiologic Profile of HIV/AIDS in Hawaii. *http://www.hawaii.gov/health/healthy-lifestyles/std-aids/aids_rep/1h2005-1.pdf.*

Heffernan, R., Mostashari, F., Das, S., Karpati, A., Kulldorff, M., and Weiss, D. (2004). Syndromic Surveillance in Public Health Practice, New York City. *Emerging Infectious Diseases* vol. 10:858–864.

Henry, J.V., Magruder, S., and Snyder, M. (2002). Comparison of Office Visit and Nurse Advice Hotline Data for Syndromic Surveillance, Baltimore–Washington, D.C., Metropolitan Area, 2002. *Morbidity and Mortality Weekly Report* vol. 53:112–116.

Hutwagner, L., Thompson, W., Seeman, G.M., and Treadwell, T. (2003). The Bioterrorism Preparedness and Response Early Aberration Reporting System (EARS). *Journal of Urban Health* vol. 80:i89–i96.

Kachur, S.P., et al. (1997). Malaria surveillance, United States. *CDC Surveillance Summaries* vol. 46:1–18.

Kaniecki, C. and Asrti, C. (2004). Massachusetts Approach. 2005, from *http://www.wetp.org/wetp/1/04meeting/thr_files/Breakout4/BO4_Kaniecki.ppt#3.*

Koo, D. and Gibson Parrish II, R. (2000). The Changing Health-Care Information Infrastructure in the United States: Opportunities for a New Approach to Public Health Surveillance. In *Principles and Practice of Public Health Surveillance.* (S.M. Teutsch and R.E. Churchill, eds.) New York: Oxford University Press.

Koo, D., Morgan, M., and Broome, C.V. (2003). New Means of data collection. In *Public Health Informatics and Information Systems.* (P.W. O'Carroll, W.A. Yasnoff, M.E. Ward, L.H. Ripp, and E.L. Martin, eds.) New York: Springer.

Koo, D. and Wetterhall, S. F. (1996). History and Current Status of the National Notifiable Diseases Surveillance System. *Journal of Public Health Management and Practice* vol. 2:4–10.

Lombardo, J.S., Burkom, H., and Pavlin, J. (2004).ESSENCE II and the Framework for Evaluating Syndromic Surveillance Systems. *Morbidity and Mortality Weekly Report* vol. 53:159–165.

Loonsk, J.W. (2004). BioSense: A National Initiative for Early Detection and Quantification of Public Health Emergencies. *Morbidity and Mortality Weekly Report* vol. 53:53–55.

Mandl, K.D., et al. (2004). Implementing syndromic surveillance: a practical guide informed by the early experience. *Journal of the American Medical Information Association* vol. 11:141–150.

Metropolitan Medical Response System. (n.d.). About MMRS. *https://www.mmrs.fema.gov/Main/About.aspx.*

Moran, G.J. and Talan, D.A. (2003). Update on Emerging Infections: News from the Centers for Disease Control and Prevention. Syndromic Surveillance for Bioterrorism following the Attacks on the World Trade Center, New York City, 2001. *Annals in Emergency Medicine* vol. 41:414–418.

National Association of County and City Health Officials. (2001). Local Public Health Agency Infrastructure: A Chartbook. *http://archive.naccho.org/documents/chartbook.html.*

National Center for Health Statistics. (2003). Revisions of the U.S. Standard Certificates of Live Birth and Death and the Fetal Death Report. *http://www.cdc.gov/nchs/vital_certs_rev.htm.*

National Committee on Vital and Health Statistics. (2002). Information for Health: A Strategy for Building the National Health Information Infrastructure. Report and Recommendations from the Work Group on the National Health Information Infrastructure. *http://nvvhs.hhs.gov.*

National Technical Information Service. (2005). Public Health Laboratory Information System. *http://www.scitechresources.gov/Results/show_result.php?rec=2601.*

Nitzkin, J.L (2001). Policy and Politics in Healthcare and Public Health Settings presented at CME Institute at the *Annual Meeting of the American College of Preventive Medicine,* Miami, FL.

Oregon Department of Human Services. (n.d.) Electronic Lab Reporting. *http://www.dhs.state.or.us/publichealth/elr/index.cfm.*

Pennsylvania Department of Health. (2005). Pennsylvania's National Electronic Disease Surveillance System (PA-NEDSS) PA-Electronic Laboratory Reporting (PA-ELR). *http://www.dsf.health.state.pa.us/health/cwp/view.asp?Q=234580.*

Powell, J. (2001). Great Thinkers on Liberty. *http://www.libertystory.net/LSTHINKACTON.html.*

Richards, E. (2005). Public Health Law Map: Disease and Injury Reporting. *http://biotech.law.lsu.edu/map/Page46.html.*

Richards, E.P. and Rathbun, K.C. (1993). *Law and the Physician: A Practical Guide.* Boston: Little Brown.

Rosenkrantz, B.G. (1972). *Public Health and the State: Changing Views in Massachusetts, 1842-1936.* Cambridge: Harvard University Press.

Ruiz, M.O. (2004). A local department of public health and the geospatial data infrastructure. *Journal of Medical Systems* Vol. 24:385–395.

Shimkus, J. (2005). H.R. 1175: To Amend the Public Health Service Act with Respect to the Shortage of Medical Laboratory Personnel. *http://www.govtrack.us/congress/billtext.xpd?bill=h109-1175.*

Texas Department of State Health Services. (2004). Electronic Pathology Laboratory Pilot Project, June 2004. *.http://www.tdh.state.tx.us/tcr/EPLP_Update0604.html.*

Thacker, S.B. (2000). State and Local Public Health Surveillance. In *Principles and Practice of Public Health Surveillance.* (S.M. Teutsch and R.E. Churchill, eds.) New York: Oxford University Press.

The Quote Garden. (2005). Quotations about Government. 2005, from *http://www.quotegarden.com/government.html.*

Townes, J.M., et al. (2004). Investigation of an Electronic Emergency Department Information System as a Data Source for Respiratory Syndrome Surveillance. *Journal of Public Health Management and Practice* vol. 10:299–307.

Trust for America's Health. (2004). Ready or Not? Protecting the Public's Health in the Age of Bioterrorism. *http://www.healthamerican.com.*

Tsui, F.C., et al. (2002). Data, Network, and Application: Technical Description of the Utah RODS Winter Olympic Biosurveillance System. *Proceedings: A Conference of the American Medical Informatics Association 815–819.*

U.S. Constitution Online. (1995). *www.usconstitution.net/const.html#Am10.*

U.S. Constitution. (1787). Article 1, Section 8. *http://www.house.gov/Constitution/Constitution.html.*

University of Washington. (2005). *http://courses.washington.edu/hserv539/vitaldata.html.*

Wagner, M.M., et al. (2004). Syndrome and Outbreak Detection using Chief-Complaint Data: Experience of the Real-Time Outbreak and Disease Surveillance Project. *Morbidity and Mortality Weekly Report* vol. 53:28–31.

Wurtz, R. and Cameron, B.J. (2005). Electronic Laboratory Reporting for the Infectious Diseases Physician and Clinical Microbiologist. *Clinical Infectious Diseases* vol. 40:1638–1643.

CHAPTER

The Healthcare System

Michael M. Wagner and William R. Hogan
RODS Laboratory, Center for Biomedical Informatics, University of Pittsburgh, Pittsburgh, Pennsylvania

Ron M. Aryel
Del Rey Technologies, Los Angeles, California

1. INTRODUCTION

The healthcare system plays an enormous role in biosurveillance. Its importance derives from a simple fact: when people contract infectious diseases, they seek medical attention.

The healthcare system comprises hospitals, doctor's offices, long-term care facilities, visiting nurse services, laboratories, dental offices, pharmacies, ambulatory and "same day" procedure facilities, and emergency medical services. All of these organizations are involved in the assessment and care of the sick.

Hundreds of thousands of highly accessible nurses, licensed nurse practitioners, and physicians work in healthcare. These frontline healthcare workers are trained to accurately observe and interpret diagnostic information. They are skilled at eliciting and recording basic elements of an epidemiological case history, and they routinely record data needed for outbreak detection and for characterization. These data include symptoms of disease, temperature, laboratory results, and diagnoses.

At present, the healthcare system plays three roles in biosurveillance. First, the healthcare system reports notifiable diseases and clusters of suspicious illness to state and local health departments. Second, the healthcare system assists outbreak investigators by providing medical records, screening services, and diagnostic work-ups of patients. Third, the healthcare system conducts biosurveillance of its own facilities (especially hospitals and long-term care facilities). Hospitals operate special divisions called *infection surveillance and control units* that monitor a facility for patients with communicable diseases and for outbreaks. Outbreaks can start or spread quickly in healthcare facilities and then spread to the community, as they frequently did during the 2003 severe acute respiratory syndrome (SARS) outbreak.

At present, the domains of medical and epidemiological practice exchange relatively little data, except in the setting of an outbreak. They exchange these data primarily by fax, mail, telephone, and e-mail and e-mail attachment. These mechanisms of communication are vulnerable to errors of omission and delays. Even when these domains use electronic transfer of data, the exchange is typically via batch transfer of data on a daily or less frequent basis, rather than via a real-time communications.

In the future, the healthcare system will provide significantly more data and services to governmental public health, and it will provide the data in real time. Indeed, this "megatrend" is already unfolding in many jurisdictions in the United States and abroad. Conversely, governmental public health will transmit case definitions and up-to-the-minute information about disease prevalence electronically to the healthcare system. The untapped potential for biosurveillance of real-time electronic communications among the healthcare system and other biosurveillance organizations, such as governmental public health, is enormous, as we will discuss.

2. ORGANIZATION OF THE U.S. HEALTHCARE SYSTEM

As every American knows, the healthcare system in the United States comprises many independent organizations, with an estimated 5,000 hospitals, 17,000 long-term care facilities, and 40,000 pharmacies. There are countless office practices, free-standing radiology practices, and commercial laboratories. There has been a trend toward consolidation of these entities over the past two decades. Some of the bigger consolidations include pharmacy chains, laboratories, the military healthcare system, the healthcare system run by the Veterans Administration (VA), and large health maintenance organizations such as the nine Kaiser Permanente health maintenance organizations. Nevertheless, the number of independent organizations remains large.

Hospitals themselves are subdivided into many somewhat independent departments, including laboratory, radiology, pharmacy, and clinical departments (e.g., infectious diseases). Each of these divisions may operate its own information systems, which is both a blessing and a curse. It is a blessing because the information systems collect data relevant to biosurveillance, and it is a curse because the sheer number of information systems imposes a barrier to data integration for biosurveillance (although as we will discuss, hospitals are

Handbook of Biosurveillance
ISBN 0-12-369378-0

motivated to integrate these systems for their own benefit and, to some extent, have already done so).

Of course, differences between healthcare in the United States and in other countries exist. Canada and the United Kingdom have national healthcare systems, which make more centralized decisions about information technology (IT). However, even in countries with national healthcare systems, the process of health care involves large numbers of individuals, facilities, organizations, and heterogeneity of information systems.

3. PERSONNEL

Most readers are quite familiar (perhaps more familiar than they would like to be) with the personnel that work in the healthcare system, such as physicians, nurse practitioners, nurses, pathologists, phlebotomists, medical technologists, radiologists, and a wide array of specialists. Readers may be less familiar with hospital epidemiologists and infection control practitioners, who are responsible for biosurveillance of the hospital. We discuss *hospital infection control* in more detail later in this chapter.

Readers may also not realize that hospitals employ a large number of specialists in IT. To function in the healthcare setting, these individuals require not only competence in their primary IT role but also an understanding of medical data and processes, which are complex. To function as part of a biosurveillance system, they similarly require a basic understanding of biosurveillance processes.

4. ROLE OF THE HEALTHCARE SYSTEM IN BIOSURVEILLANCE

Existing laws and regulations in the United States and other countries shape the role of the healthcare system in biosurveillance. As discussed in the previous chapter, health statutes in the United States require hospitals, physicians, and clinical laboratories to notify health departments whenever they encounter patients with notifiable diseases. In general, the body and spirit of American law promotes and enables the healthcare system to participate in biosurveillance. The recent Health Insurance Portability and Accountability Act (HIPAA) recognized the need of governmental public health to collect biosurveillance data from the healthcare system (and other organizations that collect personal health information). HIPAA exempts this use of personal health information from the scope of its regulations (Department of Health and Human Services, 2002).

Hospital infection control has traditionally been influenced more by scientific consensus and evidence than by laws and regulations.[1]

The informal influence takes the form of evidence-based guidelines and position papers. Many professional and governmental organizations—develop guidelines.[2]

Hospitals heed many of these guidelines based on their inherent merit. The JCAHO adopts some of them into its criteria for accreditation. JCAHO's standards have a profound influence on infection control practice and on the healthcare system in general, as we will discuss.

5. DATA COLLECTED BY THE HEALTHCARE SYSTEM

Similar to biosurveillance, the practice of medicine is information intensive. The healthcare system records many types of data for every patient encounter. Table 6.1 lists data that clinicians routinely record as part of the admission history and physical for each patient admitted to a hospital. They record similar data throughout an inpatient stay and for outpatient visits. If just these data were fully and immediately available to biosurveillance organizations for all patients seen by clinicians with possible infectious diseases, their ability to detect and characterize disease outbreaks would be enhanced considerably.

Access, however, is a significant barrier to the use of healthcare data in biosurveillance. Healthcare workers record many important data only on paper. The types of data that are most often "locked away" on paper are the very data needed for early detection and rapid characterization of an outbreak—a patient's symptoms, travel history, immunization history, history of recent foods consumed, and contacts with sick individuals or animals. This problem is especially severe in outpatient offices, which have lower levels of automation than do hospitals. This barrier to access will gradually disappear because of a number of trends, including the falling cost of IT, consolidation of the healthcare system, and federal initiatives such as the National Health Information Infrastructure (NHII) (Yasnoff et al., 2001, 2004; Rippen and Yasnoff, 2004).

Even when the healthcare system records data electronically (as is typically the case for results of laboratory tests and radiology examinations), the data are encoded in nonstandard formats that represent a barrier to regional integration of data for biosurveillance. This problem will also gradually resolve as the healthcare system adopts standard methods for representing and storing data (discussed in detail in Chapter 32, *"Information Technology Standards in Biosurveillance"*), a process that has been ongoing for several decades and is gaining momentum under NHII.

1 This situation is changing as a result of increased awareness of the societal cost of hospital-acquired infections in terms of morbidity, mortality, and economic costs. Recent JCAHO rules now hold the CEO of a healthcare organization accountable for ensuring adequate funding of infection control.

2 Influential organizations include the Society for Healthcare Epidemiology of America (SHEA), Association for Professionals in Infection Control and Epidemiology (APIC), and the Hospital Infections Control Practices Advisory Committee (HICPAC) of the CDC.

TABLE 6.1 Diagnostic and Epidemiological Data Recorded in an Admission History and Physical Examination

Type of Data	Examples of Data Relevant to Clinical Diagnosis (Case Detection), Outbreak Detection, or Characterization
Demographics	Age, gender, home and work address
History of present illness	Symptoms (cough, fever, diarrhea) and their timing; significant negatives
Physical examination	Temperature, rashes, evidence of pneumonia
Laboratory results	Blood, stool, and sputum cultures; cerebrospinal fluid analyses; examinations of stool for ova and parasites
Radiology results	Chest radiographs
Travel and exposure histories	Travel to endemic area, drinking of unboiled water, animal bites
Vaccinations	Measles, hepatitis, influenza, yellow fever vaccinations
Personal/social history	Intravenous drug use, sexual practices, occupation, household members
Past medical history	Diabetes, HIV, transfusions
Allergies	Medications, insects
Current medications	Ciprofloxacin, Tamovir
Diagnostic impression	"pneumonia, rule out anthrax"

Data are checked if they could be used in a case-control study to elucidate outbreak characteristics such as source or to determine if a patient matches a case definition.

6. INFORMATION SYSTEMS IN HEALTH CARE

Information systems are the key to solving the problem of accessing clinical data for biosurveillance. The past 3 decades have seen the emergence of electronic systems to manage almost every aspect of medical practice, including scheduling, ordering of tests, recording of clinical observations, and intensive care unit operations. Some of these systems are widely deployed; others have far less market penetration. To realize the potential of clinical data for biosurveillance, the market penetration of those systems that collect needed clinical data must increase (especially into outpatient settings) and the systems must be "biosurveillance enabled," which means they must provide certain functions and do so in a standard way.

Because of the enormous potential of data collected by the healthcare system for biosurveillance, we devote this lengthy section to a description of the information systems that the healthcare system presently uses to collect and store data. We have written this section with the needs of designers and developers of biosurveillance systems in mind. Therefore, in addition to describing the systems, we highlight data relevant to biosurveillance and any technical or administrative barriers to obtaining the data. We offer our opinion about which systems provide the most immediate and the most long-term potential for biosurveillance. We suggest questions that a biosurveillance organization should ask a healthcare system or hospital when discussing options for creating electronic data exchange. For readers that have limited time, our conclusion will be that HL7 (Health Level 7)-message routers offer the most immediate potential for bidirectional data exchange between biosurveillance organizations and the healthcare system, and that point-of-care (POC) systems represent the future of biosurveillance. Table 6.2 summarizes the systems that we will discuss—the data they contain, their market penetration, and their potential role in biosurveillance. As a general rule, the larger the hospital or healthcare organization, the more likely it will have each system.

6.1 HL7-Message Routers

Figure 6.1 illustrates the information-system architecture of a typical hospital (or multihospital system). At the heart of the architecture is an HL7-message router.[3] An HL7-message router is a communication hub that transmits information between information systems, both within the healthcare system and outside of it. HL7-message routers are commercial off-the-shelf products supplied by many vendors (there is an open-source HL7-message router called Jengine available at *http://www.jengine.org/* and an HL7 listener, which can receive messages from a hospital HL7-message router, at *openrods.sourceforge.net*). They are also called *integration engines*. Healthcare systems often employ HL7 interface engineers and network engineers to configure and maintain these systems around the clock.

The importance of the HL7-message router for biosurveillance is twofold. First, many clinical information systems send data to the HL7-message router (Figure 6.1). The data include patient chief complaints, dictations, results of laboratory tests, and results of radiological examinations. The information systems send these data to the HL7-message router in real time, and the HL7-message router can forward these data to biosurveillance organizations without delay. If a computer in a biosurveillance organization is temporarily incapable of receiving the data, the HL7-message router will queue the data for up to a week until the computer becomes available. Second, an HL7-message router can support bidirectional, real-time communication between computers in a healthcare system and computers in a biosurveillance organization by using the

3 The term *HL7* (Health Level 7) refers to the dominant messaging standard in health-care computing. Briefly, the *HL7* standard was developed in the 1980's by a coalition of information system vendors to allow their systems to more readily exchange data. We discuss the HL7 standard in detail in Chapter 32.

TABLE 6.2 Information Systems in Healthcare: Data, Relevant HL7-Message Types, U.S. Market Penetration (Estimated), and Potential Uses of the Data in Biosurveillance

			U.S. Market Penetration			
System	Data	HL7 Message Type and Event	Hospital/Health System	Office/Home Health	LTC Facility	Potential Biosurveillance Use
HL7-message router	Data from many of the systems listed below	N/A	High	Large HMOs	If owned by HS	Best current single point of integration
Registration	Chief complaints, addresses, age, sex	ADT^A04	High	?	—	Source of data for case detection, syndromic surveillance
Billing	Diagnostic and procedure codes (CPT codes, ICD-9 codes)	DFT^P03	High	High		Source of data for case detection, syndromic surveillance
Laboratory	Orders and results of tests	ORM^O01 (orders) ORU^R01 (results)	High	High	High	Culture- and test-based detection strategies
Dictation	Symptoms, signs, travel, exposures	MDM^T02	High			Source of symptom and sign data
Radiology	Orders and results of tests, images	ORM^O01 (orders) ORU^R01 (results)	High	High		Test-based strategies
Pathology	Orders and results of tests, autopsy results	ORM^O01 (orders) ORU^R01 (results)	Mod	Mod		Diagnosis-based strategies
Pharmacy	Orders for medications	RDE^O01	High			Indirect evidence
Order entry	Orders for laboratory tests and medications; admission diagnoses	Refer above message types and events	Moderate	Low	Low	Indirect evidence
Point-of-care systems	Symptoms, vital signs, signs, diagnoses, orders, epidemiological data	?	Low	Low		Data needed to satisfy case definitions; potential for decision support for physicians
Data warehouse	Data from many of the systems listed above	N/A	Low	—	—	All of the above
Call centers and patient Web portals	Symptoms, referrals, appointments	N/A	Low	—		Collect early symptom information, potential for decision support for patients

LTC indicates long-term care facility; HMO, health maintenance organization; HS, hospital system; ADT, admission discharge and transfer; ?, unknown; —, not applicable; CPT, Current Procedural Terminology; ICD-9, International Classification of Diseases 9th edition.
Adapted from Agency for Healthcare Research and Quality report.

query message type.[4] For these reasons, HL7 messaging is a core element of emerging biosurveillance IT standards, as we discuss in Chapter 32. It is worthwhile to point out that HL7 defines a communications protocol at the application layer and does not address the transport layer (e.g., TCP/IP, HTTP, RS-232). Therefore, it is possible for two systems to speak HL7 at the application layer but still not be able to physically communicate.

Healthcare organizations that own multiple hospitals and physician practices may use an HL7-message router to integrate the data systems of these geographically diverse practices. In these settings, the HL7-message router represents a single point of integration between a biosurveillance organization and scores of hospitals. For example, when the Utah Department of Health deployed the RODS system for the 2002 Winter Olympics, it created a single connection to the Intermountain Healthcare System HL7-message router, enabling it to collect registration data from nine emergency departments and 19 acute care facilities (e.g., urgent cares, instacares and now cares). It created a second connection to the University of Utah HL7-message router to add one emergency department, one urgent care, and the polyclinic located in the Olympic Village to the biosurveillance system (Gesteland et al., 2002, 2003, Tsui et al., 2002).

Regarding data exchange with a hospital or other large healthcare organization, key questions to ask about HL7-message routers are as follows: *Do you have an HL7-message router (you may have to ask the IT person and remember that it is also called an integration engine)? Which information systems send messages to it? How many hospitals and office practices are connected to it? Do you maintain it, or do you outsource its maintenance? What is your minimum lower level protocol (e.g., TCP/IP, HTTP, RS-232) for HL7 messages if you have an HL7-message router?* Note that these and questions that we suggest for other information systems may best be answered by the healthcare organization's IT staff, and in many instances, they are best answered by the organization's HIS vendor.

4 There are multiple HL7 query message types such as QBP^Q21 (query for person demographics), QRY^Q28 (query for pharmacy dispense information), VXQ^V01 (query for vaccination record).

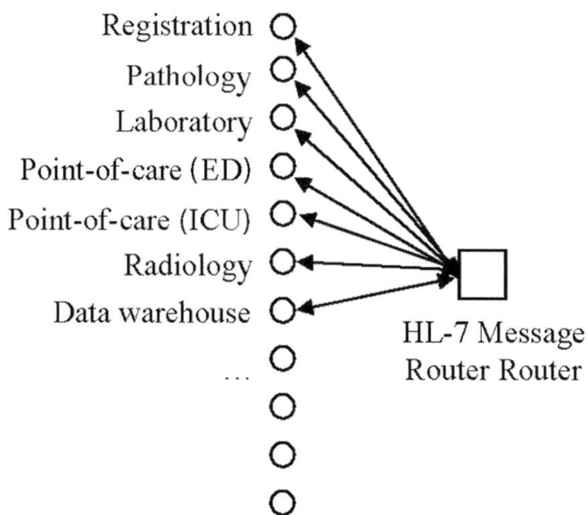

FIGURE 6.1 Representative information system architecture of a modern healthcare system. A healthcare system may also create direct connections between systems such as the laboratory information system and a point-of-care system (not shown).

6.2. Registration Systems

Clerical staff in emergency departments (and other medical facilities) register patients electronically at the time that they present for care. Electronic patient registration is nearly ubiquitous in the United States, especially in emergency departments and hospital-based or hospital-associated office practices and large health maintenance organizations. The registration clerk enters the reason for visit (also called the *chief complaint*) at the time of registration, along with the patient's age, gender, and home address. Syndromic surveillance systems analyze de-identified versions of these data (without patient name and with home address information limited to zip code) (Espino and Wagner, 2001; Lober et al., 2001; Tsui et al., 2001; Wagner et al., 2001).

The main advantage of registration data for biosurveillance is its widespread availability in the United States and inherent timeliness. If a healthcare organization has an HL7-message router, chances are good that the registration system sends messages to it and that it can de-identify and forward these data to a biosurveillance organization in real time.

6.3. Billing Systems

Billing was one of the first hospital functions to be computerized. Billing systems contain information about diagnoses and tests that were performed (but not the results). Third-party payers such as insurance companies require that a hospital encode this information by using either Current Procedural Terminology (CPT) or International Classification of Diseases, 9th edition (ICD-9) codes, a process that in many healthcare settings is done manually by professional coders. This process may introduce a several-day delay between the time the patient is seen and the diagnoses are available in a billing system. Billing data are often routed through an HL7-message router, and the combination of the widespread availability of both billing systems and HL7-message routers makes billing data highly available for biosurveillance. We note that third-party payers consolidate billing data from many hospitals and healthcare providers. Because the number of payers is typically smaller than is the number of hospitals and healthcare providers in a community, payers are a potentially more efficient source of billing data.

6.4. Laboratory Information Systems

A laboratory information system receives and stores requests for tests, and results are entered by laboratory technicians or directly from laboratory instruments (e.g., via the ASTM E-1381 protocol *Specification for Low Level Protocol to Transfer Messages between Clinical Laboratory Instruments and Computer Systems*). Results of tests are available via paper reports and electronic interfaces, both to human users and to other information systems such as a POC system (described below).

Laboratory information systems in hospitals, the animal health system, governmental public health, and commercial free-standing laboratories are virtually identical. For this reason, we devote Chapter 8 to the role of laboratories and networks of laboratories in biosurveillance.

Briefly, laboratories perform tests that include all types of cell counts, analytical chemistry, drug and toxin screening, and detection of microbes. The results of these tests are important to biosurveillance of virtually every conceivable biological agent. Fortunately, the vast majority of clinical laboratories in the United States are highly automated, using computers to perform tests, store results, and communicate results. Test results are generally available electronically very soon after the tests are performed. Results are available not only directly from the laboratory information system but may also be transmitted to an HL7-message router and to an enterprise data warehouse or a POC system (described below). There are many examples of hospitals that send laboratory data electronically to health departments, although these hospitals represent but a tiny fraction of the approximately 5,000 hospitals in the United States (Effler et al., 1999; Overhage et al., 2001; Hoffman et al., 2003).

Not only are the results of laboratory tests of value to biosurveillance, but so may be the mere fact that a particular test was ordered. The type of test ordered provides a clue to the clinician's diagnostic thinking about the nature of the illness in a patient, and it may be available well in advance of the result of the test, possibly within minutes or hours of a physician first seeing a patient.

At present, unfortunately, obtaining laboratory results from hospitals requires extensive custom interface development to translate and reformat data. Many hospitals use proprietary coding schemes to represent the names of tests and the results of tests. For example, the organism *Bacillus anthracis* may be called ANTX in one institution and BANTHRACIS in another. Dr. Clem McDonald recognized this importance of this problem in the 1980s. He promulgated the use of standards for the naming of laboratory tests and the reporting of their results (discussed in detail in Chapter 32). Until these standards are more universally used in health care, however, the construction of biosurveillance systems that collect laboratory data from hospitals will be time-consuming and expensive (the monitoring of laboratory test results and orders from national laboratory companies is far more feasible at present, as we discuss in Chapter 8).

When discussing data exchange with a hospital or other healthcare organization, the key questions to ask about the laboratory information system are as follows: *Does your laboratory information system use SNOMED and LOINC? Do you send the results to an HL7-message router, a data warehouse, or a POC system? Is the microbiology outbound feed structured (i.e., not a free-text report intended for printing or display on a computer screen only)?*

6.5. Dictation Systems

Dictation systems are a mainstay of clinical data recording in the hospital, emergency department, and outpatient settings. Clinicians often use dictation systems to record a patient's history, observations made during physical examination, progress notes, interpretations of radiological examinations, and results of postmortem examinations. Dictation systems can range in complexity from a single part-time transcriptionist working with a word processor to pools of transcriptionists using dedicated systems produced by companies such as Lanier. Although dictations are rich with clinical detail—including the patient's presenting complaint, the history of the illness, exposure information, vaccinations, vitals signs and physical findings, and diagnostic impressions—the data are recorded in English and are difficult for computers to understand. In addition, the time delay between dictation and transcription delays the availability of the data. Nevertheless, the value of the information is sufficiently high for both biosurveillance and medical applications that researchers in medical informatics have developed approaches to processing these data, which we discuss in detail in Chapter 17, *"Natural Language Processing for Biosurveillance."*

The availability of dictations for biosurveillance purposes is lower than that for laboratory data because not all transcribed dictations are stored electronically in databases and not all institutions route electronic versions of the transcriptions through an HL7-message router. Thus, even when dictations are available electronically, custom interfaces may need to be built to the database that stores the dictations.

When discussing data exchange with a hospital or other healthcare organization the key questions to ask about dictations are as follows: *Do you have a dictation system that produces electronic copy? Does it provide an outbound interface (either HL7 or proprietary)? Do you send the dictations to the HL7-message router, the data warehouse, or a POC system? Which of the many types of dictation are stored by the system, and with what time delay do they appear from the time of dictation?*

6.6. Radiology Systems

Radiology departments were early adopters of IT, and today many practices manage the reports of examinations electronically. A radiology department performs radiographic, ultrasound, computed tomography (CT), magnetic resonance imaging (MRI), and other examinations of patients. The results of many of these examinations are highly relevant to biosurveillance. Clinicians use radiological examinations to diagnose infectious diseases (e.g., pneumonias). The reports include the diagnostic impression of the radiologist (e.g., "the combination of mediastinal widening and pneumonic infiltrate is consistent with pulmonary anthrax"). Unfortunately, in the majority of practices, the radiologist dictates his report and it is subsequently transcribed; thus, the reports can be delayed a day or more by the transcription process and are transcribed in English (or another language). Researchers in medical informatics have investigated methods for processing dictated radiology reports to extract information about patient characteristics, such as presence or absence of pneumonia (see Chapter 17) (Jain et al., 1996; Knirsch et al., 1998; Chapman and Haug, 1999; Hripcsak et al., 1999; Fiszman et al., 2000a).

We note that technical advances have made it possible for radiology information systems to store the images themselves in digital form, although the market penetration of this functionality is relatively low at present with the exception of imaging modalities such as CT and MRI, which are inherently digital. Rapid access to the images themselves, especially of chest radiographs, would be of value to both hospital infection control and governmental public health, especially during outbreak investigations. There is a trend to make such images directly available to physicians in hospitals through Web browser–based interfaces, and biosurveillance organizations could negotiate to obtain permission to access such systems or otherwise use this emerging capability.

When discussing data exchange with a hospital or other healthcare organization, the key questions to ask about radiology information systems are as follows: *Do you have a radiology information system that stores reports? Does it provide an outbound interface? Do you send the reports to the HL7-message router, the data warehouse, or a POC system? Do you store images digitally (and which ones), and can your physicians access images by using the Web?*

6.7. Pathology Information Systems

Pathologists examine bodily fluids, tissue specimens, and organs. Pathology information systems are more recent additions to healthcare computing because of the image-intensive nature of pathology practice (gross and microscopic examinations). Thus, market penetration is lower than that of laboratory or radiology systems. The data that may be useful for biosurveillance include both orders (for pathology tests) and their results. Access to the data in these systems is relatively difficult, as many systems function as free-standing applications used to generate printed reports. Integration with other hospital systems has not been as critical of a design factor as in the laboratory or radiology department, so biosurveillance organizations must either develop the integration functions or influence hospitals and system designers to "biosurveillance enable" these systems.

6.8. Pharmacy Information Systems

Pharmacy information systems receive and process orders for medications such as antibiotics or antidotes for toxins, which are relevant to biosurveillance because they provide indirect evidence that the patient's illness may be caused by an infectious disease and even hint at the nature of that disease. In the vast majority of hospital pharmacies, physician orders are received on paper order forms, which pharmacists transcribe into the pharmacy information system; thus, there is a delay from the time that the physician expresses his or her understanding of the clinical problem in the form of an antibiotic order to the time that information is available electronically. The reliability of the information is extremely good, however, because pharmacists use expert knowledge and contextual information (available in the orders themselves, the pharmacy information system, and sometimes from review of the patient chart or contact with the physician) to validate the order before dispensing a medication.

6.9. Order-Entry Systems

Order-entry systems are computer systems that ward clerks or clinic staff use to communicate a physician's orders electronically to the laboratory and other departments in the hospital. If an order-entry system exists, it is a single point of access for information about orders for laboratory tests and medications, as well as orders that a patient be placed under respiratory isolation. Orders also may state a patient's diagnosis in an entry entitled "admission diagnoses."

Ward clerks or clinic staffers enter orders promptly, so the information is available without time delay, except the time lag from when the clinician writes the order on a paper order form until the time when a ward clerk transcribes the order, a delay that is measured in hours at most. The reliability and accuracy of this information is high.

In less than 10% of hospitals, physicians enter orders directly into computers, eliminating the time delay and creating an opportunity for direct interaction and decision support of the clinician (Tierney et al., 1993; Bates et al., 1994; McDonald et al., 1994; Sittig and Stead, 1994; Frost and Sullivan, 2003). We discuss the potential of direct interaction with the clinician and decision support later in this chapter.

When discussing data exchange with a hospital or other healthcare organization, the key questions to ask about order-entry systems are as follows: *Do you have an order entry system? What fraction of clinicians use it, and what fraction of orders does it capture? Does it provide an outbound interface? Do you send the orders to the HL7-message router or a data warehouse?*

6.10. Point-of-Care Systems

A POC system is a hospital (or outpatient) information system that includes bedside terminals or other devices for capturing and entering data at the location where patients receive care (Shortliffe et al., 2001). Clinicians use POC systems to record directly details of patient encounters, to review information, and to order tests and other services. POC systems replace many functions of the paper chart and, in fact, are sometimes referred to as electronic medical records (although that term is used so loosely that we recommend that it is not used). Vendors sell POC systems specialized for diverse settings, including the emergency department, physician offices, hospitals, intensive care units, long-term care facilities, and home health care. POC systems even exist for prehospital care settings. Emergency medical units may use "ruggedized" handheld computers in the field.

Depending on the POC system, a clinician may enter some subset of the data listed in Table 6.1. A clinical information system that has POC functionality has the potential to become paperless as each clinician, the laboratory, and the radiology department contribute to the collection of data about a patient. Advantages of POC systems to a hospital include quicker access to clinical information, the ability to communicate orders more quickly, elimination of the difficulties involved in reading the products of poor penmanship, and the ability to harness integrated decision-support tools such as electronic formularies, drug interaction warning databases, and electronic implementations of practice guidelines.

POC systems are the future of biosurveillance. A POC system facilitates the electronic capture of key diagnostic data (and usually in a computer-interpretable form rather than English). POC systems typically include decision-support capabilities (discussed in Chapter 13) that alert clinicians to potential drug–drug interactions and even suggest diagnoses. It is possible to program the underlying computer decision-support system to notice that a patient may have pneumonia and Gram-positive rods in a blood culture (an example of automatic case finding) and alert the clinician to consider a diagnosis of inhalational anthrax (and even report this suspicion automatically to a health department). High degrees of suspicion based on regional events can also be incorporated into the computer analysis. These capabilities are the reason that POC systems with decision support are the future of biosurveillance.

At present, most estimates of the market penetration of POC systems are in the single digits. The surgeon general offices of the nation's military services, when interviewed, were not aware of wide-spread use of POC systems in military facilities; if they are deployed, such deployments may be scattered in specific facilities. The VA, on the other hand, has high level of deployment of POC systems. Reasons for low market penetration include cost and reluctance by physicians and other providers to adopt these systems. A large multihospital organization may make a strategic decision to deploy POC system. Kaiser Permanente, in California, anticipates that its facilities will be operating an integrated POC system within five years.

In the United Kingdom, by contrast, POC systems are ubiquitous (Benson, 2002). The value of these systems for public health surveillance is illustrated by the rapidity with which the United Kingdom can potentially implement an anthrax surveillance strategy. By changing the decision support logic only, once, in a central location, the ability to detect postal workers presenting with influenza-like symptoms at the time of phone or physical presentation to any primary care physician in the country will exist.

Lazarus et al. (2002) has claimed that a POC system (a commercial product Epicare; Epic Systems Corporation, Madison, Wisconsin; *http://www.epicsys.com*) can be effective for purposes of public health reporting and bioterrorism early warning even if it serves only 5% to 10% of the population in a region being monitored. More research with POC-based surveillance is required to elucidate the relationship between the completeness of sampling of a population and the size of outbreaks of different diseases that can be detected.

6.11. Patient-Care Data Warehouses

Ralph Kimball (eminent data warehouse authority) defines a data warehouse as "a copy of transaction data specifically structured for query and analysis." Large hospital systems often build or purchase data warehouses specifically to integrate data from multiple information systems and multiple hospitals to provide clinicians with a consolidated view (often via a Web interface) of patient data (Figure 6.2). We refer to such data warehouses as *patient-care data warehouses* to distinguish them from data warehouses used for business or research purposes (Shortliffe et al., 2001).

When a patient-care data warehouse exists, it represents a point of integration of data that, similar to the HL7-message router, is a leverage point for biosurveillance. Data warehouses acquire data from other systems and transform data into a common format (e.g., data type, domain, unit of measure), and load the data into special data structures tailored for the intended use. In the case of patient-care data warehouses, the data are stored in structures that support rapid retrieval of the complete medical record of a single patient.

Transformation improves the accuracy of the data by removing duplicates from data sent to the data warehouse by laboratory or other systems. This process also may translate different hospital identification codes into a single canonical form, allowing data collected by different information systems to appear the same. Transformation is extra work that a biosurveillance organization would have to do if it were to access data directly from radiology and laboratory information systems.

If a healthcare organization has a Web-based interface to a patient-care data warehouse (as in Figure 6.2), it may be possible for a health department to negotiate with the hospital to provide its epidemiologists with access to the patient data on a need-to-know basis. Moreover, the access can be integrated within the health department's biosurveillance system. Figure 6.3 shows a sequence of screens from an early version of the RODS biosurveillance system (circa 2002) in which the user notices an increase in the number of patients with chief complaints of diarrhea, drills down to a line listing of cases, and then selects a case for which he wishes to see the patient record. After asking for authentication (a password and username issued by the healthcare system), the RODS system automatically takes him to the screen shown in Figure 6.2, which is the Web-based interface to the healthcare system's patient-care data warehouse. Note that Web browsers can remember login names and passwords, so the epidemiologist only must enter these credentials the first time he accesses the system, after which the transition to the patient-care data warehouse is seamless. During the months after the anthrax postal attacks, this function was used many times to do rapid investigations of spikes in syndrome data. A user could review approximately 40 patient charts per hour in this manner, which is several of orders of magnitude faster than conventional shoe-leather methods.

The availability of patient-care data warehouses in healthcare is low to moderate at present. Note that in many healthcare systems, a patient-care data warehouse will be a component of a more comprehensive "electronic medical record" provided by a vendor. However, it likely will still have a Web-based interface that provides a consolidated view of a patient's "chart."

In theory, it should be easy for a biosurveillance organization to interface with a data warehouse, which, if it exists, may represent a single point of integration that can provide data that have already been integrated and transformed. The key questions to ask when discussing data exchange with a hospital are as follows: *Do you have a data warehouse? Is it for clinical care or archiving and business analysis? Is it part of a more comprehensive vendor system (and which one)?* Many data warehouses now have Web-based interfaces. Although these interfaces are now being standardized, there are two competing standards. One is being promoted by Microsoft and the other by Oracle, with other vendors lining up on either side (or even both sides). So an additional question is *Does your data warehouse support either XML/A or JOLAP?*

6.12. Patient Web Portals and Call Centers

Two additional types of information systems are beginning to appear in health care. *Call centers* are facilities that receive

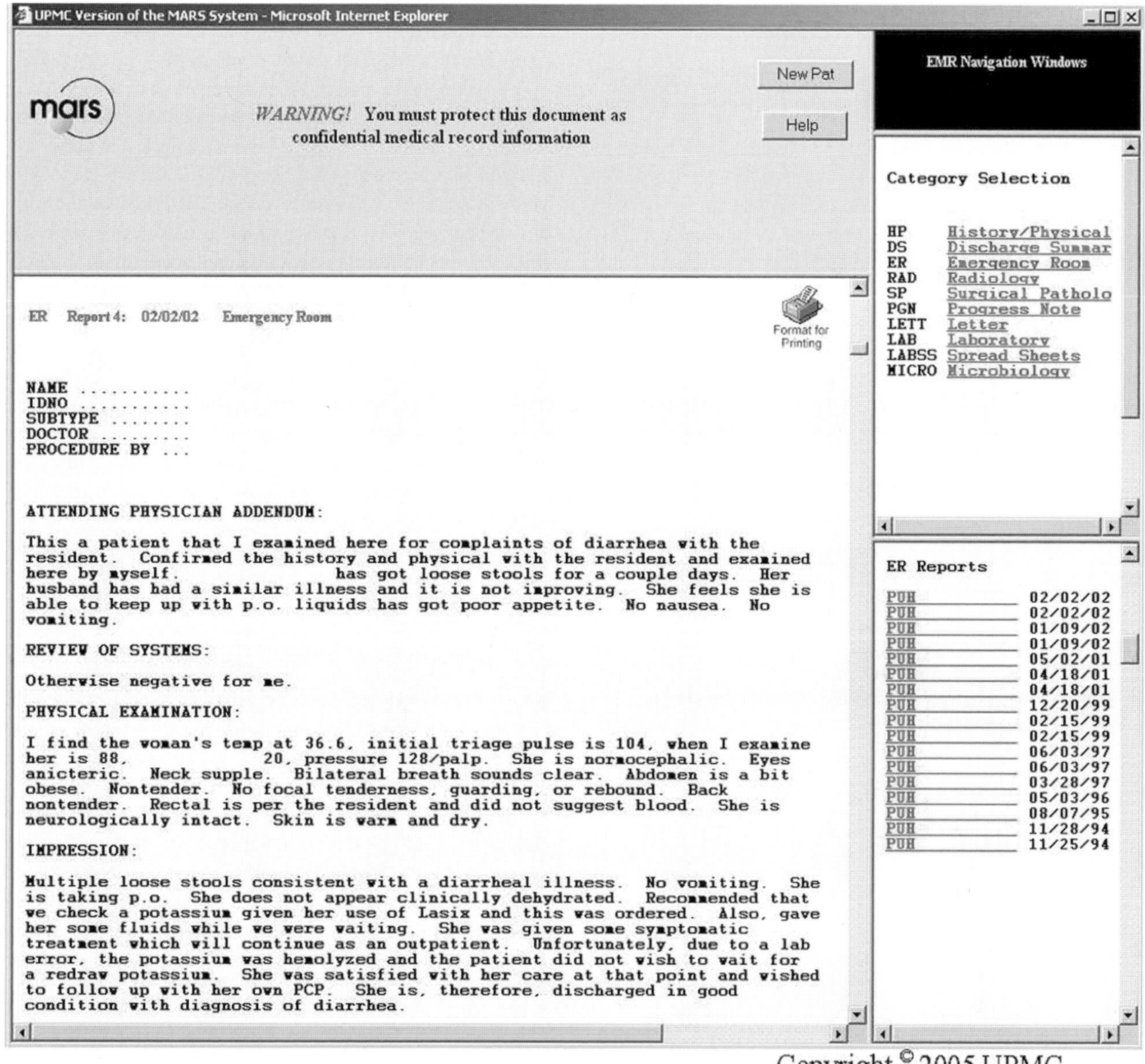

FIGURE 6.2 Web-based interface to a patient-care data warehouse. This data warehouse collects data from multiple information systems in multiple hospitals owned by the healthcare system to provide a consolidated view of a patient's medical history for a clinician. (Reproduced by permission from the University of Pittsburgh Medical Center [UPMC].)

telephone calls from sick individuals who require information, triage, appointments, or immediate assistance. The staff fielding phone calls use information systems to document the calls, typically recording diagnostic information (e.g., reason for call, symptoms, and nurse assessment). In systems for which potential use for biosurveillance has been studied (see Chapter 27), the data included the practice guideline selected by the nurse to manage the call and the reason for call with timestamps, locations, and disposition.

The most extensive use of call centers in biosurveillance is the United Kingdom's National Health Service (NHS) Direct, which is a nurse-led telephone help-line that covers the whole of England and Wales. Data on the following 10 symptoms/syndromes are received electronically from 22 call centers and are analyzed on a daily basis; cough, cold/flu, fever, diarrhea, vomiting, eye problems, double vision, difficulty breathing, rash, and lumps. Significant statistical excesses (exceedances) in calls for any of these symptoms are automatically highlighted

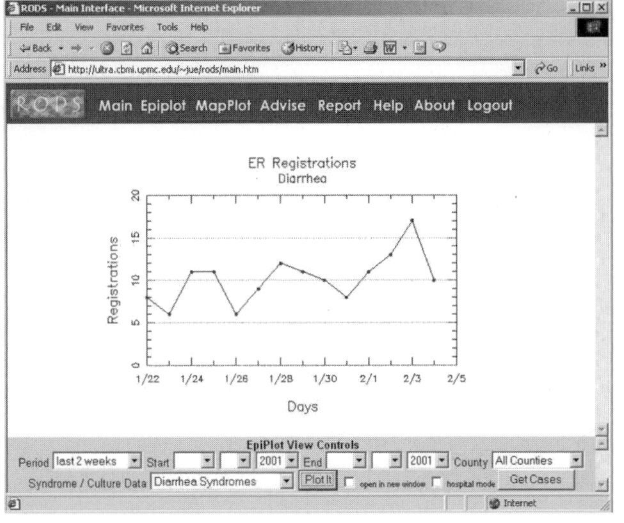

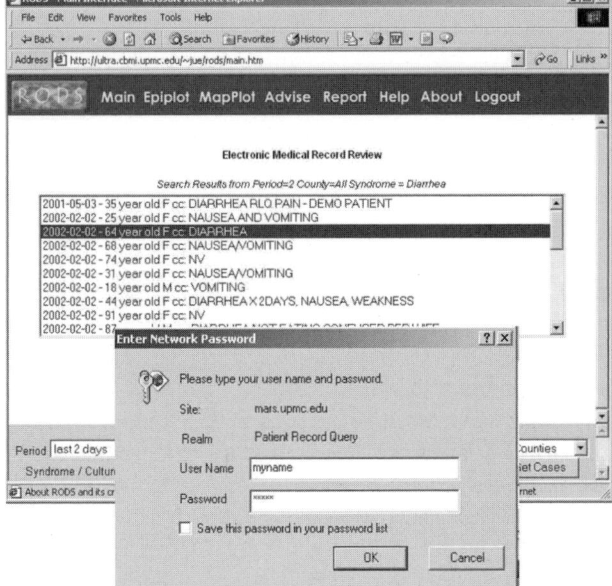

FIGURE 6.3 Sequence of screens from an early version of the RODS biosurveillance system. After noticing an increase in patients with chief complaints of diarrhea (top screen), the user drills down to a line listing of cases (bottom screen). The user selects a case to see the patient's medical record. After providing authentication (overlying dialog box), the RODS system automatically takes the user to the screen shown in Figure 6.2, which is the Web-based interface to the healthcare system's patient-care data warehouse.

and assessed by a multidisciplinary team. The aim is to identify an increase in symptoms indicative of the early stages of illness caused by the deliberate release of a biological or chemical agent, or more common infections (Harcourt et al., 2001; Cooper and Chinemana, 2004; Cooper et al., 2004a,b; Doroshenko et al., 2004; Nicoll et al., 2004).

A second large project that involves call centers is the National Bioterrorism Syndromic Surveillance Demonstration Program, coordinated by Harvard Medical School and Harvard Pilgrim Heath Care (Platt et al., 2003; Yih et al., 2004). There is no single call center for the United Sates; therefore, this project seeks to recruit and integrate the call centers for cities, regions and ultimately the entire country.

Patient Web portals provide similar functionality but are basically self-service, much like Web-based airline bookings. The types of data collected by Web portals justify their inclusion in this discussion, despite their very low market presence. Patient Web portals have the potential to collect symptom level data as early as the day of onset of illness. Call centers, if patients are encouraged to use them early rather than waiting for illness to progress, have similar potential. The reliability and availability of such data have potential to be very high, especially if such services are designed from the ground up with the needs of regional integration of data for biosurveillance purposes in mind.

7. BIOSURVEILLANCE OF THE HEALTHCARE SYSTEM

A healthcare-associated infection (HCAI) is an infection that develops in a healthcare setting such as a hospital or as a result of medical treatment. HCAIs are also known as *nosocomial infections.* HCAI is a significant problem in healthcare. In 1992, the CDC estimated that there are at least two million HCAIs in hospitalized patients alone each year in the United States, costing $4.5 billion and causing 90,000 deaths, a third of which are probably preventable (Anonymous, 1992).[5] Roughly an equal number of infections occur in long-term care facilities, dialysis centers, clinics and other settings (Martone et al., 1998).

The rate of HCAIs in hospitals has remained steady at approximately 5% of patient admissions for at least the past three decades (Haley et al., 1985a,b; Martone et al., 1998). This lack of improvement does not reflect inattention or lack of improvement in methods of prevention, but rather imperfect implementation of known measures such as hand washing, the relentless evolution of microorganisms, the severity of illness of patients, and the increasing complexity of medical treatment.

To put these statistics in perspective, Florence Nightingale, Ignaz Semmelweiss, Joseph Lister, and Oliver Wendell Holmes lived in eras in which 20% to 30% mortality from

5 These oft referenced statistics stem from data and analyses by Haley RW, Culver DH, Morgan WM, Emori TG, Munn VP, Hooton TP. *The efficacy of infection surveillance and control programs in preventing nosocomial infections in U.S. hospitals.* Am J Epidemiol 1985; 121:182-205 and Martone, W. J., Jarvis, W.R., Edwards, J.R., Culver, D.H. and Haley, R.W. (1992) In *Hospital Infections, Third Edition* (Eds, Bennett, J.V. and Brachman, P.S.) Lippincott-Raven Publishers, Philadelphia.

HCAIs was not uncommon.[6] These individuals, incidentally, pioneered methods such as hand hygiene that now form the basis of modern infection control.

Outbreaks also occur in hospitals, but they are infrequent and account for only 2% to 3.7% of HCAIs in hospitals (Wenzel et al., 1983; Haley et al., 1985b). Haley et al. (1985b) estimated that a typical community hospital has one nosocomial outbreak per year. The types of outbreaks change over time as organisms and medical technology changes. Contaminated products (e.g., blood) and medical devices are common causes of recent outbreaks investigated by the CDC (Martone et al., 1998).

7.1. Infection Surveillance Control Programs

An *infection surveillance control program* (ISCP) is a division of a healthcare organization with the mission "to identify and reduce the risks of acquiring and transmitting infections among individuals served, staff, contract service workers, volunteers, students, and visitors." (JCAHO, 2004). Among the responsibilities of an ISCP is biosurveillance of the organization.[7] In particular, an ISCP is responsible for collection, analysis and interpretation of infection control data, and the investigation and surveillance of suspected outbreaks of infection.

The origins of ISCP can be traced to a pandemic of staphylococcal infections in hospitals in the mid 1950s in the United States. In response to this problem, hospitals organized infection control committees. Over the ensuring decade, a few hospitals developed organized infection control programs, initially using physicians and then adding trained infection control nurses. By the mid 1970s, most hospitals had adopted this practice.

A typical ISCP consists of one or more doctors and nurses or medical technologists with specialized training related to epidemiology of hospital infections and disease prevention. These individuals have expertise in the recognition of disease in individual patients as well as recognition of outbreaks in the hospital population and in their prevention and control. The APIC created the Certification Board of Infection Control, which certifies infection control practitioners (Sheckler, 1998).

Approximately one-fourth to one-half of hospital HCAIs come to the attention of ISCP as a result of laboratory testing. The rest are identified by a variety of formal and informal surveillance activities. A typical ISCP identifies patients via a daily printout of "positive cultures" from an electronic laboratory system. The ISCP may also obtain a list of new prescriptions for antibiotics. A good ISCP also requires "shoe-leather" epidemiology (i.e., daily ward rounds to speak with personnel providing direct patient care) and some form of "post-discharge surveillance" to detect infections in patients who have already left the hospital. The combined result of all these processes is a list of potential patients to investigate that day. The staff reviews this list to organize and prioritize the work for the day. The staff collects additional information for each patient from hospital information systems, for example, accessing a single system or multiple systems to review radiology reports, physician dictations, medication records, and other results of laboratory testing. In addition, the staff may read the paper record of a patient or speak with physicians and nurses caring for a particular patient.

To satisfy reporting requirements, the staff may notify governmental public health when a patient with a reportable disease is found. To satisfy JCAHO requirements (discussed below), they compile periodic reports.

Prevention of infections in the healthcare setting requires cooperation of virtually all divisions and individuals. A list of related departments created by JCAHO identifies central sterile processing, environmental services, equipment maintenance personnel, facilities management (including engineering), housekeeping, information management, laboratory, medical staff, nursing, and pharmacy.

7.2. JCAHO Infection Control Guidelines

JCAHO establishes guidelines for patient safety that include guidelines for infection control. Infection control is one of JCAHO's 14 priority focus areas.

JCAHO is widely recognized for its leadership role in developing standards and performance measures and for the adaptability of its rigorous evaluation processes to emerging new forms of healthcare delivery. JCAHO evaluates and accredits more than 15,000 healthcare organizations and programs in the United States. An independent, not-for-profit organization, JCAHO is the nation's predominant

6 Ignaz Philipp Semmelweis (1818-65), a Hungarian obstetrician, introduced antiseptic prophylaxis into medicine. In the 1840s, puerperal or childbirth fever, a bacterial infection of the female genital tract after childbirth, was taking the lives of up to 30% of women who gave birth in hospitals. Women who gave birth at home remained relatively unaffected. Semmelweis observed that women examined by student doctors who had not washed their hands after leaving the autopsy room had very high death rates. When a colleague who had received a scalpel cut died of infection, Semmelweis concluded that puerperal fever was septic and contagious. He ordered students to wash their hands with chlorinated lime before examining patients; as a result, the maternal death rate in his hospital was reduced from 12% to 1% in two years. Source: funkandwagnalls.com Copyright 1999, 2000.

7 The other responsibilities of a ISCP include (1) planning, implementation and evaluation of infection prevention and control measures; (2) education of individuals about infection risk, prevention and control; (3) development and revision of infection control policies and procedures; (4) management of infection prevention and control activities; (5) provision of consultation on infection risk assessment, prevention and control strategies. Source: *http://www.cbic.org/becoming_certified.asp.*

standards-setting and accrediting body in health care. Since 1951, JCAHO has maintained state-of-the-art standards that focus on improving the quality and safety of care provided by healthcare organizations. Infection control is a critical component of safe, quality health care.

JCAHO is becoming more active, perhaps even militant, as a result of increasing awareness of the impact of HCAIs on the cost and quality of health care. Effective January 1, 2005, JCAHO established a New Patient Safety Goal (the seventh) in the area of infection control, which it promulgates as a set of standards that includes the following: accountability of the CEO of a healthcare organization for compliance and fiscal support of an ISCP; staffing and training of ISCP; communication and coordination with health departments and other community organizations; clear command and control (delegated authority); and surveillance, and monitoring of efficacy if its infection control programs (JCAHO, 2004).

JCAHO is very influential. As previously discussed, its influence derives from the Medicare Act of 1965, which included JCAHO accreditation as one basis for Medicare reimbursement. The most recent *Accreditation Manual for Hospitals* issued by JCAHO includes a set of scoring guidelines on which the compliance of a hospital will be judged. To obtain the highest score, a hospital must provide evidence of having switched from processes such as surveillance of antibiotic use and nosocomial infections as ends in themselves to measures of patient outcomes as indicators of hospital performance. To link patient outcomes such as length of stay, days or morbidity, mortality, and costs will require substantial collection of data, or integration of data, already being collected by registration systems and billing systems.

7.3. Information Systems in Infection Control

There is and will be an increasing trend to support ISCP with IT, although the barriers to comprehensive support of all functions are high.

At present, ISCPs use computers to manage surveillance data in two ways: the most common use is to store and analyze surveillance data that are collected manually. We refer to such systems as free-standing. Far less commonly, ISCPs computerize the actual collection of surveillance data. We refer to these ISCP systems as embedded because they receive data directly from clinical information systems.

7.3.1. Free-Standing ISCP Systems

An ISCP may utilize general purpose software such as Microsoft Excel, SAS, Microsoft Access, Microsoft SQL Server, and Oracle to store and analyze surveillance data. It may use software specifically designed for infection control such as AICE, NNIS-IDEAS, QLOGIC II, Epidemic Information Exchange (Epi-X), and WHOCARE

The practice of analyzing ISCP data by using free-standing computers is nearly ubiquitous because of the low cost of computers and their ability to facilitate statistical analysis and report generation. However, ISCPs still rely heavily on paper- and card-based systems. An ISCP that collects surveillance data on paper may subsequently enter the data into a computer system simply for analytic purposes. Manual collection of data with subsequent entry into computers for storage, analysis, and report generation is by far the more common use of computers in ICSP.

7.3.2. Embedded ISCP Systems

A small but growing number of health systems have deployed embedded ISCP systems, motivated by research goals and or the potential to improve the cost-efficiency and efficacy of ICSP (although the initial cost is high).

In the 1990s, a group of researchers at the University of Utah developed a program called *Antibiotic Assistant,* which was a module in the HELP clinical information system operating at LDS hospital (Evans et al., 1985, 1986, 1992, 1998; Burke et al., 1991; Evans, 1991; Gaunt, 1991; Classen et al., 1992; Rocha et al., 1994; Chizzali-Bonfadin et al., 1995; Classen and Burke, 1995; Fiszman et al., 2000b). This research demonstrated new types of hospital infection control functionality that access to clinical information systems made possible (e.g., reminders to administer preoperative antibiotics) as well as their efficacy. An offshoot of this research was Theradoc, Inc., which has commercialized this technology.

Also in the 1990s, researchers at Barnes Jewish Christian (BJC) Hospital in St. Louis developed the GermWatch and GermAlert systems (Kahn et al., 1993, 1995, 1996a,b). Similar to Antibiotic Assistant, these systems collect surveillance data automatically from clinical information systems. They use a rule-based expert system (see Chapter 13) to detect patients of interest to ISCP. These systems are still in use at BJC Health System.

Brossette and colleagues explored the use of computers to perform brute-force search through routinely collected data (also known as data mining) to detect changes in rates of infection in subpopulations (e.g., patients in intensive care units) (Brossette et al., 1998; Moser et al., 1999; Brossette et al., 2000). An offshoot of this research was MedMined, Inc., which has commercialized this technology.

Although the above systems have demonstrated the feasibility of automatic data collection, their market penetration remains low owing to their cost and lack of definitive economic data showing direct benefit to healthcare systems.

7.4. Challenges to Automating ISCP

The challenges to integrating patient data for biosurveillance in the hospital are identical or greater than are those for integrating biosurveillance data for public health surveillance. The difficulty is slightly greater because the set of diseases of concern in ISCP are a superset of those of concern in public health practice. Not only must the healthcare system report notifiable conditions to governmental public health, but it is encouraged by JCAHO to monitor for urinary tract infections, pneumonia, multiple drug-resistant organisms (e.g., methicillin-resistant *Staphylococcus aureus* and vancomycin resistant *enterococci*), surgical site infections, infections related to implanted devices, needle-stick injuries in staff, and infections within immuno-compromised patient populations. JCAHO also encourages healthcare organizations to monitor health outcomes in addition to infection control processes. To automate this type of analysis, an organization must integrate data about processes (e.g., which surgeon performed which procedure on which patient on which date) with data about outcomes (e.g., infection rate, length of stay, and hospital costs).

There are healthcare systems that have accomplished such integration. However, they are the exception rather than the rule. These healthcare systems had already achieved a high level of information system integration for other reasons. They had sufficient medical informatics expertise and grant funding to integrate the systems. They are to some degree the same organizations that we discuss in the next section on regional health information organizations (RHIOs). This overlap is not coincidental. The same IT infrastructure that is a prerequisite for automated sharing of clinical information among hospitals is required for automation of hospital infection control.

7.5. Hospital Biosurveillance as a Model of an Ideal Biosurveillance System

It is interesting to note that biosurveillance in hospitals is more intensive than in the general population. In fact, many of the ingredients of an ideal biosurveillance system (see Chapter 13) are already in place in modern hospitals: highly trained clinicians evaluate every patient every day, patient's temperatures are taken regularly, and surveillance data are available electronically in real time. The few missing ingredients include surveillance of the staff and visitors (who are part of the population) and real-time information about patterns of disease in the community, including other hospitals and long-term care facilities from which patients are transferred. Nevertheless, the ideal biosurveillance system that we will discuss in Chapter 13 has its most complete realization in the modern hospital.

8. ASPs AND RHIOs

Two important trends in clinical computing are (1) the use of application service providers (ASPs), and (2) regional integration of healthcare data for the improvement of clinical care.

8.1. Application Service Providers

ASPs are companies that are in the business of hosting computer applications in central locations. A healthcare system may contract with an ASP to outsource some or all of its data processing. Clinicians interact with the server-based applications over private or public networks. The relevance of an ASP for regional or national biosurveillance is that an ASP may, after obtaining appropriate legal and administrative permission, provide data collected by many healthcare systems. The physical colocation of hundreds of clinical information systems in a single location is helpful but it represents a 20% solution, with the residual 80% comprising unaddressed confidentiality, organizational, vocabulary, and other data integration issues.

8.2. National Health Information Infrastructure

The NHII is an initiative of the U.S. government whose goal is to promote the use of IT by the healthcare system. The government, in particular, hopes to improve the quality of medical care while also reducing its cost.[8]

Importantly, NHII understands the importance of biosurveillance (National Committee on Vital and Health Statistics, 2001, Thompson and Brailer, 2004). The objectives of NHII relevant to biosurveillance are (1) increasing the adoption of electronic medical records, (2) promoting the exchange of data among various healthcare organizations, and (3) improving public health (Thompson and Brailer, 2004).

8.2.1. Increasing Adoption of Electronic Medical Records

The relevance of the first objective to biosurveillance is that data that currently exist only on paper would become available electronically. The federal government, as part of its NHII initiative, has taken several actions to promote adoption of electronic medical records.

The Center for Medicare and Medicaid Services (CMS) announced in July 2005 that it will make available to physicians a free electronic medical record called Office-VistA, which is based on the electronic medical record, VistA, used by VA hospitals throughout the United States. Because the VA also provides outpatient care in clinics located in its facilities, VistA has significant functionality for outpatient offices. CMS is working with the VA to create Office-VistA from VistA by removing inpatient functionality and making it easy to install.

In 2004, CMS initiated a pilot program called Doctor's Office Quality (DOQ)-IT.[9] As part of the DOQ-IT pilot, four Quality Improvement Organizations (QIO)[10] in four states received contracts to assist physicians with selecting and implementing EMRs. Based on the pilot, CMS has subsequently funded the QIO in every state to assist physicians with adoption of EMRs, with the sum of all QIO contracts totaling $120 million (Monegain, 2005).

---

8 Similar initiatives exist in other countries, inculding the United Kingdom, Australia, and Canada.

9 *See http://www.doqit.org*

10 The Peer Review Improvement Act of 1982 created Quality Improvement Organizations to improve the quality of care received by Medicine beneficiaries, ensure that beneficiaries receive only medically necessary care, and handle individual beneficiary issues such as complaints about care received.

In 2004, the Secretary of Health and Human Services (HHS) created an exception to the Stark law for the development of a "community-wide health information system." The Stark law is a federal statute that prohibits physicians from referring Medicare patients to a facility with which they have a financial relationship.[11] It has been an obstacle for hospitals that wished to provide associated outpatient practices with EMRs because providing a practice with an EMR creates a financial relationship under the law. However, the Stark law includes a provision that permits the Secretary of HHS to exempt a specific financial relationship if he or she determines that the relationship does not pose a risk for abuse. Thus, the Secretary of HHS in 2004 made an exception for provision of an EMR when the EMR is necessary to connect to a community-wide health information system.[12] This action by the Secretary of HHS may encourage the development of community-wide efforts to exchange patient data from the outpatient setting and the provision of EMRs to physicians by organizations that participate in the community-wide effort.[13] We discuss some of these community-wide efforts next.

8.2.2. Promoting Exchange of Healthcare Data

Central to the NHII effort is the concept of a *RHIO*.[14] A RHIO is typically a nonprofit organization founded by a multistakeholder group in a single metropolitan region or state. Its mission is typically to develop electronic exchange of patient data (both clinical and administrative) among its member organizations. The NHII concept also includes inter-RHIO data exchange so that when patients travel or move from one region to another, their medical records are available to treating physicians and other authorized parties.

The organizations that participate in a RHIO vary, but most often they include health plans, hospitals, and physicians. Other organizations that participate less frequently include pharmacies, commercial laboratories, diagnostic imaging centers, nursing homes, and government agencies such as health departments.

The RHIO movement can be traced to the Community Health Information Network (CHIN) movement that began and largely ended in the first half of the 1990s. CHINs had similar goals as today's RHIOs: electronic exchange of health care data to support patient care. The CHIN movement largely collapsed because of lack of trust among competing organizations, concerns about privacy of data, failure of the technological approach of creating a centralized database, and the cost of technology at that time (Appleby, 1995; Starr, 1997; Payton and Ginzberg, 2001; MacDonald and Metzger, 2004).

The RHIO movement has better prospects for success because the federal government is providing incentives and addressing the aforementioned problems that CHINs encountered. At present, federal incentives to RHIOs have mostly come in the form of grant funding. The Agency for Healthcare Research and Quality (AHRQ) has provided nearly $150 million in grant funding to support healthcare data exchange. Federal efforts also include promoting healthcare IT standards (for details on federal efforts to promote the creation and adoption of IT standards necessary for data exchange, see Chapter 32).

The RHIO movement is in its infancy. Overhage et al. (2005) report the results of a recent survey of RHIOs that identified only nine operational RHIOs out of 134 that responded. A majority of RHIOs that provided information to the survey did not yet have substantial commitment from the leadership of the various organizations involved. Nearly one-third of these RHIOs had no funding. The most common technological approach was a centralized database, which was a cause of failure during the CHIN movement (MacDonald and Metzger, 2004). The report notes that a federated database is a characteristic of successful RHIOs.[15] The report judged the RHIOs' plans for implementing data exchange as overly ambitious in general.

Table 6.3 summarizes states in which a single RHIO is attempting to integrate clinical data across a whole state. These statewide RHIOs are also in a state of infancy, with only four of the RHIOs actively exchanging data (three are exchanging clinical data). Two-thirds of the RHIOs are new, having formed only in the past two years.

---

11 The rationale for this law is that physicians might refer patients to facilities with which they have a financial relationship (such as an ownership relationship) even when it is not in the best interest of the patient.

12 There are other qualifications on the exception, such as the community-wide health information system must be available to all physicians who wish to participate, the party providing the EMR cannot take referrals into account when deciding which physicians to give an EMR, and the arrangement cannot violate the Anti-Kickback Law (another law that regulates hospital-physician relationships).

13 Many observers have noted that this exception, although important, may not be sufficient to spur the provision of EMRs to physicians because it does not define "community-wide health information system" or change the Anti-Kickback Law, which is another legal barrier to hospitals providing physicians with EMRs.

14 Also known as a local or regional health information infrastructure.

15 A centralized database is a single collection of patient data from all the healthcare organizations in the community. A federated database, on the other hand, is a system of sending data about patients from one organization to another only when there is a legitimate request.

TABLE 6.3 Regional Healthcare Organizations

Year	State	RHIO	Admin*	Clinical†	Govern‡	Funded§
1993	Utah	Utah Health Information Network	X			X
1993	Wisconsin	Wisconsin Health Information Network	X	X		
1996	Indiana	Indiana Health Information Exchange	X	X		X
1997	Delaware	Delaware Health Information Network			X	
2003	Massachusetts	Massachusetts SHARE (Simplifying Healthcare Among Regional Entities)	X	X		X
2004	Colorado	Colorado Health Information Exchange				X
2004	Florida	Florida Health Information Infrastructure			X	
2004	Minnesota	Minnesota eHealth Initiative			X	
2004	Rhode Island	Rhode Island Health Improvement Initiative			X	X
2004	Tennessee	Volunteer eHealth Initiative			X	X
2005	Pennsylvania	Pennsylvania eHealth Initiative				
2005	Wyoming	WyHIO			X	

RHIO indicates regional health information organization.
*RHIO is exchanging administrative and billing data.
†RHIO is exchanging clinical data.
‡The state government played a major role in creation of RHIO, either through legislation that created the RHIO or through a government agency that convened the stakeholders.
§RHIO has received external funding as of this writing.

We next discuss three of the most well-known RHIOs (two of which are statewide) and use these examples to explore the opportunities and challenges inherent in the NHII effort.

The Indiana Health Information Exchange (IHIE) is arguably the most successful RHIO to date. IHIE was formed in 1996 as the Indianapolis Network for Patient Care when Wishard Memorial Hospital began sharing its data unilaterally with other hospitals to demonstrate the value of health-data exchange. As a result of this leadership, other organizations began sharing their data, and by 2004, the RHIO included five healthcare systems (14 hospitals total), four homeless clinics, and three hospital-affiliated physician group practices. In 2004, the scope of the RHIO expanded to statewide.

The IHIE illustrates the promise that NHII holds for biosurveillance. By 2001, IHIE was reporting notifiable diseases over a single network connection to the Marion County Health Department[16] from five clinical laboratories serving nine hospitals (Overhage et al., 2001). The IHIE had already standardized its participating laboratories' data for purposes of clinical data exchange before the biosurveillance project. It therefore reports notifiable diseases to Marion County Department of Health by using current CDC-recommended standards (HL7, LOINC, and SNOMED). As the IHIE expands and standardizes the laboratory data for additional hospitals in Indiana, the data will be available for biosurveillance as a by-product of what is fundamentally a clinical data integration project. At present, the IHIE also reports chief complaint data for biosurveillance over a single connection from all the participating hospitals.

The IHIE also illustrates the challenge of NHII. It took several years for IHIE to create the administrative and technical infrastructure necessary to integrate data from just nine laboratories. IHIE had to obtain sufficient grant funding to develop the data systems and manage the project, and the technical staff had to analyze the laboratory data and develop custom software to translate the data into standard encodings and formats. We note that the required skill set for understanding laboratory information management system data and creating translation capability is not widely available. Finally, the IHIE (as with other RHIOs) still has not achieved a business model that allows it to be self-sufficient without grant support.

The Santa Barbara County Care Data Exchange was founded in 1998 and incorporated in 1999. Its participating organizations planned to start exchanging data only in February 2005 (Anonymous, 2005). This RHIO is noteworthy for being a case study in the cost and effort to develop a RHIO. It has already spent $10 million in grant funding from the California Health Care Foundation to develop the organizational and technical infrastructure necessary for data exchange, and in 2004, it received another $400,000 in funding from the federal government (Colliver, 2005). We note that this cost and effort does not include the cost of data standardization.[17] Furthermore, data do not flow from one organization's information system to another, but instead physicians view all the data for a

16 The city of Indianapolis, IN is located in Marion Country.

17 The Santa Barbara Country Care Data Exchange makes use of HL17 Clinical Context Object Workgroup standard. However, this standard is not a data standard, but a standard way of passing patient and user informatin among clinical applications so that viewing a patient's data that resides in one application while using another application is seamless. The data viewed, however, may be (and often are) nonstandard.

patient—regardless of the organization at which the data originated—by using a Web browser.

The Utah Health Information Network (UHIN) was founded in 1993. At present, the participating organizations are hospitals, physicians, and every health plan in Utah except one. These organizations exchange administrative and billing data. By use of an AHRQ grant of $5 million awarded in 2004, it has recently begun the work to exchange clinical data among its member organizations. UHIN also illustrates the difficulty of NHII: in its 12 years of existence, it has accomplished the exchange of only nonclinical, administrative data among many, but not all health plans, hospitals, and physicians in Utah.

8.2.3. Improve Public Health

As the first two NHII goals are achieved, regional healthcare data will become increasingly available for biosurveillance. RHIOs will provide increasing coverage of the relevant organizations in a region. Instead of having to establish point-to-point data exchange with dozens of hospitals, hundreds or thousands of physicians, and numerous laboratories, pharmacies, and diagnostic imaging centers, a biosurveillance organization will establish a single relationship and technical connection to a RHIO. It is perhaps obvious, but worth stating, that ensuring that RHIOs are designed to meet the needs of biosurveillance will increase their societal value and potentially lower costs for all parties involved in the RHIO by expanding the set of organizations that might share in the development and operational costs.

8.2.4. Is the Glass Half Empty or Half Full?

Despite the current momentum of the NHII initiative, we caution that many barriers to progress exist. A recent study by Kaushal et al. (2005) suggests that the 5-year cost to achieve a model National Health Information Network[18] exceeds projected spending on IT by the healthcare system.

Even if the U.S. Congress was to authorize the additional $132 billion estimated by Kaushall et al. to create an *achievable* (as opposed to ideal) model NHII, the number of technicians required for tasks such as vocabulary mapping and system integration is likely to be rate limiting. Governmental public health departments continue to expend resources on alternative solutions such as Web-based disease reporting and direct ELR, which consumes healthcare resources. Healthcare systems have finite resources and devoting them to one project often comes at the cost of not being able to devote them to another project. The CDC has advocated for NHII but has not yet invested in its development or included it in guidelines that shape how state and local departments of health invest federal funds.

Nevertheless, the potential advantages of the NHII model and its current momentum suggest that NHII will transform significantly how the United States conducts biosurveillance in the future.

9. BARRIERS TO TIGHTER INTEGRATION BETWEEN HEALTH CARE AND GOVERNMENTAL PUBLIC HEALTH

Because of the importance of the healthcare system for the biosurveillance conducted by governmental public health, it is worth considering here, at the end of two chapters, these two pillars of biosurveillance and the current status of their integration.

Ideally, the two domains would exchange relevant data bidirectionally and in real time. The healthcare system would transmit data needed for outbreak detection, characterization, and management (e.g., bed status, treatment status of victims) to governmental public health. Governmental pubic health would transmit data needed by clinicians for case detection (e.g., case definitions and up-to-the-minute information about population health relevant to diagnosis of individual patients) to the healthcare system. The data exchanges would be secure and satisfy the ethical requirement of "minimal need to know" and a patient's rights to confidentiality.

In addition, the workflow processes related to detection and characterization of outbreaks would be distributed optimally across both domains to maximize the efficiency and speed of the biosurveillance process. Most clinicians are capable, for example, of making observations needed to complete a case-investigation form, but at present this is a task conducted by a health department. Considerable efficiency and speed-up may be possible if IT were to enable a physician to elicit and record more epidemiologically relevant data about a patient with hepatitis A at the time the patient presents in the office setting. This level of integration is many years in the future and requires a rethinking of current systems that takes into account the potential of IT to enable conceivable, but previously impossible, configurations of organizations and workflows.

The need for such integration has been understood for many years, and widely accepted since the anthrax postal attacks of 2001. That such integration is feasible technically has been proven beyond the shadow of doubt by the success of the IHIE and other RHIOs. It is therefore reasonable to ask whether the progress has been satisfactory, given the resources that have been devoted to the problem. Has the progress been exemplary, the best achievable, or less?

18 They define an achievable National Health Information Network as an information system that enables physicians to review the results of testing done in both an inpatient and outpatient setting, review and update both inpatient and outpatient medical records, order treatments including medications, verify the eligibility of patients for various services under their health insurance plans, communicate with patients securely (for example, secure email), and transmit prescriptions electronically to pharmacies.

The current status, four years after October 4, 2001 (the date of public awareness of the first case of inhalational anthrax from contaminated mail), is that the healthcare system by and large continues to report notifiable diseases to governmental public health by using forms that must be completed manually. There is an increasing trend toward the forms being online, but the workflow process for the reporting clinician is otherwise unchanged. There are many syndromic surveillance projects in which the healthcare system sends minute quantities of data (de-identified ED registration records) to health departments. A handful of the more than 5,000 hospitals in the United States send laboratory data electronically to governmental public health.

Earlier in this chapter, we identified barriers that in part explain the gap between what is and what could be. They include the recording of data on paper only (especially in outpatient settings), the sheer number of hospitals, the numerous departmental information systems within each hospital, nonstandard data formats, and the low market penetration of the types of information systems of most value.

It may be instructive to examine just the integration that should exist between laboratory information management systems and governmental public health. As we discuss in Chapter 8, laboratory test orders and results are highly automated in most hospitals. There are many firms that are in the business of creating interfaces between laboratory information systems and other information systems in hospitals. Although it is true that significant effort is required to create an interface, the effort is finite and the cost is in the neighborhood of $100,000 (Overhage et al., 2001) for a comprehensive interface (one that covers all laboratory tests). The cost would likely be lower for an interface that covered only results of interest to biosurveillance. Even were the cost to be as high as $100,000, the total cost for 10,000 interfaces (more than the number of hospitals in the United States) would be $1 billion dollars. One of the authors in fact presented this option to President Bush in February 2002. The President's DHHS advisor quickly interjected that DHHS already had a plan to send $1 billion to the states to address the problem of bioterrorism. Since then, the federal government has provided several billion dollars to the states for bioterrorism preparedness.

This example suggests that an additional barrier at present may be the direction that the government is taking. As discussed in Chapter 5, the government because of its power to tax and enact law, is essential to the creation of systems for the public's good. Such systems cannot come into being without the leadership of a government. However, the leadership must lead in the right direction.

The idea that governmental public health should provide the leadership for the integration of the healthcare system for biosurveillance may be a barrier to progress. Governmental public health organizations have different priorities at present. Their biosurveillance priority, and perhaps rightly so, is to integrate their own systems and to connect to other governmental public health organizations and governmental laboratories so that they can manage outbreaks. Their evolving plan for integrating the healthcare system at present is to connect each hospital electronically to a health department and to leverage RHIOs should they exist. The CDC is also advocating a plan to connect hospitals directly to the CDC and then route the data to health departments.

This may be the wrong idea for biosurveillance. An alternative approach would be for every hospital and healthcare provider in a region to link with each other (the RHIO model) and then create a single connection from the RHIO to governmental public health. The oldest RHIOs have been in existence for more than a decade and have solved many technical and administrative barriers to regional exchange of data. The healthcare and pharmacy organizations within the regions have far more technical capability to create such connections than does a health department. As nongovernmental organizations, they have more flexibility and in some cases more resources.

A key advantage to the RHIO model is that it avoids the construction of two separate but redundant infrastructures: one for public health purposes and one for clinical care. The alternative model of governmental public health creating a separate biosurveillance infrastructure misses the opportunity for a dual-purpose system, which would also be more mission critical to the healthcare system and therefore promoted to a higher priority for both development and long-term maintenance.

10. SUMMARY

Clinical data collected by the healthcare system are a rich source of data for biosurveillance. They include the data needed for earlier detection of cases and outbreaks and for more rapid characterization of outbreaks. Clinical data in the United States at present, however, are not highly available for biosurveillance (other than that practiced by hospital infection control). The barriers include the use of paper records, multiple departmental information systems, and nonstandard data formats. In countries with national health systems such as the United Kingdom, these barriers are less daunting.

The types of data that are available in electronic form in the United States are weighted toward data collected for administrative purposes such as patient registration and billing (and market penetration is high for such systems). Some administrative systems—registration, scheduling, and billing—have data that are of value for biosurveillance and developers of new strategies for early detection of outbreaks are using these data. The use of computers to record clinical information has lagged administrative use, and market penetration is variable depending on the type of system. Clinical information systems are widely deployed in clinical laboratories and radiology departments, and are less used in pathology departments and as POC systems. Specific data that are highly available, although difficult to access, include laboratory and radiology results.

Key gaps are symptom and sign data, which are often recorded by using English, not computer encoding. The market penetration of IT into small private practices is less than in large practices; even when small practices use IT, the sheet numbers of such practices make integration of their data into a biosurveillance network an expensive and time-consuming project.

A bright note is that the clinical computing industry has been working on the problem of interfacing and data integration for several decades, so there is a large body of work already completed toward solutions that can be applied directly to the problem of integrating clinical data into biosurveillance.

A biosurveillance organization such as a state health department that wishes to create real-time data exchange with a hospital should develop both a short-term and a long-term strategic plan. In the short term, it should work with each hospital organization to determine whether the appropriate technical approach to data exchange should focus on building an interface to an existing HL7-message router, an interface to an existing data warehouse, or a POC system (or systems). We discussed principles to guide such decisions.

In the long-term, the health department should factor two megatrends into its planning. The first is that POC systems will become commonplace in both the outpatient and inpatient settings. Unless these systems are *biosurveillance-enabled,* meaning that their manufacturers engineer these systems to be able to interoperate with biosurveillance organizations, additional work will have to be done to create such interfaces. The second is the NHII movement, which, if supported by governmental public health, may lead to the required biosurveillance enabling of clinical information systems on an accelerated time frame.

The protestant minister, when asked for his secret for giving a good sermon, responded "first I tell them what I'm gonna tell them, then I tell them, then I tell them what I told them." POC systems with decision support are the future of biosurveillance. HL7-message routers represent unique resources for the present. POC systems enable collection of symptom and sign data in coded format. Their decision support capabilities can support real-time bidirectional interactions among front-line clinicians and biosurveillance organizations. They can support computer-based case detection and case reporting. The RHIO component of the NHII movement is also important, if it is supported, to the future of biosurveillance. POC systems with decision support and RHIOs are important to the future of biosurveillance.

ADDITIONAL RESOURCES

Clinical Information Systems

Shortliffe, E.H., et al., eds. (2001). *Medical Informatics: Computer Applications in Health Care and Biomedicine.* New York: Springer. Chapters 9, 10, and 12 discuss computer-based patient record systems.

Infection Control

Journals

American Journal of Infection Control
Infection Control & Hospital Epidemiology
Morbidity and Mortality Weekly Report, CDC

Guidelines and Standards

Joint Commission on the Accreditation of Healthcare Organizations (JCAHO), including *Meeting JCAHO's Infection Control Requirements* (2004)

Occupational Health and Safety Administration (OSHA)

American Hospital Association (AHA). Position papers and guidelines.

Association for Professionals in Infection Control and Epidemiology (APIC). Guidelines on infection control

Books

Association for Professionals in Infection. (2002). *APIC Handbook of Infection Control,* 3rd ed. Washington, DC: Association for Professionals in Infection.

Association for Professionals in Infection. (2005). *APIC Text of Infection Control and Epidemiology.* Washington, DC: Association for Professionals in Infection.

Bennett, J. and Brachman, P., eds. (1998). *Hospital Infections.* Philadelphia: Lippincott-Raven Publishers. A textbook on hospital epidemiology.

Mandell, G.L., Bennett, J.E., and Dolin, R. (2003). 2003 Red Book: Report of the Committee on Infectious Diseases, 26th ed. American Academy of Pediatrics.

Wenzel, R.P. ed. (2002). *Prevention and Control of Nosocomial Infections,* 4th ed. New York: Lippincott Williams & Wilkins

ACKNOWLEDGMENTS

We thank Robert Haley, Gilad Kuperman, Tracy Gustafson, Michael Mochan, Carlene Muto, and Fu-Chiang Tsui for reading and commenting on this chapter.

REFERENCES

Anonymous. (1992). Public Health Focus: Surveillance, Prevention, and Control of Nosocomial Infections. *Morbidity and Mortality Weekly Report* vol. 41:783–787.

Anonymous. (2005). Santa Barbara County Care Data Exchange and CareScience Announce Next Step in Roll-out of Care Data Exchange Solution. *Business Wire* February 14.

Appleby, C. (1995). The trouble with CHINS (Community Health Information Networks). *Hospital Health Networks* vol. 69:42–44. *http://www.ncbi.nlm.nih.gov/entrez/query.fcgi?cmd=Retrieve&db=PubMed&dopt=Citation&list_uids=7728164.*

Bates, D.W., Boyle, D.L., and Teich, J.M. (1994). Impact of Computerized Physician Order Entry on Physician Time. *Proceedings of the Annual Symposium on Computer Applications in Medical Care* 996.

Benson, T. (2002). Why General Practitioners Use Computers and Hospital Doctors Do Not, Part 1: Incentives. *British Medical Journal* vol. 325:1086–1089.

Brossette, S.E., Sprague, A.P., Hardin, J.M., Waites, K.B., Jones, W.T., and Moser, S.A. (1998). Association Rules and Data Mining in Hospital Infection Control and Public Health Surveillance. *Journal of the American Medical Informatics Association* vol. 5:373–381.

Brossette, S.E., Sprague, A.P., Jones, W.T., and Moser, S.A. (2000). A Data Mining System for Infection Control Surveillance. *Methods of Information in Medicine* vol. 39:303–310.

Burke, J.P., Classen, D.C., Pestotnik, S.L., Evans, R.S., and Stevens, L.E. (1991). The HELP System and Its Application to Infection Control. *Journal of Hospital Infection* vol. 18:424–431.

Certification Board of Infection Control and Epidemiology. (2005). Becoming Certified. *http://www.cbic.org/Becoming_Certified.asp.*

Chapman, W. and Haug, P. (1999). Comparing Expert Systems for Identifying Chest X-Ray Reports that Support Pneumonia. *Proceedings of AMIA Annual Symposium* 216–220.

Chizzali-Bonfadin, C., Adlassnig, K.P., and Koller, W. (1995). MONI: An Intelligent Database and Monitoring System for Surveillance Of Nosocomial Infections. *Medinfo* vol. 8:1684.

Classen, D.C. and Burke, J.P. (1995). The Computer-based Patient Record: The Role of the Hospital Epidemiologist. *Infection Control and Hospital Epidemiology* vol. 16:729–736.

Classen, D.C., Evans, R.S., Pestotnik, S.L., Horn, S.D., Menlove, R.L., and Burke, J.P. (1992). The Timing of Prophylactic Administration of Antibiotics and the Risk of Surgical-Wound Infection. *New England Journal of Medicine* vol. 326:281–286.

Colliver, V. (2005). Medical Data Made Whole: Health Exchanges Hope To Offer All Patient Records in One Place. *San Francisco Chronicle* March 8.

Cooper, D. and Chinemana, F. (2004). NHS Direct Derived Data: An Exciting New Opportunity or an Epidemiological Headache? *Journal of Public Health* vol. 26:158–160.

Cooper, D., et al. (2004a). Can We Use Self-Testing To Augment Syndromic Surveillance? A Pilot Study Using Influenza. *Morbidity and Mortality Weekly Report* (Submitted).

Cooper, D., et al. (2004b). A national symptom surveillance system in the UK using calls to a telephone health advice service. *Morbidity and Mortality Weekly Report* vol. 53 (Suppl):179–83.

Department of Health and Human Services. (2002). Medical Privacy: National Standards to Protect the Privacy of Personal Health Information. *http://www.hhs.gov/ocr/hipaa/finalreg.html.*

Doroshenko, A., Cooper, D., Smith, G., Gerard, E., Chinemana, F., and Verlander, N. (2004). Evaluation of Syndromic Surveillance Based on NHS Direct Derived Data in England and Wales. *Morbidity and Mortality Weekly Report* (Submitted).

Effler, P., Ching-Lee, M., Bogard, A., Man-Cheng, L., Nekomoto, T., and Jernigan, D. (1999). Statewide System of Electronic Notifiable Disease Reporting from Clinical Laboratories: Comparing Automated Reporting with Conventional Methods. *Journal of the American Medical Association* vol. 282:1845–1850.

Espino, J. and Wagner, M. (2001). The Cccuracy of ICD-9 Coded Chief Complaints for Detection of Acute Respiratory Illness. *JAMIA Symposium Issue.*

Evans, R.S. (1991). The HELP System: A Review of Clinical Applications in Infectious Diseases and Antibiotic Use. *MD Computing* vol. 8:282–288, 315.

Evans, R.S., et al. (1985). Development of a Computerized Infectious Disease Monitor (CIDM). *Computers and Biomedical Research* vol. 18:103–113.

Evans, R.S., et al. (1986). Computer Surveillance of Hospital-Acquired Infections and Antibiotic Use. *Journal of the American Medical Association* vol. 256:1007–1011.

Evans, R.S., et al. (1992). Computerized Identification of Patients at High Risk for Hospital-Acquired Infection. *American Journal of Infection Control* vol. 20:4–10.

Evans, R.S., et al. (1998). A Computer-Assisted Management Program for Antibiotics and Other Antiinfective Agents. *New England Journal of Medicine* vol. 338:232–128. *http://www.ncbi.nlm.nih.gov/entrez/query.fcgi?cmd=Retrieve&db=PubMed&dopt=Citation&list_uids=9435330.*

Fiszman, M., Chapman, W., Aronsky, D., Evans, R., and Haug, P. (2000a). Automatic Detection of Acute Bacterial Pneumonia from Chest X-Ray Reports. *Journal of the American Medical Informatics Association* vol. 7:593–604.

Fiszman, M., Chapman, W.W., Aronsky, D., Evans, R.S., and Haug, P.J. (2000b). Automatic detection of acute bacterial pneumonia from chest X-ray reports. *Journal of American Medical Information Association* vol. 7:593–604. *http://www.jamia.org/cgi/content/full/7/6/593*

Frost & Sullivan. (2003). U.S Computerized Physician Order Entry Market, 2002. New York: Market Research. *www.marketresearch.com.*

Gaunt, P.N. (1991). Information in Infection Control. *Journal of Hospital Infection* vol. 18:397–401.

Gesteland, P.H., et al. (2002). Rapid Deployment of an Electronic Disease Surveillance System in the State of Utah for the 2002 Olympic Winter games. *Proceedings : A Conference of the American Medical Informatics Association* 285–289. *http://www.ncbi.nlm.nih.gov/entrez/query.fcgi?cmd=Retrieve&db=PubMed&dopt=Citation&list_uids=12463832.*

Gesteland, P.H., et al. (2003). Automated Syndromic Surveillance for the 2002 Winter Olympics. *Journal of the American Medical Information Association* vol. 10:547–554. *http://www.ncbi.nlm.nih.gov/entrez/query.fcgi?cmd=Retrieve&db=PubMed&dopt=Citation&list_uids=12925547.*

Haley, R.W., et al. (1985a). The Efficacy of Infection Surveillance and Control Programs in Preventing Nosocomial Infections in U.S. Hospitals. *American Journal of Epidemiology* vol. 121:182–205. *http://www.ncbi.nlm.nih.gov/entrez/query.fcgi?cmd=Retrieve&db=PubMed&dopt=Citation&list_uids=4014115.*

Haley, R.W., Tenney, J.H., Lindsey II, J.O., Garner, J.S., and Bennett, J.V. (1985b). How Frequent Are Outbreaks of Nosocomial Infection in Community Hospitals? *Infection Control* vol. 6:233–236.

http://www.ncbi.nlm.nih.gov/entrez/query.fcgi?cmd=Retrieve&db=PubMed&dopt=Citation&list_uids=3848422.

Harcourt, S., et al. (2001). Can Calls to NHS Direct Be Used for Syndromic Surveillance? A Pilot at Three Sites Using Influenza as an Example. *Communicable Disease and Public Health* vol. 4:178–182.

Hoffman, M., et al. (2003). Multijurisdictional Approach to Biosurveillance, Kansas City. *Emerging Infectious Diseases* vol. 9:1281–1286. *http://www.cdc.gov/ncidod/EID/vol9no10/03-0060.htm.*

Hripcsak, G., Kuperman, G., Friedman, C., and Heitjan, D.F. (1999). A Reliability Study for Evaluating Information Extraction from Radiology Reports. *Journal of the American Medical Information Association* vol. 6:143–150.

Jain, N.L., Knirsch, C.A., Friedman, C., and Hripcsak, G. (1996). Identification of Suspected Tuberculosis Patients Based on Natural Language Processing of Chest Radiograph Reports. *Proceedings of the AMIA Annual Fall Symposium* 542–546.

Joint Commission on Accreditation of Healthcare Organizations [JCAHO]. (2004). *Meeting JCAHO's Infection Control Requirements.* Oakbrook Terrace, IL: Joint Commission Resources, Inc.

Kahn, M.G., Steib, S.A., Fraser, V.J., and Dunagan, W.C. (1993). An Expert System for Culture-Based Infection Control Surveillance. *Proceedings of the Annual Symposium on Computer Applications in Medical Care* 171–175.

Kahn, M.G., Steib, S.A., Spitznagel, E.L., Claiborne, D.W., and Fraser, V.J. (1995). Improvement in User Performance Following Development and Routine Use of an Expert System. *Medinfo* vol. 8:1064–1067.

Kahn, M.G., Bailey, T.C., Steib, S.A., Fraser, V.J., and Dunagan, W.C. (1996a). Statistical Process Control Methods for Expert System Performance Monitoring. *Journal of the American Medical Informatics Association* vol. 3:258–269.

Kahn, M.G., Steib, S.A., Dunagan, W.C., and Fraser, V.J. (1996b). Monitoring Expert System Performance Using Continuous User Feedback. *Journal of the American Medical Informatics Association* vol. 3:216–223.

Kaushal, R., et al. (2005). The Costs of a National Health Information Network. *Annals in Internal Medicine* vol. 143: 165–173. *http://www.ncbi.nlm.nih.gov/entrez/query.fcgi?cmd=Retrieve&db=PubMed&dopt=Citation&list_uids=16061914.*

Knirsch, C., Jain, N., Pablos-Mendez, A., Friedman, C., and Hripcsak, G. (1998). Respiratory Isolation of Tuberculosis Patients Using Clinical Guidelines and an Automated Clinical Decision Support System. *Infection Control and Hospital Epidemiology* vol. 19:94–100.

Lazarus, R., et al. (2002). Use of Automated Ambulatory-Care Encounter Records for Detection of Acute Illness Clusters, Including Potential Bioterrorism Events. *Emerging Infectious Diseases* vol. 8:753–760. *http://www.ncbi.nlm.nih.gov/entrez/query.fcgi?cmd=Retrieve&db=PubMed&dopt=Citation&list_uids=12141958.*

Lober, W.B., Trigg, L.J., Bliss, D., and Brinkley, J.M. (2001). IML: An Image Markup Language. *Proceedings of a conference of the American Medical Informatics Association* 403–407. *http://www.ncbi.nlm.nih.gov/entrez/query.fcgi?cmd=Retrieve&db=PubMed&dopt=Citation&list_uids=11825219.*

MacDonald, K. and Metzger, J. (2004). *Connecting Communities: Strategies for Physician Portals and Regional Data Sharing.* New York: First Consulting Group.

Martone, W.J., Jarvis, W.R., Edwards, J.R., Culver, D.H., and Haley, R.W. (1992) In *Hospital Infections,* 3rd ed. (J.V Bennett and P.S. Brachman, eds.). Philadelpha: Lippincott-Raven Publishers.

Martone, W.J., Jarvis, W.R., Edwards, J.R., Culver, D.H., and Haley, R.W. (1998). Incidence and Nature of Endemic and Epidemic Nosocomial Infections. In *Hospital Infections,* 4th ed. (J.V. Bennett and P.S. Brachman, eds.). Philadelphia: Lippincott-Raven Publishers.

McDonald, C.J., et al. (1994). The Regenstrief Medical Record System: Experience with MD Order Entry and Community-wide Wxtensions. *Proceedings of the Annual Symposium on Computer Applications in Medical Care* 1059.

Monegain, B. (2005). Doctors' Offices Get Ready for Technology Makeovers. *http://www.healthcareitnews.com/NewsArticleView.aspx?ContentTypeID=3&ContentID=3214&Term=doq-it.*

Moser, S.A., Jones, W.T., and Brossette, S.E. (1999). Application of Data Mining to Intensive Care Unit Microbiologic Data. *Emerging Infectious Diseases* vol. 5:454–457.

National Committee on Vital and Health Statistics. (2001). Information for Health: A Strategy for Building the National Health Information Infrastructure. *http://ncvhs.hhs.gov/nhiilayo.pdf.*

Nicoll, A., Smith, G., Cooper, D., Chinemana, F., and Gerard, E. (2004). The Public Health Value of Syndromic Surveillance Data: Calls to a National Health Help-Line (NHS Direct). *European Journal of Public Health* vol. 14:69.

Overhage, J.M., Evans, L., and Marchibroda, J. (2005). Communities' Readiness for Health Information Exchange: The National Landscape in 2004. *Journal of the American Medical Information Association* vol. 12:107–112. *http://www.ncbi.nlm.nih.gov/entrez/query.fcgi?cmd=Retrieve&db=PubMed&dopt=Citation&list_uids=15561785.*

Overhage, J.M., Suico, J., and McDonald, C.J. (2001). Electronic Laboratory Reporting: Barriers, Solutions and Findings. *Journal of Public Health Management and Practice* vol. 7:60–66. *http://www.ncbi.nlm.nih.gov/entrez/query.fcgi?cmd=Retrieve&db=PubMed&dopt=Citation&list_uids=11713754.*

Payton, F.C. and Ginzberg, M.J. (2001). Interorganizational Health Care Systems Implementations: An Exploratory Study of Early Electronic Commerce Initiatives. *Health Care Management Review* vol. 26:20–32. *http://www.ncbi.nlm.nih.gov/entrez/query.fcgi?cmd=Retrieve&db=PubMed&dopt=Citation&list_uids=11293008.*

Platt, R., et al. (2003). Syndromic Surveillance Using Minimum Transfer of Identifiable Data: The Example of the National Bioterrorism Syndromic Surveillance Demonstration Program. *Journal of Urban Health* vol. 80(Suppl 1):i25–i31. *http://www.ncbi.nlm.nih.gov/entrez/query.fcgi?cmd=Retrieve&db=PubMed&dopt=Citation&list_uids=12791776.*

Rippen, H.E. and Yasnoff, W.A. (2004). Building the National Health Information Infrastructure. *Journal of AHIMA* vol. 75:20–26, quiz 29–30. *http://www.ncbi.nlm.nih.gov/entrez/query.fcgi?cmd=Retrieve&db=PubMed&dopt=Citation&list_uids=15141583.*

Rocha, B.H., Christenson, J.C., Pavia, A., Evans, R.S., and Gardner, R.M. (1994). Computerized Detection of Nosocomial Infections in Newborns. *Proceedings of the Annual Symposium on Computer Applications in Medical Care* 684–688.

Sheckler, W. (1998). The Role of Professional and Regulatory Organizations in Infection Control Programs. In *Hospital Infections,* 4th ed. (J.V. Bennett and P.S. Brachman, eds.). Philadelphia: Lippincott-Raven Publishers.

Shortliffe, E.H., Perreault, L.E., Wiederhold, G., and Fagan, L.M., eds. (2001). *Medical Informatics: Computer Applications in Health Care and Biomedicine.* New York: Springer.

Sittig, D.F. and Stead, W.W. (1994). Computer-based Physician Order Entry: The State of the Art. *Journal of the American Medical Informatics Association* vol. 1:108–123.

Starr, P. (1997). SMART Technology, Stunted Policy: Developing Health Information Networks. *Health Affairs* vol. 16:91–105. *http://www.ncbi.nlm.nih.gov/entrez/query.fcgi?cmd=Retrieve&db=PubMed&dopt=Citation&list_uids=9141326.*

Thompson, T.G. and Brailer, D.J. (2004). The Decade of Information Technology: Delivering Consumer-Centric and Information-Rich Health Care. *http://www.os.dhhs.gov/healthit/documents/hitframework.pdf.*

Tierney, W.M., Miller, M.E., Overhage, J.M., and McDonald, C.J. (1993). Physician Inpatient Order Writing on Microcomputer Workstations: Effects on Resource Utilization. *Journal of the American Medical Association* vol. 269:379–373.

Tsui, F.C., et al. (2002). Data, Network, and Application: Technical Description of the Utah RODS Winter Olympic Biosurveillance System. *Proceedings of a conference of the American Medical Informatics Association* 815–819. *http://www.ncbi.nlm.nih.gov/entrez/query.fcgi?cmd=Retrieve&db=PubMed&dopt=Citation&list_uids=12463938.*

Tsui, F.-C., Wagner, M.M., Dato, V., and Chang, C.-C.H. (2001). Value of ICD-9-Coded Chief Complaints for Detection of Epidemics. *Proceedings of a conference of the American Medical Informatics Association* 711–715.

Wagner, M.M., et al. (2001). The Emerging Science of Very Early Detection of Disease Outbreaks. *Journal of Public Health Management Practice* vol. 7:51–59. *http://www.ncbi.nlm.nih.gov/entrez/query.fcgi?cmd=Retrieve&db=PubMed&dopt=Citation&list_uids=11710168.*

Wenzel, R.P., et al. (1983). Hospital-Acquired Infections in Intensive Care Unit Patients: An Overview with Emphasis on Epidemics. *Infection Control* vol. 4:371–375. *http://www.ncbi.nlm.nih.gov/entrez/query.fcgi?cmd=Retrieve&db=PubMed&dopt=Citation&list_uids=6556158.*

Yasnoff, W.A., et al. (2004). A Consensus Action Agenda for Achieving the national Health Information Infrastructure. *Journal of the American Medical Information Association* vol. 11:332–338. *http://www.ncbi.nlm.nih.gov/entrez/query.fcgi?cmd=Retrieve&db=PubMed&dopt=Citation&list_uids=15187075.*

Yasnoff, W.A., Overhage, J.M., Humphreys, B.L., and LaVenture, M. (2001). A National Agenda for Public Health Informatics: Summarized Recommendations from the 2001 AMIA Spring Congress. *Journal of the American Medical Information Association* vol. 8:535–545. *http://www.ncbi.nlm.nih.gov/entrez/query.fcgi?cmd=Retrieve&db=PubMed&dopt=Citation&list_uids=11687561.*

Yih, W.K., et al. (2004). National Bioterrorism Syndromic Surveillance Demonstration Program. *Morbidity and Mortality Weekly Report* vol. 53:43–49. *http://www.ncbi.nlm.nih.gov/entrez/query.fcgi?cmd=Retrieve&db=PubMed&dopt=Citation&list_uids=15714626.*

7

CHAPTER

Animal Health

Richard Shephard
Australian Biosecurity Cooperative Research Centre for Emerging Infectious Disease, Brisbane, Australia

Ron M. Aryel
Del Rey Technologies, LLC, Kansas City, Missouri

Loren Shaffer
Bureau of Health Surveillance, Department of Prevention, Ohio Department of Health, Columbus, Ohio

1. INTRODUCTION

There are three ways that disease in animal populations can impact human populations. First, animals can transmit diseases to humans, so monitoring disease prevalence and distribution in animal populations may prevent outbreaks in human populations or contribute to the earlier detection of such outbreaks. Second, disease can cause sickness in both animals and humans, and, in some cases, animals are more susceptible than are humans. The classic example of this circumstance is the sensitivity of canaries to coal gases. Therefore, monitoring animals may provide an early warning of a concurrent outbreak in humans. The monitoring of mass mortality events in birds has demonstrated value as an early warning system for increased risk of West Nile virus (WNV) outbreaks in humans (Mostashari et al., 2003). WNV is an arbovirus residing within bird reservoir hosts, which is by mosquitoes and can result in infection spilling over to humans. Clinical disease in humans involves the central nervous system and can result in the occasional death. Predicting outbreak risk in humans for arbovirus diseases can be difficult because this requires knowledge of the distribution and the infection status of the reservoir and the vector. Although events such as poisoning can result in mass mortality of birds, WNV is a predominant cause of these occurrences in birds. Therefore, an increase in the expected frequency of mass mortality events in birds generally indicates increased viral activity within both reservoir hosts and the vector. Third, animals have economic value; hence, disease in animals can threaten the economic health—and in the extreme case physical health (through starvation)—of human populations.

For example, an eradication program to control an outbreak of contagious bovine pleuropneumonia in cattle in Botswana in 1996 required the slaughter of all cattle within the affected region. This mass slaughter led to a 2.3-fold increase in the risk of malnutrition in children under 5 years of age within the affected region compared with the country as a whole. The malnutrition arose from a reduction in milk and meat consumption and also from an economic decline contributed to by loss of cattle for transport and draught power (Boonstra et al., 2001).

The crossover threats, mentioned above, posed by microbes in animals are partly determined by the degree to which we share DNA with them. For example, the chimpanzee genome contains 99% of the human genome, and consequently, chimpanzees can become sick with many illnesses that also strike humans. The reverse is also true. In contrast, a human is unlikely to be affected by many pathogens that make a lion ill, with the notable exception of enteric pathogens such as salmonella. Diseases that can spread from animals to humans include cat-scratch disease, rat bite fever, rabies, tularemia, and plague. Conversely, other mammals and birds can become ill with the same viruses that affect humans, including encephalitides and WNV. Reptiles have even less in common with humans but can carry salmonella, and humans who handle the animals can become infected. Young children are especially vulnerable when they manifest an all-too-common habit of not washing their hands after handling reptiles.

2. ANIMAL HEALTH ORGANIZATIONS

Veterinarians, agribusiness (farmers), and wildlife organizations observe the health of animals. Internationally, animal disease monitoring is a major role of the World Organization for Animal Health, the Office International des Epizooties (OIE). The OIE and wildlife organizations, such as the Wildlife Conservation Society (WCS), monitor wildlife disease internationally. Nationally, animal disease monitoring is conducted by the federal and state departments of agriculture and by health departments. Organizations such as the Fish and Wildlife Service (FWS), the U.S. Geological Survey (USGS), and the American Zoo and Aquarium Association (AZA) monitor wildlife diseases as part of their work.

2.1. Office International des Epizooties

The OIE *(http://www.oie.int/eng/en_index.htm)* is the World Organization for Animal Health. It sets guidelines and

Handbook of Biosurveillance
ISBN 0-12-369378-0

provides recommendations to minimize the risk of spreading animal diseases and pests while facilitating trade between nations. The OIE lists its missions as the following:

1. To guarantee the transparency of animal disease status worldwide
2. To collect, analyze, and disseminate veterinary scientific information
3. To provide expertise and promote international solidarity for the control of animal diseases
4. To guarantee the sanitary safety of world trade by developing sanitary rules for international trade in animals and animal products

The OIE develops policy, standards, and techniques that member countries can apply to help protect themselves from animal diseases by establishing valid import barriers for the trade of certain animals or animal products. These standards are documented in the International Terrestrial Animal Health Code and the International Aquatic Animal Health Code (*http://www.oie.int/eng/publicat/en_normes.htm*). The requirements vary for individual diseases and depend on the impact of disease on animal health, agriculture, trade, and the human population.

The OIE does not support a zero-risk approach to the management of biosecurity risks—the codes document a level of protection that will provide minimal risk. If an exporting country can meet OIE code standards, then importing countries cannot prohibit access to their markets on the grounds of biosecurity concern. Therefore, the codes document the level of protection that provides an acceptably low risk to human, animal, or plant health arising from the importation of products (Biosecurity Australia, 2003). The information on animal health for any member country is available from the OIE (*www.oie.int*) or by request from the designated reporting body for that country.

2.2. U.S. Department of Agriculture

The U.S. Department of Agriculture (USDA) (*http://www.usda.gov/wps/portal/usdahome*) is a large organization concerned with helping and developing agriculture within the United States. Its divisions deal with agricultural science, human nutrition, law, marketing and trade of agricultural products, natural resource and environmental management for farmers, research, and training, and it is involved in rural community development.

The Animal and Plant Health Inspection Service (APHIS) (*http://www.aphis.usda.gov/*) is a division of the USDA that is concerned with the protection of agricultural health. This involves protection of the human population, as well as the domestic animal and certain wild animal resources and farmed and native plants, from pests and disease. APHIS has programs working in plant biosecurity, veterinary services, animal care, and welfare and wildlife monitoring.

The Veterinary Services (VS) is concerned with health, quality, and marketability of the nation's animals, animal products, and veterinary biological products. It does this primarily by preventing, monitoring and controlling, and eliminating disease within the population where possible. The VS also monitors and promotes good animal health and high animal productivity. The Center for Epidemiology and Animal Health (CEAH) is a group within VS that produces timely, factual information about animal health. CEAH collects and analyzes animal health data and reports to and works with the OIE to describe the distribution of animal disease from within the United States and internationally.

APHIS also has a wildlife services division that monitors wildlife with the objective of protecting agriculture and livestock, conserving natural resources, and managing potential sites of conflict between wildlife and people.

2.3. State Departments of Agriculture

Each state has a department of agriculture. All have purposes and objectives similar to those of the USDA; however, their focus is on state issues. Some of the issues include the provision of local animal disease diagnostic laboratories, licensing of farming premises, food safety, meat inspection, enforcement of state agricultural laws, state-based quarantine and disease control, animal movement control, animal welfare, pest management, quality assurance programs, and state-specific farm production research and extension. In general, the state departments of agriculture provide most of the institutional expertise, such as veterinarians and plant disease experts, within the local region.

Most state departments of health actively manage state-based disease or pest control programs. Some are part of national programs, but others pertain specifically to the state. Diseases that are endemic within one part of the United States, but are not present within the state, are often subject to state monitoring and control programs. Similarly, local disease eradication programs are present in many states.

All states contribute to national knowledge on animal disease status by reporting to the USDA.

2.4. Agribusiness

The aggregated livestock farming systems provide a large interface between humans and animals, and farm animals provide a large interface between themselves and wildlife (especially via insect vectors and rodents). Thus, the health of farm animals is of importance.

All commercial livestock systems follow the same template–the cost-effective production of sufficient high-quality animal product of in a regular, sustainable, and reliable manner. Profit from livestock farming is determined from the amount produced, the quality of the product, and the cost of production. Interestingly, disease within farming is simply another cost of production. Economic reasoning (the law of diminishing returns) usually precludes the complete elimination of disease from a population because the generally small economic return arising from the eradication of low prevalence disease does not justify the expense of the program. Most control programs for endemic diseases, consequently, are cost-effective prevalence reduction plans.

Product quality exerts greater influence on farm profit than it has in previous decades because consumers are demanding safety and quality. This requires capability to track food during the journey from "paddock to plate." Information about food animals and their veterinary care is available at several points during the production process. Systems for capturing this information are continually being developed in agriculture.

2.5. Wildlife, Fisheries, and Zoos

A multitude of organizations observe the health of wildlife; veterinarians within zoos, private practices, and animal refuges see only a small fraction of diseased wildlife. The OIE and Centers for Disease Control and Prevention (CDC) also have sections that are concerned with wildlife disease. The OIE wildlife group monitors diseases that can impact domestic animals and humans, and the CDC animal section monitors zoonotic diseases.

2.5.1. Fish and Wildlife Service

The FWS is a bureau within the Department of the Interior, whose mission is "working with others to conserve, protect, and enhance fish, wildlife, and plants and their habitats for the continuing benefit of the American people" (*http://www.fws.gov/*). This service manages nearly 100 million acres of land for wildlife, more than 500 wildlife refuges, thousands of wetlands, and more than 50 national fish hatcheries. The department is charged with managing wildlife laws, protecting endangered species, managing migratory birds, conserving and restoring habitats (terrestrial and aquatic), and coordinating with efforts in international conservation.

2.5.2. The U.S. Geological Survey

The USGS provides "reliable scientific information to describe and understand the earth; minimize loss of life and property from natural disasters; manage water, biological, energy, and mineral resources; and enhance and protect the quality of life" (*www.usgs.gov*). It is the largest mapping organization within the United States and offers data collection, monitoring, analysis, and interpretation services to improve the understanding of natural resource conditions, problems, and issues. The USGS has interests within the following areas (among others): atmosphere and climate, earth characteristics, ecology and environment, environmental issues, natural resources, plants, and animals.

2.5.3. The Wildlife Conservation Society

The WCS is active in the conservation of both wildlife and habitats (*www.wcs.org*) through the development and application of science, production of educational material, support for conservation programs, and management of the world's largest system of urban wildlife parks. Its overriding objective is to increase public awareness and to cultivate support for wildlife and wild habitat conservation and management.

The organization uses a science-based approach to conservation, which requires monitoring change that occurs within nature, the analysis and interpretation of data, and discussing the implication of observed change. The very nature of this work requires a population-based approach to the science. Animal population size and density, the interaction between populations (including humans), and the impact of wild resource degradation combine to produce a vision of the future.

The WCS has a Science Resource Center (SRC) that focuses on population biology, ecology, and genetics to help develop and support conservation strategies for endangered species. Animal disease is just one of many factors affecting species survival.

Organizations with expertise in wildlife need to play an increasingly greater role in disease surveillance because they primarily view disease within an ecological framework—equal emphasis applies to host, agent, and environment when describing disease. This approach allows further understanding of the factors that result in disease equilibrium and, therefore, gives further insight into the causes of disease outbreaks or the spillover of disease into new hosts.

The WCS has already played a significant role in human disease surveillance. The WCS and Programme de Conservation et Utilisation Rationelle des Ecosystémes Forestiers en Afrique Centrale combined with local Gabon and Republic of Congo authorities to undertake surveillance of wild animal carcasses for Ebola virus as a predictor of possible human Ebola disease outbreaks. This monitoring program found evidence of Ebola virus in wild animals before five separate human outbreaks and alerted human authorities before two of the human outbreaks occurred (Rouqet et al., 2005).

3. FARMING SYSTEMS

There are around 1.3 million farms in the United States that carry livestock. The majority of farms are grazing enterprises, but more than 20% of farms carry fewer than 25 grazing animals—these include hobby farms, cropping farms, and small holdings. Approximately 20% of farms house the animals—generally intensive chicken and swine systems, as well as some dairy systems (Kellog, 2002).

There are around 95 million cattle (of which around 10 million are dairy cattle) and 60 million swine currently in the United States. Broiler chicken production is around 7 billion birds per year within the United States (Figure 7.1). Almost 1 million horses are involved in the racing industry (American Horse Council; *http://www.horsecouncil.org/statistics.htm*).

We limit our discussion below to terrestrial animals, but fish (as suggested by Table 4.5 in Chapter 4) are of increasing

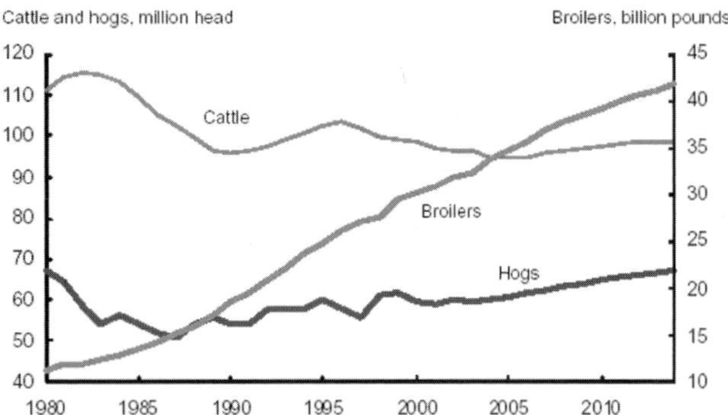

FIGURE 7.1 Livestock inventories and broiler production. (From Office of the Chief Economist and World Agricultural Outlook Board, 2005.)

importance as a food source. The unique features of the aquatic environment drive the development of new epidemiological methods and biosurveillance practices relevant to that industry (Georgiadis et al., 2001). Similarly, agricultural systems are continuously evolving, and part of the evolution is the farming of previously untried species of animal and fish.

3.1. Beef Production

The U.S. beef cattle production system usually involves several movements of animals between farms, from birth of the calf to slaughter of the finished animal. Small breeding herds produce most of the calves born in a given year. These calves may grow on the property of birth until finished, but most sell as yearling steers (castrated bulls) and heifers (cows) for further growth on larger, more extensive ranches (backgrounding). These animals aggregate from a variety of farms and a variety of regions at these establishments. Further aggregation and movement occurs when animals move for final finishing on grain at a feedlot before slaughter.

Cattle in feedlots have unique disease pressures. Animals from multiple sources are densely concentrated within the feedlot, and they enter the feedlot with varying immunological exposures and pathogen status. The alteration to diet, housing, and management can increase the risk of infectious disease. Efficient systems for sourcing, moving, managing, and feeding cattle and for controlling disease within feedlots are essential for them to be profitable.

3.2. Swine Production

The production of swine is an intensive and highly integrated industry. In contrast to beef cattle, swine live within sheds for the complete production cycle within controlled environments and eat a complete processed diet. The production cycle revolves around two often separate components: the reproductive management of sows and the growth and finishing of their progeny for sale. The key drivers of profitability in swine production are reproductive efficiency, growth rate, and the cost and utilization efficiency of the feed.

The supply of piglets to the grower facilities should be constant in order to immediately fill pen space as it becomes available. Active management of reproduction is essential to allow a constant supply of piglets. Grower pigs must also gain weight at an appropriate rate to allow early marketing at uniform weights and to maintain a uniform supply to the processors. This both reduces the cost of production and makes space available for the next grower pig. Feed conversion (the amount of feed required to produce a unit of weight gain) must be efficient to ensure production is profitable, and it is a function of the diet, feeding system, the genetics of the animal, and the effectiveness of disease control.

Management concentrates on balancing the supply of required nutrients at reasonable cost, analyzing the performance of diets, maintaining herd reproductive performance and growth rates, and controlling pathogens. The intensive nature of the production system supports the spread of infectious disease; therefore, farm biosecurity systems are becoming increasingly complex.

3.3. Poultry Production

Poultry and swine production systems are similar. For example, commercial chicken production systems can market meat birds (broilers) at around 30 days of age (broiler) and can produce around 320 eggs per year from a layer. These systems depend upon elite genetics, nutrition, and management to achieve these results. Essentially, all poultry production systems (e.g., turkeys, ducks) are similar to the chicken broiler industry described below.

Parent breeding birds produce eggs for both the broiler and layer systems; however, birds used for meat are different from those used for eggs. Thus, the broiler and layer industries are essentially separate. Workers remove the eggs from the hens to allow artificial incubation. Day-old chickens go to broiler units (meat) or for growing out to egg-laying age. Hens move into the layer shed at the point of egg production. Nutritional systems are complex and similar to the system used for intensive swine production. Complex biosecurity systems have evolved to prevent outbreaks of disease within the very large population of birds found within a commercial farm. These include the exclusion of access to wild birds, careful sourcing of replacement birds, use of batch rearing systems (all-in-all-out systems followed by disinfection of premises), and use of prophylactic vaccination and treatment programs. Veterinarians are helping to develop new ways to control disease and maintain production in commercial units housing thousands of birds as antibiotic growth stimulants fall out of favor.

Parameters that determine poultry farming profitability are similar to those for swine farming. Again, poultry veterinarians and commercial producers have systems that provide them with real-time, high-quality data on the health status of poultry within their care.

3.4. Dairy Production

Commercial dairy systems range in size from small family farms to large commercial feedlot style production systems. Farms tend to be self contained. The farmer carefully controls and times reproduction in the cows because parturition is required to initiate lactation, such that most cows produce a new calf each year. Female (heifer) calves, conceived by using artificial insemination with frozen semen, replace older cull cows, and these ensure ongoing genetic improvement to the herd.

Profitable dairy farms require careful management of reproduction, nutrition, genetics, milk harvesting, and disease. Feed is a major cost of business and is generally a combination of home-grown forages (alfalfa, pasture) and purchased supplements such as grain. Most dairy farmers work closely with nutritional advisers, agronomists, and veterinarians to ensure optimal efficiency. Dairy farmers and their advisers use data on cow production, nutrition, reproduction, and health and milk quality to manage their business.

Outbreaks of disease requiring extended quarantines are a bane of dairy operations; dairy farmers will often opt to destroy an animal rather than expend resources to quarantine animals that cannot return an income (Canon, 2005). For example, many jurisdictions around the world have control programs for bovine Johne's disease (due to infection with *Mycobacterium paratuberculosis*). These generally require younger animals to grow separated from the adults, which necessitates the dedication of land, sheds, and equipment to this specific task and can limit the ability of producers to lease land from other land holders for growing young stock. The economic impact of these requirements is generally greater than is the impact of disease within the herd.

4. ANIMAL HEALTH PROFESSIONALS

A wide range of people observe the health of animals. These include professionals such as veterinarians and scientists, as well as nonprofessionals such as animal inspectors, farm workers, animal owners, and the general public.

4.1. Veterinarians

The practice of veterinary medicine includes the examination, diagnosis, and treatment of animals—whether mammals, reptiles, or fish. Veterinarians work within multiple fields and practices that are relevant to monitoring disease in animals, including private veterinary practice, agribusiness, wildlife, and governmental public health. Each of these fields uses information systems to track disease in individual animals and populations of animals.

4.1.1. Veterinary Education

Veterinarians receive training that resembles the training given to both clinicians and epidemiologists. Similar to physicians, they become experts in the diagnosis of disease in individual animals, and similar to epidemiologists, they become experts in the recognition and characterization of *outbreaks* in populations of animals. Undergraduate training is in the basic sciences, such as anatomy, physiology, biochemistry, and histology. This allows progression into subjects such as pathology, microbiology, medicine, surgery, and epidemiology. Training in animal husbandry, nutrition, pasture agronomy, and economics allows veterinarians to help service the needs of commercial farming.

The teaching material covers all of the major species, but the increasing complexity of veterinary science within individual species exerts increasing pressure on faculties to offer course electives—these may allow students to graduate with extra skills in a given species (and therefore fewer skills in others).

There are a large number of veterinarians with expertise in the care of zoo animals, laboratory animals, or specialization training in fields such as reproduction. Many universities offer postgraduate courses in these areas. The American Veterinary Medical Association (AVMA) is the professional body for veterinarians and produces peer-reviewed journals and offers educational resources for members. The veterinarian's oath is "Being admitted to the profession of veterinary medicine, I solemnly swear to use my scientific knowledge and skills for the benefit of society through the protection of animal health, the relief of animal suffering, the conservation of animal resources, the promotion of public health, and the advancement of medical knowledge. I will practice my profession conscientiously, with dignity, and in keeping with the principles of veterinary medical ethics. I accept as a lifelong obligation the continual improvement of my professional knowledge and competence" (*http://www.avma.org/*). Veterinarians, therefore, commit to ongoing education and to the protection of human and animal health.

4.1.2. Spheres of Veterinary Practice

Although veterinary practice can be very specialized, similar to medical practice, most veterinarians work within one of five spheres: private practice caring for "companion" animals, agribusiness, institutional veterinary practice, government veterinary service, and research or industry (e.g., pharmaceuticals). Each sphere has different pursuits, responsibilities, and objectives. There are approximately 60,000 veterinarians working within the United States. Most work in private practice, but the federal government employs more than 1,100 veterinarians. Other large employers are state departments of agriculture, universities, research laboratories, agribusiness, and pharmaceutical companies, (U.S. Department of Labor Bureau of Labor Statistics, 2005).

Private Practice Caring for Companion Animals. One sphere of practice involves the care of companion animals (i.e., pets). These practitioners generally work in private clinics and hospitals and provide individual animal medicine and surgery services to clients. Many companion animal veterinarians specialize into particular fields of medicine or surgery. Often, individuals develop expertise with rarer forms of pets, including reptiles, birds, small rodents, and aquarium fish. These veterinarians provide a valuable resource in interpreting the health of these less common species. There are approximately 60 million dogs, 70 million cats, 10 million birds, and 5 million horses kept as pets in the United States (Euromonitor, 2000; American Veterinary Medical Association, 2002). Of the households in the United States, approximately 36% have dogs, 32% have cats, 5% have birds, and 2% have horses.

Livestock Farming. Veterinarians within this sphere work with production animals, which are animals kept for profit. These animals include cattle, swine, poultry, race horses, and racing dogs such as greyhounds and Alaskan "mush" sled racing dogs.

Many production animal veterinarians provide only a management consultancy service—offering nutrition, production system, disease management, etc. This is advice without also providing a basic clinical service to treat sick individuals. These veterinarians combine their knowledge of animal physiology and disease with an ability to observe change within animal populations, to gather and analyze data. For example, commercial farmers often hire veterinarians and nutritionists to help them optimize their yield of animal product (e.g., beef, eggs).

Disease diagnosis and control at the population level is the major emphasis of food animal veterinary practice. Evaluation of disease in a food animal emphasizes symptomatology. Veterinarians dealing with food animals may not undertake a complete physical examination on individual sick animals (with dairy veterinarians providing notable exceptions to this rule). There are far too many animals to survey, and it would not be cost-effective. Frequently "examination" is undertaken only to rule out a specific disease. However, the clinical examination of sick animals is undertaken in the event of evidence of an outbreak (multiple animals with similar signs are observed), or if the signs observed fit diseases of particular concern (e.g., exotic diseases such as foot-and-mouth disease). Veterinarians will readily perform necropsies on dead animals (often abbreviated). On occasion, the veterinarian and the producers will agree to sacrifice an affected animal in order to undertake necropsy.

If a carcass has an unusual appearance or the veterinarian suspects the presence of a particularly virulent pathogen, he/she will undertake a full necropsy. This may require the shipment of the body to an animal hospital of government veterinary laboratory to allow for a more extensive pathological examination than is possible in the field. A government veterinary officer is immediately contacted if an exotic disease is suspected; movements onto and off the farm are prohibited until a complete investigation has been completed.

Production animal veterinary medicine emphasizes the economics of intervention (e.g., cost/benefit analysis) to a much greater degree than does human medical practice or private companion animal veterinary practice. This is because the veterinarians are supporting food safety, economic, and animal welfare goals: growers want to maximize profit given the constraints of protecting animal welfare and the environment. Highly skilled veterinarians contribute to the profitable farming of animals by developing systems that maximize growth rates, feed conversion efficiency, reproduction and production, as well as disease control within the population. The tight profit margins common within agriculture generally limit the use of expensive veterinary skills and pharmaceuticals on the individual animal. However, systems that control or reduce the level of disease within the population generally provide a significant return on investment to the rancher when multiplied across large herds. Veterinarians typically develop these population-level disease management systems.

Institutional Veterinary Medicine. Another avenue of veterinary pursuit is institutional veterinary medicine, which includes zoos, wild animal parks, aquariums, and aviaries. Veterinarians working in these areas develop skills in health of nondomesticated animals and in ecology. These skills complement their traditional veterinary medicine and surgery skills, and the veterinarians use these skills together to maintain the health of animals within their care. The skill crossover is such that, often, these veterinarians become involved in work directed toward the preservation of species. These veterinarians also often examine injured and sick wildlife found by members of the public. They are, therefore, an important interface to animal diseases in wild populations.

Public Practice. A third sphere is public practice, essentially composed of government veterinarians working for the state and federal departments of agriculture. Veterinary duties include enforcement of the regulations to control certain stock diseases, farming practices, animal movement (interstate as well as international), animal welfare, and quarantine. They can also work to control organisms of relevance to governmental public health (e.g., birds and mosquitoes) if an animal disease of interest is involved; however, the control of specific pests, such as mosquitoes, is performed by local municipalities.

Government veterinarians' responsibilities extend to the investigation of potential exotic disease events, unusual disease events, and the gathering of disease surveillance information. They predominately work with agribusiness (cattle, swine, poultry); however, an increasing proportion of their time and resources are used for companion animals, especially hobby farm animals (such as pet cows, sheep, poultry, goats, and the increasingly popular Vietnamese pot bellied pig).

Primary responsibility for inspection of food animals at the slaughterhouse rests with the USDA. Within USDA, the Food Safety and Inspection Service (FSIS) employs well over 1,000 veterinarians for this purpose. These veterinarians perform both antemortem and postmortem examinations of animals at commercial slaughterhouses, such as those operated by the large meat processors Cargill, ConAngra, and Tyson. The USDA inspectors screen animals and divert any that raise suspicions to a holding pen, where an FSIS veterinarian examines them for evidence of disease. FSIS inspectors cover cattle, swine, sheep, and poultry; they also examine minor species such as goats, rabbits, and bison on request.

The USDA's jurisdiction includes animals intended for interstate shipment. State agriculture departments inspect food animals that are raised and processed and with food product that is sold within their respective states. The goals of state inspection programs mirror those of the USDA (California Department of Food and Agriculture, 2005; Iowa Deparment of Agriculture and Land Stewardship, 2005; New York State Department of Agriculture and Markets, 2005). Disease surveillance duties often involve trapping and monitoring of vectors. For example, culicoides insect population monitoring for blue tongue virus and mosquito monitoring for the surra trypanosome occurs in Australia (Animal Health Australia, 2004). The National Australian Arbovirus Monitoring Program uses the information from vector monitoring to predict the annual distribution of key agricultural arboviruses, such as blue tongue within northern Australia (Australian Government Department of Agriculture et al., 2003).

Nonclinical Veterinarians. Many veterinarians work outside of clinical practice in universities, research, the pharmaceutical industry, business, or their own farming enterprises. In some countries (e.g., Australia), the geographical spread of veterinarians who farm instead of practice veterinary medicine is extensive. These veterinarians provide an informal network of sentinel farms. However, there is often no system in place to capture the observations of these practitioners.

The number of veterinarians who undertake research is increasing. These professionals are in universities (where they often also have a teaching role) and research establishments. Animal health research includes areas such as ecology, diagnostic test detection, and wildlife. Many veterinarians manage the animal houses within research organizations that, for example, require expertise in rodents and primates.

Veterinarians within pharmaceutical industries generally are involved with the development, marketing, and product support for veterinary pharmaceuticals. These veterinarians observe changes that occur at an industry level: for instance, a reduction in the use of respiratory disease treatments within the beef feedlot industry, an increase in the demand for chicken vaccines, or a change in the prescribing behavior of small animal clinicians for antibiotics. Often the information that they obtain is commercially sensitive and therefore is unavailable for disease surveillance purposes.

These distinctions between spheres of veterinary pursuit are important, because each category of pursuit uses, what we will call, electronic veterinary records (EVRs) very differently.

4.2. Farmers and Farm workers

Agricultural workers possess a broad range of skills, ranging from unskilled farm laborers through to qualified workers. An increasing proportion of farmers possess a university education in agricultural science. There is increasing emphasis on agricultural workers to use continuing education resources such as specific training courses and attendance of field days and demonstrations. Many consultants and specialists who service agricultural clients have science-based degrees, including agricultural science, environmental science, horticulture, agronomy, animal science, and veterinary science.

Veterinary care delivery is mostly by farm workers. For example, feedlot ranch hands receive basic training in cattle diseases from veterinarians. These ranch hands take charge of identifying and separating sick animals from the group. Often these workers treat the sick animals by using simple treatment and management algorithms derived by veterinarians for dealing with animals with specific signs of ill health. For example, ranch hands may remove cattle with respiratory signs, such as a nasal discharge, wheezing, or coughing, from the main pen to a quarantine pen and begin treatment with a nominated antibiotic. The other commercial animal production systems have similar workers with advanced observational skill sets for the species that they work with. These farm workers develop acute observational skills for the detection of disease within the animals under their care.

4.3. Wildlife Experts

The FWS employs 7,500 people in nearly 700 locations across the United States. Employees include biologists with specialist training in wildlife, fisheries, and ecology and workers such as wildlife and refuge managers, engineers, and inspectors. Wildlife inspectors are the frontline detection agency against the illegal trade of wildlife. Volunteer and other affiliated organizations work closely with the department to observe animal populations and to assist with conservation efforts. These are often formal partnerships to deliver services. The Web site *http://www.fws.gov/partnerships/partnership_links.html* lists partner organizations, including groups concerned with wetlands, endangered species, refuges, aquatic animals, recreational boating, fishing and hunting, and farming. The department also actively encourages nonaffiliated volunteers from the general public to assist with programs. Currently, around 35,000 volunteers contribute 1.4 million hours of time assisting the department each year.

The USGS employs a wide range of experts across many disciplines. These experts combine earth and life science knowledge to address problems. The integration of diverse scientific expertise enables the USGS to address complex natural science phenomena. The USGS employs around 10,000 people—ranging from scientists to technicians—to support staff. The USGS operates from more than 400 locations throughout the United States.

The WCS currently operates nearly 300 programs in more than 50 countries around the globe. There are many biology departments within universities whose scientists observe (and frequently investigate) the health of flora and fauna of the world. The OIE has a working group (founded in 1994) that focuses on wildlife disease (*http://www.oie.int/wildlife/eng/en_wildlife.htm*).

Symposia that bring together experts from around the globe to improve understanding and recognition of the interactions and dependencies between disease in humans, domestic, and wild animals are appearing. These include the One World, One Health symposium held in 2004, which examined the current and potential movements of disease among humans, domestic animals, and wildlife populations. This meeting called for an interdisciplinary approach toward threats to the health of life on earth. The symposium developed the Manhattan principles, which call for greater coordination and cooperation in research, extension, and control measures for the increasing disease threats to humans and domestic and wild animals that most certainly will emerge during the 21st century (*http://www.oneworldonehealth.org/*).

5. ANIMAL HEALTH DATA

There are multiple organizations within a country that make up its animal disease surveillance system. They all gather various data that describe an aspect of the animal health situation within a country. A single organization within a country has responsibility to report to the OIE. In the United States, this responsibility lies with the USDA, which must gather and compile the required information from multiple sources, including state departments of agriculture, laboratories, industry groups, and internal sources. Often the centralization of information is manual and, therefore, does not occur in real time and is in summary form only. Information provided to OIE does not include individual producers. Individual states typically also monitor diseases that they are not required to report to the USDA.

The OIE does not require information on all diseases; for example, information on diseases of production animals that do not result in production loss or on diseases of wildlife with no known impact on humans or domesticated animals is generally not required. Usually, wildlife bodies collect this information but do not send it to the OIE reporting body.

It is therefore essential that all the independent organizations within a country who monitor animal health combine into an effective surveillance system to facilitate trade of animals or animal products. This combined system provides evidence to validate a country's claims of freedom from individual animal diseases. However, improvements to data collection, centralization, analysis, and reporting are certainly possible and desirable within the animal health monitoring system. A system to identify deficiencies and duplications of surveillance and a method to prioritize and fund system development are required. These developments will require coordination and consultation with many organizations involved in human, domestic animal, and wildlife health across multiple jurisdictions.

One such effort is underway. The USDA's APHIS, working together with the U.S. Animal Health Association and the American Association of Veterinary Laboratory Diagnosticians (AAVLD) launched the National Animal Health Reporting System (NAHRS), which is part of the APHIS National Surveillance Unit (NSU) Veterinary Service. NAHRS is implementing a Web-based reporting tool and database that the USDA hopes will speed reporting and make the data more useful for surveillance (Bruntz, 2004). However, we note that NAHRS-specified state reporting consists of simple yes/no assertions; that is, it merely states whether a given disease has occurred in that state. This is a significant limitation.

5.1. Electronic Veterinary Records

An EVR is defined as software running on a computer system that allows the recording of veterinary history, physical examination notes, laboratory results, surgical notes, diagnoses, and treatment. The function of an EVR parallels that of a point-of-care system in human patients. Veterinary practice is primarily a cash business; hence, third-party, payer-oriented coding schemes similar to the International Classification of Disease (ICD) codes for human disease are not used. Nevertheless, EVRs may be designed to be compatible with, and accept, a controlled vocabulary, most often the Nomenclature of Veterinary Diseases and Operations (SNOVDO). SNOMED-CT codes for veterinary medicine supersede the now obsolete SNOVDO.

5.1.1. EVR Use: Private Companion Animal Veterinary Practice

The degree of acceptance of EVRs in veterinary hospitals is about the same as the market penetration of point-of-care systems in medicine. In 2001, no more than 10% of veterinary hospitals had them installed; by the end of 2005, 20% of veterinary schools in the United States expect to have deployed EVRs. Very few private practices use them; budget constraints and a lack of familiarity with these systems have limited acceptance. These systems range from complete electronic veterinary systems with entries for history, physical examination, laboratory data, radiology data, diagnoses and procedures, to systems that merely list pedigree, vaccinations, current diagnoses, and what tests or procedures have been ordered (but not their results or follow-up). A prime example of the former is the EVR developed at the veterinary teaching hospital of the University of California at Davis; the latter, functioning more as a billing tool than a medical record, is typified by the University of Pennsylvania's veterinary hospital (as of 2005, the university's small animal hospital furnishes laboratory test results electronically). Specific elements, which form part of an EVR, have gained wider acceptance: electronic laboratory reporting and teleradiology are prime examples. Many veterinary hospitals, even those affiliated with schools of veterinary medicine, have no electronic medical record systems.

In veterinary practices that have them, one may find that some veterinarians use the system, but others refuse to do so. Some systems are extensions to practice management software systems, most commonly used for billing, with the EVR capability unused (American Animal Association, personal communication, 2005). The advantages of EVRs are similar to point-of-care systems in medical settings: veterinarians enter data in real time or near real time, and as with relational databases, the EVR allows multiple views of the data. Limitations to EVR usefulness to biosurveillance in this practice setting are also similar to those in human medical settings. The number of systems installed is relatively small. Confidentiality requires that the disclosure of information on the diagnosis and treatment of an animal by a veterinarian can only occur with permission of the pet's owner.

Private veterinary clinics frequently outsource their laboratory work. Automated equipment allows many biochemical analyses to be undertaken in-house; however, serology, microbiology, and histopathology services are usually external. This separation commonly results in failure of the EVR to capture this essential source of information. Laboratories often do not use standardized disease coding systems or data reporting systems.

5.1.2. EVR Use: Agribusiness Veterinary Practice

Most veterinarians who work with farmers use production and animal health data collected by the producer in order to tailor their advice. The intensive farm animal production systems (poultry, pigs, dairy, beef feedlots) have well-developed data capture, analysis, and interpretation systems. Extensive production systems (e.g., ranch beef) have lesser developed systems but often capture significant information on the health and productivity of livestock. These data are available to the production animal consultants (including veterinarians) but are usually from farm management software programs. These programs tend to be species and enterprise specific; therefore, they tend to vary in format and structure. The main impetus for their development is the production of a useful business management tool for the specific enterprise.

Some veterinarians in this field use specific EVR systems–especially those that also offer companion animal services. However, the systems for companion animals are best at recording individual animal records. Many records from farming systems pertain to multiple animals; consequently, the use of individual animal EVRs are not appropriate for recording animal health details. Developments toward tighter regulations for prescribing antibiotics and other restricted pharmaceuticals for use in food producing animals are promoting the modification of EVR systems to allow recording of individual animal (or pen) treatment information. These systems often allow the recording of information on the clinical condition present within the individual (or pen) as evidence to support the decision to administer medication.

5.1.3. EVR Use: Institutional Veterinary Practice

The care of these animals is not subject to the same privacy and confidentiality regulations as are privately owned pets; moreover, public zoos and parks are subject to the Freedom of Information Act, wherein interested parties may demand to inspect documents related to the care of animals at a given facility. However, this relative openness is not limitless. Zoos are understandably sensitive to disclosures, which may affect their business adversely if the public were to avoid the zoo for fear that visits to the zoo are unsafe. Zoos also bear the brunt of frequent legal assaults and harassment by animal rights advocates; as a result, zoo administrators have tried to restrict the flow of information to outside parties.

The vast majority of large- and medium-sized zoos and animal parks participate in using the International Species Identification System (ISIS), which is a standardized way of classifying and inventorying animal collections in zoos, parks, and aquariums. ISIS members include more than 450 such facilities, half of which are within the United States. ISIS developed a DOS-based program for animal cataloguing and inventory management called ARKS, which has an EVR extension called MedArks. Currently about 90% of ISIS member facilities use MedArks as their EVR, and in contrast to the situation often encountered in POC-equipped human medical facilities, the zoo or park will use MedArks to record the care of its entire animal population. MedArks is designed around a DOS-based, SQL-compliant Foxpro database. MedArks is designed to allow one institutional veterinary center to transmit its records to any other MedArks-equipped facility, as well as the ISIS central database, by e-mail or by mailing a diskette containing the desired files. Currently, ISIS members send sets of records to the ISIS central database on a monthly basis; this infrequent schedule is a result of manpower limitations and could be made arbitrarily frequent. Of course, these same transmissions could also be received by other agencies involved in biosurveillance, if appropriate arrangements were made.

A few zoos, such as the San Diego Zoo and the Denver Zoo, use other software for their EVRs. It is important to note, however, that nearly 100% of medium- to large-sized institutions use EVRs of one type or another. The distribution of institutions in most metropolitan areas of the United States (as well as many areas of the world), their lower privacy and confidentiality barriers, and their commonality of EVRs combine to offer a unique opportunity to develop an early warning sentinel system.

Use of *sentinel species* to monitor biological threats to both humans and the sentinel species offers possibilities not available within human medicine. Farms located near large cities also have the potential to provide sentinel service. For example, a herd of immunologically naïve pigs is kept at the top of Cape York in Queensland as a detector for Japanese encephalitis incursion from the Torres Straight in Australia. They are bled regularly for seroconversion (Australian Government Department of Agriculture et al., 2003; Animal Health Australia, 2004).

ISIS recognizes that the potential represented by the widespread usage of MedArks is threatened by the age of its software. There is currently no modern, Web-based, or client-server version of MedArks available to replace existing implementations, and some institutions have threatened to purchase or develop customized systems, potentially complicating compatibility across ISIS members. The National Aquarium in Baltimore and the New England Aquarium made attempts to reengineer MedArks for the Web.

ISIS is currently designing the Web-based Zoological Information Management System (ZIMS). The ZIMS project will overhaul ISIS member zoos' biological inventory systems and integrate them with an EVR. When brought on-line, ZIMS will serve 640 zoos in 71 countries, including 250 zoos in the United States (International Species Information System, 2005; J.A. Teare, personal communication, 2005). The detection spectrum of a sentinel network based on zoo monitoring, if constructed, is an interesting question. For aerosol releases of biological agents, some animal species will be more susceptible than are humans and, therefore, sicken and die early. These animals are closely watched, so if a disease is spreading through the animal population more quickly or in advance of spread through the human population, as in the case of West Nile encephalitis, the outbreak may be noticed first by surveillance of zoo-based animals. One last and unexpected role for institutional veterinarians is illustrated by the West Nile outbreak. Zoos employ keepers who inspect the grounds of their institutions and collect dead animals for examination, whether or not they originate from the institution's collection. In such cases, the institutional veterinarian will often perform a necropsy, especially when there are signs of an unusual or unexpected illness.

It was a zoo veterinarian, not a physician or public health service officer, who uncovered an epidemic of WNV in 1999 in New York, based on necropsy findings of dead birds found on zoo property.

5.1.4. EVR Use: Public Veterinary Practice

State and federal departments of agriculture provide systems for the recording of disease events within jurisdictions. Much of the electronically recorded material is from laboratory samples and information about disease from control and eradication programs (e.g., results of on-farm testing such as tuberculous skin tests), but electronic systems are being developed to capture observational data.

Public practice veterinarians use government veterinary laboratories for the diagnosis of animal disease. These laboratories undertake much of the testing of samples required for demonstrating freedom from disease for trade purposes and control and eradication programs for individual pathogens, as well as for investigating unusual disease events or suspected occurrences of exotic animal disease. Again, standardized laboratory coding systems do not exist, and little aggregation of raw data between jurisdictions is present. These factors combine to present a major hurdle to the successful aggregation of veterinary laboratory data into an integrated biosurveillance system. The USDA and the AAVLD are developing the National Animal Health Laboratory Network (NAHLN). This will allow the aggregation of data by integrating the veterinary diagnostic laboratories and standardizing the laboratory codes.

As explained in Section IV, "Animal Health Data," the NAHRS is a mechanism for state, federal, and university veterinary laboratories to report monthly on the presence or absence of disease. In addition, information/data from universities contribute to OIE reports. The newly formed NSU within the USDA CEAH is working toward improving integration and coordination of the various surveillance projects and reporting systems, as well as management of other surveillance activities, such as national eradication programs.

The Federal and state departments of agriculture are developing a register of all livestock premises and individual animals—the National Animal Identification System (NAIS). Essentially, the system identifies premises, identifies individual animals, and records movements of animals to and from premises. The purpose of centralizing this data is to allow authorities to rapidly locate animals and to identify their contacts in an outbreak, thereby reducing the risk of spread and increasing likelihood of eradication or control. Database development is occurring through consultation between government departments and industry (*http://animalid.aphis.usda.gov/nais/about/plan.shtml*).

Veterinarians involved in animal inspection at processing plants record observations on animal for slaughter. Until recently, this involved standardized paper forms. The forms from inspections at abattoirs were sent to a data entry facility in Des Moines, Iowa, at the end of each workweek. These observations include number of tumors found, condition of skin and muscle, presence of pneumonia or other infection, and presence of injuries. The forms sent to Des Moines are not complete history and physical examination records, rather they were lists of anomalies found in each animal. The forms themselves are pathology oriented. The FSIS has a database that captures all carcass (or part thereof) condemnation reasons and when the condemnation occurred. All forms were submitted over the weekend, and data from them appeared in USDA's database the following Monday. Thus, the newest data in the database were 72 hours old; the oldest data (from that week) were 10 days old. Until recently, the database itself was housed on a minicomputer running an aging, proprietary, hierarchical database, posing potential difficulties for adapting the system to public health surveillance purposes. The system's design, its age, its range of time lags, and the current absence of EVRs in the slaughterhouses tend to limit the usefulness of the USDA's legacy database in the detection of bioterrorism.

In February 2004, The USDA's FSIS began deployment of a new electronic data system operational in real time. This system is the Electronic Animal Disposition Reporting System (eADRS) for recording pre- and postmortem examination of cattle, swine, sheep, and goats. Poultry inspectors still use paper-based submission however (United States Department of Agriculture, 2005a). As these data include information regarding carcasses, it is accurate and reliable although not oriented toward infectious diseases per se. The marginal cost to the government for using FSIS data for public health purposes is modest. Given interagency cooperation, public health agencies should be able to set up their own views into the database and perform surveillance and signal detection.

Laboratory data returned from the NVSL would be available to public health, with a lag time corresponding to the time required for shipping and processing a specimen and arriving at a reportable result.

State agriculture departments are currently in various stages of reporting system deployment. For example, the New York State Department of Agriculture and Markets has deployed the New York State Animal Incident Notification and Tracking System (NYSAINTS) a Web-based reporting tool, database, and guide to reportable animal diseases (New York State Department of Agriculture and Meats, 2005). The State of California's Department of Food and Agriculture (CDFA) intends to develop and deploy such a system in the future (K. Fowler, personal communication, 2005). Other states in the midwest, including Kansas and Nebraska, are working on systems that harness the Internet to create secure communication and data networks for veterinarians who care for food animals.

5.1.5. New Quality Management Practice: Hazard Analysis, Critical Control Point

The advent of hazard analysis critical control point (HACCP) will be gradually shifting primary responsibility for identifying and addressing problems to the abattoirs and processing plants themselves. Under this scheme, rather than rely on a government inspector to report problems, a processor proactively analyzes processing steps, identifies key potential problems, addresses them, and keeps meticulous, standardized records at the plant itself. The plant's government regulator would have access to these records and would function primarily to assist the plant in effectively executing its HACCP program. Although a biosurveillance system could access them as well, the use of HACCP implies that a much more widely distributed data scheme will be the norm in the future (B. Haxton, personal communication, 2005; United States Department of Agriculture, 2005b).

5.1.6. EVR Use: Nonclinical Veterinarians

Veterinarians outside of a practice do not have need for an EVR system. These veterinarians often use other electronic systems to record the health of animals. For example, veterinary pharmaceutical companies collect industry-level data. The information on sale of individual animal health medications can provide a useful system for monitoring the health of an industry. It will be necessary to aggregate data across competing companies to control for changes in individual company market share. Companies are likely to be reluctant to make raw data available to third parties. Systems that use information brokers and can de-identify data may be needed to encourage compliance.

5.2. Data from Livestock Farming

In general, electronic data systems are highly used within the various animal production industries. But, there is little standardization of systems, formats, and the type of data captured within or between the different industries. These differences arise because the various systems have different objectives. However, all basically capture information on the general health and productive performance of farm animals. Commercial producers are generally reluctant to release their data for use by third parties, which is likely to present a major barrier to the development of systems for aggregating data for the purpose of monitoring health and disease within the total farmed animal population.

Consumers are becoming increasingly interested in understanding where their food comes from, the production system and methods used (e.g., organic), and the likely quality of the product. This interest is providing stimulus within the food animal systems for the development of electronic lifetime identification systems and centralized databases. Electronic individual animal data systems require the use of unique and electronic animal identification (usually an ear tag). The technologies, such as electronic tags, readers, automatic drafting systems, and databases, already exist. These systems theoretically allow a consumer in a supermarket to scan a bar code and obtain production information on the foodstuff. Similarly, these systems have potential use for feed contamination investigations, providing a trace-back capability to investigators. The European Union is moving toward full electronic identification systems through the cattle passport system (*http://www.defra.gov.uk/animalh/tracing/cattle/passport/pport-index.htm*). Australia is rolling out the National Livestock Identification Scheme (NLIS), which uses electronic ear tags and a central database (*http://www.mla.com.au/content.cfm?sid=131*). These pressures are leading to increased commitment from individual animal industries within the United States to adopt the NAIS.

Data collection at the farm level emphasizes aggregate and statistical reporting. Beef cows, pigs, and poultry are not assigned individual health records. Although there has been some use of individual records on a trial basis, the benefits from individual animal management are not yet apparent. However, it is likely that individual animal systems similar to those currently in place within the dairy industry will develop in beef cattle production systems in the future. The increased uptake of NAIS will encourage development of individual animal production systems. As ranchers become aware of the natural variation in individual growth rates and the impact that these have on individual animal profitability, then production systems will evolve toward the management of the individual animal. Currently, an ear tag identifies most animals, but the identification system is not electronic or national—the numbers have relevance to the producer only. Animals that move between farms often have their original tags removed, and a replacement identification with meaning to the new owner is applied (Figure 7.2).

The information collected by beef feedlots about the animals they raise can vary. A few track each animal individually, collecting data on weight, temperature, feeding, and history of illnesses and treatment. Others will collect this information on a pen or consignment level. Dairy farming relies on individual animal production monitoring. Data capture, centralization, and processing systems allow producers to measure production, regulate feeding, and monitor health and reproduction.

Sophisticated systems are present in the chicken and swine industries. Often these are client-server systems developed to integrate data from multiple production units within an enterprise. These systems can contain comprehensive benchmarking and statistical process control tools provided to allow early identification of less productive units or farms and to monitor improvements in performance. These systems are well suited for monitoring the health of the animal population. However, there are many competing computerized systems, and these have varying data formats and standards. This reduces the ability of the systems to provide data to a single centralized database for the purpose of disease surveillance.

5.3. Wildlife Data

The AZA (*www.aza.org*) is an organization of more than 200 zoos and aquariums throughout the United States. The AZA is a nonprofit organization for the advancement of wildlife conservation, education, and science through zoos

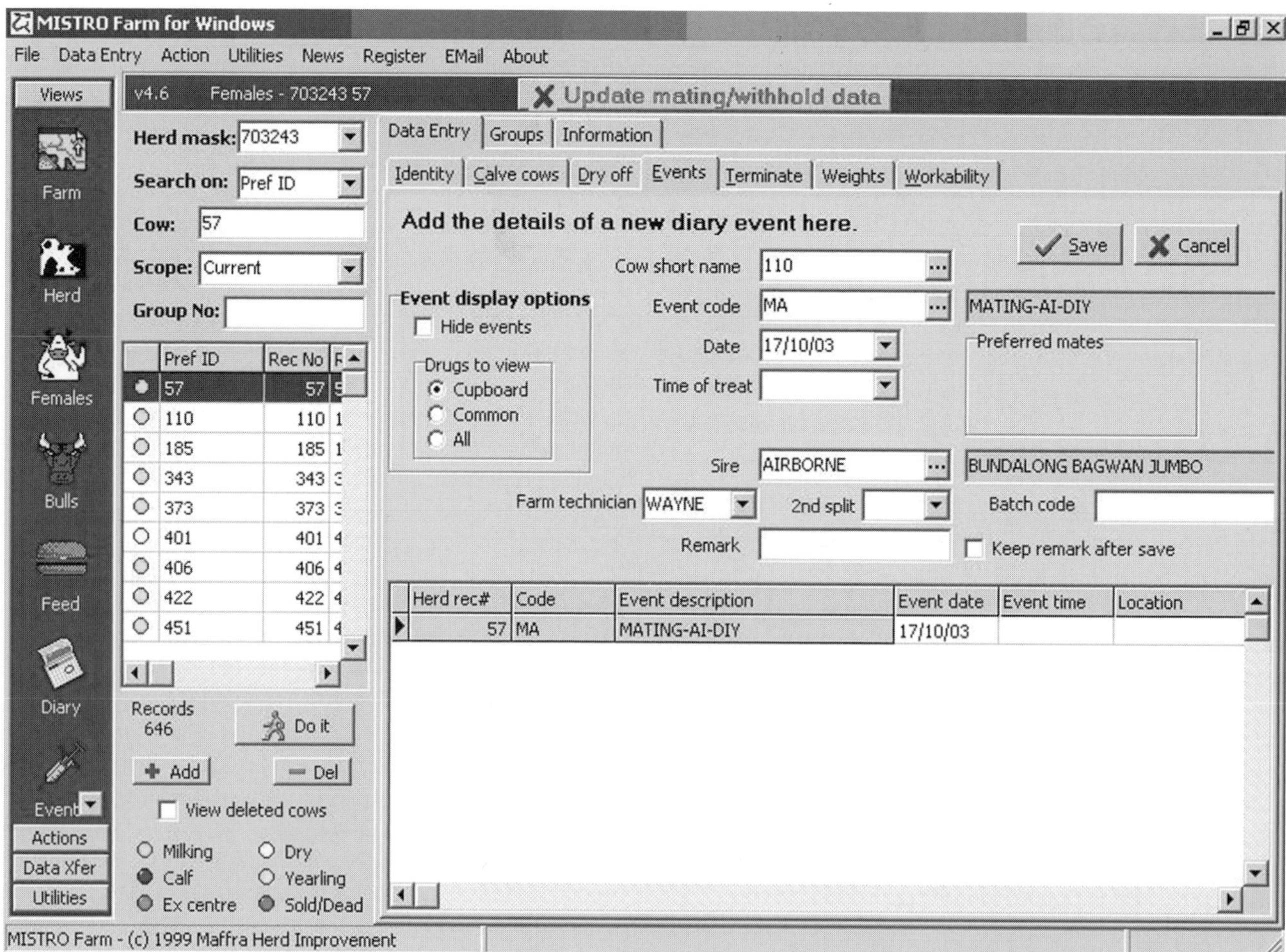

FIGURE 7.2 The animal health event screen from a dairy farm management software system (MISTRO Farm for Windows 4.0, Gippsland Herd Improvement, Victoria, Australia; *www.gippslandhi.coop*).

and aquariums. There are more than 6,000 professionals, businesses, and related organizations who are members of AZA.

The CDC provides a site for animal diseases called "Healthy Pets Healthy People" (*http://www.cdc.gov/healthypets/index.htm*). The site contains links to reports on zoonotic diseases and recent publications on zoonotic diseases for health professionals. Although this is not a database of events, it is a repository of summary information in the form of reports, publications, and other information related to zoonotic diseases.

The OIE wildlife working group, WCS, and USGS are currently looking at information sharing by using Web-based platforms and standards. Integration across species and organization is developing. Integration of data management systems is occurring and includes the ZIMS and Australian electronic Wildlife Health Information System (eWHIS) that is under development by the Australian Wildlife Health Network (AWHN) and the Canadian Cooperative Wildlife Health Centre database. Globally, all wildlife organizations have similar data collection, storage, access, and analytical needs; therefore, more uniform solutions may apply. The need for centralization and sharing of data is more obvious and accepted within the many organizations that conduct work with wildlife around the world. However, some technical issues are not resolved on subjects such as taxonomy, diagnoses, coding systems, data sharing, and integration with production animal databases.

6. ANIMAL HEALTH INFORMATION SYSTEMS

6.1. Veterinary Databases

There is no centralization of private veterinary EVR database information. Veterinarians contribute case details for notifiable diseases to their relevant authority (state or federal department of agriculture). However, the development of a central system that can capture and monitor the clinical picture of veterinary case load would be useful for detecting change to the pattern of disease presentation, similar to systems that monitor ED submissions in human medicine.

The National Animal Health Information System (NAHIS) operating within Australia is an example of a coordinated multispecies centralized animal disease database. The information within NAHIS allows Australia to report the national disease status for multiple pathogens and multiple host species to the OIE and to other trading partners. Many organizations, including state and federal departments of agriculture, quarantine services, the Animal Health Australia, and wildlife organizations, provide data to NAHIS. However, NAHIS only collects aggregated data from the various reporting bodies, and data are not provided in real time. The U.S. equivalent is the aforementioned nascent NAHRS. The USDA collects data into NAHRS from similar sources and in a similar form to the data provided to NAHIS in Australia.

The majority of university-affiliated veterinary healthcare facilities send data to the Veterinary Medicine DataBase (VMDB), a project housed at the University of Illinois. The National Cancer Institute originally established VMDB to help researchers study cancer in animals. Today, veterinary hospitals send diagnoses and procedure codes to VMDB. Submission does not occur in real time; lag times vary from a few days to months, with one participant offering an average of 2 to 3 weeks. Although VMDB is a rich storehouse of data on millions of animal health care encounters, its usefulness in the early detection of epidemics is very limited (Veterinary Medical Database, 2004; Kansas State University–Manhattan, KS, Medical Records Department, personal communication, 2005; University of Missouri–Columbia, MO, Medical Records Department, personal communication, 2005).

To encourage timely submissions and enhance the surveillance value of the database, VMDB has offered the veterinary community a free electronic submission utility that harnesses SNOMED. Submission, although still entirely voluntary, is easier now, and VMDB's staff hopes the new software will lead to significantly reduced lag times.

The Rapid Syndrome Validation Project for Animals (RSVP-A) is currently being examined to see if it offers an effective means of monitoring the health of cattle. The system required clinical veterinarians to classify cases into one of six major syndrome groupings. The data recording and transmission occurs in real time by the use of mobile computing devices. The Kansas State University study is determining the usability and utility of the system, with a primary focus on completeness of veterinary reporting and acceptance by veterinarians of the technology (De Groot et al., 2003).

6.2. Agricultural Databases

Much information above farm level has been aggregated, often because it is essential to organize payments and to monitor for quality and ensure traceability of products. For example, slaughterhouses, or beef processing plants, accept animals from feedlots for slaughtering and manufacturing of meat and meat-derived products. In the United States, three large beef processors dominate this part of the meat industry: IBP, a subsidiary of Tyson Foods; ConAgra, and Excel Corp., a subsidiary of Cargill. Of these, only ConAgra and Excel own feedlots; company-owned feedlots account for a minority of the food animals shipped to slaughterhouses. Individual processing plants collect large amounts of information on animals that they slaughter; however, the different companies record different amounts and type of data.

The dairy industry supports an industry-level system to allow efficient use of artificial insemination with semen from elite sires. These systems allow real-time identification of cows and matching with sires that are suitable for mating based on a measure of inbreeding between the two individuals. Organizations, such as the National Dairy Herd Improvement Association (*http://www.dhia.org/*), promote development of data standards and systems for integration of data across states.

There is also significant aggregation of data beyond the farm level by milk processors. Milk payment schemes use quantity and quality of milk provided to determine individual farm payments, which requires sophisticated systems to capture and aggregate production data at the processor level. Systems that monitor production in farm animals can be sensitive detectors of disease.

Similar processor-level data systems exist for commercial egg, poultry, and pig production systems.

There is little uniformity of standards and formats between species, production systems, and even processors that compete within the same industry. Systems that monitor production in farm animals can be sensitive detectors of disease. Again, variation in computer software has resulted in differences in data standards and formats between and within organizations. This will inhibit the development of systems to centralize this data for the purpose of surveillance for disease.

6.3. Wildlife Databases

ZIMS will replace existing ISIS database systems such as MedArks and will provide a more accurate and comprehensive system. ZIMS will be Web based, thus allowing users to see collections of animal data from multiple institutions in real time from any authorized computer anywhere in the world. Data will include information on animal health, laboratory accessions, genetics, disease investigations, and animal resource management.

ZIMS will enhance local care and international conservation efforts by providing better and faster access to information. This will increase efficiency, communication, productivity, and data quality of captive animals. Initially, ZIMS will provide access to information on an estimated 2 million animals from almost 700 institutions within 70 countries. ZIMS is developing robust methods for dealing with natural groupings of animals, such as schools of fish, and it will contain information on disease events and processes. Data will be stored within a central repository, allowing effective data mining and surveillance for trends. It will provide a unique level of integration among zoo, aquarium, wildlife, and environmental organizations, thus making it an extremely valuable resource for disease surveillance systems. An advantage of ZIMS will be the identification of common ground between organizations, thereby allowing the development of a prioritization system for further development and research.

The lack of uniform and consistent data recording systems across species is a major problem that prevents unification of wildlife data sources. The Australian National Information Managers Technical Group (NIMTG) was established in 2002 with the objective of encouraging the development of improved data management systems for plant and animal disease surveillance, incident response and recovery, emergency management, and the reporting national and international disease reporting compliance. This group concluded there should be one national data scheme and a set of rules for effective management of disease data. The final version is not available at present (Australian Government Department of Agriculture et al., 2003; Animal Health Australia, 2004).

7. ORGANIZATIONS THAT USE ANIMAL HEALTH DATA

A large amount of animal health data is collected, but this is generally to satisfy specific purposes, such as determining the prevalence of a disease within a target species. Organizations that use animal health data include farming enterprises, processors (such as abattoirs), farming service providers (such as veterinarians), pharmaceutical companies, and state and federal departments of agriculture. Organizations that use domestic animal health data include individual veterinary hospitals, veterinary wholesalers, and pharmaceutical companies. Personnel working in resource management, wildlife and fisheries, zoos, aquariums, and agriculture and human health have call to use wildlife data. The USDA collects information on animal health when it is a threat to agriculture. The Department of Health will collect animal health data when it is of significance to human health (e.g., salmonella, WNV vector data).

As a result of the multitude of organizations involved with collecting data and the multiple reasons for the collection of data, there exists duplication of data collection concurrent with deficiencies in data collection, little sharing of data between organizations, and therefore little secondary use of data. The development of overarching policies to guide the collection, format, sharing and analysis of all disease data is necessary to maximize information gain.

8. SUMMARY

> "Or Providence sends the pleuro, and big strong beasts slink away by themselves and stand under trees glaring savagely until death comes."–Andrew Barton (Banjo) Patterson, 1917, *The Bullock*

There is potential for disease to transfer to humans from the domesticated animals that we keep or from the wild animal population. We have seen in Chapter 2 that disease can spread from animals to humans and that new diseases of man are often zoonoses. Some animal diseases that do not directly threaten humans, nonetheless, have the potential to significantly disrupt society and threaten our food supply (e.g., foot-and-mouth disease). Therefore, systems that monitor the health of animals need to be coordinated with human disease surveillance systems.

Veterinarians monitor the health of domesticated animals. However, systems for the centralization of health data within the animal health system are not well developed. There is little crossover between systems that capture information on companion animals with those that collect information on farm animals. There is almost no compatible system operating for wild animals. The care of pet animals is through individually tailored heath services; therefore, it is directly analogous to its human counterpart. Conversely, the care of production animals in the United States involves population medicine, which refers to measures designed to maintain the health of a population as opposed to a single patient, to a greater extent than the care of people.

Although private veterinarians do not store diagnoses and treatment data electronically to any greater degree than do physicians and a pet's veterinary record is subject to the same confidentiality rules as its owner's, veterinary care in institutional (zoos, animal parks, aquariums) and animal production industry/public practice settings is well supported by computers, databases, and software. Institutions are universally equipped with EVR systems linked by e-mail into a loose but effective network.

The typical production animal veterinarian is a sophisticated user of statistical tools, more so than his/her physician counterpart, although the emphasis on population medicine has, historically, meant that EVR use for individual animals has not been a priority. This will change as animal producers and processors recognize the economic value of an animal health record. Individual animal illness and treatment records have been successfully deployed on laptop computers for field use. It is likely that agricultural software specialists will develop systems that can provide individual animal and herd data to production veterinarians.

These data are potentially highly accessible and are already very well suited for public health purposes, because of the emphasis on population medicine and a sharing of goals. However, the producers jealously guard these data. Early disease detection in animals translates into lower herd losses and higher profits for the rancher and feedlot operator. Although providing this information to public health authorities is likely to require considerable financial compensation, the industry as a whole will need education to recognize both the personal and national benefits that can arise from cooperation. The development of systems that aggregate data and allow benchmarking of disease, production, and profitability performance may encourage the sharing of data, but they systems will need to demonstrate significant management information to users to encourage large-scale uptake.

Wildlife disease is just one of the many aspects of natural system ecology. All ecological projects compete for limited funds, personnel, and resources from the many organizations involved with wildlife. Limited resources are currently allocated for the study and monitoring of wildlife disease; therefore, there are large gaps in knowledge and a limited disease surveillance system. Systems to identify and quantify risks and to prioritize funding of projects that cross multiple jurisdictions, organizations, and countries require further development. There are, potentially, thousands of diseases of wildlife; hence, surveillance for wildlife disease must be selective. Risk assessment systems are required to identify and prioritize wildlife disease surveillance. The purpose of any surveillance system must be clear. Usually, this purpose is external to wildlife ecology–wildlife disease may have significant impact on human health or trade in agricultural products. Thus, the risk assessment system that develops must be multidisciplinary and the funding system expanded to include contributions from more extensive sources (R. Wood, personal communication, 2005).

A large amount of data is collected by many different organizations working with animals; however, little information is shared or made available for other uses (such as biosurveillance). Moreover, an overarching system that monitors animal health in all species does not exist. The full elucidation of all animal health systems that are operational, identification of areas of surveillance deficit, and the cooperative development of systems to standardize the capture and sharing of animal health information, are required. This will require close cooperation between medical, veterinary, agricultural, and wildlife associations within the country. Animal monitoring is a potentially rich source of data for detection of bioterrorism and other public health threats. As in many other fields of endeavor, money, competition, and politics will greatly influence the outcomes of recent initiatives.

ADDITIONAL RESOURCES

Veterinary Epidemiology

Text books

Cameron, AR. (1999). *Survey Toolbox for Livestock Diseases. A Practical Manual and Software Package for Active Surveillance in Developing Countries.* Canberra: Australian Centre for International Agricultural Research. ACIAR Monograph No 54. This provides advice for the conduct of effective surveillance programs within countries with limited resources and infrastructure. This booklet is available free online at *http://www.ausvet.com.au/content.php?page=res_manuals.*

Dohoo, I.R., Martin, W., and Stryhn, H. (2003). *Veterinary Epidemiologic Research.* Prince Edward Island, Canada: Atlantic Veterinary College, University of Prince Edward Island. This is a new text that comprehensively covers the discipline. Extensive discussion of key statistical tools used by veterinary epidemiologists are provided.

Martin, S.W., Meek, A.H., and Willeberg, P. (1987). *Veterinary Epidemiology.* Ames, IO: Iowa State University Press.

Noordhuizen, J.P.T.M., Thrusfield, M.V., Frankena, K., and Graat, E.A.M. (2001). *Application of Quantitative Methods in Veterinary Epidemiology,* 2nd reprint. Wageningen, The Netherlands: Wageningen Pers. This text provides an excellent introduction to the most commonly used multivariate techniques used in veterinary science.

Salman, MD. (2003). *Animal Disease Surveillance and Survey Systems*. Ames, IO: Iowa State Press. This book provides an introduction to veterinary surveillance. The requirements for a functional system to detect, control, and describe disease in animal populations are provided. Relevant statistical methods are discussed.

Thrusfield, M. (1995). *Veterinary Epidemiology,* 2nd ed. Oxford: Blackwell Science. This is a good general introduction that explains basic epidemiological concepts such as causality well. Veterinary examples are used to support concepts.

Toma, B., et al. (1999). *Applied Veterinary Epidemiology and the Control of Disease in Populations.* Maisons-Alfort, France: AEEMA.

Web pages

Australian Biosecurity Cooperative Research Centre for Emerging Infectious Disease (*http://www1.abcrc.org.au/index.aspx*). This is a collection of resources and organizations in Australia. The mission is to protect Australia's public health, livestock, wildlife, and economic resources through research and education that strengthens the national capability to detect, diagnose, identify, monitor, assess, predict, and respond to emerging infectious disease threats that impact on national and regional biosecurity. This organization provides links to emerging disease, new technology, research, training opportunities, and publications on emerging diseases.

EpiVetNet (*http://www.vetschools.co.uk/EpiVetNet/*). This Web page is actively maintained providing up-to-date information within the field of veterinary epidemiology. The site provides links to texts, Web resources, software, conferences, and publications within the field of veterinary epidemiology. Members of the International Society for Veterinary Epidemiology and Economics (ISVEE) maintain the Web site.

Office International des Epizooties (OIE) (*http://www.oie.int*). The OIE is the international body for information on animal diseases around the world, providing information on the status of global animal diseases and zoonoses, scientific techniques in animal health, surveillance and risk analysis, and codes for sanitary safety and biosecurity requirements. The publications list is comprehensive with material on disease distribution, quarantine, surveillance, risk assessment, sampling, diagnostic tests, laboratory and veterinary services, verifying freedom from disease, and food contamination.

The Fish and Wildlife Service (FWS) Web page (*http://www.fws.gov/*). Information on the management of wild animals, fish, and their habitats within the United States. The site provides links to affiliated and similar organizations.

The United States Geological Survey (USGS) Web page (*http://www.usgs.gov/*). A source of scientific information on landscape resources and hazards (including disease) that threaten humanity.

The World Conservation Society contains information on conservation, education and management of wildlife resources as well as links to contributing organizations and projects conducted to protect and preserve wildlife and wild lands (*http://www.wcs.org/*).

The International Species Identification System (ISIS) Web page (*http://www.isis.org*). Information about the organization and on the developing ZIMS data management system.

AusVet Animal Health Services (*http://www.ausvet.com.au/*). A private epidemiology consulting company that specializes in (but is not limited to) animal disease within populations. The company has extensive experience in disease control and eradication programs, animal health in developing countries, and aquaculture. The site contains a comprehensive useful links page covering animal health, epidemiology, and statistics.

"Healthy Pets Healthy People" (*http://www.cdc.gov/healthypets/index.htm*). This site provided by the CDC contains information on pet care, diseases of pets, and zoonotic diseases. It also contains information on the current zoonotic disease status within the United States and relevant publications for health professionals.

(*http://www.oneworldonehealth.org/*). The One World, One Health symposium examined the difficulties of disease control within the modern world. This resulted in production of the 12 Manhattan principles for addressing the disease challenge that the world faces in the 21st century.

REFERENCES

American Veterinary Medical Association. (2002). *U.S. Pet Ownership and Demographics Sourcebook.* Schaumburg, IL: American Veterinary Medical Association.

Animal Health Australia. (2004). *http://www.aahc.com.au/.*

Australian Government Department of Agriculture, Fisheries and Forestry. (2003). Animal Health in Australia. *http://www.affa.gov.au/.*

Biosecurity Australia. (2003). *The Import Risk Analysis Handbook.* Canberra: Australian Government Department of Agriculture, Fisheries and Forestry. *http://www.affa.gov.au/biosecurityaustralia.*

Bruntz, S. (2004). The National Animal Health Reporting System. United States Department of Agriculture, Animal Plant Health Inspection Service. *http://www.aphis.usda.gov/vs/ceah/ncahs/nahrs/.*

California Department of Food and Agriculture. (2005). Animal Health and Food Safety Services. *http://www.cdfa.ca.gov/ahfss/.*

Canon, S. (2005). TB Feared in Kansas Cattle Herds. *Kansas City Star* April 16, 2005.

De Groot, B., Spire, M., Sargeant, J., and Robertson, D. (2003). Preliminary Assessment of Syndromic Surveillance for Early Detection of Foreign Animal Disease Incursion or Agri-terrorism in Beef Cattle Populations. In *Proceedings of the 10th Annual Conference of the International Society for Veterinary Epidemiology and Economics.,* November 17–21. Santiago, Chile, CD-ROM.

Euromonitor. (2000). PetFood Industry, March 2000. *http://www.euromonitor.com/Petfood.*

Georgiadis, M.P., Gardner, I.A., and Hedrick, R.P. (2001). The Role of Epidemiology in the Prevention, Diagnosis, and Control of Infectious Diseases of Fish. *Preventive Veterinary Medicine* vol. 48:287–302.

International Species Information System. (2005). Zoological Information Management System (ZIMS) Project. *http://www.isis.org/CMSHOME/.*

Iowa Department of Agriculture and Land Stewardship. (2005). Meat and Poultry Bureau. *http://www.agriculture.state.ia.us/thebasics.htm.*

Kellog, R. (2002). *Profile of Farms with Livestock in the United States: A Statistical Summary.* Washington, DC: Natural Resources Conservation Service, USDA.

Mostashari, F., Kulldorff, M., Hartman, J.J., Miller, J.R., and Kulasekera, V. (2003). Dead Bird Clusters as an Early Warning System for West Nile Virus Activity. *Emerging Infectious Diseases* vol. 9:641–646.

New York State Department of Agriculture and Meats. (2005). Animal Disease Reporting. *http://www.agmkt.state.ny.us/AI/disease_rep.html.*

New York State Department of Agriculture and Markets. (2005). Division of Animal Industry. *http://www.agmkt.state.ny.us/AI/AIHome.html.*

Office of the Chief Economist and World Agricultural Outlook Board. (2005). USDA Agricultural Baseline Projections to 2014, Baseline Report OCE-2005-1. Washington, DC: U.S. Department of Agriculture.

Rouqet, P., et al. (2005). Wild Animal Mortality Monitoring and Human Ebola Outbreaks, Gabon and Republic of Congo, 2001–2003. *Emerging Infectious Diseases* vol. 11:283–290.

U.S. Department of Labor Bureau of Labor Statistics. (2005). Occupational Outlook Handbook, 2004–05 Edition, Veterinarians. *http://www.bls.gov/oco/ocos076.htm.*

United States Department of Agriculture. (2005a). Electronic Animal Disposition Reporting System. *http://www.fsis.usda.gov/OPPDE/rdad/FSISNotices/11-04.htm.*

United States Department of Agriculture. (2005b). HACCP and Pathogen Reduction. *http://www.fsis.usda.gov/Science/hazard_analysis_&_pathogen_reduction/index.asp.*

Veterinary Medical Database. (2004). *http://www.vmdb.org/vmdb.html.*

8

CHAPTER

Laboratories

Charles Brokopp

Coordinating Office of Terrorism Preparedness and Emergency Response, Centers for Disease Control and Prevention, Atlanta, Georgia

Eric Resultan and Harvey Holmes

Bioterrorism Preparedness and Response Program, National Center for Infectious Diseases, Centers for Disease Control and Prevention, Atlanta, Georgia

Michael M. Wagner

RODS Laboratory, Center for Biomedical Informatics, University of Pittsburgh, Pittsburgh, Pennsylvania

1. INTRODUCTION

Results from laboratory testing are an important source of information for biosurveillance systems. Clinical laboratory tests are vital for the correct diagnosis and treatment of individuals. Clinical laboratories analyze blood, urine, mucus, saliva, respiratory secretions, cerebrospinal fluid, semen, vaginal secretions, sweat, feces, fluid aspirated from joints, and tissues from humans and animals. The tests performed include cell counts; analytical chemistries, including drug and toxin tests; and examinations to detect and identify microbes and markers of current and past infection. Environmental testing is critical to the detection of outbreaks, the prevention of disease, and the monitoring of the environment. Environmental laboratories analyze samples of water, food, air, soil, plant material, and unknown powders, as well as samples taken from surfaces for evidence of bacterial, viral, toxin, or chemical contamination.

Data produced by laboratories are important for biosurveillance of virtually every disease caused by biological agents, chemicals, or toxins. Data collected during the preanalytical, analytical, and postanalytical phases of testing can be captured and incorporated directly into biosurveillance systems. Preanalytical data, such as the type of test ordered and the reason for a test, can provide an early clue to the existence of an outbreak. Similarly, analytic results, such as the initial Gram stain of a cerebrospinal fluid specimen, can potentially confirm a diagnosis when combined with other clinical information, as it did in the first case of inhalational anthrax in the 2001 postal attack. The actual results of tests are obviously foundational to biosurveillance.

The range of tests offered by an individual laboratory varies significantly among laboratories in the United States. Many small laboratories perform a limited number of tests that are needed on an urgent basis or for screening purposes. Larger laboratories provide more complex confirmatory analyses. The majority of clinical laboratories in the United States are small laboratories located in physician offices; however, the larger laboratories account for a high volume of all tests performed.

Table 8.1 describes the clinical laboratory tests that contribute to the diagnosis of inhalational anthrax. Anthrax, as well as many other infectious diseases, is diagnosed after the performance of presumptive and confirmatory tests in combination with the clinical presentation. Clinical specimens, such as blood, cerebrospinal fluid, urine, sputum, throat swabs, and skin scrapings, are used to isolate a causative agent that is later subjected to further testing with confirmatory procedures to make the final identification. Preliminary tests results are sometimes released before the confirmatory tests results become available. When preliminary results are reported, the report often contains a statement about when final results will be available.

Laboratories that produce the types of data most useful for biosurveillance include clinical laboratories operated by the human or animal health systems, commercial laboratories, and governmental laboratories. Laboratories typically specialize in either human or animal testing. Commercial laboratories are free-standing laboratories that are not associated with hospitals or healthcare facilities and that often provide a broad range of services over a wide geographical area. Governmental laboratories exist at the federal, state, and local level and often provide testing that is not readily available from other laboratories. We describe each of these types of laboratories in this chapter.

Laboratories that test for biologic agents are classified as biosafety level 1, 2, 3, or 4, with biosafety level 4 providing the highest degree of protection to personnel and the environment. Most clinical laboratory work is performed at level 2. These biosafety levels combine the use of laboratory safety practices, safety equipment, and laboratory facilities to provide greater levels of safety for the more dangerous organisms. Each level is specifically appropriate for handling various biologic agents (CDC, 1999).

2. CLINICAL LABORATORIES

There are more than 186,000 clinical laboratories in the United States in which clinical laboratory scientists, pathologists,

Handbook of Biosurveillance
ISBN 0-12-369378-0

TABLE 8.1 Clinical Laboratory Tests that Contribute to the Diagnosis of Anthrax

Type of Test	Specimen	Expected Result
Nonspecific		
White blood count	Whole blood	Elevated count
Cerebrospinal fluid (CSF) analysis	CSF	Normal
Presumptive		
Growth on sheep blood	Blood, CSF, lesion	Growth within 24 hours
Colony morphology	Bacterial growth	Gray-white colonies, flat or convex, ground glass appearance
Gram stain	Bacterial growth	Large Gram-positive rods
Hemolysis	Bacterial growth	Clear hemolysis
Motility	Bacterial growth	Motile
Sporulation	Bacterial growth	Visual spores with malachite green stain
Confirmatory		
Capsular stain	Bacterial growth	Visual capsules with M'Faydean stain
Gamma phage	Bacterial growth	Lysis by gamma phage
Direct fluorescent antibody (DFA)	Bacterial growth	Positive fluorescence
Polymerase chain reaction (PCR)	Bacterial growth	Positive PCR
Time-resolved fluorescence (TRF)	Bacterial growth	Positive TRF assay
Molecular characterization	Bacterial growth	Positive match with control materials

medical technologists, and laboratory technicians perform 7 billion or more diagnostic tests annually (Centers for Medicare and Medicaid Services, 2004). The American Society for Clinical Pathology (ASCP) currently certifies more than 280,000 laboratory professionals who primarily work in clinical diagnostic and research laboratories. Clinical laboratory services in the United States are delivered either by commercial clinical laboratories or by "in-house" laboratories at healthcare facilities (hospitals, clinics, physician offices), departments of health, veterinary hospitals, and clinics. Individual veterinarians and physicians and the staff within their offices also conduct laboratory testing and produce results that are important for biosurveillance.

Professional laboratorians provide services that include simple, rapid screening tests; more advanced diagnostic tests; and complex confirmatory analyses. Clinicians use the information provided by laboratories to establish diagnoses and to make treatment decisions on virtually every patient. The demand for testing is increasing as the population ages and requires more health care, including analytical services. New tests are frequently introduced that improve diagnosis and care. The emergence of new diseases, the threat of bioterrorism, and the need for better biosurveillance systems have increased the demand for qualified laboratory professional in all fields, especially infectious disease testing. Although the demand for more laboratory professionals is increasing, the number of established laboratory professional training programs is decreasing.

The Centers for Medicare and Medicaid Services (CMS) registers all clinical laboratories in the United States that examine materials derived from the human body for diagnosis, prevention, or treatment. CMS administers the program for the Secretary of Health and Humans Services in conjunction with the Centers for Disease Control and Prevention (CDC) and the Food and Drug Administration (FDA). CMS regulates laboratories and establishes criteria for other organizations, such as state health departments, that also regulate laboratories to ensure compliance with the federal Clinical Laboratory Improvement Act (CLIA). CLIA was first enacted by Congress in 1967 and set guidelines for large independent laboratories. In 1988, Congress amended CLIA 67 to expand the type of laboratories that must comply; CLIA 88 further established quality standards for laboratories to ensure accuracy, reliability, and timeliness of test results.

In August, 2004, 186,734 laboratories were registered with the CMS (Centers for Medicare and Medicaid Services, 2004). Table 8.2 shows the distribution of these laboratories by type. More than 55% of these laboratories are located in physician offices. Skilled nursing facilities (7.9%), hospitals (4.6%) ,and home health agencies (4.4%) accounted for an additional 20% of laboratories. The remaining clinical laboratories are found in community health clinics, health maintenance organizations, blood banks, industrial facilities,

TABLE 8.2 Clinical Laboratories Registered by Center for Medicare and Medicaid Services (CMS) by Type of Facility, August 2004

Type of Laboratory	Number	Percentage
Ambulatory surgical centers	3,229	1.7
Community clinic	6,717	3.6
Comprehensive outpatient rehabilitation facility	205	0.1
Ancillary testing site in healthcare facility	2,712	1.5
End-stage renal disease dialysis facility	3,657	2.0
Health fair	482	0.3
Health maintenance organization	657	0.4
Home health agency	8,308	4.4
Hospice	1,285	0.7
Hospital	8,749	4.6
Independent	5,162	2.8
Industrial	1,647	0.9
Insurance	49	0.02
Intermediate care facility for mentally retarded	856	0.5
Mobile laboratory	1,096	0.5
Pharmacy	2,423	1.3
School/student health facility	1,771	0.9
Skilled nursing facility/nursing facility	14,792	7.9
Physician office	103,378	55.4
Other practitioner	2,239	1.2
Tissue bank/repositories	36	0.02
Blood banks	361	0.2
Rural health clinic	982	0.5
Federally qualified health center	247	0.1
Ambulance	1,993	1.1
Public health laboratories	119	0.06
Other	13,582	7.3
	186,734	100

TABLE 8.3 Clinical Laboratories Registered by the Center for Medicare and Medicaid Services (CMS) by Annual Test Volume, August 2004

Annual Test Volume	Total Number of Laboratories	Percentage of Laboratories	Physician Office Laboratories	Percentage of Physician Office Laboratories
≤2000	8,955	43.1%	6,998	51.1%
2001–10,000	6,212	29.9%	4,469	32.7%
10,001–25,000	2,347	11.4%	1,206	8.7%
25,001–50,000	1,303	6.3%	475	3.5%
50,001–75,000	643	3.1%	207	1.5%
75,001–100,000	384	1.8%	111	0.8%
100,001–500,000	789	3.8%	221	1.5%
500,001–1,000,000	74	0.4%	19	0.1%
>1,000,000	51	0.2%	1	0.1%
	20,758	100%	13,697	100%

and health departments. All of these laboratories are frequent sources of biosurveillance data.

Over 58% of the 186,734 laboratories registered with CMS only perform simple tests. These simple tests, often referred to as *waived tests,* usually are based on commercially available test kits determined by the FDA to be sufficiently simple to perform that there is little risk of operator error. Laboratories that perform waived tests must enroll in the CLIA program, pay certification fees, and follow the manufacturers' test instructions. However, laboratories that perform only waived tests do not undergo inspections or need to comply with other CLIA requirements for larger laboratories. Laboratories that perform tests that use a microscope to examine specimens that are not easily transportable during the course of a patient visit are required to enroll in a CLIA program, pay applicable fees, and maintain certain quality and administrative requirements. These laboratories, known as *provider-performed microscopy providers* (PPMPs), represent 22% of the registered laboratories and are not subject to routine inspections. The remaining clinical laboratories must either be accredited by 1 of 6 approved clinical laboratory accrediting organizations (American Association of Blood Banks, American Osteopathic Association, American Society of Histocompatibility and Immunogenetics, College of American Pathologists; Commission on Office Laboratory Accreditation, Joint Commission on Accreditation of Healthcare Organizations) or obtain a compliance certificate directly from CMS.

In August 2004, these six organizations had accredited 15,667 (8.7%) laboratories, and CMS had certified 20,758 (11.5%) laboratories. The balance of the clinical laboratories (144,022) were either waived test providers or PPMPs (Centers for Medicare and Medicaid Services, 2004). Accreditation of clinical laboratories helps ensure that laboratories meet or exceed clinical standards established by governmental and nongovernmental associations. CMS, insurance companies, and healthcare plans require laboratories to be certified or accredited by these organizations for reimbursement of laboratory services.

Table 8.3 shows the annual test volume of the 20,758 clinical laboratories that were certified by CMS in August 2004. Over 84% of these laboratories perform fewer than 25,000 tests per year. Only 125 of these laboratories performed 500,000 or more tests per year. These 125 laboratories represent only 0.6% of all laboratories, yet they perform approximately 20% of all tests.

Two state health departments have developed and currently administer state clinical laboratory improvement programs that CMS deems equivalent to the CMS program. Laboratories in Washington and New York must meet the standards of these state programs. Approximately 25 additional states have laboratory licensure programs. They receive funding from CMS to implement the federal CLIA program. In states that do not have clinical laboratory regulatory programs, the laboratories must choose between accreditation through one of the six approved organizations or submitting to inspection and certification by CMS.

3. ENVIRONMENTAL LABORATORIES

Environmental testing laboratories perform physical, chemical, and microbiological analysis of specimens collected in the environment. For example, a water sample may undergo physical testing (temperature, turbidity, odor, color), chemical testing (nitrates, sulfates, pesticides, metals), and microbiological testing (total plate counts, coliforms, *Giardia,* cryptosporium). Environmental testing laboratories provide a wide range of testing that is in many ways similar to the testing performed in clinical laboratories. Sanitarians or water quality technicians often perform basic tests (e.g., for temperature, pH, volatility, and physical appearance) at the site where samples are collected. They transmit the results of these simple tests to the laboratory along with the samples, where chemists and microbiologists perform additional presumptive and confirmatory testing. Results from the simple tests may suggest the need for more definitive testing using instruments such as atomic adsorption spectrophotometers, gas chromatographs, and mass spectrometers. Laboratories perform much of the routine environmental testing in batches of 10 to 50 samples on semi-automated or fully automated instruments. The raw analytical data are captured, processed, and reported by using software that interfaces directly with the instrument and the laboratory's data management system.

Environmental laboratories are certified by accreditation authorities recognized by the Environmental Protection Agency (EPA) as part of the National Environmental Laboratory Accreditation Program (NELAP). At least 12 states currently are recognized by the EPA as environmental laboratory accrediting authorities. These state programs apply nationally recognized standards to the laboratories that they accredit so that there is some consistency in the quality of tests performed by accredited laboratories.

4. COMMERCIAL LABORATORIES

Commercial laboratories are an important component of the medical delivery system in the United States. A commercial laboratory is a laboratory that is free-standing; that is, it is not associated with a hospital or other healthcare organization. Commercial laboratories may specialize in clinical specimens, environmental specimens, or both.

Commercial laboratories can be important partners for organizations that wish to develop biosurveillance systems because of size of the laboratories and their use of information technology. Commercial laboratories have grown significantly in size during the past decade as a result of mergers and consolidations, and they may offer tests that are not readily available. Many of these laboratories began as specialized reference laboratories, offering tests that could not be economically provided by smaller laboratories. Small laboratories merged with larger laboratories that were, in turn, purchased by large laboratory corporations. Regional consolidation of clinical laboratories that provided services to healthcare organizations led to the formation of large commercial laboratories. These commercial laboratories, such as Lab Corp, ARUP, and Quest, have developed service systems that allow them to provide clinical testing at their headquarters and at distributed sites around the country.

Large commercial laboratories have established elaborate courier systems that collect samples locally for overnight distribution to the appropriate laboratory in their network. Although a sample may be transported to a distant laboratory, the results, nevertheless, frequently become available overnight. The commercial laboratories use information systems referred to as *laboratory information management systems* (LIMSs) to track tests and results. These systems monitor test requests, capture test results, and electronically report the results, often within hours of the sample being received.

As discussed in Chapter 5, clinical laboratories are required to report notifiable diseases to local health departments. The large, multistate commercial laboratories face challenges in complying with notifiable disease reporting requirements that vary by state. Further, reporting requirements frequently change as new diseases of public health interest are identified and many states have expanded laboratory reporting requirements to include *suspected* cases of notifiable diseases. Commercial laboratories are increasingly using electronic laboratory reporting to satisfy these complex and changing requirements.

Commercial laboratories may specialize in environmental testing. These commercial environmental laboratories provide testing on a variety of samples, such as water, air, hazardous materials, dust, and soil. These laboratories often contract with governmental agencies, such as EPA, Department of Defense (DoD), and U.S. Department of Agriculture (USDA) at the federal level and with environmental and regulatory agencies at the state and local level.

5. GOVERNMENTAL LABORATORIES

Federal, state, and local government agencies, such as health departments, operate laboratories or contract with commercial laboratories for testing related to diagnosis, regulatory compliance, investigations, and environmental monitoring. Since the early 1800s, governmental laboratories have performed testing that led to the identification of outbreaks of diphtheria, cholera, smallpox, and typhoid fever. During the 20th century, these laboratories, in conjunction with academic laboratories, helped develop vaccines and contributed to the detection of polio, rubella, measles, and whooping cough. The response to West Nile fever in the United States and the release of viable *Bacillus anthracis* in 2001 required these laboratories to develop procedures quickly for diagnosis and identification of agents that they had not previously encountered. Governmental laboratories are key to the recognition of new and emerging infectious diseases and are vital to surveillance efforts.

5.1. Federal Laboratories

The Department of Health and Human Services (DHHS), USDA, Department of Energy (DOE), DoD, and Departments of Commerce and Justice, and the EPA operate or fund clinical, environmental, forensic, and research laboratories. Federal laboratories provide reference testing and are often involved with the development of new technologies, as well as the transfer of these technologies to other laboratories. Many federal laboratories collaborate with international partners and serve as reference centers for specialized testing. Federal laboratories often provide training, confirmatory testing, and reference materials for other governmental laboratories.

The DHHS laboratories that are associated with disease detection, control, and surveillance activities are found at the National Institutes of Health (NIH), CDC, and FDA. The NIH, headquartered in Bethesda, Maryland, conducts research on acute and chronic diseases, develops new therapies and immunizations, and develops new laboratory and diagnostic technologies for many infectious and noninfectious diseases and health conditions. Much of the NIH work is done through extramural grants and contracts with universities, private companies, and other governmental organizations. The CDC's laboratories focus on infectious diseases, occupational diseases, and environmental causes of diseases. The CDC specialized laboratories are in Colorado, Ohio, West Virginia, and Puerto Rico, in addition to its headquarters in Atlanta, Georgia. The CDC has led the development of rapid national laboratory reporting systems that have been successfully used to identify multistate outbreaks of diseases (Bean et al., 1992; Hutwagner et al., 1997). CDC-based scientists have developed and used new technologies to identify outbreaks of disease in cooperation with state and local public health laboratories. One such example of the successful application of a new technology is known as *PulseNet* (discussed in Chapters 3 and 5; *http://www.cdc.gov/pulsenet/*). PulseNet is a national network of public health laboratories that perform DNA "fingerprinting" of foodborne pathogens and clinical isolates to allow matching of isolates. This epidemiological typing method is the basis for detecting clusters of disease that are geographically diffuse, and for linking bacteria found in a specific food to bacteria found in one or more persons with a particular disease. PulseNet permits recognition of outbreaks that previously went undetected. A multistate outbreak of listeriosis in 2000 was identified only after the isolates of *Listeria monocytogenes* were tested by using pulsed-field gel electrophoresis and determined to all have a common PulseNet pattern (CDC, 2000).

FDA laboratories focus primarily on monitoring the food supply and ensuring the purity and potency of drugs and other pharmaceuticals. These regulatory laboratories frequently become involved in the investigation of food contamination (including ground beef, poultry) and the adulteration of drugs. FDA maintains regional laboratories in Washington, New York, Colorado, Michigan, Kansas, California, Georgia, and Arkansas, as well as specialized laboratories in Ohio, Pennsylvania, Puerto Rico, and Massachusetts. The FDA laboratories provide testing to support investigation and compliance activities. FDA's Electronic Laboratory Exchange Network (eLEXNET) is a Web-based system for real-time sharing of laboratory data derived from foods. This system allows public health officials to compare laboratory findings and to identify outbreaks earlier.

The USDA operates laboratories that support many of their regulatory, monitoring, and investigative programs. USDA laboratories conduct research on animal and plant diseases and provide testing of animals and agricultural products. The USDA's National Veterinary Services Laboratories (NVSL) located in Ames, Iowa, tests for domestic and foreign animal diseases, and function as the primary animal disease reference laboratory. The NVSL provides diagnostic support for disease control and eradication programs, import and export testing of animals, and laboratory certification for selected animal diseases. Diseases, such as anthrax, rabies, brucellosis, and bovine spongiform encephalopathy (mad cow disease), may impact both animals and humans and, therefore, constitute priorities at this laboratory. A former USDA laboratory, the Foreign Animal Disease Diagnostic Laboratory (FADDL), which is located on Plum Island off Long Island, New York, was recently transferred to the Department of Homeland Security after supporting years of research on some of the most dangerous animal pathogens.

The DOE oversees the operation of 25 DOE national laboratories, many of which were established to support the production, use, and response to nuclear materials. After the end of the Cold War, the focus of some of the DOE laboratories shifted to other projects, including the Human Genome Project and the development of technologies and assays to support homeland security initiatives. The DOE national laboratories develop new technologies for countering biologic and chemical threats, including systems for the detection, modeling, and response to terrorist attacks (see *www-ed.fnal.gov/doe/doc_labs.html*).

The DoD has established laboratories worldwide in locations such as Peru, Indonesia, Egypt, Thailand, and Kenya. DoD laboratories serve the needs of the armed forces and function as screening or sentinel laboratories for infectious diseases. The U.S. Army Medical Research Institute of Infectious Diseases (USAMRIID) has the capability to detect unusual biological agents that often require advanced testing techniques. This laboratory, located in Maryland, is a member of the Laboratory Response Network (LRN; discussed later) and one of a few laboratories worldwide that can isolate and identify the most dangerous human agents, such as Ebola, smallpox and Marburg viruses.

The EPA operates 10 regional environmental testing laboratories across the United States. These laboratories have a research and environmental monitoring mission: they analyze air, drinking water, ground water, surface water, soil, sediment, and hazardous materials for biological, chemical, and radiological materials that are toxic to the environment. EPA develops standard methods for the analysis of environmental samples. EPA maintains large databases of environmental monitoring data produced by its own laboratories and others. Its Office of Research and Development (ORD) directs laboratory activities at 12 locations, including the National Center for Environmental Assessment in Research Triangle Park, North Carolina.

Although the capability and capacity of the federal laboratories described above is large, this capacity was challenged by the volume of environmental and clinical samples generated during the 2001 anthrax postal attack. The distribution of anthrax spores in mail during October 2001 led to an unprecedented demand for quality testing throughout the United States owing to discovery of real and suspected contaminations. Although few of the more than 125,000 environmental samples tested contained *B. anthracis,* the existing network of public and commercial laboratories was barely able to meet the demands for testing, and there were significant delays caused by the sheer volume of samples. The concept of a high-throughput laboratory capable of testing thousands of biologic, chemical, or radiological samples would require the laboratory to be equipped with the latest automated instrumentation and supported with an efficient LIMS (Layne and Beugelsdijk, 2003). The establishment of high-throughput laboratories to support the nation's homeland security needs is a reasonable concept, especially if the major federal partners were to colocate resources on a national interagency homeland security laboratory campus.

5.2. State Laboratories

Approximately 200 of the more than 186,000 laboratories in the United States are classified as state public health laboratories. Included in this number are about 150 regional or branch laboratories that are administered as part of the state public health laboratory. More than 6,500 laboratory professionals are employed by state public health laboratories.

Each state and five territories operate a state public health laboratory. One major function of theses laboratories is to provide diagnostic and analytical services for surveillance of infectious, communicable, genetic, and chronic diseases. State (and local) public health laboratories provide testing to support many public health programs (tuberculosis, sexually transmitted diseases, HIV, immunizations, and newborn screening). Areas of analysis include clinical and environmental chemistry, immunology, pathogenic microbiology, virology, and parasitology. State laboratories serve as reference laboratories, and they provide confirmatory testing of specimens submitted from other laboratories. The state public health laboratories frequently measure toxicants in human samples to document exposures to chemicals found in the diet or environment. This specialized testing requires expensive equipment (mass spectrometers, chromatographs) and well-trained chemists.

The capabilities, responsibilities, and practices of the state public health laboratories vary. During recent years, many of these laboratories have received substantial federal funding to increase staff, equipment, and capabilities to respond to biologic and chemical threats, as part of the DHHS' Cooperative Agreement on Public Health Preparedness and Response for Bioterrorism. A large percentage of the state public health laboratories have used some of the federal funding to build, remodel, and upgrade facilities. Funding for state laboratories is generally a mix of state and federal funds. Many states rely on fees, reimbursements, and service contracts to carry out their mission. Some states have regional or district laboratories to provide the necessary services throughout an entire state. State public health laboratories partner with public health laboratories operated by counties or cities to meet the needs of their communities. State public health laboratories are generally better prepared to respond to an incident requiring biologic testing as opposed to an incident requiring chemical testing.

Although research is not the primary mission of public health laboratories, several state public health laboratories have developed close ties with academic institutions. The opportunity to work on surveillance projects with faculty and students from schools of public health and academic training programs for laboratory professionals has been beneficial to these laboratories. Arrangements with academic centers afford opportunities to improve laboratory services and surveillance systems. Flexible funding and other resources, such as grants, faculty, and students, help support special research projects and training opportunities within public health laboratories.

State laboratories provide services to health departments; healthcare organizations; local, state, and federal law enforcement; local hazardous materials (hazmat) teams; civil support teams; and other private and governmental laboratories. State laboratories analyze thousands of water and air samples daily. State laboratories involved in the analysis of drinking water and other environmental samples are accredited by the NELAP, certified by EPA Office of Drinking Water Programs, or accredited by state-specific accreditation programs.

State public health laboratories are the backbone of the Laboratory Response Network (LRN), which we discuss below. The LRN also includes laboratories under the jurisdiction of federal agencies discussed in this chapter. At the time of publication, 96 state and local public health laboratories make up the 146 laboratories that are members of the LRN. Of these, 72 state, territorial, and metropolitan public health laboratories are part of the LRN's chemical testing network.

5.3. Local Public Health Laboratories

More than 1,900 local public health laboratories provide support for city and county public health programs. These laboratories offer onsite testing for child health screening, tuberculosis, refugee screening, food safety, sexually transmitted diseases, and lead poisoning prevention, as well as epidemiologic investigations and environmental monitoring. Local public health laboratories often forward specimens to their state laboratories for tests that are not available locally. Approximately 40% of the local public health laboratories only perform waived tests. Another 40% perform moderate complexity testing, and 20% perform highly complex testing (Association of Public Health Laboratories, 2004).

5.4. Other State and Local Laboratories

Many state and locally operated laboratories in addition to clinical and public health laboratories produce test results that may be useful for surveillance. Forensic laboratories and toxicology laboratories are frequently associated with departments of public safety or with state or local medical examiners. These forensic laboratories may be positioned at the federal, state, or local level, depending on the jurisdiction. Medical examiners contribute to biosurveillance by elucidating unusual causes of death (see Chapter 11). Medical examiners and forensic pathologists frequently rely on laboratories to establish the exact cause of death.

6. SERVICES PROVIDED BY LABORATORIES

Readers may understand laboratories as only providing test results after the analysis of samples that are submitted.

This view of laboratories only focuses on one of their products and overlooks services provided by the professionals who work in the laboratories. Laboratory professionals provide consultation on the selection of the most useful tests for screening, diagnosis, and verification of disease. They provide information on the proper handling and transportation of samples. Development of new technologies, evaluation of technologies, and the application of technologies are other important services provided by laboratory professionals. Laboratories evaluate test kits for ease of use, storage conditions, and performance characteristics. Laboratories train individuals to perform testing and educate and inform users of laboratory services.

Laboratories may also operate laboratory-based surveillance systems for respiratory and enteric diseases. For example, several states—such as Wisconsin (*www.slh.wisc.edu/labupdates.description.php*), Minnesota (*www.health.state.mn.us/divs/idepa/diseases/flu/avain/surveillance.html*), and Nebraska (*www.hhs.state.ne.us/new/o2o5nr/flu1.htm*)—and the United Kingdom Health Protection Agency (*www.phls.co.uk/infections*) have developed laboratory-based surveillance systems that collect data on respiratory infections and make the findings available on a their Web sites daily or weekly.

7. TESTING TECHNOLOGIES

Laboratories employ many analytic technologies to provide testing for diagnosis and surveillance. These tests range from simple screening tests to presumptive diagnostic tests and confirmatory tests.

7.1. Simple Screening Tests

Simple laboratory tests are used to screen biologic samples for the presence of biological agents or other substances that might indicate a disease, an infection with a biologic agent, or a contamination with a toxin or other chemical. The classical approach to identification of biologic agents involves direct examination of stained materials by microscopic examination for the presence of agents (DHHS/CDC, n.d.; York, 2003). Microscopic examination relies on staining characteristics and the size and shape of the organisms found. With few exceptions, it is rarely possible to identify an agent based only on microscopic characteristics. Direct stains may help eliminate organisms from further consideration, but they are usually not sufficient without further testing to identify a pathogen. Wet mounts are used for the direct microscopic examination of clinical materials for fungi and parasitic agents. Additional simple assays take advantage of the ability of an organism to metabolize chemicals or produce chemical reactions that result in color changes of a substrate. Recently, many assay kits have been developed for waived tests. These waived test kits can produce reliable results in most cases; however, one must understand the limitations of these kits and the need for confirmatory testing. Field test kits, often referred to as handheld assays, have become popular with first responders who have a need to know if an unknown material contains a biologic agent or toxic chemical. Simple immunologic reactions, coupled with a colorimetric indicator, form the detection systems for many handheld assays. Validating the performance of handheld assays in comparative studies is a task generally reserved for governmental agencies or contract laboratories (Emanuel et al., 2003). The major advantage of the simple assays is that they are quick, as a result may be available in 5 to 20 minutes. The short turnaround time makes these tests ideal for use by surveillance systems provided one understands the limitations of the tests and that a method for confirmatory testing is available.

7.2. Presumptive Diagnostic Tests

Presumptive diagnostic tests are procedures that, when properly performed, may indicate the presence of a particular agent or closely related agents. Presumptive diagnostic tests for biologic and chemical agents are more complex and require more time to perform than do the simple tests described above. Most bacterial and viral agents can readily be grown in culture media or cell culture if the appropriate conditions are met. These conditions include the appropriate temperature, pH, and a source of energy (e.g., glucose). Inhibitory substances, such as antibiotics, are often placed into the growth media to prevent the growth of unwanted organisms. Growth of organisms in cell culture or artificial media takes 24 hours to 30 or more days, depending on the organism.

Once organisms are detected on or in growth media, various techniques are used to determine the genus and species of the organism. For bacteria, the differential growth on select media, metabolism of various carbohydrates and other chemicals, and the presence of selected enzymes are often used for identification. Special stains that incorporate florescent dyes coupled with antibodies to selected organisms are used for identification of bacteria and viruses.

Laboratories use immunologic assays to identify many biologic agents and for the subspecies typing that is needed for epidemiological purposes. Immunologic assays generally involve the use of a specific antibody that can attach to or react with the outer-surface structures of an agent. Other immunologic assays are used to detect antibodies in biological materials after infection with biological agents. After the isolation and identification of many organisms, drug-susceptibility testing determines the organism's susceptibility to drugs that are used to treat infections with the organism.

7.3. Definitive and Confirmatory Tests

A definitive or confirmatory test is a test that will identify with a very high degree of certainty the true identity of an agent. These tests have a very low likelihood of providing a false-positive result. Many of the definitive and confirmatory assays used today are molecular assays, which detect genetic material that is specific to a bacterium, virus, protozoa, or other organism. Nucleic-acid–based assays rely on the unique differences found in the structure of single strands of DNA and RNA. The unique pattern of bases is specific for a single organism or closely related organisms. The nucleic-acid–based assays involve the use of probes, which are strands of DNA or RNA that match distinctive DNA or RNA patterns of the organism being tested and will bind with that DNA or RNA if it is in the clinical specimen. Once the binding occurs, the binding can be detected by using electrochemical, colorimetric, and optical systems (Committee on Research and Development Needs for Improving Civilian Medical Response to Chemical and Biological Terrorism Incidents, 1999). This binding provides extreme selectivity between the known probes and the material found in clinical specimens. The use of a technique known as polymerase chain reaction (PCR) makes it is possible to amplify trace quantities of DNA or RNA in a clinical specimen to enable the detection of as few as 1,000 bacteria or viruses. The high specificity and sensitivity of molecular assays makes them especially suited for detecting minute quantities of biologic agents and toxins.

The use of genetic fingerprints has become a valuable tool for microbial forensics (Murch, 2003). By identifying distinct features of genes using sequencing techniques, it is possible to identify individual strains of organisms and to use this information as epidemiological markers. Molecular techniques allow investigators to link strains from various sources and to form associations that often unravel the mystery of disease outbreaks. Libraries of genetic patterns, or fingerprints, make it possible to trace the origin of many outbreaks. Not only does sequencing identify DNA or RNA patterns unique to a particular organism, but, in many cases, these probe technologies are simpler, faster, and less technology-dependent than are other traditional assays.

Many biotechnology companies are pursuing production of microarray systems that will test for 100 to 100,000 or more different DNA fragments simultaneously. The technology embeds the probes on a single glass or nylon substrate called a microchip. By using these arrays and various detection systems onto which the clinical specimen is applied, the individual components of the microchip will react with DNA fragments in the specimen and be detected. As this technology develops, it will become more common to use the microarray systems for screening and detection of diseases. Simultaneous PCR assays for multiple respiratory viruses now have sufficient sensitivity and specificity to be a valuable tool for diagnosis of respiratory viral infections (Hindiyeh et al., 2001). In the future, PCR assays will be able to screen a single specimen for a multitude of biologic agents.

8. LABORATORY INFORMATION MANAGEMENT SYSTEMS

A LIMS is a computer system that a laboratory uses to track specimens and manage analytical data. The function of a LIMS can perhaps best be explained by tracing a single test—a white blood count—from the time that a clinician orders the test until the time the clinician receives the result of the test.

8.1. Tracking a White Blood Count from Order to Result

A clinician requests (orders) that a laboratory perform a white blood count on a blood sample from a patient in one of several ways. The clinician writes the order on paper, gives a verbal order to a nurse, or enters the order directly into a computer system. A LIMS may receive this order in one of several ways: directly, if the LIMS provides an order-entry component that either the clinician uses directly or the nurse or ward clerk uses on her behalf; indirectly, if a paper order form accompanies the specimen; or electronically, from an order-entry system embedded in another information system (such a point-of-care system as discussed in Chapter 6).

We will trace the most automated path in which the LIMS receives the order electronically. The received order sets up a specimen-tracking process that is a central LIMS function. The LIMS (or a point-of-care system) controls a printer, which is often located in the clinical area from which the order originated. The printer generates a bar-coded label. A phlebotomist attaches the label to a "purple top" vial (containing magnesium citrate or ethylenediaminetetraacetic acid [EDTA] to prevent the blood from coagulating). The phlebotomist draws the blood after checking carefully that the identification on the labeled tube matches the patient, and sends the sample to the laboratory. Oftentimes, the labeled specimen is transported by pneumatic chute from clinical areas, such as the emergency department, or by express overnight delivery to the laboratory. A technician in the laboratory scans the barcode of the specimen with an optical scanner, which is connected to the LIMS. If the laboratory has an automated analyzer for blood counts, the technician simply places the vial into the specimen carousel of the analyzer. The analyzer recognizes the bar code, communicates with the LIMS over an internal laboratory network to determine which tests were ordered for the specimen, runs the tests, and transmits the results to the LIMS. Other types of specimens may require preparation, such as centrifugation, before being placed into the carousel of an automatic analyzer, but the information processing and communication between the analyzer and the LIMS are otherwise identical. For tests that are done by hand, the LIMS provides user-interfaces into which medical technologists register specimens and enter intermediate and final results of tests.

Depending on the laboratory and the needs of its clinical customers, the LIMS may deliver the results as paper reports, via Web-browser interfaces, or by e-mail. A LIMS typically offers all of these options. A LIMS invariably has an outbound computer-to-computer interface that can transmit results to other clinical information and public health information systems. Although our example is of a clinical test, the process is identical for environmental tests ordered by sanitarians, water quality technicians, or outbreak investigators.

The vast majority of laboratory work in the United States is highly automated in the fashion just described. With the exception of tests done in the field or in office practices, LIMSs track and manage the analytic results of most laboratory tests performed in the United States. LIMSs are designed to make test results available to clinicians and other information systems soon after they are performed. LIMSs are highly reliable and operate in real time. LIMSs are developed within the laboratory by information technology staff, purchased from commercial vendors, or a combination of both.

8.2. Use of LIMS in Biosurveillance

Because of the importance of laboratory tests in biosurveillance, biosurveillance organizations are attempting to establish connections between LIMSs and their own biosurveillance computers. At present, however, there are significant technical barriers to connecting a LIMS located in a laboratory with a computer located in a biosurveillance organization. The most difficult technical barrier is data incompatibility. Most laboratories do not use standard coding systems to identify the names and results of laboratory tests. Data standards exist, but few laboratories use them. Most laboratories have evolved their own naming or coding systems for the tests that they perform and the results of those tests. They use these proprietary codes (or free text) to identify the laboratory test, specimen type, organism, or other results of a test. As a result, each LIMS-to-biosurveillance–computer interface requires significant effort to understand the data and to create means to translate the data into a standard format so that it can be integrated with data coming from other LIMS. We discuss standard data formats in detail in Chapter 32.

Outbound communication standards, such as HL7, also exist and we discuss them in Chapter 32. Although most LIMS support these standards, the specific implementations vary. Even if both the biosurveillance computer and the LIMS use the HL7 standard, they will not be able to communicate without significant effort to understand the specifics of the messages and to create a means for extracting the data from the messages. Importantly, a new version of HL7 (version 3.0) will solve this problem, but its penetration in the LIMS market is low at present.

Because of the customization required to connect even a single LIMS to a biosurveillance organization, there are only a few regions that have integrated data from most of the LIMSs that serve their region into biosurveillance. Ultimately, these barriers will be addressed by standardization, but as we discuss in Chapter 32, it takes many years to achieve widespread standardization of any component in a biosurveillance system.

9. NETWORKS OF LABORATORIES

The organization of laboratories into collaborative networks has been ongoing for more than a decade. The concept of laboratory networks arose from the need to ensure that critical laboratory services were available throughout the country (Gilchrist, 2000). The CDC, FDA, USDA, and state governmental laboratories formed many of the original partnerships that grew into the laboratory networks that exist today.

Early networks included the National Laboratory System (NLS) of clinical, public health and federal laboratories (McDade and Hughes, 1998), and the Public Health Laboratory Information System (PHLIS). PHLIS was an early DOS-based system that involved voluntary reporting of selected laboratory tests directly to the CDC by 23 state and local public health laboratories and numerous military laboratories (see Chapter 3). PHLIS was one of the earliest laboratory-based surveillance systems that became an effective tool for the identification of outbreaks of salmonellosis (Bean et al., 1992; Hutwagner et al., 1997).

9.1. The Laboratory Response Network

The CDC's LRN was established in 1999 in compliance with a presidential directive that outlined federal agencies' counterterrorism goals and responsibilities. The mission of the LRN is "to maintain an integrated national and international network of laboratories that can respond quickly to acts of chemical or biological terrorism, emerging infectious diseases and other public health threats and emergencies." The LRN was first tasked to address state and local public health laboratory preparedness and response for bioterrorism. Since its inception, its mission has expanded to include chemical terrorism. The scope of laboratories in the LRN has expanded beyond state and local public health laboratories in order to meet national security needs (*http://www.bt.cdc.gov/lrn/*).

The LRN and its partners maintain a network of laboratories that can respond quickly to acts of chemical or biological terrorism, emerging infectious diseases, or other public health threats and emergencies. The network included 152 federal, state, and local public health; military; and international laboratories as of August 1, 2005. Figure 8.1 shows the distribution of LRN laboratories by type as of March 11, 2005. The LRN employs tests that can detect biological threat agents in clinical specimens, environmental samples, food, animals, and water. The Association of Public Health Laboratories was a partner in the early development of the LRN and has played an important role in ensuring that the LRN provides the training, standardized methods, and equipment necessary for detecting biologic agents of terrorism and other threats to public health. The Federal Bureau of Investigation (FBI) was also a key partner in establishing the network. The FBI brought its forensic expertise and evidence-gathering requirements to the program. Public health and law enforcement have overlapping approaches to their investigations that required collaboration between CDC and law enforcement to both enhance and protect the integrity of their investigations.

Three levels of laboratories are recognized within the LRN for bioterrorism agent detection. LRN laboratories responsible

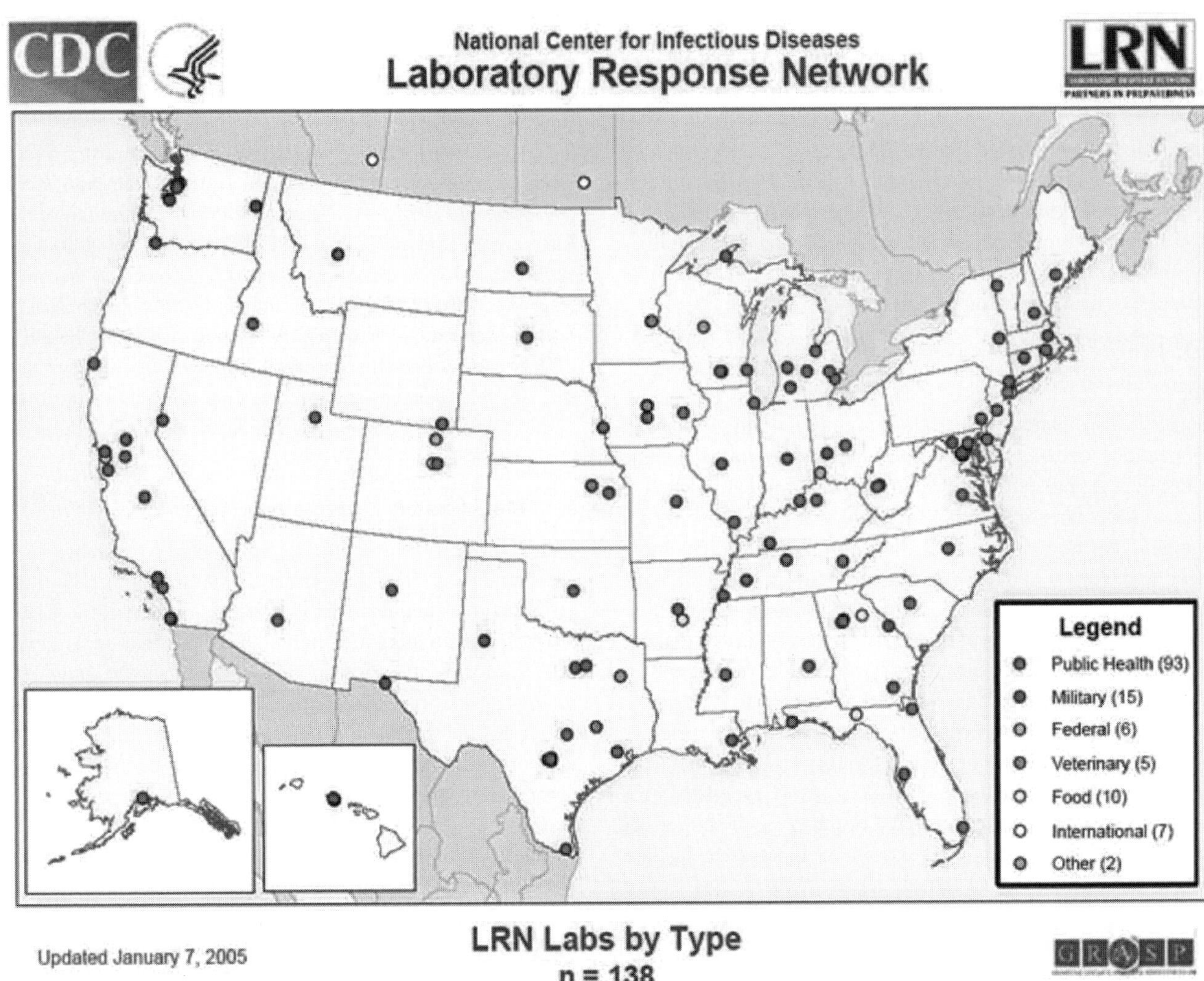

FIGURE 8.1 Map of Coverage of the Laboratory Response Network (LRN) in the United States, 2005.

for bioterrorism agent detection are designated as national, reference, or sentinel laboratories (Figure 8.2). The national laboratories include laboratories at the CDC, USAMRIID, and the Naval Medical Research Center. They have unique resources to safely identify highly infectious agents (biological safety level 4 agents) and the ability to provide definitive characterization of biologic agents. The national laboratories have developed standard tests and protocols, trained laboratory analysts, and established secure communications for the rapid sharing of laboratory results from reference laboratories.

Reference laboratories in the LRN perform tests to detect and confirm the presence of biological threat agents, such as those that cause anthrax, plague, tularemia, smallpox, or botulism. These laboratories ensure that state and local laboratories can have a timely response in the event of a bioterrorism incident or an emerging disease. Rather than having to rely on confirmation testing from national laboratories, reference laboratories are capable of producing conclusive results needed by local authorities for rapidly responding to emergencies. Specimens received by reference laboratories that contain threat agents are usually submitted to a national laboratory for definitive characterizations of the agent. State and local public health laboratories comprise most of the reference laboratories within the LRN.

Although sentinel laboratories are not counted among the LRN laboratories, they represent the thousands of clinical (human and veterinary) and environmental laboratories identified by the state LRN reference laboratory to serve as the front line of defense in detecting agents of terrorism. Qualification of the sentinel laboratory is based on the experience and competency of the laboratory staff, appropriateness of facilities, and completion of training provided by the LRN. Sentinel laboratories can often rule out potential bioterrorism agents based on a battery of simple tests. In a covert terrorist attack, a sentinel laboratory could be the first facility to identify a suspicious agent or an unusual cluster

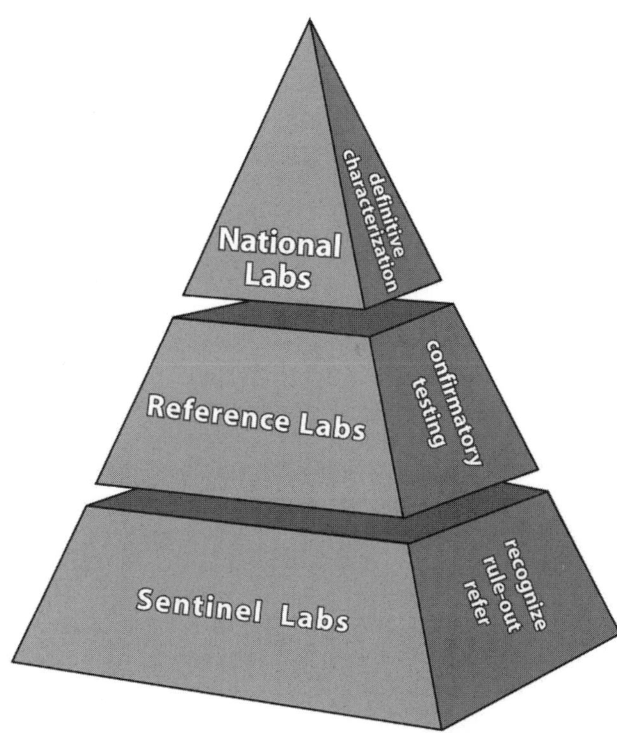

FIGURE 8.2 Laboratory Response Network (LRN) structure for bioterrorism response laboratories.

of diseases. A sentinel laboratory's responsibility is to rule out other diseases and refer a suspicious sample to an appropriate reference laboratory.

The LRN uses an array of presumptive and confirmatory assays to detect biological threat agents or chemical agents. Presumptive assays generally include traditional microbiological assays, such as growth on special media, use of special stains, and the use of rapid molecular diagnostic assays, such as real-time PCR. Confirmatory methods used by the national laboratories are considered the gold standard for detecting the target agent. Methods, such as time-resolved fluorescence (TRF) and molecular characterization, have replaced many of the more traditional confirmatory methods that were based on the growth and biochemical properties of the agent. Reference laboratories have access to LRN protocols through a secure Web site and are supplied standardized reagents needed to perform the necessary tests. Uniform procedures for use by sentinel laboratories have also been developed and are used for training of sentinel laboratory staff. All laboratories that are part of the LRN must provide a safe and secure environment for performing tests and must participate in a recognized proficiency testing program.

The chemical component of the LRN employs a more centralized structure, with only a few laboratories currently prepared to provide definitive analysis of specimens for chemical agents or their metabolites. LRN laboratories responsible for chemical agent detection are designated as level 1, 2, or 3 laboratories. Five laboratories located in California, Michigan, New Mexico, New York, and Virginia participate in level 1 activities. At this level, personnel are trained to detect exposure to an expanded number of chemicals in human blood or urine, including all level 2 laboratory analyses, plus analyses for mustard agents, nerve agents, and other toxic chemicals. Forty-one laboratories participate in the chemical LRN by providing level 2 activities. At this level, laboratory personnel are trained to detect exposure to a limited number of toxic chemical agents in human blood or urine. Analysis of cyanide and toxic metals in human samples are examples of level 2 laboratory activities. Each of the 62 chemical network members participates in level 3 activities. Level 3 laboratories are responsible for the following:

- Working with hospitals in their jurisdiction
- Knowing how to properly collect and ship clinical specimens
- Ensuring that specimens used as evidence in a criminal investigation are handled properly and chain-of-custody procedures are followed
- Being familiar with chemical agents and their health effects
- Training on anticipated clinical sample flow and shipping regulations
- Working to develop a coordinated response plan for their respective state and jurisdiction

Initial testing in a suspected chemical event will occur at CDC or 1 of 5 level 1 chemical laboratories that have been established by CDC. By use of mass spectrometry, CDC laboratories perform tests on the first 40 or more clinical specimens to measure human exposure. Results of these tests would be reported to affected states, and if needed, appropriate LRN members may be asked to test additional samples. This approach is necessary because the analytical expertise and technology resources required to respond to a chemical event is expected to be high.

The LRN supports secure communications on emerging and emergency issues, a secure mechanism for ordering reagents and testing protocols, and a system for electronically reporting test results. The LRN provided valuable testing after the release of anthrax spores in 2001, the identification of monkey pox in 2002, and the response to severe acute respiratory syndrome (SARS) in 2003.

9.2. Other Laboratory Networks

In 2001, Canada created a laboratory network, known as the Canadian Public Health Laboratory Network (CPHLN), to strengthen the linkages between federal and provincial public health laboratories. This Canadian network is modeled after the LRN in the United States. The CPHLN has been providing responses to naturally occurring infections and deliberate releases of biologic agents and toxins. The CPHLN coordinates pathogen detection and infectious disease prevention activities, as well as conducts laboratory-based surveillance and early warning systems for emerging pathogens and bioterrorism threats.

The National Animal Health Laboratory Network (NAHLN) is part of a national strategy in the United States to coordinate the nation's federal, state, and academic animal health laboratories. The USDA has taken the lead in the development of this network, which includes agriculture and animal health laboratories operated by state agricultural agencies and those associated with veterinary teaching facilities. The facilities and professional expertise of NAHLN members allows authorities to better respond to animal health emergencies that might include a bioterrorist event, the emergence of a new domestic animal disease, or the appearance of a foreign animal disease that could threaten the nation's food supply and public health. Because many of the biologic agents that cause the greatest concern as terrorist agents infect both humans and animals, the role of veterinarians in the early detection of disease is very important. The NAHLN currently consists of 44 laboratories in 37 states. An effort is underway to deploy standardized testing methods in member laboratories and to improve the information technology system used by member laboratories to track test requests and report results. The U.S. Animal Health Association (USAHA) and the American Association of Veterinary Laboratory Diagnosticians (AAVLD) have members who participate in the NAHLH and contribute expertise to protect animals and public health.

The National Plant Diagnostic Network (NPDN) is an agricultural laboratory network that provides detection, identification, and reporting of pests and pathogens that have been deliberately or accidentally introduced into agricultural systems. The primary concern of this network is food security and the economic threats to the nation's food supplies. The NPDN includes five regional centers located at Cornell University, Michigan State, Kansas State, University of Florida at Gainesville, and the University of California at Davis. The NPDN recently implemented a new database that helps with required reporting to state and national agencies. A Web-based plant diagnostic system using digital photography allows laboratories to share images with specialists in remote locations.

The Food Emergency Response Network (FERN) is a network of state and federal laboratories that are committed to analyzing food samples in the event of a biological, chemical, or radiological terrorist attack in the United States. The federal partners in the FERN include the FDA, USDA, CDC, and EPA. As of May 2005, there are 99 laboratories in FERN, representing 44 states and Puerto Rico. Twenty-six federal laboratories, 68 state laboratories, and five local laboratories are enrolled in FERN. The mission of FERN is to integrate the nation's food testing laboratories for the detection of threat agents in food while using standardized diagnostic protocols and procedures. Of the 99 laboratories in FERN, there are 64 that perform chemical tests on food, 64 that perform biological tests on food, and 25 that perform radiological tests on food. These laboratories strengthen preparedness and provide surge capacity. The FDA and USDA jointly share the leadership within FERN and have been working to obtain federal funds that can be made available to support further development of the network.

The data capture and information exchange system for FERN is the eLEXNET, is an integrated, secure system designed for use by multiple governmental agencies involved in food safety activities. Laboratories report test results and public health officials assess risks and analyze trends in foodborne diseases. eLEXNET has GIS reporting functions and uses HL7 data exchanges between laboratories. Similar to those in LRN, participants in FERN receive training on the latest equipment and are required to participate in proficiency testing programs. eLEXNET provides the necessary infrastructure for an early warning system that can identify potentially hazardous foods and share laboratory reports in a timely manner. As of January 2005, 113 federal, state, and local laboratories in all 50 states have joined the eLEXNET system. About 90 laboratories actively exchange data by using eLEXNET.

The Radiological Emergency Analytical Laboratory Network (REALnet) is a national network of radiological laboratories that are capable of responding to the needs for radiological testing after a terrorist attack. Academic, commercial, military, federal, state, and local laboratories participate in REALnet. These laboratories serve as a science and technology asset for the Department of Homeland Security. REALnet is modeled after the LRN and FERN and includes a Web-based database containing information on the capabilities, capacity, and competence of member laboratories. Information on accreditation, certification, and performance testing is also maintained. Standards, guidelines, and laboratory procedures are developed and distributed by REALnet. Gaps in the standards are being addressed by appropriate standards development organizations. Internet-based tools, such as bulletin boards and list servers, are used to promote the exchange of information. The system provides links to other resources that would be useful during an emergency caused by the release of a radioactive material.

The expansion of laboratory networks designed to produce test results in response to an act of terrorism or other public health emergency has led to increased sharing of laboratory results. Coordination and integration of the various networks has not always been a priority. Duplicate systems and overlapping missions suggest that an integrated consortium of laboratory networks could provide timely results for early detection and response to acts of terrorism. The networks need to agree on standardized tests and policies that would promote a timely response no matter which network is reporting results. An overall system to ensure that laboratory capacity will be available to test clinical (human and animal) specimens and environmental samples, including food and water, does not exist.

Laboratories that can perform screening, monitoring, and definitive testing are needed for each class of specimen. Although laboratory information systems have been developed for capturing and sharing data, the diverse systems are not fully compatible with each other. Failure to standardize nomenclatures and use recognized data transmission protocols make sharing of large quantities of data for surveillance purposes problematic.

In an effort to improve coordination among laboratory networks, the Department of Homeland Security is working to integrate these networks through the creation of the Integrated Consortium of Laboratory Networks (ICLN). The ICLN employs the LRN model for coordinating laboratory assets for terrorism among federal, state, local, and scientific partners. A memorandum of agreement was signed in 2005 by the USDA, DoD, DOE, DHHS, EPA, and the Departments of Homeland Security, State, Justice, and Interior to form the ICLN. These federal agencies will collaborate to ensure that laboratory resources available within each agency can be used to respond to terrorist events and other national emergencies.

10. SUMMARY

Clinical, commercial, and governmental laboratories are an important source of data for biosurveillance systems. Test results obtained from the analysis of human and animal specimens may be early indicators of disease within the population. Before the establishment of a definitive diagnosis, a sudden rise in the number of tests being requested by clinicians might be the first indication of an outbreak. The combination of laboratory data from environmental laboratories and the occurrence of illness in humans or animals might signal onset of an infectious disease or poisoning. The establishment of a LIMS that can rapidly capture and report laboratory data electronically will greatly contribute to the use of laboratory data in biosurveillance. Challenges still exist for integrating 190,000 laboratories into a real-time network to support biosurveillance. The multitude of laboratory networks that are being formed will enhance the analytical capabilities of laboratories as well as the ability of laboratories to distribute peak loads during emergencies and quickly and efficiently share data electronically. Enormous potential exists for the use of data that that is locked up in systems by the lack of standardization. Although much recent emphasis has been focused on building laboratory networks in preparation for a biological, chemical, or nuclear attack, many of the enhanced laboratory capabilities will assist with the response to other public health emergencies.

ADDITIONAL RESOURCES

Web Sites

American Society of Clinical Pathology (ASCP), *http://www.ascp.org.*

American Society for Clinical Laboratory Science (ASCLS), *http://www.ascls.org.*

American Society for Microbiology (ASM), *http://www.asm.org.*

Association of Public Health Laboratories, *http://www.aphl.org.*

Clinical Laboratory Management Association (CLMA), *http://www.clma.org.*

Clinical Laboratory Improvement Act (CLIA), *http://www.cms.hhs.gov/clia.*

American Association of Veterinary Laboratory Diagnosticians (AAVLD), *http//www.aavld.org.*

U.S. Animal Health Association (USAHA), *http://www.usaha.org.*

Laboratory Response Network (LRN), *http://www.bt.cdc.gov/lrn/.*

National Environmental Laboratory Accreditation Program (NELAP), *http://www.epa.gov/nerlesd1/land-sci/nelac/.*

National Animal Health Laboratory Network (NAHLN), *http://www.nahln.us.*

National Plant Diagnostic Network (NPDN), *http://www.npdn.org.*

Canadian Public Health Laboratory Network (CPHLN), *http://www.cphln.ca.*

National Laboratory System and Database (NLS), *http://www/phppo.cdc.gov/mlp/nls.aspx.*

FoodNet Surveillance, *http://www.cdc.gov/foodnet.*

U.S. EPA Laboratories, *http://www/epa/gov/OSP/tribes/sciinf/labs.htm.*

Recommended Further Reading

Cowan, D.F., ed. (2002). *Informatics for the Clinical Laboratory: A Practical Guide.* New York: Springer-Verlag.

Hinton, M. (1994). *Laboratory Information Management Systems: Development, and Implementation for a Quality Assurance Laboratory.* New York: Marcel Dekker.

Paszko, C. and Turner, E. (2001). *Laboratory Information Management Systems,* 2nd ed. New York: Marcel Dekker.

REFERENCES

Association of Public Health Laboratories. (2004). Assessing America's Local Public Health Laboratory Capacity, December 2004. *http://www.aphl.org.*

Bean, N.H., Martin, S.M., and Bradford Jr., H. (1992). PHLIS: An Electronic System for Reporting Public Health Data from Remote Sites. *American Journal of Public Health* vol. 82:1273–1276. *http://www.ncbi.nlm.nih.gov/entrez/query.fcgi?cmd=Retrieve&db=PubMed&dopt=Citation&list_uids=1323935.*

Centers for Disease Control and Prevention [CDC]. (2000). Multistate Outbreak of Listeriosis, United States, 2000. *Morbidity and Mortality Weekly Report* vol. 49:1129–1130. *http://www.ncbi.nlm.nih.gov/entrez/query.fcgi?cmd=Retrieve&db=PubMed&dopt=Citation&list_uids=11190115.*

CDC. (1999). *Biosafety in Microbiological and Biomedical Laboratories,* 4th ed. Bethesda, MD: National Institutes of Health. *http://www.cdc.gov/od/ohs.*

Centers for Medicare and Medicaid Services. (2004). CLIA Data Base. *http://www.cms.hhs.gov/clia/.*

Committee on Research and Development Needs for Improving Civilian Medical Response to Chemical and Biological

Terrorism Incidents. (1999). Detection and Measurement of Biological Agents. In *Chemical and Biological Terrorism—Research and Development to Improve Civilian Medical Response.* Washington, DC: National Academy Press.

Department of Health and Human Services [DHHS]/CDC. Bioterrorism Response Guide for Clinical Laboratories. *http://www.bt.cdc.gov/labissues/.*

Emanuel, P.A., Chue, C., Kerr, L., and Cullin, D. (2003). Validating the Performance of Biological Detection Equipment: The Role of the Federal Government. *Biosecurity and Bioterrorism: Biodefense Strategy, Practice, and Science* vol. 1:131–137. *http://www.ncbi.nlm.nih.gov/entrez/query.fcgi?cmd=Retrieve&db=PubMed&dopt=Citation&list_uids=15040191.*

Gilchrist, M.J. (2000). A National Laboratory Network for Bioterrorism: Evolution from a Prototype Network of Laboratories Performing Routine Surveillance. *Military Medicine* vol. 165(Suppl 2):28–31. *http://www.ncbi.nlm.nih.gov/entrez/query.fcgi?cmd=Retrieve&db=PubMed&dopt=Citation&list_uids=10920634.*

Hindiyeh, M., Hillyard, D.R., and Carroll, K.C. (2001). Evaluation of the Prodesse Hexaplex Multiplex PCR Assay for Direct Detection of Seven Respiratory Viruses in Clinical Specimens. *American Journal of Clinical Pathology* vol. 116:218–224. *http://www.ncbi.nlm.nih.gov/entrez/query.fcgi?cmd=Retrieve&db=PubMed&dopt=Citation&list_uids=11488068.*

Hutwagner, L.C., Maloney, E.K., Bean, N.H., Slutsker, L., and Martin, S.M. (1997). Using Laboratory-based Surveillance Data for Prevention: An Algorithm for Detecting Salmonella Outbreaks. *Emerging Infectious Diseases* vol. 3:395–400. *http://www.ncbi.nlm.nih.gov/entrez/query.fcgi?cmd=Retrieve&db=PubMed&dopt=Citation&list_uids=9284390.*

Layne, S.P. and Beugelsdijk, T.J. (2003). High-Throughput Laboratories for Homeland and National Security. *Biosecurity and Bioterrorism: Biodefense Strategy, Practice, and Science* vol. 1:123–130. *http://www.ncbi.nlm.nih.gov/entrez/query.fcgi?cmd=Retrieve&db=PubMed&dopt=Citation&list_uids=15040190.*

McDade, J. and Hughes, J. (1998). The U.S. Needs a National Laboratory System. *U.S. Medicine* vol. 34:9.

Murch, R.S. (2003). Microbial Forensics: Building a National Capacity To Investigate Bioterrorism. *Biosecurity and Bioterrorism: Biodefense Strategy, Practice, and Science* vol. 1:117–122. *http://www.ncbi.nlm.nih.gov/entrez/query.fcgi?cmd=Retrieve&db=PubMed&dopt=Citation&list_uids=15040189.*

York, M. (2003). Sentinel Bioterrorism Responders: Are Hospital Labs Ready? *Medical Laboratory Observer* vol. 35:12–17, 19; quiz 22–23. *http://www.ncbi.nlm.nih.gov/entrez/query.fcgi?cmd=Retrieve& db=PubMed&dopt=Citation&list_uids=12942658.*

9

CHAPTER

Water Suppliers

Jason R. E. Shepard, Nick M. Cirino, and Christina Egan
Biodefense Laboratory, Wadsworth Center, New York State Department of Health, Albany, New York

Ron M. Aryel
Del Rey Technologies, Los Angeles, California

1. INTRODUCTION

Biosurveillance of water supplies has been a constant area of focus in terms of environmental, clinical, and public health in the United States. The U.S. population has approximately doubled over the past 50 years, imposing new requirements and increasing demands on the infrastructure, availability, and quality of our water supply (U.S. Census Bureau, 1999; National Resource Council, 2004). As population growth and land development continue to affect the environment, new water safety issues have arisen. For example, increased agricultural development has contributed to water supply contamination from pesticides, fertilizers, and animal waste. These issues, among others, have caused new problems (most notably in the area of exposure to disease-causing microorganisms and their toxins), complicating the ability to provide potable drinking water. To counter these new problems, the tracking of illness has become more refined as communication networks and databases are developed. Even so, the prevalence of illness owing to ingestion of pathogenic microorganisms is a growing problem (Desselberger, 2000; Pedley and Pond, 2003). Further complicating the matter, many biological species are evolving, leading to the development of new pathogenic strains (Colwell, 1996) as well as antiviral-resistant (Mansky and Bernard, 2000; Gallant, 2002) and antibiotic-resistant species (Boggs, 2003; Chopra et al., 2003). The emergence of infectious biological agents is not a new issue, but their increased frequency and their effect on public health constitute growing areas of concern (Desselberger, 2000; Pedley and Pond, 2003).

Safeguarding the water supply means dealing with challenges, including damage associated with natural disasters, accidental environmental pollution, and intentional contamination. Although natural disasters and pollution are problems that have been recognized for years, the threat of bioterrorism to our water supplies is a modern problem. Over the past few years, federal government organizations–including the Federal Bureau of Investigation (FBI), the Department of Homeland Security (DHS), Environmental Protection Agency (EPA), and the Centers for Disease Control and Prevention (CDC)–have issued advisories regarding the risks to our water supply (CDC/EPA, 2003; Meinhardt, 2005). Any compromised water source, particularly one serving a large urban area, could have major public health and economic consequences. Further complicating the issue is the limited funding available for proper preparation, education, and instruction in dealing with the occurrence of water-borne disease (Meinhardt, 2005). Biosurveillance of our water supplies, as a counter-terrorism measure, is a growing necessity and obligation, especially after the U.S. Postal Service anthrax and ricin incidents (CDC, 2001, 2003a,b). Just as the postal network served as a viable means to distribute *Bacillus anthracis* (anthrax) and ricin toxin, so, too, can food and water supplies serve as a conduit for pathogen mass distribution.

There is a general consensus that the extent of water-related illnesses each year is vastly underreported, which is problematic considering that, according to World Health Organization estimates, from 1991 to 2000, healthcare providers and public health authorities could identify the causative agent in fewer than 60% of all drinking water outbreaks (Pedley and Pond, 2003). Over the past few decades, more than 175 species of infectious agents have exhibited an initial occurrence or increased incidence and, as such, are defined as emerging pathogens (not all of these pathogens represent threats to the water supply). Of those 175 emerging pathogens, more than 75% are zoonotic, or those normally associated with animals but contracted by humans. One such zoonotic parasite is linked to the worst outbreak of illness originating from water supplies in documented U.S. history. In 1993, a *Cryptosporidium parvum* outbreak occurred in Milwaukee, Wisconsin, causing more than 400,000 reported cases of illness and more than 100 deaths. The illness cost the community an estimated $96.2 million ($31.7 million in medical costs and $64.6 million in loss of productivity) (Corso et al., 2003). Analysis linked the outbreak to greater than normal parasite counts in the source water, associated with wet weather, as well as ineffective filtration and clarification processes at a municipal water treatment plant.

The New York State Department of Health in 1999 reported an *Escherichia coli* (*E. coli*) O157:H7 outbreak, which resulted in 71 hospitalizations, 14 cases of hemolytic uremic syndrome,

Handbook of Biosurveillance
ISBN 0-12-369378-0

and two deaths (Bopp et al., 2003). More than 750 suspected and confirmed cases of illness were associated with attendance at the Washington County Fair. Pulsed-field gel electrophoresis identified the water distribution system as the main source of the outbreak. The report suggested that the outbreak resulted from contamination of a local well.

During December 1989 and January 1990, extraordinarily cold weather saw two major pipes break and dozens of water meters fail in Cabool, Montana. The utility company failed to disinfect its water mains after the repairs, choosing instead to merely flush the distribution system with finished water. The result was an outbreak of *E. coli* O157H7, resulting in 240 cases of diarrhea and four deaths (EPA, 2002b). We know of other documented water supply contamination incidents, including more than 1000 cases of gastrointestinal illness traced back to the Norwalk virus found in contaminated well water (Lawson et al., 1991).

Any public exposure to a biological agent through the water system could have dire economic and medical repercussions. Substantial economic damage can result from a credible threat alone, whereas the medical consequences owing to an actual exposure would depend not only on the virulence and toxicity of the agent but also on the extent of its dissemination through the water system. The general consensus is that conventional treatment procedures—including clarification (a term that includes coagulation/flocculation, and sedimentation), filtration, and disinfection—would be able to neutralize or remove the great majority of biological threats that could appear, or be added purposefully, to source water. However, certain select agents, such as *B. anthracis* exist in forms that are small (i.e., spores of 1 nm in diameter) and more resistant to disinfectant concentrations typically used by drinking water suppliers (Burrows and Renner, 1999). Even a trace contamination of a viable biological agent with the ability to propagate inside the water distribution system could have devastating effects.

A more pronounced vulnerability to the safety of the water supply is the sheer size of the finished water supply delivery infrastructure, which includes numerous accessible points along the delivery network. These access points, often including thousands of service lines and hydrants, make it virtually impossible to definitively protect the entire water supply system. Moreover, these points are downstream of most treatment and monitoring sites, which the utilities establish on intakes and main distribution trunks; hence, the contamination would reach the public.

Researchers have used simulation exercises to achieve a better understanding of the impact that a biological release can have on the public. These exercises evaluated the overall public health response infrastructure and led to discussions of the challenges associated with such a situation (Kaufmann et al., 1997; Walden and Kaplan, 2004). The exercises have demonstrated that the healthcare system would be massively strained, and that substantial investments must be made in the public health infrastructure, vaccine and drug stockpiles, and research and development. These drills effectively modeled the benefit of rapid and effective disease control and surveillance in response to the intentional release of a biothreatening agent.

2. WATER SURVEILLANCE

Federal and state regulators require utilities to monitor water supplies continuously. Water is an important potential mode of transmission for many pathogens, including bacteria, viruses, and parasites (Table 9.1).

There are two complementary approaches to monitoring these biological contaminants. Ongoing comprehensive surveillance is intended to signal the occurrence of contaminated water before its distribution to the general public. The underlying intent of this approach is to detect contaminates to prevent illness, in terms of an early warning detection system. A secondary surveillance approach involves analysis of water supplies that is initiated once presence of a biological agent has been confirmed. This second method is a detect-to-treat strategy, intended to uncover the original contamination point and the route(s) of dispersion. A combination of the two approaches facilitates tracking of the contamination to the source, essentially activating the public health response for detection of additional exposures and the development of triage protocols.

The principal roles of water supply biosurveillance are to communicate with utility regulators and governmental public health (as well as law enforcement, as appropriate), organize emergency preparedness, and coordinate crisis management. Surveillance is crucial during situations likely to cause water contamination, such as pipe ruptures (water main ruptures reduce distribution water pressure, increasing vulnerability to contamination), shortages, and threats to the water system, in particular bioterrorism threats. Several organizations are involved in defining surveillance policy, specifically in the areas of protocol development, identification of emerging threats, protection, security, and emergency planning. These include the water utilities, the government regulators overseeing them, and law enforcement and governmental public health organizations.

2.1. Surveillance Policy

Government policy and oversight help to shape the level and extent of water surveillance. The development of any coherent policy is based on a situational assessment of hazards and general public principles. Hazard assessment covers issues such as prevalence of biological contamination of water supplies and the cost of treatment and testing. Public opinion also modifies policy, as exemplified by the recent issuance of state laws limiting smoking in public owing to heightened awareness of hazards associated with second-hand cigarette smoke. Because hazard assessment and public opinion vary geographically, policy implementation generally begins with a set of national

TABLE 9.1 Threat Potential of Biological Agents to Drinking Water

Agent	Water Threat	Stable in Water	Chlorine Tolerance	Early Clinical Presentation
Protozoa				
Cryptosporidium parvum	Yes	Stable for days	Oocysts resistant	Diarrhea
*Giardia lamblia**	Yes		Oocysts resistant	Diarrhea
*Cyclospora cayetaenesis**	Yes		Oocysts resistant	Diarrhea
*Entamoeba histolytica**	Yes		Oocysts resistant†	Diarrhea
Bacteria				
Shigella sp	Yes	2–3 days	0.05 ppm, 10 minutes	Diarrhea
Vibrio cholera	Yes	Survives well	Easily killed	Diarrhea
Salmonella sp.	Yes	8 days, fresh water	Inactivated	Diarrhea
*Escherichia coli, Pseudomonas**	Yes	Stable?		Diarrhea
*Yersinia enterocolitica**				Diarrhea
Bacillus anthracis	Yes	2 years (spores)	Spores resistant	Flulike
Brucella sp.	Probable	20–72 days	Unknown	Flulike
Clostridum perfringens	Probable	Common in sewage	Resistant	
Francisella tularensis	Yes	up to 90 days	Inactivated by 1 ppm, 5 minutes	Flulike
Legionella sp.*				Flulike
Yersinia pestis	Yes	16 days	Unknown	Flulike
Coxiella burnetti (Q fever)	Possible	Unknown	Unknown	Flulike
Psittacosis	Possible	18–24 hours, seawater	Unknown	Flulike
Viruses				
Variola (smallpox)	Possible	Unknown	Unknown	Flulike
Hepatitis A, E	Yes	Unknown	0.4 ppm, 30 minutes	Jaundice
Toxins				
Botulinum toxins	Yes	Stable	0.6 ppm, 20 minutes	Diplopia, weakness
T-2 mycotoxin	Yes	Stable	Resistant	Diverse
Aflatoxin	Yes	Probably stable	Probably tolerant	Jaundice
Ricin	Yes	Unknown	Resistant at 10 ppm	Severe N, V, D
Staphylococcus enterotoxins	Yes	Probably stable	Unknown	Flulike
Microcystins	Yes	Probably stable	Resistant at 100 ppm	
Anatoxin A	Probable	Inactivated in days	Unknown	Neurological
Tetrodotoxin	Yes	Unknown	Inactivated 0.5 ppm	N, V, D, neurological
Saxitoxin	Yes	Stable	Resistant at 10 ppm	Neurological

Agents categorized by IP-31-017 as not threats to drinking water, or as an unlikely threat: Protozoa: *Naegleria fowleri* and *Acanthamoeba* sp (meningitis), *T. gondii* (mononucleosis-like); Bacteria: *Campylobacter jejuni*†, *Campylobacter coli* and *Burkholderia mallei, Burkholderia pseudomallei* (diarrhea), *Salmonella typhi* (flulike); Viruses: encephalomyelitis and hemorrhagic fever (flulike), reoviruses and norovirus (diarrhea).
*Agents added to IP-31-017 list by authors as concerns of health departments.
†Cysts chlorine tolerant, but removed by standard sand filtration.
Data from Medical Issues Information Paper No. IP-31-017, "Biological Warfare Agents as Potable Water Threats." U.S. Army Center for Health Promotion and Preventive Medicine. Table courtesy of Michael Wagner, MD, PhD.

health standards and then incorporates regional variations that are distinct from the standards of individual municipalities. As a result, higher levels of vigilance are maintained in regions that are perceived of as more likely targets of terrorism. This description holds true for the water surveillance policy for biological agents in the United States Although the threat to our water systems has been recognized, policy making has not yet followed the rhetoric, and the current system ultimately is reactive rather than preemptive. The current state of water surveillance infrastructure serves more in a public health, detect-to-treat mode, compared with a more comprehensive preventative measure. Specifically, federal regulators impose more sophisticated treatment requirements, rather than more sophisticated monitoring requirements, on water utilities with high levels of endemic pathogens. A case in point is cryptosporidium and the enhanced surface treatment rule of 1996, which focuses primarily on filtration (EPA, 1998). Once the utility implements treatment, it focuses on post-treatment monitoring of the source waters (i.e., it monitors finished water).

2.1.1. Governmental Oversight

Government regulation of potable water use falls into two categories. The Clean Water Act (CWA), originally enacted in 1972 and subsequently amended, covers discharge of wastewater. It established the basic framework for regulation of pollutant release into U.S. waters and provided the authority to execute plans to control pollution (EPA, n.d.). This act established water pollution guidelines, particularly water quality standards for industrial wastewater.

The Safe Drinking Water Act (SDWA), passed in 1974 and amended in 1986 and 1996, is the basis for current water standards, state-based enforcement (except in Washington, DC, and Wyoming), and public notification related to drinking water (EPA, n.d.). This act defined the public water system, which now serves more than 80% of the population (EPA, 2003b).

According to the SDWA, there were approximately 170,000 public water systems in the United States in 2003 (EPA, 2000).

In addition, Congress passed a plethora of amendments, including the Total Coliform Rule, the Surface Water Treatment Rule, and the Bioterrorism Act of 2002, each addressing specific drinking water contaminants. These statutes required the EPA to establish specific rules and regulations to ensure that drinking water supplies do not pose a health risk, either acute or chronic. The EPA has established maximum contaminant levels (MCLs) for 83 contaminants known to pose a public health risk. The EPA also has acted to raise public awareness, designated laboratories that are permitted to culture for selected agents, and developed vulnerability assessments for water systems (EPA, n.d.). The combined objectives of governmental legislation are to set safety standards, develop monitoring procedures, and collect data into a national database.

Under these combined regulations, the EPA interacts closely with public utilities and other government agencies to develop clean water protocols and environmental standards. Water from the system source cannot leave the treatment plant without meeting certain health standards. The type and extent of treatment varies from source to source, depending on the initial quality of the source water, the geographic region, and the service area. Types of treatment procedures include clarification, filtration, and disinfection. Treatment practices are fairly uniform, despite minor variations related to local ordinances. As such, municipal water utilities are subject to regular water supply testing for both contaminants and microorganisms. Depending on the size of the facility and availability of testing equipment, each facility can either perform the testing in house or have its testing done through a contract laboratory. We will not discuss treatment techniques in detail.

2.2. Water Supply Organization

Water for consumption can be divided into three categories: commercially available bottled water, privately owned sources (wells and springs), and public sources. Each category has a distinct set of regulations associated with its surveillance and governmental oversight. The most important distinction is that the Food and Drug Administration (FDA) regulates bottled water as a packaged consumption item, and the EPA regulates public water suppliers as utilities. This is true whether the supplier is a municipal agency, a public corporation, or a private firm supplying water to a city. Typically, neither the federal nor state government regulates the quality of water drawn from private sources.

Bottled water is most often drawn from privately owned watersheds; however, a quarter of bottled water sold consists of reprocessed tap water (Moore, 2003). In either case this water is not substantially different from tap water supplied to residents. Bottled water undergoes treatment, such as filtration and/or ozonation, to reduce and eradicate microorganisms before its distribution to the public retail system. Although there have been documented quality-control issues with distributed bottled water found to have high bacterial counts (Illinois Department of Public Health, 1996; Pennsylvania Department of Environmental Protection, 2000), the FDA stipulates that bottled water must meet specific safety and labeling requirements, according to standards similar to those proposed by the EPA for public water systems (FDA, 2002). State and local authorities also regulate bottled water. Overall, the FDA cites a good safety record, and bottled water suppliers are given a lower inspection priority than are municipal systems (FDA, 2002). However, bottled water is not immune to contamination; in November 2003, a terrorist or terrorists (the "Aquabomber") injected bleach, acetone, and ammonia into plastic bottles of water in 20 Italian cities, causing a number of consumers to be hospitalized (Reuters, 2003), and it is clear water bottling plants could have been targeted. We will not discuss bottled water distribution and surveillance methods here.

Municipal water suppliers deliver water to public access taps from open watersheds, typically reservoirs or lakes, or from underground sources, such as wells and springs. A typical water supply and distribution chain is shown in Figure 9.1.

3. TESTING INFRASTRUCTURE

Laboratory testing on drinking water is performed for analysis of chemical, radiological, and biological contaminants. Although biological testing by definition focuses on microorganisms, there is overlap with chemical analysis because many toxins occur naturally. The existing infrastructure for chemical analysis is more established than that for biological monitoring, because water contamination owing to pesticides, pollutants, or even metals has been monitored for decades and because chemical analysis has simpler, less expensive, and more routine regulatory compliance reporting requirements. Regardless of the high numbers of naturally occurring biological contaminants, as well as the costs associated with their identification, all public water suppliers must, by law, test their water for biological organisms (L. Lindsay, personal communication). To determine the prevalence of many of these naturally occurring organisms, a study in the early 1990s tested the source waters of 66 treatment plants. The study found that 97% of the tests came back positive for *Cryptosporidium* and/or *Giardia* species.

4. WATER MONITORING TESTS

Municipal water suppliers must test and meet requirements for a number of parameters, including water pressure, chlorine residual, chemical pollutants, radiological contaminants, and biological organisms. Although this chapter is intended to focus solely on biological monitoring, note that some testing methods are complementary. Water utilities test for conditions that predispose the water supply to contamination, as well as

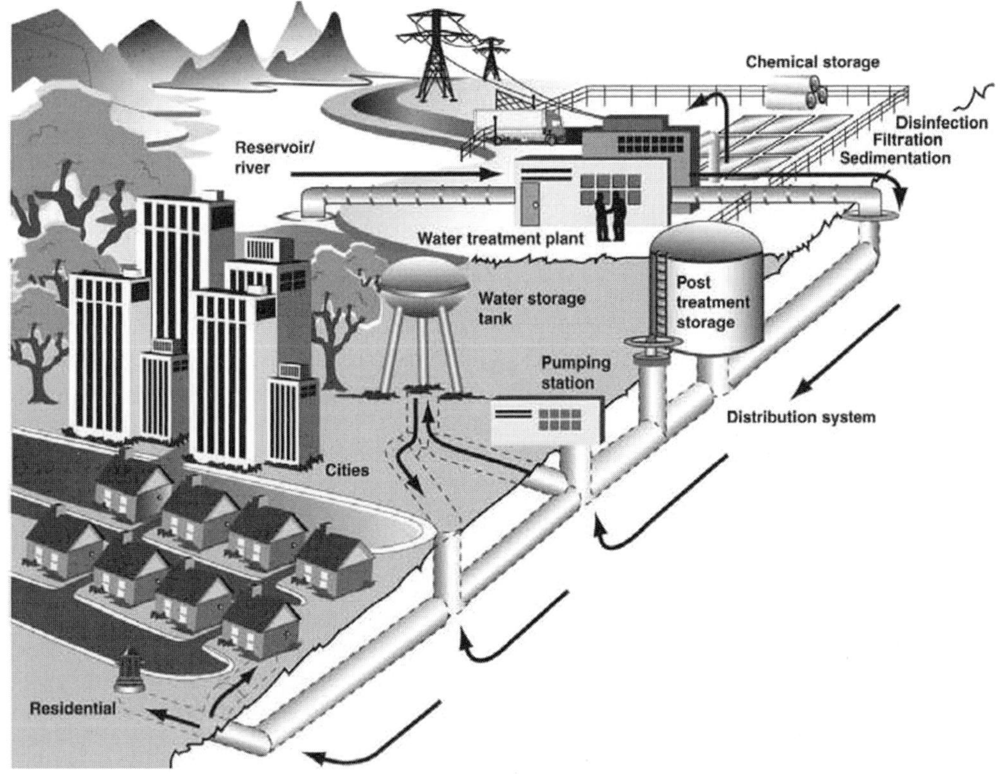

FIGURE 9.1 Schematic of the water supply distribution system. (From United States Government Accountability Office, 2004.)

for physical or chemical properties that can be viewed as proxies for contaminants.

Specifically, water utilities must maintain a water pressure of 20 pounds/inch^2 (psi) in the water mains and a chlorine residual level of 0.2 mg/L to reduce opportunities for biological contamination (L. Lindsay, personal communication). The utilities also monitor pH, turbidity, odor, pesticides, and certain other chemicals in water to be released for consumption.

Many biological agents produce toxins, such as botulinum and cholera toxins, whereas other toxins are naturally occurring, such as ricin toxin, which originates from castor beans. Implementation of surveillance for bioagents occurs primarily at the local and state levels, and is part of any comprehensive water protection plan. Security measures can range from physical barriers (deterrents for unauthorized access such as walls and fences) to cyber-safety protections (firewalls and restriction of computer access to monitoring facilities), to actual biosensor surveillance technologies. Ideally, each level of security is implemented at several checkpoints, covering multiple sites along the distribution chain, including main plants, remote sites, distribution systems, wastewater collection systems, pumping stations, treatment facilities, and water storage depots. Although the entire spectrum of security measures is needed to protect our water supplies, experts cite more comprehensive, real-time monitoring technologies as the most important concern regarding the safety of our drinking water (U.S. Government Accountability Office, 2004).

Public health authorities, both state and local, have the means to detect and identify water-associated outbreaks, and these authorities are part of a network capable of notifying the public and implementing prevention and control measures. The efficacy of these measures depends on how early a pathogen can be detected. In December 2003, the EPA issued a Response Protocol Toolbox (RPTB) (Figure 9.2) that offered specific guidance for response to a compromise of the drinking water supply (EPA, 2003a). The toolbox includes algorithms and time points designed to maximize containment and minimize harm to citizens.

Detection of a naturally occurring water-borne pathogen transpires at many places, including the source; much testing in municipal water systems occurs at the customer's tap. Routine testing procedures can detect many organisms but usually not parasites or viruses (S. States, personal communication) Current detection methods have improved substantially since 2001; most molecular assays take a few hours to perform, which should, in theory, substantially reduce the time lag in detection. However, problems remain, especially in cases in which detection still consists of a human looking through a microscope at a slide. The availability of these rapid tests is a concern, as they are often cost-prohibitive, meaning that

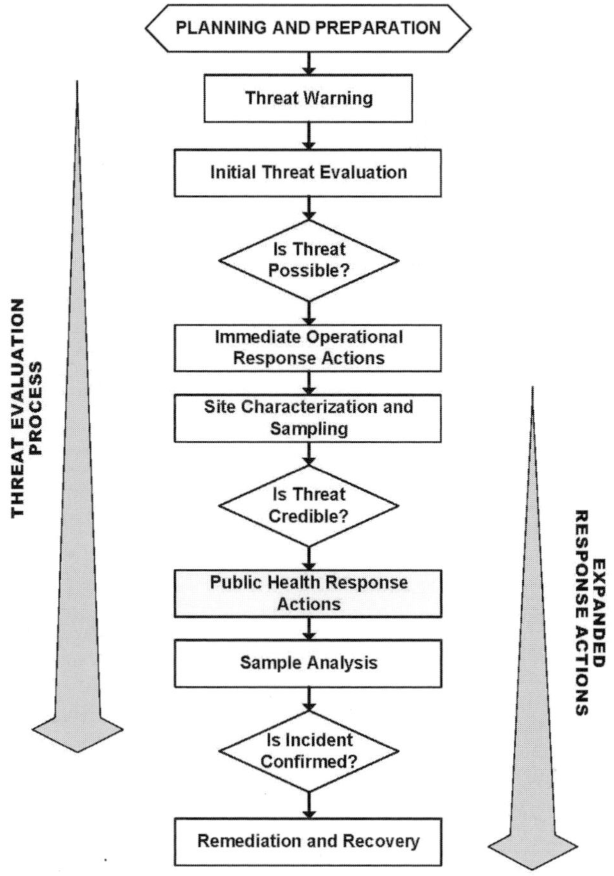

FIGURE 9.2 EPA Response Protocol Toolbox. (From *http://www.epa.gov/safewater/watersecurity/pubs/guide_response_module5.doc.)*

samples must be transferred to outside laboratories for analysis, causing delays. Utilities that store water in reservoirs retain the water, often for days, before releasing it into the distribution pipeline, thus allowing sufficient time to detect and neutralize a threat. Water volumes greater than 100 million gallons can be held in this manner. Any relevant detection methods performed for that utility's water supply should, if positive, allow the utility to cancel or delay the release of contaminated water to the general public.

There are two problems with this scenario, however. The first issue is that many water supplies are derived from combined sources. Combined source waters distributed through the pipeline have experienced various forms of treatment and monitoring. Distributed water that is a combination of reservoir source and intake (upstream from the reservoir) completely negates the rationale for testing of reservoir sources. In addition, a combined source substantially complicates any attempt to ascertain where the introduction of contaminants occurred. Utilities that supply drinking water in the United States commonly use combined source water. The second problem is that the water supply has numerous access points downstream of the reservoir, at which intentional contamination could be introduced, such as back-siphoning through a local service line. If monitoring is meant to serve as an early warning system for terrorist activities, then the safeguarding of these access points must be addressed.

Continued development of the communication infrastructure among water utility companies, regulatory agencies, and public health organizations arises from the desire to protect the public and to contain outbreaks. For example, it would be desirable if municipal water suppliers provided water sampling data to the EPA database on a frequent basis; this has not yet been accomplished. A centralized surveillance system that automatically polls water suppliers' databases and analyzes the incoming data would be an integral part in enabling early warning of potential outbreaks, but it would not, by itself, allow real-time scrutiny; efficient population of that database is also a prerequisite. These gaps in the process could severely hamper any attempt to sequester contaminated supplies before they are sent to the public. Successful surveillance depends on the testing performed by the water suppliers. Improving early detection and contamination containment calls for more frequent sampling and use of on-site rapid testing techniques to produce immediate results.

Authorities often base organism-specific tests on the prevalent threats to the local service area, such as the inherent differences between rural supplies and larger metropolitan water sources. Often, the local supplier's tests identify a natural contamination problem and correct it before any illness is reported in a network. Any problem, but particularly those related to bioterrorism, must be reported quickly if the networks are to function properly. Rapid turnaround in reporting to the network allows identification of larger issues, such as contamination of multiple supplies. Otherwise, the local or state health department would generally issue the initial alert based on patient presentation. The latter type of public health alert is a more likely situation for intentional biological contamination introduced downstream of critical surveillance points, or for agents not routinely tested. A few governmental public health agencies now perform syndromic surveillance, monitoring trends in signs and symptoms related to increased patient presentation. These signs may range from a spike in sales of a particular over-the-counter medicine to an increased incidence of emergency department patients. The number of deployed systems is slowly increasing.

The information sharing and central water-quality databases, as well as regular compliance reviews, provide a feedback loop by which we can assess the biosurveillance effectiveness. Although these databases do not operate in real time, the networks have been vastly improved since the 2001 terrorist anthrax mailings, so that, now, interdepartmental contacts, via phone and e-mail, can help streamline the notification process in the event of an actual crisis situation. The network is more likely to be effective for early detection of a wide range of possible incidents. Although a wide range of incidents is possible, each with unique circumstances, the organizational framework is set up to provide rapid turnaround on any emergency samples (e.g., clinical, environmental). Once the need for select agent identification has been determined, preliminary results for analysis can be performed within a few hours. Because testing of select agents cannot be performed at every locale, inherent delays associated with incident identification, such as sample transportation to an appropriate testing laboratory, are still a problem and can lengthen the overall response time.

In general, larger water utilities perform testing with some type of high-throughput, automated testing equipment, plus a supporting system for confirmation. At this level, they make very limited use of bioassays that employ rapid, modern genetic or immunological methods. These types of assays are for the most part, utilized by the second tier of the surveillance infrastructure, such as state public heath (clinical and environmental) laboratories or local departments of environmental protection (DEPs). The suppliers covering heavily populated areas are the most likely to employ state-of-the-art water surveillance systems to provide a more effective early-warning system for bioterrorism threats (see "Molecular Methods"). Federal response teams, such as the U.S. Army Technical Escort Unit, use mobile state-of-the-art detection techniques, but again this process is more reactive rather than proactive, given the inherent delays in traveling to the incident site. These mobile units are not practical for routine monitoring of our water systems, owing to the high cost involved in their deployments. Although the current system functions well under normal circumstances, the available surveillance methods are not amenable to heightened security, other than through increases in the number of point sources monitored or in the number of specific organisms tested.

Until recently, even the use of advanced testing methods calls for subsequent verification by standard cell culture studies, still considered the gold standard for biological identification. Confirmatory cell culture is always performed but, because the process can take days, such tests are less applicable for early detection. Indeed, laboratories still cannot culture some water-borne viruses at all. Although an initial culture may raise an alarm, triggering shut down or rerouting of water delivery, the response may well be too late to prevent significant harm, given the limitations imposed by sample collection, delivery, and processing on rapid notification and incident response.

The water utilities have responded to this limitation by searching for more rapid, single-step tests. Consider the example of *E. coli:* because *E. coli* is the most abundant organism among fecal coliforms but is not generally found in the environment, its presence is the accepted indicator of water supply contamination. The gold standard for water utilities looking for this microbe begins with the utility first determining coliform density with a filtration step, then performing a standard culture on selective media. By this method, coliforms are easy to identify, but the two steps consume precious time and resources.

Recognizing this, nearly all utilities now employ a direct test. A 100 ml water sample is incubated for 24 hours in a fluorogenic substance, 4-methylumbelliferyl-β-D-glucuronide (MUG), and the technician need only look for a simple color change. A yellow color signifies the presence of coliform bacteria; if the sample fluoresces under black light, *E. coli* is present in the sample. Although faster and cheaper, the new direct test still falls short of the ideal for early detection.

The EPA recognized this limitation when it issued its toolbox. The EPA specifies that a water utility should recognize a "possible" threat within 1 hour of receiving a warning of some sort, and be able to confirm a threat to the water supply within 2 to 8 hours of identifying a "possible" threat. Because laboratory verification can take longer, the EPA document states that a "preponderance of evidence" is sufficient to establish confirmation of a threat requiring containment measures. The EPA advises that a utility need not establish positive identification of a pathogen to execute containment measures.

To be effective, surveillance strategies must employ continuous monitoring for the agents routinely being screened. U.S. water utilities do not employ this type of surveillance routinely. For contaminants not routinely monitored, or for private wells, the sole surveillance strategy available is strictly reactive, based on the reporting of infected individuals. This will obviously involve substantial delays, depending on the laboratory testing network and efficiency of its reporting system. Outsourcing of the initial testing to commercial laboratories can also be an impediment to timely reporting, if the contract laboratory has slower turnaround times. Ideally, water monitoring systems use testing facilities adjacent to water supply or treatment sites, essentially eliminating delays associated with sample transport.

4.1. Biological Monitoring

Detection of any source of contamination is not straightforward, because many contaminants are not visible and may have no discernible odor. The majority of water supply surveillance effort is directed toward *natural* contamination at the supply sources, focusing on chemical detection compared with specific biological organisms. A good source of information on testing for drinking water contaminants is the EPA Web site (EPA, 2005b). The water delivery system is often cited as a major point of vulnerability, because multiple access sites exist in the distribution chain and are downstream of any early warning or monitoring system currently in practice (Figure 9.3) (United States Government Accountability Office, 2004). Intentional contamination at this point of the supply chain can go unnoticed until illness in the general public is identified. The reality is that without more comprehensive

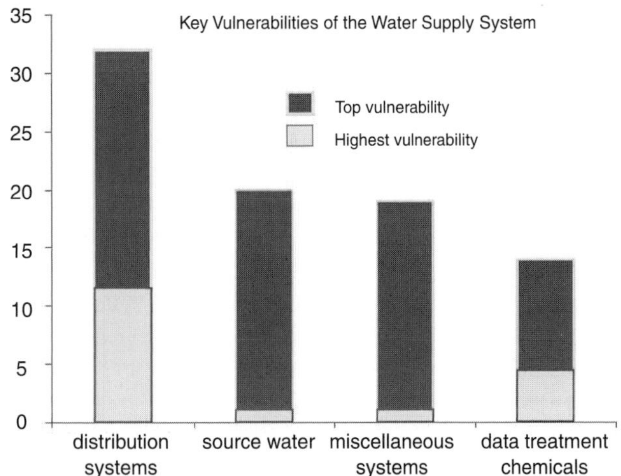

FIGURE 9.3 A survey of expert opinions regarding vulnerable areas associated with the security of our drinking water supply system. (Adapted from United States Government Accountability Office, 2004.)

biosurveillance of our water systems, the initial identification and response occur when patients begin to present in the healthcare network.

Currently, federal and state regulations only set guidelines for total fecal coliform and *E. coli* levels. Regulations dealing with standard unfiltered surface water monitoring for *Cryptosporidium* and *Giardia* species are not currently mandated, although regulations are expected to be set soon. Although areas on higher alert, such as large metropolitan regions or sites hosting major sporting events (e.g., the Olympics), may use advanced molecular methods to perform analysis of select agents, these situations are more of an exception to routine testing.

4.2. Molecular Methods

Direct culturing is the standard assay for analysis of biological agents in water supplies, except for *Cryptosporidium,* which cannot be grown. The time required for organisms to grow on culture media limits this usefulness of this method when an agency must respond quickly to a crisis. The recent progress in developing genomic sequence databases has led to the development of a number of molecular methods that are capable of more rapid, highly-specific analyses. These assays use biomolecules, such as DNA or proteins, as recognition elements based on compounds distinct for each pathogenic microorganism. Because every organism has unique biomarkers, tests have been developed for unequivocal identification of both waterborne pathogens and select agents. Knowledge of these biomarkers enables their combination with sensor schemes or detection methods. Although a number of these methods are useful in identification, issues remain concerning the ability of these methods to determine the viability of the pathogens present. Accordingly, these methods provide a "presumptive positive" result, but organism culturing is still often required for a definitive answer. The determination of viability is important, especially for ascertaining whether the normal water treatment process is capable of eliminating the threat. Broad reviews of pathogen detection methods have been published (Epstein et al., 2002a; Straub and Chandler, 2003; Cirino et al., 2004), and although detailed description is beyond the scope of this chapter, a few methods will be noted below.

4.2.1. *Concentrating Water for Testing*

Because of the huge volume of water issuing from our reservoirs, the presence of a biological microorganism in source water does not guarantee it will also appear in a sample. Hence, utilities filter samples to increase the concentration of captured pathogens. Even this is not always sufficient to ensure detection, because the total volume of water that can realistically route through such filters is limited. Thus, the utility will use both filtration and advanced molecular methods to further increase the likelihood of detecting an organism in a sample.

4.2.2. *Polymerase Chain Reaction*

For DNA analysis, polymerase chain reaction (PCR) methods exist that can selectively enrich the unique biomarker sequence of an organism. This sequence amplification allows the representative pathogen sequences to be identified by more common laboratory methods, such as mass spectrometry, optical spectroscopy, and gel and capillary electrophoresis. More specifically, optical analysis methods have been combined with PCR to form a real-time PCR assay. Real-time PCR methods allow the fluorescent response to correlate directly to the biomarker sequence amplification, which increases exponentially with the PCR process, essentially signaling the presence of the gene targets of an organism as the PCR process is ongoing. This method is now a widely adopted assay for preliminary analysis of pathogens, with results reported within a few hours. In few notable applications, this method has been used for multiplexed analysis of water-borne pathogens (Bopp et al., 2003; LaGier et al., 2004). Another novel extension of nucleic acid amplification includes a RNA amplification method, signaling the presence of pathogen and allowing determination of viability (Min and Baeumner, 2002; Baeumner et al., 2003).

4.2.3. *Microarrays*

Microarrays are analytical tools developed over the past decade that enable high-throughput multiplexed pathogen analysis. Microarray platforms employ thousands of gridded, microscopic DNA spots that are used, in parallel, to interrogate a sample, essentially screening the sample against an immobilized library of pathogen-specific DNA sequences. Real-time PCR is capable of detection of a few DNA target sequences; microarrays can address hundreds to thousands of target DNA sequences simultaneously. Conversely, although real-time PCR has been shown to detect a single copy, or only a few copies, of DNA, microarrays are generally less sensitive. Microarrays have now been developed to interrogate a broad spectrum of pathogens, including a range of food and water microbial pathogens (Chizhikov et al., 2001; Sergeev et al., 2004), as well as select agents (Wilson et al., 2002). As microarray platforms are becoming more specialized, some formats have exhibited the ability to detect DNA even when present in low abundance (Epstein et al., 2002b) and have demonstrated the ability to rapidly subtype pathogen species (Borucki et al., 2004; Shepard et al., 2005). If these methodologies are developed further, comprehensive microarray formats will eventually be able to discriminate and identify the entire spectrum of microbial species, down to the subtype level, and will be of enormous value for the analysis of emerging pathogens.

4.2.4. Immunoassays

Immunoassays constitute another practical option for pathogen detection. Rather than analyzing nucleic acids, this type of assay is directed at identification of specific proteins or toxins. This methodology is based on the manufacture of antibodies specific for an organism or toxin. The most common format resembles that of an enzyme-linked immunosorbent assay (ELISA), in which one antibody is used in a capture capacity, and a second "signal" antibody is employed to bind in a sandwich format. A common immunoassay format links the signal antibody with an optical indicator that allows measurement of the toxin concentration. A number of immunoassays have been developed to target water-borne and other pathogenic microorganisms (Shelton and Karns, 2001; McBride et al., 2003). The throughput of immunological methods can be increased when combined with techniques such as flow cytometry, which in some cases can screen millions of cells or particles per hour for specific binding events.

4.2.5. Other Advanced Methods

Other novel analytical methods have addressed the rapid detection of biological agents in drinking water. A bioluminescence method for water samples was developed as a simple method for *E. coli* analysis (Lee and Deininger, 2004). The process uses antibody binding separation steps before analysis to produce a bioluminescence response within an hour. A great deal of effort is also being put into mass spectrometric analysis of biological species, including water-borne pathogens (Magnuson et al., 2000; Fenselau and Demirev, 2001). Such analyses can target specific protein or toxins, even from intact organisms and can be used as an *in situ* test. A number of remote and/or *in situ* sampling systems are also being developed, which enable active monitoring of many basic water quality parameters, including an estimate of biological activity based on residual chlorine, dissolved oxygen, pH, and algal chlorophyll. Sampling systems allow dynamic collection in any water environment, so the autosampler can shift to regions of interest, essentially tracking water supply contamination with real-time measurements. This type of analysis could soon be implemented directly in the water main pipe, serving as a first level of warning. Such systems will likely be combined with nucleic acid and/or immunoassays for rapid, *in situ* pathogen testing of our water supplies.

Limitations. The schemes summarized above are increasingly used for biosurveillance, as costs for their implementation and maintenance decrease. These methods offer sensitive, rapid assay formats with the ability to be tailored to any pathogen with a known biomarker. As our knowledge of these organisms advances and the relevant sequence information is garnered, more thorough biothreat analysis will become routine. Because these assays are based on known genetic or other biological information, they have limited utility for identification of emerging pathogens or for pathogens for which we have limited information. Also, although a few techniques have demonstrated the capability to ascertain viability, the majority of assays provide only a preliminary positive result, and viability must be confirmed by growth in culture. One of the biggest limitations to biomonitoring is cost, and no inexpensive, reliable portable systems are yet commercially available. The cost of implementation, coupled with the enormous scale of the water supply infrastructure, makes the widespread incorporation of this type of system impractical. Also, as many of these formats are tied to specific pathogens, a single comprehensive platform is difficult to envision. For example, although microarrays have the ability to address thousands of gene targets, organisms such as *Salmonella* have hundreds to thousands of subtypes, including recent multidrug-resistant strains (Mulvey et al., 2004), making them more difficult to identify at the subtype level.

Approach to Biosurveillance. The EPA determines biosurveillance standards for treated water. The surveillance process involves three levels of action. If the routine level of screening identifies a potential problem, the second (confirmatory) level of testing is initiated, and more detailed testing is performed to validate the initial positive test. The detailed testing includes more frequent sampling, sampling of other parts of the supply and distribution chain, and an expanded regimen of agents tested. The types of tests performed are commensurate with the situation, that is, rapid, high-throughput, organism-specific tests are used when an incident is identified. The purpose of the expanded testing is to assess the public health risk and determine the need for further actions. The final stage of action seeks to determine the source of contamination. This phase of analysis guides remediation, mitigation, and policy to prevent recurrences.

Routine Monitoring. Routine monitoring most commonly entails surface water analysis for the presence of total coliform by the water suppliers. There are no federal guidelines mandating routine testing of many commonly occurring pathogens. Testing is, in practice, limited to the minimal regulatory requirements plus a few voluntary monitoring tests, such as for *Cryptosporidium* and *Giardia* species, performed with low frequency. Such monitoring does not serve in a detect-to-prevent mode and is unlikely to be useful as an early warning indicator. Testing largely focuses on the watershed and its supply chain. Utilities will maintain records on multiple indicators and baseline concentrations as a means to identify atypical levels of contaminants. As noted above, utilities will also record data from which they can infer the presence of unwanted organisms. Test results are available for real-time analysis and are often included in microbiological surveillance strategies.

Confirmation Analysis. The second level of surface drinking water source assessment requires assays for specific pathogens. Utilities perform tests most often to confirm the presence of natural pathogens, such as protozoa and enteric viruses, or common environmental contaminants, such as bacterial organisms (*E. coli* O157:H7, *Shigella, Salmonella*). Although many laboratories can perform screening assays for the above agents, the water utilities usually assign confirmatory testing to specialized laboratories under strict protocols (see "Laboratory Networks"), which involve advanced methods and an expedited response. Methods such as real-time PCR can enumerate pathogen concentrations and be used to better assess the threat level, but these methods cannot assess organism viability, which is determined via organism growth, for example, cell culturing.

Source Contamination Determination. The third level of water surveillance involves trace-back analysis, with the purpose of identifying the source(s) of microbial contamination. Although this testing has practically no early warning value, it is critical for preemptive purposes to avoid or minimize future outbreaks. This analysis provides a framework within which to establish regulations, contain the pathogen so as to limit the extent of public illness, and define the requirements for remediation. These responses are influenced by the identity of the causative pathogen, its quantity, source, and dispersal range.

The three levels of surveillance are complementary and shape the overall public health response. Surveillance is often a dynamic process, influenced by experience from past outbreaks, public policy, and ongoing data collection. Obviously, we can define the extent of an outbreak most clearly when comprehensive pre- and postexposure data are available.

Sampling and Actual Testing Methods. Sampling is a critical step in the identification of biological agents. For water sampling, a concentration step is performed because many laboratory techniques are concentration sensitive and have specific limits of detection that could compromise the result for a diluted sample. Filter sets can also serve as an initial means for unknown identification, because bacteria, viruses, or protozoa can be selectively filtered based on particle size. The order in which tests are performed is influenced by any clinical or physical evidence that suggests a particular pathogen. The EPA sets testing guidelines, such as culture protocols, for the common water-borne agents (EPA, n.d.). For bacteria, membrane filtration concentrates microorganisms from water samples, which are then spotted on selective media for culture growth, serological classification, and biochemical analysis. Concomitant with culturing, other established techniques may be performed, such as PCR and immunology-based analysis. Detection of protozoa and viruses requires analysis of much larger volumes of water than are used for bacterial analysis, because these organisms are typically present in lower concentrations. For viruses, the material from the collection filters undergoes microscopic examination for visual confirmation as well as exposure to a number of cell lines selective for particular viruses and, ultimately, PCR analyses. General parasitological analysis is via fluorescent immunologic-based separation and identification and, again, PCR assays. An important note is that although many clinical laboratories are equipped with PCR capabilities, the above methods, such as culturing, are the universally accepted measure for definitive classification. The time required to complete a real-time PCR assay is generally 2 hours for common rapid cycling instruments. The assay results depend on the sample concentration, that is, the amount of organism present. Because the assays can be interpreted in real time, an abundant biological agent can provide a positive real-time PCR result within 30 to 60 minutes of the 2-hour assay. Although the use of PCR instruments at every stage of the water testing process is still cost prohibitive, portable PCR instruments have been developed that can be deployed to problem sites as needed. Ideally, an *in situ* test would able to monitor the actual water mains, but currently this type of system is impractical, owing to both to the cost associated with such an instrument and to the dilution factor associated with water sampling. Because of this, PCR is considered the most rapid and effective measure of the presence of an organism, and organizations such as the Laboratory Response Network (LRN) have defined primer sequences and target genes that must be used for identification.

If the sample is an unknown and a select agent is suspected, a specialty laboratory receives the sample initially. The filtration/concentration process for unknowns uses ultracentrifugation with selective filters. Because bacterial and viruses differ substantially in size, filters with varied pore sizes and polarities can provide a gross discrimination of the type of microbe before culture-based and molecular analyses (immunoassays, PCR) are performed. The "Molecular Methods" section below describes the more common advanced assays for biological detection. Also, a number of specific and/or commercially available tests, including handheld assays and PCR formats are listed in a Department of Justice guide, "An Introduction to Biological Agent Detection Equipment for Emergency First Responders" (Fatah et al., 2001).

Limitations to Surveillance. One of the biggest limitations facing comprehensive biosurveillance is cost, because of the sheer size of the water system and the multiple points at which it is vulnerable to intentional contamination. A number of concern areas in need of improvement are listed in Figure 9.4. Because of the scale of the physical infrastructure, the safety policies, testing proficiency, and communication network for the nation's water supply can be thought of as a massive collective effort, organized on the federal, state, and local levels. More than $140 million in federal funding (2002–2004) has been spent on physical security, training, and vulnerability assessments, yet little mandatory implementation has taken place (U.S. Government Accountability Office, 2004). The lack of federal guidelines requiring testing of the most commonly present organisms, and the limited implementation of advanced biosurveillance techniques are issues of concern. The limited extent of drinking water surveillance in terms of biological organisms, in and of itself, can be perceived as low prioritization. Although many agents, if present, would be sufficiently eliminated by the current treatment system, several pathogens have demonstrated resistance to the concentrations of disinfectants used in common practice (Burrows and Renner, 1999). Still, comprehensive testing of all pathogens is not considered practical, and general water quality tests only determine total coliform density. The lack of comprehensive guidelines will likely serve confusion and difficulties in future incidents.

Inherent limitations to sampling and analysis exist because many water supply sources are continually fed, resulting in

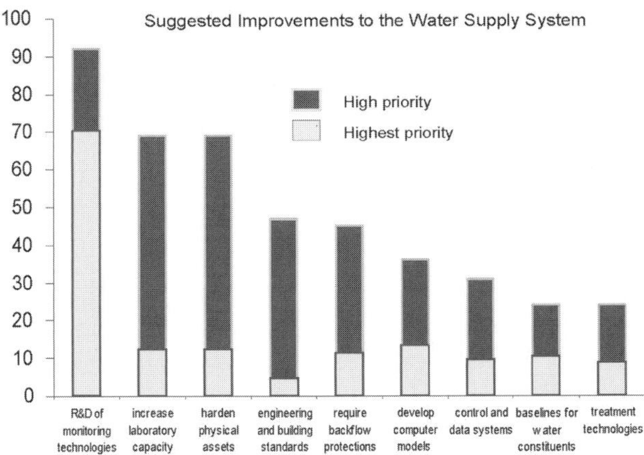

FIGURE 9.4 A survey of expert opinions regarding necessary improvements to our drinking water supply system. (Adapted from United States Government Accountability Office, 2004.)

sample dilution. Because of this, filtration and analysis protocols were developed by the EPA, typically filtering 50-liter water samples for testing, to concentrate any biological pathogen. In addition, as mentioned above, distributed water can originate from multiple sources and can undergo varied treatment measures and limited testing along the distribution route. As the basic goal of a water supplier is to provide potable water, comprehensive surveillance would require multiple check points for testing, allowing trace-back to the source of contamination, verification of the efficacy of the water treatment procedures, and rapid signaling of a problem before it develops into a public health crisis. Eventually, the current monitoring of total coliform counts, and *E. coli* specifically, will not be perceived as sufficient to meet public health needs.

5. SURVEILLANCE DATA

Water testing data are collected based on federal, state, and local regulations. The tests performed, and their frequencies, vary with the site being tested. For example, New York City has a water surveillance program (Heffernan et al., 2004), and strict guidelines for comprehensive watershed monitoring program (New York City Department of Environmental Protection, 2005a). Tests include, but are not limited to, raw fecal coliform levels and biologicals such as *Cryptosporidium* and *Giardia* species. Raw fecal coliform reservoir levels must be fewer than 20 colonies per 100 ml sampled, in greater than 90% of measurements during any 6-month period (New York City Department of Environmental Protection, 2005b). The samples are collected from various locations, including dams, intakes, mid-reservoir points, and main tributaries. The program has close to 1,000 sampling areas, and calls for more than 1,300 water samples per month from many of these sites. Public watershed information is available online, including the diagram in Figure 9.5, showing part of the New York City/Catskill Watershed.

In New York State, water samples expected to contain select agents, are sent to the Biodefense Laboratory of the State DOH for analysis. The New York State DOH performs all detection and identification of select agents, and provides extensive training to the hazardous situation first responders, such as fire and police officials, providing guidelines for sample collection, aiding in incident response, and building a feedback loop for communication. This next section details an actual timeline of select agent analysis associated with a security breach at a New York City water reservoir. Workers at a municipal water system notified the New York City DEP at 10:30 a.m. to report the presence of a suspicious bottle near the edge of the water supply. The bottle was a large plastic soda bottle, which was covered in brown tape and partially filled with a dark liquid. The bottle contained noticeable particulate matter and was found with two pieces of paper. The responding officer inspected the scene and called in the DEP

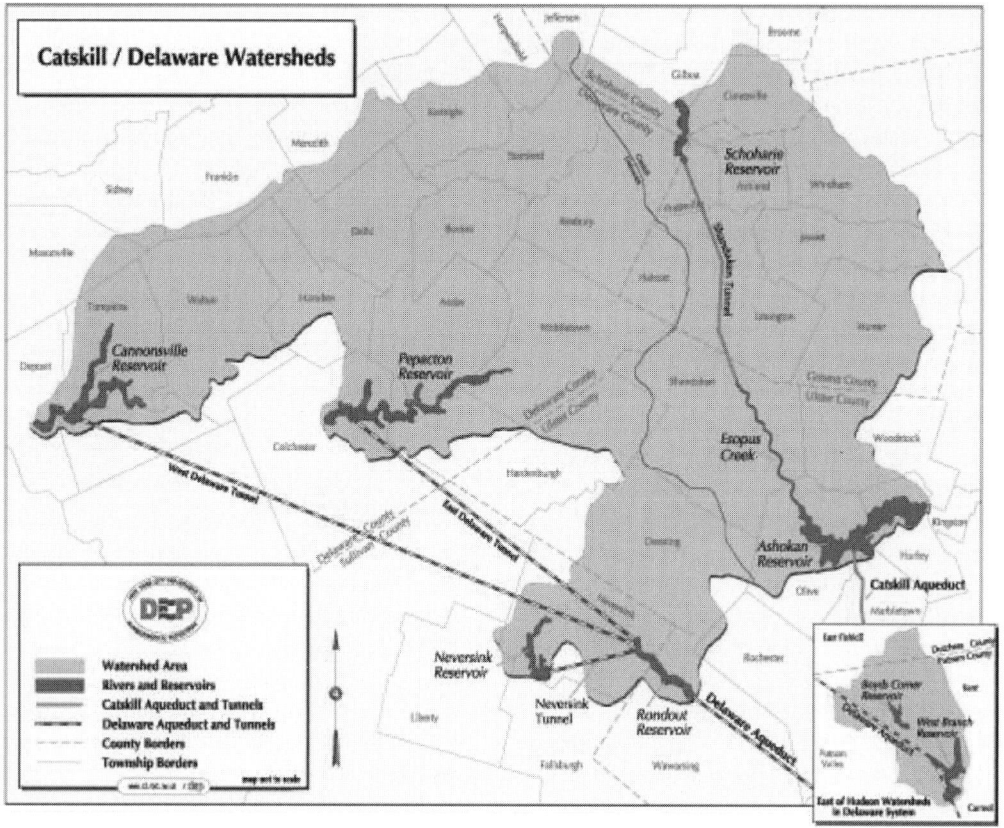

FIGURE 9.5 A depiction of the Catskill/Delaware Watersheds serving New York City. (From the New York City Department of Environmental Protection, *http://www.ci.nyc.ny.us/html/dep/html/wsmaps.html*.)

hazardous materials (hazmat) responders. The hazmat team performed a test for explosives (none identified) and notified the DOH Biodefense Laboratory of the sample at 2:30 p.m. The hazmat team transferred the material to a container for transport and chain of custody to the DOH laboratory, departing at 3:30 p.m. Material to be tested arrived at the laboratory at 6:50 p.m., and DOH preliminary testing returned negative for the presence of ricin toxin and B. *anthracis* by 11:30 p.m. Additional preliminary analysis for *Yersinia pestis, Francisella tularensis* and *Brucella* species was determined to be negative by 9:15 a.m. the following morning. Negative culture analysis for the above select agents was confirmed approximately 48 hours from receipt of sample, and this initiated the transfer of the suspect material for chemical analysis. The water in the reservoir, which was to enter the distribution chain in 1 week, was evaluated for further contamination. This incident highlighted a number of issues:

- The importance of communication between state agencies and local entities, to provide a comprehensive and rapid response to a potential bioterror incident. Although this incident was not found to involve hazardous chemicals or biological agents, it served an exercise of the response system in New York State.
- The value of further accelerating and streamlining the data collection, analysis, and response process. Although the water in this case was not scheduled for release that day, the utility's managers would have been fully justified in following EPA guidelines and isolating the water tank immediately after discovery of the plastic bottle but before tests were complete. This does not mean the utility reacted inappropriately in this case, but rather, a different water distribution schedule would have mandated a swifter containment response.

6. SYNDROMIC SURVEILLANCE CASE STUDIES

Syndromic surveillance is a public health tool that analyzes statistical data related to trends in health care as an early warning system. The basis for this type of surveillance is that outbreaks are associated with an increase in medical case loads and other tangible indicators. A number of preclinical factors, including over-the-counter remedy sales, topical Internet searches, and emergency department cases, can often be early warning signs or diagnostic precursors of an outbreak. The New York City DOH and Mental Hygiene currently operate a syndromic surveillance network in New York City (Heffernan et al., 2004). New York City has established a Water-borne Disease Risk Assessment Program to determine *Giardia* and *Cryptosporidium* species levels. This program monitors information on cases presumably linked to tap water consumption, so as to ensure rapid detection of any outbreaks. The database maintains information on gastrointestinal disease, particularly emergency department and nursing home statistics, over-the-counter sales for related medicines, and collections of relevant clinical laboratory tests performed. Detailed monitoring of these factors is expected to accelerate the public health response to a biological agent exposure. Additional examples of syndromic surveillance systems are included in two case studies below.

1. An excellent surveillance case study of gastrointestinal illness was performed via retrospective evaluation of pharmacy over-the-counter sales in two Canadian provinces (Edge et al., 2004). The study compared pharmacy sales of antinauseal and antidiarrheal medications, as well as emergency department visits from water-related outbreaks associated with *Cryptosporidium, E. coli* O157:H7, and *Campylobacter* species. The investigators compiled these data to assess the potential of a real-time link to outbreaks. Their findings showed that the over-the-counter sales spiked appreciably and correlated well to the outbreak epidemic data. Statistical analysis confirmed that the sales data could be an early indicator to the start of the outbreak periods. This study demonstrated that if public health authorities set up an automated monitoring system covering specific medical supplies, the spatial and temporal trends could be correlated with trends in community health. Such a system would be a huge asset for public health officials and would allow more rapid identification of outbreaks than is possible under the currently available laboratory-based surveillance systems alone.
2. Pascal Beaudeau et al. (1999) studied data from a syndromic system that monitored the sale of antidiarrheal medications in Le Havre, France, and they found a correlation between these sales and the failure of a drinking water treatment plant. Specifically, between April 1993 and September 1996, sales of medicine to treat gastrointestinal illness climbed dramatically between 3 and 8 days after interruption of chlorination at the water treatment plant; a sales increase also occurred in the 3 weeks after an increase in raw source water turbidity. There were several instances involving a failure to control turbidity and maintain residual chlorine levels; a significant detail here was that the turbid water still met France's microbiological standards for potable (drinkable) water. The investigators concluded that treatment plant–based monitoring may not be sufficient to consistently prevent contamination of drinking water; the study results also support the value of syndromic surveillance systems that perform signal processing on multiple and disparate data sources.

For the fiscal 2006 year, an approximate $44 million Water Sentinel Initiative has been proposed in which a pilot monitoring and surveillance study of five U.S. urban centers will be developed (Office of Management and Budget, 2005). The program will attempt to incorporate standardized syndromic surveillance systems, such as RODS and the Department of Defense's Electronic Surveillance System for the Early Notification of Community-based Epidemics (ESSENCE), in its mandate to provide an early warning system for chemical and biological contamination. This initiative is expected to ultimately result in a national surveillance system capable of greatly heightened monitoring of our water supply.

7. LABORATORY INFRASTRUCTURE

Because the release of a biological agent is intended to have devastating effects on the population, laboratory networks have been established to standardize the issues associated with bioanalysis, including sampling, testing, reporting, and sample disposal. These networks provide a hierarchical approach to identification and confirmation of biological agents, allowing water suppliers to become involved in the testing process and obviating the need for trained responders to take action in every incident on the local level. Even though local level analytical laboratories are not likely to have comprehensive biological agent monitoring systems in place, they provide initial, basic analyses. At later stages, better equipped laboratories in the networks, such as select government and academic laboratories, can perform more sophisticated analyses. In this sense, the system is set up so that laboratories are available to provide support and confirmation. Many hospital or clinic laboratories are not trained to deal with highly infectious (select) pathogens. If patients begin to present at emergency departments en masse or with unknown illnesses, involvement with a confirmatory laboratory is necessary. Recently, a number of simple handheld testing kits have been developed for select agent identification that seems well suited to nonspecialized use. However, these tests, typically antibody based, exhibit high rates of false-positive responses and low sensitivity, and they do not address organism viability. For these reasons, resulting in public health professionals have denounced their utility (Counterterrorism and Forensic Science Research Unit Hazardous Materials Response Unit/Bioterrorism Preparedness and Response Program, 2003).

8. LABORATORY NETWORKS

National networks that support biological analysis include the LRN and Emergency Laboratory Response Network (eLRN). These networks are composed of clinical hospital and public health laboratories, public and commercially owned environmental laboratories, and federal laboratories such as the CDC and U.S. Army Medical Research Institute of Infectious Diseases (USAMRIID). The LRN is a joint project of the CDC, the Association of Public Health Laboratories, and the FBI, coordinating the response to threats associated with possible bioterror incidents (see Chapter 8). The LRN is designed to improve communication between the more than 140 local, state, and federal laboratories involved in the network nationally. The LRN is organized with a chain of command structure, based on government regulations and select agent control rules, operating under the guidance of the Department of Health and Human Services. Specifically, the LRN has three levels of laboratory association: sentinel, reference, and national. This tiered analytical network provides a referral-based support system of increased proficiency and enhanced capabilities. The most common LRN elements are the sentinel laboratories, including hospitals, clinics, or first-responder networks. These laboratories typically act in the initial attempts at biological characterization, or "rule in/rule out" analysis. The middle tier comprises the reference laboratories, which provide a higher level of analytical sophistication for agent identification, often confirmatory for earlier sentinel laboratory testing. Reference laboratories typically include state DOH public health laboratories. The top tier of the network encompasses agencies such as the CDC, which are the source for definitive analysis of even the most hazardous of infectious agents. The threat level determines the LRN response, and as many naturally occurring water-borne pathogens are listed as select agents, their testing is regulated by the LRN.

The EPA has established the eLRN from a network of environmental laboratories to provide laboratory assistance during a natural disaster or a chemical, biological, radiological, or nuclear incident. The network focuses on all aspects of the protection of the nation's drinking water supply, including safety guidelines, sampling regulations, and transport documentation. This network is designed to provide laboratory support during an incident in the event that the initial jurisdictional laboratory cannot respond or requires additional assistance. An incident involving a select agent has much stricter guidelines than does one involving a natural water-borne pathogen. A laboratory, when responding to an incident may request aid and can receive assistance through either a federal EPA laboratory or state public health laboratory. In the event of an act of biological terrorism, the eLRN laboratories respond and analyze specimens by using validated protocols available through the LRN. Standardized methods are in place to ensure consistency and to avoid interlaboratory variations in reporting. Reporting laboratories are required to adhere to an overall incident response structure that requires emergency (24-hour) availability and strict guidelines for sample handling, covering analysis, custody, and sample disposal. Other similar response networks exist to respond to an attack on the nation's food supply, such as the Food Emergency Response Network (FERN), which has been established jointly by the FDA and the U.S. Department of Agriculture. The FERN has also established a laboratory system that communicates with the LRN network. Although the main focus of this network is the response to food contamination issues, the network is set up to handle water-borne events, and protocols are in place to deal with water testing. Any jurisdictional issues in the case of any hazardous materials incident are resolved under a federal mandate regarding the Incident Command System/Unified Command (ICS/UC) (U.S. Department of Labor, 2005). The ICS is an organizational system adopted by emergency responders throughout the nation, serving as a tool to coordinate efforts and structure the incident response. Any instance of a compromised water system will likely require a coordinated response from multiple agencies. For incidents necessitating cooperation among multiple federal or regulatory agencies, such as those involved in national emergencies, the UC process is further delineated so that each agency retains individual authority and accountability under the coordinated response. As of 2005, the National Incident Management System, as part of the Homeland Security Initiative, requires education and training in the use of the ICS for all incident responses down to the local level.

Broadly, there are two possible situations associated with detection of a select agent that can arise. In the first scenario, is when patients present at local hospitals or physicians' offices. Determination of the cause of illness first depends on the latency between exposure and onset of illness, which can be a matter of days. Because of this large lag time, the laboratory network system is expected to function mainly in identifying the organism responsible for illness, and not as an early warning network. The initial determination of illness could thus be determined in a few days, depending on the virulence and rate of infectivity of the agent, the number of patients presenting symptoms, etc. Any obvious warnings, such as an overwhelming number of patients at a hospital emergency department, will initiate reference and/or national laboratory responses, which in turn will dramatically speed up the identification of the organism.

In the second scenario, a network response would be invoked as the result of a discrete and verified action, such as a called-in threat or a security breach in which a terrorist has actually introduced a contaminant or is judged likely to have done so (it is unlikely that a mere telephone threat or break-in at a facility would, by itself, trigger the network response). In this type of event, the response level is elevated, likely starting with a second-tier reference laboratory, and identification of the threat is rapid, typically within hours. In the absence of a defined outbreak, the standard course of action initiates with field sampling/screening, followed by sentinel laboratory testing and, if required, reference laboratory analysis and/or confirmation. The national laboratories generally are involved when a highly infectious agent is suspected. A tentative positive response to a biological agent at any stage initiates an immediate referral to a higher level of the LRN.

Because of the complexity of testing and the large number of potential water-borne agents, analysis uses a layered procedural methodology. Biological threats are screened in order of likelihood, based on the circumstances of the case and proceeding with a systematic elimination of organisms. For example, the symptoms associated with a smallpox outbreak differ markedly from an anthrax exposure. Because an attack with weaponized anthrax is presumed more likely than an attack using smallpox, unless the symptoms showed otherwise, a test for anthrax would be performed before other rarer select agents. As the presence of certain pathogens is ruled out, sequential testing for the less common pathogens is performed.

9. SUMMARY

Concern has been growing over the state and safety of municipal water supplies. Government organizations and the water industry seem to have a good working relationship, but both federal standards and industry-wide procedures within the water supplier network are lacking. The more stringent among the existing guidelines are voluntary, and they require both greater standardization and more comprehensive development of rapid (real-time) analysis and data sharing. Currently, to determine the source of a problem, a reactive, trace-back method is the principal tool used by health authorities, which can result in major delays in the public health response. Continuous detection and monitoring strategies are becoming available and could reduce the delays in reporting, but the water supply surveillance infrastructure is currently lagging in its efforts to adopt real-time bioanalytical methods. A report by the Government Accountability Office in 2004 cites physical and technological upgrades, particularly development of real-time monitoring technologies for treated water (in the distribution system) as the area most deserving of federal support. These advances, incorporating molecular biology techniques and epidemiological investigations of water-borne illness, will continue to aid in the direction in which biological surveillance needs to follow. As suggested by the case studies and technologies described herein, methods currently in development can dramatically improve the current state of water supply surveillance. Although the current incident reporting process does not offer the kind of real-time reporting necessary for early reliable detection of bioagent exposure, a more proactive stance toward water sampling, surveillance, and reporting can be developed.

The adoption of pathogen surveillance methods is not the complete answer, as the list of potential biological contaminants is long and new pathogens continue to emerge. Development of comprehensive analysis requires continual assay development and modernization of an aging infrastructure, both of which entail substantial cost. Large numbers of vulnerable access locations exist, often in the distribution system downstream of the locations where biological agent testing is performed. Any attack introduced near the tap (i.e., toward the consumer end of the distribution system) has a greater likelihood of success because such sites are after treatment and, generally, after surveillance. An essential aspect of water surveillance is more intelligent use of the current distribution infrastructure, particularly in terms of water conservation. Water utility organizations should adopt innovative strategies, such a division of municipal water. Much of the water that is delivered through the municipal system is not directly consumed; that is, water treated for consumption is used for industrial or agricultural use, such as irrigation, and may not require the same level of treatment and surveillance. As public education, recycling, and waste reduction efforts expand, and as new detection technologies continue to be developed, the demand and burden on the existing infrastructure may be lessened. A more comprehensive approach to water regulation, management, and surveillance is slowly becoming a reality, but substantial modernization of the water supply infrastructure and policy still remains as a major problem in water supply biosurveillance.

ACKNOWLEDGMENTS

We thank Dr. Adrianna Verschoor from the New York State Department of Health (NYSDOH) Office of Research for reviewing the manuscript. For their helpful discussions, we thank Connie Schreppel, from the Mohawk Valley Water Authority, Dr. Ellen Braun-Howland from the NYSDOH; Craig Jackson Jane Thapa from NYSDOH Bureau of Water Supply Protection; Dr. Stanley States, Laboratory Director of the Pittsburgh, Pennsylvania, Water and Sewer Authority; and Lorene Lindsay, Manager of Laboratory Services, Kansas City, Missouri, Water Services Department.

REFERENCES

Baeumner, A.J., Cohen, R.N., Miksic, V., and Min, J. (2003). RNA Biosensor for the Rapid Detection of Viable *Escherichia coli* in Drinking Water. *Biosensors & Bioelectronics* vol. 18:405–413.

Beaudeau, P., et al. (1999). A Time Series Study of Antidiarrheal Drug Sales and Tap-Water Quality. *International Journal of Environmental Health Research* vol. 9:293–311.

Boggs, A.F. (2003). Antibacterial Resistance: Who Pays? *Expert Opinion on Therapeutic Patents* vol. 13:1107–1112.

Bopp, D.J., et al. (2003). Detection, Isolation, and Molecular Subtyping of *Escherichia coli* 0157:H7 and *Campylobacter jejuni* Associated with a Large Waterborne Outbreak. *Journal of Clinical Microbiology* vol. 41:174–180.

Borucki, M.K., Kim, S.H., Call, D.R., Smole, S.C., and Pagotto, F. (2004). Selective Discrimination of *Listeria monocytogenes* Epidemic Strains by a Mixed-Genome DNA Microarray Compared To Discrimination by Pulsed-Field Gel Electrophoresis, Ribotyping, and Multilocus Sequence Typing. *Journal of Clinical Microbiology* vol. 42:5270–5276.

Burrows, W.D. and Renner, S.E. (1999). Biological Warfare Agents as Threats To Potable Water. *Environmental Health Perspectives* vol. 107:975–984.

Centers for Disease Control and Prevention[CDC]. (2001). Public Health Guidance for Anthrax Threat Letters or Packages; October 9. Official Health Advisory. *http://www.phppo.cdc.gov/HAN/ArchiveSys/ViewMsgV.asp=AlertNum=00027.*

CDC. (2003a). CDC Alert on Ricin. Official Health Advisory. *http://www.phppo.cdc.gov/HAN/ArchiveSys/ViewMsgV.asp=AlertNum=00160.*

CDC. (2003b). Positive Screening Signal for *Bacillus anthracis* at the Anacostia Naval Postal Sorting Facility in the Washington Metropolitan Area; Friday, November 07. Official Health Advisory. *http://www.phppo.cdc.gov/HAN/ArchiveSys/ViewMsgV.asp=AlertNum=00162.*

CDC/Environmental Protection Agency [EPA]. (2003). CDC and EPA Water Advisory in Response to High Threat Level; Friday, March 21. *http://www.phppo.cdc.gov/HAN/ArchiveSys/ViewMsgV.asp=AlertNum=00123.*

Chizhikov, V., Rasooly, A., Chumakov, K., and Levy, D.D. (2001). Microarray Analysis of Microbial Virulence Factors. *Applied and environmental microbiology* vol. 67:3258–3263.

Chopra, I., O'neill, A.J., and Miller, K. (2003). The Role of Mutators in the Emergence of Antibiotic-Resistant Bacteria. *Drug Resistance Updates* vol. 6:137–145.

Cirino, N.M., Musser, K.A., and Egan, C. (2004). Multiplex Diagnostic Platforms for Detection of Biothreat Agents. *Expert Review of Molecular Diagnostics* vol. 4:841–857.

Colwell, R.R. (1996). Global Climate and Infectious Disease: The Cholera Paradigm. *Science* vol. 274:2025–2031.

Corso, P.S., Kramer, M.H., Blair, K.A., Addiss, D.G., Davis, J.P., and Haddix, A.C. (2003). Cost of Illness in the 1993 Waterborne Cryptosporidium Outbreak, Milwaukee, Wisconsin. *Emerging Infectious Disease* vol. 9:426–431.

Counterterrorism and Forensic Science Research Unit Hazardous Materials Response Unit and Bioterrorism Preparedness and Response Program. (2003). Preliminary Findings on the Evaluation of Hand-Held Immunoassays for *Bacillus anthracis* and *Yersinia pestis*. forensic science communication. *http://www.fbi.gov/hq/lab/fsc/backissu/jan2003/fsru.htm.*

Desselberger, U. (2000). Emerging and Re-emerging Infectious Diseases. *Journal of Infection* vol. 40:3–15.

Edge, V.L., et al. (2004). Syndromic Surveillance of Gastrointestinal Illness Using Pharmacy Over-the-counter sales. A retrospective study of Waterborne Outbreaks in Saskatchewan and Ontario. *Canadian Journal of Public Health* vol. 95:446–450.

Environmental Protection Agency [EPA]. Water trivia facts. Document 80-F-95-001. *http://www.epa.gov/ogwdw000/kids/water_trivia_facts.html.*

EPA. (1974). Safe Drinking Water Act (SDWA). *http://www.epa.gov/safewater/sdwa/index.html.*

EPA. (1998). Interim Enhanced Surface Water Treatment Rule. *http://www.epa.gov/safewater/mdbp/ieswtr.html.*

EPA. (2000). Public Drinking Water Systems: Facts and Figures. *http://www.epa.gov/safewater/pws/factoids.html.*

EPA. (2002a). Laws and Regulations, Clean Water Act. *http://www.epa.gov/R5water/cwa.htm.*

EPA. (2002b). New or Repaired Water Mains. *http://www.epa.gov/safewater/tcr/pdf/maincontam.pdf.*

EPA. (2002c). Requirements of the Public Health Security and Bioterrorism Preparedness and Response Act of 2002 (Bioterrorism Act). *http://cfpub.epa.gov/safewater/watersecurity/bioterrorism.cfm.*

EPA. (2003a). Emergency/Incident Planning. *http://cfpub.epa.gov/safewater/watersecurity/home.cfm=program_id=8.*

EPA. (2003b). Water on Tap: What You Need To Know. *http://www.epa.gov/safewater/wot/pdfs/book_waterontap_full.pdf.*

EPA. (2005a). EPA Microbiology Home page. *http://www.epa.gov/nerlcwww/.*

EPA. (2005b). Ground Water and Drinking Water: Drinking Water Contaminants. *http://www.epa.gov/safewater/hfacts.html.*

Epstein, J.R., Biran, I., and Walt, D.R. (2002a). Fluorescence-based Nucleic Acid Detection and Microarrays. *Analytica Chimica Acta* vol. 469:3–36.

Epstein, J.R., Lee, M., and Walt, D.R. (2002b). High-Density Fiber-Optic Genosensor Microsphere Array Capable of Zeptomole Detection Limits. Analytical Chemistry vol. 74:1836–1840.

Fatah, A.A., Barrett, J.A., Arcilesi Jr., R.D., Ewing, K.J., Lattin, C.H., and Moshier, T.F. (2001). An Introduction to Biological Agent Detection Equipment for Emergency First Responders. NIJ Guide 101-00. Washington, DC: National Institute of Justice, The Research, Development, and Evaluation Agency of the U.S. Department of Justice.

Fenselau, C. and Demirev, P.A. (2001). Characterization of Intact Microorganisms by MALDI Mass Spectrometry. *Mass Spectrometry Reviews* vol. 20:157–171.

Food and Drug Administration [FDA]. (2002). Bottled Water Regulation and the FDA [reprinted from *Food Safety Magazine* August/September 2002 issue]. *http://www.cfsan.fda.gov/~dms/botwatr.html.*

Gallant, J.E. (2002). Initial Therapy of HIV infection. *Journal of Clinical Virology* vol. 25:317–333.

Gesteland, P.H., et al. (2003). Automated Syndromic Surveillance for the 2002 Winter Olympics. *Journal of the American Medical Informatics Association* vol. 10:547–554.

Heffernan, R., Mostashari, F., Das, D., Karpati, A., Kulldorff, M., and Weiss, D. (2004). Syndromic Surveillance in Public Health Practice, New York City. *Emerging Infectious Disease* vol. 10:858–864.

Illinois Department of Public Health. (1996). Natural Springs Bottled Water Recall Announced. *http://www.idph.state.il.us/public/press96/water.htm.*

Kaufmann, A.F., Meltzer, M.I., and Schmid, G.P. (1997). The Economic Impact of a Bioterrorist Attack: Are Prevention and Postattack Intervention Programs Justifiable? *Emerging Infectious Disease* vol. 3:83–94.

LaGier, M.J., Joseph, L.A., Passaretti, T.A., Musser, K.A., and Cirino, N.M. (2004). A Real-Time Multiplexed PCR Assay for Detection and Discrimination of *Campylobacter jejuni* and *Campylobcater coli. Molecular and Cellular Probes* vol. 18:275–282.

Lawson, H.W., et al. (1991). Waterborne Outbreak of Norwalk Virus Gastroenteritis at a Southwest US Resort: Role of Geological Formations in Contamination of Well Water. *Lancet* vol. 337:1200–1204.

Lee, J. and Deininger, R.A. (2004). Detection of *E. coli* in Beach Water within 1 hour using Immunomagnetic Separation and ATP Bioluminescence., *Luminescence* vol. 19:31–36.

Magnuson, M.L., Owens, J.H., and Kelty, C.A. (2000). Characterization of Cryptosporidium Parvum by Matrix-Assisted Laser Desorption Ionization–Time Of Flight Mass Spectrometry. *Applied Environmental Microbiology* vol. 66:4720–4724.

Mansky, L.M. and Bernard, L.C. (2000). 3'-Azido-3'-deoxythymidine (AZT) and AZT-Resistant Reverse Transcriptase Can Increase the In Vivo Mutation Rate of Human Immunodeficiency Virus Type 1. *Journal of Virology* vol. 74:9532-9539.

McBride, M. T., et al. (2003). Multiplexed Liquid Arrays for Simultaneous Detection of Stimulants of Biological Warfare Agents. *Analytical Chemistry* vol. 75:1924–1930.

Meinhardt, P.L. (2005). Water and Bioterrorism: Preparing for the Potential Threat to U.S. Water Supplies and Public Health. *Annual Review of Public Health* vol. 26:213–237.

Min, J. and Baeumner, A. J. (2002). Highly Sensitive and Specific Detection of Viable Escherichia coli in Drinking Water. *Analytical Biochemistry* vol. 303:186–193.

Moore, M. (2003). Can Public Water Utilities Compete with Bottled Water? On Tap Online, Spring 2003. *http://www.nesc.wvu.edu/ndwc/articles/OT/SP03/BottledWater.html.*

Mulvey, M.R., Boyd, D., Cloeckaert, A., Ahmed, R., and Ng, L.K. (2004). Emergence of Multidrug-Resistant Salmonella paratyphi B dT+, Canada. *Emerging Infectious Disease* vol. 10:1307–1310.

National Resource Council. (2004). From Source Water to Drinking Water: Workshop Summary. *http://www.nap.edu/catalog/11142.html.*

New York City Department of Environmental Protection. (2005a). Cryptosporidium and Giardia Background Information and Monitoring Program. *http://www.ci.nyc.ny.us/html/dep/html/pathogen.html.*

New York City Department of Environmental Protection. (2005b). Rules and Regulations for the Protection from Contamination, Degradation and Pollution of the New York City Water Supply and Its Sources. *http://www.ci.nyc.ny.us/html/dep/html/ruleregs/finalrandr.html.*

Office of Management and Budget (2005). Fiscal Year 2006 Budget Priorities. *http://www.whitehouse.gov/omb/budget/fy2006/epa.html.*

Pedley, S. and Pond, K. (2003). Emerging Issues in Water and Infectious Disease. World Health Organization, Water Sanitation and Health. *http://www.who.int/water_sanitation_health/emerging/emerging.pdf.*

Pennsylvania Department of Environmental Protection. (2000). Pennsylvania DEP Expands Recall of Contaminated Bottled Water. *http://www.dep.state.pa.us/dep/deputate/polycomm/update/07-28-00/07280011.htm.*

Reuters. (2003). Italy on Alert for Water Poisoner. CNN.com. *http://www.cnn.com/2003/WORLD/europe/12/09/italy.water.reut/index.html.*

Sergeev, N., et al. (2004). Multipathogen oligonucleotide Microarray for Environmental and Biodefense Applications. *Biosensors & bioelectronics* vol. 20:684–698.

Shelton, D.R. and Karns, J.S. (2001). Quantitative Detection of *Escherichia coli* O157 in Surface Waters by Using Immunomagnetic Electrochemiluminescence. *Applied and Environmental Microbiology* vol. 67:2908–2915.

Shepard, J.R.E., Danin-Poleg, Y., Kashi, Y., and Walt, D. R. (2005). Array-based Binary Analysis for Bacterial typing. *Analytical Chemistry* vol. 77:319–326.

Straub, T.M. and Chandler, D.P. (2003). Towards a Unified System for Detecting Waterborne Pathogens. *Journal of Microbiological Methods* vol. 53:185–197.

U.S. Department of Labor. (2005). Incident Command System (ICS). *http://www.osha.gov/SLTC/etools/ics/index.html.*

U.S. Census Bureau. (1999). Historical National Population Estimates: July 1, 1900 to July 1, 1999. *http://www.census.gov/popest/archives/1990s/popclockest.txt.*

U.S. Government Accountability Office. (2004). Drinking Water: Experts Views on How Federal Funding Can Best Be Spent to Improve Security. Document GAO-04-1098T. *http://www.gao.gov/new.items/d041098t.pdf.*

United States Government Accountability Office. (2005). Protection of Chemical and Water Infrastructure: Federal requirements, Actions of Selected Facilities and Remaining Challenges. Document GAO-05-327. *http://www.gao.gov/new.items/d05327.pdf.*

Walden, J. and Kaplan, E. H. (2004). Estimating Time and Size of Bioterror Attack. *Emerging Infectious Disease* vol. 10: 1202–1205.

Wilson, W.J., Strout, C.L., DeSantis, T.Z., Stilwell, J.L., Carrano, A.V., and Andersen, G.L. (2002). Sequence-specific Identification of 18 Pathogenic Microorganisms Using Microarray Technology. *Molecular and Cellular Probes* vol. 16:119–127.

10

CHAPTER

Food and Pharmaceutical Industries

Ron M. Aryel
Del Rey Technologies, Los Angeles, California

T.V. (Sunil) Anklekar
Clariant Corporation, Elgin, North Carolina

1. THE FOOD INDUSTRY

Farming and food wholesaling are major businesses in the United States. In 2004, U.S. farmers harvested 279,253 bushels of barley (mostly for beer brewing), nearly 12 million bushels of corn, more than 115,000 bushels of oats, and more than 2 million bushels of wheat. In 2003, U.S. farmers produced more than 21.5 million hundred-weights of strawberries, nearly 18 million hundred-weights of dry beans, and nearly 20 million hundred-weights of broccoli (USDA, 2003). Food wholesaling accounted for $589 billion of sales in 1997. Dairy, poultry, meat, and fresh fruit and vegetable sales accounted for nearly three-fourths of this business; packaged frozen foods accounted for the rest (USDA, 2005).

The grocery and restaurant sectors are also major businesses in the United States. The National Restaurant Association (2005) estimates that Americans will spend $476 billion eating in 900,000 restaurants throughout the United States in 2005; U.S. restaurant sales will represent 46.7% of total food expenditures in this country. The Food Marketing Institute (2005), an industry trade group representing grocery stores, reported that Americans spent more than $457 billion in supermarkets in 2004.

1.1. Impact of Foodborne Illnesses

Food-borne illness is not uncommon in the United States and has significant economic impact. Food contamination causes 76 million illnesses, 325,000 hospitalizations, and 5,000 deaths every year, according to a Centers for Disease Control and Prevention (CDC) study published in 1999. The majority of deaths occur owing to unidentified agents, but 1,500 deaths per year can be attributed to *Listeria, Salmonella,* and *Toxoplasma* species (Mead et al., 1999). More than 200 different diseases may be transmitted in contaminated food, including ailments caused by bacteria, viruses, parasites, prions, bacterial toxins, and metals.

The impact of outbreaks owing to food contamination can potentially be severe. In 1951 in the French village of Pont-St.-Esprit, hundreds of villagers ate bread baked that morning from flour contaminated with ergot. The mold-infested flour caused villagers and pets fed bread scraps to hallucinate, the equivalent of a lysergic acid diethylamide (LSD) "trip." The incident caused widespread panic and many deaths, often self-inflicted as a consequence of the toxin ingestion (Fuller, 1968).

In 1994, truckers hauling ice cream premix in improperly sanitized tanker trucks, which previously held raw egg, exposed more than 220,000 people to salmonella poisoning. It was the largest single salmonella poisoning outbreak ever recorded in the United States (Hennessy et al., 1996). Although this event was a case of negligence, criminals could arrange this type of event with little trouble.

An intentional contamination might have even greater impact. Wein and Liu (2005) modeled the outcome of an intentional contamination of a milk processing facility with less than 1 gram of botulinum toxin. This model estimated that 100,000 casualties could ensue. Although some scientists have disputed the degree of vulnerability in the case of botulinum toxin, an intentional contamination of modern food distribution channels is possible.

1.2. Food Supply and Food Monitoring

In the United States, food arrives on a consumer's table via many paths of production and distribution, but the most common and typical path is as follows: growers (crop and animal cooperatives, farms, ranches, importers) sell food crops to food processors (packers, canners, or other manufacturing/processing/packaging operations). For example, grain farmers (e.g., those who grow wheat, corn, soybeans, barley, rice, sorghum, oats) bring their crops to grain elevators, which in turn ship it to makers of cereal or flour. The cereal manufacturers then sell to wholesale distributors, who in turn sell to retailers.

Individual manufacturers and wholesale distributors do not freely discuss measures they take to protect the commodities they ship from intentional contamination, beyond prohibiting nonemployees from entering warehouses, rail yards, feedlots, or grain silos. We note that interfering with the shipment of commodities by an interstate carrier, such as a trucking firm

Handbook of Biosurveillance
ISBN 0-12-369378-0

or railroad, is a federal crime. Railroads employ their own fully certified police officers to protect railroad property, but their officers, few in number, are responsible for vast territories.

2. FOOD PROCESSORS AND MANUFACTURERS

Food processors and manufacturers–who receive raw foodstuffs from farmers and can, jar, freeze, or otherwise prepare them for distribution–monitor their production processes to meet state and federal standards, as well as to ensure consistency of the prepared foods' appearance, aroma, and taste and to minimize waste and potential customer complaints. Protection against intentional contamination at present rests on physical security of the processes and not on biological monitoring of product.

The standards to which food is processed and checked are set by both companies and federal regulations (described below). Companies such as Kellogg's, General Mills, and Pillsbury impose stringent quality standards both internally and on their suppliers. For their suppliers, the quality standards cover not only contamination and spoilage but also the appearance, texture, size, and weight of the incoming foodstuffs. Their sales success depends on the degree of consistency they can achieve. For example, the largest buyer of potatoes in the United States, McDonald's Corporation, specifies to its supplier the specific strain of potatoes to be grown and its care and the size, weight, appearance, and texture that are acceptable for purchase. Every bag of French fries must look, feel, smell, and taste like every other bag.

Food processors and manufacturers generate and store information relevant to food safety and contamination, including records of pasteurization and irradiation. However, they do not routinely share data with the government except what is required for facility inspection and to comply with export requirements, as discussed below. Moreover, the industry is not vertically organized except in a few cases; hence these companies cannot track food from the farm all the way to the dinner table. Extensive records are available for review on products such as meat, seafood, low-acid canned foods, and juice that are produced under mandatory hazard analysis critical control point (HACCP) programs discussed in Chapter 3. The records required by regulations include HACCP plans, HACCP records, process authority approvals, and sanitary standard operating procedures (SSOP) records.

To convey the detail of food processing, we discuss two industry sectors in more detail: cereal grains and milk production.

2.1. Cereal Grains

When a cereal maker, such as General Mills or Kellogg's, produces cereal, it seeks to produce a product that is free of contamination and consistent in appearance, aroma, texture, and taste from box to box. To accomplish this objective, it executes purchase contracts with suppliers (usually large processors, such as Cargill, Louis Dreyfus, Archer Daniels Midland, and ConAgra) and specifies a maximum contaminant level for given contaminants, such as 4 parts per million (ppm) for vomitoxin, a mycotoxin that can affect flavors in foods and baking quality, in a five-car composite sample. (The profit margins on grain sales are measured in pennies, so testing must be reliable yet affordable; testing strategies may involve composite samples drawn from several hopper carloads.)

Before a train can pull out of the grain elevator's siding, the supplier will conduct testing to satisfy the buyer of the grain's quality. If an initial composite sample does not pass inspection, the elevator operator may reblend the load and try again or may then test individual hopper cars to find the source of the problem. If a given hopper carload fails inspection, the operator must empty and reload the car and test again (Stewart, 2005). The processor generally tests samples for the presence of bacteria, mold, fungal infestations, spores, and mycotoxins, as well as moisture, ash, fiber, protein, and carbohydrate content. Figure 10.1 is an example of a test request form used to process samples.

The processor may contract this work to a private laboratory, which will maintain these data by using a laboratory information management system (Chapter 8) as well as send the data to its client. Neither client nor laboratory shares these data with any government agency routinely. The potential value of the data to biosurveillance derives from our ability to prevent the distribution of contaminated grain, the consumption of which could lead to a range of consequences; among them is the real-life example of "St. Anthony's fire," the nightmarish scenario in the small village France described early in the chapter.

2.2. Milk Producers

Milk production in the United States in 2003 totaled 170 billion pounds, resulting from the tireless efforts of 9 million cows in more than 70,000 dairy herds (USDA, 2004b). Production is concentrated in 20 states, which house 7.8 million cows producing 147 billion pounds of milk. California is the largest milk producer, accounting for 20% of the U.S. milk supply (USDA National Agricultural Statistics Service, 2004).

Milk production and processing routines may vary slightly from state to state. In California, a farm will milk its cows and ship the raw milk by tanker truck to a silo, which can hold from 50,000 to 200,000 gallons. The milk then proceeds from the silo into a processing plant where pasteurization (heating to 170°F for 15 minutes), homogenization, and

FIGURE 10.1 CII Inc. laboratory request form. (From CII Laboratory Services, Kansas City, Missouri.)

10835 NW Ambassador Drive
Kansas City, MO 64153
Phone: 816 891-7337
Fax: 816-891-7450
e-mail: ciisvc@ciilab.com

TEST REQUEST FORM

Contact: ____________________ Pho: ____________________

Company: ____________________ Fax: ____________________

Address: ____________________ e-mail: ____________________

Customer Sample ID: ____________________

Proximate Analysis	
Moisture	
Ash	
Protein (Combustion)	
Fat (Ether Extraction)	
Fat (Acid Hydrolysis)	
Crude Fiber	
Total Dietary (Sol & Insol)	
Neutral Detergent Fiber	
Acid Detergent Fiber	
Calories (by calculation)	
Carbohydrates (by calculation)	

Physical Dough Testing	
Farinograph	
Alveograph	
Amylograph	
Standard	
Modified	
Starch (cook/cool)	
Wet Gluten	
Dry Gluten	
Gluten Index	
Mixograph	
M, A, P, FN, Farino	
M, A, P	

Bake Testing	
Sponge & Dough	
Pup Loaf	
Cake	
Cookie	
Finished Product Evaluation	

Grain and Flour Analysis	
Falling Number	
1000 Kernel Weight	
Test Weight	
Zeleny Sedimentation	
Experimental Milling	
Grade	

Grain and Flour (cont'd.)	
Starch Damage	
Total Starch	
Potassium Bromate	
Sanitation-Extraneous Matter	
Sanitation-Foreign Material	
Qualitative Enrichment	

Minerals	
Calcium	
Iron	
Phosphorus	
Potassium	
Sodium	

Physical Tests	
Granulation	
Alpine	
Ro-Tap	
pH	
Brookfield Viscosity	
Alkaline Water Retention	
Flour Color (Minolta)	
Sand and Grit	
Kernel Size Distribution	
Single Kernel Characterization	
Solvent Retention Capacity	

Mycotoxins	ELISA	HPLC	TLC
Aflatoxin			
Ochratoxin			
T2			
Vomitoxin			
Zearalenone			
Fumonisin			

Other	
GMO (Elisa PCR)	
ELISA	
Triple Test DNA	
Real Time Taqman	
Pesticide Screen	

Microbiology	
Standard Plate Count	
Coliforms	
E-coli	
Salmonella	
Staphylococcus (Coag. Positive)	
Yeast	
Mold	
Rope Spores	
Karnal Bunt	

Miscellaneous	
Semolina Speck Count	
ADA	
B_1	
B_2	
Niacin	
Cholesterol	
Vitamin A	
Vitamin C	
Saturated /Unsaturated Fat	
Folic Acid (HPLC)	
Sugar Profile (HPLC)	
Nutritional Label	
Sesame Seed Surface Oil	
Free Fatty Acids	
Peroxide Value	

***Special Notation on Low Carb**

Please note your fiber and protein sources and the use of sugar alcohols, etc. If possible, provide an estimate of these levels. This will expedite the processing of your sample. All information is strictly confidential.

Comments/Additional Information

____

____

____

____

This form can be printed from our website (www.ciilab.com) by selecting Test Request Form.

Form 003 / Revision B / October 25, 2004

vitamin fortification occurs. The processor packages the milk and places it into the distribution chain, where it heads (via wholesale distributors) to retail stores and institutional purchasers, including restaurants and catering services. The processor cleans each milk silo every 72 hours.

The U.S. Department of Agriculture (USDA) inspects milk processing plants under a voluntary program, set up under the Agricultural Marketing Act of 1946. Because compliance with USDA standards helps sell milk, participation is high. Inspections are designed to show the extent of a plant's raw milk, equipment, and procedure compliance with standards set forth in 7 CFR 58, subpart B, "General Specifications for Dairy Plants approved by USDA Inspection and Grading Service" (USDA, 2004a). The USDA will inspect processing plants under this arrangement at frequencies that depend on the plant's status. Usually, inspections are conducted every 5 to 9 months, although plants on probation will be inspected 10 days after the probationary citation.

USDA grades milk by sanitary manufacturing standards. The highest grade of milk is "A," which denotes milk that the processor has collected and cooled promptly; that has passed inspection for appearance, odor, and sediment; and that has bacterial cell counts of no more than 500,000 per milliliter. The producer retains milk production data but would send it to USDA as well.

3. RESTAURANTS AND RETAILERS

Restaurants and retailers (e.g., grocery stores) are the final point in the distribution system of foods. Both types of organization are concerned about the safety of their food as the companies' reputation as a safe source of food is important to their commercial success. Larger companies insist on stringent attention to hygiene and equipment maintenance, both at the supplier and the restaurant or store. To accomplish this goal, restaurant chains employ a quality assurance practice that collects voluminous amounts of data and seeks to reduce statistical variation as much as possible. These data can be of potential benefit to biosurveillance as they help identify the most vulnerable points in companies' operations. An increase in variation of food appearance, freshness, or amount of spoilage can be a symptom of either unintentional or intentional actions or omissions.

4. GOVERNMENTAL AGENCIES

Government agencies at the federal, state, and local level play a role in food inspection and regulation. We discussed the government regulation of food animals in Chapter 7, including the roles of the USDA and state departments of agriculture. In this chapter, we discuss the Food and Drug Administration (FDA), which has the responsibility for safety of food and drugs, and provide additional information on the role of state departments of agriculture relevant to monitoring of the food supply.

4.1. Food and Drug Administration

The FDA inspects food sold for export. FDA inspectors cannot physically inspect every bushel or hundred-weight of food crop manufacturers' export; they request documentation from the exporter and then inspect a percentage of the outgoing shipment (perhaps 10%). The federal government does not require inspections of crops grown, processed, and consumed domestically. Contractors working on behalf of the Federal Grain Inspection Service in each state conduct the mandated inspections. The inspection service records the elevator company who supplied the grain and the buyer slated to receive it. They will also record this information in the case of voluntary inspections (e.g., when a customer requests it of the supplier).

4.2. State Departments of Agriculture

Each state has an agency that oversees agricultural affairs and facility inspections. Typically, the agency is a department of agriculture, although other departments may be involved such as a health department or commercial trading regulator. The state department of agriculture's interests lie in assisting the state's agricultural producers to bring foodstuffs to market as well as regulating and inspecting facilities. Neither growers nor processors share data with the states other than what the law requires.

For example, in Nebraska, state inspectors rely on federal law 21 CFR 110, *Current Good Manufacturing Practice in Manufacturing, Packing or Holding Human Food* in enforcing minimum acceptable conditions in food processing facilities. An inspector looks for evidence of physical problems, such as improperly maintained equipment, inappropriate temperature settings, and loose screws or metal shavings that can fall into foodstuffs. The inspector also investigates for signs of biological contamination, such as rodent droppings, insect infestations (in grain silos), and mold, as well as evidence of improper hygiene or sterilization procedures.

Manufacturing or storage facilities may undergo inspection by the state only twice a year; thus, at any given time, the data for a given state-inspected facility will be an average of a few months old.

States inspect silos that store grain for export more frequently than those that store grain destined for domestic use. Countries buying grain from U.S. producers may require certification that the grain is free of contamination and may even require that the U.S. exporter fumigate the shipment (Reiman, 2005). Growers and processors have no obligation to report to the government every food safety problem that they detect occurring in the course of their business.

Some states perform rigorous inspections of produce, although this testing is not disease oriented, rather, it seeks to ensure the consistent quality and appearance of fruit and vegetable shipments. The inspections also have a sanitary function. Inspectors look for signs of rot and presence of rodents and agricultural pests; insects on the unwanted lists depend on the state and produce involved. For example, in Florida, more than 90% of citrus fruits are destined for juice extraction. The Florida Department of Agriculture and Consumer Services (FDACS) inspectors evaluate fruit for the presence of decay and maturity (defined by the ratio of sugar to acid according to a Brix scale, a method of measuring the density or concentration of sugar in a substance at a given temperature) and measure pound solids (a volumetric determination which governs how much buyers pay for the fruit). The department contracts with the USDA to inspect the extracted juice for color, foreign matter, flavor, oil, and Brix/acid ratio for purposes of grading the juice. Oranges shipped out of state for processing may fall under other organizations' inspection demands (such as the New York Board of Trade) regarding juice futures, or the USDA Commodity Purchase Program (which supplies food to school lunch programs).

FDACS stores inspection data on a database, *www.citranet.net*, and posts weekly summaries to a site where anyone can download them, *http://www.citranet.net/DownloadsCN.asp*. The state does not release producer-specific results owing to a public records exemption for trade secrets. The facilities themselves have log-ons and passwords to view their own results, as do suppliers.

Biosurveillance system operators would be interested in any data pertaining to the distribution of spoiled or contaminated fruit. However, they would want the fruit producers to submit data that are considered proprietary and to do so more frequently than weekly. Hence, at minimum, the biosurveillance system operators would have to negotiate legal agreements that protect trade secrets, as well as pay the costs involved in arranging the data transfers.

As is the case in food animal production, producers of citrus crops, reasoning that their market-driven quality standards exceed the standards set by the state of Florida, are proposing that an HACCP process replace routine full-time in-plant state inspections. The HACCP process would result in fruit processors' collecting large amounts of data, which biosurveillance system operators could try to access. Such data would facilitate environmental investigations (as discussed in Chapter 3).

States and/or local agencies inspect restaurants and food stores, grading them for safe food handling practices, cleanliness, and adherence to food safety codes. The state learns of problems through these periodic inspections, epidemiological investigations, or citizen complaints at the retail level, which are then traced back to a source.

States and/or local agencies store inspection and complaint data in a variety of ways. Some agencies maintain databases, others store records on paper, and others use combinations of paper and computer records. The computerized records may not always be in a form easily accessible for biosurveillance purposes.

4.3. Trace-Back Investigations

When individuals become ill, and suspicion falls on the food supply, governmental public health initiate a trace-back investigation, which enlists the assistance of food suppliers. As discussed in Chapter 3, trace-back investigations involve tracing a product by using shipping and purchase records at each link in the distribution chain back to common points that can explain the occurrence of illness among all or most affected individuals. Governmental public health regulations vary by jurisdiction and can provide the legal basis for trace-backs within state lines; however, in many cases, the FDA and CDC may become involved. Trace-back by the FDA involves numerous challenges, including the absence of records, the inability to require that records be maintained or provided, the existence of multiple sources of product and complex distribution systems, and the resource-intensive nature of the process, which may or may not confirm a contamination.

When consumers fall ill after eating food purchased from a retailer or restaurant, investigators will elicit a food consumption history, as well as the restaurants or retailers they patronized. The retailers or restaurants, in turn, identify the wholesalers. The wholesalers identify their suppliers, who may have difficulty identifying the farm or farms of origin because of product commingling. (The commingling of farm products is a major gap in trace-back ability.) At each point in the trace-back, inspectors observe the procedures for storing or processing the food, looking for abuses that would result in food contamination. Specimens may be taken to determine if other similarly handled food or the environment that the food was stored or processed in is contaminated. If there is evidence to suggest that contaminated food is in the distribution chain, then the manufacturer initiates a recall of that food.

Before recent molecular advances in laboratory techniques, it was not possible to know that two or more individuals who have never met and have no obvious exposure in common were, in fact, infected with the same rare strain of a virus or bacteria. With the advent of the new technology, it becomes possible to conduct case-control interviews and to determine that there is a common exposure to the same type of produce or food product. Figure 3.11 (Chapter 3) is a diagram from a FDA manual on trace-back. In this hypothetical trace-back, contaminated food infected individuals at four different point-of-service locations in different states. The trace-back allowed investigators to identify a single farm. Before these recent laboratory advancements, we would not even know that the four point-of-services events were related. Recent laws make it even easier to do trace-back, since food distributors are now required to maintain trace-back information.

4.4. Information Systems and Data

A variety of information systems store data about the source of ingredients, inspection results, and distribution of food products. Unfortunately, this information is distributed among a large number of data systems and organizations and is stored in largely incompatible formats. This situation makes gaining access to it more difficult.

4.4.1. Commercial Data

Food manufacturers and packers collect and store information regarding food (grains, fruits, vegetables) they purchase and use a variety of systems. They consider this type of information to be proprietary and highly sensitive.

A few very large vendors, such as Wal-Mart, have automated sales information and logistical systems, which are also linked to suppliers' computers, but these systems are optimized to track purchasing patterns and prevent supply bottlenecks. This design makes them potentially valuable sources of data for integrated biosurveillance systems but would not enhance detection of a discrete problem, such as a contamination incident, in the food chain per se. Commercial contract laboratories, such as CII Laboratories of Kansas City, Missouri, maintain laboratory information management systems that hold testing results.

4.4.2. Data Held by Federal and State Agencies

As noted above, the Federal Grain Inspection service retains information about grain intended for export. States and/or local agencies store inspection and complaint data in a variety of ways. Some agencies maintain databases; others store records on paper, and others use combinations of paper and computer records. The computerized records may not always be in a form easily accessible for biosurveillance purposes.

4.4.3. Food Emergency Response Network

Recognizing the need to respond quickly to foodborne disease outbreaks, the federal government created the Food Emergency Response Network (FERN), an effort to integrate the nation's food testing laboratories and improve response to the contamination of food (also discussed in Chapter 8). Roughly analogous to the CDC's Laboratory Response Network (LRN), FERN's members include federal, state, and local agencies. They include the FDA, CDC, USDA, Environmental Protection Agency (EPA), Department of Defense, Federal Bureau of Investigation (FBI), and Department of Homeland Security (DHS), state departments of public health and agriculture, and diagnostic veterinary laboratories. FERN's goals are to enhance surveillance sampling activity; improve laboratory capabilities, including "surge" capabilities (to handle a large outbreak); and help re-establish the safety of the food supply and consumer confidence. FERN improves laboratory capabilities through training, proficiency testing, method validation/development, improved surveillance, and communication. As of May 2005, 99 laboratories in 44 states and Puerto Rico operate in FERN. It is noteworthy that this laboratory network can also assist in evaluating water-borne contamination.

FERN's communication backbone is the Electronic Laboratory Exchange Network (eLEXNET), the first Internet-based food safety data network in the United States (Counter Terrorism Food Emergency Response Network, 2003). When launched in 2002, eLEXNET held data on *Escherichia coli* O157:H7, *Listeria monocytogenes, Campylobacter jejuni,* and *Salmonella* species (Levitt, 2003), but it has expanded to include thousands of analytes. eLEXNET's database of food sample and laboratory data is available online, and offers data mining and analysis tools such as On-Line Analytical Processing as well as a geographical information system. eLexNet communicates with LRN.

FERN intends to integrate three other communication and data networks under its organizational umbrella:

- FoodNet—In 1995, the FDA, USDA and CDC launched the Foodborne Disease Active Surveillance Network (FoodNet).
- PulseNet (discussed in Chapters 3, 5, and 8)—In May 1998, the federal government launched PulseNet, an Internet-based network and database linking food safety investigators at the CDC, FDA, USDA, and state public health laboratories. The network relied on pulsed-field gel electrophoresis (PFGE) to identify pathogen strains. By 2000, PulseNet linked all 50 states to a communication and laboratory data net, which reduced the time required to determine the strain of a microbe from 3 weeks to as little as 48 hours (Office of the Vice President, 1998). In 2001, PulseNet processed more than 25,000 pattern submissions (FDA, 2001). It is now possible to trace a contaminated source where, previously, (owing to the complex distribution system for food), we would not even have known that an outbreak had occurred. Agribusinesses may also detect outbreaks if they independently receive a report or reports of illness associated with a product.
- National Antimicrobial Resistance Monitoring Network (NARMS)—A joint effort by the CDC, FDA, USDA, and state health departments, the purpose of NARMS is to detect and characterize trends in bacterial resistance. Participating health departments forward every 20th non-typhi *Salmonella* isolate, every *Salmonella* typhi, every 20th *Shigella* isolate, and every 20th *E. coli* O157 isolate received at their public health laboratories to the CDC for susceptibility testing, which includes determination of the minimum inhibitory concentration (MIC) for 17 antibiotics: amikacin, ampicillin, amoxicillin-clavulanic acid, apramycin, cefoxitin, ceftiofur, ceftriaxone, cephalothin, chloramphenicol, ciprofloxacin, gentamicin, imipenem, kanamycin, nalidixic acid, streptomycin, sulfamethoxazole, tetracycline, and trimethoprim-sulfamethoxazole. The FoodNet sites (California, Colorado, Connecticut, Georgia, Maryland, Minnesota, New York, New Mexico, Oregon, and Tennessee) also send one *Campylobacter* isolate each week to the CDC. Susceptibility testing of *Campylobacter* is performed to

determine the MICs for these antibiotics: azithromycin, chloramphenicol, ciprofloxacin, clindamycin, erythromycin, gentamicin, nalidixic acid, and tetracycline (CDC, 2005).

4.5. Limitations of Current Monitoring

The food supply system is sufficiently large that direct monitoring for biological contaminants, as is being contemplated for the water system, is not feasible. Security of the food chain against accidental contamination is achieved through a combination of supplier initiatives designed to protect their businesses as well as federal and state regulations and inspections. Although the food industry's procedures work well to contain the impact of most food-related outbreaks, their limitations currently preclude the early detection advocated by the government for bioterrorism incidents. In essence, despite the potential represented by data sources we discussed earlier, current efforts to protect the population from contaminated food are limited to trace-back methods. More proactive measures are not in use because of prohibitive cost and lack of appropriate tools.

In the event of a contamination, trace-back investigation is the basis for limiting the extent of the event and preventing future events. Although the logistical aspects of a trace-back may appear straight forward to the casual observer, they are not. The information systems of the supply chains are not vertically integrated. Each entity knows only to whom they shipped or from whom they received food; even this information may be incomplete, as the wholesale warehouse will know which suppliers sell which foods but may not have lot numbers and dates available for immediate inspection.

One consequence of this situation is that, although tracing the source of a can of spoiled or contaminated green beans with a known lot number is relatively quick and easy, tracking the source or ultimate destination of foodstuffs reported by public health with unknown lot numbers remains a painstaking, tedious process. For this reason, the FDA proposed rules that would require food suppliers to be able to quickly locate food items at every level of the supply chain. The proposal met with fierce resistance, mostly related to the high cost of the requisite technology. The bar codes present today on product labels, which can be found on every American supermarket shelf, are inadequate to contain the requested data. Suppliers have concluded that enlarged bar codes would still not accomplish the main goal; instead, food suppliers are examining the value of radio-frequency identification (RFID) transponders to allow food shipments to be identified and located anywhere in the supply chain. They want to equip each case and, even individual containers, of food with transponder chips; however, they judge the hardware available today as far too expensive. Moreover, businesses at different levels of the food supply chain (growers, packers, wholesale distributors and retail stores) have not yet agreed how to divide the bill for such hardware. Without a financing plan approved at every level, or government support, the necessary investment will not occur.

In June 2002, Congress passed the Public Health Security and Bioterrorism Preparedness and Response Act of 2002 (Bioterrorism Act, Public Law 107–188), which President Bush signed into law on June 12, 2002. In October 2003, the FDA sent a report to Congress titled "Testing for Rapid Detection of Adulteration of Food." Congress wrote Title III of this law to empower the FDA to protect the food supply against terrorism and other threats (FDA, 2003).

The law's writers emphasized the FDA's authority to collect information about imported foods and to inspect incoming food shipments. Because rapid detection of adulterated foods is of high priority, Congress directed the Department of Health and Human Resources (HHS) to conduct research to determine the effectiveness of various techniques and technologies in detecting contamination in food, including biologic agents, chemical agents, and radioactive materials. Another research focus involves validating methods characterizing microbes that can work well in handheld kits.

In December 2004, to the consternation of the food distribution industry, the FDA took the first steps to requiring companies to track food distribution from one level to the next; the agency-issued regulations to interpret and administer the Bioterrorism Act. 21 CFR Parts 1 and 11 specify that persons involved in the processing, packaging, and distribution of food must maintain records regarding both the sources of the foods they receive and the recipients to whom they sell the food. The regulations apply principally to wholesale distributors, packers, importers, and others (The Federal Register, 2004).

Although the industry is complying, the thin margins inherent in the food distribution system, which includes processors, distributors, and retail stores, preclude heavy investment in new forms of information technology necessary to create a true "farm to dinner table" database. Thus, although the industry creates records to comply with federal law, these records may not provide the kind of access required for effective early detection of inadvertent or purposeful contamination.

5. THE PHARMACEUTICAL INDUSTRY

The pharmaceutical industry as we discuss it comprises manufacturers, distributors, retailers, and the FDA, which also regulates this business sector. FDA regulates active pharmaceutical ingredients (APIs), pharmaceutical excipients (substances other than the API that are used in the finished dosage form), sterile pharmaceutical products, biological products, investigational pharmaceutical products for humans, herbal medicinal products, radiopharmaceutical products, and medical devices.

The extended supply chain is vulnerable to both accidental and deliberate tampering. Tampering is the unlawful corruption or contamination of a drug product and/or its labeling and can refer to intrusion at any point in the manufacturing, shipment, and sale of a drug. The sale of counterfeit drugs can be considered a form of tampering.

5.1. Manufacturers, Distributors, and Retailers

A relatively few large, multinational firms comprise the bulk of the brand pharmaceutical manufacturing industry today. The 10 largest pharmaceutical corporations, as measured by U.S. sales, accounted for almost 60% of total U.S. sales of pharmaceuticals in 2004 (IMS Health, 2005).

The wholesale distribution industry is also closely held, having consolidated in the past 30 years, from approximately 200 distributors in 1975, to fewer than 50 in 2000. In 2004, the top three wholesale distributors accounted for almost 90% of the wholesale market.

The retail pharmacy landscape has changed from the traditional apothecary and drugstore model of decades ago. In 2004, drug store chains filled 41.2% of prescriptions, mail-order pharmacies 18.7%, other pharmacies 18.3%, supermarket pharmacies 12.3%, and mass merchants 9.6% (National Association of Chain Drug Stores, 2004).

5.2. The Food and Drug Administration

The FDA monitors and regulates the process of manufacturing the APIs, which are the key ingredients of any drug the FDA approves for marketing. The FDA holds the drug's maker responsible for producing a product that works as intended, is of consistent quality, and is free of contamination. These are necessary conditions to win the public's confidence. As noted above, the FDA regulates the manufacture of drugs: it licenses the API, approves the manufacturing process, and requires manufacturers to maintain records related to the manufacture of drugs. Any failure to comply, or deviation by the manufacturer, from the approved drug manufacturing process, can lead to the FDA ordering a shutdown of the manufacturing plant producing the API, a recall, or even withdrawal of the drug from the market. The FDA mandates the generation of data, which FDA inspects during its audits; the manufacturer, however, holds the data and it is from the manufacturer that a biosurveillance system would obtain data. Manufacturers closely guard these data as proprietary and confidential, and so governmental public health would have to obtain administrative, legal, and political agreements in order to gain access to them.

5.3. Drug Manufacturing

The manufacture of pharmaceuticals is a complex process, comprising numerous steps leading from raw materials to finished product. That finished product is the API, and it is this product, along with the necessary inert ingredients, packaging, and labeling, which the FDA licenses.

To understand what is involved in the manufacture of medicines, we should describe the basic elements of a pharmaceutical manufacturing and supply plan, which includes both obtaining raw materials and using them to create the desired product. A drug maker's plan for manufacturing a given medicine, which must pass muster with the FDA before the agency permits its distribution, contains a number of steps:

- Planning—A strategy for ordering the raw materials.
- Sourcing—Choosing the suppliers that will deliver the goods and services. Suppliers could be domestic or international.
- Making—This is the manufacturing step. Manufacturers schedule activities necessary for production, testing, packaging, and delivery. Manufacturing is also the most time-intensive portion of the supply chain. Manufacturing involves making the API and its conversion to the final dosage form, as mentioned above. More than one manufacturer may be involved, and manufacturing can take place in the United States or abroad. A pharmaceutical company may contract several chemical steps to foreign firms and then import the resulting intermediate compounds needed to complete the manufacture of the API domestically. The manufacturer can also outsource the manufacture of the API abroad; when it does so, the FDA closely monitors those plants. Conversion of API to the final dosage form may occur domestically or be "outsourced" to a foreign plant.
- Delivering—This is the "logistics" portion of the manufacturing and supply plan. Companies coordinate the orders from customers and choose transporters who carry the product to customers, such as wholesale distributors and pharmacies. Finally, the pharmacy dispenses the finished medicines to patients.
- Returning—This is the system for receiving defective and excess products back from customers and supporting customers who have problems with delivered products.

5.4. Current Good Manufacturing Practices

The FDA mandates the adoption and use of Current Good Manufacturing Practices (cGMPs) by the pharmaceutical industry (FDA, 1996). GMP manufacture applies to the following: APIs (bulk drug substances), pharmaceutical excipients (substances other than the API that are used in the finished dosage form), sterile pharmaceutical products, biological products, investigational pharmaceutical products for humans, herbal medicinal products, radiopharmaceutical products, and medical devices (FDA, 2004a).

cGMP regulations describe the methods, equipment, facilities, and controls required for producing human and veterinary products, medical devices, and processed food (FDA, 1996). cGMPs are built around safety, identity, strength, purity, and quality of the pharmaceutical product. The FDA developed GMP regulations over time in response to various public health crises caused by unsafe or ineffective medical products. FDA's charge, and the purpose of cGMPs, is to ensure that medical products are safe and have the "identity and strength and meet the quality and purity characteristics" that they are purported or represented to possess. The cGMPs are a quality assurance system that requires control of the drug (or device) production process from the time materials (excipients, actives, and packaging and labeling materials) are received at the plant, through production and testing, and shipment into the commercial marketplace. The requirements of the cGMPs, published in the Federal Register, are general and represent the minimum requirements around which a company should construct its own specific quality program. The FDA has certain expectations as to what must be covered and how matters must be handled to meet the requirements of GMP, depending on the type of product manufactured and the current state of the art in a particular industry. The FDA expects manufacturers to remain current in the state of the art of manufacture, testing, and packaging of their products, hence the name cGMP.

As noted above, the manufacture of most medicines involves many steps, which lead ultimately to the API. cGMP regulations apply to API and its final dosage form, and not necessarily to steps before the API, as explained earlier. This is the case with a non-cGMP contract manufacturer. However, most non-cGMP manufacturers comply with the International Organization for Standardization's (ISO) 9000 standards (International Organization for Standardization, 2005) or some such system in which quality of the chemical intermediates produced is subject to close monitoring and control, and the manufacturer must adhere to internationally accepted benchmarks for quality. We now present a brief description of ISO 9000 quality management.

The ISO 9000 family is primarily concerned with "quality management," which covers what the organization does to fulfill the customer's quality requirements and applicable regulatory requirements, while aiming to enhance customer satisfaction and achieve continual improvement of its performance in pursuit of these objectives. In practical terms, it means the manufacturer strives to reduce variation in the product to as close to zero as possible.

The cGMPs require thorough documentation. In FDA's eyes, if it is not documented, it did not happen or does not exist. Every procedure must be committed to writing and reflect the actual operating practices at the company, and all critical steps in the process must be controlled and recorded in a detailed manner. The FDA requires that critical processes, such as manufacturing, cleaning, and analytical testing be validated with documented evidence that they consistently perform their intended function this regulation assures that the result of the process meets predetermined quality attributes.

The FDA focuses on the company responsible for producing and packaging the drug; however, many manufacturers outsource manufacturing of several intermediates to other firms. The FDA does not oversee the manufacture of all intermediate compounds and does not regulate facilities that are not subject to its current cGMP regulations (see below). However, the final manufacturing stage, which results in API synthesis, is always subject to rigorous FDA control. Hence non-cGMP manufacturers, who synthesize intermediates and supply them to the drug maker, for the final stage, must also meticulously control their manufacturing processes. If they do not, they risk the introduction of impurities or contaminants into the pharmaceutical, which can lead to adverse effects in patients consuming the drugs.

The FDA conducts inspections of medical product manufacturers to determine if the company is operating in a state of compliance with cGMP requirements. Failure to comply with cGMP regulations results in range of consequences, from a listing of observations on an FDA form 483 and a possible warning letter, to injunction and seizure or to criminal indictment if repeated warnings are not heeded.

5.5. Counterfeiting

The FDA appointed a task force in 2003 to identify steps that the FDA, other government agencies, and the private sector could take to minimize the risks to the public from counterfeit medications. Counterfeiting, on the surface, does not appear related to biosurveillance, but it is when we consider that counterfeiters may introduce harmful substances into the drug distribution systems for consumption. The monitoring techniques being developed to address the problem of counterfeiting can potentially improve biosurveillance.

The FDA task force held discussions with the pharmaceutical industry, law enforcement officials, technology developers, and security experts and produced a report in February 2004 (FDA, 2004b). The report recommended that each drug have an electronic pedigree, which is a "secure record documenting the drug was manufactured and distributed under safe and secure conditions." A pedigree is a documented history or genealogy. In this context, a pedigree would provide the government with assurance that the product arriving in a drugstore or hospital stockroom is the genuine article—the same drug that emerged from the factory.

The FDA illustrated this method to show how the serialization and pedigree tracking process is expected to work for a drug manufactured in Virginia, which is sold to a pharmacy in Texas through a chain of wholesalers.

- A manufacturer in Virginia puts the serial number on the drug package and starts the pedigree to ship the drug to a wholesaler in South Carolina.
- The wholesaler adds to the pedigree.
- The drug is sold to a second wholesaler in Mississippi, who verifies that the pedigree has not been forged or altered. This custodian adds his certification to the pedigree.
- When a pharmacy in Texas purchases the drug, the product arrives, accompanied by its pedigree, showing its complete, authenticated history in the supply chain.

By 2006, the FDA expects the drugs vulnerable to counterfeiting to be serialized and tracked with electronic pedigrees. The National Association of Boards of Pharmacy (NABP) has published a list of the most vulnerable drugs for counterfeiting. NABP is a professional association representing the state boards of pharmacy in all 50 states in the United States, the District of Columbia, Guam, Puerto Rico, the Virgin Islands, New Zealand, eight Canadian Provinces, and three Australian states. Table 10.1 shows the 31 drugs deemed susceptible to counterfeiting.

5.6. Information Systems and Data

The pharmaceutical industry invests heavily in information technology to support manufacturing, prescription approval, postmarketing surveillance, adverse events reporting, marketing, and distribution.

5.6.1. *Manufacturing Data*

Pharmaceutical manufacturers' compliance with cGMP and ISO standards means that manufacturing is data driven. Both sets of standards require the manufacturer to measure and document many aspects of its operation. These data are both the basis for improving the manufacturing operation and safeguarding the product, and a potential source of data for a biosurveillance system that could pay attention to a pharmaceutical supply chain in ways analogous to surveillance of food supply chains.

TABLE 10.1 Drugs Vulnerable To Counterfeiting

1. Combivir (lamivudine/zidovudine)
2. Crixivan (indinavir)
3. Diflucan (fluconazole)
4. Epivir (lamivudine)
5. Epogen (epoetin alfa)
6. Gamimune (globulin, immune)
7. Gammagard (globulin, immune)
8. Immune globulin
9. Lamisil (terbinafine)
10. Lipitor (atorvastatin)
11. Lupron (leuprolide)
12. Neupogen (filgrastim)
13. Nutropin AQ (somatropin, *Escherichia coli* derived)
14. Panglobulin (globulin, immune)
15. Procrit (epoetin alfa)
16. Retrovir (zidovudine)
17. Risperdal (risperidone)
18. Rocephin (ceftriaxone)
19. Serostim (somatropin, mannalian derived)
20. Sustiva (efavirenz)
21. Trizivir (abacavir/lamivudine/zidovudine)
22. Venoglobulin (globulin, immune)
23. Videx (didanosine)
24. Viracept (nelfinavir)
25. Viramune (nevirapine)
26. Zerit (stavudine)
27. Ziagen (abacavir)
28. Zocor (simvastatin)
29. Zofran (ondansetron)
30. Zoladex (goserelin)
31. Zyprexa (olanzapine)

From National Association of Boards of Pharmacy, 2004.

Typically, a drug manufacturer's quality control department keeps track of incoming materials used in production. These records include information on suppliers and details on quantity and quality of raw materials received. For each incoming raw material, manufacturer often sets quality standards or "specifications" for the required purity of the material to be supplied and the kinds and degree of acceptable impurities.

The receiving manufacturer may not test all incoming raw materials. In most cases, a manufacturer accepts raw materials based on quality assurance provided by the supplier. A supplier will submit a "certificate of analysis" (CoA) with the materials supplied. The CoA details the quality of the raw material as analyzed by the supplier. Many raw materials arrive in several lots; each lot will arrive with its own CoA even though all are composed of the same chemical. Some suppliers will combine different lots of the same bulk chemical (chemical transported by tanker trucks or railcars) and only submit one CoA per truck or railcar. The manufacturer receiving these shipments may choose to sample the raw materials, depending on the criticality of the process. For high-value products, chemists will typically check these in-house and keep records. They may decide to test each of several lots separately or sample only a randomly selected portion of them. Once the manufacturer begins synthesizing intermediate products (which in turn become the ingredients used for synthesizing medicines), it will analyze the result of each synthetic step for yield, purity, physical properties (e.g., color), and types and amounts of impurities. The margin of error allowed in synthesizing pharmaceutical-grade chemicals is very small, as impurities in the API may retard effectiveness or cause adverse effects. Therefore, impurities in all steps leading to the API are also monitored closely and controlled to acceptable levels, and the manufacturer records detailed data about impurities.

The meticulous recording of data offers a way to detect the most vulnerable points of a manufacturing and supply chain to contamination or tampering. These data also represent a level of vertical organization to which the food industry could aspire, if only funding were available.

5.6.2. *Prescription Benefits Management/Prescription Verification*

There are about 60 pharmacy benefit management (PBM) companies in the United States; the largest players, which dominate the landscape, are Medco Health Solutions, Express Scripts, and Caremark. Pharmacies use the on-line systems operated by PBMs to verify health plan prescription coverage and complete the process of purchasing medicines by filing claims. The information recorded includes geographic and age information, the name and dosage of the drug, cost, and, in a few cases, an ICD-9 code signifying the underlying diagnosis. Prescription data enter a transactional system in real time;

periodically—often weekly—the transactional systems upload to a data warehouse designed to provide analytical "data mining" services. The data warehouse, built around modern relational technology, can be made readily accessible to public health authorities; its value is limited by the inherent lag time. Coupling warehouse access with direct access to the transactional system feeding can, potentially, produce an up-to-the minute ability to detect and interpret trends in prescription drug activity indicative of a threat to public health.

The transactional systems receive data about prescriptions as soon as they are presented to pharmacists for filling. Several companies that run such systems have been using legacy mainframe systems to store these data, and these file-based (VSAM) systems were somewhat difficult to interface with. To modernize them and facilitate compliance with the Health Insurance Portability and Accountability Act (HIPAA), the vendors introduced relational database technology to make interfaces easier. External entities, such as those charged with public health surveillance, can obtain data from this transactional system in two ways: by receiving transactions in parallel to the transactional system or launching SQL-based queries, or by using decision-support tools, against the transactional relational database.

The Veterans' Administration (VA) maintains a PBM database that could serve as a model for a source of data relevant to biosurveillance. The VA has connected to its VISTA electronic point-of-care (POC) system, in operation at all VA medical centers, a PBM database that updates monthly, extracting information from VISTA's electronic medical records. Its function is analogous to its commercial counterparts. The database stores information on both inpatient and outpatient medications (including both intravenous and oral medications); these data are specific enough to allow analysis of dispensing patterns at aggregate and geographically discrete levels or at the level of individual patients and addresses (Veterans Administration, 2001). Variations, or spikes, in dispensing patterns may potentially be of value for detecting disease outbreaks. To ensure the timeliness of these data, the VA would have to arrange for at least daily polling from VISTA, but such polling is very feasible.

Although using PBM systems' data for biosurveillance is technically feasible, it is important to note that the companies collecting these data will require payment for access to the data, as well as to offset the costs of any software modifications required to provide this new service. Biosurveillance system operators will have to deal with ethical and legal (such as confidentiality) issues, as well.

5.6.3. Postmarketing Surveillance: Adverse Event Reporting

Pharmaceutical companies and the FDA conduct postmarketing surveillance to detect problems with medicines after the FDA has approved them for distribution. These problems are similar to what can occur before a given drug is approved for sale; they include side effects, adverse interactions with other medicines, and manufacturing problems. Postmarketing surveillance, by its nature, benefits from large numbers of consumers, as opposed to the relatively limited numbers of people enrolled in premarketing studies, and can uncover issues that were statistically less likely to appear before a given medicine's commercial release. Postmarketing surveillance of drugs falls under the jurisdiction of the FDA. Although postmarketing surveillance is primarily intended to identify unexpected adverse effects of drugs, intentional contamination of drugs has occurred and adverse event reporting systems are a potential means by which such events may be detected.

For this reason, we review how pharmaceutical adverse-event reporting occurs in the United States, identify and examine related electronic systems, and consider whether additional systems are required to counter this threat.

Both pharmaceutical manufacturers and the FDA maintain adverse-event reporting, or safety, databases. The FDA's system—discussed below—contains reports that have been vetted by pharmaceutical makers but that are weeks old. Manufacturers' databases are populated with reports with actual latencies for serious or life-threatening incidents that may be as little as 48 hours, or as much as several days to weeks for minor or trivial drug reactions. In the latter cases, the patient may not have even bothered consulting a physician for care. Because monitoring for adverse events is more extensive in clinical trials involving investigational drugs, relevant information may be available earlier than 48 hours; public health authorities may use this to improve detection of events, such as drug tampering. Drug tampering can occur during such trials, but clinical trials do not represent the target of most tampering attempts.

Adverse event reports may vary in content, depending on the source of the report. A report from a physician will generally be the most complete, including a description of the event itself, the affected patient's relevant medical history, results of physical examination (if performed), and results of laboratory tests (if any). A call to an adverse drug reporting hotline by a consumer may yield a complaint that something is wrong, as well as some medical history, but will not include physical examination results or other data obtainable from a physician. A competent hotline interviewer should strive, in that case, to ensure that information relevant to most, if not all, categories of public health threats may be elicited.

FDA-Operated Adverse Event Reporting System. The FDA operates AERS, the Adverse Event Reporting System. This system resides on an Oracle database maintained by the agency. Adverse event reporting consists of a mix of voluntary and mandatory reporting. Reporting by manufacturers, healthcare providers, and patients may be accomplished in a number of ways: by telephone to a FDA hotline, by fax, or by submission of a form through the FDA's Web site. In June 1993, the FDA established Medwatch, which is specifically designed to facilitate voluntary reporting by healthcare providers.

Manufacturers are required by law to report adverse events. When they do so, names and addresses of patients suffering an adverse event are not reported to the FDA in the initial report, but the reporting manufacturer must maintain raw information used to file the report for at least 10 years.

Mandatory reporting of adverse events is governed by sections of the Code of Federal Regulations (CFR), including Part 21 sections 310.305, 312.32, 314.80, and 600.80. These regulations cover use of prescription drugs and biological products, investigational new drug (IND) safety reporting, and postmarketing reporting of adverse experiences. The FDA requires a pharmaceutical maker or distributor to report unexpected adverse events brought to its attention within 15 days. Although the regulations state such reporting should be accomplished as soon as possible, the absolute deadline is 15 days. Expected adverse events, meaning events noted during prelicensing trials, must be reported quarterly; in cases of serious or life-threatening events, the FDA requests that companies file within 15 days, a guideline that is generally accepted and adhered to. At least one company has stated it files adverse events reports with the FDA, for investigational drugs, within 7 business days. The 15-day reporting requirement reduces the timeliness and thus the potential value of AERS as a contributor to early detection of public health threats. According to FDA staff, the agency has not investigated or documented the average lag time for reporting; one staffer remarked that pharmaceutical makers, although adhering faithfully to the regulations, frequently filed reports very close to the deadline, implying that the average lag between the time of awareness of an event by a company and its report to AERS may be close to the 15-day limit. Officials at three pharmaceutical manufacturers confirmed this; although interviews with them revealed the average may be 8 to 12 days. They explained that the reason lay in a need to confirm the veracity of collected information. Moreover, this lag does not include the time between a patient's initial call or presentation for examination and the mandatory reporter's (i.e., pharmaceutical maker) becoming aware of it. This lag may be minimal, in the case of serious or life-threatening events, or it may be several days or even weeks for less serious events.

Another limitation is the misattribution of a report of illness or ill effect to another cause. Unless a physician, pharmacist, or the patient identifies a malady as a drug-related adverse event, no one will report it to AERS. Medwatch has four goals, stated on FDA's Web site. The following bullets are quoted exactly (MedWatch, 2005):

- Make it easier for healthcare providers to report serious events.
- Make it clearer to healthcare providers what types of adverse events FDA is interested in receiving.
- More widely disseminate information on the FDA's actions that have resulted from adverse event and problem product reporting.
- Increase healthcare providers' understanding and awareness of drug—and device—induced disease.

Medwatch does not demand adherence to a standard here for timeliness of reporting and inclusion of data on the AERS system.

Because of recent concerns that healthcare providers and the public are not learning about serious safety-related issues in a timely way, the FDA announced in February 2005 that it had established a new Web site, Drug Watch, to alert the agency's various constituents to problems identified after drugs are approved for marketing (FDA, 2005a). The FDA developed this Web site to address side effects, drug to drug interactions, and other safety data pertaining to the drugs themselves, not to enhance the early detection of epidemics or bioterrorism. Nonetheless, the decision is significant because it represents a new information-sharing policy. Previously, FDA consulted with pharmaceutical makers before releasing new safety-related information. Now, "we're not going to discuss it [public release] with them. They're not going to review it," acting CDER Director Steven Galson told Knight-Ridder Newspapers (Young, 2005).

How FDA Learns of Adverse Events. The FDA's staff learns of adverse pharmaceutical-related events in a number of ways. The pharmaceutical industry reports drug experience data before approval via clinical trial reports (phases I to III) submitted as part of a new drug application. After receiving marketing approval from the FDA, pharmaceutical companies send the agency the results of phase IV trials and regular postmarketing reports. These manufacturers also report data through AERS; the FDA also receives information from market research and tracking firms who work for drug makers, such as IMS or National Data Corporation.

Healthcare providers supply data to the FDA through AERS, and FDA staff regularly review published, peer-reviewed medical literature. Foreign drug regulators exchange information with the FDA on a regular basis. Figure 10.2 illustrates the various data sources available to the FDA.

Not all of these data sources can address the need for early detection. The best source available to the FDA is AERS.

Pharmaceutical Maker Adverse Event Reporting. Pharmaceutical makers maintain their own safety-related databases of adverse events, reported by healthcare providers and consumers. These database systems generate adverse event reports, as appropriate, to the FDA. The databases may be purchased from vendors or may be adaptations of software developed in-house by the manufacturer to record untoward events during clinical trials of investigational drugs. Recent installations all use relational database technology. Trained personnel,

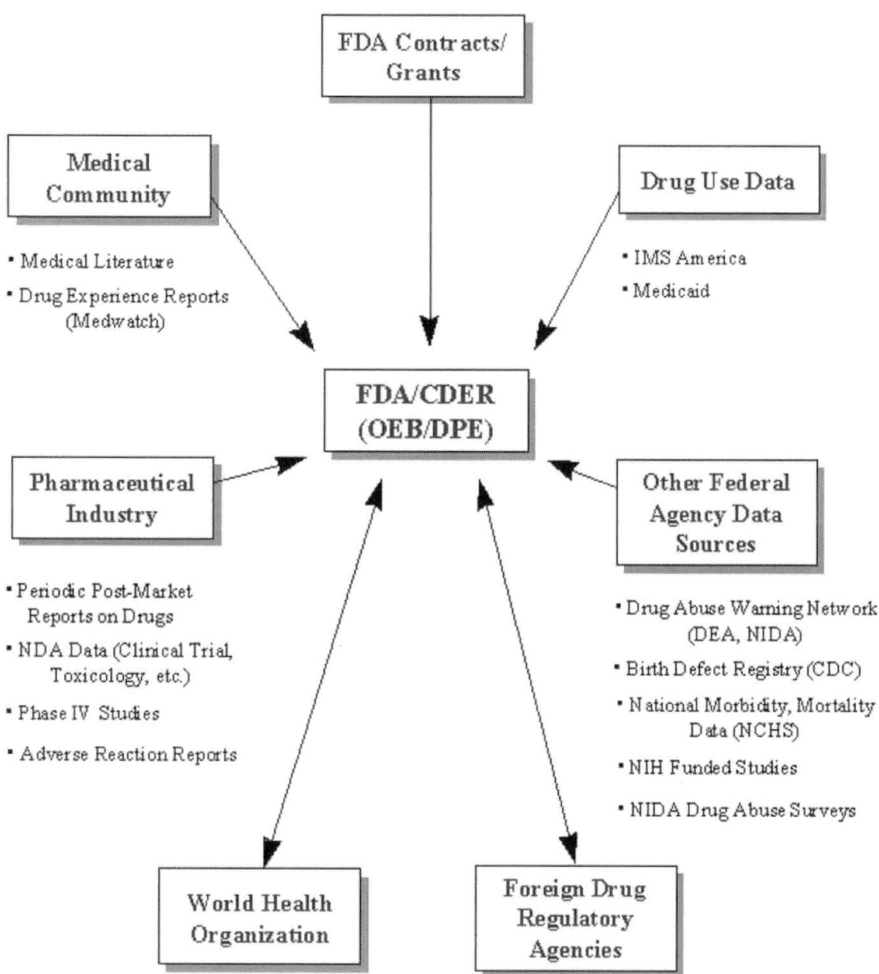

FIGURE 10.2 Postmarketing surveillance chart. (From the U.S. Food and Drug Administration Center for Drug Evaluation and Research, *http://www.fda.gov/cder/handbook/pmsinfo.htm*.)

usually registered nurses, who are familiar with the drug manufacturer's reporting requirements and protocols, record reporting data.

The time lag between receipt of report by a company and its appearance in a safety-related database is typically at least 48 hours, because the unit receiving the report must pass it on to other units that evaluate and code it before entering it in the database. In the case of serious or life-threatening events, the pharmaceutical maker receives the information very quickly, but query results from the safety database will still be at least 48 hours old. They may be even older, depending on the workload of the reporting unit and the employees' ability to "turn around" reports quickly. The following hypothetical example, based on actual events in the recent past, will illustrate how this process may be temporarily slowed: Company A's drug, intended to treat hypertension, is found to have caused a number of fatalities, and a highly publicized recall occurs. Company B markets a competing drug in the same class. Company B's consumer hotline is inundated with calls from people concerned that they are being adversely affected by Company B's drug. Each report, regardless of how improbable the actual connection to the drug may be, must be documented and followed up, resulting in an increased lag time from initial report to inclusion in Company B's database.

However, there is still some potential for safety databases' use in early warning. The company receives reports from its affiliates/programs, which in turn have stored reports only

24 hours old. In these cases, it may be possible for public health authorities to tap these databases. In addition, when drug manufacturers run clinical trials involving investigational drugs, tighter monitoring for adverse events occurs, which often enables the manufacturer's database to receive an adverse event report within 1 day. This makes the database useful for threats involving investigational drugs.

Pharmaceutical makers are very reluctant to allow outside parties, except the FDA under Part 21 of the CFR, to have access to company databases for fear of revealing proprietary information to competitors; inadvertently releasing information, out of context, which the public may misinterpret; and, of course, violating patients' right to privacy under HIPAA. Thus, the government would have to present a very compelling reason before any manufacturer or distributor would be willing to grant access to this kind of information. The manufacturers would also require the government to pay for any software or hardware development performed. Information from the safety database may be useful when coupled to the use of prescription tracking systems. One provides early, inferential warning; the other, more direct confirmation. Their complementary use may enhance detection of public health threats.

5.6.4. *Prescription Pharmaceutical Sales/Marketing Systems*

Pharmaceutical companies routinely outsource the tracking of the sales of their prescription medicines to firms that specialize in healthcare- or pharmacy-related marketing and sales analysis. These firms, such as IMS Health and National Data Corp., provide extensive sales tracking, market analysis, and customized reports to their clients. Each pharmaceutical maker is shown specific sales results in its markets, data that are not shown to competitors. These market surveillance firms will also provide data to researchers, but not at the same levels of granularity as afforded their primary clients, to ensure compliance with confidentiality agreements.

Pharmaceutical makers are provided with a range of customized reports. They detail many things, including where each drug was prescribed and sold, a comparison of one time period's sales with another, and a comparison of one sales region with another. The primary purpose of these services is to facilitate analysis of market performance for each product and to aid the pharmaceutical maker in making needed adjustments to its manufacturing, shipping, and marketing efforts. These data do not include diagnostic information or other data elements found in medical records. Their value is likely to be inferential, in that an unusual or unexpected increase in dispensing or shipments of a given drug may raise suspicion of an imminent public health threat. The data collected by the market surveillance firms arrives both electronically and on paper, with collection frequency usually ranging from weekly to monthly. Thus, raw data regarding a sales event are received at least 1 week after the drug has been sold; the resulting report package is sent to the client (pharmaceutical maker) with a lag of as much as 4 to 6 weeks. The lag is primarily related to cost, not technological obstacles.

Market surveillance firms are likely to be leery of offering public health authorities access to detailed, vendor-specific data. They decline to provide such details to researchers. But, even if a public health agency were to address such reservations and pay for software modifications with "early warning" functions, the results gained would probably be marginal. The time lag in both data collection and reporting significantly reduces the potential of pharmaceutical sales tracking systems to aid in the early detection of public health threats. They were not designed for this purpose.

5.7. New Technology and Future Developments

5.7.1. *Marketing Practices*

The FDA states that by the end of 2007, the adoption of reliable technology-based solutions for the electronic tracking, tracing, and authenticating of pharmaceuticals is expected to accomplish the goals of the Prescription Drug Marketing Act (PDMA), which was signed into law in 1988 and was modified by the Prescription Drug Amendments of 1992 (P. L. 102-353, 106 Stat. 941) on August 26, 1992 (Federal Register, 2003).

Most PDMA provisions became effective in 1988 to address some drug marketing practices that have contributed to the diversion of large quantities of drugs into a secondary market. These marketing practices include the distribution of free samples, the use of coupons redeemable for drugs at no cost or low cost, and the sale of deeply discounted drugs to hospitals and healthcare chains. These marketing practices provide a venue through which mislabeled, subpotent, adulterated, expired, and counterfeit drugs are able to enter the nation's drug distribution system. The most simple and straightforward of the acts prohibited by the PDMA is knowingly selling, purchasing, or trading, or offering to sell, purchase, or trade a prescription drug sample (U.S. House of Representatives, n.d.). Another key provision of the PDMA requires that drug wholesalers, before distributing any prescription drug, provide the purchaser with a statement of origin identifying all prior sales of the drug, including the dates of the transactions and all parties thereto (U.S. House of Representatives, n.d.). This provision, which does not apply to the drug's manufacturer and its authorized distributors, is intended to ensure that all prescription drugs in commercial channels are accompanied by a paper trail and can be traced, if need be, in the event of a recall or for any other reason. Violation of this requirement is a misdemeanor (U.S. House of Representatives, n.d.). However, if the offense is committed with the intent to defraud or mislead (including the intent to defraud or mislead the FDA) the offense is a felony.

5.7.2. New Technology

Just as food distributors mull over new technologies, and their benefits and costs, so do drug distributors. Drug distributors are looking at RFID for purposes similar to those of food distributors.

The electronic product code (EPC) is a unique number that identifies a specific item in the supply chain. EPC is a product numbering standard developed by the Uniform Code Council (UCC) that can be used to identify a variety of items using RFID technology. EPC tags were designed to identify each item manufactured, as opposed to only the manufacturer and class of products, as bar codes do today. Similar to a bar code, the EPC is divided into numbers that identify the manufacturer, product, version, and serial number. But the EPC uses an extra set of digits to identify unique items. The EPC is stored on an RFID tag. Once the EPC is retrieved from the tag, it can be associated with specific data, such as where an item originated, or the date of its production.

The goal is to track every item anywhere in the supply chain securely and in real time. RFID can dramatically reduce human error. Instead of typing information into a database or scanning the wrong bar code, goods will communicate directly with inventory systems. Readers installed in factories, distribution centers, and storerooms and on store shelves will automatically record the movement of goods from the production line to the consumer. There is no guarantee that the EPC system will catch on. But Wal-Mart, P&G, Coca-Cola, and others are implementing EPC systems. By using authenticated RFID to certify the tag and ensure chain-of-custody event information is available for supply chain software, pharmacies can confidently receive and distribute drugs that have moved through a safe and secure supply chain.

The data generated by RFID systems are likely to be very accessible, from a technical standpoint. However, the systems holding these data will be sensitive and highly guarded, and biosurveillance system access to them will require administrative, legal, and political agreements.

In November 2004, the FDA announced a new initiative to encourage use of RFID technology to protect drug supplies, and issued guidance for pilot studies using RFID (FDA, 2004c). Purdue Pharma announced it was placing RFID tags on bottles of Oxy-Contin, a powerful narcotic; in 2005, Pfizer pledged to introduce RFID for the shipment of Viagra, and Glaxo-SmithKline also announced its intention to introduce RFID tags for shipments of drugs judged likely to be counterfeited (FDA, 2005b).

As in food distribution, cost is a factor in RFID adoption. Manufacturers are focusing current efforts on specific drugs, not an across-the-board implementation. However, the largest pharmaceutical manufacturers have deeper pockets than do their counterparts in the food industry; hence, we can expect progress to be faster.

6. SUMMARY

Food, drugs, and medical devices are important routes of transmission of disease, and contamination of these products either accidentally or intentionally is of constant concern. Although there are well-practiced methods of conducting trace-back investigations to trace contaminated food shipments and procedures for reporting adverse events in our supplies of pharmaceuticals, more rapid conduct of such investigations will require further exploiting electronic sources of data, which presents fiscal, legal, and political challenges. The lack of vertical integration in food supply data poses a particularly vexing problem, but not an insurmountable one. Each level of the industry is concerned only with its immediate suppliers and customers, and not with the entire supply chain. Therefore, a manual trace-back investigation is the only method available to public health authorities to determine the source of a problem at the present time. Challenges facing investigators include a complex web of relationships between producers and consumers, limitations to governmental authority, and the limitations of existing information systems, limitations we are unlikely to overcome quickly unless we address the question of funding.

The FDA oversees various processes by which it can learn of problems with medicines. The medical community, drug industry, and consumers are all sources of information. Counterfeiting of medications is becoming a major issue, and here the FDA and the pharmaceutical industry have paired to introduce new methods and new technology, such as RFID, to protect and record information about drug shipments. These data are likely to be highly accessible from a technical standpoint, but accessing it will require legal and political agreements.

ACKNOWLEDGMENTS

We acknowledge the invaluable contribution made by Elaine Kellerman, whose meticulous research gave us a much better understanding of our food supply. We also thank Gretchen Stewart, Vice President, CII Laboratories, Kansas City, Missouri, for giving generously of her time and sage insights regarding cereal grain testing.

Clariant Corporation and its affiliated companies are not responsible for the contents of this chapter and the statements herein do not necessarily represent views or opinions of the Clariant Corporation or its affiliated companies.

REFERENCES

Aoki, N. (2001). Maker of AIDS Drug Battles Counterfeiters: Serostim's Growth Hormone May Give It Black-Market Value. *Boston Globe. http://www.sfgate.com/cgi-bin/article.cgi=file=/chronicle/archive/2001/02/05/BU170554.DTL&type=business.*

Centers for Disease Control and Prevention [CDC]. (2005). National Antimicrobial Resistance Monitoring System (NARMS): About NARMS. *http://www.cdc.gov/narms/about_narms.htm.*

Counter Terrorism Food Emergency Response Network. (2003). Counter Terrorism Food Emergency Response Network. *http://www.crcpd.org/Homeland_Security/Food_Emergency_Response_Network.pdf.*

Federal Register. (2003). Rules and Regulations. *http://www.fda.gov/CbER/rules/pdmapol013103.pdf#search='PDMA%20amendment%201992'%20(a)%2021%20U.S.C.%20%A7%20353(c)(1).*

Food Marketing Institute. (n.d). Facts and Figures. *http://www.fmi.org/facts_figs/keyfacts/grocery.htm.*

Fuller, J. (1968). *The Day of St. Anthony's Fire.* New York: MacMillan.

Hennessy, T.W., et al. (1996). A national Outbreak of *Salmonella enteritidis* Infections from Ice Cream. *New England Journal of Medicine* vol. 334:1281–1286.

IMS Health. (2005). 2004 Year-End U.S. Prescription and Sales Information and Commentary. *http://www.imshealth.com/ims/portal/front/articleC/0,2777,6599_3665_69890098,00.html.*

International Organization for Standardization. (2005). *http://www.iso.org/iso/en/ISOOnline.frontpage.*

Levitt, J. (2003). CFSAN's Program Priorities: From Food Safety to Food Security. *Food and Drug Law Journal* vol. 58:19–24. *http://www.cfsan.fda.gov/~acrobat/cfsan503.pdf.*

Mead, P., et al. (1999). Food-Related Illness and Death in the United States. *Emerging Infectious Disease* vol. 5: 607–625.

MedWatch. (2005). MedWatch Program. *http://www.fda.gov/cder/handbook/medwatch.htm.*

National Association of Boards of Pharmacy. (2004). National Specified List of Susceptible Products. *http://www.state.nj.us/health/eoh/slsdp.pdf.*

National Association of Chain Drug Stores. (2004). 2004 Community Pharmacy Results. *http://www.nacds.org/user-assets/PDF_files/2004results.PDF.*

National Restaurant Association. (2005). Restaurant Industry Fact Sheet 2005. *http://www.restaurant.org/pdfs/research/2005factsheet.pdf.*

Office of the Vice President. (1998). Vice President Gore Launches Computer Network to Fight Food-Borne Illness. *http://www.foodsafety.gov/~dms/fs-wh6.html.*

Reiman, R. (2005). Nebraska Department of Agriculture. Personal Communication.

Stewart, G. (2005). Vice President, CII Laboratories. Personal Communication, June 7.

The Federal Register. (2004). Rules and Regulations. DOCID: fr09de04-25. vol. 69:71561–71655 *http://wais.access.gpo.gov.*

U.S. Department of Agriculture [USDA]. (2004a). DA Instructions 918-PS. Instructions for Dairy Plant Surveys. *http://www.ams.usda.gov/dairy/section_1-4.pdf.*

USDA. (2004b). Licensed Dairy Herds. National Agricultural Statistics Service. *http://usda.mannlib.cornell.edu/reports/nassr/dairy/pmp-bb/specda04.txt.*

U.S. Food and Drug Administration [FDA]. (1996). 21 Code of Federal Regulations. Part 210–Current Good Manufacturing Practice in Manufacturing, Processing, Packing, or Holding of Drugs. Part 211–Current Good Manufacturing Practice for Finished Pharmaceuticals. *http://www.fda.gov/cder/dmpq/cgmpregs.htm.*

FDA. (2001). Food Safety Progress Report. Fiscal Year 200. *http://www.cfsan.fda.gov/~dms/fsirp004.html.*

FDA. (2002a). Important Drug Warning: Counterfeiting of Procrit. *http://www.fda.gov/medwatch/SAFETY/2002/Procrittamper_UD.PDF.*

FDA. (2002b). The Prescription Drug Marketing Act. Report to Congress. *http://www.fda.gov/oc/pdma/report2001/report.html#background.*

FDA. (2003). Testing for Rapid Detection of Adulteration of Food. Report to Congress. *http://www.fda.gov/oc/bioterrorism/report_congress.html#intro.*

FDA. (2004a). Code of Federal Regulations. Title 21–Food and Drugs. *http://www.accessdata.fda.gov/scripts/cdrh/cfdocs/cfcfr/CFRSearch.cfm?CFRPart?211&showFR=1.*

FDA. (2004b). Combating Counterfeit Drugs. *http://www.fda.gov/oc/initiatives/counterfeit/report02_04.html.*

FDA. (2004c). FDA Announces New Initiative to Protect the U.S. Drug Supply through the Use of Radiofrequency Identification Technology. *http://www.fda.gov/bbs/topics/news/2004/NEW01133.html.*

FDA. (2005a). FDA Improvements in Drug Safety Monitoring. *http://www.fda.gov/oc/factsheets/drugsafety.html.*

FDA. (2005b). Radiofrequency Identification Technology: Protecting the Drug Supply. *http://www.fda.gov/fdac/features/2005/205_rfid.html.*

U.S. House of Representatives. (n.d). 21 U.S.C. § 333(a)(1). *http://www4.law.cornell.edu/uscode/html/uscode21/usc_sec_21_00000333—000-.html.*

U.S. House of Representatives. (n.d). 21 U.S.C. § 353(c)(1). *www4.law.cornell.edu/uscode/html/uscode21/usc_sec_21_00000333—000-.html.*

U.S. House of Representatives. (n.d). 21 U.S.C. § 353(e)(1). *http://www4.law.cornell.edu/uscode/html/uscode21/usc_sec_21_00000333—000-.html.*

United States Department of Agriculture [USDA]. (2003). Crop Production. National Agricultural Statistics Service. *http://usda.mannlib.cornell.edu/reports/nassr/field/pcp-bb/2003/crop1203.txt.*

USDA. (2005). *http://www.usda.gov/wps/portal/usdahome.*

USDA National Agricultural Statistics Service. (2004). Milk Production. *http://usda.mannlib.cornell.edu/reports/nassr/dairy/pmp-bb/2004/mkpr1204.txt.*

Veterans Administration, 2001. *http://www.virec.research.med.va.gov/References/VirecInsights/Insights-v02n2.pdf.*

Wagner, M., et al. (2004). National Retail Data Monitor for Public Health Surveillance. *Morbidity and Mortality Weekly Report* vol. 53:40–42. *http://www.cdc.gov/mmwr/preview/mmwrhtml/su5301a9.htm.*

Wein, L.M. and Liu, Y. (2005). Analyzing a Bioterror Attack on the Food Supply: The Case of Botulinum Toxin in Milk. *Proceedings of the National Academy of Sciences* vol. 102:9984–9989. *http://www.pnas.org/cgi/content/abstract/0408526102v1.*

World Health Organization. (2003). Cumulative Number of Reported Probable Cases of SARS. *http://www.who.int/csr/sars/country/2003_06_30/en/.*

Young, A. (2005). New FDA online site issues alerts. *Kansas City Star*, May 21.

11

CHAPTER

Coroners and Medical Examiners

Ron M. Aryel
Del Rey Technologies, Los Angeles, California

Michael M. Wagner
RODS Laboratory, Center for Biomedical Informatics
University of Pittsburgh, Pittsburgh, Pennsylvania

1. INTRODUCTION

The public authorities charged with investigating causes of death and assisting with identification of deceased persons are either medical examiners or coroners. Medical examiners are appointed officials, while coroners are usually elected public servants. Medical examiners are always physicians; often they are forensic pathologists, meaning they have received specialized training in determining cause and manner of death. Coroners, on the other hand, are not required in most jurisdictions to be medical doctors. When a death did not obviously result from natural causes, a coroner conducts an inquest or investigation before a jury to determine cause of death. If the coroner is not a physician, the coroner's office appoints a licensed physician to perform the forensic medical procedures required or a coroner's office may employ a team of doctors, pathologists, and forensic pathologists. In this chapter, we use the term medical examiner to refer to all these entities.

Offices of medical examiners range in size from a single individual, who serves part-time, to departments at the city or county level that employ full-time physicians, paraprofessionals, laboratory technicians and deputies who visit crime scenes, interact with law enforcement, collect bodies while preserving evidence, and testify in court. Other specialists who may offer their services include dentists, radiologists, radiology technicians, toxicologists, genetics experts, and anthropologists. In short, many of the same competencies applicable to the care of living patients are also applicable to inquiries into causes of death.

The training of the medical examiner's staff varies by job description. A coroner need not be a physician, but will often have a medical background. A coroner may have a law enforcement background; in California, for example, coroners are also sworn peace officers who have attended a law enforcement academy (Riverside County, 2005). Universities now offer master's degree programs aimed at coroners (Philadelphia College of Osteopathic Medicine, 2005, Kezdi, 2005). They develop a mastery of skills such as interviewing, obtaining medical histories, evidence collection, and report writing.

Forensic pathologists are physicians who have completed medical school and residency training; academic centers have created fellowships in forensic pathology, designed to train pathologists to serve as medical examiners (Office of Chief Medical Examiner State of North Carolina, 2005). Forensic dentists and odontologists understand the structure and diseases of the teeth and gums and can help establish a diagnosis that is either the cause of death, contributory, or coincidental.

2. FUNCTION OF THE MEDICAL EXAMINER

Medical examiners investigate deaths due to homicide, suicide, or accidental violence, and deaths of persons unattended by a physician, or who succumbed to a contagious disease. They also intervene in cases where death occurs amid suspicious circumstances. Examples of the latter include the sudden death of persons in apparently good health, or who die while in the custody of law enforcement officers. The medical examiner is empowered to overrule family members or legal guardians who refuse permission for an autopsy; however, in cases where the cause of death becomes obvious upon preliminary review, the medical examiner, at his sole discretion, may decline the case and allow family members to claim the body without an autopsy.

Medical examiners are licensed physicians. They are most often pathologists by training, but some medical examiners, especially those for whom the job is a part-time occupation, are family practitioners or have other specialties.

The medical examiner investigating a death will conduct an autopsy. The medical examiner must carefully review and record a history and pertinent past medical history, supplied by witnesses, family members, and medical records obtained from the deceased person's healthcare providers; he/she must then perform a thorough physical examination, which includes inspection of the body, and examination, or weighing, and dissection of organs. The medical examiner may order radiological and laboratory tests as appropriate. Pathology specimens may be prepared as well. The purpose of this work is to reach a conclusion regarding the cause of death. The cause of death may be considered the deceased person's diagnosis.

The medical examiner must prepare a record for a deceased person, not unlike the medical record other physicians prepare

Handbook of Biosurveillance
ISBN 0-12-369378-0

for their living patients, which includes previous medical history and the results of the exams and procedures described above.

3. ROLE OF MEDICAL EXAMINER IN BIOSURVEILLANCE

The information collected by medical examiners in the course of their work is valuable when considering every category of lethal, or potentially lethal, public health threat. To understand what a vital resource medical examiners represent to disease surveillance, consider that our nation's examiners certify approximately 20% of all deaths in the United States.

Medical examiners must notify governmental public health whenever the deceased suffered from a reportable communicable disease, and whether that disease was the cause of death. They also report deaths resulting from environmental conditions of interest to governmental public health agencies, such as extreme heat during the summer (Young, 2005). Prompt efficient reporting by the medical examiner can assist in the early detection of an outbreak, assuming that the medical examiner did not, by happenstance, decline a relevant case (a given deceased person).

4. MEDICAL EXAMINERS' DATA: ACCESSIBILITY–USE OF ELECTRONIC RECORDS

Although medical records used by doctors and medical examiner records share common elements, there are, obviously, some aspects of medical examiner records, such as organ weights and dissection observations that are unique. Hence, information–system vendors developed electronic medical record systems especially for medical examiner use. The vast majority of medical examiners covering large U.S. cities and suburban jurisdictions has installed, or is installing now, electronic record systems. Early generation systems—developed in-house or sold by commercial vendors—used relatively inflexible data storage schemes, making their efficient utilization problematic, although they still represented an improvement over paper records. Current systems rely on relational databases and are as adaptable to new uses as modern electronic medical record systems. Moreover, their prevalence among most large medical examiner jurisdictions in the United States U.S. means they can play an important role in detecting a public health threat. This market penetration is in contrast to point-of-care systems for live patients, whose less frequent deployment reduces their potential utility in biosurveillance. (although other limitations of medical examiner data, such as the limited number of people that die and come to autopsy offsets this deployment advantage).

Of note is that medical examiner electronic record systems use laboratory information systems in the same way that point-of-care systems do for live patients. Laboratory systems can send data to a medical examiner's electronic record system. In theory, biosurveillance systems will be able to tap these data, either from the laboratory system, or the medical examiner's system.

There is no insurmountable technological barrier preventing a given biosurveillance system from accessing medical examiners' records in ways very similar to the access achievable with standard electronic medical records. However, this may require extensive modifications to existing software, depending on the system involved.

5. TIMELINESS

The postmortem medical examination, conducted only rarely, produces data about an event that is inherently late (death), using methods that are time consuming (establishing with certainty the cause of death). This implies an inherent limitation to the timeliness of medical examiner data.

The medical examiner can make available the results of the physical examination soon after its completion, thanks to the use of electronic record systems. A biosurveillance system could access the data regarding the deceased, as would be the case with any electronic medical record, as soon as the medical examiner enters them into a database (assuming that technical integration and administrative barriers have been addressed). However, some data, such as laboratory values or pathologic examination, might not be available until later. Additionally, while live patients can usually supply a past medical history immediately, the medical examiner may have to wait longer to collect such information; in some unfortunate cases, there is no one to supply it.

As a result, the time lag between time of death and a medical examiner's reporting of data regarding the deceased can be protracted. The postmortem examination takes time. If only body parts are available for autopsy, the true cause of death may take longer to establish. A heavy caseload can delay completion of autopsy reports. A mass-casualty incident, which is a rare event, can cause delays measured in days or even weeks. Unless special assistance is requested for a mass casualty incident (and such help may itself be delayed by days), the medical examiner's office must deal with whatever the daily demand is with a fixed staff. This means that, on a given day, some bodies, without obvious law enforcement or immediately discernible public health priority (higher than other bodies involved in the same mass casualty incident) may be placed in a freezer compartment until attention can be paid to them. The same may occur in cases of unrelated deaths if the medical examiner's staff is diverted to a mass casualty incident. In a recent example, an efficient and experienced mortuary team was deployed by the National Disaster Medical System to augment a county medical examiner's response to a commuter plane crash. The team required three days to complete the autopsies of 19 air-crash victims; their reporting did not include all laboratory test results, which were reported later. In another example, a county medical examiner reporting a jumbo jet crash off Long Island, New York, in 1996 did not complete the bulk of the required work for weeks; DNA-based identification of the last two victims was finally accomplished 13 months after the plane fell into the sea (New York Times, 1997). A caveat here, however, is that such data can usually be recorded and released sooner to law enforcement or governmental public health than to families.

Given an average workload, significant delays are the exception rather than the rule. Whether a medical examiner can maintain efficient operation in the face of an epidemic, or bioterrorist attack, depends on the incident's effect on the medical examiner's workload and the medical examiner's ability to quickly deploy additional staff.

The utility of medical examiner data in the context of an epidemic, whether organic or a case of bioterrorism, also depends on the degree to which the medical examiner has integrated the department's information systems and linked them to a biosurveillance system, and the degree to which the medical examiner's personnel can continue functioning efficiently.

Thus, medical examiner data seems most helpful in outbreaks that have a long window of opportunity for intervention, are fatal, and are distinctive enough or large enough to come to postmortem examination. There are many pathogens and outbreaks, for example, which have this characteristic, including Hantavirus (where there is a continuous source of exposure that will continue to produce death until it is identified), or person-to-person contagious diseases, or vector-borne illness, such as West Nile. At the other extreme, an extremely rapidly progressing outbreak, such as one caused by a terrorist's point bioaerosol release, would benefit from data that are inherently earlier.

6. RELIABILITY AND UTILITY

The reliability and accuracy of data generated by medical examiners is equivalent to that of any other medical provider. Since medical examiners are licensed physicians, they are ethically and legally obligated to maintain accurate records, the same as any other physician.

In general, the reliability of ancillary data received by the medical examiner, such as clinical laboratory data, copies of the deceased's previous medical charts and radiographs would be expected to be high. Moreover, a medical examiner's chart of the deceased is inherently more detailed regarding the physical examination than that created by physicians regarding their living patients, due to the former's ability to directly view and dissect organs and prepare extensive pathology specimens for analysis where needed.

7. SUMMARY

Offices of Medical Examiner or Coroner are public agencies charged with investigating unexpected or suspicious deaths and those where the victim was not attended by a physician. These departments have a public health reporting obligation similar to other healthcare providers. They use information systems that are similar in many respects to point-of-care systems used by physicians. Thus, we can expect reliability and data accuracy to be high. Virtually all large jurisdictions use these systems, so availability should also be high and they constitute a potentially useful source of information for biosurveillance. Post-mortem medical examinations, however, are rare and inherently late in the timeline of disease. Mass casualty incidents can cause significant delays in data entry. The utility of medical examiner data in the context of an outbreak, whether naturally occurring or a product of bioterrorism, also depends on the degree to which the information systems are integrated and linked to biosurveillance systems, and the nature of the outbreak.

8. REFERENCES

Kezdi, M. (2005). Forensic nursing one-of-a-kind in Northeast Ohio. The Cleveland Stater: A Laboratory Newspaper at Cleveland State University. *http://www.csuohio.edu/clevelandstater/highlights/highlights5.html.*

New York Times. (1997). Last two victims identified in Flight 800. Metro news briefs: New York. August 17.

Office of Chief Medical Examiner, State of North Carolina. (2005). Program Description: Forensic Pathology Fellowship. *http://www.pathology.unc.edu/fellowsp/forensic.htm.*

Philadelphia College of Osteopathic Medicine. (2005). Program Description: Master of Science in Forensic Medicine. *http://www.pcom.edu/Academic_Programs/Degree_Programs/Degree_Programs_Physician_Assi/PA_Advanced_Masters_Brochure/pa_advanced_masters_brochure.html.*

Riverside County, California, Sheriff Department. (2005). California Penal Code Section 830.35(c). *http://www.riversidesheriff.org/coroner/index.html.*

Young, T. (2005). Medical Examiner, Jackson County, Missouri Office of Medical Exminer. Personal Communication, April 15.

12

CHAPTER

Other Organizations That Conduct Biosurveillance

Michael M. Wagner
RODS Laboratory, Center for Biomedical Informatics, University of Pittsburgh, Pittsburgh, Pennsylvania

Julie Pavlin
Lieutenant Colonel, US Army, Uniformed Services, University of the Health Sciences, Washington, DC

Kenneth L. Cox
Colonel, US Air Force, Force Health Readiness, Tricare Management Activity/Deployment Health Support Directorate, Washington, DC

Nick M. Cirino
Biodefense Laboratory, Wadsworth Center, New York State Department of Health, Albany, NY

1. INTRODUCTION

This chapter completes our review of the organizations listed in Table 1.2 (Chapter 1). In particular, we explore the biosurveillance roles of the U.S. Department of Homeland Security (DHS), transportation systems (planes, trains, and ships), the U.S. Postal System (USPS) and the U.S. Department of Defense (DoD). Like the other organizations we discussed in Part II, these organizations have counterparts in most other countries. We also discuss the World Health Organization (WHO), the United Nations agency responsible for advancing health.

2. DEPARTMENT OF HOMELAND SECURITY

The DHS was established in March 2003 as a result of the terrorism events of 2001. According to Homeland Security Presidential Directive 10, the Secretary of Homeland Security is the principal federal official for domestic incident management and is responsible for coordinating domestic federal operations to prepare for, respond to, and recover from biological weapons attacks (White House, 2004). The Secretary of Homeland Security coordinates with the heads of other federal departments and agencies to accomplish this mission.

DHS plays a principal role in the monitoring of air for pathogens and has broad responsibilities for situational awareness (of which biosurveillance is but one dimension) for terrorism threats and events.

2.1. Organization of DHS

DHS is comprised of five major divisions or directorates: the Science and Technology Directorate, Border and Transportation Security, Emergency Preparedness and Response, Information Analysis and Infrastructure Protection, and Management. Besides the five directorates, several other agencies are folding into the new department or being newly created. They include the Federal Emergency Management Agency (*www.fema.gov*), Customs and Border Protection (*www.cbp.gov*), Immigration and Customs Enforcement (*www.ice.gov*), and the U.S. Citizenship and Immigration Services (*http://uscis.gov/graphics/*). The Science and Technology Directorate is active in biosurveillance. DHS has an annual budget of approximately $40 billion.

2.2. Biosurveillance Initiatives of DHS

DHS initiatives in biosurveillance include increasing the deployed level of capability at the city and state level as well as the ability to integrate data from all sources (intelligence, human, animal, and agricultural surveillance) to form a national view. Table 12.1 presents a U.S. General Accounting Office summary of DHS biosurveillance information technology initiatives in 2005.

2.2.1. BioWatch Air Monitoring

DHS has spent over $200 million on the BioWatch program (Lipton, 2005), whose goal is to detect the presence of biological agents in the air, and thus provide warning of a surreptitious release prior to victims' development of symptoms (U.S. Department of Homeland Security Science and Technology Division). The BioWatch program has deployed approximately 500 air sampling devices in 31 cities (Bridis, 2003). BioWatch devices continuously draw air through a collecting filter. The air samples are taken daily to laboratories for testing for several biological agents, including *B. anthracis, Y. pestis,* the smallpox virus, and *F. tularensis* (Bridis, 2003). DHS deploys additional sampling devices and mobile laboratory facilities during special events.

The current density of deployed devices is sparse relative to the estimated need. Although the devices are positioned to maximize the probability of detecting an attack (under assumptions about the most likely release points, agents, and quantity of material released), a detection probability of 1.0 is difficult to achieve with approximately 10 devices that exist in most cities (Lipton, 2005). In recognition of this problem, the

ISBN 0-12-369378-0

TABLE 12.1 Department of Homeland Security Biosurveillance Initiatives

Initiative	Description
Biological Warning and Incident Characterization System (BWICS)	A system that is expected to integrate data from environmental monitoring and health surveillance systems with incident characterization tools[a] in order to provide timely warning of a biological attack and to help guide an effective response. BWICS is also expected to provide secure distribution of information to different types of users.
BioNet	A cooperative program between DHS's S&T Directorate and DOD (established as a demonstration project in May 2004) that is expected to integrate civilian and military capabilities at the local level for detecting and responding to the use of biological agents. The BioNet initiative is being developed in one city. It includes the use of a syndromic surveillance system known as the Electronic Surveillance System for the Early Notification of Community-based Epidemics (ESSENCE).[b] DHS plans now call for BioNet to be terminated in fiscal year 2005 with lessons learned, tools, and capabilities transferred to the BWICS initiative.
BioWatch	An early-warning environmental monitoring system that collects air samples from high-threat cities in order to detect trace amounts of biological materials. BioWatch consists of three IT components: a sample management tracking system, a lab analysis tracking system, and an electronic reporting system. BioWatch labs use the reporting system to send data to CDC, who then sends a monthly report of negative results to DHS.
Biowatch Signal Interpretation and Integration Program (BWSIIP)	A surveillance program pilot that is intended to evaluate public data feeds for their usefulness in biomonitoring signal interpretation to provide BioWatch metropolitan areas, in the event of a signal detection, with the ongoing collection and analysis of appropriate medical information (with personally identifying information removed) that would support rapid interpretation of the signal and integration into consequence management operations. Once BWSIIP is deployed as part of BWICS, plans call for local public health agencies to use locally existing or publicly available biosurveillance tools provided by DHS, such as ESSENCE, or the Real-Time Outbreak and Disease Surveillance (RODS) software.[c]
National Biosurveillance Integration System	An effort at the federal level to combine multiple data streams from sector-specific agencies—those with medical, environmental, agricultural, and intelligence data—to give DHS situational awareness that is expected to allow earlier detection of events and to assist in response actions.

[a] Incident characterization tools are designed to integrate information from surveillance, environmental monitoring, plume hazard predictions, epidemiological forecasts, and population and critical infrastructure databases.

[b] ESSENCE is a syndromic surveillance software package available through free licensing agreements with the Johns Hopkins University Applied Physics Lab. The software is available to federal, state, and local health organizations that wish to deploy a Web-based syndromic surveillance system using their own data. DOD uses the system worldwide. The Department of Veterans Affairs and about 26 states and localities are implementing ESSENCE.

[c] RODS, developed by the University of Pittsburg, is a syndromic surveillance system used by several states that collects data from hospital emergency room visits. This system identifies patients' chief medical complaints, classifies the complaints according to syndrome, and aggregates those data in order to look for anomalous increases in certain syndromes that may reveal an infectious disease outbreak.

Source: U.S. Government Accounting Office, 2005. Information Technology: Federal Agencies Face Challenges in Implementing Initiatives to Improve Public Health Infrastructure. Report No. GAO-05-308 (http://www.gao.gov/new.items/d05308.pdf).

BioWatch program is expanding the number of devices per city (Divis, 2005), but as many as 60 may be needed for a high probability of detection (Lipton, 2005).

The Homeland Security Advanced Research Projects Agency (HSARPA) is contracting for development of "next-generation" devices (Simon, 2004) that can measure the air every three hours (as opposed to the current daily testing of samples) for up to 20 biological agents (Divis, 2005). A second next-generation device is the rapid automated biological identification system, or RABIS, which is the goal of a program to develop a detector that warns of a bioterrorist attack in less than two minutes. The intent is that RABIS units will be attached to building heating and air conditioning systems and will be able to shut down ventilation sections in the event of a release, limiting the spread of a pathogen will prevent others in the building from being exposed.

2.2.2. Integration of Surveillance Data

Despite the potential for the BioWatch program to provide early warning of an aerosol attack, it is still necessary to monitor for human and animal illness. As an event that occurred in Houston demonstrated, biological agents such as F. tularensis exist naturally in the environment and can trigger the BioWatch system (Divis, 2004, Bridis, 2003). Terrorists might deliberately release inactive agents to cause false positives and panic (Divis, 2005). At present, the BioWatch system monitors for approximately 10 biological agents (Mintz, 2003) with future systems targeted to test for 20 organisms (Divis, 2005). If terrorists released an aerosol of either a biological agent for which the system does not test or a sufficiently genetically modified strain of an organism, the release would go undetected. Thus, monitoring of human health is also necessary to distinguish between real events and Houston-like events or deliberate triggering of sensors with harmless agents from threats to human health. Monitoring of human health is also necessary to detect outbreaks that are missed due to the sparse density of the BioWatch network.

Table 12.1 also lists DHS projects whose goals include integration of biosurveillance data from military and civilian sources (BioNet), and data from animal, human health, intelligence, and agricultural surveillance (BWICS and the National Biosurveillance Integration System). We discuss BioNet in the section on the DoD. The other two programs are still in development.

3. DEPARTMENT OF DEFENSE

The primary mission of the DoD is to provide the military forces needed to deter war and to protect the security of the United States. The DoD employs 1.7 million active duty, Guard and Reserve personnel who receive health benefits. In addition, over 2.3 million family members and 4.5 million retirees and dependents of retirees receive medical care through the military system (U.S. Department of Defense, 2005e).

In order to maintain force health protection and medical readiness, the military has extensive programs to monitor the health and fitness of service members. Military installations around the globe perform health surveillance to determine the nature, magnitude, and distribution of health threats and to focus and evaluate preventive efforts. These threats include injuries, acute and chronic conditions (including infectious diseases), mental illnesses, and environmental exposures.

Protection against biological warfare has long been a part of force protection for the military. As the threat has evolved, protection against biological terrorism on domestic and overseas installations has also improved. As diseases do not respect installation boundaries, and the well-being of military retirees and family members is also of importance to the military, the DoD has greatly increased its surveillance efforts to identify natural or deliberate disease outbreaks at the earliest possible point in time. Presidential Decision Directive NSTC-7 (June 1996) stated that the United States will strengthen domestic infectious disease surveillance and response. It expanded the mission of the DoD to include support of global surveillance, training, research, and response to emerging infectious disease threats (White House, 2005). In response to this directive, DoD established the Global Emerging Infections Surveillance and Response System (DoD-GEIS). The DoD-GEIS is designed to strengthen the prevention of, surveillance of, and response to infectious diseases that pose a threat to military personnel, their families, and neighboring civilian communities, or present a risk to U.S. national security (U.S. Department of Defense, 2005b). Their efforts include sponsoring the development of syndromic surveillance systems and laboratory-based surveillance (Pavlin and Kelley, 2004, Canas et al., 2000).

3.1. Organization of DoD

The DoD is a large organization with an annual budget of over $370 billion (U.S. Department of Defense, 2005a). Its full organization is well beyond the scope of this book; interested readers should go to the military website *www.defenselink.mil* for more detailed information. Here we provide a focused overview of the organization of DoD relevant to a discussion of biosurveillance. The DoD has centralized many of its healthcare and biosurveillance functions, but there are interservice (e.g., Army, Navy) differences.

3.1.1. Centralized Functions

The Deputy Assistant Secretary of Defense for Force Health Protection and Readiness is responsible for all aspects of health care in support of military operations, military public health, and surveillance (U.S. Department of Defense Military Health System, 2005). The military health system, like the civilian healthcare system that was the topic of Chapter 6, conducts health surveillance of its own beneficiary populations. The DoD has 239 fixed military treatment facilities[1] around the world to care for beneficiaries (U.S. Department of Defense, 2005d). The DoD also monitors international disease trends that may affect force readiness. DoD shares public health information of importance with local and state health departments. Each installation has medical officers responsible for the public health of those who live and work there. The DoD aggressively identifies, characterizes, and reports on 78 reportable diseases based on Centers for Disease Control and Prevention (CDC) recommendations plus additional conditions of military importance. Reports flow not only through DoD communication channels but also to the local and state health departments within which the military treatment facility is located.

The military healthcare system is run by the TRICARE Management Activity under the Assistant Secretary of Defense for Health Affairs (U.S. Department of Defense, 2005c). The DoD has developed an electronic medical record system that records demographic information of all inpatient and outpatient visits along with diagnostic information and pharmacy, radiology, and laboratory orders. A complete electronic medical record is under deployment at the time of this writing.

3.1.2. Service-Specific Functions

The DoD is divided into Army, Navy, Marine Corps, and Air Force departments. Within each are suborganizations that play a role in biosurveillance. Surveillance is performed by each of the services for beneficiaries under their responsibility. This includes public health surveillance for naturally occurring diseases and bioterrorism.

The Army, Navy, and Air Force all have organizations dedicated to the health of their service members and families. The Army runs the U.S. Army Center for Health Promotion and Preventive Medicine (USACHPPM) located at Aberdeen Proving Ground (Edgewood), Maryland. The Army Medical Surveillance Activity (AMSA) is part of the USACHPPM, and

---

1 A fixed military treatment facility is a hospital, clinic. Unlike the civilian healthcare system, the military healthcare system has extensive capability to deploy temporary field treatment facilities.

analyzes, interprets, and disseminates information regarding the health and fitness of the Army. AMSA maintains the Defense Medical Surveillance System that contains a database of diseases and medical events, including hospitalizations, ambulatory visits, and longitudinal data on personnel and deployments. This system provides epidemiological information to policymakers, medical planners and researchers (Army Medical Surveillance Activity, 2005).

The Navy Environmental Health Center is located in Norfolk, Virginia, and its mission is to ensure Navy and Marine Corps readiness through leadership in prevention of disease and promotion of health. Its focus areas include global disease surveillance, clinical epidemiology, and the ability to detect, protect against and respond to lethal and nonlethal biological agents (Navy Environmental Health Center [NEHC], 2005).

The Air Force Institute for Operational Health (AFIOH) is located at Brooks City-Base, Texas, and its mission is to promote global health and protect Air Force warriors and communities. The AFIOH Surveillance Directorate provides field sampling, laboratory analysis, and data collection for assessment and management of risk (U.S. Air Force, 2005).

3.1.3. Research

Research, development, and acquisition of chemical, biological, radiological, and nuclear defense programs within the DoD are overseen by the Assistant to the Secretary of Defense for Nuclear, Chemical and Biological Defense Programs (Medical Chemical and Biological Defense Science and Technology Program, 2005). The Joint Program Executive Office for Chemical and Biological Defense (JPEO-CBD) is responsible for research, development, acquisition, and fielding of chemical, biological, radiological, and nuclear defense equipment and medical countermeasures (The Joint Program Executive Office for Chemical and Biological Defense, 2005f).

In addition, other laboratories in the DoD work on medical issues for biological defense. The U.S. Army Medical Research and Material Command (USAMRMC) operates laboratories responsible for ensuring the military forces are deployed in a state of optimal health and equipped to protect themselves from disease and injury (U.S. Army Medical Research and Material Command, 2005). Under USAMRMC, the U.S. Army Medical Research Institute for Infectious Diseases conducts basic and applied research on biological threats resulting in medical solutions to protect the warfighter (U.S. Army Medical Research Institute of Infectious Diseases, 2005). The Walter Reed Army Institute of Research (WRAIR) conducts biomedical research to deliver life saving products to sustain the combat effectiveness of the warrior (Walter Reed Army Institute of Research, 2005).

The Defense Advanced Research Products Agency previously sponsored the Bio-Event Advanced Leading Indicator Recognition Technology (BioALIRT) project to advance research on improving timeliness of outbreak detection. This project focused on identifying new nontraditional data sources that could detect outbreaks early and the development and evaluation of detection algorithms (Buckeridge et al., 2005).

3.2. Biosurveillance Systems

DoD conducts surveillance of both human health and of the environment, especially the air.

3.2.1. Human Health Event Surveillance

The WRAIR developed the Electronic Surveillance System for the Early Notification of Community-based Epidemics (ESSENCE) to capture and disseminate information on outpatient healthcare utilization for specific syndrome categories (Lewis et al., 2002). By monitoring this information both temporally and spatially, outbreaks of infectious disease of potential public health significance, including those caused by bioterrorism, can be quickly recognized and tracked. In collaboration with the Johns Hopkins University Applied Physics Laboratory, this system was expanded to include civilian health information in the Washington, DC region and is currently available for health departments in the District of Columbia, Virginia and Maryland (Lombardo et al., 2003). ESSENCE is the Military Health System's primary electronic method for outbreak detection.

3.2.2. Interoperation with Civilian Systems

Joint Services Installation Pilot Project. To improve the response capabilities for chemical, biological, radiological, and nuclear incidents, the Defense Threat Reduction Agency (DTRA) sponsored the Joint Services Installation Pilot Project at nine military installations in the United States. The project also used ESSENCE to integrate military and civilian data in regional systems by combining civilian data, such as emergency room chief complaints, with the military outpatient visit and pharmacy data and providing the web-based monitoring system to both military and civilian users.

BioNet. The integration of military and civilian surveillance systems was one of the key components of the BioNet project, which was sponsored by DHS and managed by DTRA. This program used San Diego as a test bed. This one-year project integrated detection and characterization capabilities of both civilian and military organizations to improve detection time and accuracy. Environmental sensor data from both military and civilian programs are combined and medical data are also provided to both military and civilian public health departments.

Military data collected during BioNet included outpatient ICD-9 codes categorizing the reason for the visit, text chief complaints from the Naval Medical Center San Diego emergency room and one ship, and outpatient pharmacy prescriptions. Civilian data included school absenteeism records, school nurse office visit data at select schools, chief complaint logs from hospital emergency rooms, over-the-counter medication sales from the National Retail Data Monitor, and pre-hospital ambulance transport data. The system combined data in the web-based Integrated Data Repository and Analysis Engine and made it available to both military and civilian users. Integration issues included difficulties using legacy systems in operation at the health departments and getting permission to install a new system. In post-use surveys, the users rated the system very high, thought having access to both civilian and military data helped in the performance of their job, and rated syndromic surveillance overall as an asset to public health work (Marsden-Haug et al., 2005).

Military outpatient data are currently used in the CDC's BioSense system, a nationwide syndromic surveillance system that combines the military data with Veteran's Administration hospitals' outpatient data and test ordering information from a nationwide commercial laboratory (Loonsk, 2004).

3.2.3. Environmental Surveillance

Under the JPEO-CBD, the Joint Project Manager for Nuclear Biological and Chemical Contamination Avoidance runs the Product Manager for Biological Detection (Joint Program Executive Office for Chemical and Biological Defense, 2005e). This area focuses on biological point detection and has a technology goal to increase detection sensitivity, lower detection thresholds, increase the specificity across threat agents, reduce false alarm rates, and integrate detectors into mapping and communication networks.

The systems currently in development or fielded include: (1) the Biological Integrated Detection System, a detection suite that is mounted on a dedicated vehicle that detects and identifies large-area biological warfare agent attacks (Joint Program Executive Office for Chemical and Biological Defense, 2005a); (2) the Joint Biological Point Detection System, a fully automated, stand-alone suite that will be capable of identifying biological warfare agents in less than 15 minutes (Joint Program Executive Office for Chemical and Biological Defense, 2005b); (3) the Joint Biological Standoff Detection System, a standoff biological detection system that will detect and track aerosol clouds out to 15 km in near real time. It will discriminate between natural and manmade aerosols (Joint Program Executive Office for Chemical and Biological Defense, 2005c); and (4) the Joint Portal Shield, an automated, modular, networked biological detection system that identifies up to eight biological warfare agents simultaneously in less than 25 minutes (Joint Program Executive Office for Chemical and Biological Defense, 2005d).

4. U.S. POSTAL SERVICE (USPS)

As a result of the anthrax-tainted envelopes of 2001, the USPS has added biosurveillance to its many core functions. Biosurveillance measures in place include the Biohazard Detection System (BDS), a device that monitors mail-sorting machines and a nationwide illness surveillance program for postal workers that can be activated in response to an event.

BDS is an air monitoring system based on a DNA polymerase chain reaction (PCR) test for *B. anthracis* (U.S. Postal Service, 2004, Military Postal Service Agency, 2004). BDS attaches to mail-sorting machines. A mail-sorting machine is both a single point through which all mail passes (except locally routed rural mail and larger packages) and a nearly ideal device for expressing spores from all but the most tightly sealed envelops (because the process compresses envelops), enabling the detection of spores by BDS.

BDS comprises an air-collection hood, a cabinet where the collection and analysis devices are housed, a local computer network connection, and a site controller (i.e., a networked computer). All the BDS processes are automated. The equipment continuously collects air samples from mail canceling equipment while the canceling operation is underway. It absorbs and concentrates airborne particles into a sterile water base. This creates a liquid sample that is injected into a cartridge. BDS then performs an automated polymerase chain reaction (PCR) on the liquid sample to detect the presence of anthrax (Bacillus anthracis). The system concentrates air samples for a one-hour period followed by the PCR test that takes approximately 30 minutes. While the PCR test is performed, the BDS is simultaneously concentrating particles for the next sample. So, although the first result requires approximately 1.5 hours, subsequent results are obtained every hour (U.S. Department of the Army, Headquarters, 2005).

According to published material, "If there is a DNA match, the BDS computer network conveys that information to the site controller computer. Local management is notified directly by on-site BDS personnel and also by multiple forms of electronic communication from the BDS site controller. The emergency action plan will be activated. The facility's building alarm will sound and everyone in the building will be evacuated. Upstream and downstream processing facilities will also be notified. An Emergency Notification Center at Postal Service headquarters will be notified as well as community first responders and the Department of Homeland Security. Once the postal employees are outside the building, supervisors will call the roll and make sure everyone in the building has been evacuated. They will explain the nature of the incident, and everyone will wait for direction from community emergency response personnel. An outside lab will perform multiple plate cultures using the BDS positive test sample and other environmental samples. Local public health officials will determine the need for any medication. The mail inside the plant will be retained until it is safe for delivery. The new mail that would normally be processed in this facility will be diverted to other mail processing facilities and delivery operations" (U.S. Department of the Army, Headquarters, 2005).

The military also deploys this type of monitoring, and it triggered the March 2005 anthrax alert at a Pentagon mail-handling facility. In this event, postal workers received postexposure prophylaxis (i.e., antibiotics) even though anthrax could not be confirmed by secondary testing. The lack of timely reporting of this incident and the laboratory's failure to use assays approved by the Federal Laboratory Response Network highlighted the disconnect between federal civilian and military surveillance programs even three-and-a-half years after 9/11/01.

5. PLANES, TRAINS, AND SHIPS

Transportation systems have a long history of transporting not only goods and people but also disease. In fact, the word quarantine (derived from the Latin word quadraginta meaning "forty") was first used between 1377 and 1403 to refer to a 40-day period of isolation enforced by Venice and the other chief maritime cities of the Mediterranean for all vessels entering their ports from areas suspected to harbor plague (Bolduan and Bolduan, 1941).

Modern transportation systems make it possible for people, vectors, and microorganisms to travel between any two cities on the earth in less than a day. This situation, however, is relatively recent. It was not until the steamship network arose in the 1870s, for example, that plague could reach seaports in Asia, the Americas, and South Africa (McNeill, 1989).

In general, biosurveillance of a transportation system can involve multiple approaches, such as monitoring of the transportation vehicle itself (e.g., its air, water, and food systems); screening of passengers and cargo for disease; post-travel monitoring of recent travelers; and investigation of outbreaks. At present, it is rare for all of these modalities to be in use together for cost or feasibility reasons, although they are complementary. In the future, we expect their combined use to be the rule rather than the exception.

5.1. Ships

Cruise ships are miniature cities with their own water systems, food systems, and medical facilities. They resemble hospitals in having many of the prerequisites for outbreaks to develop: close quarters, and steady influx of new individuals into the population.

In the year 2001, CDC's Vessel Sanitation Program, which conducts surveillance for gastrointestinal illness on cruise ships with foreign itineraries sailing into U.S. ports, received reports of 21 outbreaks of acute gastrointestinal illness on 17 cruise ships (CDC, 2002), at least half of them due to norovirus.[2] These outbreaks sickened up to 31% of passengers and occasionally recurred on subsequent trips despite sanitary efforts. Other biological agents that have caused outbreaks on cruise ships include influenza (Uyeki et al., 2003, Brotherton et al., 2003), *staphylococcal* enterotoxin (Waterman et al., 1987, CDC, 1983), E. coli (Daniels et al., 2000, Snyder et al., 1984, Lumish et al., 1980), *Legionella pneumophila* (Kobayashi et al., 2004, Regan et al., 2003, Castellani Pastoris et al., 1999, Rowbotham, 1998, Health Protection Agency, 1998, Jernigan et al., 1996), Shigella sp. (Gikas et al., 1996, CDC, 1994, Lew et al., 1991, Merson et al., 1975) and Variola minor (rubella) (Hoey, 1998, CDC, 1998). Table 12.2 presents a catalog of the gastrointestinal

TABLE 12.2 International Cruise Ship Gastrointestinal Outbreaks Reported to Centers for Disease Control and Prevention Vessel Sanitation Program, January—June 2005

Cruise Line	Ship Name	Sailing Dates	Causative Agent
Princess Cruise Line	*Sun Princess*	5/9–5/16/2005	Norovirus
Princess Cruise Line	*Dawn Princess*	4/29–5/14/2005	Norovirus
Celebrity Cruise Line	*Horizon*	4/23–4/30	Unknown
Crystal Cruise	*Crystal Symphony*	4/14–4/25	Unknown
Norwegian Cruise Line	*Norwegian Crown*	3/13–3/28	Unknown
Princess Cruise Line	*Royal Princess*	3/7–3/24	Unknown
Holland America Cruise Line	*Veendam*	3/5–3/12	Norovirus
Celebrity Cruises	*Zenith*	2/27–3/13	Norovirus
Carnival Cruise Line	*Celebration*	2/21–2/26 and 2/17–2/21	Norovirus
Radisson Seven Seas Cruises	*Seven Seas Mariner*	2/12–2/24	Salmonella
Royal Caribbean Cruise Line	*Empress of the Seas*	1/17–1/28	Norovirus
Royal Caribbean Cruise Line	*Mariner of the Seas*	1/26–1/23	Norovirus
Holland America Cruise Line	*Ryndam*	1/13–1/29	Norovirus
Princess Cruise Line	*Sun Princess*	1/8–1/18	Norovirus
Holland America Cruise Line	*Veendam*	1/3–1/15 and 1/15–1/29	Norovirus
Royal Caribbean Cruise Line	*Enchantment of the Seas*	1/3–1/8	Norovirus

Source: Centers for Disease Control and Prevention, 2005. Vessel Sanitation Program. National Center for Environmental Health (http://www.cdc.gov/nceh/vsp/surv/GIlist.htm#2005).

2 Norovirus is a family of viruses that cause vomiting and diarrhea, for which humans have little immunity, and which spread extremely efficiently between people by hand-to-mouth, environmental surfaces and even via the air.

outbreaks reported to the CDC vessel sanitation program from January through May 2005.

Cruise ships are also mobile cities. When they are in port, they become a temporary extension of larger cities. During the Olympic Games in Sydney and Athens, cruise ships served as floating hotels and biosurveillance for these special events included measures for monitoring the ship–hotels (Jorm et al., 2003, Waples et al., 2000).

5.1.2. Planes

Airplanes often carry communicable disease between cities and countries. In 1978, poliovirus was imported into Canada by people traveling from the Netherlands (White et al., 1981). In 1996, a health worker in South Africa died from Ebola transmitted by a physician who had entered the country from Gabon, which was experiencing an Ebola outbreak (WHO, 2005a). In 1987, meningococcal meningitis was spread to several continents by tourists returning from the Haj (Novelli et al., 1987) (Figure 12.1). Similarly, air travel quickly spread SARS throughout the world in 2003. There have been reports of malaria deaths in northern countries possibly related to mosquitoes transported by airplanes from endemic regions (Isaacson, 1989). Brussels, Geneva, and Oslo have all had cases of "airport" malaria (WHO, 1999).

Airplanes and airports themselves represent a risk to air travelers because of crowding, although the risk is low because contact time is limited. Diseases such as pneumonic plague, influenza, and TB, nevertheless, can spread in airports and on airplanes (Roberts, 1996, Wenzel, 1996, Ritzinger, 1965, CDC, 1995, Driver et al., 1994). In 1977, over 70% of the passengers on board an airliner grounded for several hours were infected with influenza by a fellow passenger (Moser et al., 1979). In 1994, a person with active TB is believed to have infected six fellow passengers on a flight from Chicago to Honolulu (Miller et al., 1996, Kenyon et al., 1996).

At present, there is little "dedicated" biosurveillance of air transportation systems. Rather, outbreaks are detected through conventional biosurveillance methods such as disease reporting and astute observers. The role of airports and airlines at present is largely to cooperate during outbreak investigations by providing investigators with passenger lists and cargo manifestos. There are exceptions, however. For example, air monitoring devices similar to those used in the BioWatch program were in place at the Salt Lake City airport during the 2002 Winter Olympic Games. In Taiwan and other Asian countries, the SARS outbreak of 2003 led to large scale screening of passengers arriving and departing from airports (Chapter 3, Figure 3.5). Taiwan continues to routinely screen at airports for febrile patients using infrared thermal imaging devices, followed by stages of clinician assessment and laboratory testing. Shu and colleagues reported the Taiwan experience with finding cases of dengue fever (Shu et al., 2005). Of more than 8 million inbound travelers arriving at the two international airports (Chiang Kai-Shek and Kaohsiung) between July 2003 and June 2004, infrared thermal camera and confirmation by ear temperature identified approximately 22,000 passengers with fever. After clinical screening, 3011 passengers were tested for dengue virus. Forty of 3011 serum samples tested were positive for dengue by real-time PCR and E/M-specific capture IgM and IgG ELISA.

During the SARS outbreak of 2003, the Taiwan CDC also routinely monitored for post-travel illness in recent arrivals. It linked arrival data from passport control with national health insurance claims data to identify individuals who had received medical care for selected diagnoses within a week of arrival. The Taiwan CDC has plans to automate and continue this process as there is significant business travel between Taiwan and China (approximately 20,000 arrivals from China per day). This strategy is not only a general approach for detecting an outbreak that has been imported into a country

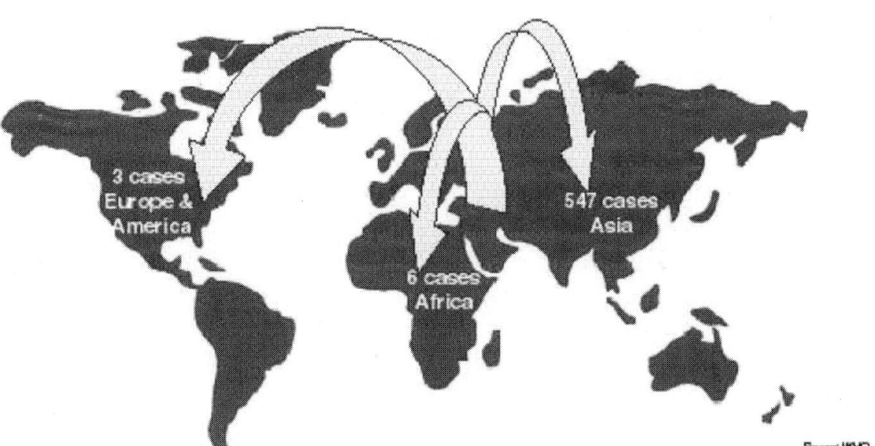

FIGURE 12.1 Spread of meningococcal meningitis by pilgrims returning from the Haj, 1987 (http://www.who.int/infectious-disease-report/pages/graph39.html). Reproduced by permission of World Health Organization (WHO).

TABLE 12.3 Passenger Traffic at 30 Major Airports in 2004

City (Airport)	Total Passengers	% Change
ATLANTA (ATL)	83,606,583	5.7
CHICAGO (ORD)	75,533,822	8.7
LONDON (LHR)	67,344,054	6.1
TOKYO (HND)	62,291,405	(0.9)
LOS ANGELES (LAX)	60,688,609	10.4
DALLAS/FT. WORTH ()	59,412,217	11.6
PARIS (CDG)	51,260,363	6.3
FRANKFURT/MAIN (FRA)	51,098,271	5.7
AMSTERDAM (AMS)	42,541,180	6.5
DENVER (DEN)	42,393,766	13.0
LAS VEGAS (LAS)	41,441,531	14.2
PHOENIX (PHX)	39,504,898	5.6
MADRID (MAD)	38,704,731	7.9
BANGKOK (BKK)	37,960,169	25.8
NEW YORK (JFK)	37,518,143	18.2
MINNEAPOLIS/ST PAUL (MSP)	36,713,173	10.6
HONG KONG (HKG)	36,711,920	35.5
HOUSTON (IAH)	36,506,116	6.9
DETROIT (DTW)	35,187,517	7.7
BEIJING (PEK)	34,883,190	43.2
SAN FRANCISCO (SFO)	32,247,746	10.0
NEWARK (EWR)	31,947,266	8.4
LONDON (LGW)	31,461,454	4.8
ORLANDO (MCO)	31,143,388	14.0
TOKYO (NRT)	31,057,252	17.0
SINGAPORE (SIN)	30,353,565	23.1
MIAMI (MIA)	30,165,197	1.9
SEATTLE (SEA)	28,804,554	7.5
TORONTO (YYZ)	28,615,709	15.7
PHILADELPHIA (PHL)	28,507,420	15.5

Notes: Airports participating in the Airports Council International annual traffic statistics collection. Total passengers enplaned and deplaned; passengers in transit counted once.

Source: Airports Council International. (2005). Passenger Traffic 2004 FINAL. Retrieved from: *http://www.aci.aero/cda/aci/display/main/aci_content.jsp?zn=aci&cp=1-5_9_2__*, with permission.

but also has potential as a method for estimating disease activity in another country. We expect the above practices to become more routine with time.

The need for biosurveillance of air travel is increasing due to the ever increasing volume of air travel, especially international travel. In 2004, 1.275 billion passengers enplaned and deplaned in just 30 large airports alone (Table 12.3). Airports Council International projects a 4% growth rate in air travel resulting in 7.4 billion passenger trips by 2020 from a current level of approximately 1.7 billion (Airports Council International, 2005). Biosurveillance personnel must consider an outbreak anywhere in the world as a threat—especially to those cities that serve as major hubs for international travel. The WHO plays a key role in disseminating information about outbreaks that pose international health, as we discuss below.

5.1.3. Trains

Public mass transit accounts for a substantial proportion of daily commuters in many cities around the world, including several metropolitan areas in the United States. The New York City subway system alone carries 5 million passengers every workday; eight of ten commuters arriving in Manhattan daily do so by train. The Washington, D.C. Metrorail is carrying nearly 700,000 commuters per day. A subway train of 10 cars operating at rush hour may carry over 2000 passengers.

Gershon and colleagues reviewed the health and safety hazards, whether natural or manmade, associated with subways (Gershon et al., 2005). The combination of high transient population density, restricted air circulation/filtration, and confined spaces creates an opportunity for terrorists to effectively expose large numbers of people to biological or chemical agents.

The vulnerability of subway infrastructure to bioterrorism has been recognized for decades. According to an unclassified U.S. Army report, "A series of trials were conducted in three major north–south subway lines in mid-Manhattan, New York City, in June 1966. A harmless simulant biological agent (*Bacillus globigii*) was disseminated within the subway tubes and from the street into the subway stations. The simulant data when translated into equivalent covert attacks with pathogenic agents during peak traffic periods indicated that large numbers of people could be exposed to infectious doses" (U.S. Department of the Army, 1977). In 1995, this credible scenario became reality when the Aum Shinrikyo successfully dispersed sarin gas on the Tokyo subway injuring 5500 and killing 12 (Okumura et al., 2003). Two other chemical agent releases in subway and railway station restrooms were thwarted that year (Okumura et al., 2003). In addition, the Aum Shinrikyo attempted to disperse anthrax in the Tokyo subway, but, fortunately, they used the vaccine strain of anthrax which does not cause disease (Keim et al., 2001).

There is little published information about air monitoring of subway systems. At present, we are aware of pilot projects in one or more U.S. subway systems that directly monitor the air for pathogens and plans to deploy this capability more widely (Telemedicine & Advanced Technology Research Center, U.S. Army Medical Research & Materiel Command, 2004).

6. WORLD HEALTH ORGANIZATION

Although WHO is the last organization that we discuss in this book, it is perhaps the most important because of the inexorable trend towards globalization of trade and travel.

WHO is the United Nations' specialized agency for health. Its objective, as stated in its constitution (WHO, 1946), is the attainment by all peoples of the highest possible level of health, defined as a state of complete physical, mental and social well-being and not merely the absence of disease or infirmity (WHO, 2005b).

The governance of WHO comprises the World Health Assembly, the Executive Board, and an executive branch called the Secretariat. The World Health Assembly is an annual meeting of representatives of all 192 member-states (countries) to approve the WHO program and the budget and to decide major policy questions. The Executive Board comprises 32 elected members who are technically qualified in the field of health. The Executive Board appoints the Director General of WHO. The Director General is the director of the Secretariat, which employs approximately 3500 professional and support staff.

WHO influences the communicable disease practices of its 192 member-states through International Health Regulations, collection and dissemination of biosurveillance information, and through direct assistance in the control of disease. The biosurveillance functions are concentrated in the Department of Communicable Disease Surveillance and Response (www.who.int/csr/).

6.1. International Health Regulations

The International Health Regulations (IHR) refers to two international legal instruments, the current International Health Regulations, IHR(1969), and the revised International Health Regulations, the IHR(2005). The IHR(1969) will be replaced by the IHR(2005) when they come into effect in June 2007.

The purpose of the IHR(1969) are to ensure maximum security against the international spread of diseases with a minimum interference with world traffic. This philosophy of striking a balance between physical and economic health is similar to that of the OIE regulations described in Chapter 17. The IHR(1969) only apply to three infectious diseases: cholera, plague and yellow fever. The experience with IHR(1969) was that many countries fail to report outbreaks out of concern for potential economic consequences due to effects on trade and tourism. Additionally, the rules were difficult to enforce.

The purpose and scope of the IHR(2005) are to prevent, protect against, control, and provide a public health response to the international spread of disease in ways that are commensurate with and restricted to public health risks, and which avoid unnecessary interference with international traffic and trade. The IHR(2005) also establish a single code of procedures and practices for routine public health measures at international airports and ports and some ground crossings (WHO, 2005c).

According to IHR(2005), the key roles of WHO (as opposed to member-states) include:

- To conduct daily global surveillance of international intelligence to detect possible risks of a public health emergency of international concern for verification with member-states.
- To determine whether a particular event notified by a state under the regulations constitutes a public health emergency of international concern.
- To provide technical assistance to states in their response to public health emergencies of international concern.
- To provide guidance to states to strengthen their existing surveillance and response capacity to contain and control public health risks and emergencies.
- To develop and recommend measures for use by states during public health emergencies of international concern, based on a consistent process of risk verification and assessment.
- To update the regulations and supporting guides as necessary to maintain scientific and regulatory validity.
- To respond to the needs of states regarding the interpretation and implementation of the regulations (WHO, 2005c).

6.2. Department of Communicable Disease Surveillance and Response

Figure 12.2 presents the organization chart of WHO's Department of Communicable Disease Surveillance and Response. Biosurveillance functions for influenza are separate from other diseases (which are under Office of Alert and Response Operations).

6.3. Global Outbreak Alert and Response Network

The Global Outbreak Alert and Response Network (GOARN) is a technical collaboration of hundreds of existing institutions and networks (administered by WHO) that pools human and technical resources for the rapid identification, confirmation, and response to outbreaks of international importance.

GOARN systematically gathers official reports and rumors of suspected outbreaks from a wide range of formal and informal sources. More than 60% of the initial outbreak reports come from unofficial informal sources, including sources other than the electronic media, which require verification.

Formal sources include ministries of health, national institutes of public health, WHO regional and country offices, WHO collaborating center, civilian and military laboratories, academic institutes, and nongovernmental organizations (NGOs), all of which submit reports to WHO.

The informal sources include global media sources, such as news wires and websites. The Global Public Health Intelligence Network (GPHIN) is one of the most important sources of informal information related to outbreaks. GPHIN was developed by Health Canada in collaboration with WHO. It is a secure Internet-based multilingual early-warning tool that continuously searches global media sources to identify information about disease outbreaks and other events of potential international public health concern (see Chapter 26).

The goal of WHO's biosurveillance is to identify outbreaks that are international threats. Figure 12.3 summarizes the process in use in 2000 to distill the large amount of biosurveillance data to identify and disseminate events of interest to international public health.

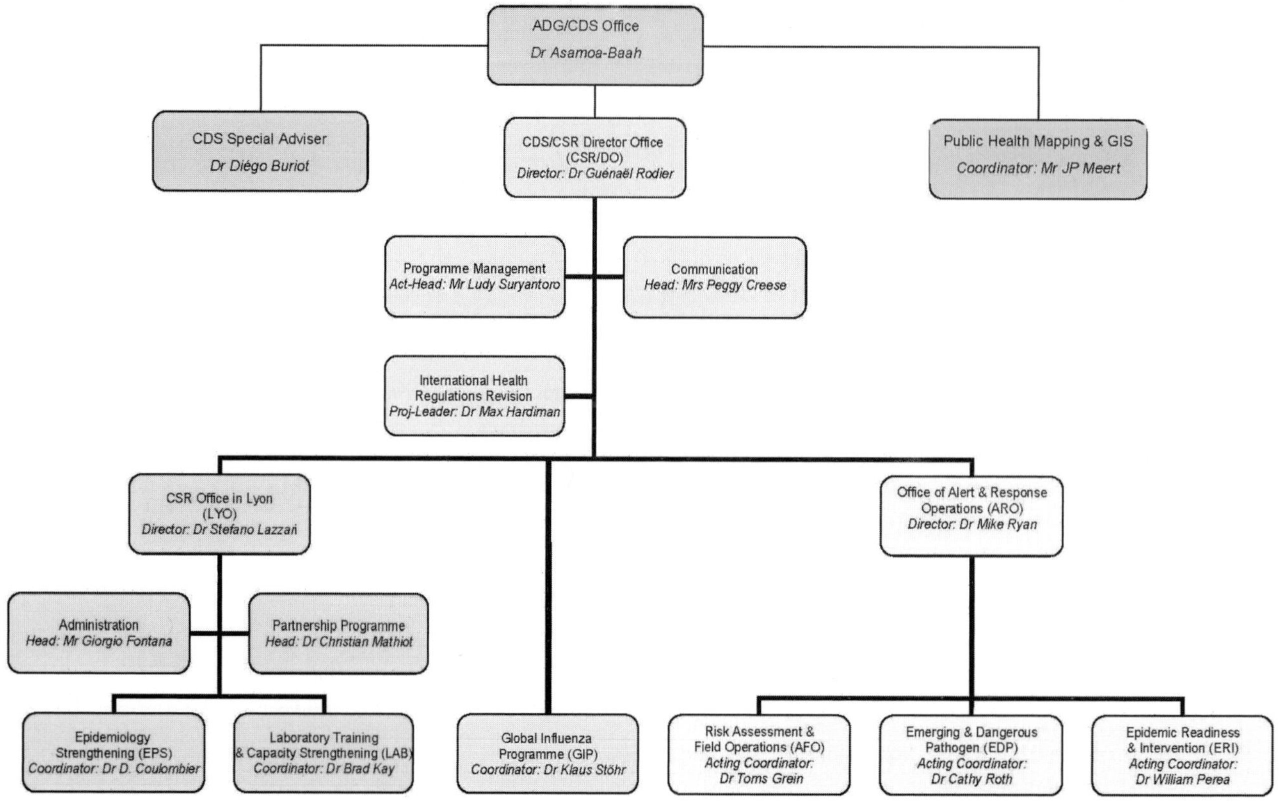

FIGURE 12.2 Organizational chart of WHO's Department of Communicable Disease Surveillance and Response.

7. SUMMARY

The U.S. Department of Homeland Security is the lead federal agency for defense against terrorism including biodefense. It plays a focused role in facilitating the development and deployment of air monitoring technology. It is playing a lead role in coordinating the federal effort to integrate intelligence (capability and intent), animal, human, and environmental data to form a coherent situational assessment.

The U.S. Department of Defense plays many biosurveillance roles. Its healthcare system can and does monitor its own bases, workforce, retirees, and dependents. It cooperates with governmental public health to provide data and detection services. It has a longstanding role in the development of sensing technology for biological agents in the environment.

Airplanes, ships, and mass transit systems are important elements in a biosurveillance system due to the concentration of individuals and their ability to quickly disseminate outbreaks between any two cities in the world. The current level of biosurveillance of these systems is low, although governments are actively developing new methods.

The principal role of the World Health Organization in biosurveillance is establishing international regulations and monitoring for disease outbreaks that pose international threats. Its biosurveillance relies on reporting by member states as well as near real time analysis of electronically available new stories and other electronic sources in multiple languages.

This chapter and the preceding seven chapters comprising Part II of this book reviewed the many organizations that participate in biosurveillance—their roles, their information systems, and the data they collect. We devoted this much time and space to this material because anyone functioning in or training for a role in biosurveillance planning/system design should have a working knowledge of these organizations. The roles and responsibilities of these organizations are still evolving; therefore, the information we presented should be understood as a snapshot in time that will likely change, driven by the need for faster and better outbreak detection and characterization as well as advances in technical methods. Additional types of organizations may emerge in the governmental or non-governmental sectors. Nevertheless, the strengths and limitations of the existing organizations and the individuals working within them are important factors to consider when thinking about the biosurveillance process and methods for its improvement.

8. ACKNOWLEDGMENTS

We thank Drs. William Hogan, Ron Aryel, Jen-Hsiang Chuang, and Chi-Ming (Kevin) Chang for providing information used in

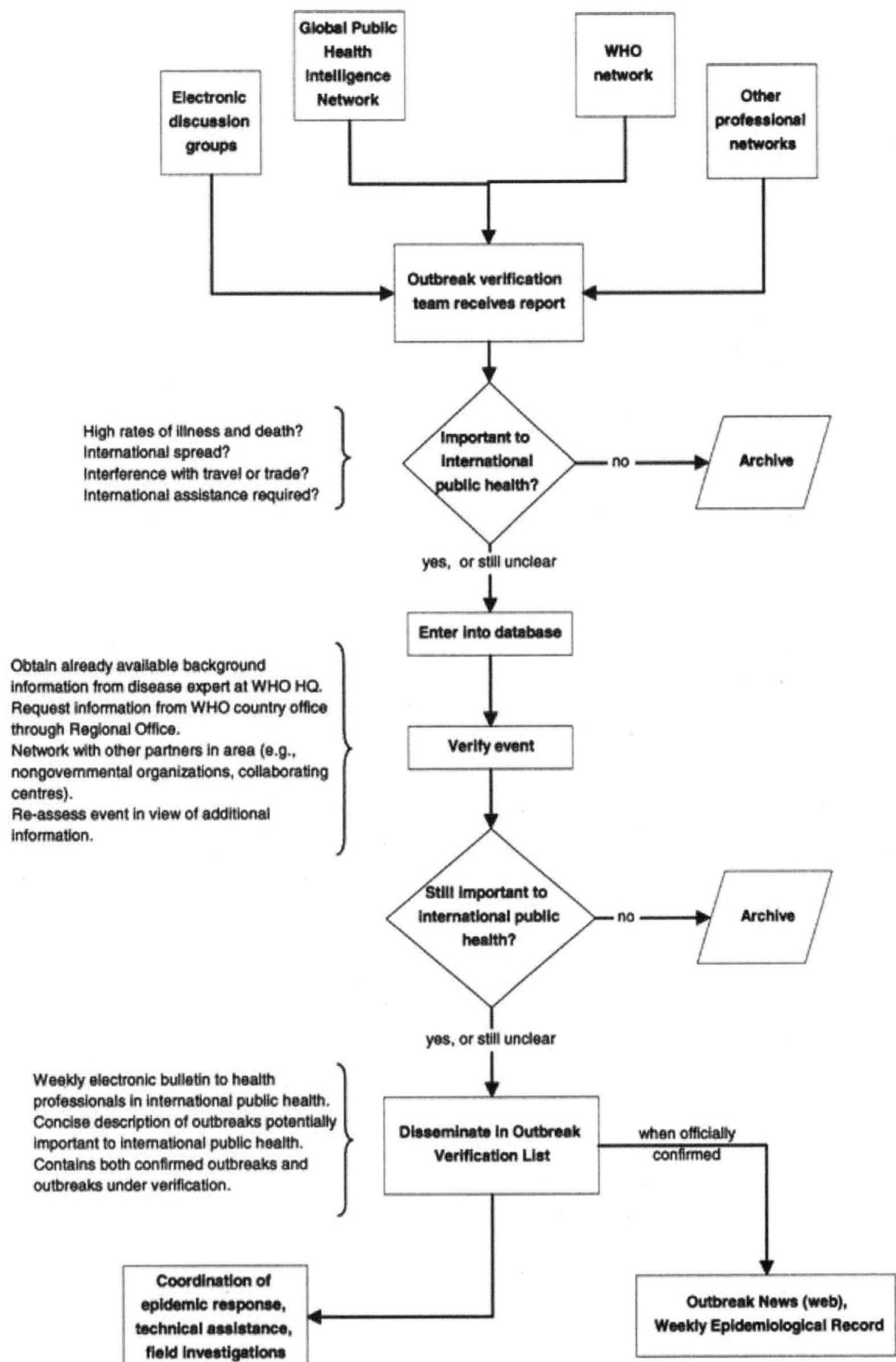

FIGURE 12.3 WHO's Outbreak Verification Process in 2000. (Source: Grein, T., Kamara, K., Rodier, G., et al. (2000). Rumors of disease in the global village: outbreak verification. Emerg Infect Dis, vol. 6, no.2, pp. 97-102 http://www.ncbi.nlm.nih.gov/entrez/query.fcgi?cmd=Retrieve&db=PubMed&dopt=Citation&list_uids=10756142.)

this chapter. We especially thank Daphne Henry for research and editing.

The opinions expressed in this chapter are those of the authors and do not represent the official position of the Department of Defense.

9. REFERENCES

Airports Council International. (2005). ACI Passenger and Freight Forecasts 2005–2020. Press release. *http://www.aci.aero/cda/aci/display/main/aci_content.jsp?zn=aci&cp=1-7-46^4549_9_2__.*

Army Medical Surveillance Activity. (2005). *http://amsa.army.mil/AMSA/amsa_home.htm.*

Bolduan C, Bolduan N. (1941). Public Health and Hygiene. Philadelphia: WB Saunders.

Bridis, T. (2003). Government discloses details of nation's bioterror sensors. *Associated Press,* November 14. *http://www.detnews.com/2003/politics/0311/16/politics-325675.htm.*

Brotherton, J. M., Delpech, V. C., Gilbert, G. L., et al. (2003). A large outbreak of influenza A and B on a cruise ship causing widespread morbidity. *Epidemiol Infect* 130:263–71. *http://ww.ncbi.nlm.nih.gov/entrez/query.fcgi?cmd=Retrieve&db=PubMed&dopt=Citation&list_uids=12729195.*

Buckeridge, D. L., Burkom, H., Campbell, M., et al. (2005). Algorithms for rapid outbreak detection: a research synthesis. *J Biomed Inform* 38:99–113. *http://www.ncbi.nlm.nih.gov/entrez/query.fcgi?cmd=Retrieve&db=PubMed&dopt=Citation&list_uids=15797000.*

Canas, L., Lohman, K., Pavlin, J., et al. (2000). The Department of Defense laboratory-based global influenza surveillance system. *Mil Med* 165(Suppl 2):52–6.

Castellani Pastoris, M., Lo Monaco, R., Goldoni, P., et al. (1999). Legionnaires' disease on a cruise ship linked to the water supply system: clinical and public health implications. *Clin Infect Dis* 28:33–8. *http://www.ncbi.nlm.nih.gov/entrez/query.fcgi?cmd=Retrieve&db=PubMed&dopt=Citation&list_uids=10028067.*

Centers for Disease Control and Prevention. (1983). Staphylococcal food poisoning on a cruise ship. *MMWR Morb Mortal Wkly Rep* 32:294–5. *http://www.ncbi.nlm.nih.gov/entrez/query.fcgi?cmd=Retrieve&db=PubMed&dopt=Citation&list_uids=6406815.*

Centers for Disease Control and Prevention. (1994). Outbreak of Shigella flexneri 2a infections on a cruise ship. *MMWR Morb Mortal Wkly Rep* 43:657. *http://www.ncbi.nlm.nih.gov/entrez/query.fcgi?cmd=Retrieve&db=PubMed&dopt=Citation&list_uids=8072477.*

Centers for Disease Control and Prevention. (1995). Exposure of passengers and flight crew to Mycobacterium tuberculosis on commercial aircraft, 1992–1995. *MMWR Morb Mortal Wkly Rep,* 44:137–40. *http://www.ncbi.nlm.nih.gov/entrez/query.fcgi?cmd=Retrieve&db=PubMed&dopt=Citation&list_uids=7862079.*

Centers for Disease Control and Prevention. (1998). Rubella among crew members of commercial cruise ships—Florida, 1997. *MMWR Morb Mortal Wkly Rep* 46:52–53, 1247–50. *http://www.ncbi.nlm.nih.gov/entrez/query.fcgi?cmd=Retrieve&db=PubMed&dopt=Citation&list_uids=9436714.*

Centers for Disease Control and Prevention. (2002). Outbreaks of gastroenteritis associated with noroviruses on cruise ships–United States, 2002. *MMWR Morb Mortal Wkly Rep* 51:1112–5.

Daniels, N. A., Neimann, J., Karpati, A., et al. (2000). Traveler's diarrhea at sea: three outbreaks of waterborne enterotoxigenic Escherichia coli on cruise ships. *J Infect Dis* 181:1491–5. *http://www.ncbi.nlm.nih.gov/entrez/query.fcgi=cmd=Retrieve&db=PubMed&dopt=Citation&list_uids=10762583.*

Divis, D. A. (2004). BioWar: BioWatch response plans lacking. United Press International, December 8. *http://www.washtimes.com/upi-breaking/20041208-092642-3814r.htm.*

Divis, D. A. (2005). BioWar: Biowatch expansion developing. United Press International, February 24.

Driver, C. R., Valway, S. E., Morgan, W. M., et al. (1994). Transmission of Mycobacterium tuberculosis associated with air travel. *JAMA* 272:1031–5. *http://www.ncbi.nlm.nih.gov/entrez/query.fcgi?cmd=Retrieve&db=PubMed&dopt=Citation&list_uids=8089885.*

Gershon, R. R., Qureshi, K. A., Barrera, M. A., et al. (2005). Health and safety hazards associated with subways: a review. *J Urban Health* 82:10–20. *http://www.ncbi.nlm.nih.gov/entrez/query.fcgi?cmd=Retrieve&db=PubMed&dopt=Citation&list_uids=15738339.*

Gikas, A., Pediaditis, J., Giti, Z., et al. (1996). Shigellosis on an Italian cruise ship. *Lancet* 348:1593–4. *http://www.ncbi.nlm.nih.gov/entrez/query.fcgi?cmd=Retrieve&db=PubMed&dopt=Citation&list_uids=8950912.*

Health Protection Agency (London). (1998). Legionella on board a cruise ship. *Commun Dis Rep CDR Wkly* 8:237. *http://www.ncbi.nlm.nih.gov/entrez/query.fcgi?cmd=Retrieve&db=PubMed&dopt=Citation&list_uids=9695289.*

Hoey, J. (1998). Rubella outbreaks on cruise ships. *CMAJ* 158:516–7. *http://www.ncbi.nlm.nih.gov/entrez/query.fcgi?cmd=Retrieve&db=PubMed&dopt=Citation&list_uids=9627566.*

Isaacson, M. (1989). Airport malaria: a review. *Bull World Health Organ* 67:737–43. *http://www.ncbi.nlm.nih.gov/entrez/query.fcgi?cmd=Retrieve&db=PubMed&dopt=Citation&list_uids=2699278.*

Jernigan, D. B., Hofmann, J., Cetron, M. S., et al. (1996). Outbreak of Legionnaires' disease among cruise ship passengers exposed to a contaminated whirlpool spa. *Lancet* 347:494–9. *http://www.ncbi.nlm.nih.gov/entrez/query.fcgi?cmd=Retrieve&db=PubMed&dopt=Citation&list_uids=8596266.*

The Joint Program Executive Office for Chemical and Biological Defense. (2005a). Biological Integrated Detection System (BIDS). *http://www.jpeocbd.osd.mil/ca_bids.htm.*

The Joint Program Executive Office for Chemical and Biological Defense. (2005b). Joint Biological Point Detection System (JBPDS). *http://www.jpeocbd.osd.mil/ca_jbpds.htm.*

The Joint Program Executive Office for Chemical and Biological Defense. (2005c). Joint Biological Standoff Detection System (JBSDS). *http://www.jpeocbd.osd.mil/ca_jbsds.htm.*

The Joint Program Executive Office for Chemical and Biological Defense. (2005d). Joint Portal Shield (JPS). *http://www.jpeocbd.osd.mil/ca_jps.htm.*

The Joint Program Executive Office for Chemical and Biological Defense. (2005e). Joint Project Manager NBC Contamination Avoidance (JPM NBC CA). *http://www.jpeocbd.osd.mil/conavoid.htm.*

The Joint Program Executive Office for Chemical and Biological Defense. (2005f). Protecting Our Warfighters. *http://www.jpeocbd.osd.mil/warfighter.htm.*

Jorm, L., Thackway, S., Churches T, et al. (2003). Watching the Games: public health surveillance for the Sydney 2000 Olympic Games. *J Epidemiol Community Health* 57:102–8.

Keim, P., Smith, K. L., Keys, C., et al. (2001). Molecular investigation of the Aum Shinrikyo anthrax release in Kameido, Japan. *J Clin Microbiol* 39:4566–7. *http://www.ncbi.nlm.nih.gov/entrez/query.fcgi?cmd=Retrieve&db=PubMed&dopt=Citation&list_uids=11724885.*

Kenyon, T. A., Valway, S. E., Ihle, W. W., et al. (1996). Transmission of multidrug-resistant Mycobacterium tuberculosis during a long airplane flight. *N Engl J Med* 334:933–8. *http://www.ncbi.nlm.nih.gov/entrez/query.fcgi?cmd=Retrieve&db=PubMed&dopt=Citation&list_uids=8596593.*

Kobayashi, A., Yamamoto, Y., Chou S, et al. (2004). Severe Legionella pneumophila pneumonia associated with the public bath on a cruise ship in Japan. *J Anesth* 18:129–31. *http://www.ncbi.nlm.nih.gov/entrez/query.fcgi=cmd=Retrieve&db=PubMed&dopt=Citation&list_uids=15127261.*

Lew, J. F., Swerdlow, D. L., Dance, M. E., et al. (1991). An outbreak of shigellosis aboard a cruise ship caused by a multiple-antibiotic-resistant strain of Shigella flexneri. *Am J Epidemiol* 134:413–20. *http://www.ncbi.nlm.nih.gov/entrez/query.fcgi?cmd=Retrieve&db=PubMed&dopt=Citation&list_uids=1652203.*

Lewis, M. D., Pavlin, J. A., Mansfield, J. L., et al. (2002). Disease outbreak detection system using syndromic data in the greater Washington DC area. *Am J Prev Med* 23:180–6. *http://www.ncbi.nlm.nih.gov/entrez/query.fcgi=cmd=Retrieve&db=PubMed&dopt=Citation&list_uids=12350450.*

Lipton, E. (2005). E.P.A. report finds lag in monitoring attacks. New York Times, March 25, p. 13.

Lombardo, J., Burkom, H., Elbert, E., et al. (2003). A systems overview of the Electronic Surveillance System for the Early Notification of Community-Based Epidemics (ESSENCE II). *J Urban Health* 80(Suppl 1):i32–42. *http://www.ncbi.nlm.nih.gov/entrez/query.fcgi=cmd=Retrieve&db=PubMed&dopt=Citation&list_uids=12791777.*

Loonsk, J. W. (2004). BioSense—a national initiative for early detection and quantification of public health emergencies. *MMWR Morb Mortal Wkly Rep* 53(Suppl):53–5. *http://www.ncbi.nlm.nih.gov/entrez/query.fcgi=cmd=Retrieve&db=PubMed&dopt=Citation&list_uids=15714629.*

Lumish, R. M., Ryder, R. W., Anderson, D. C., et al. (1980). Heat-labile enterotoxigenic Escherichia coli induced diarrhea aboard a Miami-based cruise ship. *Am J Epidemiol* 111:432–6. *http://www.ncbi.nlm.nih.gov/entrez/query.fcgi=cmd=Retrieve&db=PubMed&dopt=Citation&list_uids=6990749.*

Marsden-Haug, N., Foster V., Hakre, S., et al. (2005). Evaluation of Joint Services Installation Pilot Project and BioNet Syndromic Surveillance Systems—United States, 2004. *Morb Mortal Wkly Rep* 54(Suppl):194. *http://www.cdc.gov/mmwr/pdf/wk/mm54su01.pdf.*

McNeill, W. H. (1989). Plagues and Peoples. New York: Doubleday.

Merson, M. H., Tenney, J. H., Meyers, J. D., et al. (1975). Shigellosis at sea: an outbreak aboard a passenger cruise ship. *Am J Epidemiol* 101:165–75. *http://www.ncbi.nlm.nih.gov/entrez/query.fcgi?cmd=Retrieve&db=PubMed&dopt=Citation&list_uids=1092155.*

Military Postal Service Agency. (2004). Biohazard Detection System (BDS) Briefing Sheet. *http://hqdainet.army.mil/mpsa/nov_conf/ppt/pdf/BDS.pdf.*

Miller, M. A., Valway S., Onorato, I. M. (1996). Tuberculosis risk after exposure on airplanes. *Tuber Lung Dis* 77:414–9. *http://www.ncbi.nlm.nih.gov/entrez/query.fcgi=cmd=Retrieve&db=PubMed&dopt=Citation&list_uids=8959144.*

Mintz, J. (2003). U.S. provides a peek at air sensor program. Washington Post, November 15, p. 3.

Moser, M. R., Bender, T. R., Margolis, H. S., et al. (1979). An outbreak of influenza aboard a commercial airliner. *Am J Epidemiol* 110:1–6. *http://www.ncbi.nlm.nih.gov/entrez/query.fcgi=cmd=Retrieve&db=PubMed&dopt=Citation&list_uids=463858.*

Navy Environmental Health Center. (2005). NEHC Force Health Protection Plan. *www-nehc.med.navy.mil/NEHC_Strategic_Plan.doc.*

Novelli, V. M., Lewis, R. G., Dawood, S. T. (1987). Epidemic group A meningococcal disease in Haj pilgrims. *Lancet* 2:863. *http://www.ncbi.nlm.nih.gov/entrez/query.fcgi=cmd=Retrieve&db=PubMed&dopt=Citation&list_uids=2889067.*

Okumura, T., Ninomiya N.,Ohta, M. (2003). The Chemical Disaster Response System in Japan. *Prehosp Disaster Med* 18:189–92. *http://pdm.medicine.wisc.edu/18-3pdfs/189Okumura.pdf.*

Pavlin J., Kelley, P. (2004). Department of Defense Global Emerging Infections Systems Programs in Biodefense. In: Lindler L, Lebeda F, Korch G, eds. Infectious Diseases: Biological Weapons Defense: Infectious Diseases and Counterterrorism, 361–85, L., F., G.. Totowa, NJ: Humana Press, Totowa, NJ.

Regan, C. M., McCann, B., Syed, Q., et al. (2003). Outbreak of Legionnaires' disease on a cruise ship: lessons for international surveillance and control. *Commun Dis Public Health* 6:152–6. *http://www.ncbi.nlm.nih.gov/entrez/query.fcgi=cmd=Retrieve&db=PubMed&dopt=Citation&list_uids=12889297.*

Ritzinger, F. R. (1965). Disease transmission by aircraft. *Aeromed Rev* 4:1–10. *http://www.ncbi.nlm.nih.gov/entrez/query.fcgi?cmd=Retrieve&db=PubMed&dopt=Citation&list_uids=5825522.*

Roberts, M. (1996). Role of aviation in the transmission of disease. In: DeHart R, ed. Fundamentals of Aerospace Medicine, 2nd ed., Baltimore: Williams & Wilkins. pp. 1091.

Rowbotham, T. J. (1998). Legionellosis associated with ships: 1977 to 1997. *Commun Dis Public Health* 1:146–51. *http://www.ncbi.nlm.nih.gov/entrez/query.fcgi=cmd=Retrieve&db=PubMed&dopt=Citation&list_uids=9782626.*

Shu, P. Y., Chien, L. J., Chang, S. F., et al. (2005). Fever screening at airports and imported dengue. *Emerg Infect Dis* 11:460–2. *http://www.ncbi.nlm.nih.gov/entrez/query.fcgi=cmd=Retrieve&db=PubMed&dopt=Citation&list_uids=15757566.*

Simon, H. (2004). DHS bio-detection contracts tied to BioWatch expansion. Aviation Week's Homeland Security and Defense, February 4.

Snyder, J. D., Wells, J. G., Yashuk, J., et al. (1984). Outbreak of invasive Escherichia coli gastroenteritis on a cruise ship. *Am J Trop Med Hyg* 33:281–4. *http://www.ncbi.nlm.nih.gov/entrez/query.fcgi?cmd=Retrieve&db=PubMed&dopt=Citation&list_uids=6370005.*

Telemedicine & Advanced Technology Research Center (U.S. Army Medical Research & Materiel Command). (2004). Integrated Biosurveillance & Bioterrorism Preparedness. Advanced Briefing for Industry, December. *http://www.atmeda.org/conf/2004.SBOTS.presentations/015_Parker.ppt.*

U.S. Air Force. (2005). Air Force Institute for Operational Health (AFIOH) Fact Sheet. *http://www.brooks.af.mil/afioh/Files/2005%20AFIOH%20Fact%20Sheet.pdf.*

U.S. Army Medical Research and Material Command. (2005). About the Medical Research and Materiel Command. *https://mrmc.detrick.army.mil/aboutmrmc.asp.*

U.S. Army Medical Research and Materiel Command, Research Area Directorate IV. (2005). Medical Chemical and Biological Defense Science and Technology Program. *http://medchembio.detrick.army.mil/MedBioindex.html.*

U.S. Army Medical Research Institute of Infectious Diseases. (2005). About USAMRIID. *http://www.usamriid.army.mil/aboutpage.htm.*

Department of the Army, Headquarters. (2005). Military Postal Service. *http://hqdainet.army.mil/mpsa/nov_conf/ppt/pdf/BDS.pdf.*

U.S. Department of Defense. (2005a). DoD 101: An introductory overview of the Department of Defense. *http://www.defenselink.mil/pubs/dod101/.*

U.S. Department of Defense. (2005b). Overall Surveillance and Response System for Infectious Diseases. Global Emerging Infections Surveillance and Response System. *http://www.geis.fhp.osd.mil/aboutGEIS.asp.*

U.S. Department of Defense. (2005c). Tricare Management Activity. *http://www.tricare.osd.mil/tricarehandbook.*

U.S. Department of Defense. (2005d). Tricare Military Treatment Facility Locator. *http://www.tricare.osd.mil/mtf.*

U.S. Department of Defense. (2005e). Tricare Orientation Briefing. *http://www.tricare.osd.mil/briefings/TRICARE_101_new_brief.ppt.*

U.S. Department of Defense Military Health System. (2005). Force Health Protection and Readiness. *http://www.ha.osd.mil/FHPR/default.cfm.*

U.S. Department of the Army. (1977). US Army Activity in the US Biological Warfare Programs. Volume I, Unclassified. *http://www.gwu.edu/~nsarchiv/NSAEBB/NSAEBB58/RNCBW_USABWP.pdf.*

U.S. Postal Service. (2004). Biohazard Detection System Deployment Resumes: Statement of Azeezaly S. Jaffer, Vice President, Public Affairs and Communications. Release No. 38. *http://www.lunewsviews.com/usps/pcr.htm.*

U.S. Department of Homeland Security Science and Technology Division. Fact Sheet: BioWatch. Early Detection, Early Response. *http://www.milnet.com/wh/DoHS/BioWatchFactSheetFINAL.pdf.*

Uyeki, T. M., Zane, S. B., Bodnar, U. R., et al. (2003). Large summertime influenza A outbreak among tourists in Alaska and the Yukon Territory. *Clin Infect Dis* 36:1095–102. *http://www.ncbi.nlm.nih.gov/entrez/query.fcgi?cmd=Retrieve&db=PubMed&dopt=Citation&list_uids=12715302.*

Walter Reid Army Institute of Research. (2005). WRAIR Mission. http://wrair-*http://www.army.mil/aboutWRAIR/mission.asp.*

Waples, P., Thackway, S., Banwell, K., et al. (2000). Health surveillance on cruise ships during the Sydney 2000 Olympic and Paralympic Games. *N S W Public Health Bull* 11:150–1. *http://www.ncbi.nlm.nih.gov/entrez/query.fcgi=cmd=Retrieve&db=PubMed&dopt=Citation&list_uids=12105462.*

Waterman, S. H., Demarcus, T. A., Wells, J. G, et al. (1987). Staphylococcal food poisoning on a cruise ship. *Epidemiol Infect* 99:349–53. *http://www.ncbi.nlm.nih.gov/entrez/query.fcgi?cmd=Retrieve&db=PubMed&dopt=Citation&list_uids=3678396.*

Wenzel, R. P. (1996). Airline travel and infection. *N Engl J Med* 334:981–2. *http://www.ncbi.nlm.nih.gov/entrez/query.fcgi?cmd=Retrieve&db=PubMed&dopt=Citation&list_uids=8596601.*

White, F. M., Lacey, B. A., Constance, P. D. (1981). An outbreak of poliovirus infection in Alberta—1978. *Can J Public Health* 72:239–44. *http://www.ncbi.nlm.nih.gov/entrez/query.fcgi?cmd=Retrieve&db=PubMed&dopt=Citation&list_uids=7306906.*

The White House. (2004). HSPD-10: Biodefense for the 21st Century. Homeland Security Presidential Directive. *http://www.fas.org/irp/offdocs/nspd/hspd-10.html.*

The White House. (2005). Presidential Decision Directive NSTC-7. *http://www.geis.fhp.osd.mil/GEIS/aboutGEIS/historicaldocs/NSTC-7.asp.*

World Health Organization. (1946). Constitution of the World Health Organization. *http://whqlibdoc.who.int/hist/official_records/constitution.pdf.*

World Health Organization. (1999). Spread of meningococcal meningitis by pilgrims returning from the Haj, 1987. *http://www.who.int/infectious-disease-report/pages/graph39.html.*

World Health Organization. (2005a). 1996—Ebola haemorrhagic fever in South Africa—Update 2. *http://www.who.int/csr/don/1996_11_28b/en/.*

World Health Organization. (2005b). About WHO. *http://www.who.int/about/en/.*

World Health Organization. (2005c). Frequently asked questions about the International Health Regulations. *http://www.who.int/csr/ihr/howtheywork/faq/en/index.html#will.*

III
PART
Data Analysis

13

CHAPTER

Case Detection Algorithms

Fu-Chiang Tsui, Michael M. Wagner, and Jeremy Espino

RODS Laboratory, Center for Biomedical Informatics, University of Pittsburgh, Pittsburg, Pennsylvania

Richard Shephard

Australian Biosecurity Cooperative Research Centre for Emerging Infectious Disease, Brisbane, Australia

1. INTRODUCTION

This chapter focuses on algorithmic methods for case detection. As we discussed in Chapter 3, the objective of case detection is to notice the existence of a single case of a disease. Case detection is a core activity of biosurveillance. Detection of an outbreak usually depends on detection of individual cases, although it is also possible to detect an outbreak from data other than case data (e.g., retail sales of thermometers or satellite imagery).

For most of the 20th century, governmental public health relied almost exclusively on the astute clinician and the clinical laboratory to detect and report cases via the notifiable disease system. Algorithms for case detection did not exist, with the possible exception of case definitions discussed in Chapter 3, which are algorithms in the sense that they are formal specifications that a clinician or epidemiologist can follow to classify an individual as a suspected or confirmed case.

Around the beginning of the 21st century, emerging diseases and the threat of bioterrorism began to stress the capabilities of the notifiable disease system. Health departments endeavored to improve it by creating web-based forms for disease reporting and electronic laboratory reporting. Their goal is to increase completeness of reporting and decrease time latencies inherent in paper-based reporting.

After the anthrax postal attacks of 2001, health departments redoubled their efforts to improve disease reporting by increasing the "astuteness" of clinicians through education and training (Gerberding et al., 2002, Hughes and Gerberding, 2002). A specialist in infectious diseases had detected and reported the first case of inhalational anthrax (Kolata, 2001); thus, the conventional wisdom was that the notifiable disease system worked. For the foreseeable future, increasing the ability of front-line physicians to diagnose rare diseases through better training was our best defense against bioterrorism. The conventional wisdom was correct insofar as case detection by clinicians is important in outbreak detection. However, it was overly sanguine about the ability of training to improve the existing capability.[1]

There is a limit to which additional training can improve a clinician's ability to detect and report rare diseases to a biosurveillance organization. Humans are not perfectible in this manner, as noted first by Dr. Clem McDonald, who entitled his seminal paper on physician performance *Protocol-Based Computer Reminders, the Quality of Care and the Non-Perfectability of Man* (McDonald, 1976b).[2]

His research, conducted in the early 1970s at the Regenstrief Institute in Indiana, demonstrated that even for common conditions, such as diabetes and hypertension, and for common preventive measures, such as immunizations, physicians often failed to deliver required services. However, when reminded by a computer system that monitored electronic patient data about the need to vaccinate a *specific patient* or order a needed test, physicians complied with standards of care at twice the rate as when not reminded. When the system stopped reminding the physicians, however, their compliance rates quickly returned to baseline; thus, any "education" or "training" that the system provided had no lasting effect on compliance.

McDonald's research spawned a new line of system-level thinking in medicine that continues to this day (Kohn et al., 2000, Leavitt, 2001, Yasnoff et al., 2004). McDonald concluded this influential paper with the observation that, although man is not perfectible, *systems* of care are.

---

1 Studies of continued medical education programs also suggest that the yield is low and the cost is high (Haynes et al., 1984, Leist and Kristofco, 1990, Williamson et al., 1989, McDonald, 1983). The one study of efficacy of training (of a web-based educational program) on physicians' knowledge about diagnosis and management diseases caused by known weaponized biological agent showed no retention of information (chung et al., 2004).

2 We do not know whether the always playful Dr. McDonald and the sophisticated editors of the New England Journal of Medicine misspelled the word *perfectibility* intentionally. Also reinforcing the main conclusion of his research, it took a computer's spell checker to bring this error to our attention.

ISBN 0-12-369378-0

2. PERFECTING CASE DETECTION

The best case detection system imaginable would be one in which every individual in a community is examined every morning by the best diagnostician in the world. This diagnostician would have all the time in the world to interview and examine each person. Since she would be examining everyone every day, she would notice patterns (clusters) of early illness in a community. Her awareness of patterns would appropriately bias her diagnostic thinking (and treatment) of individual patients. Physicians are taught (and reminded incessantly) "when you hear hoof beats, don't think of zebras." This adage is an informal statement that when the evidence available about a particular patient supports equally a diagnosis of either influenza or SARS (e.g., the patient has constitutional symptoms and no history of exposure to SARS), they should conclude that the diagnosis of influenza is far more likely than SARS. This diagnostician would also never fail to report immediately each fever, early syndromic presentation, or reportable disease to governmental public health.

This ideal scenario recognizes the importance of the knowledge, judgment, and skills that clinicians bring to bear on the diagnosis of disease. The importance of expert knowledge and judgment underlies the opinion, expressed by experts in public health after the anthrax letters in 2001, that there is no substitute for an astute clinician. This ideal scenario also recognizes that knowledge of the prevalence of disease in a population influences the diagnostic work up and management of individual patients.

Of course, such a system is not feasible due to the impossibility of having every person seen by the same diagnostician every day or the alternative of cloning the best diagnostician and ensuring that the clones could instantly share information about individuals they were seeing. Nevertheless, it represents a benchmark against which we can compare other schemes that are perhaps superior to current approaches.

In the next section, we discuss diagnostic expert systems, which are computer programs (algorithms) that embody the diagnostic knowledge and diagnostic skills of expert clinicians. Diagnostic expert systems are far less expensive than clinicians, never tire, and we can clone them at will. They make it reasonable to imagine a biosurveillance system in which a highly competent diagnostician examines thousands of individuals in a community every day with consistent diagnostic quality—reporting fevers, syndromes, and reportable diseases to a health department without fail and without delay. They make it possible to imagine a biosurveillance system in which the health department analyzes highly improved case data and communicates up-to-the-minute information about patterns of illness in the community back to the diagnosticians.

3. DIAGNOSTIC EXPERT SYSTEMS

Diagnostic expert systems are computer algorithms that automate the cognitive process of medical (or veterinary) diagnosis. Many readers will be surprised to learn that researchers in the fields of artificial intelligence and medical informatics have been developing and fielding such systems since the 1960s. The first fielded system, developed by Dr. Homer Warner in Salt Lake City, provided diagnostic assistance for children with congenital heart disease (Warner et al., 1961, Warner et al., 1964). Congenital heart diseases are severe birth defects of the valves and structure of the heart. In the 1960s, the exact nature of heart malformations was very difficult to diagnose without invasive and risky angiographic procedures. Dr. F. Timothy de Dombal developed a similar system for the differential diagnosis of the acute abdomen, another high-stakes diagnostic problem. A surgeon must differentiate between conditions that require emergency surgery, such as appendicitis, and conditions, such as pancreatitis, for which surgery is relatively contraindicated (de Dombal et al., 1972, de Dombal et al., 1974, Wilson et al., 1975, de Dombal, 1975, Wilson et al., 1977, de Dombal, 1984, Adams et al., 1986, McAdam et al., 1990, de Dombal, 1990, de Dombal et al., 1993, American College of Emergency Physicians, 1994).

3.1. How Diagnostic Expert Systems Work: Data Collection

We can perhaps best explain how a diagnostic expert system works by analogy to the process that a physician uses to diagnose a patient. Like a physician, a diagnostic expert system begins by collecting information about the patient's illness—symptoms, observations from physical examination, results from laboratory tests, risk factors for disease (e.g., travel to a foreign country) and pre-existing medical conditions (e.g., diabetes). We refer to this diagnostic information collectively as the findings. Of course, the computer usually does not interview the patient and (at present) never examines the patient. Rather, a physician interviews and examines the patient after which she or an assistant enters the findings into the program (e.g., Warner and Bouhaddou, 1994, London, 1998, Buchanan and Shortliffe, 1984, Miller et al., 1986, Shwe et al., 1991, Heckerman et al., 1992). Increasingly, diagnostic expert systems acquire findings automatically from clinical information systems (Aronsky et al., 2001, Burnside et al., 2004, McDonald et al., 1991). There are also examples of diagnostic expert systems that interview patients directly to obtain their medical histories (Pynsent and Fairbank, 1989, Wald et al., 1995).

3.2. How Diagnostic Expert Systems Work: Knowledge Representation

Like the physician, the diagnostic expert system is a storehouse of medical knowledge. A diagnostic expert system stores its medical knowledge in tables of diseases and their findings. There are typically a table of prevalences for each disease (e.g., Table 13.1) and tables with every finding of every disease that the system knows about (e.g., Table 13.2). The latter tables usually represent the strength of association between diseases and findings as conditional probabilities.

TABLE 13.1 Prior Probabilities and Prior Odds of FMD and MCD

Disease	P(Disease)	Odds(Disease)
FMD	0.001	0.001001
MCD	0.001	0.001001

Notes: The notation p(FMD) actually is shorthand for a probability of an event variable called FMD that can take the values true or false. So, p(FMD) actually stands for p(foot and mouth disease is present) or p(foot and mouth disease is not present). If we to write Table 13.1 out in its full form, it would have four rows corresponding to p(FMD is present) = 0.001, p(FMD is absent) = 0.999, p(MCD is present) = 0.001, p(MCD is absent) = 0.999. The laws of probability require that the sum of the probabilities for all the event variable outcomes will be equal to 1 (e.g., p[FMD is present] + p[FMD is absent] = 1, because it is a certainty that something is either present or it is absent), so knowledge engineers usually save space by only writing one row per disease. FMD, foot and mouth disease; MCD, mad cow disease.

3.3. How Diagnostic Expert Systems Work: Differential Diagnosis

Like the physician, the computer program generates a differential diagnosis for the sick individual. A differential diagnosis is a list of diseases that are most likely to account for the findings in a patient. The diagnostic expert system typically creates its differential diagnosis by computing the posterior probability for every disease given the findings. A disease's posterior probability is the chance that the patient has the disease, given the findings.

Diagnostic expert systems use Bayes rules to compute the differential diagnosis. The relevance of Bayes rules to medical diagnosis was first introduced theoretically by Ledley and Lusted (1959) and first used in a diagnostic expert system by Homer Warner in 1962. Developers of diagnostic expert systems continue to use the same methods as did Homer Warner, as well as more complex Bayesian methods (but the original technique generally works well).

3.4. How Diagnostic Expert Systems Work: Question Generation

A differential diagnosis must be "resolved"; that is, the diseases in the list must be ruled in or ruled out. Like an expert physician, the diagnostic expert system engages in a cyclic process often referred to as "hypothesize and test" to resolve the differential diagnosis. A diagnostic expert system uses value-of-information calculations (Weinstein, 1980) to recommend to the physician user additional findings that will resolve the differential diagnosis efficiently. Value-of-information calculations identify those findings, which, if known to be present or absent, optimally discriminate among the diseases in the differential diagnosis, where optimality takes into account not only the probability of a diagnosis, but cost-benefit considerations, such as whether a diagnosis is treatable.

As new findings become available (either as a result of following recommendations from the diagnostic expert system, consultants, or the physician's own judgment), the physician user can rerun the diagnostic expert system to recompute the differential diagnosis. The most likely diagnosis from the first run may become less or more likely as the new information acts to rule in or rule out each diagnosis. The user can rerun the program whenever new findings about the individual become available during the course of the diagnostic work-up. The net result of this cyclic process is the diagnostic certainty about the diagnoses in the differential increases over time (i.e., probabilities of the diagnoses in the differential move towards zero, indicating certainty that a disease is not present, or one, indicating certainty that a disease is present) (Figure 13.1).

TABLE 13.2 Conditional Probabilities for FMD and MCD

Finding	Disease	p(Finding|Disease)[a]
Drooling of saliva present	FMD present	0.95 (sensitivity)
Drooling of saliva present	FMD absent	0.05 (1 – specificity)
Drooling of saliva present	MCD present	0.001 (sensitivity)
Drooling of saliva present	MCD absent	0.05 (1 – specificity)
More than one animal affected	FMD present	0.95 (sensitivity)
More than one animal affected	FMD absent	0.2 (1 – specificity)
More than one animal affected	MCD present	0.001 (sensitivity)
More than one animal affected	MCD absent	0.2 (1 – specificity)

[a]Similar to the previous table, the knowledge engineers have left out half of the combinations because they can be derived from those listed listed by subtraction from 1. For example, p(drooling of saliva is not present|FMD is present)= 1 – p(drooling of saliva is present|FMD is present) = 1 – 0.95 = 0.05. FMD, foot and mouth disease; MCD, mad cow disease.
Probabilities from BOVID, a cattle diagnostic program. Courtesy of Animal Information Management Pty. Ltd, Victoria, Australia, BOVID, a cattle diagnostic program.

4. EXAMPLES OF DIAGNOSTIC EXPERT SYSTEMS: BOSSS AND ILIAD

As indicated in the previous discussion, researchers have developed many diagnostic expert systems that vary in both the underlying technology as well as the domain of application. In this section, we briefly describe two systems that illustrate the key characteristics of probabilistic diagnostic expert systems. These systems—Iliad and BOSSS—are diagnostic

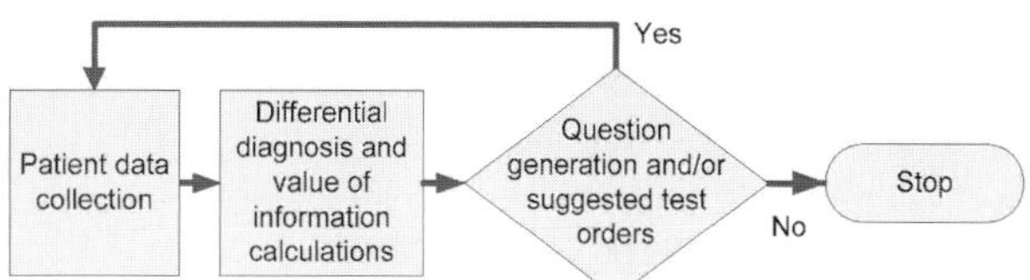

FIGURE 13.1 The process of diagnosis. Like a physician, a diagnostic expert system analyzes patient data to generate a differential diagnosis. Based on the differential diagnosis and value of information calculations, the diagnostic expert system can suggest additional questions to ask the patient, additional physical observations to make about the patient, or additional tests to consider. A physician may accept the recommendations or not. As additional information about the patient's illness arrives over time (either as a result of tests or questions suggested by the computer; or by the physician; or simply the passage of time, this user can enter this new information and rerun the computer program to generate an updated differential diagnosis). This process of data collection and analysis can be repeated frequently until the diagnosis for a patient is established with sufficient certainty and diagnostic precision that further diagnostic evaluation is not necessary.

expert systems in the domains of diseases of humans and cattle, respectively.

4.1. Iliad

Iliad was developed by Homer R. Warner, Jr. and colleagues at the University of Utah in the 1980s (Warner, 1989, Sorenson et al., 1988, Cundick et al., 1989, Warner and Bouhaddou, 1994, Warner, 1990). Iliad is a stand-alone diagnostic expert system; that is, it requires a user (physician) to enter findings into the program. Iliad uses Bayes rules to compute a differential diagnosis for the patient's illness. It uses value-of-information calculations to recommend additional tests and observations.

Iliad performs differential diagnosis of human diseases in a variety of fields: internal medicine, sports medicine, pediatrics, dermatology, psychiatry, obstetrics/gynecology, urology, peripheral vascular diseases, and sleep disorders. Iliad covers more than 900 diseases and 1500 syndromes (which means that there are tables of information about diseases and findings for approximately 2400 diseases similar to those shown in Figures 13.1 and 13.2). Iliad computes a differential diagnosis for a patient and then displays the diseases and syndromes to the user in order of posterior probability as shown in Figure 13.2. If the user asks Iliad to suggest additional tests or questions to ask the patient, the program performs value-of-information calculations to suggest findings that would discriminate among the diseases in the differential diagnosis (Figure 13.3).

Researchers have compared Iliad's diagnostic performance on real cases with that of expert physicians and found equivalent performance (Graber and VanScoy, 2003, Friedman et al., 1999, Elstein et al., 1996, Murphy et al., 1996). Interestingly, Iliad did not achieve widespread acceptance in clinical practice because the entering of patient data was too time-consuming for busy clinicians and the program is no longer in use.

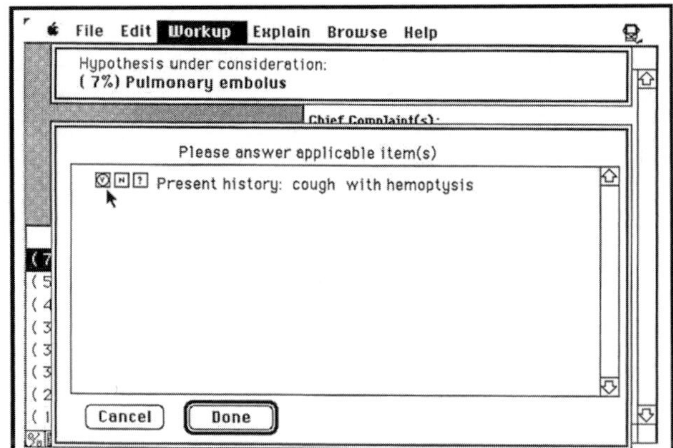

FIGURE 13.3 Iliad screen requesting additional patient information. (From Warner, H. R., Jr. (1989). Iliad: moving medical decision-making into new frontiers. Methods Inform Med 28:370–2, with permission.)

4.2. BOSSS

The Bovine Syndromic Surveillance System (BOSSS) is a web-based disease-reporting tool that incorporates a diagnostic expert system for diagnosis of diseases of cattle (Figure 13.4) (Shephard et al., 2005). BOSSS requires a user (veterinarian or cattle herd worker) to enter findings into the program (Figure 13.5). BOSSS uses Bayes rules to compute a

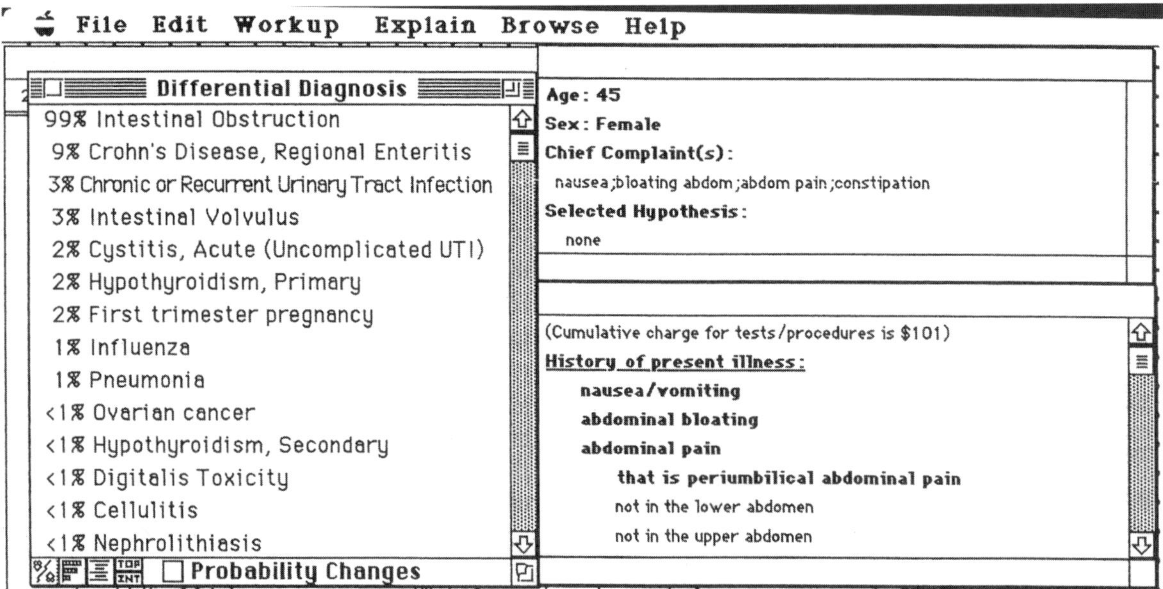

FIGURE 13.2 Consultation mode of Iliad showing the patient findings that the physician entered (lower right panel), and the differential diagnosis (panel on left) that Iliad generated from those findings. (Courtesy of LE Widman, *http://www.informaticamedica.org.br/informaticamedica/n0105/widman.htm.*)

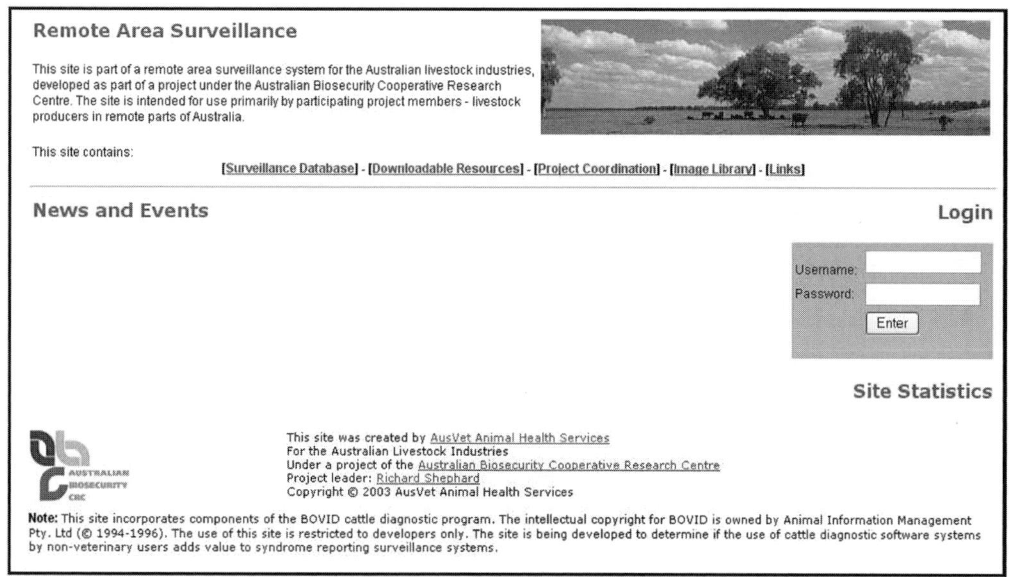

FIGURE 13.4 Log-in page of the web-based bovine syndromic surveillance system.

differential diagnosis for the patient's illness. It also uses value-of-information calculations to recommend additional tests and observations.

Shephard and colleagues developed BOSSS to meet a need to capture disease and syndrome observation data from field observations made by veterinarians, producers, and lay observers. These observations are important to population-based surveillance of cattle herds and to the business of cattle production, but are largely unrecorded and, therefore, unavailable for biosurveillance purposes. BOSSS encourages use by rewarding a user with diagnostic support in his efforts to determine the cause of illness or death in cattle.

Like Iliad, BOSSS contains information on the prevalence of diseases (for approximately 1000 diseases of cattle). It also contains the sensitivity and specificity of each finding for each disease. (We discuss sensitivity and specificity in Sect. 5.1.) The development team compiled this information from veterinary literature and the opinions of veterinary experts. Like Iliad, it

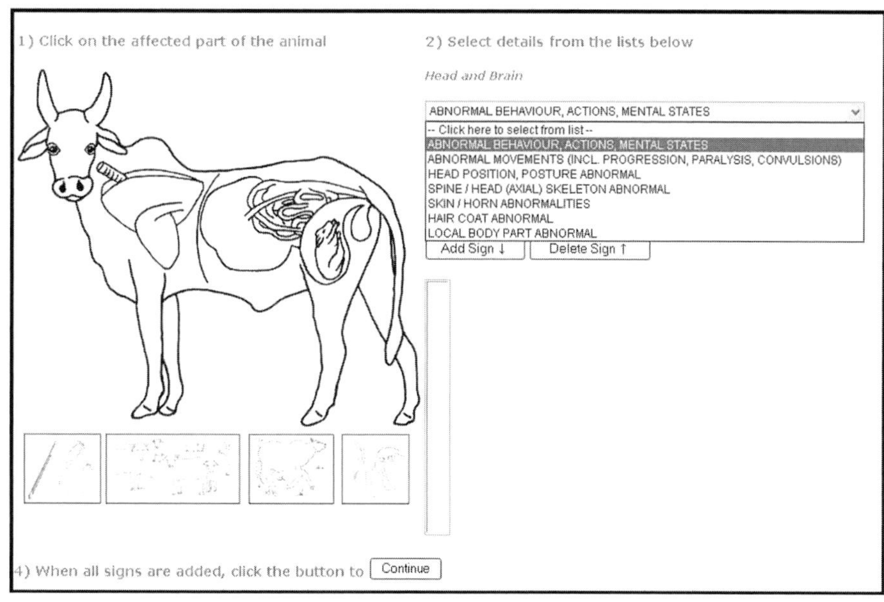

FIGURE 13.5 A graphic user interface for entering findings of a sick animal.

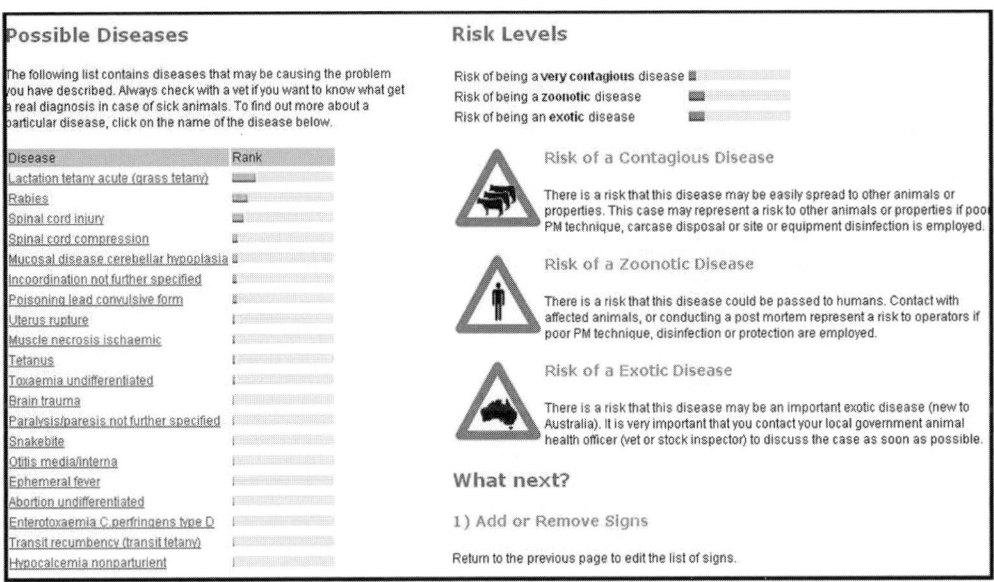

FIGURE 13.6 Differential diagnosis results ranked by posterior probabilities.

uses Bayes rules to determine the posterior probability of each disease given the findings (Figure 13.6). The program promotes the capture of extra information on each case through the use of an interrogation module that presents questions to the user (Figure 13.7). These questions are key differentiating signs for the most likely diseases identified by the system that have not already been recorded by the user. Unlike Iliad, BOSSS has an explicit biosurveillance mission. The "syndromic-surveillance" component of BOSSS is illustrated by the mapping of cases shown in Figure 13.8.

The developers of BOSSS, learning lessons from Iliad and other diagnostic expert systems about the importance of fitting into the workflow of busy clinicians, are developing a palm-computer version of BOSSS. This portable version will fit into the practice patterns of cattle veterinarians, who spend their days at different locations and most definitely do not work in an office setting. A training manual is available for download at *http://www.ausvet.com.au/bosss/resources/BOSSS_Manual.pdf.*

Guided Examination

Please indicate if the following signs were present or absent, or if you are unsure.

- Observations from a distance (may be done in the field)
- Hands-on examination (usually requires restraint)

1) Observations from a distance

Did you observe any of the following signs?

Clinical Sign	Yes	No	Not Sure
Animal details -sex, age, status, etc (signalment)			
lactating cow	○	○	⊙
calving, during first month afterwards	○	○	⊙
age adult	○	○	⊙
female	○	○	⊙
dry cow	○	○	⊙
Behaviour, actions, mental state			
excessive response to external stimuli (hyperespon	○	○	⊙
restlessness	○	○	⊙
convulsions unspecified	○	○	⊙
convulsions tetanic (rigid)	○	○	⊙

FIGURE 13.7 A graphic user interface for guided examination.

5. KNOWLEDGE REPRESENTATION AND INFERENCE IN DIAGNOSTIC EXPERT SYSTEMS

As previously discussed, a diagnostic expert system contains an internal store of facts about diseases, including (1) the prevalence of each disease, (2) their findings, and (3) the statistical relationships between the findings and the diseases. We call this collection the knowledge base. In addition to the knowledge base, a diagnostic expert system contains an inference engine, which performs diagnostic reasoning (Figure 13.9).

To demonstrate how a diagnostic expert system uses Bayes rules to compute a differential diagnosis, we created mini-BOSSS, a tiny version of BOSSS. In particular, mini-BOSSS has a knowledge base with only two diseases and two findings. The diseases are foot and mouth disease (FMD) and mad cow disease (MCD). The findings are drooling saliva and whether there is more than one cow in the herd with this symptom.

5.1. Probabilistic Knowledge Bases

We note that there are several kinds of expert systems, including the probabilistic diagnostic expert systems that we have been discussing, and rule-based expert systems. We discuss rule-based expert systems later in this chapter.

A probabilistic knowledge base uses probabilities to represent the prevalence of disease and the relationships between findings and diseases. A development team creates a knowledge

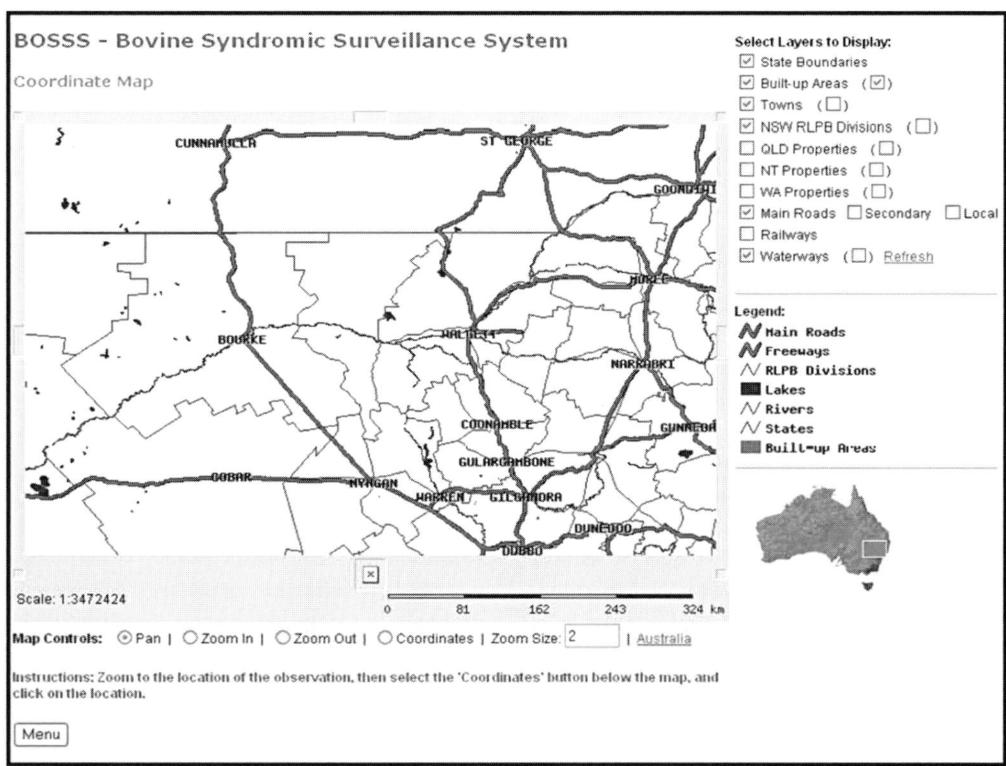

FIGURE 13.8 A mapping tool that displays geographic locations of reported cases.

base by a labor-intensive literature review supplemented by interviews with experts. Increasingly, developers use large data sets collected by hospital information systems (especially for disease prevalence) to develop knowledge bases. The process that developers use to elicit knowledge from experts and to convert information in the literature into a knowledge base is referred to as knowledge acquisition or knowledge engineering (Feigenbaum, 1977).

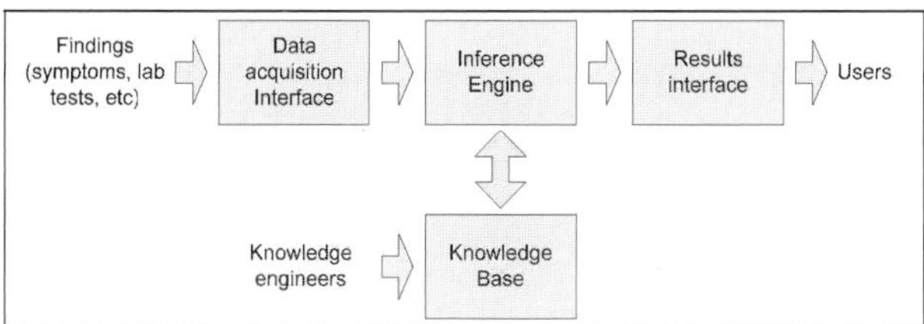

FIGURE 13.9 Components and process flow in an expert system. The inference engine uses both patient data and medical knowledge to compute a differential diagnosis and to generate suggestions for additional collection of data. The inference engine obtains data from an inbound data-acquisition interface and it outputs the differential diagnosis and suggestions for additional data through an outbound results interface. In a freestanding diagnostic expert system such as Iliad or BOSSS, both the data-acquisition and results interfaces are screens that a physician interacts with. In an embedded system like Antibiotic Assistant, the data-acquisition interface is with hospital information systems (and to some extent with the user, but only for selected items of information that the user wishes to provide). The results interface could also be with another computer system such as a point-of-care system, which would present the differential diagnosis and suggestions through its own screens.

5.1.1. *Prior Probabilities (Disease Prevalence)*

The prior probability is the prevalence of disease in the population. We represent prior probabilities using the notation P(Disease). For example, P(FMD) represents the prior probability of foot and mouth disease. Table 13.1 shows the prior probabilities of FMD and MCD that we use in our example. Table 13.1 also shows the prior odds of the diseases in mini-BOSSS. Odds are a simple mathematical transform of probabilities, which are essentially equal to probabilities when probabilities are less than 0.1. The reason that we show odds in Table 13.1 is that we use the odds-likelihood form of Bayes rules in our example. We will define odds and the odds-likelihood form of Bayes shortly.

5.1.2. *Conditional Probabilities*

A conditional probability is the chance of one event occurring given the occurrence of another event. In diagnostic expert systems, we use conditional probabilities to describe the probability of seeing a certain finding (one event) when a disease is present (another event). If the conditional probability is high, it means that the particular finding is often associated with the disease. The mathematical notation for the conditional probability of a finding, given a disease, is P(Finding|Disease). For example, P(drooling of saliva is present|FMD is present) is the probability of observing drooling of saliva in a cow with foot and mouth disease. Table 13.2 lists the conditional probabilities for the findings and diseases in mini-BOSSS.

If you are familiar with the concepts of sensitivity and specificity (discussed in detail in Chapter 20), you may recognize that the conditional probability p(drooling of saliva is present|FMD is present) is the sensitivity of the symptom *drooling of saliva* for the disease *FMD*. Similarly, p(drooling of saliva is absent|FMD is absent) is the specificity of drooling of saliva for FMD. Many readers will be quite familiar and comfortable with the concept of sensitivity and specificity of laboratory tests, but perhaps not with the idea of sensitivity and specificity of other findings. In fact, symptoms, travel history, results of physical examination, and results of laboratory tests are all nothing more than observations that we make about an individual that may help us discriminate between individuals with disease and those without disease. It is possible to measure a sensitivity, specificity (or, alternatively, likelihood ratios) for any of these diagnostic observations. Consider the diseases presented in mini-BOSSS. Cattle affected with FMD develop very severe mouth vesicles and ulcers. This mouth pain results in excessive salivation; therefore, it causes drooling of saliva. Cattle affected with MCD do not develop mouth wounds; therefore, they are not likely to be observed drooling saliva. The conditional probabilities in Table 13.2 reflect the observations above; P(drooling of saliva|FMD) is equal to 0.95 and p(drooling of saliva|MCD) is equal to 0.001.

If our knowledge base included laboratory tests for FMD and MCD, these conditional probabilities would be the sensitivities and specificities of the laboratory tests for FMD and MCD.

Researchers working in the field of diagnostic expert systems sometimes refer to conditional probabilities of findings for given diseases as textbook knowledge because these probabilities are often available in textbooks of medicine (human or veterinary). For example, a chapter in a textbook of veterinary medicine on FMD will contain many statements about the frequency with which different findings occur in animals with FMD. These frequencies are basic facts, highly relevant to diagnosis.

5.2. Probabilistic Inference Engines

A probabilistic inference engine is an algorithm that computes a differential diagnosis for a sick individual. In particular, it computes the posterior probability of each disease in its knowledge base, given the findings for a sick individual. In our example, the algorithm will compute the posterior probability p(FMD is present|drooling of saliva is present, more than one animal is affected) and the posterior probability p(MCD is present|drooling of saliva is present, more than one animal is affected).

A probabilistic inference engine uses Bayes rules to compute the posterior probability from the prior probability (or prior odds) and the sensitivities and specificities of the observed findings. Once it has computed the posterior probability of every disease in its knowledge base, the inference engine outputs a differential diagnosis, that is, a list of all diseases in the knowledge base sorted from most probable to least probable.

Bayes rules are sometimes referred to as Bayesian inversion because it inverts the conditional probability that a given finding will be observed in an individual with disease (textbook knowledge) into the probability that an individual has the disease, given that we observe a finding in that individual (diagnostic knowledge). That is, it inverts p(drooling of saliva is present|FMD is present) to p(FMD is present|drooling of saliva is present), which is exactly what a veterinarian needs to know. A veterinarian needs to know the likelihood that the cow has FMD or MCD, given the findings.

A complete discussion of the various algorithms that a probabilistic inference engine can use to compute posterior probabilities using Bayes rules would fill a book, such as the excellent textbook by Richard Neapolitan (2003). For teaching purposes, we will use a simple form of Bayes rules called the *odds-likelihood form of Bayes rules*. This simplified form rests on an assumption that findings are independent, given the disease, which is why researchers called this formulation (and similar formulations) *naive Bayes*. This simple form of Bayes rules works surprisingly well in many diagnostic expert systems, including the systems developed by Homer Warner (congenital heart advisor) and Homer Warner Jr. (Iliad), as well as the BOSSS. We will use the odds-likelihood version of Bayes rules to illustrate how BOSSS computes a differential diagnosis for two diseases, given two findings about a sick cow.

5.2.1. Definition of Odds

For clarity and convenient reference, we here provide the definition of *odds*:

$$odds = \frac{p}{1-p} \tag{1}$$

Odds are simply a rescaling of probability from a range of 0 to 1 to a range of 0 to infinity (which you can prove to yourself by substituting probabilities of zero and one into Eq. 1). For probabilities less than 0.1, probabilities and odds are roughly equal. A probability of 0.1, for example, equals an odds of $\frac{0.1}{1-0.1} = 0.11$. You can safely think "probability" whenever we use the term "odds" and vice versa, as we will be dealing with small probabilities. Upon reading Appendix D, which contains the simplest probabilistic formulation of Bayes rules, you will see why we use the odds-likelihood form in this chapter—it is simpler to learn Bayes rules using this form, and more illuminating.

5.2.2. Odds-Likelihood Form of Bayes Rule

Equation 2 is the odds-likelihood form of Bayes rules when we have only one finding.

$$Odds(D|f) = LR_{f|D} \times Odds(D) \tag{2}$$

This equation says that if we know the prior odds *Odds(D)* (read prevalence) of a disease *D* and we observe one finding *f* of that disease in an individual, we can compute the posterior odds (read probability) by multiplying the prior odds times the likelihood ratio of that finding for that disease.

Likelihood ratios are nothing more than an alternative way of expressing the sensitivity and specificity of a test or observation for a disease. In fact, the likelihood ratio is defined in terms of sensitivities and specificities. The likelihood ratio positive (positive means that the finding *drooling saliva* is known to be present) of drooling saliva for the disease FMD follows:

$$LR^{+}_{\text{drooling saliva}|\text{FMD}} = \frac{\text{p(drooling saliva present}|\text{FMD present)}}{\text{p(drooling saliva present}|\text{FMD absent)}} \tag{3}$$

In Equation 3, the numerator is sensitivity, and the denominator is the false positive rate (equal to 1 — specificity).

Note that the beauty of the likelihood ratio $LR^{+}_{\text{drooling saliva|FMD}}$ is that it is a very direct measure of how well drooling saliva discriminates between animals with FMD and animals without FMD. If drooling of saliva occurs much more frequently in animals with FMD than animals without FMD, $LR^{+}_{\text{drooling saliva|FMD}}$ will be a large number. For a finding that is pathognomic (meaning that the finding in itself is sufficient to diagnose a disease), the denominator will be zero and the LR^{+} will be infinity, meaning that no matter how small the prior odds are, the posterior odds will be infinity (which converts to a probability of 1.0, and means that the individual has the disease with certainty). If, on the other hand, drooling of saliva occurs with equal frequency in animals with FMD and animals without FMD, the numerator and the denominator will be equal and the $LR^{+}_{\text{drooling saliva|FMD}}$ will be equal to one, which when multiplied times the prior odds of FMD will result in a posterior odds that is equal to the prior odds. This result makes sense. If a finding cannot discriminate between animals with FMD and without FMD, it contains no diagnostic information for the disease FMD, and should not increase or decrease our belief that the animal has FMD.

The likelihood ratio negative (negative means that the finding *drooling saliva* is known to be absent) follows:

$$LR^{-}_{\text{drooling saliva}|\text{FMD}} = \frac{\text{p(drooling saliva absent}|\text{FMD present)}}{\text{p(drooling saliva absent}|\text{FMD absent)}} \tag{4}$$

In Eq. 4, the numerator is the false negative rate (equal to 1 – sensitivity), and the denominator is specificity. In medicine, the absence of a finding that we expect to see if the patient actually has the disease in question is useful information. There is even a term for it—*significant negative finding*. A significant negative finding, such as a negative laboratory test, helps to rule out a diagnosis. A likelihood ratio negative is always a number that is less than one (but greater or equal to zero) for findings that we expect to see more often in individuals with the disease than in individuals without the disease.

Equation 2 expresses the essence of Bayes rules. Diagnosticians use Bayes rules to update their prior belief in a diagnosis in light of new information. When a diagnostician has no information whatsoever about an individual, her belief that the individual has a disease should be the prevalence of the disease. If she makes an observation (whether positive or negative), she should update her belief in the diagnosis by multiplying the likelihood ratio for the test or observation for that disease times the prior odds of the disease (think "prevalence").

If we know nothing about a cow whatsoever, then our belief that the cow has FMD is simply the prevalence of FMD. If we subsequently observe that the cow is drooling saliva, Bayes rules instructs us to update our belief that the cow has FMD using the information in Table 13.1 and 13.2 using the following calculation:

$$\begin{aligned}
&\text{Odds(FMD | drooling saliva)}\\
&= LR^{+}_{\text{drooling saliva}|\text{FMD}} \times \text{Odds(FMD)}\\
&= \frac{\text{p(drooling saliva}|\text{FMD present)}}{\text{p(drooling saliva}|\text{FMD absent)}} \times \frac{0.001}{1-0.001}\\
&= \frac{0.95}{0.05} \times \frac{0.001}{0.999}\\
&= 19 \times 0.001\\
&= 0.019
\end{aligned} \tag{5}$$

(If we observe that the cow is not drooling saliva, we would use the likelihood ratio negative in Bayes rules.)

If we subsequently observe that a second cow is sick, we can apply Bayes rules a second time. In effect, we are treating the posterior odds from the prior calculation as the new prior odds:

$$\begin{aligned}
&\text{Odds(FMD | drooling saliva, more than one animal affected)}\\
&= LR^{+}_{\text{more than one animal affected}|\text{FMD}} \times \text{Odds(FMD|drooling saliva)}\\
&= \frac{\text{p(more than one animal affected}|\text{FMD present)}}{\text{p(more than one animal affected}|\text{FMD absent)}} \times 0.0019\\
&= \frac{0.95}{0.20} \times 0.019\\
&= 4.75 \times 0.019\\
&= 0.09
\end{aligned} \tag{6}$$

In general, if we have N diagnostic facts about a cow (or person), the odds-likelihood form of Bayes rules has the following form:

$$\text{Odds}(D|f_1,f_2,...,f_n) = LR_{f_1|D} \times LR_{f_2|D} \times ... \times LR_{f_n|D} \times \text{Odds}(D) \quad (7)$$

The result of Eq. 6, Odds(FMD|drooling saliva, more than one animal affected) + 0.09, is the odds (think probability) that the cow has FMD given that we have observed drooling of saliva in this cow and at least one other cow. The probabilistic inference engine would repeat the same calculation for the disease MCD, using the prior odds of MCD and the likelihood ratios for the two findings for MCD. The result of this calculation (not shown) is 9.8×10^{-8} . Cows with MCD stagger, but rarely drool saliva.

5.2.3. Differential Diagnosis

Table 13.3 shows the differential diagnosis that our extremely simple diagnostic expert system would show to a user. The two diseases are sorted in order of posterior odds. Note that we converted the posterior odds back to posterior probabilities using the formula $p = \frac{odds}{1+odds}$, which we obtained by solving Eq. 2 for p.

Note that the posterior odds (and probabilities) in this example are very low. The low posterior odds are also partly due to the low prior prevalence of FMD and MCD that we arbitrarily assign (note that zero prior probability is not acceptable for the likelihood ratio computation). Countries that are currently free from these diseases or within the final stage of a disease eradication program for the diseases use low prior prevalence for exotic diseases (e.g., FMD and MCD). The LR^+s for these findings for the disease FMD are only 19 and 4.75, respectively. If we had a third finding, such as blisters on feet, the LR^+ for this finding would be a third multiplier in the equation, possibly increasing the posterior odds for one or both diagnoses. If we had a positive result from a highly sensitive and specific test for FMD (very high LR^+), the posterior odds for FMD might be quite high. The key question for biosurveillance systems is at what level of certainty of a diagnosis in a single individual or a set of individuals different response actions are warranted. The answer to this question depends on treatability and other cost-benefit considerations that we discuss in Part V of this book.

TABLE 13.3 Differential Diagnosis for Cow Drooling Saliva in a Herd with At Least One Other Cow Drooling Saliva

	Posterior Odds	Posterior Probability
FMD	0.09	0.083
MCD	9.8×10^{-8}	$\sim 9.8 \times 10^{-8}$

FMD, foot and mouth disease; MCD, mad cow disease.

5.3. Computing Posterior Probabilities Using Bayesian Networks

Readers should be aware that there are other forms of Bayes rules that do not rest on an assumption of conditional independence, given a diagnosis. A developer of a probabilistic expert system may choose to use these alternative forms to improve the diagnostic accuracy of a system. Although mathematically too complex to cover in a brief introductory tutorial to Bayesian inference, the differences among these more complex forms of Bayes rules are simple to explain using a graphical representation called a Bayesian network (Figure 13.10). A Bayesian network comprises a set of nodes and arcs where each node represents a conditional probability distribution for the variable that the node represents, conditioned on its parents (the nodes from which directed arcs connect to the node). For the benefit of statisticians, a Bayesian network is a (compact) factorization of the complete joint probability distribution over all variables represented by the nodes (Neapolitan, 2003). The arcs in a Bayesian network represent the statistical dependence and independence relationships among the variables in the model. Figure 13.10 is a Bayesian network diagram for our mini-BOSSS diagnostic expert system.

The fact that there is no arc between *drooling saliva* and *more than one animal affected* indicates that the probability of observing drooling of saliva in a cow, once we know whether the cow has FMD or MCD, is not affected by knowing that other animals have these symptoms and vice versa (knowing that other animals have drooling once we know that this cow has FMD does not change the probability that this animal is drooling). If *drooling saliva* and *more than one animal affected* were not independent, given that we know that the cow has FMD, then we could add an arc between them.

Figure 13.11 shows a portion of the Bayesian network that underlies the Pathfinder system, which is a diagnostic expert system for pathologists who are interpreting biopsies of lymph

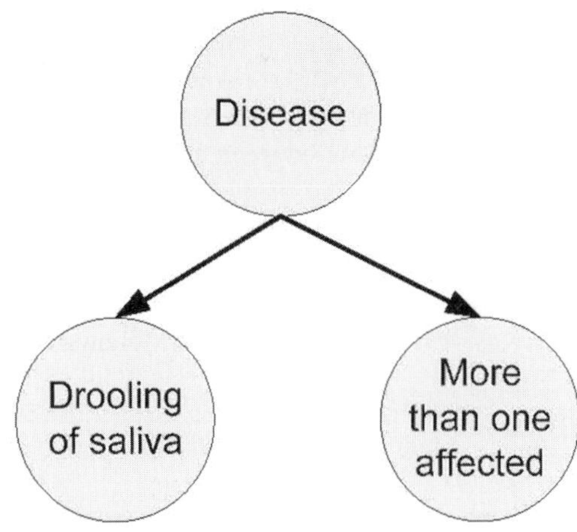

FIGURE 13.10 A Bayesian network of mini-BOSSS.

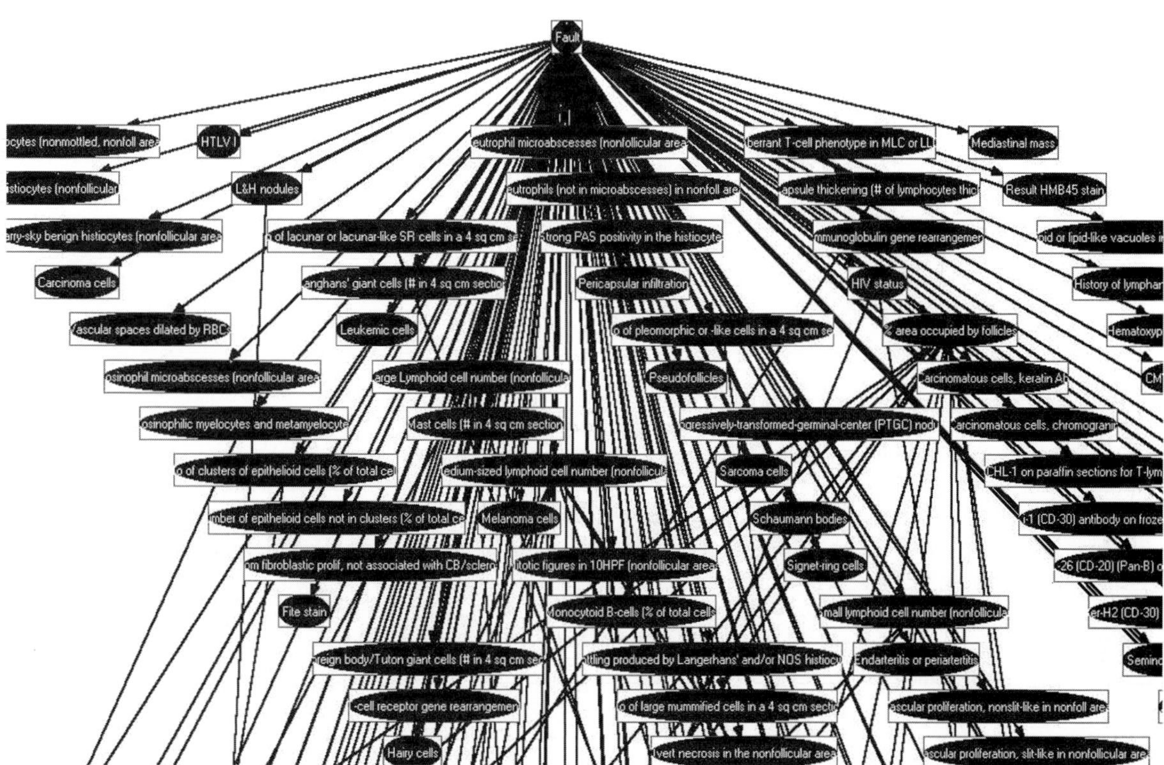

FIGURE 13.11 A portion of the Bayesian network in the Pathfinder expert system for pathology. (From Stuart Turner, lecture notes, 2001, with permission.)

nodes (Heckerman et al., 1992). The arcs between findings indicate that they are not independent, given diagnoses. Note that the Pathfinder network is a more realistic illustration of the size and complexity of the models used in diagnostic expert systems. In chapter 18, we discuss Bayesian networks in the context of Bayesian biosurveillance. That chapter discusses a really big model.

6. RULE-BASED EXPERT SYSTEMS

Rule-based expert systems may also find application in biosurveillance. A rule-based expert system (also known as a deterministic expert system) encodes medical knowledge in a set of *if–then* rules. Rule-based expert systems are simpler, conceptually, than probabilistic systems, but they are only appropriate for problems that do not involve reasoning under uncertainty. A rule-based formalism is appropriate when diagnostic or other knowledge can be represented as Boolean statements (e.g., if A and B then C). A hospital or biosurveillance organization can use a rule-based expert system to automate case detection when the findings are diagnostically precise (e.g., when case-detection is based on diagnostically precise data such as a positive tuberculosis culture [Miller et al., 1997]).

Figure 13.12 is an anthrax case detection rule from the Clinical Event Monitor, an embedded rule-based expert system developed by several of the authors (Wagner et al., 1997). This rule is in a format that can be interpreted by a rule-based inference engine called CLIPS (National Aeronautics and Space Administration). The rule contains variables that represent findings of the disease anthrax. The part of the rule before the ≥ symbol is the IF part and the part after the symbol is the THEN part. If all the conditions defined in the IF segment are true—wide mediastinum is found in a chest radiograph report and gram positive rods are reported in laboratory results within 10 days of each other—then the rule will conclude that there is a possible anthrax case. A typical rule-based expert system contains hundreds of such rules.

7. EMBEDDED EXPERT SYSTEMS

In medical applications, the requirement for manual input of data by busy clinicians is a significant barrier to the use of diagnostic expert systems (Graber and VanScoy, 2003). In the 1970s, several research groups took this barrier seriously. They began to explore methods to embed diagnostic expert systems—providing diagnostic support and, more generally, decision support—to clinicians without asking them to enter patient findings. Their research demonstrated that it is possible to obtain needed findings directly from existing clinical information systems (Evans, 1991, Gardner et al., 1999, Rind et al., 1993, Rind et al., 1995, McDonald, 1976c, McDonald et al., 1992a, McDonald, 1976b, McDonald et al., 1992b, Overhage et al., 1996,

```
(defrule detect_anthrax
  (event Radiology_report)
  (report (patient_id ?patient_id))
  (freetext ?freetext)
  (test (str-index "wide mediastinum" (lowcase ?freetext)))
  =>
  (if (> (cCheckGrams ?patient_id "GPR" 10) -1) then
    (event ANTHRAX)
    (printout t "Found gram positive rods and widened mediastinum in chest x-ray"
crlf)
    (printout t "==== Detected a possible ANTHRAX case =====" crlf)
  )
)
```

FIGURE 13.12 An anthrax-case detection rule in the Clinical Event Monitor, a rule-based expert system. The rule is written in CLIPS, a language and expert system shell developed by the National Aeronautics and Space Administration.

Wagner et al., 1997) and from alterations in work flow, such as replacing unstructured paper records with computer-generated encounter forms (from which findings can be optically scanned or manually extracted) (McDonald, 1976a, McDonald et al., 1992a, McDonald, 1976b, McDonald et al., 1992b). The research also demonstrated how diagnostic expert systems can be embedded in clinical information systems, such as physician order entry systems (Tierney et al., 1993, Dexter et al., 2001) and electronic medical records (Gardner et al., 1999, Warner et al., 1997). Several of these systems functioned in the domain of hospital infection control (Hripcsak et al., 1997, Hripcsak et al., 1999, Kahn et al., 1996a, Kahn et al., 1996b, Kahn et al., 1993).

The secret to successful deployment of a diagnostic or other expert system in medicine is designing a system in which the benefits to clinicians (from decision support and increases in efficiency) outweigh any additional effort required by clinicians to enter data by a considerable margin.

8. DIAGNOSTIC EXPERT SYSTEMS FOR BIOSURVEILLANCE

In addition to the diagnostic expert systems for hospital infection control just outlined, there are two other projects developing systems in the domain of biosurveillance. Shannon and colleagues at the Children's Hospital in Boston are developing a web-based diagnostic expert system to assist emergency room clinicians with diagnosis of approximately 20 diseases caused by biological agents known to have been weaponized or that are otherwise of concern as potential bioterrorist threats. The system has developed and evaluated on-line educational modules about the biological agents (Chung et al., 2004), and is intended to support the reporting of cases to local, state, and federal agencies (Shannon et al., 2002). Similar to stand-alone expert systems, the web-based tool requires clinicians to manually enter patient information.

The National Library of Medicine has developed a system called WISER (Wireless Information System for Emergency Responders) to assist first responders when they arrive at a hazardous material (Hazmat) incident, such as a chemical spill (http://wiser.nlm.nih.gov/). WISER operates on a personal digital assistant (PDA) and can send and receive data with a central location and other PDAs running the WISER program through a wireless network. WISER contains a diagnostic expert system that provides assistance in identification of an unknown substance and, once the substance is identified, provides guidance on actions to save lives and protect the environment. The WISER framework could be expanded to include a diagnostic expert system for the analysis of patient or animal findings.

9. PERFECTING CASE AND OUTBREAK DETECTION

Recall that the best case detection system imaginable would be one in which every individual in a community is examined every morning by the best diagnostician in the world. Since she would be examining everyone every day, she would be aware of patterns of illness in a community and this awareness would appropriately influence her diagnostic thinking (and treatment) of individual patients.[3] This diagnostician would also never fail to report immediately each fever, early syndromic presentation, or reportable disease to governmental public health.

Diagnostic expert systems are the key to building such a system. The research we reviewed in this chapter has already solved most, if not all, of the technical problems. What is needed is the will to create such a system.

3 Physicians are taught (and reminded incessantly) "when you hear hoof beats, don't think of zebras". This adage is an informal statement that when the evidence available about a particular patient supports equally a diagnosis of either influenza or SARS (e.g., the patient has constitutional symptoms and no history of exposure to SARS), they should conclude that the diagnosis of influenza is far more likely than SARS.

If and when diagnostic expert systems are embedded in the clinical information systems of every hospital (animal and human), long-term care facility, clinic, and laboratory in a region, they will be able to notify a health department or other biosurveillance organization of every fever, syndromic presentation, and reportable disease in individuals receiving medical or veterinary care. If diagnostic expert systems are made available to the public or to selected high-risk populations (e.g., postal employees or patients with preexisting conditions such as asthma or diabetes), case finding would be extended to an even larger fraction of the population, approximating the "every-patient-every-day" capabilities of an ideal case detection system.

If a biosurveillance system located in a region's health department were to receive the differential diagnosis for each individual in the region in real time (anonymously, of course, and perhaps selectively based on diseases and probability thresholds), it could compute the current incidences of these conditions. It could monitor the region for increases in incidence of findings, syndromes, and diseases of concern.

Note that the outputs of a probabilistic diagnostic expert system are posterior probabilities of diseases for one patient, so the central monitoring would be monitoring of the sums of the probabilities for all reported patients. Figure 13.13 illustrates this summation. It plots the daily sums of the posterior probabilities of "flu-like" illness of all patients seen in emergency departments on each day. If the diagnostic expert systems are well calibrated, the sums of the posterior probabilities should equal the actual number of patients with the disease in the population being evaluated by the system.

Finally, if the biosurveillance system would then communicate the current fever, syndrome, and disease incidences back to the diagnostic expert systems being used by clinicians and citizens, we would realize the "diagnosticians-are-aware-of-patterns-of-illness-in-the-community" capability of an ideal case detection system.

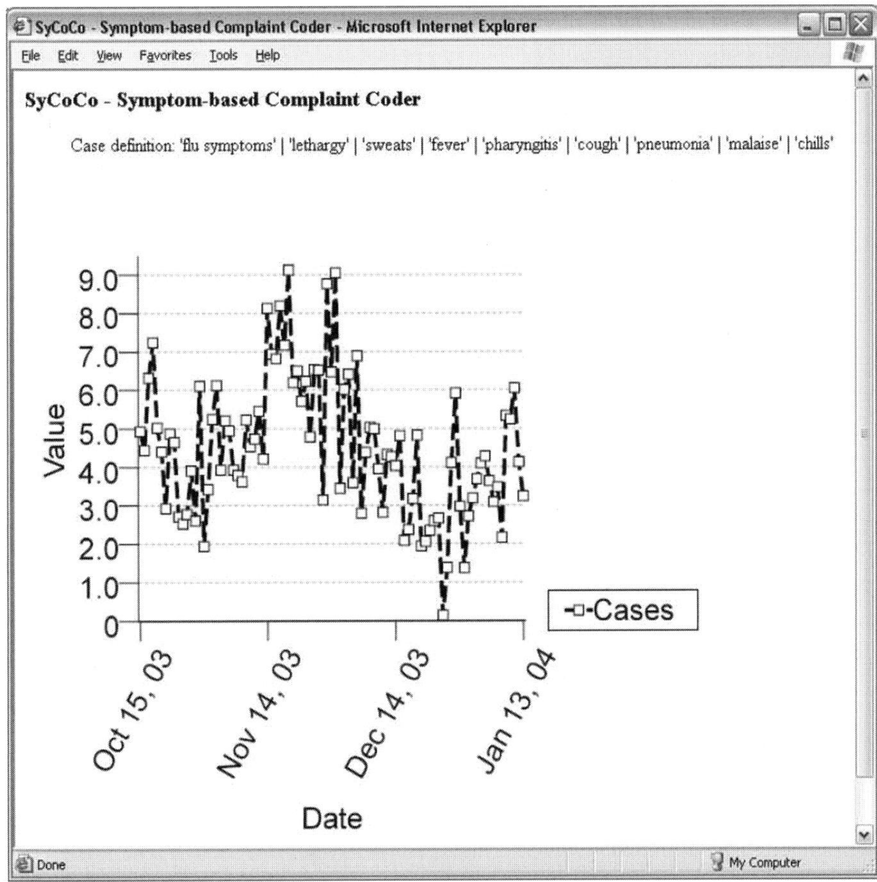

FIGURE 13.13 Daily sum of syndrome probabilities produced by SyCO². SyCO² computes the posterior probability that a patient has a flu-like illness from his chief complaint. (From Espino, J., Dara, J., Dowling J, et al. (2005). SyCo2: A Bayesian Machine Learning Method for Extracting Symptoms from Chief Complaints And Combining Them Using Probabilistic Case Definitions. Pittsburgh, PA: Center for Biomedical Informatics, University of Pittsburgh, with permission.) An outbreak-detection system would sum the posterior probabilities of flu-like illness from all patients seen in 24-hour periods to form a time series of expected daily counts of patients with respiratory illness. Readers familiar with Bayesian statistics will recognize this sum as the expectation for the number of individuals with a given diagnosis.

This ideal approach underlies the Bayesian approach to outbreak detection described in Chapter 18 Bayesian Methods for Diagnosing Outbreaks. PANDA (Population-wide ANomaly Detection and Assessment)—the research system described in that chapter—actually merges many individual diagnostic Bayesian networks into a very large network that also includes a subnetwork that draws inferences about the presence or absence of an outbreak. PANDA is pursuing this idea on a citywide scale. Conceptually, a PANDA network comprises millions of person-specific diagnostic Bayesian networks, each of whose probabilities of disease (e.g., anthrax) are being influenced by population-level observations (e.g., aggregate sales of over-the-counter medications) and population-level inferences (e.g., the likelihood that other individuals who may have inhalational anthrax are present in a population). PANDA also includes a prior probability distribution over outbreak diseases (e.g, the prior probability of inhalational anthrax, based on the national terror alert level). By integrating person-specific diagnostic submodels into a population-wide super model, approaches like PANDA are able to make inferences about the probability of a disease outbreak in the population as a whole, as well as the probability of disease in individual people (or subgroups of people) within the population.

10. COMPUTER-INTERPRETABLE CASE DEFINITIONS

An ideal case detection system would also support case finding during outbreak investigations. Case definitions (described in Chapter 3) are the basis for case–control studies and investigations of emerging diseases. A computer-interpretable case definition is a prerequisite for providing computer-support to case finding during investigations. As discussed in Chapter 3, case definitions are Boolean (logical) statement of findings (Figure 13.14).

Case definitions, as currently written, are not well suited for automation. The authors of the SARS case definition intended it for use by physicians and epidemiologists, not computers. The clause "findings of lower respiratory illness (e.g., cough, shortness of breath, difficulty breathing)" does not enumerate all findings of lower respiratory illness. A computer requires a complete enumeration of all findings that it should count as evidence of *lower respiratory illness* (e.g., cough, shortness of breath, difficulty breathing, wheezing, cyanosis, tachypnea, dullness to percussion, fremitus, whispered pectoriloquy, rales, and rhonchi). The findings would also have to be described more precisely. For example, a physician or an epidemiologist would not count chronic cough or cough associated with asthma as a finding of lower respiratory illness when applying this case definition, but a computer would (unless told otherwise). Note that it is difficult, if not impossible, to enumerate all of the possible exceptions to the counting of a finding as evidence of a disease. This difficulty is the reason that diagnostic expert systems in medicine are probabilistic. They quantify the number of exceptions to a categorical statement about the relationship between findings and disease using probabilities. For example, 70% of patients with cough have an acute lower respiratory illness, but 30% (the exceptions) have some other cause. This observation suggests that computer-interpretable case definitions will employ Bayesian networks, as illustrated by Figure 13.15.

For readers interested in the topic of knowledge representations for computer-interpretable case definitions, there is

Case definition for confirmed case of SARS-CoV disease: (early OR mild-to-moderate OR severe illness) AND laboratory confirmation

Definitions of Terms used in Case Definition:

Early illness: two or more of the following findings: fever (might be subjective), chills, rigors, myalgia, headache, diarrhea, sore throat, rhinorrhea.

Mild-to-moderate respiratory illness: Temperature of >100.4° F (>38° C) AND one or more findings of lower respiratory illness (e.g., cough, shortness of breath, difficulty breathing)

*Severe illness**

Laboratory confirmation: serum antibody to SARS-CoV by a test validated by CDC (e.g., enzyme immunoassay [EIA]), OR isolation in cell culture of SARS-CoV from a clinical specimen, OR SARS-CoV RNA by a reverse-transcription-polymerase chain reaction (RT-PCR) test validated by CDC and with subsequent confirmation in a reference laboratory (e.g., CDC).

FIGURE 13.14 Excerpt from the CDC case definition for confirmed SARS-CoV disease *(http://www.cdc.gov/ncidod/sars/guidance/b/app1.htm)*. Only the definitions relevant for classifying a patient as a probable case of SARS CoV are shown. (We omitted the definition of severe respiratory illness for clarity. The illness clause in the case definition for confirmed case of SARS-CoV disease is a disjunction, and all patients that satisfy the definitional criteria for severe disease also satisfy the criteria for mild to moderate respiratory illness.

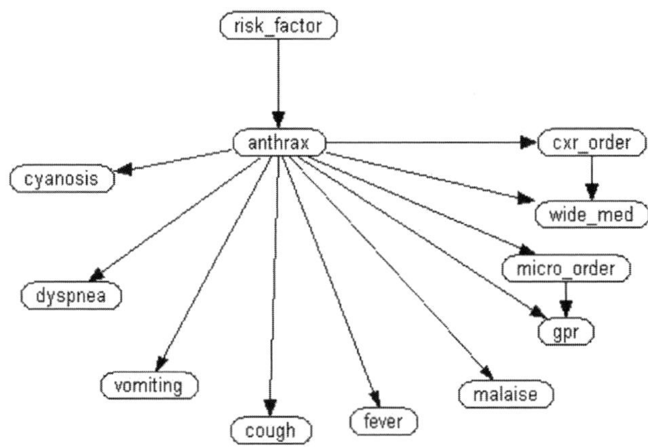

FIGURE 13.15 A Bayesian network "case definition" for anthrax. The network computes the posterior probability for a patient addmitted to the emergency department. *Cxr order*, electronic record of an order for a chest radiograph; *wide med*, chest radiograph finding of wide mediastinum; *Gpr*, gram-positive rods in blood or cerebrospinal fluid (CSF) smear; *micro order*, order for a blood or CSF culture. (From Espino J., Tsui, F.-C. (2000). A Bayesian network for detecting inhalational anthrax outbreaks. Pittsburgh, PA: Center for Biomedical Informatics, University of Pittsburgh, with permission.)

literature from research on computer-interpretable patient care guidelines that is relevant to this topic (Shiffman et al., 2004, Tu and Musen, 2001, Peleg et al., 2003, Boxwala et al., 2004, Wang et al., 2003, Seyfang et al., 2002, Fox et al., 1997, Terenziani et al., 2003, de Clercq et al., 2004, Johnson et al., 2001, Ciccarese et al., 2004, Parker et al., 2004).

11. SUMMARY

Despite the unfamiliarity of most readers with diagnostic expert systems, we chose to begin our discussion of algorithms for biosurveillance with this topic because case detection provides the case data needed for outbreak detection. Outbreak detection cannot function without case detection (unless the surveillance data are aggregated data, such as daily sales of over-the-counter thermometers).

Diagnostic expert systems have the potential to improve the quality and completeness of the case data available to analytic methods designed to detect and characterize outbreaks, which we discuss in the following chapters. They can profoundly improve the reporting of syndromes. Improvements in case detection will translate directly into improvements in the earliness of outbreak detection and characterization.

McDonald did not conclude in his seminal paper that the solution to the non-perfectibility of man was to admit only women to medical schools. Rather, he stated, "Thus, I conclude that though the individual physician is not perfectible, the system of care is, and that the computer will play a major part in the perfection of future care systems." His point—that technology can be used to create a system (involving both humans and computers) that can then be perfected—also seems to apply to biosurveillance systems.

12. ADDITIONAL RESOURCES

12.1. Expert Systems in Plant Pathology

For readers who are interested in diagnostic expert systems for diagnosing plant diseases, Travis and Latin (1991) briefly reviewed several diagnostic expert systems in plant pathology, including PLANT/ed, Apple Pest and Disease Diagnosis, CALEX/Peaches, Muskmelon disorder management system, and Penn State Apple Orchard Consultant, in "Development, Implementation, and Adoption of Expert Systems in Plant Pathology."

13. ACKNOWLEDGMENTS

This work was supported in part by grants from the National Science Foundation (IIS-0325581), the Defense Advanced Research Projects Agency (F30602-01-2-0550), and the Pennsylvania Department of Health (ME-01-737). We thank William Hogan and Gregory Cooper for helpful discussions and comments.

14. REFERENCES

Adams, I. D., Chan, M., Clifford, P. C., et al. (1986). Computer aided diagnosis of acute abdominal pain: a multicentre study. *BMJ* (Clin Res Ed) 293:800–4.

American College of Emergency Physicians. (1994). Clinical policy for the initial approach to patients presenting with a chief complaint of nontraumatic acute abdominal pain. American College of Emergency Physicians. *Ann Emerg Med* 23:906–22.

Aronsky, D., Fiszman, M., Chapman, W. W., et al. (2001). Combining decision support methodologies to diagnose pneumonia. In: Proceedings of American Medical Informatics Association Annual Fall Symposium, 12–16.

Boxwala, A. A., Peleg, M., Tu, S., et al. (2004). GLIF3: a representation format for sharable computer-interpretable clinical practice guidelines. *J Biomed Inform* 37:147–61.

Buchanan, B. G., Shortliffe, E. H. (1984). Rule-based expert systems: the MYCIN experiments of the Stanford Heuristic Programming Project. Reading, MA: Addison-Wesley.

Burnside, E. S., Rubin, D. L., Shachter, R. D., et al. (2004). A probabilistic expert system that provides automated mammographic-histologic correlation: initial experience. *AJR Am J Roentgenol* 182:481–8.

Chung, S., Mandl, K. D., Shannon, M., et al. (2004). Efficacy of an educational web site for educating physicians about bioterrorism. *Acad Emerg Med* 11:143–8.

Ciccarese, P., Caffi, E., Boiocchi, L., et al. (2004). A guideline management system. *Medinfo* 11:28–32.

Cundick, R., et al. (1989). Iliad as a patient care simulator to teach medical problem solving. Paper presented at 13th Symposium on Computer Applications in Medical Care, Washington, DC.

de Clercq, P. A., Blom, J. A., Korsten, H. H., et al. (2004). Approaches for creating computer-interpretable guidelines that facilitate decision support. *Artif Intell Med* 31:1–27.

de Dombal, F., de Baere, H., van Elk, P., et al. (1993). Objective medical decision making: acute abdominal pain. In: Beneken J, Thevenin V, eds. *Advances in Biomedical Engineering: Results of the 4th EC Medical and Health Research Programme.* Vol. 7 of Studies in Health Technology and Informatics, 65–87. Burke, VA: IOS Press.

de Dombal, F., Leaper, D., Staniland, J., et al. (1972). Computer-aided diagnosis of acute abdominal pain. *BMJ* 2:9–13.

de Dombal, F. T. (1975). Computer-aided diagnosis and decision-making in the acute abdomen. *J R Coll Physicians Lond* 9:211–8.

de Dombal, F. T. (1984). Computer-aided diagnosis of acute abdominal pain. The British experience. *Rev Epidemiol Sante Publique* 32:50–6.

de Dombal, F. T. (1990). Computer-aided decision support in acute abdominal pain, with special reference to the EC concerted action. *Int J Biomed Comput* 26:183–8.

de Dombal, F. T., Leaper, D. J., Horrocks, J. C., et al. (1974). Human and computer-aided diagnosis of abdominal pain: further report with emphasis on performance of clinicians. *BMJ* 1:376–80.

Dexter, P. R., Perkins, S., Overhage, J. M., et al. (2001). A computerized reminder system to increase the use of preventive care for hospitalized patients. *N Engl J Med* 345:965–70.

Elstein, A. S., Friedman, C. P., Wolf, F. M., et al. (1996). Effects of a decision support system on the diagnostic accuracy of users: a preliminary report. *J Am Med Inform Assoc* 3:422–8.

Espino, J., Dara, J., Dowling, J., et al. (2005). SyCo2: A Bayesian Machine Learning Method for Extracting Symptoms from Chief Complaints And Combining Them Using Probabilistic Case Definitions. Pittsburgh, PA: Center for Biomedical Informatics, University of Pittsburgh.

Espino J., Tsui, F.-C. (2000). A Bayesian network for detecting inhalational anthrax outbreaks. Pittsburgh, PA: Center for Biomedical Informatics, University of Pittsburgh.

Evans, R. S. (1991). The HELP system: a review of clinical applications in infectious diseases and antibiotic use. *MD Computing* 8:282–8, 315.

Feigenbaum, E. A. (1977). The Art of Artificial Intelligence: Themes and Case Studies of Knowledge Engineering. Paper presented at 5th International Joint Conference on Artificial Intelligence.

Fox, J., Johns, N., Lyons, C., et al. (1997). PROforma: a general technology for clinical decision support systems. *Comput Methods Programs Biomed* 54:59–67.

Friedman, C. P., Elstein, A. S., Wolf, F. M., et al. (1999). Enhancement of clinicians' diagnostic reasoning by computer-based consultation: a multisite study of 2 systems. *JAMA* 282:1851–6.

Gardner, R. M., Pryor, T. A., Warner, H. R. (1999). The HELP hospital information system: update 1998. *Int J Med Inform* 54:169–82.

Gerberding, J. L., Hughes, J. M., Koplan, J. P. (2002). Bioterrorism preparedness and response: clinicians and public health agencies as essential partners. *JAMA* 287:898–900.

Graber, M. A., VanScoy, D. (2003). How well does decision support software perform in the emergency department? *Emerg Med J* 20:426–8.

Haynes, R. B., Davis, D. A., McKibbon, A., et al. P. (1984). A critical appraisal of the efficacy of continuing medical education. *JAMA* 251:61–4.

Heckerman, D. E., Horvitz, E. J., Nathwani, B. N. (1992). Toward normative expert systems. Part I. The Pathfinder project. *Methods Inform Med* 31:90–105.

Hripcsak, G., Knirsch, C., Jain, N., et al., (1999). A health Information network for managing innercity tuberculosis: bridging clinical care, public health, and home care. *Computers Biomed Res* 32:67–76.

Hripcsak, G., Knirsch, C. A., Jain, N. L., et al. (1997). Automated tuberculosis detection. *J Am Med Inform Assoc* 4:376–81.

Hughes, J. M., Gerberding, J. L. (2002). Anthrax bioterrorism: lessons learned and future directions. *Emerg Infect Dis* 8:1013–4.

Johnson, P., Tu S., Jones, N. (2001). Achieving reuse of computable guideline systems. *Medinfo* 10:99–103.

Kahn, M. G., Bailey, T. C., Steib, S. A., et al. (1996a). Statistical process control methods for expert system performance monitoring. *J Am Med Inform Assoc* 3:258–69.

Kahn, M. G., Steib, S. A., Dunagan, W. C., et al. (1996b). Monitoring expert system performance using continuous user feedback. *J Am Med Inform Assoc* 3:216–23.

Kahn, M. G., Steib, S. A., Fraser, V. J., et al. (1993). An expert system for culture-based infection control surveillance. In: *Proceedings of Annual Symposium on Computer Applications in Medical Care,* 171–5.

Kohn, L. T., Corrigan, J., Donaldson, M. S., et al. (2000). To Err Is Human: Building a Safer Health System. Washington, DC: National Academy Press.

Kolata, G. (2001). Florida Man Is Hospitalized with Pulmonary Anthrax. *New York Times,* October 5.

Leavitt, M. (2001). Medscape's response to the Institute of Medicine Report: crossing the quality chasm: a new health system for the 21st century. *Med Gen Med* 3:2.

Ledley, R. S., Lusted, L. B. (1959). Reasoning foundation of medical diagnosis: symbolic logic, probability, and value theory aid our understanding of how physicians reason. *Science* 130:9–12.

Leist, J. C., Kristofco, R. E. (1990). The changing paradigm for continuing medical education: impact of information on the teachable moment. *Bull Med Libr Assoc* 78:173–9.

London, S. (1998). DXplain: a web-based diagnostic decision support system for medical students. *Med Ref Serv* Q 17:17–28.

McAdam, W. A., Brock, B. M., Armitage, T., et al. (1990). Twelve years' experience of computer-aided diagnosis in a district general hospital. *Ann R Coll Surg Engl* 72:140–6.

McDonald, C. (1976a). Use of a computer to detect and respond to clinical events: its effect on clinician behavior. *Ann Intern Med* 84:162–7.

McDonald, C., Tierney, W., Overhage, J. (1991). H.I.S. and the physician: direct inpatient order entry by physicians through medical gopher workstations—problems and promises. In: *Proceedings IMIA Working Conference on Trends in Modern Hospital Informatics Systems, Gottingen,* Germany, 71–80.

McDonald, C., Tierney, W., Overhage, J., et al. (1992a). The Regenstrief Medical Record System: 20 years of experience in hospitals, clinics, and neighborhood health centers. *MD Computing* 9:206–17.

McDonald, C. J. (1976b). Protocol-based computer reminders, the quality of care and the non-perfectability of man. *N Engl J Med* 295:1351–5.

McDonald, C. J. (1976c). Use of a computer to detect and respond to clinical events: Its effect on clinician behavior. *Ann Intern Med* 84:162–7.

McDonald, C. J. (1983). Computer technology and continuing medical education: a scenario for the future. *Conn Med* 47:159–62.

McDonald, C. J., Hui, S. L., Tierney, W. M. (1992b). Effects of computer reminders for influenza vaccination on morbidity during influenza epidemics. *MD Computing* 9:304–12.

Miller, P. L., Frawley, S. J., Sayward, F. G., et al. (1997). Combining tabular, rule-based, and procedural knowledge in computer-based guidelines for childhood immunization. *Comput Biomed Res* 30(3):211–31.

Miller, R., Masarie, F. E., Myers, J. D. (1986). Quick medical reference (QMR) for diagnostic assistance. *MD Computing* 3:34–48.

Murphy, G. C., Friedman, C. P., Elstein, A. S., et al. (1996). The influence of a decision support system on the differential diagnosis of medical practitioners at three levels of training. In: *Proceedings of American Medical Informatics Association Annual Fall Symposium,* 219–23.

National Aeronautics and Space Administration A. C language integrated production system (CLIPS). *http://www.www.ghg.net/clips/WhatIsCLIPS.html.*

Neapolitan, R. (2003). Learning Bayesian Networks. Upper Saddle River, NJ: Pearson Prentice Hall.

Overhage, J. M., Tierney, W. M., McDonald, C. J. (1996). Computer reminders to implement preventive care guidelines for hospitalized patients. *Arch Intern Med* 156:1551–6.

Parker, C. G., Rocha, R. A., Campbell, J. R., et al. (2004). Detailed clinical models for sharable, executable guidelines. *Medinfo* 11:145–8.

Peleg, M., Tu, S., Bury, J., et al. (2003). Comparing computer-interpretable guideline models: a case-study approach. *J Am Med Inform Assoc* 10:52–68.

Pynsent, P. B., Fairbank, J. C. (1989). Computer interview system for patients with back pain. *J Biomed Eng* 11:25–9.

Rind, D. M., Davis, R., Safran, C. (1995). Designing studies of computer-based alerts and reminders. MD. Computing 12:122–6.

Rind, D. M., Safran, C., Russell, S. P., et al. (1993). The effect of computer-based reminders on the management of hospitalized patients with worsening renal function. In: *Proceedings of the 15th Annual Symposium on Computer Applications in Medical Care,* Washington, D.C., 28–32.

Seyfang, A., Miksch S., Marcos, M. (2002). Combining diagnosis and treatment using ASBRU. *Int J Med Inform* 68:49–57.

Shannon, M., Burstein, J., Mandl K, et al. (2002). Usage of a web-based decision support tool for bioterrorism detection. *Am J Emerg Med* 20:384–5.

Shephard, R., Cameron, A., Toribio J., et al. (2005). Bovine syndromic surveillance system. *http://www.www.ausvet.com.au/bosss/.*

Shiffman, R. N., Michel, G., Essaihi A., et al. (2004). Bridging the guideline implementation gap: a systematic, document-centered approach to guideline implementation. *J Am Med Inform Assoc* 11:418–26.

Shwe, M. A., Middleton, B., Heckerman, D. E., et al. (1991). Probabilistic diagnosis using a reformulation of the INTERNIST–1/QMR knowledge base. I. The probabilistic model and inference algorithms. *Methods Inform Med* 30:241–55.

Sorenson D, et al. (1988). Generation and maintenance of ILIAD medical knowledge in a hypercard environment. In: *Proceedings of 12th Symposium on Computer Applications in Medical Care,* Washington, D.C.

Terenziani, P., Montani, S., Torchio, M., et al. (2003). Temporal consistency checking in clinical guidelines acquisition and execution: the GLARE's approach. In: *Proceedings of American Medical Informatics Association Annual Symposium,* 659–63.

Tierney, W. M., Miller, M. E., Overhage, J. M., et al. (1993). Physician inpatient order writing on microcomputer workstations. Effects on resource utilization. *JAMA* 269:379–83.

Travis J., Latin, R. (1991). Development, implementation, and adoption of expert systems in plant pathology. *Ann Rev Phytopathol* 29:343–360.

Tu, S. W., Musen, M. A. (2001). Modeling data and knowledge in the EON guideline architecture. *Medinfo* 10:280–4.

Turner, S. (2001). Bayesian Networks & More. *http://www.www.medInformatics.org/mdi207/lab3.html.*

Wagner, M. M., Pankaskie, M., Hogan, W., et al. (1997). Clinical Event Monitoring at the University of Pittsburgh. In: *Proceedings of American Medical Informatics Association Annual Fall Symposium,* 188–92.

Wald, J. S., Rind, D., Safran, C., et al. (1995). Patient entries in the electronic medical record: an interactive interview used in primary care. In: *Proceedings of Annual Symposium on Computer Applications in Medical Care,* 147–51.

Wang, D., Peleg, M., Bu, D., et al. (2003). GESDOR—a generic execution model for sharing of computer-interpretable clinical practice guidelines. In: *American Medical Informatics Association Annual Symposium,* 694–8.

Warner, H., Jr., Blue, S. R., Sorenson, D., et al. (1997). New computer-based tools for empiric antibiotic decision support. In: *Proceedings of American Medical Informatics Association Annual Symposium,* 238–42.

Warner, H. R., Jr. (1989). Iliad: moving medical decision-making into new frontiers. *Methods Inform Med* 28:370–2.

Warner, H. R., Jr. (1990). Iliad: diagnostic tools for general medicine. Interview by Bill W. Childs. *Healthcare Inform* 7:38.

Warner, H. R., Jr., Bouhaddou, O. (1994). Innovation review: Iliad—a medical diagnostic support program. *Top Health Inform Manag* 14:51–8.

Warner, H. R., Toronto, A. F., Veasey, G. (1964). Experience with Bayes theorem for computer analysis of congenital heart disease. *Ann N Y Acad Sci* 115:558–67.

Warner, H. R., Toronto, A. F., Veasey, G., et al. (1961). A mathematical approach to medical diagnosis: application to congenital heart disease. *JAMA* 177:177–83.

Weinstein, M. C. (1980). The value of information. In: *Clinical Decision Analysis,* 131–67. Philadelpha: Saunders.

Williamson, J. W., German, P. S., Weiss, R., et al. (1989). Health science Information management and continuing education of physicians: a survey of U.S. primary care practitioners and their opinion leaders. *Ann Intern Med* 110:151–60.

Wilson, D. H., Wilson, P. D., Walmsley, R. G., et al. (1977). Diagnosis of acute abdominal pain in the accident and emergency department. *Br J Surg* 64:250–4.

Wilson, P. D., Horrocks, J. C., Lyndon, P. J., et al. (1975). Simplified computer-aided diagnosis of acute abdominal pain. *BMJ* 2:73–5.

Yasnoff, W. A., Humphreys, B. L., Overhage, J. M., et al. (2004). A consensus action agenda for achieving the national health Information infrastructure. *J Am Med Inform Assoc* 11:332–8.

14

CHAPTER

Classical Time-Series Methods for Biosurveillance

Weng-Keen Wong

School of Electrical Engineering and Computer Science, Oregon State University, Corvallis, Oregon

Andrew W. Moore

School of Computer Science, Carnegie Mellon University, Pittsburgh, Pennsylvania

1. INTRODUCTION

Healthcare data are commonly represented as a sequence of observations recorded at regular time intervals. For instance, these observations could be the number of emergency department (ED) cases per hour, the number of ED cases involving respiratory symptoms per day, or the number of thermometers sold per day. This sequence of observations is called a univariate time series. A time series is a sequence of observations made over time from a random process. The term univariate refers to the fact that we only get one piece of information per time step, such as the number of patients with respiratory symptoms appearing at an ED. Multivariate data consist of records with more than one variable associated with each record. An example of multivariate data would be patient records containing information about the patient's age, gender, home zip code, work zip code, and symptoms. In this chapter, we will discuss algorithms for detecting unexpected increases in a univariate time series that might indicate an outbreak. We will present multivariate detection algorithms in the next chapter.

The purpose of this chapter is a tour of a nonexhaustive set of approaches that, along the way, introduce the reader to some of the issues faced when choosing or implementing a time-series based method. We keep the discussion at a high level, with plenty of examples, and without going into the many statistical details. Our intent is that anyone who wishes to skip over the equations we *do* provide will still (primarily through the worked examples) understand the central issues in this kind of analysis.

Figure 14.1 contains an example of a univariate time series in which we plot the number of physician visits in which the patient's recorded complaint was of a respiratory problem in a simulated town for each day between August 12, 2004 and September 30, 2004. This graph also illustrates the effects of a simulated outbreak on the data. One of the key assumptions of many detection algorithms is that the observed data are assumed to be the sum of cases from regular background activity plus any cases from an outbreak. We produced this graph according to this assumption: outbreak cases were injected into the background activity.

One simple way to detect anomalies in a univariate time series is to look for an obvious spike. However, by visually inspecting Figure 14.1, it is difficult to determine exactly when the outbreak begins: it is clearly happening by Monday, September 27, but could it have started before? The answer is revealed in Figure 14.2 the dark gray line indicates a "ramp outbreak" that starts on Saturday, September 25. This injection simulates a situation in which on Saturday there were 10 extra visits on top of the initial series; 20 additional cases on Sunday; 30 additional cases on Monday, 40 on Tuesday and 50 on Wednesday. By inspecting Figure 14.2 carefully, it is possible to notice day-of-week effects in the weeks leading up to the simulated outbreak. Most noticeably, Sundays tend to have much lower counts, and Mondays often have slightly higher counts than the surrounding days. Interestingly, if we take day-of-week effects into account, the increment on Sunday, September 26 is noticeable. It is the first Sunday we have seen with counts higher than the previous Friday. It is arguable whether the small rise on Saturday is in any way detectable.

Let us now consider how to develop automated systems to notice rises such as these in time series data. If we could model the background activity, which represents the conditions under non-outbreak periods, we could then quantify the degree of departure from the predicted count for the day. We could then detect outbreaks by looking for severe deviations from the background activity. The majority of univariate algorithms discussed in this chapter work exactly in this manner. First, the univariate algorithm method estimates the background activity according to various assumptions, such as requiring the background activity to have a normal distribution. This first step is the main difference between the various algorithms we will present. Some algorithms are able to characterize the background data more accurately by taking the temporal fluctuations into account. In the second step, limits are imposed in order to determine the amount of deviation from the background that will be tolerated before an alert is raised.

How can we tell if one detection algorithm is better than another? This is a very large and important question, described in detail in Chapter 20. For the purposes of this chapter,

Handbook of Biosurveillance
ISBN 0-12-369378-0

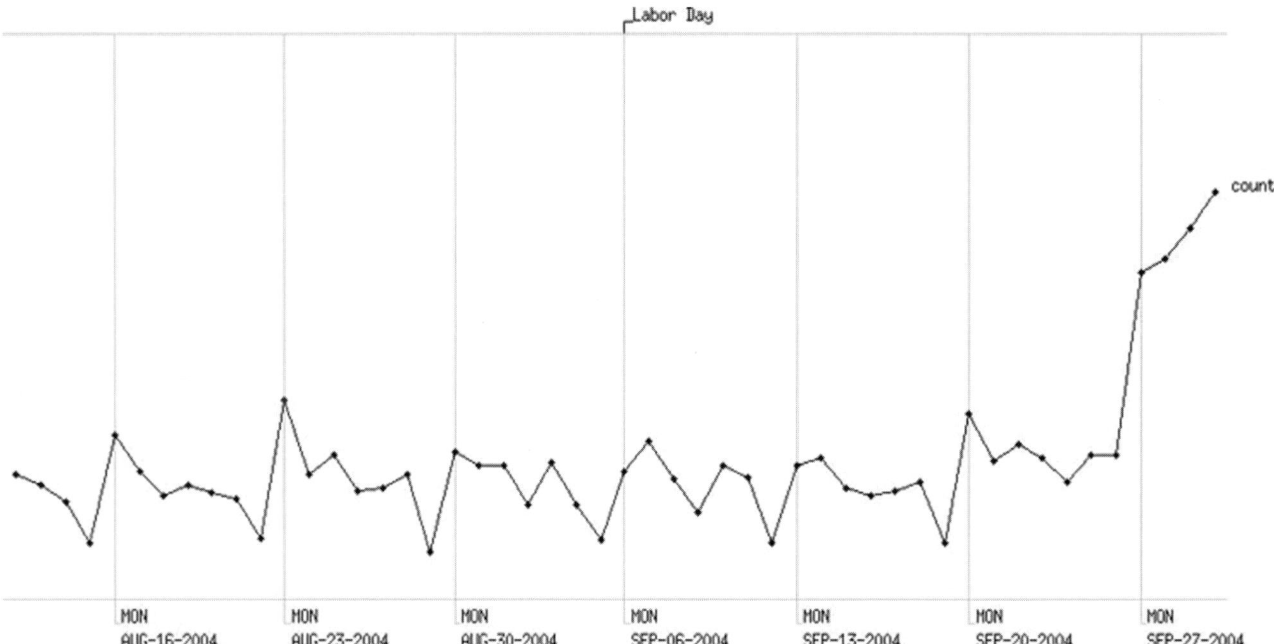

FIGURE 14.1 A time series of physician visits over time, with a simulated outbreak added. A puzzle for the reader: when did the simulated outbreak start? (See answer in the text.)

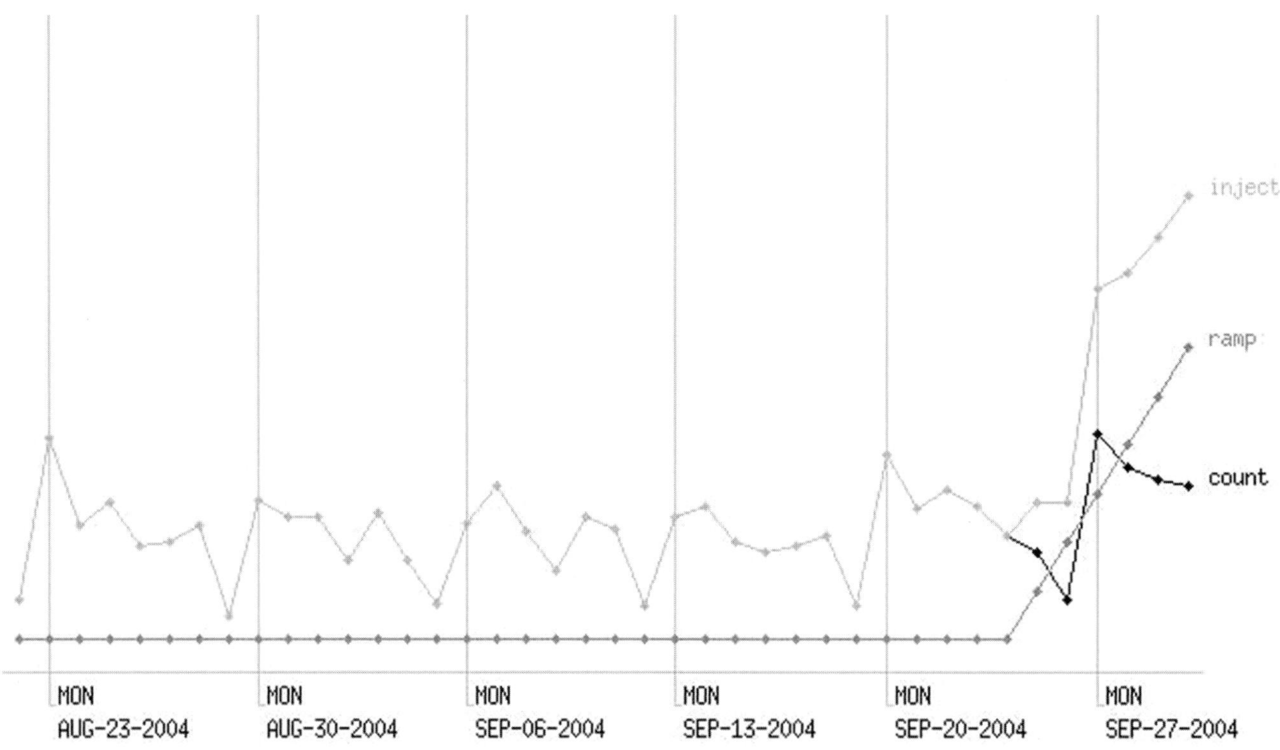

FIGURE 14.2 Light gray ("inject") is the simulated data. It is the sum of two time series: the black baseline ("count"), which is mostly covered by "inject," and the simulated outbreak (labeled ramp) (dark gray).

we will evaluate detection algorithms based on the tradeoff between the detection time and the false-positive rate. The detection time is defined as the number of time steps needed to detect the outbreak after it begins. A false positive is defined as an alert that is raised in the absence of an outbreak. The false-positive rate is the number of false positives divided by the total number of time steps. The detection time/false positive tradeoff is best explained by considering the two extremes. At one extreme, we have a "boy who cried wolf" algorithm that raises an alert extremely frequently. Due to the frequency of these alarms, however, the time to detect an outbreak will be very short. At the other extreme, we have a very insensitive algorithm that is reluctant to raise an alarm unless it is very sure an outbreak is occurring. In this case, the false-positive rate will be lower at the expense of a longer detection time. The ideal detection algorithm finds just the right compromise between the detection time and the false-positive rate.

In subsequent sections of this chapter we will describe some illustrative time series detection algorithms. Before that we will briefly discuss in more detail the simulated data we will be using.

2. SYNTHESIZING HEALTHCARE UTILIZATION DATA

Consider healthcare utilization records in which respiratory symptoms are a part of the recorded chief complaint. This is an example of an important data stream to monitor because there are several severe outbreak diseases, such as anthrax, which might be revealed by a blip in respiratory cases. It is also an interesting data stream to model because, even without any abnormal outbreak, respiratory cases vary considerably throughout the year and between years, according to the timing, severity and length of the annual influenza outbreak. In order to simulate such data we begin with a smoothly varying (and entirely fictitious) level of influenza in the population over a three-year period, shown in Figure 14.3.

Sadly, real healthcare utilization data rarely looks so smooth. This is partly because counts of physician and emergency department cases are drawn from a small population of people sufficiently ill to seek health care. On any given day, the expected count of people sick enough to decide to seek medical attention varies according to the smooth count. But the actual count will have random variation. One reasonable way to simulate the actual counts is to draw them from a Poisson distribution with mean proportional to expected count (why this is reasonable is a technical detail that we will not pursue any further here). When we simulate counts of people wishing to utilize health-care services, we get Figure 14.4.

For many sources of data, Figure 14.4 is, unfortunately, not the end of the story. For a variety of reasons, the actual number of healthcare visits is affected by the day of the week. For example, on the weekend, and particularly on Sundays, patients are more likely to wait until early the following week to seek care. Holidays, such as Thanksgiving and July 4th, have a similar effect. When these effects are simulated, we see the data in Figure 14.5.

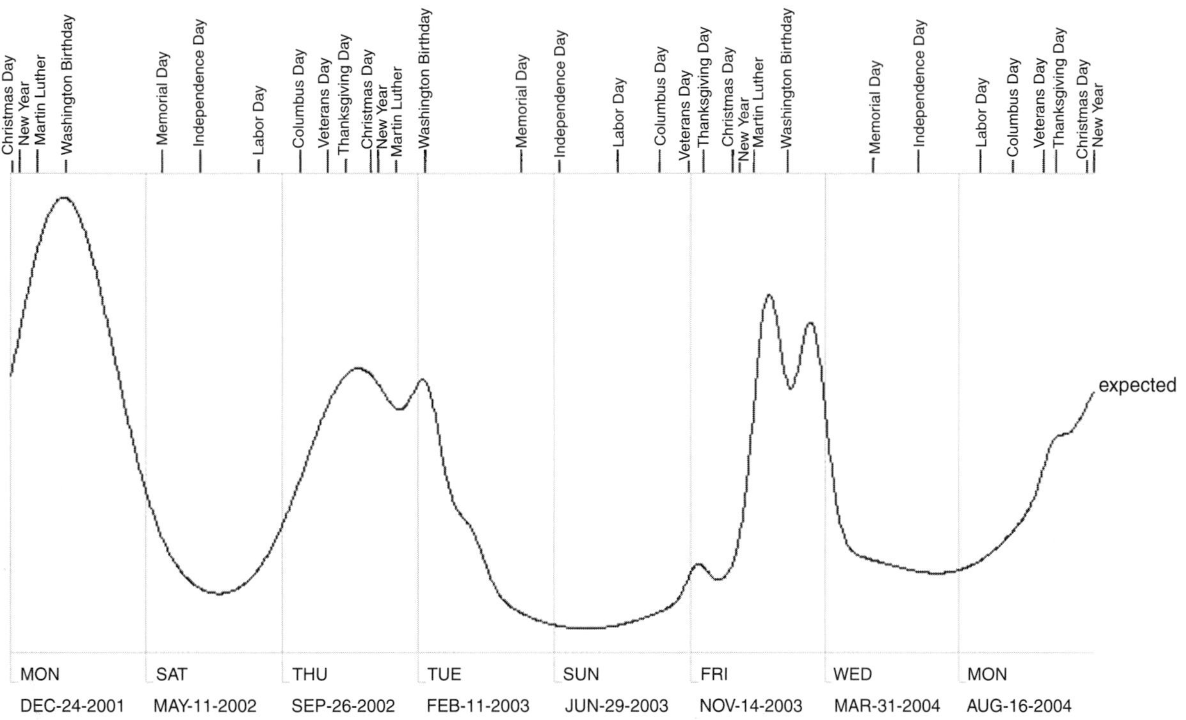

FIGURE 14.3 A smoothly varying (and entirely fictitious) level of influenza in the population over a three-year period. The (sadly overcrowded) holiday labels will be relevant when we simulate the effects of holidays.

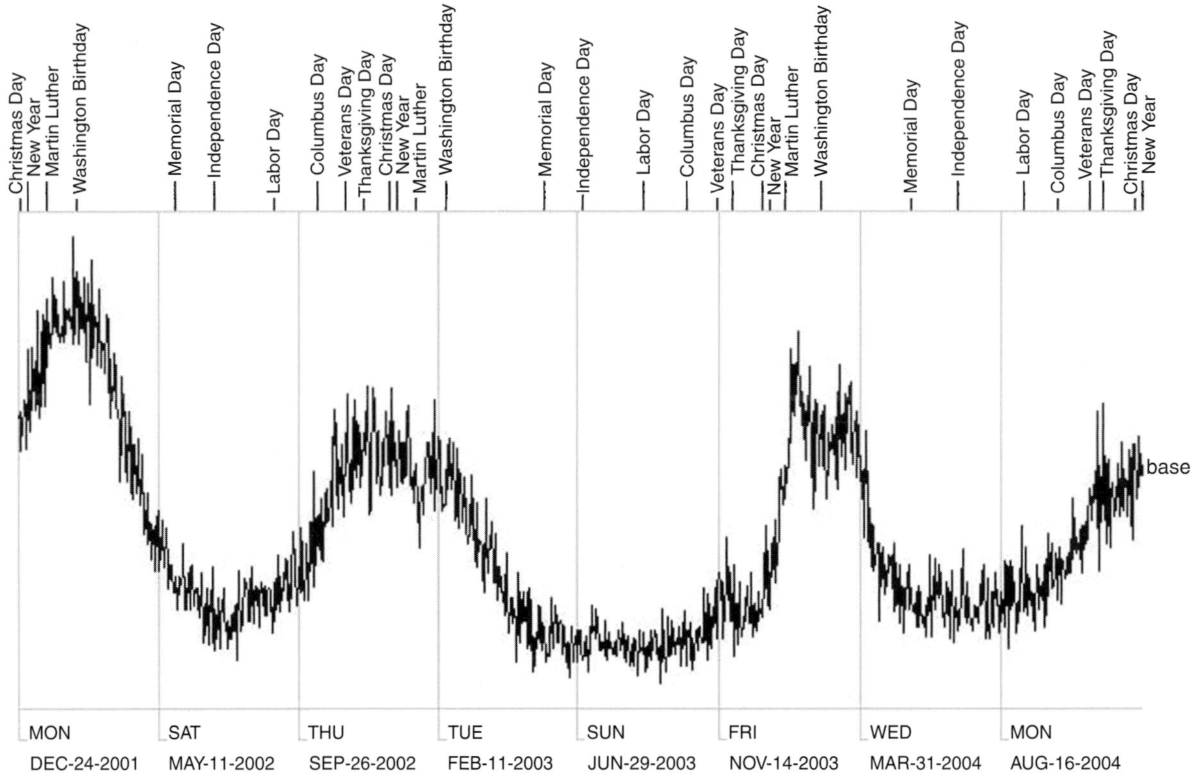

FIGURE 14.4 Simulating the number of people with respiratory complaints who seek health care.

Figure 14.5 is the baseline data to which our ramp outbreak was added in Figure 14.2. It has both the coarse and fine-grain properties of real univariate biosurveillance data.

Why use simulated data in this chapter? The answer is a take-home lesson in one of the issues of data privacy. Since we will be looking at specific numbers in the data in detail, and since we will make the data publicly available for readers who would like to manipulate it them themselves, it would be a violation of the spirit and letter of data use agreements to provide a real time series.

3. CONTROL CHARTS

Many of the univariate algorithms used in biosurveillance are techniques taken from statistical quality control, which is a field concerned with monitoring the quality of a production process. One of the simplest statistical quality control methods that we can apply to surveillance is the control chart. A control chart sets an upper control limit (UCL) and a lower control limit (LCL) on the time series being monitored. If the daily counts remain between the UCL and the LCL, then the process is said to be *in control.* However, if the daily counts exceed the UCL or go below the LCL, then the process is said to be *out of control.* When monitoring health-related activity, we are usually concerned with a high number of counts in data. As a result, we signal an alarm only if the process exceeds the UCL.

In order to determine the control limits, we need to measure the behavior of the process under normal conditions. Suppose that we have N observations $X_1, X_2, \ldots, X_N$ from the background activity of the process being monitored. Assuming that the daily counts in the background activity follow a normal distribution, we can estimate the sample mean $\hat{\mu}$ and standard deviation $\hat{\sigma}$ of the normal from the N observations. The formulas are shown in Eqs. 1 and 2.

$$\hat{\mu} = \frac{1}{N}\sum_{i=1}^{N} X_i \quad (1)$$

$$\hat{\sigma} = \sqrt{\frac{1}{N-1}\sum_{i=1}^{N}(X_i - \hat{\mu})^2} \quad (2)$$

Once we have the sample mean and standard deviation, we can calculate the upper control limit as shown in Eq. 3. We can signal an alarm if the actual count exceeds the upper control limit, and the larger the amount by which we exceed the upper control limit, the greater the severity of the alarm. The alarm level is defined as the probability of seeing the count X_i or a lesser count under the assumption that the observations are distributed according to a normal distribution with mean $\hat{\mu}$ and standard deviation $\hat{\sigma}$. In Eq. 4, the symbol Φ refers to the

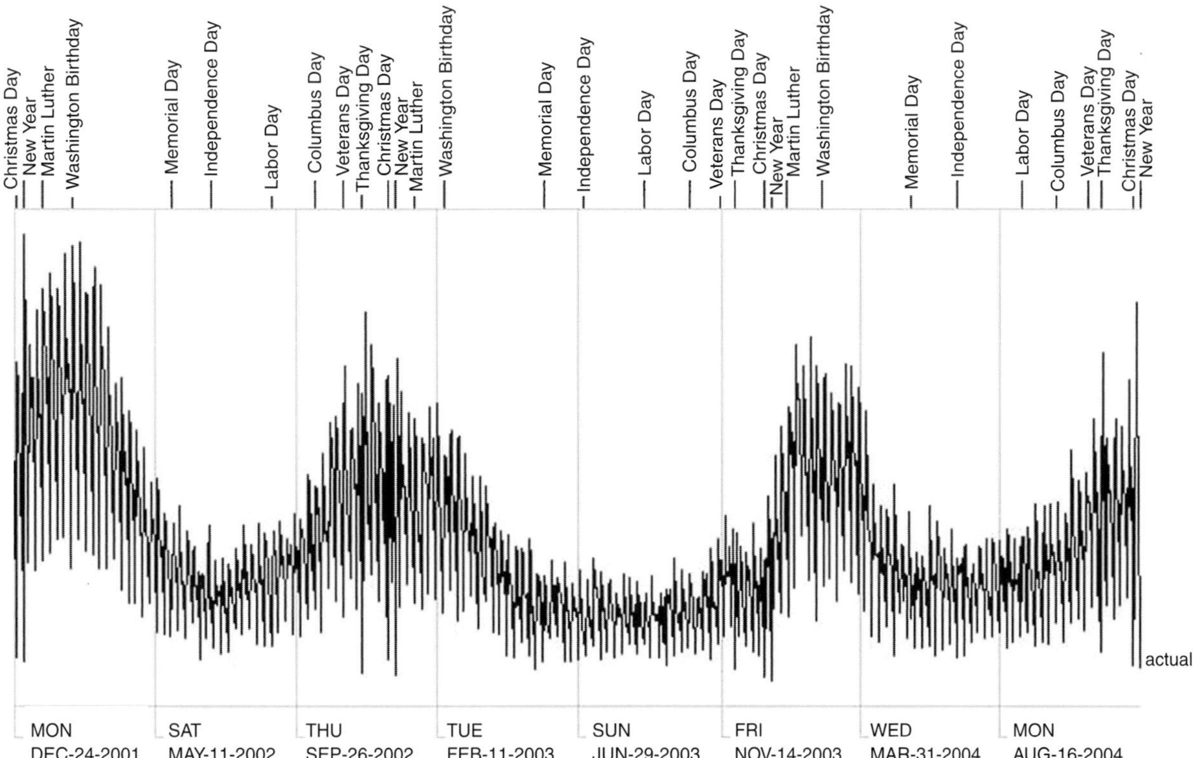

FIGURE 14.5 Synthetic data that account for seasonality, noise from a finite sample, day-of-week effects, and the effects of holidays.

cumulative distribution function for a standard normal–that is, a normal distribution with mean 0 and variance 1. The max term in Eq. 4 ensures that only counts greater than the mean will have an alarm value greater than 0.5.

$$\text{UCL} = \hat{\mu} + k\hat{\sigma} \tag{3}$$

$$\text{Alarm Level} = \Phi\left(\frac{\max(0, X_i - \hat{\mu})}{\hat{\sigma}}\right) \tag{4}$$

The constant k determines how sensitive the control chart will be. With a lower value of k, the control chart becomes more sensitive as it will trigger an alarm with less of a deviation from the mean of the process. Two typical values used in industrial applications of k are 2 or 3. Setting the constant k to be 2 corresponds to a 5% chance that an in control process is mistakenly said to be out of control while setting k to be 3 corresponds to a 1% chance. These percentages are obtained from the assumption that the observations X_1, X_2, ..., X_N follow a normal distribution.

Figure 14.6 shows the performance of a control chart on our example from Figure 14.1. Figure 14.7 shows the same data (with the same injected ramp) on a large scale. This figure shows many interesting phenomena. The height of the dark gray "predict" line at time t shows the historical average of the count based on all data previous to time t. The light gray line shows the upper limit of the control chart. The good news is that the counts do indeed exceed the upper limit during the synthetic outbreak. But the bad news is that there are many other occasions where this also happens, especially during the peaks in fall 2003 and winter 2004. The major problem with the control chart in this form is that it does not anticipate the seasonal effects, and when it is in the middle of a winter peak, it issues an alarm frequently. The next section alleviates this problem.

4. CHANGES FROM YESTERDAY

We can also transform the original time series $X = \{X_1, X_2, ..., X_N\}$ into a new time series that indicates the increase or decrease in count from the previous time step, which in our example is a day. Let us define this new time series by $Y_t = X_t - X_{t-1}$ as the changes from the previous day. We can then use a control chart algorithm on the Y time series to detect the onset of an outbreak.

Figures 14.8 and 14.9 graph the results of applying the "changes-from-yesterday" transformation to the data from Figure 14.5. When we compare the control chart in Figure 14.5 to the "changes-from-yesterday" algorithm in Figures 14.8 and 14.9, we notice that the control chart can be too insensitive to recent changes while the yesterday approach can be too easily influenced by recent changes. We would like an algorithm that is a happy medium between these two approaches, such

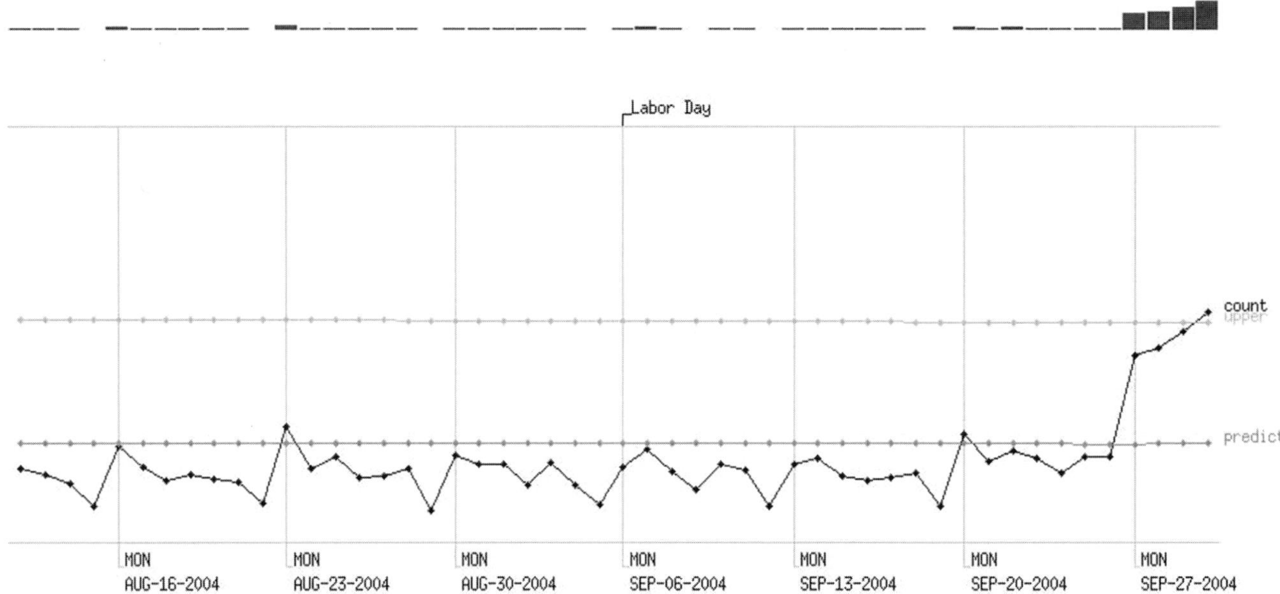

FIGURE 14.6 The black line ("count") shows the data. The dark gray line ("predict") shows the historical mean value, and the light gray line ("upper") is the 2-sigma upper level. The dark bars on the top show the degree of alarm: the larger the bar, the more surprised is the control chart (formally, the height of a bar is proportional to the log of the reciprocal of the alarm level from Eq. 4). We see that the control chart only became very excited on Monday, September 27, 2004.

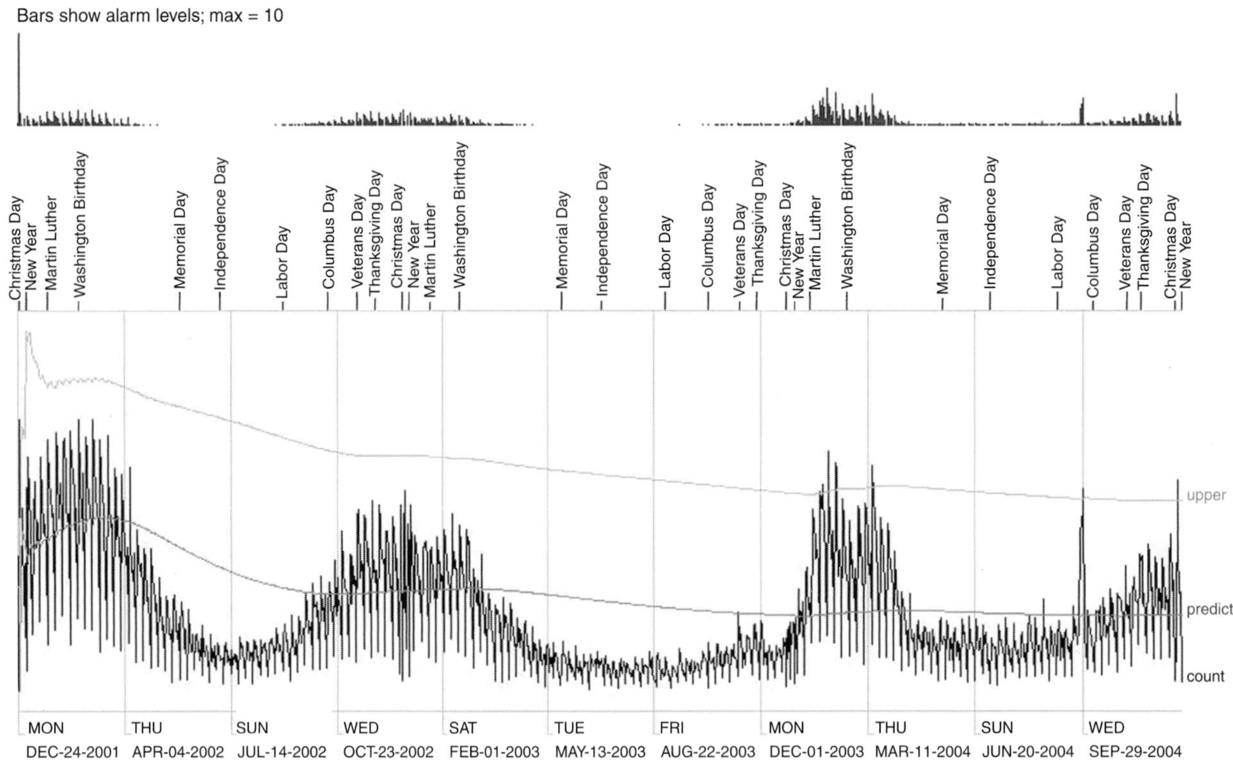

FIGURE 14.7 This shows the same information as Figure 14.6, but shows it over the entire three-year time period. Note the large number of strong alarms (indicated by the fine bar chart on the top), and how many have equal strength as the alarm corresponding the simulated outbreak at the end of September 2004.

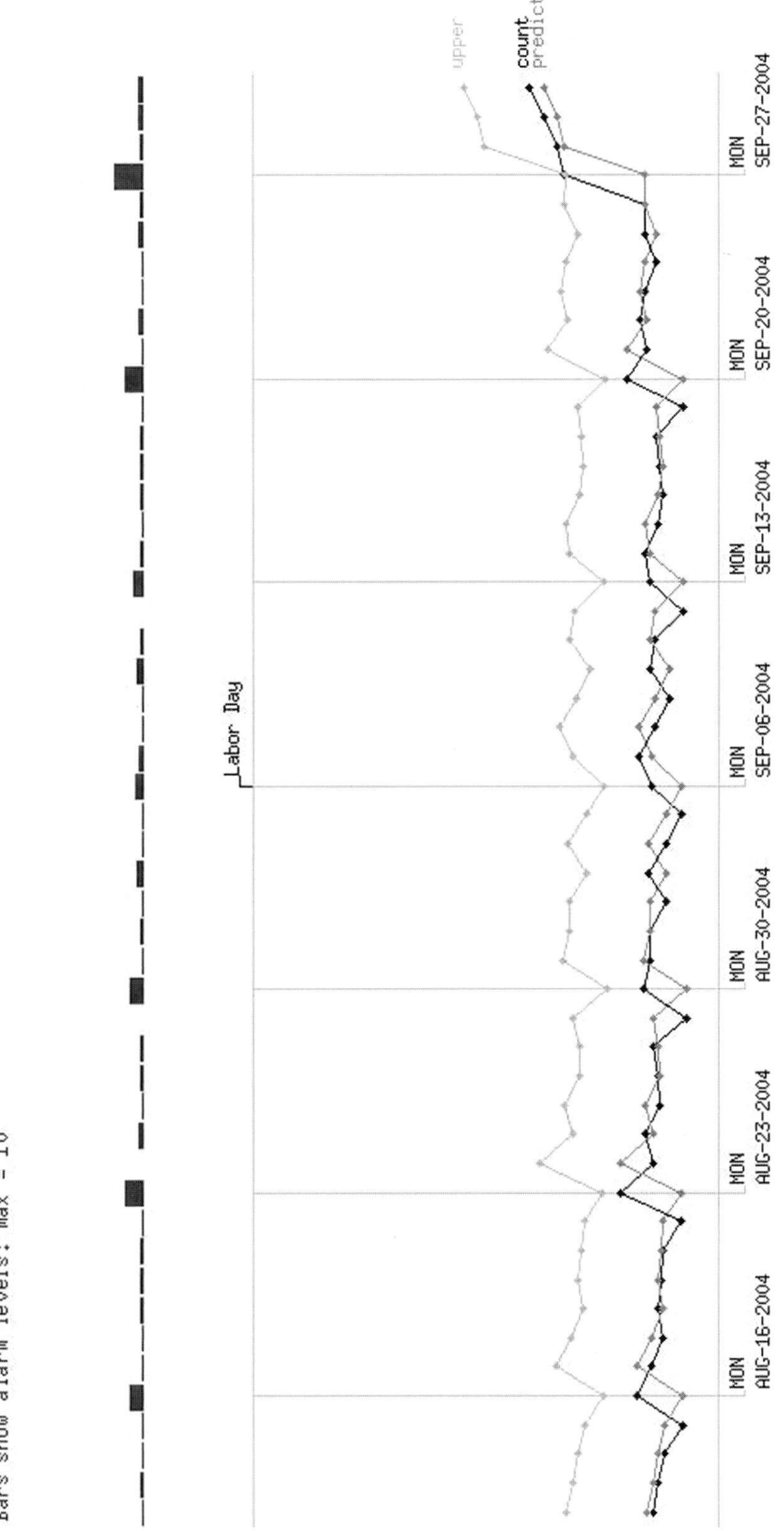

FIGURE 14.8 Alerts based on changes from yesterday. The algorithm successfully detects the increase on the Monday of the outbreak, but it also signals an increase on earlier (non-outbreak) Mondays, since all Mondays tend to have a jump in physician visits compared with the day before.

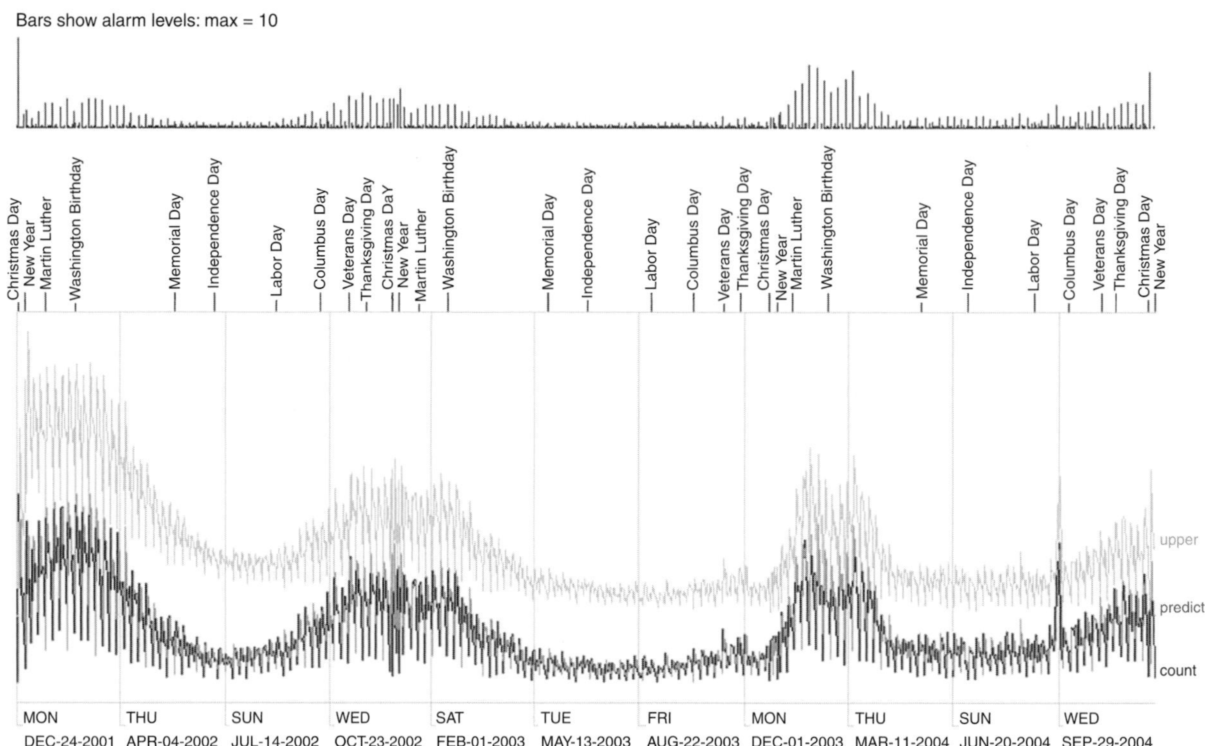

FIGURE 14.9 The same information as Figure 14.8, but over a longer period. It shows alerts based on changes from yesterday on the timescale of several years. The "comb" shape of the alarm levels indicates that the algorithm almost always signals an alarm on Mondays.

as the moving average algorithm, which we will describe in the next section.

Using yesterday's count as the baseline has at least two problems, however. First, the baseline itself will be subject to random fluctuation, so if day t happens to have a low count, then day t+1 will automatically look more dangerous. Second, day-of-week effects are a disaster: Mondays will always signal an alert since they follow the very low-counting Sundays.

5. MOVING AVERAGE

The moving average algorithm predicts the value for the next time step based on an average of the values in the last W time steps, where W is the window size.

More precisely, a moving average algorithm predicts the following:

$$X_{t+1} = \frac{1}{W}(X_t + X_{t-1} + \ldots + X_{t-W-1})$$

In order to compute the alarm level for the moving average algorithm, we first fit a Gaussian distribution to the counts from the recent W days. We estimate the mean and standard deviation for this Gaussian from the W observations within the window using Eqs. 1 and 2. We then calculate the alarm level using the cumulative distribution function of this fitted Gaussian as in Eq. 4.

On the physician data, moving average avoids both the problems of control charts and the problems of comparison against yesterday. Figure 14.10 shows the algorithm nicely mimicking the human eye, and signaling surprise on the Monday with the large rise. Notice how the upper limit is much lower than in the control chart approach of Figures 14.6 and 14.7. This reduction is because the estimate of σ (the standard deviation) is now based only on recent data, instead of all data throughout history, and the recent data does not contain the variation caused by seasonality. As a result, the Monday jump is well above the upper limit, leading to a strong alert signal. It is interesting to note that although the ramp outbreak gets more substantial after Monday, the alarm strength begins to decrease (indicated by the drop of the bars on the top right). This phenomenon reveals a weakness of this and several other methods: the predictions (the "predict" line) become accustomed to increased counts during an outbreak. Since expected counts are higher, so is the upper limit and alarms become less substantial. This problem is a prime motivation for the CUSUM approach in the next section.

Figure 14.11 shows moving average in action over the entire three-year period. It shows how the baseline nicely follows the recent average of the data, and so the upper limit adapts to seasonal variation. As a result, over the entire three-year period, by far the largest alarm level occurs during the simulated outbreak.

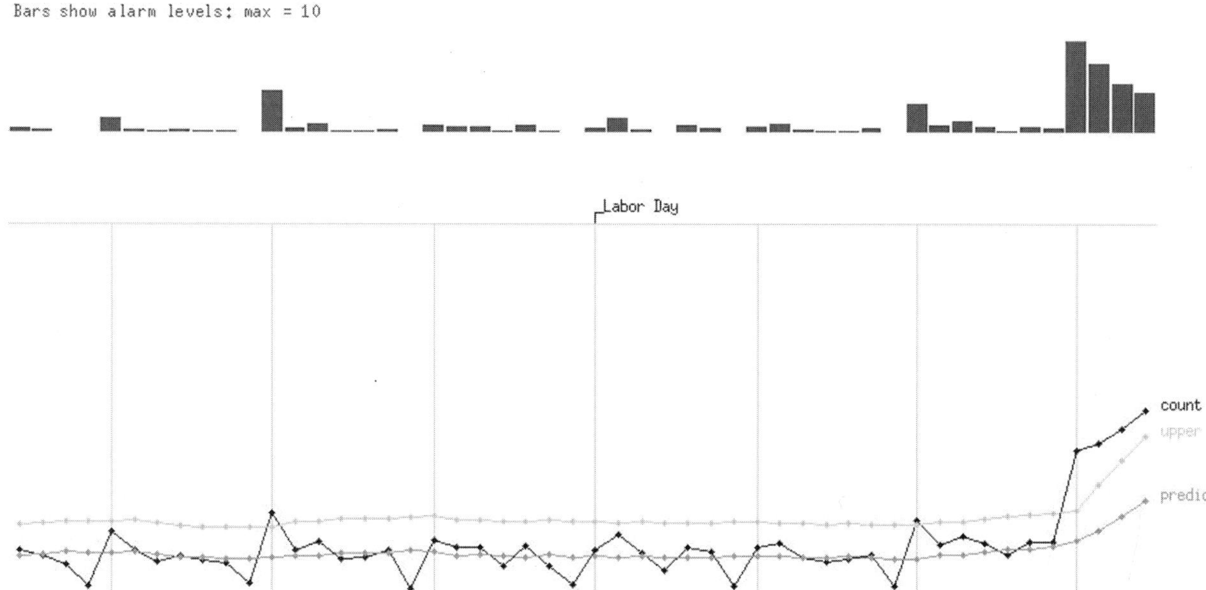

FIGURE 14.10 Alerts based on moving average (with a seven-day window to estimate predicted mean ♦ and 28-day window to estimate standard deviation ●).

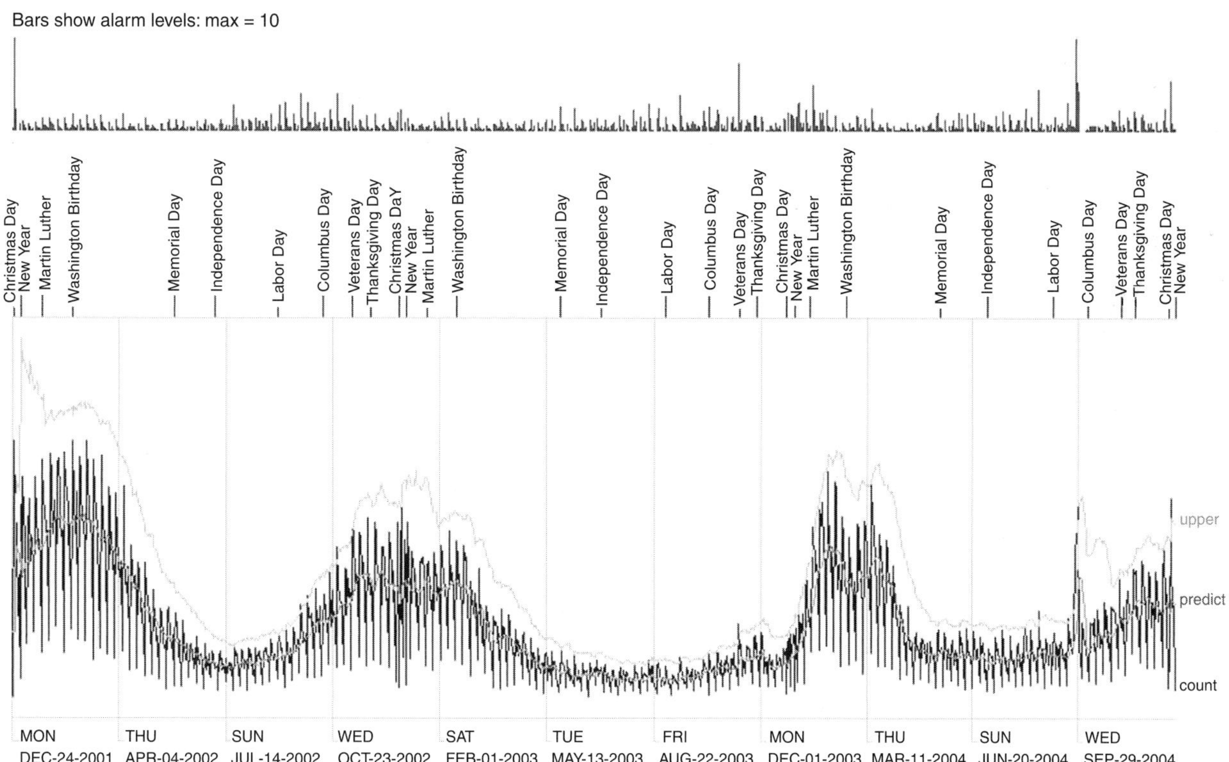

FIGURE 14.11 Expanded view of Figure 14.10: moving average alerts over three years.

6. CUSUM

Another detection algorithm taken from statistical quality control is CUSUM (Page, 1954), which stands for cumulative sum. The CUSUM algorithm is able to detect small shifts from the mean more quickly than a control chart. As its name suggests, CUSUM maintains a cumulative sum of deviations from a reference value r, which is usually the mean of the process. Let X_i be the ith observation and let S_i be the ith cumulative sum. The calculation of the cumulative sum is as follows:

$$S_1 = X_1 - r$$
$$S_2 = (X_2 - r) + (X_1 - r) = (X_2 - r) + S_1$$
$$\vdots$$
$$S_k = \sum_{i=1}^{k} (X_i - r) = (X_k - r) + S_{k-1}$$

If the observations X_i are close to the mean, then the cumulative sums S_i will be around zero. However, once a shift from the mean occurs, the S_i values will either increase or decrease quickly. Since we are usually only concerned with an increase in counts in biosurveillance, we can rewrite the formula for S_i to keep the cumulative sums positive. The max term in Eq. 5 ensures that the cumulative sum never goes below 0.

$$S_k = \max(0, (X_k - r) + S_{k-1}) \qquad (5)$$

In the equations above, we have used the mean as the reference value r. Typically, the reference value consists of the mean plus or minus a slack value or allowance variable called K. We can rewrite Eq. 5 as:

$$S_k = \max(0, X_m - (r + K) + S_{m-1}) \qquad (6)$$

In Eq. 6, any value within K units of r is effectively ignored. The value of K is typically set to be the midpoint between the in control process mean and the out-of-control process mean μ_i expressed in terms of the standard deviation σ of the in-control process:

$$K = \frac{|\mu_1 - \mu_0|}{2}$$

Alerts are raised whenever S_i exceeds the threshold H; once an alert is raised, we reset the cumulative sum value to be 0. The value of H is determined from the value of K and the average run length (ARL) of the in-control process. The ARL is the average number of time steps before an alert is raised. Whenever the process is operating as expected, the ARL should be large. However, when a shift in the mean occurs, the ARL should be small in order to detect the shift as quickly as possible. The ARL can be determined through a variety of approximating equations, which are summarized in (Montgomery, 2001). According to (Montgomery, 2001), a reasonable value for H is 50, which is five times the standard deviation of the in-control process.

7. COMPARING THE UNIVARIATE ALGORITHMS

Let us take stock of the methods described up to this point. How do they rank against each other? We performed a test in which 2000 synthetic data sets were created. Each data set is a copy of the original synthetic data of Figure 14.5, except that each data set also involves a simulated inject outbreak. Half of the tests involved a "spike pattern" in which one day had an increase of between 10% to 100% over the true count. The other 1000 tests involved a "ramp pattern" similar to Figure 14.2 in which five days in a row had an increased count, with the increment for each outbreak day being larger than its predecessor. Noise was simulated in these ramps, and the ramp size was randomly chosen between 10% and 100% of the true count. Thus, there are a variety of outbreak patterns. Some are so easy to detect that any self-respecting algorithm should identify them; others are trickier, and some might be virtually impossible to detect. The results are shown in Table 14.1. For each algorithm, we measure the following four characteristics:

- What fraction of spike outbreaks are detected if alarm thresholds are set at a level that produces one false alarm every two weeks?

TABLE 14.1 Experiments with Univariate Methods

	Average Performance During 1000 Simulated Outbreaks			
	Fraction of Spike Outbreaks Detected		Number of Days Until Ramp Outbreak Detected	
Method	1 FP every two weeks	1 FP every 10 weeks	1 FP every two weeks	1 FP every 10 weeks
Standard control chart	0.33	0.14	4.15	4.74
Using yesterday	0.42	0.08	3.53	4.76
Moving average W=3	0.67	0.51	2.46	3.64
Moving average W=7	0.75	0.61	1.96	2.86
Moving average W=14	0.70	0.58	2.04	3.07
Moving average W=28	0.65	0.53	2.34	3.34
Moving average W=56	0.49	0.32	2.89	3.76
Best CUSUM	0.46	0.12	3.71	4.64

Notes: Experiments are described in the text. The CUSUM results are the best results out of CUSUM runs with all values of the H parameter between 1 and 100. A more complete version is provided in Table 14.2.

- What fraction of spike outbreaks are detected if alarm thresholds are set at a level that produces one false alarm every ten weeks? This performance must be inferior (i.e., a lower fraction detected) to the two-week case because the alarm threshold must be higher.
- How many days on average does it take until a ramp outbreak is detected if alarm thresholds are set at a level that produces one false alarm every two weeks? If the ramp is undetected we score it as requiring five days.
- How many days on average does it take until a ramp outbreak is detected if alarm thresholds are set at a level that produces one false alarm every 10 weeks?

From the table we see that simple control charts perform poorly (because of seasonal problems, and because of serious day-of-week effects). When we use "yesterday" we do not suffer from the seasonality-induced problems but day-of-week effects are even worse. Moving average is more effective (although methods described later in this chapter will thoroughly resolve it), and CUSUM is, on these examples, relatively insensitive. The performance of MA seems to be little affected by the window size *W*: it is best at seven days and degrades only slightly if it is halved or doubled.

8. EXPONENTIALLY WEIGHTED MOVING AVERAGE

Another statistical quality control technique that is similar to CUSUM is the exponentially weighted moving average (EWMA) algorithm (Roberts, 1959). As its name suggests, the EWMA algorithm is a variation of the moving average algorithm in which we assign the current and prior observations weights such that observations that are further in the past receive an exponentially decreasing amount of weight. The parameter λ, which is in the range $0<\lambda\leq 1$, controls the weight we assign to the observations. EWMA is better at detecting small shifts in a process than a control chart, with its performance being similar to that of the CUSUM algorithm (Montgomery, 2001).

Let X_i be a measurement at time *i*. The EWMA statistic Z_i is defined as:

$$Z_i = \lambda X_i + (1-\lambda)Z_{i-1} \tag{7}$$

The starting value Z_0 is set to the desired process mean μ_0 or to the average of some initial data ie. $Z_0 = \bar{X}$. If we unravel the recursion in Eq. 7 as shown below, we can see that observations further in the past are weighted more lightly.

$$\begin{aligned} Z_i &= \lambda X_i + (1-\lambda)[\lambda X_{i-1} + (1-\lambda)Z_{i-2}] \\ &= \lambda X_i + (1-\lambda)\lambda X_{i-1} + (1-\lambda)^2 Z_{i-2} \\ &\vdots \\ &= \lambda \sum_{j=0}^{i-1} (1-\lambda)^j X_{i-j} + (1-\lambda)^i Z_0 \end{aligned}$$

Assuming that the X_is are independent random variables with variance σ^2, the variance of Z_i can be computed as:

$$\sigma_{Z_i}^2 = \sigma^2 \left(\frac{\lambda}{2-\lambda}\right)[1-(1-\lambda)^{2i}]$$

With this variance, we can establish an upper control limit for an EWMA chart as shown below:

$$UCL = \mu_0 + L\sigma_{Z_i}$$

The parameters L and λ are chosen according the desired ARL. In practice, values of $0.050 \leq \lambda \leq 0.25$ and L = 3 work well (Montgomery, 2001).

9. REGRESSION

As we discussed previously, the background activity in actual healthcare utilization data is never a clean signal. There are noisy fluctuations in the data due to seasonal, day-of-week, and holiday effects. The detection algorithms described up to this point have a difficult time working with such data because the background activity is so difficult to model. If the background cases are estimated from a time period with a high count due to seasonality, then the detection algorithm becomes very insensitive to high counts in future data. Conversely, if the background is estimated during a valley, then the algorithm becomes too sensitive. In this section we describe how a simple and familiar approach—linear regression—can help with these kinds of problems. First, we will very briefly recall what linear regression does. Next, we will see how it can help model seasonal effects and then other effects, such as the day of week.

9.1. What Regression Does

Regression is a statistical method used to determine the relationship between an output *Y* and a given set of inputs *X*. In linear regression, *Y* is assumed to be a linear combination of the set of inputs *X*, that is, $Y=\beta_0+\beta_1X_1+\ldots+\beta_mX_m$. The output *Y*, however, is assumed to be shifted by some noise, where the noise is assumed to be generated from a Gaussian random variable with mean 0 and variance σ^2. We will refer to the variable *Y* as the dependent variable and the set as the independent variables.

We will illustrate linear regression using a simple example in which the set X_1 consists of only variable X_1. A typical regression problem begins with a set of *n* observed data points of the form (X_i, y_i) such as a set of two dimensional coordinates shown in Figure 14.12. From these data, we would like to find a line of the form $Y=\beta_1X+\beta_0$ that best fits this data. The best fitting line is defined as the line that minimizes the sum of squared residuals, where the residual is defined as the

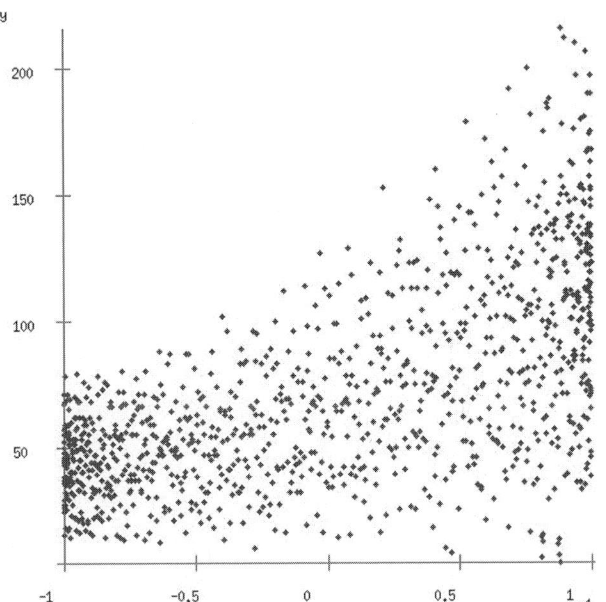

FIGURE 14.12 A scatterplot of 1100 (x,y) pairs used to illustrate linear regression.

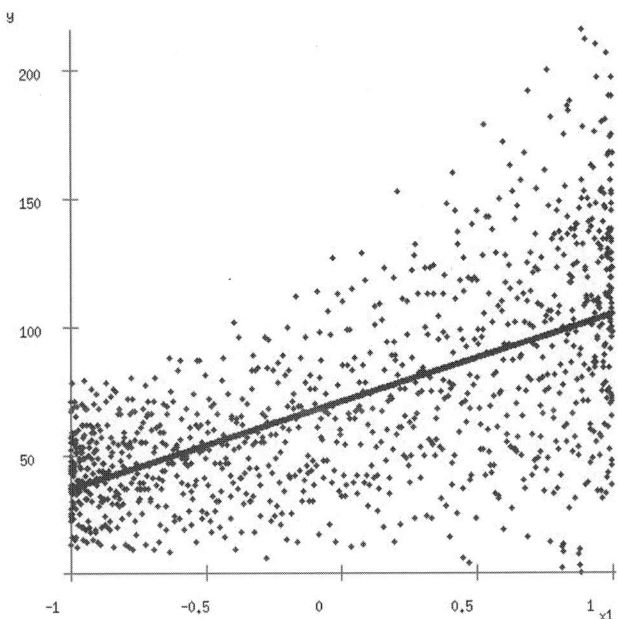

FIGURE 14.13 The best-fit linear regression for the data.

difference between the actual value and the predicted value, that is, $y_1 - \beta_1 X_1 - \beta_0$. The method of least squares picks the values of β_0 and β_1 that minimize the formula:

$$\text{Residual Sum of Squares} = \sum_{i=1}^{n} (y_i - \beta_0 - \beta_1 x_1)^2 \tag{8}$$

If we perform a linear regression on the points in Figure 14.12, the optimal values of β_0 and β_1 turn out to be $\beta_0 = 70.97$ and $\beta_1 = 34.01$. The resulting predictions are $\hat{Y} = 34.01 \times X_1 + 70.97$, shown in Figure 14.13.

9.2. Using Regression to Account for Seasonality

In order to handle seasonal effects in the data, we can use the part of the year that we are currently in as a hint to the method that predicts the expected count for today. For example, we can define the term *hours_of_daylight* as:

$$\sin\left(2\pi * \left(\frac{\text{Number of days since July 31}}{365.25}\right) - \frac{\pi}{2}\right)$$

This expression does not really model hours of daylight, but it provides a numerical value that is low (–1) in summer when we expect little influenza activity and is high (+1) in winter when we expect more influenza. How can we do the best job of predicting the expected count for today based on this hours-of-daylight feature and a table of historical data? The answer is, of course, to use linear regression, where the X-values are hours-of-daylight values from all previous days in our historical data, and the Y-values are the corresponding counts in our historical data. For example, if we wish to predict the expected count for January 1, 2005, from the data in Figure 14.5 in this manner, the regression we must perform is shown in Figure 14.14. It turns out that our example data sets in the regression tutorial were derived in exactly this way.

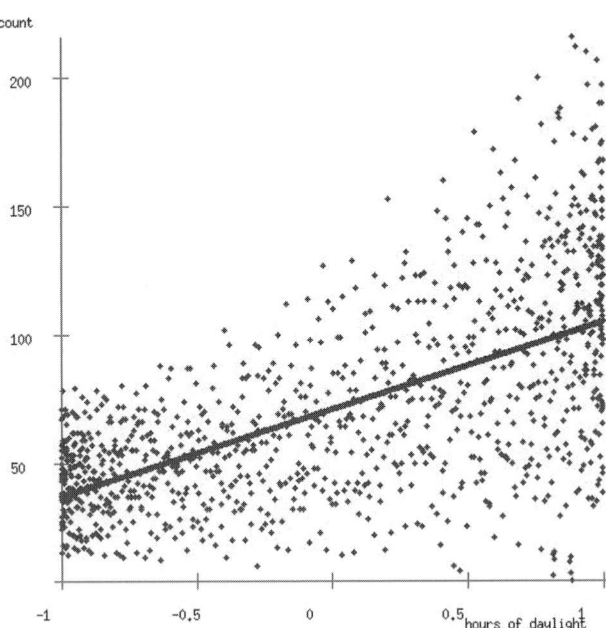

FIGURE 14.14 A total of 1104 data points corresponding to the historical data in Figure 14.5, and the best linear fit for predicting expected counts based on hours of daylight.

Using this approach, the predicted counts can be modeled as shown in Figure 14.15. The upper limit can be derived using a well-known technique called regression analysis, which we will not describe here. Note how in Figure 14.15 the expected counts and upper limits loosely model our expectations based on seasonality. As a result, there is lower estimated standard deviation in the noise compared with the control chart of Figure 14.5, and so the upper limit can stay closer to the expectation. This means the alarms can be more sensitive, and you can notice a far stronger reaction to the synthetic ramp outbreak.

There is an interesting feature in Figure 14.15 which might have caught the eye of a reader familiar with time-series regression. Why do the magnitudes of the sinusoids change over time? The reason is that at each date, the regression method can be trained only on data from before the current date, and so the training data set is varying as time goes by, which is why the linear relationship varies, and, in turn, causes the expected counts and upper limits to follow a varying pattern.

In addition to seasonal effects, regression can also deal with day-of-week effects. For example, if we know that on Mondays, the count for health utilization cases tends to differ relative to the surrounding days, we can add a binary term *is_monday* to the regression. With this extra term, the independent variable X no longer consists of a single value. Instead, X is now a tuple, meaning that each data point consists of the values ($hours\_of\_daylight_i$, is_monday_i, y_i) where *is_Monday* takes the value 0 for all days which are not Monday, and value 1 on all days that are Monday. Linear regression then learns the following relationship:

expected_count
$= \beta_0 + \beta_1 \times \text{hours\_of\_daylight}$ if today is not a Monday

expected_count
$= \beta_0 + \beta_1 \times \text{hours\_of\_daylight} + \beta_2$ if today *is* a Monday

Thus, Mondays can be given a little "bump" in their expectation, and historical data combined with linear regression can determine the best value for that bump. A close-up of the resulting predictions is shown in detail in Figure 14.16 and across all 3 years in Figure 14.17. Note that the pattern of alarm gets stronger and stronger during the ramp outbreak. Monday's alarm is substantial, but less substantial than in the earlier method because we were already anticipating the Monday "bump." Note in Figure 14.17 how our injected spike now has a far stronger alarm than any other period in the three years.

Suppose we would also like to account for the fact that a date falls on a Tuesday. We can simply add another binary term called *is_tuesday* to the regression. We can even consider all of the days of the week by adding the six binary terms *is_monday*, *is_tuesday*, *is_wednesday*, *is_thursday*, *is_friday*, and *is_saturday*. Note that a term for *is_sunday* is not needed because a value of 0 for all six binary terms would indicate

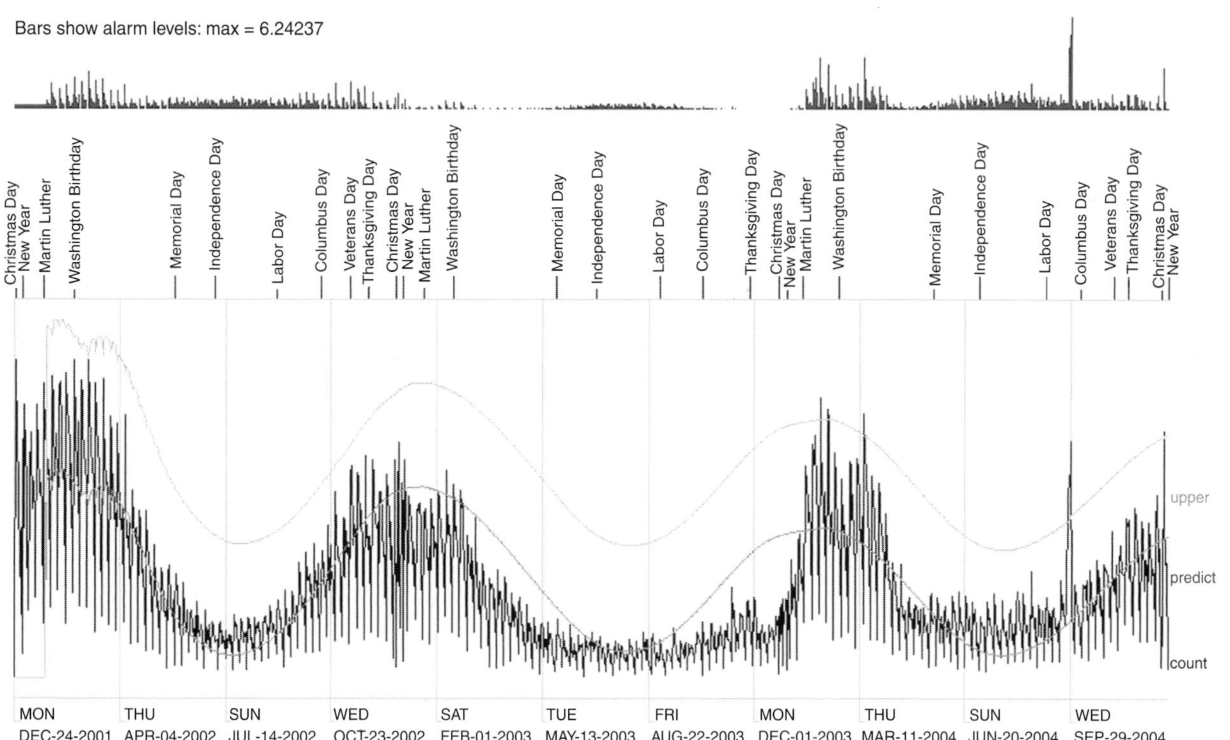

FIGURE 14.15 Performance of a regression method which uses only one feature: hours_of_daylight.

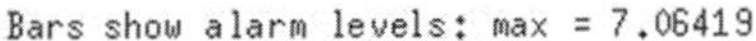

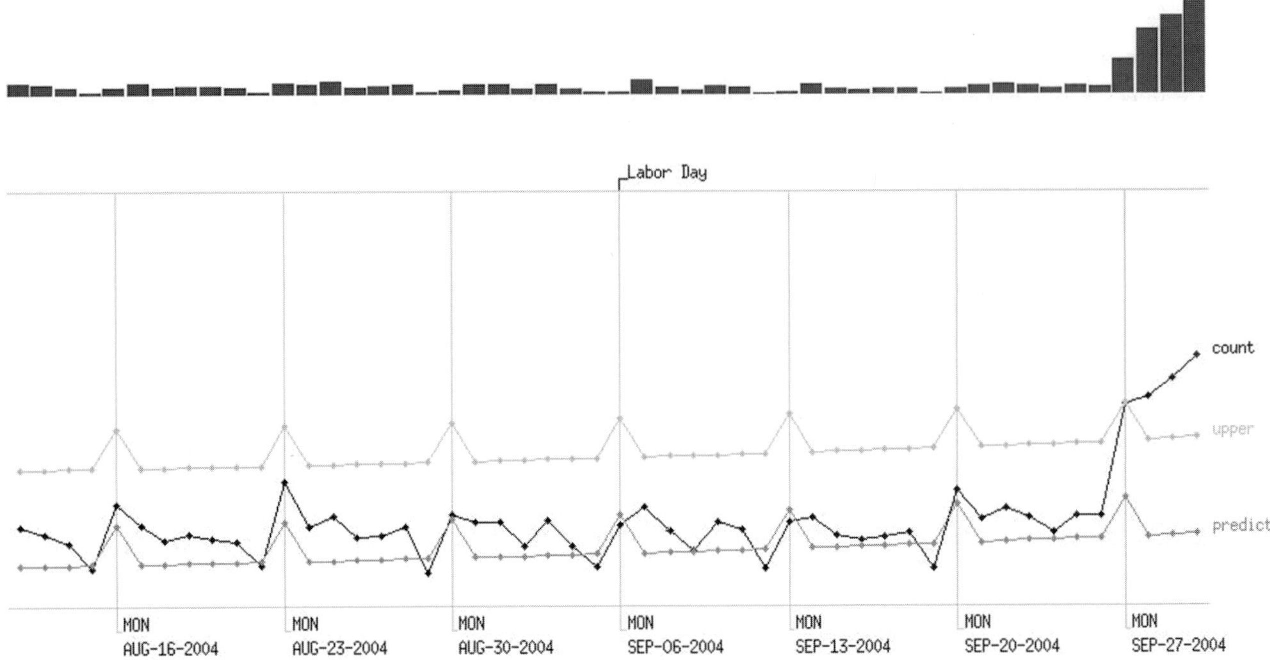

FIGURE 14.16 Predictions and alarms from a regression method that uses hours_of_daylight and is_monday as features.

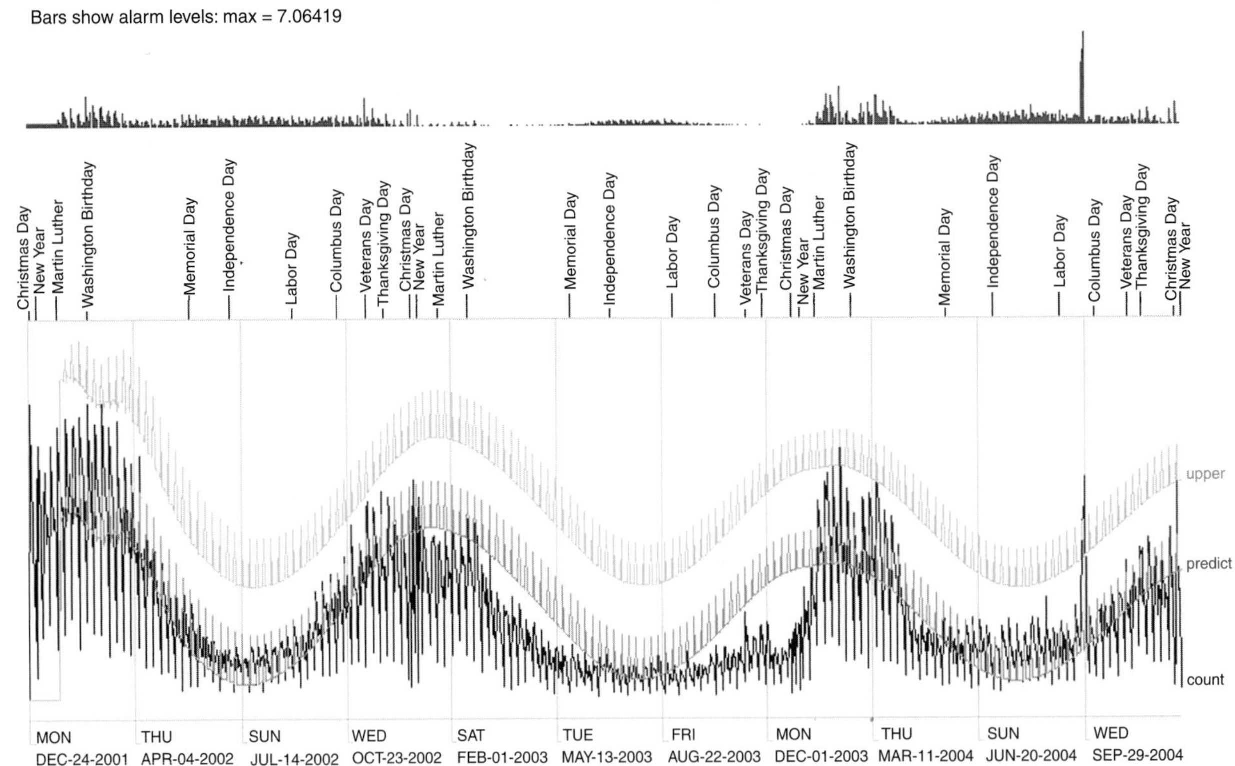

FIGURE 14.17 Over the 3-year period, predictions and alarms from a regression method that uses hours_of_daylight and is_monday as predictive features.

that the date is a Sunday. It is also possible to add local terms, such as the mean count over the previous seven days as an additional feature to allow the method to adapt better to recent local events. These additions can be very helpful, and as we shall see at the end of the chapter, regression methods that include these extra terms perform better than moving average (which had been our favorite method) on our data.

10. SICKNESS AVAILABILITY

Another way to deal with day-of-week variations is to use the sickness availability method to smooth the time series by removing noise due to the day-of-week effect. This algorithm transforms the daily counts in the time series into a daily sickness value, which is defined as the number of people getting sick every day irrespective of whether they seek health care or not. The term availability refers to the probability that a patient will seek health care during a specific day of the week; hence, there are a total of seven values of availability, one for each day of the week. The availability of a day can be thought of as the fraction of a weeks-worth of visits that get assigned to the given day. The sickness availability method is based on the intuitive assumptions that the expected count is the product of the true amount of sickness and the current day's availability.

We can estimate the expected availability for a specific day of week (dow) using the average of the availabilities on that day for the past m weeks. We can calculate the expected availability A_{dow} as:

$$A_{dow} = \frac{\sum_{i=1}^{m} (C_{i(dow)} / \sum_{dow=0}^{6} C_{i(dow)})}{m} \tag{9}$$

In Eq. 9, A_{dow} refers to the expected availability for the day of week specified by the variable *dow*, which takes on seven different values ranging from 0 to 6. The term $C_{\text{i(dow)}}$ is the actual number of patients that visited the ED on the particular day of week *dow* during the *i*th week in the past. Since national holidays affect the number of patients visiting the ED, weeks containing holidays are ignored completely in the availability calculations. Finally, the parameter *m* controls the smoothness of the sickness curve.

Since sickness is defined as the total number of people in the city getting sick on a particular day, we can calculate it as follows:

$$S_{today} = \frac{C_{today}}{A_{today}} \tag{10}$$

The term S_{today} in Eq. 10 refers to the number of people who are sick on the current day while the term C_{today} refers to the number of people who visited the ED on the current day. A_{today} is the probability that a patient will visit the ED for the day of week specified by *today*. Figure 14.18 shows the availability estimated in the period leading up to the synthetic ramp outbreak: it shows a consistent picture that we usually see far more patients on Mondays than Sundays, and then the visits taper off during the rest of the week. Figure 14.19 shows both the original count and the estimated sickness (count/availability). Sickness is much more stable than the original count because day of week effects have been greatly reduced. It is now possible to run a time series algorithm on the sickness values. In this case, we chose to use moving average with a window of seven days. The resulting alarms show something that was not achieved in any

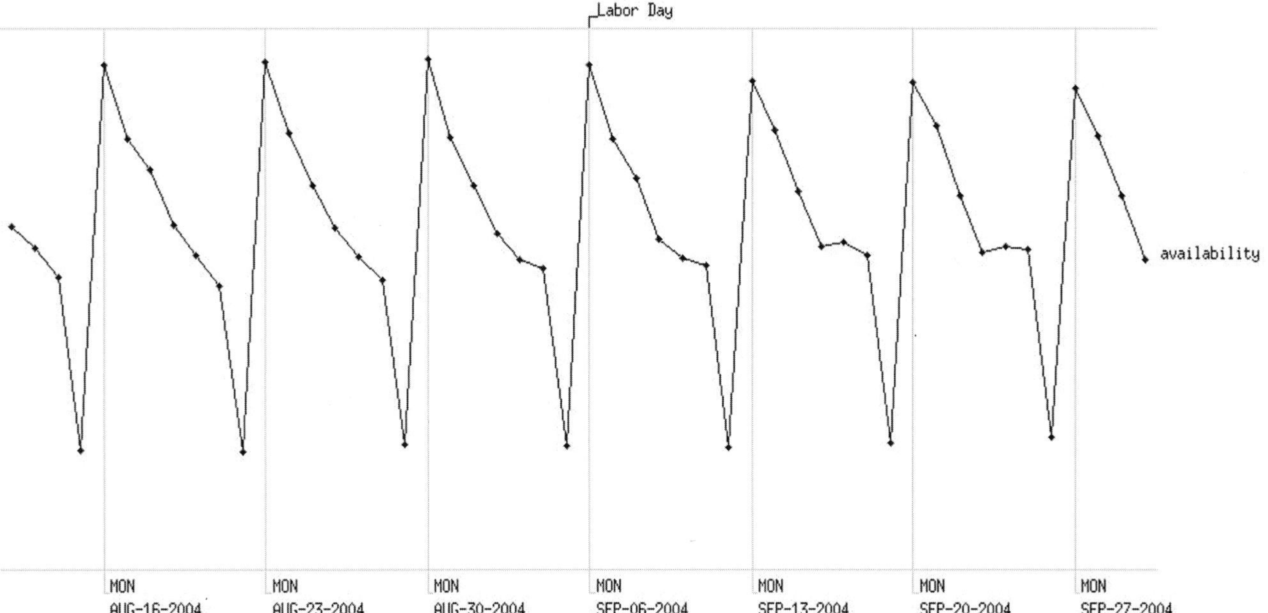

FIGURE 14.18 The availability of days of the week using the sickness availability method described in the text.

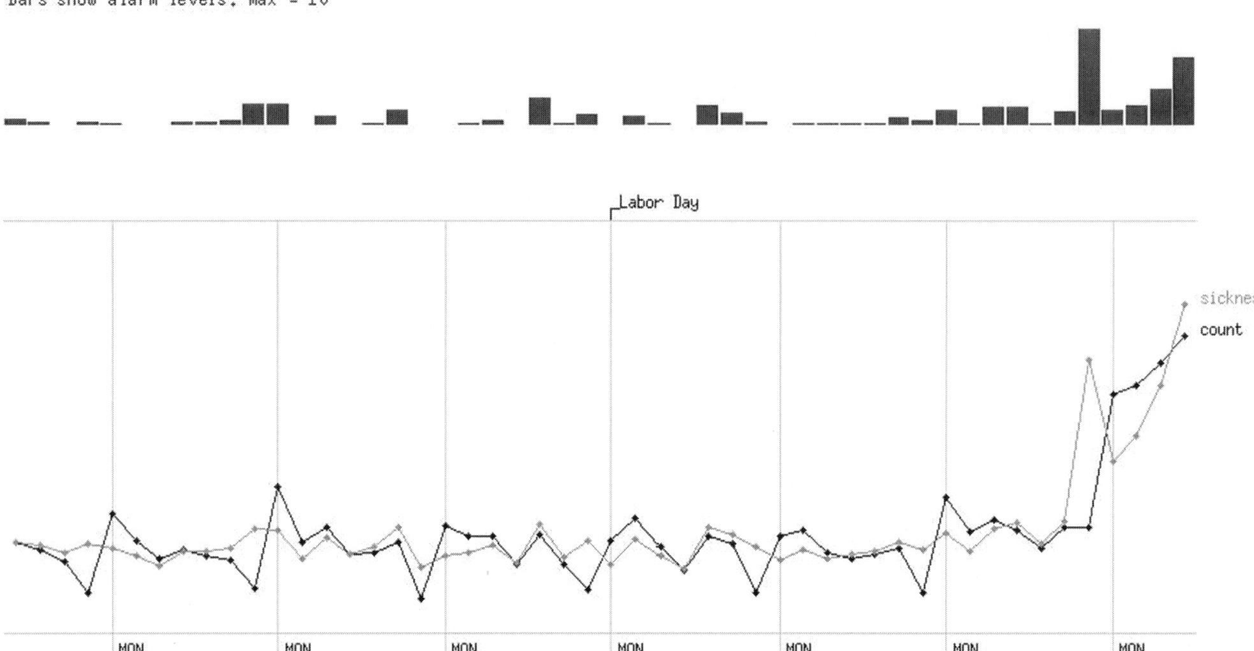

FIGURE 14.19 The black "count" line shows the raw data. The gray "sickness" line shows the sickness after the count has been divided by the corresponding availability value from Figure 14.18. Note how the sickness time series is now smoother than the original data and decorrelated with day of week. The alarm levels are derived from moving average applied to the sickness time series.

of the previous illustrations: a strong alarm resulting from the higher-than-expected counts for the Sunday.

We would like to emphasize that the sickness availability method only smoothes the time series. It is not a detection algorithm by itself. Instead, a detection algorithm, such as the control chart, moving average, or CUSUM algorithms should be used on the smoothed data.

11. FURTHER COMPARISON OF THE UNIVARIATE ALGORITHMS

We insert another copy of Table 14.1, with some of the above methods added for comparison. The newly evaluated algorithms follow:

- Regression using two features: the mean count over the past week and hours of daylight. This allows the algorithm to account for seasonal variation by putting a negative coefficient in front of hours of daylight.
- Regression using the additional feature *is_Monday*, which is set to 1 if today is Monday and 0 otherwise. This allows the algorithm to anticipate the Monday bump in physician visits and so be less prone to false positives.
- Regression using indicator variables for all days of the week except for Sunday (which would be redundant), and additionally, *hours_of_daylight* and *mean_count_over_previous_seven_days*.
- Using sickness/availability to compensate day-of-week effects, and then using the approach of comparing against yesterday. This method thus looks for jumps in the day-of-week-adjusted counts.

For this data set, with its seasonal components and day-of-week components, we see that sickness availability (to cope with day-of-week effects) combined with moving average (to cope with seasonal trends) performs well. Some of the regression methods perform almost equally well. We should note that this does not mean these methods are best in general: individual properties of individual data sets mean different approaches can be stronger for different data sets. Our only general advice is that in our experience relatively simple methods usually work at least as well as complex approaches. A second important note is that the numbers in Table 14.2 cannot be used as an estimate of how quickly real outbreaks are expected to be detected: the numbers are a function of many things, including the simulated magnitude of the outbreaks and the simulated noise levels.

12. ADDITIONAL METHODS

A complete set of approaches would require a very thick book, such as Hamilton (1994). The purpose of this chapter has been a tour of a sufficient variety of approaches to introduce the reader to some of the issues faced when choosing or implementing a time-series–based method, without going into the

TABLE 14.2 Experiments with Univariate Methods

	Average Performance Over 1000 Simulated Outbreaks			
	Fraction of Spike Outbreaks Detected		Number of Days Until Ramp Outbreak Detected	
Method	1 FP every two Weeks	1 FP every ten weeks	1 FP every two weeks	1 FP every ten weeks
Standard control chart	0.33	0.14	4.15	4.74
Using yesterday	0.42	0.08	3.53	4.76
Moving average 7	0.75	0.61	1.96	2.86
Moving average 28	0.65	0.53	2.34	3.34
Moving average 56	0.49	0.32	2.89	3.76
Moving average 14	0.70	0.58	2.04	3.07
Moving average 3	0.67	0.51	2.46	3.64
Regress with hours_of_daylight (HOD)	0.59	0.33	3.32	4.17
Regress: HOD + is_mon	0.61	0.33	3.10	4.34
Regress: HOD + is_mon is_tue	0.60	0.39	3.07	4.07
Regress: HOD + is_mon ... is_sat	0.86	0.62	1.84	3.41
Best CUSUM	0.46	0.12	3.71	4.64
Sick/availability with moving average W=1	**0.87**	0.63	1.70	2.97
Sick/availability with moving average W=7	0.84	**0.64**	**1.17**	2.42
Sick/availability with moving average W=14	0.80	0.62	1.30	2.37
Sick/availability with moving average W=28	0.76	0.60	1.45	**2.34**
Sick/availability with moving average W=56	0.63	0.52	1.76	2.47

Notes: Experiments are described in the text. In each column, the winning method is highlighted.

many interesting statistical details. Here, we provide pointers to other popular or promising approaches.

- Wavelets are another popular approach to smoothing time series data, and wavelets of varying resolutions can be used to model data properties at both broad and fine time resolutions. Examples used in biosurveillance include Zhang et al. (2003) and Goldenberg et al. (2002).
- Changepoint statistics (Carlstein, 1988, Buckeridge et al., 2005, Baron, 2002) are increasingly popular in the statistics literature for noticing when an underlying process changes.
- Kalman filters (Hamilton, 1994) or hidden Markov models (Rabiner, 1989) would be appropriate for biosurveillance time-series modeling. Madigan (2005) contains an overview of the use of hidden Markov models in surveillance.
- Much of this chapter used a Gaussian model for deriving confidence intervals and thus alarm levels. It is common to use different distributions such as a Poisson model.

ADDITIONAL RESOURCES

Montgomery (2001) is an excellent textbook on statistical quality control. Two useful references on time series methods are Chatfield (1989) and Hamilton (1994). Other surveys of methods for the early detection of disease outbreaks can be found in LeStrat (2005), Farrington and Andrews (2004), Sonesson and Bock (2003), Moore et al. (2003), and Wong (2004).

ACKNOWLEDGMENTS

Thanks to Robin Sabhnani for his help in developing this chapter.

REFERENCES

Baron, M. I. (2002). Bayes and asymptotically pointwise optimal stopping rules for the detection of influenza epidemics. In: Gastonis C, Kass RE, Carriquiry A, et al., eds. *Case Studies in Bayesian Statistics*, Vol. 6, 153–63. New York: Springer-Verlag.

Buckeridge, D. L., Burkom, H., Campbell, M., et al. (2005). Algorithms for rapid outbreak detection: a research synthesis. *J Biomed Inform* 38:99–113.

Carlstein, E. (1988). Nonparametric change-point estimation. *Ann Stats* 16:188–97.

Chatfield, C. (1989). *The Analysis of Time Series: An Introduction.* London: Chapman and Hall.

Farrington P., Andrews, N. (2004). Outbreak detection: application to infectious disease surveillance. In: Stroup, R. B., eds. *Monitoring the Health of Populations*, 203–31. New York: Oxford University Press.

Goldenberg, A., Shmueli, G., Caruana, R. A., et al. (2002). Early statistical detection of anthrax outbreaks by tracking over-the-counter medication sales. *Proc Natl Acad Sci U S A* 99:5237–40.

Hamilton, J. D. (1994). *Time Series Analysis*. Princeton, NJ: Princeton University Press.

LeStrat, Y. (2005). Overview of temporal surveillance. In: Lawson, A. B., Kleinman, K. eds. *Spatial and Syndromic Surveillance for Public Health*, 13–8. Chichester: John Wiley & Sons.

Madigan, D. (2005). Bayesian data mining for health surveillance. In: Lawson, A. B., Kleinman, K., eds. *Spatial and Syndromic Surveillance for Public Health*, 203–1. Chichester: John Wiley & Sons.

Montgomery, D. C. (2001). *Introduction to Statistical Quality Control.* New York: John Wiley and Sons.

Moore, A., Cooper, G., Tsui R., et al. (2003). *Summary of Biosurveillance-Relevant Technologies*. Pittsburgh,

PA: Realtime Outbreak and Disease Surveillance Laboratory, University of Pittsburgh.

Page, E. S. (1954). Continuous Inspection Schemes. *Biometrika* 41:100–15.

Rabiner, L. R. (1989). A tutorial on hidden Markov models and selected applications in speech recognition. *Proc IEEE* 77:257–85.

Roberts, S. W. (1959). Control-charts-tests based on geometric moving averages. *Technometrics* 1:239–50.

Sonesson C., Bock, D. (2003). A review and discussion of prospective statistical surveillance in public health. *J R Stat Soc* 166:5–2.

Wong, W.-K. (2004). Data Mining for Early Disease Outbreak Detection. Doctoral dissertation. Carnegie Mellon University.

Zhang, J., Tsui, F.-C., Wagner, M. M, et al. (2003). Detection of Outbreaks from Time Series Data Using Wavelet Transform. In: *Proceedings of American Medical Informatics Association Annual Symposium*, 748–52.

15

CHAPTER

Combining Multiple Signals for Biosurveillance

Andrew W. Moore
School of Computer Science, Carnegie Mellon University, Pittsburgh, Pennsylvania

Brigham Anderson
Robotics Institute, Carnegie Mellon University, Pittsburgh, Pennsylvania

Kaustav Das
Center for Automated Learning & Discovery, Carnegie Mellon University, Pittsburgh, Pennsylvania

Weng-Keen Wong
Center for Biomedical Informatics, University of Pittsburgh, Pittsburgh, Pennsylvania

1. INTRODUCTION

This chapter begins with two simple illustrations of the power of combining multiple data sources for surveillance. We then survey four representative multivariate approaches. The first two (multiple regression and the Hotelling T-squared test) are conventional and time-honored statistical approaches. Next, we describe how a famous probabilistic approach called hidden Markov models can be used for outbreak diagnosis from many data streams. Finally, we discuss WSARE, a system that searches for anomalous subsets of multivariate records.

2. THE IMPORTANCE OF MULTIPLE SOURCES OF DATA

Multiple independent sources of information can increase both sensitivity and specificity. Let us begin with a drastically oversimplified example. Suppose Sensor A has a daily 90% chance of signaling an attack (event SIGA) if one occurs, and a 1% chance of signaling an attack when there is none. Suppose sensor B monitors an independent data source, and has a daily 90% chance of signaling an attack (event SIGB) if one occurs, and a 1% chance of signaling an attack when there is none. Writing ATT as the event of attack, we have:

P(SIGA|~ATT) = 0.01
P(SIGA|ATT) = 0.9
P(SIGB|~ATT) = 0.01
P(SIGB|ATT) = 0.9

Then there is now a 99% chance that at least one detector will signal if there is an attack, and there is only a probability of 1 in 10,000 that both detectors will signal if there is no attack. Thus, in many situations (both the [SIGA and SIGB] case and the [~SIGA and ~SIGB] case), the operational decision is much clearer. Even in the case of inconsistent signals, the analysis task can be better informed.

Another compelling example concerns time series analysis. Figure 15.1 shows two times series, for daily sales of two fictional products in a fictional city. Apart from a general upward trend, no serious anomalies stand out.

If, however, we look at the same data in a different way, February 20 stands out as somewhat anomalous. The new view is a scatterplot. For each date, it plots one data point, with the x-coordinate denoting sales of Product A, and the y-coordinate denoting sales of Product B. There is a general correlation in sales, but February 20 is atypical because B sales are high, *taking into account A's sales.*

In this chapter, we briefly survey methods which can notice effects that are revealed by inspecting more than one time series at a time.

3. COMBINING MULTIPLE TIME SERIES USING REGRESSION ANALYSIS

Multiple regression is a statistical technique that can be used to analyze the relationship between a single dependent variable and several independent variables. The objective of multiple regression analysis is to use the independent variables whose values are known to predict the value of the single dependent value. Each predictor value is weighed, the weights denoting their relative contribution to the overall prediction.

$$Y = a + b_1X_1 + b_2X_3 + \ldots + b_nX_n \quad (1)$$

Here Y is the dependent variable, and $X_1,\ldots,X_n$ are the *n* independent variables. In calculating the weights, a, $b_1,\ldots,b_n$, regression analysis ensures maximal prediction of the dependent variable from the set of independent variables. This is usually done by least squares estimation.

This approach can be applied to analyze multivariate time series data when one of the variables is dependent on a set of other variables. We can model the dependent variable Y on the set of independent variables. At any time instant when we are given the values of the independent variables, we can predict the value of Y from Eq. 1.

ISBN 0-12-369378-0

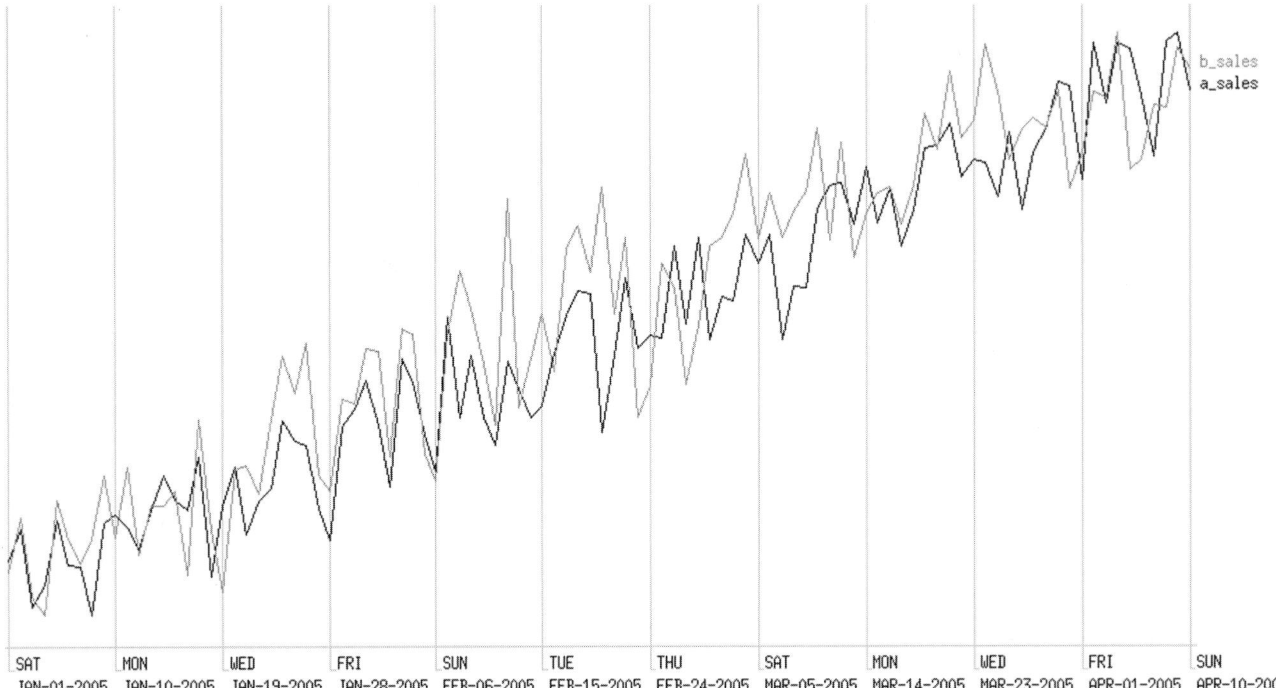

FIGURE 15.1 Synthetic time series for sales of two products.

In time series analysis, it is possible to do regression analysis against a set of past values of the variables. This is known as autoregression (AR). Let us consider *n* variables. We have a time series corresponding to each variable. At time *t*, the vector Z_t represents the values of the *n* variables. The general autoregressive model assumes that Z_t can be represented as:

$$Z_t = A_1 Z_{t-1} + A_2 Z_{t-2} + \dots + A_p Z_{t-p} + E_t \tag{2}$$

where each A_i (an n × n matrix) is the autoregression coefficient. Z_t is the column vector of length *n*, denoting the values of the time series variables at time *t*. *p* is the order of the filter which is generally much less than the length of the series. The noise term or residual, E_t, is almost always assumed to be Gaussian white noise.

In a more general case, we can consider that the values $Z_{t-1}, Z_{t-2} \dots Z_{t-p}$ are themselves noisy values. Adding the noise values to Eq. 2, we get the ARMA (autoregressive moving average) equation:

$$\begin{aligned} Z_t = {} & A_1 Z_{t-1} + A_2 Z_{t-2} + \dots + A_p Z_{t-p} \\ & + E_t - B_1 E_{t-1} - B_2 E_{t-2} \dots - B_p E_{t-p} \end{aligned} \tag{3}$$

Here $B_1 \dots B_p$, (each an n × n matrix), are the MA coefficients. These coefficients can be determined using the standard Box-Jenkins methodology (Box et al., 1994).

AR and ARMA provide a nice way to predict what should be happening to all of the time series simultaneously, and they allow one time series to use other series to increase their accuracy.

4. COMBINING MULTIPLE TIME SERIES USING THE HOTELLING STATISTIC

The Hotelling T^2 test provides a statistical test of the deviation of the recent mean of a set of signals from the current time period relative to their expected values, accounting for known correlation between the signals. Let us consider a data set with *p* variables and *n* samples from the recent past. Let X_i, a vector of length *p*, denote the *i*th sample. The Hotelling T^2 statistic is defined as:

$$T^2 = n(\bar{\mathbf{X}} - \boldsymbol{\mu}_0)\ S^{-1}(\bar{\mathbf{X}} - \boldsymbol{\mu}_0) \tag{4}$$

where $\bar{\mathbf{X}}$ is the sample mean vector of the *n* recent samples, S is the sample variance—covariance matrix and μ_0 is the expected mean vector. For example, in Figure 15.2, February 20 would be the data point with the largest T^2 value.

It is well known that T^2 is distributed as $\frac{p(n-1)}{n-p} F_{(p,n-p)}$, where F represents the Fisher F distribution.

The Hotelling T^2 test can be applied to a multivariate time series to detect whether a *new* datapoint has a substantial deviation from an expected value. The expected value μ_0 and the covariance matrix S can be computed from historical data. For example, we can take the mean and covariance of all the data seen so far. A more sophisticated approach, which

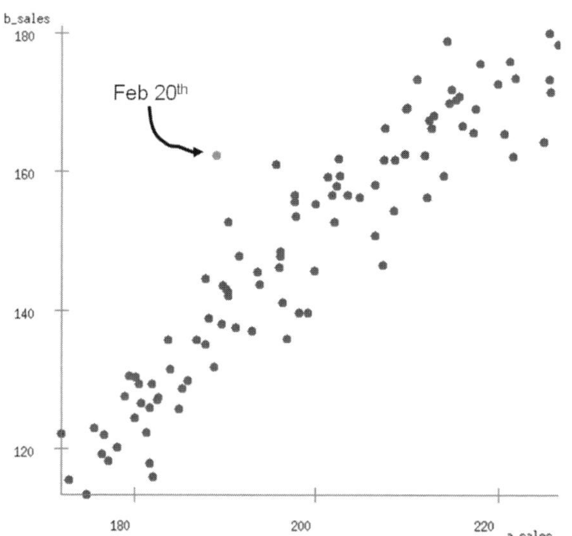

FIGURE 15.2 A scatterplot of sales reveals an outlier.

accounts for gradual drift in the process is to model the time series (using AR or MA as mentioned before) and compute μ_0 and S from this model.

At any time t, we consider the past n values of the p variables. Let $\bar{\mathbf{X}}$, a vector of size p, be the sample mean of these n values. And let μ_0 be the expected mean vector. We can test the null hypothesis that $\bar{\mathbf{X}} = \mu_0$ by computing T^2 (from Eq. 4), and comparing with $\frac{p(n-1)}{n-p} F_{\alpha(p,n-p)}$ for a suitably chosen α. If the T^2 statistic exceeds the value of $\frac{p(n-1)}{n-p} F_{\alpha(p,n-p)}$, we can reject the null hypothesis with a confidence of (1-α) and signal an alarm.

Several biosurveillance research groups have made significant successful use of Hotelling methods. Good examples, within the ESSENCE framework, are described in Burkom et al. (2004, 2005).

5. COMBINING MULTIPLE TIME SERIES USING PROBABILITY

Hidden Markov models (HMMs) are fundamental tools in fields such as bioinformatics, signal processing, and machine learning. In biosurveillance, Le Strat and Carrat (1999) used HMMs to monitor influenza-like illnesses and poliomyelitis using a Gaussian observation model. Rath et al. (2003) improved their model by, among other things, replacing the Gaussian observation model with an exponential distribution. Other applications include Cooper and Lipsitch (2004) modeling hospital infections with HMMs. Madigan (2005) reviewed the literature and discussed issues such as model selection and random observation-time HMM. He also noted the excellent fit between the capabilities of HMMs and the requirements of multivariate time series.

Figure 15.3 shows an HMM. Arrows indicate causality, shaded variables are observed, and the unshaded state variables

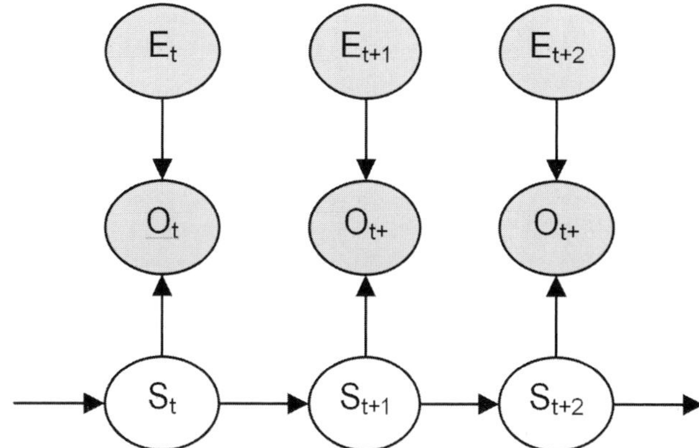

FIGURE 15.3 Hidden Markov model structure.

are unobserved. At time t, the underlying disease state is S_t, which has N possible values. The observation O_t is a vector of K observations, which can assume count values, such as specific over-the-counter (OTC) sales, ER visits, and school absenteeism. The third variable type is for environmental influences, E_t, on the observations. The environment variable is a vector of categorical variables, and has no effect on the disease state.

5.1. States

The most common epidemiologic HMM is binary; it has two states: disease and no-disease. It is, however, quite easy to track the stages of multiple diseases by assigning one state to every possible combination of disease. Thus, if we are tracking four diseases with three stages each, the HMM will have $3^4 + 1 = 82$ states. The number of states is much reduced if we assume that especially rare diseases will not co-occur.

5.2. Transitions

The word "Markov" in "hidden Markov model" indicates the assumption that state at time t depends only on the state at time t–1. The probability of transitioning from one state to another is thus governed by the transition model $P(S_{t+1}|S_t)$, which is an $N \times N$ matrix derived analytically from estimates of prior probabilities and estimates of disease stage lengths.

5.3. Observations

At a high level, the observation model described how we expect the signals to be affected by the disease state and environmental conditions. For example, if the observation vector includes cough/cold sales as one of its components, then the observation model might predict sales of one cough product per household per six months under normality, but one per household per two months during an influenza outbreak.

5.4. Environment

The counts in the observation vectors are driven by more than the disease state. Additional observed information, such as day of week, weather, promotions (in the case of retail data), vacations (for school absenteeism), and other external factors are expected to influence the observed counts. For example, the HMM could expect, with high probability, to observe zero counts for school absenteeism on public holidays, and a 20% increase in OTC sales at stores with promotional activity.

The following simulation illustrates the benefits of fusion of multiple data streams. For details of such an HMM approach, see the references above. The data in the first three panels of Figure 15.4 were generated from a vastly oversimplified HMM where there are three possible disease states: "none," "influenza," and "allergy." The simulated data streams are ED visits, school absenteeism, and OTC antihistamine sales. In this example, the ED visits have a strong weekend effect, while weekend data are missing for school absenteeism. "Influenza" is assumed to double the rates of all three data streams. "Allergy" also doubles OTC antihistamines. The population is in state "allergy" during time steps 10–20, and in "influenza" for time steps 30 to 40. A three-state HMM was applied to the data and the inference results are shown in the bottom panel of Figure 15.4. Inference involves computing the disease state probabilities (which are not in the observed data) from the counts and environmental factors (which *are* in the observed data). Note that the HMM is not told when allergy is present.

In Figure 15.4, note that the model was successful in explaining the jump in OTC sales and missing absenteeism data did not noticeably impair inference. Individually, these data streams would be inconclusive, or worse, misleading, but the proper statistical model effectively fuses them into a robust and highly informative resource. This example is *very simplified*, but illustrates several of the components of a realistic multivariate system.

6. COMBINING FIELDS IN EVENT DATA: THE "WHAT'S STRANGE ABOUT RECENT EVENTS" APPROACH

The what's strange about recent events (WSARE) algorithm (Wong et al., 2002, Wong and Moore, 2002, Wong et al., 2003) is a rule-based anomaly pattern detector that operates on discrete, multidimensional data sets with a temporal component. This algorithm compares recent data against a baseline distribution with the aim of finding rules that summarize significant patterns of anomalies. Each rule is made up of components of the form $X_i = V_i^j$, where X_i is the ith feature and V_i^j is the jth value of that feature. Multiple components are joined together by a logical AND. For example, a

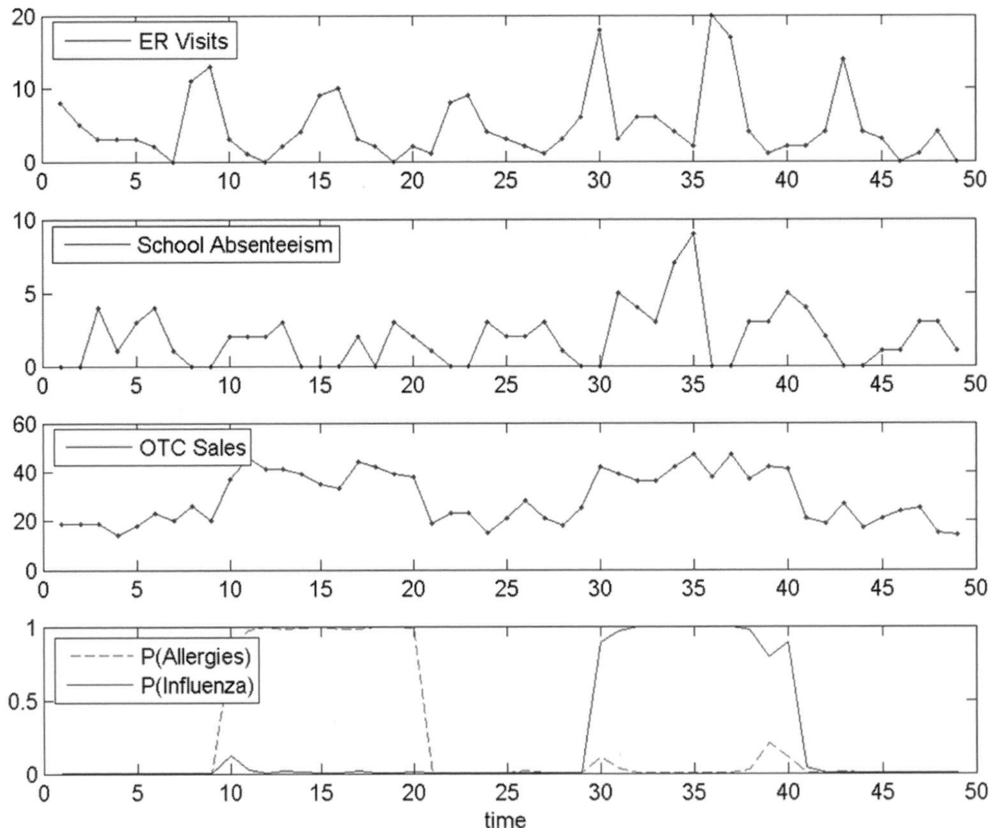

FIGURE 15.4 Synthetic time series and diagnostic probabilities described in the text.

two-component rule would be *Gender = Male AND Home Location = NW*. These rules should not be interpreted as rules from a logic-based system in which the rules have an antecedent and a consequent. Rather, these rules can be thought of as SQL SELECT queries because they identify a subset of the data having records with attributes that match the components of the rule. WSARE finds these subsets whose proportions have changed the most between recent data and the baseline.

We will overview versions 2.0 and 3.0 of the WSARE algorithm. These two algorithms only differ in how they create the baseline distribution; all other steps in the WSARE framework remain identical. WSARE 2.0 uses raw historical data from selected days as the baseline while WSARE 3.0 models the baseline distribution using a Bayesian network.

6.1. What's Strange About Recent Events

The basic question asked by anomaly detection systems is whether anything strange has occurred in recent events. This question requires defining what it means to be recent and what it means to be strange. The WSARE algorithm (Wong and Moore, 2002, Wong et al., 2002, Wong et al., 2003) is one example of an algorithm that attempts to do this. It defines all patient records falling on the current day under evaluation to be recent events. Note that this definition of recent is not restrictive—the approach is fully general and "recent" can be defined to include all events within some other time period such as over the last six hours. In order to define an anomaly, we need to establish the concept of something being normal. In WSARE version 2.0, baseline behavior is assumed to be captured by raw historical data from the same day of week in order to avoid environmental effects such as weekend versus weekday differences in the number of ED cases. This baseline period must be chosen from a time period similar to the current day. This can be achieved by being close enough to the current day to capture any seasonal or recent trends. On the other hand, the baseline period must also be sufficiently distant from the current day. This distance is required in case an outbreak happens on the current day but it remains undetected. If the baseline period is too close to the current day, the baseline period will quickly incorporate the outbreak cases as time progresses. In the description of WSARE 2.0 below, we assume that baseline behavior is captured by records that are 35, 42, 49, and 56 days prior to the day under consideration. We would like to emphasize that this baseline period is only used as an example; it can be easily modified to another time period without major changes to the algorithm. In a later section, we will illustrate how version 3.0 of WSARE automatically generates the baseline using a Bayesian network.

We will refer to the events that fit a certain rule for the current day as C_{recent}. Similarly, the number of cases matching the same rule from the baseline period will be called $C_{baseline}$. As an example, suppose the current day is Tuesday

November 2003

S	M	Tu	W	Th	F	S
						1
2	3	4	5	6	7	8
9	10	11	12	13	14	15
16	17	18	19	20	21	22
23	24	25	26	27	28	29
30						

FIGURE 15.5 The baseline for WSARE 2.0 when the current day is December 30, 2003.

December 30, 2003. The baseline used for WSARE 2.0 will then be November 4, 11, 18 and 25 of 2003 as seen in Figure 15.5. These dates are all from Tuesdays in order to avoid day of week variation.

6.2. One-Component Rules in WSARE

In order to illustrate this algorithm, suppose we have a large database of one million ED records over a two-year span. This database contains roughly 1370 records a day. Suppose we treat all records within the last 24 hours as "recent" events. In addition, we can build a baseline data set out of all cases from exactly 35, 42, 49, and 56 days prior to the current day. We then combine the recent and baseline data to form a record subset called *DB*, which will have approximately 5000 records. The algorithm proceeds as follows. We first consider all possible one-component rules. For every possible feature—value combination, obtain the counts C_{recent} and $C_{baseline}$ from the data set *DB*. As an example, suppose the

TABLE 15.1 Sample 2 × 2 Contingency Table

	C_{recent}	$C_{baseline}$
Home location = NW	6	496
Home location ≠ NW	40	9504

feature under consideration is the *Home_Location* for the ED case. Imagine (for simplicity) that there are four possible *Home_Location* values: NE, NW, SE, SW. We start with the rule *Home_Location* = *NW* and count the number of cases for the current day that have *Home_Location* = *NW* and those that have *Home_Location* ≠ *NW*. The cases from five to eight weeks ago are subsequently examined to obtain the counts for the cases matching the rule and those not matching the rule. The four values form a 2 × 2 contingency table such as the one shown in Table 15.1.

Each contingency table is scored according to whether there is a significant increase in the fraction of records matching the rule in the recent data compared with the historical data. There many possible scoring functions: these are discussed in (Wong et al., 2003). By default, WSARE uses the Fisher exact test, described in the same paper.

WSARE searches over all possible rules that can be made out of attributes of the database records. In a typical ED data set the rules would thus include both general and specific examples, for example: HomeLocation=NW, Gender=Male, AgeGroup=Over60, Syndrome=GI, InternalBleeding=True, HomeZip=12345.

There are many anomalous scenarios that would not be detected by single-component rules. For example, there might be a slight increase in pediatric cases throughout the city and a slight increase in cases from zip code 54321, but a dramatic increase in pediatric cases from 54321. For this reason, WSARE also searches for rules made up of multiple components.

6.3. Finding the *p*-Value for a WSARE Rule

When reporting the significance of the highest scoring rule, it is essential that we take into account the intensity of WSARE's search for anomalies. Even if surveillance data were generated randomly, entirely from a null distribution, we can expect that the best rule would be alarmingly high-scoring if we had searched over 1000 possible rules. In order to illustrate this point, suppose we follow the standard practice of rejecting the null hypothesis when the *p*-value is $>\alpha$, where $\alpha = 0.05$. In the case of a single hypothesis test, the probability of a false positive under the null hypothesis would be α, which equals 0.05. On the other hand, if we perform 1000 hypothesis tests, one for each possible rule under consideration, then the probability of a false positive could be as bad as $1 - (1 - 0.05)^{1000} \approx 1$, which is much greater than 0.05 (Miller et al., 2005). Thus, if the algorithm returns a large score, we cannot accept it at face value without adding an adjustment for the multiple hypothesis tests we performed. This problem can be addressed using a Bonferroni correction (Bonferroni, 1936), but this approach would be unnecessarily conservative. Instead, we turn to a randomization test in which the date and each ED case features are assumed to be independent. In this test, the case features in the data set *DB* remain the same for each record but the date field is shuffled between records from the current day and records from five to eight weeks ago. The full method for the randomization test is shown below.

Let *UCS* = Uncompensated score: the score of the best scoring rule *BR* as defined above.

For $j = 1$ to 1000:

Let $DB^{(j)}$ = newly randomized data set
Let $BR^{(j)}$ = Best rule on $DB^{(j)}$
Let $UCS^{(j)}$ = Uncompensated *p*-value of $BR^{(j)}$ on $DB^{(j)}$

Let the compensated *p*-value of *BR* be

$$CPV = \frac{\text{\# of Randomized Tests in which } UCP^{(j)} < UCP}{\text{\# Randomized Tests}}$$

CPV is an estimate of the chance that we would have seen an uncompensated score as large as *UCS* if in fact there was no relationship between date and case features. Note that we do not use the uncompensated score *UCS* after the randomization test. Instead, the compensated *p*-value *CPV* is used do determine the level of anomaly.

6.4. WSARE 3.0

Many detection algorithms (Goldenberg et al., 2002, Zhang et al., 2003, Fawcett and Provost, 1997) assume that the observed data consist of cases from background activity, which we will refer to as the baseline, plus any cases from irregular behavior. Under this assumption, detection algorithms operate by subtracting away the baseline from recent data and raising an alarm if the deviations from the baseline are significant. The challenge facing all such systems is to estimate the baseline distribution using data from historical data. In general, determining this distribution is extremely difficult due to the different trends present in surveillance data. Seasonal variations in weather and temperature can dramatically alter the distribution of surveillance data. For example, influenza season typically occurs during mid-winter, resulting in an increase in ED cases involving respiratory problems. Disease outbreak detectors intended to detect epidemics such as severe acute respiratory syndrome (SARS), West Nile virus and anthrax are not interested in detecting the onset of flu season and would be confused by influenza. Day-of-week variations make up another periodic trend.

In WSARE 2.0, we made the baseline distribution to be raw data obtained from selected historical days. WSARE 3.0 instead learns the baseline distribution from historical data in order to better account for observed environmental factors for the current day (e.g., public holiday, day of week, weather). WSARE 3.0 takes all records prior to the past 24 hours and builds a Bayesian network from this subset.

Bayesian networks are overviewed in Chapters 13 and 18, and an introductory tutorial is available at *www.cs.cmu.edu/~awm/tutorials/ bayesnet.html*. During the structure learning, we differentiate between *environmental attributes*, which are features such as the season and the day of week that cause trends in the data and *response attributes*, which are the remaining features. The environmental attributes are specified by the user based on the user's knowledge of the problem domain.

6.5. WSARE Results

Our evaluation criteria examines algorithm performance on an AMOC curve (Fawcett and Provost, 1997). AMOC curves are described in more detail in Chapter 20. These results were generated from 100 synthetic data sets described in the WSARE paper (Wong et al., 2003), and are available at *www.cs.cmu.edu/~awm/wsare-data*. The AMOC curve in Figure 15.6 plots detection time versus false positives over alarm thresholds ranging from 0 to 0.2 in 0.001 increments. The lower alarm thresholds yield lower false positives and higher detection times while the converse is true with higher alarm thresholds. The best possible detection time is one day, as shown by the dotted line at the bottom of the graph. We add a one-day delay to all detection times to simulate reality where current data is only available after a 24-hour delay. Any alert occurring before the start of the simulated anthrax attack is treated as a false positive. Detection time is calculated as the first alert raised after the release date. If no alerts are raised after the release, the detection time is set to 14 days.

The Israel Center for Disease Control evaluated WSARE 3.0 retrospectively using an unusual outbreak of influenza type B that occurred in an elementary school in central Israel (Kaufman et al., 2004). WSARE 3.0 was applied to patient visits to community clinics between the dates of May 24, 2004 to June 11, 2004. The attributes in this data set included the visit date, area code, ICD-9 code, age category, and day of week. The day of week was used as the only environmental attribute. WSARE 3.0 reported two rules with p-values at 0.002 and five other rules with p-values below 0.0001. Two of the five anomalous patterns with p-values below 0.0001 corresponded to the influenza outbreak in the data. The rules that characterized the two anomalous patterns consisted of the same three attributes of ICD-9 code, area code and age category, indicating that an anomalous pattern was found involving children aged 6 to 14 having viral symptoms within a specific geographic area. WSARE 3.0 successfully detected the outbreak on the second day from its onset.

6.6. WSARE Conclusions

The WSARE algorithms approach the problem of early outbreak detection on multivariate surveillance data using two key components. In WSARE 2.0, the main component is an association rule search, which is used to find anomalous patterns between a recent data set and a baseline data set.

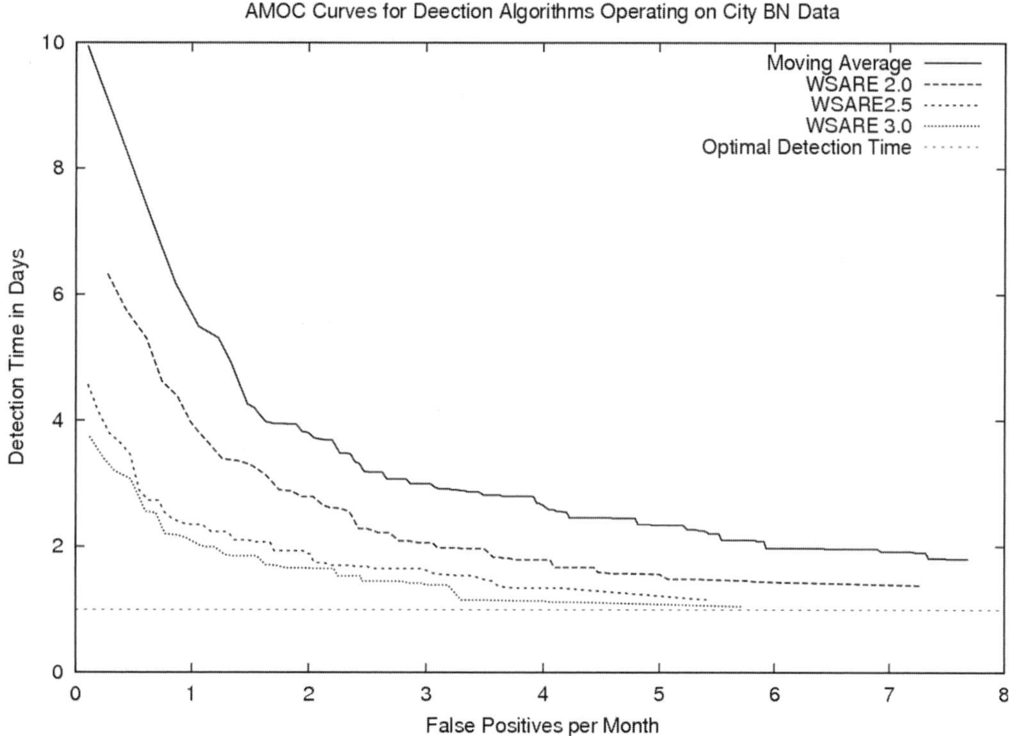

FIGURE 15.6 Asymptotic behavior of algorithms for simulated data.

The contribution of this rule search is best seen by considering the alternate approach of monitoring a univariate signal. If an attribute or combination of attributes is known to be an effective signal for the presence of a certain disease, then a univariate detector that monitors this signal will be an effective early warning system for that specific disease. However, if such a signal is not known a priori, then the association rule search will determine which attributes are of interest. We intend WSARE to be a general purpose safety net to be used in combination with a suite of specific disease detectors. The key to this safety net is to perform nonspecific disease detection and notice any unexpected patterns.

7. FURTHER MULTIVARIATE APPROACHES

A number of other multivariate methods for syndromic surveillance have been proposed. Burkom et al. (2004, 2005) discuss many such methods, including multivariate versions of the exponentially weighted moving average and cumulative sum methods. These methods are compared to methods that combine the output of multiple univariate detectors, for example using a Bayesian network.

ADDITIONAL RESOURCES

WSARE and additional biosurveillance software are available from *www.autonlab.org*. Example synthetic data sets are available from *www.cs.cmu.edu/~awm/wsare-data*.

REFERENCES

Bonferroni, C. E. (1936). *Pubblicazioni del R Istituto Superiore di Scienze Economiche e Commerciali di Firenze* 8:3–62.

Box, G., Jenkins, G. M., Reinsel, G. (1994). *Time Series Analysis: Forecasting and Control*. Englewood Cliffs, NJ: Prentice Hall.

Burkom, H. S., et al. (2005). *Public health monitoring tools for multiple data streams.* MMWR Morb Mortal Wkly Rep Aug 26;54 (Suppl): 55–62.

Burkom, H. S., Elbert, Y., Feldman A., et al. (2004). *The role of data aggregation in biosurveillance detection strategies with applications from ESSENCE.* MMWR Morb Mortal Wkly Rep 53 (Suppl):67–73.

Cooper, B. and M. Lipsitch (2004). *The analysis of hospital infection data using hidden Markov models.* Biostatistics 5, 223–237.

Cooper, G.F., D. H. Dash, J. D. Levander, W-K Wong, W. R. Hogan, and M.M. Wagner (2004). *Bayesian Biosurveillance of Disease Outbreaks*. In *Proceedings of the Conference of Uncertainty in Artificial Intelligence,* 94–103.

Fawcett, T., Provost, F. (1997). Adaptive fraud detection. Data Mining and Knowledge Discovery. 1:291–316.

Goldenberg, A., Shmueli, G., Caruana, RA, Fienberg SE. (2002). Early statistical detection of anthrax outbreaks by tracking over-the-counter medication sales. Proceedings of the National Academy of Sciences. 99 (8):5237–40.

Kaufman, Z., Cohen, E., Peled-Leviatan, T., et al. (2004). *Evaluation of Syndromic Surveillance for Early Detection of Bioterrorism Using a Localized, Summer Outbreak of Influenza.* In Proceedings of the Third National Syndromic Surveillance Conference.

Le Strat, Y. and Carrat, F. (1999). *Monitoring epidemiologic surveillance data using hidden Markov models.* Statistics in Medicine. 18: 3463–3478.

Madigan, D. (2005). *Bayesian Data Mining for Health Surveillance, in Spatial and Syndromic Surveillance for Public Health* (A.B. Lawson and K. Kleinman, Editors). John Wiley & Sons Ltd, West Sussex, England, pp. 203–221.

Miller, C. J., Genovese, C., Nichol, R. C., et al. (2001). Controlling the false discovery rate in astrophysical data analysis. The Astronomical Journal, 122(6):3492–3505.

Moore A, Wong W-K (2003). Optimal reinsertion: A new search operator for accelerated and more accurate Bayesian network structure learning. In Proceedings of the 20th International Conference of Machine Learning. Menlo Park, CA:AAAI Press; 552–9.

Rath, T. M., Carreras, M., Sebastiani, P. (2003). *Automated detection of influenza epidemics with hidden Markov models, in Advances in intelligent Data Analysis V* (M. R. Berthold, H.-J. Lenz, E. Bradley, R. Kruse, C. Borgelt and F. Pfenning, Editors) Springer-Veriag: Berlin. p 521–531.

Wong, W. K. *Data Mining for Early Disease Outbreak Detection.* (2004). Doctoral dissertation. (School of Computer Science) Carnegie Mellon University, Pittsburgh.

Wong, W. K. and A. Moore (2004). *Efficient Algorithms for Non-Parametric Clustering with Clutter*, in *Proceedings of the 34th Interface Symposium.*

Wong, W-K, Moore A, Cooper G, Wagner M (2002). Rule-based anomaly pattern detection for detecting disease outbreaks. In Proceedings of the 18th National Conference on Artificial Intelligence. Menlo Park, CA: AAAI Press; 217–23.

Wong W-K, Moore A, Cooper G, Wagner M. (2003). Bayesian network anomaly pattern detection for disease outbreaks. In Proceedings of the 20th International Conference on Machine Learning. Menlo Park, CA: AAAI Press; 2003. 808–15

Zhang J, Tsui F-C, Wagner MM, Hogan WR (2003). Detection of outbreaks from time series data using wavelet transform. In Proceedings of AMIA Fall Symposium: Omni Press CS; 748–52.

16

CHAPTER

Methods for Detecting Spatial and Spatio-Temporal Clusters

Daniel B. Neill and Andrew W. Moore

School of Computer Science, Carnegie Mellon University, Pittsburgh, Pennsylvania

1. INTRODUCTION

This chapter focuses on the detection of spatial clusters of disease, with the goal of rapidly detecting emerging outbreaks by prospective surveillance. Spatial cluster detection has two main goals: identifying the locations, shapes and sizes of potential disease clusters, and determining whether each of these potential clusters is due to a genuine outbreak or due to chance fluctuations in case counts. In other words, we want to know whether anything unexpected is going on, and if so, where? This task can be broken down into two parts: first, figuring out what we expect to see, based on populations or on expected counts inferred from historical data, and then determining which regions deviate significantly from our expectations. In this chapter, we present an overview of spatial cluster detection, and then discuss a number of cluster detection methods, focusing on the spatial scan statistic. In addition to presenting the standard spatial scan framework, we consider a number of extensions to this framework, including generalization of the scan statistic to situations in which baselines must be inferred from historical data, computational methods for fast spatial scanning, and extensions to spatio-temporal cluster detection. This chapter does three things. First, it provides the basic statistical and computational tools to make the spatial scan applicable and useful for analyzing large real-world data sets. Second, it motivates cluster detection approaches and third, it compares them to other outbreak detection methods.

Epidemiologists have been analyzing biosurveillance data spatially since the seminal work of John Snow on cholera (Snow, 1855). In the 1970s, researchers automated the map creation aspect of spatial analysis. The results of this work—geographic information systems—are now in widespread use in health departments. Over the past decade, researchers have developed spatial scan statistics, which automate the pattern recognition component of spatial analysis. This advance enables a biosurveillance organization to analyze spatial data far more exhaustively than ever before.

2. OVERVIEW OF SPATIAL CLUSTER DETECTION

In this chapter, we focus on the task of *spatial cluster detection*: finding spatial areas where some monitored quantity is significantly higher than expected. For example, we may want to monitor the observed number of cases of influenza, or some other specific type of disease, and find any regions where the number of cases is abnormally high. The spatial cluster detection techniques that we describe are disease independent; that is, they are capable of detecting clusters of any type of disease including those of previously unknown diseases.

At present, health departments are using automatic spatial cluster detection primarily on syndromic data with a goal of detecting regions in a city or even a country with abnormally high case counts of some syndrome (e.g., respiratory), based on observed quantities such as the number of emergency department visits or sales of over-the-counter cough and cold medication. The detected clusters of disease may be indicative of a naturally occurring outbreak, a bioterrorist attack (e.g., anthrax release) or an environmental hazard.

The main goals of spatial scanning are to identify the locations, shapes, and sizes of potential clusters (i.e., pinpointing those areas which are most relevant), and to determine whether each of these potential clusters is more likely to be a "true" cluster (requiring further investigation by public health officials) or simply a chance occurrence (which can safely be ignored).

As mentioned previously, the spatial cluster detection task involves the two questions: Is anything unexpected going on? If so, where? In order to answer these questions, we must first have some idea of what we expect to see. We typically take one of two approaches, illustrated in Figure 16.1. In the *population-based* approach, we estimate an at-risk population for each area (e.g., zip code); this population can either be estimated simply from census data, or can be adjusted for a variety of covariates (patients' age and gender, seasonal and day of week effects, etc.). Then the expected number of cases in an area is assumed to be proportional to its at-risk population, and thus clusters are areas where the disease rate (number of cases per unit population) is significantly higher inside the region than outside. The *expectation-based* approach directly estimates the number of cases we expect to see in each area; typically by fitting a model based on past data (e.g., the number of cases in each area on each previous day). Wagner et al. (2003) describes such an approach. Using an expectation-based method, clusters are areas where the number of cases is significantly greater than its expectation. One important difference between these two approaches is their response to a global increase (i.e., where the number of cases increases in the entire area being monitored). The expectation-based approach

Handbook of Biosurveillance
ISBN 0-12-369378-0

Population-based

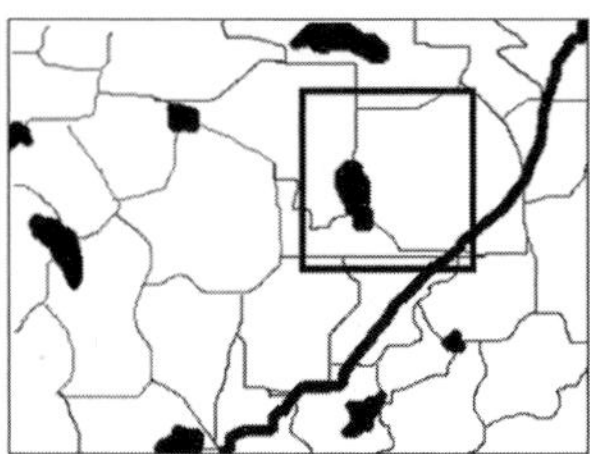

In region:
Population = 50000
# of cases = 150
Disease rate = 0.003

Out of region:
Population = 1 million
# of cases = 1000
Disease rate = 0.001

Expectation-based

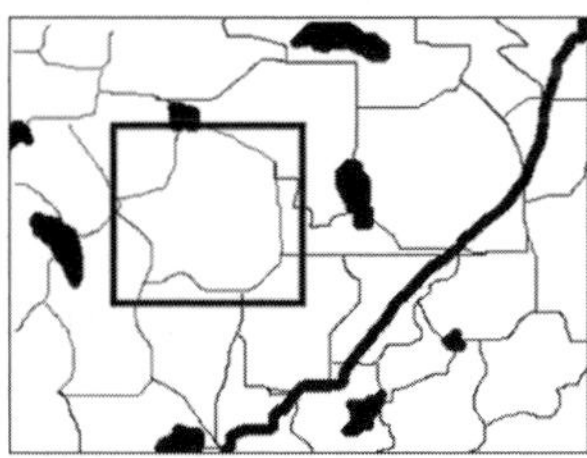

In region:
# of cases = 200
Expected cases = 100

FIGURE 16.1 Population-based versus expectation-based scan statistics.

will find this increase very significant because the counts everywhere are higher than expected. However, the population-based approach will only find the increase significant if there is spatial variation in the amount of increase; otherwise the ratio of disease rate inside the region to disease rate outside the region remains constant, so no significant increase is detected. We discuss these approaches in more detail below.

For now, let us consider the expectation-based approach, assuming that we are given the observed number of cases c, as well as an estimated mean μ and standard deviation σ, for each zip code. How can we tell whether any zip code has a number of cases that is significantly higher than expected? One simple possibility would be to perform a separate statistical test for each zip code: for example, we might want to detect all zip codes with observed count more than three standard deviations above the mean. However, there are two main problems with this simple method. First, treating each zip code separately prevents us from using information about the spatial proximity of adjacent zip codes. For instance, while a single zip code with count two standard deviations above the mean might not be sufficiently surprising to trigger an alarm, we would probably be interested in detecting a cluster of four adjacent zip codes each with count two standard deviations above the mean. Thus, the first problem with performing separate statistical tests for each zip code is reduced power to detect clusters spanning multiple zip codes: we cannot detect such increases unless the amount of increase is so large as to make each zip code individually significant. A second, and somewhat more subtle, problem is that of *multiple hypothesis testing*. We typically perform statistical tests to determine if an area is significant at some fixed level α, such as $\alpha = 0.05$, which means that if there is no abnormality in that area (i.e., the "null hypothesis" of no clusters is true) our probability of a false alarm is at most α. A lower value of α results in less false alarms, but also reduces our chance of detecting a true disease cluster. Now let us imagine that we are searching for clusters in a large area containing 1000 zip codes, and that there happen to be no outbreaks today, so any areas we detect are false alarms. If we perform a separate significance test for each zip code, we expect each test to trigger an alarm with probability $\alpha = 0.05$. But because we are doing 1000 separate tests, our expected number of false alarms is $1000 \times 0.05 = 50$. Moreover, if these 1000 tests were independent, we would expect to get at least one false alarm with probability $1-(1-0.05)^{1000} \approx 1$. Of course, counts of adjacent zip codes are likely to be correlated, so the assumption of independent tests is not usually correct. The main point here, however, is that we are almost certain to get false alarms every day, and the number of such false alarms is proportional to the number of tests performed. One way to correct for multiple tests is the *Bonferroni correction* (Bonferroni, 1935): If we want to ensure that our probability of getting any false alarms is at most α, we report only those regions which are significant at level α/N, where N is the number of tests. The problem with the Bonferroni correction is that it is too conservative, thus reducing the power of the test to detect true outbreaks. In our example, with $\alpha = 0.05$ and $N = 1000$, we only signal an alarm if a region's statistical significance (p-value) is less than 0.00005, and thus only very obvious outbreaks can be detected.

As an alternative to this simple method, we can choose a set of regions to search over, where each region consists of a set of one or more zip codes. We can define the set of regions based on what we know about the size and shape of potential outbreaks; we can either fix the region shape and size, or let these vary as desired. We can then do a separate test for each region rather than for each zip code. This resolves the first problem of the previous method: assuming we have chosen the set of regions well, we can now detect attacks whether they affect a single zip code, a large number of zip codes, or anything in between. However, the disadvantage of this method is that it makes the multiple hypothesis testing problem even worse: the number of regions searched, and thus the number of tests performed, is typically much larger than the number of zip codes. In principle, the number of regions could be as high as 2^Z, where Z is the number of zip codes, but in practice the number of regions searched is much smaller (because we want to enforce constraints on the connectedness, size, and shape of regions). For example, if we consider circular regions centered at the centroid of some zip code, with continually varying radius (assuming that a region contains all zip codes with centroids inside the circle), the number of distinct regions is proportional to Z^2. For the example above, this would give us one million regions to search, creating a huge multiple hypothesis testing problem; less restrictive constraints (such as testing ellipses rather than circles) would require testing an even larger number of regions.

This method of searching over regions, without adjusting for multiple hypothesis testing, was first used by Openshaw et al. (1988) in their geographical analysis machine (GAM). Openshaw et al. test a large number of overlapping circles of fixed radius, and draw all of the significant circles on a map; Figure 16.2 gives an example of what the output of the GAM might look like. Because we expect a large number of circles to be drawn even if there are no outbreaks present, the presence of detected clusters is not sufficient to conclude that there is an outbreak. Instead, the GAM can be used as a descriptive tool for outbreak detection: whether any outbreaks are present, and the location of such outbreaks, must be inferred manually from the number and spatial distribution of detected clusters. For example, in Figure 16.2 the large number of overlapping circles in the upper right of the figure may indicate an outbreak, while the other circles might be due to chance. The problem is that we have no way of determining whether any given circle or set of circles is statistically significant, or whether they are due to chance and multiple testing; it is also difficult to precisely locate those clusters which are most likely to correspond to true outbreaks. Besag and Newell (1991) propose a related approach, where the search is performed over circles containing a fixed number of cases; this approach also suffers from the multiple hypothesis testing problem, but again is valuable as a descriptive method for visualizing potential clusters.

The *scan statistic* was first proposed by Naus (1965) as a solution to the multiple hypothesis testing problem. Let us assume we have a score of some sort for each region, such as the *Z-score*, $Z = (c - \mu)/\sigma$. The Z-score is the number of standard deviations that the observed count c is higher than the expected count μ; a large Z-score indicates that the observed number of cases is much higher than expected. Rather than triggering an alarm if *any* region has Z-score higher than some fixed threshold, we instead find the distribution of the *maximum* score of all regions under the null hypothesis of no outbreaks. This distribution tells us what we should expect the most alarming score to be when the system is executed on data in which there is no outbreak. Then we compare the score of the highest-scoring (most significant) region on our data against this distribution to determine its statistical significance (or p-value). In other words, the scan statistic attempts to answer the question, "If there were no outbreaks, and we searched over all of these regions, how likely would we be to find *any* regions that score at least this high?" If the analysis shows that we would be very unlikely to find any such regions under the null hypothesis, we can conclude that the discovered region is a significant cluster. The main advantage of the scan statistics approach is that we can adjust correctly for multiple hypothesis testing: we can fix a significance level α, and ensure that the probability of having any false alarms on a given day is at most α, regardless of the number of regions searched. Moreover, because the scan statistic accounts for the fact that our tests are not independent, it will typically have much higher power to detect than a Bonferroni-corrected method. In some applications, the scan statistic results in a *most powerful* statistical test (see Kulldorff, 1997 for more details).

FIGURE 16.2 Example of a set of significant regions.

Although the scan statistic focuses on finding the single most significant region, it can also be used to find multiple regions: secondary clusters can be examined, and their significance found, though the test is typically somewhat conservative for these. The technical difficulty, however, is finding the distribution of the maximum region score under the null hypothesis. Turnbull et al. (1990) solved this problem for circular regions of fixed population, using the maximum number of cases in a circle as the test statistic, and using the method of randomization testing (discussed below) to find the statistical significance of discovered regions. The disadvantage of this approach is that it requires a fixed population size circle, and thus a multiple hypothesis testing problem still exists if we want to search over regions of multiple sizes or shapes. Kulldorff and Nagarwalla (1995) and Kulldorff (1997) solved the problem for variable size regions using a *likelihood ratio test*: The test statistic is the maximum of the likelihood ratio under the alternative and null hypotheses, where the alternative hypothesis represents clustering in that region and the null hypothesis assumes no clusters. We discuss their method, the "spatial scan statistic," in the following section.

3. THE SPATIAL SCAN STATISTIC

The *spatial scan statistic* (Kulldorff and Nagarwalla, 1995, Kulldorff, 1997) is a powerful and general method for spatial cluster detection. It is in common use by epidemiologists for finding significant spatial clusters of disease cases, which may be indicative of an outbreak. In this section, we present the spatial scan statistic as originally described by Kulldorff, along with a number of generalizations, extensions, and variants which extend the scope and applicability of this method.

In its original formulation, Kulldorff's statistic assumes that we have a set of spatial locations s_i, and are given a count c_i and a population p_i corresponding to each location. For example, each s_i may represent the centroid of a census tract, the corresponding count c_i may represent the number of respiratory emergency department visits in that census tract, and the corresponding population p_i may represent the "at-risk population" of that census tract, derived from census population and possibly adjusted for covariates. The statistic makes the assumption that each observed count c_i is drawn randomly from a Poisson distribution with mean $q_i p_i$, where p_i is the (known) at-risk population of that area, and q_i is the (unknown) *risk*, or *underlying disease rate*, of that area. The risk is the expected number of cases per unit population; that is, we expect to see a number of cases equal to the product of the population and the risk, but the observed number of cases may be more or less than this expectation due to chance. Thus, our goal is to determine whether observed increases in count in a region are due to increased risk, or chance fluctuations. The Poisson distribution is commonly used in epidemiology to model the underlying randomness of observed case counts, making the assumption that the variance is equal to the mean. If this assumption is not reasonable (i.e., counts are "overdispersed" with variance greater than the mean, or "underdispersed" with variance less than the mean), we should instead use a distribution which separately models mean and variance, such as the normal or negative binomial distributions. We also assume that each count c_i is drawn independently, although the model can be extended to account for spatial correlations between nearby locations.

3.1. Detailed Description of the Spatial Scan Statistic

As discussed above, Kulldorff's spatial scan statistic attempts to detect spatial regions where the underlying disease rates q_i are significantly higher inside the region than outside the region. Thus, we wish to test the null hypothesis H_0 ("the underlying disease rate is spatially uniform") against the set of alternative hypotheses $H_1(S)$: "The underlying disease rate is higher inside region S than outside region S." More precisely, we have:

H_0: $c_i \sim \text{Poisson}(q_{all}\, p_i)$ for all locations s_i, for some constant q_{all}.

$H_1(S)$: $c_i \sim \text{Poisson}(q_{in}\, p_i)$ for all locations s_i in S, and $c_i \sim \text{Poisson}(q_{out}\, p_i)$ for all locations s_i outside S, for some constants $q_{in} > q_{out}$.

The test statistic that we use is the likelihood ratio, that is, the likelihood (denoted by Pr) of the data under the alternative hypothesis $H_1(S)$ divided by the likelihood of the data under the null hypothesis H_0. This gives us, for any region S, a *score function*:

$$D(S) = \frac{\Pr(Data \mid H_1(S))}{\Pr(Data \mid H_0)}.$$

For Kulldorff's statistic, we obtain

$$D(S) = \left(\frac{C_{in}}{P_{in}}\right)^{C_{in}} \left(\frac{C_{out}}{P_{out}}\right)^{C_{out}} \left(\frac{C_{in}+C_{out}}{P_{in}+P_{out}}\right)^{-(C_{in}+C_{out})}, \text{ if } \frac{C_{in}}{P_{in}} > \frac{C_{out}}{P_{out}},$$

and

$$D(S) = 1 \quad \text{otherwise;}$$

see Kulldorff (1997) for a derivation. In this equation, C_{in} and C_{out} represent the aggregate count $\sum c_i$ inside and outside region S, and P_{in} and P_{out} represent the aggregate population $\sum p_i$ inside and outside region S, respectively. See Figure 16.3 for an example of the evaluation of $D(S)$ for a region. Kulldorff (1997) proved that this likelihood ratio statistic is *individually most powerful* for finding a single region of elevated disease rate: For the given model assumptions (H_0 and H_1), for a fixed false alarm rate, and for a given set of regions searched, it is more likely to detect the cluster than any other test statistic.

Given the above test statistic $D(S)$, the spatial scan statistic method can be easily applied by choosing a set of regions S, calculating the score function $D(S)$ for each of these regions, and obtaining the highest scoring region S^* and its score $D^* = D(S^*)$. We can imagine this procedure as moving a "spatial window" (like the rectangle drawn in Figure 16.3) all around

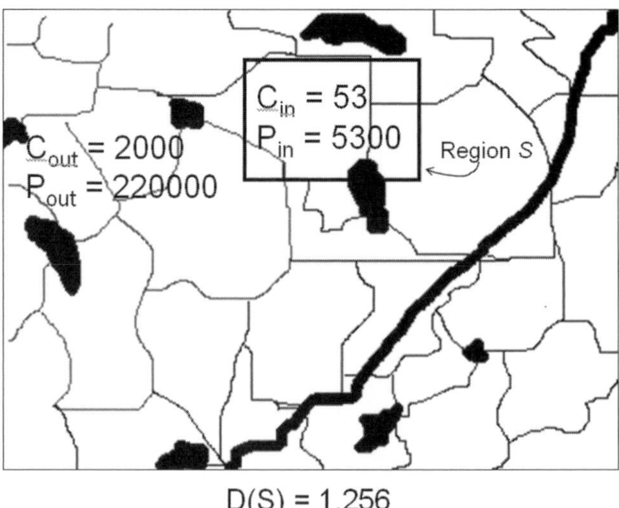

FIGURE 16.3 Counts inside and outside rectangular regions.

the search area, changing the size and shape of the window as desired, and finding the window which gives the highest score $D(S)$. Even though there are an infinite number of possible window positions, sizes, and shapes, we only need to evaluate the score function a finite number of times, since any two regions containing the same set of spatial locations s_i will have the same score. The region with the highest score $D(S)$ is the "most significant region," that is, the region that is most likely to have been generated under the alternative hypothesis rather than the null hypothesis, and thus the region most likely to be a cluster. We typically search over the set of all "spatial windows" of a given shape and varying size; for example, Kulldorff et al. (1997) search over circular regions, Neill and Moore (2004a) search over square regions, and Neill and Moore (2004b) search over rectangular regions. Searching over a set of regions which includes both compact and elongated regions (e.g., rectangles or ellipses) has the advantage of higher power to detect elongated clusters resulting from wind dispersal of pathogens, but because the number of regions to search is increased, this also makes the scan statistic more difficult to compute. We discuss computational issues in more detail below. Chapter 19 describes more accurate modeling of windborne dispersion patterns.

Once we have found the regions with the highest scores $D(S)$, we must still determine which of these "potential clusters" are likely to be "true clusters" resulting from a disease outbreak, and which are likely to be due to chance. To do so, we calculate the statistical significance (p-value) of each potential cluster, and all clusters with p-value less than some fixed significance level α are reported. Because of the multiple hypothesis testing problem discussed above, we cannot simply compute separately whether each region score $D(S)$ is significant, because we would obtain a large number of false positives, proportional to the number of regions searched. Instead, for each region S, we ask the question, "If this data set were generated under the null hypothesis H_0, how likely would we be to find any regions with scores higher than $D(S)$?" To answer this question, we use the method known as *randomization testing*: we randomly generate a large number of "replicas" under the null hypothesis, and compute the maximum score $D^* = \max_S D(S)$ of each replica. More precisely, each replica is a copy of the original search area that has the same population values p_i as the original, but has each value c_i randomly drawn from a Poisson distribution with mean $\frac{C_{all}}{P_{all}} p_i$, where C_{all} and P_{all} are respectively the total number of cases and the total population for the original search area. Once we have obtained D^* for each replica, we can compute the statistical significance of any region S by comparing $D(S)$ to these replica values of D^*, as shown in Figure 16.4. The p-value of region S can be computed as $\frac{R_{beat}+1}{R+1}$, where R is the total number of replicas created, and R_{beat} is the number of replicas with D^* greater than $D(S)$. If this p-value is less than our significance level α, we conclude that the region is significant (likely to be a true cluster); if the p-value is greater than α, we conclude that the region is not significant (likely to be due to chance). We typically start from the most significant region S^* and test regions in order of decreasing $D(S)$, since if a region S is not significant, no region with lower $D(S)$ will be significant. We note that the randomization testing approach given here has the benefit of bounding the overall false positive rate: regardless of the number of regions searched, the probability of any false alarms is bounded by the significance level α. Also, the more replications performed (i.e., the larger the value of R), the more precise the p-value we obtain; a typical value would be $R = 1000$. However, since the run time is proportional to the number of replications performed, this dramatically increases the amount of computation necessary. Finally, we note that spatial scan software is available at *www.satscan.org* and *www.autonlab.org*. The former is the very widely used SaTScan software of (Kulldorff and Information Management Services Inc., 2002). The latter is prototype software that is discussed below.

3.2. Generalizing the Spatial Scan Statistic

In this subsection, we consider a general statistical framework for the spatial scan statistic, extending it to allow for a large class of underlying models and thus a wide variety of application domains. As above, we wish to test a null hypothesis H_0 (a model of how the data is generated, assuming there are no clusters of interest) against the set of alternative hypotheses $H_1(S)$, each of which represents a relevant cluster in some region S of space. Assuming that the null hypothesis and each alternative hypothesis are point hypotheses (with no free parameters), we can use the likelihood ratio $D(S) = \frac{\Pr(Data \mid H_1(S))}{\Pr(Data \mid H_0)}$ as our test statistic. A more interesting question is what to do when each hypothesis has some parameter space Θ: let $\theta_1(S) \in \Theta_1(S)$ denote parameters for the alternative hypothesis $H_1(S)$, and let $\theta_0 \in \Theta_0$ denote parameters for the null hypothesis H_0. There are two possible answers to this question. In the more typical *maximum likelihood framework*, we use the estimates of each set of parameters that maximize the likelihood of the data: $D(S) = \frac{\max_{\theta_1(S)\in\Theta_1(S)} \Pr(Data \mid H_1(S), \theta_1(S))}{\max_{\theta_0\in\Theta_0} \Pr(Data \mid H_0, \theta_0)}$. We then perform randomization testing using the maximum likelihood estimates of the parameters under the null hypothesis. In the *marginal likelihood framework*, we instead average over the possible values of each parameter:

$$D(S) = \frac{\int_{\theta_1(S)\in\Theta_1(S)} \Pr(Data \mid H_1(S), \theta_1(S)) \Pr(\theta_1(S))}{\int_{\theta_0\in\Theta_0} \Pr(Data \mid H_0, \theta_0) \Pr(\theta_0)}.$$

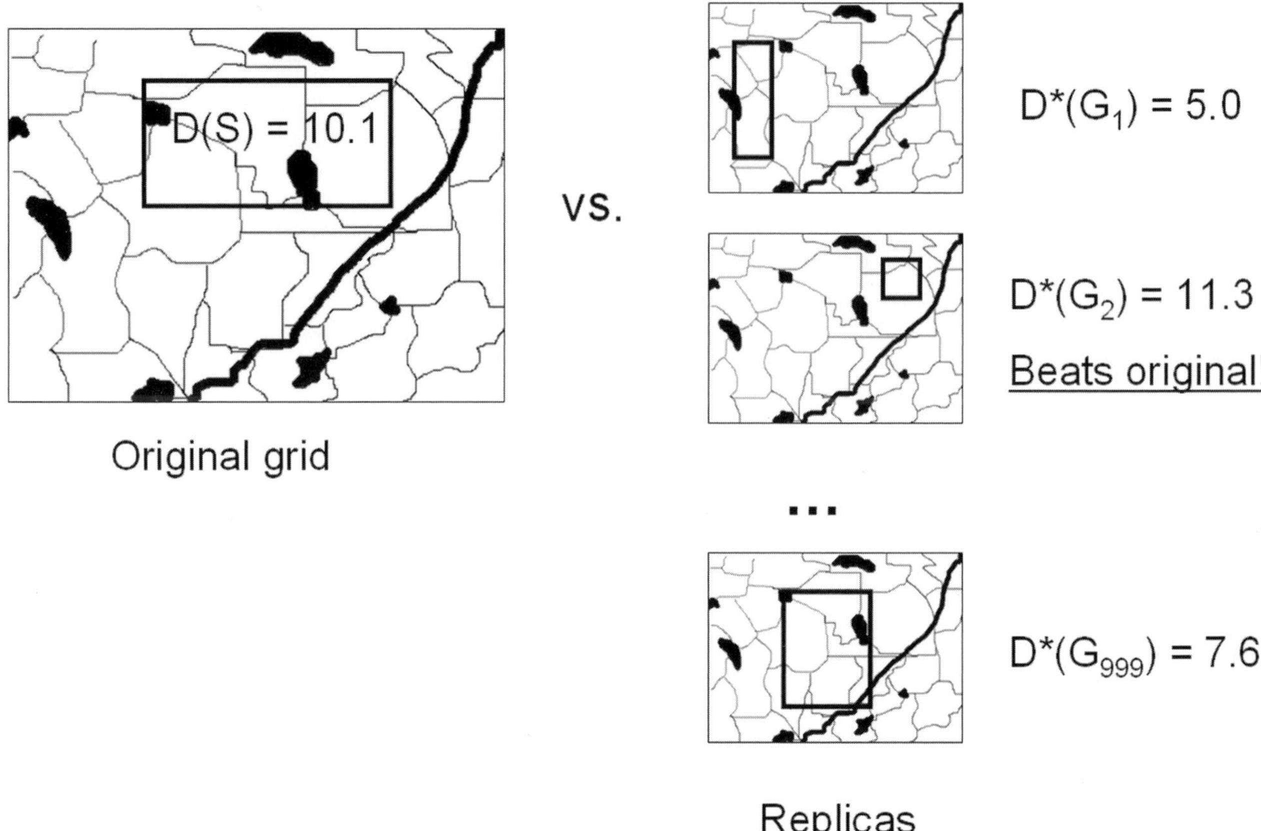

FIGURE 16.4 Calculating the largest region score 1000 times. On the left it is calculated on the real data. On the right it is calculated 999 times on randomized data.

Neill et al. (2005c) present a Bayesian variant of Kulldorff's spatial scan statistic using the marginal likelihood framework; here we focus on the simpler, maximum likelihood approach, and give an example of how new scan statistics can be derived.

Our first step is to choose the null hypothesis and set of alternative hypotheses that we are interested in testing. Here we consider the expectation-based scan statistic discussed above, where we are given the *baseline* (or expected count) b_i and the observed count c_i for each spatial location s_i, and our goal is to determine if any spatial location s_i has c_i significantly greater than b_i. We test the null hypothesis H_0 against the set of alternative hypotheses $H_1(S)$, where:

> H_0: $c_i \sim \text{Poisson}(b_i)$ for all spatial locations s_i.
> $H_1(S)$: $c_i \sim \text{Poisson}(q\, b_i)$ for all spatial locations s_i in S, and $c_i \sim \text{Poisson}(b_i)$ for all spatial locations s_i outside S, for some constant $q > 1$.

Here, the alternative hypothesis $H_1(S)$ has one parameter, q (the relative risk in region S), and the null hypothesis H_0 has no parameters. Computing the likelihood ratio, and using the maximum likelihood estimate for our parameter q, we obtain the following expression for $D(S)$:

$$D(S) = \frac{\max_{q>1} \prod_{s_i \in S} \Pr(c_i \sim \text{Poisson}(qb_i)) \prod_{s_i \notin S} \Pr(c_i \sim \text{Poisson}(b_i))}{\prod_{s_i} \Pr(c_i \sim \text{Poisson}(b_i))}$$

We find that the value of q that maximizes the numerator is $q = \max(1, C/B)$, where C and B are the total count $\sum c_i$ and total baseline $\sum b_i$ of region S respectively. Plugging in this value of q, and working through some algebra, we obtain: $D(S) = \left(\frac{C}{B}\right)^C \exp(B - C)$, if $C > B$, and $D(S) = 1$ otherwise. Then the most significant region S^* is the one with the highest value of $D(S)$, as above. We can calculate the statistical significance (p-value) of this region by randomization testing as above, where replicas are generated under the null hypothesis $c_i \sim \text{Poisson}(b_i)$.

For the population-based method, a very similar derivation can be used to obtain Kulldorff's statistic: we compute the maximum likelihood parameter estimates $q_{in} = C_{in}/P_{in}$, $q_{out} = C_{out}/P_{out}$, and $q_{all} = C_{all}/P_{all}$. We have also used this general framework to derive scan statistics assuming that counts c_i are generated from normal distributions with mean (i.e., expected count) μ_i and variance σ_i^2; these statistics are useful if counts might be overdispersed or underdispersed. Many other likelihood ratio scan statistics are possible, including models with simultaneous attacks in multiple regions and models with spatially varying (rather than uniform) disease rates. We believe that some of these more complex model specifications may have more power to detect relevant and interesting clusters, while excluding those potential clusters which are not epidemiologically relevant.

3.3. Computational Considerations

While considering deployment of spatial methods, it may be necessary to account for the computational time of an algorithm if it is to be run over a very large area or use many randomizations. In this section we return to the question of what set of regions to search over, and discuss how to perform this search efficiently. First, we note that the run time of the spatial scan can be approximated by the product of three factors: the number of replications R, the average number of regions searched per replication $|S|$, and the average time to search a region t. The number of replications R is typically fixed in advance, but we can stop early if many replicas beat the original search area (i.e., the maximum region scores D^* of the replicas are higher than the maximum region score D^* of the original). If this happens, it is clear that no significant clusters are present. The other two factors $|S|$ and t depend on both the set of regions to be searched and the algorithm used to search these regions. For a set of M distinct spatial locations in two dimensions, the number of circular or axis-aligned square regions (assuming that the size of the circle or square can vary) is proportional to M^3, while the number of axis-aligned rectangular regions (assuming that both dimensions of the rectangle can vary) is proportional to M^4. For non—axis-aligned squares or rectangles, we must also multiply this number by the number of different orientations searched. However, most algorithms only search a subset of these regions: for example, Kulldorff (1999) algorithm searches only circles centered at one of the M spatial locations, and the number of such regions is proportional to M^2, not M^3. Another possibility is to aggregate the spatial locations to a grid, either uniform or based on the distinct spatial coordinates of the data points. For a two-dimensional $N \times N$ grid, the number of axis-aligned square regions is proportional to N^3, and the number of axis-aligned rectangular regions is proportional to N^4. Whatever set of regions we choose, the simplest possible implementation of the scan statistic is to search each of these regions by stepping through the M spatial locations, determining which locations are inside and outside the region, computing the aggregate populations/baselines and counts, and applying the score function. Thus, in this approach, we have $|S|$ (number of regions searched per replication) equal to the total number of distinct regions, and t (time to search a region) proportional to the number of spatial locations M.

There are several possible ways to improve on the runtime of this naïve approach. First, we can reduce the time to search a region t, making this search time independent of the number of spatial locations M. We consider two possible methods for searching a region in constant time. The first method, which we call "incremental addition," assumes that we want to search over all regions of a given type: for example, in the approach of Kulldorff (1999), we want to search all distinct circular regions centered at one of the spatial locations. To do so, we increase the region's size incrementally, such that one new spatial location at a time enters the region; for each new location, we can add that location's count and population/baseline to the aggregates, and recompute the score function. For example, in Kulldorff's method, for each location s_i we keep a list of the other locations, sorted from closest to furthest away. Then we can search over the M distinct circular regions centered at s_i by adding the other points one at a time in order. Because the sorting only has to be done once (and does not have to be repeated for each replication), this results in constant search time per region. In other words, Kulldorff's method requires time proportional to M^2 to search over all M^2 regions. This must be done for each of the R replications, giving total search time proportional to RM^2. The second method assumes that points have been aggregated to an $N \times N$ grid, and that we are searching over squares or rectangles. We can use the well-known "cumulative counts" technique to search in constant time per region; see Neill and Moore (2004b) for more details. As a result, we can perform the scan statistic for gridded square or rectangular regions in time proportional to R times the number of regions, that is, RN^3 or RN^4 for square or rectangular regions, respectively.

Even if we can search in constant time per region, the spatial scan statistic is still extremely computationally expensive, because of the large number of regions searched. For example, to search over all rectangular regions on a 256×256 grid, and perform randomization testing (assuming $R = 1000$ replications), we must search a total of 1.1 trillion regions, which would take 14 to 45 days on our test systems. This is clearly far too slow for real-time detection of emerging outbreaks. While one option is to simply search fewer regions, this reduces our power to detect disease clusters. A better option is provided by the *fast spatial scan* algorithms of Neill and Moore (2004a, 2004b) and Neill et al. (2005a), which allow us to reduce the number of regions searched, but without losing any accuracy. The idea is that, since we only care about the most significant regions, that is, those with the highest scores $D(S)$, we do not need to search a region S if we know that it will not have a high score. Thus, we start by examining large regions S, and if we can show that none of the smaller regions contained in S can have high scores, we do not need to actually search each of these regions. Thus, we can achieve the same result as if we had searched all possible regions, but by only searching a small fraction of these. Further speedups are gained by the use of multiresolution data structures, which allow us to efficiently move between searching at coarse and fine resolutions. A detailed description of these structures is beyond the scope of this chapter, but this has been an active area of research in computer science and biomedical informatics. In these algorithms, we apply multiresolution techniques in order to accelerate the spatial scan statistic for searching square regions (Neill et al. 2005a) or rectangular regions (Neill and Moore, 2004b). We have also extended these techniques to rotated rectangular regions and multidimensional data sets (Neill et al., 2005a). These methods are able to search hundreds or thousands of times faster than an exhaustive search, without any loss of accuracy (i.e., the fast spatial scan finds exactly the same region and p-value as exhaustive search). As a result, these methods have enabled us to perform spatial scans on data sets such as nationwide over-the-counter sales data, from over 20,000 stores in near real time, searching for disease clusters in minutes or hours rather than days or weeks. In Figure 16.5, we show a screen shot from the current version of our spatial scan software, available from our websites *www.autonlab.org* (Auton Laboratory) and *rods.health. pitt.edu* (RODS Laboratory).

3.4. Calculation of Populations/Baselines

In the discussion of the spatial scan methods above, we have paid relatively little attention to the question of how the underlying populations or baselines are obtained. In the population-based methods, we often start from census data, which gives an unadjusted population p_i corresponding to each spatial location s_i. This population can then be adjusted for covariates such as the distribution of patient age and gender, giving an estimated "at-risk population" for each spatial location. In a recent paper, Kleinman et al. (2005) suggest two additional model-based adjustments to the population estimates. First, they present a method for temporal adjustment (accounting for day of week, month of the year, and holidays),

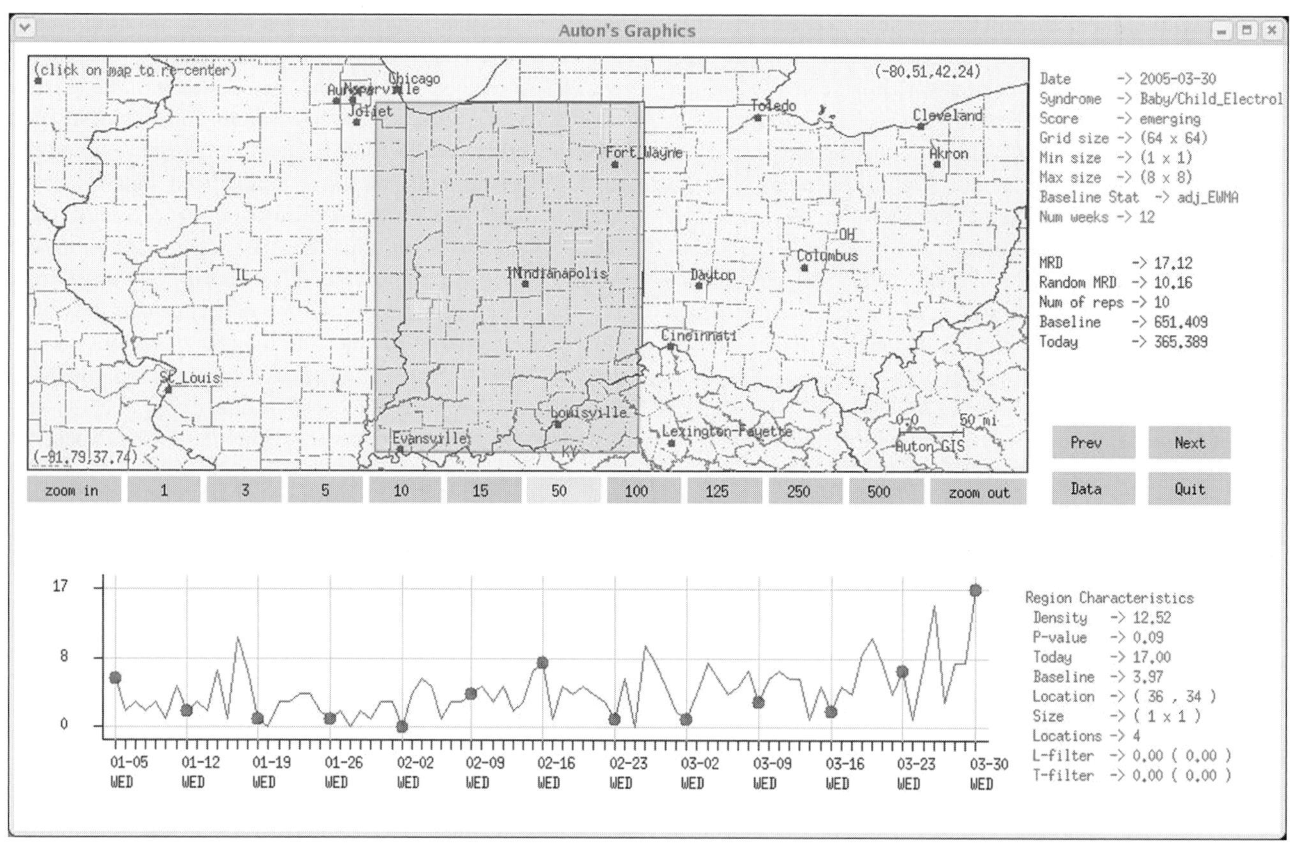

FIGURE 16.5 A screen shot from the current RODS spatial scan software. The bottom time series is the recent history of electrolyte sales in a region north of Indianapolis that was determined as the most significant region in the state on that day (the actual region is hidden in order to avoid providing information that might reveal the data providers).

making the populations larger on days when more visits are likely (e.g., Mondays during influenza season) and smaller on days when fewer visits are likely (e.g., Sundays and holidays). Second, they apply a "generalized linear mixed models" (GLMM) approach, first presented in Kleinman et al. (2004), to adjust for the differing baseline risk in each census tract. This makes the adjusted population larger in tracts that have a larger baseline risk, which makes sense since a given number of observed cases should not be as significant if the observed counts in that region are consistently high. These baseline risks are computed from historical data, that is, the time series of past counts in each census tract, using the GLMM version of logistic regression to fit the model; see Kleinman et al. (2004) for details.

In the expectation-based methods, we also make use of the historical data, but for these methods the goal is to directly estimate the number of cases we expect to see in each area. Thus, we must predict the expected number of cases b_i for each spatial location s_i based on the history of past counts at that location (and optionally, considering spatial correlation of counts at nearby locations). This becomes, in essence, a univariate time series analysis problem, and many of the techniques discussed in Chapter 14 can be used. For example, to adjust for day of week effects, we can either stratify by day of week (i.e., predict Tuesday's expected count by using only prior Tuesdays) or adjust for day of week using the sickness availability method (see Chapter 14). For data sets without strong seasonal effects, simple mean or exponentially weighted moving average methods (e.g., estimating the expected value of today's count as the mean of the counts 7, 14, 21, and 28 days ago) can be sufficient, but for data sets with strong seasonality, these methods will lag behind the seasonal trend, resulting in numerous false positives for increasing trends (e.g., sales of cough and cold medication at the start of winter) or false negatives for decreasing trends (e.g., cough and cold sales at the end of winter). To account for these trends, we recommend the use of regression methods (either weighted linear regression or nonlinear regression depending on the data) to extrapolate the current counts; see Neill et al. (2005b) for more details of this expectation-based approach. Another possibility is to make the assumption of independence of space and time, as in Kulldorff et al. (2005); this means that the expected count in a given region is equal to the total count of the entire area under surveillance, multiplied by the historical proportion of counts in that region. This approach is successful in detecting very localized outbreaks, but loses power to detect more widespread outbreaks (Neill et al., 2005b). The reason for this is that a widespread outbreak will increase the total count significantly, thus increasing the expected count in the outbreak region, and hence making the observed increase in counts seem less significant. In the worst scenario, a massive outbreak which causes a constant, multiplicative increase in counts across the entire area under surveillance would be totally ignored by this approach; this is also true for many of the population-based methods, since they only detect spatial variation in disease rate, not an overall increase in counts. If these methods are used, we recommend using a purely temporal method in parallel to ensure that large-scale outbreaks (as well as localized outbreaks) can be detected. Either way, the accurate inference of expected counts from historical data is still an open problem, with different methods performing well for different data sets and outbreak types. See Neill et al. (2005b) for empirical testing of the various time series methods and further discussion.

3.5. From Space to Space-Time

In this subsection, we briefly consider extensions of the spatial scan statistic to spatio-temporal cluster detection. The space-time scan statistic was first proposed by Kulldorff et al. (1998), and a variant was applied to prospective disease surveillance by Kulldorff (2001). The goal of the space-time scan statistic is a straightforward extension of the purely spatial scan: to detect regions of space-time where the counts are significantly higher than expected. Let us assume that we have a discrete set of time steps $t = 1...T$ (e.g., daily observations for T days), and for each spatial location s_i, we have counts c_i^t representing the observed number of cases in the given area on each time step. There are two very simple ways of extending the spatial scan to space-time: to run a separate spatial scan for each time step t, or to treat time as an extra dimension and thus run a single multidimensional spatial scan in space-time (for example, we could search over three-dimensional "hyper-rectangles" which represent a given rectangular region of space during a given time interval). The problem with the first method is that, by only examining one day of data at a time, we may fail to detect more slowly emerging outbreaks. The problem with the second method is that we tend to find less relevant clusters: for prospective disease surveillance, we want to detect newly emerging clusters of disease, not those that have persisted for a long time. Thus, in order to achieve better methods for space-time cluster detection, we must consider the question, "How is the time dimension different from space?" In Neill et al. (2005b), we argue that there are three main distinctions:

1. The concept of "now." In the time dimension, the present is an important point of reference: we are typically only interested in disease clusters that are still "active" at the present time, and that have emerged within the recent past (e.g., within a few days or a week). We do not want to detect clusters that have persisted for months or years, and we are also not interested in those clusters which have already come and gone. The exception to this, of course, is if we are performing a *retrospective analysis*, attempting to detect all space-time clusters regardless of how long ago they occurred. The space-time scan statistic for retrospective analysis was first presented in Kulldorff et al. (1998),

and the space-time scan statistic for prospective analysis was first presented in Kulldorff (2001). In brief, the retrospective statistic searches over time intervals $t_{min}...t_{max}$, where $1 \leq t_{min} \leq t_{min} \leq T$, while the prospective statistic searches over time intervals $t_{min} \ldots T$, where $1 \leq t_{min} \leq T$, adjusting correctly for multiple hypothesis testing in each case. We focus here on prospective analysis, since this is more relevant for our typical disease surveillance task.

2. "Learning from the past." In the space-time cluster detection task, we often do not have reliable denominator data (i.e., populations), so we must infer the expected counts b_i^t of recent days from the time series of previous counts c_i^t, taking into account effects such as seasonality and day of week. Some methods for inferring these expected counts were discussed in the previous section; see Neill et al. (2005b) for further discussion.
3. The "arrow of time." Time has a fixed directionality, moving from the past, through the present, to the future. We typically expect disease clusters to *emerge* in time: For example, a disease may start out having only minor impact on the affected population, then increase its impact (and thus the observed symptom counts) either gradually or rapidly until it peaks. Based on this observation, we propose a variant of the scan statistic designed for more rapid detection of emerging outbreaks (Neill et al., 2005b). The idea is that rather than assuming (as in the standard, "persistent" space-time scan statistic) that the disease rate q remains constant over the course of an epidemic, we expect the disease rate to *increase* over time, and thus we fit a model which assumes a monotonically increasing sequence of disease rates q_t at each affected time step t in the affected region. In Neill et al. (2005b), we show that this "emerging cluster" space-time scan statistic often outperforms the standard "persistent cluster" approach. We note that Iyengar (2005) accounts for a different aspect of the arrow of time: this method searches over truncated pyramid shapes in space-time, allowing detection of spatial clusters that move, grow, or shrink linearly with time.

Taking these factors into account, the prospective space-time scan statistic has two main parts: inferring (based on past counts) what we expect the recent counts to be, and finding regions where the observed recent counts are significantly higher than expected. More precisely, given a "temporal window size" W, we wish to know whether any space-time cluster within the last W days has counts c_i^t higher than expected. To do so, we first infer the expected counts $b_i^t = E[c_i^t]$ for all spatial locations on each recent day t, $T - W < t \leq T$. See Neill et al. (2005b), Kulldorff et al. (2005), and Kleinman et al. (2005) for methods of inferring these expected counts; earlier methods such as Kulldorff et al. (1998) and Kulldorff (2001) instead use at-risk populations determined from census data. Next, we choose the models H_0 and $H_1(S, t_{min})$, where the null hypothesis H_0 assumes no clusters and the alternative hypothesis $H_1(S, t_{min})$ represents a cluster in spatial region S starting at time t_{min} and continuing to the present time T. Neill et al. (2005b) gives two such models, one for persistent clusters and one for emerging clusters. From our model, we can derive the corresponding score function $D(S, t_{min})$ using the likelihood ratio statistic, and then find the space-time cluster (S^*, t_{min}^*) which maximizes the score function D. Finally, we can compute the statistical significance (p-value) of this space-time cluster by randomization testing, as above. More details of the space-time method described here, as well as empirical tests on several semi-synthetic outbreak data sets, are given in Neill et al. (2005b). We also refer the reader to Kulldorff et al. (1998, 2005), Kulldorff (2001), and Kleinman et al. (2005) for other useful perspectives on space-time cluster detection.

3.6. When to Apply Spatial Scan Approaches

Within epidemiology, scan statistics are a well-used and thriving analytic method. As a result, they have been incorporated into several experimental biosurveillance systems such as Heffernan et al. (2004), Lombardo et al. (2003), and Yih et al. (2004). We recommend caution when using scan statistics on new kinds of data. For example, in our early experiences of applying scan statistics to over-the-counter retail pharmacy data, it was immediately clear that simplistic assumptions in the underlying model can lead to false alarms: there are dozens of nondisease-related reasons for clusters of over-the-counter medication purchases to occur. Conversely, with the wrong data, even a sophisticated model will fail. For example, if home zipcodes are the only data in an emergency department's records then an attack on a downtown office location might not appear as a spatial cluster (although it is possible that appropriate use of commuting statistics can help in this case) (Buckeridge et al., 2003, Duczmal and Buckeridge, 2005). We believe that careful modeling is needed in order to overcome these effects on novel sources of data, and there is considerable ongoing work in the area, such as Kleinman et al. (2004, 2005).

4. RELATED METHODS

In this chapter, we have discussed those methods that can be used to solve the two problems of cluster detection: determining whether any significant clusters exist, and pinpointing the spatial location and extent of clusters. Many other spatial methods can be found in the literature on spatial epidemiology and spatial statistics, although most such methods either do not find specific clusters, or do not evaluate the statistical significance of discovered clusters. More general overviews of the literature on spatial statistical methods can be found in Lawson (2001) and Elliott et al. (2000). In addition to the spatial cluster detection methods discussed here, these methods include *general* and *focused* clustering methods, *disease mapping* approaches, and *spatial cluster modeling*.

General clustering methods are hypothesis testing methods that test for a general tendency of the data to cluster; in other words, they attempt to answer the question, "Is this data set more spatially clustered than we would expect?" Such methods do not identify specific clusters, but instead give a single result of "spatially clustered" or "not spatially clustered." These methods are useful if we want to know whether anything unexpected is going on, but do not care about the specific locations of unexpected events. Examples of such methods include Whittemore et al. (1987), Cuzick and Edwards (1990), and Tango (1995); see Lawson (2001) and Elliott et al. (2000) for more details. We also refer the interested reader to two general tests for space-time clustering: Knox (1964) and Mantel (1967).

Focused clustering methods are hypothesis testing methods that, given a prespecified spatial location, attempt to answer the question, "Is there an increase in risk in areas near this location?" These methods can be used to examine potential environmental hazards, such as testing for an increased risk of lung cancer near a coal-burning power plant. Since the locations are specified in advance, these methods cannot be used to identify specific cluster locations, but are instead used to test locations that have been identified by other means. Examples of such methods include Stone (1988), Besag and Newell (1991), and Lawson (1993); see Lawson (2001) and Elliott et al. (2000) for more details.

Disease mapping approaches have the goal of producing a spatially smoothed map of the variation in disease risk. For example, a very simple disease mapping approach might plot the observed disease rate (number of observed cases per unit population) in each area; more advanced approaches use a variety of Bayesian models and other spatial smoothing techniques to estimate the underlying risk of disease in each area. These methods do not explicitly identify cluster locations, but disease clusters may be inferred manually by identifying high-risk areas on the resulting map. Nevertheless, no hypothesis testing is typically done, so we cannot draw statistical conclusions as to whether these high-risk areas have resulted from true disease clusters or from chance fluctuations. Examples of such methods include Clayton and Kaldor (1987), Besag et al. (1991), and Clayton and Bernardinelli (1992); see Lawson, (2001) and Elliott et al. (2000) for more details.

Finally, *spatial cluster modeling* methods attempt to combine the benefits of disease mapping and spatial cluster detection, by constructing a probabilistic model in which the underlying clusters of disease are explicitly represented. A typical approach is to assume that cases are generated by some underlying process model which depends on a set of cluster centers, where the number and locations of cluster centers are unknown. Then we attempt to simultaneously infer all the parameters of the model, including the cluster centers and the disease risks in each area, using a simulation method such as reversible jump Markov chain Monte Carlo (Green, 1995). Thus, precise cluster locations are inferred, and while no formal significance testing is done, the method is able to compare models with different numbers of cluster centers, giving an indication of both whether there are any clusters and where each cluster is located. One typical disadvantage of such methods is computational: the underlying models rarely have closed-form solutions, and the Markov chain Monte Carlo methods used to approximate the model parameters are often computationally intensive. Examples of such methods include Lawson (1995), Lawson and Clark (1999), and Gangnon and Clayton (2000). For a more detailed discussion of spatial cluster modeling, see Lawson and Denison (2002).

SUMMARY

In this chapter, we have presented an overview of methods for spatial cluster detection, focusing on spatial scan statistics. The statistical and computational techniques described here have extended the spatial scan framework by increasing the generality of the underlying models, as well as making the spatial scan computationally feasible even for very large sources of data.

ADDITIONAL RESOURCES

- *http://www.autonlab.org* (Auton Laboratory website includes Fast Spatial Scan software, available for download, and spatial scan papers by Neill, Moore, and others.)
- *http://www.satscan.org* (SaTScan website includes latest version of Martin Kulldorff's SaTScan software, available for download, and spatial scan papers by Kulldorff and others.)

REFERENCES

Besag, J., Newell, J. (1991). The detection of clusters in rare diseases. *J R Stat Soc A* 154:143–55.

Besag, J., York, J., Mollie, A. (1991). Bayesian image restoration with two applications in spatial statistics. *Ann Inst Stat Math* 43:1–59.

Bonferroni, C. E. (1935). Il calcolo delle assicurazioni su gruppi di teste. In: *Studi in Onore del Professore Salvatore Ortu Carboni*, 13–60.

Buckeridge, D. L., Musen, M. A., Switzer, P., et al. (2003). An analytic framework for space-time aberrancy detection in public health surveillance data. In: *Proceedings of American Medical Informatics Association Annual Symposium*, 120–4.

Clayton, D. G., Bernardinelli, L. (1992). Bayesian methods for mapping disease risk. In: Elliot, P., Cuzick, J., English, D., et al., eds. *Geographical and Environmental Epidemiology: Methods for Small Area Studies*, 205–20. Oxford: Oxford University Press.

Clayton, D. G., Kaldor, J. (1987). Empirical Bayes estimates of age-standardized relative risks for use in disease mapping. *Biometrics* 43:671–81.

Cuzick, J., Edwards, R. (1990). Spatial clustering for inhomogeneous populations. *J R Stat Soc B* 52:73–104.

Duczmal, L., Buckeridge, D. (2005). Using modified spatial scan statistics to improve detection of disease out breaks when exposure occurs in the workplace. *MMWR Morb Mortal Wkly Rep* 54(suppl): 187.

Elliott, P., Wakefield, J. C., Best, N. G., et al., eds. (2000). *Spatial Epidemiology: Methods and Applications*. Oxford: Oxford University Press.

Gangnon, R. E., Clayton, M. K. (2000). Bayesian detection and modeling of spatial disease clustering. *Biometrics* 56:922–35.

Green, P. J. (1995). Reversible jump Markov chain Monte Carlo computation and Bayesian model determination. *Biometrika* 82:711–32.

Heffernan, R., Mostashari, F., Das, D., et al. (2004). Syndromic surveillance in public health practice. *Emerg Infect Dis* 10:858–64.

Kleinman, K., Abrams, A., Kulldorff, M., et al. (2005). A model-adjusted space-time scan statistic with an application to syndromic surveillance. In: *Epidemiology and Infection*. 133:409–19.

Kleinman, K., Lazarus, R., Platt, R. (2004). A generalized linear mixed models approach for detecting incident clusters of disease in small areas, with an application to biological terrorism. *Am J Epidemiol* 159:217–24.

Knox, E. G. (1964). The detection of space-time interactions. *Appl Stat* 13:25–29.

Kulldorff, M. (1999). Spatial scan statistics: models, calculations, and applications. In: Glaz, J., Balakrishnan, N., eds. *Scan Statistics and Applications*, 303–22. Boston: Birkhauser.

Kulldorff, M. (2001). Prospective time periodic geographical disease surveillance using a scan statistic. *J R Stat Soc* 164:61–72.

Kulldorff, M., Athas, W. F., Feurer, E. J., et al. (1998). Evaluating cluster alarms: a space-time scan statistic and brain cancer in Los Alamos, New Mexico. *Am J Public Health* 88:1377–80.

Kulldorff, M., Feuer, E. J., Miller, B. A., et al. (1997). Breast cancer clusters in the northeast United States: a geographic analysis. *Am J Epidemiol* 146:161–70.

Kulldorff, M., Heffernan, R., Hartman, J., et al. (2005). A space-time permutation scan statistic for the early detection of disease outbreaks. *PLOS Med* 2:216–24.

Kulldorff, M., Information Management Services Inc. (2002). *SaTScan v. 3.1: Software for the spatial and space-time scan statistics*.

Kulldorff, M., Nagarwalla, N. (1995). Spatial disease clusters: detection and inference. *Stat Med* 14:799–810.

Lawson, A. B. (1993). On the analysis of mortality events around a prespecified fixed point. *J R Stat Soc A* 156:363–77.

Lawson, A. B. (1995). Markov Chain Monte Carlo techniques for putative pollution source problems in environmental epidemiology. *Stats Med* 14:2473–86.

Lawson, A. B. (2001). *Statistical Methods in Spatial Epidemiology*. Chichester: John Wiley & Sons.

Lawson, A. B., Clark, A. (1999). Markov chain Monte Carlo methods for putative sources of hazard and general clustering. In: Lawson, A.B., ed. *Disease Mapping and Risk Assessment for Public Health*, Chichester: John Wiley & Sons.

Lawson, A. B., Denison, D. G. T., eds. (2002). *Spatial Cluster Modelling*. Boca Raton, FL: Chapman & Hall/CRC.

Lombardo, J., Burkom, H., Elbert, E. (2003). A systems overview of the Electronic Surveillance System for the Early Notification of Community-Based Epidemics (ESSENCE II). *J Urb Health* 80(Suppl 1):i32–42.

Mantel, N. (1967). The detection of cancer clustering and the generalized regression approach. *Cancer Res* 27:209–20.

Naus, J. I. (1965). The distribution of the size of the maximum cluster of points on the line. *J Am Stat Assoc* 60:532–8.

Neill, D. B., Cooper, G. F., Moore, A. W. (2005c). A Bayesian scan statistic for spatial cluster detection. In *Advances in Disease Surveillance*, in press.

Neill, D. B., Moore, A. W. (2004a). A fast multi-resolution method for detection of significant spatial disease clusters. *Adv Neural Information Processing Syst* 16:651–8.

Neill, D. B., Moore, A. W. (2004b). Rapid detection of significant spatial clusters. In proc. 10th ACM SI6KOO Conference on Knowledge Discovery and Data Mining, 256–65.

Neill, D. B., Moore, A. W., Pereira, F., et al. (2005a). Detecting significant multidimensional spatial clusters. *Adv Neural Information Processing Syst* 17:969–976.

Neill, D. B., Moore, A. W., Sabhnani, M., et al. (2005b). Detection of emerging space-time clusters. In proc. 11th ACM SI6KOO Conference on Knowledge Discovery and Data Mining, 218–227.

Openshaw, S., Charlton, M., Craft, A., et al. (1988). Investigation of leukemia clusters by use of a geographical analysis machine. *Lancet* 1:272–3.

Snow, J. (1855). *On the Mode of Communication of Cholera*. London: John Churchill.

Stone, R. A. (1988). Investigations of excess environmental risks around putative sources: statistical problems and a proposed test. *Stats Med* 7:649–660.

Tango, T. (1995). A class of tests for detecting "general" and "focused" clustering of rare diseases. *Stats Med* 14:2323–34.

Turnbull, B. W., Iwano, E. J., Burnett, W. S., et al. (1990). Monitoring for clusters of disease: application to leukemia incidence in upstate New York. *Am J Epidemiol* 132:s136–43.

Wagner, M., Robinson, J., Tsui, F., et al. (2003). Design of a national retail data monitor for public health surveillance. *J Am Med Inform Assoc* 10:409–18.

Whittemore, A., Friend, N., Brown, B., et al. (1987). A test to detect clusters of disease. *Biometrika* 74:631–5.

Yih, W. K., Caldwell, B., Harmon, R. (2004). The National Bioterrorism Syndromic Surveillance Demonstration Program. *MMWR Morb Mortal Wkly Rep* 53(Suppl):43–6.

17
CHAPTER

Natural Language Processing for Biosurveillance

Wendy W. Chapman
Center for Biomedical Informatics
University of Pittsburgh, Pittsburg, Pennsylvania

1. INTRODUCTION

Data useful for biosurveillance are often only available in a free-text format that can be easily read and understood by a human but not by a computer. Natural language processing (NLP) refers to automated methods for converting free-text data into computer-understandable format (Allen, 1995). This conversion is necessary so that information stored in free-text format can contribute to detection and characterization of outbreaks.

2. THE ROLE OF NLP IN BIOSURVEILLANCE

Detection algorithms count the number of occurrences of a variable in a given spatial location over a given time period to look for anomalous patterns. Detection algorithms require structured data, that is, data in a format that can be interpreted by a computer. By far the most common structured data formats are relational database tables. An example of a structured data element is the number of units of cold and cough medicine sold over the last 24 hours in a particular county. Many other examples of structured data elements were described in earlier chapters and in Part IV.

Much data that could potentially be useful in biosurveillance are unstructured. These include symptoms reported by a patient when presenting at an emergency facility, physical and radiological findings recorded by a physician, and queries to healthcare related web sites. In order to use these data for biosurveillance, the information must be converted. We focus our discussion on the use of NLP to encode information from textual patient records for input to outbreak detection algorithms.

3. EXAMPLE USE OF NLP

As an example, assume we want to develop an expert system that generates the probability a patient presenting to an emergency department (ED) has severe acute respiratory syndrome (SARS) given their free-text medical records. According to the World Health Organization and Centers for Disease Control and Prevention case definitions of SARS, the required input variables for diagnosing SARS are whether the patient (1) has an acute respiratory finding (Respiratory Fx), (2) is febrile (Fever), (3) has an abnormal chest radiograph consistent with consolidation or pneumonia (CXR), and (4) has recently traveled to a country currently affected by SARS (Travel). Values for the first three variables are often described in electronic textual patient records. Figure 17.1 shows excerpts from a patient's medical record generated during an ED visit, including the triage chief complaint, history and physical exam, and chest radiograph report. A human physician reading these textual records could easily determine the correct values for the first three variables. Information from the fourth variable (Travel) may not be accessible anywhere in the patient's medical record unless the dictating physician had been concerned about SARS and had dictated the travel history. Retrieving the value for the Travel variable may require a semi-automatic technique in which patients with a high probability of SARS based on the three clinical values could be interviewed regarding their travel history.

A simple expert system may only use the variable Respiratory Fx, monitoring the number of patients presenting to the ED with respiratory complaints. A physician may be able to determine the true value of the Respiratory Fx variable from information in the triage chief complaint. A more complex expert system may use all three variables. According to the medical record in Figure 17.1, a physician would assign values to the variables in a more complex expert system as follows. Respiratory Fx: yes, because the patient had a chief complaint of cough, he complained to the ED physician of productive cough and shortness of breath, and the shortness of breath was probably not cardiac in nature, given that the patient did not have a history of CHF and denied chest pain; Fever: yes, because the chief complaint and the ED report said the patient was febrile; and CXR: yes, because the radiograph report described an opacity consistent with pneumonia.

It would be impractical to hire physicians to read medical records and extract values for the variables required by our expert system. Therefore, if we want to know the variables' values, we must determine them automatically using natural language processing.

4. HOW HARD IS NLP?

NLP can be relatively easy or difficult depending on how complex the text is and on what variables you want to extract. For example, it is relatively easy to extract symptoms from free-text chief complaints using simple methods, because chief complaints are short phrases describing why the patient came

Handbook of Biosurveillance
ISBN 0-12-369378-0

(a)

Fever/cough

(b)

HISTORY OF PRESENT ILLNESS:
CHIEF COMPLAINT: FEVER.

This is a **AGE[in 60s]-year-old white male who presented to the Emergency Department with a one-day history of fever to 102 degrees Fahrenheit, as well as minimally productive cough. The patient also complains of shortness of breath that occurs both with rest and with exertion over the past 24 hours. No past history of CHF. Denies chest pain. He has minimal congestion and denies sore throat. The patient was evaluated by his PCP, Dr. **NAME[ZZZ UUU], and was placed on decongestant, as well as given one dose of Augmentin today, which he took. Despite this, patient continued to have fevers through the day, cough, and shortness of breath, which prompted his visit to the Emergency Department. He denies sick contacts at home. ...

(c)

There is a subtle opacity in the lower portion of the right lung that may be a pneumonia. The patient has normal heart size. No pleural or mediastinal abnormalities noted.

IMPRESSION: SUBTLE OPACITY IN THE LOWER PORTION OF THE RIGHT LUNG IS CONSISTENT WITH PNEUMONIA.

FIGURE 17.1 De-identified excerpts from a patient's electronic medical record. (a) Chief complaint recorded by a triage nurse. (b) First paragraph of a history and physical exam dictated by the ED physician. (c) Impression section of a transcribed chest radiograph report.

to the ED. It is not possible to extract diagnoses from chief complaints, because information in a chief complaint is recorded before the patient even sees a physician. Once a patient is examined by a physician, the patient's diagnosis may be recorded in a dictated report. Extracting information from dictated reports is much more difficult, because a report tells a complex story about the patient involving references to time and negation of symptoms that are not present in chief complaints.

There are many types of technologies used in NLP. In general, the selection of technology depends on the linguistic characteristics of the text. There are some linguistic characteristics that are so difficult to process that effective NLP methods do not exist for them. For example, few NLP systems can accurately extract information that is being conveyed by use of a metaphor. Fortunately, metaphor is not a frequent characteristic in the data sources of potential value in biosurveillance.

In the remainder of this chapter we will discuss (1) the linguistic characteristics of clinical texts that should be considered when implementing NLP for biosurveillance, (2) the types of NLP technologies researchers are using to successfully model information in text, (3) evaluation methods for determining how successful an NLP application is in the domain of outbreak and disease surveillance, and (4) the feasibility of using NLP to encode information for biosurveillance expert systems.

5. LINGUISTIC CHARACTERISTICS OF CLINICAL TEXT—WHAT MAKES NLP HARD?

According to Zelig Harris (Friedman et al., 2002), the informational content and structure of a domain form a specialized language called a sublanguage. The sublanguage of patient medical records exhibits linguistic characteristics that influence an NLP system's ability to extract information from the text. When a physician reads a patient's medical reports, she understands the linguistic characteristics of the text and can make reasonable inferences from the record. For instance, a physician will not assign the Respiratory Fx a value of yes if the respiratory finding described in the report is described as occurring in the patient's past history. For an NLP application to determine the values of clinical variables from patient records the same way a physician would, the application must account for or model the linguistic characteristics of the clinical text. Some important linguistic characteristics of the sublanguage of patient reports are (1) linguistic variation, (2) polysemy, (3) negation, (4) contextual information, (5) finding validation, (6) implication, and (7) co-reference.

5.1. Linguistic Variation

Natural language provides us with freedom to express the same ideas in different ways. Humans are generally capable of understanding the meaning of a natural language expression in spite of such variation; however, the freedom that accompanies natural language makes computerized understanding of the language difficult. In patient reports, a patient's clinical state can be expressed differently due to the linguistic characteristics of derivation, inflection, and synonymy.

Derivation and inflection change the form of a word (the word's morphology) while retaining the underlying meaning of the word. The adjective "mediastinal" can be derived from the noun "mediastinum" by exchanging the suffix -um for the suffix -al. Similar rules can be used to derive the adjective "laryngeal" from the noun "larynx" or to derive the noun "transportation" from the verb "transport."

There are other forms of linguistic variation to contend with. The two most important are inflectional rules (which change a word's form, such as by pluralization of a noun or tense change of a verb) and synonymy (in which different words or phrases mean the same thing).

Physicians reading reports are seldom confused by derivation, inflection, or synonymous expressions. An NLP application attempting to determine whether a patient has shortness of breath, for example, must account for linguistic variation in order to identify "dyspnea," "short of breath," or "dyspneic" as evidence of shortness of breath.

5.2. Biomedical Polysemy

Terms that have the identical linguistic form but different meanings are polysemous. Biomedical polysemy manifests itself in different ways (Roth and Hole, 2000, Liu et al., 2001). Some words in clinical texts have different biomedical meanings or word senses. For instance, the word "discharge" has two word senses—one word sense meaning a procedure for being released from the hospital, as in "prior to discharge," and one word sense meaning a substance that is emitted from the body, as in "purulent discharge."

Acronyms and abbreviations with more than one meaning may be the most frequently occurring type of biomedical polysemy. A striking example of this is the acronym "APC," which has more than thirty unique biomedical definitions, including activated protein c, adenomatosis polyposis coli, antigen-presenting cell, aerobic plate count, advanced pancreatic cancer, age period cohort, and alfalfa protein concentrated. According to one study (Wren and Garner, 2002), 36% of the acronyms in MEDLINE are associated with more than one definition. The number of unique acronyms in MEDLINE is increasing at the rate of 11,000 per year, and the number of definitions associated with unique acronyms is increasing at 44,000 per year. In the sublanguage of patient reports, the type of report is helpful in disambiguating the correct meaning of an acronym or abbreviation, because the report type indicates the type of medical specialty. In this way, "APC" in a microbiology lab report is more likely to mean aerobic plate count, whereas "APC" in a discharge summary may be referring to advanced pancreatic cancer.

Triage chief complaints are full of abbreviations created by clerks and triage nurses to keep the complaint short (Travers and Haas, 2003). Some of the abbreviations are standard and are easily understood by physicians, such as "rt" for "right" and "h/a" for headache. But many abbreviations in chief complaints are unique to the sublanguage of chief complaints or perhaps even to a single hospital or registration clerk. For example, "appy" is commonly used to describe an "appendectomy," and in one hospital "gx" indicates the patient came to the ED by ground transportation.

Depending on the particular clinical variables that we want to extract or encode from text, understanding the meaning or word sense of polysemous words in the patient reports can be critical to success.

5.3. Negation

One of the primary goals in differential diagnosis is to definitively rule out as many hypotheses as possible in order to concentrate on the most probable set of diagnoses. One study (Chapman et al., 2001a) estimated that between 40% and 80% of all findings were explicitly negated in ten different report types, with surgical pathology and operative notes demonstrating the least amount of negation and mammograms and chest radiograph reports demonstrating the most. Explicit negations are indicated by negation terms such as "no," "without," and "denies." Findings can also be implicitly negated. For example, "The lungs are clear upon auscultation" indicates that rales/crackles, rhonchi, and wheezing are all absent. We focus on explicit negation, which is the most common type of negation in patient reports.

In most cases, a physician can easily determine from a report whether a finding is negated in the text. In the sentence, "The patient denies chest pain but has experienced shortness of breath," a physician would assign the clinical variable chest pain the value of no and the variable shortness of breath the value of yes. The types of information a human uses to identify explicitly negated findings include (1) negation terms, (2) scope of the negation term, and (3) expressions of uncertainty.

5.3.1. Negation Terms

Explicit negations are triggered by negation terms that may precede the finding being negated, as in "The chest x-ray revealed no abnormalities," or may follow the observation, as in "The patient is tumor free." Consistent with Zipf's law (Manning and Schutze, 1999), which states that there exist a few very common words, a middling number of medium-frequency words, and many low-frequency words, very few negation phrases account for the majority of negation in patient reports. Two studies on automated negation (Mutalik et al., 2001, Chapman et al., 2001a) found that a few negation phrases accounted for approximately 90% of negation in different report types: "no," "denies/denied," "without," and "not."

The other 10% of negated observations are triggered by a potentially huge number of low-frequency negation phrases.

Once a human identifies a negation term, he must decide whether a relevant finding in the sentence is being negated by that term, that is, whether the finding is within the scope of the negation term. For example, in sentences (1) and (2) the words "source" and "change" are being negated by "not" instead of the findings "infection" and "pain."

(1) This is not the source of the *infection*.
(2) There has not been much change in her *pain*.

5.3.2. *Expressions of Uncertainty*

Unfortunately, differential diagnosis is not a clear-cut science in which physicians are completely confident in what findings or diseases a patient has, and the language used in patient reports expresses the dictating physician's uncertainty on a continuum ranging from certain absence to certain presence. Consider the implications of sentences (5) to (12). The first sentence expresses certainty that pneumonia is absent, whereas the last sentence expresses certainty that pneumonia is present. The intervening sentences express different amounts of uncertainty about a diagnosis of pneumonia. A sophisticated expert system may try to incorporate uncertainty of the variables into its decision making. A simpler expert system may only allow variables to be present or absent. In that case, determining whether pneumonia is negated in the sentences below depends on the goal of the expert system. An expert system designed to be especially sensitive may accept a finding with uncertainty to be present and may set the value of pneumonia to yes for all but the first two sentences. An expert system designed to be specific may consider uncertainty about the variable an indication of negation and may only set the value of pneumonia to yes for the last two.

(5) The chest x-ray ruled out *pneumonia*.
(6) We performed a chest x-ray to rule out *pneumonia*.
(7) Cannot rule out *pneumonia*.
(8) It is not clear whether the opacity is atelectasis or *pneumonia*.
(9) Radiographic findings may be consistent with *pneumonia*.
(10) Discharge diagnosis: possible *pneumonia*.
(11) The patient has *pneumonia*.
(12) He did have sputum that grew out klebsiella *pneumonia* during his admission.

5.4. Contextual Information

Information contained in a single word or phrase is not always sufficient for understanding the value of a clinical variable; the context around the phrase is often essential in understanding the patient's clinical state. Among other things, contextual information is important for determining when the finding occurred and what anatomic location was involved.

Any expert system attempting to increase timeliness in outbreak detection must distinguish between findings that occurred in past history and current problems. For example, one of the variables in our SARS detector is whether the patient has an acute respiratory finding. The definition of acute is not straightforward. However, at the least, an NLP application attempting to determine the value of this variable should be able to accurately assign the value yes to pleuritic chest pain in sentence 13 and no to pneumonia in sentence 14.

(13) The patient presents today with *pleuritic chest pain*.
(14) She has a past history significant for *pneumonia*.

A physician reading a report uses contextual clues like the structure of a report to discriminate between acute or current findings and those in the past history. For example, a finding described in an ED report within a section that is titled "Past Medical History" is probably a historical finding. A human may also use linguistic cues within sentences to determine whether a finding is current. For instance, in sentence 15, a physician would know that myocardial infarction occurred in the past history but that chest pain is a current finding.

(15) He has a past history significant for *myocardial infarction*, and presents to the ED today with *chest pain*.

Determining what findings are described in a patient report also entails discriminating current findings from future or hypothetical findings. In sentence 16, the instance of fever is described as a hypothetical finding, but shortness of breath is described as a finding that probably occurred at the current hospital visit.

(16) She should return for *fever* or exacerbation of her *shortness of breath*.

Some findings can occur with multiple anatomic locations. For detection of SARS, our expert system needs to know whether the edema described in sentence 17 was found in the lung or in the skin.

(17) Chest is *edematous*.

Sometimes the anatomic location is explicitly stated, as in sentence 18. Other times, the anatomic location is not explicitly stated (e.g., sentence 19). The context around the finding is important for disambiguating the anatomic location—even when a location is not explicitly stated.

(18) The *lump* on her back has not changed.
(19) Chest x-ray showed no *mass*.

5.5. Finding Validation

Not all terms representing findings or diseases in a patient report are actual findings in the patient; some findings must have a particular value in order to be considered positive. The variable of oxygen desaturation may be useful in our SARS detector, but a physician may not describe oxygen desaturation with those words. Instead he may say "the patient's O_2 saturation is low" or "the patient is satting at 85% on room air." The qualitative value of "low" in the first example and the quantitative value of "85%" are what let the reader know the patient has oxygen desaturation. Similarly, the presence of the word "temperature" does not inform the reader of whether the patient has a fever–the variable together with its value provide the requisite information to the reader.

5.6. Implication

The main audience of patient reports consists of other physicians. For this reason, understanding what is said in a dictated medical report is difficult for a human reader without domain knowledge. Researchers compared laypeople against physicians at reading chest radiograph reports and judging whether the report described radiological evidence of acute bacterial pneumonia (Fiszman et al., 1999). Not surprisingly, laypeople performed much worse than physicians. As long as the report stated explicitly that the findings were consistent with pneumonia, the laypeople agreed with the physicians in their judgment, but pneumonia was mentioned in only one-third of the positive reports. In the remaining two-thirds of the reports, the evidence for pneumonia was inferred by the physicians and missed by the laypeople.

Implication in medical reports can occur at the sentence level and at the report level. A simple example is the sentence, "The patient had her influenza vaccine." If our SARS expert system had a variable for influenza, even a layperson reading the previous sentence could determine that the value for the variable would probably be no, because the patient was vaccinated. This inference requires domain knowledge that a vaccine generally prevents the target disease. In the radiology study reported above, evidence for pneumonia in positive reports was not always explicitly stated by the radiologist. Instead, the radiologist described "hazy opacities" or "ill-defined densities" in the lobes of the lung, which can be inferred to mean localized infiltrates. Once the inference at the sentence level has been correctly made, a physician reading the radiology report can integrate the findings described throughout the entire report and can infer that because the chest x-ray shows localized infiltrates not explained by other causes, there is evidence for acute bacterial pneumonia. Domain knowledge about words, combinations of words, and combinations of findings make it possible for a physician to make inferences from reports that a lay person—or an expert system—may not be able to make without training in knowledge of the domain.

5.7. Coreference

As described above, sometimes information across sentences must be combined to truly understand the patient's clinical state. A single entity (which could be a finding, a person, or some other object mentioned in a report) may be referred to in more than one sentence. True to the human inclination towards conciseness, once an entity has been evoked, we can refer to the entity with shortened phrases, including pronouns (e.g., "it," "he," or "she") or definite noun phrases (e.g., "the finding," or "her mother"). When two expressions refer to the same entity, they corefer. Determining which referring expressions refer to which referent is important in understanding a clinical report.

5.8. Summary of Linguistic Issues

We have described some of the linguistic characteristics of the sublanguage of patient medical records, including linguistic variation, polysemy, negation, contextual information, finding validation, implication, and coreference. If we want to automatically determine an individual patient's values for the variables used in our expert system, we must address these linguistic characteristics, using the types of information a physician uses to understand the meaning of the words and sentences in the reports. Below we describe some of the techniques current natural language processing research employs for extracting information from clinical texts.

6. TECHNOLOGIES FOR NATURAL LANGUAGE PROCESSING

NLP techniques fall into two broad classes: statistical and symbolic. Statistical techniques use information from the frequency distribution of words within a text to classify or extract information. Symbolic techniques use information from the structure of the language (syntax) and the domain of interest (semantics) to interpret the text to the extent necessary for encoding the text into targeted categories. Although some NLP applications exclusively use one or the other technique, many applications use both statistical and symbolic techniques. In this section, we give a brief background of NLP research in the medical domain and describe some statistical and symbolic NLP techniques used for classifying, extracting, and encoding information from biomedical texts, focusing on techniques useful for addressing the linguistic characteristics of patient medical reports described in the previous section.

6.1. Brief Background of NLP in Medicine

Over the last few decades researchers have actively applied NLP techniques to the medical domain (Friedman and Hripcsak, 1999, Spyns, 1996). NLP techniques have been used for a variety of applications, including quality assessment in radiology (Fiszman et al., 1998, Chapman et al., 2001b); identification of structures in radiology images (Sinha et al., 2001a, Sinha et al., 2001b); facilitation of structured reporting (Morioka et al., 2002, Sinha et al., 2000) and order entry (Wilcox et al., 2002, Lovis et al., 2001); encoding variables required by automated decision-support systems such as guidelines (Fiszman and Haug, 2000), diagnostic systems (Aronsky et al., 2001), and antibiotic therapy alarms (Fiszman et al., 2000); detecting patients with suspected tuberculosis (Jain et al., 1996, Knirsch et al., 1998, Hripcsak et al., 1999); identifying findings suspicious for breast cancer (Jain and Friedman, 1997), stroke (Elkins et al., 2000), and community acquired pneumonia (Friedman et al., 1999b); and deriving comorbidities from text (Chuang et al., 2002).

Probably the most widely used and evaluated NLP system in the medical domain is MedLEE, which was created at Columbia Presbyterian Medical Center (Friedman, 2000, Friedman et al., 1994, 1998, 1999a). MedLEE extracts clinical information from several types of radiology reports, discharge summaries, visit notes, electrocardiography, echocardiography, and pathology notes. MedLEE has been shown to be as accurate as physicians at extracting clinical concepts from chest radiograph reports (Hripcsak et al., 1995, 2002).

NLP has only recently been applied to the domain of outbreak and disease surveillance, and most of the research has focused on processing free-text chief complaints recorded in the ED (Olszewski, 2003, Ivanov et al., 2002, Ivanov et al., 2003, Travers et al., 2003, Travers and Haas, 2003, Chapman et al., 2005a).

Below we describe some of the statistical and symbolic NLP techniques implemented in the medical domain.

6.2. Statistical NLP Techniques

Statistical text classification techniques use the frequency distribution of words to automatically classify a set of documents or text fragments into one of a discrete set of predefined categories (Mitchell, 1997). For example, a text classification application may classify MEDLINE abstracts into one of many possible MeSH categories or may classify websites by topic. Various statistical models have been applied to the problem of text classification, including regression models, Bayesian belief networks, nearest neighbor algorithms, neural networks, decision trees, and support vector machines. The basic element in all text classification algorithms is the frequency distribution of the words in the text. Applications of text classification of free-text patient medical records include retrieving records of interest to a specific research query (Aronis et al., 1999, Cooper et al., 1998), assigning ICD-9 admission diagnoses to chief complaints (Gundersen et al., 1996), and retrieving medical images with specific abnormalities (Hersh et al., 2001). In the domain of biosurveillance, text classification techniques have been applied to triage chief complaints and chest radiograph reports. CoCo (Olszewski, 2003) is a naive Bayesian text classification application that classifies free-text triage chief complaints into syndromic categories, such as respiratory, gastrointestinal, or neurological, based on the frequency distribution of the words in the chief complaints. For example, the chief complaint "cough" would be assigned a higher probability of being respiratory than of being gastrointestinal or neurological, because chief complaints in the training corpus that contained the word "cough" were classified most frequently as respiratory. The IPS system (Aronis et al., 1999, Cooper et al., 1998) was used to create a query for retrieving chest radiograph reports describing mediastinal findings consistent with inhalational anthrax (Chapman et al., 2003). The IPS system uses likelihood ratios to identify words that discriminate between relevant and not relevant documents.

Statistical NLP techniques have been applied to the problem of biomedical polysemy. Given a word or phrase with multiple meanings, the statistical distribution of the neighboring words in the document could be helpful in disambiguating the correct meaning or sense of the word. As an example, consider the word "discharge," which has two word senses: a procedure for being released from the hospital (Disch1) and a substance emitted from the body (Disch2). If we applied a statistical learning technique to text containing the word "discharge," we may learn that Disch1 occurs significantly more often with the neighboring words "prescription," "upon," "home," "today," and "instructions," and that Disch2 occurs more often with the words "purulent," "rashes," "swelling," and "wound."

Beyond text classification, statistical techniques can be used for complex NLP tasks. For instance, Taira and Soderland (1999) have developed an NLP system for radiology reports that uses mainly statistical techniques to encode detailed information about radiology findings and diseases, including the finding, whether it was present or absent, and its anatomic location.

Because of the complexity of patient medical reports, purely statistical techniques that only rely on words and their frequencies are less common than hybrid or purely symbolic techniques that leverage knowledge about the structure or meaning of the words in the text in order to classify, extract, or encode information in clinical documents.

6.3. Symbolic NLP Techniques

Linguistics is the study of the nature and structure of language, including the pronunciation of words (phonetics), the way words are built up from smaller units (morphology), the way words are arranged together (syntax), the meaning of linguistic utterances (semantics), and the relation between language and context of use (pragmatics) including the relationships of groups of sentences (discourse). As humans, we combine all of this linguistic knowledge with knowledge of the world to understand natural language. Symbolic NLP techniques also utilize this linguistic information in attempting to interpret free-text. Below we describe NLP techniques that take advantage of syntactic, semantic, and discourse knowledge in order to address the linguistic characteristics of clinical texts.

6.3.1. Syntax: The Way Words Are Arranged Together

Every word in a language has at least one part of speech. The most common parts of speech in English are noun (e.g., "tuberculosis," "heart"), verb (e.g., "see," "prescribe"), adjective (e.g., "severe," "red"), adverb (e.g., "quickly," "carefully"), determiner (e.g., "the," "some"), preposition (e.g., "of," "in"), participle (e.g., "up," "out"), and conjunction (e.g., "and," "but"). The difficulty in automatically assigning a part of speech to words in a sentence is that some words can have more than one part of speech. For example, the word "discharge" can be a verb or a noun. Automated part-of-speech taggers use either rules or probability distributions learned from hand-tagged training sets to assign parts of speech and perform with an accuracy of 96–97% on general English texts, such as newspaper articles, scientific journals, and books. Part-of-speech distribution in patient reports is different than that of nonclinical texts. For example, discharge summaries contain more nouns and past tense verbs and fewer proper nouns (e.g., people and company names) and present tense verbs (Campbell and Johnson, 2001). Not surprisingly, training a part-of-speech tagger on medical texts improves its accuracy when assigning parts of speech to patient reports (Campbell and Johnson, 2001, Coden et al., 2005). Publicly available part-of-speech taggers trained on medical documents are just beginning to become available (Smith et al., 2004).

A word's part of speech can sometimes be helpful in understanding which word sense is being used in a sentence. Returning to the example of the word "discharge," a statistical analysis of the distribution of "discharge" in patient reports may show that if "discharge" is being used as a verb, the word sense is more likely Disch1 (release from hospital).

Syntactic rules use the part of speech to combine words into phrases and phrases into sentences. For instance, an adjective followed by a noun is a noun phrase, an auxiliary verb followed by a verb is a verb phrase, and a preposition followed by a noun phrase is a prepositional phrase. Phrases can be combined so that a noun phrase followed by a prepositional phrase creates another noun phrase and a noun phrase followed by a verb phrase creates a sentence. This process of breaking down a sentence into its constituent parts is called parsing. Automated parsers employ a grammar consisting of rules or probability distributions for generating combinations of words and a lexicon listing the possible parts of speech for the words. Automated parsers may attempt to produce a deep parse that connects all the words and phrases together into a sentence (Figure 17.2[a]) or a partial parse (also called a shallow parse), which combines words into noun phrases, verb phrases, and prepositional phrases but does not attempt to link the phrases together (Figure 17.2[b]). A deep parse gives you more information about the relationships among the phrases in the sentence but is more prone to error. A partial parse is easier to compute without errors and may be sufficient for some tasks.

As with part-of-speech tagging, the syntactic characteristics of patient reports differ from those of nonclinical texts (Campbell and Johnson, 2001). A publicly available parser trained on medical texts does not yet exist. Szolovitz (2003) showed that the Link Grammar Parser (available at *www.link.cs.cmu.edu/link/*) only recognized 38% of the words in a large sample of ED reports. For this reason, he adapted the SPECIALIST Lexicon distributed by the National Library of Medicine to the format required for the Link Grammar Parser

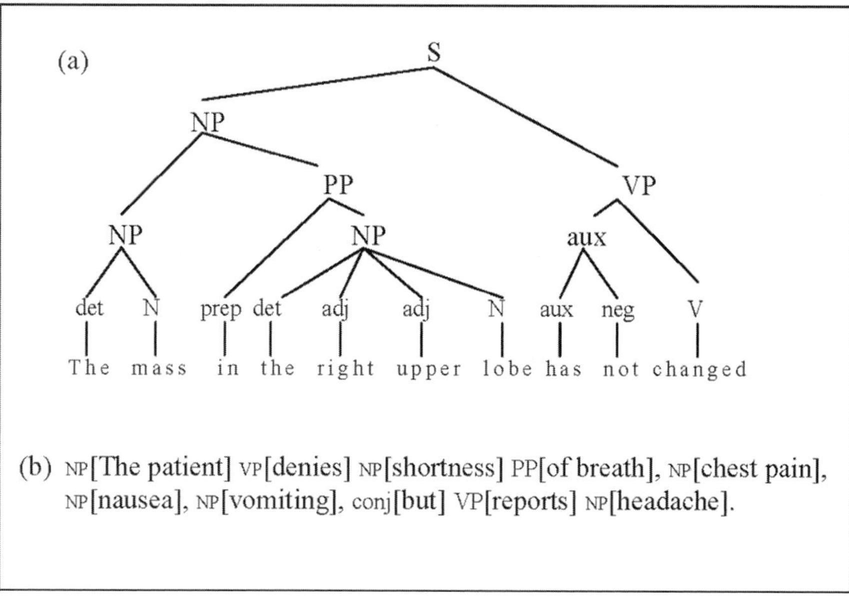

FIGURE 17.2 (a) The tree structure of a deep parse in which words are combined into phrases and phrases are combined into a sentence. det, determiner; N, noun; prep, preposition; adj, adjective; aux, auxiliary verb; v, verb. (b) A partial parse that only labels simple phrases and conjunctions (conj) without linking the phrases together.

and provided over 200,000 new entries for the Link Grammar Lexicon, quintupling the size of the original Lexicon (*available at www.medg.lcs.mit.edu/projects/text/*).

The syntactic structure of a sentence can provide information about the semantic relationships among the words. For example, in Figure 17.2(a) a relationship between the mass and the right upper lobe is indicated by the fact that the prepositional phrase "in the right upper lobe" is attached to the noun phrase "the mass." Statistical methods that rely on whether or not a word or phrase occurs in the sentence without requiring a syntactic relation between the constituents may mistakenly infer a location relation between a noun phrase and prepositional phrase. For instance, in sentence 22, the noun "mass" and the prepositional phrase "in the right upper lobe" both occur in the sentence, but without syntactic knowledge there is no way to know the phrases are actually unrelated.

(22) There is no change in the mass, but the infiltrate in the right upper lobe has increased.

6.3.2. Semantics: The Meaning of Linguistic Utterances

Understanding the syntactic relationships among words in a sentence does not assure understanding the meaning of the sentence. Sentence (23) shows a famous example by Noam Chomsky, the father of modern linguistics, of a perfectly grammatical sentence that has no meaning.

(23) Colorless green ideas sleep furiously.

Understanding a patient report requires not only knowledge of the syntactic relation of the words but also knowledge of the meaning of the words in the report, knowledge of semantic relations between the words, and knowledge of the relationships between the words and the ideas they represent.

Lexical Semantics Refers to Meaning of Words. Understanding the meaning of the words used in a patient medical report is the first step to understanding what is wrong with the patient. The National Library of Medicine's (NLM) Unified Medical Language System® (UMLS®) has created several resources to "facilitate the development of computer systems that behave as if they 'understand' the meaning of the language of biomedicine and health" (*www.nlm.nih.gov/research/umls/about_umls.html*). The NLM freely distributes three UMLS knowledge sources: the Metathesaurus®, the Semantic Network, and the SPECIALIST Lexicon. The three knowledge sources can assist NLP applications in understanding the meaning of the words in clinical reports.

The Metathesaurus is a vocabulary database of biomedical and health related concepts containing over 900,000 concepts compiled from more than 60 different source vocabularies. The Metathesaurus integrates existing vocabularies (such as SnoMed and ICD-9), which provide terms and sometimes hierarchies relating the terms. The Metathesaurus organizes the terms into concepts, organizes the concepts into hierarchies, and relates concepts to each other. If a concept from a new source vocabulary already exists in the Metathesaurus, the concept is added as a synonym. The Metathesaurus is the most complete collection of biomedical concepts and their synonyms.

The Semantic Network provides a consistent categorization of all concepts represented in the Metathesaurus, which are the nodes in the network, and provides a useful set of relations among these concepts, which are the arcs in the network. Every concept in the Metathesaurus is assigned at least one of 135 different semantic types (e.g., finding, anatomical structure, pathologic function, etc.). The Semantic Network contains 54 relationships among the semantic types, such as "part of," "is-a," and "caused by."

An NLP application for our SARS detector may find the phrase "shortness of breath" in a patient report, which is a synonym for the Metathesaurus concept Dyspnea. Other synonyms for the concept Dyspnea are "difficulty breathing," "SOB," and "breathlessness." The concept Dyspnea has the semantic type of Sign or Symptom and has children like "hypoventilation," "paroxysmal dyspnea," and "respiratory insufficiency." A knowledge base with synonyms and semantic information can be helpful in identifying variables and their values from text.

The SPECIALIST Lexicon is a general English lexicon that includes many of the biomedical terms in the Metathesaurus together with the most commonly occurring English words. As of 2003, the SPECIALIST contained almost 300,000 entries. A lexical entry for each word or term records information about spelling variants, derivation, inflection, and syntax. Using the SPECIALIST, we could know, for example, that the term "mediastinal" is the adjectival form of the noun "mediastinum" and that the following phrases are equivalent: "mediastinal widening," "widened mediastinum," and "wide mediastinum."

The Semantic Relationships Among Words Are Also Important. An NLP technique with a syntactic model of a sentence and a semantic model of the words in a sentence has a better chance of understanding relationships among the words in the sentence. For example, a noun phrase comprising an adjective followed by a noun signifies a relationship between the noun, which is the head of the phrase, and the adjective, which is the modifier. Precisely what that relationship is depends on the meaning of the words in the phrases. Consider the phrase "atrial fibrillation." The UMLS semantic type for "atrial" is Body Part, Organ, or Organ Component, and the semantic types of "fibrillation" are Disease or Syndrome and Sign or Symptom. An NLP application that modeled both the syntactic and the semantic information in this phrase could have a rule that stated: *If a syntactic modifier has the semantic type Body Part, Organ, or Organ Component, and the head has the semantic type Disease or Syndrome or Sign or Symptom, the semantic relationship is* Head-*has-location*-Modifier.

An application that validated whether a term mentioned in a report actually is a finding could benefit from modeling semantic and syntactic relationships. For instance, the NLM system FindX (Sneiderman et al., 1996) contains rules based on the semantic type of the words in a modifier-head relation to validate the finding. For example, one rule states: *An abnormality or anatomical site modified by a SNOMED adjective is a finding*, validating "*chest* clear to auscultation" as a finding. Another rule says: *A diagnostic or laboratory procedure modified by a SNOMED adjective or a numeric value is a finding.* This rule correctly validates arterial blood gas as a finding in (24) and invalidates it in (25).

(24) *Arterial blood gas* 7.41/42/43/27
(25) We suggest *arterial blood gas* preoperatively.

Semantic modeling of syntactically related words can also be useful in understanding implicit information in a report. MPLUS (Medical Probabilistic Language Understanding System) is an NLP system that uses Bayesian networks to model the relationship between the words in a report and the ideas or concepts the words represent (Christensen et al., 2002). Figure 17.3 shows a simplified network for radiological findings. The syntactic parse helps determine which words in the sentence should be slotted together into the Bayesian network (i.e., which words are syntactically related). When a new phrase or sentence is slotted into the network, MPLUS can make inferences about the meaning of the words in a sentence in spite of the different combinations of words that can be used to describe the same concept. For example, the phrases "hazy opacity in the left lower lobe" and "ill-defined densities in the lower lobes" both indicate localized infiltrates—even though the word "infiltrate" was not used by the radiologist.

6.4. Discourse: Relationships Among Sentences

Sentences in a patient report are not meant to stand alone–they often convey a story about the differential diagnosis and treatment process for a patient. Some of the variables our example SARS expert system would need cannot be obtained without integrating and disambiguating information from the entire report. Once the individual variables have been located in a report, some type of discourse processing must integrate values for the variables to answer questions such as: (1) Were the relevant findings reported for the patient or for someone else (e.g., a family member, as in "patient's mother died at the age of 48 with an MI")? (2) Did the relevant findings occur at the current hospital visit (versus past history or hypothetical findings)? (3) Is it likely the patient has a respiratory disease or disorder? Three discourse techniques that may help answer these questions are section identification, co-reference resolution, and diagnostic modeling.

Patient reports are semistructured, depending on the type of report and the institution from which the report is generated. For instance, ED reports may contain sections for chief complaint, past history, history of present illness, physical exam, radiologic or lab findings, hospital course, discharge diagnosis, and plan. The section in which a finding is described can provide information important to understanding the meaning of the report. For example, our SARS detector may have a variable for pneumonia history, a variable for radiological

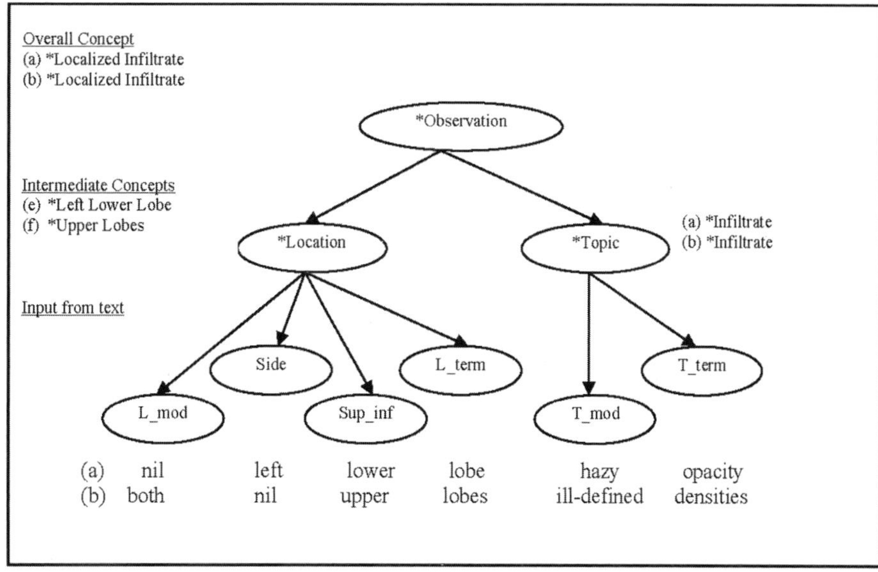

FIGURE 17.3 Partial Bayesian network for radiological findings. Words in leaf nodes come directly from text, concepts in other nodes (shown with asterisks[*]) are inferred based on training examples. Two sentences are slotted in the network: (a) There is a hazy opacity in the left lower lobe. (b) Both upper lobes show ill-defined densities.

evidence of pneumonia, and a variable for a pneumonia diagnosis. An instance of pneumonia described in the radiological findings section of a report is likely to provide radiological evidence of pneumonia, whereas an instance of pneumonia in the discharge diagnosis section probably indicates a diagnosis. Report section identification can also assist in understanding whether the finding occurred in the past history, the current visit, or as a hypothetical finding (e.g., a finding described in the plan section is more likely to be a hypothetical finding), can identify findings for family members (e.g., a finding in the social history section may not be the patient's finding), and can provide insight regarding the anatomic location of an ambiguous finding (e.g., a mass described in the radiology finding section is probably a pulmonary mass).

Patient reports tell a story involving various findings, physicians, patients, family members, medications, and treatments that are often referred to more than once in the text. Identifying which expressions are really referring to the same entity is important in integrating information about that entity. Useful discourse clues for identifying coreferring expressions include how close the expressions are within the text (e.g., a referring expression is more likely to refer to a referent in the previous sentence than to a referent five sentences back), overlapping words (e.g. "the pain" is more likely to refer to "chest pain" than to "atelectasis"), and the semantic type of the entities (e.g., "she" can only refer to a human entity, not to a finding or disease).

Integrating the clinical information within a report to determine the clinical state of the patient (e.g., the likelihood the patient has SARS) requires a diagnostic model relating the individual variables or findings to the diagnosis. Many diagnostic models have been used in medicine, including rule sets, decision trees, neural networks, and Bayesian networks. Diagnostic models are also helpful for determining the values of individual variables. For example, a Bayesian network can model which radiological findings occur with which diseases. With this type of semantic model, even if a report did not mention pneumonia, for example, the model could infer that acute bacterial pneumonia is probable given the radiologic finding of a localized infiltrate (Chapman et al., 2001c).

None of the NLP techniques we have described perform perfectly, but some of the techniques described in this section are easier to address than others. For instance, automatic part-of-speech taggers perform similarly to human taggers. The ability to perform inference on information in a report as a physician does is more complex, entailing both semantic and discourse modeling.

Although the task is difficult, developing NLP techniques for classifying, extracting, and encoding individual variables from patient medical reports is feasible and has been accomplished to different extents by many groups. Successful extraction of variables in spite of imperfect syntactic and semantic techniques can occur for many reasons, including access to the UMLS databases and tools, structure and repetition within reports, and modeling a limited domain. NLP research over the years has revealed that NLP techniques perform better in narrower domains. For instance, modeling the lexical semantics of the biomedical domain is easier than modeling the lexical semantics of all scientific domains, and modeling the lexical semantics of patient reports related to SARS would be easier than modeling all clinical findings in patient reports.

Most of the studies in NLP have focused on the ability of the technology to extract and encode individual variables from the reports. Fewer studies have integrated NLP variables from an entire report to diagnose patients or have evaluated whether an NLP-based expert system can improve patient care. Below we discuss different levels of evaluation of NLP technology related to biosurveillance.

7. EVALUATION METHODS FOR NLP IN BIOSURVEILLANCE

The first step in evaluating an NLP application is to validate its ability to classify, extract, or encode features from text (feature detection). Most evaluations of NLP technology in the biomedical domain have focused on this phase of evaluation. Once we validate feature detection performance, we can evaluate the ability of the encoded features to diagnose individual cases of interest (case detection). Finally, we can perform summative evaluations addressing the ability to detect epidemics (epidemic detection). Figure 17.4 shows how the three levels of evaluation relate to one another, using the diagnostic system for SARS as an example.

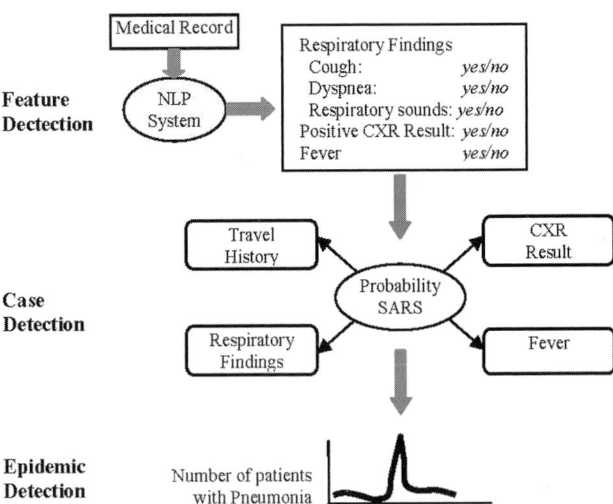

FIGURE 17.4 Relationship between the three levels of evaluation for biosurveillance. Evaluations of feature detection quantify how well variables and their values are automatically encoded from text. Evaluations of case detection quantify the ability to accurately diagnose a single patient from the variables encoded from text, which may or may not be combined with other variables. Evaluations in epidemic detection quantify whether the variable being monitored by detection algorithms can detect outbreaks.

7.1. Feature Detection

The first type of NLP evaluation should measure the application's ability to detect features from text. The question being addressed when quantifying the performance of feature detection for the domain of biosurveillance is: *How well does the NLP application determine the values to the variables of interest from text?* For our SARS detector, examples of feature detection evaluations include how well the NLP application can determine whether a patient has a respiratory-related chief complaint, whether an ED report describes fever in a patient, or whether a patient has radiological evidence of pneumonia in a radiograph report.

Figure 17.5 illustrates the evaluation process for feature detection. Studies of feature detection do not evaluate the truth of the feature in relation to the patient–that is, whether the patient actually had the finding of interest–but only evaluate how well the technique interpreted the text in relation to the feature. Therefore the reference standard for an evaluation of feature detection is generated by experts who read the same text processed by the NLP application and assign values to the same variables. If the reference standard and the NLP application both believe the chest radiograph report describes the possibility of pneumonia, the NLP system is considered correct–even if the patient turned out to not have pneumonia.

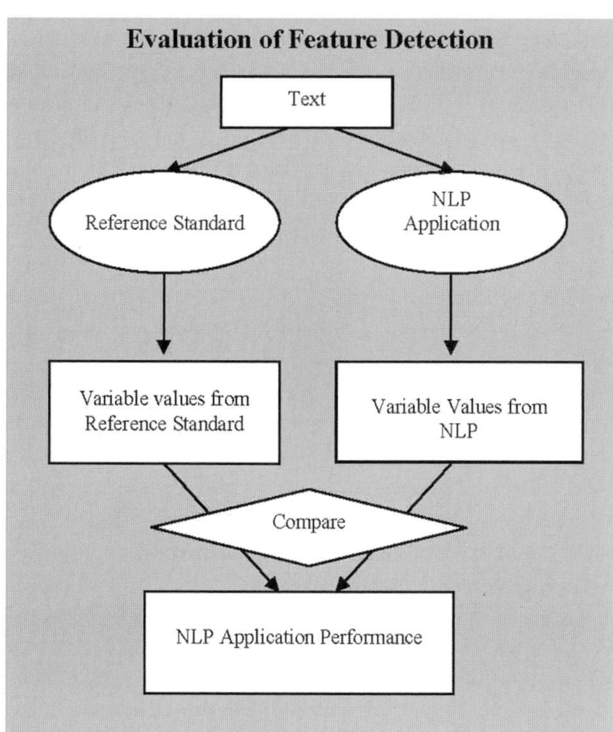

FIGURE 17.5 In an evaluation of feature detection, the NLP application and the reference standard independently extract the relevant variable values from the same text. Performance metrics are calculated by comparing the NLP output against that of the reference standard.

Several studies have evaluated how well NLP applications can encode findings and diseases, such as atelectasis, pleural effusions, CHF, stroke, and pneumonia from radiograph reports (Hripcsak et al., 1995, Friedman et al., 2004, Fiszman et al., 2000, Elkins et al., 2000). The reference standard for these studies was physician encodings of the variables, and the studies showed that the NLP applications performed similarly to physicians. One study (Chapman et al., 2004) evaluated how well the variable fever could be automatically identified in chief complaints and ED reports compared to a reference standard of physician judgment from the ED report. The application identified fever from chief complaints with 100% sensitivity and 100% specificity, and from ED reports with 98% sensitivity and 89% specificity.

Other studies have evaluated how well NLP technology can classify chief complaints into syndromic categories (e.g., respiratory, gastrointestinal, neurological, rash, etc.). Olszewski (2003) evaluated CoCo, a naive Bayesian classifier (Mitchell, 1997) that classifies chief complaints into one of eight syndromic categories. Chapman et al. (2005a) evaluated a chief complaint classifier (MPLUS [Christensen et al., 2002]) that used syntactic and semantic information to classify the chief complaints into syndromic categories. The reference standard for both studies was a physician reading the chief complaints and classifying them into the same syndromic categories. Performance of the NLP applications was measured with the area under the receiver operating characteristic (ROC) curve (areas under curve [AUC]), with AUCs ranging from 0.80 to 0.97 for CoCo and 0.95 to 1.0 for MPLUS, suggesting that NLP technology is quite good at classifying chief complaints into syndromes.

Studies of feature detection do not make claims about whether the NLP technology can accurately diagnose patients with the target findings, syndromes, or diseases. The conclusions only relate to the application's ability to determine the correct values for the variables given the relevant input text. Once feature detection has been validated, the next step is to apply the technology to the problem of diagnosing the patients and evaluate the technology's accuracy at case detection.

7.2. Case Detection

The question being addressed when measuring the case detection ability of an NLP application for the domain of biosurveillance is: *How well does the NLP application identify relevant patients from textual data?* For our SARS detector, examples of case detection evaluations include how well the NLP application can determine whether a patient has a respiratory syndrome, whether a patient has a fever, whether a patient has radiological evidence of pneumonia, or whether a patient has SARS.

Figure 17.6 illustrates the evaluation process for a study on case detection. The reference standard is generated by expert diagnosis of the patients. The source of the expert diagnosis

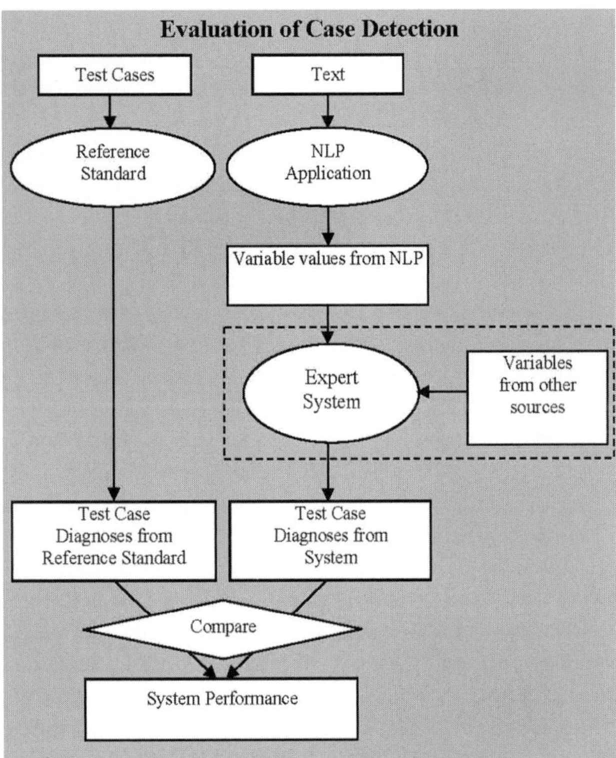

FIGURE 17.6 In an evaluation of case detection, the NLP application extracts relevant variable values from text, which may or may not be combined with variables from other sources (dashed box) to diagnose patients. The reference standard reviews test cases independently and generates a reference diagnosis. Performance metrics are calculated by comparing the diagnoses generated in part or in whole by the NLP application against that of the reference standard.

depends on the finding, syndrome, or disease being diagnosed, and may comprise review of textual patient reports or complete medical records, results of laboratory tests, autopsy results, and so on.

One of the first case detection studies involving an NLP-based system evaluated the ability of a computerized protocol to detect patients suspicious for Tuberculosis with data stored in electronic medical records (Hripcsak et al., 1997, Knirsch et al., 1998). In a prospective study, the system correctly identified 30 of 43 patients with TB. The computerized system also identified four positive patients not identified by clinicians. Aronsky et al. (2001) showed that a Bayesian network for diagnosing patients with pneumonia performed significantly better with information from the chest radiograph encoded with an NLP system than it did without that information (AUC 88% without NLP vs. 92% with NLP).

Several studies have evaluated how well automatically classified chief complaints can classify patients into syndromic categories (Espino and Wagner, 2001, Ivanov et al., 2002, Beitel et al., 2004, Chapman et al., 2005b, Gesteland et al., 2004). The studies used either ICD-9 discharge diagnoses or physician judgment from medical record review as the reference standard for the syndromic categories. The majority of the studies have focused on more prevalent syndromes—respiratory and gastrointestinal—but a few studies have evaluated classification into more rarely occurring syndromes, such as hemorrhagic and botulinic. Results suggest that syndromic surveillance from free-text chief complaints can diagnose patients into most syndromic categories with sensitivities between 40% and 77%, in spite of the limited nature of chief complaints.

In the section on feature detection, we described a study that evaluated the ability of an NLP application to determine whether chief complaints and ED reports described fever (Chapman et al., 2004). The fever study also measured the case detection accuracy of fever diagnosis from chief complaints and ED reports. The NLP application for identifying fever in chief complaints performed with perfect sensitivity and specificity in the feature detection evaluation. However, when quantifying how well the automatically extracted variable of fever from chief complaints identified patients who had a true fever based on reference standard judgment from the ED report, the chief complaint fever detector only performed with a sensitivity of 61%. The specificity remained at 100%. On the one hand, whenever a chief complaint mentioned fever, the patient actually had a fever, so there were no false-positive diagnoses from chief complaints. On the other hand, despite the fact that the NLP technology did not make any mistakes in determining if fever was described in a chief complaint, the chief complaints themselves did not always mention fever when the patient was febrile in the ED. As demonstrated by this study, coupling evaluations on feature detection with evaluations on case detection can inform us about the source of diagnostic errors, which could be the NLP technology, the input data itself, or a combination of the two.

7.3. Epidemic Detection

The question being addressed when measuring the epidemic detection performance of an NLP application in the domain of biosurveillance is *How well does the NLP application contribute to detection of an outbreak?* Evaluating epidemic detection is difficult. The first requirement for an epidemic detection study is reference standard identification of an outbreak. Outbreaks of respiratory and GI illnesses, such as influenza, pneumonia, and gastroenteritis, occur yearly throughout the country. Outbreaks of other infectious or otherwise concerning diseases, such as anthrax, West Nile virus, hemorrhagic fever, or SARS, rarely occur in the United States. Once an outbreak is identified, the next requirement for an epidemic detection evaluation is having access to textual data for an adequate sample of patients living in the geographical area of the outbreak.

One example of an evaluation of epidemic detection involving NLP technology was performed by Ivanov et al. (Ivanov et al., 2003). The evaluation used ICD-9 discharge diagnoses to define retrospective outbreaks of pediatric respiratory and gastrointestinal syndromes over a five year period (1998–2001) in four contiguous counties in Utah. Outcome measures were reported for correlation between chief complaint classifications and ICD-9 classifications and for timeliness of detection. Figure 17.7 from the Ivanov publication shows the time series plot of respiratory illness admissions (reference standard) and chief complaints. It is evident from the plot that chief complaints generated the same type of signal that the reference standard generated. Chief complaint classification detected three respiratory outbreaks with 100% sensitivity and specificity, and time series of chief complaints correlated with hospital admissions and preceded them by an average of 10.3 days.

A study by Irvin (Irvin et al., 2003) showed that numeric chief complaints could correctly detect an influenza outbreak between 1999 and 2000 with one false positive alarm. Although the chief complaints were numeric instead of textual, the same study design could be applied to free-text chief complaint classification for known outbreaks.

Evaluating feature detection is an important first step in evaluation of NLP techniques to ensure that the technology is working as expected. However, to truly understand the impact of NLP in outbreak and disease surveillance, evaluations of case detection and epidemic detection must also be performed.

8. SUMMARY

Natural language processing techniques are far from perfect. However, the question is not whether the techniques perform perfectly but whether the performance is good enough to contribute to disease and outbreak detection. For instance, a few errors in part-of-speech tagging or negation identification may not substantially decrease the ability of an NLP application to determine whether a patient has a fever. Evaluation studies of NLP in biosurveillance are still young, but we have learned a few things about how variables extracted from free-text medical records with NLP can contribute to outbreak detection.

First, we have learned that automated classification of free-text chief complaints, while not perfect, is sufficient to detect one-third to two-thirds of positive syndromic cases. Moreover, chief complaints for pediatric patients are accurate and timely at detecting respiratory and gastrointestinal outbreaks. Second, we have learned that ED reports can provide more detailed information about the state of a patient than chief complaints. For example, we can detect 40% more patients with fever from ED reports than from chief complaints (Chapman et al., 2004). Third, several researchers have shown that identification of radiological variables required for detection of many public health threats (including SARS and inhalational anthrax) from chest radiograph reports is feasible with NLP techniques.

This chapter has focused on applying NLP techniques to variable extraction from patient medical records, but other types of free-text documents contain information that may be useful for biosurveillance, including web queries, transcripts from call centers, and autopsy reports. Regardless of the type of free-text data, we suggest three questions to consider when deciding whether application of NLP techniques to textual data is feasible for disease and outbreak detection: (1) How complex is the text? The simple phrases in chief complaints are much easier to understand than complex discourses contained in ED reports. Textual data that require coreference resolution, domain modeling for inference, and other more difficult techniques required to identify values for the variables of interest will be more challenging to process and will be more prone to error. (2) What is the goal of the NLP technique? If the goal is to understand all temporal, anatomic, and diagnostic relations described in the text as well as a physician could, you may be in for a lifetime of hard but interesting work. Extraction of a single variable, such as fever, or encoding temporal, anatomic, and diagnostic relations for a finite set of findings, such as all respiratory findings, is more feasible. (3) Can the detection algorithms that will use the variables extracted with NLP handle noise? Detecting small outbreaks requires more accuracy in the input variables.

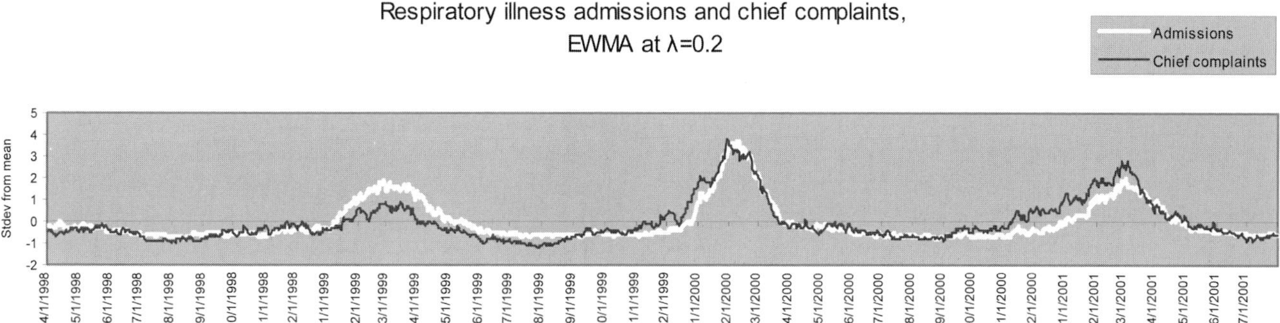

FIGURE 17.7 Time series plot of chief complaint syndromic classifications against ICD-9 discharge diagnoses for admissions of patients with respiratory illnesses including pneumonia, influenza, and bronchiolitis.

As an extreme example, some diseases such as inhalational anthrax require only a single case to be considered a threatening outbreak. If the NLP-based expert system did not correctly detect that case, then the detection system would have failed. However, in detecting an outbreak of a gastrointestinal illness, for example, if the NLP-based expert system only detected two-thirds of the true cases, there may still be enough positive patients to detect a moderate to large-sized outbreak. In addition, the consistent stream of false positive cases identified by the NLP-based expert system would comprise a noisy baseline that may not prevent the algorithm from detecting a significant increase in gastrointestinal cases but would require a larger increase to detect the outbreak. Consideration of these three questions can help determine the feasibility of using NLP for outbreak and disease surveillance.

NLP techniques can be applied to determine the values of predefined variables that may be useful in detecting outbreaks. The linguistic structure of the textual data being processed and the nature of the variables being used for surveillance determine the feasibility of applying NLP techniques to the problem. Characteristics such as linguistic variation, polysemy, negation, contextual information, finding validation, implication, and coreference must be accounted for to understand the information within patient medical reports as well as a physician does. However, because many of the variables helpful in biosurveillance do not require complete understanding of the text, NLP techniques may successfully extract variables useful for outbreak detection. In fact, evaluations of feature detection, case detection, and epidemic detection of NLP techniques have begun to demonstrate the utility of NLP techniques in this new field. More research in NLP techniques and more evaluation studies of the effectiveness of NLP will not only increase our understanding of how to extract information from text but will also help us continue to learn what types of data provide the most timely and accurate information for detecting outbreaks.

ADDITIONAL RESOURCES

Natural Language Processing Textbooks

- Allen, J. (1995). *Natural Language Understanding*. Redwood City, CA: Benjamin/Cummings Publishing Company.
- Charniak, E. (1993). *Statistical Language Learning*. Cambridge, MA: MIT Press.
- Jurafsky, D., Martin, J. H. (2000). *Speech and Language Processing*. Upper Saddle River, NJ: Prentice Hall.
- Manning, C. D., Schutze, H. (1999). *Foundations of Statistical Natural Language Processing*. Cambridge, MA: MIT Press.

Data Mining from Biomedical Texts

- Aronis, J. M., Cooper, G. F., Kayaalp, M., et al. (1999). Identifying patient subgroups with simple Bayes'. In: *Proceedings of American Medical Informatics Association Annual Symposium*, 658–62.
- Cooper, G. F., Buchanan, B. G., Kayaalp, M., et al. (1998). Using computer modeling to help identify patient subgroups in clinical data repositories. In: *Proceedings of American Medical Informatics Association Annual Symposium*, 180–4.
- Friedman, C., Kra, P., Yu, H., et al. (2001). GENIES: a natural-language processing system for the extraction of molecular pathways from journal articles. *BioInformatics* 17 (Suppl 1): S74–82.
- Johnson, D. B., Chu, W. W., Dionisio, J. D., et al. (1999). Creating and indexing teaching files from free-text patient reports. In: *Proceedings of American Medical Informatics Association Annual Symposium*, 814–8.
- Krauthammer, M., Rzhetsky, A., Morozov, P., et al. (2000). Using BLAST for identifying gene and protein names in journal articles. *Gene* 259:1–2, 245–52.
- Liu, H., Friedman, C. (2003). Mining terminological knowledge in large biomedical corpora. *Pacific Symposium on Biocomputing*, 415–26.
- Lussier, Y. A., Shagina, L., Friedman, C. (2001). Automating SNOMED coding using medical language understanding: a feasibility study. In: *Proceedings of American Medical Informatics Association Annual Symposium*, 418–22.
- McCray, A. T. (1991). Extending a natural language parser with UMLS knowledge. In: *Proceedings of Annual Symposium on Computer Applications in Medical Care*, 194–8.
- McCray, A. T. (2000). Digital library research and application. *Stud Health Technol Inform* 76:51–62.
- McCray, A. T., Nelson, S. J. (1995). The representation of meaning in the UMLS. *Methods Inform Med* 34:1–2, 193–201.
- McCray, A. T., Razi, A. M., Bangalore, A. K., et al. (1996). The UMLS Knowledge Source Server: a versatile Internet-based research tool. In: *Proceedings of American Medical Informatics Association Fall Symposium*, 164–8.
- McCray, A. T., Soponsler, J., Brylawski, B., et al. (1987). The role of lexical knowledge in biomedical text understanding. In: *Symposium on Computer Applications in Medical Care* 87:103–7.
- McCray, A. T., Srinivasan, S., Browne, A. C. (1994). Lexical methods for managing variation in biomedical terminologies. In: *Symposium on Computer Applications in Medical Care*, 235–9.
- Mendonca, E. A., Cimino, J. J., Johnson, S. B., et al. (2001). Accessing heterogeneous sources of evidence to answer clinical questions. *J Biomed Inform* 34:85–98.
- Rzhetsky, A., Koike, T., Kalachikov, S., et al. (2000). A knowledge model for analysis and simulation of regulatory networks. *BioInformatics* 16:1120–8.
- Yu, H., Hatzivassiloglou V., Friedman, C., et al. (2002). Automatic extraction of gene and protein synonyms from MEDLINE and journal articles. In: *Proceedings of American Medical Informatics Association Symposium*, 919–23.

ACKNOWLEDGMENTS

We thank Thomas C. Rindflesch for his helpful comments.

REFERENCES

Allen, J. (1995). *Natural Language Understanding*. Redwood City, CA: Benjamin/Cummings Publishing Company.

Aronis, J. M., Cooper, G. F., Kayaalp, M., et al. (1999). Identifying patient subgroups with simple Bayes'. In: *Proceedings of American Medical Informatics Association Symposium*, 658–62.

Aronsky, D., Fiszman, M., Chapman, W. W., et al. (2001). Combining decision support methodologies to diagnose pneumonia. In: *Proceedings of American Medical Informatics Association Symposium*, 12–6.

Beitel, A. J., Olson, K. L., Reis, B. Y., et al. (2004). Use of emergency department chief complaint and diagnostic codes for identifying respiratory illness in a pediatric population. *Pediatr Emerg Care* 20:355–60.

Campbell, D. A., Johnson, S. B. (2001). Comparing syntactic complexity in medical and non-medical corpora. In: *Proceedings of American Medical Informatics Association Symposium*, 90–4.

Chapman, W. W., Bridewell, W., Hanbury, P., et al. (2001a). Evaluation of negation phrases in narrative clinical reports. In: *Proceedings of American Medical Informatics Association Symposium*, 105–9.

Chapman, W. W., Christensen, L. M., Wagner, M. M., et al. (2005a). Classifying free-text triage chief complaints into syndromic categories with natural language processing. *Artif Intell Med* 33:31–40.

Chapman, W. W., Cooper, G. F., Hanbury, P., et al. (2003). Creating a text classifier to detect radiology reports describing mediastinal findings associated with inhalational anthrax and other disorders. *J Am Med Inform Assoc* 10: 494–503.

Chapman, W. W., Dowling, J. N., Wagner, M. M. (2004). Fever detection from free-text clinical records for biosurveillance. *J Biomed Inform* 37:120–7.

Chapman, W. W., Dowling, J. N., Wagner, M. M. (2005b). Classification of emergency department chief complaints into seven syndromes: a retrospective analysis of 527,228 patients. *Ann Emerg Med*. 46(5):445–55.

Chapman, W. W., Fiszman, M., Frederick, P. R., et al. (2001b). Quantifying the characteristics of unambiguous chest radiography reports in the context of pneumonia. *Acad Radiol* 8:57–66.

Chapman, W. W., Fizman, M., Chapman, B. E., et al. (2001c). A comparison of classification algorithms to automatically identify chest X-ray reports that support pneumonia. *J Biomed Inform* 34:4–14.

Christensen, L., Haug, P. J., Fiszman, M. (2002). MPLUS: a probabilistic medical language understanding system. In: *Proceedings of Workshop on Natural Language Processing in the Biomedical Domain*, 29–36.

Chuang, J. H., Friedman, C., Hripcsak, G. (2002). A comparison of the Charlson comorbidities derived from medical language processing and administrative data. In: *Proceedings of American Medical Informatics Association Symposium*, 160–4.

Coden, A. R., Pakhomov, S. V., Ando, R. K., et al. (2005). Domain-specific language models and lexicons for tagging. *J Biomed Inform*. 38(6):422–30.

Cooper, G. F., Buchanan, B. G., Kayaalp, M., et al. (1998). Using computer modeling to help identify patient subgroups in clinical data repositories. In: *Proceedings of American Medical Informatics Association Symposium*, 180–4.

Elkins, J. S., Friedman, C., Boden-Albala, B., et al. (2000). Coding neuroradiology reports for the Northern Manhattan Stroke Study: a comparison of natural language processing and manual review. *Comput Biomed Res* 33:1–10.

Espino, J. U., Wagner, M. M. (2001). Accuracy of ICD-9-coded chief complaints and diagnoses for the detection of acute respiratory illness. In: *Proceedings of American Medical Informatics Association Symposium*, 164–8.

Fiszman, M., Chapman, W. W., Aronsky, D., et al. (2000). Automatic detection of acute bacterial pneumonia from chest X-ray reports. *J Am Med Inform Assoc* 7:593–604.

Fiszman, M., Chapman, W. W., Evans, S. R., et al. (1999). Automatic identification of pneumonia related concepts on chest x-ray reports. In: *Proceedings of American Medical Informatics Association Symposium*, 67–71.

Fiszman, M., Haug, P. J. (2000). Using medical language processing to support real-time evaluation of pneumonia guidelines. In: *Proceedings of American Medical Informatics Association Symposium*, 235–9.

Fiszman, M., Haug, P. J., Frederick, P. R. (1998). Automatic extraction of PIOPED interpretations from ventilation/perfusion lung scan reports. In: *Proceedings of American Medical Informatics Association Symposium*, 860–4.

Friedman, C. (2000). A broad-coverage natural language processing system. In: *Proceedings of American Medical Informatics Association Symposium*, 270–4.

Friedman, C., Alderson, P. O., Austin, J. H., et al. (1994). A general natural-language text processor for clinical radiology. *J Am Med Inform Assoc* 1:2, 161–74.

Friedman, C., Hripcsak, G. (1999). Natural language processing and its future in medicine. *Acad Med* 74:890–5.

Friedman, C., Hripcsak, G., Shablinsky, I. (1998). An evaluation of natural language processing methodologies. In: *Proceedings of American Medical Informatics Association Symposium*, 855–9.

Friedman, C., Hripcsak, G., Shagina, L., et al. (1999a). Representing Information in patient reports using natural language processing and the extensible markup language. *J Am Med Inform Assoc* 6:76–87.

Friedman, C., Knirsch, C., Shagina, L., et al. (1999b). Automating a severity score guideline for community-acquired pneumonia employing medical language processing of discharge summaries. In: *Proceedings of American Medical Informatics Association Symposium*, 256–60.

Friedman, C., Kra, P., Rzhetsky, A. (2002). Two biomedical sublanguages: a description based on the theories of Zellig Harris. *J Biomed Inform* 35:222–35.

Friedman, C., Shagina, L., Lussier, Y., et al. (2004). Automated encoding of clinical documents based on natural language processing. *J Am Med Inform Assoc*. 11(5):392–402.

Gesteland, P. H., Wagner, M. M., Gardner, R. M., et al. (2004). Surveillance of Syndromes during the Salt Lake 2002 Winter Olympic Games: An Evaluation of a Naive Bayes Chief Complaint Coder. In preparation.

Gundersen, M. L., Haug, P. J., Pryor, T. A., et al. (1996). Development and evaluation of a computerized admission diagnoses encoding system. *Comput Biomed Res* 29:351–72.

Hersh, W., Mailhot, M., Arnott-Smith, C., et al. (2001). Selective automated indexing of findings and diagnoses in radiology reports. *J Biomed Inform* 34:262–73.

Hripcsak, G., Austin, J. H., Alderson, P. O., et al. (2002). Use of natural language processing to translate clinical Information from a database of 889,921 chest radiographic reports. *Radiology* 224:157–63.

Hripcsak, G., Friedman, C., Alderson, P. O., et al. (1995). Unlocking clinical data from narrative reports: a study of natural language processing. *Ann Intern Med* 122:681–8.

Hripcsak, G., Knirsch, C. A., Jain, N. L., et al. (1997). Automated tuberculosis detection. *J Am Med Inform Assoc* 4:376–81.

Hripcsak, G., Knirsch, C. A., Jain, N. L., et al. (1999). A health Information network for managing innercity tuberculosis: bridging clinical care, public health, and home care. *Comput Biomed Res* 32:67–76.

Hripcsak, G., Wilcox, A. (2002). Reference standards, judges, and comparison subjects: roles for experts in evaluating system performance. *J Am Med Inform Assoc* 9:1–15.

Irvin, C. B., Nouhan, P. P., Rice, K. (2003). Syndromic analysis of computerized emergency department patients' chief complaints: an opportunity for bioterrorism and influenza surveillance. *Ann Emerg Med* 41:447–52.

Ivanov, O., Gesteland, P., Hogan, W., et al. (2003). Detection of pediatric respiratory and gastrointestinal outbreaks from free-text chief complaints. In: *Proceedings of American Medical Informatics Association Fall Symposium*, 318–22.

Ivanov, O., Wagner, M. M., Chapman, W. W., et al. (2002). Accuracy of three classifiers of acute gastrointestinal syndrome for syndromic surveillance. In: *Proceedings of American Medical Informatics Association Symposium*, 345–9.

Jain, N. L., Friedman, C. (1997). Identification of findings suspicious for breast cancer based on natural language processing of mammogram reports. In: *Proceedings of American Medical Informatics Association Fall Symposium*, 829–33.

Jain, N. L., Knirsch, C. A., Friedman, C., et al. (1996). Identification of suspected tuberculosis patients based on natural language processing of chest radiograph reports. In: *Proceedings of American Medical Informatics Association Fall Symposium*, 542–6.

Knirsch, C. A., Jain, N. L., Pablos-Mendez, A., et al. (1998). Respiratory isolation of tuberculosis patients using clinical guidelines and an automated clinical decision support system. *Infect Control Hosp Epidemiol* 19:94–100.

Liu, H., Lussier, Y. A., Friedman, C. (2001). Disambiguating ambiguous biomedical terms in biomedical narrative text: an unsupervised method. *J Biomed Inform* 34:249–61.

Lovis, C., Chapko, M. K., Martin, D. P., et al. (2001). Evaluation of a command-line parser-based order entry pathway for the Department of Veterans Affairs electronic patient record. *J Am Med Inform Assoc* 8:5, 486–98.

Manning, C. D., Schutze, H. (1999). *Foundations of Statistical Natural Language Processing*. Cambridge, MA: MIT Press.

Mitchell, T. M. (1997). *Machine Learning*. Boston: McGraw-Hill.

Morioka, C. A., Sinha, U., Taira, R., et al. (2002). Structured reporting in neuroradiology. *Ann N Y Acad Sci* 980:259–66.

Mutalik, P. G., Deshpande, A., Nadkarni, P. M. (2001). Use of general-purpose negation detection to augment concept indexing of medical documents: a quantitative study using the UMLS. *J Am Med Inform Assoc* 8:598–609.

Olszewski, R. T. (2003). Bayesian classification of triage diagnoses for the early detection of epidemics. In: *Proceedings of FLAIRS Conference*, 412–6.

Pakhomov, S. V., Ruggieri, A., Chute, C. G. (2002). Maximum entropy modeling for mining patient medication status from free text. In: *Proceedings of American Medical Informatics Association Symposium*, 587–91.

Roth, L., Hole, W. T. (2000). Managing name ambiguity in the UMLS metathesaurus. In: *Proceedings of American Medical Informatics Association Fall Symposium*, 1124.

Sinha, U., Dai, B., Johnson, D. B., et al. (2000). Interactive software for generation and visualization of structured findings in radiology reports. *AJR Am J Roentgenol* 175:609–12.

Sinha, U., Taira, R., Kangarloo, H. (2001a). Structure localization in brain images: application to relevant image selection. In: *Proceedings of American Medical Informatics Association Symposium*, 622–6.

Sinha, U., Ton, A., Yaghmai, A., et al. (2001b). Image content extraction: application to MR images of the brain. *Radiographics* 21:535–47.

Smith, L., Rindflesch, T., Wilbur, W. J. (2004). MedPost: a part-of-speech tagger for bioMedical text. *BioInformatics* 20: 2320–1.

Sneiderman, C. A., Rindflesch, T. C., Aronson, A. R. (1996). Finding the findings: identification of findings in medical literature using restricted natural language processing. In: *Proceedings of American Medical Informatics Association Fall Symposium*, 239–43.

Spyns, P. (1996). Natural language processing in medicine: an overview. *Methods Inform Med* 35:4–5, 285–301.

Szolovits, P. (2003). Adding a medical lexicon to an english parser. In: *Proceedings of American Medical Informatics Association Symposium*, 639–43.

Taira, R. K., Soderland, S. G. (1999). A statistical natural language processor for medical reports. In: *Proceedings of American Medical Informatics Association Symposium*, 970–4.

Travers, D. A., Haas, S. W. (2003). Using nurses' natural language entries to build a concept-oriented terminology for patients' chief complaints in the emergency department. *J Biomed Inform* 36:4–5, 260–70.

Travers, D. A., Waller, A., Haas, S. W., et al. (2003). Emergency department data for bioterrorism surveillance: electronic data availability, timeliness, sources and standards. In: *Proceedings of American Medical Informatics Association Symposium*, 664–8.

Wilcox, A. B., Narus, S. P., Bowes, W. A., 3rd. (2002). Using natural language processing to analyze physician modifications to data entry templates. In: *Proceedings of American Medical Informatics Association Symposium*, 899–903.

Wren, J. D., Garner, H. R. (2002). Heuristics for identification of acronym-definition patterns within text: towards an automated construction of comprehensive acronym-definition dictionaries. *Methods Inform Med* 41:426–34.

18

CHAPTER

Bayesian Methods for Diagnosing Outbreaks

Gregory F. Cooper, Denver H. Dash, John D. Levander, Weng-Keen Wong, William R. Hogan, and Michael M. Wagner

RODS Laboratory, Intel Research, Santa Clara, California

1. INTRODUCTION

Early, reliable detection of outbreaks of disease, whether natural (e.g., West Nile virus and SARS) or bioterrorist induced (e.g., anthrax and smallpox), is a critically important problem today. We need to detect outbreaks as early as possible in order to provide the best response and treatment, as well as improve the chances of identifying the source.

Outbreaks often present signals that are weak and noisy early in the event. If we hope to achieve rapid and reliable detection, it likely will be necessary to integrate multiple weak signals that together provide a relatively stronger signal of an outbreak. Combining spatial and temporal data is an important instance of such integration. For example, even though the number of patients in a given city with fever, who were seen in emergency departments (EDs) in the past 24 hours, may not be noticeably higher than average, nonetheless, for the past 12 hours, it may be significantly higher for a given neighborhood of the city.

Because of the noise in signals early in the event, early detection is almost always detection under uncertainty. In the research reported here, we use probability as a measure of uncertainty. A well-organized probabilistic approach allows for the rational combination of multiple, small indicators into a big picture. Since the interrelationships among risk factors, diseases, and symptoms often are causal, a causal representation provides a natural approach to modeling. This chapter concentrates on causal modeling of how the data could be produced by various hypothesized outbreak diseases. The more we know about the causal relationships among the risk factors, the outbreak disease etiologies, and the clinical presentations of those outbreak diseases, the more compelling is the causal modeling approach.

This chapter focuses on the use of causal Bayesian networks to represent causal relationships under uncertainty. The chapter starts with a brief overview of Bayesian modeling in general and then causal Bayesian networks in particular. In order to illustrate Bayesian modeling using Bayesian networks, the remainder of the chapter describes a detailed example in which a causal Bayesian network is used to model an entire population of people (Cooper et al., 2004).[1]

2. BACKGROUND

This section provides a brief background on Bayesian modeling and inference (Bernardo and Smith, 1994) and on Bayesian networks (Neapolitan, 2004).

2.1. Bayesian Modeling and Inference

Let H be some hypothesis of interest (e.g., the disease state of a patient) and let E denote available evidence (e.g., a patient's symptoms). We often are interested in knowing the probability of H in light of E, that is, $P(H|E)$.

In order to derive $P(H|E)$, we need a model that relates the evidence to the hypothesis. It often is easier to model how the hypothesis might lead to the evidence, namely $P(E|H)$, than to model directly how the evidence implicates the hypothesis, namely $P(H|E)$. When the hypotheses being modeled are causally influencing the evidence, it is natural to model in the causal direction from hypotheses to evidence. For example, we could model the probability distribution of *cough* in an individual with *influenza* (i.e., P ([*cough* = present | *influenza* = present])), and alternatively, the probability of *cough* given that there is no *influenza* (i.e., P(*cough* = present | *influenza* = absent)).

We cannot derive $P(H|E)$ just from $P(E|H)$; we also need to know the probabilities $P(H)$ and $P(E)$, and then combine them as follows:

$$P(H|E) = \frac{P(E|H) \cdot P(H)}{P(E)}.$$

By replacing $P(E)$ with an equivalent expression, we obtain the following equation, which is called Bayes rule:

$$P(H|E) = \frac{P(E|H) \cdot P(H)}{\sum_{H'} P(E|H') \cdot P(H')},$$

where the sum is taken over all the hypotheses H' being modeled (e.g., "influenza" and "no influenza"). The terms in Bayes rules are referred to as follows: $P(H)$ is the *prior probability* of H, $P(E|H)$ is the likelihood of E given H, and $P(H|E)$ is the *posterior probability* of H given E.

1 This chapter is based on a paper by Cooper [2004].

Handbook of Biosurveillance
ISBN 0-12-369378-0

Modeling the likelihood function, $P(E|H)$, involves specifying parameters within the model. For example, in the simplest case that E and H each consist of one binary variable, then two parameters are needed to specify the likelihood function. $P(H)$ represents the prior probability of the hypothesis itself (e.g., the probability of *influenza*). The parameters and hypothesis priors may be estimated in a variety of ways, including the use of previously available training data, the literature, expert opinion, and their combination. Section 4.1 illustrates how priors are estimated from a combination of information sources.

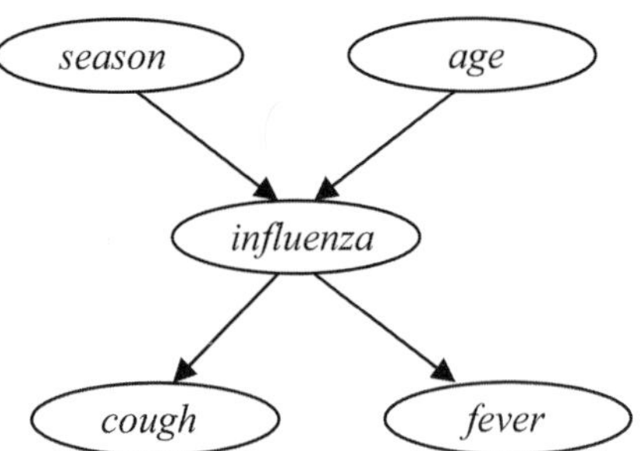

FIGURE 18.1 An example of a Bayesian network structure.

2.2. Bayesian Networks

A Bayesian network (BN) model represents the joint distribution of a set of variables of interest. For example, over all possible joint states of the variables in E and H, a BN would represent $P(E, H)$. From such a joint distribution, we can derive any probability of interest, such as $P(H|E)$.

A BN has two parts: a structure and a set of parameters. The structure contains nodes, which represent model variables,[2] as well as arcs, which connect nodes that are related probabilistically. The resulting graph is required to contain no directed cycles, meaning that it is not possible to start at some node and follow a path of arcs that leads back to that same node. The parents of a node (variable) X_i are those nodes that have arcs into X_i. A descendant of Xi is a node X_j for which there is a directed path from X_i to X_j. The following Markov condition specifies the independence that is expressed in a BN: *A node is independent of its nondescendants, given just the state of its parents.*

The Markov condition is the most important key to understanding Bayesian networks. It tells us that if we want to predict the value of some variable in the Bayesian network, and we already know the values of all of its parents, then no variable (except possibly some of our node's descendents) could possibly give us any useful predictive information beyond that supplied by its parents. The BN structure in Figure 18.1 contains five nodes. The node season could be modeled as having the values *spring*, *summer*, *fall*, and *winter*. *Age* represents a patient's age, perhaps discretized into ranges of years. The nodes *influenza*, *cough*, and *fever* could be modeled as being *present* or *absent*.

As an example of the BN Markov condition, we see that the structure in Figure 18.1 specifies that *cough* is independent of *season* given the state of *influenza*. If the designer of the network had decided that this independence assumption was unreasonable they could have added an arc from *season* to *cough*.

Someone who designs a Bayesian network first chooses the structure. But more work is needed after that. The network must be populated with numerical parameters. For each node X_i, there is a probability distribution $P(X_i \mid \text{parents}(X_i))$, where parents($Xi$) are the parents of X_i in the BN. For example, we might have $P(cough = \text{present} \mid influenza = \text{present}) = 0.90$. If X_i contains no parents, then the probability $P(X_i)$ is specified. The BN Markov condition implies that we can factor the joint distribution of the variables in the model as follows (Neapolitan, 2004):

$$P(X_1, X_2, \ldots, X_n) = \prod_{i=1}^{n} P(X_i | parents(X_i)) \quad (1)$$

The fewer the number of parents per node, the fewer the number of parameters needed to specify each conditional probability $P(X_i \mid \text{parents}(X_i))$, and thus, the fewer the number of parameters in the model. Therefore, a BN with relatively fewer arcs will require relatively fewer model parameters. A BN can thereby represent parsimoniously the joint distribution over n variables.

If the arcs are interpreted as being direct causal relationships (relative to the variables being modeled) then the model is called a *causal Bayesian network*. For a causal BN, the Markov condition states (in part) that effects are independent of distant causes, given the state of all the proximal causes. For example, *fever* is independent of the distant influence of *season* and *age*, given that we know the state (present or absent) of *influenza*, which is the proximal cause of *fever* in the model in Figure 18.1.

Equation 1 specifies a complete joint probability distribution over all n variables in the BN model. From such a joint distribution, we can derive the probability of any subset of nodes conditioned of the state of another subset of nodes. Thus, for the example BN, we could derive $P(influenza \mid cough = \text{present}, season = \text{winter}, fever = \text{present})$. Note that information about the patient's age is missing in this conditional probability; in general, we need only condition on a subset of the variables in the model.

Researchers have developed exact BN inference algorithms that take advantage of the independence among the variables that follows from the BN Markov condition when some arcs are missing. These algorithms are often able to derive conditional probabilities relatively efficiently (Neapolitan, 2004). When exact inference would require too much computation time, approximate algorithms are available (Neapolitan, 2004).

2 We use the terms node and variable interchangeably.

3. BIOSURVEILLANCE USING BAYESIAN NETWORKS

As an example of the use of Bayesian networks for biosurveillance, this section presents an approach to modeling and detecting non-contagious outbreak diseases, such as disease due to airborne-released anthrax. In particular, the approach models each individual in the population being monitored for an outbreak. Modeling an entire population of people in just one city-wide area leads to a Bayesian network model with millions of nodes. For example, the model reported here contains approximately 20 million nodes. Each individual in the population is represented by a 14-node subnetwork, which captures important syndromic information that is commonly available for health surveillance (such as ED chief complaints), while avoiding any information that could personally identify the individual (e.g., name, social security number, and home street address).

Given current data about individuals in the population, we use a Bayesian network to infer the posterior probabilities of outbreak diseases in the population. To provide timely detection, inference needs to be performed in real time, such that the biosurveillance system "keeps up" with the data streaming in. Once the probability of an outbreak exceeds a particular threshold, an alert is generated by the Bayesian-network–based biosurveillance system; this alert can serve to warn public health officials.

Using such a large Bayesian network presents both modeling and inference challenges. To help make modeling more tractable in terms of computational space, we use the following approach: if some groups of people are indistinguishable, according to the data being captured, we model them with a single subpopulation subnetwork. To speed up inference, we use a method that need only update the network state based on new information about individuals in the population (such as newly available clinical information that is based on people who have recently visited EDs in seeking care).

A key contribution of this section of the chapter is the explication of assumptions and techniques that are sufficient to allow the scaling of Bayesian network modeling and inference to millions of nodes for real-time surveillance applications, thus providing a proof of concept that Bayesian networks can serve as the foundation of a system that effectively performs Bayesian biosurveillance of disease outbreaks. With this foundation in place, many extensions are possible, and we outline several of them in the final section of the chapter.

In the remainder of this section, we first outline our general approach for using causal Bayesian networks to represent noncontagious diseases that can cause outbreaks of disease. Next, we introduce the specific network we have constructed to monitor for an outbreak caused by the outdoor release of anthrax spores. We then describe an experiment that involves injecting simulated cases of patients with anthrax (which were generated from a separate model) onto background data of real cases of patients who visited EDs during a period when there were no known outbreaks of disease occurring. We measure how long it takes the Bayesian network system to detect such simulated outbreaks. Finally, we discuss these results and suggest directions for future research.

3.1. Modeling

Our methodology uses Bayesian networks to explicitly model an entire population of *individuals*. Since, in this chapter, we are specifically interested in disease outbreak detection from syndromic information, we will refer to models of these individuals as *person models*, although obviously the same ideas could be applied to model other entities that might provide information about disease outbreaks, such as biosensors and livestock.

We explicitly model each person in the population, and thus in our BN there will exist (at least conceptually) a subnetwork (or object) $\boldsymbol{P}_i$ for each person. Each such subnetwork is in essence a diagnostic expert system applied to a given person. By connecting those subnetworks appropriately, we obtain a population diagnostic model that is built from person-specific diagnostic components. An example of a complete model for four people is shown in Figure 18.2, where each person in the population is represented with a simple six-node subnetwork structure. In this particular example, there is only one person model (class), but the methodology can allow for more.

In this chapter, we restrict the methodology to model non-contagious diseases. We partition all the nodes $\boldsymbol{X}$ in the network into three parts:

1. A set of global nodes $\boldsymbol{G}$,
2. A set of interface nodes, $\boldsymbol{I}$, and
3. A set of person subnetworks $\boldsymbol{P} = \{\boldsymbol{P}_1, \boldsymbol{P}_0, \ldots, \boldsymbol{P}_n\}$.

The set $\boldsymbol{G}$, defined as $\boldsymbol{G} = \boldsymbol{X} \setminus \{\boldsymbol{I} \cup \boldsymbol{P}\}$, contains nodes that represent global features common to all people. For the example in Figure 18.2, $\boldsymbol{G}$ consists of two nodes: *Terror Alert Level* (having states *Green*, *White*, *Yellow*, *Orange*, and *Red*), and *Anthrax Release* (having states *Yes* and *No*). Set $\boldsymbol{I}$ contains factors that directly influence the status of the outbreak of disease in people in the population. Each $\boldsymbol{P}_i$ subnetwork (object) represents a person in the population.

Structurally, we make the following two assumptions. Both assumptions use the notion of d-separation, which is described fully in (Neapolitan, 2004). In brief, *d-separation* is a graphical condition on Bayesian network structures that implies conditional independence. Thus, if for some BN the nodes in set A d-separate each node in set B from each node in set C, then it follows from the BN Markov condition that each node in B is independent of each node in C, given the states of the nodes in A.

Assumption 1: The interface nodes, $\boldsymbol{I}$, d-separate the person subnetworks from each other, and any arc between a node I in $\boldsymbol{I}$ and a node X in some person subnetwork $\boldsymbol{P}_i$ is oriented from I to X.

Thus, we do not allow arcs between the person models.

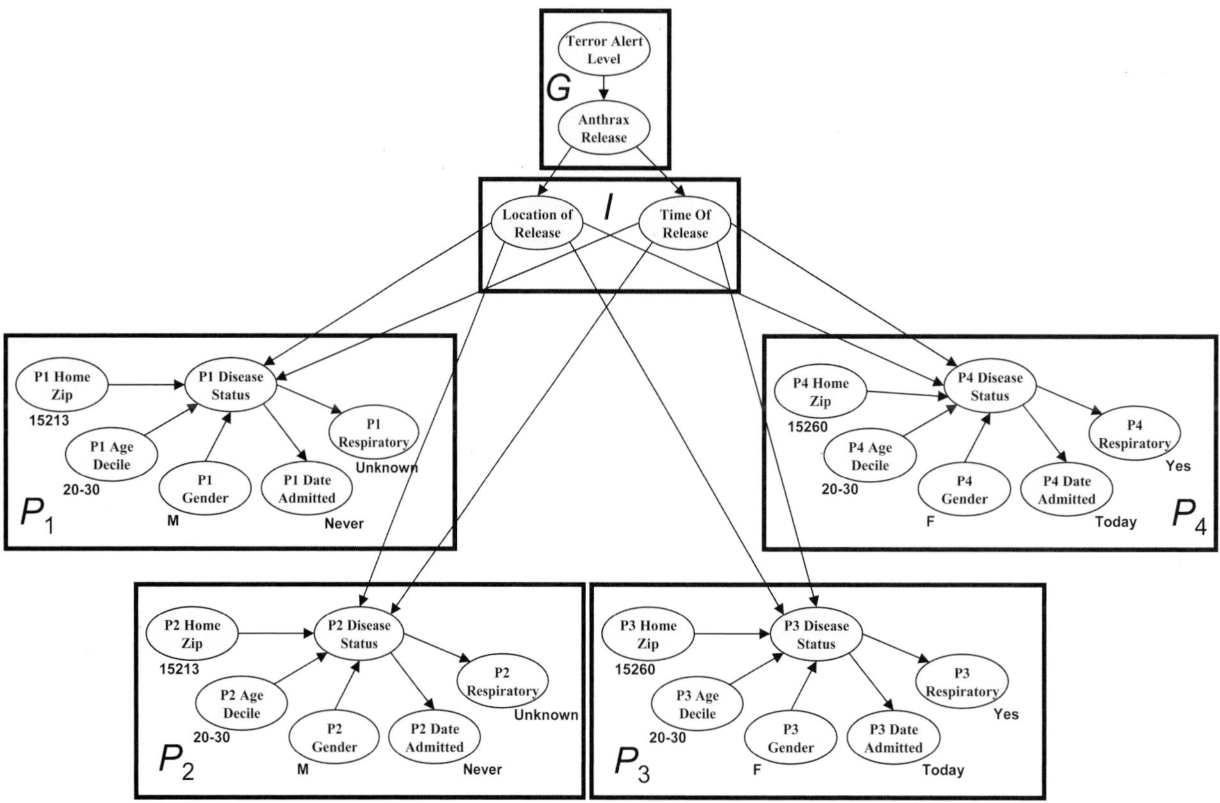

FIGURE 18.2 A simplified four-person model for detecting an outbreak of anthrax. Each person $\boldsymbol{P}_i$ in the population is represented explicitly by a six-node subnetwork.

Assumption 2: The interface nodes, ***I***, d-separate the nodes in ***G*** from the nodes in ***P***, and any arc between a node *G* in ***G*** and a node *I* in ***I*** is oriented from *G* to *I*.

Figure 18.3 presents the above two assumptions in diagrammatic form.

For noncontagious diseases that may cause outbreaks, Assumptions 1 and 2 are reasonable when ***I*** contains all of the factors that significantly influence the status of an outbreak disease in individuals in the population. In the case of bioterrorist-released bioagents, for example, such information includes the time and place of release of the agent. Key characteristics of nodes in ***I*** are that they have arcs to the nodes in one or more person models, and they induce the conditional independence relationships described in Assumptions 1 and 2. Often the variables in ***I*** will be unmeasured. It is legitimate, however, to have measured variables in ***I***. For example, the regional smog level (not shown in Figure 18.2) might be a measured variable that influences the disease status of people in the population, and thus it would be located in ***I***.

Let *T* be a variable in ***G*** that represents a disease outbreak. In Figure 18.2, *T* is the node *Anthrax Release*. The goal of our biosurveillance method is to continually derive an updated posterior probability over the states of *T* as data about the population streams in.

We consider spatio-temporal data in deriving the posterior probability of *T*. For example, we consider information about when patient cases appear at the ED, as well as the home location (at the level of zip codes) of those patients. In our current implementation, spatio-temporal information is explicitly

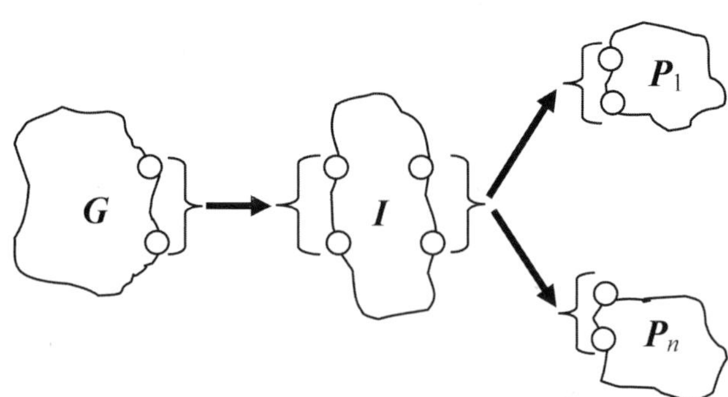

FIGURE 18.3 The closed regions represent Bayesian subnetworks. The circles on the edges of the subnetworks denote nodes that are connected by arcs that bridge the subnetworks. Only two such "I/O" nodes are shown per subnetwork, but in general there could be any number. The arrows between subnetworks show the direction in which the Bayesian-network arcs are oriented between the subnetworks. The braces show which nodes can (possibly) be connected by arcs. In subnetwork *I*, the I/O nodes on the left and those on the right are not necessarily distinct.

represented by nodes in the network, such as *Location of Release*, *Time of Release*, and patient *Home Zip*. We note that in Figure 18.2, the *Disease Status* nodes contain values that indicate when the disease started (if ever) and ended. This temporal representation has the advantage (over, for example, dynamic Bayesian networks (DBNs) [Neapolitan, 2004]) of being relatively compact. The method allows us to create a network with fewer parameters than the corresponding DBN, and simplifies our method for performing real-time inference.

3.2. Inference

When performing inference for biosurveillance, our goal is to continuously monitor target variable T by deriving its updated posterior probability as new data arrives. At any given time, there are two general sources of evidence we consider:

1. General evidence at the global level: $\boldsymbol{g} = \{G = g: G \in \boldsymbol{G}\}$
2. The collective set of evidence $\boldsymbol{e}$ that we observe from the population of people: $\boldsymbol{e} = \{X = e: X \in \boldsymbol{P}_i, \boldsymbol{P}_i \in \boldsymbol{P}\}$

In our application, $\boldsymbol{g}$ might consist of the observation that the *Terror Alert Level* = *Orange*, and $\boldsymbol{e}$ might include information about the patients who have visited EDs in the region in recent days, as well as demographic information (e.g., age, gender, and home zip code) for the people in the region who have not recently visited the ED.

Given $\boldsymbol{e}$ and $\boldsymbol{g}$, our goal is to calculate the following:

$$P(T|\boldsymbol{e},\boldsymbol{g}) = k \cdot P(\boldsymbol{e}|T,\boldsymbol{g}) \cdot P(T|\boldsymbol{g}), \tag{2}$$

where the proportionality constant is

$$k = 1/\sum_T P(\boldsymbol{e}|T,\boldsymbol{g}) \cdot P(T|\boldsymbol{g}).$$

Since T and $\boldsymbol{g}$ are in $\boldsymbol{G}$, it follows from Assumptions 1 and 2 that the term $P(T|\boldsymbol{g})$ in Eq. 2 can be calculated using Bayesian network inference on just the portion of the model that includes $\boldsymbol{G}$. Performing BN inference over just the nodes in $\boldsymbol{G}$ is much preferable to inference over all the nodes in $\boldsymbol{X}$, because in the model we evaluated the number of nodes in $\boldsymbol{X}$ is approximately 10^7.

The term $P(\boldsymbol{e}|T, \boldsymbol{g})$ in Eq. 2 can be derived as follows:

$$P(\boldsymbol{e}|T,\boldsymbol{g}) = \sum_i P(\boldsymbol{e}|\boldsymbol{I} = \boldsymbol{i}) \cdot P(\boldsymbol{I} = \boldsymbol{i}|T,\boldsymbol{g}),$$

because by Assumption 2 the set $\boldsymbol{I}$ renders the nodes in $\boldsymbol{P}$ (including $\boldsymbol{e}$) independent from the nodes in $\boldsymbol{G}$ (including T and $\boldsymbol{g}$). The above summation can be very demanding computationally, because $\boldsymbol{e}$ usually contains many nodes; therefore, we next discuss its computation in greater detail.

We first show an example from Figure 18.2. Here we are modeling exactly four people in the population. The two on the left have identical attributes, as do the two on the right. We want to calculate the probability of this configuration of evidence, given the interface nodes. For this example, we have two distinct sets of evidence, $\boldsymbol{e}_1$ = {*Home Zip*=15213, *Age*=20–30, *Gender*=M, *Date Admitted*=never, *Respiratory symptoms*=unknown} and $\boldsymbol{e}_2$ = {*Home Zip*=15260, *Age*=20–30, *Gender*=F, *Date Admitted*=today, *Respiratory symptoms*=yes}.[3] We need to calculate:

$$P(\boldsymbol{e}|\boldsymbol{I}) = P(\boldsymbol{P}_1 = \boldsymbol{e}_1, \boldsymbol{P}_2 = \boldsymbol{e}_1, \boldsymbol{P}_3 = \boldsymbol{e}_2, \boldsymbol{P}_4 = \boldsymbol{e}_2|\boldsymbol{I}).$$

By Assumption 1, $\boldsymbol{I}$ d-separates each person model from each other, so this equation can be factored as follows:

$$P(\boldsymbol{e}|\boldsymbol{I}) = P(\boldsymbol{P}_1 = \boldsymbol{e}_1|\boldsymbol{I}) \cdot P(\boldsymbol{P}_2 = \boldsymbol{e}_1|\boldsymbol{I}) \cdot P(\boldsymbol{P}_3 = \boldsymbol{e}_2|\boldsymbol{I}) \cdot P(\boldsymbol{P}_4 = \boldsymbol{e}_2|\boldsymbol{I}). \tag{3}$$

It follows from Assumptions 1 and 2 that we can derive each quantity $P(\boldsymbol{P}_i = \boldsymbol{e}_j|\boldsymbol{I} = \boldsymbol{i})$ via BN inference using just the model fragment defined over the set of nodes in $\boldsymbol{P}_i \cup \boldsymbol{I}$. However, this quantity must be calculated for all configurations $\boldsymbol{I} = \boldsymbol{i}$ of the interface nodes. Performing this calculation for each of millions of person models would not be feasible within the time limits required for real-time biosurveillance. We could cache these conditional probability tables so that at run-time they amount to a constant-time table lookup. This technique is problematic, however, because it requires caching of a conditional probability table for all configurations of $\boldsymbol{I}$ and for all possible states of evidence $\boldsymbol{e}_j$. Such a table would be very large. As described in the next two sections, we use two techniques to deal with the large size of the inference problem: equivalence classes and incremental updating. Using equivalence classes saves both space and reduces inference time. Using incremental updating also reduces inference time, often dramatically so.

3.2.1. Equivalence Classes

If some person subnetworks are identical in structure and parameters, and they are instantiated to the same evidence, then fewer calls to the inference engine are needed. Eq. 3 can be written as:

$$P(\boldsymbol{e}|\boldsymbol{I}) = P(\boldsymbol{P}_1 = \boldsymbol{e}_1|\boldsymbol{I})^2 \cdot P(\boldsymbol{P}_3 = \boldsymbol{e}_2|\boldsymbol{I})^2,$$

because P_1 and P_2 share the same evidence, as do P_3 and P_4.

We define an *equivalence class* Q_i^j as a pair $Q_i^j = \langle \boldsymbol{P}_i, \boldsymbol{e}_j \rangle$, where $\boldsymbol{P}_i$ is a person model and $\boldsymbol{e}_j$ is a (possibly incomplete) set

3 "never" means the person is in the population at large and has not recently been admitted to the ED.

of evidence over the variables in $\boldsymbol{P}_i$.[4] A given evidence state $\boldsymbol{e}$ for the entire population corresponds to a unique set of equivalence classes and the instance count of each class. Using this set, the general expression for the quantity $P(\boldsymbol{e}\,|\,\boldsymbol{I})$ is as follows:

$$P(\boldsymbol{e}|\boldsymbol{I}) = \prod_{Q_i^j \in \Omega} P(E_j = e_j | \boldsymbol{I})^{N_{ij}}, \tag{4}$$

where N_{ij} is the instance count of equivalence class Q_i^j, that is, it is the number of people for whom we model with person model $\boldsymbol{P}_i$, and for which the evidence is $\boldsymbol{E}_j = \boldsymbol{e}_j$.

If the person model is relatively simple, then there could be many fewer equivalence classes than there are people in the population. For our example, since all person models are identical, the number of equivalence classes is equal to the number of possible ways to instantiate the variables of the person model (including not observing the state for those variables that sometimes have missing values). In this case, we would have at most (101 zip codes) × (2 genders) × (10 age ranges) × (4 relative dates of admission) × (3 respiratory states) = 24,240 states. In practice, the actual number of equivalence classes present at any one time is usually smaller, since rarer equivalence classes often do not appear.

In previous work, object-oriented Bayesian networks (Koller and Pfeffer, 1997) and related work (Srinivas, 1994, Xiang and Jensen, 1999) have been used to improve the efficiency of BN inference. The method we have described in this section takes advantage of those computational savings, as well as the savings that accrue from performing inference only once for objects of a given class that share the same evidence.

3.2.2. *Incremental Updating*

When we apply this technique to a population of millions of people, calculating $P(\boldsymbol{e}\,|\,\boldsymbol{I})$ will be a time-consuming task, even with the savings that results from using equivalence classes. Since we would like to perform this calculation very frequently (e.g., every hour as ED patients are coming into EDs throughout the modeled region), it is important to avoid re-calculating $P(\boldsymbol{e}\,|\,\boldsymbol{I})$ for the entire population every hour. To do so, we use the fact that as a single person moves from one equivalence class Q_i^j to another Q_i^k, $P(\boldsymbol{e}|\boldsymbol{I})$ can be updated to $P(\boldsymbol{e}'|\boldsymbol{I})$ as follows:

$$P(\boldsymbol{e}'|\boldsymbol{I}) = P(\boldsymbol{e}|\boldsymbol{I}) \frac{P(Q_i^k|\boldsymbol{I})}{P(Q_i^j|\boldsymbol{I})} \tag{5}$$

When biosurveillance monitoring is begun, the set of evidence $\boldsymbol{e}$ represents background information about the population. Currently, as background information, we use U.S. Census data to provide the age, gender, and home zip code information for the people in the region being monitored for disease outbreaks.

After $P(\boldsymbol{e}|\boldsymbol{I})$ is calculated once for the entire population, we can apply Eq. 5 and update this quantity incrementally as we observe people enter the ED in real time. As we observe a person from equivalence class Q_i^k enter the ED, we find the class Q_i^j that this person must have originated from in the background population. For example, if we observe a patient in the ED with the following attributes: Q_i^k={*Home Zip*=15260, *Age*=20–30, *Gender*=F, *Date Admitted*=today, *Respiratory symptoms* =yes}, then we know that she originated from the background class Q_i^j= {*Home Zip*=15260, *Age*=20–30, *Gender*=F, *Date Admitted*=never, *Respiratory symptoms*= unknown}.

Applying the incremental updating rule allows us to reduce the number of updates that need to be processed each hour to dozens (= rate of patient visits to all the EDs in the region) rather than the millions (= the number of people in the regional population).

By caching equivalence classes and applying incremental updating, we can process an hour's worth of ED patient cases (about 26 cases) from a region of 1.4 million people in only 11 seconds using a standard Pentium III PC and the Hugin BN inference engine v6.2 (Hugin, 2004). Thus, there is enough computing reserve to "keep ahead" of the real-time data, even when in the future we extend our model to be considerably richer in detail, and we widen the geographic region being monitored for a disease outbreak.

4. EMPIRICAL EVALUATION

This section describes the detection model that we evaluated. The model represents a preliminary prototype for use in detecting disease outbreaks that would result from an outdoor, airborne release of anthrax spores. The model plus the inference algorithms constitute a Bayesian biosurveillance system that we call PANDA (Population-wide ANomaly Detection and Assessment). We conclude this section with a description of the method we used to evaluate PANDA, as well as the results of that evaluation.

4.1. Model for Outbreak Detection

In our empirical tests, we use a model similar to the example model shown in Figure 18.2, with two primary differences: (1) we do not use the *Terror Alert Level* node, and (2) we use a more complex person model. Figure 18.4 shows the person model we use.

The meanings of the nodes are listed below. For each underlined variable, its conditional probability table was estimated from a training set consisting of one year's worth of ED patient data from the year 2000. The variables in boldface were estimated from U.S. Census data about the region.

4 We abuse notation somewhat by using Pi here to denote a class, whereas previously it has been used to denote an object.

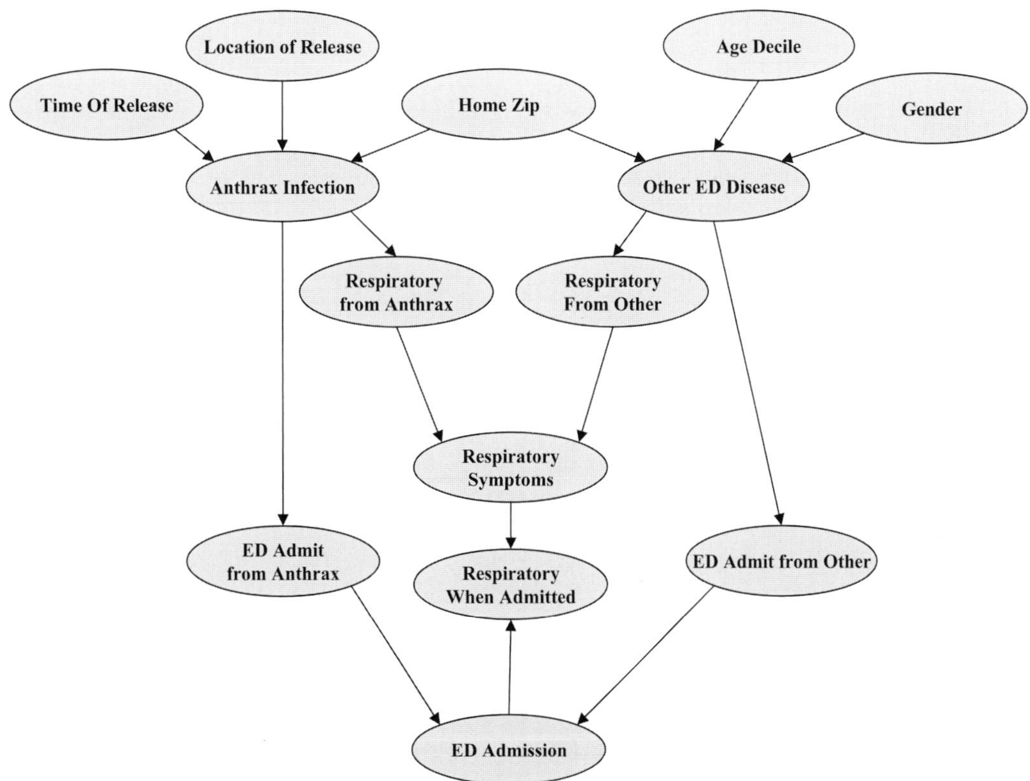

FIGURE 18.4 The person model used in the evaluation. We used Hugin software (v6.2, *www.hugin.com*) to implement this model.

The remaining variables had their respective probabilities (whether prior or conditional) assessed subjectively; these assessments were informed by the literature and by general knowledge about infectious diseases.

Time of Release: This is the hypothesized day that anthrax was released, if ever. It has the states *never, today, yesterday*, and *day before yesterday*.

Location of Release: This is the hypothesized location at which the anthrax was released, if released anywhere. It has the states: *nowhere*, and *one state for each of about 100 zip codes being covered by the model.* In the current model, we assume only a single point of release.

Home Zip: This node represents the location of the person's home zip code; it can take on one of about 100 zip codes in Allegheny County, Pennsylvania, which is the region being modeled. There is currently a "catch-all" zip code called *other* that represents patients who do not live in Allegheny County, but who are seen in EDs there. In the initial prototype being described here, we make the very simplifying assumption that only individuals with a home zip code in the zip code of release will become infected with anthrax. In Section 4.5 we relax this assumption.

Age Decile: This node represents the individual's age, which can take one of 9 values: 0, 1 ... 8 corresponding to (0–10 years), (10–20 years), ..., (>80 years), respectively.

Gender: This represents the gender of the individual, taking values *female* and *male*.

Anthrax Infection: This node represents whether or not the individual has been infected with a respiratory anthrax infection within the past 72 hours. This node takes the following states: *AAA* (indicating that anthrax was absent for the past three days), *AAI* (indicating that within the past 24 hours the patient was infected with anthrax and still is infected), *AII* (indicating that the patient was infected with anthrax between 24 and 48 hours ago and is still infected now), and finally, *III* (indicating that the patient was infected between 48 and 72 hours ago and continues to be infected now). There are in principle four other states that this node could have (*IAA, IIA, IAI*, and *AIA*); however, we make the assumption that once a person gets anthrax, he or she maintains the disease for at least three days, so these other states have probability zero. In future work, we plan to extend the *Anthrax Infection* variable (as well as other temporal variables described here) to model over more than three days.

<u>Other ED Disease</u>: This variable is conceptually similar to *Anthrax Infection*, but it denotes instead some other disease or disorder, which by definition is sufficient to cause the individual to go into the ED, but is not an anthrax infection. This node has the same type of states as *Anthrax Infection*.

Respiratory from Anthrax: Indicates that the individual is showing respiratory symptoms (e.g., *cough*) due to anthrax. It has states similar to those of *Anthrax Infection*.

Respiratory from Other: Respiratory symptoms from ED disease other than anthrax.

Respiratory Symptoms: Node indicating whether or not the patient exhibits respiratory symptoms. It is a "logical OR" function of *Respiratory from Anthrax* and *Respiratory from Other*.

Respiratory When Admitted: This node represents whether the person has respiratory symptoms when he or she is seen in the ED. It has states *True*, *False*, and *Unknown*. If the person was admitted to the ED today, then we typically know whether the value is *True* or *False*; otherwise the value is *Unknown*. For those patients admitted to the ED, the value of this variable is based on a patient's chief complaint as given to the triage nurse in the ED.

ED Admit from Anthrax: Indicates that the person was admitted to the ED due an anthrax infection.

ED Admit from Other: Indicates that the person was admitted to the ED due to a disease other than an anthrax infection.

ED Admission: Indicates the day (if any) that the person was admitted to the ED within the past 72 hours. It is a "logical OR" function of *ED Admit from Anthrax* and *ED Admit from Other*. We currently do not model the possibility that a person could be admitted more than once. To do so, the de-identified data that we receive on each patient could be extended to include a unique integer for the patient that does not reveal the patient's personal identity.

We emphasize that the current model, which is presented here for illustration, is an initial prototype, which we are refining and extending.

4.2. Simulation Model

We evaluated the performance of PANDA on data sets produced by injecting simulated ED cases into a background of actual ED cases obtained from several hospitals in Allegheny County. In accordance with (HIPAA) regulations, all personal identifying information was removed from these actual ED cases. The simulated cases of anthrax were produced by the BARD simulator (see Chapter 19) that models the effects of an airborne anthrax release using an independently developed Gaussian plume model of atmospheric dispersion of anthrax spores (Hanna et al., 1982).

Given weather conditions and parameters for the location, height, and amount of the airborne anthrax release, the Gaussian plume model derives the concentration of anthrax spores that are estimated to exist in each zip code. The extent of an outbreak for a zip code is influenced by the spore concentration in the zip code and the number of people living in the zip code. The output from the simulator consists of a list of anthrax cases, where each case consists of a date-time field and a zip code. The full details of the model are in Chapter 19. For our experiments, we selected historical meteorological conditions (e.g., wind direction and speed) for Allegheny County from a random date as the meteorological input to the simulator. The height of the simulated release was sampled from a prior distribution, created using expert judgment (see Chapter 19). This distribution was skewed towards heights less than 1500 feet. Finally, the release locations were sampled from a prior distribution, which favors release locations that would be expected to infect large numbers of people given the current meteorological conditions.

The output of the BARD simulator cannot be used directly by PANDA because a full evidence vector for a case includes information about the patient's age and gender. As a result, we took the partially complete patient cases produced by the simulator and probabilistically assigned the age and gender fields using the person model Bayesian network. The age of the patient is sampled from the conditional distribution of age given the home zip code of the patient and given the fact that the patient had respiratory symptoms when admitted. We use a similar procedure for determining the gender.

The anthrax release simulator that we used generally generates multiple downwind cases of anthrax that span several zip codes. The simulator also includes a minimum incubation period of 24 hours after the release during which no cases of anthrax are generated.[5] Beyond that minimum period, the incubation period varies, with greater airborne concentrations of anthrax leading to a shorter incubation period, in general, than lesser concentrations.

In order to evaluate the detection capability of PANDA, we generated data sets corresponding to simulated releases of anthrax of the following amounts: 1.0, 0.5, 0.125, and 0.015625.[6] For each release amount, we created 96 data sets, each with a unique release location. For each month in 2002, we chose eight random release dates and times to use with the simulator, thus

---

5 Arguably this minimum could be larger. The results in this chapter, however, emphasize detection time that is relative to the incubation time, rather than detection time from the point of the simulated release. Thus, the exact choice of an incubation time is less critical.

6 The units of concentration are not reported here in order to avoid providing results that could pose some degree of security risk. The concentration of 0.015625 was included for technical reasons.

producing a total of 96 different anthrax release data sets. We used only 91 data sets for the 0.015625 concentration because five of the data sets generated had no reported anthrax cases. PANDA was applied to monitor the data from each data set, starting on midnight of January 4, 2001 and extending through to six days after the simulated anthrax release occurred.

We measured the performance of PANDA using an AMOC curve, introduced by (Fawcett 1999) and described in Chapter 20. It plots time to detection as a function of false alarms per week. The points on the AMOC curve are generated by determining the false-positive alarm rate (ranging from 0 to 1) and detection time of the algorithm over a range of alarm thresholds, where an alarm is considered to be raised if the posterior probability of an *Anthrax Release* = yes exceeds the given alarm threshold. Since no known releases of anthrax have ever occurred in the region under study, the false-positive alarm rate was measured by determining the fraction of monitored hours that the release probability exceeded the alarm threshold under consideration for the period starting on January 4, 2001 and continuing until the simulated release date for the particular data set. In order to measure the timeliness of detection, we counted the number of hours that passed between the time of the simulated anthrax release and the first time the posterior probability of *Anthrax Release* = yes (as produced by PANDA) exceeded the alarm threshold. If no alarms were raised within six days after the simulated release point, the detection time was set to be 144 hours.

4.3. Baseline Results

Figure 18.5 illustrates the activity monitoring operating characteristic (AMOC) curve for PANDA over the four anthrax concentrations. Since the incubation period of the simulation is set at a minimum of 24 hours, the earliest possible detection time is shown with a dotted line at the 24-hour mark. As expected, the detection time decreases as the simulated release amount increases, since a larger release is more easily detected. In particular, at zero false positives, the detection time is approximately 132, 84, 58, and 46 hours for respective simulated release concentrations of 0.015625, 0.125, 0.5, and 1.0. The maximum widths of the 95% confidence intervals for the detection times at concentrations of 0.015625, 0.125, 0.5, and 1.0, are ±5.21, 4.00, 2.67, and 1.68 hours, respectively.

The majority of false positives with a release probability of over 50% occurred during a 10-hour period from January 20, 2002, 11:00 PM to January 21, 2002, 9:00 AM and also during a 17-hour period from midnight August 18, 2002 to August 19, 2002, 5:00 PM.

We note that the model parameters were based in part on data from the year 2000, whereas the evaluation was based on using test data from 2002. So, some false positives may have been due to a lack of synchronization between the model and the test data. When tested on data from the year 2001, there were no false alarms above the 50% level.

4.4. Optimizing Model Parameters

We performed a set of experiments in which we considered the BARD simulation model as a gold standard, and then optimized the PANDA detection model based on the BARD model. Optimizing the PANDA model required a change to both the structure of the person Bayesian network and to its parameters. The structure of the optimized person model is shown in Figure 18.6.

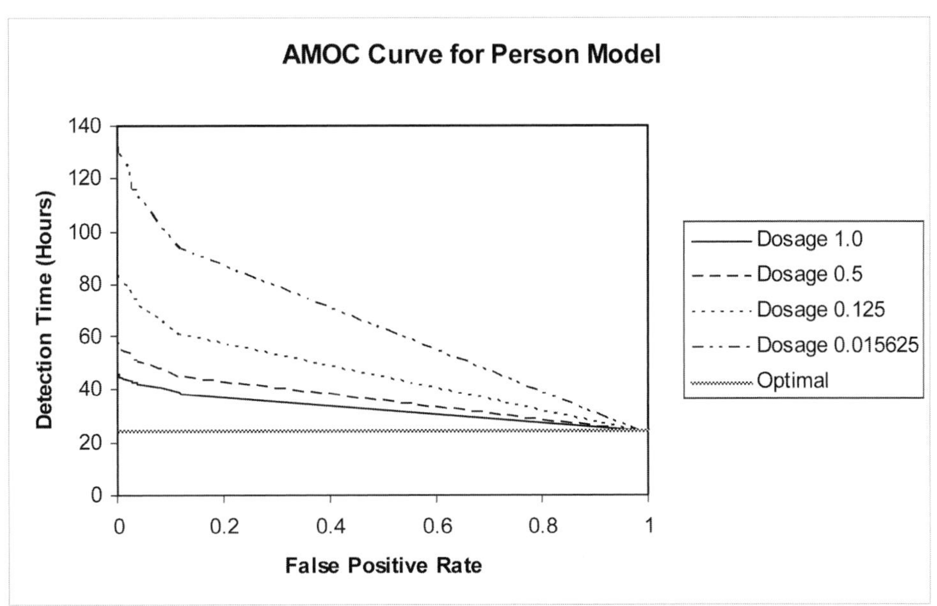

FIGURE 18.5 An AMOC curve showing the detection capabilities of PANDA over different anthrax concentrations.

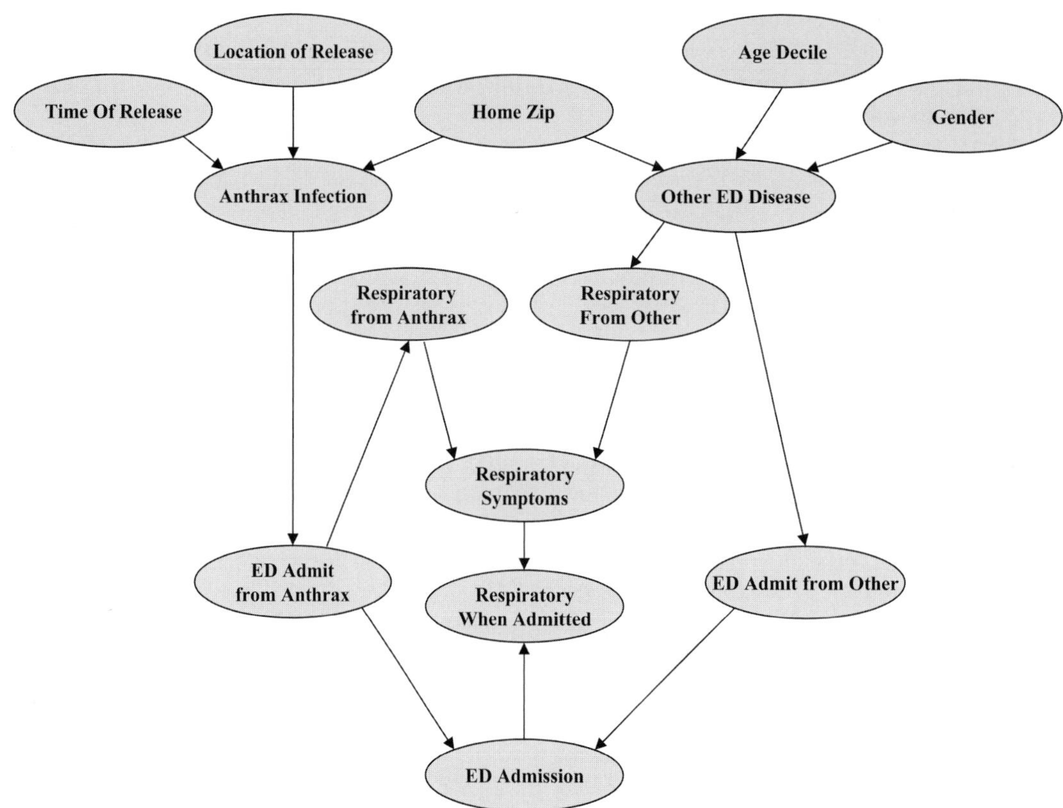

FIGURE 18.6 The optimized person model.

One of the main differences between the optimized model and the original model is the removal of the previously existing arc from the *Anthrax Infection* node to the *Respiratory from Anthrax* node. We also add an arc from the *ED Admit from Anthrax* node to the *Respiratory from Anthrax* node. These changes are intended to reflect the behavior of the BARD simulator, which assumes that all patients infected with respiratory anthrax will exhibit respiratory symptoms. Therefore, all patients who are admitted to the ED due to a respiratory anthrax infection are assumed to exhibit respiratory symptoms.

We estimated the optimal parameters for the *Anthrax Infection* node and the *ED Admit from Anthrax* node over a training set obtained from the BARD simulator that was separate from the test set used to evaluate the performance of the PANDA algorithm. The training set consisted of 96 anthrax-release data sets for each of the four spore dosages. The parameters were generated by averaging the probabilities obtained from files of a specific dosage in the training set. Once these probabilities specific to a dosage were obtained, the probabilities were averaged across the four different dosages because, currently, PANDA does not model the dosage.

The results of using the optimized person model on the data sets described in Section 4.2 are shown in Figure 18.7. As expected, the optimized model significantly improves on the detection time of the original model. At zero false positives, the average detection time is approximately 127, 63, 44, and 38 hours for the corresponding simulated attack concentrations of 0.015625, 0.125, 0.5, and 1.0. The average improvement in detection time over the original model is 5, 21, 14, and 8 hours, respectively, for the four attack concentrations given above. The maximum widths of the 95% confidence intervals for the detection times at concentrations 0.015625, 0.125, 0.5, and 1.0 are ± 4.38, 4.07, 2.15, and 1.62, respectively.

4.5. Incorporating the Spatial Distribution of Cases into the Model

In this section, we describe changes to the PANDA model to account for the situation in which an anthrax release infects people in more than one zip code. In particular, we added a new interface node, called *Angle of Release*, which describes the orientation of the airborne anthrax release and takes on the eight possible values of *N, NE, E, SE, S, SW, W*, or *NW*, as shown in Figure 18.8. Figure 18.9 depicts the modified person model, which is based on the original nonoptimized model shown in Figure 18.4. For computational reasons, Figure 18.9 shows that we decomposed the previous *Anthrax Infection*

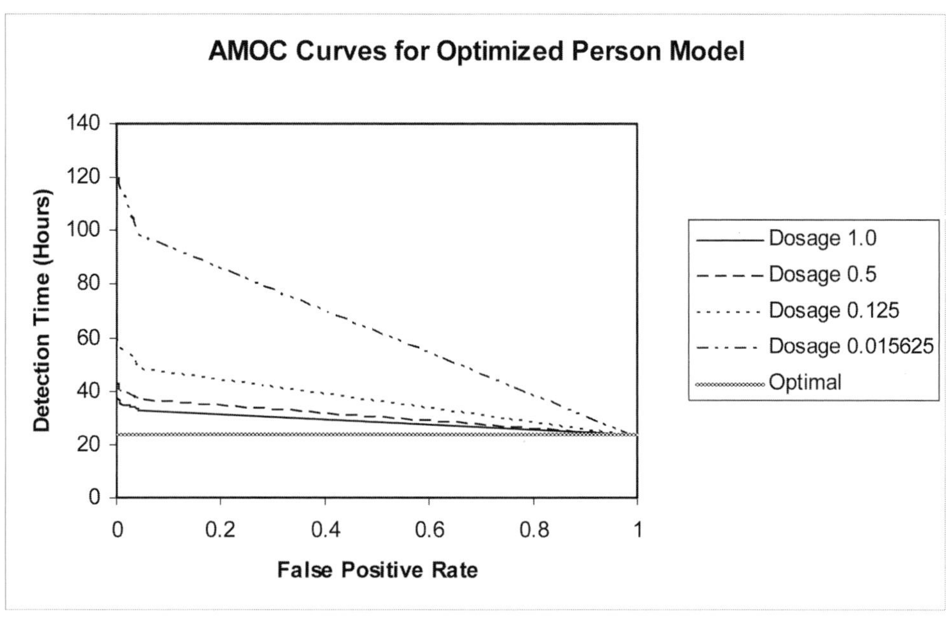

FIGURE 18.7 The AMOC curves for the optimized person model for the four anthrax dosages.

node in Figure 18.4 into an *Exposed to Anthrax* node and a new *Anthrax Infection* node. We will refer to this modified version of PANDA as the spatial model and the previous version will be referred to as the non-spatial model.

The *Exposed to Anthrax* node represents the probability that the person is exposed to anthrax during a release, given the zip code of the home location of the person, the location of the release, and the angle of the release. The spatial model assigns a probability of 1.0 for *Exposed to Anthrax* to anyone who has the same home zip code as the zip code of the hypothesized anthrax release, regardless of the angle of release. For people outside of the release zip code, we consider them to be potentially exposed to anthrax if their home zip code is within a rectangular region that originates at the centroid of the hypothesized release zip code and is oriented according to the angle of the release variable. As an example, suppose the release occurs in 15213. The two small circles in Figure 18.8 represent zip code centroids. The circle in the center of the figure is for zip code 15213. The circle in the bottom right is for zip code 15132. There are eight rectangular regions centered at the centroid of zip code 15213. If a person has a home zip in 15132, and the angle of release is SE, then we would consider that person to be potentially exposed to anthrax. The actual probability of being exposed to anthrax is computed by decaying the value 1.0 by a half for every three miles of distance between the release zip code's centroid and the person's home zip code centroid. The distance of three miles was obtained by manually tuning the model over data sets produced by the simulator; these data sets were distinct from the data sets that we used to evaluate PANDA. The width of the rectangle is set to be approximately three miles, which was chosen by calculating the average area per zip code in Allegheny County, determining the diameter of a circle with this average area, and then assigning that diameter as the width. The length of the rectangle is assumed to extend to infinity, as shown by the arrows in Figure 18.8.

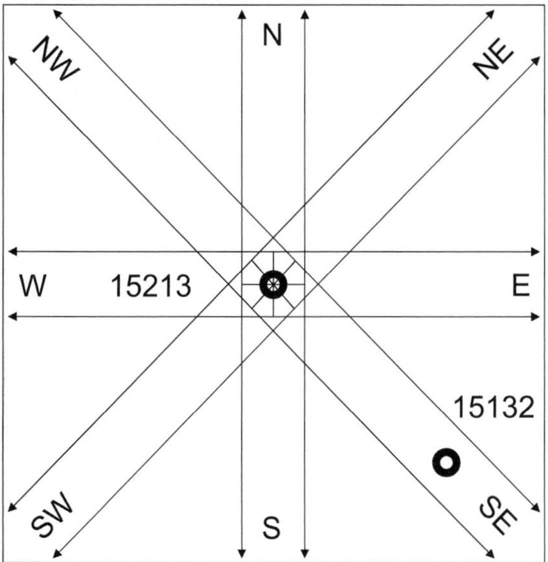

FIGURE 18.8 The rotating rectangular regions centered at (for example) the centroid of zip code 15213 that are used to determine if a person is exposed to anthrax.

FIGURE 18.9 The person model modified to incorporate spatial information.

Figure 18.10 illustrates another variation on the spatial model, which we will call the spatial model with *temporal fluctuation nodes*. This version of the spatial model contains three additional nodes for *Season, Day of Week*, and *Time of Day*, which are intended to capture the fluctuations in the number of ED cases due to these three factors. The *Season* node takes on the values of *Spring, Summer, Fall*, or *Winter*. The *Day of Week* node can be assigned one of the possible seven names of the day of the week. We allow the *Time of Day* node to have three possible discrete values of 12:00 midnight to 8:00 AM, 8:00 AM to 4:00 PM, and 4:00 PM to 12:00 midnight.

We evaluated both spatial models over the 96 simulated test data sets for the 1.0 concentration that were previously used. The false-positive rate was measured over the period of January 4, 2002 until the start of the simulated release for the particular data set. The results are shown in Figure 18.11. Adding spatial information improves the detection time significantly. Adding the temporal fluctuation nodes in addition to the spatial information produces only a slight improvement over the spatial model. The largest difference in detection time between the nonspatial and either of the spatial models is approximately 9.7 hours. The maximum widths of the 95% confidence intervals for the nonspatial, spatial, and spatial with temporal fluctuation nodes models are ± 1.64, 1.68, and 1.65 hours, respectively.

4.6. Optimizing the Spatial Model Parameters

The spatial model was also optimized using the process described in Section 4.4. Figure 18.12 illustrates the structure of the optimized spatial person model, which is based on the optimized person model in Figure 18.6. We will refer to this model as the optimized spatial model.

Figure 18.13 illustrates the AMOC curve for the optimized spatial models when compared to the nonoptimized spatial model. The optimized spatial model improves the detection time of the non-optimized spatial model by nearly five hours. Optimizing the decay rate seems to have minimal effect on the detection time. At zero false positives, the detection times are 39.7 and 34.7 for the spatial model and the optimized spatial model, respectively. The corresponding maximum 95% confidence interval widths for the two models are ± 1.55 and 1.37 hours, respectively.

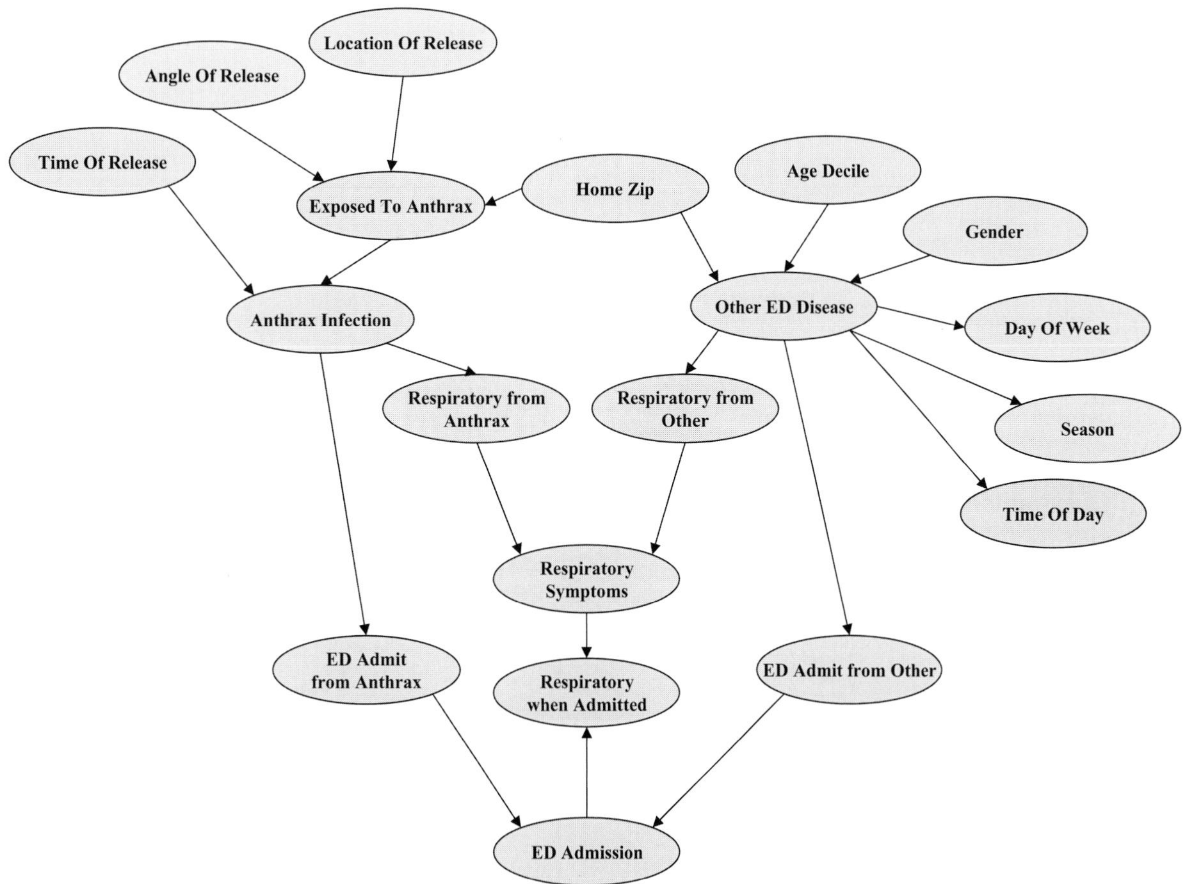

FIGURE 18.10 The person model modified to incorporate spatial information as well as temporal fluctuations due to day of week, season, and time of day.

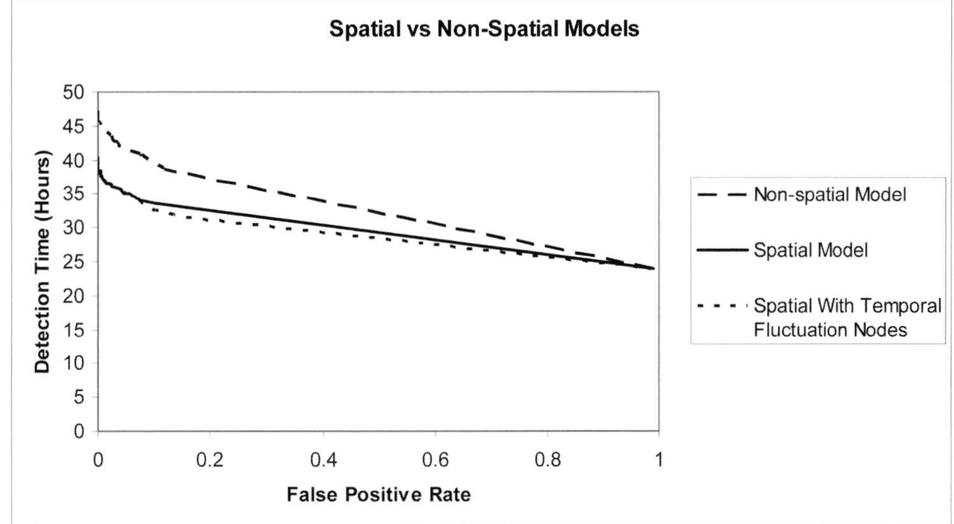

FIGURE 18.11 An AMOC curve comparing the non-spatial model results with those from the spatial model and the spatial with temporal fluctuation nodes model.

FIGURE 18.12 The optimized spatial person model.

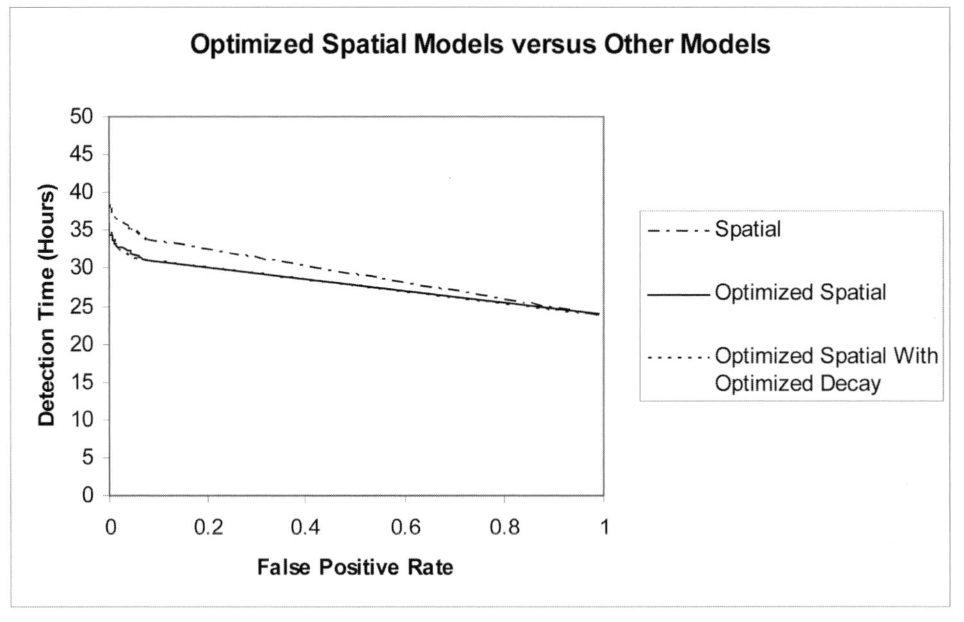

FIGURE 18.13 AMOC curves for the optimized spatial models versus the nonoptimized spatial model.

5. SUMMARY

This chapter introduced a biosurveillance method that uses causal Bayesian networks to model noncontagious diseases in a population. By making two independence assumptions in the model, both of which appear plausible for noncontagious diseases, and by performing inference using equivalence classes and incremental updating, it is possible to achieve tractable Bayesian biosurveillance in a region with 1.4 million people. We implemented and evaluated an outbreak detection system called PANDA. Overall, the run-time results and the detection performance of this initial evaluation are encouraging, although additional studies are needed and are in process.

6. EXTENSIONS

There are several possible extensions to PANDA that are straightforward to implement, including (1) increasing the number of days being modeled, (2) modeling on an hourly basis, rather than a daily one, and (3) adding nodes to the model that represent *prevailing wind direction* and wind *speed.*

A more fundamental extension involves modeling a set of variables that represent the amount of over-the-counter (OTC) medication sales of a particular type (e.g., cough medication sales) per subregion (e.g., zip code) per day. Preliminary work indicates that it is feasible to develop a causal Bayesian network model that incorporates both ED data and OTC data (Wong et al., 2005).

Within the current framework, additional non-contagious outbreak diseases can be modeled, as well as non-outbreak diseases that might be easily confused with outbreak diseases. A more ambitious goal is to causally model contagious diseases, where there is much less independence among the individuals being modeled than in noncontagious diseases. Beyond the modeling issues, which are substantial, developing inference algorithms that are fast enough to permit real-time biosurveillance of contagious diseases will also be challenging. Approximate inference algorithms may be required.

Finally, much additional testing is needed of the run-time and detection performance of Bayesian biosurveillance methods, including their relative performance to other detection methods.

ADDITIONAL RESOURCES

A good introduction to Bayesian networks is Neapolitan (2004). A useful source of information about research on Bayesian networks and other graphical models is available at *http://www.auai.org*. Additional information about PANDA and Bayesian biosurveillance is available at *http://www.cbmi.pitt.edu/panda*.

ACKNOWLEDGMENTS

We thank Andrew Moore and the other members of the Pitt-CMU Detection Group for helpful discussions. This research was supported in part by grants from the National Science Foundation (IIS-0325581), the Defense Advanced Research Projects Agency (F30602-01-2-0550), and the Pennsylvania Department of Health (ME-01-737).

REFERENCES

Bernardo, J., Smith, A. (1994). *Bayesian Theory*. New York: John Wiley & Sons.

Buckeridge, D. L., Burkom, H., Moore, A., et al. (2004). Evaluation of syndromic surveillance systems: design of an epidemic simulation model. *MMWR Morb Mortal Wkly Rep* 53(Suppl):137–43.

Carlstein, E. (1988). Nonparametric change-point estimation. *Ann Stats* 16:188–97.

Cooper, G., Dash, D., Levander, J., et al. (2004). Bayesian biosurveillance of disease outbreaks. In: *Proceedings of the Conference on Uncertainty in Artificial Intelligence* 94–103.

Fawcett, T., Provost, F. (1999). Activity monitoring: noticing interesting changes in behavior. In: *Proceedings of Fifth International Conference on Knowledge Discovery and Data Mining*, 53–62.

Hamilton, J. (1994). *Time Series Analysis*. Princeton, NJ: Princeton University Press.

Hanna, S., Briggs, G., Hosker, J. (1982). *Handbook of Atmospheric Diffusion*. DOE/TIC-11223. Washington, DC: U.S. Department of Energy.

Hugin Expert A/S. (2004). Hugin, Version 6.2. *http://www.hugin.com*.

Koller, D., Pfeffer, A. (1997). Object-Oriented Bayesian networks. In: *Proceedings of the Conference on Uncertainty in Artificial Intelligence*, 302–13.

Kuldorff, M. (2001). Prospective time periodic geographical disease surveillance using a scan statistic. *J R Stat Soc* 164:61–72.

Kulldorff, M. (1997). A spatial scan statistic. *Commun Stats Theory Methods* 26:1481–96.

Moore, A., Cooper, G., Tsui, R., et al. (2003). Summary of biosurveillance-relevant technologies. Pittsburgh, PA: Computer Science Department, Carnegie Mellon University. *http://www-2.cs.cmu.edu/~awm/biosurv-methods.pdf.*

Neapolitan, R. E. (2004). *Learning Bayesian Networks*. Upper Saddle River, NJ: Prentice Hall.

Neill, D., Moore, A. (2003). A fast multiresolution method for detection of significant spatial disease clusters. *Adv Neural Info Processing Syst.*

Rabiner, L. (1989). A tutorial on hidden Markov models and selected applications in speech recognition. *Proc IEEE* 77:257–85.

Reis, B., Mandl, K. (2003). Time series modeling for syndromic surveillance. *BMC Med Inform Decision Making* 3:2.

Serfling, R. (1963). Methods for current statistical analysis of excess pneumonia-influenza deaths. *Public Health Rep* 78:494–506.

Srinivas, S. (1994). A probabilistic approach to hierarchical model-based diagnosis. In: *Proceedings of the Conference on Uncertainty in Artificial Intelligence*, 538–45.

Tsui, F., Wagner, M., Dato, V., et al. (2001). Value of ICD-9-coded chief complaints for detection of epidemics. In: *Proceedings of Fall Symposium of American Medical Informatics Association*, 711–715.

Williamson, G., Hudson, G. (1999). A monitoring system for detecting aberrations in public health surveillance reports. *Stats Med* 18:3283–98.

Wong, W. (2004). *Data Mining for Early Disease Outbreak Detection.* Doctoral dissertation. Carnegie Mellon University.

Wong, W., Cooper, G., Dash, D., et al. (2005). Bayesian biosurveillance using multiple data streams. *MMWR Morb Mortal Wkly Rep* (Suppl) 54:63–9.

Wong, W., Moore, A., Cooper, G., et al. (2003). Bayesian network anomaly pattern detection for disease outbreaks. In: *Proceedings of the Conference on Machine Learning*, 808–15.

Xiang, Y., Jensen, F. (1999). Inference in multiply sectioned Bayesian networks with extended Shafer-Shenoy and lazy propagation. In: *Proceedings of the Conference on Uncertainty in Artificial Intelligence*, 680–7.

19

CHAPTER

Atmospheric Dispersion Modeling in Biosurveillance

William R. Hogan

RODS, Laboratory, Center for Biomedical Informatics

University of Pittsburgh, Pittsburgh, Pennsylvania

1. INTRODUCTION

The motivation for using models of atmospheric dispersion in the analysis of biosurveillance data is the basic physics of an aerosol release. When a biological agent is released outdoors into the atmosphere as an aerosol cloud, wind and other weather conditions determine the locations to which the cloud travels and its density when it reaches a particular location. These factors in turn influence the spatial pattern of disease in a community.

An atmospheric dispersion model is a mathematical description of how wind and other weather conditions determine the spread of substances such as biological agents through the air. An aerosol release of a biological agent into the atmosphere is the most feared route of a biological attack. Windborne distribution of a biological agent is an efficient means of (nearly) simultaneously exposing as many as hundreds of thousands to millions of people to an agent. In addition to the terrorist threat, wind can also spread several naturally occurring diseases of concern to biosurveillance over distances as large as 100 km.

One naturally occurring disease for which the wind is a route of transmission is foot and mouth disease (FMD), a highly contagious infectious disease of cloven-hoofed animals such as cattle, pigs, and sheep. Researchers have studied the windborne spread of FMD extensively because FMD has significant economic consequences to nations that are currently free of the disease (through trade embargoes that disease-free nations may impose). Additionally, experts consider FMD as the most likely terrorist threat to agribusiness (Breeze, 2004, Cupp et al., 2004). When managing an outbreak of FMD, many governments use atmospheric dispersion models specialized for FMD to identify farms at risk of acquiring FMD from infected farms and to make decisions about restrictions on movement of animals and vaccination strategies.

Besides the FMD virus, biological agents of concern for aerosol dissemination include, but are not limited to, *Bacillus anthracis* (anthrax), *Francisella tularensis* (tularemia), *Yersinia pestis* (plague), the Smallpox virus, and botulinum toxin (see Table 3.2 in Chapter 3). Outbreaks of other naturally occurring diseases known or suspected to be the result of windborne spread include *Q* fever in humans, due to windborne spread of *Coxiella burnetii* spores generated by sheep husbandry (Hawker et al., 1998); bluetongue in sheep, due to windborne spread of the Culicoides midge that is the insect vector of the disease (Sellers et al., 1978, Sellers et al., 1979, Sellers and Pedgley, 1985); and Western equine encephalitis in horses, due to windborne spread of its mosquito vector (Sellers and Maarouf, 1988). Windborne spread of plant pathogens is also common (Davis, 1987).

We begin with an example of how researchers have used atmospheric dispersion models and weather data in the past to identify that an outbreak resulted from windborne dissemination of a biological agent. We then explore the basics of atmospheric dispersion modeling and contrast several specific atmospheric dispersion models. We then discuss more generally the role of atmospheric dispersion models in the analysis of biosurveillance data. We discuss additional uses of atmospheric dispersion models and weather data—a key input parameter of an atmospheric dispersion model—at the end of the chapter.

2. EXAMPLE OF USING AN ATMOSPHERIC DISPERSION MODEL TO PROVE WIND AS ROUTE OF OUTBREAK TRANSMISSION

Meselson et al. (1994) demonstrated—using the Gaussian plume model of atmospheric dispersion as a critical component of their analysis—that the 1979 outbreak of anthrax in Sverdlovsk, Soviet Union, was due to an aerosol release of *B. anthracis* spores from a military microbiology facility (Meselson et al., 1994). Furthermore, they determined the time of the release of spores to within 2.5 hours (Guillemin, 2000) and the quantity of the release (Meselson, 2001, Meselson et al., 1994). Their analysis, when combined with the following two facts established that the route of transmission of this outbreak was the wind. During the course of their investigation, Russian president Boris Yeltsin admitted officially that the outbreak was related to military production of biological weapons (The Record, 1992). Prior to their study, analysis of autopsy specimens determined that the victims suffered from inhalational anthrax and not gastrointestinal anthrax as was the official Soviet explanation for over a decade (Abramova et al., 1993).

The analytic procedure that Meselson et al. (1994) used follows. After first obtaining as complete a list of cases as possible, they interviewed surviving victims and friends and relatives of deceased victims. They determined the whereabouts of each

Handbook of Biosurveillance
ISBN 0-12-369378-0

case in the days prior to the onset of illness and dates of symptom onset, entry into the healthcare system, and death. Of 77 victims identified, they were able to determine the home and work locations for 66. They mapped the daytime locations during the week of April 2, 1979 for 57 victims (Figure 19.1). The area encompassing the 57 locations was a narrow, 4-km zone extending south-southeast from a Soviet military microbiology facility. Of the nine patients whose daytime whereabouts were unknown, three lived inside this zone and another three, although they lived and worked outside the zone, had occupations (truck driver, pipe layer, and telephone worker) that might have taken them inside the zone.

They also identified and mapped animal populations that experienced anthrax outbreaks at the same time. They identified six villages that experienced outbreaks of anthrax in animals at the same time that humans were ill in Sverdlovsk. On April 2, 1979, these villages lay downwind of Sverdlovsk along the axis (when extended to 50 km) of the high-risk zone for humans (Figure 19.2). The compass bearing of the centerline of human and animal anthrax outbreaks was 330° ± 10°.[1]

They obtained historical weather data from the Koltsovo airport (located 10 km east of the high-risk zone). Records of surface observations every three hours showed that the only time the wind direction was consistent with both the known locations of victims and the compass bearing of 330° was April 2, when the wind direction ranged from 320° to 350° from 4:00 AM to 5:00 PM local time (Figure 19.3). The date of April 2 and the time delay to onset of illness in the victims were consistent with the incubation period of inhalational anthrax.

Finally, Meselson et al. (1994) used the Gaussian plume model in combination with data about breathing rates to show that the meteorological data and spatial distribution of cases were entirely consistent with an aerosol release of spores from the military microbiology facility as the cause of the outbreak. The ellipse-like lines in Figures 19.1 and 19.2 show contours of constant dosage (airborne concentration times the breathing rate) computed from the Gaussian plume model with the microbiology facility as the release location.

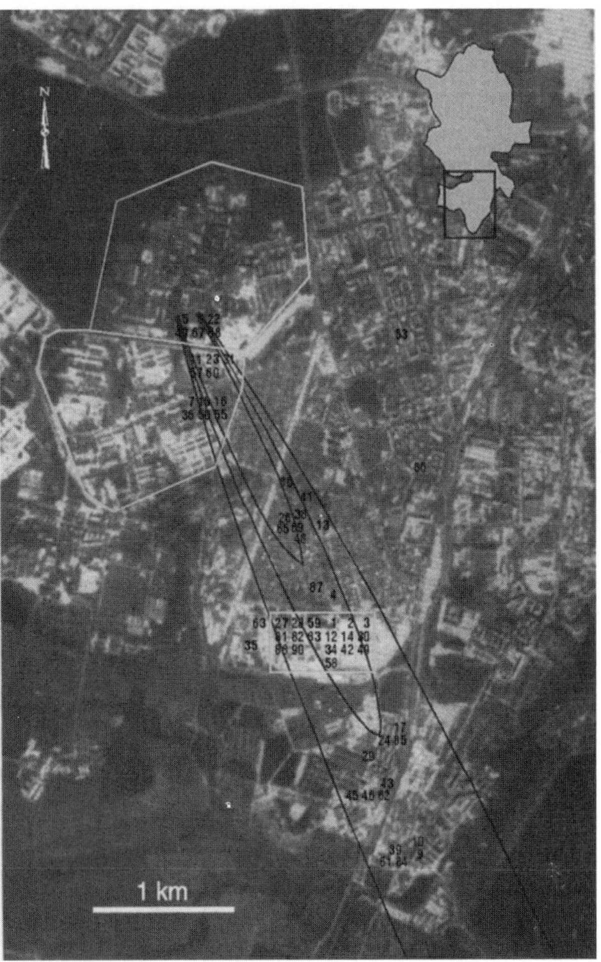

FIGURE 19.1 The high-risk zone for humans for the Sverdlovsk outbreak of inhalational anthrax in 1979. The ellipse-like lines are lines of equal concentration produced by the Gaussian plume model. The numbers are the daytime locations of victims. (From Meselson, M., Guillemin, J., Hugh-Jones, M., et al. (1994). The Sverdlovsk anthrax outbreak of 1979. *Science* 266:1202–8, with permission.)

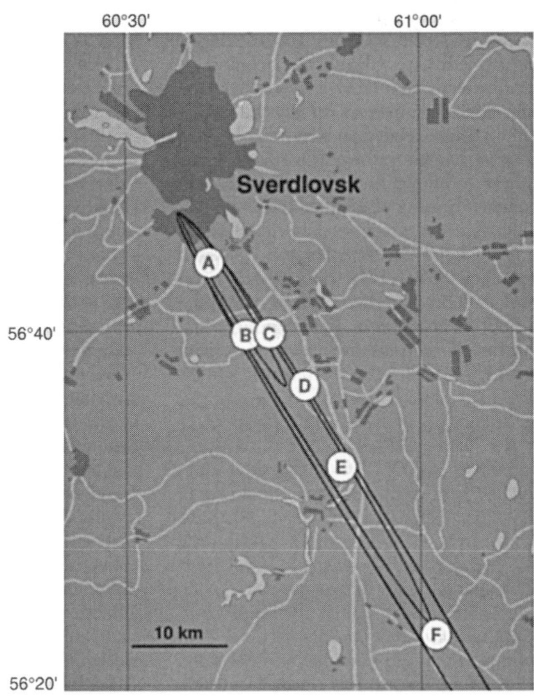

FIGURE 19.2 The high-risk zone for animals, which is an extension of the high-risk zone for humans to 50 km along the same axis. (From Meselson, M., Guillemin, J., Hugh-Jones, M., et al. (1994). The Sverdlovsk anthrax outbreak of 1979. *Science* 266:1202–8, with permission.)

1 The meteorological convention for wind directions is to report the direction from which the wind is blowing, and 0°/360° is due North.

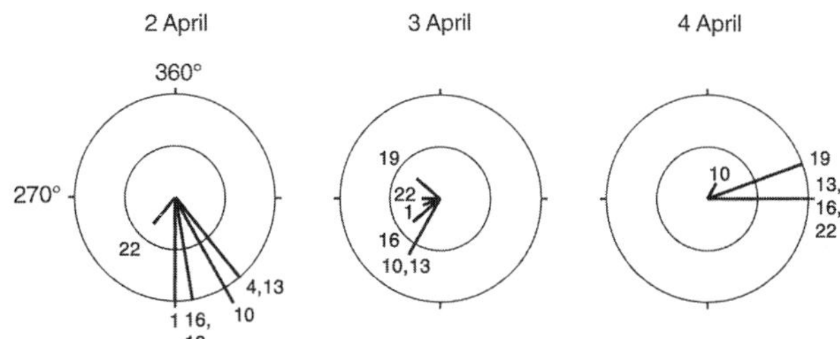

FIGURE 19.3 Weather data from April 2–4, 1979. Wind directions are reported as the direction from which the wind was blowing. Only on April 2 is the wind direction in alignment with the daytime locations of victims. Reprinted with permission from (From Meselson, M., Guillemin, J., Hugh-Jones, M., et al. (1994). The Sverdlovsk anthrax outbreak of 1979. *Science* 266:1202–8, with permission.)

Using the Gaussian plume model, Meselson (2001) estimated the number of spores released. Given the data available about the victims, he calculated that the release amount ranged from a few milligrams to a gram, depending on assumptions about the LD_{50}[2] of anthrax spores and a difficult-to-estimate meteorological parameter used by many atmospheric dispersion models, the atmospheric stability class (Meselson, 2001). This quantity of release is consistent with one explanation for the outbreak put forward by at least two former Soviet officials: workers in the military microbiology facility mismanaged a ventilation system with subsequent discharge of *B. anthracis* spores into the air outside the facility (Guillemin, 2000).

In summary, Meselson reasoned from (1) the spatial and temporal distribution of cases of anthrax and (2) meteorological conditions present in Sverdlovsk prior to the outbreak, to the fact that an aerosol release had occurred and the location, time, and quantity of release that caused the outbreak.

Meselson's analysis, however, took months to complete, and therefore would not provide timely results for the prospective detection and characterization of an outbreak. In the following sections, we provide background about atmospheric dispersion models and then a method for automating Meselson's analysis to enable prospective detection of outbreaks.

3. ATMOSPHERIC DISPERSION MODELS

A model of atmospheric dispersion is an algorithm that predicts the downwind concentrations of a substance that result when a given quantity of the substance is released into the air. Atmospheric dispersion models require as input (at a minimum) the quantity of substance released, the release location, and weather conditions. Many atmospheric dispersion models exist. They have been developed for pollution control and modeling the spread of radioactive material from accidents at nuclear power plants (both hypothetical and actual). Not all are suitable for all applications.

To select a model for a particular application, it is necessary to know how models differ. After discussing the ways in which models differ, we discuss the Gaussian plume model in detail. Although it is among the simpler models, it is sufficient for many applications and—because it is computationally efficient—it is ideal for quickly obtaining estimates, which may be adequate or can be further refined using other models. We compare and contrast the features of four other aerosol dispersion models.

3.1. Attributes of Atmospheric Dispersion Models

Atmospheric dispersion models differ in several ways. One major difference is whether the model computes concentrations that result from a continuous release of a substance into the air over time (e.g., a smokestack) or from an episodic release (limited in time to a few seconds to a few hours). Models of a continuous release are known as plume models and models of an episodic release are called puff models. Plume models estimate the steady-state airborne concentrations at locations downwind from the source. Puff models estimate the concentration at a given downwind location as function of time since the release. Some models are both puff and plume models. They can estimate downwind concent rations for either continuous or episodic releases.

Atmospheric dispersion models also differ in whether they represent the release as a point, a line, an area, or a volume. Nearly every model can represent a point release. Many models also represent line releases (e.g., a release from an airplane in flight or dispersion of vehicle pollutants from highways), area releases (e.g., a burning field), and volume releases (e.g., dispersion after the source initially explodes). Some models are capable of representing all these geometries of the release as well as plume and puff (continuous or short term) variants for each of them.

Atmospheric dispersion models also differ in the assumptions they make to simplify the problem of computing downwind concentrations. In general, simpler models take less time to compute concentrations. So there is a tradeoff between accuracy of prediction and computational time that is especially relevant to uses of atmospheric dispersion models for prospective surveillance applications. To simplify computation, model developers may make assumptions such as the terrain is flat, the substance does not settle out of the air, the substance behaves as a gas, weather conditions are the same at every location, and weather conditions do not change with time.

Models that attempt to improve accuracy by making fewer simplifying assumptions typically require additional input parameters about weather, the substance itself, and/or terrain. The simplest atmospheric dispersion models require as meteorological input only wind speed, wind direction, and a measurement of atmospheric turbulence such as the stability class of the atmosphere, and do not require input data about terrain or the substance.[3]

2 The LD_{50} is the dose that is lethal for half the population.

3 One can compute stability classes from meteorological data that are widely available such as wind speed and cloud cover.

Finally, we note that plume models are useful in biosurveillance even when the release is episodic. We want to know the total quantity of biological agent that individuals would inhale—the *inhaled dose*—because it determines the probability that an individual will develop an infection. Because puff models compute the airborne concentration of spores at a given instant following the release, they enable computation of the inhaled dose as a function of time. To compute the total inhaled dose over the entire release however, we must sum the quantity inhaled at each instant after the release over all such instants. If we assume a constant breathing rate, we can multiply it by the sum of the concentrations from the puff model over all instants in time to obtain the inhaled dose. Some plume models compute this very sum of concentrations, because they are the time-integrated (or time-summed) form of a puff model: they assume that a continuous release of spores is equivalent to the situation where a puff release occurs at every instant. The concentrations that this type of plume model computes can be interpreted in two ways: (1) steady-state airborne concentrations that result from a continuous release, or (2) the sum over time of concentrations from a puff release.

3.2. The Gaussian Plume Model

The Gaussian plume model is a fundamental and widely used model of atmospheric dispersion. Mathematically, the Gaussian plume model is the integration over time of the Gaussian puff model (Barrat, 2001). It is both a puff and a plume model. You can use it to compute steady-state concentrations from a continuous release or to compute the total quantity of substance that passes some geographic point after a puff release. A number of researchers have used the Gaussian plume model to model releases of B. anthracis spores, (Wein et al., 2003, Meselson et al., 1994, Buckeridge et al., 2004) including Meselson (1994) in his famous study of Sverdlovsk discussed above.

The Gaussian plume model makes several simplifying assumptions: (1) the source of the substance is a point, (2) weather conditions do not vary over time or location, (3) the released material behaves as a gas (that has the same density as air), (4) the material does not settle out of the air or otherwise decay (such as might occur due to chemical reactions with air or sunlight), and (5) the terrain is flat. Of the five assumptions, the ones least likely to hold in a bioterrorism scenario are (1), (2), and (5), but these assumptions did not affect the conclusions of Meselson and colleagues. Assumptions (3) and (4) are reasonable. The size of particles that most effectively enter the lung to cause infection is 1 to 5 microns, a particle size that behaves like a gas with respect to atmospheric dispersion (Office of Technology Assessment, 1993). For *B. anthracis* spores, ultraviolet light has an insignificant effect on the infectivity of spores as they travel through the atmosphere (World Health Organization, 1970).

Hanna et al. (1982) cite several reasons for the popularity of the Gaussian plume model, including:

1. It produces results that agree with experimental data as well as any model.
2. It is fairly easy to perform mathematical operations on this equation.
3. It is appealing conceptually.
4. It is consistent with the random nature of turbulence.

The Gaussian plume model is given by the following equation, which computes the atmospheric concentration C at a given point (x, y, z) downwind from the release point:

$$C = \frac{Q}{2\pi\sigma_y(x)\sigma_z(x)u} e^{\frac{-y^2}{2\sigma_y(x)^2}} \left[e^{\frac{-(z-h)^2}{2\sigma_z(x)^2}} + e^{\frac{-(z+h)^2}{2\sigma_z(x)^2}} \right]$$

where

C = concentration of the substance at location x, y, z in μg/m³

Q = rate that the substance is released into atmosphere (in μg/second)

x = number of meters downwind from release point at which C is measured

y = number of meters cross wind from release point at which C is measured

z = number of meters above ground at which C is measured

u = wind speed in meters/second

h = height above ground in meters at which the substance was released

$\sigma_y(x)$ = standard deviation of the distribution of the substance in the crosswind (y) direction, as a function of x

$\sigma_z(x)$ = standard deviation of the distribution of the substance in the vertical (z) direction, as a function of x

The parameters $\sigma_y(x)$ and $\sigma_z(x)$ are the standard deviations of the distribution of C in the y and z directions. They are functions of the downwind distance x and atmospheric turbulence (dispersion is greater in more turbulent air). Stability classes are commonly used as a measurement of turbulence because they can be derived from inexpensive and easy meteorological measurements. There are a number of stability classification schemes, but most atmospheric dispersion models use Pasquill's scheme in conjunction with Brigg's formulas (Gifford, 1976) for calculating $\sigma_y(x)$ and $\sigma_z(x)$ for each stability class (Table 19.1).

The Gaussian plume model produces a pattern of concentrations that is longer than it is wide, with the highest concentrations occurring at locations where $y = 0$. Although the Gaussian plume equation is the integration over time of the Gaussian puff equation, and therefore can be used to estimate total inhaled dose, different dispersion parameters $\sigma_y(x)$ and $\sigma_z(x)$ are necessary for puff versus plume scenarios and

TABLE 19.1 Briggs Formulas for Computing $\sigma_y(x)$ and $\sigma_z(x)$, 10 m< x <10 km

Pasquill Stability Class	$\sigma_y(x)$	$\sigma_z(x)$
Urban		
A-B	$0.32x(1+0.0004x)^{-0.5}$	$0.24x(1+0.001x)^{0.5}$
C	$0.22x(1+0.0004x)^{-0.5}$	$0.20x$
D	$0.16x(1+0.0004x)^{-0.5}$	$0.14x(1+0.0003x)^{-0.5}$
E-F	$0.11x(1+0.0004x)^{-0.5}$	$0.08x(1+0.0015x)^{-0.5}$
Open country		
A	$0.22x(1+0.0001x)^{-0.5}$	$0.20x$
B	$0.16x(1+0.0001x)^{-0.5}$	$0.12x$
C	$0.11x(1+0.0001x)^{-0.5}$	$0.08x(1+0.0002x)^{-0.5}$
D	$0.08x(1+0.0001x)^{-0.5}$	$0.06x(1+0.0015x)^{-0.5}$
E	$0.06x(1+0.0001x)^{-0.5}$	$0.03x(1+0.0003x)^{-1}$
F	$0.04x(1+0.0001x)^{-0.5}$	$0.016x(1+0.0003x)^{-1}$

From Gifford, F. (1976). Turbulent diffusion-typing schemes: a review. *Nucl Saf* 17:68–86, with permission.

puff parameters such as those from Turner (1994) should be used (Table 19.2).

3.3. More Complex Models of Atmospheric Dispersion

Relative to the Gaussian plume model, almost all other dispersion models make fewer simplifying assumptions and require more input parameters. They thus account for additional properties of the atmosphere, weather, the substance released, and/or the terrain (Table 19.3). Examples of such properties include changes in wind direction and/or other meteorological parameters over time and space, non-flat terrain, air flow around buildings, settling of particles from the atmosphere, washout of the substance from the air by precipitation, temperature inversions, and decay or deactivation of the substance (e.g., some chemicals react with air and become inert quickly relative to the time it takes to disperse to locations of interest).[4]

TABLE 19.2 Dispersion Parameters for Puff Releases, $\sigma_y(x) = ax^b$ and $\sigma_z(x) = cx^d$

Pasquill Stability Class	a	b	c	d
A	0.18	0.92	0.72	0.76
B	0.14	0.92	0.53	0.73
C	0.1	0.92	0.34	0.72
D	0.06	0.92	0.15	0.70
E	0.045	0.91	0.12	0.67
F	0.03	0.90	0.08	0.64

TABLE 19.3 Characteristics of Dispersion Models

Dispersion Model	Plume or Puff?	Meteorological Factors	Substance-Related Factors	Terrain-Related Factors	Other Features
Gaussian plume model	Plume	Equation can be easily modified to include temperature inversion effects, but height of inversion must be known	None	None	None
Industrial Source Complex 3 (ISC3)[a]	Plume	Adjust windspeed by elevation above ground, temperature inversion effects	Gravitational settling, dry deposition, precipitation washout, decay	Building effects, non-flat terrain	Line, area, volume, open pit sources
AERMOD[a]	Plume	Adjust windspeed by elevation above ground, temperature inversion effects	Gravitational settling, dry deposition, precipitation washout, decay	Building effects, non-flat terrain	Line, area, volume, open pit sources
SCIPUFF/HPAC[b]	Both	Wind shear, temperature inversion effects, advanced method for handling turbulence, can make use of upper air meteorological observations and observations at multiple locations	Gravitational settling, dry deposition, precipitation washout, decay	Building effects, non-flat terrain	Line, area, volume, open pit, explosive sources Outputs probability distribution over concentrations Can model dispersion up to continental scale (other models limited to <100 km)
CALPUFF[c]	Puff	Temperature inversion effects, can make use of upper air meteorological observations and observations at multiple locations	Gravitational settling, dry deposition, precipitation washout, decay	Building effects, non-flat terrain	Models unique weather conditions of coastal areas such as sea breeze effects

[a]ISC and AERMOD are available at: *www.weblakes.com/lakeepa3.html.*
[b]SCIPUFF is available at: *www.titan.com/products-services/336/download_scipuff.html.*
[c]CALPUFF is available at: *www.src.com/calpuff/calpuff1.htm.*

4 The temperature of air normally decreases the further above ground one goes and thus a temperature inversion occurs when a warmer layer of air exists above a cooler one. A temperature inversion influences dispersion because the layer of warmer air prevents dispersion of substances from the cooler air and vice versa. The net effect of a temperature inversion is to increase ground-level concentrations when the release occurs below the inversion and to decrease them when the release occurs above the inversion. The effect can be dramatic, increasing or decreasing concentrations by several fold.

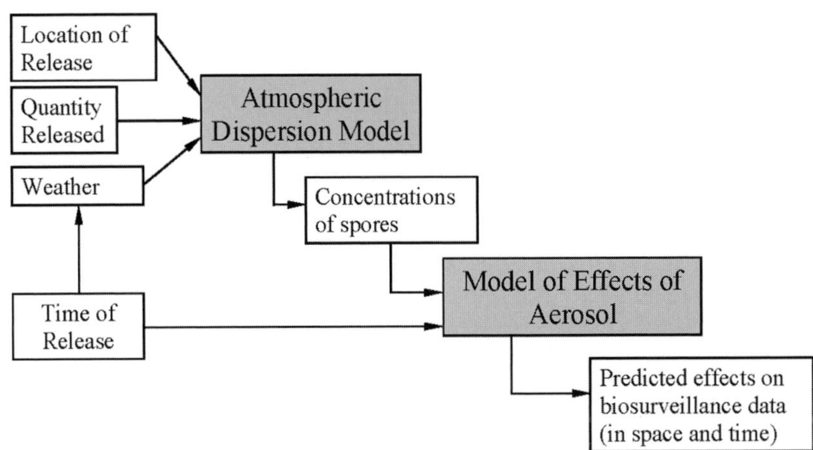

FIGURE 19.4 The more typical use of atmospheric dispersion models is in the simulation of a windborne outbreak. They compute downwind concentrations, then a separate model computes the effects of those concentrations on biosurveillance data.

There are far too many models to discuss in a chapter. Instead, we describe four models that are likely to be of value in biosurveillance, well known, and freely available: (1) the third version of the Industrial Source Complex model (ISC3), (2) AERMOD[5] (Lakes Environmental, 2005), (3) the Second-order Closure Integrated Puff model (SCIPUFF), and (4) CALPUFF.[6] Table 19.3 contrasts these models based on the additional aspects of weather, substance, and terrain that they take into account. We also note any unique features.

The ISC3 model (U.S. Environmental Protection Agency [EPA], 1995) is one of several models that the EPA recommends for regulatory purposes with respect to emissions of pollutants into the air by industry. It is based on the Gaussian plume model but makes adjustments for temperature inversions, dry and wet deposition of the substance, terrain, and building effects. The program, its source code (written in the FORTRAN programming language), and a user's guide are all freely available on the web (see website address in Table 19.3).

The AERMOD model is a proposed replacement for ISC3 for regulatory use. Its capabilities appear identical to ISC3 and it is also based on the Gaussian plume model, but it incorporates results from dispersion research that occurred subsequent to the development of ISC3. It includes improved modeling of atmospheric turbulence, dispersion in the layer of the atmosphere called the convective boundary layer, and the effects of terrain. Neither AERMOD nor ISC3 model weather conditions as changing with time, such as changes in wind speed and direction. As with ISC3, AERMOD, its source code (also in the FORTRAN programming language), and (a draft of) a user's guide are available for free on the web (see Table 19.3 for website address).

The SCIPUFF model is a model that accounts for changing weather conditions over time and location (Sykes et al., 1998). It is also the only model we mention here that is capable of modeling long-range dispersion up to the scale of a continent. The other models are limited to distances of 100 km or less. Despite its name, it can also estimate concentrations that result from continuous releases. SCIPUFF, its source code, and a user's guide are all available for free download (see Table 19.3 for website address).

Note that SCIPUFF is part of the Hazard Prediction and Assessment Capability (HPAC) software that the Defense Threat Reduction Agency makes available to first responders for a license fee. HPAC adds geographic information system features to SCIPUFF, enabling the user to visualize the dispersion of a substance over time on a map. HPAC also has the capability to predict the consequences of outdoor releases of various chemical and biological agents.

The CALPUFF model is a puff model. We include it here because it models over-water dispersion and other unique meteorological features that occur along coastlines such as sea breeze effects[7] (Scire et al., 2000). Given the proximity of many population centers to coastal areas, its use may be preferred in these areas. Like SCIPUFF, it also handles weather conditions that vary by location. It is also freely available for download (see Table 19.3 for website address).

5 AERMOD stands for the American Meteorological Society/Environmental Protection Agency Regulatory Model.

6 CALPUFF presumably stands for the California Puff model, as its developers originally created it for the California Air Resources Board.

7 For the curious, "sea breeze" is the term for the phenomenon of the wind blowing from the ocean to shore during the day (and from the shore to the ocean at night) because of a temperature differential between the air over land and over water.

4. ATMOSPHERIC DISPERSION MODELS AND THE ANALYSIS OF BIOSURVEILLANCE DATA

Because the atmospheric dispersion of a biological agent that is released into the air will determine in large part the spatial distribution of cases, it is logical to construct an outbreak detection algorithm that employs atmospheric dispersion modeling. The additional information provided by the model and weather data may enable earlier and more specific detection of a windborne outbreak. The potential increase in specificity results from the ability to distinguish between a disease pattern that is consistent with known weather conditions and the knowledge of how biological agents disperse in the atmosphere that is embodied in an atmospheric dispersion model. Such algorithms may be less prone to false alarms due to non-windborne outbreaks because other modes of transmission of disease produce different spatial patterns in data.

To incorporate an atmospheric dispersion model into a detection algorithm, we must invert it. An atmospheric dispersion model takes as input a known release of biologic agent and projects its downwind effects. To detect an unknown release, we must take observations about downwind effects (e.g., sick people or environmental samples such as the number of spores estimated from a BioWatch or other environmental monitor) and work backwards to infer the location, amount, and time of the release of the biologic agent (Figures 19.4 and 19.5). We now discuss this inversion problem and examples of algorithms and the methods they use to solve it.

4.1. Model Inversion

We can invert either the dispersion model alone or a combined model that includes the dispersion model and a model of the effects of aerosol cloud of biological agent (Figure 19.5). We refer to the former inversion as two-stage inversion because we must invert the aerosol effects model and the dispersion model separately. The inverted aerosol-effects model—using biosurveillance data as input—estimates downwind concentrations for input into the inverted dispersion model. We refer to the inversion of the combined model (that includes a dispersion model) as one-stage inversion. It does not require computing downwind concentrations, but instead involves estimation of release parameters directly from biosurveillance data.

There are two possible approaches to inverting a model, whether it is the dispersion model, the aerosol effects (or disease) model, or a model that is a combination of the two. The first—to invert them algebraically by solving a system of equations with release parameters as unknowns—is usually not possible. Another method of inversion is to employ a search: try thousands of combinations of values for the release parameters and output the parameters that best explain either downwind concentrations or the biosurveillance data. This device requires a measurement of how well a set of parameters explain downwind concentrations or biosurveillance data. All but one of the approaches we describe in this section use probability as a measurement.

We next describe examples of two-stage and one-stage inverted models. The work is preliminary and the improvement in outbreak detection performance (if any) that the models provide over other outbreak detection algorithms is not yet known.

4.2. Two-Stage Inverted Models

Hogan et al. (2004a) describe a two-stage inverted model called the Bayesian Aerosol Release Detector (BARD). This model takes as input weather data and counts of visits to the emergency department for respiratory complaints. It outputs

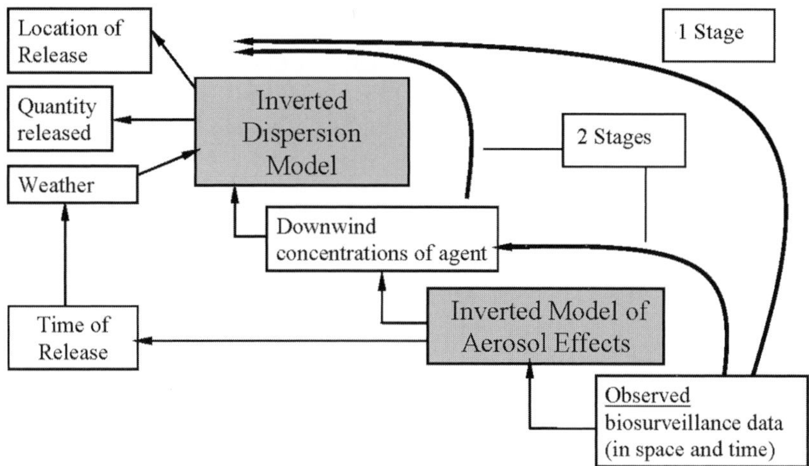

FIGURE 19.5 The inversion process required to construct outbreak-detection algorithms with atmospheric dispersion models. In two-stage inversion, the inverted aerosol effects model computes downwind concentrations for input into the inverted dispersion model. The inverted dispersion model then computes release parameters from downwind concentrations. In one-stage inversion, the inversion of the combined model computes release parameters directly from biosurveillance data–the intervening step of computing downwind concentrations from biosurveillance data is not required.

the likelihood ratio $P(data|H_1)/P(data|H_0)$, where data are counts of visits to emergency departments by zip code (i.e., the biosurveillance data), H_1 is the hypothesis that both an anthrax release (of a particular amount at a particular location and time) and usual respiratory disease occurred, and H_0 is the hypothesis that usual respiratory disease alon occurred. If the likelihood ratio is greater than a threshold value, BARD also outputs the release location, timing, and quantity.

The inverted dispersion model is an inverted Gaussian plume model (IGPM) that they created. The IGPM uses search to solve the Gaussian plume model equation with the coordinates of, and concentration at, four downwind locations called sensor locations (Hogan et al., 2004b). It does not use probability as a measurement. The measurement it does use is too complicated to describe in detail; essentially, it is the difference between predicted and actual sensor locations. When given exact downwind concentrations produced by the Gaussian plume model, IGPM was able to find the release location to within 1 meter and the release quantity to within 0.18 kilograms. With increasing levels of noise introduced into the downwind concentrations however, the accuracy of the procedure degraded rapidly, off by as much as 7 km with an amount of noise equal to approximately 20% of the actual concentrations.

The second stage (the inverted model of aerosol effects) is a Bayesian model of inhalational anthrax and usual respiratory disease (Hogan et al., 2004a). It models the probability of presenting to the ED for a respiratory complaint due to inhalational anthrax as a function of the number of spores inhaled and time since release. BARD uses Bayesian inversion (i.e., it uses probability as a measurement) of the anthrax model to compute an expectation over the number of spores inhaled and time since release. We discussed Bayesian inversion in Chapter 13 and Bayesian methods in greater detail in Chapter 18. BARD inputs the four zip codes with the highest expected number of inhaled spores and the expected time of release into IGPM (the time since release determines the weather data that the IGPM uses for analysis). BARD then uses the release location and quantity that IGPM estimates to compute the likelihood ratio.

BARD may also support more diagnostic precision at the time of outbreak detection. Current outbreak-detection algorithms cannot distinguish between a windborne outbreak and one resulting from a different mode of transmission. They detect an anomalous level of ED visits for "respiratory" complaints, for example, and include a line listing of the verbatim chief complaints. However, the differential diagnosis of a putative "respiratory" outbreak is quite large. Knowing that the mode of transmission is windborne reduces the differential diagnosis of a respiratory outbreak to fewer than 11 agents (Table 3.2 in Chapter 3).

Another benefit of BARD is that it estimates the likely location and date and time of the release. This information could help focus an environmental and criminal investigation, as well as the response effort.

BARD can also simulate windborne outbreaks of inhalational anthrax. We discuss the importance of outbreak simulations in biosurveillance below. Note that the ability to operate a model in either a forward direction for simulation or inverse direction for detection is general to Bayesian models and not unique to BARD.

4.3. One-Stage Inverted Models

The Population-wide ANomaly Detection and Assessment (PANDA) algorithm contains a causal model of a release of *B. anthracis* spores into the atmosphere and inverts the model using Bayesian inversion. It does not use a model of a model of atmospheric dispersion, however. Chapter 18 describes PANDA and its spatial model that is based on simulations produced from BARD.

A newer version of BARD is a one-stage model (Hogan et al., 2005). The motivation for this version of BARD was the inaccuracy of the estimate of the release location that the previous version computed. The source of much of the inaccuracy was the IGPM. Preliminary studies show that the one-stage version of BARD estimates release location to within 3.5 km. The previous version estimated it to within 35 km. Research is ongoing into whether the estimate can be improved further.

5. OTHER USES OF DISPERSION MODELS IN BIOSURVEILLANCE

Besides their role in analysis of biosurveillance data, atmospheric dispersion models are useful in biosurveillance both before and after outbreaks are detected. Once an outbreak has been detected, dispersion models can help to estimate the geographical extent of contamination with the biological agent. This information is critical to the response to an outbreak and decisions to perform enhanced human and environmental surveillance in particular areas.

We discuss these uses and the National Atmospheric Release Advisory Center, an atmospheric dispersion modeling resource that biosurveillance organizations and first responders may wish to use either for planning and/or for assistance during an actual event.

5.1. Simulation

Before an outbreak is detected, researchers and system builders use dispersion models to simulate outbreaks to understand the most likely tactics of a terrorist or military opponent. They also generate simulated outbreak data for use in measuring the outbreak detection performance of algorithms and systems.

Terrorists or military adversaries who wish to create an outbreak with maximal impact are likely to release a biological agent from locations or under weather conditions that maximize the lethality of the attack. Simulating releases from various locations and under various weather conditions can identify scenarios with maximal morbidity and mortality. Planners and system builders can then direct resources at preventing, detecting, and preparing responses to these scenarios. For example, one could design detection algorithms with heightened sensitivity to these scenarios. The BioWatch program conducts analyses about where to place air samplers to maximize the probability of detecting an aerosol release of biological agent. Atmospheric dispersion modeling is a critical component of these analyses.

A number of researchers have used the Gaussian plume model to simulate windborne outbreaks of inhalational anthrax. For example, Buckeridge et al. describe a model of anthrax attacks that generates simulated office visit data (Buckeridge et al., 2004). It uses the Gaussian plume model to simulate the windborne spread of *B. anthracis* spores. Evaluators can add the simulated outbreak data to historical baseline data and measure the false alarm rate, sensitivity, and timeliness of outbreak detection (Chapter 20 discusses evaluating algorithms).

5.2. Estimating Downwind Contamination

Once an outbreak has been detected, responders can use dispersion models to further characterize the outbreak. A key aspect of characterization is identification of exposed populations who are not yet symptomatic so that prophylactic treatment can be directed toward them. Using estimates of release location and time—perhaps those obtained from the use of dispersion models in the analysis of biosurveillance data—responders can "simulate" the outbreak to inform decisions about prophylactic treatment of populations.

An example of how authorities have used dispersion models in the past to respond to known outbreaks is a FMD outbreak that started in northern France. On March 4, 1981, French veterinary authorities and the World Organization for Animal Health notified the Ministry of Agriculture in the United Kingdom (UK) of outbreaks of FMD in northern France (Donaldson et al., 1982). Some previous outbreaks of FMD in southern Great Britain were the result of windborne spread of FMD virus across the English Channel from France. Thus, the Ministry of Agriculture on March 6 and March 12 asked the Meteorological Office of the UK about the likelihood of spread of FMD across the English Channel to the UK.

The Meteorological Office of the UK advised the Ministry of Agriculture that conditions favorable for windborne spread had occurred. Given recent wind directions and other weather conditions, the Meteorological Office predicted that airborne concentrations of virus at two islands in the English Channel—Jersey and the Isle of Wight—were high enough to cause infection in animals. The risk was low for southern England.

Subsequent to the receipt of this advice, the Ministry of Agriculture received notification of FMD on Jersey and the Isle of Wight. No outbreaks developed in southern England. Additional research into the outbreaks confirmed that the strain of FMD virus on the two islands and the strain in northern France were identical. This research also ruled out other modes of transmission of FMD virus such as movement of animals or equipment.

The study makes no mention of any heightened surveillance or preventive action (such as vaccination) based on the information about a threat to the islands. However on the same day that the first farmer reported symptoms in his cattle (on the Isle of Wight), veterinarians diagnosed the outbreak and imposed restrictions on the movement of animals. Prior to these outbreaks of FMD, the United Kingdom had a plan in place to deploy to the site of any FMD outbreak a team of experts. These experts included individuals from the Animal Virus Research Institute, the Ministry of Agriculture's state veterinary service, and other appropriate experts such as meteorologists and statisticians. Such a team deployed to the Isle of Wight after the outbreak was detected and collected the epidemiological data reported in the study.

5.3. The National Atmospheric Release Advisory Center

The National Atmospheric Release Advisory Center (NARAC) is a set of services and tools of the Lawrence Livermore National Laboratory (LLNL) (Lawrence Livermore National Laboratory). NARAC allows authorized emergency managers from local, state, and federal levels of government to request assistance in the event of a release of a radiological, chemical, or biological agent into the atmosphere. NARAC uses state-of-the art dispersion modeling tools developed at LLNL to predict downwind concentrations and even the health consequences of releases. It displays downwind concentrations graphically on maps and even aerial photographs so that responders can view the downwind concentrations at various landmarks such as roads, buildings, and stadiums. It staffs an emergency hotline 24 hours a day, seven days a week for emergency managers to call for assistance. NARAC also has other tools and services such as software for emergency managers to perform modeling locally, training on that software, and support for conducting drills and exercises.

6. WEATHER DATA

Data about weather conditions are essential for the use of both usual and inverted models of atmospheric dispersion. Fortunately, weather data from the United States and around the world are available because data are collected in real time, in standard formats, and integrated in a central location that is publicly available without any technical or administrative barriers (*http://weather.noaa.gov/*).

Automated weather stations located at civilian and military airports around the world produce minute-by-minute observations of wind direction, wind speed, temperature, dewpoint temperature, visibility, precipitation, cloud cover, and other variables. These stations transmit hourly weather observations to the National Weather Service, which then makes available for free download the previous 24 hours' data. In the United States, the weather stations are part of the Automated Surface Observation System (ASOS) (*www.nws.noaa.gov/asos/index.html*), a joint effort between the National Weather Service, the Federal Aviation Administration, and the Department of Defense. The stations code data using a standard format called METAR (see Chapter 32). The *Federal Meteorological Handbook No. 1* (Office of the Federal Coordinator for Meteorological Services and Supporting Research, 1995) explains the meaning of the codes, enabling one to write a parser to extract the data contained within. Approximately 1700 stations in the United States report data.

Given the importance to aviation of accurate data about current weather conditions, a human operator usually reviews and corrects the data. The National Weather Service also performs quality control functions on data they receive (as part of the ASOS program).

The ASOS data do not include measurements of atmospheric turbulence such as atmospheric stability class. However, they do contain measurements that you can use in combination with other measurements to compute a Turner stability class (see WebMET, 2002). Turner stability classes have direct mappings to the commonly used Pasquill stability classes. The RODS Laboratory has been downloading and storing METAR data since 2003, including computing and storing the atmospheric stability class in this manner. The RODS Laboratory makes these data available to organizations that conduct biosurveillance.

Relative humidity affects survivability of FMDV and other biological agents as they travel long distances in the atmosphere, and thus measurements of relative humidity of atmosphere are needed for modeling the spread of these agents. Fortunately, computing relative humidity from measurements of temperature and dewpoint temperature is straightforward, and both these temperature measurements exist in the ASOS data sets.

7. SUMMARY

Atmospheric dispersion models describe how wind disperses biological agents from the source of an aerosol release. They have several important uses in biosurveillance, including configuration of biosurveillance systems before an outbreak, detection and characterization of an outbreak, and focused surveillance in areas suspected to have been exposed but where individuals are not yet ill after outbreak detection.

For atmospheric dispersion models to be applied to outbreak detection, they must be inverted. Most research on the use of inverted atmospheric dispersion models for outbreak detection has used Bayesian methods. Inverted models for outbreak detection have the potential to narrow down the differential diagnosis of outbreaks by inferring that outbreaks are windborne (which limits the possible organisms) early in the course of an outbreak. They can also estimate the location, quantity, and timing of the release of biological agent, information critical to directing response efforts.

ADDITIONAL RESOURCES

National Atmospheric Release Advisory Center (NARAC) website: *http://narac.llnl.gov/index.html*. This is NARAC website, which is discussed in this chapter.

Environmental Protection Agency Support Center for Regulatory Air Models website: *www.epa.gov/ttn/scram/*. This website lists several models of atmospheric dispersion that the EPA recommends for use in regulatory applications.

Hazard Prediction and Assessment Capability (HPAC) website:*www.dtra.mil/Toolbox/Directorates/td/programs/acec/hpac.cfm*. This is the HPAC website, which is discussed in this chapter.

ACKNOWLEDGMENTS

I would like to thank Matthew Meselson and the American Association for the Advancement of Science for permission to reproduce figures from Dr. Meselson's paper on the Sverdlovsk outbreak and Michael Wagner for reading and commenting on this chapter.

REFERENCES

Abramova, F. A., Grinberg, L. M., Yampolskaya, O. V., et al. (1993). Pathology of inhalational anthrax in 42 cases from the Sverdlovsk outbreak of 1979. *Proc Natl Acad Sci U S A* 90:2291–4.

Barrat, R. (2001). *Atmospheric Dispersion Modeling: An Introduction To Practical Applications*. Sterling, VA: Earthscan Publications.

Breeze, R. (2004). Agroterrorism: betting far more than the farm. *Biosecur Bioterror* 2:251–64.

Buckeridge, D. L., Burkom, H., Moore, A., et al. (2004). Evaluation of syndromic surveillance systems—design of an epidemic simulation model. *MMWR Morb Mortal Wkly Rep* 53(Suppl):137–43.

Cupp, O. S., Walker, D. E. 2nd, Hillison, J. (2004). Agroterrorism in the U.S.: key security challenge for the 21st century. *Biosecur Bioterror* 2:97–105.

Davis, J. M. (1987). Modeling the long-range transport of plant pathogens in the atmosphere. *Ann Rev Phytopathol* 25:169–88.

Defense Threat Reduction Agency. Hazard Prediction and Assessment Capability (HPAC). *http://www.dtra.mil/Toolbox/Directorates/td/programs/acec/hpac.cfm*.

Donaldson, A. I., Gloster, J., Harvey, L. D., et al. (1982). Use of prediction models to forecast and analyse airborne spread during the foot-and-mouth disease outbreaks in Brittany, Jersey and the Isle of Wight in 1981. *Vet Rec* 110:53–7.

Gifford, F. (1976). Turbulent diffusion-typing schemes: a review. *Nucl Safety* 17:68–86.

Guillemin, J. (2000). Anthrax: the investigation of a deadly outbreak. *N Engl J Med* 343:1198.

Hanna, S., Briggs, G., Hosker, J. (1982). *Handbook of Atmospheric Diffusion*. DOE/TIC-11223. Washington, DC: U.S. Department of Energy.

Hawker, J. I., Ayres, J. G., Blair, I., et al. (1998). A large outbreak of *Q* fever in the West Midlands: windborne spread into a metropolitan area? *Commun Dis Public Health* 1:180–7.

Hogan, W., Cooper, G., Wallstrom, G. L., et al. (2004a). *A Bayesian Anthrax Aerosol Release Detector*. Pittsburgh, PA: Real-Time Outbreak and Disease Surveillance Laboratory, Center for Biomedical Informatics, University of Pittsburgh.

Hogan, W., Cooper, G., Wallstrom, G. L., et al. (2004b). *An Inverted Gaussian Plume Model for Estimating the Location and Amount of Release of Airborne Agents from Downwind Atmospheric Concentrations*. Pittsburgh, PA: Real-Time Outbreak and Disease Surveillance Laboratory, Center for Biomedical Informatics, University of Pittsburgh.

Hogan, W., Cooper, G., Wallstrom, G. L., et al. (2005). *An Improved Bayesian Anthrax Aerosol Release Detector*. Pittsburgh, PA: Real-Time Outbreak and Disease Surveillance Laboratory, Center for Biomedical Informatics, University of Pittsburgh.

Lakes Environmental. (2005). AERMOD Tech Guide. *http://www.weblakes.com/aermodvol1/tableofcontents.html*.

Lawrence Livermore National Laboratory. (2005). National Atmospheric Release Advisory Center. *http://narac.llnl.gov/*.

Meselson, M. (2001). Note regarding source strength. ASA Newsletter 87:1, 10–12.

Meselson, M., Guillemin, J., Hugh-Jones, M., et al. (1994). The Sverdlovsk anthrax outbreak of 1979. *Science* 266:1202–8.

Office of Technology Assessment. (1993). *Technologies Underlying Weapons of Mass Destruction*. OTA-BP-ISC-115. Washington, DC: Office of Technology Assessment, U.S. Congress.

Office of the Federal Coordinator for Meteorological Services and Supporting Research. (1995). *Federal Meteorological Handbook No. 1—Surface Weather Observations and Reports. http://www.ofcm.gov/fmh-1/pdf/fmh1.pdf*.

Scire, J., Strimaitis, D., Yamartino, R. et al. (2000). *A User's Guide for the CALPUFF Dispersion Model (Version 5)*. Concord, MA: Earth Tech, Inc.

The Record. (1992). Germ warfare caused epidemic, Yeltsin says. *The Record* (Kitchener-Waterloo, Ontario), Moscow, p. A4.

Sellers, R. F., Gibbs, E. P., Herniman, K. A., et al. (1979). Possible origin of the bluetongue epidemic in Cyprus, August 1977. *J Hyg (Lond)* 83:547–55.

Sellers, R. F., Maarouf, A. R. (1988). Impact of climate on western equine encephalitis in Manitoba, Minnesota and North Dakota, 1980–1983. *Epidemiol Infect* 101:511–35.

Sellers, R. F., Pedgley, D. E. (1985). Possible windborne spread to western Turkey of bluetongue virus in 1977 and of Akabane virus in 1979. *J Hyg (Lond)* 95:149–58.

Sellers, R. F., Pedgley, D. E., Tucker, M. R. (1978). Possible windborne spread of bluetongue to Portugal, June-July 1956. *J Hyg (Lond)* 81:189–96.

Sykes, R. I., Parker, S. F., Henn, D. S., et al. (1998). *PC-SCIPUFF Version 1.2PD Technical Documentation*. ARAP Report No. 718. Princeton, NJ: Titan Corporation.

Turner, D. B. (1994). *Workbook of Atmospheric Dispersion Estimates: An Introduction to Dispersion Modeling*. Boca Raton, FL: Lewis Publishers.

U.S. Environmental Protection Agency. (1995). *User's Guide for the Industrial Source Complex (Isc3) Dispersion Models: Volume II—Description of Model Algorithms*. EPA report EPA-454/B-95-003b. Research Triangle Park, NC: U.S. Environmental Protection Agency.

WebMET. (2002). Turner's Method. *http://www.webmet.com/met_monitoring/641.html*.

Wein, L. M., Craft, D. L., Kaplan, E. H. (2003). Emergency response to an anthrax attack. *Proc Natl Acad Sci U S A* 100:4346–51.

World Health Organization. (1970). *Health Aspects of Chemical and Biological Weapons*. Geneva: World Health Organization.

20

CHAPTER

Methods for Algorithm Evaluation

Michael M. Wagner and Garrick Wallstrom
RODS Laboratory, Center for Biomedical Informatics
University of Pittsburgh, Pittsburgh, Pennsylvania

1. INTRODUCTION

Our objective when evaluating a biosurveillance detection algorithm is to measure its accuracy (sensitivity and false alarm rate) and time to detection (e.g., from the time of infection in an individual case or from the start date of the outbreak). If the algorithm is capable of inferring other outbreak characteristics, such as size, route of transmission, or infectivity, we are also interested in its accuracy for these parameters and the time during the course of the outbreak when the algorithm is capable of estimating these parameters.

Researchers typically measure these algorithm characteristics in the laboratory using surveillance data collected during outbreaks or synthetic data that resembles real outbreak data. However, it is also important to know how well an algorithm or analytic method works when incorporated into a fielded biosurveillance system. Therefore, evaluators also conduct field testing to determine whether the performance in the field is similar to that found in the laboratory. Field testing also assesses other important factors, such as what amount of training is required for professionals working in the field to use the algorithm effectively.

In this chapter, we discuss laboratory evaluation of algorithms. We discuss methods for field testing of algorithms in Chapter 37.

2. GOALS OF ALGORITHM EVALUATION

A laboratory evaluation of an algorithm may address one or more of the following questions:

- With what sensitivity and with what error rate can a *case* detection algorithm detect individuals with different syndromes or diseases?
- With what sensitivity and with what error rate can an *outbreak* detection algorithm detect outbreaks of different sizes and types?
- What is the smallest size outbreak that can be detected?
- When are outbreaks of different sizes detected?
- How accurately can an *outbreak* detection algorithm identify other characteristics of an outbreak, such as geographic location of cases, source, route of transmission, infectivity, and other individuals potentially exposed?
- Is a new or modified algorithm an improvement over existing algorithms?
- How can the performance of an algorithm be improved?

Although an algorithm may be capable of detecting many diseases or types of outbreaks, evaluators typically address these questions in reference to one disease or one type of outbreak, or for a limited set of diseases or outbreaks.

The general experimental approach to obtaining answers to these questions involves running the algorithm on surveillance data that contains known cases or outbreaks and determining whether the algorithm detected each case or outbreak, on what date, and whether there were any false alarms. Although the general approaches for evaluating algorithms for case detection, outbreak detection, and outbreak characterization are similar, there are sufficient differences to warrant separating the discussions into separate sections.

3. EVALUATING ALGORITHMS FOR CASE DETECTION

To evaluate an algorithm designed to detect a single individual with disease (a case), an evaluator employs well-developed methods for measuring *sensitivity* and *specificity* of classification (defined below). These methods are also used to evaluate new laboratory tests and biosensors—and many other systems and devices that are fundamentally classifiers (e.g., a machine that distinguishes bad ball bearings from good ones).

3.1. Standard Evaluation Method for Evaluating a Classifier

A case detection algorithm is an example of a classifier. At the most general conceptual level, it classifies individuals into the categories "sick" or "not sick." In practice, these categories are typically more diagnostically precise ranging from, for example, "patient has *E. coli* 0157" to "patient has fever." Similar to a diagnostic laboratory test, a case detection algorithm outputs a numeric value that is then compared to a threshold to determine whether to classify the individual as "sick" or "not sick."

To evaluate a case detection algorithm, the evaluator assembles two sets of individuals: a set of individuals with the disease or condition of interest (called *cases*) and a set of individuals that do not have the disease (called *controls*). The evaluator can find the cases from the records of a known outbreak, by a

Handbook of Biosurveillance
ISBN 0-12-369378-0

review of laboratory test results (e.g., microbiological cultures), or by asking experts to review medical or veterinarian records. The evaluator assembles the set of control patients by random selection. The control patients could be healthy individuals, or they could be individuals whose malady resembles the condition of interest, such as patients that have influenza-like illness, but do not have SARS. The choice depends on the intended use of the algorithm. If the algorithm will be used, for example, in an emergency department (ED) setting to automatically differentiate patients presenting with influenza-like symptoms into "has SARS" and "does not have SARS," the evaluator would select controls from the set of all patients presenting with influenza-like symptoms determined not to have SARS. If the intended use is to differentiate SARS patients automatically from all other patients presenting to the ED, the evaluator would draw the control group from the set of all patients presenting to the ED known not to have SARS.

To illustrate the evaluation method concretely, let us suppose an evaluator has assembled a set of 100 patients with SARS and a control group of 100 individuals drawn from the set of patients coming to the ED who were subsequently proven not to have SARS. The evaluator's goal is to measure the accuracy of a case detection algorithm that classifies patients into *SARS* or *not SARS* based on patient data routinely collected in an ED. To measure the accuracy of the case detection algorithm, the evaluator would run the case detection algorithm on the patient data for each case and each control, recording whether the algorithm classifies each case and each control as "SARS" or "not SARS." The evaluator summarizes the results of this experiment in a 2×2 table, such as the one depicted in Table 20.1.

Table 20.1 shows the results of this hypothetical experiment. Of the 100 individuals with SARS, the case detection algorithm classified 90 correctly (*true positives*) and 10 incorrectly (*false negatives*). Of the 100 controls, the case detection algorithm classified five incorrectly as having SARS (*false positives*), and it classified 95 correctly as not having SARS (*true negatives*).

3.2. Sensitivity and Specificity

To summarize the accuracy of the case detection algorithm, the evaluator computes two simple ratios from the numbers in Table 20.1: sensitivity and specificity. *Sensitivity* is the ratio of the number of cases of SARS that the case detection algorithm classified correctly (90) to the total number of patients with SARS (100), or 0.9. Sensitivity is the probability that the case detection algorithm will "detect" a patient with SARS. *Specificity* is the ratio of individuals without SARS that the case detection algorithm classified correctly as not having SARS (95) to the total number of individuals without SARS (100), or 0.95. The evaluator can also compute the ratio of individuals without SARS (5) who were incorrectly labeled by the case detection algorithms as SARS (5) over the total number of patients without SARS (100), or 0.05. This quantity is called the false positive rate (or false alarm rate), and it is the probability that a case detection algorithm will incorrectly classify a patient as SARS-infected. Note that specificity and false positive rate (also known as the false alarm rate) sum to one because those two rates account for all of the patients that did not have SARS (the ones who were classified negative for SARS by the algorithm and the ones who were classified positive for SARS). For this reason, specificity and false positive rate (false alarm rate) are really two sides of the same coin, and, since each conveys the same information about the classifier, evaluators will use them interchangeably depending on whether they want to stress the number of false alarms that the case detection algorithm will generate, or its specificity. Sensitivity and specificity of classification are fundamental properties of a classifier for a condition of interest (in this case the disease SARS). These two numbers completely characterize the detection characteristics of the case detection algorithm. (Note that these methods are also the basis for measuring data accuracy, which we discuss in Chapter 37, and for studying case definitions, which can be understood as a type of classifier that is used for case detection and during outbreak investigations. Case definitions are discussed in Chapters 3 and 13.)

TABLE 20.1 Results of Hypothetical Evaluation of SARS Case Detection Algorithm

	100 Individuals with SARS (cases)	100 Individuals without SARS (controls)	Total Classifications
Algorithm positive for SARS	90	5	95
Algorithm negative for SARS	10	95	105
Total patients	**100**	**100**	**200**

Notes: sensitivity=0.9; specificity=0.95.

3.3. Predictive Value Positive and Negative

Students are often confused by two other ratios that an evaluator can compute from the data in Table 20.1. The world would be a far less confusing place if these ratios did not exist at all. We will not discuss them further because they are rarely useful, but for the record, they are as follows:

Predictive value positive (PVP): the fraction of the total number of cases that were classified as SARS that truly were SARS.

Predictive value negative (PVN): the fraction of the total number of cases that were classified as non-SARS that truly were non-SARS.

Readers who wish to learn more about PVP and PVN and the reason why we discourage their use can refer to Appendix B.

3.4. Using Bayes Theorem to Compute Posterior Probability of Disease

Biosurveillance personnel want to know how well a case detection algorithm can *predict* the actual diagnosis of a patient from biosurveillance data about that patient, because the real-world purpose of such an algorithm would be to screen individuals for SARS. The preferred method (and standard practice) has two parts. First, the evaluator measures and reports only sensitivity and specificity. Second, consumers of this information (system developers or biosurveillance personnel) use Bayes theorem to compute predictive probabilities from the reported sensitivity and specificity of the case detection algorithm and the current estimates of prevalence (also known as prior probability) of disease in their country or locality. We illustrate this use of Bayes theorem below.

$$P(SARS \mid Algorithm\ Pos) = \frac{P(Algorithm\ Pos \mid SARS)P(SARS)}{P(Algorithm\ Pos \mid SARS)P(SARS) + P(Algorithm\ Pos \mid No\ SARS)P(No\ SARS)}$$

$$P(No\ SARS \mid Algorithm\ Neg) = \frac{P(Algorithm\ Neg \mid No\ SARS)P(No\ SARS)}{P(Algorithm\ Neg \mid SARS)P(SARS) + P(Algorithm\ Neg \mid No\ SARS)P(No\ SARS)}$$

If the background prevalence of SARS in the population is one case per 100,000 people, the predictive probabilities of disease status given the test result are:

$$P(SARS \mid Alg\ Pos) = \frac{0.90 * 0.00001}{0.90 * 0.00001 + 0.05 * 0.99999} = 0.00018$$

$$P(No\ SARS \mid Alg\ Neg) = \frac{0.95 * 0.99999}{0.10 * 0.00001 + 0.95 * 0.99999} = 1 - 1.053 \times 10^{-6}$$

These numbers, although derived by somewhat dry, boring mathematics, are quite useful to doctors, veterinarians, and personnel involved in outbreak detection and investigation. The quantity $1–1.053\times10^{-6}$ (essentially 1.0) is the probability that the individual who classifies negative does not have SARS. That is reassuring for outbreak investigators and for the patient. On the other hand, the number 0.00018, the probability that the individual has SARS given that the algorithm classified him as positive, does not help. Although 0.00018 is 18 times greater than the prevalence of disease, meaning that the chance that this individual has SARS is higher than before the classification, it is not high enough to change the management of the patient. The evaluator would conclude that the performance of the case detection algorithm with 90% sensitivity and 95% specificity is not good enough for this purpose.

3.5. ROC Curve Analysis

It is important to note that the preceding table was produced by a case detection algorithm that was operating at a preselected and fixed "detection threshold." Classification algorithms, such as a case detection algorithm, typically produce continuous numeric outputs over some range of values. An evaluator selects a numerical value in this range as the cut-off point above which patients are classified as having disease and below which they are classified as not having disease.

Figure 20.1 illustrates the raw numeric results for a hypothetical classification algorithm that produces as an intermediate result a number between 0 and 15 for most cases and controls.

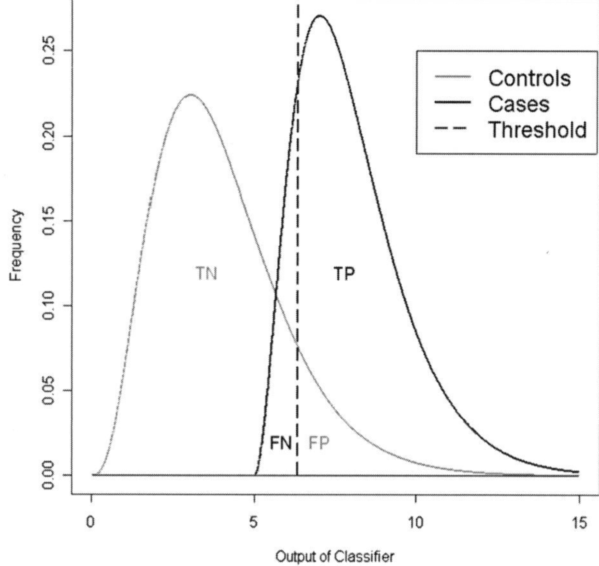

FIGURE 20.1 Classifier discrimination. Subjects are classified as positive when the output from the classifier exceeds a threshold, and negative otherwise. TP, true positive; FN, false negative; FP, false positive; TP, true positive.

The left-most curve in Figure 20.1 is the distribution of numeric test results for control patients, and the right-most curve is the distribution of numeric results for cases. We can understand the numeric results produced by a classification algorithm as a measurement of some property of the individual patients (e.g., temperature) that we are hoping will be useful for discriminating between individuals with disease and individuals without disease. The fact that the two curves in Figure 20.1 overlap indicates that the classifier cannot perfectly discriminate between cases and controls.

Since most classification algorithms produce numeric results, the evaluator must pick some threshold to convert the numeric output of a case detection algorithm (or diagnostic test) into a determination that an individual "has disease" or "does not have disease." Figure 20.1 shows an arbitrarily selected detection threshold of six (dashed line).

The results of the experiment—including the counts in the 2×2 table above and the measured sensitivity and specificities—are highly dependent on the threshold selected by the evaluator. In fact, the areas under the curves labeled TN, FN, TP, and FP in Figure 20.1 correspond to the numbers that go into each of the four cells in the above table.

Note that if the evaluator were to change the detection threshold in Figure 20.1 by moving it to the left or to the right, the areas under the curves corresponding to TN, FN, FP and TP would all change, as would the numbers in the table and the sensitivity and specificity. As she increases the threshold (by moving it to the right), the sensitivity (fraction of the disease cases that are correctly classified) decreases, but the specificity (fraction of controls that are correctly classified) increases. Since both sensitivity and specificity are desirable properties of a classifier, there is obviously a tradeoff between sensitivity and specificity involved when setting the threshold for a classifier. Since an evaluator does not know whether a future user of a classifier will prefer sensitivity to specificity, she explores a range of thresholds spanning the overlap between the two curves (thresholds from five to approximately 12 in Figure 20.1), recording the sensitivity and specificity at each threshold and plotting a graph of sensitivity versus specificity. This type of graph is called a *receiver operator characteristic* (ROC) curve (Egan, 1975, Metz, 1978, Lusted, 1971). Note that most evaluators plot 1 – specificity, which as we previously discussed is the *false alarm rate*, on the horizontal axis. Their reason is that users of classifiers are most interested in the tradeoff between the value of detecting cases (sensitivity) and the cost and other consequences of misdiagnosing healthy people.

In Figure 20.2, an evaluator has plotted two ROC curves, corresponding to the results of evaluations of two different case detection algorithms. The evaluator would conclude that Algorithm 1 is a better classifier than Algorithm 2 for the hypothetical disease in question because at every value of false-alarm rate, Algorithm 1 has better sensitivity than Algorithm 2.

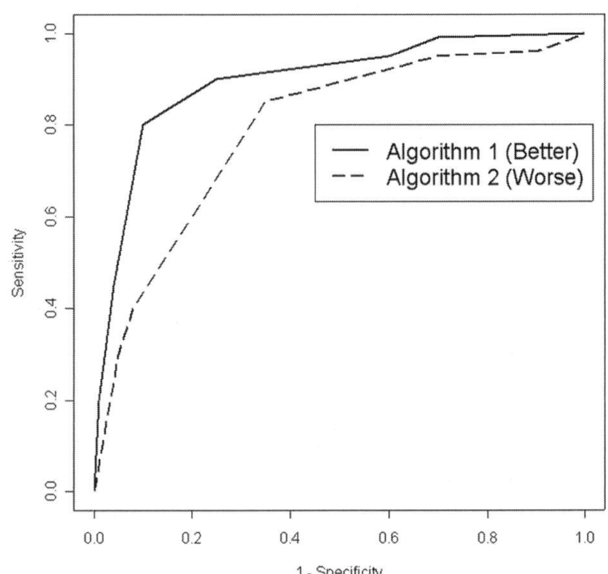

FIGURE 20.2 ROC curves from a hypothetical experiment comparing two algorithms.

Evaluators use the area under an ROC curve as an overall measure of the classification accuracy of a classifier. The technical term they use is area under (the ROC) curve (AUC). AUC has a very natural interpretation: if the evaluator produced one positive patient and one negative patient and asked the classifier to compare the two as to which is more likely to be infected with SARS, then the AUC is the probability that the classifier would correctly rank the positive patient higher than the negative patient. A perfect classifier has an AUC of 1.0, and a classifier that guesses at random has an AUC of 0.5, which corresponds to a diagonal ROC curve from the point (0,0) to the point (1,1) in Figure 20.2.

3.6. Measuring Timeliness of Case Detection

Timeliness is not a fundamental characteristic of a case detection algorithm. Timeliness is the time between the initial infection of a patient until the case is classified correctly. A case detection algorithm's time of detection depends on when the algorithm is applied to patient data, which depends on when the case detection algorithm is used in a field application, which we discuss in Chapter 37. An evaluator may measure timeliness of detection in reference to time points other than the time of initial infection, such as the date the diagnosis was established by a clinician or a definitive laboratory test.

4. EVALUATING ALGORITHMS FOR OUTBREAK DETECTION

To evaluate an outbreak detection algorithm, an evaluator uses methods similar to the methods we just discussed for evaluating case detection algorithms. There are several technical and practical differences however. A technical difference is that timeliness of detection of an outbreak correlates to sensitivity and specificity and is a fundamental property of outbreak detection algorithms. As the evaluator changes the threshold setting for an outbreak detection algorithm, not only do sensitivity and specificity change, but so does time of detection. This difference requires evaluators to use a form of ROC curve analysis that is generalized to a third dimension—time—which we will discuss. A key practicality is that it is extremely difficult for an evaluator to assemble a set of 100 outbreaks for study. This difference induces evaluators to resort to a variety of creative approaches to using simulated or partly simulated data in the evaluation of detection algorithms.

4.1. Standard Evaluation Method for an Outbreak-Detection Algorithm

Biosurveillance systems use outbreak detection algorithms to analyze surveillance data and search for signs of an outbreak. As discussed in previous chapters, the surveillance data are typically formed into time series of daily or weekly counts prior to analysis. For each unit in the time series, an outbreak detection algorithm calculates a value that is a measure of how unusual one or more recent counts are, when compared to historical counts. If the degree of anomaly is above some threshold, an outbreak is considered to be present and the algorithm generates an alarm. Although for concreteness, we will discuss methods for evaluating outbreak detection algorithms for algorithms that analyze time series data, these methods also apply to algorithms, such as PANDA (described in Chapter 18) or WSARE (described in Chapter 15), that do not aggregate data into time series.

To evaluate an outbreak detection algorithm, the evaluator requires (1) surveillance data representing a sufficiently large number of outbreaks that he can measure the algorithm's sensitivity and average timeliness, and (2) surveillance data recorded during outbreak-free intervals with which to measure the algorithm's specificity (false alarm rate).

4.1.1. Measuring Timeliness of Detection

The evaluator first determines the *start date*, defined as the date the outbreak began. When the evaluator simulates outbreak data, the start date is known because it is a parameter of the simulation. When using real outbreak data, however, the start date may be difficult to establish. The question of when any particular influenza outbreak began would generate a healthy discussion among epidemiologists. For purposes of research, the date can be established by a voting procedure or some other method to achieve expert consensus (Buckeridge et al., 2005). The evaluator also establishes an end date using similar methods.

The *date of detection* is the first date on or after the start date that the algorithm generated a true alarm, defined as an alarm that occurred after the start date and before the end date. Figure 20.3 illustrates a hypothetical outbreak for which an evaluator established a start date of February 10. The algorithm generated alarms, indicated by "X" marks, on January 26, February 13, and February 15. In this example, the date of the first true alarm was February 13, which the evaluator would consider the date of detection.

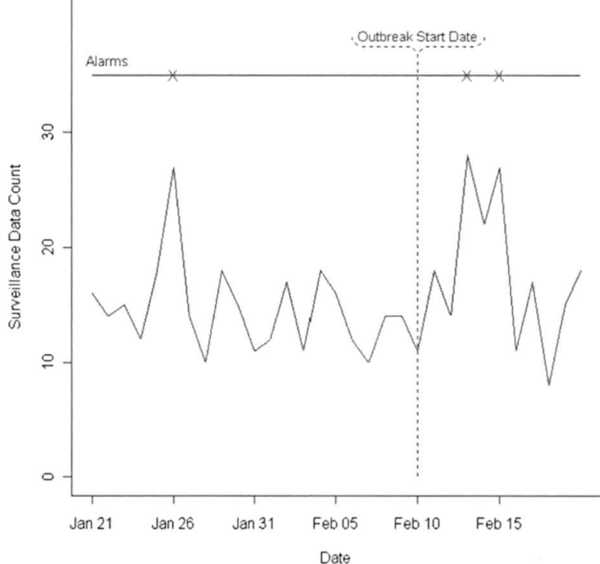

FIGURE 20.3 Alarms generated by a detection algorithm running on surveillance data organized into a time series of daily counts. The date of detection is February 13, the date of the first alarm after the start date of the outbreak (February 10). The alarm on January 26 is a false alarm.

Timeliness of detection is the difference between the date of the first true alarm and a reference date. In Figure 20.3, the timeliness is three days. If the algorithm does not detect the outbreak at all, evaluators typically set the timeliness of detection to a fixed large value that approximates when the outbreak would most certainly be detected by conventional means (e.g., the duration of the entire outbreak).

Note that for studies that compare the performance of two algorithms, the accuracy or choice of the reference date is not critical, although it is most useful for interpretation purposes if it corresponds to a date that has decision importance, such as the date that the outbreak was actually detected by a health department. However, it is important that the reference date is defined objectively so that it can be consistently calculated for each outbreak in the sample or used by other researchers in future studies whose results collectively would then be more amenable to meta analysis.

4.1.2. Measuring Sensitivity

When evaluating a *case* detection algorithm, sensitivity is the proportion of cases that the algorithm classified correctly. When evaluating an *outbreak* detection algorithm, sensitivity is the proportion of *outbreaks* the algorithm detected. Since this measure ignores the time of detection, we refer to it as the *overall sensitivity*. Because time is important, this measure is not very informative about performance except for outbreaks that might go undetected by current methods because they are small or too diffuse to detect.

A more informative measure is sensitivity as a function of timeliness. For example, we can measure the proportion of outbreaks that an algorithm detects within three days from the start of the outbreak, while holding the false positive rate fixed. More generally, we define the sensitivity function S as follows: $S(x)$ = proportion of outbreaks that were detected within x days from the reference date.

An evaluator can plot this sensitivity function against timeliness to produce a sensitivity-timeliness curve (Wallstrom et al., 2004). Figure 20.4 is an example of a sensitivity–timeliness plot for two algorithms. Algorithm 1 has a higher probability of detecting the outbreak until day five of the outbreak, but after that Algorithm 2 has a higher probability of having detected the outbreak. If very early detection is critical for this particular type of outbreak, then Algorithm 1 is preferable to Algorithm 2.

4.1.3. False Alarm Rate

Evaluators measure the false-alarm rate by running an outbreak detection algorithm on surveillance data that do not contain outbreaks. The false-alarm rate is the proportion of non-outbreak days (or weeks, depending on the organization of the time series) on which the algorithm signals an alarm. In Figure 20.3, the false-alarm rate is 1/20, or 0.05 false alarms per day of monitoring, or one false alarm every 20 days. (The expression *one false alarm every 20 days* is called the *recurrence interval*. It is an easy-to-understand way to present the false alarm rate that you may encounter in the literature.)

4.1.4 Generalized ROC Curves

The previous three sections described how an evaluator measures timeliness of detection, sensitivity, and false alarm rate for a detection algorithm with a fixed threshold. As with case detection algorithms, if the evaluator alters the algorithm's threshold level, the measurements all change.

When studying an outbreak detection algorithm, an evaluator also constructs an ROC curve to understand the tradeoff between its sensitivity and false alarm rate. To understand the relationship between its false alarm rate and timeliness, an evaluator plots timeliness against the false-alarm rate. This type of plot is called an activity monitoring operating characteristic (AMOC) curve (Fawcett and Provost, 1999). From the AMOC curve, an evaluator can easily see the tradeoff between the false alarm rate and timeliness. For example, in Figure 20.5, we see that if we were to increase the detection threshold to reduce the false alarm rate from four per year to two per year, the average day of detection would increase from 2.6 days to 4.4 days. The price we would pay for a lower rate of false alarms is a delay in detection of 1.8 days.

4.1.5. Measuring Detectability as a Function of Outbreak Size

One question of particular interest to evaluators is how small of an outbreak is the algorithm capable of detecting. We refer to analyses that pursue this question as *detectability analyses*. A detectability analysis is only possible if the evaluator can quantify the size of the outbreaks under study in some manner (e.g., as the fraction of people in a region who were sickened during an outbreak). She can study the effect of outbreak size

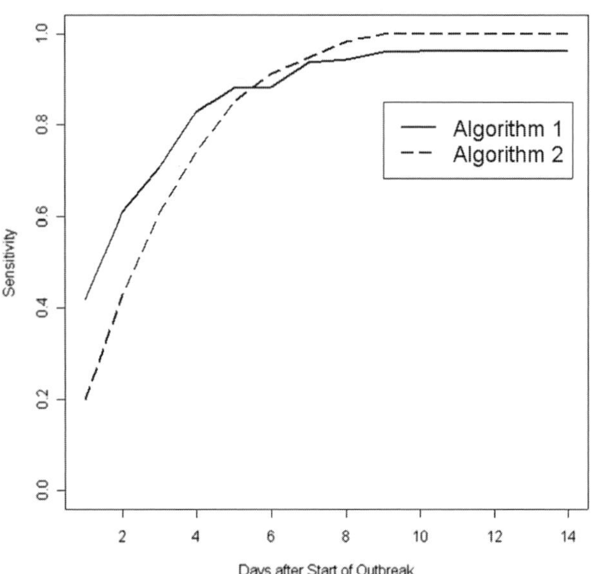

FIGURE 20.4 Sensitivity versus timeliness for two detection algorithms, computed from a hypothetical experiment.

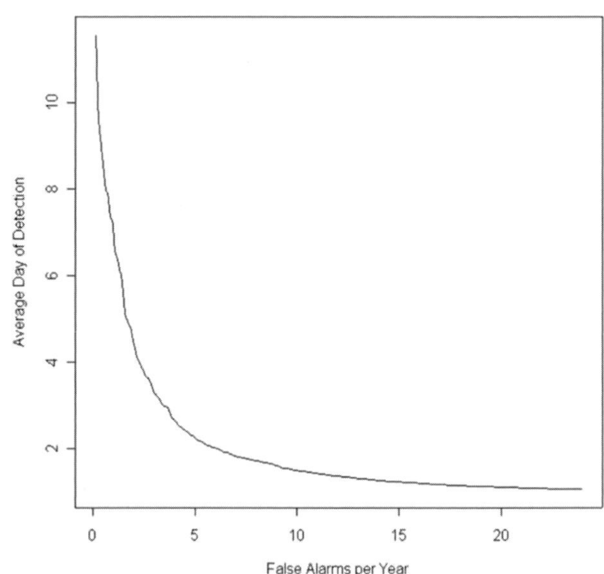

FIGURE 20.5 An AMOC curve from a hypothetical experiment. The curve shows the tradeoff involved in manipulating the detection threshold to improve either timeliness or false alarm rate.

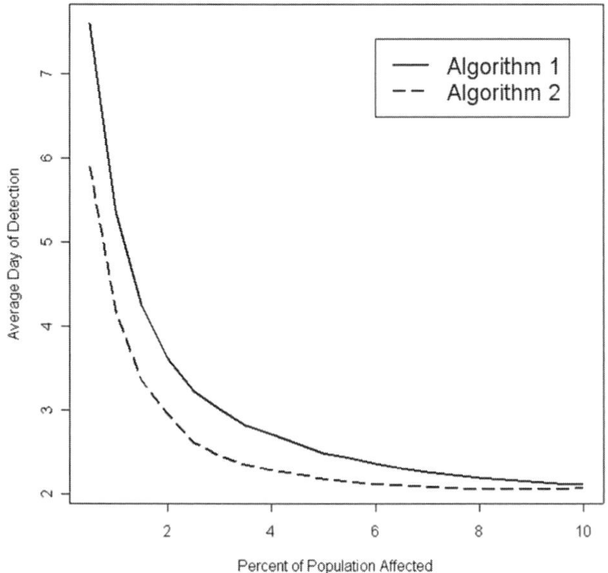

FIGURE 20.6 Timeliness versus outbreak size for two algorithms. The curves shows that large outbreaks are detected earlier and it provides insight about the smallest outbreak that is detectable by this algorithm (using a particular type of surveillance data).

on its detectability graphically by plotting timeliness against outbreak size, while holding the false alarm rate at a constant level. Figure 20.6 is a timeliness versus outbreak size plot that compares two algorithms.

4.2. Obtaining Surveillance Data for Evaluation

The biggest challenge for an evaluator who wishes to study an outbreak detection algorithm is obtaining surveillance data for a sufficiently large number of outbreaks with which to measure sensitivity and timeliness. The exact number of outbreaks required depends on the tightness of the statistical error bounds on these measurements that he desires, but as a rough approximation, 10 outbreaks are required as a bare minimum.

To have the greatest validity, an evaluator tests an algorithm using surveillance data collected from real outbreaks. However, with the exception of influenza, rotavirus, adenovirus, and some foodborne illnesses, obtaining surveillance data for even 10 outbreaks is difficult, at best. For some diseases of great concern as bioterrorist threats, outbreaks are non-existent. For outbreaks to be useful to the evaluator, they must have occurred in regions that collected biosurveillance data. Although the availability of suitable data is rapidly improving, the present lack of surveillance data is a significant barrier to research using real data. Indeed, there are few published evaluations of outbreak detection algorithms using real data that also have had sufficient sample size to compute confidence intervals on their measures of sensitivity and timeliness. At the time of this writing, we are aware only of studies by Hogan et al. (2003), Ivanov et al. (2003), and Campbell et al. (2004).

Fortunately, much can be learned about outbreak detection algorithms using synthetic data whose characteristics resemble those of real data, at least for those characteristics relevant to the algorithm's performance. Evaluators have used three types of synthetic surveillance data in published studies: fully synthetic, semi-synthetic, and high-fidelity synthetic.

4.2.1. Fully Synthetic Test Data

An evaluator can use a computer simulator to generate data with which to evaluate an outbreak detection algorithm. The most available simulator to use might be the algorithm itself, run backwards to generate simulated data. This type of evaluation tests whether the algorithm performs well using data that match the algorithm's modeling assumptions exactly. WSARE (Wong et al., 2003) and BARD (Chapter 19) are two examples of algorithms that were evaluated by using their own models run in a reverse direction. At the other end of the spectrum, an evaluator could use an agent-based simulator, a program—such as the popular SimCity game—comprising many individual entities (or agents) that represent people (or animals) who exhibit behavioral characteristics of sick individuals. The evaluator introduces an infectious disease into the simulator and runs the simulator forward in time, resulting in a growing number of sick individuals who may be located in different neighborhoods and who may buy thermometers or visit physicians when falling ill. By summing the number of individuals who visit physicians or purchase thermometers on any given day in the simulation, the evaluator creates a time series with which to test the detection algorithm.

The key problem with fully synthetic data is validity; the simulator embodies many assumptions about what people (or animals) do, what their test results show when they are sick and when these behaviors or test results appear in data. At present, the body of evidence regarding the behavior of sick individuals (used to estimate the parameters required by simulators) is very small.

The advantages of using fully synthetic data are its availability and the control that the evaluator has over the properties of the simulation. The evaluator can change the infectivity of an organism, the size of the outbreak, the geographic distribution and many other characteristics. Because of the problem with validity, however, evaluators restrict their use of fully synthetic data to early evaluations of an algorithm designed to test whether the algorithm is working as intended and whether its performance can be improved. They also use fully synthetic data when comparing an algorithm against alternative algorithms (which could be an earlier version of the same algorithm).

4.2.2. Semisynthetic Test Data

Evaluators can generate *semisynthetic data* by injecting geometrically shaped spikes into real surveillance data collected during non-outbreak periods (Goldenberg et al., 2002,

Zhang et al., 2003, Reis et al., 2003, Reis and Mandl, 2003). This technique was used for illustration in Chapter 14. The advantage of the semisynthetic approach is that it allows the evaluator to manipulate the spike size to find the smallest spike that the algorithm can detect above the background noise in real surveillance data. The evaluator can manipulate the shape of the spike and its time course. He can inject the spike repeatedly on every single day of the year to explore the effect on detectability of seasonal and day-of-week variations in the surveillance data.

The key problem with using semisynthetic data is that the resulting measures of sensitivity, false alarm rate, and timeliness of detection are for spike detection, not outbreak detection. For example, a semi-synthetic analysis may determine that an algorithm can detect an outbreak that causes a 10-unit increase in sales of cough products. However, unless the evaluator already knows how often sick individuals purchase cough products, the conclusions an evaluator can draw about the *detectability* properties of an algorithm (defined above as the relationship between outbreak size and sensitivity, specificity, and timeliness) are limited and rest on assumptions about the purchasing behaviors of sick individuals. The technique is, nevertheless, useful for early studies of algorithms and has greater validity than fully synthetic data because it employs real baseline data.

There is one other aspect of the semisynthetic approach that affects its validity. Geometric shapes are noise-free and, thus, underestimate the noise one expects to find in real surveillance data, a bias that may result in the overestimation of an algorithm's sensitivity and timeliness.

4.2.3. *High-Fidelity Injection*

The high-fidelity detectability experiments (HiFIDE) technique extends the semisynthetic method by forming injections whose shapes and noise levels are derived from surveillance data collected during an actual outbreak (Wallstrom et al., 2004, Wallstrom et al., 2005). The technique also scales the height of the inject in a way that preserves the known relationship between the magnitude of the real outbreak and the strength of the signal in the surveillance data collected during the real outbreak. The scaling adjusts for differences in population size and in *data completeness* (defined as the proportion of data that is available in a region). This technique estimates the effect of an outbreak on surveillance data in one region and allows the evaluator to determine the detectability of that type of outbreak in a second region (where the two regions may differ in population size, population density, completeness of surveillance data, and sizes of outbreaks).

The mechanics of a HiFIDE analysis are similar to those of a semi-synthetic analysis. An evaluator creates multiple injects, varying in outbreak size and surveillance data completeness. Each inject is then combined with real surveillance data to form a time series. An evaluator runs detection algorithms on each injected time series, varying the detection threshold, and summarizes the detectability characteristics of the algorithms using AMOC curves and other generalized ROC curves that display the relationships between timeliness of detection, sensitivity, false alarm rate, outbreak size, and data completeness.

A software package, also called HiFIDE, is available from *www.hifide.org* that automates the analysis. The software can be freely downloaded for noncommercial use. Note that HiFIDE is limited to synthetic additions to univariate data. The problem of adding synthetic additions to multivariate data is an ongoing area of research. It is a complex problem because one would need to simulate correctly the interdependencies between the variables. Buckeridge et al. (2004) are developing an approach to injecting simulated anthrax records into a set of relational tables.

5. EVALUATING ALGORITHMS FOR OUTBREAK CHARACTERIZATION

The purpose of a biosurveillance system is not limited to case and outbreak detection. It should also support outbreak investigation and characterization. Many of the algorithms discussed in Part II can provide information about key outbreak characteristics, including the following:

- Set of affected individuals
- Geographic scope
- Time course
- Causative organism or toxin
- Incubation period
- Source if other than a person
- Route of transmission
- Host characteristics that explain why some individuals are more susceptible
- Relevant environmental factors

Each characteristic is an uncertain quantity (at least early during an outbreak) that algorithms can potentially infer from biosurveillance data. The algorithms that perform this inference can be evaluated using the same techniques described above for measuring sensitivity, specificity, and timeliness. To our knowledge, no such evaluations have been conducted. Therefore, many questions remain open for future research to determine what kind of data sets could be used to gain insight into the performance of such algorithms, with what accuracy can they infer, automatically, each quantity for different types of outbreaks, and what additional surveillance data or other types of data would improve the accuracy of their inferences.

6. DETERMINING WHETHER BAYESIAN ALGORITHMS ARE WELL CALIBRATED

For those algorithms that compute posterior probabilities (of a case, an outbreak, or an outbreak characteristic) from surveillance data, it is important to know whether the posterior probabilities are accurate; that is, do they in the long run reflect the actual frequency of cases or outbreaks as determined by a gold standard. We refer to an algorithm that satisfies this requirement as *well calibrated*. For example, if an evaluator runs a case detection algorithm on a sample of 100 patients of which 10 patients have a disease in question and 90 do not, the sum of the posterior probabilities of a well-calibrated case detection algorithm for the 100 patient would be 10.

We note that an algorithm can trivially meet this requirement by simply outputting the prior probability of an outbreak for each day; therefore, such an evaluation represents a necessary but not sufficient test of the algorithm. As illustrated in Figure 13.13 (Chapter 13), the RODS system uses the case detection algorithm SyCO2 to compute a time series of daily counts of "respiratory syndrome" by summing the posterior probability of "respiratory syndrome" produced by SyCO2 for all patients visiting an emergency room on each day. If SyCO2 is well-calibrated, the daily sums should equal the actual number of patients with respiratory illness each day, which could be verified by examination of patient charts or some other gold-standard ascertainment method.

The importance of accuracy calibration is that these algorithms may be used within a decision-analytic framework to compute the expected utility of a decision (Part V), which requires that the posterior probabilities are accurate.

As with other evaluations of case and outbreak detection algorithms, case data required to measure the calibration of a case detection algorithm are far more available than outbreak data required to evaluate an outbreak detection algorithm. To evaluate the calibration of an outbreak detection algorithm, an evaluator would assemble a library of time series consisting of outbreak (simulated or real) data and non-outbreak data. He might then compute the posterior probability of an ongoing outbreak for each day in each of the time series. The posterior is well calibrated if the sum of the posterior probabilities across all time series equals the total number of outbreak days in the time series. Again, the evaluator would have to ensure that the outbreak detection algorithm is not "gaming the system" by always outputting a posterior probability that is equal to the prior probability.

7. EVALUATING THE COMBINATION OF ALGORITHM AND DATA

The timeliness, sensitivity, and specificity of an algorithm depend not only on its ability to extract information from surveillance data but also on the information content of surveillance data that are available to the algorithm. When studying an algorithm, an evaluator—like any good scientist—will tend to hold the data available to the algorithm "constant" to avoid introducing a confounding variable into the experiment. However, the objective of algorithm development is to improve detection relative to current "best practice," and improvement in the quantity and quality of surveillance data will likely improve overall performance more than any algorithm improvement. Thus, holding data constant is not always appropriate in evaluation.

Although we discuss algorithm evaluation and data evaluation separately in this book (and it is worthwhile scientifically to study their contributions separately), there are important scientific questions for which the appropriate method involves treating the combination of data and algorithm as the object of a study. Perhaps the best way to think of this is that, early in the course of algorithm development, it is appropriate to focus on the algorithm alone; but ultimately, to confirm that a new algorithmic approach is superior to current "best practice," the object of study becomes the combination of surveillance data and algorithm.

8. DIAGNOSTIC PRECISION AND THE QUESTION OF "HOW GOOD IS GOOD ENOUGH?"

If a study compares the combination of *algorithm plus data* with current "best practice," the results of the study will directly answer the question of whether the algorithm is "good enough." Otherwise, the question of whether the algorithm's sensitivity, timeliness, and false alarm rate are good enough must be addressed indirectly, based on consideration of whether the algorithm's users will take actions based on anomalies identified by the algorithm. When discussing the significance of the results from an algorithm evaluation, one factor to keep in mind is that diagnostic precision (discussed in Chapter 3) is a strong determinant of an algorithm's value. If a case detection algorithm, for example, informs a clinician that a patient has a probability of "respiratory syndrome" of 0.2, in most cases, this information will not influence the management of the patient. In contrast, a probability of inhalational anthrax of 0.2 would be very influential if the clinician were not already considering this diagnosis. Similarly, if an outbreak detection algorithm informed an epidemiologist that the probability there was an ongoing "respiratory" outbreak was 0.2, the assessment would be less influential than if the algorithm were to suggest an outbreak due to inhalational anthrax with the same level of confidence.

9. SUMMARY

Algorithms for case detection, outbreak detection, and outbreak characterization are critical components in a biosurveillance system. Although these components ultimately must be tested under field conditions, this level of evaluation is expensive and difficult, especially for outbreak detection and characterization due to the rarity of real outbreaks. Field testing is appropriate for algorithms whose potential has been demonstrated by laboratory testing. The methods for measuring sensitivity, specificity, and timeliness discussed in this chapter are general methods that can be applied both to laboratory and field evaluations.

Evaluators employ well-developed methods for evaluating classifiers when they evaluate case detection algorithms. These existing methods measure sensitivity and specificity of case detection. They summarize results using ROC curve analysis.

Researchers have extended these methods to enable the evaluation of algorithms for outbreak detection and characterization. In addition to measuring sensitivity and specificity, these evaluations always measure timeliness. Since sensitivity, specificity, and timeliness are correlated, evaluators have extended ROC curve analysis to include the dimension of time.

Researchers have further extended these methods to allow analysis of the relationship between the size of outbreak and detectability. Since outbreak size is also correlated with sensitivity, specificity, and timeliness, they have further extended ROC curve analysis to a fourth dimension—outbreak size. We refer to studies of this type as detectability analyses because their goal is to understand the smallest outbreak that an algorithm can detect.

Diagnostic precision is an important determinant of the utility of a case or outbreak detection algorithm, although the surveillance data available to the algorithm are the primary determinant of diagnostic precision. Diagnostic precision, sensitivity, specificity, timeliness, and outbreak size are all correlated. Thus, we expect that future researchers will extend ROC analysis into this fifth dimension.

ADDITIONAL RESOURCES

For a description of more sophisticated approaches to semi-synthetic studies, see Buckeridge et al. (2004).

REFERENCES

Buckeridge, D. L., Burkom, H., Moore, A., et al. (2004). Evaluation of syndromic surveillance systems: design of an epidemic simulation model. *MMWR Morb Mortal Wkly Rep* 53(Suppl):137–43.

Campbell, M., Li, C.-S., et al. (2004). *An Evaluation of Over-the-Counter Medication Sales for Syndromic Surveillance.*

Egan, J. (1975). *Signal Detection Theory and ROC Analysis*. New York: Academic Press.

Fawcett, T., Provost, F. (1999). Activity monitoring: Noticing interesting changes in behavior. *Proceedings of the Fifth International Conference on Knowledge Discovery and Data Mining*, 53–62.

Goldenberg, A., Shmueli, G., Caruana, R. A., et al. (2002). Early statistical detection of anthrax outbreaks by tracking over-the-counter medication sales. *Proc Natl Acad Sci U S A* 99:5237–40.

Hogan, W. R., Tsui, F.-C., Ivanov, O., et al. (2003). Early detection of pediatric respiratory and diarrheal outbreaks from retail sales of electrolyte products. *J Am Med Inform Assoc* 10:555–62.

Ivanov, O., Gesteland, P. H., Hogan, W., et al. (2003). Detection of pediatric respiratory and gastrointestinal outbreaks from free-text chief complaints. In: *Proceedings of American Medical Informatics Association Symposium*, 318–22.

Lusted, L. (1971). Signal detectability and medical decision making. *Science* 171:1217–1219.

Metz, C. (1978). Basic principles of ROC analysis. *Semin Nucl Med* 8:283–298.

Reis, B., Pagano, M., Mandl, K. (2003). Using temporal context to improve biosurveillance. *Proc Natl Acad Sci U S A* 100:1961–1965.

Reis, B. Y., Mandl, K. D. (2003). Time series modeling for syndromic surveillance. *BMC Med Inform Decision Making* 3:2.

Wallstrom, G. L., Wagner, M. M., Hogan, W. R. (2004). Using Surveillance Data from Actual Outbreaks to Characterize Detectability. Rods Laboratory Technical Report. Center for Biomedical Informatics, University of Pittsburgh, Pittsburgh, Pennsylvania.

Wallstrom, G. L., Wagner, M. M., Hogan, W. R. (2005). High-Fidelity Injection Detectability Experiments: A Tool for Evaluation of Syndromic Surveillance Systems.

Wong, W., Moore, A., Cooper, G., et al. (2003). Bayesian network anomaly pattern detection for disease outbreaks. In: *Proceedings of the Twentieth International Conference on Machine Learning*, Menlo Park, CA, 808–15.

Zhang, J., Tsui, F.-C., Wagner, M., et al. (2003). Detection of outbreaks from time series data using wavelet transform. In: *Proceedings of Fall Symposium of American Medical Informatics Association,* 748–52.

IV
PART
Newer Types of Surveillance Data

21

CHAPTER

Methods for Evaluating Surveillance Data

Michael M. Wagner
RODS Laboratory, Center for Biomedical Informatics
University of Pittsburgh, Pittsburgh, Pennsylvania

1. INTRODUCTION

We begin Part IV, "Newer Types of Surveillance Data," with this chapter on methods for evaluating surveillance data. The other chapters in Part IV discuss the results of experiments using the methods described in this chapter.

The goal of data evaluation is to understand (1) the ability of data to contribute to the earlier detection of outbreaks (or of individuals with disease); (2) their availability; and (3) the cost to obtain the data. A successful evaluation will characterize these factors in a manner that they can be compared against each other to form an overall understanding of the cost and benefit of the data.

The questions that data evaluation seeks to understand about surveillance data include:

- Can the surveillance data contribute to the detection of disease or outbreaks (and which ones)?
- How much earlier can outbreaks or cases of disease be detected using the new data?
- What level of diagnostic precision of case or outbreak detection can the data support?
- How difficult are the data to obtain? At what cost?
- How available are the data for a region of interest?

The primary application for the methods described in this chapter is the study of new types of surveillance data. Examples of such data include chief complaints of patients, absenteeism rates, telephone calls to appointment lines and advice centers, 911 calls, and orders for laboratory tests. There are many types of data that may be of value in biosurveillance. These techniques form the basis for a research agenda whose goal is to screen them for possible inclusion in biosurveillance systems.

We note that the importance to biosurveillance of traditional surveillance data—reports of confirmed cases, results of diagnostic laboratory tests, and the results of routine environmental testing—is beyond question: these data have proven their value in the field. Thus, an evaluator would only be motivated to study their costs and benefits in order to compare them with newer types of data that might provide the same information, but perhaps be easier to obtain, at lower cost, or have the potential to improve the timeliness, sensitivity, or diagnostic precision of outbreak detection and characterization.

We begin the chapter with a discussion of the characteristics of surveillance data that an evaluation should elucidate–informational value, availability, and cost to obtain. We follow with a review of experimental techniques that evaluators have used in published research to measure informational value. Since evaluators use different experimental methods to measure the informational value of data for case detection and for outbreak detection, we discuss these methods in separate sections. The scope of this chapter is limited to methods appropriate to laboratory evaluations of surveillance data. We discuss methods for field testing of surveillance data in Chapter 37.

2. ATTRIBUTES OF SURVEILLANCE DATA RELEVANT TO ACQUISITION DECISIONS

The benefits of surveillance data derive from their ability to contribute to the timely detection and characterization of disease outbreaks (or cases). Unless data can contribute to the detection or characterization of at least one disease or type of outbreak, they are of no value.

The costs of surveillance data derive from the money, time, and effort that a biosurveillance organization must invest to develop a systematic method to obtain the data on a routine basis. We note that additional factors, such as the cost of analysis and the cost of false alarms, may be significant as well.

2.1. Informational Value of Surveillance Data

The most fundamental question about surveillance data is whether they contain information that can facilitate detection or characterization of some specific outbreak (or disease).

To facilitate detection and characterization, surveillance data must be both sufficiently timely and indicate either the presence of disease in a population (or individual) or contribute to the analytic power (as in the case of weather or water supply data, which on its own cannot help detect outbreaks, but in combination with clinical data may enable earlier detection). Note that evaluators always ask these questions with respect to a specific disease or type of outbreak (or a group of similar diseases). Evaluators consider data that are both sufficiently early and indicate disease as having informational value.

There are a variety of methods that evaluators may employ to measure informational value. These methods vary in the effort required to conduct the study and the validity of the results.

Handbook of Biosurveillance
ISBN 0-12-369378-0

They are often used in combination. Many of the methods that we discuss have been used in published studies, which we will cite as examples. The methods range from surveys of sick individuals to value-of-information calculations, which attempt to measure the information value of surveillance data in the same units (dollars) as the costs of a system to enable direct comparisons.

The types of measurements that are most feasible at present include the following:

- Measures of the improvement in sensitivity, specificity, and timeliness of *case* detection attributable to the surveillance data
- Signal-to-noise ratios (for surveillance data intended for use in outbreak detection)
- Correlation analysis (for surveillance data intended for use in outbreak detection)
- Measures of the improvement in sensitivity, specificity, and timeliness of *outbreak* detection attributable to the surveillance data
- Linked record analysis
- Surveys of the sick

2.2. Availability

Although an evaluator may be able to obtain a sample of data for research purposes, it may be difficult or expensive for a biosurveillance organization to obtain the same data on a routine basis. If routine collection is not feasible, it may not be worth studying informational value (or at least this knowledge may help to set research priorities). Availability of surveillance data is often far easier to assess than informational value, suggesting that data availability may be a more efficient criteria for screening potential types of data than informational value (Wagner et al., 2001).

An evaluator can estimate availability as the number of organizations in a biosurveillance region that own the data, the capabilities of their information systems to provide the data, and any legal and proprietary business concerns that influence the willingness of the owners of the data to provide them for biosurveillance purposes. At one extreme, there may be organizations that already aggregate the data for other purposes at the regional level (e.g., the data aggregators mentioned earlier), or the data may come from a highly consolidated industry, such as the retail industry, as discussed in the next chapter. At the other extreme, the data may be owned by a large number of small physician practices and hospitals.

2.3. Cost of Data

The specific costs associated with obtaining data may include fees that the owners of the data charge and the cost of building and operating a data-collection system. It may be difficult to estimate these costs prior to building a system. Therefore, we discuss costs in detail in Chapter 37 where we recommend that evaluations of fielded systems measure costs and publish their results for the benefit of other organizations.

3. METHODS FOR ESTIMATING INFORMATIONAL VALUE OF SURVEILLANCE DATA FOR CASE DETECTION

To evaluate the value of surveillance data for detecting disease in a single patient, an evaluator employs methods identical to those described in Chapter 20.

The evaluator assembles a set of individuals known to have the disease (termed *cases*) and a set of individuals without the disease to serve as controls. The cases can be established by reviewing medical records, veterinarian records, or lab results. The evaluator uses a case-detection method that is appropriate to the data and computes the sensitivity and specificity of detection and the area under the receiver operator characteristic (ROC) curve, as described in Chapter 20. The evaluator uses the area under the ROC curve as a measure of the data's informational value for detecting cases of the disease in question. Since the area under the ROC curve is not an absolute measure of value, these evaluations typically compare the area under the ROC curve with that obtained from an identical analysis using some other type of surveillance data. Alternatively, evaluators rely on knowledge about what levels of sensitivity and specificity are required by a biosurveillance system. Some example published studies are (Chapman et al., 2003b, Espino and Wagner, 2001, Ivanov et al., 2003).

If the evaluation fails to show that the data can discriminate between patients with the disease and without the disease (i.e., the area under the ROC curve equals 0.5) or that the performance is not sufficient for a particular application, the evaluator should question whether a different case-detection algorithm might produce better results. Even if discrimination ability is found, an evaluator may experiment with different case-detection algorithms to establish the best possible performance that can be achieved with the data.

4. METHODS FOR ESTIMATING INFORMATIONAL VALUE OF SURVEILLANCE DATA FOR OUTBREAK DETECTION/ CHARACTERIZATION

A prerequisite for most studies of the informational value of surveillance data for outbreak detection is an adequate sample of surveillance data. This prerequisite can be difficult to satisfy. The surveillance data must be from an outbreak. The surveillance data should be a nearly complete sample from the outbreak region, otherwise a negative result from the study will be difficult to interpret. It will either indicate that the data are of no value or that the sampling may not have been sufficient.

To obtain an adequate sample, an evaluator may have to obtain data from multiple hospitals, physicians' offices, veterinarians, laboratories, or retail stores. Occasionally, it is possible to find a data aggregator, such as Information

Resources, Inc. or ACNielsen, or a health department that can provide a suitable data set. Note that, unlike evaluations of detection algorithms, there is no role for synthetic data in evaluation of surveillance data; the properties of real data are of primary interest.

4.1. Signal-to-Noise Ratio

If the surveillance data can be formed into a time series, as illustrated by Figure 21.1, evaluators can use a variety of methods to understand their informational value for detecting outbreaks of disease.

The evaluator will first examine the time series visually to determine whether an outbreak effect is present. The evaluator can transform the daily or weekly counts into units of "standard deviations from the mean" (y axis), which enables him to measure the signal-to-noise ratio. This transformation is a standard way of normalizing time-series data in the field of signal analysis (e.g., Lobanov [1971] used this method to compare two time series in the domain of speech analysis). The normalization also allows an evaluator to plot multiple types of surveillance data, which may vary in scaling considerably, on the same graph.

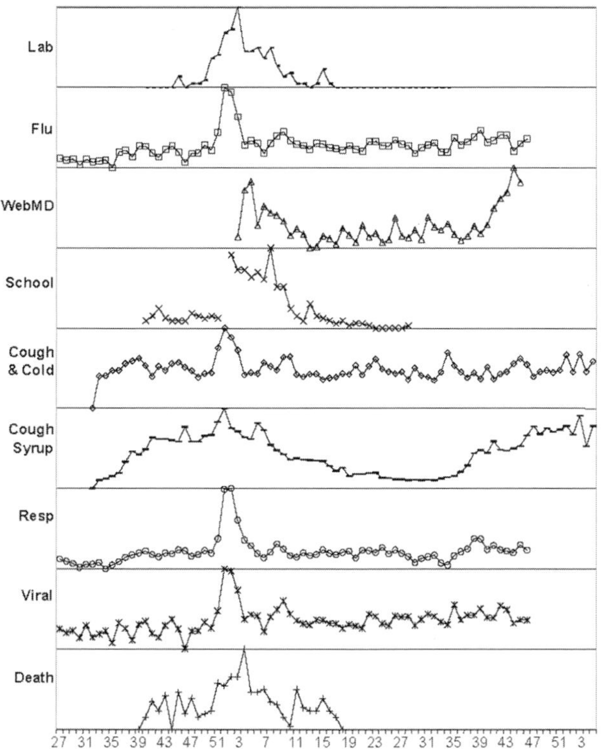

FIGURE 21.1 Weekly counts for several types of routinely collected data for various time periods around the December 1999 influenza outbreak in Pittsburgh. Each data type is plotted on a normalized scale. Lab, influenza cultures from the University of Pittsburgh Medical Center Health System; WebMD, counts of queries to a national web health site using words such as "cold" and "flu"; Cough and cold and cough syrup, grocery chain point of purchase counts; School, school nurse influenza reporting; Resp. and Viral, categories of emergency department ICD-9-coded chief complaints.

The *signal-to-noise ratio* is the strength of signal in surveillance data during an outbreak, divided by the level of the signal at baseline in the absence of an outbreak. For example, Hogan plotted the mean and standard deviation of weekly sales of pediatric electrolytes for a four year period (Hogan et al., 2003). The signal-to-noise ratio of sales of pediatric electroytes for winter outbreaks in children was 18.3 standard deviations.

The signal-to-noise ratio is easy to calculate and is informative about the informational value of surveillance data. If the evaluator finds no signal in the surveillance data, there is little point in further study.[1] An evaluator can compare the signal-to-noise ratios of multiple types of data to determine which one will provide the most sensitive outbreak detection.

This method cannot provide quantitative answers to the questions of whether and how early an outbreak can be detected. It only answers the question of whether there is any outbreak effect in the data and suggests how strong and early the effect is.

4.2. Correlation Analysis

If the evaluator can obtain a second time series of data from the outbreak (e.g., pneumonia and influenza deaths), he can use the correlation function to compare the two time series. The correlation function finds the time lag at which the correlation between two time series is maximized as well as the strength of the correlation (Figure 21.2).

The second time series must have face validity; that is, it must already be known to reflect outbreak activity.

FIGURE 21.2 Correlation analysis. Two time series (A and B) show outbreak effect. The correlation function finds the time shift that will bring the two time series into maximal alignment. An evaluator uses this difference (hypothetically given as two weeks in the example) as a measure of the relative earliness of two types of surveillance data.

1 It may be possible to bring signal out of noise by focusing the analysis more precisely on the subset of patients affected by the outbreak. For example, if all of the affected individuals live in the same zip code, the evaluator will create a time series for that zip code. If all of the affected individuals are also children, the evaluator will create a time series of data for children in that zip code.

Pneumonia and influenza deaths, for example, have been used for many years by researchers to identify the existence of influenza outbreaks and, therefore, have face validity.

The evaluator uses the time lag as a relative indication of earliness of detection (relative to the second time series). He uses the strength of correlation as an indication of the degree to which the perturbations in the surveillance data are caused by outbreak versus non-outbreak effects.

The conclusions the evaluator can draw about timeliness of surveillance data for the outbreak in question using correlation analysis are limited. The correlation function finds the time lag at which the peaks of the signals in the time series data (which occur mid-outbreak, typically) are most highly correlated. The evaluator is interested in knowing the time latency between the initial upticks in the two time series, which correlation analysis cannot measure. Although the time difference between peaks and the time difference between initial upticks may be the same, it will only be so if the shape of the outbreak effect is identical in the two signals. For this reason, most evaluators also use the detection algorithm method, described in the next section.

The idea of using the correlation function to evaluate biosurveillance data was first demonstrated in 1979 by Welliver et al. (1979). Other studies of biosurveillance data that have used the correlation function include (Hogan et al., 2003, Magruder and Florio, 2003, Espino et al., 2003, Campbell et al., 2004, Johnson et al., 2004).

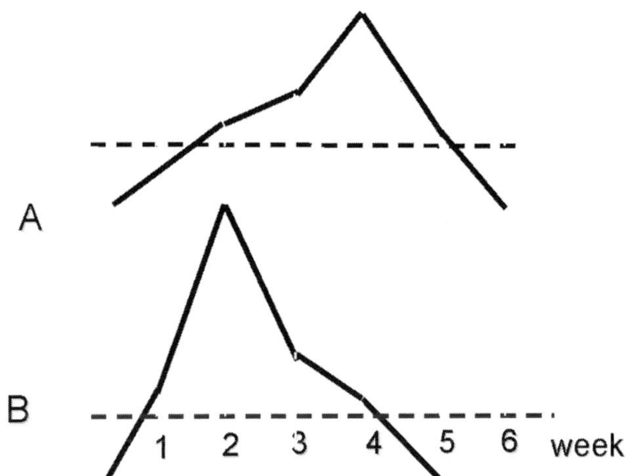

FIGURE 21.3 Detection algorithm method. Two time series (A and B) show outbreak effect. The detection algorithm method finds the dates on which a detection algorithm (whose threshold is indicated by the dashed lines) first detects the increase in the surveillance data. An evaluator uses this difference as a measure of the relatively earliness of two types of surveillance data.

4.3. Detection Algorithm Method

The *detection algorithm method* determines the date that a statistical detection algorithm first detects an anomaly in surveillance data. The evaluator compares this date with a gold standard reference date. The evaluator may establish the gold standard reference date by convening a panel of experts to review the outbreak, or he may determine the date by running a detection algorithm on a second time series, one with face validity for the outbreak (Figure 21.3).

If the evaluator obtains the reference date from a second time series, he may elect to use the same detection algorithm on both time series, or different algorithms. It is important that his choice not bias the results, so that in general, an evaluator will use the same detection algorithm unless the time series have different statistical properties (e.g., one has weekly periodic effects and the other is nonstationary) (See Chapter 14 for a discussion of periodic effects in surveillance data). The use of two different algorithms creates an experimental situation in which two variables have been manipulated, thus the evaluator must use care in the selection of the algorithms so that the differences observed can be reasonably attributed to underlying differences in the data, not a bias that was introduced by his selection of algorithms.

The detection algorithm method was first introduced by Quenel et al. (1994) in a survey of novel types of surveillance data for detecting influenza. It was next used by Tsui et al. (2001) for a study of chief complaints coded with the ICD-9-CM (International Classification of Diseases, 9th Revision, Clinical Modification). We discuss other examples of its use in the next chapter (Hogan et al., 2003, Campbell et al., 2004, Ivanov et al., 2003).

4.4. Linked Record Analysis

If an evaluator does not have access to time series surveillance data collected during an outbreak, there are still methods that he can use to study the informational value of surveillance data for outbreak detection. *Linked record analysis* requires, as a prerequisite, only a set of patients with the disease of interest. The name "linked record" refers to the fact that the patients of interest are identified by links between records in the surveillance data and other data that are a gold standard for the presence of disease in the individual.

With this method, an evaluator can measure the fraction of individuals with a disease that appear in the surveillance data. An evaluator can also measure the average time that an individual is encountered in the surveillance data, relative to the onset of disease, or some other reference data.

Examples of linked record analysis can be found in Henry et al. (2004), Chapman et al. (2003a), Chapman et al. (2004a), Chapman et al. (2005), and Gesteland et al. (2005).

4.5. Surveys

A method that does not require any surveillance data whatsoever is a survey of sick (or recently sick) individuals to determine when they "emitted" behaviors that may show up in surveillance data.

A recent survey of pediatric patients is an example of a survey designed specifically to understand the informational value of biosurveillance data (Johnson et al., 2005). In this study, Johnson interviewed caregivers of children who were being seen in emergency rooms for fevers, respiratory ailments, and gastrointestinal symptoms. She asked the children's caregivers whether they had purchased over-the-counter medications (and which ones and when), visited physicians, called for appointments and several other actions that would leave footprints in routinely collected data. The survey instruments used in that study are available at *http://rods.health.pitt.edu/PedsEDStudy.htm*. We discuss this study and one other survey explicitly designed to understand the informational value of biosurveillance data in detail in the next chapter.

There are several surveys that were conducted for other purposes, but remain of interest both for the results they obtained relevant to understanding biosurveillance data as well as for providing methodologies for conducting such studies. We discuss the results of the studies in the next chapter because they were studies of use of over-the-counter medications.

Note that surveys do not directly measure the effect of disease outbreaks on surveillance data. Their interpretation relies on assumptions that the behavior reported actually occurred and that it was recorded in an information system. Surveys, however, can provide insights that other methods cannot. For example, the results of surveys can be used to estimate the fraction of sick individuals in a population that will appear in surveillance data, thus enabling evaluators to understand the relationship between the size of a spike in surveillance data and the size of an outbreak (Johnson et al., 2005). A single survey can produce these estimates for many types of surveillance data; thus, it is a highly efficient research design. Surveys are complementary to other studies, confirming the measurements of timing and strength of a signal that can be expected in different outbreak situations.

5. VALUE OF INFORMATION

The ideal method to measure the informational value of surveillance data is a *value-of-information study* (Friedman and Wyatt, 1997), although it is time consuming and difficult to conduct. There are no examples of such analyses of biosurveillance data as of this writing.

A value-of-information measurement is a general technique that can be used to quantify the value of data for any purpose, including purposes other than biosurveillance. When applying this technique to assess biosurveillance data, an evaluator would compare the ability of a biosurveillance system to detect outbreaks or cases with and without the data in question using the detection algorithm method previously described. She would further estimate, using models, the reduction in morbidity and mortality and economic impact that could be attributed to the improvement in earliness of detection. For example, if the analysis found that outbreaks could be detected 24 hours earlier if the data were available, the evaluator would use a model to translate the 24-hour improvement in speed of detection into, for example, an additional 100 lives saved and the removal of 1000 people from the expensive "requires treatment with ventilator" category to "requires prophylactic treatment with antibiotics" category. The number of false alarms expected might be one per year. The evaluator could then compare the benefit of the data denominated in lives saved or sickness averted against the cost and effort required to build and maintain systems for their routine collection, as well as the cost of the false alarms.

Value-of-information analyses can also be taken to the level of economic analyses, which allow more direct comparisons of benefits and costs. Briefly, an economic analysis would translate "lives saved" or "sickness averted" into dollars. We discuss these transformations further in Chapter 31.

6. GOLD STANDARDS FOR EVALUATION OF SURVEILLANCE DATA

We conclude this chapter with a brief overview of gold standards. We use the term gold standard to refer to data and methods that an evaluator can use to establish "ground truth" about the existence and timing of an outbreak or of an individual case. An example of gold standard data is pneumonia and influenza surveillance data. An example of a gold standard method is ascertainment of disease status of individual patients by manual review of patient charts.

If an outbreak has been investigated and documented by a biosurveillance organization, the documentation will serve as an excellent gold standard for research. Otherwise, an evaluator will have to develop a gold standard if he wishes to conduct many of the above studies.

Examples of gold standards that have been used in biosurveillance research include:

- ICD-9—coded *hospital* discharge diagnoses
- ICD-9—coded *outpatient* billing diagnoses
- Influenza surveillance data
- Manual review of hospital charts

6.1. ICD-9 Hospital Discharge Diagnoses

Many state health departments (approximately 60%) compile diagnoses of patients discharged from hospitals located within the state. These *hospital discharge data sets* include diagnoses encoded using the ICD-9-CM coding system (discussed in Chapter 32), dates of admission and discharge, home zip code, hospital zip code, and patient age. Evaluators can use the diagnoses to construct reference epidemiological curves for diseases such as influenza, salmonella, respiratory syncytial virus, and rotavirus. Using the location information, evaluators can construct these curves for different zip codes, cities, counties, or states. Examples of hospital discharge data sets include the Pennsylvania Health Care Cost Containment Council (PHC4) data set, which comprises all discharges for all hospitals in

Pennsylvania, and the Utah Hospital Discharge Database, which comprises all discharges for hospitals in Utah except Intermountain Shriners Hospital, which is exempt from reporting requirements because it is a charity hospital.

As an example of how these hospital discharge diagnosis data sets are used, Hogan and colleagues defined four sets of ICD-9-CM codes: pneumonia and influenza (P&I), bronchiolitis due to respiratory syncytial virus (RSV), rotavirus gastroenteritis, and pediatric gastroenteritis due to all causes (Hogan et al., 2003). They used these sets of codes to form time series against which to compare sales of pediatric electrolytes using the correlation function and detection algorithm methods.

6.2. ICD-9–Coded Outpatient Billing Diagnoses

Outpatient billing data sets are compiled from physician outpatient bills that are submitted electronically to insurers. These data sets are important for research in biosurveillance because a majority of patient interactions with the healthcare system, for both acute and chronic disease, occur as visits to physician outpatient offices, rather than hospitals; therefore, they potentially can identify smaller outbreaks or outbreaks of diseases that do not often cause hospitalization. These data were used in an evaluation conducted by Siegrist et al. (2004).

Claims data include date of service, patient and physician location, patient age and gender, all diagnoses for the visit by ICD-9 code and all procedures performed by CPT code. Although the source data contains specific street address information for both the physician office and the patient home that potentially could be available for public health operations purposes, data are available to researchers only at the five-digit zip code level for patients.

6.3. Influenza Surveillance Data

Influenza surveillance data sets are readily available from the Centers for Disease Control and Prevention (CDC) website at *www.cdc.gov/ncidod/diseases/flu/*. They include the following:

- Weekly state epidemiologist reports of influenza activity.
- Weekly regional number of influenza-like illness[2] (ILI) cases from the U.S. Influenza Sentinel Physicians Surveillance Network.
- Weekly regional number of positive influenza tests.

State health department estimates are a second source of influenza data. State health departments report the *estimated* level of influenza activity in their states each week using four levels:

- *No Activity*: No cases of influenza or influenza-like illnesses (ILI) reported.
- *Sporadic*: Cases of influenza or ILI are reported, but reports of outbreaks in places such as schools, nursing homes, and other institutional settings have not been received.
- *Regional*: Outbreaks of influenza or ILI are occurring in geographic areas containing less that 50% of the state's population. A geographic area could be a city, county, or district.
- *Widespread*: Outbreaks are occurring in geographic areas representing more than 50% of the state's population.

6.4. Manual Review of Hospital Charts

Gold standards for evaluating surveillance data for case detection are often established by manual review of hospital charts. The following papers provide methods for the development of such gold standards and the conduct of such studies (Chapman and Haug, 1999, Chapman et al., 2004a, Chapman et al., 2004b)

SUMMARY

Data are the foundation of biosurveillance. The range of data that are potentially useful in biosurveillance is broad and continues to expand as information technology enables more detailed and timely collection of information about individuals, populations, and the environment. There are well-developed methods for studying the informational value of surveillance data. The methods can be applied to data that are collected for both the purpose of detecting cases, as well as data that are collected for the purpose of detecting or characterizing outbreaks. For those data that have value, evaluators should assess the availability of data, and the cost and effort to obtain the data, so that the decision makers can judge whether to incorporate the data into a biosurveillance scheme.

There are additional aspects of data, such as data quality and sampling biases, which we discuss in Chapter 37. Data quality and sampling bias influence the informational value of data, but the methods that we discuss in this chapter by and large measure the overall value of data, which includes these effects.

The research discussed in the next chapters is representative of the research that should be conducted to search for new types of data that can contribute to the goal of improving biosurveillance.

ADDITIONAL RESOURCES

Friedman, C. and J. Wyatt, eds. (1997). *Evaluation Methods in Medical Informatics. Computers in Medicine*. New York: Springer. A comprehensive monograph on the design, conduct and interpretation of evaluations of information systems in medicine.

2 The CDC defines influenza like illness as fever (temperature of >100°F) plus either a cough or a sore throat.

Wagner, M., J. Pavlin, Brillman, J. et al. (2004). *Synthesis of Research on the Value of Unconventional Data for Early Detection of Disease Outbreaks*. Washington, DC: Defense Advanced Research Projects Agency. This report summarizes results from the Biological Advanced Leading Indicator Recognition Technology program (Bio-ALIRT), which was funded by the Defense Advanced Research Projects Agency from 2001 to 2003. The overall goal of the program was to develop technology for early detection of a covert biological event through statistical analysis of non traditional data sources. This report summarizes the work performed to identify and evaluate non traditional data sources.

Wagner, M. M., R. Aryel, Dato, V. et al. (2001). *Availability and Comparative Value of Data Elements Required for an Effective Bioterrorism Detection System*. Washington, DC: Agency for Healthcare Research and Quality. A report on the methods and results of a 1-year project to estimate the availability and potential of new types of surveillance data.

ACKNOWLEDGMENTS

We thank Dr. Fu-Chiang Tsui for providing illustrations of the correlation function and the detection algorithm method.

REFERENCES

Campbell, M., Li, C.-S., Aggarwal, C and et al. (2004). An Evaluation of Over-the-Counter Medication Sales for Syndromic Surveillance. IBM Research Technical Report. IBM Research Division, Thomas J. Watson Research Center, Yorktown Heights, New York. Retrieved: from http://domino.research.ibm.com/library/cyberdig.nsf/papers/5C83AC4CCAE8A5EA852570130067B870/$File/rc23368.pdf.

Chapman, W., Haug, P. (1999). Comparing expert systems for identifying chest x-ray reports that support pneumonia. In: *Proceedings of American Medical Informatics Association Symposium*, 216–20.

Chapman, W., Wagner, M., Cooper, G., et al. (2003a). Creating a text classifier to detect chest radiograph reports consistent with features of inhalational anthrax. *J Am Med Inform Assoc* 10:494–503.

Chapman, W. W., Dowling, J. N., Wagner, M. M. (2004a). Fever detection from free-text clinical records for biosurveillance. *J Biomed Inform* 37:120–7.

Chapman, W. W., Dowling, J. N., Wagner, M. M. Classification of emergency department chief complaints into 7 syndromes: a retrospective analysis of 527,228 patients. Ann Emerg Med. Nov 2005; 46(5): 445–55.

Chapman, W. W., Espino, J. U., Dowling, J. N., et al. (2003b). *Detection of Multiple Symptoms from Chief Complaints*. Technical Report, Center for Biomedical Informatics, University of Pittsburgh, Pittsburgh, Pennsylvania.

Chapman, W. W., Fiszman, M., Dowling, J. N., et al. (2004b). Identifying respiratory findings in emergency department reports for biosurveillance using MetaMap. *Medinfo* 2004:487–91.

Espino, J., Wagner, M. (2001). The accuracy of ICD-9 coded chief complaints for detection of acute respiratory illness. In: *Proceedings of American Medical Informatics Association Symposium*, 164–8.

Espino, J. U., Hogan, W., Wagner, M. M. (2003). Telephone triage: A timely data source for surveillance of influenza-like diseases. In: *Proceedings of American Medical Informatics Association Symposium*, 215–9.

Friedman, C., Wyatt, J., eds. (1997). *Evaluation Methods in Medical Informatics*. New York: Springer.

Gesteland, P., Gardner, R., Rolfs, R., et al. (2005). Accuracy of a naive bayes chief complaint classifier for conducting automated emergency department based syndromic surveillance. Technical Report. School of Medicine, University of Utah, Salt Lake City, Utah.

Henry, J. V., Magruder, S. and Snyder, M (2004). Comparison of office visit and nurse advice hotline data for syndromic surveillance–Baltimore-Washington, D. C., metropolitan area, 2002. MMWR Morb Mortal Wkly Rep, 53(suppl):112–6. *http://www.ncbi.nlm.nih.gov/entrez/query.fcqi?cmd=Retrieve&db=PubMed&dopt=Citation&list uids=15714639.*

Hogan, W. R., Tsui, F.-C., et al. (2003). Early detection of pediatric respiratory and diarrheal outbreaks from retail sales of electrolyte products. *J Am Med Inform Assoc* 10:555–62.

Ivanov, O., Gesteland, P. H., Hogan, W., et al. (2003). Detection of pediatric respiratory and gastrointestinal outbreaks from free-text chief complaints. In: *Proceedings of American Medical Informatics Association Symposium*, 318–22.

Johnson, H. A., Wagner, M. M., Hogan, W. R., Chapman, W., Olszewski, R. T., Dowling, J et al. (2004). Analysis of Web access logs for surveillance of influenza. Medinfo, vol. 11, no. Pt 2, pp. 1202–6.

Johnson HA, Wagner MM, Saladino RA. (2005) A New Methods for Investigating Non-traditional Biosurveillance Data: Studying Behaviors Prior to Emergency Department Visits. RODS Laboratory Technical Report. Center for Biomedical informatics, University of Pittsburgh, Pittsburgh, Pennsylvania.

Lobanov, B. (1971). Classification of Russian vowels spoken by different speakers. *J Acoustical Soc Am* 49:606–8.

Magruder, S. (2003). Evaluation of over-the-counter pharmaceutical sales as a possible early warning indicator of human disease. *Johns Hopkins University Applied Physics Laboratory Technical Digest,* 24:349–353.

Quenel, P., Dab, W., Hannoun, C., et al. (1994). Sensitivity, specificity and predictive values of health service based indicators for the surveillance of influenza A epidemics. *Int J Epidemiol* 23:849–55.

Siegrist, D., Pavlin, J. (2004). Bio-ALIRT biosurveillance detection algorithm evaluation. *MMWR Morb Mortal Wkly Rep* 53(Suppl):152–8.

Tsui, F.-C., Wagner, M. M., Dato, V., et al. (2001). Value of ICD-9-coded chief complaints for detection of epidemics. In: *Proceedings of American Medical Informatics Association Symposium*, 711–5.

Wagner, M. M., Aryel, R., Dato, V. (2001). *Availability and Comparative Value of Data Elements Required for an Effective Bioterrorism Detection System*. Washington, DC: Agency for Healthcare Research and Quality.

Welliver, R. C., Cherry, J. D., Boyer, K. M., et al. (1979). Sales of nonprescription cold remedies: a unique method of influenza surveillance. *Pediatr Res* 13:1015–7.

22
CHAPTER

Sales of Over-the-Counter Healthcare Products

William R. Hogan and Michael M. Wagner
RODS Laboratory, Center for Biomedical Informatics
University of Pittsburgh, Pittsburgh, Pennsylvania

1. INTRODUCTION

Sales of over-the-counter (OTC) healthcare products have significant potential as early indicators of disease outbreaks. Sick individuals treat themselves with nonprescription cough syrups, flu remedies, and diarrhea remedies. In addition to medications, they also purchase thermometers and other items for their illness, such as tissues, orange juice, and chicken soup. They frequently make purchases before seeking medical care or even instead of seeking medical care. People purchase these products both for themselves and for their children.

The effort that a biosurveillance organization must expend to obtain these data from stores is low in comparison to other types of surveillance data. Pharmacies, grocery stores, and other retail sources of such products collect sales data electronically. In the United States, the effort required on the part of a biosurveillance organization to obtain daily data on sales of 18 OTC product categories at the level of zip code has been reduced to zero by the National Retail Data Monitor project, which collects OTC sales data nationwide for purposes of biosurveillance (Wagner et al., 2004, Wagner et al., 2003).

2. DESCRIPTION OF OTC DATA

The retail industry has built significant infrastructure to capture data about product sales for business purposes. Retailers analyze sales data for supply chain management and to make decisions, such as which product lines to carry in which stores, how and when to promote products, which stores to close, and where to open new stores.

The vast majority of retailers in the United States collect sales data using optical cash registers. Clerks scan the familiar barcode imprinted on every product (Figure 22.1) at the point of purchase. The barcode encodes a standard product identifier called the *Global Trade Item Number* or GTIN.[1]

The cash register either records sales in the store's computer in real time or stores the records for periodic automatic transfer to the store's central computer. The cash register and store computer record all products purchased in a single transaction, which is sometimes referred to as a *market basket*. If the customer presents a loyalty card, then the cash register and store computer also record the customer number associated with the card. On a daily or more frequent basis, the store computer transmits these sales data to a central data warehouse operated by the retail chain. Thousands of stores may send data to a single data warehouse.

The data that retailers collect are quite detailed. They include a record of each check-out, including the GTIN for each item in the set of items purchased (the market basket), the quantity of each item, the time of purchase, and—if the customer used a loyalty card or credit card when making the purchase—some indication of the identity of the purchaser. Retailers collect these data at the moment of purchase. Retailers also have electronic records of which items were being promoted at the time of the sale, which may be useful in distinguishing a spike in sales due to disease from a spike in sales due to promotion.

3. AVAILABILITY OF OVER-THE-COUNTER SALES DATA

At present, OTC sales data available for biosurveillance are less detailed and timely than those available within a store due to technical and privacy considerations. At present, OTC sales data are available for biosurveillance as daily counts of sales for each store and for each GTIN. The daily sales figures become available anywhere from 6 AM to 3 PM the following day.

At present, market basket and loyalty card indexing of purchases are not available for biosurveillance. These data have immense potential for improving biosurveillance due to their ability to distinguish between purchases made for treatment of infectious disease and purchases made for other reasons (Fienberg and Shmueli, 2005). For example, someone who

FIGURE 22.1 This barcode represents the GTIN 00001234567895. The two leading zeros are not explicitly encoded. As discussed in Chapter 32, two leading zeros are added to the Universal Product Code to convert it to a GTIN.

1 The Global Trade Item Number is a new name for the Universal Product Code. See Chapter 32.

Handbook of Biosurveillance
ISBN 0-12-369378-0

bought cough syrup and tissues is more likely to have an acute respiratory illness (or an acquaintance or child with respiratory illness) than someone who purchased cough syrup and a pack of cigarettes. The major barrier to obtaining these additional data is the reluctance of retailers to provide the data due to concerns about the volume of data they would have to process on behalf of a biosurveillance organization, proprietary business concerns, and that participation in loyalty card programs may be adversely affected.

4. THE INFORMATIONAL VALUE OF OTC SALES DATA

There have been many studies that have measured the informational value of OTC sales for the detection of outbreaks. These studies utilized experimental methods that we discussed in Chapter 21.

4.1. Surveys of Sick Individuals about their Use of OTC Healthcare Products

In surveys of individuals who have had recent upper respiratory infections, the majority of respondents recall self treatment with OTC healthcare products (Labrie, 2001, McIsaac et al., 1998, Metzger et al., 2004). Importantly, fewer individuals recall seeking medical attention and those that did seek medical attention rarely did so as their first response to their illness.

In a random-digit-dialing survey of 1505 adults, 77% of those who reported an illness in the past six months had self-treated with an OTC medication, in contrast to 43% who said they visited a physician for their illness (Labrie, 2001). Of those reporting headache symptoms, 81% self-treated; of those reporting cold, cough, flu, or sore throat symptoms, 72% self-treated.

This survey did not determine the time between OTC use and first contact with the healthcare system. However, it did ascertain whether OTC use was an individual's first action after the onset of symptoms. Of the individuals reporting headache symptoms, 54% said their first course of action was to take an OTC medication (34% said their first course of action was to wait and see if the symptoms would go away, and only 4% said their first course of action was to consult a physician). For individuals with cold, cough, flu, and sore throat symptoms, the first course of action was self-treatment with an OTC product in 42%, watchful waiting in 34%, using a dietary supplement in 9%, and consulting a physician in less than 9%.

Another survey found that 76% of individuals who reported an upper respiratory tract infection in the last two weeks self-treated with an OTC medication; only 14% visited a physician (McIsaac et al., 1998). This survey did not assess the timing of OTC use relative to the timing of visits to physicians, but it did assess the timing of visits to physicians in the course of illness. The study found that nearly two-thirds of sick individuals who visited a physician did not do so until after the second day of illness and one-third did not do so until after the fourth day of illness. This survey was extensive. The researchers interviewed 42,223 adults 20 years of age or older in Canada over a period of 11 months.

Corso et al. (2003) surveyed usage of diarrhea remedies during a 1993 Milwaukee outbreak of cryptosporiosis by retrospective review of medical records of approximately 2000 patients. In the records of 378 patients who met their definition of moderate or severe cryptosporidiosis, they found evidence of self-treatment with OTC medications in approximately 30%.

A limitation of the above surveys is that they did not ascertain whether individuals *purchased* an OTC product or used one that was already in the medicine cabinet at home. Thus, these studies simply suggest the potential of monitoring of OTC medications for detection of outbreaks. They leave unanswered the question of how many unit sales might occur per sick individual during an outbreak and when these purchases might occur relative to the onset of the patient's symptoms.

A survey that Metzger et al. (2004) conducted in New York City partially addressed one of these limitations: they asked about OTC purchase as opposed to use. However, they did not determine how many OTC products each individual purchased or the relative timing of OTC purchase with respect to onset of illness.

Of the 2433 residents of New York City that Metzger and colleagues interviewed, 460 (18.9%) reported having a flu-like illness in 30 days prior to the interview. Of the 460 who reported a flu-like illness, 53.2% purchased one or more OTC products. OTC purchase was the most common action: 32.6% missed work or school, 29.1% visited a physician, 21.4% called a physician, 8.8% visited an emergency department, and 3.8% called a nurse or health advice line. Of the 108 individuals who took two or more of these actions, 36.6% purchased an OTC product as their first action in response to their symptoms. This was the most common first action: 30.3% missed work or school first, 16.2% visited a physician first, 11.8% called a physician first, 3.3% visited the emergency department first, and 0.7% called a nurse or health advice line first.

A study by Johnson et al. (2005) measured purchase (versus use) of an OTC product and the timing of purchase. In a survey of the caregivers for 78 children visiting an emergency department (ED) for an acute infectious illness, they found that caregivers purchased an OTC product for 32 (41%) of the children prior to the visit. On average, the caregivers made the purchase 1.88 (95% confidence interval [CI], 1.4–2.3) days prior to bringing their child to the ED. Of 29 children with cough and fever, caregivers purchased an OTC medication for 18 (62.1%). Of 30 children with vomiting or diarrhea with fever, caregivers purchased an OTC medication for 19 (63.3%). A limitation of this study is that it examined only children who presented to an emergency department, as opposed to the other surveys we have discussed, which conducted telephone interviews of a more representative sample of the population. The Walter Reed Army Institute of Research is conducting a similar study in adults (Pavlin, 2005).

4.2. Retrospective Studies of Sales during Known Outbreaks

Researchers have studied the sales of many types of OTC healthcare products during known outbreaks (Hogan et al., 2003, Campbell et al., 2004, Ohkusa et al., 2005, Magruder et al., 2004, Das et al., 2005, Welliver et al., 1979, Proctor et al., 1998, Stirling et al., 2001, Edge et al., 2004, Rodman et al., 1997, Beaudeau et al., 1999). These studies found strong and oftentimes early correlations between OTC sales and the epidemic curve for the outbreak. Two studies found that rises in sales of OTC healthcare products preceded detection of outbreaks by governmental public health (Proctor et al., 1998, Stirling et al., 2001). Other studies demonstrated that the increase in sales preceded the increase in traditional surveillance data used in public health surveillance. These studies collectively suggest that certain types of outbreaks can be detected earlier by monitoring of sales of OTC products than by current methods.

4.2.1. Pediatric Electrolytes

Hogan et al. (2003) used correlation analysis and the detection algorithm method to study the sales of pediatric electrolyte products during outbreaks in children. Pediatric electrolyte products are solutions of salts and water that are indicated for the treatment of dehydration in children ages five years and under. They studied the sales of pediatric electrolytes during annual winter outbreaks due to diseases such as rotavirus gastroenteritis and influenza in children ages five and under. Their study involved six urban regions (which we refer to as *cities*) over the course of three winters.

The researchers obtained a historical data set of sales of pediatric electrolytes from Information Resources, Incorporated. The data set contained weekly sales of pediatric electrolyte products sold for the six cities for the three-year study period. The sales data represented nearly 100% of all sales. They formed the sales data into time series for each city. To create a gold-standard reference time series, they obtained all hospital ICD-9–coded discharge diagnoses for the same time period for each of the cities, grouped hospital ICD-9 diagnoses into respiratory and diarrheal groups, and created time series of weekly counts (Figure 22.2). When forming the weekly counts, they used the date of admission of the patient to hospital (the earliest date at which a biosurveillance organization could possibly detect an outbreak from hospital data) to bias the

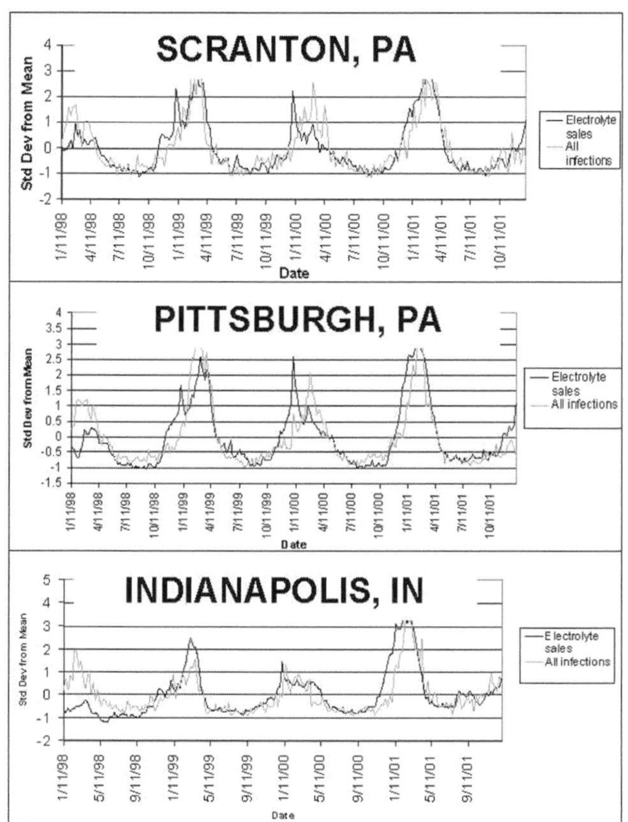

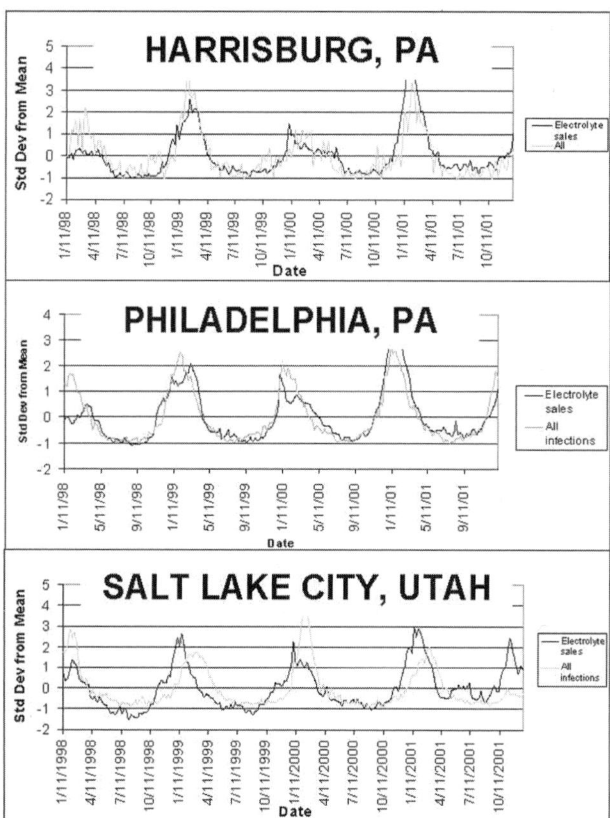

FIGURE 22.2 Correlation of weekly sales of electrolyte products and hospital diagnoses of respiratory or diarrheal illness in children aged 0 to 5. (Data courtesy of IRI, Utah Department of Health, Indianapolis Network for Patient Care, and PA HC4. From Hogan, W. R., Tsui, F.-C., Ivanov, O., et al. (2003). Early detection of pediatric respiratory and diarrheal outbreaks from retail sales of electrolyte products. *J Am Med Inform Assoc* 10:555–62, with permission.)

analysis against finding that detection from OTC monitoring was earlier than detection from more traditional data used in public health surveillance. They performed both correlation and detection algorithm analyses on the time series for each city. The hospital diagnosis data showed a total of three winter respiratory/diarrhea outbreaks in children age five and under for each of the six cities; thus, there were a total of 18 outbreaks to study.

The results of the correlation analysis showed a consistently high correlation between sales of pediatric electrolytes and hospital diagnoses of respiratory and gastrointestinal illness, with a mean correlation of 0.90 (95% CI, 0.87–0.93) for the 18 outbreaks. They found that the time lag between sales and diagnoses at which the correlation was maximal was on average 1.7 weeks earlier for sales of pediatric electrolytes than hospital admissions (95% CI, 0.5–2.9). The results by outbreak ranged from an OTC lead of eight weeks to an OTC lag of one week relative to the date of hospital admission.

The results of the detection-algorithm analysis were similar. The researchers used the exponentially weighted moving average algorithm (see Chapter 14) to detect the 18 outbreaks from sales of pediatric electrolytes and from hospital admissions. Detection from pediatric electrolytes occurred on average 2.4 weeks earlier than detection from hospital diagnoses. There was, however, significant variability in the timing among cities and years. The authors used regression analysis to estimate that 25% of the variation was explained by differences in signal strength (the ratio of the signal strength of pediatric electrolytes to the signal strength of hospital diagnoses varied across winters and cities) caused by differences in the severities of outbreaks and differences in the underlying variability in how people use pediatric electrolytes and medical services in different cities and different years. They hypothesized that the remaining variability might be explained by nosocomial outbreaks of pediatric diseases, which would appear only in the hospital diagnosis data set. The organisms causing these outbreaks—rotavirus, respiratory syncytial virus, adenovirus, and influenza virus—are important causes of nosocomial infections in pediatric hospitals. They were not able to control for this factor because the hospital discharge data sets did not specify which diagnoses were nosocomial. They called for additional studies using outpatient diagnoses to control for this potential confounder.

This study identified pediatric electrolytes as an important OTC category to monitor for outbreaks in pediatric populations. This study was the first study of the information value of biosurveillance data to include a sufficient number of outbreaks to compute a confidence interval on the measurement of timeliness of detection. By studying a sufficiently large number of similar outbreaks, this study revealed that there is variability from year to year and city to city in when outbreaks may be detected, and it sounded a cautionary note on the interpretation of studies of single outbreaks.

4.2.2. Cough, Cold, Sinus, and Allergy Medications

Campbell et al. (2004) used the detection algorithm method to study sales of cough, cold, sinus, and allergy products during outbreaks of respiratory diseases. They used the detection algorithm method to assess the lead time of sales of these products relative to a gold standard formed from billing data from physician office visits. They studied ten major U.S. cities for a period of three years (2000–2002). Similar to Hogan and colleagues, they formed time series by aggregating the OTC sales data into weekly counts for each of eight OTC product categories (adult and pediatric versions of cough, cold, sinus, and allergy). They formed gold-standard reference time series by aggregating the billing data (which comprised 22.5 million anonymized records corresponding to visits with respiratory-related ICD-9 codes for the 10 cities during the 2000–2002 period) into weekly counts.

They identified the beginning of the seasonal rise of respiratory ICD9 codes by inspection for each year and for each city. This method identified a total of 30 outbreaks for study. They used several simple detection algorithms (such as autoregression) to identify the first date of detection from both the OTC and office visit time series.

They found that for a number of categories (with the notable exception of allergy medications), detection of seasonal outbreaks of respiratory disease from OTC sales preceded detection from physician office visits. The lead time for the best categories (sinus and pediatric cough) was one to seven days.

Ohkusa et al. (2005) used correlation analysis to study the relationship between sales of OTC cold remedies and influenza activity in Japan. They found a poor correlation, with the adjusted r^2 never rising above 0.50. However, they performed the analysis at the national level with respect to geography, as opposed to other studies discussed in this chapter, which analyzed data at the metropolitan level. Also, the study included herbal remedies in the cold remedy category, a type of medication not included in other analyses. Finally, the authors caution that their results are preliminary, and that they plan to conduct additional analyses.

4.2.3. Flu Remedies and Chest Rubs

Magruder (2003) used correlation analysis to study sales of two categories of OTC products: medications used to treat "flu" symptoms (which they called "flu remedies"), and chest rubs. Similar to Campbell and colleagues, Magruder used billing data from physician office visits as a gold standard. He studied the National Capital Area that includes Washington DC and its surrounding suburbs for a 19-month period. He found that the peak correlation of chest rubs with physician office visits was 0.86, and that the correlation was maximal when sales of chest rubs preceded office visits by seven days. Similarly, the peak correlation of flu remedies with physician office visits was 0.89 when flu remedies preceded visits by 12 days. He was unable to compute a confidence interval on these measurements because he computed a single correlation for each OTC category for the entire National Capital Area. After accounting for holiday effects in the physician visit data and dividing the National Capital Area into six subregions, he found that the peak correlation of flu remedies with office visits averaged 0.895 (95% CI, 0.87–0.92) over the six subregions, and flu remedies preceded office visits by 2.8 days (95% CI, 0.09–5.58).

Das et al. (2005) used correlation analysis and the detection algorithm method to study OTC products for the treatment of influenza-like illness or ILI in New York City. They found that the ratio of sales of ILI products (a synonym for flu remedies) to sales of pain-relief products (the OTC ratio) was highly correlated with the ratio of ED visits for ILI to ED visits for other syndromes (the ED ratio), with an r^2 of 0.60. This correlation occurred when neither time series was lagged relative to the other one. They also measured the correlation when OTC sales were lagged by –14, –7, 7, and 14 days, and found lower correlations at these lags, and thus a lag of zero produced the highest correlation (of the lags they studied). In their detection algorithm analysis, they were able to detect the 2003–2004 influenza outbreak approximately two to three weeks earlier from the ED ratio than from the OTC ratio using a cyclical regression model. Similarly, they were able to detect the 2004–2005 influenza outbreak approximately six days earlier from the ED ratio.

Of historical interest, a study by Welliver et al. (1979) was the first to provide a quantitative estimate of the lead time of OTC sales data over data collected during physician office visits. Welliver and colleagues observed a strong peak in sales of cold remedies (another synonym for flu remedies) just before the rise in physician encounters with patients subsequently diagnosed as having influenza B virus, and one week before the peak in those encounters. They also observed an earlier rise in sales of cold remedies approximately coincident with the early winter rise in non-influenza respiratory virus activity as evidenced by laboratory test results. They found no correlation between sales of antifever medications, such as aspirin (either adult or pediatric), and influenza. Welliver's study was based on two outbreaks in a single city, and utilized OTC data that were aggregated at the weekly level.

4.2.4. Diarrhea Remedies and Outbreaks of Cryptosporidiosis

There have been four retrospective studies of the sales of diarrhea remedies during waterborne outbreaks of cryptosporidiosis (Proctor et al., 1998, Stirling et al., 2001, Edge et al., 2004, Rodman et al., 1997). All of these studies used similar sets of products (e.g., Kaopectate®, Imodium®, and Pepto Bismol®), and all found that sales of diarrhea remedies increased during outbreaks of cryptosporidiosis, and that the daily or weekly sales of diarrhea remedies rose and fell in proportion to the epidemic curve for the outbreak (although offset in time). Importantly, two studies found that significant increases in sales of diarrhea remedies preceded detection of the outbreak by the responsible public health organizations (Proctor et al., 1998, Stirling et al., 2001).

Proctor and colleagues studied sales of diarrhea remedies during the massive outbreak of cryptosporidiosis in Milwaukee in 1993 (Proctor et al., 1998). They obtained monthly counts of sales of diarrhea remedies from a single pharmacy in Milwaukee from March 1993—just prior to detection of the outbreak by health authorities—through July of 1995. They found that sales in March were approximately 600 units and sales in April were approximately 800 units. Monthly sales never exceeded 300 units in the ensuing two years after the outbreak. Given that the outbreak started in late March and public health became aware of the outbreak on April 5 (Mac Kenzie et al., 1994), even monthly sales data from a single pharmacy (for March 1993) may have provided four days earlier detection.

Stirling et al. (2001) studied sales of diarrhea remedies by three pharmacies during an outbreak of cryptosporidiosis in North Battleford, Saskatchewan. They found that weekly sales of diarrhea remedies increased seven-fold approximately three weeks prior to public health awareness of the outbreak and issuance of a drinking water advisory.

Edge et al. (2004) studied the relationship between sales of diarrhea remedies during the North Battleford outbreak (first studied by Stirling et al. [2001]) and data about ED visits for gastrointestinal illness during the outbreak. They found that OTC sales data for this outbreak were a more sensitive and consistent indicator of disease activity than visits to the ED for gastrointestinal syndrome. They concluded that monitoring of sales of OTC diarrhea remedies is particularly important for diseases, such as cryptosporidiosis, that have relatively mild symptoms that individuals may self-treat rather than seek care at EDs or physician offices.

Rodman et al. (1997) attempted to synthesize information about sales of OTC diarrhea remedies for five waterborne outbreaks of cryptosporidiosis in the United States and Canada, including the Milwaukee outbreak. To our knowledge, this is the first study that attempted to analyze OTC sales data for multiple outbreaks. Unfortunately, they were only able to obtain sales of diarrhea remedies for one of the five outbreaks. During an outbreak of cryptosporidiosis that occurred in Collingwood, Ontario in 1996, one pharmacy in Collingwood saw a 26.4-fold increase in sales of diarrhea remedies during the first month of the outbreak relative to the same month in the previous year and 5.6 times higher in the second month. Their paper is also interesting because it summarizes interviews they conducted with pharmacists. During outbreaks that occurred in the cities of Kelowna and Cranbrook, in British Columbia, 10 to 12 pharmacies in each city related that sales of diarrhea remedies increased "significantly" during the outbreaks, but could not provide sales figures.

4.2.5. Diarrhea Remedies and Other Infectious Diarrheal Diseases

There have been two studies that have analyzed sales of diarrhea remedies during outbreaks of diarrheal diseases other than cryptosporidiosis (Edge et al., 2004, Angulo et al., 1997).

Edge and colleagues studied sales of diarrhea remedies during an outbreak of diarrhea in Walkerton, Ontario caused by contamination of the city's water supply with the bacterium *E. coli* O157:H7 (Edge et al., 2004). They found that weekly counts of sales of diarrhea remedies at one pharmacy increased twofold coincident with the rapid rise in cases as reflected by the epidemic curve for the outbreak. They found that the increase in OTC sales preceded an increase in emergency department visits for gastrointestinal illness. They used the detection-algorithm method to study the date that the outbreaks would have been detected by routine monitoring. The two detection algorithms that they studied both alarmed on weekly sales counts for the week ending May 26. The local health department had issued a boil-water advisory on May 21. Their study used a false alarm rate of one per year. They did not determine the false alarm rate at which a detection system using these algorithms would have detected the outbreak on the week ending May 19.

Angulo et al. (1997) studied a waterborne outbreak due to contamination of the water supply of the city of Gideon MO with *Salmonella typhimurium*. They found that sales of diarrhea remedies increased 600% during the outbreak relative to historical sales levels. The increase occurred in early December 1993 before health authorities issued a boil-water advisory on December 18. The paper does not provide details about the sales data they collected (number of pharmacies, weekly vs. daily counts, and specific dates of significant increase), and we mention this study for completeness.

Das et al. (2005) performed a correlation analysis of OTC diarrhea remedies in New York City. They found a poor correlation between sales of OTC diarrhea remedies and ED visits for diarrhea, with an r^2 of 0.24. However, they did find an increase in sales of OTC diarrhea remedies during known outbreaks of gastrointestinal illness due to norovirus. They also noted an increase in sales of OTC diarrhea remedies during the electrical blackout that occurred in New York City in August 2003, suggesting that loss of refrigeration during the blackout led to contamination of food and a subsequent increase in gastrointestinal illness.

4.2.6. Diarrhea Remedies and Drinking Water Turbidity

There has been one study of the correlation of sales of diarrhea remedies with turbidity of drinking water (Beaudeau et al., 1999). The goal of this study was to assess whether drinking water turbidity (which water companies measure routinely) correlates with gastrointestinal illness in the community served by the water supply. Beaudeau and colleagues analyzed the correlation of daily water turbidity measurements with the daily sales of diarrhea remedies for a three-year period in Le Havre, France. There were no known outbreaks of cryptosporidiosis or other gastrointestinal illnesses during the period of the study.

They found that there were two 3-week periods of increases in sales of diarrhea remedies following two increases in water turbidity. Additionally, they found a statistically significant increase in sales of diarrhea remedies three to eight days after an interruption in the chlorination of water. They estimated that approximately 10% of annual cases of gastrointestinal illness were due to consumption of tap water. We note that they studied a combination of both prescription and OTC diarrhea remedies. However, only 20% of sales were of prescription diarrhea remedies, a proportion unlikely to have accounted for the entire effect.

Although it is possible that the correlation they observed was coincidental, other studies have analyzed the relationship between water turbidity and health indicators (other than OTC sales) and found similar correlations (Schwartz et al., 2000, Schwartz et al., 1997, Morris et al., 1996). A randomized, prospective epidemiological study found that individuals drinking tap water had a 14–19% higher incidence of gastrointestinal illness than individuals drinking purified bottled water (Payment et al., 1997), lending additional support to a real association between turbidity (presumably reflecting inadequate processing of source water that is a reservoir of one or more pathogens) and disease.

4.2.7. Case Studies of the National Retail Data Monitor

There is a systematic effort to study the effects of outbreaks on OTC sales data that is part of the National Retail Data Monitor (NRDM) project. (We discuss the project itself in greater detail in Section 5 of this chapter.) Each case study describes the effect of a single outbreak or other public health event, such as low air quality due to forest fires, on OTC sales (and other types of data available for the event). At present, these case studies are available only to authorized public health users of the NRDM system because of legal agreements between the NRDM and retailers.

One case study found that sales of bronchial remedies increased dramatically on the same day that air quality in southern California deteriorated significantly due to smoke from wildfires.

Two case studies of influenza outbreaks found that sales of thermometers increased significantly and early during the course of the 2003–2004 influenza outbreak. One of these case studies also found that sales of pediatric cough syrups and pediatric antifever medications were also a timely indicator of the influenza outbreak.

TABLE 22.1 Summary of Studies of Informational Value of Over-the-Counter Sales by Product Category

Category	Positive	Negative	References
Pediatric electrolytes[a]	Can detect winter diarrheal/respiratory outbreaks in children ages 5 and under earlier than hospital data		Hogan 2003
Flu remedies	For acute respiratory illnesses, precedes office diagnoses		Magruder 2004, Welliver 1979
Chest rubs	For acute respiratory illnesses, precedes office diagnoses		Magruder 2004
Diarrhea remedies[b]	Can detect large cryptosporidium outbreaks as much as 3 weeks before conventional methods. Can also detect other waterborne illness such as that due to *E. coli* and Salmonella		Proctor 1998, Stirling 2001, Edge 2004, Rodman 1997, Corso 2003, Beaudeau 1999
Cold	Winter respiratory outbreaks		Campbell 2004
Sinus	Winter respiratory outbreaks/bronchial diagnoses		Campbell 2004
Allergy		Winter respiratory outbreaks (later than office records)	Campbell 2004
Bronchial remedies	California wildfire showed effect on day that air quality deteriorated	No seasonal effects due to winter respiratory outbreaks	NRDM case study
Thermometers	Influenza outbreaks		NRDM case study
Pediatric cough	Winter respiratory outbreaks		NRDM case study
Adult and pediatric antifever	Increase in pediatric sales in one influenza outbreak	No increase in either during another influenza outbreak	NRDM case study, Welliver 1979

[a]Variability of effect noted from year to year and city.
[b]Literature review.
NRDM, National Retail Data Monitor.

4.3. Summary of Research on Information Value

Table 22.1 summarizes the results of the above studies. The categories of OTC healthcare products that show the most promise for the early detection of disease outbreaks include pediatric electrolytes for large outbreaks of respiratory and gastrointestinal disease in children (including disease caused by influenza virus, rotavirus, and respiratory syncytial virus); flu remedies, chest rubs, and "cold, cough, sinus, and allergy medications" for large outbreaks of respiratory disease; diarrhea remedies for waterborne outbreaks of cryptosporidiosis and other gastrointestinal pathogens; and thermometers, pediatric cough syrups, and pediatric antifever medications for outbreaks of influenza.

5. ANALYTICAL ISSUES IN MONITORING SALES OF OTC PRODUCTS

Monitoring of OTC sales data involves specific analytical issues. These issues include the collection of additional data to enable interpretation of anomalous levels of sales, the sources of variability in OTC sales data, and the requirement to monitor product category sales versus individual product sales.

5.1. Data Needed to Interpret Anomalous Levels of Sales

The National Retail Data Monitor project has identified additional data about OTC sales that can help biosurveillance organizations interpret anomalies in OTC sales data. Because an increase in OTC sales might be due to a promotion (such as a discount or a special display at the front of the store), it is helpful when attempting to differentiate between an outbreak and a promotion to know the count of promoted sales and the count of unpromoted sales of a product or product category.

Users of the NRDM have also found that the number of customer transactions that produced the count of sales of a particular product on a given day can help with the interpretation of sales anomalies. For example, if a store sold 15 bottles of Robitussin® DM cough syrup today, whereas it typically sells one bottle per day, the fact that one customer bought the 15 bottles is less suggestive of an outbreak than had 15 different customers purchased one bottle each.

5.2. Sources of Variability in OTC Sales Data

There are many non-disease factors that influence the level of sales of a given OTC product or product category. Sources of variability include day of week, season, holidays, severe weather, and promotions.

For many products and product categories, the daily sales exhibit day-of-week effects; that is, they vary by day of the week (Figure 22.3). For example, abuse of dextromethorphan-containing cough syrups is a well-known phenomenon (Murray and Brewerton, 1993), and purchases for the purpose of abuse are likely to reflect the opportunity to buy the product (e.g., store hours), which varies by day of week. In general, products with uses other than treatment of the symptoms of acute infectious illnesses exhibit day-of-week effects.

In addition to day-of-week effects, sales of some products exhibit seasonal variation (Figure 22.4). Sales are higher in winter because there is a higher level of infectious respiratory and gastrointestinal illnesses in winter. Certain allergens are also at their highest airborne concentrations at particular times of the year and cause seasonal variations in sales of OTC allergy medications (Magruder, 2003).

Besides day of week and season, holidays and severe weather, such as major snow storms, affect OTC sales. Specifically, these events cause decreases in sales that can be dramatic (e.g., there a steep decline in sales on December 25 in Figure 22.4). Oftentimes, the decrease in sales is associated with higher-than-normal sales before or after the holiday or weather event.

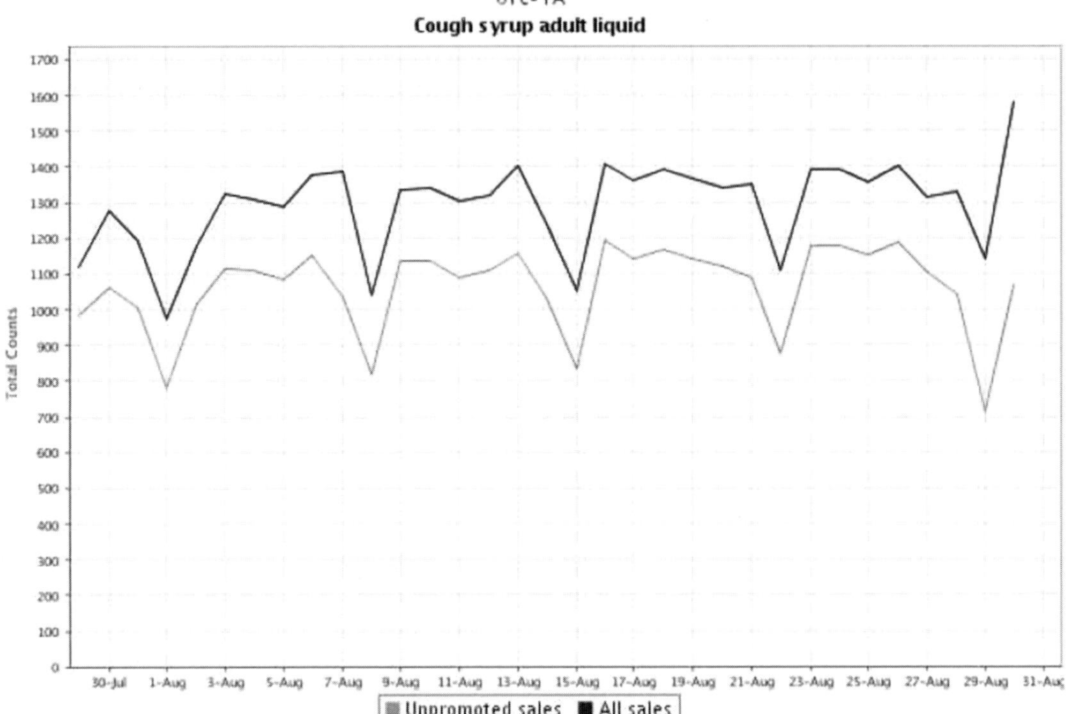

FIGURE 22.3 Day of week effect in OTC sales data.

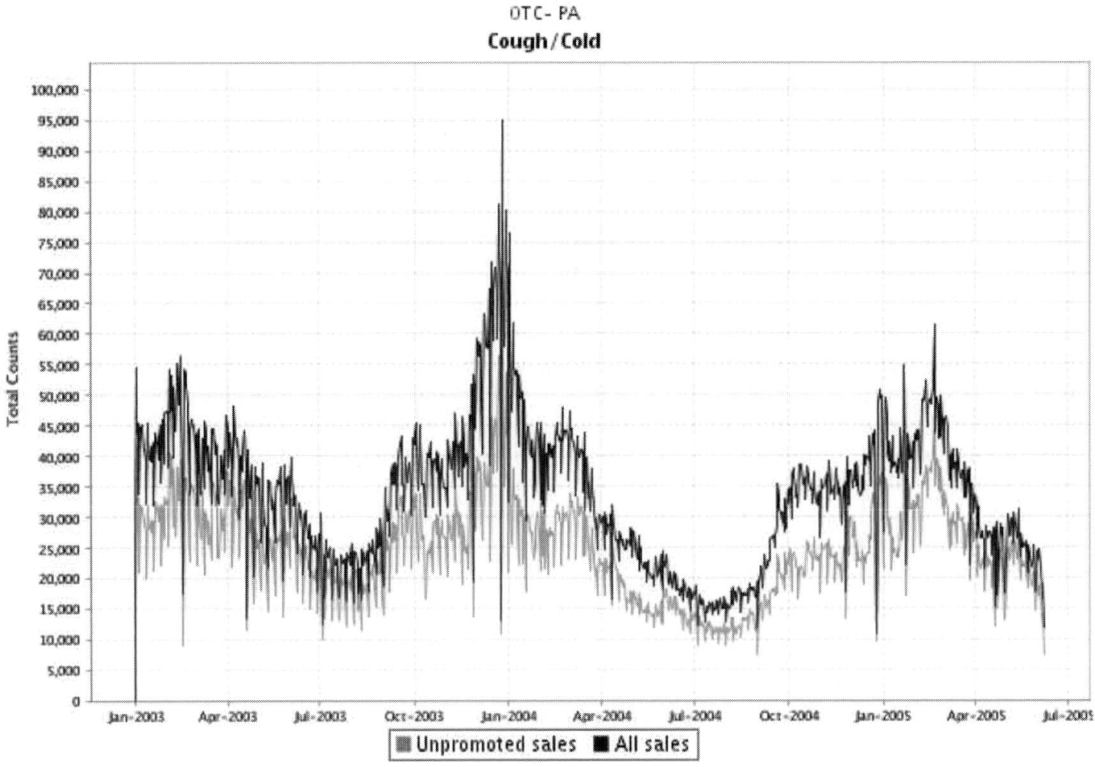

FIGURE 22.4 Seasonal and holiday effects in 2.5 years of OTC sales data. Large increases of sales occur in colder months of the year and when schools are in session. The large peak in December 2003 and January 2004 is due to a relatively severe influenza outbreak. Significant dips in sales occur on Christmas, Easter, and Thanksgiving annually.

These events are rare, however. Sales typically drop significantly on only three holidays per year, when many stores are closed or have reduced hours. In the United States, Christmas Day has the lowest counts (approximately one-tenth of usual). The drop in sales is less dramatic on Thanksgiving and Easter.

As we have discussed, promotions have the potential to increase sales, although as Wagner et al. (2003) note, the retail industry is aware that promotions, for the most part, shift sales from one product to another within a product category; thus, the overall effect on many categories is negligible.

5.3. Product Categories

At present, all biosurveillance systems that monitor OTC sales data monitor category data, not individual products. There are at least 9000 individual products intended for self-treatment of the symptoms of infectious disease, and many distinctions among these products, such as flavor or bottle size, are not relevant to biosurveillance. Aggregating products with similar uses into analytic categories reduces variability in sales data due to promotions and other causes.

A prerequisite to monitoring of sales at the product category level is a set of categories. The GTIN standard facilitates the creation and maintenance of product categories. A product-category mapping is an assignment of each GTIN code to a category.

How to best aggregate GTINs into categories for biosurveillance is, ultimately, a research question. Studies about the informational value of OTC sales data, such as the ones we have already discussed, are the best way to address this question. One study, however, grouped products into categories with a statistical analysis of historical OTC data sales alone (Magruder et al., 2004). A limitation of this approach is that it finds some associations between sub categories that are coincidental. It also does not provide information about which categories are best to monitor for which disease outbreaks.

6. A "DATA UTILITY" MODEL: NRDM

The National Retail Data Monitor is an intermediary organization positioned between the retailers and the organizations that conduct biosurveillance—a data utility for collection, analysis and distribution of OTC sales data to these organizations (Wagner et al., 2003, Wagner et al.). The NRDM as data utility saves both biosurveillance organizations and retailers' time and effort. Rather than 3000 local, state, and federal biosurveillance organizations negotiating with five or more retailers each and creating computer-to-computer interfaces ($3000 \times 5 = 15{,}000$ interfaces potentially), the retailers interact with the data utility (20 interfaces), and the biosurveillance organizations interact with the data utility (an additional 3000 interfaces in the worst case). The NRDM has enlisted participation of 15 retail chains that have 22,000 stores and roughly 50% of market share. At present, eight chains with nearly 20,000 stores send data; the remaining seven chains are in the process of connecting to the NRDM.

The retail chains transmit data on a daily or more frequent basis to the NRDM. They transmit their total daily sales (or the total sales since the last transmission for more frequent transmissions) for each store and each GTIN.[2] If a store is promoting a product, most retailers indicate that sales of the product were promotional. The NRDM aggregates the data into product categories and large geographical units such as zip codes, counties, and states.

The NRDM project makes OTC sales data available in three ways. First, individuals in biosurveillance organizations may obtain a user account to view OTC sales data in the NRDM's user interface. At present, 577 individuals working for biosurveillance organizations in 48 states, the District of Columbia, and Puerto Rico have user accounts for the NRDM. Second, a biosurveillance organization may obtain a daily file of OTC sales data for its jurisdiction. The NRDM provides 13 biosurveillance organizations OTC sales data in this manner. Third, a biosurveillance organization may configure its biosurveillance system to obtain OTC sales data using web services protocols.[3] One biosurveillance organization acquires OTC sales data from the NRDM in this manner.

The NRDM is a model for the collection and distribution of biosurveillance data of national scope. It is easy to envision similar data utilities for other national sources of biosurveillance data, such as laboratory data from national commercial laboratories and weather data.[4]

7. SUMMARY

Sales of OTC products are among the most highly available types of biosurveillance data that we discuss in this section of the book. At present, OTC sales data are available as daily counts of sales for each GTIN by store, with a one-day delay from time of sale. We expect the one-day delay to gradually improve as more retailers migrate their national data warehouses to real time. For example, Wal-Mart has developed real-time sales analysis and communication with its suppliers

2 As discussed in detail in Chapter 32, the GTIN product coding standard greatly facilitated construction of the NRDM and its ongoing maintenance.

3 See Chapter 33 *Architecture* for more information on web services.

4 We note that the CDC's BioSense system collects national data from two laboratory companies, the military healthcare (within the U.S.), and the Veteran's Administration healthcare system. BioSense at present provides these data only through the BioSense interface.

to minimize inventory costs (Kalakota and Robinson, 2003). Market basket and customer loyalty-card data have great potential for biosurveillance, but are not available at present.

Sales of OTC products are also among the most heavily studied types of biosurveillance data that we discuss in this section of the book. The available evidence strongly suggests that OTC sales are an early indicator of outbreaks of influenza, other annually recurring infectious respiratory diseases, cryptosporidiosis, and gastroenteritis due to *E. coli*, Salmonella, and rotavirus. The most promising categories include pediatric electrolytes, flu remedies, chest rubs, "cough, cold, sinus, and allergy" medications, diarrhea remedies, thermometers, pediatric cough syrup, and pediatric antifever medications.

The National Retail Data Monitor is a data utility for OTC sales data. It facilitates the communication of OTC sales data from retailers to biosurveillance organizations. It is one model for collecting biosurveillance data of national scope and providing them to biosurveillance organizations.

ADDITIONAL RESOURCES

National Retail Data Monitor website: *http://rods.health.pitt.edu/NRDM.htm*. This website is the home page of the NRDM project.

ACKNOWLEDGMENTS

We thank the retail industry, AC Nielsen, and Information Resources Inc. for assistance in developing both the science and practice of OTC sales monitoring for biosurveillance. The NRDM project and our research on OTC sales have been funded by Pennsylvania Department of Health Bioinformatics Grant ME-01-737; contract F30602-01-2-0550 from the Defense Advanced Research Projects Agency managed by Rome Laboratory; contract 2003-6-19 from the Sloan Foundation; the Passaic Water Commission; the departments of health of the states of New York, Ohio, Pennsylvania, Utah, and Washington; contract 290-00-0009 from the Agency for Healthcare Research and Quality; and award number 1 R21 LM008278-01 from the National Library of Medicine.

REFERENCES

Angulo, F. J., Tippen, S., Sharp, D. J., et al. (1997). A community waterborne outbreak of salmonellosis and the effectiveness of a boil water order. *Am J Public Health* 87:580–4.

Beaudeau, P., Payment, P., Bourderont, D., et al. (1999). A time series study of anti-diarrheal drug sales and tap-water quality. *Int J Environ Health Res* 9:293–311.

Campbell, M., Li, C.-S., et al. (2004). *An Evaluation of Over-the-Counter Medication Sales for Syndromic Surveillance.*

Corso, P. S., Kramer, M. H., Blair, K. A., et al. (2003). Cost of illness in the 1993 waterborne Cryptosporidium outbreak, Milwaukee, Wisconsin. *Emerg Infect Dis* 9:426–31.

Das, D., Metzger, K., Heffernan, R., et al. (2005). Monitoring over-the-counter medication sales for early detection of disease outbreaks–New York City. *MMWR Morb Mortal Wkly Rep* 54(Suppl):41–46.

Edge, V. L., Pollari, F., Lim, G., et al. (2004). Syndromic surveillance of gastrointestinal illness using pharmacy over-the-counter sales. A retrospective study of waterborne outbreaks in Saskatchewan and Ontario. *Can J Public Health* 95:446–50.

Fienberg, S. E., Shmueli, G. (2005). Statistical issues and challenges associated with rapid detection of bio-terrorist attacks. *Stat Med* 24:513–29.

Hogan, W. R., Tsui, F.-C., Ivanov, O., et al. (2003). Early detection of pediatric respiratory and diarrheal outbreaks from retail sales of electrolyte products. *J Am Med Inform Assoc* 10:6, 555–62.

Johnson, H. A., Wagner, M. M., Saladino, R. A. (2005). *A New Method for Investigating Non-Traditional Biosurveillance Data: Studying Behaviors Prior to Emergency Department Visits.* RODS Laboratory Technical Report. Pittsburgh, PA: Real-Time Outbreak and Disease Surveillance Laboratory, Center for Biomedical Informatics, University of Pittsburgh.

Kalakota, R., Robinson, M. (2003). From e-business to services: why and why now? In: *Services Blueprint: Roadmap for Execution.* Boston, MA: Addison-Wesley.

Labrie, J. (2001). *Self-care in the New Millenium: American Attitudes Towards Maintaining Personal Health.* Washington, DC: Consumer Healthcare Products Association.

Mac Kenzie, W. R., Hoxie, N. J., Proctor, M. E., et al. (1994). A massive outbreak in Milwaukee of cryptosporidium infection transmitted through the public water supply. *N Engl J Med* 331:161–7.

Magruder, S. (2003). Evaluation of over-the-counter pharmaceutical sales as a possible early warning indicator of human disease. *Johns Hopkins University Applied Physics Laboratory Technical Digest* 24:349–53.

Magruder, S. F., Lewis, S. H., Najmi, A., et al. (2004). Progress in understanding and using over-the-counter pharmaceuticals for syndromic surveillance. *MMWR Morb Mortal Wkly Rep* 53(Suppl):117–22.

McIsaac, W. J., Levine, N., Goel, V. (1998). Visits by adults to family physicians for the common cold. *J Fam Pract* 47:366–9.

Metzger, K. B., Hajat, A., Crawford, M., et al. (2004). How many illnesses does one emergency department visit represent? Using a population-based telephone survey to estimate the syndromic multiplier. *MMWR Morb Mortal Wkly Rep* 53(Suppl):106–11

Morris, R. D., Naumova, E. N., Levin, R., et al. (1996). Temporal variation in drinking water turbidity and diagnosed gastroenteritis in Milwaukee. *Am J Public Health* 86:237–9.

Murray, S., Brewerton, T. (1993). Abuse of over-the-counter dextromethorphan by teenagers. *South Med J* 86:1151–3.

Ohkusa, Y., Shigematsu, M., Taniguchi, K., et al. (2005). Experimental surveillance using data on sales of over-the-counter medications–Japan, November 2003–April 2004. *MMWR Morb Mortal Wkly Rep* 54(Suppl):47–52.

Pavlin, J. (2005). Personal communication.

Payment, P., Siemiatycki, J., Richardson, L., et al. (1997). A prospective epidemiological study of the gastrointestinal health effects due to the consumption of drinking water. *Int J Environ Health Res* 7:5–31.

Proctor, M. E., Blair, K. A., Davis, J. P. (1998). Surveillance data for waterborne illness detection: an assessment following a massive waterborne outbreak of Cryptosporidium infection. *Epidemiol Infect* 120:43–54.

Rodman, J., Frost, F., Davis-Burchat, L., et al. (1997). Pharmaceutical sales: a method of disease surveillance? *J Environ Health*:8–14.

Schwartz, J., Levin, R., Goldstein, R. (2000). Drinking water turbidity and gastrointestinal illness in the elderly of Philadelphia. *J Epidemiol Community Health* 54:45–51.

Schwartz, J., Levin, R., Hodge, K. (1997). Drinking water turbidity and pediatric hospital use for gastrointestinal illness in Philadelphia. *Epidemiology* 8:615–20.

Stirling, R., Aramini, J., Ellis, A., et al. (2001). Waterborne cryptosporidiosis outbreak, North Battleford, Saskatchewan, Spring 2001. *Can Commun Dis Rep* 27:185–92.

Wagner, M., Espino, J., Hersh, J., et al. (2004). National Retail surveillance. *MMWR Data Monitor for public health Morb Mortal Wkly Rep* 53(Suppl):40–42.

Wagner, M., Robinson, J., Tsui, F., et al. (2003). Design of a national retail data monitor for public health surveillance. *J Am Med Inform Assoc* 10:409–18.

Welliver, R. C., Cherry, J. D., Boyer, K. M., et al. (1979). Sales of nonprescription cold remedies: a unique method of influenza surveillance. *Pediatr Res* 13:1015–7.

23

CHAPTER

Chief Complaints and ICD Codes

Michael M. Wagner and William R. Hogan
RODS Laboratory, Center for Biomedical Informatics
University of Pittsburgh, Pittsburgh, Pennsylvania

Wendy W. Chapman
Center for Biomedical Informatics
University of Pittsburgh, Pittsburgh, Pennsylvania

Per H. Gesteland
Inpatient Medicine Division, Department of Pediatrics
University of Utah School of Medicine, Salt Lake City, Utah

1. INTRODUCTION

A *chief complaint* is a concise statement in English or other natural language of the symptoms that caused a patient to seek medical care. A triage nurse or registration clerk records a patient's chief complaint at the very beginning of the medical care process (Figure 23.1).

In contrast, an ICD code is a number (e.g., 558.9) that a clinician or professional coder uses to represent a medical diagnosis, syndrome, or symptom–usually for the purpose of billing. The ICD-coding system allows physicians and professional coders to express their diagnostic impression of a patient at different levels of diagnostic precision, ranging from very precise (e.g., ICD code 022.1 for inhalational anthrax) to syndrome (e.g., ICD code 079.99 for viral syndrome) to symptom (e.g., ICD code 780.6 for fever). The diagnosis may be a working diagnosis (a provisional diagnosis) or a definitive diagnosis, although ICD does not allow the clinician or coder to indicate this distinction. A clinician or professional coder may record an ICD code early in the process of medical care. Professional coders, not clinicians, invariably encode hospital discharge diagnoses, which are not available until after a patient is discharged from a hospital. The important points to remember about ICD coding are the heterogeneity in diagnostic precision, who does the encoding, and when the encoding is done.

Chief complaints and ICD codes are used ubiquitously in medical care in the United States in both the civilian and military healthcare systems. Medicare and other third party payers require these data for billing and claims. As a result, the healthcare industry has built significant electronic infrastructure to capture chief complaints and ICD codes.

Over the past six years, researchers have studied methods to obtain and analyze patient chief complaints and ICD codes for the purpose of early detection of outbreaks. The intensity of research on these data has been motivated in part by their availability. The objective of research is to test hypotheses that these data can be used either alone or in conjunction with other data to improve the timeliness of outbreak detection (Wagner et al., 2001). As a result of the research, many health departments are now routinely monitoring chief complaints and ICD codes.

For clarity of exposition, we discuss chief complaints and ICD codes separately in this chapter. However, we do not wish to reinforce a somewhat prevalent impression that they are competing alternatives. Both types of data contain information that is useful in biosurveillance and together they are complementary. In the future, we expect that biosurveillance systems will collect both types of data routinely. They will link these data to other data about a patient to support more accurate inference about a patient's true disease state. We explore the future roles of chief complaints and ICD codes and their synergies in the final section of this chapter.

2. CHIEF COMPLAINTS

The concept of a *chief complaint* is important in medicine. It is a statement of the reason that a patient seeks medical care. Medical and nursing schools teach future clinicians to begin their verbal presentations of patient cases with a statement of the chief complaint. They teach them to record the chief complaint using the patient's words and to avoid replacing the patient's words with their diagnostic interpretation. It is considered bad form to proffer a diagnostic impression in a chief complaint.[1]

1 The reasons for this practice are myriad. One reason is that other clinicians such as consultants and supervisory clinicians in academic medical centers read a patient's chart like detectives: They wish to form an independent diagnostic impression. They are interested in knowing 'just the facts.'

Handbook of Biosurveillance
ISBN 0-12-369378-0

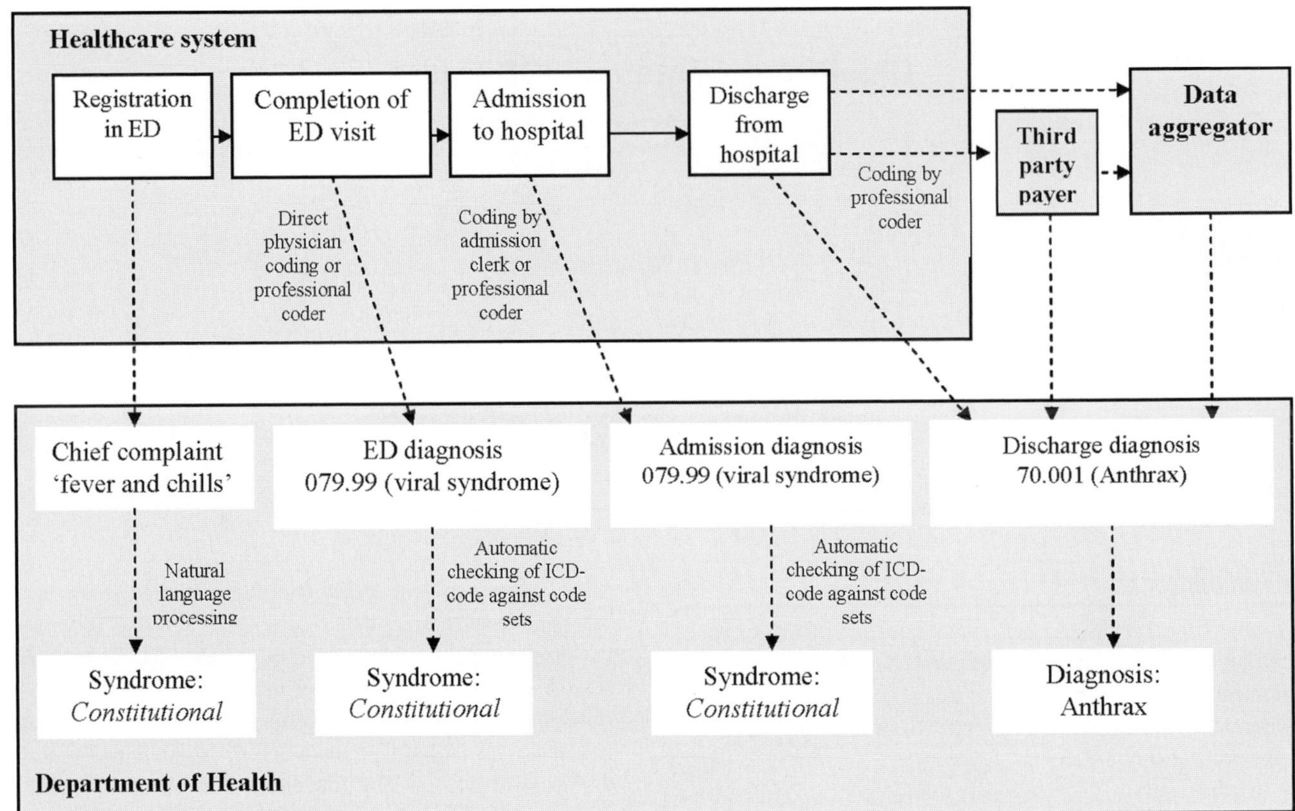

FIGURE 23.1 Points in the healthcare process at which chief complaints and ICD codes are recorded and transmitted to a health department or other biosurveillance organization. This figure illustrates a hypothetical patient with *anthrax* who seeks care at an emergency department (ED) and is subsequently admitted to a hospital. The patient's chief complaint is recorded at the time of registration and transmitted immediately to a health department via a HL7 message router. When the patient is discharged from the ED and admitted to the hospital, a professional coder reads the patient's ED chart and assigns ICD codes for billing purposes. The delay in transmission to a health department is indicated by a slanted arrow. Another ICD code may be assigned at the time of hospital admission. Finally, ICD codes are assigned by professional coders at the time of hospital discharge. These codes are transmitted to third party payers, who may submit them to data aggregators (e.g., commercial companies that analyze healthcare trends or health departments that assemble hospital discharge data sets for statistical purposes). In general, the diagnostic precision of the data available to a health department increases over time (moving from left to right in the figure). Note that there is variability from healthcare system to healthcare system. In some settings, the chief complaints are coded directly into ICD codes by physicians at the time of service (e.g., U.S. military).

As a result, the chief complaint usually states the key symptoms that a patient is experiencing.

During the process of medical care, a patient's chief complaint is recorded many times. Triage nurses and registration clerks create the first record at the time of initial registration for service at a clinic or emergency department (ED). Clinicians also record chief complaints in daily progress notes and discharge, transfer, and patient acceptance summary notes.

The research that we will discuss has shown that chief complaints contain information that may be very useful in biosurveillance. This result is not surprising. If a patient is ill with an infectious disease and presents to a physician, we would expect her chief complaint to reflect the nature of the illness.

2.1. Description of Chief Complaint Data Used in Biosurveillance

The recorded chief complaint of most interest to biosurveillance is the one recorded at the time a patient initially presents for medical care.[2] This chief complaint is often recorded directly into a registration computer by triage nurses or registration clerks and is highly available for biosurveillance purposes.

Table 23.1 is a sample of chief complaints from a registration computer in an ED. The chief complaints are more terse (four or five words) than those recorded in physician notes. They also contain misspellings, unconventional abbreviations, and unorthodox punctuation. Only two of these chief complaints contain diagnoses (*finger lac* and *uti*, which are abbreviations

2 Chief complaints are also recorded by call centers, as we discuss in Chapter 28. The methods and results discussed in this chapter apply equally to chief complaints obtained in these settings.

TABLE 23.1 Examples of Chief Complaints Recorded in an Emergency Department

Chief complaint	CoCo Syndrome
diff breathing	*Respiratory*
chest pain	*Other*
abd pain nausea vomiting	*Gastrointestinal*
Finger lac	*Other*
resp dist	*Respiratory*
Fever	*Constitutional*
nausea diarhea chest tightness sob	*Gastrointestinal, Respiratory*
chest pain vomiting	*Gastrointestinal*
r side pain	*Other*
rectal bleeding walkin	*Hemorrhagic*
chest pain	*Other*
Uti	*Other*
urinary problems	*Other*
abd pain	*Gastrointestinal*

Notes: These 14 examples come from a file used to train a Bayesian natural language processing (NLP) program called CoCo (described in Chapter 17). The second column shows the syndromes that a physician assigned to the chief complaints. For clarity, we adopt a typographical convention of italicizing syndromes.

for *finger laceration* and *urinary tract infection*, respectively). The rest describe the patient's symptoms. The second column of the table shows the syndromes that a human expert assigned to the patient for purposes of training a Bayesian natural language processor. We will discuss syndromes shortly.

2.1.1. Natural Language Processing of Free Text Chief Complaints

Before chief complaints can be analyzed by computerized biosurveillance systems, they must be converted from English (or other natural language) into computer-interpretable format. Biosurveillance systems typically use natural language processing (NLP) to convert chief complaints into computer-interpretable format. We are aware of one system that takes advantage of a routine translation of chief complaints into computer-interpretable form (Beitel et al., 2004).

There are two basic NLP methods for converting free-text chief complaints into computer-interpretable format—keyword parsing and probabilistic. We discussed these methods in Chapter 17 and will not repeat the discussion here.

The NLP component of a biosurveillance system analyzes a recorded chief complaint to classify a patient into a syndrome category. Some biosurveillance systems use NLP to identify syndromes directly and others use NLP to identify *symptoms* in the chief complaint and subsequently use Boolean (AND, OR, NOT) or probabilistic combinations of symptoms to assign a syndrome.

The subsequent (non-NLP) analysis performed by a biosurveillance system searches for clusters of syndromes in space, time, and/or demographic strata of a population, as discussed in Part III.

2.1.2. Syndromes

The concept of a syndrome is important in medical care (and in epidemiology). A syndrome is a constellation of symptoms, possibly combined with risk factors and demographic characteristics of patients (e.g., age and gender). Familiar examples of syndromes are *SARS* (*severe acute respiratory syndrome*) and *AIDS* (*acquired immune deficiency syndrome*). A syndrome plays the same role as a diagnosis in medical care—it guides the physician in selection of treatments for patients.

In this chapter, we will be discussing syndromes such as *respiratory* that are far less diagnostically precise than SARS or AIDS. The syndromes used in automated analysis of chief complaints and ICD codes are diagnostically imprecise by intent. The developers of these syndromes recognize that chief complaints (and ICD-coded diagnoses obtained close to the time of admission) in general do not contain sufficient diagnostic information to classify a patient as having SARS or other more diagnostically precise syndrome. They create syndrome definitions that are sufficiently precise to be useful, but not so precise that few if any patients will be assigned to them automatically, based solely on information contained in a four- or five-word chief complaint (or ICD code assigned early during the process of medical care).

Table 23.2 is the set of syndromes used by the RODS system. The table shows each syndrome name (which is just a convenient handle to reference the syndrome definition) and its definition. The RODS system uses the syndrome definitions in two ways. First, it makes them available to epidemiologists and other users of the system to assist in interpreting time series and maps of chief complaint data that have been aggregated by syndrome. If the user sees an increase in a syndrome such as *respiratory*, his interpretation of the increase should be that it could be due to any disease that is consistent with the definition of the *respiratory* syndrome. Second, RODS provides the definitions to individuals who are developing training sets (discussed in Chapter 17) for the CoCo parser.[3]

Tables 23.3 and 23.4 are syndrome classification systems used by Maryland Department of Hygiene and Mental Health and the New York City Department of Health and Mental Hygiene, respectively.

A final subtle point about the definition of syndromes that applies to both chief complaint syndrome definitions and ICD-code sets described later: The field of artificial intelligence distinguishes between intensional and extensional definitions. Tables 23.2 and 23.4 are intensional definition of syndrome categories. They express the intent of the system designer.

3 In practice, we train the CoCo parser using a set of approximately 10,000 chief complaints that a human has manually classified into one of these eight categories. Note that there is nothing domain specific about a Bayesian classifier as it is a mathematical formalism. CoCo could just as easily be trained to recognize a set of injury-related chief complaints.

TABLE 23.2 CoCo Syndrome Definitions

CoCo Syndrome	Definition
Gastrointestinal	Includes pain or cramps anywhere in the abdomen, nausea vomiting, diarrhea and abdominal distension or swelling.
Constitutional	Is made up of nonlocalized, systemic problems including fever, chills, body aches, flu symptoms (viral syndrome), weakness, fatigue, anorexia, malaise, irritability, weight loss, lethargy, sweating (diaphoresis), light headedness, faintness and fussiness. Shaking (not chills) is not constitutional but is other. Includes all of the "vaguely unwell" terms: doesn't feel well, feels ill, feeling sick or sick feeling, feels bad all over, not feeling right, sick, in pain, poor vital signs. Shaking or shaky or trembling (not chills) are not constitutional but are other (8). However, tremor(s) is neurological (7). Note: cold usually means a URI (cold symptoms; 3), not chills. Weakness, especially localized, is often neurological (7), rather than constitutional.
Respiratory	Includes the nose (coryza) and throat (pharyngitis), as well as the lungs. Examples of respiratory include congestion, sore throat, tonsillitis, sinusitis, cold symptoms, bronchitis, cough, shortness of breath, asthma, chronic obstructive pulmonary disease (COPD), pneumonia, hoarseness, aspiration, throat swelling, pulmonary edema (by itself; if combined with congestive heart failure, it is 8). If both cold symptoms and flu symptoms are present, the syndrome is respiratory. *Note: "Sore throat trouble swallowing" is respiratory, not respiratory and botulinic. That is, the difficulty in swallowing is assumed to be an aspect of the sore throat.*
Rash	Includes any description of a rash, such as macular, papular, vesicular, petechial, purpuric or hives. Ulcerations are not normally considered a rash unless consistent with cutaneous anthrax (an ulcer with a black eschar). *Note: Itch or itchy by itself is not a rash.*
Hemorrhagic	Is bleeding from any site *except the central nervous system*, e.g., vomiting blood (hematemesis), nose bleed (epistaxis), hematuria, gastrointestinal bleeding (site unspecified), rectal bleeding and vaginal bleeding. Bleeding from a site for which we have a syndrome should be classified as hemorrhagic and as the relevant syndrome (e.g., Hematochesia is gastrointestinal and hemorrhagic; hemoptysis is respiratory and hemorrhagic). Bleeding from a site for which we have a syndrome should be classified as hemorrhagic only without reference to the relevant syndrome, *except* hematochesia… hemoptysis. *Note: "Spitting up blood" is assumed to be hemoptysis.*
Botulinic	Includes ocular abnormalities (diplopia, blurred vision, photophobia), difficulty speaking (dysphonia, dysarthria, slurred speech) and difficulty swallowing (dysphagia).
Neurological	Covers nonpsychiatric complaints which relate to brain function. Included are headache, head pain, migraine, facial pain or numbness, seizure, tremor, convulsion, loss of consciousness, syncope, fainting, ataxia, confusion, disorientation, altered mental status, vertigo, concussion, meningitis, stiff neck, tingling, numbness, cerebrovascular accident (CVA; cerebral bleed), tremor(s), vision loss or blindness (but changed or blurred vision or vision problem is botulinic). Dizziness is constitutional and neurological. *Note: headache can be constitutional is some contexts, for example, "headache cold sxs achey" or "headache flu sxs."*
Other	Is a pain or process in a system or area we are not monitoring. For example, flank pain most likely arises from the genitourinary system, which we are not modeling, and would be considered other. Chest pain with no mention of the source of the pain is considered other (e.g., chest pain [other] versus pleuritic chest pain [respiratory]). Earache or ear pain is other. Trauma is other. Hepatic encephalopathy (not neurological), dehydration (not constitutional), difficulty sleeping or inability to sleep (not constitutional), constipation (not constitutional), and choking (but aspiration is respiratory) are all other.

Note: A physician or other medical expert refers to these definitions when creating training examples for the CoCo Bayesian classifier (described in Chapter 17).

The extensional definition of each syndrome is the actual set of chief complaints that NLP parsers, trained or configured based on the intensional definitions, assign to the categories.

2.1.3. Symptoms

Some researchers have divided the one-step chief-complaint-to-syndrome assignment process into two steps. Rather than using NLP to assign a syndrome directly to a patient based on the chief complaint, they use NLP to find all of the symptoms embedded in the chief complaint. They define a syndrome as a Boolean or probabilistic combination of symptoms. The two-step process then is: (1) NLP extracts symptoms from the chief complaint, and (2) a non NLP process determines whether the symptoms satisfy a Boolean or probabilistic syndrome definition. As an example, consider the chief complaint "n/v/d." With a two-step process, an NLP system would extract three symptoms from the chief complaint: nausea, vomiting, and diarrhea. Any syndrome definition that required nausea, vomiting, or diarrhea would be satisfied by this chief complaint. Biosurveillance systems operated by Washington State, New York State, and those using the CoCo classifier translate free text directly to syndromes; ESSENCE, SyCo2, and MPLUS classify to symptoms first.

TABLE 23.3 Syndrome Classification System Used by Maryland Department of Hygiene and Mental Health

Syndrome
Death
Gastrointestinal
Neurologic
Rash
Respiratory
Sepsis
Unspecified
Other

From Sniegoski, C. (2004). Automated syndromic classification of chief complaint records. *Johns Hopkins University APL Technical Digest* 25:68–75, with permission.

The two-step approach has several potential advantages. It is more natural for epidemiologists and physicians, who conceive of a syndrome as a combination of symptoms. In fact, *case definitions* (discussed in Chapter 3) are Boolean combinations of symptoms. Additionally, it is possible to create a new syndrome definition "on the fly" without retraining a Bayesian classifier or restructuring lists of keywords. To do this, one simply defines a Boolean or probabilistic combination

TABLE 23.4 Syndrome Classification System Used by New York City Department of Health and Mental Hygiene

Syndrome	Includes	Excludes
Common cold	Nasal drip, congestion, stuffiness	Chest congestion, sore throat
Sepsis	Sepsis, cardiac arrest, unresponsive, unconscious, dead on arrival	
Respiratory	Cough, shortness of breath, difficulty breathing, croup, dyspnea, bronchitis, pneumonia, hypoxia, upper respiratory illness, chest congestion	Cold
Diarrhea	Diarrhea, enteritis, gastroenteritis, stomach virus	
Fever	Fever, chills, flu, viral syndrome, body ache and pain, malaise	Hay fever
Rash	Vesicles, chicken pox, folliculitis, herpes, shingles	Thrush, diaper and genital rash
Asthma	Asthma, wheezing, reactive airway, chronic obstructive airway disease	
Vomiting	Vomiting, food poisoning	

Heffernan, R., Mostashari, F., Das, D., et al. (2004b). Syndromic surveillance in public health practice, New York City. *Emerg Infect Dis* 10:858–64. *http://www.ncbi.nlm.nih.gov/entrez/query.fcgi?cmd=Retrieve&db=PubMed&dopt=Citation&list_uids=15200820* with permission.

of symptoms. This advantage is important because new syndromes emerge with regularity, so it is important to be able to create a new syndrome definition quickly. Human biology changes very slowly, so new symptoms do not occur and the NLP conversion from free-text to symptom will be relatively stable, except as the language patients and triage nurses use to record chief complaints slowly evolves (e.g., the first time a patient uttered "I think I have SARS").

A limitation of the two-step approach is that it has not been validated. A real concern is that users of such systems can define syndromes for which the system's sensitivity and specificity may be extremely poor. A user may create a syndromic definition that is rational from an epidemiological standpoint but is not well-suited to the input data being classified. Without deep knowledge of the underlying processing method and its assumptions, a user will be completely unaware of this phenomenon and may be falsely reassured by the absence of disease activity when the newly created syndrome is put into operational use. As an extreme example of this problem, consider that a user might define a syndrome as a Boolean conjunction (AND statement) of five symptoms. Since the average registration chief complaint comprises four words, it is almost inconceivable that any patient would match such a syndrome definition.[4]

2.2. Availability and Time Latencies of Registration Chief Complaints

The chief complaints recorded at the time of registration are among the earliest data available electronically from a patient's interaction with the healthcare system. They are typically recorded before a doctor sees a patient. If the ED or a clinic is busy, many hours may pass before a registered patient is seen by a clinician.

The time latency between recording of a chief complaint and its availability to a biosurveillance organization can range from seconds to days, depending on whether the data collection system utilizes the HL7-messaging capability of a healthcare system for real-time communication or batch transfer of files (Figure 23.2). Hospitals are frequently capable of real-time transmission whereas office practices are not—unless they are associated with a larger organization (e.g., the Veterans Administration, U.S. military, or a large healthcare system).

Real-time transmission is possible when a healthcare system has a pre-existing Health Level 7 (HL7) messaging capability. Several publications describe the technical approach to HL7-based data collection and chief-complaint processing (Tsui et al., 2002, 2003, Gesteland et al., 2002, 2003, Olszewski, 2003a). Briefly, when a patient registers for care at an ED, a triage nurse or registration clerk enters the chief complaint into a registration system. This step is part of normal workflow in many U.S. hospitals (Travers et al., 2003). The registration system almost always transmits chief-complaints in the form of HL7 messages (Tsui et al., 2003) to an HL7-message router located in the healthcare system. To transmit these data to a biosurveillance organization, the healthcare system would configure the HL7-message router to de-identify these messages and transmit them via the Internet to a biosurveillance organization as they are received from the registration system. This configuration process is a native capability of commercial HL7-message routers and it is a routine task for an HL7 engineer or other information technology staff working in or for a healthcare system.

Batch transfer can either be automatic or manual. Automatic means that a computer program periodically queries the registration computer (or other system in which the chief complaint data are stored) for recent registrations, writes a file, and transmits the file to the biosurveillance organization via

4 Febrile syndromes provide a more realistic and common example of this problem. Many of the infectious diseases that represent threats to the public's health produce a febrile response in affected individuals early in the course of illness. From an epidemiological standpoint, monitoring febrile syndromes, such as Febrile Respiratory, will increase the chance that a positive patient actually has an infectious disease and will decrease the number of false positives. However, chief complaints—being terse–rarely describe both a syndromic symptom and fever. Of 610 patients who actually had febrile syndromes (Febrile Respiratory, Febrile GI, Febrile Neurological, Febrile Hemorrhagic, or Febrile Rash), only 5% of the chief complaints described both fever and the symptoms related to the organ (Chapman and Dowling, 2005).

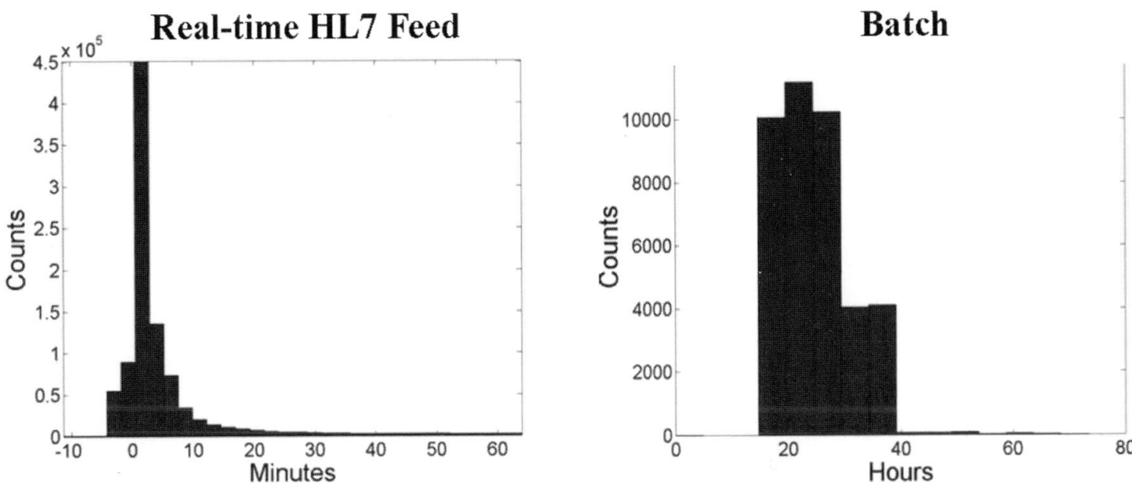

FIGURE 23.2 Comparison of time latencies of real-time and batch feeds. The negative time latencies associated with the real-time feed are due to slight clock differences between the biosurveillance system and the ED registration system.

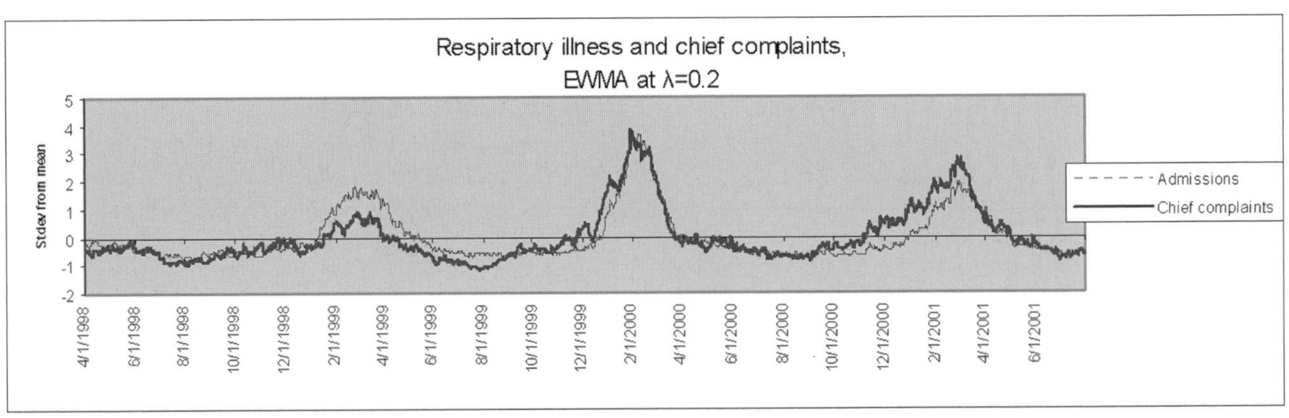

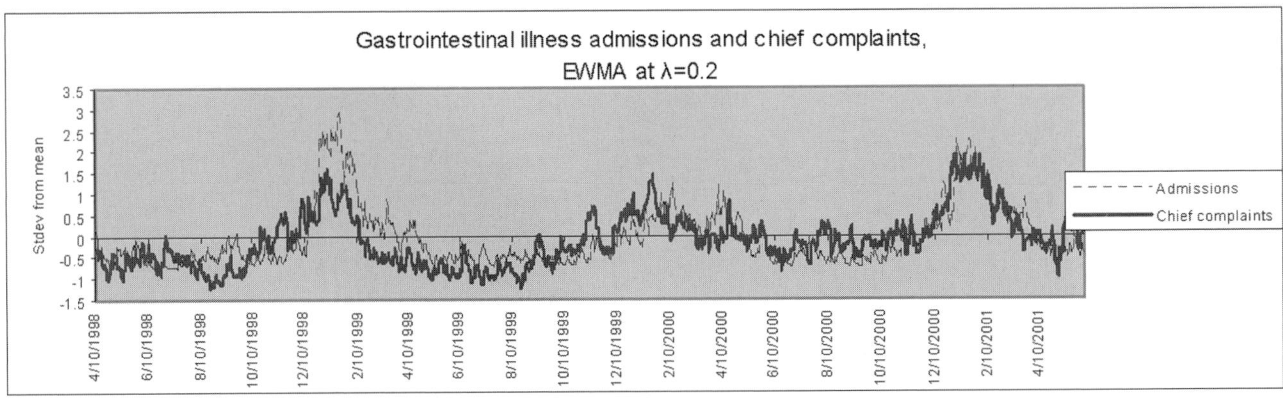

FIGURE 23.3 Detection of outbreaks in pediatric population from chief-complaint analysis, Utah 1998–2001.

the Internet. The transmission may use a secure file transfer protocol, non secure file transfer protocol, a web transfer protocol, or PHIN MS.[5] Manual means that someone working in the healthcare system must run a query and attached the results of the query to an email to the biosurveillance organization, or upload the file to a computer in the biosurveillance organization. In the past, manual data transfer often involved faxing of paper log files.

Tsui et al. (2005) studied the time latencies, data loss, and reliability associated with real-time HL7 feeds and batch feeds. Figure 23.3 compares the distribution of time delays between the time that a chief complaint was recorded during registration and receipt of that chief complaint by a biosurveillance system. The median time delay for a real-time feed was 0.033 hours and for batch was 23.95 hours.

The proportion of U.S. hospitals that are capable of sending a real-time HL7 feed appears to be approximately 84% based on our experience. Table 23.5 summarizes our experience with hospitals in the United States, suggesting that many hospitals have this capability.

2.3. Studies of Informational Value of Chief Complaints

Researchers have studied the ability of algorithms to detect syndromes and outbreaks of disease from chief complaints.

TABLE 23.5 Numbers of Hospitals Using Real-Time Versus Batch Connections to the RODS System, September 2005

	Healthcare Facilities		
Jurisdiction (project inception)	Real time	Batch	Location of Server
Pennsylvania (1999)	123	2	
Utah (2002)	26	0	
Ohio (2004)	50	14	Solaris/Linux/Oracle (Running at the RODS Public Health Data Center, University of Pittsburgh)
Kentucky (2005)	0	5	
Nevada (2005)	3	0	
Atlantic City NJ (2004)	3	0	
California (2005)	1	1	
Illinois (2005)	2	0	
Kentucky (2005)	0	5	
Michigan (2005)	2	0	
Los Angeles (2004)	1	4	Los Angeles County DOH
Houston TX (2004)	14	2	Houston DOH
Dallas Forth Worth Area TX (2004)	16	14	Tarrant County DOH
El Paso TX (2004)	2	0	El Paso
Totals	243	47	

Note: Many of these projects are statewide or citywide efforts that have an objective to connect every hospital to the health department. We provide the year of project inception to indicate the rate at which biosurveillance organizations are able to develop chief complaint data feeds from hospitals.
DOH, Department of Health; RODS, Real-Time Outbreak and Disease Surveillance Laboratory, Center for Biomedical Informatics, University of Pittsburgh.

These studies contribute to our understanding of the informational value of chief complaints for the detection of cases and of outbreaks. The studies utilized experimental methods that we discussed in Chapter 21. In this section, we review these studies and discuss how they address the following three hypotheses of interest:

Hypothesis 1: A chief complaint can discriminate between whether a patient has *syndrome* or disease X or not (stated as the null hypothesis: the chief complaint cannot discriminate).

Hypothesis 2: When aggregated with the chief complaints of other patients in a region, chief complaints can discriminate between whether there is an *outbreak* of type Y or not.

Hypothesis 3: When aggregated with the chief complaints of other patients in a region, algorithmic monitoring of chief complaints can detect an *outbreak* of type Y earlier than current best practice (or some other reference method).

It is important to note that the experiments that we will discuss differ in many details. They differ in the hypothesis being tested; the syndrome or type of outbreak studied; the NLP method; the detection algorithm; and the reference standard used in the experiment. Thus, achieving a "meta-analytic" synthesis about the informational value and role of chief complaints in biosurveillance requires that we pay attention to these distinctions. The one thing these studies share in common, however, is that they are all studies of the informational value of chief complaints.

2.3.1. *Detection of Cases*

Table 23.6 summarizes the results of studies that are informative about Hypothesis 1: A chief complaint can discriminate between whether a patient has syndrome or disease X or not.

Methodologically, the studies measure the sensitivity and specificity with which different NLP methods (which we refer to as "classifiers") identify patients with a variety of syndromes using only the recorded chief complaints. The reference syndrome ("gold standard") for patients in these studies was developed by physician review of narrative medical records, such as ED reports, or automatically from ICD-9 primary discharge diagnoses. Most studies have evaluated detection of syndromes in adults, whereas a single study examined detection of syndromes in pediatric patients (Beitel et al., 2004).

Table 23.6 groups the experiments by syndrome because many experiments studied the same or similar syndromes. Each row in Table 23.6 reports the sensitivity and specificity of a classifier for a particular syndrome, and the likelihood ratio

5 Note that time latencies associated with automatic batch transfer cannot necessarily be decreased by decreasing the periodicity of the batch transfer. The query to the healthcare information system that generates the batch file may be resource intensive and a healthcare system may only be willing to schedule the query during non peak load periods (e.g., midnight).

TABLE 23.6 Performance of Bayesian and other classifiers in detecting syndromes

Classifier Being Tested	Reference Standard for Comparison	Sensitivity (95% CI)	Specificity (95% CI)	Positive Likelihood Ratio (95% CI)	Negative Likelihood Ratio (95% CI)
Respiratory Syndrome					
Chief Complaint Bayesian Classifier *Respiratory* (CCBC)[a]	Utah Department of Health (UDOH) *Respiratory with fever*	0.52 (0.51–0.54)	0.89 (0.89–0.90)	5.0 (4.74–5.22)	0.54 (0.52–0.56)
CCBC[b]	Human review of ED reports	0.77 (0.59–0.88)	0.90 (0.88–0.92)	7.9 (5.8–10.8)	0.26 (0.13–0.49)
CCBC[a]	Utah ICD-9 list	0.60 (0.59–0.62)	0.94 (0.94–0.95)	10.45 (9.99–10.96)	0.25 (0.13–0.49)
Manual Assignment[c]	Human review of ED reports for *Pediatric respiratory illness*	0.47 (0.38–0.55)	0.99 (0.97–0.99)	56.73 (18.12–177.59)	0.53 (0.46–0.63)
CCBC[d]	ICD-9 list	0.63 (0.63–0.64)	0.94 (0.94–0.94)	11.14 (11.00–11.30)	0.39 (0.39–0.40)
CCBC[e]	Human review of ED reports	0.34 (0.30–0.38)	0.98 (0.97–0.99)	18.0 (11.24–28.82)	0.67 (0.63–0.71)
CCBC *Respiratory* and Keyword *Fever*[e]	Human review of ED reports for *Febrile respiratory*	0.02 (0.01–0.04)	0.99 (0.99–1.0)	20.83 (2.18–199.28)	0.99 (0.97–1.0)
Gastrointestinal (GI) Syndrome					
CCBC *GI*[a]	UDOH *Gastroenteritis without blood*	0.71 (0.69–0.74)	0.90 (0.90–0.90)	7.34 (6.98–7.72)	0.32 (0.29–0.35)
CCBC *GI*[f]	Human review of ED reports for *Acute infectious GI*	0.63 (0.35–0.85)	0.94 (0.92–0.96)	7.77 (4.77–12.65)	0.40 (0.20–0.80)
CCBC[a]	Utah ICD-9 list	0.74 (0.72–0.76)	0.92 (0.92–0.92)	9.5 (9.04–9.94)	0.28 (0.26–0.30)
CCBC[d]	ICD-9 list	0.69 (0.68–0.70)	0.96 (0.96–0.96)	15.70 (15.46–15.95)	0.32 (0.32–0.33)
CCBC[e]	Human review of ED reports	0.22 (0.16–0.29)	0.90 (0.88–0.91)	2.09 (1.51–2.91)	0.87 (0.80–0.95)
CCBC *GI* and Keyword *Fever*[e]	Human review of ED reports for *Febrile GI*	0.04 (0.02–0.08)	0.99 (0.99–1.0)	60.82 (7.65–483.45)	0.96 (0.93–0.99)
Neurologic/Encephalitic Syndrome					
CCBC Neurologic[a]	UDOH Meningitis/Encephalitis	0.47 (0.32–0.63)	0.93 (0.93–0.94)	4383.26[g] (1394.21–13780.56)	0.53 (0.39–0.72)
CCBC[a]	Utah ICD-9 list	0.72 (0.69–0.76)	0.95 (0.94–0.95)	13.5 (12.57–14.41)	0.29 (0.26–0.33)
CCBC[d]	ICD-9 list	0.68 (0.67–0.69)	0.93 (0.93–0.93)	9.25 (9.076–9.418)	0.35 (0.34–0.36)
CCBC[e]	Human review of ED reports	0.31 (0.27–0.35)	0.97 (0.95–0.98)	8.89 (6.25–12.65)	0.72 (0.68–0.76)
CCBC *Neurologic* and Keyword *Fever*[e]	Human review of ED reports for *Febrile Neurologic*	0.03 (0.01–0.07)	0.99 (0.99–1.0)	12.79 (2.89–56.59)	0.98 (0.95–1.0)
Constitutional Syndrome					
CCBC[d]	ICD-9 list	0.46 (0.45–0.47)	0.97 (0.97–0.98)	13.65 (13.30–14.0)	0.56 (0.55–0.57)
CCBC[e]	Human review of ED reports	0.27 (0.23–0.32)	0.95 (0.93–0.96)	5.12 (3.82–6.85)	0.77 (0.72–0.82)
Rash syndrome					
CCBC *Rash*[a]	UDOH *Febrile illness with rash*	0.50 (0.40–0.59)	0.99 (0.99–0.99)	55.6 (44.25–69.91)	0.51 (0.42–0.61)
CCBC[a]	Utah ICD-9	0.60 (0.52–0.67)	0.99 (0.99–0.99)	80.9 (67.43–97.07)	0.40 (0.33–0.49)
CCBC[d]	ICD-9 list	0.47 (0.45–0.49)	0.99 (0.99–0.99)	65.25 (61.79–68.90)	0.54 (0.52–0.56)
CCBC[e]	Human review of ED reports	0.31 (0.24–0.39)	0.99 (0.98–1.0)	34.01 (18.76–61.68)	0.70 (0.62–0.78)
CCBC *Rash* and Keyword *Fever*[e]	Human review of ED reports for *Febrile rash*	0.12 (0.05–0.27)	1.0 (0.99–1.0)	[h]	0.88 (0.78–1.0)
Hemorrhagic Syndrome					
CCBC[d]	ICD-9 list	0.75 (0.74–0.76)	0.99 (0.98–0.99)	49.01 (47.79–50.25)	0.25 (0.24–0.26)
CCBC[e]	Human review of ED reports	0.39 (0.34–0.44)	0.99 (0.98–0.99)	36.61 (20.96–63.93)	0.62 (0.57–0.68)
CCBC *Hemorrhagic* and Keyword *Fever*[e]	Human review of ED reports for *Febrile hemorrhagic*	0.0 (0–0.07)	1.0 (1.0–1.0)	[h]	1.0 (1.0–1.0)

Continued

TABLE 23.6 Performance of Bayesian and other classifiers in detecting syndromes—Cont'd

Classifier Being Tested	Reference Standard for Comparison	Sensitivity (95% CI)	Specificity (95% CI)	Positive Likelihood Ratio (95% CI)	Negative Likelihood Ratio (95% CI)
Botulinic syndrome					
CCBC[a]	UDOH *Botulism-like*	0.17 (0.05–0.45)	0.998 (0.998–0.999)	104.45 (28.57–381.86)	0.83 (0.64–1.07)
CCBC[a]	Utah ICD-9 list	0.22 (0.13–0.36)	0.999 (0.998–0.999)	166.94 (89.07–312.90)	0.78 (0.67–0.91)
CCBC[d]	ICD-9 list	0.30 (0.28–0.32)	0.99 (0.99–0.99)	44.26 (41.06–47.70)	0.70 (0.68–0.72)
CCBC[e]	Human review of ED reports	0.10 (0.06–0.17)	0.99 (0.99–1.0)	10.96 (5.11–23.48)	0.91 (0.86–0.97)
Fever					
Keyword[g]	Human review of ED reports	0.61 (0.51–0.69)	1.0 (0.96–1.0)	[h]	0.39 (0.31–0.49)

[a]From Wagner, M., Espino, J., Tsui, F.-C., et al. (2004). Syndrome and outbreak detection from chief complaints: the experience of the Real-Time Outbreak and Disease Surveillance Project. *MMWR Morb Mortal Wkly Rep* 53(Suppl):28–31, with permission.
[b]From Chapman, W. W., Espino, J. U., Dowling, J. N., et al. (2003). *Detection of Acute Lower Respiratory Syndrome from Chief Complaints and ICD-9 Codes.* Technical Report, CBMI Report Series 2003. Pittsburgh, PA: Center for Biomedical Informatics, University of Pittsburgh, with permission.
[c]From Beitel, A. J., Olson, K. L., Reis, B. Y., et al. (2004). Use of emergency department chief complaint and diagnostic codes for identifying respiratory illness in a pediatric population. *Pediatr Emerg Care* 20:355–60, with permission.
[d]From Chapman, W. W., Dowling, J. N., and Wagner, M. M. (2005). Classification of emergency department chief complaints into seven syndromes: a retrospective analysis of 527,228 patients. *Ann Emerg Med* 46(5):445–455.
[e]From Chapman WW, unpublished results.
[f]From Ivanov, O., Wagner, M. M., Chapman, W. W., et al. (2002). Accuracy of three classifiers of acute gastrointestinal syndrome for syndromic surveillance. In: *Proceedings of American Medical Informatics Association Symposium*, 345–9, with permission.
[g]From Chapman, W. W., Dowling, J. N., Wagner, M. M. (2004). Fever detection from free-text clinical records for biosurveillance. *J Biomed Inform* 2004;120–7, with permission.
[g]Large positive likelihood ratio due to specificity of 0.9999.
[h]Not able to calculate (denominator is zero).
CCBC, Chief complaint Bayesian classifier CI, confidence interval.

positive and negative. The likelihood ratio positive is the purest measure of the informational content of a chief complaint for detecting a syndrome (i.e., its ability to discriminate between a person with the syndrome and one without the syndrome). In a Bayesian analysis, it is a number that indicates the degree to which a system should update its belief that a patient has the syndrome, given the chief complaint (see Chapter 13).

The gold standard used in these studies varied. The most valid standard used was classification based on review of patients' ED reports using random selection of patients. The earliest studies evaluating the ability of chief complaints to identify syndromes were able to use this gold standard because they studied common syndromes, such as *respiratory* (Chapman et al., 2003, Beitel et al., 2004) or *gastrointestinal* (Ivanov et al., 2002). When a syndrome is common, a pool of randomly selected patients will produce a sufficient sample of actual respiratory cases.

Later studies examined less common syndromes. To obtain a sufficient sample of patients with uncommon syndromes, researchers searched ICD-9 discharge diagnoses to find cases (Wagner et al., 2004, Chapman et al., 2005b, Mundorff et al., 2004). Using a patient's discharge diagnosis as the gold standard enabled these studies to acquire large numbers of patients—even for rare syndromes, such as *botulinic*. Chart review, however, probably provides more accurate gold standard classifications than ICD-9 codes (Chang et al., 2005).

A few recent studies have used chart review as the gold standard for evaluating a variety of syndromes, including syndromes of low prevalence (Chang et al., 2005, Chapman et al., 2005c). One study compared chief complaint classification during the 2002 Winter Olympic Games against gold-standard classification of potentially positive cases selected by Utah Department of Health employees who performed drop-in surveillance (Wagner et al., 2004).

Chapman et al. (2005c) used ICD-9 searching to find a set of patients with discharge diagnoses of concern in biosurveillance. Physicians then reviewed ED reports for each of the cases to finalize a reference syndrome assignment. Using ICD-9 codes to select patients made it possible to use chart review on a fairly small sample of patients while still acquiring a reasonably sized set of patients for seven different syndromes.

An important issue is whether the same classification accuracy observed in a study of chief complaints from hospital X will be observed for chief complaints from hospital Y. Levy et al. (2005) showed that classification accuracy of a keyword-based parser differed from hospital to hospital for *gastrointestinal* syndrome. Chapman et al. (2005b), however, showed that the classification accuracy of a Bayesian chief complaint classifier was no different when it was used on a set of chief complaints from a geographic region other than the one that it had been trained on.

There are a number of studies in the literature that we did not include in Table 23.6 because they measured the sensitivity and specificity of an NLP program's syndrome assignment *relative to a physician who is classifying a patient only from the chief complaint* (Chapman et al., 2005a, Olszewski, 2003a, Sniegoski, 2004). These studies report much higher sensitivities and specificities than those in Table 23.6. These studies represent formative studies of NLP algorithms. The accuracy of syndrome classification should always be measured relative to the actual syndrome of the patient as determined by a method at least as rigorous as medical record review or discharge diagnoses when accepting or rejecting Hypothesis 1 for a syndrome under study.

In summary, the experiments in Table 23.6, although somewhat heterogeneous methodologically, are similar enough to be considered meta-analytically. They made the same measurements (sensitivity and specificity), studied similar syndromes, used simple techniques for classifying chief complaints into syndromic categories, and used similar gold standards.

With respect to Hypothesis 1, these experiments demonstrate that:

1. Chief complaint data contain information about syndromic presentations of patients and various NLP techniques including a naïve Bayesian classifier and keyword methods can extract that information.
2. For syndromes that are at the level of diagnostic precision of *respiratory* or *gastrointestinal* it is possible to automatically classify ED patients (both pediatric and adult) from chief complaints with a sensitivity of approximately 0.60 and a specificity of approximately 0.95.
3. Sensitivity of classification is better for some syndromes than for others.
4. When syndromes are more diagnostically precise (e.g., *respiratory with fever*), the discrimination ability declines quickly.
5. The specificity of syndrome classification from chief complaints is less than 100%, meaning that daily aggregate counts will have false positives among them due to falsely classified patients.[6]

2.3.2. Detection of Outbreaks

This section describes studies that address Hypotheses 2 and 3, which we reproduce here for convenient reference:

Hypothesis 2: When aggregated with the chief complaints of other patients in a region, chief complaints can discriminate between whether there is an *outbreak* of type Y or not.

Hypothesis 3: When aggregated with the chief complaints of other patients in a region, algorithmic monitoring of chief complaints can detect an *outbreak* of type Y earlier than current best practice (or some other reference method).

As discussed in Chapter 20, studies of *outbreak* detection are more difficult to conduct than studies of syndrome (case) detection. These studies require chief complaint data collected from multiple healthcare facilities in a region affected by an outbreak. Research groups often expend significant time and effort building biosurveillance systems to collect the chief complaint data needed to conduct this type of research. Because outbreaks are rare events, many of the studies we will discuss are of common diseases such as *influenza* or of seasonal outbreaks (winter) that may be caused by multiple organisms.

Adequate sample size (of outbreaks) is difficult to achieve in studies of outbreak detection. Only one of the studies computed a confidence interval on its measurements of sensitivity or time of outbreak detection. The scientific importance of adequate sample size cannot be overstated. There are two possible approaches to increasing sample size: (1) conduct research in biosurveillance systems that span sufficiently large geographic regions that they are expected to encounter sufficient numbers of outbreaks, or (2) meta-analysis. To emphasize the importance of adequate sample size, we divide this section into studies of multiple outbreaks (labeled N>1), prospective studies, and studies of single outbreaks (N=1).

N>1 Studies. Ivanov and colleagues used the detection-algorithm method and correlation analysis to study detection of six seasonal outbreaks in children from CoCo classifications of patients into *respiratory* and *gastrointestinal* based on chief complaints (Ivanov et al., 2003). They studied the daily visits to ED of a pediatric hospital during annual winter outbreaks due to diseases such as rotavirus, gastroenteritis and influenza for the four-year period 1998–2001.

The researchers identified outbreaks for study using the following procedure: They created two ICD-9 code sets, corresponding to infectious diseases of children that are respiratory or gastrointestinal, and used them to create two reference time series from ICD9-coded hospital discharge diagnoses. Figure 23.3 (top) from the publication is a plot of the daily *respiratory* time series and the reference time series of respiratory disease created from hospital discharge diagnoses of children under age five. Figure 23.3 (bottom) is a similar comparison of *gastrointestinal* time series and the reference time series of infectious gastrointestinal conditions.

The detection algorithm (exponentially weighted moving average) identified three respiratory and three gastrointestinal outbreaks in the hospital discharge data (the reference standard

6 Even were the specificity to be 100%, the diagnostic precision of the syndrome categories studied is low. As a result, the syndromes used in research have many causes other than acute infectious diseases, which are the diseases of interest in biosurveillance. When monitoring aggregate daily counts, even a syndrome with 100% specificity will show baseline levels of counts in the absence of an outbreak due to the presence of these chronic and sporadic conditions.

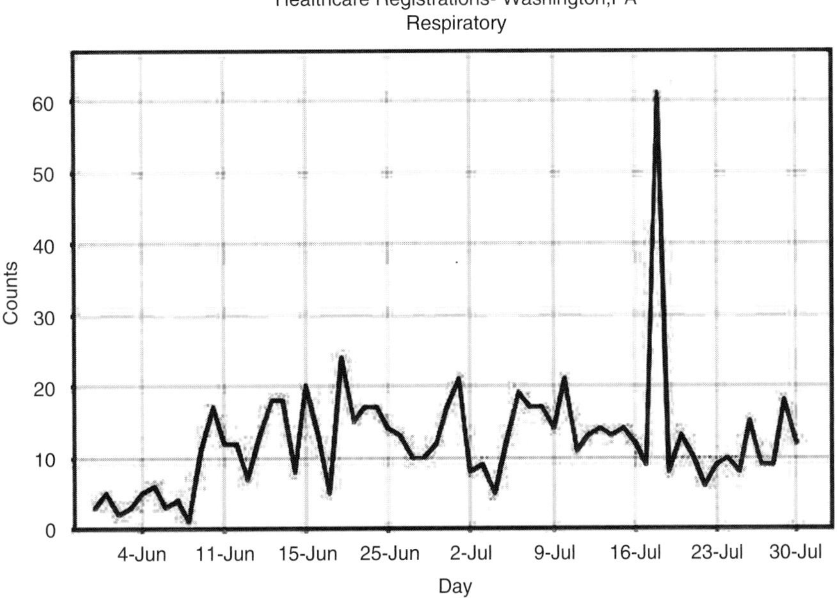

FIGURE 23.4 Daily counts of *respiratory* cases, Washington County, Pennsylvania, June–July 2003. The small increase in early June 2003 corresponds to new hospitals being added to the surveillance system.

for outbreaks). The detection from chief complaints preceded detection from automatic analysis of hospital discharge diagnoses by a mean of 10.3 days (95% confidence interval [CI], 15.15–35.5) for respiratory outbreaks and 29 days (95% CI, 4.23–53.7) for gastrointestinal outbreaks (Table 23.7). The researchers used the date of admission rather than the date of discharge in constructing the reference time series.

The correlation analysis of three respiratory outbreaks showed that on average the chief complaint time series was 7.4 days earlier (95% CI, 8.34–43.3), although the 95% CI included zero. For the three gastrointestinal outbreaks, the chief complaint time series was 17.6 days earlier (95% CI, 3.4–46.7).

Prospective Studies. Prospective studies are field evaluations of a biosurveillance system. In a prospective evaluation, the detection algorithms are operated at a fixed detection threshold for an extended period and the ability of the biosurveillance system to detect known outbreaks or to identify new outbreaks is measured.

Heffernan et al. (2004b) used the detection-algorithm method prospectively to study *respiratory and fever* syndrome monitoring in New York City (Heffernan et al., 2004b). They studied the New York City Department of Health and Mental Hygiene (DOHMH) syndromic system for the one-year period November 2001-November 2002. Note they also report the DOHMH one-year experience monitoring *diarrhea* and *vomiting*, however, the paper by Balter, which we discuss next, included that year in a three-year analysis, so we do not discuss it here.

In New York City, EDs transmit chief complaints to the DOHMH on a daily basis as email attachments or via FTP. The researchers estimated that the DOHMH system received chief complaint data for approximately 75% of ED visits in New York City. The NLP program was a keyword-based system that assigned each patient to exactly one syndrome from the set: *common cold*, *sepsis/dead on arrival*, *respiratory*, *diarrhea*, *fever*, *rash*, *asthma*, *vomiting*, and *other* (Table 23.4). The NLP program was greedy, which means that the algorithm assigned a patient to the first syndrome from the list of syndromes whose definition was satisfied and did not attempt further assignment.

DOHMH used the detection-algorithm method to identify potential outbreaks from daily counts of *respiratory* and *fever*. They used a univariate detection algorithm on data aggregated for the entire city (*citywide*), and spatial scanning for data aggregated by patient home zip code and by hospital (separate analyses).

The citywide monitoring of *respiratory* found 22 above-threshold anomalies (called *signals*), of which the researchers

TABLE 23.7 Detection Algorithm Analysis of Timeliness of Detection from Chief Complaints

Syndrome	Gold Standard Outbreak	Sensitivity	Specificity	Timeliness (95% CI)
Respiratory	Seasonal outbreaks of pediatric respiratory illness (bronchiolitis, P&I)	100%	100%	10 days (−15–35)
Gastrointestinal	Seasonal outbreaks of pediatric gastrointestinal illness (rotavirus gastroenteritis)	100%	100%	29 days (4–53)

CI, confidence interval; P&I, pneumonia and influenza.

stated that 14 (64%) *occurred during periods of peak influenza activity*. The first citywide signal occurred in December 2001 and it was followed by additional signals in both *respiratory* and *fever* signals on the six successive days. The authors commented that these signals coincided with a sharp increase in positive influenza test results, but did not report a correlation analysis. They also commented that the reports of influenza-like illness (ILI) from the existing sentinel physician ILI system showed increases three weeks after the first signal. Three other *respiratory* signals occurred during periods of known increases in asthma activity. The remaining five signals occurred during periods of increasing visits for respiratory disease. Thus, there were no signals that could not be attributed to known disease activity.

The citywide monitoring of *fever* generated 22 signals, of which 21 (95%) occurred during periods of peak influenza activity.

The hospital monitoring of *respiratory* and *fever* produced 25 signals. The home zip code monitoring of these two syndromes produced 18 signals. Investigations of these 43 (25+18) signals found no associated increases in disease activity.

Balter and colleagues analyzed the DOHMH three-year experience (November 2001–August 2004) monitoring *diarrhea* and *vomiting* using the same biosurveillance system as described in the previous paragraphs (Balter et al., 2005). The authors estimate that by the end of the study period, the monitoring system received data for approximately 90% of ED visits in New York City.

During the three years, the DOHMH system signaled 236 times (98 citywide and 138 hospital or zip code) for *diarrhea* or *vomiting*. Of 98 citywide signals, 73 (75%) occurred during what the authors referred to as "seasonal" outbreaks likely due to norovirus (fall and winter) and rotavirus (spring). One citywide signal after the August 2003 blackout was *believed to have represented a true increase in diarrheal illness*. Their investigations of the 138 hospital or zip code signals found no increased disease activity.

During the same period, DOHMH investigated 49 GI outbreaks involving ten or more cases; none of which were detected by monitoring of *diarrhea* or *vomiting*. In 36 of these outbreaks, few or no patients went to EDs. In two outbreaks, the victims were visitors to New York City who returned to their homes before onset of symptoms. In three outbreaks, victims visited EDs not participating in the monitoring system. In three outbreaks, victims visited EDs over a "days or weeks" (the algorithms used by DOHMH were sensitive to rapid increases, not gradual increases in daily counts of syndromes). In two outbreaks, the victims presented to the ED as a group and their chief complaints were recorded by reference to the group (e.g., "school incident"). In two outbreaks, *a combination of the above causes explained the failure*.

N=1 Studies. Irvin and colleagues (Irvin et al., 2003) used the detection-algorithm method to retrospectively study the ability of their *anthrax* syndrome to detect a single influenza outbreak. The paper is not explicit about the *anthrax* syndrome, but states, "The presence of any of the following symptoms were sufficient to categorize a patient into *anthrax*: cough, dyspnea, fever, lethargy, pleuritic chest pain, vomiting, generalized abdominal pain, or headache," suggesting that the researcher included symptoms with which pulmonary anthrax may present. They studied an atypical monitoring system based on numeric chief-complaint codes from a commercial ED charting system. This charting system, called E/Map (Lynx Medical Systems, Bellevue WA, *http://www.lynxmed.com*), offers clinicians charting templates for approximately 800 chief complaints. Each template has a numerical code. A clinician's selection of charting template reflects the patient's chief complaint. The detection algorithm used a fixed detection threshold set at two standard deviations from a recent two-month mean. The algorithm signaled when two of the previous three days exceeded the threshold. The reference standard was the Centers for Disease Control and Prevention (CDC) defined peak week of influenza activity. The system signaled one week prior to the CDC peak and signaled one false positive.

Yuan et al. (2004) used the detection-algorithm method to study the timeliness of detection of one influenza outbreak in southeastern Virginia. They manually assigned chief complaints to seven syndromes (*fever*, *respiratory distress*, *vomiting*, *diarrhea*, *rash*, *disorientation*, and *sepsis*). The detection algorithm was CUSUM, operated at three different moving averages (7-day window, window days 3–9, and 3-day window) and set at a threshold of 3 S.D. They reported that the CUSUM algorithm detected trends in *fever* and *respiratory* in one hospital that preceded the local sentinel influenza surveillance system by one week.

A key limitation of N=1 studies is that any correlation found may be spurious. Meta analysis could address this problem if differences among analytic and reporting methods used by studies were reduced so that studies of single outbreaks could be merged analytically. In 2003, the RODS Laboratory developed a case-study series that encourages the use of a standard method of studying single outbreaks that would enable the application of uniform analytic methods across outbreaks (or alerts) occurring in different regions (Rizzo et al., 2005). The objectives of the case report series are to: (1) ensure complete description of outbreak and analytic methods, and (2) collect the raw surveillance data and information about the outbreak in a way that future re-analyses are possible.

Each case study describes the effect of a single outbreak or other public health event, such as low air quality due to forest fires, on surveillance of data available for the event. At present, these case studies are available only to authorized public health users of the NRDM system because of legal agreements with organizations that provide surveillance data (employees of governmental public health organizations can access case studies through the RODS interface by sending e-mail to nrdmaccounts@cbmi.pitt.edu).

Of the 15 case studies developed to date, eight are examples of outbreaks considered "detectable" from available surveillance data; six were not detectable. These case studies include outbreaks of influenza, salmonella, norovirus, rotavirus, shigella, and hepatitis A. One case study describes a false alarm investigation that resulted from a retailer recording error.

Figure 23.4 is taken from a case study of a large spike in CoCo *respiratory* cases in a single county outside Pittsburgh that resulted in an alert being sent automatically on Friday July 18, 2003 at 8 PM to an on-call epidemiologist. Normally, daily counts of respiratory cases numbered 10, but on that day they numbered 60 by 8 PM. The epidemiologist logged into the RODS web interface, reviewed the verbatim chief complaints of affected patients and discovered that the cases were related to carbon-monoxide exposure, which a phone call to an ED revealed to be related to a faulty furnace at a day-care center.

The case studies include three studies of the effect of influenza on emergency room visits for the CoCo *constitutional* and *respiratory* syndromes. Figure 23.5 illustrates the size of the influenza effect in 2003–2004 in Utah (middle spike) on *constitutional*, and *respiratory*, as well as sales of thermometers by pharmacies participating in the National Retail Data Monitor.

These case studies add to the previously described studies the following: Influenza has a strong early effect on free text chief complaints in the *constitutional* and *respiratory* categories. Air pollution and small carbon monoxide events may have marked effects on chief complaints in the *respiratory* category. The results for gastrointestinal outbreaks have been negative for relatively small, protracted outbreaks of Norovirus and Shigella.

Summary of Studies of Outbreak Detection from Chief Complaints. With respect to Hypotheses 2 and 3, the studies we reviewed demonstrate that:

1. Some large outbreaks causing respiratory, constitutional, or gastrointestinal symptoms can be detected from aggregate analysis of chief complaints. Small outbreaks of gastrointestinal illness generally cannot (Hypothesis 2).
2. Research to date is suggestive but not conclusive that influenza can be detected earlier by chief complaint monitoring than current best practice (Hypothesis 3).
3. The false-alarm rates associated with such monitoring can be low. In New York City for city-wide monitoring of *respiratory* and *fever*, there were few signals that did not correspond to disease activity. There were more signals that did not correspond to disease activity from monitoring of *diarrhea* and *vomiting*. Conversely, all of the signals from spatial monitoring of hospital or zip code were not correlated with known disease activity.
4. The methodological weaknesses in the studies included failure to describe or measure time latencies involved in data collection. Some studies did not report sampling completeness, the method by which chief complaints are parsed, or details of the syndrome categories.
5. In general, the number of published studies is small, perhaps due to the fact that chief complaint monitoring systems are still being constructed. We expect more studies to be published in the near future.

The answers to Hypotheses 2 and 3 for surveillance of chief complaints in isolation may not be as important long term as the question of whether chief complaints contain diagnostic information (Hypothesis 1). The reason is that chief complaints can be used with other surveillance data to detect outbreaks, either through linking at the level of the individual patient or as a second source of evidence. Nevertheless, because of their availability and earliness, and the threat of bioterrorism and large outbreaks, it is important to understand the ability to detect outbreaks solely from this type of data.

3. ICD CODES

The International Classification of Diseases, 9th Revision, Clinical Modification (ICD) is a standard vocabulary for diagnoses, symptoms, and syndromes (see Chapter 32). ICD has a code for each class of diagnoses, syndromes, and symptoms that it covers. For example, the ICD code 034.0 *Streptococcal sore throat* includes tonsillitis, pharyngitis, and laryngitis caused by any species of *Streptococcus* bacteria. There are more than

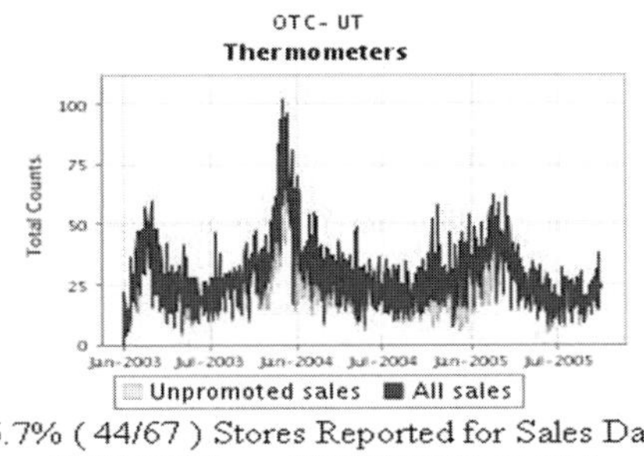

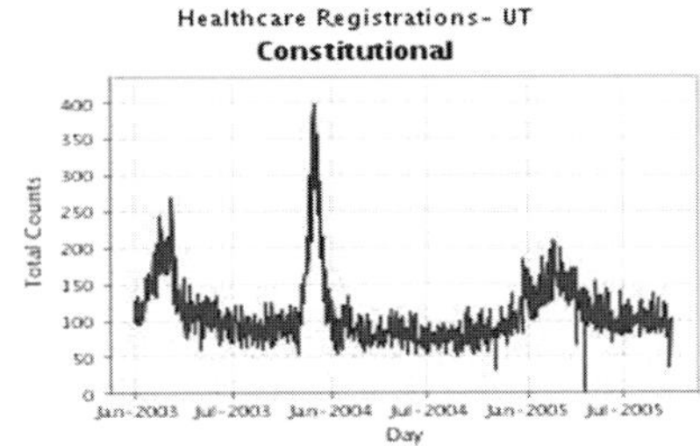

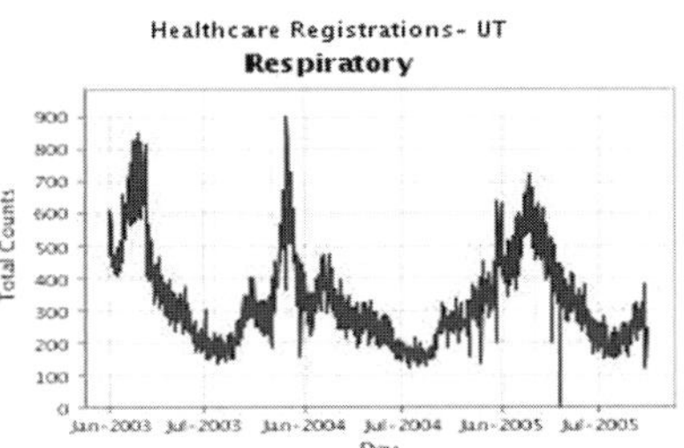

FIGURE 23.5 Daily counts of *constitutional* and *respiratory* syndrome. January 2003–September 2005. The largest spikes correspond to the 2003–2004 influenza outbreak, which was more severe (involving more people) than the previous or following year's outbreak.

12,000 ICD codes. Internationally, some countries use the 10th revision of the International Classification of Diseases, or a modification of it.

Data encoded using ICD are widely available in the United States. Most visits to physicians or other healthcare providers and hospitalizations result in one or more ICD codes. The reason is that healthcare insurance corporations require providers of care to use ICD codes when submitting insurance claims to receive reimbursement for their services.

ICD codes range in diagnostic precision from the very imprecise level of symptoms to very precise diagnoses. There are precise codes for infectious diseases, specifying both the causative organism and the disease process (e.g., 481 *Pneumococcal pneumonia*). However, there are less precise codes that providers can use if the organism is unknown or not documented (e.g., 486 *Pneumonia, organism unspecified*). There are also ICD codes for syndromes (e.g., 079.99 is the code for *Viral syndrome*) and even for symptoms (e.g., 786.2 *Cough* and 780.6 *Fever*).

ICD codes may be assigned at different times during the course of care (Figure 23.1). As you go from left to right in Figure 23.1, who assigns the ICD code, how, and when vary. Physicians, when they do assign ICD codes to office or ED visits, usually do so during or within hours to days of the visit. They either enter ICD codes into a point-of-care system or record them on an encounter form (also sometimes known as a "superbill"). Professional coders usually assign the final, billing ICD codes for ED and office visits days later. They also assign ICD codes to hospital discharge diagnoses, typically days to weeks after the patient leaves the hospital. Professional coders often enter ICD codes into specialized billing software. ICD-coded data from organizations that collect large volumes of insurance claims data (we discuss these "data aggregators" in more detail below) are usually not available for months after visits or hospital stays.

The diagnostic precision of ICD codes generally increases with time, as you go from left to right in Figure 23.1. The reason that discharge diagnoses generally have higher diagnostic precision relative to visit diagnoses is that discharge diagnoses typically represent the outcome of a greater amount of diagnostic testing that leads to greater diagnostic certainty (i.e., providers are more likely to order—and have the results available from—laboratory tests, microbiology cultures, x-rays, and so on).

Health services researchers have established that the accuracy of ICD-coded data is highly variable and often only moderately high (O'Malley K et al., 2005, Hsia et al., 1988). They have identified several causes for inaccuracy (Peabody et al., 2004, O'Malley et al., 2005). One cause is that two different, highly trained, experienced coders may assign different codes to the same hospitalization (Fisher et al., 1992, Lloyd and Rissing, 1985, MacIntyre et al., 1997). One reason is that coders work from the patient chart, which is an imperfect representation of the patient's true medical history and is subject to variable interpretation. Professional coders are typically not physicians or nurses, so their level of understanding of the medical process is imperfect. Finally, the rules for assigning codes are complex and change at least annually.[7]

The problem of correct assignment of ICD codes is compounded when clinicians encode the diagnoses (Yao et al., 1999). Clinicians rarely have formal training in the rules for assigning codes. They typically have little time to ensure that the codes they assign are accurate. They often view the assignment of ICD codes to patient encounters as a distraction from patient care. To address these problems, physicians often use preprinted encounter forms that have check boxes for an extremely small subset of commonly used codes. Although these forms typically include a blank space to write in additional ICD codes, clinicians are extremely busy so an open question is how often they use the space and how accurate are the ICD codes that are hand entered. Another question is whether busy clinicians, who do not use the data on the encounter form for subsequent patient care, completely code all diagnoses made during a patient visit. One study found that, during patient visits, physicians addressed an average 3.05 patient problems but documented only 1.97 on billing forms (they documented nearly as many problems in the paper record as they addressed) (Beasley et al., 2004).

ICD codes, because of their inaccuracy and the fact that their primary use and purpose is billing, are likely to be less than ideal when used for other purposes. One study found low accuracy of billing data about cardiac diseases relative to a clinical research database (Jollis et al., 1993). Another study found that one-third of patients who received an ICD code that indicated the presence of a notifiable disease did not truly have the notifiable disease (Campos-Outcalt, 1990). A third study found that data about prescriptions identified patients with tuberculosis more accurately than all 60 ICD codes for tuberculosis combined (Yokoe et al., 1999).

ICD codes might be less ideal for biosurveillance than other coding systems such as SNOMED-CT. The designers of ICD did not design it with biosurveillance requirements in mind. One study found that SNOMED-CT was superior to ICD for coding ED chief complaints (McClay and Campbell, 2002). SNOMED-CT had a term that was a precise match for 93% of chief complaints; ICD had a precise match for only 40% of chief complaints.

In summary, billing ICD codes from insurance claims and hospital discharge data sets are widely available, but at long time latencies (weeks to months). ICD codes at shorter time latencies ICD (within 24 hours of ED or office visit) are less available. Who assigns ICD codes and when and how influence the time latency, diagnostic precision, and accuracy of ICD-coded data. Thus, it is essential that studies describe the process that generated the ICD codes and measure time latency.

7 Hence the need for professionals to do the coding in the first place.

3.1. Categories of ICD Codes ("Code Sets")

Despite the potential for high diagnostic precision, biosurveillance researchers and developers group ICD codes into categories ("code sets") such as "respiratory illness." The set of all 60 ICD codes for tuberculosis mentioned above is an example of an ICD code set. It is not necessary to group ICD codes into codes sets, although it is almost always done. For example, we could monitor for the single ICD code for inhalational anthrax (022.1). Creators of code sets usually group codes of diseases and syndromes that share similar early presentations to form syndrome code sets. *Respiratory*, *gastrointestinal*, *neurological*, *rash* and *febrile illnesses* are representative of code sets in common use.

A key reason that developers create code sets is to improve the sensitivity of case detection, because patients with the same disease may be assigned different ICD codes. This variability may be due to variability in how coders assign codes or that the patients are at different stages in their diagnostic work-ups. For example, a patient with influenza who has not yet undergone definitive testing may be coded as 780.6 *Fever* (or any of a number of other ICD codes for symptoms of influenza), 079.99 *viral syndrome*, 465.9 *Acute upper respiratory infection NOS*, or 486 *Pneumonia, organism unspecified*.

The most difficult and "art-more-than-science" aspect of ICD-code monitoring is development of code sets. The next sections describe several code sets used in biosurveillance. Our purpose is to illustrate how a code set is developed. It is important to note that one code set could be superior (i.e., have better case detection and/or outbreak detection performance) to others for the detection of one disease (e.g., influenza), but inferior to other code sets for the detection of another disease (e.g., bronchiolitis due to respiratory syncytial virus). To date, only one study (Tsui et al., 2001) has compared the accuracy of two alternative code sets to determine their differential ability to detect the same set of cases or outbreaks. That study lacks generalizability because no other research groups have found the data they studied—ICD codes for chief complaints assigned by registration clerks—to be available at other institutions.[8] Therefore, which code sets are better than others for the detection of various outbreaks such as influenza or cryptosporidiosis remains unknown.

3.1.1. ESSENCE Code Sets

The Department of Defense (DoD) developed the first ICD code sets for use in biosurveillance as part of its Global Emerging Infections System (DoD-GEIS). The ESSENCE biosurveillance system[9] uses these code sets (DoD-GEIS, 2005) to aggregate ambulatory visits at DoD outpatient clinics into seven syndromes (Table 23.8). A number of researchers and system developers have used these code sets, or slight modifications of them, in their work (Lazarus et al., 2002, Reis and Mandl, 2004, Lazarus et al., 2001, Lewis et al., 2002, Mocny et al., 2003, Magruder et al., 2005, Buckeridge et al., 2005, Henry et al., 2004).

TABLE 23.8 The Seven ESSENCE ICD Code Sets

Code Set	Number of ICD Codes in Set
Coma/Shock/Death	10
Dermatologic–Hemorrhagic	9
Dermatologic–Infectious	18
Fever/Malaise/Sepsis	26
Gastrointestinal	118
Neurologic	72
Respiratory	175

Note: The code sets are available at the U.S. Department of Defense, Global Emerging Infections System website, *http://www.geis.fhp.osd.mil/GEIS/SurveillanceActivities/ESSENCE/ESSENCE.asp#ICD9*.

3.1.2. CDC/DoD Categories

Representatives from the CDC and other stakeholders formed a working group to develop a suggested list of ICD code sets (CDC, 2003). These representatives reviewed all 12,000 ICD codes for possible inclusion in one of twelve ICD code sets (Table 23.9). They also looked at the frequency of code usage in both a DoD outpatient visit data set and a civilian ED data set. Based on the frequency of code usage and their knowledge of symptoms associated with diseases represented by ICD codes, they assigned ICD codes to syndrome code sets. They further specified three levels of membership in the categories, which they somewhat confusingly labeled *categories 1, 2*, and *3*:

> *Category 1: …codes that reflect general symptoms of the syndrome group and also … the bioterrorism diseases of highest concern or those diseases highly approximating them.*

TABLE 23.9 CDC/DoD ICD Code Sets

Code Set	Number of ICD Codes in Set
Botulism-like	41
Fever	87
Gastrointestinal, All	125
Gastrointestinal, Lower	108
Gastrointestinal, Upper	17
Hemorrhagic Illness	33
Lesion	71
Lymphadenitis	26
Neurological	95
Rash	44
Respiratory	171
Severe Illness or Death	14

Note: The code sets are available at the Centers for Disease Control and Prevention website, *http://www.bt.cdc.gov/surveillance/syndromedef/index.asp*. DoD, U.S. Department of Defense.

8 And even the health system that Tsui and colleagues studied no longer collects this data.

9 ESSENCE was first part of DoD-GEIS and now civilian health departments in a number of jurisdictions use it also.

Category 2: ...codes that might normally be placed in the syndrome group, but daily volume could overwhelm or otherwise detract from the signal generated from the Category 1 code set alone.

Category 3: ...specific diagnoses that fit into the syndrome category but occur infrequently or have very few counts. These codes may be excluded to simplify syndrome category code sets.

The end result was 12 ICD code sets, each with three categories of codes. The code sets are not mutually exclusive (some ICD codes appear in more than one code set). The report of the working group does not describe the procedure they followed to reach a consensus on—or resolve conflicts over—which ICD codes to include in a code set.

This effort made use of intensional definitions for each code set (Table 23.10). The extensional definition of the *Fever* ICD code set is given in Table 23.11.

As with the ESSENCE code sets, researchers and system developers have used the CDC/DoD code sets in their work (Yih et al., 2004, 2005).

3.1.3. Other Code Sets

No other code sets other than the DoD-GEIS and CDC/DoD code sets appear to be in use in a functioning biosurveillance system. At present, DoD-GEIS code sets are in use in the ESSENCE biosurveillance system (Lombardo et al., 2003) and the CDC/DoD code sets are in use in the National Bioterrorism Syndromic Surveillance Demonstration Program (Yih et al., 2004). The other code sets that researchers describe in the literature were created for research studies.

TABLE 23.10 Two Examples of Intensional Definitions of CDC/DoD ICD Code Sets

Syndrome	Definition	Category A Condition
Botulism-like	ACUTE condition that may represent exposure to botulinum toxin ACUTE paralytic conditions consistent with botulism: cranial nerve VI (lateral rectus) palsy, ptosis, dilated pupils, decreased gag reflex, media rectus palsy ACUTE descending motor paralysis (including muscles of respiration) ACUTE symptoms consistent with botulism: diplopia, dry mouth, dysphagia, difficulty focusing to a near point	Botulism
Fever	ACUTE potentially febrile illness of origin not specified INCLUDES fever and septicemia not otherwise specified INCLUDES unspecified viral illness even though unknown if fever is present EXCLUDE entry in this syndrome category if more specific diagnostic code is present allowing same patient visit to be categorized as respiratory, neurological or gastrointestinal illness syndrome	Not applicable

CDC, Centers for Disease Control and Prevention; DoD, U.S. Department of Defense.

For example, Lazarus et al. (2002) used a slight modification of ESSENCE code sets and also created an *influenza-like illness* (*ILI*) code set. They based this code set on the CDC sentinel surveillance definition of *ILI*, which is fever (body temperature measured in the office >37.8°C) plus cough or sore throat without a known cause. Miller and colleagues created an *ILI* code set that includes 31 ICD codes (Miller et al., 2004). Most of the codes in the set are symptoms of influenza. Ritzwoller et al. (2005) created an *ILI* code set. This code set included codes for respiratory illness plus fever (780.6). Espino and Wagner (2001a) created a *respiratory* code set that includes 64 ICD codes. To create this code set, experts reviewed all ICD codes assigned as ED chief complaints over a three-year period. Tsui et al. (2001) created an *influenza* code set from this *respiratory* code set and viral illness ICD codes (Tsui et al., 2001). Ivanov created a *gastrointestinal* code set that includes 16 ICD codes (Ivanov et al., 2002). As with Espino and Wagner (2001a), experts reviewed all ICD codes assigned as ED chief complaints over a three-year period. Beitel et al. (2004) created *respiratory*, *lower respiratory*, and *upper respiratory* code sets. To create the *respiratory* code set, they merged ICD codes 460.00–519.xx (encompasses all "diseases of the respiratory system"), 786.xx (encompasses all "symptoms involving the respiratory system and other chest symptoms"), and the ESSENCE *respiratory* code set except ICD code 079.99 (*Viral syndrome*). They then divided *respiratory* into *lower respiratory* and *upper respiratory* based on the part of the respiratory system implied by the code. When a code did not imply any anatomy or implied anatomy that spanned the upper and lower parts of the respiratory system, they added it to the *upper respiratory* set.

3.1.4. Standardizing Code Sets

We note that some authors have referred to either the DoD-GEIS or the CDC/DoD code sets as "standard" ICD code sets (Heffernan et al., 2004a, Mandl et al., 2004). However it would be extremely premature to standardize on these code sets other than for research reporting purposes. Given the variability in diseases, ICD assignment by clinicians and coders, and diagnostic precision, it is unlikely that a one-size-fits-all ICD code set will be optimal for all settings and purposes. Moreover, it is trivial for computer systems to aggregate raw ICD-coded data using any desired code set in real time and as needed.[10]

We note that identifying an optimal code set is not trivial. A computer scientist would, in fact, call it an intractable problem, which means that it cannot be done. The reason: there are

10 A single microchip cannibalized from a discarded DVD player likely has sufficient processing power to keep up with all the ICD-to-code set translations required for biosurveillance by the entire world.

TABLE 23.11 Extensional Definition of the CDC/DoD *Fever* ICD Code Set

ICD9CM	ICD9DESCR	Consensus	ICD9CM	ICD9DESCR	Consensus
020.2	PLAGUE, SEPTICEMIC	1	054.5	HERPETIC SEPTICEMIA	3
020.8	OTHER TYPES OF PLAGUE	1	060.0	YELLOW FEVER, SYLVATIC	3
020.9	PLAGUE NOS	1	060.1	YELLOW FEVER, URBAN	3
021.8	TULAREMIA NEC	1	060.9	YELLOW FEVER, UNSPEC	3
021.9	TULAREMIA NOS	1	066.0	PHLEBOTOMUS FEVER	3
022.3	ANTHRAX, SEPTICEMIA	1	066.1	TICK-BORNE FEVER	3
022.8	ANTHRAX, OTHER SPECIFIED	1	066.2	VENEZUELAN EQUINE FEVER	3
022.9	ANTHRAX, UNSPECIFIED	1	066.3	MOSQUITO-BORNE FEVER NEC	3
038.3	ANAEROBES SEPTICEMIA	1	066.8	ARTHROPOD VIRUS NEC	3
038.40	GRAM-NEGATIVE ORGANISM UN	1	066.9	ARTHROPOD VIRUS NOS	3
038.49	SEPTICEMIA, OTHER GRAM-NEG	1	078.2	SWEATING FEVER	3
038.8	SEPTICEMIAS, OTHER SPECIF	1	080	LOUSE-BORNE TYPHUS	3
038.9	SEPTICEMIA, NOS	1	081.0	MURINE TYPHUS	3
079.89	VIRAL INFECTION OTHER S	1	081.1	BRILL'S DISEASE	3
079.99	VIRAL INFECTIONS UNSPECIFIED	1	081.2	SCRUB TYPHUS	3
780.31	FEBRILE CONVULSIONS	1	081.9	TYPHUS NOS	3
780.6	FEVER	1	082.8	TICK-BORNE RICKETTS NEC	3
790.7	BACTEREMIA	1	082.9	TICK-BORNE RICKETTS NOS	3
790.8	VIREMIA NOS	1	083.0	Q FEVER	3
002.0	TYPHOID FEVER	3	083.1	TRENCH FEVER	3
002.1	PARATYPHOID FEVER A	3	083.2	RICKETTSIALPOX	3
002.2	PARATYPHOID FEVER B	3	083.8	RICKETTSIOSES NEC	3
002.3	PARATYPHOID FEVER C	3	083.9	RICKETTSIOSIS	3
002.9	PARATYPHOID FEVER NOS	3	084.0	MALARIA, FALCIPARUM	3
003.1	SALMONELLA SEPTICEMIA	3	084.1	VIVAX MALARIA	3
023.0	BRUCELLA MELITENSIS	3	084.2	QUARTAN MALARIA	3
023.1	BRUCELLA ABORTUS	3	084.3	OVALE MALARIA	3
023.2	BRUCELLA SUIS	3	084.5	MIXED MALARIA	3
023.3	BRUCELLA CANIS	3	084.6	MALARIA UNSPECIFIED	3
023.8	BRUCELLOSIS NEC	3	086.2	CHAGA'S DISEASE WITHOUT MENTION OF ORG	3
023.9	BRUCELLOSIS, UNSPECIFIED	3			
024	GLANDERS	3	086.3	TRYPANOSOMIASIS, GAMBIAN	3
025	MELOIDOSIS	3	086.4	TRYPANOSOMIASIS, RHODESIAN	3
027.0	LISTERIOSIS	3	086.5	TRYPANOSOMIASIS, AFRICAN	3
034.1	SCARLET FEVER	3	086.9	TRYPANOSOMIASIS, UNSPEC	3
038.0	SEPTICEMIA STAPHYLOCOCCAL	3	087.0	LOUSE-BORNE RELAPS FEVER	3
038.10	SEPTICEMIA STAPHYLOCOCCAL	3	087.1	TICK-BORNE RELAPS FEVER	3
038.11	SEPTICEMIA STAPHYLOC. AUR	3	087.9	RELAPSING FEVER NOS	3
038.19	SEPTICEMIA STAPHYLOCOCCAL	3	088.0	BARTONELLOSIS	3
038.2	PNEUMOCOCCAL SEPTICEMIA	3	088.81	LYME DISEASE	3
038.41	SEPTICEMIA, HEMOPHILUS INFLUENZAE	3	088.82	BABESIOSIS	3
038.42	SEPTICEMIA, E. COLI	3	088.89	OTHER ARTHROPOD-BORNE	3
038.43	PSEUDOMONAS	3	088.9	ARTHROPOD-BORNE DIS NOS	3
038.44	SEPTICEMIA SERRATIA	3	100.82	LEPTOSPIROSIS, OTHER	3

CDC, Centers for Disease Control and Prevention; DoD, U.S. Department of Defense.

approximately 12,000 ICD-9 codes and therefore $2^{12,000}$ possible ICD code sets. This number is far greater than the total number of atoms in the universe or the number of milliseconds since the beginning of time. Thus, it is not possible for a computer, let alone a human, to try every code set to find the best one. That is not to say that it is not possible to determine whether Code Set A is superior to Code Set B for some specific biosurveillance application, just that it is not possible to establish that Code Set A is superior to every other code set for all ICD-coded data.

3.2. Availability and Time Latencies of ICD-coded Diagnoses

The availability of ICD codes, by which we mean the proportion of patients being seen in a region for which ICD codes are available, in general is poorly understood. ICD coding of chief complaints is uncommon. ICD coding of ED diagnoses by clinicians using clinical information systems is not universal. Nationwide, only 17% of physician practices and 31% of EDs have even adopted point-of-care systems (Burt and Hing, 2005). ICD codes from healthcare-insurance claims data are widely available, but the time latency is too long for many biosurveillance applications.

We discuss the empirical data about time latency from published reports shortly. But it is worthwhile to review the steps in the process of assigning ICD codes and transmitting them to a biosurveillance system (Figure 23.1), and how these steps may contribute to time latency. First, an individual clinician or coder must acquire enough information about a patient to assign an ICD code. This individual may be a triage nurse who obtains the patient's chief complaint within minutes of the patient's arrival to the ED, or a professional coder who studies a four-inch thick medical record weeks after the patient left the hospital. Next, someone must enter an ICD code into a computer system.[11] For physician-assigned ICD codes, that may occur within minutes after she sees the patient (which in turn may be minutes to hours after the patient arrived at the office or ED) or even hours or days after the patient leaves the office or ED. Finally, the computer system must transmit the codes to a biosurveillance system or other computers that in turn transmit them to a biosurveillance system. Systems may either transmit ICD codes in real time (the exception) or in batch mode (the rule). Ideally, studies of ICD codes would measure and report the contribution of each step to overall time latency.

Studies to date—with one exception we discuss below—have not *measured* time latencies involved in ICD-code monitoring. Instead, authors provide general statements such as *most records were received by the next day*.

Lewis et al. (2000) report a one- to three-day time latency for diagnoses encoded with ICD by physicians working at DoD clinics. They state that this latency is from the time of the patient visit to the time data are available for viewing and analysis in the ESSENCE biosurveillance system. They note two factors that affect this time latency: time delays prior to physicians assigning codes and frequency of data transmission to ESSENCE from the DoD's Ambulatory Data System.

Note that the CDC's BioSense biosurveillance system receives the same ICD codes that Lewis and colleagues studied. Anecdotally, the time latency of ICD codes obtained by BioSense is not short enough to meet the needs of some state and local public health officials (U.S. Government Accountability Office, 2005). Such anecdotes highlight the importance of accurate *measurements* of each contribution to time latency and detailed descriptions of who assigns ICD codes and how and when.

Reports by the National Bioterrorism Syndromic Surveillance Demonstration Program suggest that the time latency of ICD codes assigned by physicians using a point-of-care system in an outpatient setting are "usually" less than 24 hours (Yih et al., 2004, Lazarus et al., 2002, Miller et al., 2004). This program involves multiple healthcare providers and health plans; thus, each healthcare system may have different latencies. Lazarus and colleagues report that at Harvard Vanguard Medical Associates in Boston, Massachusetts, ICD-coded diagnoses are available for "essentially all episodes by the end of the same day on which care is given" (Lazarus et al., 2002). Miller and colleagues report that at HealthPartners Medical Group in the Minneapolis-St. Paul region of Minnesota, ICD-coded diagnoses are available "within approximately 24 hours of a patient's initial visit" (Miller et al., 2004). Yih and colleagues report that, for the Demonstration Program as a whole, providers usually enter ICD-coded diagnoses into a point-of-care system on the same day as the patient visit, and that the systems put in place by the Program extract these data on a nightly basis (Yih et al., 2004).

Beitel and colleagues report that ED physicians at Boston Children's' Hospital code diagnoses with ICD within "hours" of the visit (Beitel et al., 2004). They do not describe how available these data are for biosurveillance nor whether they are transmitted in real time or via a daily batch-mode process. They also report that billing administrators assign "the final ICD-9 code to all charts, usually within 48 to 72 hours."

Suyama and colleagues report that billing ICD codes at the ED of University Hospital (which is part of the University of Cincinnati Medical Center) are available within 12 hours from the time the patient leaves the ED (Suyama et al., 2003). They do not state who assigns the ICD codes to each visit.

Begier and colleagues report that two community hospitals in the National Capital Region do not assign any ICD codes to patient visits within 24 hours of the patient leaving the ED (Begier et al., 2003). Whether ICD codes are available after 24 hours of the patient leaving the ED, they do not say.

Espino and colleagues conducted the only study that measured and reported time latencies for ICD codes (Espino and Wagner, 2001b). It is important to note that this study was of an atypical health system that expended effort to make physician-assigned ICD codes available to a biosurveillance system *in real time*, instead of using daily batch-mode extraction of ICD codes as other studies have described. Espino found that ICD codes for diagnoses (assigned by ED physicians) arrived at the RODS system on average 7.5 hours *after* ICD codes for chief complaints (assigned by ED registration clerks). The maximum time delay was 80.6 hours. Because the ICD codes for chief complaints and diagnoses were transmitted in real time, this measurement of time latency includes only a negligible transmission delay. Thus, nearly all of the 7.5 hours comprises patient waiting time, time for the physician to see the patient, and time for the physician to assign ICD codes to the visit using a point-of-care system.

In summary, a pervasive methodological limitation of reported studies is failure to measure and report the distribution of time latencies between patient presentation to the healthcare system and appearance of an ICD code in data systems. The single study that measured time latency was conducted in a best case—and non representative—situation. In the absence of definitive studies, our best assessment of time latencies and availability—based on the literature and our knowledge of biosurveillance systems and the healthcare system—is summarized in Table 23.12.

11 This someone may or may not be the person who originally elicited the information from the patient. For example, a triage nurse may elicit the chief complaint and write it on paper, and then the registration clerk, using the chief complaint on paper, enters an ICD code into a computer.

TABLE 23.12 Estimated Time Latencies and Availability of ICD Codes Originating from Different Points in Healthcare Process and Different Practice Settings

Point in Process of Care	Latency	Availability	Population Coverage
Registration (chief complaint)	Negligible if sent as HL7 ADT	Very low	Few EDs
Conclusion of ED or office visit (billing diagnosis)	VA: days DoD: days Civilian: days to months	VA: high DoD: high Civilian: unknown	Most EDs and offices
Admission (admission diagnosis)	Similar to ED diagnosis; or months if part of discharge data set	Unknown	All hospitalized patients
Discharge from hospital (discharge diagnosis)	Months	Very high	All hospitalized patients

DoD, U.S. Department of Defense; ED, emergency department; VA, Veterans Administration.

3.3. Studies of Informational Value of ICD Codes

There have been several studies that have measured the informational value of ICD codes for chief complaints and diagnoses. As with free-text chief complaints, these studies utilized experimental methods that we discussed in Chapter 21 and address the three hypotheses of interest (restated for ICD codes):

Hypothesis 1: An individual ICD code or a code set can discriminate between whether a patient has *syndrome* or disease X or not.

Hypothesis 2: ICD code sets can discriminate between whether there is an *outbreak* of type Y or not.

Hypothesis 3: Algorithmic monitoring of ICD code sets can detect an *outbreak* of type Y earlier than current best practice (or some other reference method).

3.3.1. Detection of Cases

Table 23.13 summarizes the results of studies of case detection using only ICD-coded data. The hypothesis underlying these studies is Hypothesis 1: *An individual ICD code or a code set can discriminate between whether a patient has syndrome or disease X or not.* We have grouped the studies in Table 23.13 by syndrome. Each row reports the sensitivity and specificity of an ICD code or code set for a particular syndrome or diagnosis. For convenient reference, we also computed the likelihood ratio positive and negative.

Unlike chief complaints, in which the syndromes studied are at coarse level of diagnostic precision (e.g., *respiratory*), research on ICD code sets has also studied code sets that are diagnostically precise. For this reason, we have grouped the published studies into those using diagnostically less precise and diagnostically more precise code sets.

Detection of Less Diagnostically Precise Syndromes. Four studies measured the case detection accuracy of less diagnostically precise ICD code sets. Only one study measured the case detection accuracy of a DoD-GEIS or CDC/DoD code set.

Espino and Wagner conducted the first study of case detection accuracy of an ICD code set for a diagnostically imprecise syndrome (Espino and Wagner, 2001). They studied the ability of ICD codes for ED chief complaints and ICD codes for ED diagnoses to detect patients with acute respiratory syndrome (defined as symptom duration of five days or less). In this ED, registration clerks assigned ICD codes to the chief complaints and ED physicians assigned ICD codes to the diagnoses at the time of the patient's visit. Two internists created the gold standard by reviewing the dictated ED visit note and assigning patients to *acute respiratory* if the duration of illness was five days or less and the patient had respiratory symptoms, abnormal pulmonary examination, or radiological evidence of pneumonia. The sensitivity and specificity of a *respiratory* ICD code set (that contained 64 ICD codes) for detecting patients with acute respiratory illness from chief complaints were 0.44 and 0.97, respectively. The sensitivity and specificity of the same *respiratory* code set for detecting patients with acute respiratory illness from diagnoses were 0.43 and 0.97, respectively. The accuracy of ED diagnoses was not significantly different from the accuracy of ED chief complaints.

Betancourt (2003) studied the case detection accuracy of three DoD-GEIS ICD code sets. He used manual review of medical records by two primary care physicians as the gold standard. He found that the sensitivity of the *respiratory*, *fever*, and *gastrointestinal* code sets was 0.65, 0.71, and 0.90, respectively. All three code sets had a specificity of 0.94.

Ivanov et al. (2002) studied the case detection accuracy of ICD codes for the detection of acute gastrointestinal illness (duration of illness two weeks or less) in EDs. ED physicians assigned ICD codes for diagnoses after seeing a patient. Review of the transcribed emergency-department report by two internists was the gold standard. As in the study by Espino and Wagner, the internists were instructed to label a patient as acute gastrointestinal if symptoms were of duration two weeks or less. They found that an ICD code set (that contained 16 ICD codes) had a sensitivity and specificity of 0.32 and 0.99, respectively, for the detection of *acute gastrointestinal* syndrome.

Beitel et al. (2004) studied the case detection accuracy of ICD codes for the detection of *respiratory*, *lower respiratory*, and *upper respiratory* syndromes in a pediatric ED. They do not state whether physicians or billing administrators assigned the ICD codes that they studied (but mention that both groups assign ICD codes to ED visits). They used single-physician review of the medical record as the gold standard. The sensitivity and specificity of their *respiratory* ICD code set was 0.70 and 0.98, respectively, for detection of *respiratory* syndrome. Similarly, the sensitivity and specificity of their *upper respiratory* ICD code set was 0.56 and 0.98 for *upper*

TABLE 23.13 Performance of ICD Codes and Code Sets in Detecting Syndromes and Diseases

ICD Code or Code Set Being Tested	Reference Standard for Comparison	Sensitivity	Specificity	Positive Likelihood Ratio	Negative Likelihood Ratio
Respiratory Syndrome					
Respiratory code set, ICD codes for ED chief complaints[a]	Two internist-review of dictated ED note	0.44	0.97	14.7	0.58
Respiratory code set, ICD codes for ED diagnoses[a]	Two internist-review of dictated ED note	0.43	0.97	14.3	0.59
Respiratory code set[b]	Two physician-review of medical record	0.65	0.94	10.8	0.37
Respiratory code set[c]	Single physician review of medical record	0.70	0.98	35.0	0.31
Combined chief complaint and respiratory code set[c]	Single physician review of medical record	0.72	0.97	24.0	0.29
Lower respiratory syndrome					
Lower respiratory code set[c]	Single physician review of medical record	0.87	0.99	87.0	0.13
Upper respiratory syndrome					
Lower respiratory code set[c]	Single physician review of medical record	0.56	0.98	28.0	0.45
Gastrointestinal syndrome					
Gastrointestinal code set[d]	Two internist-review of dictated ED note	0.32	0.99	32.0	0.69
Gastrointestinal code set[b]	Two physician-review of medical record	0.90	0.94	15.0	0.11
Fever					
Fever code set[b]	Two physician-review of medical record	0.71	0.94	11.8	0.31
Pneumococcal pneumonia					
038.00 Streptococcal septicemia[e]	Microbiology cultures	0.20	0.999	50.0	0.80
038.20 Pneumococcal septicemia[e]	Microbiology cultures	0.19	>0.999	190.0	0.81
481 Pneumococcal pneumonia (or lobar pneumonia, organism unspecified)[e]	Microbiology cultures	0.58	0.98	23.2	0.43
482.30 Pneumonia due to *Streptococcus*[e]	Microbiology cultures	0.11	0.99	11.0	0.90
486 Pneumonia, organism unspecified[e]	Microbiology cultures	0.14	0.48	0.3	1.78
518.81 Respiratory failure[e]	Microbiology cultures	0.15	0.92	1.8	0.93
Pneumococcal pneumonia code set (038.00, 038.20, 481, 482.30, 486, 518.81)[e]	Microbiology cultures	0.89	0.44	1.6	0.25
Pneumococcal pneumonia code set (038.00, 038.20, 481, 482.30)[e]	Microbiology cultures	0.81	0.96	20.3	0.20
Tuberculosis					
Tuberculosis code set (all ICD codes with tuberculosis in title)[f]	Review of medical record	0.57	0.75	2.3	0.57
HIV					
HIV/AIDS code set (042, 043, 044, 795.8)[g]	Review of medical record	0.98	0.89	8.9	0.02
AIDS					
HIV/AIDS code set (042, 043, 044, 795.8)[g]	Review of medical record	0.97	0.58	2.3	0.05

[a]From Espino, J., Wagner, M. (2001a). The accuracy of ICD-9 coded chief complaints for detection of acute respiratory illness. In: *Proceedings of American Medical Informatics Association Symposium*, 164–8, with permission.

[b]From Betancourt, J. A. (2003). An evaluation of the Electronic Surveillance System for the Early Notification of Community-based Epidemics (ESSENCE). Doctoral dissertation, George Washington University, with permission.

[c]From Beitel, A. J., Olson, K. L., Reis, B. Y., et al. (2004). Use of emergency department chief complaint and diagnostic codes for identifying respiratory illness in a pediatric population. *Pediatr Emerg Care* 20:355–60, with permission.

[d]From Ivanov, O., Wagner, M. M., Chapman, W. W., et al. (2002). Accuracy of three classifiers of acute gastrointestinal syndrome for syndromic surveillance. In: *Proceedings of American Medical Informatics Association Symposium*, 345–9, with permission.

[e]From Guevara, R. E., Butler, J. C., Marston, B. J., et al. (1999). Accuracy of ICD-9-CM codes in detecting community-acquired pneumococcal pneumonia for incidence and vaccine efficacy studies. *Am J Epidemiol* 149:282–9, with permission.

[f]From San Gabriel, P., Saiman, L., Kaye, K., et al. (2003). Completeness of pediatric TB reporting in New York City. *Public Health Rep* 118:144–53, with permission.

[g]From Rosenblum, L., Buehler, J. W., Morgan, M. W., et al. (1993). HIV infection in hospitalized patients and Medicaid enrollees: the accuracy of medical record coding. *Am J Public Health* 83:1457–9, with permission.

respiratory syndrome, respectively, and for their lower respiratory ICD code set was 0.87 and 0.99 for *lower respiratory* syndrome, respectively.

Notably, Beitel and colleagues studied the combined information present in both chief complaints and ICD codes. They assigned patients to *respiratory* if they had either a *respiratory* chief complaint or an ICD code in the *respiratory* code set. The sensitivity and specificity of the combined definition for *respiratory* were 0.72 and 0.97, respectively. These results vary only slightly from the sensitivity and specificity of the *respiratory* ICD code set alone.

Detection of More Diagnostically Precise Syndromes. Several studies have measured the case detection accuracy of ICD codes available only after a patient has left the hospital. We discuss three examples of such studies here.

Guevara et al. (1999) studied the sensitivity and specificity of ICD codes and code sets for the detection of the disease *pneumococcal pneumonia*. The ICD codes were part of data collected during a study of community-acquired pneumonia. They used the results of microbiology cultures as a gold standard. They found that ICD code 481 had the highest sensitivity—58.3%—for *pneumococcal pneumonia* (out of all six ICD codes studied) and a specificity of 97.5% (Table 23.13). Of the six ICD code sets they studied, an ICD code set with four ICD codes had the highest sensitivity and specificity–81.2% and 96.0%, respectively. One note of caution in interpreting these results is they limited the study to patients who met the inclusion criteria for community-acquired pneumonia. Thus the sensitivity and specificity results reported are specifically for discriminating among pneumococcal and other causes of community-acquired pneumonia in hospital patients.

San Gabriel et al. (2003) studied the accuracy of hospital discharge ICD codes for the detection *tuberculosis* in children. The gold standard against which they compared ICD codes was medical record review. An ICD code set that contained all ICD codes with "tuberculosis" in their definition had a sensitivity and specificity of 57% and 75%, respectively, for the detection of *tuberculosis* in children.

Rosenblum et al. (1993) studied the accuracy of hospital discharge ICD codes for the detection of human immunovirus (*HIV*) infection and acquired immunodeficiency syndrome (*AIDS*) (Rosenblum et al., 1993). They found that the sensitivity and specificity of an ICD code set (that contained four ICD codes for HIV infection and AIDS) for *HIV* was 98% and 89%,[12] respectively. Similarly, the sensitivity and specificity of the ICD code set for *AIDS* were 97% and 58%, respectively.[13]

Summary of Studies of Case Detection from ICD Codes. With respect to Hypothesis 1, the literature on accuracy of both disease-detection and syndrome-detection from ICD code analysis suggest that:

1. ICD-coded diagnoses and chief complaints contain information about syndromic presentations and diagnoses of patients and existing code sets can extract that information.
2. For syndromes that are at the level of diagnostic precision of *respiratory* or *gastrointestinal*, it is possible to automatically classify ED and office patients (both pediatric and adult) from ICD codes with a sensitivity of approximately 0.65 and a specificity of approximately 0.95. This level of accuracy is similar to that achievable with chief complaints.
3. For diagnoses that are at the level of diagnostic precision of disease (e.g., pneumonia) or disease and organism (e.g., pneumococcal pneumonia) it is possible to automatically classify patients from hospital discharge data.
4. Sensitivity of classification is better for some syndromes and diagnoses than for others.
5. The specificity of case detection from ICD codes and code sets is less than 100%, meaning that daily aggregate counts will have false positives among them due to falsely classified patients.

3.3.2. Detection of Outbreaks

This section describes studies that address Hypotheses 2 and 3, which we reproduce here for convenient reference:

Hypothesis 2: ICD code sets can discriminate between whether there is an *outbreak* of type Y or not.

Hypothesis 3: Algorithmic monitoring of ICD code sets can detect an *outbreak* of type Y earlier than current best practice (or some other reference method).

As with chief complaints, challenges in studying the outbreak detection performance of ICD code sets include collecting data from multiple healthcare facilities in the region affected by an outbreak and achieving adequate sample size of outbreaks for study. To emphasize the importance of sample size, we divide this section into studies of multiple outbreaks (N>1) and studies of single outbreaks (N=1). Unlike chief complaints, no published prospective studies exist.

N>1 Studies. Yih et al. (2005) used the detection algorithm method to study retrospectively 110 gastrointestinal disease outbreaks in Minnesota. They studied daily counts of the CDC/DoD *Gastrointestinal, All* code set. The detection algorithm they used was a space-time scan statistic algorithm

12 They computed a PPV of 97%. We analyzed the data reported in the paper to compute this specificity result.

13 They computed a PPV of 65%; We analyzed the data reported in the paper to compute this specificity result.

(Kleinman et al., 2005). The ICD code data they studied included 8% of the population. The outbreaks ranged in size from 2 to 720 cases with a median outbreak size of 7. Half of the outbreaks were foodborne and approximately half were caused by viruses. They defined a true alarm as a cluster of disease identified by the detection algorithm that was within 5 km of the outbreak, and that occurred any time from one week prior to the start of the outbreak to one week after the outbreak investigation began. The sensitivity of *Gastrointestinal, All* for the detection of small gastrointestinal outbreaks at a false alarm rate of 1 every 2 years was 1%. They also measured sensitivity at false alarm rates of one per year (sensitivity = 2%), 1 per 6 months (3%), 1 per 2 months (5%), 1 per month (8%), and 1 per 2 weeks (13%). They did not measure timeliness of detection. The low percentage of population covered may partly explain the low sensitivity.

N=1 Studies. Tsui et al. (2001) conducted the first study of outbreak detection accuracy and timeliness of ICD codes available soon after a patient visit. It is also the only study to compare the outbreak-detection performance of two ICD code sets for the same outbreak. They used correlation analysis and the detection-algorithm method to study the accuracy and timeliness of an *ILI* and a *respiratory* ICD code set for the detection of an influenza outbreak.[14] Pneumonia and influenza (P&I) deaths were used as the gold-standard determination of the outbreak. The correlation analysis showed a two-week lead time for both ICD code sets relative to P&I deaths.[15] The detection algorithm analysis used the Serfling algorithm. Both ICD code sets had a sensitivity of 100% (1/1). The false alarm rate for the *ILI* code set was one per year and for the *respiratory* code set was one per three months. Detection of the influenza outbreak occurred one week earlier from both code sets relative to detection from P&I deaths at these false alarm rates and measurements of sensitivity.

Miller et al. (2004) used correlation analysis and the detection algorithm method to study the ability of ICD data obtained from a large healthcare organization in Minnesota to detect an influenza outbreak. Using P&I deaths as a gold standard, Miller demonstrated a significant Pearson correlation of 0.41 of an *ILI* ICD code set with P&I deaths. When they lagged the *ILI* time series by one week, the Pearson correlation remained 0.41. For the detection-algorithm method, they used a CUSUM algorithm. CUSUM detected the outbreak from *ILI* (sensitivity of 1/1) one day earlier than the date the health department confirmed the first positive influenza isolate. They did not report the false alarm rate of the CUSUM algorithm for this detection sensitivity and timeliness.

Lazarus et al. (2002) used correlation analysis to study the ability of ICD codes to detect an influenza outbreak (Lazarus et al., 2002). They obtained the ICD codes from a point-of-care system used by a large multispecialty physician group practice. Clinicians assigned ICD-coded diagnoses at the time of either an in-person or phone consultation. They used hospital discharge data as a gold standard. They created two time series using the same *lower respiratory* ICD code set (based on the DoD-GEIS *respiratory* code set): one from their point-of-care data and one from the hospital discharge data. The maximum correlation of these time series was 0.92 when point-of-care ICD codes preceded hospital-discharge ICD codes by two weeks.

Lewis et al. (2002) used correlation analysis to study the ability of ICD codes to detect an outbreak of influenza. They used CDC sentinel physician data for influenza as a gold standard. Physicians assigned ICD codes at outpatient DoD facilities. The correlation of the DoD-GEIS *respiratory* code set with the CDC sentinel physician data was 0.7. They did not measure the correlation at time lags other than zero.

Suyama et al. (2003) used correlation analysis to study whether ICD codes contain a signal of notifiable diseases reported to the health department. They used health department data about notifiable diseases as a gold standard. Billing administrators assigned ICD codes to ED visits. They created time series from both data sets using the same ICD code sets: *gastrointestinal*, *pulmonary*, *central nervous system*, *skin*, *fever*, and *viral*. They found statistically significant correlations between *gastrointestinal*, *pulmonary*, and *central nervous system* code sets and notifiable diseases. Conversely, the *skin*, *fever*, and *viral* syndromes did not have a statistically significant correlation with notifiable diseases.

Summary of Studies of Outbreak Detection from ICD Codes. With respect to Hypotheses 2 and 3, the studies we reviewed demonstrate that:

1. Some large outbreaks of respiratory and ILI can be detected from aggregate analysis of ICD codes. Small outbreaks of gastrointestinal illness generally cannot. There were no studies of large outbreaks of gastrointestinal illness (Hypothesis 2).
2. Research to date is suggestive but not conclusive that outbreaks of influenza can be detected earlier than current surveillance methods (Hypothesis 3).
3. The methodological weaknesses of studies included:

   a. Not measuring all three characteristics of outbreak-detection performance (false alarm rate, sensitivity, and timeliness).

14 They used the same same *Respiratory* ICD code set as Espino and colleagues (Espino and Wagner, 2001).

15 They do not report the correlation measurement, but in one figure in the paper, the correlation of the *Respiratory* code set appears close to 1.0 at a time lag of two weeks.

b. Not reporting either the correlation or the time lag at which correlation was maximal.
c. Measuring correlation at too few time lags (sometimes at just one or two values).
d. Not reporting sampling completeness.

4. In general, the number of published studies is small, perhaps due to the fact that ICD monitoring systems are relatively recent. We expect more studies to be published in the near future.

4. USING CHIEF COMPLAINTS AND ICD-CODED DIAGNOSES IN BIOSURVEILLANCE

Although there is the occasional perception that a biosurveillance organization must choose between collecting chief complaints or ICD codes, this is a false issue. They each have strengths and limitations. They vary in availability by healthcare organization, time latencies, and the accuracy with which they can discriminate among different syndromes and diagnoses. Outbreaks vary in their rapidity, early syndromic presentation, and whether they cause people to seek hospital care versus outpatient care. It is likely that future research will show that chief complaints are superior to ICD codes for monitoring for some outbreaks and ICD codes are superior for other outbreaks.

At the present time, the strongest statement about their respective roles is that chief complaints—because of the low time latency—may prove to be more useful for detecting sudden events. ICD codes—because of their ability to support more diagnostically precise syndromes (or disease categories)—may prove to be more useful for detecting smaller outbreaks.

A biosurveillance organization should collect and analyze both. Algorithms for case detection can utilize both types of data (Chapter 13). Because chief complaints and ICD-codes are associated with a specific patient, it is possible to link them (and other data) to improve the sensitivity and specificity of case detection.

5. SUMMARY

Chief complaints and ICD-coded diagnoses are among the most highly available types of biosurveillance data that we discuss in this section of the book and they are being monitored by many biosurveillance organizations.

Chief complaints are routinely collected by the healthcare system. They are available early in the course of the clinical care process and they contain information about a patient's symptoms. Research to date has demonstrated that they can be readily obtained from emergency departments and hospitals, more often than not in real time as HL7 messages. They are less accessible in the outpatient setting. Chief complaints must be subject to NLP to extract information for biosurveillance. The two basic approaches are keyword matching and Bayesian text processing. The NLP assigns a chief complaint either directly to a syndrome category, or extracts symptoms that are then further processed in a second step to determine whether the patient matches a Boolean case definition. Research on the accuracy of automatic syndrome assignment from chief complaints shows that automatic classification from chief complaint data into syndromes such as *respiratory* and *gastrointestinal* can be accomplished with moderately good sensitivity and specificity. This accuracy is not sufficient to support detection of small outbreaks. When researchers have attempted to automatically assign patients to more diagnostically precise syndromes such as *febrile respiratory*, sensitivity declines quickly—chief complaints simply do not contain sufficient information to support such automatic assignments. Research has also demonstrated that detection algorithms can use daily counts of syndromes derived in this manner to detect large respiratory or diarrheal outbreaks such as those due to influenza and rotavirus. It is likely that large outbreaks of diseases that initially present with other syndromes monitored for would also be detected, although studies of such outbreaks do not yet exist, so this point remains a matter of opinion.

ICD codes are also a type of data routinely collected by the healthcare system. They are far more heterogeneous in meaning than chief complaints as the ICD coding system contains codes for symptoms, syndromes, and diagnoses at different levels of diagnostic precision. Additionally, codes are collected at multiple points during the course of a patients illness and the accuracy and diagnostic precision of coding may vary across these points. ICD codes become available later in the course of the clinical care than chief complaints. From the perspective of biosurveillance, they contain indirect information about a patient's symptoms: the reasoning process that developers use when they include an ICD code in a "code set" is "would a patient with disease X have respiratory symptoms? If so, let's include the code for disease X in a code set for respiratory." Research to date has demonstrated that ICD codes can be readily obtained from the healthcare system. The military fortuitously uses ICD to encode all services (and doctors do the encoding). There are several examples of ICD code sets developed for biosurveillance. A surprisingly small amount of work has been done on developing very specific ICD codes sets to automate the surveillance for conditions such as pneumonia. Research on the accuracy of automatic syndrome assignment shows results similar to chief complaints. As with chief complaints, this accuracy is not sufficient to support detection of small outbreaks. Research on the accuracy of automatic disease assignment using hospital discharge diagnoses shows higher accuracy. Research has also demonstrated that detection algorithms can use daily counts of syndromes derived in this manner to detect large respiratory or diarrheal outbreaks such as those due to influenza and rotavirus. It is likely that large outbreaks of diseases that initially present with other syndromes monitored for would also be detected, although there are no natural occurring examples, so this point remains a matter of opinion.

The results suggest a few immediate applications such as influenza monitoring and early warning of cohort exposures. They also carry an important implication: users of systems should not be overly reassured about the absence of a spike of activity as these data when used alone are not likely to detect a small outbreak. Monitoring of chief complaints and ICD-coded diagnoses alone simply does not have high sensitivity for small outbreaks (unless they are tightly clustered in space and time and possibly demographic strata and the detection system is capable of exploring those dimensions).

It is unfortunate that some authors over-interpret these limited results to conclude that "syndromic surveillance" does not work. A more accurate summarization of the state-of-the-art might be that surveillance of diagnostically imprecise syndrome categories is not capable of detecting small outbreaks because of the background level of patients satisfying the broad syndrome definition. When the availability of additional clinical data (e.g., temperatures and laboratory test orders or results) allows monitoring of more diagnostically precise syndromes, we expect that smaller outbreaks will be detectable and the time of detection of larger outbreaks will improve. This type of syndromic surveillance has been practiced for quite some time (e.g., polio, AIDS), albeit manually due to the lack of automatic access to required surveillance data.

ADDITIONAL RESOURCES

- CDC site with annotated bibliography that will include future publications on this topic: *http://www.cdc.gov/epo/ dphsi/syndromic/*
- Syndromic.org "latest studies" site: *http://www.syndromic. org/ index.php*

ACKNOWLEDGMENTS

We thank the many hospitals and health departments with whom we have collaborated to develop the science and practice of real-time disease surveillance. The RODS project and our research have been funded by PA Bioinformatics Grant ME-01-737, contract F30602-01-2-0550 from the Defense Advanced Research Projects Agency managed by Rome Laboratory, and contract 290-00-0009 from the Agency for Healthcare Research and Quality.

REFERENCES

Balter, S., Weiss, D., Hanson, et al. (2005). Three years of emergency department gastrointestinal syndromic surveillance in New York City: what have we found? *MMWR Morb Mortal Wkly Rep* 54(Suppl):175–80. *http://www.ncbi.nlm.nih.gov/entrez/query.fcgi?cmd=Retrieve&db=PubMed&dopt=Citation&list_uids=16177711.*

Beasley, J. W., Hankey, T. H., Erickson, R., et al. (2004). How many problems do family physicians manage at each encounter? A WReN study. *Ann Fam Med* 2:405–10. *http://www.ncbi.nlm.nih.gov/entrez/query.fcgi?cmd=Retrieve&db=PubMed&dopt=Citation&list_uids=15506571.*

Begier, E. M., Sockwell, D., Branch, L. M., et al. (2003). The National Capitol Region's Emergency Department syndromic surveillance system: do chief complaint and discharge diagnosis yield different results, *Emerg Infect Dis* 9:39–6. *http://www.ncbi.nlm.nih.gov/entrez/query.fcgi?cmd=Retrieve&db=PubMed&dopt=Citation&list_uids=12643841.*

Beitel, A. J., Olson, K. L., Reis, B. Y., et al. (2004). Use of emergency department chief complaint and diagnostic codes for identifying respiratory illness in a pediatric population. *Pediatr Emerg Care* 20:355–60. *http://www.ncbi.nlm.nih.gov/entrez/query.fcgi?cmd=Retrieve&db=PubMed&dopt=Citation&list_uids=15179142.*

Betancourt, J. A. (2003). An Evaluation of the Electronic Surveillance System for the Early Notification of Community-Based Epidemics (ESSENCE). Doctoral dissertation, George Washington University.

Buckeridge, D. L., Switzer, P., Owens, D., et al. (2005). An evaluation model for syndromic surveillance: assessing the performance of a temporal algorithm. *MMWR Morb Mortal Wkly Rep* 54(Suppl): 109–15. *http://www.ncbi.nlm.nih.gov/entrez/query.fcgi?cmd=Retrieve&db=PubMed&dopt=Citation&list_uids=16177701.*

Burt, C., Hing, E. (2005). *Use of Computerized Clinical Support Systems in Medical Settings: UnitedStates, 2001–03*. Atlanta, GA: Centers for Disease Control and Prevention.

Campos-Outcalt, D. E. (1990). Accuracy of ICD-9-CM codes in identifying reportable communicable diseases. *Qual Assur Util Rev* 5:86–9. *http://www.ncbi.nlm.nih.gov/entrez/query.fcgi?cmd=Retrieve&db=PubMed&dopt=Citation&list_uids=2136670.*

Centers for Disease Control and Prevention. (2003). Syndrome Definitions for Diseases Associated with Critical Bioterrorism-Associated Agents. *http://www.bt.cdc.gov/surveillance/syndromedef/index.asp.*

Chang, H., Cochrane, D., Tserenpuntsag, B., et al. (2005). ICD9 as a surrogate for chart review in the validation of a chief complaint syndromic surveillance system. In: *Proceedings of Syndromic Surveillance Conference*, Seattle, WA.

Chapman, W. W., Christensen, L. M., Wagner, M. M., et al. (2005a). Classifying free-text triage chief complaints into syndromic categories with natural language processing. *Artif Intell Med* 33:31–40. *http://www.ncbi.nlm.nih.gov/entrez/query.fcgi?cmd=Retrieve&db=PubMed&dopt=Citation&list_uidsg=15617980.*

Chapman, W. W., Dowling, J. N., Wagner, M. M. (2005b). Classification of emergency department chief complaints into seven syndromes: a retrospective analysis of 527,228 patients. *Ann Emerg Med*. vol. 46, no. 5, pp. 445–55.

Chapman, W. W., Dowling, J. N., Wagner, M. W. (2005c). Generating a reliable reference standard set for syndromic case classification. *J Am Med Inform Assoc*. vo. 12, no. 6, pp. 618–29 http://www.ncbi.nlm.nih.gov/entrez/query.fcgi?cmd=Retrieve&db=PubMed&dopt=Citation&list_uids=16049227.

Chapman, W. W., Espino, J. U., Dowling, et al. (2003). *Detection of Multiple Symptoms from Chief Complaints.* Technical Report, CBMI Report Series. Pittsburgh, PA: Center for Biomedical Informatics, University of Pittsburgh.

Espino, J., Wagner, M. (2001a). The accuracy of ICD-9 coded chief complaints for detection of acute respiratory illness. In: *Proceedings of American Medical Informatics Association Symposium*, 164–8.

Espino, J. U., Wagner, M. M. (2001b). Accuracy of ICD-9-coded chief complaints and diagnoses for the detection of acute respiratory illness. In: *Proceedings of American Medical Informatics Association Symposium*, 164–8. *http://www.ncbi.nlm.nih.gov/entrez/query.fcgi?cmd=Retrieve&db=PubMed&dopt=Citation&list_uids=11833477.*

Fisher, E. S., Whaley, F. S., Krushat, et al. (1992). The accuracy of Medicare's hospital claims data: progress has been made, but problems remain. *Am J Public Health* 82:243–8. *http://www.ncbi.nlm.nih.gov/entrez/query.fcgi?cmd=Retrieve&db=PubMed&dopt=Citation&list_uids=1739155.*

Gesteland, P. H., Gardner, R. M., Tsui, F. C., et al. (2003). Automated Syndromic Surveillance for the 2002 Winter Olympics. *J Am Med Inform Assoc. http://www.ncbi.nlm.nih.gov/entrez/query.fcgi?cmd=Retrieve&db=PubMed&dopt=Citation&list_uids=12925547.*

Gesteland, P. H., Wagner, M. M., Chapman, W. W., et al. (2002). Rapid deployment of an electronic disease surveillance system in the State of Utah for the 2002 Olympic Winter games. In: *Proceedings of American Medical Informatics Association Symposium*, 285–9. *http://www.ncbi.nlm.nih.gov/entrez/query.fcgi?cmd=Retrieve&db=PubMed&dopt=Citation&list_uids=12463832.*

Guevara, R. E., Butler, J. C., Marston, B. J., et al. (1999). Accuracy of ICD-9-CM codes in detecting community-acquired pneumococcal pneumonia for incidence and vaccine efficacy studies. *Am J Epidemiol* 149:282–9. *http://www.ncbi.nlm.nih.gov/entrez/query.fcgi?cmd=Retrieve&db=PubMed&dopt=Citation&list_uids=9927225.*

Heffernan, R., Mostashari, F., Das, D., et al. (2004a). Syndromic surveillance in public health practice, New York City. *Emerg Infect Dis* 10:858–64. *http://www.ncbi.nlm.nih.gov/entrez/query.fcgi?cmd=Retrieve&db=PubMed&dopt=Citation&list_uids=15200820.*

Heffernan, R., Mostashari, F., Das, D., et al. (2004b). Syndromic surveillance in public health practice, New York City. *Emerg Infect Dis* 10:858–64. *http://www.ncbi.nlm.nih.gov/entrez/query.fcgi?cmd=Retrieve&db=PubMed&dopt=Citation&list_uids=15200820.*

Henry, J. V., Magruder, S., Snyder, M. (2004). Comparison of office visit and nurse advice hotline data for syndromic surveillance—Baltimore-Washington, D.C., metropolitan area, 2002. *MMWR Morb Mortal Wkly Rep* 53(Suppl):112–6. *http://www.ncbi.nlm.nih.gov/entrez/query.fcgi?cmd=Retrieve&db=PubMed&dopt=Citation&list_uids=15714639.*

Hsia, D. C., Krushat, W. M., Fagan, A. B., et al. (1988). Accuracy of diagnostic coding for Medicare patients under the prospective-payment system. *N Engl J Med* 318:352–5. *http://www.ncbi.nlm.nih.gov/entrez/query.fcgi?cmd=Retrieve&db=PubMed&dopt=Citation&list_uids=3123929.*

Irvin, C. B., Nouhan, P. P., Rice, K. (2003). Syndromic analysis of computerized emergency department patients' chief complaints: an opportunity for bioterrorism and influenza surveillance. *Ann Emerg Med* 41:447–52. *http://www.ncbi.nlm.nih.gov/entrez/query.fcgi?cmd=Retrieve&db=PubMed&dopt=Citation&list_uids=12658241.*

Ivanov, O., Gesteland, P. H., Hogan, W., et al. (2003). Detection of pediatric respiratory and gastrointestinal outbreaks from free-text chief complaints. In: *Proceedings of American Medical Informatics Association Symposium*, 318–22. *http://www.ncbi.nlm.nih.gov/entrez/query.fcgi?cmd=Retrieve&db=PubMed&dopt=Citation&list_uids=14728186.*

Ivanov, O., Wagner, M. M., Chapman, W. W., et al. (2002). Accuracy of three classifiers of acute gastrointestinal syndrome for syndromic surveillance. In: *Proceedings of American Medical Informatics Association Symposium*, 345–9. *http://www.ncbi.nlm.nih.gov/entrez/query.fcgi?cmd=Retrieve&db=PubMed&dopt=Citation&list_uids=12463844.*

Jollis, J. G., Ancukiewicz, M., DeLong, E. R., et al. (1993). Discordance of databases designed for claims payment versus clinical Information systems. Implications for outcomes research. *Ann Intern Med* 119:844–50. *http://www.ncbi.nlm.nih.gov/entrez/query.fcgi?cmd=Retrieve&db=PubMed&dopt=Citation&list_uids=8018127.*

Lazarus, R., Kleinman, K., Dashevsky, I., et al. (2002). Use of automated ambulatory-care encounter records for detection of acute illness clusters, including potential bioterrorism events. *Emerg Infect Dis* 8:753–60. *http://www.ncbi.nlm.nih.gov/entrez/query.fcgi?cmd=Retrieve&db=PubMed&dopt=Citation&list_uids=12141958.*

Lazarus, R., Kleinman, K. P., Dashevsky, I., et al. (2001). Using automated medical records for rapid identification of illness syndromes (syndromic surveillance): the example of lower respiratory infection. *BMC Public Health* 1:9. *http://www.ncbi.nlm.nih.gov/entrez/query.fcgi?cmd=Retrieve&db=PubMed&dopt=Citation&list_uids=11722798.*

Levy, R., Cochrane, D., Chang, H., et al. (2005). Components of Interhospital variability in chief complaint classification of gastrointestinal syndrome. In: *Syndromic Surveillance Conference*, Seattle, WA.

Lewis, M. D., Pavlin, J. A., Mansfield, J. L., et al. (2002). Disease outbreak detection system using syndromic data in the greater Washington, DC area. *Am J Prev Med* 23:180–6. *http://www.ncbi.nlm.nih.gov/entrez/query.fcgi?cmd=Retrieve&db=PubMed&dopt=Citation&list_uids=12350450.*

Lloyd, S. S., Rissing, J. P. (1985). Physician and coding errors in patient records. *JAMA* 254:1330–6. *http://www.ncbi.nlm.nih.gov/entrez/query.fcgi?cmd=Retrieve&db=PubMed&dopt=Citation&list_uids=3927014.*

Lombardo, J., Burkom, H., Elbert, E., et al. (2003). A systems overview of the Electronic Surveillance System for the Early Notification of Community-Based Epidemics (ESSENCE II). *J Urban Health* 80(Suppl 1):i32–42. *http://www.ncbi.nlm.nih.gov/entrez/query.fcgi?cmd=Retrieve&db=PubMed&dopt=Citation&list_uids=12791777.*

MacIntyre, C. R., Ackland, M. J., Chandraraj, E. J., et al. (1997). Accuracy of ICD-9-CM codes in hospital morbidity data, Victoria: implications for public health research. *Aust N Z J Public Health* 21:477–82. *http://www.ncbi.nlm.nih.gov/entrez/query.fcgi?cmd=Retrieve&db=PubMed&dopt=Citation&list_uids=9343891.*

Magruder, S. F., Henry, J., Snyde, M. (2005). Linked analysis for definition of nurse advice line syndrome groups, and comparison to encounters. *MMWR Morb Mortal Wkly Rep* 54(Suppl):93–7.

http://www.ncbi.nlm.nih.gov/entrez/query.fcgi?cmd=Retrieve&db=PubMed&dopt=Citation&list_uids=16177699.

Mandl, K., Reis, B., Cassa, C. (2004). Measuring outbreak-detection performance by using controlled feature set simulations. *MMWR Morb Mortal Wkly Rep* 53(Suppl):130–6.

McClay, J. C., Campbell, J. (2002). Improved coding of the primary reason for visit to the emergency department using SNOMED. In: *Proceedings of American Medical Informatics Association Symposium*, 499–503. *http://www.ncbi.nlm.nih.gov/entrez/query.fcgi?cmd=etrieve&db=PubMed&dopt=Citation&list_uids=12463874.*

Miller, B., Kassenborg, H., Dunsmuir, W., et al. (2004). Syndromic surveillance for influenzalike illness in ambulatory care network. *Emerg Infect Dis* 10:1806–11. *http://www.ncbi.nlm.nih.gov/entrez/query.fcgi?cmd=Retrieve&db=PubMed&dopt=Citation&list_uids=15504267.*

Mocny, M., Cochrane, D. G., Allegra, J. R., et al. (2003). A comparison of two methods for biosurveillance of respiratory disease in the emergency department: chief complaint vs ICD9 diagnosis code. *Acad Emerg Med* 10:513. *http://www.ncbi.nlm.nih.gov/entrez/query.fcgi?cmd=Retrieve&db=PubMed&dopt=Citation&list_uids=12720909.*

Mundorff, M. B., Gesteland, P., Haddad, M., et al. (2004). Syndromic surveillance using chief complaints from urgent-care facilities during the Salt Lake 2002 Olympic Winter Games. *MMWR Morb Mortal Wkly Rep* 53(Suppl):254.

Olszewski, R. (2003a). Bayesian classification of triage diagnoses for the early detection of epidemics. In: *Recent Advances in Artificial Intelligence: Proceedings of Sixteenth International FLAIRS Conference*, 412–16.

O'Malley K, J., Cook, K. F., Price, M. D., et al. (2005). Measuring diagnoses: ICD code accuracy. *Health Serv Res* 40:20–39. *http://www.ncbi.nlm.nih.gov/entrez/query.fcgi?cmd=Retrieve&db=PubMed&dopt=Citation&list_uids=16178999.*

Peabody, J. W., Luck, J., Jain, S., et al. (2004). Assessing the accuracy of administrative data in health Information systems. *Med Care* 42:1066–72. *http://www.ncbi.nlm.nih.gov/entrez/query.fcgi?cmd=Retrieve&db=PubMed&dopt=Citation&list_uids=15586833.*

Reis, B. Y., Mandl, K. D. (2004). Syndromic surveillance: the effects of syndrome grouping on model accuracy and outbreak detection. *Ann Emerg Med* 44:235–41. *http://www.ncbi.nlm.nih.gov/entrez/query.fcgi?cmd=Retrieve&db=PubMed&dopt=Citation&list_uids=15332065.*

Ritzwoller, D. P., Kleinman, K., Palen, T., et al. (2005). Comparison of syndromic surveillance and a sentinel provider system in detecting an influenza outbreak—Denver, Colorado, 2003. *MMWR Morb Mortal Wkly Rep* 54(Suppl):151–6. *http://www.ncbi.nlm.nih.gov/entrez/query.fcgi?cmd=Retrieve&db=PubMed&dopt=Citation&list_uids?16177707.*

Rizzo, S. L., Grigoryan, V. V., Wallstrom, G. L., et al. (2005). *Using Case Studies to Evaluate Syndromic Surveillance.* RODS Technical Report. Pittsburg, PA: Real-Time Outbreak and Disease Surveillance Laboratory, Center for Biomedical Informatics, University of Pittsburgh.

Rosenblum, L., Buehler, J. W., Morgan, M. W., et al. (1993). HIV infection in hospitalized patients and Medicaid enrollees: the accuracy of medical record coding. *Am J Public Health* 83:1457–9. *http://www.ncbi.nlm.nih.gov/entrez/query.fcgi?cmd=Retrieve&db=PubMed&dopt=Citation&list_uids=8214239.*

San Gabriel, P., Saiman, L., Kaye, K., et al. (2003). Completeness of pediatric TB reporting in New York City. *Public Health Rep* 118:144–53. *http://www.ncbi.nlm.nih.gov/entrez/query.fcgi?cmd=Retrieve&db=PubMed&dopt=Citation&list_uids=12690068.*

Sniegoski, C. (2004). Automated syndromic classification of chief complaint records. *Johns Hopkins University APL Technical Digest* 25:68–75.

Suyama, J., Sztajnkrycer, M., Lindsell, C., et al. (2003). Surveillance of infectious disease occurrences in the community: an analysis of symptom presentation in the emergency department. *Acad Emerg Med* 10:753–63. *http://www.ncbi.nlm.nih.gov/entrez/ query.fcgi?cmd=Retrieve&db=PubMed&dopt=Citation&list_uids=12837650.*

Travers, D. A., Waller, A., Haas, S. W., et al. (2003). Emergency department data for bioterrorism surveillance: electronic data availability, timeliness, sources and standards. In: *Proceedings of American Medical Informatics Association Symposium*, 664–8. *http://www.ncbi.nlm.nih.gov/entrez/query.fcgi?cmd=Retrieve&db=PubMed&dopt=Citation&list_uids=14728256.*

Tsui, F., Espino, J., Wagner, M. (2005). *The Timeliness, Reliability, and Cost of Real-time and Batch Chief Complaint Data.* RODS Laboratory Technical Report. Pittsburg, PA: Real-Time Outbreak and Disease Surveillance Laboratory, Center for Biomedical Informatics, University of Pittsburgh.

Tsui, F., Wagner, M., Dato, V., et al. (2001). Value of ICD-9–coded chief complaints for detection of epidemics. In: *Proceedings of the Fall Symposium of the American Medical Informatics Association*, 711–5.

Tsui, F.-C., Espino, J. U., Dato, V. M., et al. (2003). Technical description of RODS: A real-time public health surveillance system. *J Am Med Inform Assoc* 10:399–408. *http://www.jamia.org/preprints.shtml.*

Tsui, F. C., Espino, J. U., Wagner, M. M., et al. (2002). Data, network, and application: technical description of the Utah RODS Winter Olympic Biosurveillance System. In: *Proceedings of American Medical Informatics Association Symposium*, 815–9. *http://www.ncbi.nlm.nih.gov/entrez/query.fcgi?cmd=Retrieve&db=PubMed&dopt=Citation&list_uids=12463938.*

U.S. Government Accounting Office. (2005). *Information Technology: Federal Agencies Face Challenges in Implementing Initiatives to Improve Public Health Infrastructure.* GAO-308. Washington, DC: Government Accounting Office. *http://www.gao.gov/new.items/d05308.pdf.*

U.S. Department of Defense, Global Emerging Infections System. (2005). ICD9 Code Sets Used in ESSENCE. *http://www.geis.fhp.osd.mil/GEIS/SurveillanceActivities/ESSENCE/ESSENCE.asp#ICD9.*

Wagner, M., Espino, J., Tsui, F.-C., et al. (2004). Syndrome and outbreak detection from chief complaints: the experience of the Real-Time Outbreak and Disease Surveillance Project. *MMWR Morb Mortal Wkly Rep* 53(Suppl):28–31.

Wagner, M. M., Tsui, F. C., Espino, J. U., et al. (2001). The emerging science of very early detection of disease outbreaks. *J Public Health Manag Pract* 7:51–9. *http://www.ncbi.nlm.nih.gov/entrez/query.fcgi?cmd=Retrieve&db=PubMed&dopt=Citation&list_uids=11710168.*

Yao, P., Wiggs, B. R., Gregor, C., et al. (1999). Discordance between physicians and coders in assignment of diagnoses. *Int J Qual Health Care* 11:147–53. *http://www.ncbi.nlm.nih.gov/entrez/query.fcgi?cmd=Retrieve&db=PubMed&dopt=Citation&list_uids=10442845.*

Yih, K. W., Abrams, A., Danila, R., et al. (2005). Ambulatory-care diagnoses as potential indicators of outbreaks of gastrointestinal illness—Minnesota. *MMWR Morb Mortal Wkly Rep* 54(Suppl):157–62. *http://www.ncbi.nlm.nih.gov/entrez/query.fcgi?cmd=Retrieve&db=PubMed&dopt=Citation&list_uids=16177708.*

Yih, W. K., Caldwell, B., Harmon, R., et al. (2004). National Bioterrorism Syndromic Surveillance Demonstration Program. *MMWR Morb Mortal Wkly Rep* 53(Suppl):43–9. *http://www.ncbi.nlm.nih.gov/entrez/query.fcgi?cmd=Retrieve&db=PubMed&dopt=Citation&list_uids=15714626.*

Yokoe, D. S., Subramanyan, G. S., Nardell, E., et al. (1999). Supplementing tuberculosis surveillance with automated data from health maintenance organizations. *Emerg Infect Dis* 5:779–87. *http://www.ncbi.nlm.nih.gov/entrez/query.fcgi?cmd=Retrieve&db=PubMed&dopt=Citation&list_uids=10603211.*

Yuan, C. M., Love, S., Wilson, M. (2004). Syndromic surveillance at hospital emergency departments—southeastern Virginia. *MMWR Morb Mortal Wkly Rep* 53(Suppl):56–8. *http://www.ncbi.nlm.nih.gov/entrez/query.fcgi?cmd=Retrieve&db=PubMed&dopt=Citation&list_uids=15714630.*

24

CHAPTER

Absenteeism

Leslie Lenert and David Kirsh
University of California, San Diego
San Diego, California

Jeffrey Johnson
Health & Human Services Agency, County of San Diego
San Diego, California

Ron M. Aryel
Del Rey Technologies
Los Angeles, California

1. INTRODUCTION

Absenteeism has been of interest as a potential early indicator of outbreaks in a population. Among the earliest studies demonstrating this potential was work by Costagliola et al. (1991) on modeling of data surrounding French influenza outbreaks. They found that increased absenteeism historically preceded influenza outbreaks. Later, Quenel et al. (1998) confirmed this finding and compared absenteeism to other data sources, identifying it as a leading (temporal) indicator of influenza outbreaks. About this time, Lenaway and Ambler (1995) also noted links between school absenteeism and influenza outbreaks.

Absenteeism is also of interest because it offers windows into symptoms in particular populations that may be at special risk for outbreaks. If the outbreak is related to some process or toxin in the workplace or the learning institution, absenteeism provide a very early sign that an outbreak is occurring in that population. For example, absenteeism among postal workers might be an early sign of some mail-borne outbreak. Both the time and the subpopulation focus of absenteeism data make them one of the more potentially important and interesting types of data for health surveillance.

One of the primary reasons to monitor absenteeism data is that many individuals with illnesses avoid contact with the healthcare system. Johnson et al. (2005) studied the antecedent health behaviors of pediatric patients reporting to the emergency room with viral-like syndromes. In syndromes with fever, absenteeism preceded emergency room visits in 44–50% of patients by a mean of 1.3 days. This study is somewhat limited by its focus on the emergency room. Metzger et al. (2004) examined the relationship between emergency room visits and illnesses in the community using a telephone survey. They found that every emergency department visit for an influenza-like illness represented approximately 60 cases of such illnesses in the community. Of these, approximately 32.6% were accompanied by absenteeism from work or school. This extrapolates to 20 potentially measurable absences for every emergency room visit. The only activity that occurred more frequently in this population was a purchase of over-the-counter medication (53.2%).

Because of the desirable properties of absenteeism as an early indicator in many monitoring systems, including ESSENCE II in the nation's capital region (Lombardo et al., 2003), the New York City Public Health Department systems (Heffernan et al., 2004), and the Bionet system in San Diego, all use absenteeism data for health monitoring. This chapter will describe how systems for absenteeism work, review the present data for evidence of the effectiveness of this information source, and speculate about future models for surveillance that use improved methods to take advantage of the social structures that create absenteeism data.

2. PUBLIC/PRIVATE SCHOOLS

Among the entities that track attendance most closely in American society are primary and secondary schools. School districts track attendance because of statutory responsibility. Virtually all states mandate school attendance. In California, the attendance tracking must be supplemented by some effort to ascertain if an illness caused the absence or it is unexcused. State Education Code 46011 reads, "Absences due to illness or quarantine shall be verified by the district or the county superintendent of schools in such manner as the Superintendent of Public Instruction may provide." As a result, in California and most other states, schools make routine queries of parents into the causes of absensteeism.

Schools also track absenteeism because of both fiduciary and financial responsibilities. Absenteeism is correlated with poor academic performance, social difficulties in school, and dropping out from high school prior to graduation (Weitzman et al., 1986).

Handbook of Biosurveillance
ISBN 0-12-369378-0

A percentage of absenteeism is known to be due to either chronic illness or severe personal or family life problems confronting students. At a certain point, students who miss an excessive amount of school become difficult to educate (six to 10 days per quarter), and those missing more require extraordinary efforts (Weitzman et al., 1986). Therefore, concern for students' welfare is also an important factor. From a financial perspective, states compensate many districts on the basis of average daily attendance; as a result, school districts have direct financial incentive to ensure students attend classes. As a result, most school districts place a high priority on detection of high rates of absenteeism and expend considerable effort determining if a child's absence is illness related or unexcused.

Attendance management in schools is a complex process that depends on the organizational preferences of districts and even of schools. Every school and district with will vary to some extent. In the Chula Vista Elementary School District (a school district in southern California that is typical of size and funding levels of districts in the state), the processes for management of attendance are focused around the attendance clerk. This individual, who probably has a high school education and earns near minimum wage, is responsible for producing, at the end of the day, and updating, as the week unfolds, a report on the number of absences, including whether a particular student's absence is due to illness or medical necessity.

Figure 24.1 illustrates the process. Absence data initially comes to the attendance clerk from parents by telephone, by personal communications from parents dropping off other children at school, and by notes sent to school with siblings. The percentage of children absent without any parent contact varies from 20% to 60% of absences depending on the school.

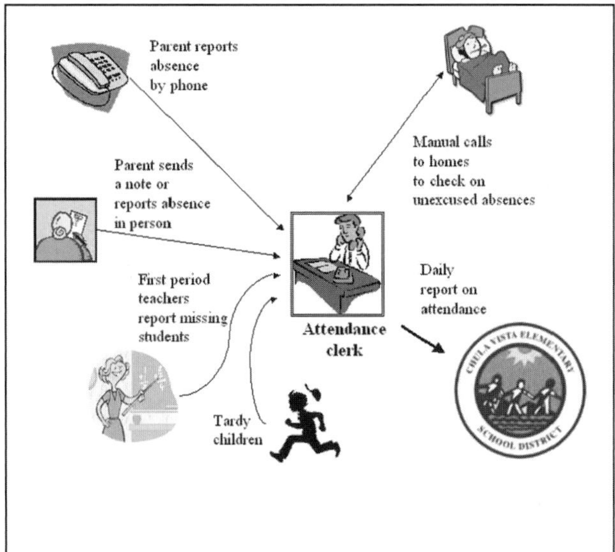

FIGURE 24.1 Current work flow for attendance reporting in the Chula Vista schools.

School clerks merge these data with attendance records from teachers of the first period class. The clerks attempt to determine which students have excused absences. As children who are late for legitimate and other reasons arrive, clerks then edit the absentee list. Attendance clerks also may attempt to determine the causes for unexplained absences by calling parents at home. Some districts use automated calling systems to inform parents of unexcused absences during evening hours. The final rate of unexplained absences varies among schools in the district depending on the motivation of the clerk and the time he or she has available for following up on the unexplained absences. At the end of each business day, clerks manually enter the absence data they have collected into the district's computer database. Clerks update the absence records for each day throughout the school week, as they collect data or receive information from parents documenting that absences are due to illnesses or medical appointments or other excusable causes.

A parent or guardian typically supplies the reason for an absence. There are no standards for documentation of illnesses and most systems assume parents are cooperative participants who are working with the district to insure appropriate attendance. Of course, in some circumstances, absenteeism may be due to the personal problems or preferences of a parent for holiday or leave time. As argued by Weitzman et al. (1986) and later by Alberg et al. (2003), school absenteeism patterns result from complex interactions between multiple factors. While child health is a major contributor, parent or sibling health or parent preferences for vacation time may be the cause.

Many school districts use computer technology to track student attendance. There are no published data on the rates of use of software systems. Attendance software may be a stand alone product focused just on this activity or may be embedded in a district wide student management system that manages grading, health records, and other information. The technology for attendance management is rapidly advancing. Recent developments include the use of personal digital assistants and either wireless or wired synchronization technologies that allow teachers to report attendance and automatically integrate reports. However, as with any information technology, the range of technologies used within a region may vary greatly. Furthermore, technicians have not designed these systems for interoperability.

Software for tracking attendance in many districts is supplemented by digital telephony communication systems. These systems can be programmed to call the homes of students with a recorded message, and it is not an uncommon practice to call the home of absent students who were not excused for illness to inform parents or guardians. The current objective of such systems is to inform parents of absenteeism, but it might be possible to expand the role to collect better data on the cause of absenteeism by using these same systems to collect data through interactive questionnaires.

Absenteeism offers only a partial picture of the effects of outbreaks on student health. Other data sources, such as information about students sent home with illness and the symptoms they are experiencing, can help complete the picture. If the primary objective is subpopulation health monitoring, these other sources of data may be invaluable in understanding events.

Public health officials attempting to integrate attendance results from the schools and school districts in a region face problems akin to those encountered with the use of medical data from hospitals for outbreak detection. There are numerous disparate and incompatible data sources and few financial incentives for districts to collaborate with public health officials. While there are significant technical issues for integration of data, probably the greatest issues are legal and political. Technical issues include connectivity and strategies for merging data and forwarding reports to public health officials in a timely way. Legal and political issues focus upon privacy considerations similar to those raised by compliance with the Health Insurance Portability and Accountability Act (HIPAA). In personal interviews conducted by one of the authors, one large school district's CIO indicated that the preference would be for "off-line" access. Similar to the approach used for hospital data in ESSENCE (Lombardo et al., 2004), school districts would need to configure attendance systems to report anonymous attendance data to public health authorities at regular intervals. Schools without computing infrastructure for attendance reporting might need to use a "drop-in" style system where clerks used a website or a telephone computer system to report overall absenteeism figures.

2.1. Linkages Between Outbreaks and School Attendance

While it is clear that illnesses often cause absences, there is also substantial evidence suggesting that illnesses in student populations can be detected using absenteeism as an indicator of health status (Weitzman et al., 1986). Simple absenteeism rates have been correlated with environments with known environmental toxins by Houghton et al. (2003). Perhaps the most elegant and forward-looking work has been performed by Gilliland et al. (2001, 2003). These researchers sought to overcome the problems with the imprecision of absenteeism data by collecting additional information using telephone interviews to define the symptoms that students were experiencing that caused their illness. Working backward from attendance records, these researchers called the parents of children and determined if absences were due to respiratory or other syndromes. They then correlated respiratory illnesses with daily ozone, respirable particle and nitric oxide concentrations in the atmosphere around the regions of the schools. The results of this analysis were remarkable. High concentrations of ozone during school hours were correlated with significant increases (up to 16%) in the rates of absenteeism due to respiratory illnesses with a two-week delay period (Figure 24.2). Gilliland et al. (2003) have also gone on to link indoor air pollution due to passive cigarette smoking with more frequent absences due to respiratory illnesses. These researchers' studies conclusively show that absenteeism data, when combined with information on the types of symptoms being experienced, provides a highly precise view into the health of a school-age population.

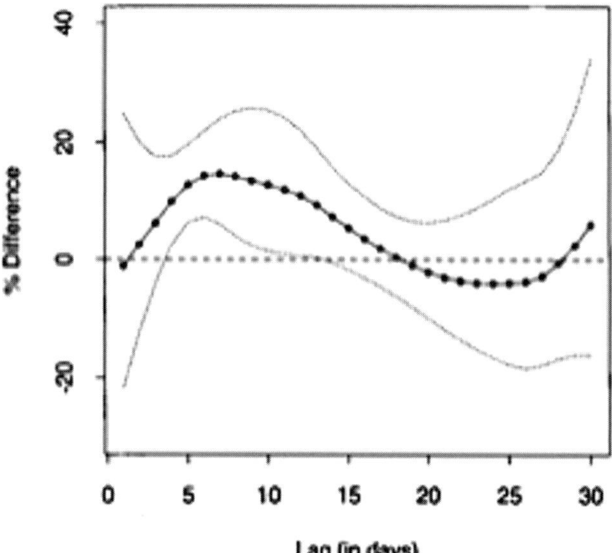

FIGURE 24.2 Effects of ozone levels of greater than 20 parts per billion in air between 10 AM and 6 PM on future absenteeism.

Other work has looked at the association between school absenteeism and influenza outbreaks. Overall, monitoring systems using school absenteeism appear to have considerable power to detect influenza outbreaks. Lenaway and coauthors described a surveillance system for Boulder Valley School District in which schools, on a weekly basis, reported rates of absences due to illnesses (Lenaway and Ambler, 1995). These rates were compared to the rates of influenza A and B observed in sentinel clinical practices. Over a five-year period, reports of absenteeism increases paralleled (two of five years) or preceeded by one week (three of five years) rates of influenza-like illnesses observed in the sentinel medical care system.

Bescuilides et al. (2005) attempted to replicate this experience in New York City using absenteeism reports from city schools. This was a more difficult environment for monitoring than suburban Boulder, Colorado, with high rates of absenteeism from non-illness–related causes and poor follow-up. Their study noted differences in the quality of elementary school and high school data, with many more unexplained absences in high schools. Absenteeism was slightly higher on Mondays and Fridays and nearly twice as high on days with parent teacher conferences, state exams, or school half-days. Analyses that controlled for temporal variability in absenteeism found statistically increased absenteeism in three of four influenza A outbreaks studied. Satscan statistics applied to elementary school absenteeism data frequently detected clusters of schools with increased absenteeism—up to three clusters per day with a median of five schools in each cluster. Similar findings occurred in high schools. These results suggest that spatial analyses may be over sensitive and at the minimum produce more "hits" in New York City than the Department of Health could investigate.

In Japan, Takahashi et al. (2001) developed an outbreak detection system for elementary schools. This system differs from the prior two in that it did not use attendance estimates. Rather, teachers and nurses use the system to report daily counts of children with influenza-like illness in classes. Counts of these illnesses were well correlated with cases identified by an influenza surveillance system (r=0.85) with good sensitivity and specificity for outbreaks of both influenza A and B. This finding provides further evidence that researchers may need to combine absenteeism data with information on symptoms that cause absenteeism to achieve adequate performance.

2.2. Case Study of a School Surveillance System

Project SHARE (School Health and Absenteeism Reporting Exchange) began as an idea to link schools and public health for early event detection for bioterrorism events, as well as identification and management of both large- and small-scale outbreaks. The vision was to create an electronic surveillance system that had minimal workload impact on the school staff, provided regular updated data to public health agencies for analysis, and built an informal network of collaboration between front-line school nurses and public health. The project is a collaboration between the County of San Diego, the state's local county Office of Education, and school districts throughout San Diego County. Funding was made available from the Centers for Disease Control and Prevention through bioterrorism preparedness funds.

The project began formally with Board of Supervisors approval and authorization for the expenditure of project funds in June 2003. The County of San Diego contracted with Voxiva, a transnational company specializing in information solutions, to develop and customize Project SHARE as a dual web-telephony application. The objective was to allow higher and lower technology schools and school districts to collaborate on data entry by supporting data entry through both digital phone applications and Internet applications. Data collected through a password protected website or toll-free telephone portals are stored directly into a database. No confidential or student identifying information is collected. Rather, information is collected on a daily basis about overall attendance and the symptoms students experienced leading to health office visits. Reporting requires a minimal time commitment. The users of Project Share data (school nurses and health administrators) can monitor the information in real time through web-based interfaces. Figure 24.3 shows an illustration of a sample report from the system. Users can also extract data from the web-based system into easy-to-use Excel spreadsheets. Data are also accessed and analyzed by the county's Community Epidemiology Branch within the Health and Human Services Agency, which looks for trend information and identifies and investigates abnormal levels of activity, in collaboration with school health practitioners.

The Community Epidemiology Branch worked closely with the Office of Education in designing the Project SHARE system. The Office of Education also played a role in identifying selected schools that would serve as pilot schools during the first year of implementation. The pilot schools were selected based upon a geographic distribution and for school nurses who expressed interest in participating in the pilot project. To insure the success of the system, the Office solicited extensive school nurse input in the design and implementation of Project SHARE.

The pilot year of Project SHARE was evaluated by an external group from University of California at San Diego led by David Kirsh. This evaluation took the following factors into consideration:

- System usage (e.g., data entry and data retrieval)
- Cost of the program
- Value of the system to schools and their staff

2.3. Use of the System

Pilot data reflected the diversity of users of the Project SHARE system: users included 35 nurses (which encompasses school site and district nurses with both licensed vocational nursing and registered nursing qualifications), 27 health clerks, 16 attendance clerks (includes secretaries), six analysts, and four principals and administrators. The total number of users was 88. Users were from 35 different schools, representing 11 different San Diego school districts. Twenty elementary schools (average enrollment of 688), two middle schools (average enrollment of 1112), and 12 high schools (average enrollment of 1181) participated in the pilot.

Log-in data between December 1, 2003 and April 22, 2004 revealed that 97 users with passwords logged in a total of 10,560 times or an average of 108.9 log-ins per user. Eighty-five percent of potential users accessed the system to either view screens or submit an Attendance or Health Report.

The timeliness of data submission can be a problem with "drop-in" style voluntary reporting systems because these systems ask users to add reporting tasks to their usual work load. Using system logs, the evaluators determined the percentage of reports submitted on time by clerks and nurses in the pilot testing group. Results show mean delays of about 2.7 days in submission of attendance reports and delays of 1.7 days in submission of health reports. Table 24.1 summarizes the reporting data.

2.4. Cost of the Program

The cost to develop the Project SHARE system was $107,754. This figure includes costs for web customization and

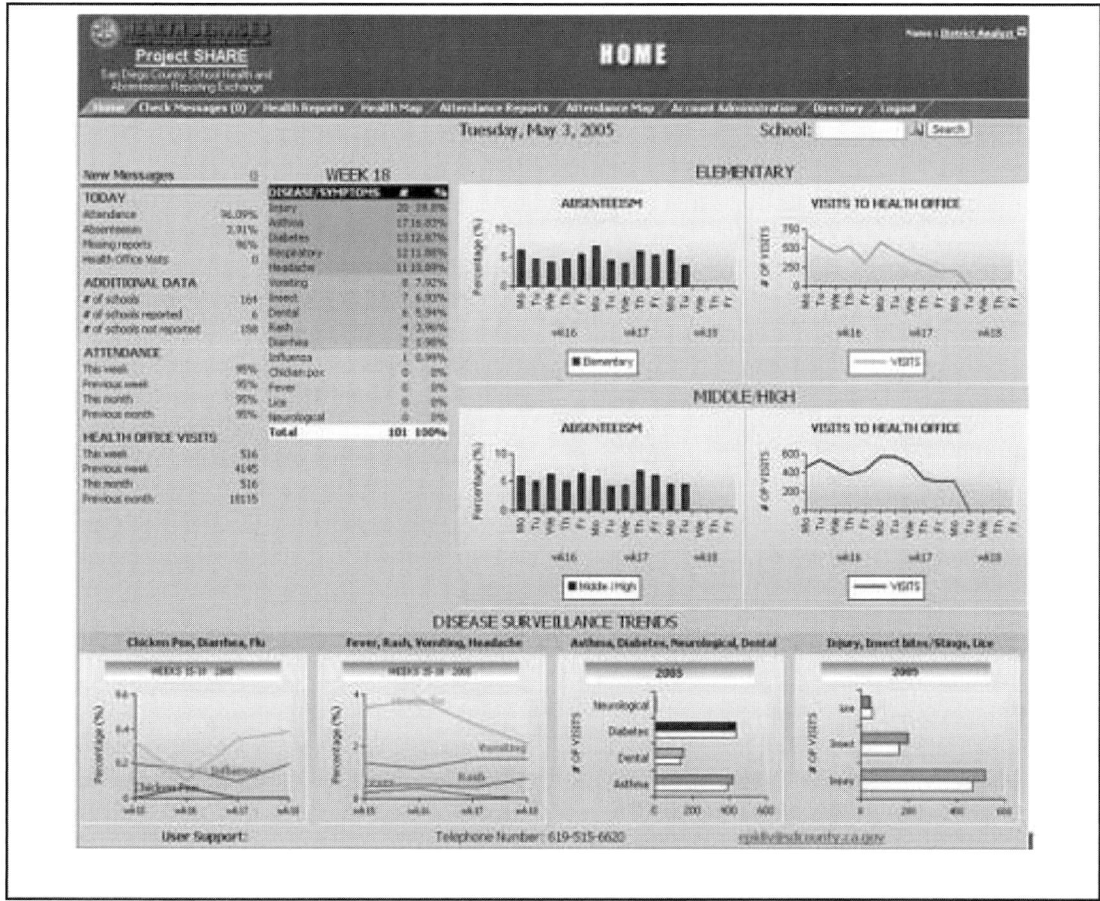

FIGURE 24.3 Project SHARE report on school absenteeism and symptoms.

development, systems resources, a test environment and system training, and one year of software maintenance and support. In addition, the county contracted with the Office of Education to liaison with the school districts in the selection, recruitment, and formalization of participating school sites. The Office of Education also facilitated the pilot year project evaluation and conducted two regional meetings for SHARE users. The contract cost was $25,000 for the pilot year and $15,000 for the second project year.

A combination of public health and education staff maintain the system on an as-needed basis. Staff responsibilities include responding to school nurse inquiries, supporting systems, managing user accounts, and analyzing data for syndromic surveillance purposes. Costs to support the system on an annual basis are about $75,000 a year.

2.5. Results/Success of the Program

Subjective user response to the system was good. In terms of the perceived value of the system to the user, 56.0% of the users responding believed the system to be "very valuable." The SHARE system itself was widely praised in comments as user-friendly, and 66.7% of users reported that they need

TABLE 24.1 Submission of Reports During the Pilot Study

Attendance Office	Elementary School	Middle/High School	Countywide
Total reports	1324	831	2155
Mean length of time to submit report	1.3 days	5.1 days	2.7 days
% Reports submitted within two days	86.3%	62.9%	77.3%
Health Office	**Elementary School**	**Middle/High School**	**Countywide**
Total reports	1236	918	2154
Mean length of time to submit report	1.4 days	2.2 days	1.7 days
% Reports submitted within two days	85.0%	74.4%	80.5%

only 10 minutes or less to enter their daily data. Users also found it easier to prepare reports with SHARE.

County officials felt that Project SHARE had met its objectives in enhanced monitoring of school health data. The collaborative approach between public health, Office of Education, and school districts proved valuable in connecting various front line medical professionals involved in student health. The Community Epidemiology Branch also found Project Share was useful during the severe influenza season of 2003–2004, allowing the agency to assess the severity of an outbreak in the school systems. An unexpected benefit to schools was additional financial support for the districts participating in the project. California law allows school districts to reclaim funds lost due to absenteeism when it is due to an outbreak or natural disaster. Several districts used project SHARE data, together with other public information, to document claims of lost attendance due to the annual influenza outbreak and recover lost funds.

2.6. Conclusions

Project SHARE illustrates a successful school health monitoring system that combines both absenteeism data and nurse reporting of student symptoms. This combination provides a relatively complete picture of student health. Absenteeism alone was not an adequate data source, but provided a foundation, which when combined with reports from school nurses provided a highly valued source of information about the health of the school populations. Obstacles, such as a lack of interoperability among attendance systems and the lack of information technology, were overcome by use of web and telephone reporting systems.

3. EMPLOYER AND MILITARY ATTENDANCE REPORTING

In contrast to the situation in schools, business and financial needs drive the design and operation of employer attendance systems. Further, successively higher levels of the business hierarchy are concerned primarily with operational status. They ask, "Is our business unit functioning properly today?," and not "Who is absent today and why are they absent?" In general, tracking attendance serves human resource and compensation needs, and departments that track attendance design their work processes and information systems around those needs, which do not include the real-time absence or health status of a given employee.

An employer-run attendance system must cope with two classes of employees: essential and nonessential employees. All employers have essential employees, regardless of their size. Essential employees are those who contribute to activities where staffing must be maintained at specific levels for safety and efficiency. The law requires some kinds of facilities to adhere to specific minimum staffing regulations, especially where the facility cannot operate without the presence of a licensed or certified employee. Examples include procedure-certified nurses in hospital intensive care units, licensed reactor operators in nuclear power plants, and licensed brokers in a real estate office. Other employees are nonessential, and a given business unit may continue to operate without them.

Intuitively, the presence or absence of essential employees represents an opportunity for a biosurveillance system to monitor attendance; in practice, this opportunity is often lost because the organization's hierarchy is focused on performance, not individuals. This principle extends to the U.S. military's current attendance-taking policy. A record of absence because of illness is registered when a member of the service presents to a military health facility; those records may be electronic or manual. Personnel records, however, will reflect simply whether a service member is on duty or on leave and may not be completed in a timely way. The reasoning is quite straightforward: The captain of a U.S. Navy warship needs to know and must be able to report to superiors the state of readiness of that warship. For this purpose, he/she needs to know the state of readiness of all the ship's departments. The identities or even (to a point) number of service members absent is a secondary matter, so long as all departments are functioning properly, the ship is fully fueled and armed, and all sensors and weapons systems are ready for combat (Medina, 2005).

Employer-run attendance reporting at most large employers is computerized, with a wide variety of systems in place. Newer systems use relational databases, facilitating the creation of different views of these data. Moreover, vendors such as Peoplesoft and SAP have created web-based systems that can take attendance daily and make lists of absentees available in real time to authorized users. Whether these systems are actually used to maintain daily attendance records varies from employer to employer based on business practices. Other clients using older systems with proprietary networking or hierarchical database technologies may pose additional challenges.

For many employers, attendance is a more local problem. Employees call in sick to a supervisor who may not formally track absences. This reporting would typically be supplemented by payroll-based attendance recording using a time card. Employers generally do not enter absenteeism information into the electronic database until the end of the pay period, to determine monthly salary and eligibility for benefits. This practice might result in delays of days to weeks between events and data entry into computer systems. Many employers use a category known as "PT," or personal time (or a similar classification), which represents an absence on the employee's request for any reason, including a short-term illness, to identify periods of absence.

Even though as much as 10–15% of all absenteeism is due to influenza outbreaks (O'Reilly and Stevens, 2002), there is minimal evidence for the effectiveness of employment-based efforts for absenteeism surveillance. Early work by Quenel and coauthors demonstrated linkages between sick leave requests and influenza outbreaks; however, subsequent work has not confirmed or expanded this relationship. For example, Nguyen-Van-Tam et al. (1999) were not able to find associations between outbreaks of influenza and hospital employee absenteeism.

4. LIMITATIONS TO ABSENTEEISM DATA

Besculides et al. (2005) detail in their recent paper many of the limitations of absenteeism data. Absenteeism data has greater periodicity than many other sources and is sometimes completely unavailable. Time periods prior to holidays often have higher absenteeism. In analyses, the authors had to retrospectively remove certain days from analysis (e.g., Halloween). Signals are absent during holiday periods and weekends. Holiday periods can be prolonged in schools. Issues with absenteeism data arise largely from lack of specificity. Health-related absenteeism is often difficult to distinguish from other causes.

Gilliland et al. (2001) point the way toward approaches to improve the specificity of absenteeism data. If data are collected on the symptoms that led to absences, absenteeism becomes a sensitive and precise signal of health problems. How might we achieve this? Weedn et al. (2004) have suggested developing self-administered questionnaires for web or interactive voice response systems on telephones to help members of the public determine if they have symptoms consistent with an ongoing outbreak. If self-report measurements could be integrated with work processes for collecting absenteeism data, the result might be low-cost timely data on symptoms experienced in a population.

Figure 24.4 illustrates how self-report systems might be integrated within a school absenteeism system. Parents would either call a dedicated telephone number or access a website to report children will be absent. During this call, they would confidentially report the cause of the absence and the symptoms their child was experiencing. Children absent without an excuse would be detected by the system using current workflow processes. An outbound call to the home would not only inform parents of the absence (as is currently a common practice) but would also attempt to elicit associated symptoms. To complete the picture of school health, districts might add voluntary reporting systems for types of symptoms experienced in children with health problems during school hours, using a system similar to project share.

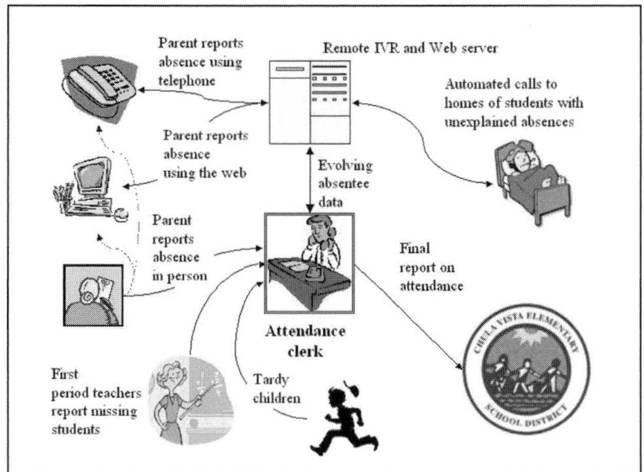

FIGURE 24.4 Proposed workflow for reporting of absenteeism using electronic questionnaire to capture symptoms.

4.1. Alternatives to Attendance-Based Systems

Attendance-keeping systems are not the only ways to determine absenteeism. We can imagine existing data sources that could be used to derive whether an individual is at work, school or home. Such signals might include utilization of various resources at the place of work (or at home) especially those that require self-identification. Examples include logging into a computer network, utilization of computer and network resources, and entry into facilities (buildings, parking garages, elevators) that require pass cards. Even phone activity (not content, just time of placement and minutes) may provide a useful indicator. These approaches do not address issues of the lack of specificity of attendance data.

SUMMARY

Although absenteeism represents relatively early behavior of sick individuals, and there are good epidemiological data linking absenteeism with outbreaks, absenteeism data has significant problems as a tool for detection. Data are imprecise (reflecting events other than illness) and are unavailable for certain periods. In some circumstances, there may be long delays in reporting. In most situations, substantial investments in information technology may be required. One of the most successful regional models, Project Share, avoids this problem by using a "drop-in" style web and telephone system to collate data from diverse systems. Absenteeism, however, might (with careful investment) evolve into a high-quality data source on illness in specific populations of interest. To achieve these ends, attendance monitoring would need to be combined with self-report technologies to allow the accurate assessment of symptoms and syndromes. Indirect measures of absenteeism may also provide low-cost real-time approaches for detection of events.

REFERENCES

Alberg, A. J., Diette, G. B., Ford, J. G. (2003). Invited commentary: Attendance and absence as markers of health status—the example of active and passive cigarette smoking. *Am J Epidemiol* 157:870–3. *http://www.ncbi.nlm.nih.gov/entrez/query.fcgi?cmd=Retrieve&db=PubMed&dopt=Citation&list_uids=12746238.*

Besculides, M., Heffernan, R., Mostashari, F., et al. (2005). Evaluation of school absenteeism data for early outbreak detection, New York City. *BMC Public Health* 5:105.

http://www.ncbi.nlm.nih.gov/entrez/query.fcgi?cmd=Retrieve&db=PubMed&dopt=Citation&list_uids=16212669.

Costagliola, D., Flahault, A., Galinec, D., et al. (1991). A routine tool for detection and assessment of epidemics of influenza-like syndromes in France. *Am J Public Health*, 81:97–99. *http://www.ncbi.nlm.nih.gov/entrez/query.fcgi?cmd=Retrieve&db=PubMed&dopt=Citation&list_uids?=1983924.*

Gilliland, F. D., Berhane, K., Islam, T., et al. (2003). Environmental tobacco smoke and absenteeism related to respiratory illness in school children. *Am J Epidemiol* 157:861–9. *http://www.ncbi.nlm.nih.gov/entrez/query.fcgi?cmd=Retrieve&db=PubMed&dopt=Citation&list_uids=12746237.*

Gilliland, F. D., Berhane, K., Rappaport, et al. (2001). The effects of ambient air pollution on school absenteeism due to respiratory illnesses. *Epidemiology* 12:43–54. *http://www.ncbi.nlm.nih.gov/entrez/query.fcgi?cmd=Retrieve&db=PubMed&dopt=Citation&list_uids=11138819.*

Heffernan, R., Mostashari, F., Das, D., et al. (2004). New York City syndromic surveillance systems. *MMWR Morb Mortal Wkly Rep* 53(Suppl):23–7. *http://www.ncbi.nlm.nih.gov/entrez/query.fcgi?cmd=Retrieve&db=PubMed&dopt=Citation&list_uids=15714622.*

Houghton, F., Gleeson, M., Kelleher, K. (2003). The use of primary/national school absenteeism as a proxy retrospective child health status measure in an environmental pollution investigation. *Public Health* 117:417–23. *http://www.ncbi.nlm.nih.gov/entrez/query.fcgi?cmd=Retrieve&db=PubMed&dopt=Citation&list_uids=14522157.*

Johnson, H. A., Saladino, R. A., Wagner, M. M. (2005). Investigating Non-Traditional Biosurveillance Data by Studying Behaviors Prior to Emergency Room Visits. Master's Practicuum, Biomedical Informatics, University of Pittsburgh, Pittsburg, PA.

Lenaway, D. D., Ambler, A. (1995). Evaluation of a school-based influenza surveillance system. *Public Health Rep* 110:333–7. *http://www.ncbi.nlm.nih.gov/entrez/query.fcgi?cmd=Retrieve&db=PubMed&dopt=Citation&list_uids=7610226.*

Lombardo, J., Burkom, H., Elbert, E., et al. (2003). A systems overview of the Electronic Surveillance System for the Early Notification of Community-Based Epidemics (ESSENCE II). *J Urban Health* 80(Suppl 1):i32–42. *http://www.ncbi.nlm.nih.gov/entrez/query.fcgi?cmd=Retrieve&db=PubMed&dopt=Citation&list_uids=12791777.*

Lombardo, J. S., Burkom, H., Pavlin, J. (2004). ESSENCE II and the framework for evaluating syndromic surveillance systems. *MMWR Morb Mortal Wkly Rep* 53(Suppl):159–65. *http://www.ncbi.nlm.nih.gov/entrez/query.fcgi?cmd=Retrieve&db=PubMed&dopt=Citation&list_uids=15714646.*

Medina, J. (2005). Captain, U.S. Airforce Media Outreach Office. Personal Communication, September 27, 2005.

Metzger, K., Hajat, M., Crawford, F. M. (2004). How many illnesses does one emergency department visit represent? Using a population-based telephone survey to estimate a syndromic multiplier. *MMWR Morb Mortal Wkly Rep* 53(Suppl):106–11.

Nguyen-Van-Tam, J., Granfield, R., Pearson, J., et al. (1999). Do influenza epidemics affect patterns of sickness absence among British hospital staff. *Infect Control Hosp Epidemiol* 20:691–4. *http://www.ncbi.nlm.nih.gov/entrez/query.fcgi?cmd=Retrieve&db=PubMed&dopt=Citation&list_uids=10530649.*

O'Reilly, F. W., Stevens, A. B. (2002). Sickness absence due to influenza. *Occup Med (Lond)* 52:265–9. *http://www.ncbi.nlm.nih.gov/entrez/query.fcgi?cmd=Retrieve&db=PubMed&dopt=Citation&list_uids=12181375.*

Quenel, P., Dab, W. (1998). Influenza A and B epidemic criteria based on time-series analysis of health services surveillance data. *Eur J Epidemiol* 14:275–85. *http://www.ncbi.nlm.nih.gov/entrez/query.fcgi?cmd=Retrieve&db=PubMed&dopt=Citation&list_uids=9663521.*

Takahashi, H., Fujii, H., Shindo, N., et al. (2001). Evaluation of a Japanese school health surveillance system for influenza. *Jpn J Infect Dis* 54:27–30.

Weedn, V. W., McDonald, M. D., Locke, S. E., et al. (2004). Managing the community response to bioterrorist threats. *IEEE Eng Med Biol Mag* 23:162–70. *http://www.ncbi.nlm.nih.gov/entrez/query.fcgi?cmd=Retrieve&db=PubMed&dopt=Citation&list_uids=15154273.*

Weitzman, M., Walker, D. K., Gortmaker, S. (1986). Chronic illness, psychosocial problems, and school absences. Results of a survey of one county. *Clin Pediatr (Phila)* 25:137–41. *http://www.ncbi.nlm.nih.gov/entrez/query.fcgi?cmd=Retrieve&db=PubMed&dopt=Citation&list_uids=3948456.*

25

CHAPTER

Emergency Call Centers

Ron M. Aryel

Del Rey Technologies, Los Angeles, California

Michael M. Wagner

RODS Laboratory, Center for Biomedical Informatics
University of Pittsburgh, Pittsburgh, Pennsylvania

1. INTRODUCTION

Emergency call centers receive telephone calls from people requiring immediate assistance to prevent the imminent loss of life or property, to prevent injury, or to address a potentially hazardous situation requiring immediate attention. Examples of emergency call centers include 911 call centers, military call centers; commercial call centers such as General Motors' "On-Star"®, and transport company (truck or rail) control centers. Government agencies or private contractors may operate these facilities. People call these centers about sick and deceased people and animals, chemical spills or hazardous materials. These calls may provide an early warning of a natural outbreak or bioterrorism incident.

2. 911 CALL CENTER/DISPATCH COMPUTER SYSTEMS

911 systems were developed to allow a person to contact emergency services by dialing an easily recalled, short string of digits. In the vast majority of jurisdictions, law enforcement personnel operate the primary service answering point (PSAP), the center receiving the calls and coordinating the dispatch of emergency personnel; this is because most of the call volume over 911 systems is either law-enforcement related or may require the involvement of a peace officer.

2.1. How 911 Systems Work

When a person dials 911, the telephone company's central office routes the call automatically via dedicated lines to a PSAP. In some systems, especially in smaller communities, the PSAP operator will handle all callers, regardless of the nature of their emergency needs. PSAPs in cities will typically route the caller to a secondary service answering point (SSAP), operated by the appropriate fire or emergency medical services (EMS). While the PSAP operator talks to the caller; he/she enters information pertinent to the call using a computer keyboard. The computer then routes this information to emergency service workers. The operator(s) can also speak to emergency personnel by radio.

911 systems are widely available in the United States. In 1976, only 17% of the U.S. population had access to 911. By contrast, a 1997 Bureau of Justice Statistics survey found that Enhanced 911 (E911, defined below) was operational at 83% of all American law enforcement agencies. The National Emergency Number Association (NENA) estimates that 85% of jurisdictions (which serve 93% of the U.S. population) are equipped with some form of 911, and the vast majority have access to the technologically advanced E911.

There are still jurisdictions, however, which do not offer this service. In those areas, callers must dial specific, separate 7-digit (or 10-digit) telephone numbers to reach law enforcement, fire, or EMS.

Basic 911 systems, deployed decades ago, cannot provide an address to the PSAP operator. Basic 911 may be equipped to provide the caller's telephone number to the operator, and the majority of 911 systems have this capability. E911 adds Automatic Number Identification and Automatic Location Identification to Basic 911. In E911, Automatic Number Identification is used to retrieve a street address or precise location from an Automatic Location Identification database. (Automatic Location Identification is actually a stand-alone database in its own right, and could be accessed by any Basic 911 system if the operator merely had an additional computer terminal in front of him/her.)

Emergency call centers in most medium-sized and virtually all large cities use computer-aided dispatch (CAD) systems. CAD stores and retrieves for the operator call histories for a given location or caller (e.g., it will note previous calls reporting altercations with police officers, previous calls reporting fire, previous calls for cardiac arrest, or previous calls requiring "HazMat" dispatches); the location of desired response resources, such as patrol officers, fire trucks, ambulances and special response teams; and additional notes received from responders in the field. CAD systems help the 911 operators to efficiently manage the call and resources deployed in response to the call. The operator can locate and dispatch the

ISBN 0-12-369378-0

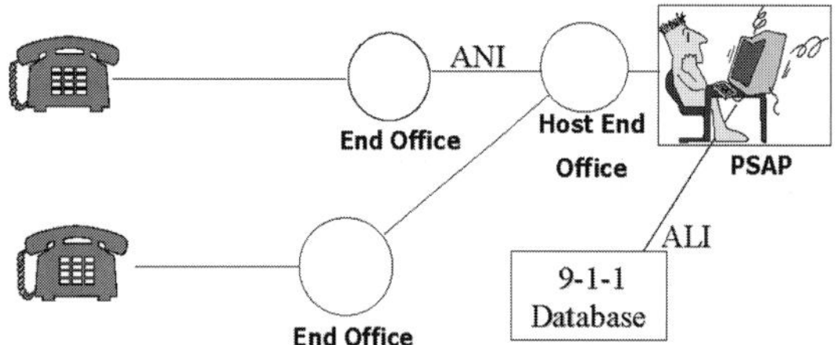

FIGURE 25.1 Call flow process, Basic 911. End Office, telephone company switching office local to caller; Host End Office, switching office operating 911 PSAP. (Courtesy of National Emergency Number Association.)

most appropriate responders to a call. CAD systems may also be linked to geographic information systems (GIS).

Figure 25.1 illustrates the architecture of a Basic 911 system. Industry professionals estimate that roughly half of 911 systems in the U.S. are equipped with an integrated CAD system. Figure 25.2 shows a typical E911 setup.

In 2000, callers placed 30% of the 150 million calls to 911 in the United States from cellular phones, according to NENA, and the organization estimates that in 2005, more 911 callers will use cellular phones than conventional land lines. Despite this, the infrastructure required for cellular E911 systems is only now in place. Currently, a PSAP operator receiving a 911 call can view the cellular phone number and the location of the cellular tower handling the call; this satisfies the Federal Communications Commission's (FCC) E911 Phase I requirement. Cellular tower transmission ranges vary from under 3 miles in dense urban settings, to 4 miles in suburbs, to up to 16 miles in rural settings. Taking advantage of overlapping

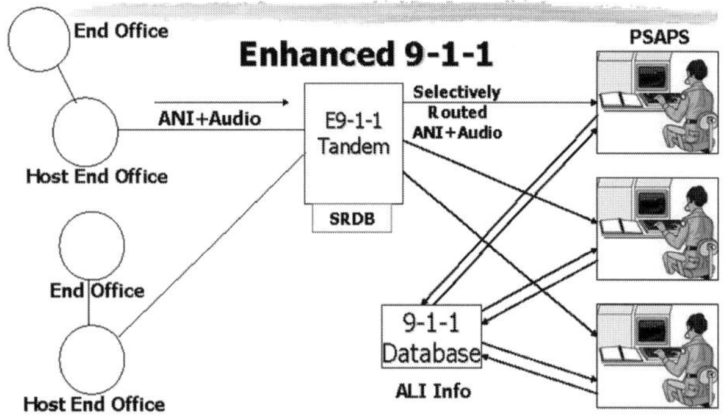

FIGURE 25.2 Call flow process, Enhanced 911. End Office, telephone company switching office local to caller; Host End Office, switching office operating 911 PSAP. (Courtesy of National Emergency Number Association.)

tower coverage and previous call handoffs (in cases where the caller is traveling), the PSAP can place an individual within about 2 to 3 square miles in a city. If only one tower had contact with a stationary caller, the known area encompassing the caller's location may be 20 square miles or more.

The FCC, in its E911 Phase II standard, requires cellular carriers to locate a caller automatically to within 300 meters (Federal Communications Commission, 2005b). Triangulation of a signal using cellular towers is one way to accomplish this. T-Mobile and Cingular Wireless support this scheme. One drawback is that, in many places, the phone may not be able to communicate with three towers due to topography or to coverage gaps.

The use of global positioning system (GPS) satellites is a more precise way to locate callers. The latter technique is already used by many commercial transport companies, which equip their trucks with GPS receivers. New cellular phones manufactured by Motorola are GPS-ready; the carriers who sell them to the public must place software in the phones which will access the GPS interface (Motorola Inc., 2001, Rivera, 2005). Once this occurs, the PSAP answering the 911 call will, in theory, be able to ascertain a caller's location within a few square meters; the FCC wants GPS-equipped phones to provide a caller's location with a maximum error of 400 feet.

Nextel has very recently enabled this function within its service areas in the United States, and as of this writing, Sprint and Verizon are initiating similar services, although it is unclear whether they are able to meet the FCC's requirements. The caller must be talking on a late-model cellular phone equipped with the requisite interface and software.

The FCC wants wireless carriers to complete the Phase II roll-out by December 2005, but the carriers have encountered numerous technical obstacles to meeting the FCC's Phase II requirements, and the agency has already issued a number of waivers. As of the first quarter of 2005, PSAPS in 35% of the 3,100 counties in the United States have Activated Phase II systems, those that should be able to determine the precise location of a wireless 911 caller (Hixson, 2005). However, tests to determine compliance have seen mixed results, which was not unexpected given the new technologies and methods involved.

Another limitation is that the caller may not be in the PSAP's jurisdiction; that is, the call may not be routed to the PSAP, which would have handled a call from a land line at the caller's location. In this case, the caller must supply the PSAP with the address and city requiring an emergency response; often, this requires the caller to be transferred to another PSAP that is empowered to dispatch the appropriate agency's personnel. If the caller specifies the address where emergency response is needed and the jurisdiction has CAD installed, the operator will be able to retrieve and display the reasons for previous calls from and dispatches to the address.

Lastly, Internet-based phones (voice-over Internet protocol, or VoIP) have made their debut. These phones often cannot be used to reach a PSAP directly; instead, the caller relies on operator relay. Several tragedies and near tragedies involving people in distress trying to call 911 on VoIP phones prompted the FCC to take action and in the spring of 2005, the FCC ordered all VoIP service providers to provide direct 911 connections to PSAPS; the providers must obtain the physical address of the phone's user and display this information and a call-back number to the PSAP. The FCC expects VoIP service providers to comply with this rule by the end of 2005. The degree to which the providers will be able to comply remains to be seen (Federal Communications Commission, 2005a).

If the subject of interest is a private vehicle, useful systems for locating them include Lo-Jack®, which, when activated, allows police to track its movements.

2.2. Implications for Biosurveillance

A 911 system is actually a set of components that work together to provide communications and data storage functions. The component with the most immediate potential for biosurveillance is the CAD database, which collects, in real time, in addition to the information listed above (the problem, previous call history from the location), information from emergency responders at the location. This information includes clinical information regarding patients and sick animals, contaminants and human/animal exposure to them. Although this information overlaps with information that will be collected in emergency facilities, should the patient be transported, some 911 callers may refuse transport, not require transport, decline to follow advice to seek medical care, or in large-scale outbreak situations, may be referred or routed to locations that do not have clinical information systems that are part of biosurveillance networks.

2.3. Field Use of Online Databases

One very important aspect of police dispatch, directly related to data accessibility, is the recent introduction of CDPD (cellular digital packet data). CDPD is a wireless encrypted data network that employs 128-bit (DES) encryption. CDPD coverage is very widespread, and although it operated at a low rate of data transfer (19.2 k) at roll-out, it was anticipated to increase to 28.8 within months.

Following the Los Angeles riots of 1992, supporters have touted CDPD as a strong supplemental system assisting police agencies in the dissemination and accessing of data through police units employing MDTs (mobile data terminals). CDPD allow police to transmit confidential information by wireless.

Linking MDTs with CDPD enables law enforcement officers in the field to readily access their record management systems (RMS) and other databases. Although many of these databases, such as the National Crime Information Center (NCIC) and/or State Crime Information Center (SCIC), are clearly outside the scope of this book, CDPD technology also makes feasible access to databases of direct interest to public health authorities. MDTs are not limited to laptops; recent advances are also allowing local law enforcement to consider wireless palm pilot services or BlackBerry® handheld computers for similar applications. New Jersey State Police Officers have begun using these devices while patrolling commuter trains. The World-Wide Web can be used for this purpose; virtually any Web-enabled device can participate.

CDPD/wireless web applications must be kept simple because of screen size limitations, with image transfers or large data files kept to a minimum. Data distributive systems using MDT/CDPD could also be invaluable during major catastrophes (such as the bombing of the World Trade Center) because they are supported by the inherent robustness and flexibility of the Internet; however, there still exists the potential for failure if the telephone system itself fails. CDPD is available in virtually all telephone company service areas in the United States, except for a few remote rural areas. Not every 911 call center in a CDPD-capable area harnesses it, however.

3. MILITARY CALL CENTER/DISPATCH

Although military systems serve only a small portion of the population, they are relevant to biosurveillance because outbreaks can be centered in military facilities. Most military bases have their own 911 call centers; some are tied into civilian systems. For example, Fort Belvoir, Virginia is served by Fairfax County's 911 system. A caller who is located on the base dials 911 and reaches a Fairfax County PSAP operator. That operator acts as the base' dispatcher, alerting military police, fire or EMS workers as needed (U.S. Army Military Police, 2005).

The National Emergency Number Association considers military bases to be underserved by 911 when compared to civilian jurisdictions. The majority of emergency services on military bases in the United States are equipped with at least Basic 911, but there are still many locations, which require callers to dial seven or ten digit numbers to reach emergency services. The 911-equipped centers are not uniformly equipped with CAD technology. A particular 911-equipped base may have E911 with CAD, or it may have Basic 911 with Automatic Number Identification and Automatic Location Identification, or Automatic Number Identification only. When callers must dial all 7 to 10 digits, the dispatcher may be able to display the originating telephone number, and may also be able to manually retrieve an address (but this information would not be available online). Often, base services are all dispatched centrally, meaning an incoming call is handled by one PSAP, and the same dispatcher will initiate and coordinate the response of military or DoD police, fire, and medical services as needed. As is the case in the civilian world, E911 and CAD functions are not yet available for callers using cellular phones. However, some military and Department of Defense personnel can also reach 911 via the satellite phone network operated by Iridium.

Iridium's phones, built by Motorola, use a constellation of 66 satellites in low orbit. The phones offer automatic 911 localization via satellite-based triangulation. Iridium® phones are not compatible with cellular systems, and must rely on a terrestrial wireless switch to connect a satellite phone user with landlines or cellular callers. The location of an Iridium user cannot be determined by cellular tower triangulation. This means that locating military callers in real time (and placing this information in Automatic Location Identification) using Iridium is more difficult than locating civilian callers using satellite phones when they can access the cellular phone network.

Military call centers can potentially play a sentinel role in biosurveillance if sufficient numbers of them can store information and provide the same access that civilian centers offer. For the moment, military centers remain less immediately useful, especially in regard to early detection.

4. COMMERCIAL ASSISTANCE CALL CENTERS

Commercial transport companies such as truckers and railroads, do not participate in 911; however, they operate control centers that stay in contact with their trucks and trains and these systems, along with advanced logistics systems central to their business, collect information that may be of value for biosurveillance. Larger enterprises use GPS to track the location of their trains and trucks. Their data systems contain information about what substances their trains and trucks are carrying and collect en-route information from train engineers and truck drivers who are in direct contact with control centers that dispatch assistance in emergencies. This information can include sudden changes in the engineer's, or driver's, ability to function, as would be caused by an illness or other mishap of interest (such as a spill) to biosurveillance. In addition, engineers or drivers may report incidents or sightings, such as dead animals or birds..

Railroads assign grade crossings location codes and usually post them, along with a toll-free telephone number, on the familiar red-light equipped "cross-bucks" sign. This is to enable passersby to notify the railroad if they witness a dangerous situation, such as an accident, a derailment, or a hazardous material spill.

The systems currently employed by trucking companies can automatically track not only where a truck is located, but also whether the trailer is empty or full, or disconnected from the tractor and thus located elsewhere. Drivers can reach a control center operator at any time, as needed, and so the company's call center will often become aware of a problem before a public 911 center will.

These companies, with fleets of thousands of tractors and trailers, also employ advanced, computerized logistical systems allowing a shipped item to be tracked continuously from shipping dock to receiving dock. Since some are licensed to carry hazardous materials, pharmaceuticals, biological samples and other items of interest to a bioterrorist, the companies' tracking systems and call centers are important sources of data to consider. Since these modern systems use relational databases, they will generally be technically amenable to modifications for purposes of biosurveillance. Access to these databases is as valuable to public health authorities as access to CAD databases, described earlier (JB Hunt Transport Services Inc., 2001, Schneider National Inc, 2005, Werner Enterprises, 2001, CRST International, 2005, CSX Corporation, 2005, Norfolk Southern, 2001).

5. POISON INFORMATION CENTERS

Dissemination of information to the public and clinicians is the primary role of poison information centers. The information is provided by specialists in poison information (nurses and pharmacists) and clinical toxicologists. The poison control specialists document each patient interaction electronically; these interactions form the basis for the poison center medical record. Additionally, if a physician treating a patient calls the center, he or she records the center's advice on the patient's medical record in the clinical office or emergency department.

The poison control center's record contains the standard personal and demographic information about the patient as well as specific information that are categorized by substance (the poison), treatment, patient symptoms, route of exposure, and laboratory values. Each of these areas constitutes a fully searchable section of the medical record. The values within each section are standardized and coded so that data from all poison centers in the United States can be incorporated into a single database. There is also a free-text documentation section. The free-text section is not searchable as part of the standard poison center medical record, and is therefore not included in the national database on poisoning exposures.

Currently, participating poison control centers submit data to the American Association of Poison Control Centers Toxic Exposure Surveillance System (AAPCC TESS) on a monthly, quarterly, or semiannual basis. Sixty-four poison centers submitted data in 2003 and these data reflect all 50 states and the Commonwealth of Puerto Rico (Watson et al., 2003). Participation in AAPCC TESS and submission of all data is mandatory for certified regional poison information centers and voluntary for noncertified centers. Data are reviewed and published on an annual basis approximately nine months after the end of the calendar year. Since data submission is intermittent and partially voluntary, there is no real-time surveillance of national poison center data for toxidromes that are consistent with exposure to biological and chemical terrorism agents. Furthermore, most individual centers do not conduct real-time toxico-surveillance. Figure 25.3 shows the type of surveillance that is feasible using poison center data.

However, poison information centers collect a large volume of information on exposures involving both humans and animals. For example, the Pittsburgh Poison Center responded to

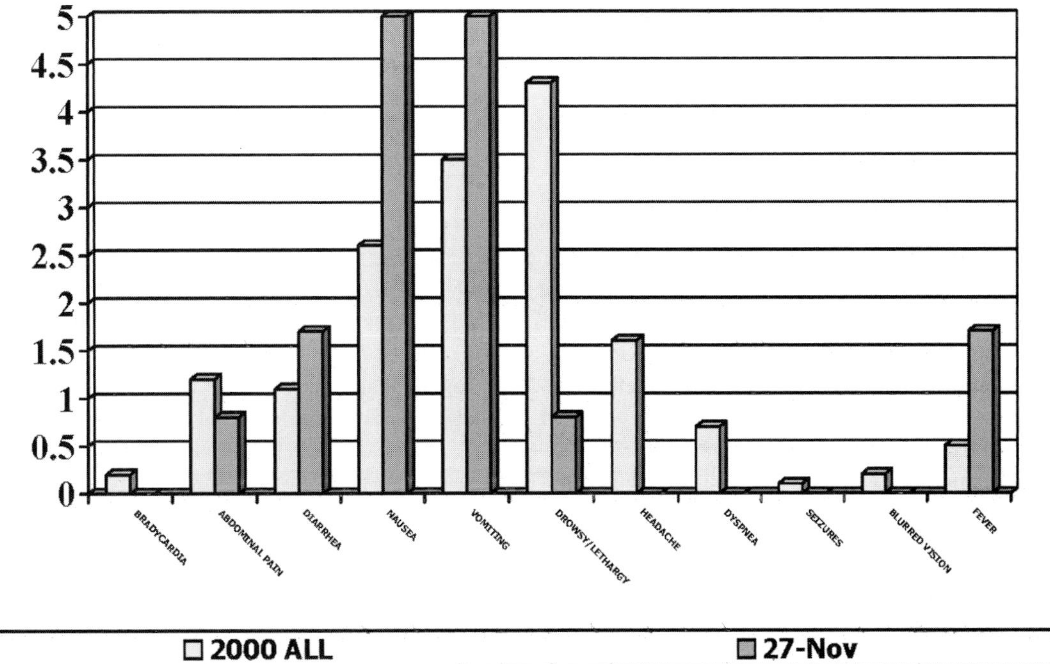

FIGURE 25.3 Daily Surveillance report from Pittsburgh Poison Center for November 27, 2001. Abscissa is the percent of calls normalized by total calls. Note the increased proportion of calls for this date relative to the average for diarrhea, nausea, and vomiting.

approximately 78,000 inquiries in 2000. Nearly 5000 of the inquiries involved animal exposures. This reflects the public awareness of poison information services. This call volume, the active data reservoir, and the reliance upon the poison center by the public and medical professionals make the poison center an ideal active surveillance center to identify sentinel events and to profile medically and demographically the nature of a biological or chemical terrorist event.

Software to facilitate the identification of sentinel events or toxidromes through the surveillance of real-time poison center data has not yet been widely deployed, though at least one vendor offers software which claims to tap into these data (Stout Solutions, LLC, 2005a). Currently, the only way to conduct analysis of data is through the identification of toxidromes based on expert opinion and then the tedious analysis of data to identify the presence of those toxidromes. The data are available readily and are accessible easily through the use of any database management system. Most poison centers document cases in real-time, which make the data fields available for immediate analysis. It would not be a challenging task to conduct real-time background surveillance of a single poison center's medical record database. However, poison centers within most states and throughout the nation are not linked in real-time and all national data analysis is conducted retrospectively. The challenge is to network U.S. poison centers so that data are contributed in real time. Artificial intelligence may help to identify sentinel events that are occurring, below the established threshold of recognition. The collaboration of data that includes poison center data, nonprescription and prescription pharmaceutical sales, emergency department diagnoses, veterinary clinic data on companion and large animal problems, and so on could help to identify problems in the early stages of their evolution.

6. USE OF CALL CENTERS IN BIOSURVEILLANCE

Biosurveillance systems have only recently begun to tap into call center data; these projects focus on 911 data in particular. Its value is not yet clearly established. The types of applications fall into two classes: stand alone systems that analyze only 911 data (perhaps implemented as part of a 911 system), and more general biosurveillance systems that collect 911 data as one of several types of data.

FirstWatch Biosurveillance®, a software suite produced by Stout Solutions, LLC, is an example of a stand-alone system used in Kansas City MO, Las Vegas, Plano TX, Tulsa OK, and Des Moines IA, and is being deployed in other areas of the Midwest. The vendor claims that its software will provide early warning during flu season by monitoring 911 data for spikes in respiratory and abdominal symptoms (Stout Solutions, LLC, 2005b). These systems' utility is under evaluation. Governmental public health spokespersons have noted that the FirstWatch alarm system does indeed become busier during flu season, but they question its true value; it has not justified any change in response strategy and its false positive rate is far too high for the system to be of any value, by itself, to detecting an unusual epidemic or bioterrorism event (Archer, 2005). A study conducted by New York City's Department of Health and Mental Hygiene (DOHMH) found that an EMS data collection alarm could be tuned to fairly high sensitivity for a community-wide respiratory outbreak (Mostashari F et al., 2003), but another found that the use of EMS data by itself achieved 58% sensitivity and a positive predictive value of 22% for "influenza-like illness" (Greenko et al., 2003). We note that any system that focuses on a single category of data collection arrives with the limitations inherent in its respective category. For example, an upper respiratory ailment is usually first seen by a primary care physician or nurse practitioner, not by an ambulance crew. Similarly, a seemingly innocuous skin rash that is caused by a bioterrorism incident is not likely to be reported through an emergency call center.

These results and the paucity of peer-reviewed research regarding use of emergency call center data suggest that we will need to study this subject more closely before we know how to best use these kinds of systems for biosurveillance purposes.

7. SUMMARY

Emergency call centers collect and distribute data likely to be relevant to biosurveillance. Railroad and trucking call and dispatch centers hold data regarding the contents of shipments and their status, including adverse events such as spills. Among the databases used by 911 call centers, CAD databases are the most useful and immediately relevant to public health authorities, because they can provide information collected in real time from the scene of an incident, and if appropriately linked to GIS, geographic information. The data often include observations by trained observers and so are generally reliable. These systems cover a wide area of the United States, since E911 with CAD is widely deployed in this country. Systems currently being deployed use modern database technology, so access does not present unduly difficult technical challenges.

Creating a technical connection to 911 and other databases access is feasible and not very difficult, as recent installations demonstrate, but there are legal and cost considerations. The budgets of the organizations that operate these systems do not provide funding for developing interfaces to outside entities, so governments must be prepared to support the necessary software and hardware modifications. Other systems that may contribute information of use to biosurveillance are commercial emergency call centers. Poison information centers collect large amounts of information from callers about poisoning incidents, and this information can be readily extracted from their databases. However, truly useful nationwide biosurveillance using poison center data awaits the linking of these centers in a network, as well as new developments in automating toxidromes surveillance and recognition.

REFERENCES

Archer, R. (2005). Director of Health, Kansas City, MO, Health Department. Personal Communication, May 11.

CRST International. (2005). *http://www.crst.com.*

CSX Corporation. (2005). *http://www.csx.com.*

Federal Communications Commission. (2005a). *http://www.fcc.gov/cgb/consumerfacts/voip911.html.*

Federal Communications Commission. (2005b). Enhanced 911—Wireless Services. *http://www.fcc.gov/911/enhanced/.*

Greenko, J., Mostashari, F., Fine, A., et al. (2003). Clinical evaluation of the Emergency Medical Services (EMS) ambulance dispatch-based syndromic surveillance system, New York City. *J Urban Health* 80(Suppl 1):i50–6.

Hixson, R. (2005). Technical Issues Director, National Emergency Number Association (NENA). Personal Communication, March 11.

JB Hunt Transport Services Inc. (2001). Personal Communication.

Mostashari F, Fine A, Das D, et al. (2003). Use of ambulance dispatch data as an early warning system for communitywide influenzalike illness, New York City. *J Urban Health* 80(Suppl 1):i43–9.

Motorola Inc. (2001). *http://www.motorola.com/home/.*

Norfolk Southern. (2001). Personal Communication.

Rivera, M. (2005). CEO Airsoft Wireless, Santa Monica CA. Personal Communication.

Schneider National Inc. (2005). Personal Communication.

Stout Solutions, LLC. (2005a). FirstWatch® Real-Time Early Warning System. Retrieved from: *www.firstwatch.net/.*

Stout Solutions, LLC. (2005b). New Computer Network Will Provide Early Flu Warnings in Central U.S. States. Press Release, January 27. *http://www.firstwatch.net/fwflu05.pdf.*

U.S. Army Military Police. (2005). Headquarters, Fort Belvoir, VA. Personal Communication.

Watson, W., Litovitz, T. L., Klein-Schwartz, W., et al. (2003). 2003 Annual Report of the American Association of Poison Control Centers Toxic Exposure Surveillance System. *Am J Emerg Med* 22:335–404.

Werner Enterprises. (2001). Personal Communication.

26

CHAPTER

The Internet as Sentinel

Michael M. Wagner and Heather A. Johnson
RODS Laboratory, Center for Biomedical Informatics
University of Pittsburgh, Pittsburgh, Pennsylvania

Virginia Dato
Center for Public Health Practice, University of Pittsburgh, Graduate School of Public Health, Pittsburgh, Pennsylvania

1. INTRODUCTION

The Internet is revolutionizing biosurveillance.[1] It is already the electronic network over which people and biosurveillance systems located anywhere on the planet communicate and exchange data. Modern biosurveillance systems (e.g., Real-Time Outbreak and Disease Surveillance [RODS], National Retail Data Monitor [NRDM], web-based disease reporting systems, BOSSS, ESSENCE, and BioSense) would not be possible without the Internet. Hospitals, laboratories, schools, retailers, and individuals contribute data to these systems via the Internet, and end users log in to these systems using Internet browsers. The Internet has made it possible not only to conceive of biosurveillance systems of global scope, but also to construct them, and to do so quickly.

The Internet facilitates communication among the many individuals involved in biosurveillance. Like the telegraph (1844) and telephone (1876), the Internet quickly and profoundly changed how professionals communicate (Kravitz, 1969). Colleagues use e-mail to exchange text, images, video, sound and (password-protected and encrypted) data. Newsgroups, blogs, and chat rooms support more structured or topical communication among self-defining communities of individuals. These communications, although inherently asynchronous, have become nearly synchronous due to the use of wireless devices such as BlackBerry® handheld computers. Increasingly, the Internet supports synchronous communication such as telephony, conference calls, video conferencing, instant messaging, telemedicine, and remote diagnosis.

The Internet has become a real-time encyclopedia for epidemiologists, infection control practitioners, researchers, and outbreak investigators. These individuals quickly "Google" to journal articles, technical reports, newspapers, books, and video. They find both highly reliable information (e.g., peer-reviewed papers, technical reports, books, ProMED-mail items, treatment guidelines, World Health Organization [WHO] alerts), and less reliable but still useful information such as newswire stories, unmoderated e-mail lists, chat rooms, and online newspapers.[2]

The Internet provides new tools for outbreak investigators. An investigator need only type an address into an internet browser to see a map, satellite image, or both of the location of an outbreak or a building of interest (Figure 26.1). When shoe-leather epidemiology is required, the same website can provide driving directions to the location. An investigator may use a search engine to obtain a restaurant's menu (e.g., when investigating a Salmonella outbreak) or use Internet-based "people locators" such as *http://www.anywho.com* to obtain contact information for individuals with communicable disease or their contacts.

Perhaps of greatest importance, the massive amount of information on the Internet represents a new form of *biosurveillance data*. A newswire story, for example, is input data for computer programs such as those operated by the Global Public Health Intelligence Network (GPHIN), which we discuss in detail in this chapter. These programs use search engines and text-filtering programs to bring information from websites, newswires, and other Internet resources to the attention of analysts. The idea that computers could search and filter large document collections was conceived soon after the birth of digital computers (Luhn, 1958, Luhn, 1961, Connor, 1967, IBM, 1962); but its application in biosurveillance required the invention of the Internet.

1 Many people use the term *Internet* loosely to refer not only to the physical network (what an information technologist would consider the Internet), but also the information that is available on the World Wide Web and applications such as email that utilize the network. In this chapter, we also use the term *Internet* loosely.

2 We note that the Internet has become an important means for public health education. The Center for Disease Control and Prevention's Public Health Training Network http://www2a.cdc.gov/phtnonline/, and the Supercourse, http://www.pitt.edu/~super1/ are just two of the many examples of readily available free training.

Handbook of Biosurveillance
ISBN 0-12-369378-0

FIGURE 26.1 MSN Virtual Earth hybrid map. Visual inspection of this map may suggest hypotheses about the cause of illness (e.g., proximity to vegetation, industrial plants, and special events) that a more conventional map might not suggest. The oblong object is the stadium at the University of Pittsburgh. Microsoft product screen shot reprinted with permission from Microsoft Corporation.

In this chapter, we focus on the Internet as a source of biosurveillance data and begin with a discussion of ProMED-mail—a worldwide biosurveillance system crafted from standard e-mail-list software. We then provide a brief primer on the Internet as background for the subsequent discussion of GPHIN—a system that automatically analyzes newswire and other resources available on the Internet—and futuristic biosurveillance systems that may search spatial and temporal patterns of Internet utilization that may provide an early indication of an outbreak.

2. INTERNET AS SENTINEL I: PROMED-MAIL

ProMED is the Program for Monitoring Emerging Diseases. As described on its website, ProMED-mail is "an Internet-based reporting system dedicated to rapid global dissemination of information on outbreaks of infectious diseases and acute exposures to toxins that affect human health, including those in animals and in plants grown for food or animal feed." ProMED-mail is developed and maintained by the International Society for Infectious Diseases and the Harvard School of Public Health, and is available at *http://www.promedmail.org*.

ProMED-mail is an e-mail list. Briefly, an e-mail list is a computer program that can accept subscribers (to the list), accept e-mails from subscribers, allow an operator of the software to decide whether to *post* an e-mail to the list, and for the operator to assign a posting to a topical category (called a *thread*). The primary function of an e-mail list is to disseminate new postings by e-mail to subscribers in a selective manner, based on their indicated interest in specific threads and their preferences for receiving new information immediately or periodically. We discuss e-mail lists in detail in Section 3.

ProMED-mail is a moderated e-mail list, which means that list moderators screen the reports submitted to ProMED-mail before posting them to a thread (and adding them to the archives on the ProMED website). The postings in ProMED-mail originate from a number of sources including ProMED-mail subscribers and the moderators of the threads, who may add information from media, government, or other sources. ProMED-mail has been in existence for 12 years and has 30,000 subscribers in 150 countries. Its subscribers include public health physicians, epidemiologists, animal health experts, reporters, and many other professionals and lay people. A subscription to ProMED is free.

ProMED-mail demonstrated its value during the 2003 outbreak of SARS. On February 10, 2003, ProMED-mail disseminated a posting entitled "Pneumonia–China (Guandong): RFI" asking for information on an outbreak of pneumonia in the Guandong province of China (ProMED-mail, 2003e). Figure 26.2 reproduces this and the second posting in what was subsequently renamed the "SARS thread" on ProMED-mail.

On the following day (February 11), ProMED-mail disseminated a WHO Routine Communicable Disease Surveillance Report about an outbreak of acute respiratory syndrome with 300 cases and five deaths in Guangdong Province (World Health Organization, 2003), along with a related Associated Press article (ProMED-mail, 2003c). ProMED-mail disseminated additional information from a *British Medical Journal* article on February 21 (ProMED-mail, 2003d).

There were 15 additional postings between February 10 and March 14. On March 14, ProMED-mail disseminated five reports of outbreaks of deadly or severe respiratory illness in four countries in East Asia (Singapore, Taiwan, Vietnam, and China), many in hospitals, and all with no known etiology (ProMED-mail, 2003b). That same day, ProMED-mail disseminated a Government of Ontario Press Release about four cases of atypical pneumonia in Canada (Toronto and British Columbia), and the editor noted an unconfirmed report that the British Columbia case had a travel history to Hong Kong (ProMED-mail, 2003a). On March 15, 2003, ProMED-mail followed up with a formal WHO report of a new entity named SARS (ProMED-mail, 2003f).

This brief chronology illustrates the role of ProMED-mail in keeping the multinational infectious disease community abreast of information. No single country or group of researchers could have easily accessed the wide variety of information submitted to and distributed by ProMED-mail about SARS in February and March 2003. Readers interested in the rest of the SARS story may refer to Chapter 2 or they can search ProMED-mail with the keyword "SARS" and read the story as it unfolded.

ProMED-mail is an example of a biosurveillance system in which computers are used for data collection, archiving, and selective dissemination of new information, but not for analysis. People (list moderators) analyze the data, deciding whether to post the information to the list, and if so, with what title and framing comments to add a submission to a thread.

3. INTERNET PRIMER

People use the word "Internet" loosely to refer to both the physical Internet as well as the software applications that run on it. The physical Internet comprises the wires, optical fiber, satellites, protocols, and routing computers (what a technologist would consider the Internet). Examples of the software applications include e-mail programs, web servers, search engines, instant messaging programs, and file transfer programs. In this chapter, we use the term "Internet" to refer to both the physical Internet and the software applications that run on it.

3.1. The Physical Internet

The physical Internet is the "network of networks," a set of linked electronic networks–much like the telephone system–that computers worldwide use to communicate with each other. Computers transmit information across the Internet using several computer languages, known as *protocols*. The two most important protocols are IP (Internet protocol) and TCP (transmission control protocol). IP specifies a standard form of packets of data transmitted across the Internet through a network of intelligent routers. It is a globally agreed upon format where bytes in a packet of data specify the destination address, the source address, the lifetime of the packet, and many other pieces of bookkeeping information. TCP is a set of standards for how complex communications between machines are broken up into many packets at one end, and reassembled at the other end, with built in redundancy to cope intelligently with lost packets, duplicate packets, and automatic request for resending of packets.

3.2. The World Wide Web

The World Wide Web (web) is a hypertext-based, distributed information system originally created by researchers at CERN, the European Laboratory for Particle Physics, to facilitate sharing research information. The web presents the user with documents, called web pages, full of links to other documents or information systems. Selecting one of these links, the user can access more information about a particular topic. Web pages include text as well as multimedia (images, video, animation, sound).

URLs (uniform resource locators) are the address system used by the Internet to locate resources such as web sites. URLs specify the type of resource being accessed such as gopher or hypertext, the address of the server, and the location of the file on the server. For example, the complete URL for the RODS Laboratory website is *http://rods.health.pitt.edu/index.htm*. The *http://* indicates the access method as "hyper text transfer protocol." The *rods.health.pitt.edu* is the address of the server, and *index.htm* is the initial page of the website. Web browsers will assume *http://* and *index.html*, so you can simply use *rods.health.pitt.edu* as the URL. When authors of web pages embed URLs into HTML documents (an authoring language for web pages), they are creating *hyperlinks*, which a web browser uses to jump to another page that may be located on the same computer or on a computer on the other side of

Archive Number 20030210.0357
Published Date 10-FEB-2003
Subject PRO/EDR> Pneumonia - China (Guangdong): RFI

PNEUMONIA - CHINA (GUANGDONG): RFI
****
A ProMED-mail post
<http://www.promedmail.org>
ProMED-mail is a program of the
International Society for Infectious Diseases
<http://www.isid.org>

[1]
Date: 10 Feb 2003
From: Stephen O. Cunnion, MD, PhD, MPH <cunnion@erols.com>

This morning I received this e-mail and then searched your archives and found nothing that pertained to it. Does anyone know anything about this problem?

"Have you heard of an epidemic in Guangzhou? An acquaintance of mine from a teacher's chat room lives there and reports that the hospitals there have been closed and people are dying."

--
Stephen O. Cunnion, MD, PhD, MPH
International Consultants in Health, Inc
Member ASTM&H, ISTM
<cunnion@erols.com>

******
[2]
Date: 10 Feb 2003
From: Jack Soo <kyip@gmx.net>
Source: Hong Kong's Information Services Department [edited]
<http://www.news.gov.hk/en/category/healthandcommunity/030210/html/030210en05017.htm>

Take precautions when traveling abroad

The public should take precautions when traveling abroad and tell doctors if there are signs of fever or infections that do not abate and are unusual, says Secretary for Health, Welfare & Food Dr Yeoh Eng-kiong.

Commenting on the problem of pneumonia on the Mainland, Dr Yeoh said the Department of Health has already touched base with the Guangdong authorities to learn more about the type of infection prevalent there. The department will also determine whether there is any particular risk of that infection coming to Hong Kong.

He assured the public that the Government is always on the alert, as the Department of Health has a very good communicable disease surveillance system.

Coupled with the network of reporting sources both from the public and private sectors, as well as communication channels with authorities on the Mainland and Macau, the Government is informed of any infections that may spread to Hong Kong.

He called on the public not to be unduly concerned.

"We'll certainly be doing our part as the health authorities, but individuals should always take precautions when they travel aboard," he added.
--
Jack Soo
<kyip@gmx.net>

[ProMED-mail appreciates the preliminary information above from Jack Soo and would be grateful for any additional information. The etiology and extent of this apparent outbreak of pneumonia are unclear, as is whether the outbreak is secondary to influenza. - Mod.LM]
...............lm/jw/pg/lm
##
ProMED-mail makes every effort to verify the reports that are posted, but the accuracy and completeness of the information, and of any statements or opinions based thereon, are not guaranteed. The reader assumes all risks in using information posted or archived by ProMED-mail. ISID and its associated service providers shall not be held responsible for errors or omissions or held liable for any damages incurred as a result of use or reliance upon posted or archived material.
****
Visit ProMED-mail's web site at <http://www.promedmail.org>. Send all items for posting to: promed@promedmail.org (NOT to an individual moderator). If you do not give your full name and affiliation, it may not be posted. Send commands to subscribe/unsubscribe, get archives, help, etc. to: majordomo@promedmail.org. For assistance from a human being send mail to: owner-majordomo@promedmail.org.
##
##

FIGURE 26.2 Beginning of the ProMED-mail SARS thread.

the world. These hyperlinks are responsible for the Web's web-like structure and its name (as well as the name of another type of program called a "web spider," which we discuss shortly).

3.3. Websites

A *website* is a related collection of web pages and files (*resources*) that are managed by a single organization usually, but not necessarily, on a single computer or at a single physical site. The address (URL) for a website takes you to the initial (home) page. From the home page you can go to all the other pages on the website.

3.4. Search Engines

An Internet search engines is a computer system that (1) locates and indexes web pages, and (2) processes queries from users who are searching for information on the web. The most common way people find information on the Internet is through a search engine (PEW Internet & American Life Project, 2004).

A search engine comprises three components: a *web spider*, a *database*, and one or more *information retrieval algorithms*. The web spider (also known as a "web crawler") searches the Internet for new web pages (Gordon and Pathak, 1999). It systematically follows hyperlinks found on known pages. If the spider comes upon a web page it has not previously encountered, it sends this page to the information retrieval algorithms for indexing and storage in the database (Kirsanov, 1997). The indexing enables the search engine to retrieve the URL of the web page from the database based on query terms entered into the search engine by users of the search engine.

The information retrieval algorithms create the indexing from the content of the web page (words, phrases, whether there are images), as well as whatever other clues the algorithm developer can exploit (e.g., the popularity of the page, the nature of pages that hyperlink to it).

Information retrieval algorithms are the subject of study of a subfield of information science called "document retrieval," and there are many books on the topic (e.g., Pao, 1989, van Rijsbergen, 1986, Salton and McGill, 1983) Briefly, there are three basic approaches to document indexing and retrieval: Boolean, vector space, and probabilistic. These approaches are distinguished partly by their use of different retrieval strategies. Boolean systems retrieve the subset of documents whose indexing terms match exactly the query generated by the user (Salton, 1989). Many electronic library catalogs use Boolean retrieval algorithms. Vector space and probabilistic systems impose a partial ordering on the document collection, according to a document score. The vector space algorithms compute a heuristic score, typically the cosine of the query and document's index term vectors (Salton, 1989). Probabilistic systems order documents by the probability of relevance, which they learn from a training set of documents that developers of the system or end users have judged relevant or not relevant. All Internet search engines use one of the latter two approaches, as does the GPHIN system that we discuss in the next section (its information retrieval algorithm is proprietary). GPHIN computes document scores (called "relevance scores"), and uses them to select documents to disseminate to subscribers.

Table 26.1 displays the current market share of the 10 most frequently used search engines, as measured by number of searches (as of March 2005) (Nielsen//NetRatings MegaView Search, 2005). These systems differ according to the methods by which their web spiders search the Internet and in their information retrieval algorithms, which accounts for differences in the information they retrieve in response to the same user query. At present, Google®, Yahoo®, and MSN® are the most popular search engines (Sullivan, 2005).

3.5. E-Mail

E-mail—for which the Internet plays the role of postal service by receiving, sorting, and delivering the mail—is the most popular use of the Internet (Pew Internet and American Life Project, 2003). Approximately 58 million American adults check their e-mail via the Internet on any given day (Pew Internet and American Life Project, 2005). E-mail is packaged and routed through the Internet via a communication protocol called "simple mail transfer protocol" (SMTP).

3.6. E-Mail Lists

An e-mail list is mechanism to distribute an e-mail to a set of people who have indicated an interest in receiving e-mails from this source. An organization sets up an e-mail list by installing software (e.g., LISTSERV® or Majordomo) on its own hardware or by contracting with another organization to install and maintain the software and hardware.

Individuals then may subscribe to the list, which usually means providing at a minimum their e-mail addresses and usually some preferences about the types of e-mails they wish to receive and the periodicity. A subscriber can elect to receive e-mails on an immediate message-by-message basis or in a daily/weekly digest format. Most e-mail lists are free, but

TABLE 26.1 Top 10 Search Providers Ranked by Searches, March 2005

Search Engine	Searches (Thousands)	Share of Total Searches (%)
Google Search	2,057,897	47%
Yahoo! Search	907,751	21%
MSN Search	592,153	14%
AOL Search	195,130	4%
Ask Jeeves Search	78,758	2%
My Way Search	73,162	2%
Netscape Search	63,736	1%
iWon Search	44,039	1%
Earthlink Search	39,993	1%
My Search Search	36,194	1%

From Nielsen//NetRatings MegaView Search. (2005). *Top 10 Search Providers Ranked by Searches.*

some require payment of a subscription fee. Usually, each list has a general topical area, meaning (theoretically) that all posting to the list will relate to that topic. An e-mail list can be moderated (meaning that submissions are reviewed by an editor prior to posting) or unmoderated.

ProMED-mail (discussed in detail above) is a moderated e-mail list whose topic is threats to the public's health. If a ProMED-mail subscriber such as a public health physician or a veterinarian wishes to be notified immediately of information about animal, zoonotic, or vector-borne disease outbreaks, she would subscribe to the *ProMED-ahead list*, which forwards e-mail on those topics as soon as an e-mail is posted to the list. There is also *ProMED-ahead-digest*, which sends a summary e-mail daily containing all the day's postings. E-mail lists typically provide an online archive of all past postings to the list, which a subscriber can search for specific topics or words.

3.7. Internet as Sentinel II: The Global Public Health Intelligence Network

The Global Public Health Intelligence Network (GPHIN) is a new type of biosurveillance system that continuously monitors global media sources such as newswires and websites for articles about disease outbreaks and other events of international public health concern (World Health Organization, 2005, Mawudeku and Blench, 2005, Heymann and Rodier, 2001). GPHIN disseminates this information to subscribers such as the WHO, governments, and non-governmental organization. Health Canada in collaboration with the WHO developed the initial prototype of GPHIN in 1997. The annual subscription fee for GPHIN in 2004 ranged from $30,000 Canadian (for a university) to $250,000 (for a country).

The *input data* for GPHIN comprise news service items, newspaper articles, and reports from approximately 10,000 primary sources. Two news aggregators—Factiva and Al Bawaba—account for many of the sources.[3] Al Bawaba is the Arabic language content provider to GPHIN.

ProMED-mail is a GPHIN data source. GPHIN also uses a web spider to search selected websites. GPHIN selects articles for dissemination to its subscribers using a two-step process (Figure 26.3). First, a proprietary information retrieval algorithm computes a relevance score for each article for each of eight topics of interest: *animal diseases, human diseases, plant diseases, biologics, natural disasters, chemical incidents, radioactive incidents*, and *unsafe products*. If an article has a sufficiently high relevance score (X in Figure 26.3), GPHIN disseminates it to subscribers immediately. If an article scores

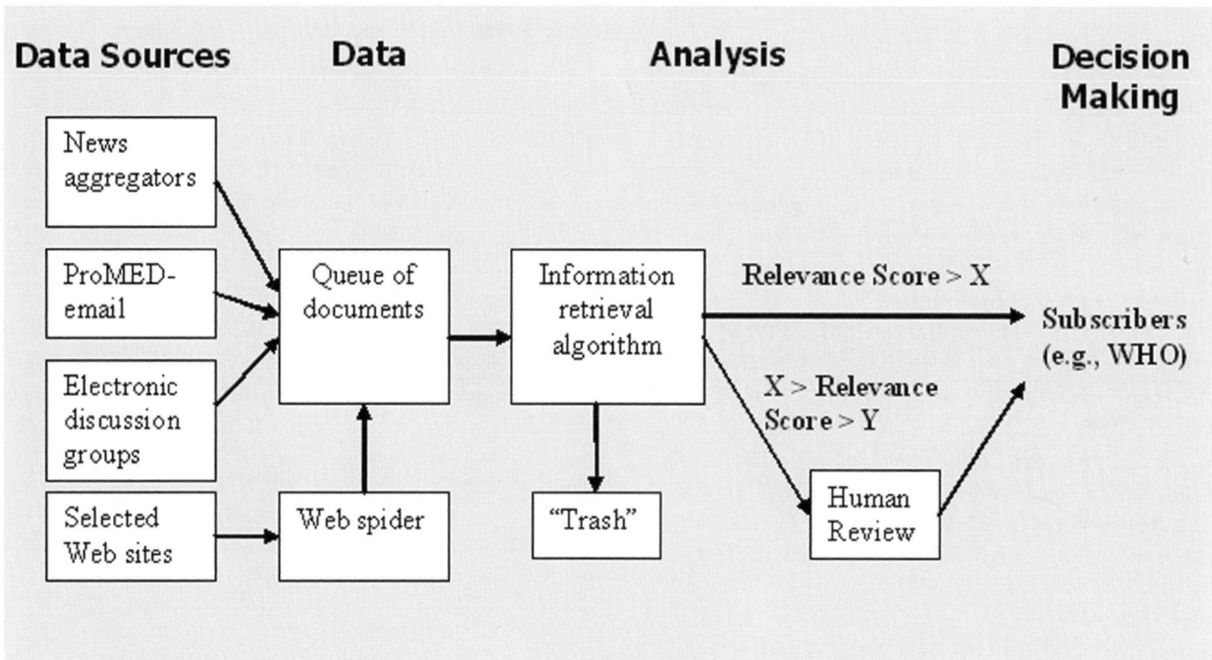

FIGURE 26.3 The GPHIN Biosurveillance System. An information retrieval algorithm computes relevancy scores for a large number of documents as they arrive from news aggregators, ProMED e-mail, GPHIN's web spider and other sources. If a document's relevance score is greater than X (high relevance), it is disseminated immediately. If the score is between Y (lower relevance) and X, it is queued for human review. If the document scores lower than Y, it is assigned to "Trash" but also reviewed, but not with the same level of effort or immediacy. (Components not shown: machine translation, human editing of machine translations, and archiving.)

3 Factiva(R) aggregates news from approximately 9,000 primary sources in 152 countries and 22 languages. A1 Bawaba is the Arabic language content provider to GPHIN.

below a certain threshold (Y in Figure 26.3), it is automatically "trashed." The system places articles that score between the two thresholds in a queue to await human analysis (the second step). The net result of this computer filtering followed by human review is approximately 100–150 articles disseminated per day.

Subscribers then decide whether to take action based on the information. WHO, for example, contacts about four countries per week to request verification of information found in GPHIN articles.

GPHIN is multilingual. It analyzes articles in eight languages: Arabic, Chinese (Simplified), and Chinese (Traditional), English, Farsi, French, Russian, and Spanish. The automatic filtering of articles occurs in the original language. Natural language programs translate English articles into the other seven languages and non-English articles into English. Human analysts edit the translations performed by the translation software prior to distribution.

Although Health Canada describes GPHIN as a near real-time system, the reference point is the time that an outbreak or other event is described in one of the sources being monitored (i.e., after it is detected by some other means and written about). Nevertheless, GPHIN often provides an earlier warning of events of interest to the international community than other methods. Of the 578 outbreaks verified by WHO between July 1998 and August 2001, 56% were initially picked up by GPHIN (Heymann and Rodier, 2001).

The U.S. Department of Agriculture's (USDA) Animal and Plant Health Inspection Services, Veterinary Services, Center for Emerging Issues (CEI) also filters and analyzes information gleaned from the Internet for indications of the emergence and/or spread of animal diseases in the United States and abroad. CEI uses an electronic scanning methodology that analyzes large amounts of Internet-sourced text in a relatively short time. The system evaluates the text against predefined queries to extract articles of possible interest, thus quickly identifying information requiring further review. CEI staff construct queries to find records containing information about specific animal diseases, disease situations for which a specific disease is not identified, outbreaks of a novel disease, and changes in nations' veterinary and regulatory activities that may indicate disease outbreak. For example, one query finds records that may indicate an animal disease outbreak of unknown or undiagnosed cause. In this case, the query includes specific animal names in association with terms such as dead, dying, sick, ill, outbreak, and so on. CEI periodically reviews and updates the search queries.

Internet data sources such as listserves, industry websites, and articles from worldwide electronic news sites are loaded into the system on a daily basis. CEI analysts review articles that are tagged by the queries to determine if the information is of value. Information that is deemed useful is then stored in the system's file manager.

Typically, CEI's querying and filtering tool analyzes 900 to 1000 records per day, of which about 190 to 200 are tagged and then manually reviewed. Of the reviewed records, on average, about two or three per day are deemed of value and are stored. CEI then tracks these disease situations or potential disease situations.

Both GPHIN and the CEI system use computers for not only data collection, archiving, and selective dissemination of new information, but also for finding documents and filtering them for relevance. The latter uses distinguish them from ProMED-mail, in which list moderators do all the filtering, and there is no automatic searching of the web.

4. INTERNET AS SENTINEL III: MONITORING USAGE OF HEALTH WEBSITES AND HEALTH-RELATED QUERIES TO SEARCH ENGINES

The goal of ProMED-e-mail and GPHIN are similar: to detect an outbreak or unusual events close to the time that astute observers or news media report them. In this section, we explore the potential of analyzing patterns of Internet utilization by sick individuals (or their caregivers) to detect events even earlier.

This area of research is predicated on the assumption that sick people (or their caregivers or doctors) turn to the Internet early during the course of an illness. At present, it is known only that individuals frequently turn to the Internet for health information, but not that they turn to the Internet early in their illnesses. According to a 2004 Harris Interactive Poll, almost 33% of American adults say they "often" or "very often" search for health information online (Harris Interactive, 2004). A survey by Pew Internet found that on any day as many as 7 million American adults look for health information online. There have been no surveys of the timing of searches relative to illness onset or that distinguished between chronic and acute diseases.

The conventional industry measure of Internet utilization is the number of "hits" to a website from some region for some period (or the number of searches received by a search engine). To detect increases in Internet utilization by sick individuals against a background of extremely high levels of utilization for other purposes, more specific measures of Internet utilization will likely be necessary—measures such as the number of requests to a health-related website for documents about influenza or the number of queries to Internet search engines that include the word "fever."

4.1. Web Page Accesses

Many organizations operate websites designed for patients or for healthcare providers. For example, the National Library of Medicine at the National Institutes of Health operates MedlinePlus (*http://www.medlineplus.gov*), a nationally recognized website that contains health information for patients and caregivers on virtually every medical topic of interest. WebMD, Medscape, Embase, the Cochrane Library of Databases, and Medline/PubMed provide a wide range of information from very general information on diseases to peer-reviewed, evidence-based practice guidelines. The WHO, CDC, and state and local health departments maintain websites targeted to the public, medical professionals, and public health professionals.

These websites maintain *web access logs* that contain a variety of information about users and the information that users accessed. For each access to the site, the log contains a record of:

- IP address—The IP address of the computer making the HTTP request.
- RFC—A field that is used to identify the person requesting information from the website. (This field will rarely be available and only from websites that register users.)
- Auth—A field included to list the authenticated user, if required for website access.
- Timestamp—Date and time of request.
- Action—The action requested (e.g., request for a specific article or for a search).
- Status—Whether the user experienced a success, redirect, failure, or server error when they tried to access the web page.
- Transfer volume—How many bytes were transferred to the requester.
- Referring URL—The URL of any website that the requester was using just before coming to the present site.

Johnson et al. (2004) used correlation analysis to study the relationship between the volume of web page accesses to a health-related website for articles on the topic of influenza and the number of influenza-like illness (ILI) cases reported by the U.S. Influenza Sentinel Physicians Surveillance Network. They obtained 12 web access logs (one for each month of the 2001 calendar year) from Healthlink, a consumer health information website developed and maintained by the Office of Clinical Informatics at the Medical College of Wisconsin. The web access logs contain the identification number of documents that persons retrieved and any free-text search terms that they may have entered into the search engine that brought them to the documents. The study was limited to an analysis of the documents retrieved, not the search terms.

Although Healthlink receives queries from many countries, the study was limited to requests from the United States. Influenza activity varies by location, and the influenza season is inverted in the Southern Hemisphere. They removed requests made by users outside of the United States using GeoIP Country, an open source Perl module developed by MaxMind™ to analyze the IP address of each record and assign (when possible) the country of the user. We discuss "geolocation" (determining a requester's locations from an IP addresses) in detail shortly.

Figure 26.4 shows the weekly accesses of 17 documents about influenza diagnosis/treatment and vaccination versus the ILI reference standard. Both time series are normalized into units of standard deviations from the mean. The correlation analysis was limited to those periods for which ILI reference data were available; namely, portions of the 2000–2001 and 2001–2002 influenza seasons (weeks 1–20 and 40–52 of the year 2001, respectively).

The correlation between the article-access time series and the ILI reference standard was 0.78 for weeks 1–20 of 2001 and 0.76 for weeks 40–52. The time lag at which the correlation was maximal was zero.

Although these correlations are interesting, we note that sick individuals have a wide choice of health-related websites, including those located in other countries. Therefore, a monitoring system based on analysis of website access would face a monumental organizational task in achieving high coverage of all sick individuals from a particular region who accessed websites.

We also note that, at present, the potential of monitoring websites designed for clinicians is not high. In addition to the number of such sites, physicians and nurse practitioners tend to consult each other, textbooks, and reference manuals rather than websites when they have patient-specific questions that they want to answer. A recent survey of 13 faculty members and 25 residents found that "the group sought immediate answers to 66% of questions [that arose in routine practice]," but that they "most commonly use another person or a pocket reference" when doing so (Ramos et al., 2003). In the future, it may be that clinicians (or patients) will consult new types of websites designed to provide diagnostic assistance—similar to the BOSSS system for cattle diseases described in Chapter 13—and that such websites will be few in number, or linked.

4.2. Queries to Search Engines

About half of people who use the Internet to access health information online do so via a search engine; thus, monitoring the queries received by search engines is a potential biosurveillance strategy. A rapid increase in the number of Google searches containing the word "fever" would be of concern in the absence of a known outbreak or other explanation. In contrast to website monitoring, monitoring of queries to the three most popular search engines would catch nearly 80% of the health-related searches issued over the Internet, assuming that people do not switch to less commonly used search engines for health-related searches.

Privacy policies of the search engines (and websites) are, however, a barrier to developing a system to monitor query data from search engines. Organizations that operate search engines (and websites) respect the rights of individuals to confidentiality and have strict policies concerning the distribution and use of personal information. There are "privacy-protecting" techniques that might allow society to benefit from this information (should research prove its value), while protecting an individual's right to confidentiality. For example, an organization operating a search engine could run biosurveillance algorithms within its computing facility. It could send only the results (e.g., daily or hourly counts of queries that matched predefined criteria from a city or county) to governmental public health.

4.3. Geolocation

Because of the importance of spatial analysis in biosurveillance, the precision with which the spatial location of an Internet user can be determined affects the value of Internet utilization data for biosurveillance. There are numerous geolocation methods based on the IP address associated with queries.[4] However, none of the methods claims 100% accuracy and their accuracy degrades as one tries to determine precise location information for reasons that include the use of dynamic IP addressing. Many companies that offer geolocation products will report accuracy rates for the country or city, but not for smaller geographic regions. Digital Envoy™, one of the leaders in geolocation, reports country accuracy rates of 99.4% and city accuracy rates of up to 94% (Digital Envoy). Digital Envoy also reports that they can obtain zip codes for IP addresses, but do not provide accuracy rates for this information, stating only "Accuracy rates vary greatly" and "Accuracy rates are higher at the 3-digit zip code level."

5. INTERNET AS SENTINEL IV: SELF-REPORTING

Figure 26.5 shows several screen snaps from a prototype disease self-reporting system called *Diarrhea Detective,* developed by Jeremy Espino. In this prototype, an individual who suspects she is ill from exposure to food would access the system to (1) register her illness, (2) record the restaurants at which she had recently dined, and (3) receive feedback about whether other individuals who had eaten at those locations were self-reporting at a higher rate than usual. Although this Web-based application was never field tested, it illustrates a fourth potential use of the Internet, as a vehicle that a citizenry could use to elucidate collaboratively a point-source outbreak. We note that for such a scheme to

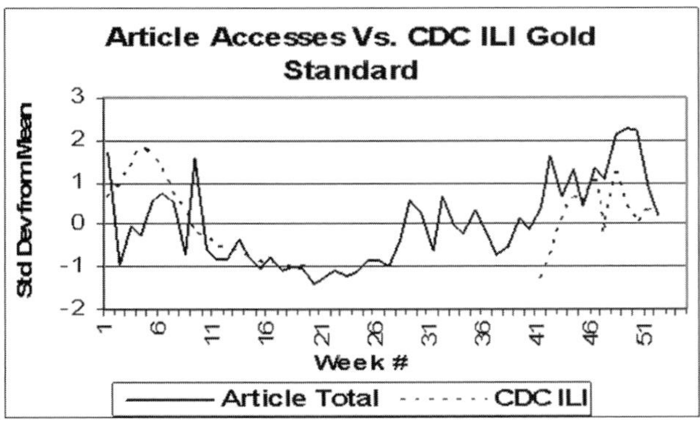

FIGURE 26.4 Correlation of influenza article accesses on the web and influenza activity as measured by standard surveillance data sets, 2001.

Diarrhea Detective

Copyright University of Pittsburgh, 1999. All rights reserved.

The purpose of the Diarrhea Detective is to identify local sources of infectious diarrhea. Your participation in this website will assist your community health department in preventing future illness and outbreaks. All personal information is kept confidential and will only be used to contact you if needed.

Step 1 of 2: Identification (fields marked with * are optional)

*First Name
*Last Name
*Address
Zip Code (required)
*Phone Number
*Email Address

Submit Identification | Reset Form

Step 3 of 3: Confirm Eatery Address

Confirm the eatery by selecting it's address. You may want to go back to the previous page and retype your eateries if there are no addresses available.

Address:
not listed
not listed

Get me a map!

Submit Confirmation | Reset Form

Address services provided by Yahoo and InfoUSA/Map services provided by MapBlast

Step 2 of 3: Questions

did your symptoms begin (diarrhea, vomiting, stomach pain)?

Jan 01 1999

Where did you eat in the last 24 hours prior to symptoms? (You may identify up to four places where you ate. The pulldown list identifies current suspected eateries. If you do not see the place where you ate, fill in the name in the textbox.)

Eatery 1: or
Eatery 2: not listed or
Eatery 3: not listed or
Eatery 4: not listed or

Did other people in your family have similar symptoms and ate at the above places?

yes no no comment

What did you eat?

Submit Answers | Reset Form

FIGURE 26.5 Screen snaps from Diarrhea Detective, a prototype self-reporting tool.

4 An *IP address* is a 32-bit identifier for a computer that is its address on the Internet. It is comprised of four numbers separated by periods (e.g., 123.456.789.1).

work, the public would have to know it existed and find the feedback provided by the system sufficiently valuable to use it.

6. SUMMARY

The Internet plays many roles in biosurveillance. It serves as the electronic network over which people and biosurveillance systems located anywhere on the planet communicate. It is the platform for the web, which is a medium for rapid dissemination of new information and an electronic encyclopedia for the storage and retrieval of vast quantities of older information. It provides a variety of programs such as e-mail, maps, driving directions, and person locators that epidemiologists and other biosurveillance personnel can use to increase their efficiency.

Organizations such as ProMED, Health Canada, and the USDA's Veterinary Services, CEI, have constructed novel biosurveillance systems that use general Internet software tools such as e-mail lists and search engines to bring the existence of known outbreaks or clusters of cases to the attention of the world's public health and animal health communities. In the future, we expect that new forms of biosurveillance may emerge based on analysis of Internet utilization, websites that offer diagnostic assistance, or websites that encourage self-reporting of illness.

ADDITIONAL RESOURCES

ProMED-mail

- *http://www.promedmail.org*. An e-mail list that disseminates information about events of interest to human and animal health surveillance. Subscription is free.

Websites for Patients and Clinicians

- MedlinePlus– *http://www.medlineplus.gov*
- healthfinder®– *http://www.healthfinder.gov*
- Health Information from the National Institutes of Health–*http://www.health.nih.gov*
- First Gov for Consumers– *http://www.consumers.gov*
- Veterans Consumer Health Information– *http://www.va.gov/vbs/health/consumers.htm*
- Food and Drug Administration– *http://www.fda.gov*
- MedlinePlus (*http://www.medlineplus.gov*)
- WebMD (*http://www.webmd.com*)
- MedHunt from Health on the Net (*http://www.medhunt.com*)
- OmniMedicalSearch.com (*http://www.omnimedicalsearch.com*)
- MedicineNet.com (*http://www.medicinenet.com*)

ACKNOWLEDGMENTS

Grant T15 LM/DE07059 from the National Library of Medicine supported Heather Johnson. We thank the USDA, Centers for Epidemiology and Animal Health, Center for Emerging Issues for providing a description of their text filtering system; William Hogan and Andrew Moore for reading and commenting on the chapter; Ron Aryel for suggesting the chapter title; and Jeremy Espino for permission to use the screen snaps from Diarrhea Detective.

REFERENCES

Connor, J. H. (1967). Selective dissemination of information: a review of the literature and the issues. *Library Quarterly* 373–91.

Digital Envoy. *http://www.digitalenvoy.net/*.

Gordon, M. D., Pathak, P. (1999). Finding information on the World Wide Web: the retrieval effectiveness of search engines. *Info Processing Manag* 35:141–80.

Harris Interactive. (2004). Email, Research, News and Weather, Information about Hobbies or Special Interests Top the List of How People Use the Internet as it Continues to Grow: Biggest Increases Since Last Year: Travel Planning and Health. *http://www.harrisinteractive.com/harris_poll/index.asp?PID=527*.

Heymann, D. L., Rodier, G. R. (2001). Hot spots in a wired world: WHO surveillance of emerging and re-emerging infectious diseases. *Lancet Infect Dis* 1:345–53.

IBM. (1962). *Storage, Retrieval and Dissemination of Information*. White Plains, NY: IBM.

Johnson, H. A., Wagner, M. M., Hogan, W. R., et al. (2004). Analysis of Web access logs for surveillance of influenza. *Medinfo* 11:1202–6.

Kirsanov, D. (1997). HTML Unleashed PRE: Strategies for Indexing and Search Engines. How Search Engines Work. *http://www.webreference.com/dlab/books/html-pre/43-1.html*.

Kravitz, H. (1969). The telephone and the practice of medicine. *Illinois Med J* 135:268–9.

Luhn, H. P. (1958). A business intelligence system. *IBM J* 2:314–9.

Luhn, H. P. (1961). Selective dissemination of new scientific information with the aid of electronic processing equipment. *American Documentation*, April, pp. 131–8.

Mawudeku, A., Blench, M. (2005). *Global Public Health Intelligence Network (GPHIN)*. Ottowa, Canada: Global Public Health Intelligence Network.

Nielsen//NetRatings MegaView Search. (2005). *Top 10 Search Providers Ranked by Searches*.

Pao,. (1989). *Concepts of Information Retrieval*. Englewood, CO: Libraries Unlimited.

PEW Internet & American Life Project. (2004). Data Memo: The Popularity and Importance of Search Engines. *http://www.pewinternet.org/pdfs/PIP_Data_Memo_Searchengines.pdf*.

Pew Internet and American Life Project. (2003). Internet Health Resources: Health Searches and Email Have Become More Commonplace, But There Is Room for Improvement in Searches and Overall Internet Access. *http://www.pewinternet.org/reports/toc.asp?Report=95*.

Pew Internet and American Life Project. (2005). Internet Evolution: How the Internet has Woven Itself into American Life. *http://www.pewinternet.org/pdfs/Internet_Status_2005.pdf*.

ProMED-mail. (2003a). Acute Respiratory Syndrome–Canada (Ontario). *http://www.promedmail.org/pls/promed/f?p=2400:1001:::NO::F2400_P1001_BACK_PAGE,F2400_P1001_PUB_MAIL_ID:1000%2C20958.*

ProMED-mail. (2003b). Acute Respiratory Syndrome–East Asia. *http://www.promedmail.org/pls/promed/f?p=2400:1001:::NO::F2400_P1001_BACK_PAGE,F2400_P1001_PUB_MAIL_ID:1000%2C20957.*

ProMED-mail. (2003c). Pneumonia–China (Guangdong) (02). *http://www.promedmail.org/pls/promed/f?p=2400:1202:7217457626651181122::NO::F2400_P1202_CHECK_DISPLAY,F2400_P1202_PUB_MAIL_ID:X,20670.*

ProMED-mail. (2003d). Pneumonia–China (Guangdong) (07). *http://www.promedmail.org/pls/promed/f?p=2400:1001:::NO::F2400_P1001_BACK_PAGE,F2400_P1001_PUB_MAIL_ID:1000%2C20761.*

ProMED-mail. (2003e). Pneumonia–China (Guangdong): RFI. *http://www.promedmail.org/pls/promed/f?p?2400:1202:463072846851417654::NO::F2400_P1202_CHECK_DISPLAY,F2400_P1202_PUB_MAIL_ID:X,20658.*

ProMED-mail. (2003f). Severe Acute Respiratory Syndrome–Worldwide: Alert. *http://www.promedmail.org/pls/promed/f?p=2400:1202:15294495450593885846::NO::F2400_P1202_CHECK_DISPLAY,F2400_P1202_PUB_MAIL_ID:X,20964.*

Ramos, K., Linscheid, R., Schafer, S. (2003). Real-time information-seeking behavior of residency physicians. *Fam Med* 35:257–60.

Salton, G. (1989). *Automatic Text Processing: The Transformation, Analysis, and Retrieval of Information by Computer*. Reading, MA: Addison-Wesley.

Salton, G., McGill, M. (1983). *Introduction to Modern Information Retrieval*. New York: McGraw-Hill.

Sullivan, D. (2005). *Nielsen NetRatings Search Engine Ratings*.

van Rijsbergen, C. J. (1986). *Information Retrieval*. London: Butterworths.

World Health Organization. (2003). Acute Respiratory Syndrome in China, 11 February 2003. *http://www.who.int/csr/don/2003_02_11/en/.*

World Health Organization. (2005). *http://www.who.int/en/.*

27

CHAPTER

Physiologic and Space-Based Sensors

Ron M. Aryel
Del Rey Technologies, Los Angeles, California

1. INTRODUCTION

Sensors are devices that detect, measure, and record physical phenomena. Sensors are useful to biosurveillance, because they can monitor people, animals, the water, air, the ground, and even the mail system. They are the basis for systems that collect data about weather and climate, as well as other conditions that predispose to disease outbreaks. Sensor technology is continuously evolving. At present, the technologies utilized include such disparate tools as microchips, DNA polymerase amplification; various imaging techniques, including photographic analysis in various frequency ranges (normal, infrared); Doppler radar; various physiological monitoring techniques, such as EKG recording; and mass spectrometry. These technologies can be deployed in orbiting satellites, wearable devices, air/water monitoring devices, and portable or laboratory-based analyzers that process tissue specimens. The value they represent, however, is often tied to human analysts who are limited in their ability to process large amounts of data.

2. SENSORS FOR MONITORING THE HEALTH STATUS OF PEOPLE OR ANIMALS

2.1. Physiological Monitors

Physiological monitoring systems assess aspects of human performance and either respond with a corrective action, or alert a human being to do so. Physiologic monitors can measure and record temperature, heart rate, respiration, alertness, and activity.

We will consider the potential of these systems to assist in biosurveillance by equipping a "sentinel population." A sentinel population is any group of people who are willing to participate in a monitoring scheme. Security professionals, for example, could act as the proverbial canary in the mineshaft in a variety of settings. Examples of this approach could include advance members of the president's Secret Service detail, or other executive protection forces, or security agents patrolling special events (e.g., the Republican National Convention). In the latter case, we imagine that devices worn unobtrusively by the sentinels would provide early warning of spiking fevers, tachycardia or other signs arising in a fraction of the sentinel population. The information transmitted automatically from these sentinels would be timely.

To offer this capability, the equipment worn must be able to carry out accurate measurements while withstanding, and compensating for, the activities of an active subject. For example, a thermometer worn on the body must be able to compensate for clothing, weather, and exercise.

2.2. Remote Physiological Monitoring

Remote physiological monitoring, in one form or another, has a long, established history in aviation and space applications.

Gear intended to fit under or on clothing to measure various parameters of bodily function and environmental conditions has existed since the late 1940s. Beginning soon after World War II, the U.S. Air Force, the National Advisory Committee on Aeronautics (NACA), and its successor, the National Aeronautics and Space Administration (NASA), built a series of rocket-powered experimental aircraft to explore flight at very high speeds and altitudes. In 1947, Charles Yeager flew the X-1 past mach 1; in 1956, the X-2 flew at nearly mach 3. Two years later an X-2 test pilot set a world altitude record of nearly 126,000 feet. By 1967 a pilot flying North American Aviation's X-15 had reached Mach 6.7, soaring 60 miles above the Earth's surface. Test pilots flying in the late 1950s and after needed special suits, helmets, and oxygen supply systems to stay alive, and scientists needed to collect data on human performance in their cockpits. Physiologic data was not collected consistently throughout the test flight programs, due to resistance by test pilots to becoming "guinea pigs." When collected, this information was generally recorded in the aircraft, and collected for study after landing.

Collecting data on human bodily function was also a vital part of research in the space program. In 1958, the Mercury program ushered in the first group of physicians specifically assigned to care for astronauts. NASA decided that monitoring astronaut physiology in real time was of critical importance; therefore physiologic data collected by sensors was included in spacecraft telemetry from the beginning of the space program. These data included heart rate and rhythm, respiration and core and skin temperature. In the 1960s, Hamilton-Standard, now part of United Technologies Corp., developed a series of space suits designed to accept NASA-developed biometric-measuring gear. These suits were used in the Apollo, Skylab, and Space Shuttle programs. Data from these

Handbook of Biosurveillance
ISBN 0-12-369378-0

systems was recorded during flight and added to other telemetry sent to the ground. To these systems the Russians added (on the space station, not the suits), a blood analyzer, built by Daimler-Benz in Germany, which automatically sent blood chemistry results to ground control from the Mir space station.

Medical researchers have successfully demonstrated, on a small scale, the potential of physiological monitoring technology to aid in the remote monitoring of personnel functioning in extremely inhospitable environments. Examples of this include trials of biosensors designed and built by FitSense Technology Corporation Inc.

Working at the NASA-funded Commercial Space Center of Yale University's Department of Surgery, researchers led by Dr. Richard Satava organized climbing expeditions on Mount Everest, during the climbing seasons of 1998–1999, in order to demonstrate the reliability of FitSense biometrics gear and its value to the health maintenance of people in remote environments. Three climbers wore sensors measuring heart rate, skin temperature, core body temperature, and activity level, as well as a global positioning system (GPS) receiver to determine location. Data from these devices were transmitted to the Everest Base Camp and relayed to Yale University. The outcome measures were correlated via time-stamp identification. Sensor availability ranged from 78% to 100%. The researchers concluded that this application of biosensor technology was feasible, but that improvements in reliability and robustness were needed.

FitSense also equipped 16 Marines at Quantico VA, and nine U.S. Army Rangers at Fort Benning GA, with biosensors during 10-day war fighting exercises in 1998, 1999, and 2000. The sensors were taped on or otherwise worn by soldiers wearing standard-issue battle dress uniforms (BDUs). The Rangers' biosensors transmitted data in real time to a command post; the Marine's sensor data was recorded by a device worn by each soldier. The company reported 100% availability of sensors, with reporting accuracy of 97% (Blackadar, 2001). When interviewed, the company's CEO noted that pulse oximeters proved problematic at first, and the company expended a good deal of effort to improve them.

As shown by these trials, these types of sensors are not intended exclusively for the military. Law enforcement, medical surveillance, search and rescue, and public health applications offer possible uses. The idea common to all these applications is to enable a control center to determine whether and how severely people under its control are injured or otherwise disabled. Possible scenarios this technology can be applied to include a military command post watching over soldiers embarking on a reconnaissance mission, a police unit commander watching over officers who are dispersed over a wide area, or a fire department battalion chief keeping a close watch on fire fighters advancing into a burning building. A photo showing the transceivers appears in Figure 27.1.

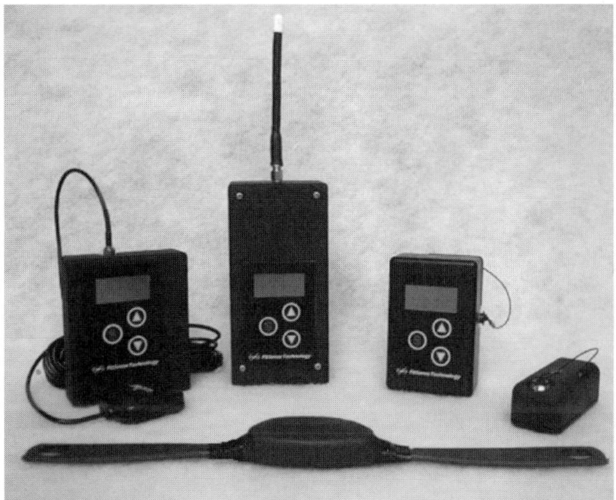

FIGURE 27.1 Transceivers. (Courtesy of FitSense Technologies.)

One drawback of this equipment is its expense. It costs thousands of dollars to outfit a soldier with this type of equipment.

2.3. Monitoring the Body's Vital Functions

Monitoring such parameters of bodily functions as heart rate, respiration, temperature, blood glucose, and even the heart's electrical activity (an EKG rhythm strip) does not require a very large bandwidth. A transmission capability as low as 8 bits per second, and an 8-bit microprocessor can accomplish this. FitSense created a low-power (microwatt) local area network (LAN) to handle multiple wireless sensors attached to clothing or jewelry (FitSense Technology Corporation Inc., 2005).

There are several devices currently available that monitor heart function and transmit data in real time to a monitoring station. These device include both externally worn and implantable monitors. EKG monitors consisting of one to three leads that are no larger than a music tape cassette can monitor heart rhythm and transmit information by landline, cellular, or satellite phone. Small, light, battery-powered Holter monitors are also available. Implantable cardiac pacemakers employ low-power sensors to measure heart rhythm and deliver a counter shock when a dysrhythmia is noted. These devices have been limited to employing 8-bit microprocessors in order to minimize heat generation during operation within the body. Conceivably, these devices could be adapted to store and forward physiological data for public health purposes (Schiller AG, 2005).

Pulse oximeters measure the degree of blood oxygenation noninvasively. Models currently available weigh a few ounces, are battery powered, and feature digital signal processing. Adding a transmitter to such devices is technically quite feasible (Nellcor Inc., 2005).

Commercial vendors sell a variety of wearable devices that are intended for monitoring human performance in clinical trial settings or exercise settings. Wrist and elbow devices manufactured by Pittsburgh-based BodyMedia, for example, measures movement, heat flow, skin temperature, ambient temperature, and galvanic skin response. A transceiver mounted on each unit allows the instrument to both transmit data in real-time to a workstation, and to receive data from another, third-party device.

Biometric gear offers great potential, but is expensive and evaluated only with small groups of volunteers thus far. Government agencies have yet to adopt these systems as standard issue.

2.4. Chemical Sniffing

During the Vietnam War, the U.S. military developed sensors that detect the presence of ammonia from mammalian waste. These sensors, which could be inserted by helicopter teams or dropped by aircraft into the jungle, could not distinguish between humans and other mammals, so their correct use required an accompanying terrain analysis. For example, knowing at which elevation an elephant, predator cat, or primate would be very unlikely to appear meant the detection of ammonia was much more likely to correlate with human activity. However, the North Vietnamese and their allies learned to defeat these sensors by spraying any they could find with animal urine, thus hiding their own presence amid the false-positive alarms.

However, this limitation, while important in warfare, may not be as important in biosurveillance. If we could use such monitoring to establish a baseline activity map of a heavily wooded area, which is difficult to survey by other means, then sudden, unexpected changes in such activity could in theory signal an event worthy of greater scrutiny by other appropriate means. For example, if a system detected a sudden absence of activity in the region, this could possibly signify a catastrophic event. In less extreme circumstances, its best use is likely to occur when paired with other types of biosurveillance.

2.5. Bioethics

The widespread use of sentinel populations, wearing physiological monitors, will require the development of ethical principles governing their use. If a sentinel's monitor alerts a control center to changes in physiological functioning, under what conditions should the sentinel be withdrawn, and under what conditions should he/she remain in place? How will such policies balance, on the one hand, maximizing the sensitivity and specificity of detection, and on the other, the sentinel's safety?

3. IMAGE ANALYSIS

Imaging helps us detect many different kinds of phenomena or events, from tracking storms to detecting troop movements and rocket launches. To use images, we need first to detect the image, classify the image (and determine its relevance to our purpose), interpret the image, and, finally, decide what our response to this image should be.

The number of surveillance systems, of all types, in use in the United States is truly staggering. These systems produce vast quantities of images, quantities that are orders of magnitude above the number we can analyze and interpret. Culling the trivial and retaining the significant comprise the biggest challenge to using images in biosurveillance.

Reliable automated image recognition and classification comprise a field still in its infancy; accurately interpreting images and communicating their significance are tasks for highly trained professional analysts. A competent image analyst boasts a good eye, relevant domain knowledge, and an ability to accurately translate what he/she is seeing into information that decision makers can use. Further, analysts specialize within domains and professions. For example, a radiologist is a highly skilled image analyst who understands how his instruments (x-ray, fluoroscope, CT scanner, MRI scanner) prepare images, the scale of those images, correlations between the images and the human body systems they portray, and the needs of the physician who requested the imaging. He can also aim the instruments so that the resulting images are more likely to reveal details relevant to the patient's case, such as a tumor in the spinal cord or an abscess in the liver.

Similarly, a government intelligence analyst understands how a reconnaissance satellite or aircraft's camera, radar or other sensor prepares images, what scale to use while viewing them, the characteristics of objects likely to appear in those images, and what each type of object conveys in terms of a target country's military or economic capabilities. The analyst can order an adjustment to the sensor's aim in order to optimize the value of the resulting images, for example, adjusting an orbital track and camera angle to provide greater detail about a missile installation or ground vegetation.

However, a radiologist's domain knowledge will not allow her to competently conclude whether a shape photographed from 200 miles away is part of an oil refinery or a facility for manufacturing biological agents, just as a military image analyst cannot competently diagnose a bowel obstruction by looking at an x-ray picture.

4. SATELLITES

Space-based observation is well established as a means for collecting data for a wide variety of purposes. These missions include military reconnaissance, weather forecasting, agricultural analysis, commercial surveying, and search and rescue.

Satellites obtain images via optical cameras, infrared sensors, and radar (especially using synthetic aperture technique). Each modality has advantages and disadvantages, situations where it is optimal, and situations where its use is impossible. Often the combination of sensors (i.e., obtaining a multispectral image) optimizes the chances of detecting a given object or phenomenon.

Space-based observation can detect the presence of deceased humans and animals, or degraded crops. A reconnaissance satellite photographs a scene of interest, and transmits its images in digital form to an interpretation center on the ground, either directly, or relayed by a communications satellite. The value of the images depends on the type of image, its resolution, and the time elapsed between the occurrence of an event and receipt of imagery by a trained analyst.

4.1. Satellite Imaging Quality and Access

Satellite image quality depends on the equipment available and the aiming decisions executed at control centers. The resolution of photography varies among platforms. The best nonmilitary satellites offer very good imagery: Quickbird can resolve objects as small as 0.6 m across, while Ikonos offers images with a resolution of 1 m per pixel (4 m per pixel for multispectral images). Older spacecraft, such as the French SPOT, offer resolution of 10 m per pixel, while Landsat can resolve objects 30 m across.

The resolution of military reconnaissance satellites is classified, but civilian experts, such as those at the Federation of American Scientists, have opined that 10 cm per pixel would not be unrealistic. In addition, these satellites' sensors can also take pictures in a night sky. Access to such data is restricted; however, if human intelligence ("humint") suggested that some kind of incident would occur, governmental public health and/or law enforcement could seek access to these "birds."

Photographs of use in detecting stricken people or animals are most likely taken using optical cameras, infrared sensors, and perhaps radar antennas using synthetic-aperture techniques, assuming that a high enough resolution is achieved.

The data are available in real-time and reliable. Degree of coverage depends on how many satellites of a particular type are available to cover the area being monitored at any given time. Temporary gaps in coverage have been known to occur when malfunctioning satellites, or those reaching the end of their useful lives, have not been replaced in time.

When an imaging customer (government or commercial user) wants to study something of interest, a satellite specialist can recommend what type(s) of imaging would be best and, therefore, which "birds" would be most useful. After the satellite has taken photos, the analyst interprets the photos and may recommend additional views of the target in question, or perhaps other sites. It may be desirable to tilt the camera to do such tasks as photograph the side of a building or other geographic feature rather than the top of it. Satellites are equipped with maneuvering thrusters allowing their controllers to adjust their orbits and quickly bring their cameras to bear on scenes of interest; the camera's angle is adjustable, which allows the camera to change from an overhead view to an oblique view. Should an image taken provoke curiosity or concern, the satellite's operator can maneuver it to pay special attention to the location where the image was taken. The interpretation center will process and enhance the images retrieved from satellites by using sophisticated software.

Unfortunately, this software cannot interpret the images for us. Image analysts are highly skilled professionals that the government calls upon to detect a developing catastrophe before it harms us; they are in very short supply.

4.2. Image Recognition Technology

Image recognition technology is relatively young. To quote a timeworn cliché, our true state of the art in image recognition is the "Mark One eyeball," a joking reference to military weapon naming conventions.

Researchers are pursuing automated recognition via a number of techniques, such as pattern matching, use of "primitives," and others. The military has demonstrated its utility; in 1991, cruise missiles fired from aircraft and naval vessels employed image recognition technology during the final homing phase of their flight to strike their targets. One key to their success was the relative simplicity of the images that their seeker heads sought, such as a building window. Current weapons utilize improved software; the exact capability of that software is classified.

This simplicity can, potentially, be found in public health-related tasks as well. For example, a dead cow assumes a profile quite distinct from a sleeping cow, or one that is contentedly munching on alfalfa. If a computer program can be designed to recognize the first case, the latter two cases should provide more than enough variance to convince the program to reject them, given sufficient image resolution. A large group of dead cows would be spotted in the same way.

Humans could present a more difficult problem. Our bodies are smaller than cattle, and discrimination between death and sleep, or even death and any stationary pose, is more challenging. One way of making this distinction is by using infrared sensors. Infrared imaging, making use of the portion of the electromagnetic spectrum just above visible light, traces its practical origins to the discovery in the 1940s that exposing lead sulfide to heat energy reduces its electrical resistance, a property known as photoconductivity. Thus, a lead sulfide photocell generates an electric current when in the presence of a heat source (Westrum, 1999). The military took advantage of this beginning in the early 1950s.

By using a combination of optical and infrared photography, a surveillance analyst should be able to spot groups of human beings, and determine whether they are alive or dead. Other indirect means may include monitoring the level of radio traffic, which is currently easily accomplished by military satellites.

However, significant changes in the condition or activities of groups of humans are readily documented by satellite data; fully automated detection of desired signals, however, may require technologies that, if they exist, are classified. It is clear, however, that if satellites offering very high-resolution photography, coupled to appropriate enhancement and interpretive or detection software, were to be available to public health authorities, they could add to the U.S. ability to detect a significant disease outbreak or bioterrorism event.

4.3. Limitations

The biggest limitation, bar none, to the effective use of satellite data is twofold: an astronomical amount of data flows to us from the skies, and trained analysts available to sift through it are few. Moreover, the number of analysts with domain knowledge to perform biosurveillance constitutes a subset of these analysts. Currently, the probability that space-based surveillance would miss the early signs of an epidemic or bioterrorism incident are high if only because a small hillock of relevant images would be buried in mountain ranges of other pictures.

Aircraft and satellites are affected by weather. Heavy cloud cover hides scenes of interest from cameras; storm activity can prevent aircraft or drones from flying their assigned missions. This is a factor in manmade epidemics. If bioterrorists understand that satellites are being used for surveillance, they are likely to attempt to strike when weather conditions most favor them, or to try as much as possible to conceal their attacks for as long as possible.

In addition to the technical limitations related to human bodies discussed above, other issues that would confront public health authorities are ethical (especially those related to privacy), and financial (we must pay for analyst time and spacecraft observation time). Indeed, it is likely that certain advocacy groups may protest the use of high-resolution satellite imagery for observation of people.

Important limitations to the use of military assets include legal restrictions that regulate military deployments on domestic soil and bureaucratic difficulties within federal agencies. The domestic use of military assets is restricted by many Executive Orders, Presidential Directives, Supreme Court decisions, and laws, such as the Posse Comitatus Act (U.S. Coast Guard, 2005). These legal directives, laws, and precedents exist to protect citizens from intrusion by the military and to ensure that military units remain subordinate to civilian control.

Aerospace assets are not controlled by one agency, but rather by a number of agencies; the different lines of authority tend to hinder efficient collection and dissemination of information. For example, the National Image and Mapping Agency is responsible for image intelligence products, while the National Security Agency collects signal intelligence.

Hence, the prompt distribution of information related to epidemics or bioterrorism, collected by such assets, to domestic agencies, is likely to be significantly hampered without the following:

- Anticipatory guidelines and policies consistent with federal law and able to withstand judicial scrutiny.
- Strong leadership in the federal government committed to cooperation between agencies that collect intelligence.
- Commercial satellite data are readily obtained; limitations are related primarily to the cost of their analysis and interpretation, and secondarily to the cost of data acquisition.

Putting spacecraft and aircraft to good use in biosurveillance will require significant investments in training experts who can interpret and analyze data, as well as research that could one day automate some of this process.

5. MONITORING WEATHER AND ENVIRONMENTAL CONDITIONS

Weather surveillance can serve biosurveillance in a number of ways. Predicting the spread of a bioaerosol could help governmental public health contain an organism's release in air. Changes in weather can improve or hamper the reproduction and spread of virulent organisms in the environment.

Weather and climate data currently used in epidemiological analysis include temperature, wind direction and speed (for bioaerosol related analyses), and precipitation. Of possible value, but not used according to our experts, might be barometric pressure and ultraviolet exposure.

In the United States, weather data are already highly available. The specific data available include temperature, wind speed, wind direction, and precipitation. Up-to-the-minute information for the entire nation is available because data are collected in real time, in standard formats, and integrated in a central location that is publicly available without any technical or administrative barriers (*http://weather.noaa.gov/*). We note that the weather system—in addition to a mature source of data for public health surveillance and early warning of bioterrorism—represents a good case study in how to integrate data from many sources and independent monitoring stations using communication networks and standards to achieve a national surveillance capability.

The space-based weather satellites most interesting and relevant to public health authorities are those that measure water temperature and ecological conditions, such as floods, fires, and conditions of vegetation, which have been used for epidemic prediction. These ecological factors are often predictive of changes in intermediate host and vector activity that are, in turn, predictive of outbreaks of human disease.

In the United States, satellite data are already highly available. As with reconnaissance satellites, data for the entire U.S. are usually constantly available, so long as enough satellites are in orbit. There have been occasions when the United States has not launched a satellite in time to replace a failing "bird," resulting in temporary gaps in coverage.

The raw data are of various types ranging from numeric to image. We have begun to develop the ability to extract useful information for public health purposes from these images; researchers at Johns Hopkins used satellite imagery (medium-resolution spectroradiometer) to assess a Canadian forest fire's impact on Baltimore's particulate air pollution (Sapkota et al., 2005). Another example is a joint venture formed by the U.S. Department of Defense and the NASA Goddard Space Flight Center, called the GEIS Rift Valley Information System. This project uses satellite near-real-time data, such as vegetation measurements, cloud cover measurements and sea surface temperatures to predict the appearance of Rift Valley fever in Kenya; a virus in the bunyaviridae family causes this disease, which affects primarily sheep and cattle but can also afflict humans who work with these animals. Rift Valley fever is actually endemic in many African countries, including Sudan, Egypt, Mauritania, and Senegal. The outbreaks are usually preceded by heavy and persistent rainfall in geographic regions accustomed to relative drought, and can result in heavy losses of livestock, leading to famine conditions (U.S. Department of Defense, 2005). We have yet to uncover other factors that determine availability and utility, such as the extent to which they are available in standard formats, and integrated in a central location that is publicly available without any barriers.

The National Oceanographic and Atmospheric Administration (NOAA) operates geostationary environmental satellites, parked in either equatorial or polar orbits 22,300 miles above the Earth's surface; these satellites observe weather patterns, measure winds, detect ash plumes from volcanic eruptions, help assess ground moisture, and provide 99% of the observations that NOAA requires to make climate predictions (Ray, 2005). The most recent program is called the National Polar-Orbiting Environmental Satellite System (NPOESS), a new family of spacecraft managed jointly by the Departments of Defense and Commerce and by NASA, which collect and disseminate data on weather, atmosphere, bodies of water and land (National Oceanographic and Atmospheric Administration, 2005). Government, agricultural, and other uses will access high-quality real-time data.

Spacecraft, such as Ikonos, as well as government spacecraft, can image the ground and help interpreters assess the condition of vegetation, moisture, the presence of fires, and so on, with the help of multispectral imaging as well as false color overlays.

The European Space Agency's Remote Sensing Satellite (ERS) measures atmospheric and surface properties using radar. Measurements include sea-surface wind-speed/direction, sea-surface height, wave height, sea-surface temperature, cloud-top temperature, and atmospheric water vapor; these are in addition to photography offering 30-m resolution (U.S. Air Force 2005).

6. SUMMARY

We can use a variety of sensor technologies, used both in intimate contact with humans or remotely from space, to collect data relevant to the presence of a disease outbreak. Physiologic sensors can tell us how well the subjects are functioning; remote sensors, such as those on aircraft or spacecraft, can monitor a wide variety of potential indicators. The primary obstacle to the latter's effective use is the shortage of trained analysts who can turn data into useful timely information, upon which we can act.

ACKNOWLEDGMENTS

We thank Dana Moreland, Senior Software Engineer, Raytheon Co, Denver CO, for his important insights regarding the usefulness of space-based sensors and the role of analysts.

REFERENCES

Blackadar, T. (2001). CEO, FitSense. Personal Communication.

FitSense Technology Corporation Inc. (2005). *http://www.fitsense.com/b/BodyLAN.asp*.

National Oceanographic and Atmospheric Administration. (2005). *http://www.noaa.gov*.

Nellcor Inc. (2005). Retrieved from: *www.nellcor.com*.

Ray, J. (2005). Delta 2 Rocket to Launch Earth Weather Probe. *SpaceflightNow.com. http://www.space.com/missionlaunches/sfn_noaa_prelaunch_050509.html*.

Sapkota, A., Symons, J. M., Kleissl, J., et al. (2005). Impact of the 2002 Canadian forest fires on particulate matter air quality in Baltimore city. *Environ Sci Technol* 39:24–32. *http://www.ncbi.nlm.nih.gov/entrez/query.fcgi?cmd=Retrieve&db=PubMed&dopt=Citation&list_uids=15667071*.

Schiller AG. (2005). *http://www.schiller.ch/*.

U.S. Coast Guard. (2005). *http://www.uscg.mil/hq/g-cp/comrel/factfile/Factcards/PosseComitatus.html*.

U.S. Department of Defense. (2005). Global Emerging Infections System. *http://www.geis.fhp.osd.mil/GEIS/SurveillanceActivities/RVFWeb/indexRVF.asp*.

Westrum, R. (1999). *Sidewinder: Creative Missile Development at China Lake*. Annapolis, MD: Naval Institute Press.

28

CHAPTER

Data NOS (Not Otherwise Specified)

Michael M. Wagner
RODS Laboratory, Center for Biomedical Informatics
University of Pittsburgh, Pittsburgh, Pennsylvania

Murray Campbell
IBM T.J. Watson Research Center, Yorktown Heights, New York

1. INTRODUCTION

This chapter explores types of biosurveillance data not already discussed in Part IV and elsewhere in this book. We discuss data for which research shows significant potential, as well as data for which value is suggested by early research or conjectured, but awaits confirmation by definitive studies. We also mention data that have been studied and have not panned out. We conclude the chapter with a discussion of monitoring in permissive environments. Permissive environments are situations in which more 'privacy-invasive' biosurveillance may be possible through consent of those being monitored.

We group the data into clinical, preclinical, and presymptomatic categories. *Preclinical data* are data that are the result of illness behavior that precedes seeking of clinical care. *Presymptomatic data* are data that might be obtained about a person during the incubation period of a disease.

We use the clinical–preclinical–presymptomatic distinction to group data in this chapter because it indicates the inherent earliness of data for detection of outbreaks. Although the inherent earliness of data is not the only factor that determines when an outbreak will be detected, it does define a theoretical limit to the earliness that can be achieved.

Clinical data, for example, become available about a patient relatively late after the onset of illness, but this delay is potentially offset by better diagnostic precision of the data. Moreover, clinical data typically are associated with a patient identifier, allowing a biosurveillance system to link the data with other observations about a sick individual to improve the accuracy and precision with which the patient is classified into a syndrome or disease category.

Preclinical data are inherently earlier than clinical data, but often cannot be linked at the patient level. Preclinical data are especially helpful for outbreaks in which people do not seek medical advice or care, or do not do so early in the illness.

Presymptomatic data such as antibody or antigen levels detected in blood samples are theoretically the earliest data—dramatically earlier than preclinical or clinical data for diseases with very long incubation periods such as HIV—but, presymptomatic data are rarely collected at present.

For convenient reference, Table 28.1 summarizes the surveillance data already discussed in this book and provides chapter numbers. Table 28.2 lists data discussed in this chapter.

2. CLINICAL DATA

This section discusses data from clinical call centers (e.g., nurse call centers), orders for laboratory tests, and prescriptions for drugs. These types of data are frequently mentioned in lists of data with potential for biosurveillance (Henning, 2004, Wagner et al., 2001).

There are existing biosurveillance systems, which we will discuss shortly, that collect and monitor these data routinely. These projects were initiated in advance of scientific evidence of value of the data and their goals include investigating the value of these data. Given the existing threat and the likelihood that these data will be of value, parallel system development and research may be optimal policy. As with most clinical data, these data contain diagnostic information and therefore have "face validity" for detection of cases and outbreaks (i.e., biosurveillance). Nevertheless, quantitative studies are important. Designers of biosurveillance systems will benefit from knowing the strength of association between particular clinical data and particular diseases and outbreaks, as well as the effort and time latencies associated with their collection and the fraction of any population from which they may be collected.

2.1. Data Collected by Clinical Call Centers

Clinical call centers (also known as nurse triage and nurse hotlines) are facilities that receive telephone calls from sick individuals or individual seeking advice or appointments.

Clinical call centers capture rich data about a person's illness and they capture this data before a physician encounter. The staff—usually registered nurses—document calls electronically and in real time. They use electronic documentation systems and problem-specific (e.g., *fever* and *cough*) charting templates. The nurse (or other call center employee) selects a charting template that is specific to the nature of the call. The choice of charting template itself represents a diagnostic interpretation that is available for analysis.

Handbook of Biosurveillance
ISBN 0-12-369378-0

TABLE 28.1 Data Already Discussed in this Book with Chapter References

Data	Chapter
Notifiable disease reports and laboratory test results	5, 8
Chief complaints (free text)	17, 23
ICD codes	23
Over-the-counter medications	22
Internet resources and usage	26
Absenteeism	24
Weather	19
Water quality and water system	9
Air monitoring	19
Remote imaging and physiologic	27

Charting by template results in detailed data about symptoms that are computer interpretable (not free text). The charting template is often based on a diagnostic/problem management algorithm, which is a tree structure that suggests key facts about the illness (e.g., presence of fever) for the nurse to elicit during the call. The resulting data include the reason for the call, symptoms, severity, date of onset, self-measured temperature of the patient—and a numerical code that corresponds to the template selected by the nurse for documentation of the call (which reflects the nurse's professional judgment). The data also include a time-stamp and the location of the caller.

2.1.1. Syndromes

At present, biosurveillance systems that monitor call-center data classify each patient into a syndrome category prior to analyzing the data for clusters of illness that might indicate the presence of an outbreak. The United Kingdom's NHS (National Health Service) Direct syndromic surveillance system (also discussed in Chapter 6) uses 10 syndromes: *cold/influenza, cough, diarrhea, difficulty breathing, double vision, eye problems, lumps, fever, rash*, and *vomiting* (Doroshenko et al., 2004). The ESSENCE II system uses seven syndromes: *coma, gastrointestinal, respiratory, neurologic, hemorrhagic, infectious dermatologic*, and *fever* (Henry et al., 2004b). The syndrome assignment may be based on the template selected by the nurse for documenting the call or other data collected during the call.

TABLE 28.2 Data Discussed or Mentioned in this Chapter

Type of data	Value of Information
Clinical	
Call center data	Probably high
Test orders	Unknown
Prescription drugs	Probably high
Preclinical	
Weekly polling by email	One study
Volumetric telephone usage	One study
Bus ridership	Unknown
Sales of food items in cafeterias or restaurants	Unknown
Counts of users of medical facility parking structure	Low, due to poor diagnostic precision
Waste water monitoring	Low, due to dilution (no effect seen in Milwaukee)
Missed appointments	Low, due to high background rates of cancellations for non disease-related reasons

2.1.2. Studies of the Informational Value of Clinical Call Center Data

Espino et al. (2003) used correlation analysis to test a hypothesis that call center data could provide more timely detection of influenza than reference data collected by the Centers for Disease Control and Prevention (CDC). They studied data from two types of call centers—emergency room telephone triage and after hours telephone triage—and one outbreak of influenza. In emergency telephone triage, an insured person calls a telephone triage center to obtain authorization to visit an emergency room. In after-hours telephone triage, a person calls a triage center instead of his or her physician.

The data sets obtained from the call centers included the following information:

1. Start date/time of initial call.
2. Acuity. Acute or nonacute.
3. Inclination of the caller. There were 12 unique reasons why a call could be made (e.g., call physician, research).
4. Disposition assigned to the call by the nurse. There were 23 unique dispositions (e.g., home care, see physician within 24 hours, call primary care physician within 24 hours, emergency room immediately).
5. Call outcome. This element used the same 23 possible values used in the disposition.
6. Primary symptoms expressed in free text. Examples are "stomach bug—six loose bowel movements throughout the day" and "feels hot and some pain in throat."
7. Five-digit home zip code.
8. Age.
9. Gender.
10. Treatment guideline used. There were 380 unique guidelines used in the data sets. A single guideline is recorded for each call. Examples include *diarrhea, earache, breathing difficulty, vomiting*, and *asthma attack*.

They studied the treatment guideline (item 10) used to handle the call. They assigned 21 of the specific guidelines used for the calls to *respiratory* syndrome and four to *constitutional* syndrome. They combined the calls for *respiratory* and for *constitutional* and studied the weekly counts of the combined categories.

Reference standards for influenza included three types of influenza surveillance data from the CDC:

Weekly state epidemiologist reports of influenza activity (CDC-SI).

TABLE 28.3 Maximum Correlation and Time Lag (in Weeks) Between *Influenza* Calls to Two Call Centers and Reference Influenza Time Series Data

	CDC-SI	CDC-RI	CDC-RC
Emergency room telephone triage	–1 (0.773)	–4 (0.771)	0 (0.876)
After hours telephone triage	1 (0.474)	1 (0.749)	1 (0.760)

Notes: Maximum correlation is shown in parentheses. Correlation values are re-scaled from those reported in the paper to a more conventional range of 0 to 1.0. Adapted from Espino, J. U., Hogan, W., Wagner, M. M. (2003). Telephone triage: a timely data source for surveillance of influenza-like diseases. In: *Proceedings of American Medical Informatics Association Symposium*, 215–9, with permission.

Weekly regional number of influenza-like illness (ILI) cases from the United States.[1]

Influenza Sentinel Physicians Surveillance Network (CDC-RI).

Weekly regional number of positive influenza tests (CDC-RC).

They found a strong correlation between calls to emergency room telephone triage for respiratory problems and influenza (Table 28.3). The time latency was variable and no strong conclusion about earliness can be drawn from this study.

Henry et al. (2004b) used linked analysis (Chapter 21) to study the accuracy with which nurse advice hotline data predict the physician diagnosis (mapped by the researchers to a coarser "syndrome" level of diagnostic precision) made during a subsequent office visit (Henry et al., 2004b). They also studied the time elapsed between each nurse advice call and the linked subsequent office visit.

They studied nurse advice hotline data from Kaiser Permanente of the Mid-Atlantic States for the period January—December 2002. In this clinical call center, nurses select from 586 nurse advice clinical practice protocols (e.g., diarrhea, adult). The researchers reviewed these protocols and assigned 68 to ESSENCE II syndromes based on their names and presumed usage (Figure 28.1).[2]

They studied the set of callers for whom a nurse had selected one or more of these 68 protocols and for whom they could also establish a link between the call and a subsequent office visit. They required that the link have the following characteristics: (1) the patient identification numbers for the hotline call and the office visit were identical, (2) the date of the hotline call was the same as the date the patient called for an appointment,

BOX. Classification for coding of nurse advice guidelines — Kaiser Permanente of the Mid-Atlantic States (KPMAS)

GI (gastrointestinal)
Abdominal pain, adult
Abdominal pain, pediatric
Acute GI, gastroenteritis, adult
Diarrhea, 0–24 months, pediatric
Diarrhea, >2 years, pediatric
Diarrhea, adult
Diarrhea, long-term care
Diarrhea, pediatric
Diarrhea, prenatal, OBGYN
HIV diarrhea, adult
Nausea, adult
Nausea, pediatric
Vomiting and hyperemesis, adult
Vomiting, long-term care
Vomiting, pediatric

DERMHEM (hemorrhagic manifestations)
Bruise/hematoma, adult

NEURO (neurologic)
Headache, adult
Headache, long-term care
Headache, pediatric
HIV headache, adult
HIV mental status changes, adult
Meningitis, adult
Meningitis, neonates
Meningitis, pediatric, 3 months–2 years
Meningitis, pediatric, children/young adults
Meningitis, pediatric, infants
Meningitis, pediatric, >2 years

FEVER
Fever, adult
Fever, long-term care
Fever, neonatal
Fever, pediatric
HIV fever, adult

RESP (respiratory infection)
Asthma, adult
Asthma, pediatric
Bronchiolitis, pediatric
Bronchitis, acute, adult
Croup, pediatric
Earache, pediatric
HIV dyspnea, adult
HIV pneumonia, adult
Influenza, adult
Influenza, pediatric
Laryngitis, adult
Respiratory distress, adult
Shortness of breath, long-term care
Sore throat, adult
Sore throat, pediatric
Throat culture, positive
Upper respiratory infection, adult
Upper respiratory infection, long-term care
Upper respiratory infection, pediatric

DERMINF (dermatologic, infectious)
Chicken pox, adult
Chicken pox, pediatric
Chicken pox, prenatal, OBGYN
Fifth disease, pediatric
Hand, foot, mouth disease, pediatric
Herpes zoster, adult
Herpes zoster/shingles, adult
Measles, pediatric
Rash, adult
Rash/fungal infection, adult
Rashes, pediatric
Rash, prenatal, OBGYN
Roseola, pediatric
Shingles
Smallpox

FIGURE 28.1 Six syndrome definitions based on 68 nurse advice protocols. (From Henry, with permission.)

1 The CDC defines influenza like illness as fever (temperature of >100°F) plus either a cough or a sore throat.

2 Magruder and collegues developed and evaluated an empirical method for creating syndrome groupings that they applied to call center data (Magruder et al., 2004).

TABLE 28.4 Number of Nurse Advice Hotline Calls and Outpatient Office Visits, by Syndrome Group–Kaiser Permanente of Mid-Atlantic States, 2002

Syndrome Group	Number of Hotline Calls[a]	Number of Outpatient Office Visits[a]
Gastrointestinal	72,107	26,521
Respiratory infection	242,785	201,402
Neurologic	22,957	144
Dermatologic, infectious	20,117	440
Fever	17,866	5,230
Hemorrhagic manifestation	1,580	456

[a]Multiple calls or multiple visits by a given patient in a single day for the same syndrome are counted only once.

From Henry, J., Magruder, S., Snyder, M. (2004a). Using Kaiser Permanente Nurse Advice Hotline Data for syndromic surveillance in the national capital area. *MMWR Morb Mortal Wkly Rep* Under review.

TABLE 28.5 Sensitivity, Specificity, Positive Predictive Value, and Timeliness of Nurse Advice Hotline Calls Relative to Office Visits by Syndrome Group–Kaiser Permanente of Mid-Atlantic States, 2002

Syndrome	Sensitivity	Specificity	Positive Predictive Value	Mean Time (Hours)
Gastrointestinal	0.72	0.96	0.37	12.0
Respiratory infection	0.75	0.89	0.64	12.3
Neurologic	0.27	0.98	0.01	23.6
Dermatologic	0.09	0.98	0.02	8.2
Fever	0.33	0.96	0.12	8.3
Hemorrhagic manifestations	0.13	0.999	0.02	25

From Henry, J., Magruder, S., Snyder, M. (2004a). Using Kaiser Permanente Nurse Advice Hotline Data for syndromic surveillance in the national capital area. *MMWR Morb Mortal Wkly Rep* Under review.

and (3) the time of the hotline call was earlier than or simultaneous with the time of the call for the appointment.

They assigned zero, one, or more than one syndrome to the office visit based on the ICD-9 codes assigned for the visit, using the existing ESSENCE II ICD-9-to-syndrome mappings (Chapter 23, Table 23.3). For each syndrome, they calculated the sensitivity, specificity, and positive predictive value, using office visits as a diagnostic standard.

Table 28.4 shows the number of hotline calls and outpatient office visits by syndrome. Calls categorized as *respiratory* were most likely to result in a (linkable) subsequent office visit, and calls for *neurologic* least likely.

Of the syndrome groups studied, the sensitivity of hotline calls for *respiratory* was highest (74.7%), followed by hotline calls for *gastrointestinal* (72%). Sensitivity for *hemorrhagic* was lowest (13.3%). The specificity ranged from 88.9% to 99.9% (Table 28.5). The mean time lag between the nurse advice call and the corresponding physician encounter by syndrome ranged from 8.2 to 25 hours. Note that the actual time difference experienced by a biosurveillance system will likely be greater as the physician diagnosis is recorded after the encounter, whereas the nurse selected protocol is available very early in the telephone call.

Doroshenko et al. (2004) used correlation analysis to study calls to the United Kingdom's Nurse Direct helpline and one outbreak of influenza. NHS Direct is a nurse-led telephone helpline that provides health information and advice to callers with symptoms. Nurses at the 22 NHS Direct sites use a computerized clinical decision support system containing approximately 200 clinical algorithms, each with a series of questions relating to symptoms. The NHS Direct syndromic surveillance system automatically assigns a patient into one of 10 syndromes (listed above).

The evaluation measured the Spearman rank correlation coefficient between weekly counts of *cold/influenza* syndrome, as determined by the NHS Direct syndromic surveillance system with weekly counts the Royal College of General Practitioner's Weekly Return Service (WRS) (Figure 28.2). WRS is an established sentinel clinician surveillance system for ILI. The Spearman correlation of 0.85 is a measure of the

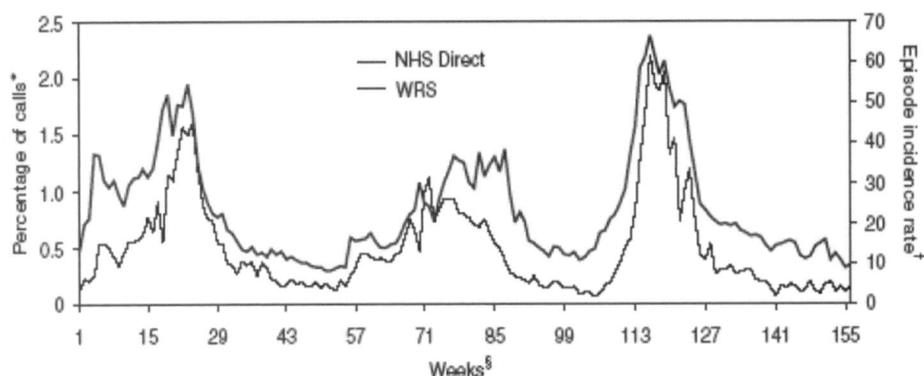

FIGURE 28.2 Time series of weekly counts of cold/influenza from the NHS (National Health Service) Direct syndromic surveillance system (upper line) and the Royal College of General Practitioner's Weekly Return Service (lower line). These data are from England and Wales for the period August 2001–August 2004. (From Doroshenko, A., Cooper, D., Smith, G., et al. (2004). Evaluation of syndromic surveillance based on NHS Direct derived data in England and Wales. *MMWR Morb Mortal Wkly Rep* 54(Suppl):117–22, with permission.)

correlation without a time offset, and the result indicates that the two time series are strongly correlated.[3]

Doroshenko et al. (2004) also reported that, in structured interviews with stakeholders, the majority of respondents indicated that the NHS Direct syndromic surveillance system registered an increase in calls about diarrhea and vomiting at the times when traditional public health surveillance systems indicated a national increase in norovirus.

Platt et al. (2003) operate the National Syndromic Surveillance Demonstration Project. Included among its data sources are telephone triage data from Optum, a company that operates triage call centers that serve healthcare facilities in 50 states (Yih et al., 2004). There have been no published studies as of the time of this writing of the value of the call center data collected.[4]

2.1.3. Availability of Call Center Data

In the United States, the proportion of a population in any geographic region that has access to a call center varies by region. The Optum plan, which participates in the National Syndromic Surveillance Demonstration Project, covers 7% of the U.S. population that is unevenly distributed (Yih et al., 2004). Over time, the use of call centers by managed care organizations will increase (Galimi, 1999). Through better allocation of resources, they can lead to a significant reduction in costs (Sabin, 1998), and there is often a high degree of patient satisfaction with call centers (Delichatsios et al., 1998).

Even in the United Kingdom, which has national health care and provides access to a clinical call center to all patients, not every sick individual uses a call center. The United Kingdom's NHS Direct system covers all of England and Wales; however, it reports that the calling volume is 1/40th of the volume of visits to primary care physicians (Cooper et al., 2005).

2.2. Prescription Drugs

Prescriptions (either filled or presented to pharmacists) are often included in lists of data with potential value in biosurveillance. There have been studies of the use of prescriptions to detect cases of disease (especially tuberculosis), and one study of the effect of a public health intervention for a pertussis outbreak on Medicaid claims for macrolide antibiotics, but none of outbreaks.

It is anecdotally reported that at the time that AIDS was first uncovered in 1981, the CDC's investigation drug unit was receiving unusual numbers of requests for the drug pentamidine for treatment of young men (CDC, 2001). The CDC's investigation drug unit was the sole distributor of pentamidine at the time, so these still rare requests were funneled to a single entity and the increase therefore noticeable. This experience nevertheless suggests the potential of prescription monitoring.

Maggini et al. (1991) studied the use of Italy's National Health Service pharmacy dispensing information to identify TB cases in Rome. They found that pharmacy screening detected seven times more new TB cases than routine public health reporting.

Yokoe et al. (1999) studied simple case-detection logic that defined a screening criteria for a new case of tuberculosis as two or more antituberculosis drugs. This simple rule identified 43 incident cases of TB from analysis of electronic records. Of these, seven (16%) had not been identified by the health department.

Chen et al. (2004) used the detection algorithm method to study retrospectively the sales of prescription macrolide antibiotics during an outbreak of pertussis. The New York State Syndromic Surveillance Project receives daily summaries from the New York State Medicaid program, which provides healthcare coverage for 4–20% of the non-NYC regions of the state, of claims for prescription medications in 18 medication groups. The summary counts are reported by zip code, age, and gender. The retrospective analysis used the CUSUM statistic on data aggregated at the county level. The reference date for the outbreak was July 21, 2004. CUSUM signaled on the countywide counts for the affected county on July 23, which was the day of treatment/prophylaxis of the first case and its contacts.

Approximately 300 individuals received prophylactic antibiotics. The total number of excess macrolide prescriptions (over baseline) received by the New York State Syndromic Surveillance Project over baseline was approximately 110 (estimated from graphs in the paper).

The increased prescriptions detected in this study were the result of a public health intervention after an outbreak was detected, so this study is not directly informative about detection of an outbreak. However, this case study illustrates the sensitivity of prescription monitoring to increases in prescription medications. This sensitivity, coupled with the fact that the researchers estimated that approximately 95% of the records that New York State system receives are for prescriptions filled within the preceding 48 hours, suggest the potential of prescription monitoring for outbreak detection and need for additional research (Table 28.6).

3 Doroshenko and collegues also measured the correlation using a second method, in which they first removed seasonal and other trends from both time series by fitting ARIMA models to the data and then computed the Pearson correlation coefficient (because the modeling yielded normally distributed data) at different time lags. Correlations at zero to four week lags were statistically significant, but small. The maximum correlation of 0.224 occurred at a lag of one week. The lack of strong correlation is not surprising since the ARIMA modeling removed influenza effect.

4 This system's detection performance from ambulatory care data (not call data) has been reported for influenza and outbreaks of gastrointestinal disease. We discuss these results in Chapter 23.

TABLE 28.6 Medicaid Prescription Drug Categories Included in New York State's Syndromic Surveillance System

Analgesics (narcotic)
Analgesics (non-narcotic)
Antacids
Antiasthmatic medications
Antibiotics
Cephalosporins, first and second generation
Cephalosporins, third and fourth generation
Fluoroqinolones
Macrolide antibiotics
Penicillin G and ampicillins
Penicillinase-resistant, extended spectrum, and penicillin combinations
Tetracyclines
Antidiarrheal medications
Antihistamines
Cough, cold, and allergy medications
Electrolyte mixtures
Herpes Agents
Influenza agents
Systemic and topical nasal products

From Chen, J.-H., Schmit, K., Chang, H., et al. (2004). Use of medicaid prescription data for syndromic surveillance—New York. *MMWR Morb Mortal Wkly Rep* 54(Suppl):31–4, with permission.

2.3. Orders for Laboratory Tests

Orders for laboratory tests are also often included in discussions of data of value for biosurveillance. The data themselves contain the test ordered, reason for the order encoded in ICD-9, patient gender, birth year, time stamp, and zip code (Ma et al., 2004). Like prescriptions, they have "face validity," and in fact are being collected routinely by CDC's Biosense system (Ma et al., 2004). However, there are as yet no published studies on their informational value.

Chotani and Lewis of Johns Hopkins Applied Physicians Laboratory compared orders received by health departments for influenza tests, results of those tests, and onset of influenza season (unpublished results are summarized in Wagner et al. [2004]). They found high correlation between influenza testing and the onset of the influenza during four seasons. There was, however, no lead time for test requests compared to patterns of influenza-like illnesses ICD-9 diagnoses in physician office visit records. This result suggests that either source can be used for monitoring and the preference between them depends on whether the data can be obtained without time delay and differences in cost and effort to build the data collection system. A caveat to this research is that the data source investigated was limited to tests ordered from the state laboratory whereas influenza testing is also done by private laboratories, which may produce an earlier or stronger signal, due to increased sampling.

3. PRECLINICAL DATA

DARPA's Bio-ALIRT program was a three-year program (2001–2003) with a goal of developing technology to reduce the delay in detection of outbreaks. It focused on development of detection algorithms (Buckeridge et al., 2005) and evaluation of novel surveillance data (Wagner et al., 2004). The Bio-ALIRT program contributed results about many types of data described in this book. The program especially encouraged study of preclinical data.

This section summarizes results about routine polling of a population about its health status, volumetric telephone usage, and food and beverage sales at cafeterias and restaurants. The interested reader can find details about the other types of data studied by Bio-ALIRT (such as bus ridership volume, parking lot utilization, automatic cough detection, waste water monitoring, and even missed appointments to orthodontists), which were largely limitative or negative results, in the report from that program (listed under Additional Resources).

3.1. Polling and Self-Reporting

The earliest possible detection from surveillance of human health is detection on the day that someone becomes symptomatic (ignoring tests capable of detecting incubating infections). Such detection requires self reporting, which could be obtained through frequent polling or methods to train individuals to report on the day they wake up ill.

Li and Aggarwal (2003) (see also Wagner et al., 2004) studied the responses of employees of the IBM T.J. Watson Research Center to a weekly e-mail asking about their health. Of approximately 2200 employees, 397 employees agreed to participate in the project. During the four-month period—January 25, 2002 through May 31, 2002—each participant was sent e-mail once a week at 7 AM on a randomly chosen work day. Participants were asked to self-assess their health for the day the e-mail was sent, choosing from one of the following categories: As usual, worse than usual (sick), or much worse than usual (very sick). If the participant reported being sick or very sick, there were a series of additional questions about the illness, including a question about symptoms (fever, headache, muscle ache, fatigue, cough, stuffy/runny nose, etc.). During the course of the survey, there was minimal dropout (<2%), and responses were prompt: 73% by noon and 92% on the same day as the e-mailing.

The researchers studied the informational value of polling data using correlation analysis with a reference time series that reflected disease activity in Westchester County. They constructed a reference time series from insurance claims data for patients seen in physician's office for respiratory complaints (ICD-9 codes 460–519). The correlation between the self-reporting obtained through polling and the physician office visits for respiratory illness was high, with the maximum correlation of 0.82 occurring with polling responses leading office visits by three days.

Although this result is promising, there are a number of factors that may limit the routine application of polling. Addressing privacy concerns will be critical in encouraging participation and obtaining reliable results. Liability issues must be fully understood. The responding process must be simple and non-intrusive enough to discourage dropouts. It may be that polling is most appropriate during situations of heightened alert or in permissive environments (see Section 5 in this chapter).

3.2. Volumetric Telephone Usage

Telephone calling patterns could provide a very early indication of a change in health status of a population. For example, an increase in telephone call volume emanating from a given neighborhood or apartment building—especially in the very early hours of the morning or late at night when levels are typically low—may indicate an emerging infection within that location. An increase in calls from one region of a city or building to medical clinics or pharmacies might also provide an early signal that requires further investigation.

Li and colleagues used correlation analysis to study the relationship between daily telephone call volume from the Watson Research Center to medical facilities and pharmacies and the same reference time series of respiratory illness as described in the previous section (Li and Aggarwal, 2003, Wagner et al., 2004). They compiled a list of approximately 5000 phone numbers that corresponded to medical offices in the area surrounding the Watson Research Center. They obtained these numbers from various sources including both online and hardcopy phone directories. The Watson Research Center tracks outgoing phone calls for billing purposes. They obtained a daily report of the total number of outgoing calls to the numbers on the compiled list of medical offices, as well as the number of callers (extensions) making those calls.

They studied call daily volumes during the period October 2001 through May 2002. The analysis ignored weekends, as call volumes dropped dramatically on non-working days. The maximum correlation between the call volume and disease activity was 0.72, which occurred when telephone calls led office visits by four days.

There have been no other studies of volumetric telephone usage. In our attempts to obtain volumetric data from phone companies, we have encountered concerns that even highly de-identified monitoring may be perceived as a "big brother" invasion of privacy by phone company subscribers. It would perhaps be easier to enlist the phone companies' participation in such monitoring if there were additional studies showing ability to detect outbreaks, but without their participation, it will be difficult to further develop the scientific evidence. The most promising setting for future studies will likely be that of permissive environments (see Section 5 of this chapter).

3.3. Sales of Food Items in Cafeterias or Restaurants

People suffering symptoms of various illnesses may alter their patterns of purchase of foods in locations where some observation is possible, such as at cafeterias. For example, respiratory illness may cause an increase in sales of soups or beverages, both foods commonly considered helpful or soothing in such cases. Such a change in pattern could be a timely indicator that the population suffers symptoms not yet reported to clinicians.

Campbell and colleagues use correlation analysis to examine the purchases of beverages during the first three months of 2002 in the cafeterias at the IBM T.J. Watson Research Center, located in Yorktown Heights NY and Hawthorne NY (Li and Aggarwal, 2003, Wagner et al., 2004). They compared total beverage sales data with the same reference standard as described in previous sections for the period January–March 2002. Weekends were ignored because the cafeterias were closed on these days, and Mondays were ignored because beverages were free and sales not recorded. The maximum correlation was 0.66, with a lead time of 19 days. The investigators considered this preliminary study of cafeteria data to be inconclusive due to the biologically implausible 19-day lead time, but suggested that there appears to be sufficient evidence of potential to warrant further investigation.

4. PRESYMPTOMATIC DATA

There is potential for early detection of disease using samples such as blood serum obtained during testing conducted during physical examinations or medical treatment for chronic conditions. High-throughput proteomic mass spectral technologies could potentially be incorporated into automated analyzers in existing laboratories. There are barriers both administrative and scientific before this potential can be realized. We mention this last data source because of its potential to detect disease during the incubation period (prior to symptoms).

5. PERMISSIVE ENVIRONMENT

Certain types of "permissive" environments may allow collection of data that are not easily available in more general contexts. Examples of permissive environments include some types of work sites, university campuses, or military bases. They are typically characterized by a geographically constrained location (a "site") and some centralized control of information technology (IT) and telecommunications infrastructure, as well as a shared sense of community and trust. The centralized control of IT/telecommunications allows electronic tracking of data (e.g., outgoing phone calls, Internet accesses) that are difficult to obtain on, say, a citywide basis. The shared sense of community and trust makes it easier to gather personal information, as there is a belief that these data will not be used to violate privacy rights.

Surveillance in permissive environments can have two purposes. The site itself may be sufficiently large or of sufficiently high value to justify surveillance for its own sake. Alternatively, if the population of a site draws upon the surrounding community for its members (e.g., work site), it may act as a high-sensitivity sentinel for disease outbreak in the surrounding area.

6. SUMMARY

Of the data types discussed in this chapter, clinical call centers have perhaps the greatest potential because of the interaction between a patient, a nurse, and a decision support system (the documentation system based on clinical guidelines) relatively early in the course of illness. This combination not only leads to a high level of clinical detail elicited by highly trained nurse but opens up the opportunity for adaptive questioning driven by Bayesian priors computed from probabilistic output of regional biosurveillance systems. Moreover, there is a long-term trend towards increased use of call centers in healthcare because of potential cost savings.

Prescriptions and test orders simply have not been studied, although we expect this situation to change quickly as existing biosurveillance systems are collecting these data and are being studied.

Of the preclinical data we discussed, telephone call volume has high potential because of the earliness suggested by the described study and that telephone systems are already so highly automated. However, call volume is extraordinarily non specific (caller may be sick), so its role seems to be limited to detection of sudden outbreaks that affect a large fraction of a region. Much more study of the potential of volumetric telephone calling is needed.

With polling and self-reporting, the big question is sustainability and feasibility. The one existing study of polling was for a limited duration in a permissive environment, and further work is needed to assess the costs and benefits for routine polling of larger populations. Restaurant sales do not have strong evidence supporting their use, and are generally difficult to obtain. The most likely use of restaurant sales would be for site-based surveillance in a permissive environment. The potential value of presymptomatic data awaits development of technology—it hasn't been studied at all.

Permissive environments allow enhanced surveillance, and therefore could represent a promising sentinel population for a surrounding community. Further study is warranted to investigate this question.

As with all of the data discussed in Part IV of this book, each of the data types described here may have a potential role in an all-threat biosurveillance system. It is inconceivable (at least for the foreseeable future) that there will be a one-size-fits all type of data as there was in the days of notifiable disease reporting. People with the same disease behave differently (not everyone sees a clinician), and people with different diseases behave quite differently. Data type X may show the earliest signal for outbreak Y, but show no signal whatsoever for outbreak Z.

We expect that a great deal more will become known in the next few year now that research methods have been worked out by the studies described in Part IV. There is at least one biosurveillance systems that is routinely collecting nearly every one of these type of data and investigators are poised to seek data retrospectively when outbreaks occur.

A great deal more research will be needed as the computerization of the world progresses and new types of data become available for study. The speed at which research will progress is limited by funding, availability of data, and occurrence of outbreaks.

ADDITIONAL RESOURCES

Wagner, M., Pavlin, J., Brillman, J., et al. (2004). *Synthesis of Research on the Value of Unconventional Data for Early Detection of Disease Outbreaks*. Washington, DC: Defense Advanced Research Projects Agency. Available at: *http://rods.health.pitt.edu/*. This report details research results from DARPA's BioALIRT program related to novel types of biosurveillance data.

ACKNOWLEDGMENTS

The authors were funded in part by the Air Force Research Laboratory (AFRL)/Defense Advanced Research Projects Agency (DARPA) under AFRL Contract No. F30602-01-C-0184; and by Contract No. F30602-01–2-0550 from DARPA.

REFERENCES

Buckeridge, D. L., Burkom, H., Campbell, M., et al. (2005). Algorithms for rapid outbreak detection: a research synthesis. *J Biomed Inform* 38:99–113. *http://www.ncbi.nlm.nih.gov/entrez/query.fcgi?cmd=Retrieve&db=PubMed&dopt=Citation&list_uids=15797000.*

Centers for Disease Control and Prevention. (2001). First Report of AIDS. *MMWR Morb Mortal Wkly Rep* 50:429. *http://www.cdc.gov/mmwr/PDF/wk/mm5021.pdf.*

Chen, J.-H., Schmit, K., Chang, H., et al. (2004). Use of medicaid prescription data for syndromic surveillance—New York. *MMWR Morb Mortal Wkly Rep* 54(Suppl):31–4.

Cooper DL, Verlander NQ, Smith GE, et al. (2006). Can syndromic surveillance data detect local outbreaks of communicable disease? A model using a historical cryptosporidiosis outbreak. *Epidemiol Infect* vol. 134, no. 1, pp. 13–20.

Delichatsios, H., Callahan, M., Charlson, M. (1998). Outcomes of telephone medical care. J Gen Intern Med 13:9, 579–85. *http://www.ncbi.nlm.nih.gov/entrez/query.fcgi?cmd=Retrieve&db=PubMed&dopt=Citation&list_uids=9754512.*

Doroshenko, A., Cooper, D., Smith, G., et al. (2004). Evaluation of syndromic surveillance based on NHS Direct derived data in England and Wales. *MMWR Morb Mortal Wkly Rep* 54(Suppl):117–22.

Espino, J. U., Hogan, W., Wagner, M. M. (2003). Telephone triage: A timely data source for surveillance of influenza-like diseases. In: *Proceedings of American Medical Informatics Association Symposium*, 215–9.

Galimi, J. (1999). The Market for Medical Call Center Applications: Gartner Group.

Henning, K. J. (2004). What is syndromic surveillance? *MMWR Morb Mortal Wkly Rep* 53(Suppl):5–11. *http://www.ncbi.nlm.nih.gov/entrez/query.fcgi?cmd=Retrieve&db=PubMed&dopt=Citation&list_uids=15714620.*

Henry, J. V., Magruder, S. and Snyder, M. (2004). Comparison of office visit and nurse advice hotline data for syndromic surveillance–Baltimore-Washington, D. C., metropolitan area, 2002. MMWR Morb Mortal Wkly Rep, vol. 53 Suppl, no., pp. 112–6. *http://www.ncbi.nlm.nih.gov/entrez/query.fcqi cmd=Retrieve&db=PubMed&dopt=Citation&list uids=15714639.*

Henry, J. V., Magruder, S., Snyder, M. (2004b). Comparison of office visit and nurse advice hotline data for syndromic surveillance—Baltimore-Washington, D.C., metropolitan area, 2002. *MMWR Morb Mortal Wkly Rep* 53(Suppl):112–6. *http://www.ncbi.nlm.nih.gov/entrez/query.fcgi?cmd=Retrieve&db=PubMed&dopt=Citation&list_uids=15714639.*

Li, C.-S., Aggarwal, C. (2003). *Epi-SPIRE: A System for Environmental and Public Health Monitoring*. Paper presented at IEEE International Conference on Multimedia & Expo, July.

Ma, H., Rolka, H., Mandl, K., et al. (2004). Implementation of laboratory order data in BioSense early event detection and situation awareness system. *MMWR Morb Mortal Wkly Rep* 54(Suppl):27–30.

Maggini, M., Salmaso, S., Alegiani, S. S., et al. (1991). Epidemiological use of drug prescriptions as markers of disease frequency: an Italian experience. *J Clin Epidemiol* 44:1299–307. *http://www.ncbi.nlm.nih.gov/entrez/query.fcgi?cmd=Retrieve&db=PubMed&dopt=Citation&list_uids=1753261.*

Magruder, S. F., Henry, J., Snyder, M. (2004). Linked analysis for definition of nurse advice line syndrome groups, and comparison to encouters. *MMWR Morb Mortal Wkly Rep* 54(Suppl):93–97.

Pavlin, J., Murdock, P., Elbert, E., et al. (2004). Conducting population behavioral health surveillance using automated diagnostic and pharmacy systems. *MMWR Morb Mortal Wkly Rep* 53 (Suppl): 166–72.

Platt, R., Bocchino, C., Caldwell, B., et al. (2003). Syndromic surveillance using minimum transfer of identifiable data: the example of the National Bioterrorism Syndromic Surveillance Demonstration Program. *J Urban Health* 80(Suppl 1):i25–31. *http://www.ncbi.nlm.nih.gov/entrez/query.fcgi?cmd=Retrieve&db=PubMed&dopt=Citation&list_uids=12791776.*

Sabin, M. (1998). Telephone triage improves demand management effectiveness. *Healthcare Financ Manag* 52:49–51. *http://www.ncbi.nlm.nih.gov/entrez/query.fcgi?cmd=Retrieve&db=PubMed&dopt=Citation&list_uids=10182276.*

Wagner, M., Pavlin, J., Brillman, J., et al. (2004). *Synthesis of Research on the Value of Unconventional Data for Early Detection of Disease Outbreaks*. Washington, DC: Defense Advanced Research Projects Agency. *http://rods.health.pitt.edu/.*

Wagner, M. M., Aryel, R., Dato, V. (2001). *Availability and Comparative Value of Data Elements Required for an Effective Bioterrorism Detection System*. Washington, DC: Agency for Healthcare Research and Quality.

Yih, W. K., Caldwell, B., Harmon, R., et al. (2004). National Bioterrorism Syndromic Surveillance Demonstration Program. *MMWR Morb Mortal Wkly Rep* 53(Suppl):43–9. *http://www.ncbi.nlm.nih.gov/entrez/query.fcgi?cmd=Retrieve&db=PubMed&dopt=Citation&list_uids=15714626.*

Yokoe, D. S., Subramanyan, G. S., Nardell, E., et al. (1999). Supplementing tuberculosis surveillance with automated data from health maintenance organizations. *Emerg Infect Dis* 5:779–87. *http://www.ncbi.nlm.nih.gov/entrez/query.fcgi?cmd=Retrieve&db=PubMed&dopt=Citation&list_uids=10603211.*

V
PART
Decision Making

29
CHAPTER

Decision Analysis

Agnieszka Onisko, Garrick Wallstrom and Michael M. Wagner
RODS Laboratory, Center for Biomedical Informatics, University of Pittsburgh, Pittsburgh, Pennsylvania

1. INTRODUCTION

Decision making is part of the daily routine of biosurveillance. When a healthcare provider or a citizen notifies a health department about a case or an unusual cluster of illness, the staff at the health department must decide whether and how to further investigate. Hospital infection control practitioners, veterinarians, or nursing home directors face similar decisions when they receive results of testing or discover anomalous numbers of cases of disease. During an outbreak, these individuals face many decisions that they must make quickly and often under uncertainty and resource limitations.

Recently, the number of routine decisions faced by biosurveillance staff has increased. Now, not only must staff decide whether to investigate reports of confirmed cases from physicians and laboratories, and the less frequent reports of clusters of illnesses, but they must also decide whether to react to surveillance data collected from schools, 911 call centers, doctor's offices, pharmacies, and emergency departments. These data are associated with higher degrees of uncertainty than are reportable disease data, which increases the difficulty of the decisions.

Biosurveillance organizations also make decisions. They decide how to set the alarm thresholds of detection algorithms. These decisions are closely related to the above decisions about how to react to anomalies in biosurveillance data. A biosurveillance organization also decides how to invest finite resources in biosurveillance systems.

In this chapter, we focus on the decisions elicited by anomalies in biosurveillance data. We review the science of decision making, which includes decision theory and decision analysis. Decision analysis is a technique that people and organizations can use to improve the quality of their decisions. Organizations may use decision analysis to develop policies to guide frontline personnel or even incorporate the mathematical models created by these techniques directly into biosurveillance systems.

1.1. Examples of Decisions in Biosurveillance

Some decisions in biosurveillance are not difficult. When a staff member of a health department receives a report of a case of measles, for example, the appropriate response may be so straightforward that the staff member may not even perceive it as a decision. An easy decision is one that an individual makes regularly, that an individual has been trained for, or that does not involve a great deal of uncertainty.

Some decisions can be quite difficult. A difficult decision is one in which the best course of action is not clear. Examples include decisions about what actions to take in response to anomalies in surveillance data that are suggestive, but not conclusive, of the existence of an outbreak; decisions such as

Should I wait another day to see if the trend in the surveillance data continues?
Should I ask the city's emergency departments to collect more data?
Should I initiate a formal investigation?
Should I request assistance from experts?
Should I begin to mobilize treatment resources such as the national strategic stockpile?
Should I issue a public advisory?

Difficult decisions involve tradeoffs between the costs and benefits of action versus inaction. The most difficult decisions are those in which the costs and benefits are denominated in units of human health and lives. Thompson et al. (2005) classified different categories of decisions that have to be made in a short period of time by various individuals and organizations.

To make this discussion concrete, we focus on the decisions faced by a biosurveillance organization when faced with evidence of an environmental contamination or of a potential disease outbreak. *Cryptosporidium* outbreaks have occurred in the recent past or have been suspected from environmental data. We focus our discussion to decisions that have arisen around this disease sufficient published information about this problem is available to support a decision analysis.

2. THE DECISION TO ISSUE A BOIL-WATER ADVISORY (GLASGOW, 2002)

To elucidate the decisions that surround a suspected contamination of a water supply with *Cryptosporidia,* we consider the events that took place in Glasgow, Scotland, in the summer of 2002. Table 29.1 summarizes the chronology of events as related by the Incident Management Team (IMT) in an "after-action" report about that event (IMT, 2003).

On July 31, 2002, the day after an unusually heavy rain, Scottish Water (the water department supplying the greater Glasgow area) noted a doubling of turbidity in water from Loch Katrine. Because the community had experienced a

Handbook of Biosurveillance
ISBN 0-12-369378-0

Cryptosporidium outbreak in 2000 from contaminated water, Scottish Water promptly drew samples of water for microscopic analysis.

On August 1, Scottish Water reported oocysts resembling *Cryptosporidia* to Glasgow Public Health. *Cryptosporidia* is a genus of parasites that includes several species that may cause gastrointestinal infection in humans. The concentration of oocysts was 0.07 per 10 liters of water (approximately one oocyst per 143 liters of water). Oocyst concentrations in water do not correlate well with levels of disease in human populations. Therefore, Glasgow Public Health *decided* to await the results from additional samples of water.

On August 2, Scottish Water notified Glasgow Public Health of a higher measurement of oocysts in water. Glasgow Public Health *decided* to transmit a fax message to physicians heightening suspicion for cryptosporidiosis in patients with gastrointestinal complaints.

On August 3, Scottish Water notified Glasgow Public Health of an even higher reading from a second water source. By this time, Scottish Water had already decided to reconfigure the water system to reduce the population exposed to water from Loch Katrine from 416,000 to 160,000 people. Glasgow Public Health convened its Policy Advisory Group, which promptly *decided* to call together its IMT. Within two hours of convening, the IMT *decided* to issue a boil-water advisory. A boil-water advisory is a general notification to the public to boil tap water before consumption. This decision resulted in considerable activity to disseminate the advisory to the public. The actions included press releases, recorded messages on the water company's call-in lines, loudspeaker trucks, direct contact with large facilities such as hospitals, and postings on Web sites. The notification took approximately three days to complete.

TABLE 29.1 Timeline for Glasgow *Cryptosporidium* Incident

Date	Time	New Information, Decision, or Action
July 30, 2002		Heavy rain
July 31, 2002		SW detects doubling of turbidity
		SW draws water samples for crypto analysis
Aug 1, 2002		Glasgow Public Health Protection Unit (Glasgow) receives notice from SW of doubling of turbidity
		Glasgow receives report from SW of *Cryptosporidium* -like oocysts at 0.07 per 10 liters
Aug 2, 2002	16:25	Glasgow receives report from SW of 0.275 oocysts per 10 liters
	17:15	Glasgow notifies medical practitioners by fax of deterioration of water quality
Aug 3, 2002	9:50	Glasgow receives report from SW of 0.353 oocysts per 10 liters
	12:15	Glasgow receives report from SW of 11 oocysts per 10 liters
	15:00	PAG convened
		SW finishes rerouting water to reduce exposed population to 160,000
	15:40	PAG decides to form IMT
	17:00	IMT meets
	18:45	IMT decides to issue boil-water advisory
	22:00	First notification action completed (helpline activated)
Aug 4, 2002	0:15	Web site notice posted
Aug 7, 2002	14:00	Dissemination of boil-water advisory completed
		Water rerouting completed (water safe)
Weeks later		Molecular typing returns showing nonpathogenic strains

SW indicates Scottish Water; PAG, policy advisory group; and IMT, incident management team.

Several weeks later, molecular analysis of the oocysts revealed that they were predominantly species of *Cryptosporidia* that do not cause disease in humans. In retrospect, the boiling of water was unnecessary.

We cite this example to make two key points about decision making in biosurveillance. First, Glasgow Public Health had considerable difficulty interpreting the available surveillance data (oocyst levels) because the data did not provide a definitive answer to the pivotal question: *Were people being exposed to* C. parvum *or another strain of* Cryptosporidium *that causes disease in humans?* Glasgow Public Health did not know the answer to this question until several weeks later. Second, the IMT did not know the cost or the benefit of a boil-water advisory. The IMT discussed the potential for scalding injuries owing to increased boiling of water over the usual amount of water boiling for the making of tea and coffee, but it had no way to estimate the magnitude of the risk. The IMT did not know the benefit of a boil-water advisory; it did not know its efficacy (the number of cases that might be averted) or the benefit of preventing a case. Corso et al.'s (2003) economic study of the Milwaukee outbreak, estimating a cost of $239 per sick individual in 1993 U.S. dollars, had not yet been published.

Cryptosporidiosis is one of the mildest diseases of concern to a health department. Corso et al.'s estimate of $239 per sick individual reflects the mildness of the disease (although the disease can be disproportionately fatal in immunocompromised patients, as was the case for 85% of the 54 fatalities in the 1993 in Milwaukee outbreak) (Hoxie et al., 1997). Cryptosporidiosis is not a reportable disease in most jurisdictions. However, because this problem could potentially have sickened 400,000 people (as it did in Milwaukee), the mere possibility of it disrupted the city of Glasgow for a week.

Although the boil-water advisory, in retrospect, was unnecessary, the appropriate question to ask is whether the decision was optimal. It remains unknown to this day whether it was optimal and what is—were a heavy rain in Glasgow to lead to a repetition of the above events—the optimal decision. It is similarly unknown what the optimal decision should be if evidence—suggestive, but not conclusive—of a large-scale aerosol release of a biological agent over a city were to present itself tomorrow. There are several accounts of BioWatch signals (DNA-polymerase–based air sampling systems) in Houston and San Diego that raised the specter of a tularemia or other biological agent release. IMTs were convened that considered high stakes decisions under uncertainty and time pressure (Roos, 2003, Houston Department of Health and Human Services, 2003).

Because of the rapid improvement in the world's ability to monitor water, air, and the health of populations, we expect such situations demanding rapid action despite limited information to occur with increasing regularity.

3. THE SCIENCE OF DECISION MAKING

Scientists have studied decision making for centuries. In the 16th century, the Italian Gerolamo Cardano (1501–1576) studied betting strategies for gambling. Before his time, the outcomes of gambling were viewed as the will of the gods. His book *The Book on Games of Chance,* written in the 1560s but published only in 1663 after his death, contains the first systematic treatment of probability. In the 17th and 18th centuries, Pierre Fermat (1601–1665), Blaise Pascal (1623–1662), and Pierre Laplace (1749–1827) developed probability theory (Laplace, 1812). Daniel Bernoulli (1700–1782), first introduced the subjective (nonfrequentist) view of uncertainty. After this idea, Thomas Bayes (1702–1761) and Laplace introduced a concept, called a *subjective probability,* defined as the degree of a person's belief in some proposition. This view of probability enabled Bayes to formulate a rule for updating subjective probabilities in the light of new evidence. Bayes' view of probability theory was published posthumously in 1763 (Bayes, 1958). The concept of subjective probability and Bayes' rule greatly expanded the range of decision problems that scientists could analyze. Subjective probability and Bayes rule have become cornerstones of medical diagnosis (Cooper, 1984) and of modern biosurveillance (Wagner et al., 2001; Cooper et al., 2004).

Daniel Bernoulli introduced an economic theory that accounted for discrepancies between the recommendations of probabilistic models and how people make decisions and choices in every aspect of life (Bernstein, 1996). One well known discrepancy at the time was the St. Petersburg paradox, which Bernoulli addressed. He was the author in 1738 of the *Exposition of a New Theory on the Measurement of Risk* (Bernoulli, 1954 [1738]; Stanford Encyclopedia of Philosophy, 2005). His solution to the St. Petersburg paradox formed the basis for future economic theories of risk aversion, risk premium, and utility.

Behavioral psychologists study how humans make decisions from an observational perspective. Their research demonstrates that decisions made by experts can be inferior to those made by even the simplest of mathematical models (Dawes, 1979). They have found that humans, when making decisions, use heuristics (rules of thumb) that lead to systematic errors and suboptimal decisions (Tversky and Kahneman, 1974), and they have demonstrated an inverse relationship between time pressure and decision quality (Edland and Svenson, 1993). This long-standing interest of psychologists in human decision making has evolved into a branch of psychology known as behavioral decision theory (Tversky and Kahneman, 1974). Recently, a prominent behavioral psychologist, Daniel Kahneman, received a Nobel Prize for his contribution to our understanding of human judgment under uncertainty.

3.1. Decision Theory and Decision Analysis

In the 20th century, John Von Neumann and Oskar Morgenstern reintroduced, formalized, and extended the work of Daniel Bernoulli, founding what is now referred to as decision theory (von Neumann and Morgenstern, 1944).

Decision theory is an extension of probability theory, adding adds axioms and theorems to probability theory that enable mathematicians to represent decision alternatives, costs, and benefits. Leonard J. Savage (1954) further developed decision theory into its modern form.

The field of decision analysis emerged in the 1960s, as engineers and economists applied decision theory to the modeling of actual decision problems (Raiffa, 1968). Decision analysis is sometimes referred to as the art of decision theory, because creativity is required to translate a complex real-world situation into a mathematical model (Henrion et al., 1991). Decision analysis is a large field. Many individuals working in both the government and the private sector specialize in decision analysis. We list a sample of textbooks, journals, and professional societies under "Additional Resources" at the end of this chapter.

3.2. The Principle of Maximum Expected Utility

Decision theory derives from a simple, common-sense premise: a rational person should make a decision that maximizes his expected benefit. This principle is called *the principle of maximum expected utility,* and its origins can be traced to Daniel Bernoulli.

Decision theory makes it possible to associate a real number between zero and one with each possible consequence of a decision. This number— termed a *utility*—is the value of that consequence to the decision maker. In the theory, the number is always expressed on a scale from zero to one, where zero typically means the worst possible outcome and one the best possible outcome. In practice, the scale can be other than zero to one (but, technically, it is not called a utility).

The *expected utility* of a decision is simply the sum of the utilities of each of the possible consequences of the decision, multiplied by their probabilities. For example, if I am considering placing a \$10 wager on a horse (in the hope of receiving a \$20 payout from the track) and I believe the horse has a 0.3 probability of winning, the expected utility of my decision to place the bet is the sum of $0.3 \times (\$20 - \$10)$ (I win my bet and take home \$10 profit) and $0.7 \times -\$10$ (I lose my \$10 wager). The expected value of a decision to place the bet is $-\$4$ ($0.3 \times \$10 + 0.7 \times -\$10 = -\$4$), and the expected value of a decision not to bet is \$0. Of course, if I ascribe to the maximum expected value principle, I would not place this bet.

The alternative of not placing the bet is $0 (I neither gain nor lose money), which is greater than –$4. In general, one should only take bets with a positive expected utility.[1]

4. DECISION ANALYSIS: ISSUE A BOIL-WATER ADVISORY OR WAIT FOR RESULTS OF DEFINITIVE TESTING

Of course, most gamblers do not have the training (or the inclination) to apply the principle of maximum expected utility in such a formal manner. Part of the reason is that, in most games, the prize is the same (money) and its interaction with probability of winning is relatively simple. However, there are many real-world games of chance in which the prizes are far more complex. The prizes may be denominated in lives saved or lost, economic loss, and disruption to a city. Each prize can involve a mixture of lives saved or lost, money, and disruption. In these situations, decision makers often turn to professionals who have the training to assist with such decisions.

In this section, we present an analysis of the decision to issue a boil-water advisory or to wait for the results of definitive testing. We use this example to explain the process of decision analysis. Note that, our example decision analysis is similar but not identical to the decision problem faced by the IMT in Glasgow. The IMT was faced with evidence of contamination of the water. In our example, the decision maker is faced with early evidence of an outbreak of cryptosporidiosis in people—evidence that is suggestive but not conclusive of an outbreak. In particular, the decision maker is confronted with an increase in sales of diarrhea remedies (Figure 29.1). We analyze a decision problem related to the Glasgow incident, but to simplify the analysis, we consider a variant in which the evidence of a potential health threat comes from sales of diarrhea remedies instead of water quality. Our research group conducted a decision analysis of this particular situation, hence, it is available to present as an example (Wagner et al., 2005).

We considered the Chicago metropolitan area to make our study concrete. We chose Chicago because we had previously studied detectability of *Cryptosporidium* outbreaks from sales of diarrhea remedies in Chicago. Chicago is served by two water treatment plants, and we assumed that each treatment plant serves half of the city. We model the situation in which there is a biosurveillance system that analyzes the sales data for the area served by one of the plants. Every day, the biosurveillance system determines whether the level of sales exceeds a threshold and, if so, alerts the biosurveillance staff. The staff then faces a decision to *act now* (issue a boil-water advisory) or *wait* for the results of definitive testing of individuals with diarrhea or definitive testing of the water.

Our example is representative of many decisions health departments face as a result of increased monitoring of time-series data to achieve earlier detection of outbreaks. This example, incidentally, is the only decision analysis of a biosurveillance decision that exists, to our knowledge.

4.1. Overview of the Decision-Analytic Process

Decision analysis is a process in which a decision analyst and domain experts work together to construct a mathematical model of a decision. The model is called a *decision model.* A decision analyst, who is in some ways a psychoanalyst for people with decision problems, uses a set of techniques to assist an organization or an individual to construct the decision model. The decision analyst and the experts build a decision model and then use it to compute the expected utilities of the choices (two or more) that are available to the decision maker.

A decision analysis in biosurveillance or any domain is based on substantial information about the domain that can only be obtained from experts or literature review. In biosurveillance, the information basis of a decision analysis comprises knowledge about diseases, epidemics, and costs of treatments. The decision analyst must elicit this knowledge from experts and from published sources.

Figure 29.2 depicts the steps that a decision maker and a decision analyst follow to construct a decision model and then to analyze the model to produce insight about the decision.

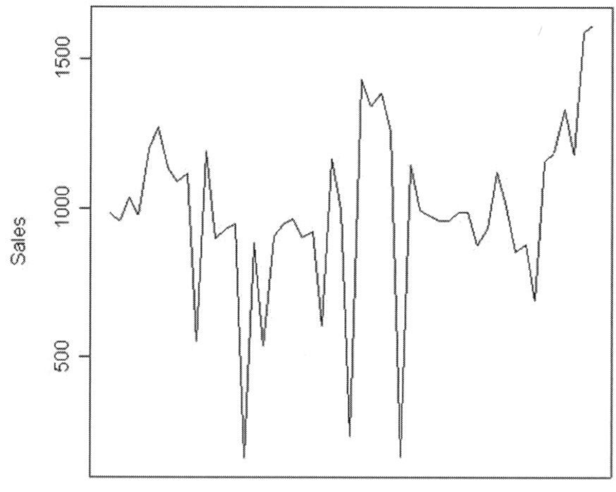

FIGURE 29.1 Rising sales of diarrhea remedies that triggered an alert (detection threshold set to one false-alarm per year).

1 Experts on decision theory know that dollar amounts do not map linearly onto a utility scale. For small dollar amounts such as a $10 bet with a $20 payout, however, they often do.

Building a decision model is an iterative process. The analyst and experts construct a first draft, examine it, and then refine it based on new insights about the factors influencing the decision. Usually, regular, short consultations work best. Between these consultations, the decision analyst refines the model and prepares questions for the decision maker. The decision analyst may read textbooks to become familiar with the terminology and the important factors relevant to the decision. A decision analyst may tape record the sessions with the decision maker because it is often hard to process all the knowledge that the decision maker provides during a meeting. The decision analyst may organize brainstorming sessions with experts who are not directly involved in building the model to confirm that no aspect of the decision problem has been overlooked.

The products of a decision analysis are (1) a base-case analysis, which is the decision model configured using the best estimates of the various parameters that go into the model, and (2) a sensitivity analysis of the effects of varying parameters over a range that spans the uncertainty that the decision maker has about the exact values of those parameters. The sensitivity analysis demonstrates how robust (or fragile) the decision recommended by the model is to the imprecision of knowledge about the domain (e.g., the uncertainty about the cost of a boil-water advisory).

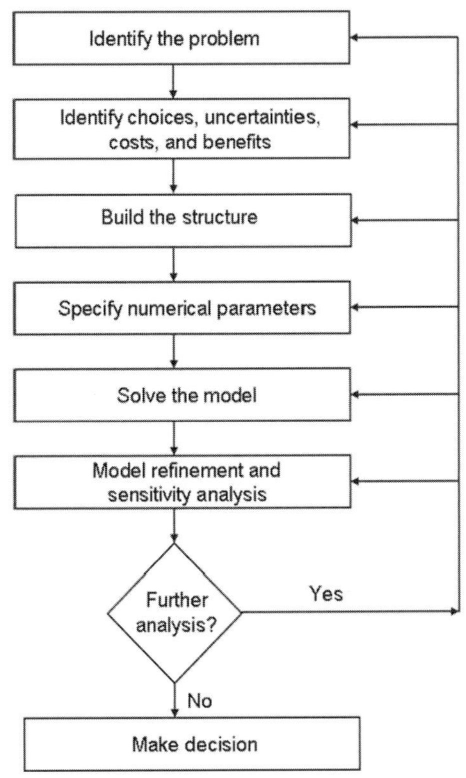

FIGURE 29.2 Building the decision model.

The next sections describe how we used a standard approach to decision analysis (Clemen and Reilly, 2001) to build a model of the boil-water decision. Each section focuses on one step of the process depicted in Figure 29.2.

4.2. Identifying the Problem

An organization may commission a decision analysis, or a research group may choose a decision problem for study. In our example, we (a research group) decided to study the decision to issue a boil-water advisory.

If this had been a study commissioned by a biosurveillance organization, the decision analyst and the organization would discuss and clarify the objectives of the analysis. In particular, they would clarify the question the analysis was intended to address, and what form the final result should take (e.g., a policy, a publication, or a set of guidelines for staff or IMTs).

4.3. Identifying Choices, Uncertainties, Costs, and Benefits

The analyst then works with experts to identify the decision alternatives (choices), uncertainties, costs, and benefits. In our simple example, the choices available to a decision maker include issuing a boil-water advisory now or ordering tests (of water and any sick individuals that can be found) and waiting for the results. The uncertainty in this example is whether the surveillance data indicate a *Cryptosporidium* outbreak or some other phenomenon. The costs and benefits associated with each possible outcome of the decision include the cost of a boil-water advisory and the benefit in reduction of the number of sick individuals, should the increase in sales of diarrhea remedies actually prove to be a result of an outbreak.

One of the techniques used in building decision models is the *clarity test* (Howard, 1988). For each element of the model (i.e., choices, uncertainties, costs, and benefits), the analyst checks whether it has been defined in an unambiguous manner. Ambiguous definitions of model elements will generally backfire at some modeling stage, especially when it comes to elicitation of numerical parameters. The definition of an outbreak, as discussed in Chapter 2, is an example of an ambiguous definition that would not pass a clarity test. The phrases *higher levels than normal levels* and *localized increases* are not precise. A decision analyst would ask the experts to provide an operational definition for the term *outbreak* were the experts to indicate that the model include a variable called *outbreak*.

4.4. Building the Structure

The analyst then represents the choices, uncertainties, and outcomes that the expert identified in a graphical structure by using a decision tree (Figure 29.3) or some other structure described below. By using the standard symbols for building decision trees (Raiffa, 1968), the decision analyst draws a rectangle, called a *decision node,* to represent the decision (the node *issue BWA* in Figure 29.3) and two branches (lines) to represent the two possible choices: *act now* and issue a

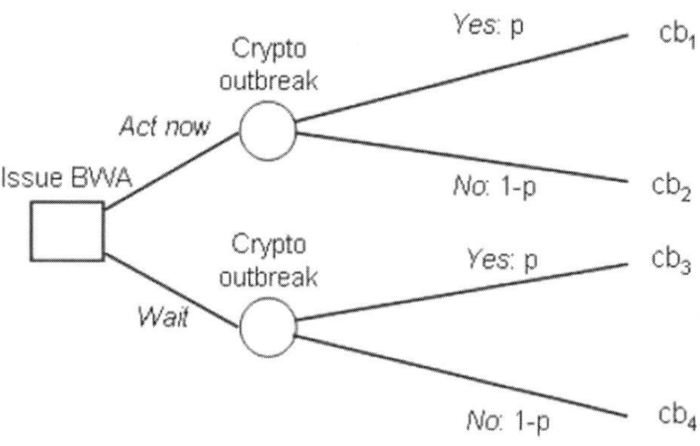

FIGURE 29.3 Decision tree model.

boil-water advisory, or *wait* 72 hours for the results of testing of sick individuals and testing of water.[2]

The decision analyst draws circles, called *chance nodes,* to represent uncertain propositions (the chance node *Crypto outbreak* represents the proposition that there is a *Cryptosporidium* outbreak). The decision analyst draws two chance nodes because both choices are affected by uncertainty. These chance nodes are identical because both choices are affected by the same uncertainty. Two branches emanate from each chance node to represent the two values of the proposition, i.e., *yes,* it is true that there is a *Cryptosporidium* outbreak, and *no,* it is not true. This decision tree has four outcomes. Each outcome represents a different future scenario. For example, the topmost outcome, described by the variable cb_1, represents the outcome of *acting now* and subsequently discovering that a *Cryptosporidium* outbreak indeed was in progress.

One can understand a decision tree as a model of what might happen in the future based on a decision that we make in the present. The chance nodes represent fate. No matter what the decision maker decides, only fate will determine whether he is a hero or a goat. If the decision maker decides to issue a boil-water advisory and is right (i.e., three days later the laboratory confirms his wisdom), he is a hero. If he decides to issue a boil-water advisory and he is wrong (the second end point of the model), he is a goat. He could decide to await further testing and also either be a hero or a goat (the fourth and the third end points of the decision tree). However, it is important to note that a bad outcome (goat) does not necessarily mean a bad decision. As we will show when we fold back the decision tree, the model will recommend a decision that in the long run (i.e., if, hypothetically, the decision maker makes the same decision the next 1000 times he faces this situation), will be most beneficial to him (or the organization that he represents).

It is worth emphasizing that a key benefit of a decision analysis is that it establishes rational criteria for taking difficult decisions under uncertainty. If a decision maker makes the decision recommended by the model, he can defend his decision by reference to the principle of maximum expected utility. The psychology of decision making is such that decision makers tend to be overly conservative about taking actions for which they may have remorse (Tversky and Kahneman, 1974). If the decision maker makes a decision that has been carefully analyzed in advance, taking into account the values of his community as expressed by the costs used in the model, as well as all available information expressed as chance nodes in the model, then it is more likely he will feel comfortable that he will be perceived as a responsible decision maker, whatever the ultimate outcome of a particular decision is.

For completeness, we shall briefly digress in this paragraph to explain that decision analysts may use alternative graphical models, such as influence diagrams (Howard and Matheson, 1984b). Readers simply need to be aware that there are graphical models other than decision trees. Figure 29.4 is an influence diagram representation of the boil-water decision. A decision analyst may use influence diagrams when modeling decision problems with more complex probabilistic structure. He may use dynamic influence diagrams (Tatman and Shachter, 1990) or the influence view formalism (Leong, 1994) when modeling a decision that is revised frequently as new information accumulates. The decision analyst may use a Markov decision process (Beck and Pauker, 1983; Dean and Wellmann, 1991), or combination of Markov decision process and the influence view formalism (Magni et al., 2000) to model a series of different decisions that are made over time (e.g., decisions made during an outbreak investigation).

4.5. Specifying Numerical Parameters

Once the structure of the model is completed, the decision analyst quantifies the model. This process involves the elicitation of numerical parameters for the probabilities and costs in the model. Parameters of the model from experts and published studies are elicited (Raiffa, 1968; Howard and Matheson, 1984a).

Our example model requires the following numerical parameters: the probability p that there is an ongoing *Cryptosporidium* outbreak and values for the variables cb_1, cb_2, cb_3, and cb_4 that represent costs and benefits incurred for each outcome of the model (Figure 29.3).

4.5.1. Probability of Outbreak, Given Surveillance Data

Although newer Bayesian detection algorithms, such as PANDA and BARD (described in Chapters 18 and 19),

2 In our model, we assume that testing of individuals with diarrhea will confirm the existence of an outbreak, if it exists, in three days. We note that direct testing of water in Glasgow did not produce a definitive answer for weeks. This process can be accelerated but at present it is unclear whether in practice it can occur in a more rapid time frame.

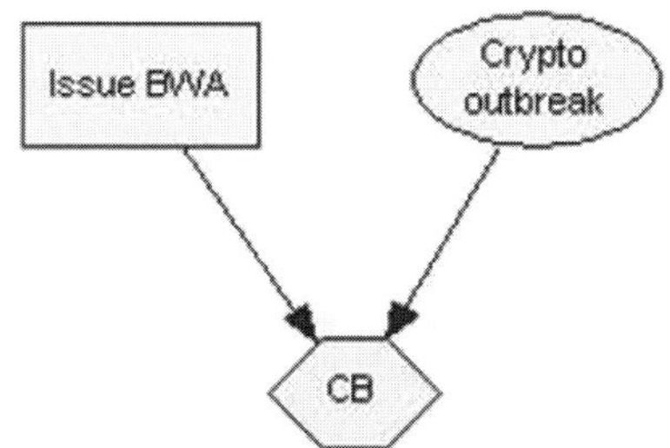

FIGURE 29.4 Influence diagram model for boil-water advisory decision. CB indicates costs and benefits of the possible outcomes of the model.

compute the posterior probability of an outbreak from surveillance data directly, older methods do not.[3] In the next chapter, we discuss methods to estimate the probability of an outbreak from surveillance data when newer detection algorithms are not available (by far the more common situation). For our discussion here, we simply use the numerical result that we obtain in that chapter: $p = 0.0410$.

4.5.2. Costs and Benefits

The variables cb_1, cb_2, cb_3, and cb_4 represent both the costs and benefits of each of the outcomes of the decision model. In our simple model, we consider only two costs and benefits: cost of a boil-water advisory and the benefit of averting illness. In general, the costs and benefits associated with outcomes in biosurveillance are some combination of financial costs (e.g., treatment), and effects on numbers of sick and dead individuals.

Efficacy of a Boil-Water Advisory. We assume that a boil-water advisory is heeded by every person in the half of the city where the water contamination is suspected; thus, there are no further infections and the total number of sick individuals equals the number of individuals infected up to the date of the advisory. We realize that not every individual in a community will heed a boil-water advisory. We make this assumption because there is little data about compliance with boil-water advisories. We could explore other efficacies in the sensitivity analyses phase, as discussed below.

Incubation Period of the Disease. Cases infected before the advisory may become symptomatic after the issuance of the boil-water advisory. The average incubation period of cryptosporidiosis is often stated to be seven days. For the base case, we assume that the incubation period for all cases is exactly seven days.

Number of Sick Individuals. Figure 29.5 depicts a portion of the epidemic curve that we used in our model. It is a best-estimate curve that is derived from the surveillance data in Figure 29.1 and sales of diarrhea remedies during a known water-borne *Cryptosporidium* outbreak that occurred in North Battleford in 2001 (Stirling et al., 2001). We describe the methods for its estimation in Chapter 30.

The area under the epidemic curve up to a given date is the total number of individuals who became sick by that date. Based on our assumption of a constant incubation period of seven days, we computed the total number of *infected* individuals on the date of intervention by finding the cumulative number of symptomatic individuals seven days after the intervention.

Cost of a Boil-Water Advisory. We assume that the cost of a boil-water advisory is the cost of additional bottled water consumed. We note that the decision-makers in Glasgow did not consider this cost but rather the risk of increased scalding injuries. For the base case, we make an assumption that biases the decision against issuing a boil-water advisory without confirmatory testing. We assume the advisory motivates half the population of Chicago[4] to consume one extra liter of bottled water at $1 per bottle per day. The population of Chicago

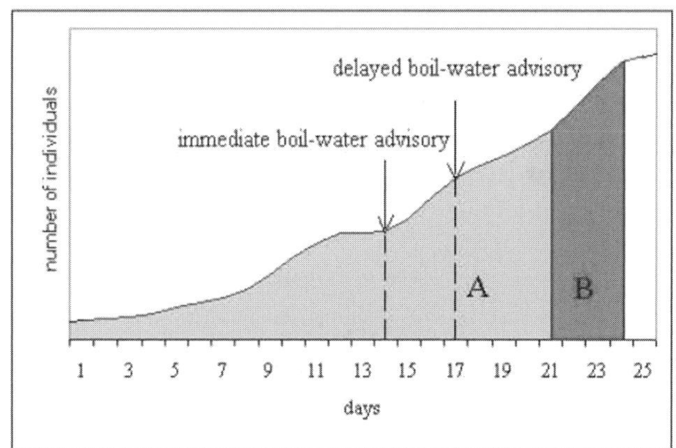

FIGURE 29.5 The estimated epidemic curve by date of onset of symptoms. The alarm occurred on day 14. If a boil-water advisory is issued on that day, the total number of individuals who will develop symptomatic cryptosporidiosis is the area under the curve marked A. If the boil-water advisory is issued three days later after confirmation, the total number of individuals who will develop symptomatic cryptosporidiosis is the sum of the areas under the curve marked *A* and *B*.

3 Note that the BARD and PANDA models are large networks of nodes similar to the chance nodes in Figure 29.3. These networks have a single node which represents the probability of outbreak given surveillance data. At a conceptual level, these detection systems can be linked directly to a decision tree via that node.

4 As we previously assumed, we model the situation in which there is a biosurveillance system that analyzes the sales data for the area served by one of the plants which provides the water for half of the city.

is 5.35 million, so the cost of a boil-water advisory for half of the city is \$2.675 million per day. We believe this is a high estimate.

Benefit of Preventing Sickness. To quantify the benefit of a boil-water advisory, which depends both on the number of cases averted and the benefit per averted case, we use the average cost per case developed by Corso et al. (2003) for the Milwaukee 1993 *Cryptosporidium* outbreak of \$239, adjusted from 1993 dollars to 2005 dollars. Our model ignores the possibility that immunocompromised individuals may die from *Cryptosporidium* infections.

cb_1, cb_2, cb_3, and cb_4. The variable cb_4 describes the outcome in which the decision maker did not issue a boil-water advisory, and there was no *Cryptosporidium* found on testing. There are no sick individuals, and there is no cost incurred because no boil-water advisory was issued. Therefore, the cost and benefit for this outcome cb_4 is zero.

The variable cb_1 represents the outcome in which the decision maker issued a boil-water advisory and the *Cryptosporidium* outbreak occurred (true alarm). We assume that the problem in the water supply is resolved in five days; thus, the boil-water advisory will be in effect for five days at a cost of \$2.675 million per day, for a total of \$13.375 million. In Figure 29.5, the area under the epidemic curve marked A is the expected number of sick individuals for the outbreak in the case that the boil-water advisory is issued immediately. The expected number of sick individuals is equal to 235,110 individuals. The cost per sick individual is \$323[5] so the total cost of sickness is \$75,940,780.49.[6] Therefore, the value of cb_1 is negative \$89,315,780.49 (\$13,375,000 + \$75,940,780.49).

The variable cb_2 describes the outcome in which the decision maker issued a boil-water advisory and the *Cryptosporidium* outbreak did not occur (false alarm). In this case there are no sick individuals; therefore, cb_2 equals the total cost of the boil-water advisory. We assume the advisory is issued for three days (after this time, we know the results of testing the water). The cost of the boil-water advisory for a total of three days is \$8.025 million (\$2.675 million per day multiplied by three).

The variable cb_3 represents the outcome in which the decision maker waited three days before issuing a boil-water advisory during an actual *Cryptosporidium* outbreak. The cost of the boil-water advisory is the same as cb_1, \$13.375 million. The total number of sick individuals will be the area $A + B$, or 316,034. The cost per sick individual is \$323 so the total cost of sickness is \$102,079,219.5. Therefore, the value of cb_3 is negative \$115,454,219.5 (\$13,375,000 + \$102,079,219.5).

4.6. Solving (Folding Back) the Decision Tree

Once the decision analyst has elicited all the required numerical parameters, he can mathematically "solve" the decision tree.

Figure 29.6 depicts the solved decision tree. For each of the choices, the decision analyst calculates the expected value according to the principle of maximum expected utility. The expected value of *act now* is $p \cdot cb_1 + (1-p) \cdot cb_2 = 0.041 \times -\$89{,}315{,}780.49 + 0.959 \times -\$8{,}025{,}000 = -\$11{,}357{,}922$. The expected value of *wait* is $p \cdot cb_3 + (1-p) \cdot cb_4 = 0.041 \times -\$115{,}454{,}219.5 + 0.959 \times \$0 = \$4{,}733{,}623$ (these equations are identical in form to those we used in analyzing the \$10 bet on a horse race earlier). Keep in mind that this decision model is a very simple one. More elaborate models may have additional chance and decision nodes. In such models, the decision analyst repeats the above calculations for each chance node, starting from the rightmost chance nodes. Once the value of a chance node is calculated, he substitutes the value for the node, allowing us to compute the value of the parent node. The repetitive nature of this right-to-left process is why the process of evaluating a decision tree is sometimes called folding back the tree.

The decision analyst and the expert now have the results of the base-case analysis. When the surveillance data indicate a posterior probability of *Cryptosporidium* equal to 0.041, the expected cost of issuing an immediate boil-water advisory is –\$11,357,922, whereas the expected cost of waiting for confirmation is –\$4,733,623. Hence, the optimal decision is to wait three more days until laboratory confirmation of the outbreak before issuing a boil-water advisory (under the assumptions of our example analysis, which are not sufficiently realistic to guide real-world decision making without refinement).

The results of the base-case analysis suggesting testing to the point of definitive diagnosis were not surprising. The probability that there was a *Cryptosporidium* outbreak was low, and the cost of illness was low. We set the cost of issuing a boil-water advisory high (we set the cost of a boil-water advisory \$1 per person per day).

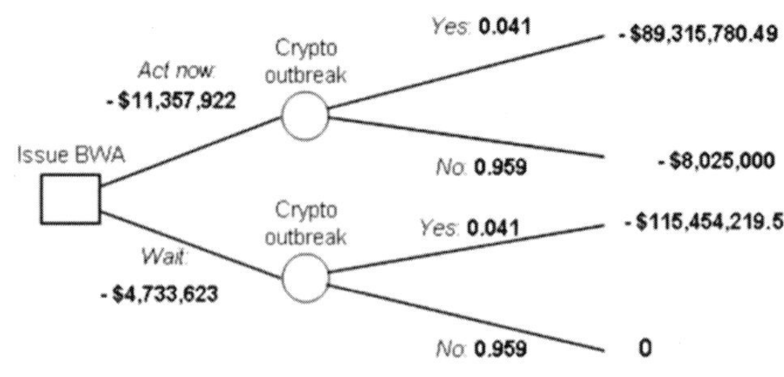

FIGURE 29.6 Solved decision tree.

5 The value of \$238 was adjusted from 1993 dollars to 2005 dollars.

6 Please note, that the cost of sickness depends on what day the outbreak is detected.

Cryptosporidium is a mild disease that in many cases does not require a patient to seek medical attention. If we were to conduct a similar analysis for a disease with a high mortality rate, such as anthrax, and assumed a three-day delay for a confirmation, the set of circumstances would be far narrower in which waiting instead of acting to control the disease would be the optimal decision. The maximum time for waiting for confirmatory testing might be measured in hours. We expect decision analyses of such situations to produce recommendations for action that may differ significantly from current practice.

4.7. Performing Sensitivity Analyses

Readers who are not familiar with decision analysis may wonder that with so many assumptions and estimates of parameters in even this simple example, how a decision analysis can produce a valid conclusion. The process of sensitivity analysis addresses the issue of validity by exploring the effects of the assumptions on the decision in a systematic fashion.

Sensitivity analysis answers the question: *How sensitive is the recommended decision to the assumptions that we made in structuring and assigning numerical values to the model?* In particular, the goal of sensitivity analysis is to understand whether changing the assumptions changes the decision alternative recommended by the model.

The decision analyst systematically explores the effect of changing the numerical parameters of a model. After each change, the analyst re-resolves the decision tree. In our example, the analyst could vary the values for the probability p or any of the parameters that underlie the values for the variables cb_1, cb_2, cb_3, or cb_4.

The decision analyst plots the results of sensitivity analyses on graphs. The simplest sensitivity analysis is a *one-way sensitivity analysis.* Figure 29.7 shows the effect of varying the cost of the boil-water advisory over the range of $0 per person per day to the level of $1 per person per day used in the base-case analysis. The graph plots the expected total costs (cost of water and of sickness) for the choices *act now* and *wait.* This one-way sensitivity analysis reveals that if the cost of bottled water consumed during a boil-water advisory is less than 13.92 cents per person per day (to the left of the intersection of the two lines depicted in Figure 29.7), then the decision should be to issue a boil-water advisory immediately because the expected costs (including both the cost of bottled water as well as the benefit of averted illness) are lower than the expected cost of waiting.

We also derived the threshold value for p at which the decision changes. For $p=0.2349$, the decision *act now* and issue boil-water advisory is associated with smaller costs than the decision *wait* and collect more data.

A decision analyst can also perform two- and three-way sensitivity analysis. In a two-way sensitivity analysis, the results are plotted on a three-dimensional graph. One approach to perform a multivariate analysis is to assign distributions to the parameters, assume the distributions are independent, and then randomly sample the parameters from their respective distributions to generate a distribution over the expected utility of each decision option. For more detail description of these techniques, see von Winterfeldt and Edwards (1988).

The decision analysis process is iterative. The sensitivity analysis might identify model parameters, such as the cost estimate of a boil-water advisory, for which better estimates would be highly desirable. This discovery may lead to a deeper literature search, consultation with experts to refine the estimates or new research. Once the decision analyst and the decision maker agree that the process of model building is finished, then the next step is to make the final decision.

4.8. Acting on the Decision

A decision analysis may influence decisions in different ways. It may simply be published, and readers will consider the analysis in forming policies and making future decisions. If the analysis recommends a strategy that differs from current "best practice," then, perhaps, what is considered "best practice" will gradually change in response to the published decision analysis. Practice will change more quickly if professional organizations endorse the decision analysis or incorporate its results into their recommendations (Plante et al., 1986; Pauker and Kassirer, 1987). If the analysis was commissioned by an organization to serve as a basis for policy formulation, the change in practice may be more immediate. The impact of a decision analysis is the extent to which a particular decision or class of decisions come into alignment with the recommendations of the analysis.

The process of creating a decision model often identifies parameters that experts feel are critical for optimal decision

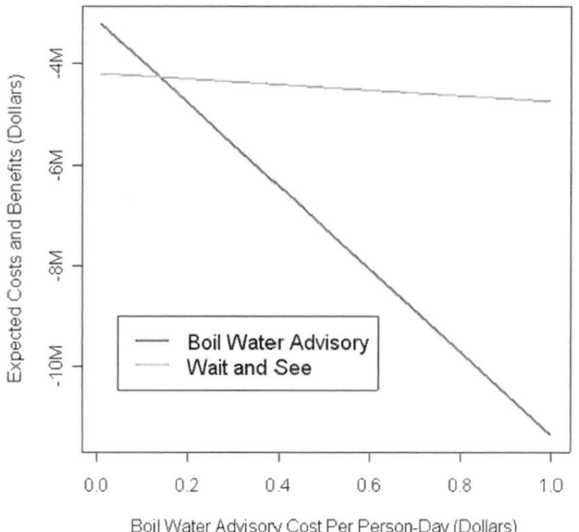

FIGURE 29.7 The effect of varying the cost of boiling water advisory on the decision.

making but for which existing information is either incomplete or completely unavailable. In our example, the recommended decision was sensitive to the cost of a boil-water advisory and the efficacy of an advisory. The identification of gaps in existing knowledge may influence funding agencies or organizations to sponsor research to fills the gaps.

5. COMPUTATIONAL DECISION ANALYSIS

A decision model can be embedded in a biosurveillance system. The field that designs and evaluates embedded decision models is called *computational decision analysis*.

The types of decisions that benefit from embedded models are decisions for which it is not possible to write a policy that covers all possible variations. For example, the decision about what piece of information to collect next in a medical diagnostic workup of where to focus investigation energy during an outbreak investigation cannot be expressed on paper or even in a textbook of medicine or epidemiology. There are simply too many conditionals. For this reason, many of the diagnostic expert systems discussed in Chapter 13 incorporate decision models. Decision models underlie the value-of-information–driven questioning that we describe in Chapter 13. These decision models are typically represented by using "cost of misdiagnosis" tables (Drummond et al., 1997).

Note that when a decision model is embedded in a biosurveillance system, a decision analyst is not available to help the user understand or manipulate the model. Researchers have developed methods, called *explanation modules,* to automate the function normally played by the decision analyst. Studies of explanation modules show that they can significantly increase insight into system's recommendations and, effectively, increase the quality of the decision made by the user of the embedded system (Clancey, 1984; Wallis and Shortliffe, 1984). Frontline decision makers are reluctant to accept a decision-support system's advice if they are not able to understand how it reached its conclusions (Teach and Shortliffe, 1984).

6. SUMMARY

Decision making is ubiquitous in biosurveillance. Decisions link biosurveillance systems to response systems. Decision makers must be skilled in both interpretation of biosurveillance data and in weighing the costs and benefits of their decisions. They, ideally, must have access to probabilistic interpretations of biosurveillance data as well as objective information about the costs and benefits of their actions. At present, the assistance available to frontline personnel in taking decisions is the training that they receive in school, on the job, experts on whom they can call, and policies established by their organizations about these decisions. The biosurveillance systems that they use at present support rapid collection and analysis of patient data and epidemiological information, but with rare exception, they do not provide probabilistic output needed for decision making (the exceptions are BARD and PANDA described in Chapters 19 and 18, respectively).[7] To date, there have been few economic analyses (see Kaufmann et al., 1997; Meltzer et al., 1999) and only one decision analysis of a biosurveillance problem (Wagner et al., 2005). Potentially, such analyses will be helpful for all major threats. Achieving this level of analysis is a long-term goal.

Decision analysis will be of increasing importance in biosurveillance as a result of the new requirement to detect outbreaks as early as possible. Biosurveillance organizations can use the techniques of decision analysis to build models of decision problems and use these models to determine policy. Computational decision analysis can provide direct support to frontline personnel as well as incident managers.

The next chapter discusses how to wrap existing biosurveillance algorithms in a Bayesian wrapper so that they can output the probabilities required by modelers (and decision makers). Chapter 31 discusses the state of the art in cost-benefit analysis.

ADDITIONAL RESOURCES

Journals

Journal of Operational Research (http://or.pubs.informs.org/)
European Journal of Operational Research
Decision Analysis Journal (http://da.pubs.informs.org/)
Journal of Multi-Criteria Decision Analysis
Journal of Behavioral Decision Making
Medical Decision Making
Theory and Decision
Journal of Risk and Uncertainty
Decision Support Systems

Electronic Journals (Statistics and Probability)

http://www.statsci.org/online.html

Textbooks

Clemen, R.T. (1996). *Making Hard Decisions: An Introduction to Decision Analysis,* 2nd ed. Pacific Grove, CA: Duxbury Press. This text is a modern introduction to decision analysis.

7 In 2002, Bravata *et al.* (Bravata et al., 2004) conducted an evidence based study of 55 detection systems and 23 diagnostic decision support systems that could potentially be used for biosurveillance. Although this study was premature in that many of the systems studied were in early phases of development, the conclusions are still valid. The researchers used an explicit decision analytic focus and found that these systems were not in their then present levels of development capable of producing actionable information. In particular, their sensitivity and specificities had not been measured. None of them had cost models.

Cooke, R.M. (1991). *Experts in Uncertainty: Opinion and Subjective Probability in Science.* Oxford University Press, New York, NY. This book discusses in detail the elicitation and use of expert opinion in science.

DeGroot, M.H. (1970) *Optimal Statistical Decisions.* New York: McGraw-Hill Publishing. This book contains axiomatic development of subjective probability theory and utility theory.

Hammond, J.S, Keeney, R.L., and Raiffa, H. (1998). *Smart Choices: A Practical Guide to Making Better Decisions.* New York: Broadway Books.

Hunink, M., et al (2001). *Decision Making in Health and Medicine.* Cambridge: Cambridge University Press.

Lindley, D.V. (1988). *Making Decision,* 2nd ed. London: John Wiley & Sons. This classic introduces decision theory using only elementary mathematics.

Savage, L.J. (1972). *The Foundations of Statistics,* 2nd ed. New York: Dover Publications. This classic book by a pioneer in Bayesian statistics describes the foundations of subjective probability and utility.

von Winterfeldt, D. and Edwards, W. (1988). *Decision Analysis and Behavioral Research.* Cambridge: Cambridge University Press, 1988.

Weinstein, M. C. and Fineberg, H. V. (1980). *Clinical Decision Analysis.* W.B. Saunders, Philadelphia, PA.

Societies

Decision Analysis Society *(http://faculty.fuqua.duke.edu/daweb/).* This society is a subdivision of the Institute for Operations Research and the Management Sciences (INFORMS), the world's largest organization of operations researchers and management scientists, with more than 12,000 members; the Decision Analysis Society is one of its largest and most active subdivisions, with more than 700 members worldwide.

Society for Medical Decision Making *(http://www.smdm.org/).*

Decision Analysis Mailing List *(http://faculty.fuqua.duke.edu/daweb/dalist.htm).*

ACKNOWLEDGMENTS

We would like to thank Greg Cooper for his helpful comments and suggestions on this chapter.

This work was supported by Pennsylvania Department of Health Award number ME-01-737 and grant F 30602-01-2-0550 from the Defense Advanced Research Projects Agency, Contract 290-00-0009. Dr. Onisko is supported by an NSF-NATO Postdoctoral Fellowship.

REFERENCES

Bayes, T. (1958). Essay Towards Solving a Problem in the Doctrine of Chances. *Biometrika* vol. 46:293–298.

Beck, J. and Pauker, S. (1983). The Markov Process in Medical Prognosis. *Medical Decision Making* vol. 3:419–458.

Bernoulli, D. (1954 [1738]). Exposition of a New Theory on the Measurement of Risk. *Econometrica* vol. 22:23–36.

Bernstein, P. (1996). *Against the Gods: The Remarkable Story of Risk.* John Wiley & Sons.

Clancey, W.J. (1984). Use of MYCIN's Rules for Tutoring. In *Rule-Based Expert Systems: The MYCIN Experiments of the Stanford Heuristic Programming Project,* Vol. 26. Reading, MA: Addison-Wesley.

Clemen, R.T. and Reilly, T. (2001). *Making Hard Decisions.* Belmont, CA: Duxbury Press.

Cooper, F., Dash, D., Levander, J., Wong, W., Hogan, W., and Wagner, M. (2004). Bayesian Biosurveillance of Disease Outbreak. In proceedings of the 20th Annual Conference on Uncertainty in Artificial Intelligence (UAI-04), pp. 94–103.

Cooper, G.F. (1984). *NESTOR: A Computer-based Medical Diagnostic Aid that Integrates Causal and Probabilistic Knowledge* [PhD dissertation]. Stanford, CA: Stanford University, Computer Science Department.

Corso, P., Kramer, M., Blair, K., Addiss, D., Davis, J., and Haddix, A. (2003). Cost of Illness in the 1993 Waterborne *Cryptosporidium* Outbreak, Milwaukee, Wisconsin. *Research* vol. 9:426–431.

Dawes, R.M. (1979). The Robust Beauty of Improper Linear Models in Decision Making. *American Psychologist* vol. 34:571–582.

Dean, T. and Wellmann, M. (1991). *Planning and Control.* San Mateo, CA: Morgan Kaufman.

Drummond, M., O'Brien, B., Stoddart, G., and Torrance, G. (1997). *Methods for the Economic Evaluation of Health Care Programs.* Oxford: Oxford University Press.

Edland, A. and Svenson, O. (1993). Judgment and Decision Making under Time Pressure: Studies and Findings. In *Time Pressure and Stress in Human Judgment and Decision Making,* Vol. 2 (O. Svenson and A.J. Maule, eds.). New York: Plenum Press.

Elstein, A. and Schwarz, A. (2002). Clinical Problem Solving and Diagnostic Decision Making: Selective Review of the Cognitive Literature. *British Medical Journal* vol. 324:729–732.

Henrion, M., Breese, J.S., and Horvitz, E.J. (1991). Decision Analysis and Expert Systems. *AI Magazine* vol. 12:64–91.

Howard, R.A. (1988). Decision Analysis: Practice and Promise. *Management Science* vol. 34:679–695.

Howard, R.A. and Matheson, J. (1984a). *Readings on the Principles and Applications of Decision Analysis.* Menlo Park, CA: Strategic Decisions Group.

Howard, R.A. and Matheson, J.E. (1984b). Influence Diagrams. In *The Principles and Applications of Decision Analysis* (R.A. Howard and J.E. Matheson, eds.). Menlo Park, CA: Strategic Decisions Group.

Hoxie, N., Davis, J., Vergeront, J., Nashold, R., and Blair, K. (1997). Cryptosporidiosis-associated Mortality Following a Massive Waterborne Outbreak in Milwaukee, Wisconsin. *American Journal of Public Health* vol. 87:2032–2035.

Houston Department of Health and Human Services. (2003). Officials Following Up on Bacteria Detection. Houston, TX: Houston Department of Health and Human Services.

Incident Management Team [IMT]. (2003). *Cryptosporidium* Contamination of the Drinking Water Supply from Milngavie (Mugdock) Water Treatment Works, August 2002. Glasgow: IMT.

Kaufmann, A.F., Meltzer, M.I., and Schmid, G.P. (1997). The Economic Impact of a Bioterrorist Attack: Are Prevention and Postattack Intervention Programs Justifiable? *Emerging Infectious Diseases* vol. 3:83–94.

Laplace, P.S. (1998). *Philosophical Essays on Probabilities. Translated by A.I. Dale from the fifth French edition of 1825 (Sources in the History of Mathematics and Physical Sciences).* Springer-Verlag, New York.

Leong, T. (1994). *An Integrated Approach to Dynamic Decision Making under Uncertainty.* [PhD dissertation]. Cambridge: Massachusetts Institute of Technology.

Magni, P., Quaglini, S., Marchetti, M., and Barosi, G. (2000). Deciding When To Intervene: A Markov Decision Process Approach. *International Journal of Medical Informatics* vol. 60:237–253.

Meltzer, M.I., Cox, N.J., and Fukuda, K. (1999). The Economic Impact of Pandemic Influenza in the United States: Priorities and Intervention. *Emerging Infectious Diseases* vol. 5:659–671.

Pauker, S. and Kassirer, J. (1987). Medical Progress: Decision Analysis. *New England Journal of Medicine* vol. 316:250–258.

Plante, D., Kassirer, J., Zarin, D., and Pauker, S. (1986). Clinical Decision Consultation Service. *American Journal of Medicine* vol. 80:1169–1176.

Raiffa, H. (1968). *Decision Analysis. Introductory Lectures on Choices under Uncertainty.* New York: McGraw-Hill Book Company.

Roos, R. (2003). Signs of Tularemia Agent Detected in Houston Air. *CIDRAP News. http://www.cidrap.umn.edu/cidrap/content/bt/tularemia/news/oct1003biowatch.html.*

Savage, L.J. (1954). *The Foundations of Statistics.* New York: John Wiley.

Stanford Encyclopedia of Philosophy. (2005). *http://setis.library.usyd.edu.au/stanford/archives/win2000/entries/paradox-stpetersburg/.*

Stirling, R., et al. (2001). Waterborne Cryptosporidiosis Outbreak, North Battleford, Saskatchewan, Spring 2001. *Canadian Communicable Disease Report* vol. 27:185–192.

Tatman, J. and Shachter, R.D. (1990). Dynamic Programming and Influence Diagrams. *IEEE Transactions on Systems, Man, and Cybernetics* vol. 20:365–379.

Teach, R.L. and Shortliffe, E.H. (1984). An Analysis of Physicians' Attitudes. In *Rule-Based Expert Systems: The MYCIN Experiments of the Stanford Heuristic Programming Project,* Vol. 34. Reading, MA: Addison-Wesley

Thompson, K.M., Armstrong, R.E., and Thompson, D.F. (2005). Bayes, Bugs, and Bioterrorists. Lessons Learned from Anthrax Attacks. Washington, DC: Center for Technology and National Security Policy, National Defense University.

Tversky, A. and Kahneman, D. (1974). Judgment Under Uncertainty: Heuristics and Biases. *Science* vol. 185:1124–1131.

von Neumann, J. and Morgenstern, O. (1944). *Theory of Games and Economic Behavior.* Princeton, NJ: Princeton University Press.

von Winterfeldt, D. and Edwards, W. (1988). *Decision Analysis and Behavioral Research.* Cambridge: Cambridge University Press,.

Wagner, M., et al. (2001). The Emerging Science of Very Early Detection of Disease Outbreaks. *Journal of Public Health Management Practice* vol. 7:51–59.

Wagner, M., Wallstrom, G.L., and Onisko, A. (2005). Issue a Boil-Water Advisory or Wait for Definitive Information? A Decision Analysis. In: proceedings of the AMIA 2005 Annual Fall Symposium (October 22–26, 2005, Washington DC), pp. 774–78.

Wallis, J.W. and Shortliffe, E.H. (1984). Customized Explanations Using Causal Knowledge. In *Rule-Based Expert Systems: The MYCIN Experiments of the Stanford Heuristic Programming Project,* Vol. 20. Reading, MA: Addison-Wesley.

30

CHAPTER

Probabilistic Interpretation of Surveillance Data

Garrick Wallstrom, Michael M. Wagner, and Agnieszka Onisko
RODS Laboratory, Center for Biomedical Informatics, University of Pittsburgh, Pittsburgh, Pennsylvania

1. INTRODUCTION

As illustrated by the postincident report from the Glasgow Health Department (summarized in the previous chapter), uncertainty about whether there was a threat to human health was a key topic in the deliberations of the Incident Management Team. In particular, the team had difficulty estimating the probability that people might become infected based on the available water surveillance data. The problem that Glasgow faced is increasingly common owing to the widespread adoption of new methods of biosurveillance.

In this chapter, we continue our example of a decision analysis of a boil-water advisory decision, focusing on techniques for estimating the probability of an outbreak from surveillance data. Figure 30.1 (reproduced from the previous chapter) is an example of a spike in surveillance data that raises questions of whether there is an outbreak, and, if so, what is the biological agent, how big is it, how long has it been ongoing, how much longer will it last, and how aggressively should it be investigated.

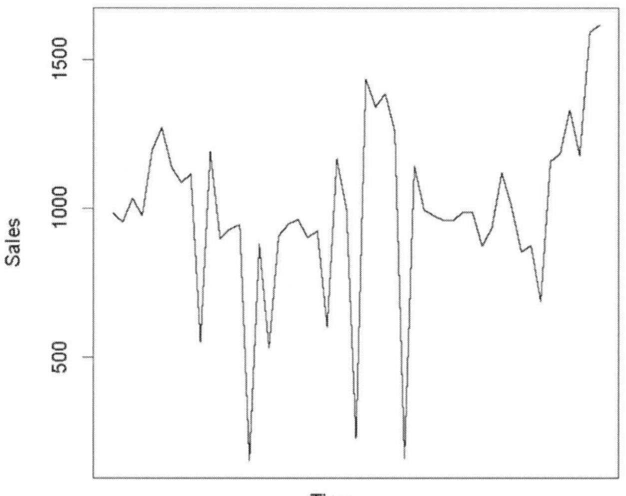

FIGURE 30.1 Rising sales of diarrhea remedies that triggered an alert (detection threshold set to one false-alarm per year).

We generally cannot answer questions about the existence and nature of an outbreak with certainty from imprecise surveillance data, such as sales of diarrhea remedies. However, we can interpret the surveillance data probabilistically; that is, we can express our uncertainty about the presence of an outbreak as a probability, which we can then use in combination with cost and benefit considerations to make a decision. For example, we may estimate that the probability that there is a *Cryptosporidium* outbreak is 0.80, which strongly suggests that *some* action is in order. Alternatively, we may estimate that the probability is only 0.001, and therefore, no action is required.

In fact, there are a small number of algorithms that do compute the probability of an outbreak directly. We call them *Bayesian* detection algorithms because they use Bayes' theorem. PANDA and BARD (discussed in Chapters 18 and 19, respectively) are two examples of Bayesian detection algorithms that are capable of computing the probability of an outbreak. The far more common situation in current practice is different, however.

2. CURRENT METHODS FOR INTERPRETING BIOSURVEILLANCE DATA

In current practice, most biosurveillance systems do not use Bayesian algorithms to compute the posterior probability of an outbreak (or of its characteristics). Currently available systems output alerts when daily counts exceed a threshold, which is set in an ad hoc manner to limit the number of false alarms. When interpreting an alert, the decision maker has available only (1) a display of the time series data, and (2) the false-alarm rate (or recurrence interval, which is the average time between false alarms in the absence of an outbreak) of the system. The decision maker must interpret the time series knowing only the false-alarm rate of the system.[1]

Decision makers do not interpret surveillance data probabilistically, although they could. It is instructive to consider

1 In some cases, the situation is even worse. Some detection systems under-estimate the false-alarm rate because they calculate it from an assumed distribution (e.g., normal or Poisson) rather than measuring it empirically. This practice makes the signal interpretation problem even more difficult if the actual distribution differs from the assumed parametric form.

ISBN 0-12-369378-0

what can be discovered about interpretation of biosurveillance data if they were to interpret the output of their systems probabilistically. To make this exercise concrete, we use the signal shown in Figure 30.1 and a prior probability of a *Cryptosporidium* outbreak equal to 0.0035 per day (we describe how we obtained this prior probability later in this chapter).

Equation 30.1 is Bayes' theorem applied to the interpretation of biosurveillance data. Equation 30.1 allows a decision maker to compute the posterior probability of outbreak, given an alarm provided that three quantities are known: (1) the prior probability of outbreak, *P*(Outbreak) (2) the sensitivity of the detection system, *P*(Alarm | Outbreak) and (3) the false-alarm rate, *P*(Alarm | No Outbreak), of the detection system. In current practice, a decision maker typically knows the false-alarm rate and can estimate the prior probability (e.g., influenza occurs once per year, so the probability of an influenza outbreak being ongoing on any given day is roughly the duration of influenza outbreaks divided by 365). Typically the decision maker does not know the sensitivity of an outbreak-detection system but can, nevertheless, compute upper and lower bounds as we will now demonstrate.

A decision maker can use Equation 30.1 to compute upper and lower bounds on the posterior probability by assuming that the sensitivity of the system lies within some range. If the sensitivity of the system is unknown, the worst and best possible sensitivities conceivable would be used to obtain bounds on what the actual posterior probability could be. In particular, the decision maker would use the best possible sensitivity of 1.0 to computer an upper bound, and the worst sensitivity possible to compute a lower bound. The worst possible sensitivity equals the false-alarm rate, as we will now prove: Figure 30.2 is an receiver operating characteristic (ROC) curve (see Chapter 20) for a system with the worst possible sensitivity. The ROC curve for this system is a straight line with slope equal to one. The mathematical equation for the slope is 1 = sensitivity/false-alarm rate; after trivial algebraic manipulation, we obtain sensitivity = false-alarm rate. *QED* (end of proof).

$$P(\text{Outbreak} \mid \text{Alarm}) = \frac{P(\text{Alarm} \mid \text{Outbreak})P(\text{Outbreak})}{P(\text{Alarm} \mid \text{Outbreak})P(\text{Outbreak}) + P(\text{Alarm} \mid \text{No Outbreak})\left[1 - P(\text{Outbreak})\right]} \tag{30.1}$$

Equation 30.2 shows how the posterior probability for the worst-case detection system equals the prior probability. We simply use the result that we just obtained showing that sensitivity is always equal to the false-alarm rate, and algebraically simplify the substituted expression to obtain this result. This result is exactly what a Bayesian would expect. A detection system that is incapable of discriminating between outbreaks and nonoutbreaks should not change our belief (prior probability) that an outbreak is present.

$$\begin{aligned}
P(\text{Outbreak} \mid \text{Alarm}) &= \frac{P(\text{Alarm} \mid \text{Outbreak})P(\text{Outbreak})}{P(\text{Alarm} \mid \text{Outbreak})P(\text{Outbreak}) + P(\text{Alarm} \mid \text{No Outbreak})\left[1 - P(\text{Outbreak})\right]} \\
&\geq \frac{\{\text{Sensitivity of Worst System}\}P(\text{Outbreak})}{\left\{\begin{matrix}\text{Sensitivity of} \\ \text{Worst System}\end{matrix}\right\}P(\text{Outbreak}) + P(\text{Alarm}|\text{No Outbreak})\left[1 - P(\text{Outbreak})\right]} \\
&= \frac{\{\text{False-AlarmRate}\}P(\text{Outbreak})}{\{\text{False-AlarmRate}\}P(\text{Outbreak}) + P(\text{Alarm} \mid \text{No Outbreak})\left[1 - P(\text{Outbreak})\right]} \\
&= \frac{P(\text{Alarm}|\text{No Outbreak})P(\text{Outbreak})}{\{P(\text{Alarm} \mid \text{No Outbreak})\}P(\text{Outbreak}) + P(\text{Alarm} \mid \text{No Outbreak})\left[1 - P(\text{Outbreak})\right]} \\
&= \frac{\{P(\text{Alarm} \mid \text{No Outbreak})\}P(\text{Outbreak})}{P(\text{Alarm} \mid \text{No Outbreak})} \\
&= P(\text{Outbreak})
\end{aligned} \tag{30.2}$$

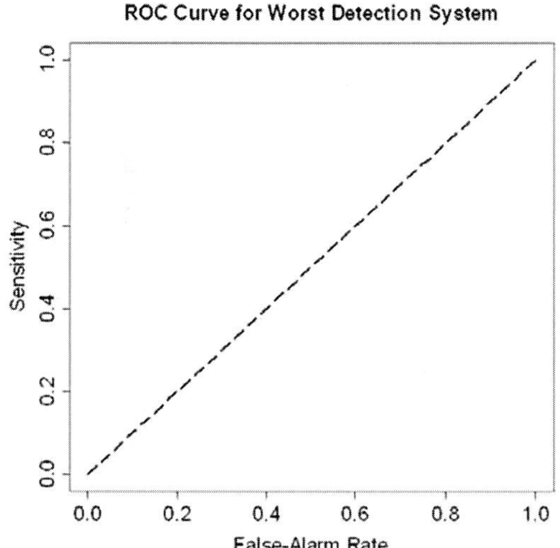

FIGURE 30.2 Receiver operating characteristic (ROC) curve for worst detection system.

Equation 30.3 shows the calculation for the best case. We use a false-alarm rate of one per year, or 1/365 days = 0.0027 per day (from our *Cryptosporidium* example). We use a prior probability of an outbreak of 0.0035 per day, also from our *Cryptosporidium* example (we describe how we obtained this probability later in this chapter).

Thus, a decision maker faced with the output of our example system (that has a false-alarm rate of one per year) and a situation in which it is believed that the prior probability of an outbreak is equal to 0.0035, only knows that the posterior probability of an outbreak is between 0.0035 and 0.5654. For decision-making purposes, this is a very broad range. The optimal action to take when the probability of a *Cryptosporidium* outbreak is equal to 0.5654 is quite different from when it is 0.0035. In fact, the interpretation method that we next discuss shows that an estimate for the posterior probability given this anomaly in surveillance data is 0.0410.

3. BAYESIAN WRAPPER METHOD

In this section, we use the *Cryptosporidium* example to demonstrate how to compute the probability of an outbreak after observing a spike in the surveillance data when using a non-Bayesian detection algorithm. We refer to this approach as creating a *Bayesian wrapper* for detection systems because the approach uses Bayes' theorem to incorporate the prior probability of the outbreak into the interpretation of an alert from a non-Bayesian system. To make the example as concrete as possible, we demonstrate this technique for a specific detection system and decision scenario. We used the results of these calculations in the decision analysis that we presented in the previous chapter.

$$\begin{aligned} P(\text{Outbreak} \mid \text{Alarm}) &= \frac{P(\text{Alarm} \mid \text{Outbreak})P(\text{Outbreak})}{P(\text{Alarm} \mid \text{Outbreak})P(\text{Outbreak}) + P(\text{Alarm} \mid \text{No Outbreak})[1 - P(\text{Outbreak})]} \\ &\leq \frac{\{\text{Sensitivity of Best System}\}P(\text{Outbreak})}{\left\{\begin{matrix}\text{Sensitivity of}\\ \text{Best System}\end{matrix}\right\}P(\text{Outbreak}) + P(\text{Alarm} \mid \text{No Outbreak})[1 - P(\text{Outbreak})]} \\ &= \frac{\{1.0\}P(\text{Outbreak})}{\{1.0\}P(\text{Outbreak}) + P(\text{Alarm} \mid \text{No Outbreak})[1 - P(\text{Outbreak})]} \\ &= \frac{\{1.0\}0.0035}{\{1.0\}0.0035 + 0.0027[1 - 0.0035]} \\ &= 0.5654 \end{aligned} \tag{30.3}$$

3.1. Detection System

As discussed in Chapter 29, we created a hypothetical surveillance system that monitors the area supplied by a single water treatment plant for a spike in over-the-counter (OTC) sales of diarrhea remedies. We use the Chicago metropolitan area for convenience because we previously studied detectability of *Cryptosporidium* from OTC data in Chicago. Chicago is served by two water treatment plants. We, therefore, assumed that each treatment plant serves half of the city, and that our detection system receives aggregated sales for the region served by a single plant.

The detection system uses as a detector the CuSum algorithm with a moving average component that accounts for day-of-week effects in the surveillance data. In biosurveillance, CuSum algorithms are used typically to monitor for outbreaks in which people become sick over several days or weeks (Page, 1954; Hutwagner et al., 1997; Morton et al., 2001; Sonesson and Bock, 2003; Rogerson and Yamada, 2004), as is historically the case with *Cryptosporidium* outbreaks caused by water contamination. The moving average component enables the algorithm to adapt to short- and long-term trends in the baseline data while still detecting outbreaks of moderate duration. The CuSum algorithm uses a fixed detection threshold that we set to limit false alarms to a rate of one per year. Every day, the moving average component forecasts the daily sales of diarrhea remedies by using an analysis of variance (ANOVA) model with a day-of-week factor that is fit to the sales data for the previous 28 days. The biosurveillance system then computes the CuSum statistic based on the standardized forecast error; determines whether the statistic exceeds the detection threshold; and, if so, alerts the biosurveillance staff.

3.2. Decision Scenario

We reproduce Figure 29.3 (as Figure 30.3) to show a graphical representation of the decision model. The base model consists of a decision node *act now* with two possible actions: *yes/no* that corresponds to issuing an immediate boil-water advisory or testing and waiting for the results, and chance nodes labelled *Crypto outbreak.* We represent the costs and benefits associated with each combination of action and chance event by cb_1, cb_2, cb_3, and cb_4. p represents the posterior probability that there is an ongoing *Cryptosporidium* outbreak.

We assume that the decision to *act now* or *wait* is made by the biosurveillance staff in response to an alert received from the detection system. After receiving the alert, the biosurveillance staff reviews the logs from the surveillance system for the previous five days and finds that there were no alerts over this period. The probability p in Figure 30.3 is, therefore, the posterior probability that there is an ongoing *Cryptosporidium* outbreak given that there was an alert today and no alerts in the previous five days.

3.3. Bayes' Theorem

We formalize the calculation by letting C denote the event that there is an ongoing water-borne *Cryptosporidium* outbreak. We refer to the information that there was an alert today and no alerts over the past five days as the *alert history* and denote

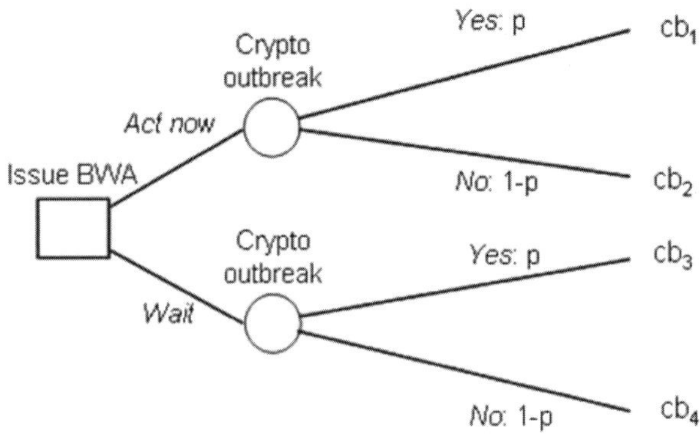

FIGURE 30.3 Decision tree model.

it by A. With this notation, the posterior probability that there is an ongoing water-borne *Cryptosporidium* outbreak given the alert history can be written as We use Bayes' theorem to calculate this probability:

$$P(C \mid A) = \frac{P(A \mid C)\, P(C)}{P(A \mid C)\, P(C) + P(A \mid \bar{C})\, P(\bar{C})}. \quad (30.4)$$

In the equation above, $P(C)$ is the prior probability that there is an ongoing *Cryptosporidium* outbreak, and $P(\bar{C}) = 1 - P(C)$ is the prior probability that there is *not* a *Cryptosporidium* outbreak. The conditional probability $P(A \mid C)$ is the probability that the system generates the alert history given that an outbreak is ongoing. Similarly, $P(A \mid \bar{C})$ is the conditional probability of the alert history when there is not an outbreak of *Cryptosporidium.*

The calculation of $P(A \mid C)$ requires additional notation for characteristics of an outbreak and to account for time. We denote the *outbreak size* (defined as the proportion of the population that is affected by an outbreak) by S, the duration of an outbreak by D, and the number of days that an ongoing outbreak has been in progress by Y. We calculate $P(A \mid C)$ by finding the *expectation* (average value) of $P(A \mid S,D,Y,C)$ under the joint conditional distribution of S, D, and Y given C: $P(A \mid C) = E_{S,D,Y \mid C}[P(A \mid S,D,Y,C)]$, where $P(A \mid S,D,Y,C)$ is the conditional probability that the system generates the alert history given that a *Cryptosporidium* outbreak of size S and duration D has been ongoing for Y days.

We need two types of probabilities to compute the posterior probability of a *Cryptosporidium* outbreak. One type consists of prior probabilities. These include $P(C)$, which arises in the application of Bayes' theorem above, and the joint conditional distribution of S, D, and Y given C, which is used to calculate $P(A \mid C)$. The other type of probabilities consists of the conditional probabilities of the alert history: $P(A \mid S, D, Y, C)$, and $P(A \mid \bar{C})$

3.4. Priors

The priors are subjective probabilities, elicited in the absence of the alert history. In a real analysis, we would elicit prior probabilities from domain experts and from literature review. In this analysis, we simply chose priors that appeared reasonable to us in order to illustrate the approach.

We need to specify $P(C)$, as well as the joint conditional distribution of S, D, and Y given C. It is difficult to directly produce these priors owing to possible relationships among S, D, Y, and C. Instead, we specify several simpler priors that we then use to derive the required priors.

We suppose that the outbreak size S has a uniform distribution between 0% and 50%, that the duration D is uniformly distributed from 21 to 56 days, and that S and D are independent. We also suppose that Y is uniformly distributed from one to D days given D and C. Finally, we suppose that there might be one outbreak every 30 years on average; that is, the probability that an outbreak *starts* on any given day is $1/(30 \times 365)$. We use these probability statements to derive the prior probability that there is an ongoing outbreak on any given day: $P(C) = 38.5/30 \times 365) = 0.0035$. We also use these priors to find the joint conditional distribution of *S, D,* and *Y* given *C.*

3.5. Conditional Probabilities

We then only need the conditional probabilities $P(A|S, D, Y, C)$, and $P(A|\bar{C})$

3.5.1. $P(A|\bar{C})$

This is the probability of observing the alert history in a period during which there is no *Cryptosporidium* outbreak. One approach to computing this probability is to formulate a model for the surveillance data under nonoutbreak conditions. By using the model and the detection algorithm, we can compute the probability either analytically or using simulation.

An alternative approach is to compute this probability empirically by obtaining a sample of surveillance data for a period during which there are no outbreaks and computing the fraction of windows (with length equal to the length of the alert history) within the time series with alert histories equal to *A*. Unless there are several years of surveillance data available for analysis, this approach may fail because it is possible that the alert history *A* may not have arisen in the historical data, and therefore, we would calculate $P(A \mid \bar{C}) = 0$ and ultimately find $P(C \mid A) = 1$.

A third approach, and the one that we use in this analysis, is to supplement a sample of nonoutbreak surveillance data with a theoretical assumption about the surveillance data and detection algorithm. Specifically, we assume that false alarms occur approximately independently of each other. In this case, we compute the probability as follows: $P(A \mid \bar{C}) = (1 - P(False\ Alarm))^{A^-} \times P(False\ Alarm)^{A^+}$, where A^- is the number of days in the alert history without an alert and A^+ is the number of days with an alert. Under the independence assumption, we only need to know the false-alarm rate of the detection system to compute $P(A \mid \bar{C})$ for these scenarios. We compute the false-alarm rate of the system empirically by running the detection algorithm on our sample of nonoutbreak surveillance data and calculating the fraction of days with alerts.

3.5.2. $P(A|s, d, y, C)$

Ideally, we would use real outbreaks of *Cryptosporidium* that have occurred in the region under study to compute the conditional probability of the alert history given an outbreak. However, the rarity of real outbreaks makes this strategy generally infeasible. The alternative approaches require either simulation or modeling of the surveillance data that would arise during an outbreak.

The HiFIDE methodology (Wallstrom et al., 2004) uses surveillance data collected during a real outbreak to construct a Poisson generalized linear model (McCullagh and Nelder, 1983) for the effect of the outbreak on surveillance data. Under our model, the average effect is a smooth function of time that we estimate by using the EM algorithm (Dempster et al., 1977) in conjunction with Bayesian adaptive regression splines (Dimatteo et al., 2001).

We use the HiFIDE approach to model the effect of a *Cryptosporidium* outbreak on sales of diarrhea remedies using published sales data during the North Battleford *Cryptosporidium* outbreak (Stirling et al., 2001a,b). We then simulate the effect of an outbreak from this model and inject that effect into OTC data for Chicago. HiFIDE accounts for the population difference between Chicago and North Battleford as well as differences in OTC data quality owing to retailer market share (Wallstrom et al., 2004). This process enables us to parameterize the known outbreak effect on OTC sales in North Battleford in a way that we can vary the size *(S)* and duration *(D)* of the outbreak in Chicago. We use this approach to construct multiple outbreak data sets. Each data set contains the simulated daily sales of diarrhea remedies during a *Cryptosporidium* outbreak of size *S* and duration *D.* For each outbreak data set and each day within the outbreak, the detection system determines whether or not to send an alert. We store these results and use them to compute $P(A \mid S,D,Y,C)$

We use Monte Carlo integration (Gilks et al., 1996; Robert and Casella, 2004) and the equation $P(A \mid C) = E_{S,D,Y|C}[P(A \mid S,D,Y,C)]$ to compute the probability $P(A \mid C)$ from the values of $P(A \mid S,D,Y,C)$. Specifically, we sampled 10,000 values of *S, D,* and *Y* from their joint conditional distribution given *C.* For each set of values, we compute $P(A \mid s,d,y,C)$. The probability $P(A \mid C)$ is then computed by averaging these values of $P(A \mid s,d,y,C)$.

3.6. Posterior Probability

We used the above Bayesian wrapper method to compute *p,* the posterior probability that there is an ongoing *Cryptosporidium* outbreak given that there was an alert today and no alerts in the previous five days. We found $p = 0.0410$. Therefore, after observing the spike in sales of diarrhea remedies, we conclude that there is approximately a one out of 25 chance that there is a *Cryptosporidium* outbreak.

4. PROBABILITIES AND COSTS

In some situations, it is sufficient for a biosurveillance organization simply to know the posterior probability of an ongoing *Cryptosporidium* outbreak to make a decision, but in other situations (as in Glasgow), the costs and benefits of the available actions must be considered. If the posterior probability is near one, it is clear that action is required. If, the posterior probability is near zero, action is likely not warranted. A more formal decision analysis that includes costs and benefits of actions is needed to clarify what is the best action in the broad range of uncertainty between zero and one. If our goal is early detection, we will typically be operating in this gray area and require an understanding of the explicit tradeoffs between waiting and acting.

5. GENERALIZATION OF THE BAYESIAN WRAPPER METHOD

We illustrated the use of the Bayesian wrapper method to compute the posterior probability of a *Cryptosporidium* outbreak given a spike in sales of diarrhea remedies, $P(C|A)$. In fact, we can find the expected characteristics of the possible outbreak by another application of Bayes' theorem. Specifically, we can compute the posterior distribution of *S, D,* and *Y* given *A* and *C,* and use it to find the expected size, duration, and start date of the outbreak. Even outside of a formal decision analysis, this information can provide insight into the source of the outbreak and guide management of the outbreak.

We can also use the Bayesian wrapper method when the differential diagnosis contains more than one disease. For example, we could consider that there are two possible causes for a spike in diarrhea remedies that is geographically mapped to a water distribution system: *Cryptosporidium* contamination and *Giardia* contamination. We could let C_0 denote no outbreak, C_1 denote *Cryptosporidium* contamination, and C_2 denote *Giardia* contamination. We would specify prior probabilities for C_0, C_1, and C_2. We would then compute $P(A|C_0)$, $P(A|C_1)$, and $P(A|C_2)$ as before, using nonoutbreak data and simulation. The set of outbreak characteristics and the priors for those characteristics could be different for the two diseases. Finally, we would use Bayes' theorem to calculate $P(C_0|A)$, $P(C_1|A)$, and $P(C_2|A)$.

The example illustrated how to use the Bayesian wrapper when the surveillance data consists of a single time series. Application of the method to multiple time series is more difficult because of the need to calculate $P(A|C)$ and $P(A|\bar{C})$. To calculate $P(A|C)$, we need to model how an outbreak simultaneously affects multiple time series. Calculation of $P(A|\bar{C})$ requires either modeling of multiple time series collected during nonoutbreak periods, or a substantial amount of surveillance data to enable empirical calculation of false-alarm rates.

6. SUMMARY

The probabilistic interpretation of surveillance data is required for a formal analysis of a biosurveillance decision, and provides guidance for making decisions even outside of a formal decision analysis. Bayesian detection algorithms, such as BARD and PANDA, can provide this interpretation directly. However, most algorithms used in surveillance systems are non-Bayesian and do not provide posterior probabilities.

In this chapter we introduced the Bayesian wrapper method for computing posterior probabilities from the output of conventional (non-Bayesian) detection algorithms. We used the method to compute the posterior probability of a *Cryptosporidium* outbreak after seeing a spike in sales of diarrhea remedies. This probabilistic interpretation of surveillance data is one of the critical items that the Glasgow Incident Management Team was lacking, in addition to the costs and benefits of possible outcomes. Both of these items are needed to conduct a formal decision analysis. The following chapter discusses methods for estimating and modeling costs and benefits.

ADDITIONAL RESOURCES

List of useful Web sites and recommended further reading. HiFIDE software may be obtained from *http://www.hifide.org.*

ACKNOWLEDGMENTS

This work was supported by Pennsylvania Department of Health Award number ME-01-737 and grant F 30602-01-2-0550 from the Department of Homeland Security. Dr. Onisko is supported by an NSF-NATO Postdoctoral Fellowship.

REFERENCES

Dempster, A.P., Laird, N.M., and Rubin, D.B. (1977). Maximum Likelihood from Incomplete Data Via the EM Algorithm. *Journal of the Royal Statistical Society, Series B,* vol. 39:1–38.

Dimatteo, I., Genovese, C.R., and Kass, R.E. (2001). *Bayesian Curve-Fitting with Free-Knot Splines.* Biometrika vol. 88:1055–1071.

Gilks, W.R., Richardson, S., and Spiegelhalter, D. (1996). In *Markov Chain Monte Carlo in Practice.* (W.R. Gilks, S. Richardson, and D. Spiegelhalter, eds.). Boca Raton: Chapman & Hill/CRC.

Hutwagner, L.C., Maloney, E.K., Bean, N.H., Slutsker, L., and Martin, S.M. (1997). Using Laboratory-based Surveillance Data for Prevention: An Algorithm for Detecting *Salmonella* Outbreaks. *Emerging Infectious Diseases* vol. 3:395–400.

McCullagh, P. and Nelder, J. A. (1983). *Generalized Linear Models.* London: Chapman and Hall.

Morton, A.P., et al. (2001). The Application of Statistical Process Control Charts to the Detection and Monitoring of Hospital-Acquired Infections. *Journal of Quality Clinical Practice* vol. 21:112–117.

Page, E.S. (1954). Continuous Inspection Schemes. *Biometrika* vol. 41:100–115.

Robert, C.P. and Casella, G. (2004). *Monte Carlo Statistical Methods.* New York: Springer.

Rogerson, P.A. and Yamada, I. (2004). Approaches to Syndromic Surveillance When Data Consist of Small Regional Counts. *Mortality and Morbidity Weekly Report* vol. 53:79–85.

Sonesson, C. and Bock, D. (2003). A Review and Discussion of Prospective Statistical Surveillance in Public Health. *Journal of the Royal Statistical Society, Series A,* vol. 166:5–21.

Stirling, R., et al. (2001a). North Battleford, Spring 2001 Waterborne Cryptosporidium Outbreak. *Health Canada Report, Ottawa, Ontario: Health Canada.*

Stirling, R., Aramini, J., Ellis, A., Lim, G., Meyers, R., Fleury, M. and Werker, D. (2001b). Waterborne Cryptosporidiosis Outbreak, North Battleford, Saskatchewan, Spring 2001. *Canadian Communicable Disease Report* vol. 27:185–192.

Wallstrom, G.L., Wagner, M.M., and Hogan, W.R. (2004). *Using Surveillance Data from Actual Outbreaks to Characterize Detectability.* RODS Laboratory Technical Report. Center for Biomedical Informatics, University of Pittsburgh, Pittsburgh, Pennsylvania.

31 CHAPTER

Economic Studies in Biosurveillance

Bruce Y. Lee
Section of Decision Sciences and Clinical Systems Modeling, Division of General Internal Medicine University of Pittsburgh, Pittsburgh, Pennsylvania

Michael M. Wagner, Agnieszka Onisko and Vahan Grigoryan
RODS Laboratory, Center for Biomedical Informatics, University of Pittsburgh, Pittsburgh, Pennsylvania

1. INTRODUCTION

Economic considerations often influence the selection, use, and even success of biosurveillance systems. Organizations, when acquiring or planning biosurveillance systems, take into account the considerable financial resources required to develop, use, and maintain such systems, as well as the feasibility and costs of recruiting and retaining capable personnel. Organizations consider the effects of false alarms, which result in the unnecessary mobilization of multiple resources. False alarms can be expensive and in the extreme case may lead to a boy-who-cried-wolf effect in which users of a system begin to ignore a system either partially or completely. Without a proper understanding of the tradeoffs between action and inaction discussed in the previous two chapters, users of a biosurveillance system may also under-react or delay response to early warning signs of an outbreak, potentially resulting in higher levels of injury, loss of life, and damage to the psychology, operations, and infrastructure of the affected region.

Economic studies can contribute to the rational selection, optimal use, and success of biosurveillance systems. An economic study is a formal scientific analysis of different choices that individuals, organizations, or societies have to make when resources are scarce. Economic studies can assist an organization in all phases of biosurveillance system operation: from acquisition to setting of alarm thresholds, to decision making in novel situations. In particular, economic studies of such decisions make explicit the choices available in each situation and the potential costs and rewards of each choice. Economic studies can provide guidance to organizations about how much and where to invest in biosurveillance and what level and type of biosurveillance to develop and maintain. They can provide guidance about whether to develop and configure a biosurveillance system that can catch subtle early warning signs but potentially generate many false alarms, or one that only looks for very suspicious warning signs but may miss insidious cases.

Although the field of economics is mature with many well-established techniques for conducting economic studies, researchers have only rarely applied these techniques to problems in biosurveillance. The economic subfield of biosurveillance is still in its infancy and is relatively uncharted territory. We expect this subfield to grow. For this reason, this chapter will review the strengths and limitations of existing economic techniques and provide examples of these techniques applied to the analysis of decision problems in biosurveillance and, in particular, the modeling of the consequences of outbreaks and bioterrorism. It will also discuss the challenges faced in applying economics in biosurveillance as well as potential future directions.

2. DEFINITIONS AND BASIC CONCEPTS

Economic studies come in a variety of forms, and economists have applied them to almost every sphere of human activity. An economic study can range in complexity from a simple "back-of-the envelope" calculation to a sophisticated model that requires substantial computer power for its evaluation. An economic study may focus on a single possibility or compare multiple alternatives. Despite the diversity of technique and domains found in the published economic literature, economic studies share several basic attributes.

2.1. Perspective

One of the most important attributes of an economic study is its *perspective,* which is the person or organization whose point of view or interests determine the costs and benefits considered in the study. For example, a publicly traded company may be primarily interested in protecting shareholder value, and therefore, an economic study commissioned by the company about the impact of a bioterrorist event might include only elements directly relevant to shareholder value and exclude costs of treating sick individuals other than those employed by the company.

The perspective of an economic study is important because available choices, costs, and benefits vary significantly depending on whom and where you are. For example, people in densely populated areas that are vulnerable to bioterrorist attacks may benefit more from biosurveillance than would people in remote rural areas who are less likely to experience such attacks. If the residents in both areas have to pay the same

Handbook of Biosurveillance
ISBN 0-12-369378-0

amount of taxes (i.e., shoulder equal burdens of paying for the biosurveillance system), the rural residents may be less interested in such a system.

Every economic study should state clearly its perspective: whether it is taking the perspective of an individual, a particular institution or organization, a government body, or society in general. As you might imagine, changing the perspective of a study can drastically alter its composition and results. For any given decision situation, the optimal decision for one individual or organization may differ from that for another individual or organization. The perspective of the economic study should match that of the decision maker.

2.2. Retrospective, Prospective, and Model-Based Analyses

An organization or individual can perform an economic study of an event that has already transpired (retrospective analysis), will soon occur (prospective analysis), or could occur in the future (theoretical or predictive analysis). Each type of study has strengths and limitations and differs in feasibility. Retrospective studies are useful, because the past often repeats itself or helps predict the future. However, current and future situations may not mirror the past, and reconstructing past events can be difficult, especially without accurate and comprehensive data. Prospective studies, which involve collecting data while natural or created situations occur, give an analyst much more control over the situations and the information collected. However, prospective studies can be difficult and expensive to perform and only generate results representing specific situations. In the case of outbreaks, prospective studies may be nearly impossible, because the onset and timing of events are unpredictable and creating such an event would be unethical (and quite damaging to one's career). A limitation shared by both retrospective and prospective studies is that the studies are only feasible if the event has occurred or is likely to occur. Predictive studies overcome this limitation because the analyst builds a mathematical model or computer simulation of hypothetical situations. Because most outbreaks are uncommon and many types have never occurred, many biosurveillance-related economic analyses are at least partially predictive. A predictive analysis is always feasible. Moreover, it provides the analyst more flexibility in manipulating the situation that is being modeled to produce insights about a range of potential situations. The key limitation of a predictive study is that it rests on many modeling assumptions.

Often, answering a question requires a series of economic studies or an economic study that involves retrospective, prospective, and predictive elements. For example, in deciding whether to administer a certain vaccine, retrospective study of previous outbreaks and experience with vaccination programs can provide important quantitative data for a predictive economic model. Performing a prospective study, such as vaccinating a small representative sample of the population and tracking the ensuing costs and rewards, can provide additional estimates that facilitate projections about what would happen if a strategy was applied on a larger scale (e.g., the entire population). Retrospective and prospective analyses often provide data for a predictive analysis, such as a computer model of different potential outbreaks and the effects of vaccination programs.

2.3. Time Horizon

When conducting an economic study, an analyst should choose an appropriate time horizon or period that adequately captures the immediate and longer-term consequences of a decision or action, but does not make the study unrealistic to perform. For example, a study that only measures costs up to one month after a bioterrorist attack may seriously underestimate the impact of that attack because diseases and injuries can have long-term effects. Conversely, if the study attempted to measure the economic impact of that event 300 years into the future, it would clearly run aground on the banks of infeasibility, owing to the many uncertainties whose cumulative effect would be impossible to model. Analysts frequently face a tension between the benefit of increased fidelity from extending the time horizon of an economic study and the difficulty in constructing the model. The ideal balance between the two is often obvious from the specifics of the problem being modeled; nevertheless, it is the rule rather than the exception that economic modelers revise the time horizon as they develop a study.

2.4. Costs

One of the two major data inputs for an economic study is costs, which can be complicated to measure and very difficult to obtain. A cost is the cash value of money, time, and labor spent for goods or services. Although some costs (e.g., purchasing a stretcher) are relatively straightforward, others (e.g., the cost of caring for a tularemia patient) can include many component costs, such as nursing and physician time, diagnostic testing, medications, and emergency and hospital room occupancy. Each of these "subcosts" may be difficult to quantify and subject to error or variability.

2.4.1. Direct and Indirect Costs

Costs can be subtle or hidden. Even seemingly small events can have immediate and long-term effects on many different people and organizations. People do not even have to be present at the time and place of an event for it to affect them. For instance, a death or severe injury may influence the victim's family, friends, and workplace. Therefore, you must carefully account for every person and organization that may be reasonably involved and affected by an event. For example, a single successful small-scale bioterrorist contamination of a commercial building can lead to many direct costs, including the cost of diagnosing and treating the victims and the cost of decontaminating the building. However, it can also entail indirect costs to the company or companies occupying the building, including lost worker productivity, the cost of finding replacement employees, and damage to the company's reputation and worker morale. When summed, these "hidden costs" can be considerable and even outweigh the more obvious direct costs.

2.4.2. Methods of Estimating Costs

Once you have identified the causes and sources of the costs, you must quantify them, which can be challenging. Rarely do items and services have clear price tags. Often, you must do a fair amount of sleuthing to determine what an item or service actually costs. There are many methods of gathering or estimating costs, each with its relative advantages and disadvantages:

Charges. In some cases, the *charges* for a service or visit found on hospital, clinic, and insurance bills can serve as reasonable proxies for costs. Of course, these charges usually exceed the actual costs; therefore, an economic modeler will convert them to costs by using established cost-to-charge ratios or conversion factors. Moreover, bills may not break down the charges to the level of detail needed. For example, an emergency department visit charge may aggregate many components (e.g., placing an intravenous line, transporting the patient to different locations) of the visit but not identify what fraction of the charge is associated with each component. Finally, charges often do not always accurately reflect the resources consumed or services provided, as some items and services are not billable and some items on a bill may not have been consumed.

Microcosting. *Microcosting* is typically more accurate than is using charge information, but microcosting is usually expensive and time-consuming to perform. Microcosting identifies every resource used during an event and then assigns a cost to each resource. Time-and-motion studies frequently help microcost. A time-and-motion study may involve following a patient (e.g., tularemia patient) during a given event (e.g., stay in the emergency department) and counting every item used (e.g., medications, catheters, saline, gauze, radiology film) and every service performed (e.g., 30 minutes of a nurse's time, 10 minutes of a patient transporter's time). An analyst would then assign a cost to each item and each fraction of personnel time and sum the costs to compute an overall cost. The use of microcosting tends to be limited to simple and well-defined events.

Resource-Unit Use. Another approach, *resource-unit use* or *health-service-resource use,* involves measuring resources that are more readily measured (e.g., number of hospitalizations, length of hospital stay, number of radiology procedures) and assigning unit costs (e.g., cost per hospitalization, cost per hospital day, cost per radiology procedure) to each resource. For example, if an anthrax attack resulted in a patient staying 30 days in the hospital, then multiplying the cost of a hospital day by 30 could estimate the per-hospitalized-patient cost of the attack. Of course, the resource-unit use method provides gross estimates, may not account for significant variability (e.g., fluctuation in cost of a day in the hospital), and assumes that the resource unit accurately reflects everything that is being done.

Other Resources for Costing. An analyst may obtain costs from the medical literature, insurance reports, or other publications. Before using these numbers in an economic study, one should ascertain whether the source is credible and the source's circumstances are comparable to the study at hand. For example, the cost of hospitalizing a patient for a simple uncomplicated case of diarrhea may not be applicable to a case of bioterrorist agent-induced diarrhea. If there is more than one source for a particular cost and the numbers vary significantly among the sources, the analyst may use either a simple or a weighted average of the costs.

2.5. Benefits

The other major data inputs for an economic study are the benefits of a successful medical, public health, or policy intervention, such as money saved, lives saved, quality-of-life improvements, productivity increases, suffering prevented, or adverse events avoided. An economic study should include benefits relevant to the interventions or actions under study. For example, measuring the number of lives saved by an acne cream would not be a useful measure of benefit, whereas improvement in quality of life (or number of dates saved) would.

Over the years, analysts have developed many measures of benefits, and new measures will emerge to fit the needs of different kinds of studies. The earliest economic studies expressed all benefit in purely monetary terms, converting all potential benefits of an intervention into dollars, pounds, yen, francs, etc. However, researchers soon found that they could not easily express all benefits in monetary terms. For example, they could not use monetary terms to capture completely the value of saving a life (e.g., a person contributes to society in many nonmonetary ways, such as providing emotional and psychological support to friends and families), so researchers began using "life-years saved" as a reward for some interventions. However, life-years saved could not adequately represent the benefits of some quality-of-life-improving interventions (e.g., pain medications, walking devices) that do not save lives or the suffering caused by non-life-threatening diseases. As a result, researchers developed quality-adjusted life years (QALYs) as a health status measure, with one QALY representing a year of perfect health and less than one QALY representing a year of impaired health. Researchers also have developed and used many other reward measures specific to particular classes of interventions, such as the number of bypass operations prevented to measure the success of cardiac medications.

2.6. Discounting

If the time horizon of an analysis is longer than a year, an analyst will discount future costs and benefits. The practice of discounting recognizes that inflation and opportunity costs (i.e., the value of the next best alternative that you must forego when you make a choice) make a dollar (or any other cost or reward) in the future worth less than a dollar today. An analyst uses discount rates to adjust future costs and rewards to present day values. A discount rate, denoted typically by *r*, is the rate used to convert future costs or benefits to their present value.

For example, if C_n represents a cost *n* years from now and *r* is the discount rate, C_0 (i.e., the current or net present value of C_n) is $C_n/(1+r)^n$. Although typically researchers use a discount rate between 3% to 5%, there is still considerable debate over the exact appropriate rate (van Hout, 1998; Gravelle and Smith, 2001; Brouwer and van Exel, 2004). By using a 3% discount rate in the above formula, an intervention that earns a \$100 ten years from now will be worth $\$100/(1+0.03)^{10} = \$74.41$ in today's dollars.

3. TYPES OF ECONOMIC ANALYSES

Economic studies relevant to biosurveillance can be of several types, which we summarize in Table 31.1 and discuss in this section.

The question often dictates the appropriate method. For example, do we want to know the magnitude of a problem to help decide the course of action? Are we unsure about how much money to invest in a given strategy? If a strategy has different "settings" or "calibrations," how should they be set? An analyst selects the appropriate method for a study after first defining the question. In many cases, one analytic method will not be enough, and only a progression of different methods will answer a question.

3.1. Cost-of-Illness

A *cost-of-illness* study can quantify the magnitude of a problem. It can quantify a disease's total monetary effect, including all the resulting medical costs and, if necessary, loss of productivity. A well-performed cost-of-illness study will estimate not only the total cost but also different categories of cost, such as the amount spent on medications, hospitalizations, emergency care, and days off from work, allowing one to target the areas of greatest economic burden. Often, the first step in tackling a new and unfamiliar problem is a cost-of-illness study to "map out" the problem.

3.2. Cost-of-Intervention or Treatment

After quantifying the magnitude of a problem (sometimes referred to as *profiling* the problem), a *cost-of-intervention* or *cost-of-treatment* analysis can estimate the cost of possible solutions. These studies calculate all the monetary costs associated with executing a solution. Such studies should include and clearly identify every important fixed and variable cost. Running multiple scenarios may show how variable costs change with different situations. Such studies can help decision makers allocate an appropriate level of funds and identify particularly costly aspects of the solution that may be targets of cost reduction.

3.3. Cost-Minimization Analysis

After one profiles the possible solutions, a *cost-minimization analysis (CMA), cost-benefit analysis (CBA), cost-effective analysis (CEA)* or *cost-utility analysis (CUA)* can help choose among multiple alternative solutions. The type of problem guides the choice of analysis.

If all alternatives yield identical rewards, a CMA, which focuses only on costs, can help choose the least costly possibility. For example, if medication A and medication B have the same success rate in treating a disease, a CMA might find that medication A should be used because it costs \$200 less. This type of analysis seems to apply to only a small number of not-too-difficult decisions in biosurveillance, such as which of two identical surveillance systems to purchase. We note that formally asking and conducting such a study will have the benefit of requiring clarity about whether two systems are equally effective.

3.4. Cost-Benefit Analysis

A CBA is suitable when the potential benefits are different but easily translate to monetary terms (e.g., dollars, yen, pounds).

TABLE 31.1 Types of Economic Studies Relevant To Biosurveillance

Type of Study	Typical Cost Units	Typical Benefit Units	Types of Biosurveillance Decision That Can Be Analyzed
Cost-of-illness	Dollars	N/A	Quantify a disease or outbreak's total monetary effect
Cost-of-intervention/treatment	Dollars	N/A	Profile the costs of interventions and treatments
Cost-minimization analysis	Dollars	Benefits must be identical	Choose between alternatives (systems, response strategies) that have the same benefits but different costs
Cost-benefit analysis	Dollars	Dollars	Choose between alternatives in which all the costs and benefits can be expressed in monetary terms
Cost-effectiveness analysis	Dollars	Clinical measures such as *life years saved* or *deaths averted*	Choose between alternatives in which benefits are expressed in clinical measures
Cost-utility analysis	Dollars	Health status measure such as *quality-adjusted life years* (QALYs)	Choose between alternatives in which benefits are expressed in health status measures

So, for instance, a CBA may compare building a \$20 million wall with building a \$10 million wall to protect a \$100 million building. Let us assume that the \$20 million wall would save the entire building during an explosion while the \$10 million wall would only save half of the building, or \$50 million. Then, a CBA would find the \$20 million wall (which would provide a net benefit of \$80 million = \$100 million – \$20 million) to be more favorable than the \$10 million wall (which would offer only a net benefit of \$40 million). The analysis in Chapter 29 is a CBA.

3.5. CEA and CUA

However, if all the potential rewards do not translate easily into pure monetary terms, a CEA (which measures rewards in simple clinical units such as life years saved, deaths avoided, or operations avoided) or a CUA (which measures rewards in health status measures such as QALYs or utilities) is more useful. Because it can be difficult to quantify the economic value of saving a single life or avoiding a medical procedure, a CEA and CUA will measure the costs and benefits of each alternative separately and compare the alternatives by using incremental cost-effectiveness (or cost-utility) ratios, described in the next section.

3.6. Marginal and Incremental Analyses

Incremental analyses quantify the resulting differences in choosing one alternative over others. An incremental analysis can tell you whether strategy A is more preferable than B, but will not tell you in absolute terms whether either strategy is better than doing nothing.

In CBAs, the *incremental cost* indicates the change in cost when moving from one alternative to another. For example, if cb_x and cb_y are the net costs and rewards of strategies X and Y, respectively, then the incremental cost of using strategy Y instead of X is $cb_y - cb_x$. A negative incremental cost suggests that strategy Y is preferable over X, whereas a positive one favors X.

Similarly in CEAs, an *incremental cost-effectiveness ratio (ICER)* is the change in cost per change in effectiveness when shifting from one alternative to another, and in CUAs, an *incremental cost-utility ratio* is the change in cost per change in health status when shifting from one alternative to another. For example, if C_A and C_B are the costs of strategies A and B, respectively, and E_A and E_B are the resulting effectiveness (benefit) of A and B, respectively, then the ICER is $(C_B - C_A)/(E_B - E_A)$. Interpreting this ratio is somewhat more complicated than is interpreting an incremental cost. If the ICER is negative, then strategy B is favorable or dominant to strategy A. If the ICER is positive, then the magnitude of the ICER matters. If for instance, the ICER is \$10 per life year saved, then choosing strategy B requires only \$10 more for each life year saved. Most, except for the most penurious, would view this as a worthwhile investment and choose B. However, if the ICER were \$100,000 per life year saved, then decision makers would have to debate over whether this reward is worth the investment. There is extensive literature debating the appropriate threshold dollar value per life year (Gold et al., 1996; Neumann et al., 2000; Hershey et al., 2003; Ubel et al., 2003).

Marginal cost and marginal cost-effectiveness studies can help reveal the implications of changing a certain parameter (e.g., number of medications given, dollars invested, items used, people employed) by a single unit. For example, in a CBA, to measure the added cost of giving every patient an extra day of a medication, if C_N represents the net monetary value of giving a medication for N days and C_{N+I} is the net monetary value for giving it $N+1$ days, then the marginal cost would be $C_{N+1} - C_N$. In a CEA, if C_N and E_N represents the cost and effectiveness, respectively, of giving a medication N days and C_{N+1} and E_{N+1} represent the cost and effectiveness, respectively, of giving the medication $N + 1$ days, the marginal cost-effectiveness of an additional day of medication is $(C_{N+1} - C_N)/(E_{N+1} - E_N)$. A similar calculation would yield a marginal cost-utility in a CUA.

3.7. Decision Analysis and CEA and CUA

The procedure to perform a CEA or CUA is similar to procedure described in Chapter 29, except that each branch in the decision tree has two sets of outcomes (i.e., costs and effectiveness measures or costs and utility measures) instead of just one (i.e., the net costs and benefits denominated in dollars). Therefore, you will need to fold back costs separately and then effectiveness or utilities separately before combining them in ICERs. By using the example in Chapter 29, Figure 31.1 shows the same decision tree structure as Figure 29.3 with one difference: each branch has a cost outcome (C_1,C_2,C_3,C_4) and an effectiveness outcome (E_1,E_2,E_3,E_4). So if C_1 was equal to cb_1 or –\$89,315,780.49, C_2 was equal to cb_2 or –\$8,025,000, C_3 was equal to cb_3 or –\$115,454,219.50, and C_4 was equal to cb_4 or –\$115,454,219.50, then folding back the costs would yield an expected cost value of *act now* of –\$11,357,922 and an expected cost value of *wait* of –\$4,733,623. Now, if both E_1 was equal to 10,000 life years saved, E_3 was equal to 10 life years saved, and both E_2 and E_3 were equal to 0 life years saved, then the expected life years saved for *act now* would be $p \cdot E_1 + (1-p) \cdot E_1 = 0.041 \times 10{,}000$ life years saved $+ 0.959 \times 0$ life years saved = 4.1 life years saved. The expected value of *wait* is $p \cdot E_3 + (1-p) \cdot E_4 = 0.041 \times 10$ life years saved $+ 0.959 \times 0$ life years saved = 0.41 life years saved. The ICER of *act now* would then be [Expected Costs$_{\text{actnow}}$ – Expected Costs$_{\text{wait}}$]/[Expected Effectiveness$_{\text{actnow}}$ – Expected Effectiveness$_{\text{wait}}$] or [–\$11,357,922 – (–\$4,733,623)]/[10,000 – 10] = \$16,173 per life year saved.

4. SENSITIVITY ANALYSES

Because it is impossible to collect perfect data, an analyst frequently has to make a series of assumptions based on findings from prior studies, expert opinion, educated guesses from

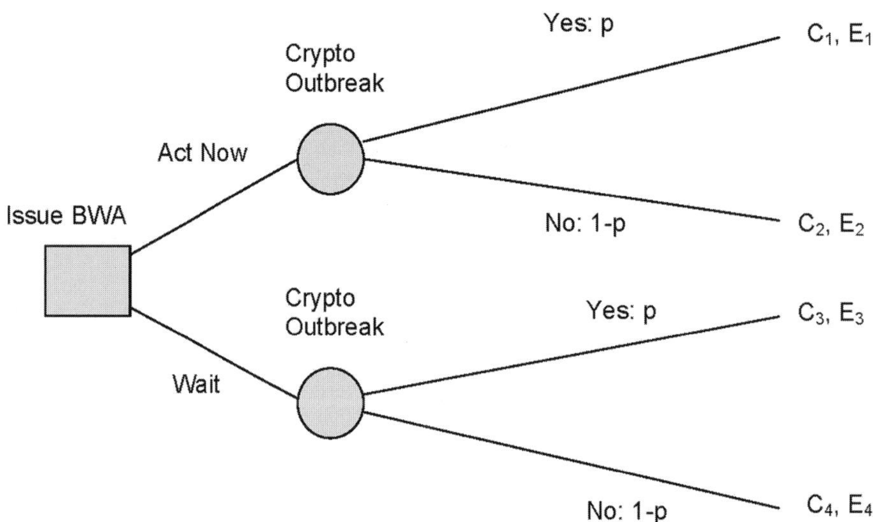

FIGURE 31.1 Decision tree for a cost-effectiveness analysis. This tree is identical to that for the cost-benefit analysis in Chapter 29 (Figure 29.3), except the leafs of the tree are pairs of numbers corresponding to dollar costs *(C)* and a measure of effectiveness *(E)* instead of a single number representing net benefit in dollars (*cb* in Figure 29.3).

personal experience, or, in some cases, truly random guesses. The study assumes that available information about costs and other model parameters is accurate, and make estimates for model parameters for which published data are not available. As a result, an economic study is only as strong as the information on which it rests, including its assumptions. For this reason, an organization or other consumer of an economic study must be aware of all assumptions and their accompanying reasons. Although some assumptions may be minor and have little bearing on the study results, others may be very controversial and dramatically influence study conclusions. Sensitivity analyses can help ascertain the impact of these assumptions.

Sensitivity analyses involves changing important variables along a range of different values and measuring the consequent effects on the results. For example, what would happen to the results if the discount rate varied from 2% to 6%, the cost of a specific medication ranged from $100 to $500, the percentage of people receiving a certain test changed from 40% to 60%, or the study excluded certain costs that were previously included? Running these different scenarios will not only identify the variables that have an important impact on the results but also demonstrate the credibility of the economic study. An economic study that does not change significantly is considered "robust"; i.e., most analysts would consider its results definitive. However, an economic study with results that fluctuate significantly during sensitivity analyses is not necessarily useless. The sensitivity analyses can help target the items and issues that are most responsible for the costs and rewards of a situation. If, for example, if the results of a study depend heavily on medication costs, then one may make extra efforts to either reduce the cost of medications or find alternative treatments.

5. EXAMPLES OF ECONOMIC ANALYSES IN BIOSURVEILLANCE

The following examples demonstrate how economic studies can address many issues important to biosurveillance. In each case, understanding of the above principles of economic analysis will help a reader (a consumer of these published studies) to catch more subtle implications from the study. Rather than accept or reject a final conclusion, one should fully discern the set of steps that led to that conclusion. Even if the results are not applicable to a decision maker's specific situation, components of the analysis may be. In fact, many times the greatest value of a study is not in the answers it provides, but instead in the questions that it raises. A study such as the boil-water CBA in Chapter 29 may focus decision makers on the key parameters or the crux of the matter. Economic studies may identify the need for additional future studies. In addition, there may be multiple approaches in dealing with a given decision or problem. Therefore, for each issue, although the selected examples offer important teaching points, they are not necessarily the best or most definitive studies.

5.1. How Significant Is the Threat?

Corso et al. (2003) conducted a retrospective cost-of-illness study of the 1993 Milwaukee *Cryptosporidium* outbreak. This study is a good example of a study that used multiple disparate sources to profile the economic impact of an outbreak. The analysts conducted a telephone survey of a random sample of

Milwaukee residents to estimate the total number of people affected by the outbreak, the percentages who sought different levels of medical care, the length of illness for those who did not seek medical care, and the total number of work days missed. The investigators also selected a representative sample of patients who sought medical care and reviewed each patient's medical chart to determine the medical resources (e.g., medications, diagnostic tests, emergency medical services) each patient consumed. They then used billing records to obtain the associated hospital charges. They used the Wisconsin 1993 average urban hospital and emergency department cost-to-charge ratios (0.70 and 0.67, respectively) to convert charges to costs.

As with most economic studies, the investigators made assumptions about information that was not readily available and used standard industry and government sources to estimate some costs. For example, they assumed patients with mild illness would self-medicate themselves with either loperamide or oral rehydration solution for 50% of the duration of illness. They used 1993 retail drug prices to estimate the unit cost of each medication. The City of Milwaukee Health Department provided data on the average cost of a single outpatient physician visit.

The study demonstrated not only that the outbreak had a substantial economic impact (in Milwaukee, $96.2 million) but that decreased worker productivity represented the largest economic consequence of the outbreak ($64.6 million), a finding confirmed by studies of outbreaks of other diseases such as severe acute respiratory syndrome (SARS) (Achonu et al., 2005) and hepatitis A (Sansom et al., 2003). One implication of these studies for biosurveillance economic analyses is that any cost-of-illness study that fails to include productivity losses may seriously underestimate the potential impact of an outbreak. Outbreaks can be devastating to not only the health care system and directly affected individuals but also many businesses and the economy. This result, if further developed, may perhaps persuade individuals and organizations initially uninterested in biosurveillance to reconsider their stance. In fact, the Milwaukee study actually underestimated productivity costs by not accounting for the lost lifetime productivity of those who died from the outbreak, the degree to which the outbreak diverted companies and the government from daily normal operations, and the damage the outbreak had on consumer and public confidence.

5.2. What Investigation and Response Options Are Available?

Cost-of-investigation/response studies can identify which potential investigation and response options are economically feasible. No matter how effective, a prohibitively expensive response option may not be possible. In addition, this type of analysis can guide the structure of more detailed analyses. An analyst may exclude relatively expensive investigation and response options such as mobilizing the Strategic National Stockpile from a study of the appropriate initial actions for a low probability alert.

Costs-of-investigations/responses have not been well studied; in some cases, cost-of-investigation/response analyses are included almost as afterthoughts in overall cost estimates that focus mainly on cost of illness. In some of these studies, the cost-of-investigation appeared to be relatively small compared with the overall economic burden of an outbreak, accounting for less than 5% of the total costs in a *Salmonella* outbreak (Cohen et al., 1978) and less than 3% in a New Mexico botulism outbreak (Mann et al., 1983). In other studies, such as Zhorabian et al. (2004) study of the 2002 Louisiana West Nile virus outbreak, the cost-of-response is a sizable percentage of the overall costs ($9.2 million of the $20.1 million total in the West Nile virus study). However, depending on retrospective cost-of-illness studies to determine costs-of-investigation is fraught with problems. The type and degree of response depend on the severity, nature, and location of the problem. Moreover, there is considerable regional and potentially temporal (e.g., varying by time of day, day of week, and month of year) variation in response mechanisms. In addition, some locations have antiquated accounting systems (Roberts et al., 1989; Bownds et al., 2003), making it difficult to accurately capture all costs. Therefore, there is a need for predictive studies.

An example of a predictive study is the evaluation by Gupta et al. (2005) of whether mass quarantine during a SARS epidemic would be cost saving, life saving, or both. They estimated the direct costs of SARS by the following formula:

$$\begin{aligned}&\text{Direct Cost of SARS/person} =\\&\quad[(\text{probability of hospitalization})\times\\&\quad(\text{Average Hospital Length of Stay})\times\\&\quad(\text{Cost per hospital day})]\\&\quad+\\&\quad[(\text{probability of an intensive care unit stay})\times\\&\quad(\text{Average ICU Length of Stay})\times\\&\quad(\text{Cost per ICU day})]\end{aligned}$$

They calculated the indirect costs or the lost productivity using the following:

$$\begin{aligned}&\text{Indirect Cost of SARS/person} =\\&\quad[(\text{Average Hospital Length of Stay})\times\\&\quad(\text{Average Daily Wage})]\\&\quad+\\&\quad[(\text{Probability of Death})\times\\&\quad(\text{Years of Potential Life Lost})\times\\&\quad(\text{Annual Salary})]\end{aligned}$$

In the second formula (and in many economic studies), the cost of a death is equivalent to the total value of the victim's potential future earnings. (This means that people are only worth as much as they can potentially earn in their lifetimes. Of course, some may take issue with this contention, but that is a discussion best left for another time.) The following formula generated the cost of quarantine:

Total Cost of Quarantine =
[(Number of People Coming into Contact with SARS) × (Incubation period of SARS) × (Average Daily Wage)]
+
Fixed Administrative Costs

As can be seen, lost worker productivity is one component cost associated with quarantine. In the 2003 Toronto outbreak, the fixed administrative costs associated with quarantine were around $12 million. The component owing to lost productivity was $1140 per person quarantined, for a total of $0.2 million for the primary wave, $1 million for the second wave, and $5 million for the tertiary wave of the Toronto outbreak. The following equation yielded the net savings from quarantine:

Net Savings =
[(Total Cost of SARS/person) ×
(Total Number of Infections-number infected before Quarantine)]
−
Total Cost of Quarantine

In other words, the net savings is the difference between the number of SARS cases prevented times the total cost of SARS per person and the total cost of quarantine. The investigators found that mass quarantine would not only save lives but also costs: $279 million during the primary wave of a SARS epidemic, $274 million during a secondary wave, and $232 million during a tertiary wave.

5.3. What is the Value of Rapid Response?

As is often said, time is money, especially in biosurveillance, in which delays in response can have significant consequences. Early action can lead to effective containment of a threat, and prompt measures can minimize disruptions in government and business operations. Many treatment options are only effective very early in an outbreak. Because biological agents used by terrorists would be expected to kill quickly, a tardy response can result in substantial morbidity and mortality. A predictive CBA, such as Kaufmann, Meltzer and Schmid's simulation of three types of bioterrorist attacks (with *Bacillus anthracis, Brucella melitensis,* and *Francisella tularensis*) over a major city suburb of 100,000 people, can help quantify the cost of such delays.

Because outbreaks of *B. anthracis, B. melitensis,* and *F. tularensis* have been rare, the investigators in this study had to make a number of assumptions to create an economic model involving extrapolations from available data. First, the investigators assumed that the spread (e.g., weather conditions would be ideal and the agents would travel with prevailing winds), physical and biologic decay (minimal decay), and infectivity of the agents (ID_{50} = infectious dose 50% = the number of infectious particles (spores, viral particles, bacteria, etc.) needed to cause disease in 50% of the people who are exposed of 20,000 spores of *B. anthracis* or 1,000 cells of *B. melitensis* or *F. tularensis*) would be uncomplicated. Then, data from the few previous outbreaks of these agents helped forecast how soon patients would develop symptoms and die after exposure. Next, extrapolations from published laboratory and clinical experimental data supplied the clinical efficacy of administering different antibiotic interventions to the exposed population. The investigators also postulated that 90% of the exposed population would participate in the treatments.

The analysts obtained cost estimates from several different sources and used the following formula to calculate the potential economic benefit of each antibiotic intervention program:

Net savings =
Number of Deaths Averted × Present Value of Expected Future Earnings
+
(Number of Days of Hospitalization Averted) × (Cost of Hospitalization)
+
(Number of Outpatient Visits Averted × Cost of Outpatient Visits)
−
Cost of Intervention

Once again, the cost of a death was equivalent to the present value of the victim's potential future earnings. The investigators derived age- and sex-specific salary data from the U.S. census and adjusted these numbers to match the age and sex distribution of their theoretical suburban population. They used discount rates of 3% (in one set of scenarios) and 5% (in another set) to express all future earnings in current dollars. The cost per day of hospitalization came from multiplying an average single hospital day charge ($875 in 1993 from the National Center for Health Statistics) by the April 1994 New York State cost-to-charge ratio (0.635) and adding a cost of $65 per day for missing work (lost productivity to society), a figure frequently used by health economists. To tabulate outpatient costs, they surmised the number of outpatient visits that victims would require and then used Medicare average allowance data to derive the cost per outpatient visit. The 1996 *Drug Topics*

Redbook provided prices to calculate the costs of the antibiotic interventions.

The study included multiple scenarios that varied in the period between the initial release of biological agent and antibiotic intervention, and found that a rapid response was the single most important means of preventing significant mortality, morbidity, and accompanying costs. In the absence of any intervention (antibiotic or vaccine), *B. anthracis* was most costly to society (ranging from $18 billion to more than $26 billion), followed by *F. tularensis* ($3.8 billion to more than $5.4 billion) and a *B. melitensis* ($477 million to more than $579 million). The potential cost savings of antibiotic interventions may be huge if initiated during the first day after the attack, but savings exponentially shrink with each passing day. For example, for an anthrax attack, antibiotic prophylaxis could save somewhere between $14 billion and $22 billion if administered on the day of the attack, $12 billion to $20 billion when administered the day after the attack, but only $5 billion to $8 billion when administered three days after the attack.

5.4. How Much Should One Pay for "Insurance" against an Attack or Outbreak?

Investing in biosurveillance (and other types of emergency preparedness) is analogous to taking out an insurance policy for protection against accidents, disability, death, or natural disasters. Similar to an accident, an outbreak or attack may occur at any moment or location. Although on most days, carrying an "insurance policy" may feel like paying a cost without obvious rewards, it is actually protection against that uncommon but potentially catastrophic occurrence. Therefore, to realize what "premium" is fair to pay for such "insurance," one must factor in the risk and the cost of the occurrence as well as the risk reduction that the "insurance policy" provides. CBAs, such as the analysis of Meltzer et al. (1999) of vaccination responses to a simulated U.S. influenza epidemic, can help ascertain the fair premium for such insurance.

In this economic study, investigators used prior clinical studies, charge data, and salary information to evaluate the economic benefit of employing different mass vaccination strategies to curtail a theoretical influenza pandemic. The analysts identified the diagnosis codes (International Classification of Diseases, Ninth Edition [ICD-9]) associated with each possible clinical sequela of influenza, such as pneumonia, bronchitis, and exacerbations of pre-existing conditions (e.g., heart disease), and searched health insurance claims data to calculate the average charges associated with each code. Previous clinical studies furnished the risk of each outcome for each age and risk category (high risk versus not high risk for contracting influenza). Age- and sex-weighted average wage data helped estimate lost productivity for each outcome. As before, the economic cost of a death was equal to the present value of how much the victim would have earned in his or her remaining lifetime. By use of this procedure, the investigators estimated that without large-scale immunization, an influenza epidemic would cost somewhere between $71.3 billion and $166.5 billion. The majority of these costs would come from deaths, suggesting that vaccine strategies should target those patients most likely to die from influenza.

The investigators then computed the economic value of different vaccination strategies (ranging from vaccinating specific populations to vaccinating the entire population), while varying the influenza attack rates, vaccine effectiveness, the rates of people vaccinated (i.e., compliance), and vaccine costs ($21 and $62). The investigators calculated the net returns of vaccinating each different age and risk category by the following formula:

$$\text{Net Returns}_{\text{age, risk group}} = \text{Savings from Outcomes Averted in Population}_{\text{age, risk group}} - \text{Cost of Vaccination of Population}_{\text{age, risk group}}$$

The "Savings from Outcomes Averted" for each age and risk group came from

$$\text{Savings from Outcomes Averted}_{\text{age, risk group}} = \text{Number with outcomes before intervention}_{\text{age, risk group}} \times \text{Compliance} \times \text{Vaccine Effectiveness}_{\text{Outcomes}} \times \text{Value of Outcome Prevented}$$

The "Cost of Vaccination of Population" for each age and risk group came from

$$\text{Cost of Vaccination of Population}_{\text{age, risk group}} = \$\text{Cost/Vaccine} \times \text{Population}_{\text{age, risk group}} \times \text{Compliance}_{\text{age, risk group}}$$

The cost per vaccinated person included the cost of the vaccine, the distribution and administrative costs, patient travel, time lost from work, and side effects, including Guillain-Barre syndrome.

According to the study, the amount of "insurance premium" to spend on maintaining proper influenza preparedness ranges from $48 million to $2,184 million annually. The investigators calculated this premium by using the following formula:

$$\text{Annual Insurance Premium} = \text{Net returns from an intervention} \times \text{Annual probability of a pandemic}$$

The results of this study suggest that the United States should be willing to spend somewhere between $48 million and $2 billion per year to prevent an influenza pandemic, depending on which assumptions one uses. Moreover, although the influenza attack rate, vaccine effectiveness, and compliance all affect this premium, the probability of the pandemic was the most important driving factor. This implies that ongoing monitoring and threat assessment is important, as determining the risk of pandemic will help determine the appropriate level of vaccination.

5.5. What Is the Economic Value of an Intervention?

Often, the economic benefits and penalties of an intervention are not necessarily obvious, and economic studies can better elucidate the true value of the intervention. For example, the CUA by Khan et al. (2005) compared different response strategies to a hypothetical SARS outbreak in New York City. Because SARS can be very difficult to distinguish from other illnesses (e.g., caused by influenza, respiratory syncytial virus, *Bordetella pertussis, Legionella pneumophilia*) that cause respiratory symptoms and fever, i.e., febrile respiratory illnesses (FRI), the investigators wanted to see the value of home isolation versus testing (for SARS or other common diseases) of patients with FRI. Their analysis included a variety of costs (such as transportation, laboratory tests, influenza vaccination, antimicrobial agents, hospitalization, public health investigation, and patient time) and used the Health Utilities Index Mark 3 (HUI) to estimate the changes in health-related quality of life from different situations such as home isolation. The study revealed that using a test (multiplex polymerase chain reaction [PCR] assays) to diagnose other common respiratory infections[1] would save $79 million and 8,474 quality-adjusted life-years over home isolation.[1] Adding SARS testing to the multiplex PCR assays would actually cost $87 million more and decrease utility. The explanation for this less-testing-is-better result is that causes of FRIs other than SARS are much more common than is SARS; therefore, SARS testing would generate false positives, resulting in more patients without SARS erroneously isolated. This study is an excellent example of how additional information provided by testing could actually be suboptimal.

5.6. What Factors Affect the Economic Value of an Intervention?

Because the right choice in some situations can be the wrong choice in others, economic studies can help determine what factors affect the relative values of different interventions. An example is a CUA conducted by Fowler et al. (2005) that evaluated the incremental cost-utility of four different postanthrax attack strategies (doing nothing, vaccination, administering antibiotics, and administering both antibiotics and vaccinations) and two preanthrax attack strategies (vaccination versus no vaccination). The investigators created a hypothetical cohort of a large metropolitan population with a similar age and sex distribution to New York City and obtained costs, probabilities, and rewards from a variety of sources, including the published literature, Centers for Medicare and Medicaid Services data, and the 1998 Statistical Abstract of the United States. The analyses showed that administering vaccine and antibiotics offered more utility (21.36 QALYs) and cost less ($46,099) than did the other three postattack strategies. Of the two preattack strategies, no vaccination was less expensive and resulted in higher QALYs gained per person when the annual risk for attack was 1% and during an attack 10% of the population was infected. However, sensitivity analyses revealed an interesting finding: if the probability of an individual being exposed (i.e., the risk for an attack multiplied by the probability of exposure given an attack) is less than one in 200, then the ICER drops below $50,000 per QALY. In health economics, an ICER of $50,000 per QALY is often used as an arbitrary threshold, as researchers consider anything below this threshold cost-effective. These findings imply that probability of exposure is pivotal in deciding whether to mass vaccinate a population preemptively, another important implication for biosurveillance.

6. CURRENT LIMITATIONS AND FUTURE DIRECTIONS

Although the current body of published literature addresses some critical issues and raises important questions, it is still limited in both the range of problems that have been studied and in the technical approaches used in the studies. A lack of funding and interest may partly explain the current state of the art. However, as interest in biosurveillance grows, so will the sophistication and use of economic studies of biosurveillance. Future methods and studies will have to address some of the following technical limitations:

6.1. Current Measures May Not Be Adequate

It remains to be seen whether the traditional cost and reward measures (such as dollars, life years, and QALYs) in their present forms are appropriate or if researchers need to modify current measures or develop new ones to match the unique aspects of bioterrorist attacks and epidemics. After all, many of these current measures originally arose in the context of more well circumscribed medical events, such as individual acute and chronic diseases. Such measures may not capture the complex scientific, economic, and social interactions that occur when the ambient environment is threatened and changed. For example, how does surrounding panic or loss of faith in daily business operations affect quality of life? Will existing measures adequately represent psychological distress? What is the cost of losing or damaging the life of a person, such as a healthcare worker, who is essential to mounting an adequate response to the outbreak? Do potential future earnings fully represent costs from a death? Because different measures may lead to different results and

1 This diagnostic strategy is an example of excluding SARS as a diagnosis by ruling-in another cause for a patient's illness.

different optimal choices, it will be important to use and develop measures germane to decision makers.

6.2. Studies Are Not Capturing All of the Effects

Most existing studies likely underestimate the impact of an outbreak or attack and do not account for all of the short-term and long-term effects. Outbreaks can shake the foundation of businesses, governments, and other organizations. Depending on who becomes ill, an attack can impede or disrupt vital services, such as transportation, health care, law enforcement, and food distribution, further compounding problems. For example, a sudden massive influx of victims into the healthcare system would divert resources and attention from other patients with more "traditional" but still urgent medical conditions, such as heart disease and stroke. Losing healthcare workers to death or quarantine would decrease an already limited response capacity. Setting up areas to place or quarantine victims would disrupt hospital workflow and reduce overall available space.

Existing studies also frequently overlook the psychological consequences of an outbreak, such as fear, hysteria, loss of confidence, and depression, which in sum could be substantial. Studies have shown that stress (Manning et al., 1996; Bejean and Sultan-Taieb, 2005), post-traumatic stress disorder (Frayne et al., 2004), and depression (Greenberg et al., 2003) are extremely costly ailments with insidious long-term consequences. Fear and hysteria can result in injury and bodily harm, as well as hinder response. A decline in consumer confidence could be very detrimental to businesses and the overall economy.

6.3. Studies Should Look at Other Scenarios

Many studies include only a limited number of scenarios, when, in fact, there is tremendous variability in where an outbreak can arise, how an agent may spread, and how a public health response may proceed. Although a number of studies have focused on very large cities such as New York City, attacks and outbreaks can occur almost anywhere. Conclusions from a New York City scenario may not be applicable to other cities and locations. A plethora of factors, including weather and climate conditions, geography, social structure and interactions, and transportation systems, can influence the pattern of spread, detection, and the ensuing response. Many other human and economic factors can alter the response. In addition, the response may not be efficient, especially if the event occurs during the weekends, holidays, or other concomitant crises.

6.4. Studies Should Take Other Perspectives

Most studies take the societal perspective, which is not necessarily the ideal perspective for all decision makers. The societal perspective may seem too abstract and inapplicable to many organizations and businesses that are busy addressing competing concerns that affect their daily operations. As a result, they may not make the time or effort to draw the link between the impact on society and the impact on their own situations. Therefore, taking other perspectives to show specifically how outbreak and bioterrorist attacks will harm their own interests may be helpful for planning, lobbying, and funding purposes.

6.5. Costs and Rewards Are Not Necessarily Linear

Many of the studies assume that costs and rewards change linearly, which is not always the case in real life. For example, in many analyses, doubling the number of people killed by an attack will double the productivity losses. However, in reality, the cost of losing two million people presumably will not be exactly twice the cost of losing one million people. Similarly, doubling the death toll from seven to 14 is not the same as doubling it from 700,000 to 1.4 million. The relationship between costs and deaths is probably much more complicated and shifts at different thresholds.

6.6. There Is a Need for More Data

Because the current poverty of data forces researchers to make many assumptions, future studies should further assess the validity of these assumptions and acquire more data to improve existing and future economic studies. Multidimensional sensitivity analyses, i.e., sensitivity analyses that vary more than one variable at a time, can test these assumptions. Complex simulation studies can measure how these assumptions may behave in a variety of conditions. Researchers can see how these assumptions fare when applied to other better-characterized diseases and problems. At the same time, organizational and policy changes can help data collection. Collecting and generating necessary data requires adequate accounting systems, cooperation from correspondent authorities, alleviation of administrative barriers, appropriately trained personnel, and, in some cases, innovative research methods.

6.7. Current Analytic Methods, Benchmarks, and Resources May Not Be Enough

Because the nature, scale, and impact of bioterrorist attacks and outbreaks are so different from many other medical and health problems, established health economic analytic methods, benchmarks (such as $50,000 QALY) and resources (such as the HUI) may not be applicable or enough to tackle important biosurveillance questions. For instance, is it appropriate to label a biosurveillance measure as not cost-effective if its cost-utility exceeds $50,000 QALY? Is it reasonable to rely on quality-of-life data derived from people who were not in the midst of an epidemic or attack? Can researchers use other more advanced economic methods from other industries? These are just some of the questions researchers and decision makers will struggle with in the near future.

7. SUMMARY

Economics has and will continue to play a significant role in biosurveillance. Economic studies can provide insight about some of the most challenging decisions related to biosurveillance, including what level of investment is justified by the threat, how best to invest available resources, and how to react to anomalies in surveillance data. The available methods of economic study include cost-of-illness, cost-of-intervention, and a set of techniques, such as CBA, that allow decision makers to explore the ever present tradeoff between cost and benefit in a world in which resources are finite. Although the field of economics is well developed in many areas, the economic study of biosurveillance is still in its early stages. Analysts have applied the existing set of techniques to but a handful of important problems. It is likely that this domain, as have many domains to which economics has been applied, will require the development of additional methods. There remain a large number of pressing problems, especially in the area of bioterrorism preparedness, still to be explored.

ADDITIONAL RESOURCES

Harvard CEA Registry (*www.hsph.harvard.edu/ceareqistry*). Provides electronic access to a database of cost-effectiveness ratios in the published literature.

Russell, L.B., Gold, M.R., Siegel, J.E., Daniels, N., and Weinstein, M.C. (1996). The Role of Cost-Effectiveness Analysis in Health and Medicine. Panel on Cost-Effectiveness in Health and Medicine. *Journal of the American Medical Association* vol. 276:1172–1177. First in a series of three articles on conducting cost-effectiveness analysis.

Siegel, J.E., Weinstein, M.C., Russell, L.B., and Gold, M.R. (1996). Recommendations for Reporting Cost-Effectiveness Analyses. Panel on Cost-Effectiveness in Health and Medicine. *Journal of the American Medical Association* vol. 276:1339–1341. Third in a series of articles on conducting cost-effectiveness analysis.

Weinstein, M.C., Siegel, J.E., Gold, M.R., Kamlet, M.S., and Russell, L.B. (1996). Recommendations of the Panel on Cost-Effectiveness in Health and Medicine. *Journal of the American Medical Association* vol. 276:1253–1258. Second in a series of articles on conducting cost-effectiveness analysis.

ACKNOWLEDGMENTS

The Pennsylvania Department of Health supported this work under award ME-01-737, and the Department of Homeland Security supported this work under grant F 30602-01-2-0550. Dr. Onisko is an NSF-NATO postdoctoral fellow. Dr. Grigoryan is a postdoctoral associate.

REFERENCES

Achonu, C., Laporte, A., and Gardam, M.A. (2005). The Financial Impact of Controlling a Respiratory Virus Outbreak in a Teaching Hospital: Lessons Learned from SARS. *Canadian Journal of Public Health* vol. 96:52–54.

Bejean, S. and Sultan-Taieb, H. (2005). Modeling the Economic Burden of Diseases Imputable to Stress at Work. *European Journal Health Economics* vol. 6:16–23.

Bownds, L., Lindekugel, R., and Stepak, P. (2003). Economic Impact of a Hepatitis A Epidemic in a Mid-Sized Urban Community: The Case of Spokane, Washington. *Journal of Community Health* vol. 28:233–246.

Brouwer, W.B. and van Exel, N.J. (2004). Discounting in Decision Making: The Consistency Argument Revisited Empirically. *Health Policy* vol. 67:187–194.

Cohen, M.L., Fontaine, R.E., Pollard, R.A., VonAllmen, S.D., Vernon, T.M., and Gangarosa, E.J. (1978). An Assessment of Patient-Related Economic Costs in an Outbreak of Salmonellosis. *New England Journal of Medicine* vol. 299:459–460.

Corso, P.S., Kramer, M.H., Blair, K.A., Addiss, D.G., Davis, J.P., and Haddix, A.C. (2003). Cost of Illness in the 1993 Waterborne *Cryptosporidium* Outbreak, Milwaukee, Wisconsin. *Emerging Infectious Diseases* vol. 9:426–431.

Fowler, R.A., et al. (2005). Cost-Effectiveness of Defending against Bioterrorism: A Comparison of Vaccination and Antibiotic Prophylaxis against Anthrax. *Annals of Internal Medicine* vol. 142:601–610.

Frayne, S.M., et al. (2004). Burden of Medical Illness in Women with Depression and Posttraumatic Stress Disorder. *Archives of Internal Medine* vol. 164:1306–1312.

Gold, M., Siegel, J., Russell, L., and Weinstein, M. (1996). *Cost-Effectiveness in Health and Medicine.* New York, NY: Oxford University Press.

Gravelle, H. and Smith, D. (2001). Discounting for Health Effects in Cost-Benefit and Cost-Effectiveness Analysis. *Health Economics* vol. 10:587–599.

Greenberg, P.E., et al. (2003). The Economic Burden of Depression in the United States: How Did It Change between 1990 and 2000? *Journal of Clinical Psychiatry* vol. 64:1465–1475.

Gupta, A.G., Moyer, C.A., and Stern, D.T. (2005). The Economic Impact of Quarantine: SARS in Toronto as a Case Study. *The Journal of Infection* vol. 50:386–393.

Hershey, J.C., Asch, D.A., Jepson, C., Baron, J., and Ubel, P.A. (2003). Incremental and Average Cost-Effectiveness Ratios: Will Physicians Make a Distinction? *Risk Analysis* vol. 23: 81–89.

Kaufmann AF, Meltzer MI, Schmid GP. The economic impact of a bioterrorist attack: are prevention and postattack intervention programs justifiable? Emerg Infect Dis. 1997 Apr-Jun; 3(2): 83–94.

Khan, K., Muennig, P., Gardam, M., and Zivin, J.G. (2005). Managing Febrile Respiratory Illnesses during a Hypothetical SARS Outbreak. *Emerging Infectious Diseases* vol. 11:191–200.

Mann, J.M., Lathrop, G.D., and Bannerman, J.A. (1983). Economic Impact of a Botulism Outbreak: Importance of the Legal Component in Food-Borne Disease. *Journal of the American Medical Association* vol.249:1299–1301.

Manning, M.R., Jackson, C.N., and Fusilier, M.R. (1996). Occupational Stress, Social Support, and the Costs of Health Care. *Academy of Management Journal* vol. 39:738–750.

Meltzer, M.I., Cox, N.J., and Fukuda, K. (1999). The Economic Impact of Pandemic Influenza in the United States: Priorities for Intervention. *Emerging Infectious Diseases* vol. 5:659–671.

Neumann, P.J., Sandberg, E.A., Bell, C.M., Stone, P.W., and Chapman, R.H. (2000). Are pharmaceuticals cost-effective? A review of the Evidence. *Health Affairs (Millwood)* vol.19:92–109.

Roberts, J.A., Sockett, P.N., and Gill, O.N. (1989). Economic Impact of a Nationwide Outbreak of Salmonellosis: Cost-Benefit of Early Intervention. *British Medical Journal* vol. 298:1227–1230.

Sansom, S.L., et al. (2003). Costs of a Hepatitis A Outbreak Affecting Homosexual Men: Franklin County, Ohio, 1999. *American Journal of Preventive Medicine* vol. 25:343–346.

Ubel, P.A., Hirth, R.A., Chernew, M.E., and Fendrick, A.M. (2003). What Is the Price of Life and Why Doesn't It Increase at the Rate of inflation? *Archives of Internal Medicine* vol. 163: 1637–1641.

van Hout, B.A. (1998). Discounting Costs and Effects: A Reconsideration. *Health Economics* vol. 7:581–594.

Zohrabian, A, et al. (2004). West Nile Virus Economic Impact, Louisiana, 2002. *Emerging Infectious Diseases* vol. 10:1736–1744.

VI
PART
Building and Field Testing Biosurveillance Systems

32
CHAPTER

Information Technology Standards in Biosurveillance

William R. Hogan and Michael M. Wagner
RODS Laboratory, Center for Biomedical Informatics, University of Pittsburgh, Pittsburgh, Pennsylvania

1. INTRODUCTION

A standard is a specification devised to facilitate construction of complex systems from component parts. In the field of biosurveillance, the component parts of particular interest are information systems owned and operated by the myriad organizations that participate in biosurveillance.

The role of standards in information technology is identical to their role in more mundane fields, such as bicycle manufacturing or house plumbing. Standards allow the building of bicycles and plumbing systems ("complex systems") from components, such as nuts, bolts, and pipes. Standards ensure that the parts fit. In these fields, it is a sure sign that standards have fallen short of ideal when you find yourself driving around town looking for the right "adaptor."

Thousands of information technology standards exist. Standards are at every level of an information system, from microchips through graphical user interfaces. National or international bodies endorse some standards, and others are de facto standards that have emerged based on usage and acceptance. Standardization has made it possible to create something as complex as the World Wide Web.

Fortunately, if you are building a biosurveillance system, you need to know about only a fraction of the vast number of existing information technology standards (although biosurveillance benefits from virtually all of them). The standards that are most relevant to your task are those required for components, such as hospital information systems and computers in health departments, to exchange data using "the same language." We define what we mean by language below.

In this chapter, we first highlight the importance of standards by describing several biosurveillance projects, which varied in the extent to which language standards facilitated or inhibited their construction. We then review existing language standards and other standards that you will need to be familiar with in your work. Finally, we discuss organized efforts to promote adoption of standards. Standards are a prerequisite for building the kinds of systems that you will ultimately need to meet the existing threats. It may be necessary for a biosurveillance organization to act strategically over a prolonged period to bring needed standards into existence and use.

2. THE GOOD, THE BAD, AND THE UGLY

Whether you are building a bicycle or a biosurveillance system, the greater the number of nonstandard components that you must assemble, the harder it will be. To emphasize the difference that standards can make, we discuss three systems. The first is the National Retail Data Monitor (NRDM; the Good). The second is a biosurveillance system for electronic laboratory reporting (ELR) from hospitals (the Bad). The third is a hypothetical biosurveillance system that is more comprehensive than are systems currently being built (the Ugly).

2.1. The National Retail Data Monitor (the Good)

In Chapter 22, we described the NRDM project in detail. Briefly, this project connects data warehouses owned by large retail corporations to a biosurveillance system. The retailers' data warehouses collect and store sales data from their individual stores in a central location. The retailers' systems that enable this centralized collection are complex. They include optical scanners in stores, network connections from each store to the data warehouse, and the central data warehouse itself.

The construction of the NRDM was facilitated by a standard—the global trade item number (GTIN), which is perhaps more familiar to readers by its former name the *universal product code (UPC)*. Retailers use GTINs to identify products. For example, the GTIN 00030045044943 stands for the 50-caplet bottle of extra-strength Tylenol, and the GTIN for the 100-caplet bottle is 00030045044972. Product manufacturers print the GTIN on the outside of each product as a barcode. Checkout scanners read the bar-coded GTIN optically.

Because, in large part, of the widespread use of the GTIN standard, a small team built the initial version of the NRDM in two months. The initial version collected data from 3,000 stores. Two years into the project, the NRDM collects data from 20,000 stores. The project team's effort to build the NRDM was limited to negotiating with retailers for permission to obtain the data and building a system that could store and analyze the data.

Note that the entire NRDM system, if viewed by aliens visiting from space, would appear to be a network that connects hundreds of thousands of optical scanners in stores

ISBN 0-12-369378-0

into a single biosurveillance system. Fortunately, the retail industry had already made 200,000 (20,000 stores times approximately 10 optical scanners per store) of the 200,008 connections[1], otherwise the NRDM project would have required significantly more work. The GTIN standard enabled the initial 200,000 connections as well as the final 8.

2.2. Hospital-Based ELR (the Bad)

In 2000, the same team began building an ELR system. At the onset, the team's goal was to connect 20 hospitals in Allegheny County, Pennsylvania to a biosurveillance system. There were no legal barriers. The hospitals already stored data in electronic form, and there were very good reasons (and still are) to create such connections: the benefits of ELR are dramatic in terms of completeness and timeliness of reporting (Effler et al., 1999; Overhage et al., 2001; Panackal et al., 2001).

Unlike the retail industry, however, hospital laboratories do not use GTIN or any uniform method to identify their "products," which are the results of laboratory tests. Standards exist, but very few hospitals use them. Instead, hospitals invent their own ways to identify the tests that they conduct and the results of those tests. They use proprietary codes or free text to identify the laboratory test, specimen type, organism, and other results of a test.

The same team that created NRDM in a few months has been working on the ELR project for five years. At present, the ELR project receives data from just 10 hospitals, all owned by a single health system that had integrated its laboratory information systems before the beginning of the ELR project. These 10 hospitals all used the same "vocabulary" of codes internally for microbiology cultures, organisms, and specimen source, but they do not use SNOMED or LOINC, which are current standards for biosurveillance. The team, therefore, had to create a custom adaptor to convert these codes to standard SNOMED and LOINC codes.

In addition to the custom adaptor that translates nonstandard hospital codes to standard ones, the team also had to build an adaptor that allows the biosurveillance system to interpret accurately the results coming from the laboratory information system. Although the hospitals used the HL7 standard to transmit microbiology results, the HL7 standard, unfortunately, is not strict. It allows *optionality,* permitting data elements to appear in different locations in messages. In addition, the laboratories update microbiology results by using a series of HL7 messages that are not standardized. The adaptor has to remember the entire sequence of laboratory messages about a specific specimen (sometimes spanning weeks) to correctly interpret the latest message. Creating this adaptor was the most time-consuming task, requiring several person-months of effort.

2.3. The Ugly

If the same team were to attempt to develop a comprehensive biosurveillance system—one capable of satisfying requirements for early detection and rapid characterization of many types of outbreaks as discussed in Chapter 4—or a biosurveillance system that could meet those requirements for outbreaks that cross national borders, they would encounter "the Ugly." It would be impossible to connect that many incompatible systems without first standardizing them.

3. TYPES OF INFORMATION TECHNOLOGY STANDARDS

Hundreds of information technology standards exist that facilitate the construction of complex information systems, such as the World Wide Web or the NRDM. These standards specify everything from how hardware components can be assembled into computers to how computers communicate with one another over networks and the Internet, to what data they communicate and how they encode data so that programs can use the data after receiving them (Table 32.1). Many of these standards are mature, representing the culmination of work that began soon after computers were invented in the 1950s. Other standards are still evolving. Without the existence of the majority of the standards in Table 32.1, we could not have built the NRDM, nor for that matter could the retailers have built optical-scanning cash registers or national data warehouses.

The number of standards is daunting. Fortunately, if you are working in the field of biosurveillance, you can benefit from many of these standards without knowing a thing about them. You can take them for granted.

There are standards, however, that you cannot take for granted. In general, you cannot take language standards for granted. As with a human language, such as English, language standards in information technology are shared conventions about how two or more entities communicate. Two people who wish to communicate in English employ a common vocabulary of words, an agreed on grammatical structure, and other rules about English usage that fall under the rubric of semantics. Computers in biosurveillance systems must also talk to each other using a common vocabulary, grammar, and semantics. There are information technology standards that play the same roles as vocabularies, grammars, and semantics.

We focus the remainder of the chapter on the language standards with which you need to have a working knowledge (dark-gray areas in Table 32.1) and other standards with which you need passing familiarity (light-gray areas in Table 32.1). The standards in the core group are all language standards. The "passing familiarity" group includes standards for data communication, networking, and representing knowledge.

1 These numbers are estimates with the exception of the number eight, which is the number of connections that the NRDM project had to build.

TABLE 32.1 Information Technology Standards

"Architectural" Level	Thing Requiring Standardization	Example Standards
Hardware	Memory chips and CPUs	Socket7, Slot A, Socket 423, Socket T, PAC611, SDRAM, DDR2
	Power supply to components	Motherboard connector, disk drive connector
	Motherboard size	ATX, AT, baby AT
	Disc drives	ATAPI, SCSI, ATA, SATA
	Expansion slots for network cards, video cards, etc.	ISA, PCI, VLB, AGP
	Peripheral connections	Serial/parallel port, USB, Firewire
	Video monitor connections	HD15, DVI, S-Video, Composite video
	Speaker connections	Stereo mini jack, RCA
	Computers	PC, Apple, Alpha, Sparc
	Physical connections between computers	RJ-45, Category 5 cable, coaxial cable, coaxial connectors, routers, switches
Software	Communications across physical connections	TCP/IP, http, ftp, 100 BaseT Ethernet, Gigabit Ethernet
	Operating systems	Unix/Linux, Windows
	Databases	SQL
	Programming languages	Java, C, C++
	Web browsers, web servers	HTML, XML
	Application servers	J2EE
Networks		VPN, sftp, SSL
Language	Vocabulary	LOINC, SNOMED, HL7 tables, ICD-9-CM, CPT, UPC/GTIN, NDC, RxNorm, UMLS, FIPS, METAR
	Message format (Grammar)	HL7, ASC X12, NDPDP, METAR
	Semantics	HL7 RIM, PHIN Logical Data Model, METAR
Data Access		JDBC/ODBC, GIS file formats
Knowledge		Notifiable Condition Mapping Tables

4. LANGUAGE STANDARDS

Without language standards, information systems must employ expensive translators that may or may not translate accurately. Building translation capability is both time-consuming and expensive. A common or standard language greatly facilitates construction of biosurveillance systems.

You should have a working knowledge of language standards because (1) you may need to talk with hospitals or other organizations about sharing their data; (2) you may need to evangelize at the local and national level to encourage hospitals and other organizations (including perhaps the organization where you work) to use standards so you can eventually build the types of systems you need; and (3) because the components you are using to build your biosurveillance system will be nonstandard for the foreseeable future, and you will need to create and/or purchase adaptors.

4.1. Vocabulary Standards

Standard vocabularies provide names for things, actions, and states of being (henceforth referred to collectively as *concepts*). Standard vocabularies assign a unique code to each concept. For example, the code 11389007 means the disease *respiratory anthrax* in "SNOMED-CT-ese."

Because code numbers are meaningless to humans, standard vocabularies also include one or more English labels for each concept. For example, SNOMED-CT associates "respiratory anthrax" and "Woolsorters' disease" with concept 11389007. The computers use the codes, and people tend to use the English labels.

There are many standard vocabularies, and none strive to be comprehensive. Each vocabulary fills a niche. For example, LOINC provides codes for laboratory tests and clinical observations such as vital signs. This was a deliberate choice on the part of the creators of LOINC, which they designed from the outset to be a vocabulary of laboratory tests and clinical observations.

Because vocabularies specialize, you need to know which standard vocabularies provide codes for which types of data. Table 32.2 lists many types of data used in biosurveillance and standard vocabularies that provide codes for each. Table 32.3 lists the standard vocabularies that we discuss in this chapter and, for each, tells whether it requires a license fee for use, the organization that created and maintains it, and the year in which the organization first created it.

4.1.1. LOINC

LOINC is a standard vocabulary for clinical laboratory tests and clinical observations, such as electrocardiogram measurements and vital signs (Huff et al., 1998). LOINC stands for "Logical Observation Identifiers, Names, and Codes."

The Regenstrief Institute for Health Care created and maintains LOINC. In particular, the LOINC committee—composed of representatives from various stakeholder groups, such as clinical laboratories and vendors of laboratory information systems—maintains LOINC. The meetings of the LOINC committee are open to the public. No license fee is necessary to use LOINC.

LOINC has broad coverage of laboratory tests done for both human and animal diseases. Note that LOINC only covers observations and tests, not their results. For example, the proper way to code the result of a microbiology culture is to use a LOINC code to identify the test (e.g., 17928-3 for

TABLE 32.2 Vocabulary Standards for Biosurveillance

Data element	Standard Widely Implemented	Possible Standards for Implementation (List Is Not Exhaustive)
Laboratory test names		LOINC
Organism names (viruses, bacteria, protozoa)		SNOMED-CT
Specimen types		SNOMED-CT, HL7
Organism susceptibility to antibiotics		LOINC
Radiology procedures		CPT, SNOMED-CT, LOINC
Occupation		SNOMED-CT
Signs, symptoms, diseases, diagnoses	ICD-9-CM (claims data)	ICD-9-CM, SNOMED-CT
Procedures	CPT (physician billing), ICD-9-CM (hospital billing)	CPT, SNOMED, ICD-9-CM
Drugs		NDCs, RxNorm, SNOMED-CT
Over-the-counter healthcare products	UPCs/GTIN	
Animal species/breeds		SNOMED-CT
Plant species		SNOMED-CT
Geopolitical entities (countries, states, etc.)	FIPS 5-2, 6-4, and 10-4, ISO 3166	
Weather observations	METAR	

CPT indicates Current Procedural Terminology; ICD, International Classification of Diseases; NDC, National Drug Code; UPC, universal product code; GTIN, global trade item number; FIPS, Federal Information Processing Standards; and ISO, International Organization for Standardization.

aerobic blood culture) and a SNOMED-CT code to identify the organism that grew in the culture (e.g., 21927003 for *Bacillus anthracis*).

In biosurveillance, LOINC is used at present for ELR from large commercial laboratories, such as Quest and LabCorp. ELR involves a subset of LOINC codes: those for immunology tests that indicate the presence of infection (e.g., hepatitis B surface antigen), microbiology cultures (e.g., blood culture), and susceptibility tests of culture isolates (e.g., the test that determines an organism's sensitivity to oxacillin).

It will be some time before clinical information systems in *hospitals* are capable of sending LOINC-encoded data to a biosurveillance organization. Until then, you will have to create or purchase an adaptor that maps each hospital's homegrown codes to LOINC if you wish to integrate laboratory data from multiple hospitals in a region.

TABLE 32.3 Source and Cost of Vocabulary Standards

Standard	Free?	Standards Developing Organization	Created
LOINC	Y	LOINC Committee	1998
SNOMED-CT	Y*	SNOMED International	2002
HL7 Tables	Y	Health Level Seven, Inc	1987
ICD-9-CM	Y	National Center for Health Statistics	1979
CPT	N	American Medical Association	1964
GTIN	N	GS1	1972
NDC	Y	Food and Drug Administration	1968
RxNorm	Y	National Library of Medicine	2001
UMLS	Y†	National Library of Medicine	1993
FIPS	Y	National Institute of Standards and Technology	Varies‡
ISO 3166	Y§	International Standards Organization	
METAR	Y	World Meteorological Organization	1995¶

The standards developing organization is the organization that currently maintains the standard. CPT indicates Current Procedural Terminology; GTIN, global trade item number; NDC, National Drug Code; UMLS, Unified Medical Language System; FIPS, Federal Information Processing Standards.
*The National Library of Medicine distributes SNOMED-CT for free only as part of the UMLS.
†The UMLS requires a signed license agreement to use.
‡FIPS 5-2 was created in 1987, FIPS 6-4 in 1990, and FIPS 10-4 in 1995.
§Only the two-character country codes in ISO 3166-1 are free. The rest of ISO 3166 (including the three-character and three-digit country codes in ISO 3166-1, ISO 3166-2, and ISO 3166-3) require a license fee.
¶METAR existed before 1995 but took its current form and became a U.S. standard that year.

4.1.2. SNOMED-CT

The Systematized Nomenclature of Medicine, Clinical Terms (SNOMED-CT) is a standard vocabulary for diseases, symptoms, signs, specimen types, living organisms, procedures (includes diagnostic, surgical, and nursing procedures), chemicals (includes biological chemicals such as enzymes and proteins and compounds used in drug preparations), drugs, anatomy, physiological processes and functions, occupations, and social contexts (e.g., religion and ethnic group). SNOMED-CT has codes for more than 360,000 concepts in human and veterinary medicine. SNOMED-CT provides significant coverage of the data elements in Table 32.2.

SNOMED International created SNOMED-CT by merging an earlier version of SNOMED called SNOMED-RT with the Read Codes, a similar standard vocabulary developed in the United Kingdom. SNOMED-CT codes for veterinary medicine supersede the now obsolete Standard Nomenclature of Veterinary Diseases and Operations (SNOVDO).

SNOMED International has an editorial board composed of representatives from several stakeholder groups, including physicians, nurses, and medical informatics specialists. No license fee is necessary to use SNOMED-CT in the United States or the United Kingdom because the governments of those countries have contracted with SNOMED International to make SNOMED-CT freely available (SNOMED International, 2004).

It will still be some time, however, before any clinical information systems send SNOMED-CT encoded data because SNOMED-CT is a relatively new terminology (it debuted in 2002) and became available for free in the United States only in 2004. Vendors are just beginning to incorporate it into their products, either to support new functionality or to replace of nonstandard vocabularies and standard vocabularies such as ICD-9-CM that are insufficient for clinical use. Once incorporated in new versions of vendors' products, SNOMED-CT encoded data will still not be available from healthcare organizations for biosurveillance purposes until those organizations buy the new products and perhaps do substantial work internally to convert proprietary codes or free text into SNOMED-CT codes. Finally, as the developers of SNOMED-CT have recognized, SNOMED-CT by itself may be insufficient for the entry of diagnoses by clinicians into clinical information systems (Spackman et al., 1997). For example, they note that clinical information systems may need lists of frequently used diagnoses and abbreviations to facilitate input (Spackman et al., 1997). Because the interim period until healthcare organizations widely implement SNOMED-CT may be a decade or longer, developers of biosurveillance systems will face a choice of waiting or creating adaptors that map proprietary codes (or parse free text) to SNOMED-CT codes for diagnoses, signs, symptoms, anatomy, and living organisms such as bacteria and viruses. As with mapping proprietary laboratory test codes to LOINC, mapping proprietary codes for organisms to SNOMED-CT is resource intensive, requiring personnel with expertise with SNOMED-CT and personnel familiar with the laboratory's testing procedures and its use of proprietary codes or free text.

4.1.3. HL7 Tables

We discuss HL7 as a grammar in the next section. Here we discuss standard vocabularies developed by the HL7 organization. HL7 has defined nearly 300 tables of codes that have become de facto vocabulary standards for encoding such data elements as a patient's gender, marital status, race, religion, and the order status and specimen source of laboratory tests. For the most part, these codes do not overlap with SNOMED-CT, although there is overlap of codes for specimen source. Given the widespread adoption of HL7 by healthcare information technology vendors, it is likely that the specimen source code set of HL7 will be the de facto standard for some time.

Clinical information systems that send data in HL7 format are likely to use HL7 codes for those data elements for which an HL7 table exists. However, because of the optionality allowed by HL7 even for basic data elements such as the gender of a patient, the encodings may exhibit variability from hospital to hospital. In our experience, registration systems in hospitals use the HL7 codes "M" (male) and "F" (female) in the "administrative sex" field of HL7 messages consistently, but these systems inconsistently use codes for other "genders," using a blank field versus "U" (unknown), "O" (other), "A" (ambiguous), and "N" (not applicable). The inconsistent use of this field, with just six possible values in HL7 2.x standards, highlights the optionality in HL7 that leads to language incompatibilities even among systems that use the HL7 standards.

4.1.4. International Classification of Diseases

The International Classification of Diseases (ICD) is a standard vocabulary for diseases, health status, types of patient visits to doctors and other health providers, and external causes of injuries. The purpose of ICD is to enable comparison of causes of death among nations. The latest version is ICD-10 (released in 1989), which the United States has used since 1999 to report mortality statistics to the World Health Organization.

The World Health Organization creates and maintains ICD (since 1948), but countries modify ICD for other purposes. In the United States, the National Center for Health Statistics (NCHS) created a "clinical modification" of ICD-9 (ICD-9-CM) in 1979 for the purpose of coding diagnoses and procedures when submitting health care insurance claims to the government. Private payers have since also adopted ICD-9-CM for the same purpose.

There are two major parts of ICD-9-CM: a classification of diseases and a classification of procedures. The NCHS maintains the classification of diseases, and the Center for Medicare and Medicaid Services (CMS) maintains the classification of procedures. NCHS and CMS have added and deleted codes every year since 1986.

You may encounter ICD-9-CM–encoded diagnoses and procedures when obtaining data from hospital information systems. Healthcare providers (e.g., physicians and hospitals) encode diagnoses and procedures (hospitals only) with ICD-9-CM for the purpose of billing. Because providers do not get paid if they do not use standard billing vocabularies to submit claims, ICD-9-CM is a widely implemented vocabulary for the purposes of billing and reimbursement.

If you encounter nonstandard diagnosis or procedure codes (or free text), then you should convert them to SNOMED-CT and not ICD-9-CM. The reason is that SNOMED-CT is the emerging standard for encoding diseases, symptoms, and signs for purposes other than billing and reimbursement. If you encounter a mix of proprietary codes and ICD-9-CM, you can convert the ICD-9-CM codes to SNOMED-CT codes by using an adaptor created by the Unified Medical Language System (UMLS, which we discuss below). The Unified Medial Language System also provides a version of ICD-9-CM that computers can use; the NCHS releases ICD-9-CM in a text document from which it is extremely difficult to create automatically a database of ICD-9-CM codes.

4.1.5. Current Procedural Terminology

Current Procedural Terminology (CPT) is a standard vocabulary for surgical procedures, minor procedures that physicians perform in the office, radiology tests, and a small number of laboratory tests (approximately 1,000). Whereas hospitals use ICD-9-CM for billing, physicians use CPT to bill for their services. Thus, CPT covers laboratory tests that physicians and/or their staff perform in office settings.

The American Medical Association (AMA) created the first version of CPT in 1966 and until 1984 released new versions every 4 years. Since 1984 it has released a new version annually. CPT requires a license fee for its use.

Because the purpose of CPT is billing, distinctions among codes often relate to the level of effort typically required to perform a procedure. For example, codes 11620 through 11624 and 11626 (six codes total) all refer to *Excision, malignant lesion, except skin tag (unless listed elsewhere), scalp, neck, hands, feet, genitalia.* The difference is that the codes refer to different size lesions; presumably larger lesions require more effort to remove and thus provide greater reimbursement.

You may encounter CPT-encoded procedures when obtaining claims data. If you are building or purchasing an adaptor, it should map proprietary laboratory test codes to LOINC, as LOINC is the standard for laboratory test codes. The LOINC committee, with the support of the AMA, is creating a mapping from CPT laboratory test codes to LOINC with funding from the National Library of Medicine (NLM) (Anonymous, 2004).

4.1.6. Global Trade Item Numbers

As mentioned at the beginning of this chapter, the UPC standard and its successor standard, the GTIN system, are standard vocabularies for manufactured products. GTIN facilitated building the NRDM because virtually every manufacturer and retailer has implemented GTINs in their systems. Every product sold—including over-the-counter healthcare products—in every major retail chain has a GTIN barcode that enables scanning of products at checkout.

The Uniform Code Council (UCC) and EAN International (the European counterpart to the UCC) created the GTIN system by merging the UPC and European Article Numbering (EAN) systems. They have also merged organizationally, forming a standards-developing organization of global scope called GS1. The UCC is a member organization of GS1 and oversees the creation and maintenance of GTINs in the United States. Similarly, other countries also have organizations that are members of GS1 and assign GTINs in their respective countries.

Each manufacturer in the United States who wishes to print a GTIN barcode on its products must first pay a fee to join the UCC. The UCC then assigns the manufacturer its own manufacturer code. The manufacturer then assigns unique product codes to each product and creates GTINs by concatenating its manufacturer code with the product codes and computing a check digit.

GTINs uniquely identify products down to the level of packaging, thus there is a different GTIN for different flavors of the otherwise same cough syrup and for two-bottle packs of cough syrup versus single bottles. Because GTINs are so specific, their use in biosurveillance requires grouping them into categories that are meaningful for surveillance. For example, the NRDM has pediatric electrolyte, antidiarrheal, and pediatric cough syrup categories.

Unlike the other vocabularies we discuss here, no easily accessible database of GTINs exists. You must purchase a listing of GTINs from organizations such as AC Nielsen that analyze market share of the retail industry and thus collect data on GTINs. Even then, some GTINs will be missing because some retailers do not provide data to companies such as AC Nielsen, and thus the GTINs of products sold only by those retailers will not appear. This situation occurs, for example, when a manufacturer creates special packaging of its products for a single retail chain, resulting in the assignment of a different GTIN.

Most retailers have converted from UPCs to GTINs already. However, should you encounter both UPCs and GTINs in your work, the conversion of a UPC to a GTIN is simple: pad the UPC with two leading zeros (Uniform Code Council, 2004).

4.1.7. The National Drug Code Directory

The National Drug Code (NDC) directory is a vocabulary for prescription drug products in the United States. Other countries have similar vocabularies for prescription drugs (e.g., the Canadian Drug Identification Number). NDCs are similar to GTINs in that the code identifies products down to the level of the packaging. Thus a 50-tablet bottle of Cipro XR 500 mg has NDC 0026-8889-50, whereas the 100-tablet bottle has NDC 0026-8889-51.

The U.S. Food and Drug Administration (FDA) has maintained the NDC Directory since 1968 (Anonymous, 1969). The Drug Listing Act of 1972 mandated that all manufacturers of prescription drugs register with the FDA and provide on an ongoing basis a listing of all the prescription drug products they manufacture. This listing must include the NDC for each drug.

The FDA releases NDCs on its Web site as 10-digit numbers, but automated pharmacies almost always use 11-digit NDCs. Thus, if you are going to use NDCs in a biosurveillance system, it is important to know these two different formats and how to convert between them. Ten-digit NDCs consist of a four- or five-digit labeler code (analogous to the GTIN manufacturer code), a three- or four-digit product code, and a one- or two-digit package code, leading to the following possible configurations of the 10 digits: 5-4-1, 5-3-2, and 4-4-2, respectively. The de facto standard is the 11-digit NDC, where four-digit labeler codes, three-digit product codes, and one-digit package codes are padded with a leading zero to produce a uniform 5-4-2 configuration with no hyphens. For example, the two NDCs above for Cipro XR 500 mg translate into the following 11-digit NDCs, respectively: 00026888950 and 00026888951.

In addition to the issue of 10-digit versus 11-digit NDCs, the format of NDCs is likely to change. The FDA has plans to revise the NDC as part of its regulatory initiative to require barcodes on all prescription and some over-the-counter drug products (Food and Drug Administration, 2004). The FDA does not intend to require new NDCs for existing products, and thus current NDCs will continue to be valid. Similar to the UPC-to-GTIN conversion, however, converting existing NDCs to the new format might involve, for example, increasing the number of digits in an NDC.

If one or more of the components in your biosurveillance system is a pharmacy information system, it is likely that you will encounter NDCs. As with GTINs, however, NDCs specify products at a level of detail too specific for biosurveillance, and thus you will need to group NDCs into categories for monitoring for various public health events. No standard categories of NDCs exist for this purpose.

4.1.8. RxNorm

RxNorm is a second vocabulary for prescription drugs. RxNorm provides a set of codes for *clinical drugs,* which are the combination of active ingredients, dose form, and strength of a drug. For example, the RxNorm code for *ciprofloxacin 500 mg 24-hour extended-release tablet* (the generic name for Cipro XR 500 mg) is RX10359383, regardless of brand or packaging.

The NLM, in consultation with the FDA and the HL7 Vocabulary Technical Committee, created RxNorm in 2001 to facilitate development of electronic medical records.[2] RxNorm does not require a license fee to use.

Although prescription drug information encoded with RxNorm would be far easier for a biosurveillance organization to process than are NDCs (unless it needed package-level information for trace back), RxNorm is a relatively new vocabulary that is not widely in use, and its purpose is managing clinical drug information in electronic health records, which are still relatively rare applications in hospitals (13%) and physician practices (14% to 28%). Vendors in general have not incorporated RxNorm into their products. Therefore, at present, developers of biosurveillance systems are unlikely to encounter systems that encode prescription drug information with RxNorm.

RxNorm is available as part of the UMLS (described next). If you need to use both NDCs and RxNorm, RxNorm includes a many-to-one mapping from NDCs to its clinical drug concepts, enabling you to build an adaptor that converts from NDC codes to RxNorm codes.

4.1.9. Unified Medical Language System

The UMLS is an amalgamation of preexisting vocabularies. It includes as "source" vocabularies the code sets described here (except GTINs/UPCs and HL7 tables) and more than 80 other vocabularies.

The NLM created and maintains the UMLS to *facilitate the development of computer systems that behave as if they "understand" the meaning of the language of biomedicine and health* (National Library of Medicine, 2004). One must sign a license agreement to use the UMLS, but there is no fee.

With respect to biosurveillance, we mention the UMLS because (1) it is the vehicle by which the NLM is distributing SNOMED-CT for free use, (2) it is the distribution vehicle for RxNorm, (3) it contains a machine-readable version of ICD-9-CM, and (4) when vocabularies overlap in their coverage (such as the case in which two vocabularies both provide codes for diagnoses), it provides mappings among them. Future versions will soon include HL7 tables from 2.x versions of HL7 and vocabularies that are part of HL7 version 3.0. UMLS is a one-stop-shopping source for many vocabularies.

4.1.10. Federal Information Processing Standards

The Federal Information Processing Standards (FIPS) are vocabularies for geopolitical entities around the world, such as countries and their political subdivisions (e.g., states, provinces, and counties). FIPS 5-2 provides standard codes for the 50 states in the United States, the District of Columbia, and the various territories of the United States such as Puerto Rico and the Marshall Islands. FIPS 6-4 provides standard codes for the counties in each U.S. state and territory, and FIPS 10-4 provides standard codes for nations around the world and their principal administrative divisions (analogous to U.S. states).

In the United States, the National Institute of Standards and Technology (NIST) develops FIPS when no existing industry standard exists to meet the needs of federal government computer systems. NIST created FIPS 5-2 in 1987, FIPS 6-4 in 1990, and FIPS 10-4 in 1995. You can use FIPS for free.

Because biosurveillance data are inherently spatial, standard methods for identifying geographical entities such as counties and states is critical to representing biosurveillance data. FIPS standards provide a standard vocabulary that covers many geopolitical entities worldwide.

4.1.11. ISO 3166

ISO 3166 is another standard vocabulary for countries and their subdivisions. ISO is short for the International

2 Because NDCs identify products down to the level of packaging, they are not suitable for use in physician order entry or decision support applications. The reason is that doctors prescribe, for example, "Cipro 500 milligrams three times a day." They do not specify whether the tablets come from a 15 tablet or a 20 tablet bottle for example, and pharmacies may substitute an equivalent generic or brand-name drug.

Organization for Standardization, the standards-developing organization that created and maintains ISO 3166.

There are three parts to ISO 3166: ISO 3166-1, ISO-3166-2, and ISO 3166-3. ISO 3166-1 contains codes for existing countries. ISO 3166-1 itself is composed of three parts: two-character country codes[3], three-character country codes (not available for free), and three-digit numeric country codes (not available for free). ISO 3166-2 (not available for free) contains codes for political subdivisions of countries such as states (such as the 50 U.S. states) and provinces (such as Canadian provinces). ISO 3166-3 contains codes for countries no longer in existence, owing either to a name change[4] or political changes such as subdivision or merging of countries[5].

Despite the existence of FIPS, the defacto standard on the Internet is the two-character country codes that are part of ISO 3166. HL7, during the ongoing development of version 3.0, initially standardized on ISO 3166. However, they removed ISO 3166 because of concerns that ISO 3166 was not available for free. The ISO clarified its position, announced that its two-character country codes are free, and made these codes available for free download from its Web site. There has of yet been no final decision by HL7 with respect to which country-code standard it will specify in version 3.0 of HL7.

If you are developing a biosurveillance system, you will have to choose between FIPS 10-4 and ISO 3166-1 for countries and their political subdivisions. Given that the Internet has adopted ISO 3166-1 two-character country codes and that these codes are now freely available, it makes sense to use ISO 3166-1 codes, except when exchanging data with U.S. federal government systems that require FIPS 10-4. The FIPS 5-2 codes for the 50 states and other territories of the United States are the standard in the United States and are free. The remaining decision is whether to use FIPS 10-4 or ISO 3166-2 for the political subdivisions of countries besides the United States. We are unaware of any reason to believe that either one is widely used or more likely than the other to become standard. Thus, based on consideration of cost alone, FIPS 10-4 is the better choice.

4.1.12. METAR

The METAR code is an international vocabulary, message format, and semantic standard for encoding surface weather observations. Vocabulary elements in METAR include codes for existing weather stations, units of measurement, cloud cover and type, and types of precipitation.

The World Meteorological Organization created and maintains METAR. The Office of the Federal Coordinator for Meteorological Services and Supporting Research created a U.S. implementation of METAR, a description of which appears in Federal Meteorological Handbook Number 1 (Office of the Federal Coordinator for Meteorological Services and Supporting Research, 1995). The first version of this handbook appeared in 1995. No license fee is necessary to use METAR.

If you are incorporating weather data into a biosurveillance system, you should know about METAR. One of the most widely available, free sources of weather data is the National Oceanic and Atmospheric Administration, which encodes weather data by using METAR. We discussed meteorological data in Chapter 19.

4.2. "Grammar" (Message Format) Standards

A standard message format, similar to the grammar of a human language, specifies the position of vocabulary elements in longer texts (also referred to as "messages"). A message format also specifies a set of characters, called delimiters, analogous to punctuation marks and blank spaces in English, that separate individual vocabulary items or groups of vocabulary items. Table 32.4 lists standard message formats that are relevant to biosurveillance.

4.2.1. HL7 Clinical Messaging Standards

HL7 is a standard message format used in health care. It is a widely recognized standard, and almost every company that

TABLE 32.4 Message Format Standards for Biosurveillance

Type of Data	Standard Widely Implemented	Possible Standards for Implementation
Healthcare/clinical	HL7	
Over-the-counter healthcare products		
Weather	METAR	
Water authority		
Veterinary/other animal		HL7
911		HL7
Poison control		HL7
Emergency medical services/fire/police/military dispatch		
Public health		HL7
Food processing and manufacturing		
Absenteeism reporting		
Claims/billing	ASC X12, NCPDP	
Vital statistics		HL7
Sentinel population/biometrics		

3 Note that the two-character codes used in Internet URLs (such as the "uk" in the URL of web site of the government of the United Kingdom: http://www.direct.gov.uk) are ISO 3166-1 two-character country codes.

4 For example, in 1989 Burma changed its name to Myanmar and the ISO 3166-1 two-character code changed from BU to MM.

5 For example, Czechloslovakia (former ISO 3166-1 code CS) split in 1993 into the Czech Republic (ISO 3166-1 code CZ) and Slovakia (ISO 3166-1 code SK).

builds clinical information systems has implemented its understanding of the standard in its products.

Health Level Seven, Inc. created the first version of HL7 in 1987. Since then, it has continued to develop and maintain the HL7 message standard. Health Level Seven, Inc., became an American National Standards Institute (ANSI)-accredited standards-developing organization in 1994. It has 26 international affiliates, meaning that organizations in 26 countries are using the standard and contributing to its ongoing development. The American Veterinary Medical Association endorses HL7 for use in veterinary information systems. HL7 is not free; one purchases documents that specify the standards (members of HL7 receive a discount).

HL7 is event based, meaning that as soon as an information system completes a task, it generates and transmits an HL7 message to other information systems. For example, after a registration clerk registers a patient in the emergency department (ED), the registration computer transmits a message to other systems such as the laboratory and radiology information systems (to prepare them to receive orders or specimens for the patient). Figure 32.1 shows a typical de-identified ED registration message that hospitals send to the RODS biosurveillance system. Similarly, laboratory and radiology information systems transmit messages about test results as soon as the results become available. HL7 defines as "trigger events" various tasks or transactions, including patient registration, orders, and results. HL7 defines nearly 90 types of messages. For each type of trigger event, HL7 specifies which data elements should be in a message, where they may be located in the message, and whether they are required versus optional.

Health Level Seven, Inc., began work on version 3 of HL7 in the late 1990s to remove the optionality present in the 2.x versions of HL7. The goal of version 3 is to improve interoperability of clinical information systems. Version 3 HL7 also includes a semantic model called the *HL7 Reference Information Model* or RIM (see below). HL7 has formally approved and released portions of the version 3 standard, including the HL7 RIM.

If you are building a biosurveillance system that connects to hospitals, it is a certainty that you will encounter HL7. It is the most widely implemented language standard in health care.

As with SNOMED-CT and LOINC, however, it will be several years before you encounter healthcare organizations that are capable of transmitting data by using version 3.0 of HL7. Vendors only recently have begun implementing HL7 version 3.0 in new versions of their products, and healthcare organizations still must then upgrade to those new versions and use version 3.0 of HL7 in place of version 2.x.

4.2.2. Accredited Standards Committee X12 Insurance Claims Standards

Standards of the *Accredited Standards Committee X12* (ASC X12) are message format standards for information systems in a number of industries. The X12N subcommittee of ASC X12 creates standards for insurance transactions. ASC X12 standards are not free; one must purchase the documents that specify the standards.

We mention ASC X12 here because the Department of Health and Human Services has mandated ASC X12 standards for the submission of healthcare insurance claims as part of the administrative simplification provisions of the Health Insurance Portability and Accountability Act. Developers of biosurveillance systems may thus encounter ASC X12 when using claims data from physician offices or hospitals for surveillance.

ASC X12 has signed memoranda with Health Level Seven, Inc., and the National Council on Prescription Drug Programs (see below) to ensure that its standards do not overlap and/or are compatible with the standards of those organizations.

4.2.3. Pharmacy Claims Standards

The standards of the National Council for Prescription Drug Programs (NCPDP) are message format standards for transmitting data as part of commercial pharmacy operations. The Batch Transaction and Telecommunications standards are message formats for submitting claims to healthcare insurance

```
MSH|^~\&||xxx||RODS|200202241715||ADT^A04|2002022XXXXXXXX|P|2.3<C
R>
PID||||||^020|M|||^^^^84204||||<CR>
PV1||E|||||||||||||98765432|||||||||||||||||||200202XXXXXXX||<CR>
DG1||||CARBON MONOXIDE EXPOSURE<CR>
IN1|||||||||||||||||||||||||||||||||||^^^^84056<CR>
<ETX>
```

FIGURE 32.1 A typical HL7 message for transmitting chief complaint data to the RODS biosurveillance system. The message contains date of service (February 24, 2002 at 17:15 hours), message type (ADT 04 standing for emergency department registration), patient age (20), gender (M), chief complaint (CARBON MONOXIDE EXPOSURE), and home zip code of patient (84056).

companies on behalf of individuals with prescription-drug coverage. The SCRIPT standard is a message format standard for transmitting electronically prescriptions from a physician's office to a pharmacy.

The NCPDP has created and maintained pharmacy transaction standards since 1977. Similar to HL7 standards, NCPDP standards are not free. You must either be a member of NCPDP or purchase the documents (which come with a membership to NCPDP) that specify the standards.

You may encounter NCPDP standards if you are using pharmacy claims data for biosurveillance because the Department of Health and Human Services mandated the Batch and Telecommunication Standards for use with pharmacy claims submission as part of regulations it developed under the Health Insurance Portability and Accountability Act.

4.2.4. METAR Meteorological Observations Message Standards

METAR refers to the METAR message-format standard—as well as its vocabulary and semantic standard—for hourly observations of surface weather. It specifies a strict ordering of the data elements in the message, as well as which data elements must appear and which are optional. For example, the weather station code is mandatory and always appears first in the message. See the discussion of METAR under "Standard Vocabularies" for more details about METAR and its role in biosurveillance.

4.3. Semantics Standards

Semantic standards specify additional rules for constructing messages. In human languages such as English, semantics refers to a set of rules that humans intuitively understand, and therefore (unless you are a linguist), you will not have encountered any semantic rules in your life (except indirectly when you learned English as a very young child or as a second language). The following sentence adheres to English grammar and vocabulary rules but violates semantic rules: *Blue mathematics eat televisions.* A speaker of English realizes that the verb *eat* cannot have blue mathematics as a subject or a television as its direct object.

Because computers do not have such intuitions, language standards must specify that when the verb is *eat,* only a limited number of subjects and direct objects are legitimate. In the clinical domain of laboratory testing, for example, you must tell the computer that laboratories "test" specimens from patients. Only laboratories can "test," and the direct object of the "testing" act is a specimen. You must then also tell the computer which vocabulary codes are for laboratories, which are for patients, and which refer to specimens. This latter function is accomplished by the vocabularies, but the relationship between the action verb test and the legitimate classes of entities is the job of semantics.

A standard information (or data) model is a semantic standard. It specifies for a domain the set of verbs or *acts,* the set of allowable actors for each act, and the set of allowable direct objects for each act. It also specifies the other data elements (analogous to adjectives, adverbs, and prepositional phrases) that qualify each act, such as when (date/time) and where (location) it occurred.

4.3.1. HL7 RIM

The HL7 RIM is a standard information model (semantic standard) for medicine. It specifies the acts, actors, and entities acted upon and the data elements associated with each that are necessary for clinical information systems to communicate data. The significance of the HL7 RIM is that it is the key to eliminating ambiguity in HL7 caused by the lack of semantic standards; thereby eliminating a source of incompatibility between systems. Health Level Seven, Inc., approved and released the HL7 RIM in 2003.

Because HL7 RIM is the core of version 3.0 HL7 standards, it is necessary to understand the RIM to create and to process HL7 version 3.0 messages. However, as we discussed previously, it will likely be some time before you encounter clinical information systems capable of sending version 3.0 messages.

4.3.2. Public Health Information Network Logical Data Model

The Public Health Information Network (PHIN) Logical Data Model (LDM) is a standard information model of public health. It is an extension of the HL7 RIM to accommodate public health "acts" such as case detection and outbreak detection. The Centers for Disease Control and Prevention (CDC) released version 1.0 of the PHIN LDM in 2004.

The PHIN LDM is still in relatively early stages of development. Because the development of standard data models is as much an art as it is a science (and moreover is a highly iterative process), until developers use the PHIN LDM in biosurveillance applications, the validity of the modeling decisions and their ability to facilitate data sharing will not be known.

Nevertheless, the PHIN LDM may serve as a useful starting point in creating a database to store biosurveillance data (provided that the PHIN LDM covers the acts, entities, roles, and data elements you need to store the data you are collecting). Wilson et al. (2001) describe in detail how they derived a LDM from a predecessor of the PHIN LDM.

4.3.3. METAR Semantics

METAR is the semantic standard for hourly observations of surface weather (as well as the vocabulary and message format standard). It specifies various rules for constructing messages about weather observations. For example, it specifies that messages about observations made at a particular time of day must include certain data elements. For example, messages about observations made during the 12:00 UTC hour must contain the amount of rainfall that occurred in the preceding 24 hours. See the discussion of METAR under "Standard Vocabularies" for more details about METAR and its role in biosurveillance.

5. STANDARDS WITH WHICH YOU NEED PASSING FAMILIARITY

As discussed previously, the standards for which you need passing familiarity include standards for data communication, networking, and knowledge.

5.1. Standards for Transferring Data to Data Analytic Programs

There are several standards for transferring data from a database to a data analytic program such as SAS and Crystal Reports or to geographical information system (GIS) packages such as ESRI's ArcGIS software.

5.1.1. Open Data Base Connectivity/Java Data Base Connectivity

Open Data Base Connectivity (ODBC) and Java Data Base Connectivity (JDBC) are standards that enable analytic (and any other) programs to retrieve data from relational databases. In general, each relational database vendor creates an ODBC and/or JDBC "adaptor" for its database product. Programs compliant with these standards can access databases that are also compliant.

The relevance of these standards to biosurveillance is that many epidemiologists and other public-health personnel use data-analytic programs such as SAS and Crystal Reports to analyze biosurveillance data. If your biosurveillance database is ODBC compliant, users can access data from these programs. One note of caution is that some data requests that users submit via these programs may be so resource intensive to execute in terms of the processor, memory, and/or disk that they severely decrease the performance of a production biosurveillance system, hindering access and use by other users.

5.1.2. ArcGIS File Formats

ArcGIS file formats are standard file formats that GIS programs use to store geospatial data. Geospatial data are often used to draw maps. Two important standard file formats that GIS systems typically have the ability to read and write include ESRI ArcImport and Shape files. Viewing biosurveillance data in GIS programs, such as ESRI's ArcGIS package, requires that the biosurveillance system format geographical data in a standard format understood by the GIS program.

5.2. Networking Standards

Networking standards are standards for establishing a communication channel between computers. Networking standards are sufficiently mature and widely deployed that you only need to know that the hardware and software you use to create networks will be using standard protocols such as transmission control protocol/internet protocol (TCP/IP). One networking standard with which you may need to be familiar is virtual private networking (VPN), because you will have to make a conscious decision whether to use it and you will likely need special hardware and software to use it.

VPN is a standard method for establishing secure communications among applications over the Internet. The advantage of VPN is that it does not require any customized software additions to applications to enable them to encrypt data or to implement Internet protocols such as hypertext transfer protocol (http). VPN does this. VPN is useful, for example, when connecting a hospital's HL7 message router to a biosurveillance system located outside the hospitals' computer network. No modifications need to be made to the message router or its software to send data over a VPN.

You can choose between using VPN software or using dedicated VPN hardware. In our experience, dedicated VPN hardware provides a more reliable connection than does using VPN software on devices that also perform other non-VPN functions (such as a firewall). Many hospitals already operate a VPN device, making VPN the ideal method for connecting to such hospitals.

We discuss other methods for secure transmission of data over the Internet in Chapter 33 because the discussion is more relevant to the architecture of a system than to standards.

5.3. Knowledge Standards

There are two types of knowledge standards relevant to biosurveillance: standard knowledge bases and standard knowledge representations. A standard knowledge base is an agreed upon set of facts or knowledge about the world represented in a way that a computer use it. A standard knowledge representation is an agreed upon way of representing facts and knowledge about the world in a knowledge base. Standard knowledge bases facilitate your work because you do not have to recreate the knowledge; you can just use it. A standard knowledge representation facilitates sharing of knowledge among knowledge bases. None of the standard knowledge bases we discuss here use a standard knowledge representation, thus the remainder of this section discusses knowledge bases that are either defacto standards or could become so through more widespread use.

5.3.1. Notifiable Condition Mapping Tables

The Notifiable Condition Mapping Tables (formerly the "Dwyer tables") are a standard set of facts about which tests and results indicate that a patient has a notifiable disease. The Notifiable Condition Mapping Tables facilitate implementation of ELR by providing the "diagnostic" knowledge that laboratory information systems need when determining which results to send to a biosurveillance system. For example, one table maps from LOINC codes to notifiable diseases. It maps LOINC code 22863-5—a test for antibodies to *B. anthracis* in serum—to the notifiable disease anthrax. Another table maps from SNOMED-CT codes to notifiable diseases. For example, it maps the SNOMED-CT code 21927003—the *B. anthracis* bacterium—to anthrax. The idea is that if the 22863-5 test is positive or a microbiology culture grows 21927003, then the laboratory information system should transmit the test and its results to public health.

As notifiable diseases vary from state to state, a laboratory would utilize the knowledge about only those diseases on a particular state's notifiable disease list. Specifically, the laboratory can pull the LOINC and SNOMED-CT codes that are associated with each disease that they must report. The laboratory can then filter its test results for the appropriate laboratory test data about notifiable diseases. These tables do not require a license fee for use.

5.3.2. Standard Ontologies

A standard ontology is a standard specification of concepts and the relationships between concepts. The relationships between concepts encode knowledge, and thus, an ontology is a knowledge base. Many of the standard vocabularies we have discussed in this chapter are also ontologies because they provide relationships between concepts as well as concepts. Specifically, LOINC, SNOMED-CT, ICD, the NDC directory, RxNorm, and the UMLS all contain relationships between concepts and thus are ontologies. For example, SNOMED-CT encodes the following knowledge: the disease respiratory anthrax (11389007) has an "associated etiology" (116675007) relationship with *B. anthracis* (21927003), meaning that *B. anthracis* causes respiratory anthrax. The code 116675007 uniquely identifies the "associated etiology" relationship of SNOMED-CT.

The most common relationship between concepts in ontologies is the "is a" relationship, which indicates that one concept is a type of another concept. For example, SNOMED-CT specifies that the disease respiratory anthrax (11389007) "is a" (116680003) anthrax (17540007), meaning that respiratory anthrax is a type of anthrax.

This type of knowledge can facilitate creation and maintenance of the logical rules that a system follows to perform certain tasks. For example, if an ELR system needs to report all cultures that grow an organism of the genus *Salmonella,* then it is easier to write a rule about the SNOMED-CT code for *Salmonella*—27268008—and have the system check each organism code to see if it is a type of *Salmonella* than to write a rule with more than 100 codes for each species and strain of *Salmonella.* This approach also facilitates system maintenance: it requires no modification to the logic of your system when a new strain of *Salmonella* is added to SNOMED-CT.

With respect to biosurveillance, the utility of ontological knowledge about diseases, symptoms, drugs, and organisms depends on the application. At a minimum, when inferences about class membership (one thing being a type of another) are important, a standard ontology can greatly facilitate your work.

6. ADOPTION OF STANDARDS

You need to be aware of efforts to promote the adoption of standards to know in advance which standards will ultimately be adopted and thus which standards to use in your system.

Ultimately, users of information systems drive adoption of standards. They adopt a standard when there is a strong incentive to do so. In the retail industry, manufacturers, distributors, and retailers developed and implemented the GTIN standard to reduce the costs of conducting business. By contrast, hospitals, health insurance companies, and physicians have not put standards high on their priority list because until recently they have not had sufficient incentive to adopt standards.

In the United States and other countries such as Canada, the United Kingdom, and Australia, national efforts to promote exchange of healthcare data to improve decision making in health care (and the desired side effect of reduced healthcare costs) are changing the dynamic. Because standards can greatly reduce the effort required to exchange data, standards are a key focus of these national healthcare information infrastructure efforts.

In the United States, the federal government is actively promoting the adoption of standards. The Consolidated Health Informatics initiative seeks to ensure that all healthcare-related information systems owned by the federal government are compliant with industry standards. CHI has named many of the standards discussed here as standards with which systems should be compliant. The idea behind this initiative is that the private sector often adopts standards used by the federal government to reduce the difficulty of communicating data to government systems. The hope is that the CHI effort drives adoption of healthcare information technology standards such as LOINC and SNOMED-CT.

The Medicare Modernization Act of 2003 mandates that the Secretary of Health and Human Services create the Commission on Systemic Interoperability. The act charges this commission with developing a strategy for promoting the adoption and implementation of healthcare information technology standards. The commission held its first meeting in January 2005, and its final report is due in October 2005.

The private sector is collaborating with the federal government to identify and promote important standards for adoption. The Certification Commission for Healthcare Information Technology is a voluntary, private-sector initiative to certify that healthcare information technology products comply with standards. It began meeting in September 2004 and released for public comment its first work products, which include an assessment of existing standards and the state of the industry in adopting them.

The CDC have been recommending standards for adoption by public health information systems, including those that perform surveillance. The CDC terms this standards effort the PHIN. We have already discussed the PHIN data model standards, but the CDC is also recommending standards that enable surveillance systems to report notifiable diseases to CDC information systems. The implementation guides for these standards all specify the use of standards that we have already discussed, including HL7 version 3.0. To our knowledge, these implementation guides are the only existing standards for communication of data from one biosurveillance system to another.

7. SUMMARY

Standards facilitate the construction of complex systems from components. The road to widespread implementation of a standard is a long and arduous one. First, an organization must create a standard. Then the community at large must agree on the standard, after which companies must incorporate the standard into their new products or new versions of their products.[6] Finally, customers must either implement *de novo* the new products or upgrade their existing systems.

As healthcare organizations increasingly communicate patient data as part of regional health information organizations that in turn are part of the larger national health information infrastructure, the source of much biosurveillance data—hospitals, physician's offices, clinical laboratories, and diagnostic imaging centers—will increasingly implement standards, facilitating the construction of biosurveillance systems that use those data.

In addition, biosurveillance systems must also adopt the same standards, enabling them to communicate with other biosurveillance systems as well as computers operated by myriad organizations.

In the next chapter we discuss architecture, which is not a standard per se, but is also an important determinant of your ability to add or modify components in a growing biosurveillance system. Architecture resembles standards from the perspective of facilitating construction of complex systems.

ADDITIONAL RESOURCES

The SNOMED International Web site *(http://www.snomed.org)*. This Web site is the home of the SNOMED International organization, which developed and maintains the SNOMED-CT vocabulary.

The LOINC Web site *(http://www.regenstrief.org/loinc)*. This Web site is the location where you can download the latest version of LOINC.

The Health Level Seven Web site *(http://www.hl7.org)*. This Web site is the home of Health Level Seven, Inc., and describes the various standards-developing efforts of that organization.

The GS1 Web site *(http://www.gs1.org)*. This Web site is the home of GS1, the organization that oversees the Global Trade Item Number standard.

The National Drug Code Directory *(http://www.fda.gov/cder/ndc)*. This Web site is where you can download the latest version of the National Drug Code Directory.

The RxNorm Web site *(http://www.nlm.nih.gov/research/umls/rxnorm_main.html)*. This Web site provides information about the RxNorm vocabulary. You can also download a program that enables you to browse and search RxNorm from this site.

The UMLS Web site *(http://www.nlm.nih.gov/research/umls)*. This site provides information about the UMLS. To download the UMLS, you must first register (registration is free).

The Federal Information Processing Standards Web site *(http://www.itl.nist.gov/fipspubs/)*. This site describes various Federal Information Processing Standards. You can download FIPS 5-2, FIPS 6-4, and FIPS 10-4 (described in the text) from this site.

The Web site for ISO 3166 *(http://www.iso.org/iso/en/prods-services/iso3166ma/index.html)*. This site describes ISO 3166 and is also the place to download the two-character country codes that are part of ISO 3166-1 and purchase the remainder of the ISO 3166 standards.

The National Oceanic and Atmospheric Administration's Web site about METAR *(http://metar.noaa.gov/)*. This site describes the METAR standard. You can find Federal Meteorological Handbook No. 1 (discussed in the text) here.

The Accredited Standards Committee X12 Web site *(http://www.x12.org/)*. This is the official Web site of the ASC X12 organization, where you can find more information about ASC X12 standards.

The National Council on Prescription Drug Programs Web site *(http://www.ncpdp.org/)*. This is the official Web site of the National Council on Prescription Drug Programs (NCPDP), where you can find more information about NCPDP standards.

The Public Health Information Network Web site *(http://www.cdc.gov/phin)*. This Web site describes the Public Health Information Network effort and the standards it recommends.

REFERENCES

Anonymous. (1969). National drug code directory. *Hospitals* vol. 43:80–1.

Anonymous. (2004). Minutes: Laboratory LOINC Meeting. *http://www.regenstrief.org/loinc/meetings/20040824/20040824/20040824_Lab_committee_minutes.*

Effler, P., Ching-Lee, M., Bogard, A., Man-Cheng, L., Nekomoto, T., and Jernigan, D. (1999). Statewide System of Electronic Notifiable Disease Reporting from Clinical Laboratories: Comparing Automated Reporting with Conventional Methods. *Journal of the American Medical Association* vol. 282: 1845–1850.

Food and Drug Administration. (2004). Bar Code Label Requirements for Human Drug Products and

6 These steps are likely to overlap, because IT vendors are often members of standard-developing organizations and part of the community that agrees on standards.

Biological Products; Final Rule. *The Federal Register* vol. 69:9119–9171.

Huff, S.M., et al. (1998). Development of the Logical Observation Identifier Names and Codes (LOINC) vocabulary. *Journal of the American Medical Informatics Association* vol. 5:276–292.

National Library of Medicine. (2004). About the UMLS Resources. *http://www.nlm.nih.gov/research/umls/about_umls.html.*

Office of the Federal Coordinator for Meteorological Services and Supporting Research. (1995). The Federal Meteorological Handbook Number 1. *http://www.ofcm.gov/fmh-1/pdf/fmh1.pdf.*

Overhage, J.M., Suico, J., and McDonald, C.J. (2001). Electronic laboratory reporting: barriers, solutions and findings. *Journal of Public Health Management Practice* vol. 7:60–66.

Panackal, A.A., et al. (2001). Automatic Electronic Laboratory-Based Reporting of Notifiable Infectious Diseases. *Emerging Infectious Diseases* vol. 8:685–691.

SNOMED International. (2004). HHS Secretary Tommy G. Thompson announces access to SNOMED CT through the National Library of Medicine. *http://www.snomed.org/news/documents/050404_E_NLMPressRelease_Final_001.pdf.*

Spackman, K.A., Campbell, K.E., and Cote, R.A. (1997). SNOMED RT: A Reference Terminology for Health Care. *Proceedings of a conference of the American Medical Informatics Association* 640–644.

Uniform Code Council. (2004). Global trade item number (GTIN) implementation guide. *www.uc-council.org/ean_ucc_system/pdf/GTIN.pdf.*

Wilson, T., Tsui, R., Espino, J., and Wagner, M. (2001). Logical and Physical Data Models for a Health-System-Resident Component in a national Disease Surveillance System. Pittsburgh: Center for Biomedical Informatics, University of Pittsburgh.

33
CHAPTER

Architecture

Jeremy Espino and Michael M. Wagner
RODS Laboratory, Center for Biomedical Informatics, University of Pittsburgh, Pittsburgh, Pennsylvania

1. INTRODUCTION

The word *architecture* carries many meanings, but in this chapter, we use it to refer to the design of an information system. An architecture for an information system is a description of the system's components (e.g., communications network[s], servers, applications, and databases) and how they interact. The description should specify the components, what they do, which components communicate, how the components communicate, and what functions and services the components provide to each other.

The idea that there should be a grand design or architecture for biosurveillance systems has gained acceptance as a result of evangelizing by two projects of the Centers for Disease Control and Prevention (CDC): National Electronic Disease Surveillance System (NEDSS) and Public Health Information Network (PHIN). The argument for having an architecture is quite simple: without an architecture, the situation that existed in governmental public health pre-2000—in which each health department used dozens of information systems that could not exchange data—would be impossible to resolve.

Although the importance of architecture for biosurveillance is now widely appreciated, no publications address the topic in a general way. The available information about architecture from NEDSS and PHIN (see *http://www.cdc.gov/phin*) is useful but is too fine grained and dispersed over multiple documents for most people to comprehend. Thus, our goal is to provide, in one chapter, a general discussion of architectures for biosurveillance systems.

To make the discussion as concrete as possible, we discuss architecture from the perspective of an architect—a person given the task of designing a system for a biosurveillance organization such as a state department of health. We will frequently draw on an analogy to the design of buildings. We will discuss how an architect would approach the design of two types of systems—an *enterprise information system,* which is an information system that is intended to support the function of a single organization, such a state department of health, and a *pan-enterprise information system,* which is designed to support the interoperation of many organizations. The reason that we discuss both enterprise and pan-enterprise architectures in this chapter is that the design of these systems influences the cost and effort to build, maintain, expand, and modify biosurveillance systems.

2. DEFINITIONS

The word *architecture* may be a source of confusion rather than clarity in this field because it has multiple senses and people use the word loosely. For example, you may have seen speakers point to a diagram and say, "This is the architecture." You may have encountered Web pages entitled *Architecture* that contain text descriptions of standards and specifications. You may also have encountered the term in various phrases, such as *multitiered architecture* and *enterprise architecture.*

To avoid adding to the confusion, we here define the terms we use in this chapter:

Architectural style is a set of general design principles (rules) for designing an information system.[1] These rules usually serve as constraints on (1) how we design the functions of each component (e.g., functions are stateless, functions operate independently), (2) how we arrange the components (e.g., in layers, in a star pattern, in a chain), and (3) how the components communicate with each other. We will be discussing architectural styles suitable for constructing enterprise information systems (e.g., a health department's information systems) and for constructing pan-enterprise systems (e.g., a worldwide biosurveillance system).

A *blueprint* is a set of specifications in sufficient detail that a diverse group of entities (e.g., network engineers, database administrators, programmers) can work collaboratively to construct a system.[2] A blueprint comprises

1 A commonly cited work about architectural styles is Roy Thomas Fielding's dissertation *Architectural Styles and the Design of Network-based Software Architectures.* In this work, Fielding comprehensively reviews the architectural styles used in the architectures of systems that must operate in a distributed manner over a network.

2 We introduce the term *blueprint* for expository purposes. This term is not used by architects of information systems.

Handbook of Biosurveillance
ISBN 0-12-369378-0

1. A data model that describes the data elements, vocabularies, grammar, and semantics of the system
2. The set of functional components of the system (e.g., database, user interface)
3. A diagram that shows how the components are connected

Architectural style and blueprint are at opposite ends of a design process. An architectural style is a general approach that exists because of previous experience and that seems to have merit for a current project. A blueprint is the result of casting that style into a concrete plan for construction that allows multiple individuals and/or organizations to build the envisioned system.

An *architect* is a designer of a system. Similar to an architect of a building, an architect of an information system brings a great deal of knowledge to a project. She is familiar with designs used in other successful projects, what prefabricated (pre-fab) components are available, costs, and local requirements (e.g., the building code). An architect must combine her knowledge about the architectural style (e.g., the layered-client server [LCS] style or, in the case of buildings, multiunit apartment building), building materials (data elements and vocabularies or, in the case of buildings, concrete and steel), the specific requirements (types of functions or, in the case of building, no more than $105 per square foot), and the local regulations (laws and regulations about confidentiality, or in the case of building, the building codes) to produce as a final product a set of blueprints.

3. ENTERPRISE AND PAN-ENTERPRISE ARCHITECTURES

Just as the architect for a building may be tasked with the design of a single family house or a skyscraper, an information technology (IT) architect can design an information system that serves a single organization or one that serves multiple organizations. We (and others) refer to an architectural style for a system that serves a single organization as an *enterprise architecture.* We use the term *pan-enterprise architecture* to refer to the architectural style for a system that comprises more than one enterprise. We discuss enterprise and pan-enterprise architectures in detail in the next sections.

Pan-enterprise architecture is of particular importance because biosurveillance is a distributed activity involving many organizations. Although these organizations operate computer systems and these systems often contain data that would be useful to other organizations, data exchange occurs primarily via person-to-person communication or, at best, via batch transfers of data. As we will discuss, an appropriate pan-enterprise architecture can accelerate progress toward the goal of computer-to-computer interoperability.

4. AN ENTERPRISE ARCHITECTURE FOR BIOSURVEILLANCE SYSTEMS

This section describes how an architect tasked with the design of an information system for a single organization would approach the design task. Such an organization might be a city health department or a larger organization, such as the CDC. IT professionals understand the principles of enterprise architecture very well. This section is a primer on methods in current use in many biosurveillance projects.

4.1. Architectural Style

The architect would begin by selecting an architectural style, which today would be LCS architecture.[3] LCS is an architectural style that follows the following principles (rules): (1) some components (the servers) provide functions or data at the request of other components (the clients), and (2) components must be grouped into layers, with components of one layer accessing the functions of components in the layer below it (i.e., the upper components are clients of the lower layer components) (Figure 33.1). RODS, Electronic Surveillance System for the Early Notification of Community-based Epidemics (ESSENCE), Biosense, and virtually all modern enterprise systems use LCS architecture. Vendors of commercial off-the-shelf software design their products to fit within a LCS architecture For example, database management systems (DBMSs; e.g., Oracle, Microsoft SQL Server) and geographic information systems (GISs; e.g., ESRI's ArcGIS software) are all designed to function as servers in an LCS architecture.[4]

4.2. Blueprint

To complete the specification of the system, the architect would specify a data model, a set of components and a set of diagrams that specify how the components are connected.

4.2.1. Data Model

A data model for any information system will specify the *data elements, vocabulary, grammar, and semantics* of the system. Data elements are the basic units of information in a system, for example, in a biosurveillance system, the type of data collected (e.g., emergency department registration data, over-the-counter (OTC) medication sales, or microbiology results). Microbiology results will include data elements such as the species and genotype of the organism, antibiotic sensitivities, and specimen source. PHIN specifies the data elements required in a section called *Implementation Guides.* (CDC, 2005b)

3 You will also find LCS architectures referenced as two-tiered, three-tiered, or multi-tiered architectures.

4 Although the current PHIN documentation does not explicitly state that PHIN employs a layered client-server architecture style, previous technical specifications of NEDSS used this term and an information technologist reading the PHIN documentation would recognize PHIN as a layered client server design.

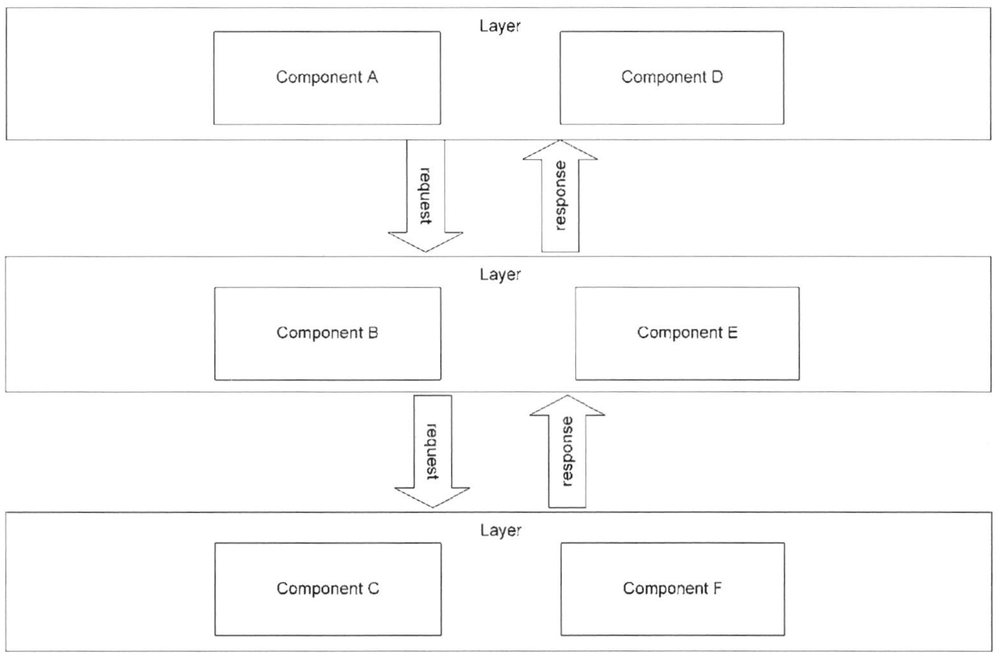

FIGURE 33.1 Layered-client server architecture. In a layered-client server architecture, the architect groups components into layers, and components in one layer make requests to the layer below it but do not jump across layers. Note that a component can be both a client and a server (component B is a client of component C and a server to component A).

The system encodes the data elements by using standard vocabularies, grammar, and/or semantics dictated by the type of data that are collected. We described the common vocabularies, grammar, and semantics appropriate for many data types in Chapter 32 of this book. For example, prescribed medications should use RxNorm as the vocabulary, National Council for Prescription Drug Programs (NCPDP) as the grammar, and some form of the PHIN Logical Data Model (LDM) as the semantic model. PHIN specifies the use of LOINC, SNOMED, International Classification of Diseases Ninth Edition (ICD-9), and Current Procedural Terminology (CPT) for vocabularies; HL7 and X12 for grammar; and HL7 RIM and PHIN LDM for semantics.

4.2.2. Set of Components

The architect must specify the set of components necessary to carry out the biosurveillance processes for the planned system. Almost all enterprise systems will include components that provide network, directory, and security functions. We describe these components first in this section. Most systems will also include many components that support directly the biosurveillance process—data collection, data storage, case detection, outbreak detection, outbreak characterization, user interface, and notification. Figure 33.3 is a diagram that shows how we lay out these components in relation to each other.

Network. The network allows multiple computers and the applications that run on them to communicate with each other. In the past, architects had to choose among a wide range of network technology. Today, the most pervasive network technology is the Internet Protocol (IP), and a biosurveillance system should run on an IP-based network.

Directory Component. The directory component comprises one or more applications that contain information about the resources—users, computers, applications, and servers—available in the system. For example, PHIN recommends that a biosurveillance system store information about the users of the system on a lightweight directory access protocol (LDAP)-compatible application, such as Microsoft's Active Directory, Sun's Directory Server, or the open source OpenLDAP. The LDAP application maintains a list of users along with their contact information, passwords, etc. This information is necessary for the system to ensure secure access to the system and notify personnel of a possible outbreak. PHIN makes recommendations on what data elements should be included in the directory in the PHINDIR specification (CDC, 2005e) and specifies public health exchange directories by using directory service markup language (DSML) and ebXML messaging.

Security Component. The purpose of the security component is to perform authentication and authorization. Authentication refers to the process that determines whether the people accessing the system are who they say they are. Authorization refers to the process that determines which data and functions of the systems a user may access.

The architect might specify that an authentication component use one or more of the following:

1. Something the user knows (e.g., their name and password)
2. Something the user has (e.g., a security token, such as an RSA Secure identification card or a digital certificate)
3. Something the user is (e.g., biometrics such as a fingerprint or iris pattern)

A biosurveillance system, particularly those that contain identifiable data (e.g., persons with HIV) should use at least two of these methods to ensure that data do not get into the wrong hands. Information technologists regard the use of more than one authentication method as *strong authentication*. PHIN recommends the use of strong authentication. (CDC, 2005c)

For each user, a biosurveillance system maintains a list of accessible functions and data types. To perform authorization, a system compares a user's request with this list and allows or rejects a user's request.

Data Collection Component. A biosurveillance system must be able to obtain data from a variety of different systems, share its data with other systems, and make requests for additional data. IT professionals refer to the process of collecting and sharing data across multiple existing systems as *data integration*. Biosurveillance systems such as RODS include a data integration component that collects and shares data from/with multiple outside systems. Commercially available software for data integration includes Sun Microsystems SeeBeyond, Microsoft's Biztalk, IBM's Websphere Business Application, and the open source Business Integration Engine.

The process of data collection can be sorted into four subprocesses—extraction, transport, transformation, and loading (Figure 33.2). Extraction is the process for pulling data out of another computer system, and takes place within that system (usually at a computer located in a facility owned by an organization other than the organization for whom the architect is designing the system). Transport is the transfer of the extracted data to the data recipient. Transformation is the process of manipulating the data (i.e., changes in data format, mapping local terms to standard terms, ignoring certain data elements, and generating new data elements from existing elements) so that they matches the data model of the data recipient's systems. Loading is the process of inserting the data into the storage component. PHIN does not specify how to extract data from a provider system but does specify the use of the PHIN Messaging System standard (a secure method of transferring data from one computer to another) to transport the data and XML for data transformation.

We note that the data collection component is a major component in a biosurveillance system that requires proper configuration and maintenance. We discuss support of a data collection component in the next chapter.

Data Storage Component. The purposes of the data storage component in a biosurveillance system are to provide local storage and efficient retrieval mechanisms for biosurveillance data. An architect must consider two major types of data that need to be stored in a biosurveillance system—transactional data and cached data.

Transactional data are raw data collected from data providers, such as users, hospitals, laboratories, water supply systems, and retailers, as well as the results from analyses of the data. Examples of transactional data include manually entered case information (e.g., name, contact information, partners), an electronic patient registration from an emergency department (e.g., date/time of the visit, age, sex, chief complaint, home zip code and work zip code of the patient), and outbreak alerts (e.g., date/time of the alert, data types analyzed, severity of the alert).

Cached data are either cached query results or Online Analytic Processing (OLAP) data cubes. Cached query results are the results from previous requests for data from a client to

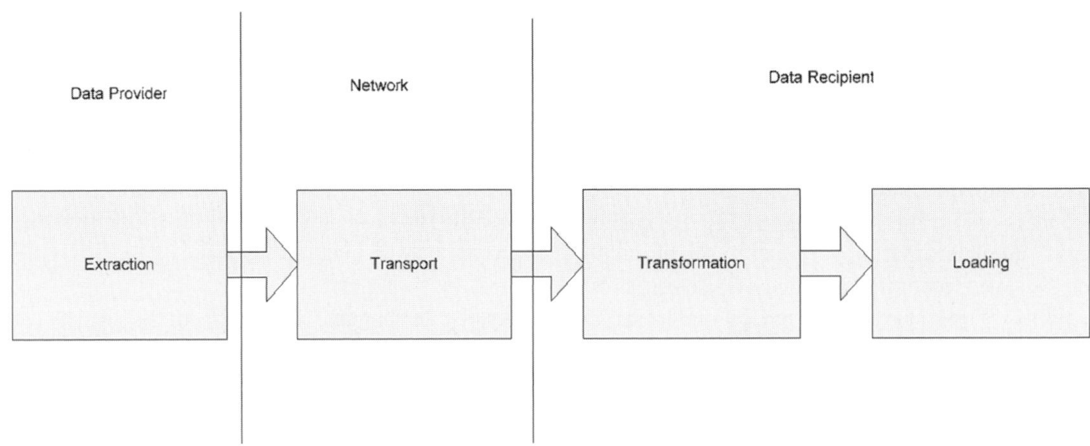

FIGURE 33.2 Data collection comprises four subprocesses—extraction, transport, transformation, and loading. Boundary lines indicate where each of these processes take place.

the storage component. These cached query results improve efficiency when users make multiple requests for the same data because the storage system does not have to search and assemble the results repeatedly. OLAP data cubes are a transformation of the transactional data to support efficient analysis. An example of data found in an OLAP cube is *the number* of patients who live in a particular zip code under the age of 18 with respiratory complaints each day from January 1, 2004, to January 1, 2005. Although a biosurveillance system can obtain such information without an OLAP data cube by retrieving all of the transactional records that meet the criteria and counting the number of records obtained, such a process takes an inordinate amount of time (1000 times slower).

An architect should use a DBMS designed for an enterprise information system, such as Oracle and Microsoft SQL server, for the storage component. The DBMS should support transactional data, cached query results, and OLAP data cubes. We discuss these configuration and maintenance issues of a DBMS in the next chapter.

Detection and Characterization Components. The detection and characterization components are the brains of a biosurveillance system. They use natural language processing tools to extract features from free-text data, diagnostic expert systems to detect possible cases of disease, and spatial and/or temporal analytic methods to detect and characterize possible outbreak of disease. The detection and characterization components store the results of their analyses—cases, outbreaks, and outbreak characteristics—in the storage component. Chapters 3, 13, and 17 provide detailed discussions of case detection, outbreak detection, and the details of natural language processing. The detection and characterization components may use a general statistical analytic tool, such as SAS, or specialized analytic software, such as the algorithms discussed in Part III of this book.

User Interface Component. An architect will include interfaces for the users of a biosurveillance system to facilitate review of the data. The predominant mechanism for interacting with users today is a computer display, keyboard, and mouse. Such a user interface may be Web-based or desktop-based. In a desktop-based system, a program that runs on the user's desktop computer connects to a server that accesses the data or provides the functions of the system. In a Web-based system, a Web browser runs on the user's desktop computer that connects to a Web server that accesses the data or provides the functions of the system.

An architect chooses a Web-based or desktop-based user interface based on several factors that include the needs of the users and the ability of the organization to maintain desktop applications. Desktop-based user interfaces are faster, more interactive, and integrate well with other desktop applications but require installation onto a user's desktop computer. Web-based user interfaces are universally accessible from any Web browser but lack the interactivity and integration that a desktop application provides. PHIN specifies Web-based user interfaces for some parts of the system. For example, the section of PHIN that discusses manual entry of data into an early event detection system says that, "Web browser-based data systems should be developed using commercial application server technology as part of a multitiered web development system using open-platform web servers" (CDC, 2005d).

Epidemiologists commonly view surveillance data as a set of tables (e.g., line listings), graphs (e.g., epidemic curves), or maps. In a biosurveillance system, the user interface component displays data as tables and graphs generated by a graphing tool or statistics package, such as SAS. The user interface component generates maps from a GIS, such as ESRI's ArcGIS software. PHIN recommends that the system display data as line listings, graphs, and maps. (CDC, 2005a)

Notification Component. The last component in a biosurveillance system is the notification component. This component sends alerts to users in the form of an e-mail, automated voice to a phone, page, or fax and tracks the responses to these alerts. The alerts contain information or links to information about possible outbreaks or cases of disease.

PHIN recommends the use of the Common Alerting Protocol (CAP) (Jones and Botterell, 2005) for this purpose. CAP is an open, nonproprietary digital format for emergency alerts and notifications.

4.2.3. Layout of the Components

We show the layout of the components in Figure 33.3.

5. A PAN-ENTERPRISE ARCHITECTURE FOR BIOSURVEILLANCE

This section describes how an architect might approach the design of an information system that interoperates with many organizations that have information systems with little or no native ability to interoperate. When constructing pan-enterprise systems, multiple technical construction teams may be located in different organizations (e.g., hospitals, laboratories, governmental public health, water companies). In both industry and biosurveillance, these teams may be difficult to coordinate unless the architectural style allows them to work relatively independently.

An appropriate architectural style for a system that requires interoperability between organizations but maintains their independence is a service-oriented architecture (SOA). We will discuss SOA, how Amazon.com used SOA to handle more than 10 million computer-to-computer exchanges of data per day with the computer systems of independent organizations in under three years (Roush, 2005), and how SOA could accelerate the development of worldwide biosurveillance systems.

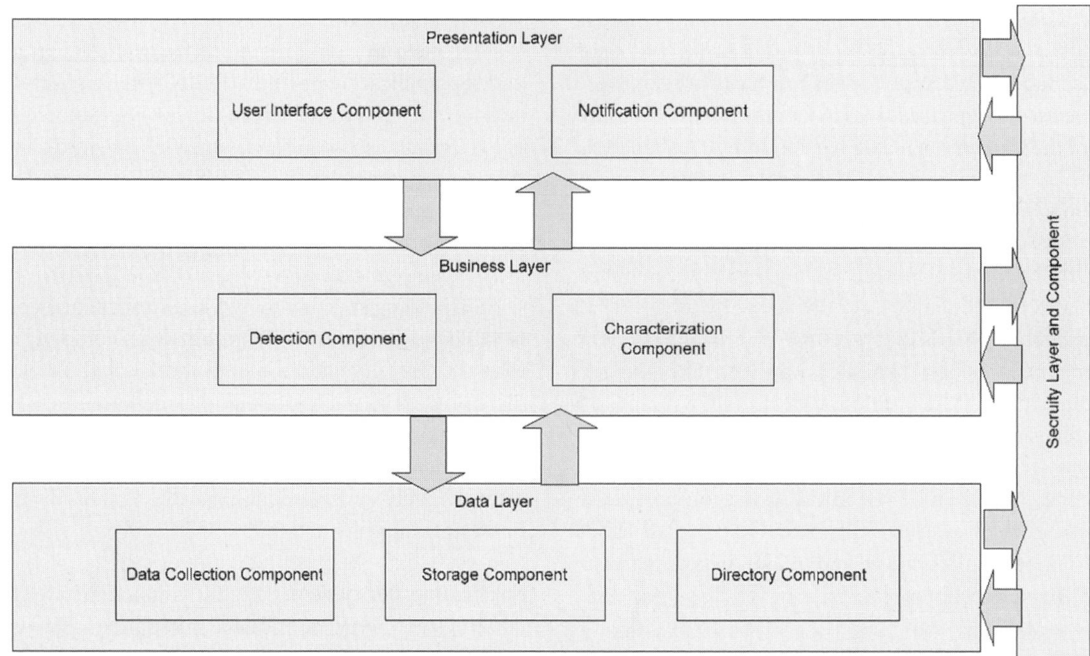

FIGURE 33.3 There are four layers in this enterprise architecture for biosurveillance—a presentation layer, a business layer, a data layer, and a security layer. The presentation layer accesses the business layer. The business layer accesses the data layer. All layers access the security layer. The presentation layer comprises the user interface component and notification components. The business layer comprises the detection and characterization components. The data layer comprises the data collection, storage, and directory components.

5.1. What is SOA?

The architectural principles of SOA start with those of client-server architecture (i.e., client components make requests to server components that return appropriate responses) but add additional principles that facilitate independent construction of the system's components. It is these rules that make SOA ideal for a pan-enterprise architecture for biosurveillance.

Services are the equivalent of components in other architectures (e.g., a database is a common component in a client-server layered architecture). When designing a pan-enterprise system in the SOA style, the building blocks that an architect thinks about are the services that systems owned by different organizations will provide to each other. According to Erl, (2005) services in an SOA adhere to the following principles:

- *Loose coupling*—Services maintain a relationship that minimizes dependencies and only requires that they retain an awareness of each other.
- *Service contract*—Services adhere to a communications agreement, as defined collectively by one or more service descriptions and related documents.
- *Autonomy*—Services have control over the logic they encapsulate.
- *Abstraction*—Beyond what is described in the service contract, services hide logic from the outside world.
- *Reusability*—Logic is divided into services with the intention of promoting reuse.
- *Composability*—Collections of services can be coordinated and assembled to form composite services.
- *Statelessness*—Services minimize retaining information specific to an activity.
- *Discoverability*—Services are designed to be outwardly descriptive so that they can be found and assessed via available delivery mechanisms.

The most important principles for interoperability between organizations are the principles of loose coupling, service contracts, and discoverability. Loose coupling allows organizations to build services independently. Service contracts are a set of documents that completely describe how to access the service. Service providers usually write service contracts by using a language that a computer can understand, such as XML (e.g., the Web Service Description Language [WSDL]). Discoverability allows organizations that need to interoperate to find required services easily via a registry (e.g., white pages for a computer).

5.2. The Example of Amazon.com

Amazon.com is an instructive case study of SOA, especially for biosurveillance, because it is a company that depends on real-time interoperation with the information systems of other organizations (enterprises).

In 2002, Amazon.com had a problem. Their *sales associates* (organizations that sell Amazon.com products through their own Web sites) had a need for real-time, up-to-date product information in order to market and sell Amazon.com products. According to Amazon.com's director of web services, Jeff Barr:

> For a long, long time people have been asking for a more structured way to access our data and, in fact, for many years people have been writing robots that would go to our site and download various kinds of information. And we were generally actually pretty friendly to robots, allowing them to proceed whereas a lot of other sites would just kind of block them off. But we talked to these folks, kind of, trying to understand what they wanted, and we tried to give them something a bit better. (Kaye, 2003)

Before coming up with a solution to the preceding problem, Amazon.com took the time to understand the needs of their sales associates and their capabilities. The sales associates needed a way for their information systems to query the Amazon.com database for specific product information (Figure 33.4). The solution, known as Amazon.com's E-commerce service (ECS), was the first service in Amazon's SOA. As we will see, the ECS had benefits that have gone beyond the initial needs of their sales associates.

A key benefit to SOA is the speed at which a large number of organizations can begin to interoperate. In the three years since Amazon.com launched its ECS, more than 140,000 organizations (or individuals) have signed up for the service, and experts estimate that these organizations make more than 10 million requests for data each day (Roush, 2005). In contrast, in the four years since the anthrax postal attacks in fall 2001, only a small fraction of the 3,000 health departments, 160,000 laboratories, 7,000 hospitals, and myriad other organizations with roles in biosurveillance in the United States have begun to interoperate at the computer-to-computer level in an effective manner.

An unexpected result (and benefit to Amazon.com) was novel applications that developers created by using the ECS. Examples of these applications include a program that lets students order textbooks from Amazon.com directly from an academic course management Web site *(http://www.concord-usa.com/amzcourseiteminfo.htm)*, comparison engines that let sellers compare their prices with the prices of other sellers *(http://www.monsoonworks.com/mw.main/index.asp)*, and a

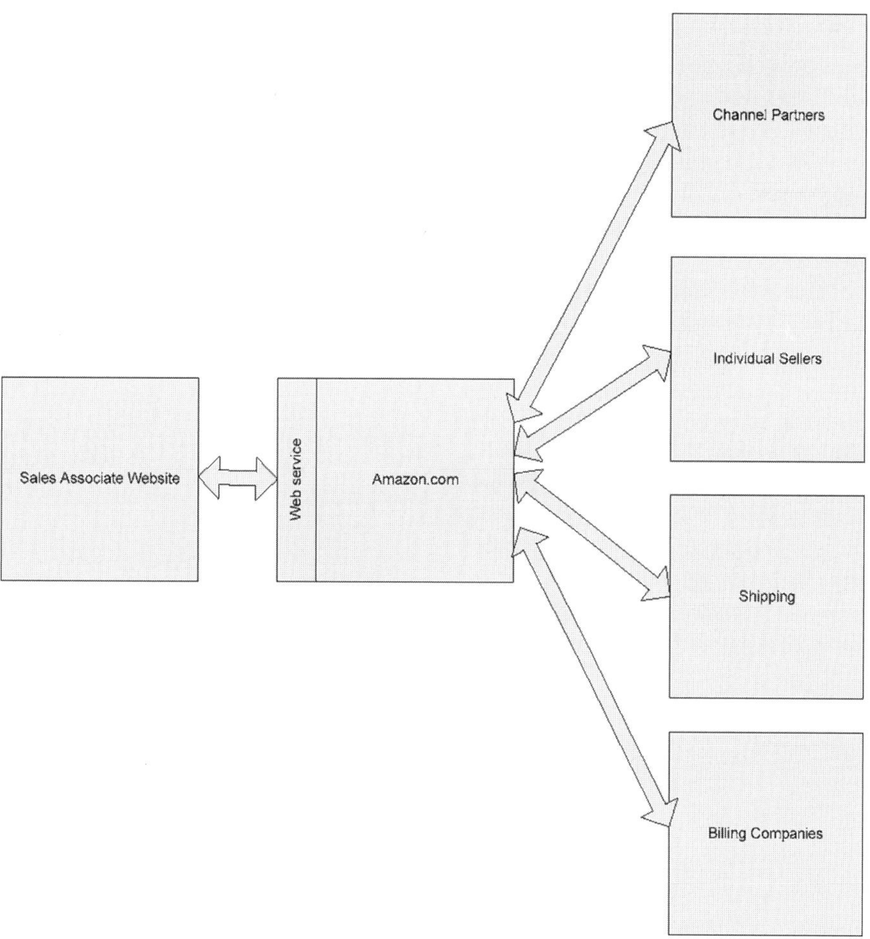

FIGURE 33.4 By use of the Amazon.com Web services, the Web sites of sales associates have access to product information, an electronic shopping cart, inventory information, and billing services.

program that lets users browse through books, music, and movies as an interactive "pile of books" *(http://amaztype.tha.jp/)*.

ECS was the beginning of Amazon.com's investment in SOA. They view it as a fundamental aspect of the company's IT strategy. Since the launch of the service in 2002, Amazon.com has continued to develop their Web-service capability, producing four revisions of the ECS and adding additional complementary services, including the Amazon.com Historical Pricing Service (three years of historical pricing information for all of Amazon.com's products), Alexa Web Information Service (a Web search engine), and the Amazon.com Simple Queue Service (an electronic version of a message box that enables separate computer applications to interact asynchronously). They also plan to utilize Web services to interact with their other partners, such as the retailer Target (Puget Sound Business Journal, 2003).

Amazon.com is not alone in focusing on SOA. A 2004 survey of 473 enterprise buyers by the Yankee Group of Boston revealed that in the next 12 months, 75% plan on investing in the technology and staffing necessary to create a SOA. Seventy-one percent of health-related companies intend on using SOAs (The Yankee Group, 2005).

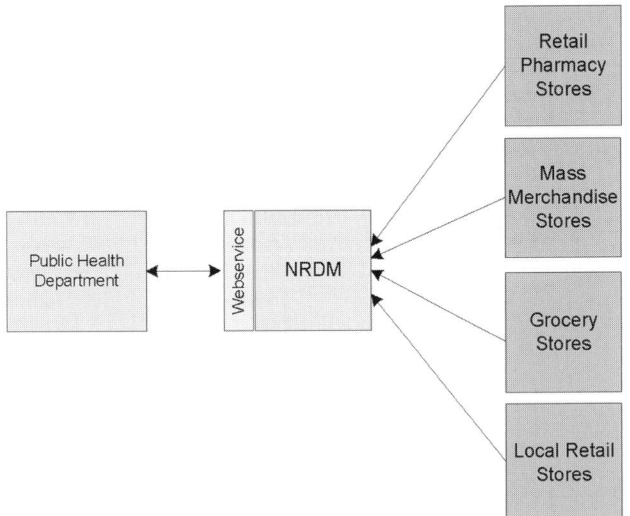

FIGURE 33.5 National Retail Data Monitor (NRDM) Web Service. Similar to the Amazon.com Web services, the NRDM Web Service provides access to the sales information of approximately 20,000 retail locations. This example is an extremely simple example of the use of Web services by two organizations that wish to exchange data or services.

5.3. Applying SOA to Biosurveillance

We now show how an architect could apply the principles of SOA to facilitate pan-enterprise biosurveillance. First, we describe an extremely simple, but real and functioning biosurveillance service. Then, we build on this example to discuss the design of a more comprehensive system.

5.3.1. The National Retail Data Monitor Web Service

The Web service offered by the National Retail Data Monitor (NRDM) is an example of a service in a biosurveillance SOA. The NRDM (discussed in Chapter 32) is a system that collects sales data for selected OTC healthcare products.

Before the NRDM created a Web service, it transmitted its data to public health departments either through its own user interface or by daily file transfer. The latter approach met the needs of health departments that wished to incorporate the data into their own biosurveillance systems. But, the file transfer approach had several limitations for both the NRDM and the health departments, which led the NRDM to develop a Web service (Figure 33.5).[5]

In the Web-services method of data distribution, computers in health departments query the NRDM for data whenever they need the data (Figure 33.6). For example, a computer in a health department might send a request to the NRDM Web

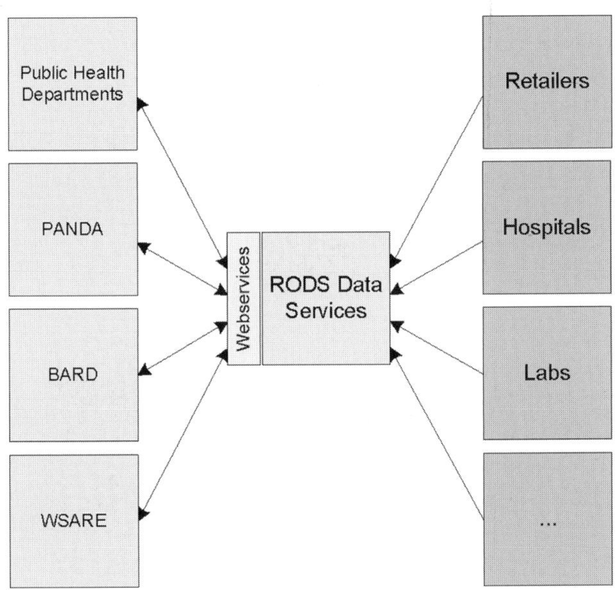

FIGURE 33.6 Future RODS Data Services. In the future, RODS laboratory will provide biosurveillance data via Web services to health departments and other biosurveillance applications. Presently, these Web services include the National Retail Data Monitor (NRDM) Web service, which distributes over-the-counter medication sales data. In the future, the RODS Laboratory will have chief complaint Web service for distributing chief complaint data.

5 These limitations included: (1) a health department had to store a local copy of the NRDM data, which are voluminous, (2) the health department needed to create processes that parse the data files and update the data. (3) even though the NRDM continually processes OTC sales data received from retailers and makes it available via a Web-based user interface, the local public health department receives updates on OTC sales only once a day.

service to "send me the daily sales of diarrhea remedies for zip code 70808 between August 1 and September 1 and the number of stores in that zip code that are reporting these sales."

The work required by the RODS Laboratory to create the NRDM Web Service was not extensive. Because the NRDM already made OTC data available via a user interface, it already had the needed query capability to the NRDM data storage component. The RODS Laboratory IT staff exposed this query capability as a Web service that adhered to SOA principles.

The NRDM Web Service is just one of potentially thousands or hundreds of thousands of services that could exist in a biosurveillance SOA. Biosurveillance will realize the full potential of SOA when more organizations offer services to each other.

Figure 33.7 illustrates the set of organizations that might interoperate during an aerosol release of anthrax in a large city such as Dallas or Fort Worth. In this figure we consider the needs of just one of the "business" partners—the health department—and we assume that the health department is responsible for coordinating the overall health response to the event, including assessment of risk, promulgating guidelines for prophylactic treatment, and declaring areas of the city unsafe for inhabitation. As in the 2001 postal attacks, such a scenario would generate an enormous volume of laboratory testing. Investigators would presumably distribute the specimens and requests for tests to the local hospital laboratories and the laboratory networks discussed in Chapter 8 (and depicted in Figure 33.7 either individually or via their participation in laboratory networks).

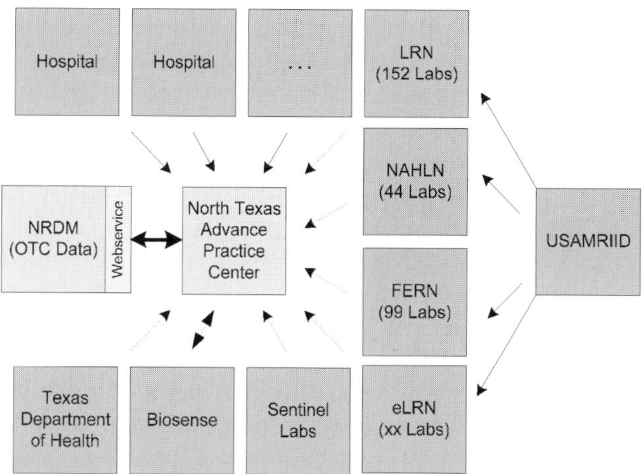

FIGURE 33.7 Pan-enterprise architecture from the perspective of a health department. Dotted arrows denote nonautomated access to data. Solid unidirectional arrows denote automated one-way reporting that does not support data requests. The bidirectional arrow between the National Retail Data Monitor (NRDM) and North Texas Advanced Practice Center denote support for data requests. LRN indicates Laboratory Response Network; NAHLN, National Animal Health Laboratory Network; FERN, Food Emergency Response Network, eLRN, Environmental Laboratory Response Network; and USAMRIID, United States Army Medical Research Institute for Infectious Diseases.

We assume that the health department will require access to individual results (e.g., the result of testing a clinical sample taken from patient P or the environmental samples taken from building X 2 days ago) and that these samples may have been processed at any of the organizations depicted in the figure. If each of these laboratories or networks shared data via a service, the investigation process would be more efficient.

At present, the level of direct computer-to-computer exchange of data between the health department and these myriad organizations is low, as depicted by dotted lines between entities. Where exchange of data does exist, it is often a one-way flow of information (e.g., electronic laboratory reporting) that does not support data queries. Although it is true that some of the laboratory networks provide a Web site that public health entities may access, they involve different passwords and data formats, and the end-user must have considerable knowledge of where a specimen was processed or likely to have been processed in order to find it. You may recall from earlier chapters that an essential part of the biosurveillance process is the collection of additional data. Services in an SOA that distribute biosurveillance data through a query mechanism (e.g., the NRDM Web Service) support the collection of additional data.

SOA has considerable potential to accelerate progress toward the goal of interoperability among organizations involved in biosurveillance because it allows each organization to articulate its needs in terms of data and services and negotiate with other organizations to provide those services. These negotiations—when framed in terms of data and services—are eminently understandable to the chief executive officers, who must decide matters of cost and prioritization of effort. Once the parties agree on the terms of a service, IT staff can take these service terms and work independently to externalize the organization's existing capabilities by using available tools and standard technologies for building a service.

5.3.2. *Adding Web Services to an Existing Information System*

In this section, we show how to add a Web services component to an organization's existing information systems. Adding services to an existing architecture is not as difficult as previous technology upgrades, such as the movement from mainframes to personal computers. The Yankee Group acknowledges that "embracing SOA is cheaper than other types of IT overhauls because the technology is based on the Web. That's interesting to a lot of large enterprises, especially the ones involved with numerous partners and acquisitions" (Stansberry, 2005).

Services externalize the existing capabilities and resources of an organization. For example, a typical health department collects surveillance data, such as notifiable disease reports and possibly patient chief complaints, from emergency departments for its region. The health department can externalize part of these data for the benefit of health departments in adjacent jurisdictions (or anywhere in the world) by adding a service as

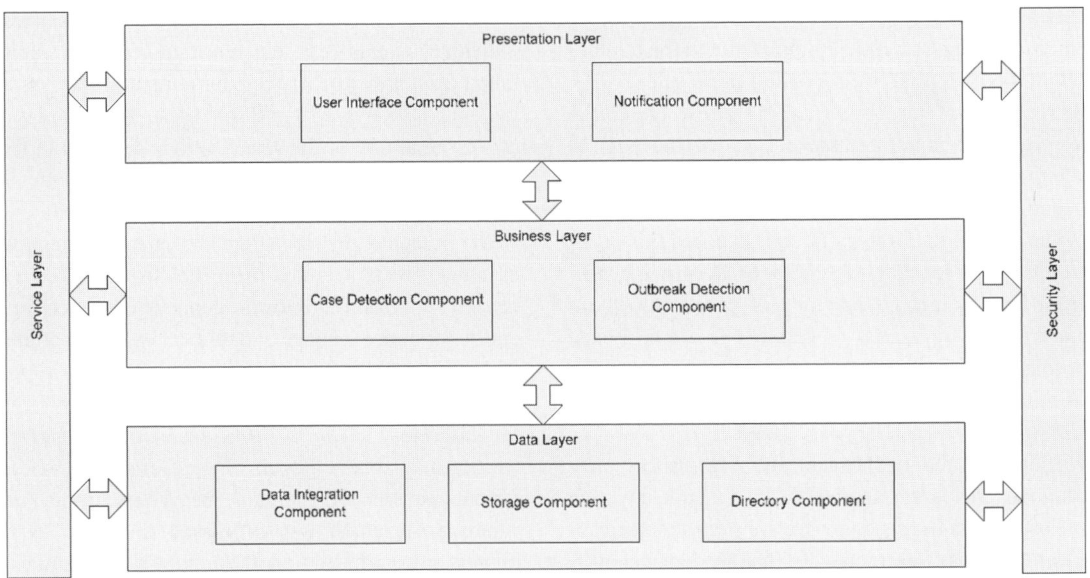

FIGURE 33.8 A service layer added to the previously discussed enterprise architecture for biosurveillance (Figure 33.3).

shown in Figure 33.8. These services also function as a bridge between biosurveillance systems and frontline personnel (doctors, nurses, veterinarians, water supply managers, the Federal Bureau of Investigation, military field commanders). These people work everyday with possible sources of disease (humans, animals, water, food, and bioweapons). We know that an astute clinician is the frontline defense for detecting disease outbreaks. An astute clinician who has up-to-the-minute additional information about the health status of those in their community and neighboring communities is likely to be "more astute." For example, doctors who have knowledge of bacterial meningitis cases in a neighboring county will be on the lookout for patients with meningitis symptoms in their own county. SOA can augment PHIN so that public health information is available to not only public health personnel but also frontline personnel. The potential for reciprocity exists as outside groups make their data and functionality available as services for use by public health.

SOA not only facilitates data sharing, it is also facilitates sharing of functions (e.g., analysis of data). Government, health departments, and biosurveillance research groups all participate in biosurveillance. Undoubtedly, some organizations will have better capabilities than do others (e.g., in the physics community, some groups have the capability to carry out more sophisticated particle collision experiments, and some groups have the capability to carry out analysis of the data from these experiments; but as a community, physicists share their resources to advance science). Services will allow the more widespread use of the capabilities of these groups, advancing the mission of biosurveillance.

6. SUMMARY

The architecture of a biosurveillance system determines the cost and effort to construct it as well as the cost and effort to expand and evolve it over time to meet the changing requirements of biosurveillance and take advantage of new technology. The commercial world has evolved and tried many approaches to architecture of both enterprise systems and pan-enterprise systems.

The LCS architecture is a well-established architectural style for enterprise systems. Its essential properties are the use of components that we organize into layers. The layering serves to separate components so that each component interacts with only a few components, thus minimizing the effort required to replace that component with an improved component in the future. Some of the components perform generic functions (e.g., databases) and are available as commercial off-the-shelf products (or open source products) that an architect can use to drastically reduce construction costs. Many PHIN specifications reflect the influence of the LCS architectural style.

The SOA is emerging as the preferred architectural style for constructing pan-enterprise systems. Its essential property is the *service*, which is a structured way that one organization can offer (externalize) some of its data, computing services, or more complex business process (e.g., delivery of a book or testing of a patient). As suggested by the example of Amazon.com, its main advantage is the speed at which independent organizations can begin to interoperate. The key underlying reason is that organizations can work relatively independently to externalize their capabilities, based on an understanding of the needs of their business partners. This architectural style has significant potential to accelerate the movement toward interoperability among the myriad organizations that participate in biosurveillance. The PHIN specifications at present stop short of endorsing SOA, although IT professionals see messaging methods (e.g., ebXML and PHIN MS) as useful adjuncts to Web service implementations of SOAs.

ADDITIONAL RESOURCES

Public Health Information Network Web site *(http://www.cdc.gov/phin).*

Fielding, R.T. (2000). Architectural Styles and the Design of Network-based Software Architectures [PhD dissertation]. Irvine, CA: University of California, Irvine. An overview of architectural styles can be found here.

Erl, T. (2005). *Service-Oriented Architecture: Concepts, Technology, and Design.* Indianapolis, IN: Prentice Hall Professional Technical Reference. A complete description of service-oriented architecture can be found here.

ACKNOWLEDGMENTS

Jeremy Espino is president of General Biodefense, LLC, a software and IT consulting group specializing in biosurveillance.

REFERENCES

Centers for Disease Control and Prevention [CDC]. (2005a). PHIN: Analysis and Visualization. *http://www.cdc.gov/phin/architecture/standards/Analysis_Visualization.html.*

CDC. (2005b). PHIN: Implementation Guides. *http://www.cdc.gov/phin/architecture/implementation_guides/.*

CDC. (2005c). PHIN: IT Security and Critical Infrastructure Protection. *http://www.cdc.gov/phin/architecture/standards/IT_Security.html.*

CDC. (2005d). PHIN: Manual Data Entry for Event Detection. *http://www.cdc.gov/phin/architecture/standards/Manual_Data_Entry.html.*

CDC. (2005e). Public Health Directory Schema. *http://www.cdc.gov/phin/architecture/schema.doc.*

Erl, T. (2005). *Service-oriented architecture: Concepts, Technology, and Design.* Indianapolis, IN: Prentice Hall Professional Technical Reference.

Jones, E. and Botterell, A. (2005). Common alerting protocol, 1.1. Billerica: MA: OASIS. *http://www.oasis-open.org/committees/download.php/14759/emergency-CAPv1.1.pdf.*

Kaye, D. (2003). Jeff Barr: Amazon Web Services. Kentfield: CA: IT Conversations. *http://www.itconversations.com/shows/detail31.html.*

Puget Sound Business Journal. (2003). Amazon.com will continue Web services for Target. *Puget Sound Business Journal* August 8, 2003. *http://seattle.bizjournals.com/seattle/stories/2003/08/11/daily21.html.*

Roush, W. (2005). Amazon: Giving Away the Store. Cambridge, MA: Technology Review.com. *http://www.technologyreview.com/articles/05/01/issue/roush0105.4.asp.*

Stansberry, M. (2005). Yankee Group: SOA Everywhere by 2006. Needham, MA: SearchDatacenter.com.

The Yankee Group. (2005). US Enterprise Service Oriented Architecture Survey. Boston, MA: The Yankee Group. *http://www.yankeegroup.com/public/products/survey/brochures/2005USServiceOrientedArchitectureSurvey.pdf.*

34

CHAPTER

Advancing Organizational Integration: Negotiation, Data Use Agreements, Law and Ethics

M. Cleat Szczepaniak, Michael M. Wagner, Judith Hutman and Sherry Daswani
RODS Laboratory, Center for Biomedical Informatics, University of Pittsburgh, Pittsburgh, Pennsylvania

Kenneth W. Goodman
University of Miami, Miami, Florida

1. INTRODUCTION

Many organizations are involved in biosurveillance; they are encouraged to cooperate by laws, regulations, money, altruism, and mutual benefit (or self-interest). For example, state, local, and federal health departments have a long tradition of voluntary and regulated interoperation. The healthcare system interoperates with governmental public health for reasons that include regulatory compliance and altruism.

As in most spheres of human activity, laws and regulations tend to lag behind real-world needs. This situation is especially true of late in biosurveillance because of the rapid changes in the types of data that are now perceived as necessary for early detection of outbreaks. Little precedent, less regulation, and yet less law, are available to guide organizations that are attempting to exchange newer types of biosurveillance data. Until regulatory and legal frameworks catch up with emerging requirements, the ability of biosurveillance organizations to implement desired new practices will depend largely on their ability to negotiate agreements with other organizations

In this chapter, we first concentrate on negotiating data use agreements (DUAs). Documents of this type can be called by other names than "data use agreements," but regardless of the name, they are documents that legally govern use, care, and confidentiality of data. DUA negotiations are set in the general context of existing laws and regulations (which may not explicitly require the desired data or service); thus, negotiators oftentimes find themselves considering fundamental ethical principles related to privacy and confidentiality in reaching agreements. For this reason, we discuss ethical principles that constrain negotiations. We provide a brief discussion at the end of this chapter on *service agreements,* which are expanded DUAs that may specify reliability of provisioning of data and services such as testing of patients.

2. AGREEMENTS TO PROVIDE OR EXCHANGE DATA

Biosurveillance depends on data about the health of people and animals, as well as the microbiological status of the environment. It also depends on data related to the production of food, drugs, drinkable water, and the movement of people.

Many organizations hold the requisite data. The number of organizations in any region that hold these data is large, and existing legal or customary practices only govern a fraction of the data and organizations. Any other exchanges of data are subject to mutual voluntarily agreement of both parties, as well as prevailing community standards of privacy and confidentiality, general law and regulations. As discussed in Chapter 5, in the United States each state has its own provisions about disease reporting, and health departments must therefore develop their surveillance systems with regard to the data they can collect in accordance with their particular state's provisions (Broome et al., 2003).

For clarity, we will use the term *data requester* to refer to an organization that is asking another organization to provide data. We will similarly use the term *data provider* to refer to the organization that is being asked to provide data. Our discussion will be from the perspective of the data requester. We recognize that the goal of a negotiation may be a mutual exchange of data and the principles that we discuss apply to that situation as well.

This section discusses general principles, specific strategies, and documents that we have found useful when negotiating agreements for data exchange. In our discussion, we assume that a decision is already made to pursue a specific type of data (for the types of data that might be useful in biosurveillance and what organizations have the data, see Parts II and IV). We address the issues that come up in negotiations and how they can be addressed. We emphasize the need for flexibility in handling the issues, which means that the data requester must be willing to adjust to and accommodate the needs of a data provider.

2.1. General Principles

A DUA is the end result of a negotiation that begins with an initial approach by one organization to another. Some general

Handbook of Biosurveillance
ISBN 0-12-369378-0

principles apply to the entire process of negotiation, which begins with an initial contact and ideally results in an exchange of data.

2.1.1. *Enlist Relevant Governmental Organizations*

The support of governmental organizations that may have legal authority to acquire, hold, or receive the data in question is essential in most negotiations. Not only will these organizations utilize the data, but also most potential data providers, such as hospitals, recognize the authority of these organizations and likely have existing relationships with them.

Departments of health are quite active in both supporting and initiating data exchanges. Many employ a "biosurveillance coordinator" or "director of biosurveillance" who is charged to develop the biosurveillance capabilities of the department. These individuals understand the need for data, and their involvement in data use negotiations is highly desirable.

Governmental and other organizations may have been approached, are working with, or may approach the same data provider during your negotiation. Knowing both the history of previous and concurrent approaches is important. A data provider will be confused if your request overlaps or is inconsistent with other approaches. If another organization is unaware of your request and approaches the data provider during your negotiation, it may undermine or slow your discussion. Multiple uncoordinated requests are a recipe for failure.

2.1.2. *Consider the Needs of Both Organizations*

Although an agreement must meet the needs of both parties, it is especially important to consider the needs of the data provider. Flexibility on the part of the data requester is a key to eliciting participation. A data requester must understand that it is generally not the primary mission of many data providers to supply data to biosurveillance organizations. A hospital's mission, for example, is to treat its patients; a laboratory's mission is to process and report the results of tests that will facilitate the proper treatment of patients; and the mission of a commercial company is to achieve profitability for the company and its shareholders.

2.1.3. *Handle Problems Expeditiously*

Minor problems and questions frequently arise during a negotiation. There may be questions and disagreements over the language in a DUA, unanticipated problems in designing the technical mechanism for data transfer, or costs involved in setup. Each step in the process must be closely monitored on a daily or weekly basis to ensure success and all problems are satisfactorily and quickly resolved.

2.1.4. *Recognize the Importance of Education*

Establishing data exchange for biosurveillance is a rather complex process, involving technical, business, ethical, and legal dimensions. The parties involved in a negotiation must have a working understanding of relevant technology, laws, regulations, and work flows in both organizations. A negotiation typically involves individuals with different backgrounds, which may include medicine, technology, law, and business, and typically some individuals who do not understand some of the relevant dimensions. This lack of shared understanding of the problem will be a barrier to progress unless one can educate all individuals involved in the negotiation.

Early in the negotiation, the data requester should provide two fact sheets. One fact sheet should provide basic information about the biosurveillance system and convey its purpose (Figure 34.1). Another should "make the business case," which means to convey why the organization should provide the requested data and describe the benefits to the data provider for participating in the project (Figure 34.2). The business case should be presented from the perspective of the potential data provider and make selling points that will resonate with the organization. These fact sheets should address the information needs of high-level decisions makers. It is likely that the organization will pass the documents up and down a chain of command.

The data requester should also offer a document describing the information technology (IT) specifications for the system (e.g., Appendix F). This document must be detailed enough to allow the IT staff (oftentimes the chief information officer) of the data provider to estimate the time, effort, and cost required to make any IT modifications necessary to transfer data. The IT-specifications document must clearly describe what is required of the data provider, provide specific step-by-step instructions, and use illustrations and graphics wherever possible. Making the process as clear, simple, and efficient as possible demonstrates professionalism of the organization requesting the data.

Concerns about data security arise frequently. The data requester should have available a document devoted exclusively to the topic of data security, including how the data are protected, if they are encrypted, who will use the data, and how the organization guarantees secure handling and use of the data (Figure 34.3).

2.1.5. *Develop and State the Scientific Case*

The decision makers at potential data providers organizations are highly educated and will weigh the costs and benefits of participating in any project. In many cases (especially with hospital and pharmacies), the decision makers have extensive domain knowledge and general familiarity with the goals of public health surveillance or biosurveillance. They often want scientific evidence that the data you request is of value. One or more scientific publications showing that the data are capable of contributing to biosurveillance is perhaps the most efficient way to satisfy the decision makers. We often provide decisions makers with a research publication on syndrome

NRDM

National Retail Data Monitor

a public health surveillance tool

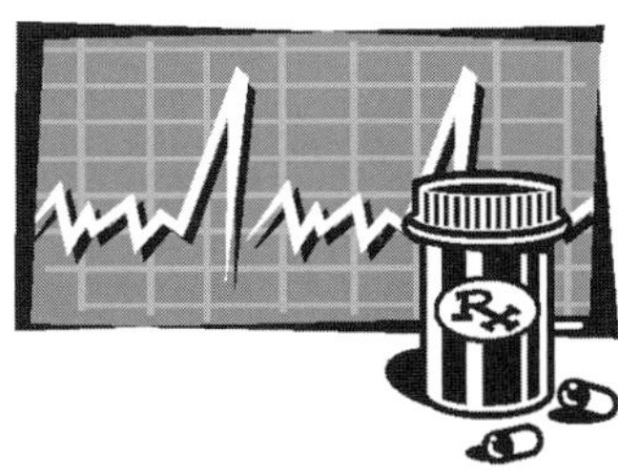

The National Retailer Data Monitor

The National Retail Data Monitor (NRDM) is a public health surveillance tool that collects and analyzes daily over-the-counter point of sale data to rapidly identify disease outbreaks, either natural or as a result of bioterrorism. It was built by the Real-time Outbreak and Disease Surveillance (RODS) Laboratory at the University of Pittsburgh in collaboration with the food and drug retail industry and state and local health departments. Began in 2002 with seed funding by the Commonwealth of Pennsylvania, the NRDM has grown into a national priority. Funding also is provided by the Department of Homeland Security, Alfred P. Sloan Foundation, Passaic Water Commission of New Jersey, and a consortium of states.

As of August 2005, there are 13 major food, drug and mass merchandise chains representing over 20,000 stores that are participating across the nation and over 675 system users in health departments across 49 states, the District of Columbia, and Puerto Rico who have accounts to review daily aggregate data in maps and graphs. Access to the system is free to public health officials through a secure web-based user interface.

The immediate project goals are:

- to increase the market share coverage to 70% nationally
- see that data are reviewed by public health on a daily basis
- report case studies of outbreaks to public health
- create a sustainable organizational model

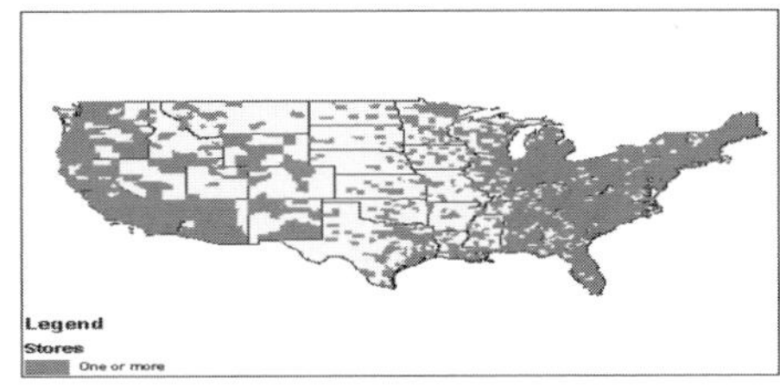

Current Coverage by County

The role of the retail industry

Industry participation involves contributing point-of-sale data to the NRDM for public health surveillance. The system looks at sales of over-the-counter (OTC) healthcare products and analyzes them for anomalies indicative of a disease outbreak whether naturally occurring or as a result of a bioterrorist attack. Many of the nation's top retailer's are providing POS data, which are transmitted via a secure FTP link with a delay of less than a few hours from point-of-sale. Once received, the data are merged to show total sales for a given geography and to de-identify the specific retailer. The output is available to public health officials via a secure extranet which displays time series charts and maps for public health surveillance.

A national effort

Cities, counties and states across the country are working with urgency to deploy methods to identify a disease outbreak that would result from a bioterrorism attack. The National Retail Data Monitor is the first attempt to centralize collection and distribution across the nation and thus eliminate the need for retailers to work with multiple public health departments across the country.

The significance of retail OTC data

Sales of OTC healthcare products are of great interest to early-warning bioterrorism surveillance. When people get sick, they often purchase OTC products prior to visiting a health care provider. By analyzing many years of market data across many cities and states, scientists have been able to correlate certain disease outbreaks with significant OTC sales increases. In some cases it is possible to identify an outbreak up to two weeks earlier with OTC sales than by monitoring clinical data.

OTC Sales peak Jan 22

OTC Medication Sales vs. Hospital Admissions

Hospital admissions peak Feb 4

5, 4, 3, 2, 1, 0, -1

12/24/00, 12/31/00, 1/7/01, 1/14/01, 1/21/01, 1/28/01, 2/4/01, 2/11/01, 2/18/01, 2/25/01, 3/4/01, 3/11/01, 3/18/01, 3/25/01

Date

OTC Sales
Hosp Adm

Privacy

The NRDM does not collect any personally identifiable information. The system simply looks for abnormal spikes in unit sales and is interested in how many aggregated products sell across a zip code or larger geography, not what individual people buy or what an individual retailer sells. These data are similar to what retailers routinely evaluate and share with trading partners and syndicated analysts such as IRI and ACNielsen. No retailer names are identified to public health officials reading the screens, and there is no breakdown of sales numbers by retailer.

Real-time Outbreak and Disease Surveillance (RODS) Laboratory • University of Pittsburgh
Suite 550 • 100 Technology Drive • Pittsburgh PA 15219 • USA
Tel: 412-383-8126 • Fax: 412-383-8135 • http://rods.health.pitt.edu/

FIGURE 34.1 National Retail Data Monitor fact sheet (August 2005).

NRDM

National Retail Data Monitor

a public health surveillance tool

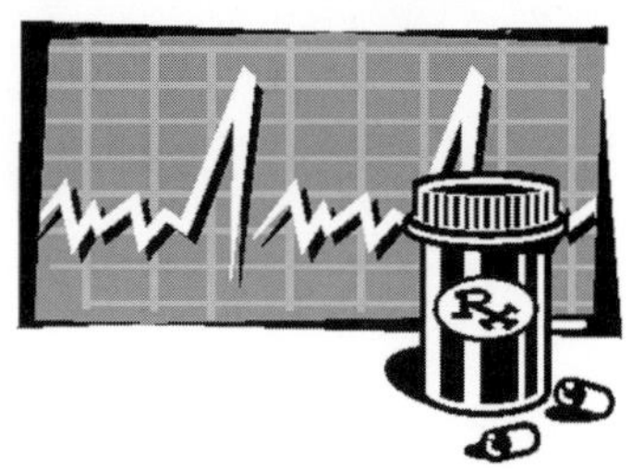

The Business Case

For Participation in the

National Retail Data Monitor

The National Retail Data Monitor (NRDM) is a public health surveillance tool that collects daily sales data for selected over-the-counter (OTC) products from retailers nationwide for the purpose of detecting outbreaks of disease, either naturally occurring or as a result of bioterrorism. The potential benefits to a retail corporation that participates in the NRDM are listed here.

- *Mitigate the impact of a bioterrorism attack.* The worldwide economic impact of a successful bioterrorism attack would be enormous and would likely affect your business by causing economic recession in the same way that SARS produced recession in Asia. That impact can only be mitigated through earlier detection and thereby earlier control.

- *Protect your employees and communities.* The direct impact of an attack that actually involved stores and employees of the firm could be even greater through lost productivity, insurance claims, and the cost of medical care.

- *Address multiple requests from health departments for data.* Many health departments have been approaching retailers directly for OTC sales data. These requests are received by representatives of your firm at the national, regional, or local level. These requests can be handled by the NRDM rather than by your firm's personnel.

- *Enhance your prestiage with local communities, physicians, public health.* Historically, the retail industry has risen to meet challenges during wartime and other national emergencies and these civic contributions pay indirect benefits in the form of prestige and national recognition.

- *Minimize cost to comply with reporting legislation.* As a result of the threat of bioterrorism, new legislation is being proposed (and has passed in Michigan, Wisconsin and Utah) that mandates that pharmacies report unusual increases in prescription and OTC sales volume. For OTC sales, the NRDM is a powerful reporting tool that is already in place that can satisfy these new requirements. The NRDM is happy to assist retailers in satisfying requirements relating to prescriptions as well.

Working Group

Pennsylvania
Joel A. Hersh, M.Ed. M.P.A.
Ohio
Robert J. Campbell, PhD
Georgia
Michael A. Coletta, MPH
New York
Carolyn J. Stetson, PhD
New Hampshire
Stephanie Miller, MPH

Founding Sponsor

Commonwealth of Pennsylvania

Supporters

Florida
Idaho
Indiana
Massachusetts
Michigan
Montana
New Hampshire
New Jersey
New York State
Ohio
Pennsylvania
Tennessee
Texas
Utah
Washington State
West Virginia
Wisconsin
Los Angeles County, CA
Louisville, KY
New York City, NY
San Diego, CA
Alfred P. Sloan Foundation

Real-time Outbreak and Disease Surveillance (RODS) Laboratory • University of Pittsburgh
Suite 550 • 100 Technology Drive • Pittsburgh PA 15219 • USA
Tel: 00+1-412-647-7113 • Fax: 00+1-412-383-8135 • http://www.health.pitt.edu/rods

FIGURE 34.2 National Retail Data Monitor (NRDM) business case.

RODS Data Security and Confidentiality

I. Data Security

I.1 Data Transmission

RODS provides a secure file server (ftp.rods.pitt.edu) that allows data providers to upload their batch data through secure file transfer protocol (SFTP) or secure socket layer (SSL) connections. The file server is located in a demilitarized zone (DMZ) and is protected by a firewall that consists of rules -- allowing only specific remote hosts (IPs) to reach it -- maintained by the network security group at the University of Pittsburgh. The firewall plays the role of a security guard that blocks all external unauthorized connections.

The firewall rules protecting the secure file server only permit access through a certain port (*e.g.*, port 22) for connections from remote hosts. An authorized person from a provider may request the addition of a new rule to allow an external host to connect to the secure file server. This request requires the approval of the internal change management team prior to being implemented. Once a request is reviewed and approved, it is submitted to the network group to add a firewall rule to allow the new host to connect to the server from their given IP address.

The RODS IT group issues a unique cryptographic DSA public key to all new data providers for SFTP or SSL connections during the setup stage. Each data provider has its own unique public key to encrypt their data before sending and only the corresponding DSA private key on the RODS server can decrypt the message.

The RODS secure file server is configured with secure packages to guarantee each account on the server is quarantined to its home directory. Upon authentication each user's account is set to chroot (send) them directly to their home directory and not allow them to access or view any other directory on the server. This process protects each user's data by making the user's home directory look like the root directory, thus the user cannot see, nor can they access, any other directory structure on the server. The process also limits access to this area to just the user and the system administrator.

I.2 Data Relocation

The RODS data loading application retrieves the data files sitting in the data provider's local directory through a secure network file system (SNFS) link. The data loading process, running on the OTC database server cluster located in the database network zone, loads the data files from the file server into an Oracle data warehouse, and then moves the data files from the file server to the database server. The database server cluster resides in the same VLAN as the file server, thus ensuring that the transfer happens locally within one network segment. There is also a rule set in place between the production server cluster and the file server to allow only this traffic to pass between them and to block all other traffic.

I.3 Data Archival

A scheduled task on the backup application server moves the raw data files on the database server to tape once a month. Once the data is archived, a RODS system engineer moves the tapes off-site to a secure location where only a select number of authorized personnel in the RODS Laboratory can access them. At no point during this archival and storage process does a person outside of the RODS Lab have access to the data.

II. Data Confidentiality

During the period where the data files are stored on the database server, only a select list of RODS lab personnel has access to them. These personnel are the system administrators and are responsible for maintaining the security and integrity of each machine. No other lab personnel or external parties have access to this data.

Access to the raw UPC data in a database is only limited to RODS database administrators who are responsible for database backups and maintenance. All the users who need to access aggregate OTC sales at the product level need to sign a data use agreement (see attached document). All the public health users can only retrieve aggregate data at the product category level. No individual UPC data are accessible by public health users. RODS researchers who need to study OTC sales at UPC product level will not be able to see vendor information.

FIGURE 34.3 National Retail Data Monitor (NRDM) data security and confidentiality.

and outbreak detection from chief complaint data when requesting emergency room chief complaints from hospitals (see Wagner et al., 2004b). Similarly, when attempting to acquire over-the-counter medications sales data from retailers, we provide decision makers with publications about the National Retail Data Monitor (NRDM) (Wagner et al., 2004a, Hogan et al., 2005). In the future, we will likely provide Chapter 22 from this book. We also develop PowerPoint presentations of the same materials when we are planning to personally present our case to data providers.

2.1.6. Provide Draft Legal Documents

Data providers must have assurances that the receiving organization will process and store the data securely. To that end, it is advisable to execute ethically optimized and legally binding documents that detail the measures you employ to safeguard the data. These documents are usually DUAs, which can also be called by other names, e.g., memorandum of understanding (in this chapter, we refer to all such documents as DUAs in order to avoid confusion). These agreements must be reviewed and approved by each organization's legal counsel—which include the data provider, the data requester, and possibly other organizations involved in processing the data, for example, biosurveillance service or data users. In addition, when such agreements involve hospitals, the hospital may choose to have its ethics committee or Health Insurance Portability and Accountability Act) HIPAA-compliance officer review the agreement. We discuss DUAs in detail shortly.

In addition to the primary agreement between the data provider and requester, a health department or other biosurveillance organization may have existing documents that it requires end-users of the biosurveillance system to sign. These documents cover the proper use of passwords and handling of data. Because these agreements are an element of the overall approach to preserving confidentiality and security of data, their existence is germane to discussions. A data requester should be prepared to discuss or provide end-user agreements to the potential data provider. We also discuss end-user agreements later in this chapter.

2.1.7. Getting to Yes

"Getting to yes" means identifying the person(s) in an organization who can make a decision or a commitment on behalf of the organization, getting their verbal agreement to participate, obtaining signatures on a DUA (or other legal contract or agreement), and finally getting to the point at which the IT work begins.

Identifying the person or persons who can make the decision may take time because a request for data for biosurveillance is outside the usual business of most organizations. More often than not, such a request will require the attention of the chief executive officer (CEO). CEOs are busy, so it may be worth suggesting that the CEO consider delegating decision making to an official lower in the chain of command. In some organizations, more than one person must give approval, and arranging a conference call to discuss a request with all decision makers may facilitate getting to yes. For certain types of data, most notably clinical data, there may be a need for more than two organizations to sign an agreement or for multiple agreements to be executed pairwise among organizations. For example, if a health department's surveillance system is not located in, or operated by, that health department but rather by another organization, there may be need for two agreements. One agreement will be between the health department and each hospital within the health department's jurisdiction that will be providing clinical data, and another will be between the health department and the organization operating the biosurveillance system being used by the health department.

The process of obtaining legal approvals on DUAs, service contracts, or other binding agreements can introduce delays or even be a reason for failure of a negotiation. Often, an organization's attorney is not fully apprised of a project and may introduce a delay, or attempt to block participation altogether, simply because he or she lacks full understanding of the project. Procrastination may be the result of a judgment about workload priorities within the legal office. A data requester can minimize such delays by maintaining regular contact with the decision maker in the organization to ascertain the status of the legal process and, if unknown, asking him/her to investigate if there is a delay.

There may be specific aspects of the required IT work that are difficult for an organization to satisfy. For example, the organization may not have specific IT infrastructure or staff required for a project. It will also be problematic if a programmer within an organization does not have enough expertise or is simply not dependable in bringing the project to fruition. In these circumstances, regular contact with the person who approved the project is valuable in bringing to his or her attention a possible internal problem that is delaying completion of the required IT work.

In summary, getting to yes requires tenacity and follow-up on the part of the data requester. He or she must maintain regular contact with the people assigned the various tasks involved in getting to yes, including the decision maker, the person assigned the task of directing the organization's steps toward participation, the legal counsel who will approve and/or sign the DUA, and the IT staff assigned to a project.

A data requester should never consider a negative reply to a request as final decision, but rather as a temporary barrier. Decision makers within an organization change positions or take positions with other organizations, legal representatives change, and IT infrastructure within organizations usually improves with time. As change occurs regularly within organizations, data requesters should make repeated attempts to change negative decisions.

2.2. Data Use Agreements

A DUA is a contractual agreement between two or more organizations, involving either a one-way or two-way exchange of data. One can also call a DUA a *data use and confidentiality agreement*.[1] A DUA affords protection to both organizations providing the data. The DUA is a means for protecting the data

1 The term 'Data Use Agreement' has a specific technical meaning in connection with The American Health Insurance Portability and Accountability Act (HIPAA) Privacy Rule that should not be confused with the term as we use it here. We discuss the HIPAA Privacy Rule later in this chapter.

provider's interests, including privacy, security, and liability concerns. The process of crafting it and ratifying it in both organizations will entail considerable discussion of important issues that will reduce the possibility of misunderstandings in the future.

Appendix G of this book is an example DUA for use between hospitals and health departments. We have executed this agreement many times, and it contains clauses that cover many issues that arise in such negotiations. The clauses in the agreement address the following: identification of all parties to the agreement; legal basis for providing the data; definitions used in the agreement; the data specifically requested; definition of authorized users of the data; assurances regarding security, confidentiality, and data management; and, finally, standard disclaimers, warrantees, compliance, term of agreement, and termination clauses. It may also include specific stipulations concerning audits, options to publish, options for research, right to terminate, reports, or institutional review board (IRB) approvals.

The DUA should describe the types of data that the two organizations anticipate exchanging and include specific examples of those types of data. It is wise to include types of data that the organizations anticipate exchanging in the future, even if the intent is to initially only share a subset of the data described. This approach will obviate the need to re-execute or modify the agreement. The DUA should clearly define the scope of the project

If possible, the data requester should provide a previously successful DUA as a suggested starting point. Lawyers and decision makers have previously vetted successful agreements; thus, such agreements are more likely to address satisfactorily the concerns of the present organization. The data requester should explain to the prospective data provider that the document is the result of previous negotiations with similar data providers. This approach may save considerable time as most legal counsel will be concerned with similar issues and will also value an agreement that has precedence associated with it. Data providers, however, must have the option to revise the DUA based on the input of their legal counsel. Legal counsel on both sides must approve any revisions.

Once both parties agree to the final wording of the document, the authorized representatives of both organizations should sign two copies of the document, with each organization having an original signed document. Some data providers will demand ratification of an agreement before permitting any technical work to begin. Others will allow a concurrent process; that is, they will allow the data connection and transmission work to proceed while negotiating the specific terms of the DUA.

2.3. End-User Agreements

Anyone who will see confidential data coming from any source must sign an agreement to acknowledge use of the data and for what purpose, to guarantee confidentiality of the data, and correct handling and maintenance of the data as per legally-binding clauses set forth in all DUAs with data providers. These agreements can have various names, for example, authorized access agreement, account access agreement, or acknowledgement of DUA. Titles of agreements can also indicate the specific type of data to which the user will have access. There are basically two types of "end-users" of data that must sign a "user" agreement.

2.3.1. End-Users of the Biosurveillance System or Its Data

The end-users of a biosurveillance system should sign an agreement that includes a description of the data to which they will have access. This agreement should cover the provisions of the DUAs negotiated with all data providers (end-users typically see data obtained from multiple sources). They should include the identity/title and affiliation of the individual, the reason for using the data, and any restrictions on the use of the data.

Appendix H and Appendix I provide two examples of end-user agreements.

2.3.2. Technical Developers or Operators of Biosurveillance Systems

All technical personnel who have access to the data (i.e., personnel working on maintenance of or improvements to the biosurveillance system) should sign an agreement to acknowledge data use and confidentiality requirements. Every employee (current and new) who has or will have access to the data should sign such an agreement. The document should indicate that the signer is aware of the content of DUAs with data providers, and that he or she agrees to abide by the provisions of those agreements. The agreement must require the signer to agree to the following additional provisions: that he or she may only access the data for the purposes of development of the biosurveillance system for which the data has been acquired, and that he or she will not disseminate the information to a third party. It is also advisable to include a clause stating that the user agrees to submit any publications based on the data for review and approval to ensure that no one inadvertently discloses confidential data. The agreement should include a list of all data sources and data types provided to the biosurveillance system (for an example, see Appendix J, "Acknowledgement of Data Use and Confidentiality Agreement").

It is good practice to hold an annual (or even twice yearly) internal security meeting in which employees sign this agreement yearly, and the agreement should be revised to list new data sources and types of data added that year. This practice serves to reinforce the importance of confidentiality and security

In addition, all independent contractors working on any project that uses confidential data must also sign an agreement that includes the provisions noted above. This type of agreement can be some altered form of the example given in Appendix J, which lists only those data providers whose data will become knowledge of independent contractors.

2.4. Negotiations with Healthcare Organizations

In this section, we focus on the specifics of a negotiation between a health department and a healthcare system within that health department's jurisdiction. The relationship between these two types of organizations is arguably the most important one in biosurveillance. Although health statutes require disease reporting, there are other potentially desired data and services that a health department may request from a healthcare organization that require negotiation.

In negotiating for data from a healthcare system, the key to success is education. Hospitals and healthcare systems are businesses, and they will base their decisions to participate in a project on a consideration of cost and benefit. That analysis will require input from various departments and key personnel within the organization (see subsequent paragraphs that identify departments and key personnel). Each of these departments or individuals will need to know how this new project is going to affect them, and the decision maker, in turn, may ask each of them for input on the decision.

In getting to yes, it is vital to educate each of the key personnel whom the decision maker is going to turn to for input on the decision and to tailor that education to address the concerns coming from each of those personnel or departments.

We have tried three approaches to negotiations with a healthcare system: (1) a "top down" approach, in which we first contact the CEO or president about participating in the project; (2) a "bottom up" approach, in which we first contact other key personnel within the organization to champion the project up the chain of command; and (3) a two-pronged method, in which we contact both the CEO and key personnel simultaneously. We have had successes with all three of these approaches. Choosing your method will depend on existing relationships within the hospital. If there is an enthusiastic supporter for biosurveillance of this kind at any level within the organization, you may wish to start with that contact. If you have no existing relationships and are cold calling, you may wish to opt for the two-pronged method. Whatever the approach, we have found that the key personnel and departments within a hospital or healthcare system include the following:

1. The CEO or president will be the decision maker. The CEO or president may turn to his/her chief financial officer (CFO), legal counsel or HIPAA privacy officer, IT director, and possibly others for input on how participation in the project will affect the organization. In our experience, we have seen different levels of interest and involvement by the CEO. Some CEOs, when approached directly, may be very taken with the project and direct staff to meet with you and get it implemented. Other CEOs may elect to participate in the project based solely on recommendations from within the organization (e.g., from an emergency planning director or emergency department [ED] director) and simply sign off on a DUA. However, CEOs may want input from all relevant departments before agreeing to participate. In negotiations with the CEO or president of a healthcare system or hospital, be prepared with clear, bullet-pointed answers to questions related to all areas of operation. Examples of questions you may hear include the following: *How much will this cost us? How much IT work will this entail? What will legal have to say about it? What is in it for us?* Your answers will need to be consistent with the answers the CEO will get if he or she asks the same questions internally. For example, if you tell the CEO the project will require 10 person-hours from the IT department, the CEO had better hear the same answer from the IT department.
2. The CFO will assess the cost to the organization. The CFO will want a cost analysis that includes person-hours, hardware, and other costs to the hospital. If the health department requires participation in the project, be prepared with information about how the health department is going to work with the hospital to cover its internal personnel, hardware, or other costs.
3. The legal counsel, who the CEO will consult about the legal ramifications of participating in the project and to approve a DUA. These discussions may also include the HIPAA privacy officer. When negotiating, be thoroughly familiar with HIPAA issues related to the data you are requesting. As with any data provider, you want to ensure the comfort level of the hospital with privacy, security, liability, and other issues. Be prepared to be flexible on language in the DUA and, as much as possible, defer to the needs of the hospital. Offer a template of a successful DUA from past implementations. Getting to yes is not possible without the approval of a hospital's legal counsel.
4. The IT director will assess feasibility for the implementation, including hardware requirements, availability of personnel, and other technical issues. You may encounter a variety of scenarios on the IT side: The hospital may not have the appropriate personnel in-house for the implementation; the hospital may outsource some or all of its IT work; the hospital's IT vendor may demand thousands of dollars to create an interface; or the IT department may claim other priorities. In some cases, the IT director may not know details of the systems his or her own department is running, and you will need to speak directly with network and interface engineers. You may also find that an IT director may say the implementation is not possible because the hospital does not meet technical requirements when, in fact, network and interface engineers will tell you there is no problem at all. If an IT director is holding up approval of the project, it will be helpful to enlist the support of the engineers who will actually do the work and who will be able to give their IT director an accurate assessment of the actual time or cost involved.
5. An ED chair or infection control practitioner, although not directly involved in an implementation, may be the biggest

champion for the project. Most often ED personnel are very enthusiastic, as they understand the importance of biosurveillance. Enlist the aid of the ED chair in selling your project at all levels of the organization.

At all levels within a healthcare system or hospital, you may encounter the objection that a department has other priorities. A CEO who is well educated about the time and financial costs of the project will be better able to direct his or her departments to cooperate with the implementation.

General recommendations are the following: (1) educate the key personnel before the CEO talks to them, and their roles and the scope of the project will become clear and your implementation will move along faster if they are prepared with their support of the project when the CEO asks for their input; (2) tailor the discussion to your audience because, for example, infection control does not need an in-depth discussion of the legal issues; (3) be prepared with specifics such as the anticipated number of person-hours that will be required or precisely how much money the project will (or will not) cost the hospital; and (4) prepare brief hard-copy documents that addresses how the project will impact each department.

To illustrate more concretely some of the types of issues one might need to address, following are examples of negotiations.

2.5. Case Study

We requested chief-complaint data from two healthcare systems before the February 2002 Olympic Winter Games in Utah. Both healthcare systems were technically capable of providing the requested data in an electronic format and in real time. Together, they accounted for 19 urgent care centers, 10 EDs, and the polyclinic located in the Olympic Village.

This negotiation involved the healthcare systems, the Utah Department of Health (UDOH), six local health departments, and the University of Pittsburgh. Specific individuals included the state epidemiologist, an Olympic surveillance coordinator, a vice president for medical research, an associate CIO, a medical informatics department chair, the director of the RODS Laboratory, and lawyers representing the healthcare systems, UDOH, and the University of Pittsburgh. The negotiation lasted seven weeks.

Several issues framed this negotiation:

- UDOH and the Salt Lake Olympics Committees had spent years planning and preparing enhanced biosurveillance systems for the 2002 Olympic Winter Games. The biosurveillance systems included daily polling of selected physicians, pharmacies, veterinarians, the poison control center staffs, and the forensic pathologists (sentinel surveillance); 24-hour batch-mode syndromic surveillance for Intermountain Health Care's urgent care facilities (an urgent care facility is a clinic that sees patients without appointments); and a manual ED syndromic surveillance system employing daily review of encounter logs and the medical records of syndromic patients. Thus, healthcare systems in Utah were already responding to a legal mandate to share data and allow intrusive chart reviews during the 2002 Winter Olympic Games.
- UDOH had established administrative rules providing the legal framework for these systems months earlier; however, the RODS system was sufficiently different in that it was not covered by the rule.[2]
- One health system classified RODS as a "research project" because it was not the planned-on and agreed-to surveillance methodology. Consequently, before allowing the implementation of RODS, compliance with federal privacy law (HIPAA) and the IRB approval were required. Although the HIPAA privacy rule did not go into effect until April 14, 2003, the healthcare system chose to adopt their understanding of the then current draft privacy rule. The proposed HIPAA regulations permitted disclosure for public health surveillance activities but did not specifically address the sharing of data for public health surveillance research.

For each healthcare system, our goal was to execute a trilateral DUA signed by the healthcare system, UDOH, and the University of Pittsburgh. We had used this type of agreement successfully in Pennsylvania. After a seven-week negotiation,

---

2 Prior to the Salt Lake Olympics, UDOH put into effect an administrative reporting rule designed to establish legal authority for syndromic surveillance. Administrative rules are statements written by state agencies that have the effect of law. Under this rule, designated emergency centers were to report syndromic information for patients from the preceding day's encounters, either by reporting it themselves or by allowing UDOH representatives to gather the data. Encounters required a report if and when diagnostic information indicated the presence of one of 11 syndromes defined by the UDOH (i.e., *Respiratory Tract Infection with Fever, Gastroenteritis without Blood, Bloody Diarrhea, Febrile Illness with Rash, Suspected Viral Hepatitis, Meningitis/Encephalitis or Unexplained Acute Encephalopathy/Delirium, Sepsis or Unexplained Shock, Unexplained Death with History of Fever, Botulism-like Syndrome, Lymphadenitis with Fever, or Illicit-drug-related Episode)*. The report included the following protected health information: the reason for visit, chief complaint, presenting diagnosis, final diagnosis (when available), facility, date/time of visit, patient demographics (age, gender, residential zip code), the syndrome detected, admission status, and a record identifier for follow-up investigation. RODS did not quite fit this rule because the seven syndromes used in RODS were more general.

we executed one trilateral agreement and one bilateral agreement (signed by one healthcare system and the University of Pittsburgh). This outcome reflects the different views held by the health systems on sharing data with the UDOH. Because the one healthcare system received partial funding from the state, the working relationship between the two entities (at least from a data provisioning perspective) had always been strong, making data sharing for collaborative public health projects relatively commonplace and painless. The second healthcare system was a large not-for-profit healthcare organization that directly competed with the first. According to a senior vice president, it had, on occasion, borne the brunt of state mandates to provide clinical data for public health purposes because of its dominant market share and advanced data systems. In addition, its data often became property of the state. The final agreement permitted UDOH to only view aggregate time-trended and spatially displayed data (e.g., absent the free-text chief complaint and with age reported in five-year ranges) and could only receive more detailed data in the event of a public health emergency. The solution also entailed that the data were located on servers at the University of Pittsburgh.

In this project, organizational issues outweighed technical issues. The factors that increased the chances of success in this negotiation were that all of the stake-holding organizations were accustomed to working together because of the planning and preparation for the 2002 Olympic Winter Games. Our ability to customize the RODS application quickly to satisfy the needs that arose during negotiation was also helpful. We also note the potential role of serendipity. Ratification of the agreement with one of the healthcare systems may not have occurred if it were not for the decision by President Bush to visit the RODS Laboratory three days before the opening ceremony for a demonstration of the RODS system, which would have only contained data from one healthcare system without ratification.

More detail about the administrative and technical and aspects of this project is available in a pair of articles in the *Journal of the American Medical Informatics Association* (Gesteland et al., 2003, Tsui et al., 2003).

2.6. Negotiations with Commercial Firms

A variety of commercial firms collect data of potential or established value to biosurveillance. These firms include retail pharmacies, drug manufacturers, food producers, grocers, credit card companies, and other entities discussed in Parts II and IV of this book. Although there are some regulations that encourage their participation in biosurveillance, their willingness to provide data or services largely hinges on altruism.

To illustrate the issues that one may encounter in negotiation with commercial firms, we describe the NRDM's experience with getting to yes. The goal of this project is to collect sales data for over-the-counter medications from retail pharmacies, grocers, and mass merchandising stores (see Chapter 22). In the course of building the NRDM, we formulated a number of tactics for approaching retailers for data. These methods may be relevant to negotiations with other commercial entities and industries.

Our approach involved working with a consultant who had worked in the industry and was familiar with both management and IT in the industry. This individual was successful in recruiting two initial retailers for the NRDM based on his credibility with his contacts and also by appealing to the retailers' sense of community responsibility.

The consultant, based on the early feedback from the industry, identified the need to develop the scientific case, which we did rapidly through literature review and conducting research on historical data (Hogan et al., 2003).[3] The consultant also identified the need for governmental public health to make the "ask" in the form of an official letter from an as highly placed individual as possible. We first attempted to obtain a request letter from the White House and then from Department of Health and Human Services (DHHS) and from senators. This process took considerable time and lobbying through several federal agencies, but after several months the director of the Centers for Disease Control and Prevention sent a request letter on our behalf to the CEOs of approximately 10 large retail corporations.

During this time, we also formed the NRDM Working Group, consisting of health department leaders in the Commonwealth of Pennsylvania, Ohio, New York, Georgia, and New Hampshire. The group enlisted the support of other state health departments and sent additional "ask" letters that introduced retailers and mass merchandisers to the program and requested their participation. The letters included fact sheets describing the purpose of the NRDM and current statistics about use and participation, reasons they should participate, an IT specifications document, and a data security document.

Very few retailers responded to the letters from the CDC director and the NRDM Working Group. The senior executives who received the letters often did not answer follow-up calls. Retailers receive many requests from organizations and vendors to participate in public benefit projects. We learned that it was going to take more than just a letter to get the attention of a retailer.

We researched the retailers and learned their chains of command. We started cold calling personnel in key positions at the retailers, such as vice presidents of pharmacy or operations, in an attempt to generate interest in our efforts. Often an executive shunted us from one vice president to another before we could pique someone's interest.

3 This literature review identified the existing literature on the correlation of sales of diarrhea remedies with cryptosporidium outbreaks that has now become part of the biosurveillance literature (see Chapter 22).

We marketed to the industry as a whole, enlisting the assistance of a national industry organization that gave us a speaking slot to present at their convention. We staffed a booth on the convention floor (Figure 34.4). We enlisted the assistance of one of the industries' data integrators, and we received a speaking slot at one of their industry meetings. We also kept a log on all of the people from retailers and mass merchandisers we met, including business cards, and if no card was available, we filled out a contact information sheet via our discussions with individuals.

After this meeting, we did follow-up with people we met. The issues that frequently arose were the status of the NRDM (profit, nonprofit, relationship to public health) and concerns about competitors gaining access to the data. We were sometimes able to address these concerns in person by flying to the retailer's headquarter and giving a presentation on the NRDM and a demonstration of the system. In some instances, public health officials from the headquarter city met with retailers and/or invited them to visit the local health department to show executives the system and explain how the health department used the data for the purpose of disease detection.

One of the most effective tools in our recruitment efforts was the development of the "business case" (Figure 34.2). The business case listed reasons for the retailer to participate in this project. In one page, it made a compelling case for participation, including helping to mitigate the effects of a bioterrorism attack, protecting the retailer's employees and communities, addressing multiple requests for data from health departments in a single interface, enhancing prestige with the health community, and minimizing cost of compliance with reporting legislation.

In some cases, public health officials led the follow-up effort by contacting retailers to ask for their participation. Sometimes a cold call to a retailer resulted in an agreement to participate without the assistance of public health officials.

Once a retailer verbally approved participating in the NRDM, a DUA was executed (Appendix K, "Data Use Agreement with Commercial Data Provider").

Our negotiations with retailers have not always been successful. In some cases, the necessary IT infrastructure was not in place for data collection. Many larger retailers operate multiple "banners" (i.e., smaller chains) and do not consolidate their data collection systems. In some of these cases, we will go back to the retailer after they have upgraded their systems. We were unable to make progress with another retailer because the health department in the state where the retailer had its headquarters made the request and took charge of the effort. The person at the health department with whom we were working had many other priorities and was unable to follow through on the request in a timely manner. The largest retailer simply turned down our request for data, despite our making a joint visit to their headquarters together with the health department of the state where the retailer had headquarters. There simply was no requirement for them

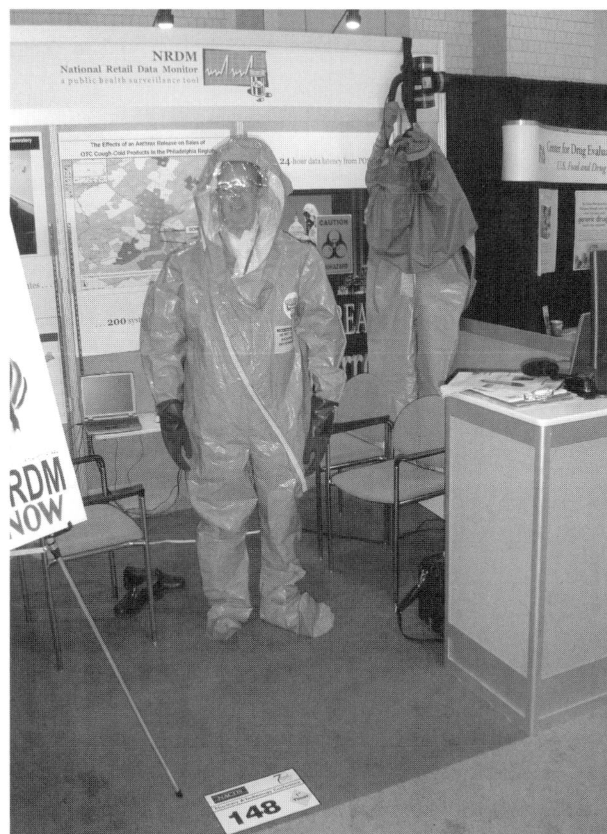

FIGURE 34.4 National Retail Data Monitor Booth at the National Association of Chain Drug Stores Annual Meeting, Philadelphia, Pennsylvania, August 2003.

to provide data; they had a corporate policy of never sharing data, and they were unwilling to make an exception. They were unwilling to consider technical solutions that would allow them to analyze their data and provide deviations from expected sales that our algorithms could still combine with data from other retailers in a region to form an overall indication of anomaly.

2.7. Negotiations with Utilities

A variety of utilities collect data of potential or established value to biosurveillance, including water companies, transit authorities, and telephone companies. These utilities may be privately or government owned and are often regulated by government agencies such as the Food and Drug Administration (FDA), U.S. Department of Agriculture (USDA), and boards of public utilities. Utilities may be more motivated to provide data or services for biosurveillance than are commercial companies because of their quasi public nature. Water companies are also motivated because they have a primary biosurveillance role and can benefit directly from better public health surveillance. Deregulation in the United States has moved several types of companies such as phone companies out of public realm, although in other countries these activities continue as utilities or governmental services.

In 2005, we approached a water company as a data requester, requesting both water quality data and information about the water distribution system. As with any data negotiation, sensitivity to the needs of the data provider was paramount. The water company was interested in participating in the project but insisted on anonymity for protection of the company and its customers. This water company was motivated to participate in a biosurveillance system for several reasons: (1) possible incentives for future projects and funding from the Environmental Protection Agency (EPA), (2) its altruistic desire to participate in protecting public health, and (3) the company's desire to improve its relationship with the health department by participating in this project.

We approached the water utility in conjunction with the EPA, which regulates water quality and the local department of health. We worked with the water quality manager at the utility to get to "yes" from the CEO.

The DUA that we negotiated included a requirement for the RODS Laboratory to delete reference to the city where the water company is located in publications and reports about the project. The water company was also concerned for the security of the water distribution maps they would give us and concern that contributing daily water quality data to a biosurveillance system will cause health departments to more readily attribute public health trends to problems in the drinking water.

The RODS Laboratory's excellent record for data security gave this water company the confidence to share details of its water distribution systems with us. To address the concern for undue attention because of anomalies detected by the public health surveillance system, we agreed to develop an alerting mechanism to let the water company know when the health department receives a water quality alert from the system. That way, the water company could do its own investigation into the possible causes for the anomaly that generated the alert and be prepared for discussion with the health department and possibly allay its concerns before a more extensive public health investigation.

3. LEGAL AND ETHICAL CHALLENGES

We discuss laws and regulations that promote data exchange in Part II of this book. In this section, we discuss the HIPAA Privacy Rule. Although the Privacy Rule was crafted to allow exchanges of data for purposes of public health surveillance (it exempts these exchanges explicitly from its scope), this rule often comes up during negotiations with holders of personal health information (PHI) (insurers, pharmacists, and healthcare systems). We also discuss the field of ethics, with principles related to privacy, confidentiality, and the tradeoff between public and individual interests that underlie many discussions of newer types of data.

3.1. HIPAA Privacy Rule

The HIPAA Privacy Rule, issued by the secretary of the DHHS in 2003, specifies regulations for protecting the privacy of PHI. Health plans, healthcare clearinghouses, and healthcare providers ("covered entities") that electronically transmit personal health information for specific reasons must follow these rules (CDC, 2003).

The Privacy Rule permits covered entities to disclose PHI to public health authorities. The Privacy Rule states that covered entities may disclose PHI without individual authorization to a public health authority legally authorized to collect or receive the information for the purpose of preventing or controlling disease, injury, or disability.

It also states, "De-identified data (e.g., aggregate statistical data or data stripped of individual identifiers) require no *individual* privacy protections and are not covered by the Privacy Rule." The Privacy Rule specifically indicates,

> De-identifying can be conducted through statistical de-identification—a properly qualified statistician using accepted analytic techniques concludes the risk is substantially limited that the information might be used, alone or in combination with other reasonably available information, to identify the subject of the information; or the safe-harbor method—a covered entity or its business associate de-identifies information by removing 18 identifiers and the covered entity does not have actual knowledge that the remaining information can be used alone or in combination with other data to identify the subject. (45 CFR 164.514[b]) (Table 34.1)

After and satisfaction of the second condition, one calls the data set a *limited data set* (Table 34.2) HIPAA also requires that the covered entity execute a DUA with the organization or individual to whom it is releasing the data. The DUA must indicate,

> who is permitted to use or receive the limited data set, and provide that the recipient will not use or disclose the information other than as permitted by the agreement or as otherwise required by law; use appropriate safeguards to prevent uses or disclosures of the information that are inconsistent with the DUA; report to the covered entity any use or disclosure of the information, in violation of the agreement which it becomes aware; ensure that any agents to whom it provides the limited data set agree to the same restrictions and conditions that apply to the limited data set recipient with respect to such information; and not attempt to re-identify the information or contact the individual.

3.2. Ethics

Morality precedes the law. The reason HIPAA makes violations of confidentiality illegal, for instance, is that it is wrong to violate confidentiality, not the other way around. It would

TABLE 34.1 Eighteen Identifiers to be Removed for De-Identified Data

To de-identify using this method, the following identifiers of the individual—or of relatives, employers, or household members of the individual— are removed:

1. Names.
2. All geographic subdivisions smaller than a state, including street address, city, county, precinct, zip code*, and their equivalent geocodes, except for the initial three digits of a zip code if, according to the current publicly available data from the Bureau of the Census the following apply:
   A. The geographic unit formed by combining all zip codes with the same three initial digits contains more than 20,000 people; and
   B. The initial three digits of a zip code for all such geographic units containing 20,000 or fewer people is changed to "000."
   C. Currently, 036, 059, 063, 102, 203, 556, 592, 790, 821, 823, 830, 831, 878, 879, 884, 890, and 893 are all recorded as "000."
3. All elements of dates (except year) for dates directly related to an individual, including birth date, admission date, discharge date, date of death; and all ages over 89 and all elements of dates (including year) indicative of such age, except that such ages and elements may be aggregated into a single category of age 90 or older.
4. Telephone numbers.
5. Fax numbers.
6. Electronic mail addresses.
7. Social security numbers.
8. Medical record numbers.
9. Health plan beneficiary numbers.
10. Account numbers.
11. Certificate/license numbers.
12. Vehicle identifiers and serial numbers, including license plate numbers.
13. Device identifiers and serial numbers.
14. Web Universal Resource Locators (URLs).
15. Internet Protocol (IP) address numbers.
16. Biometric identifiers, including finger and voice prints.
17. Full-face photographic images and any comparable images.
18. Any other unique identifying number, characteristic, or code—except covered identities may under certain circumstances—that assigns a code or other means of record identification that allows de-identified information to be re-identified.

*The first three digits of a zip code are excluded from the protected health information (PHI) list if the geographic unit formed by combining all zip codes with the same first three digits contains more than 20,000 persons.

TABLE 34.2 HIPAA-Allowable Limited Data Set

A limited data set must have all direct identifiers removed, including the follwing:

- Name and social security number
- Street address, e-mail address, telephone and fax numbers
- Certificate/license numbers
- Vehicle identifiers and serial numbers
- URLs and IP addresses*
- Full-face photos and any other comparable images
- Medical record numbers, health plan beneficiary numbers, and other account numbers
- Device identifiers and serial numbers
- Biometric identifiers, including finger and voiceprints

A limited data set could include the following (potentially identifying) information:

- Admission, discharge, and service dates
- Dates of birth and, if applicable, death
- Age (including age 90 or over)
- Five-digit zip code or any other geographic subdivision—such as state, county, city, or precinct—and its equivalent geocodes (except street address)

*URL indicates Web Universal Resource Locators; IP, Internet Protocol.

be wrong to violate confidentiality even if there were no HIPAA, no state statutes, and no litigators. Ethics, a branch of philosophy, is the study of morality.

It is by doing ethics, therefore, that we might discover that "violating confidentiality" is not always wrong. There might be reasons, at least on occasion, to place greater emphasis on other values. Public health, life and death, and national security might be examples of such values. This is not to say that public health or national security (terms that are rather vague if left at that) should always trump confidentiality; it is only to say that there will be circumstances in which morality permits the elevation of public health and other values over confidentiality. Applied ethics is the discipline that sorts these things out.

Biosurveillance systems pose many difficult and interesting ethical issues. Fortunately, we have some precedents for examining these issues, at least to the extent that they arise when computers are used in ordinary health care (Goodman, 1998), emergency response (Goodman, 2003), and data mining (Goodman, 2006). Here, we briefly survey, but do not resolve, a few of the ethical issues that arise in the use of biosurveillance systems. Any attempt at a comprehensive organizational integration for biosurveillance must attend at the outset to ethical issues raised by such systems.

3.2.1. *Privacy and Confidentiality*

It has long been known that privacy (generally, the right to be free of various intrusions) and confidentiality (generally, the right to be free of inappropriate acquisition of information about an individual) must be balanced against other rights or expectations. It would be perverse for a patient seeking medical care to suggest, for example, that her physician should not have access to her medical record because that would violate confidentiality. As a practical matter, personal health information, whether stored in stone tablets or electronic data bases, poses the following challenge: How can appropriate access be made easy, and inappropriate access made difficult or impossible?

To be effective, a biosurveillance system must acquire and analyze vast amounts of personal information. This information might be trivial, or it might be deeply personal. In a normal environment, one might be able to opt out of such a system. One can direct not to disclose or share personal information in contexts ranging from banking and financial services to health care and education, and the law permits or in some cases even requires that such an opt-out opportunity be available. In the case of biosurveillance, however, opt-outs will tend to degrade or weaken the system; if there are too many of them, the system is more likely to fail.

There are several ways a community might balance demands for privacy against the benefits reckoned to accrue from biosurveillance. They include various levels of anonymization of the data collected; strict controls on access to the data at any level of anonymization; and, most generally, buy-in of the community to be protected by the biosurveillance system in the first place. To the extent that citizens in a free society view a biosurveillance system as an extension of familiar—and trusted—public health surveillance, that system will be more acceptable. However, if one sees a biosurveillance system as intrusive and its operators or applications as not trustworthy, the system is more likely disdained.

3.2.2. *Appropriate Uses and Users*

The importance of assessing and determining appropriate use and appropriate users of clinical information systems has been apparent for decades (Miller et al., 1985). Determining appropriateness is not always a straightforward matter. One must ask several questions in any such determination. These questions include decisions about (1) who gets to make the determination, (2) what values one should embody in the determination and (3) what is to be done in cases of *in*appropriate use or *in*appropriate users.

One can answer the question as to who is responsible for vetting appropriateness by appealing to traditions of trust. Ideally, the world would be so that citizens in a democracy would, at least by tacit agreement, determine an appropriate user of a biosurveillance system. Then, public health authorities—and not law enforcement agencies or manufacturers of over-the-counter drugs, say—might come to be regarded as one class of appropriate users. This arrangement largely governs current public health and epidemiologic data collection. A distinction must be made between health and law enforcement or health services and corporate pharmacologies to make clear that not all users have the same privilege. (Much of the debate over the U.S. Patriot Act is about the extent to which law enforcement entities should have access to information that had previously been either private or available to entities other than the police.)

The values that underlie determinations of appropriate use and user must explicit and public. Because we value the lives of the many over the rights of a few, we, for instance, countenance quarantines, vaccination programs, and the like. In the context of biosurveillance, we similarly should be able to make explicit that we value the lives to be saved from disease outbreaks or bioterrorists at least as highly as we value being able to buy an inhaler at the local pharmacy without being eavesdropped. We encourage caution, however: we will usually be able to say we are putting life over privacy, over the right to move about freely, over liberty. Only when those purported to be saving the lives are trusted and only when the uses to which they put biosurveillance systems are reasonable and controlled will the incorporation of values be credible.

By "control" we mean a system or mechanism for forbidding or punishing an inappropriate use or an inappropriate user. This will require a mechanism for oversight. "Inappropriate" is a vague term, and its scope and limits should perhaps be negotiated in advance as part of any DUA and be subjected to ongoing review. Procedures for evaluating the propriety of a particular decision in a specific case must be—and be seen to be—loyal to guiding values and impartial, as well as flexible enough to accommodate judgment calls in cases in which reasonable people might disagree.

What should be clear about determining appropriate users and enforcing appropriate uses is that these processes cannot be undertaken in isolation from the community we earlier insisted had to trust those operating or managing a biosurveillance system. The public health model has served us well, with everything from vital statistics to HIV surveillance to diet education programs. That biosurveillance systems enjoy such precedents is a happy development for a community that seeks simultaneously to prepare for the worst while enjoying the best of what open societies have to offer.

3.2.3. *Risk Communication*

An ethically balanced and trusted biosurveillance system is the beginning. Such trust is, for better or worse, open to revision if not revocation. It is not enough to balance privacy and safety, and it is not enough to ensure appropriate uses and users. There is the subsequent and awesome task of making sure the system does what it is intended to do. Put differently, it will not do to develop an ethically optimized system that generates false alarms. In addition, it will be a disappointment (at the least) if our biosurveillance system fails to inform the kind of early warning system it is hoped or designed to support. Negotiating policies and procedures for disclosure of a public health risk is arguably one of the greatest challenges in crafting DUAs.

Indeed, the challenges of risk communication are among the most interesting and difficult for any form of public safety surveillance or tracking. This is especially true if we are to rely on or collaborate with the news media to warn biosurveillance stakeholders of risks (Friedman and Dunwoody., 1999). One might profitably compare the risk communication challenges faced by tropical storm forecasters to those faced by biosurveillance system operators. Indeed, many physicians and nurses find themselves needing to make sure that they "titrate" their duty to warn of risks with any number of probabilistic disclaimers.

If one sounds the alarm early and the risk does not materialize, one erodes credibility for future alarms. If there is too much harm done and one sounds the alarm too late, the system has failed at its prime purpose. Yet it is exquisitely difficult to get it just right. Data providers are likely to have loyalty duties to stakeholders with different needs and expectations. A water department, for instance, serves a community in different ways than does a pharmacy. Yet the collection and mining of data from these and other entities will require information disclosure protocols that take into account these disparate duties and harmonize them or make them congruent for the sake of public health and safety.

A number of approaches are available to address this challenge, and one should evaluate all approaches. These include negotiating the creation of entities analogous to the data safety and monitoring boards (DSMBs) that review and monitor clinical trials. These boards have the duty to interpret data and help decide if investigators should tell human subjects about new risks or dangers that might have emerged after the experiment began; in extreme cases, DSMBs have the responsibility to decide whether to stop a study altogether. (Stopping a trial early means the data collected so far will be useless or less useful than they would have been if the study were completed. However, if one completes the study, it might be at the expense of exposing subjects to unacceptable risks.) Another potentially useful entity would resemble a clinical or hospital ethics committee, a group of health professionals, and lay or community members who provide nonbinding advice to clinicians and others. A biosurveillance ethics committee should be able to take probabilistic data and weigh its disclosure against the hazards posed by the Scylla of too-early warning and the Charybdis of tardiness. We might even hypothesize that a combination DSMB/ethics committee *qua* risk communication special weapons and tactics (SWAT) team would be useful.

To be sure, the creation and operation of any such entity will also be of use in the negotiation of DUAs and other legal/contractual documents governing data sharing. The clinical ethics committee model generally includes capacity to offer education, write and revise policy, and provide a consultation service. One should consider therefore requiring the process of organizational integration to include an ethics component to carry out these three functions.

What is called for throughout, if ethics is to be included in the development and use of biosurveillance systems, is that one informs this public service by a public process, that one identifies core values and incorporate them into policies that guide and govern operations, and that one protects and fosters the trust earned by public servants in democracies. There will be a great deal to learn along the way, but this is often the case with new enterprises with very high stakes.

4. SERVICE AGREEMENTS

As discussed in the previous chapter on architecture, we expect that organizations will increasingly enter into service agreements that cover more than exchange of data. We define a *service agreement* as one in which an organization agrees to provide a service. For example, a hospital may agree to test a certain number of patients for anthrax for a certain price on request by the health department (or even an automated system running in the health department, as would be feasible if it were simply a case of conducting additional testing on samples already obtained from patients).

Current DUAs (the subject of a previous section) include a description of the data that will be exchanged and perhaps the frequency of transmission, but not the reliability of the data provisioning. In our experience, many organizations are unwilling to agree to provide more than "best effort" reliability at data provisioning. Reliable and timely exchange of data is important; therefore, we expect it to be an increasingly important subject in negotiations of data use/service agreement.

As a concrete example of other services that may be the subject of negotiations, consider the efforts of the New York City Department of Health and Mental Hygiene (DOHMH) to obtain investigation services from the healthcare system. As discussed in Chapter 23, DOHMH operates a syndromic surveillance system that monitors ED visits for *gastrointestinal* and other syndromes. When this system finds an anomalous number of cases, DOHMH investigates. Balter et al. (2005) reported that efforts to persuade EDs to augment their specimen collections have not succeeded because these laboratory studies typically do not affect clinical care and incur added effort, cost, and burden of tracking results. They also report having piloted sending specimen collection kits to five outpatient clinics with supplies for testing children with chief complaints of vomiting or diarrhea. The results of the data collection were incomplete. For such mechanisms to work, the basic workflows and understandings of duties and responsibilities of the healthcare system must change, and for such changes to occur, high-level buy-in and agreement must be obtained through negotiation.

The above example is but a hint of the types of services that organizations involved in biosurveillance could provide to each other. A more expansive list of services might include testing patient P for tuberculosis; providing electronic access to patient charts (as in the example in Chapter 6); obtaining a travel history for patient Q; testing water supply for cryptospordium, providing a cargo manifesto and/or passenger lists; providing the production history of a medicine, an animal, or a foodstuff. Laboratories, water companies, the animal healthcare system, food production systems, and hospital infection control could all contractually agree to prearranged services that might improve the speed at which outbreaks are detected and characterized.

5. SUMMARY

Diverse organizations that are not yet required by law or regulation to share data hold many types of data now needed for early detection and characterization of outbreak. For this reason, the art of negotiation will be important in biosurveillance for the near term. Some general principles to improve the likelihood of getting to yes include involving relevant governmental organizations; identifying the decision makers; paying exquisite attention to the needs of the data provider; handling problems expeditiously; tracking the process closely; using fact sheets to educate the decision makers about the legal, workflow, and technical dimension of the request; making the business case; making the scientific case; and providing a draft DUA.

We expect that in the near future, negotiations will begin to expand to include services as well as data. The general principles apply equally to these agreements.

The primary constraint on what is feasible is primarily ethical, in particular, the ethical balance between an individual right to privacy and confidentiality versus the expected benefit to society of sharing data. Most projects today are pioneering and exploring both the best technical and ethical means to maximize the benefit to society by better sharing of data. The results of these pioneering projects will likely influence the laws and regulations that will obviate the need for negotiation and facilitate a more rapid realization of improved systems of biosurveillance.

ADDITIONAL RESOURCES

The U.S. Department of Health and Human Services Web site *(http://www.dhhs.gov/)*. This site provides information on the regulations pertaining to the responsible and ethical handling of data in research for funded projects through grants from the DHHS. See also the Public Health Service, Office of Research Integrity Web site *(http://ori.dhhs.gov/)*.

American Medical Informatics Association's Ethical, Legal and Social Issues (ELSI) Working Group *(http://www.amia.org/mbrcenter/wg/elsi/)*.

University of Miami Privacy and Data Protection Project *(http://privacy.med.miami.edu)*.

ACKNOWLEDGMENTS

We thank the many individuals in health departments and universities with whom we have collaborated to develop these practices.

REFERENCES

Balter, S., Weiss, D., Hanson, H., Reddy, V., Das, D., and Heffernan, R. (2005). Three Years of Emergency Department Gastrointestinal Syndromic Surveillance in New York City: What Have We Found? *Morbidity and Mortality Weekly Report* vol. 54:175–180. *http://www.ncbi.nlm.nih.gov/entrez/query.fcgi?cmd=Retrieve&db=PubMed&dopt=Citation&list_uids=16177711.*

Broome, C.V., Horton, H.H., Tress, D., Lucido, S.J., and Koo, D. (2003). Statutory Basis for Public Health Reporting beyond Specific Diseases. *Journal of Urban Health* vol. 80(Suppl):i14–i22. *http://www.ncbi.nlm.nih.gov/entrez/query.fcgi?cmd=Retrieve&db=PubMed&dopt=Citation&list_uids=12791774.*

Centers for Disease Control and Prevention [CDC]. (2003). HIPAA Privacy Rule and Public Health Guidance from CDC and the U.S. Department of Health and Human Services. *Morbidity and Mortality Weekly Report* vol. 5:1–17, 19–20. *http://www.ncbi.nlm.nih.gov/entrez/query.fcgi?cmd=Retrieve&db=PubMed&dopt=Citation&list_uids=12741579.*

Friedman, S.M., and Dunwoody, C.L.R. (Eds.) (1999). *Communicating Uncertainty: Media Coverage of New and Controversial Science.* Mahwah, NJ: Lawrence Erlbaum Associates.

Gesteland, P.H., et al. (2003). Automated Syndromic Surveillance for the 2002 Winter Olympics. *Journal of the American Medical Informatics Association* vol. 10:547–554 *http://www.ncbi.nlm.nih.gov/entrez/query.fcgi?cmd=Retrieve&db=PubMed&dopt=Citation&list_uids=12925547.*

Goodman, K.W. (Ed.) (1998). *Ethics, Computing and Medicine: Informatics and the Transformation of Health Care.* Cambridge: Cambridge University Press.

Goodman, K.W. (2003). Ethics, Information Technology and Public Health: Duties and Challenges in Computational Epidemiology. In *Public Health Informatics and Information Systems* (P.W. O'Carroll, W.A. Yasnoff, M.E. Ward, L.H. Ripp, and E.L. Martin, eds.). New York: Springer-Verlag.

Goodman, K.W. (2006). Moral Foundations of Data Mining. In *Encyclopedia of Data Mining*. Hershey, PA: IDEA Group Reference.

Hogan, W.R., et al. (2003). Detection of Pediatric Respiratory and Diarrheal Outbreaks from Sales of Over-the-Counter Electrolyte Products. *Journal of the American Medical Informatics Association* vol. 10:555–562. *http://www.ncbi.nlm.nih.gov/entrez/query.fcgi?cmd=Retrieve&db=PubMed&dopt=Citation&list_uids=12925542.*

Hogan, W.R., Wallstrom, G.L., and Wagner, M.M. (2005). An Evaluation of Three Policies for Updating Product Categories in the National Detail Data Monitor. In: Proceedings of the AMIA 2005 Annual Fall Symposium (October 22-26, 2005, Washington DC), 325–29.

Miller, R.A., Schaffner, K.F., and Meisel, A. (1985). Ethical and Legal Issues Related To the Use of Computer Programs in Clinical Medicine. *Annals of Internal Medicine* vol. 102:529–537. *http://www.ncbi.nlm.nih.gov/entrez/query.fcgi?cmd=Retrieve&db=PubMed&dopt=Citation&list_uids=3883869.*

Tsui, F.-C., Espino, J.U., Dato, V.M., Gesteland, P.H., Hutman, J., and Wagner, M.M. (2003). Technical Description of RODS: A Real-Time Public Health Surveillance System. *Journal of the American Medical Informatics Association* vol. 10:399–408. *http://www.jamia.org/preprints.shtml.*

Wagner, M., et al. (2004a). National Retail Data Monitor for Public Health Surveillance. *Morbidity and Mortality Weekly Report* vol. 53(Suppl):40–42.

Wagner, M., et al. (2004b). Syndrome and Outbreak Detection from Chief Complaints: The Experience of the Real-Time Outbreak and Disease Surveillance Project. *Morbidity and Mortality Weekly Report* vol. 53(Suppl):28–31.

35

CHAPTER

Other Design and Implementation Issues

Steve DeFrancesco, Kevin Hutchison, and Michael M. Wagner

RODS Laboratory, Center for Biomedical Informatics, University of Pittsburgh, Pittsburgh, Pennsylvania

1. INTRODUCTION

An organization, when designing, developing, or acquiring a biosurveillance system, faces many decisions that range from the design of the physical facility that will house the system, to what type of software to use for the system, to methods for data collection and storage. Understanding of the full range of technical options and their strengths and limitations is essential for making informed decisions. The purpose of this chapter is to explore a set of key decisions not previously discussed in the chapters on standards and architecture. In particular, we will discuss the physical facility, software, functionality, methods of data transmission, and the use of data utilities. An organization should make most if not all of these decisions before acquisition or implementation. The immediate and long-term process of building and operating a biosurveillance system will be less costly and disruptive to the organization if the issues we discuss in this chapter are considered in advance of acquisition or implementation.

2. HOSTING FACILITY

Installing and maintaining a biosurveillance system is a large undertaking that will consume significant resources both initially and in the future. Because there are companies that can operate biosurveillance systems off site, an organization needs to weigh the costs and benefits of operating a system in-house or outsourcing its operation (or joining with similar organizations to share the cost and effort). The decision has significant ramifications for the organization. It will influence how the organization's staff spend their time: installing and maintaining a biosurveillance system, or using it as a tool in their daily activities. In some respects, this decision resembles the decision whether to develop a phone system in-house or use external services, except that the widespread availability of excellent telephone companies makes the latter decision quite straightforward.

For some organizations, this decision may be a very quick one: If an appropriate facility does not exist or resources are not sufficient to develop and maintain a facility—including funding, hardware, and seasoned maintenance personnel, the organization should not attempt to install and maintain a biosurveillance system. Building an adequate facility can be quite expensive. Each situation is different, but a rough estimate is a few hundred thousand dollars to a few million dollars in initial capital investment alone. In addition, there will be recurring costs owing to yearly maintenance contracts for equipment, the cost of utilities, and salaries for support personnel. In general, local hosting is expensive and time-consuming.

An alternative to the local installation, and perhaps a better use of funds, is to outsource the entire system to an application service provider (ASP). Using an ASP to install and maintain a system at its location can eliminate or reduce many of the costs associated with building, operating, and maintaining a hosting facility. The ideal situation is when the provider of the biosurveillance software (commercial vendor, government organization, university) also offers its product as an ASP service. Using a provider who offers an ASP service will not only be more cost-effective than a local installation but will also provide the comfort of knowing that the biosurveillance software provider is managing all aspects of the actual biosurveillance service for the organization.

Whether building a local system or using an ASP, it is important to review the considerations outlined in the following section to ensure the hosting facility is suitable for your biosurveillance system.

2.1. Building and Environment

To achieve the goal of maintaining a highly available system (one that is available 24 hours a day, seven days a week), whether it be for Web-based disease reporting, hospital infection control, electronic laboratory reporting, or other purposes, you must consider the physical and environmental characteristics of the facility housing the system. In particular, you must consider location, power, cooling, security, cabling/connectivity, monitoring/management, and service/maintenance contracts. There are entire books devoted to many of these factors. In this section, we provide a brief overview of each in order to give you a base understanding. We suggest that you skim these sections at this time and note them for future reference when you are ready to start your design process.

2.1.1. Location

When assessing the suitability of a potential hosting facility, whether it be for the primary location or a backup location,

ISBN 0-12-369378-0

take into consideration how prone the site is to a natural disaster, be it an earthquake, hurricane, tornado, volcano, landslide, or flooding. Evaluate the surrounding area; talk to local utility companies and to neighboring landowners to determine if the site is susceptible to other types of interruptions. Local landowners can provide insights to problems that may not be available from formal sources. Examples of this type of information might include knowing that the local power grid is prone to outages during bad weather or that new construction is about to begin in the surrounding area, which may translate to backhoe season. With any proposed nearby construction, a concern to be aware of is the digging up of water lines and communication cables. Also, ask if the local storm sewer system ever overflows during heavy rains. Good site selection and proper planning, design, and maintenance can help avoid main outages. Evaluate the site completely internally and externally.

2.1.2. Power

Stable power is the most important resource to any system. Take it away and your system will come to a grinding halt (and may not restart). When reviewing a facility's electrical system, pay particular attention to the size, number, and location of feeds servicing the building; the internal division of power to different circuit panels; grounding; and, of course, the size and type of uninterruptible power supply (UPS) and generator(s) that are in place in the event that the building looses power.

The location of the electrical service coming into the building is critical. Ideally, the electrical service to a building will be redundant, with two service lines entering the facility from opposite sides of the facility. This configuration reduces the chance of the facility losing power owing to any construction or external mishaps.

A central in-line UPS system is the key to preventing downtime caused by short power outages, surges, or brown outs. The UPS will have the building's main power feed(s) connected to its input, will condition the power, and then will supply the conditioned power to the attached systems within the facility. The UPS should be able to maintain this supply for an average of 10 to 20 minutes (Liebert Corporation, 2003).

There should be a diesel or gasoline generator to provide sustained electrical backup to the UPS system. There should be one or more generator units, depending on the level of reliability required and financial constraints. The unit(s) serves as the main source of power for the facility in the event that the main power feed(s) go down. During a power failure, the UPS system will automatically take the load of the attached systems until the generators can start and reach certain operating parameters. The generator will automatically assume the load of the UPS and continue to supply power to the systems until the main power feed(s) is re-established. The period that the generator can supply power to the systems depends on the power draw from the systems and the amount of fuel on site. Typically, an on-site fuel tank will hold 1000 gallons or more and support the operation of the generator(s) at maximum capacity for eight hours or more. If the facility is located in an area that is prone to harsh weather, such as large snow falls or flooding, it is highly recommended to have two different fuel suppliers in order to minimize the possibility of running out of fuel owing to one supplier not being able to access the facility from their location.

2.1.3. Cooling

Computers produce heat. To maintain normal system performance and to allow people to service the machines, the average operating temperature in a hosting facility should be between 64°F and 70°F.

Today's servers produce between 1600 to 3400 British thermal units (BTU) per hour, and a typical cabinet enclosure can house up to 42 of these units (a typical room air conditioner generates 5000 BTU per hour of cooling). The facility's cooling system needs to be sized appropriately to the load, and the backup power system needs to be able to sustain the cooling systems in the event of a building power loss. If the cooling systems go off-line for even a short period, the ambient air temperature of the server room could quickly increase to the point of causing machine failure.

2.1.4. Security

The most secure facility has most, if not all, of the following security measures: fencing, gates, guard stations, cameras, and motion detection systems to reduce intruder access. Inside the facility, additional provisions exist, including mantraps in lobbies and common areas, card and/or biometric readers, more cameras, and a fire alerting and suppression system.

The fire suppression system in a facility will typically be a very early smoke detection and annunciation system that notifies staff of a problem before smoke or fire are seen. In the event of a fire, a preaction sprinkler system activates and controls the areas and amount of water release. FM200 gaseous suppression systems work in conjunction with a preaction sprinkler system. The facility should utilize zoned systems, so that a fire extinguished in one area will not affect equipment and personnel in other areas.

2.1.5. Cabling

A structured cabling system is critical to ensure reliable data transmission within a facility as well as provide a reliable path for data access outside the facility. There are many different structured cabling system designs and types, but all should follow the Electronic Industries Alliance/Telecommunications Industry Association (EIA/TIA) transmission and installation standards (TIA, 2005).

2.2. Network Connectivity

Once the facility's physical and environmental characteristics are deemed adequate, examine the network connectivity. Analyze the entire network, starting from the point where the system connects to the local area network (LAN) all the way to the outbound Internet connections. The goal is to identify any single points of failure (i.e., any single device or connection that, when shut off, will disable part or all of the biosurveillance system), for example, one network connection from the system to the LAN, one network switch that all incoming and outgoing traffic must traverse, or only having one connection out to the Internet (or two Internet connections taking the same path out of the building). While you assess the network topology, take note of any physical vulnerabilities that may be present, such as unmarked network cabling being run through open spaces where it can be confused with other types of cabling, network devices not being securely located, or diverse Internet connections traversing the same physical path outside of the building.

2.3. Monitoring

A hosting facility should have management tools that the operators of the facility use to monitor, measure, and manage the performance and availability of systems and applications. This service needs to have a tie-in to all of the key components of the facility, including, but not limited to, generators, fuel tanks, cooling units, humidity and water sensors, fire protection, security and network services, and your system. The monitoring system will have a user interface for operators that provides a clear and comprehensive view into all of these key systems so operators can monitor performance, spot incipient problems, and predict system growth needs (e.g., the need for more disk capacity).

2.4. Service and Maintenance Contracts

A server facility has many electrical and mechanical systems, including diesel generators, air conditioners, computers, and disk and tape systems. Having appropriate maintenance contracts in place for regular service of all facility equipment is necessary to maintain high availability. The importance of maintenance contracts is easy to overlook, but they should receive the same attention as other factors to reduce down time and prolong the life of the equipment. Each system or piece of equipment may have a different service period, so work with the equipment manufacturers to establish proper service intervals.

Figure 35.1 summarizes factors to consider when evaluating a hosting facility.

2.5. Recovery from System or Facility Failure

Once the server facility is selected or designed, it is time to decide what means you will use to prevent interruption of service or loss of data. Designers use the concept of mean time to repair (MTTR) when weighing outage related decisions.

MTTR is the average amount of time required to restore a system to a normal working condition from a failure. The designers should establish the acceptable MTTR for the system (i.e., the acceptable "downtime").

It is possible to develop or acquire a recovery strategy that has zero MTTR, which we refer to as "Wall street" reliability, because financial transactions are protected in this manner. To accomplish this, you must have a second mirrored site (commonly referred to as a *disaster recovery site*). The secondary site takes over all functionality of the primary site in the event the primary site goes offline. The site is called *mirrored* because it receives a mirror copy of data received by the primary site. This recovery strategy may come with a large

1. No single point of failure
2. Alternate communication and electrical paths
3. Backup systems respond automatically without human intervention
4. Power system to support equipment with dual power supplies from different sources; All equipment should be properly grounded
5. Air-conditioning should be available in sufficient quantities to allow all systems and network equipment to operate within their published environmental conditions
6. Alarm and fire protection systems isolated and zoned to not affect the services outside of any area of concern
7. Preventative maintenance contracts in place for all major systems avoiding unnecessary service interruptions
8. Emergency response procedures in place, including documented change management, equipment manuals, network architecture diagrams, system schematics, labeling conventions and personnel / vendor problem escalation charts

FIGURE 35.1 Data center quick checklist. Summary checklist to determine a facility's readiness to be classified as server grade and support a highly available system.

price tag, however, because it requires a duplicate set of systems and network connections to data sources. The cost can be reduced by using a commercial disaster-recovery service because it already exists and multiple customers share its cost.

If for some reason, such as cost, you decide not to use a disaster-recovery service, then the next level of reliability comes from operating a mirrored system at your primary facility, which will be as reliable and data-loss proof as using a disaster-recovery service except in the event of loss of the entire facility (assuming that your facility satisfies all of the power, cooling, network, and other criteria described above).

Still less expensive (and riskier) is to rely solely on periodic (typically daily) back up of data to tape or disk. The risk is that if the primary data storage system fails completely, you will lose all data for the period between the last backup and the failure. Perhaps more importantly, some or all of the biosurveillance system functions will be unavailable to users until the system is replaced and data reloaded.

3. SOFTWARE

Software is at the heart of a biosurveillance system. Software underlies the collection, storage, algorithmic analysis, and display of data. Although biosurveillance systems will always include manual elements, the long-term trend will be to automate as much of the process as possible to increase reliability and decrease time latencies. Given the centrality of software in biosurveillance systems, it is important to understand the basics of software.

Modern software is designed and built from existing software components, similar to a modern house in which the windows and door assemblies come prehung. Examples of pre-fabricated (pre-fab) components are database management systems (e.g., Oracle, Microsoft SQL Server, MySQL), Web Servers (e.g., Microsoft IIS, Apache), and application servers (e.g., WebSphere, JBoss, Tomcat). Just as in the construction of house, the use of these pre-fab components increases quality of the overall system and decreases its cost. Pre-fab components are used wherever possible with as little custom programming as necessary to knit them together.

Development of software is costly. It is so costly that humans systematically fail to appreciate how costly it is. On average 71% (Standish Group, 2004) of software development projects experience problems that result in delays and cost overruns. Nearly 20% of projects completely fail.[1]

The most effective way to mitigate this risk is to keep software projects small, as the effort and cost is not linearly related to project size. Small tightly defined software projects have significantly fewer problems. In stark contrast, the development of a biosurveillance system from scratch is a very large, poorly defined task. Building a biosurveillance system from components partly addresses this risk, but a far less risky approach is to acquire an existing biosurveillance system and to extend and customize the system to meet the organization's needs.

3.1. Software Selection

Organizations must select software. They must select the pre-fab components from which to create a system and possibly the entire system. The key criteria include the software provider's track record and the functionality and technical evaluations of the software. Neglecting to conduct due diligence on these criteria can result in significant expense and lost time. In the worst case, the result may prove useless to the organization.

The first criterion is the provider's track record with the software. References from those who use the software are the best way to assess a provider's track record. The goal of checking references is to determine whether users are satisfied with the software. When asking for references, the organization should also obtain the number of customers using the software. A large, well-satisfied user base is the best indicator that the provider's software performs well. It is important to ascertain the financial position of the company or division providing the software. Large corporations frequently request financial statements from small software providers to demonstrate that the provider has the financial strength to continue to improve the software, fix defects, and provide support.

It is almost inevitable that an organization will work with a small provider offering new software. In this situation, there is risk that the provider may not be in business to support the product in the future. Many organizations require a code escrow arrangement, which places a copy of the source code of the software into the hands of an escrow company. A code escrow provides protection for the organization by ensuring a copy of the source code is available in case the provider goes out of business or prematurely terminates ongoing support for the product.

Evaluation of software functionality includes a detailed assessment of the organization's needs against what the software provides. An evaluation should be done on a working product to demonstrate that the needs are truly met. It is insufficient to evaluate the behavior of sample or prototype software. This evaluation may demonstrate how the software *could* work, but not how it *actually does* work. The original target market for the software must also be included in an evaluation of functionality. As markets change, software providers adapt by redirecting their software to target newer, emergent markets. When providers try to adapt their existing software to a new market, they may overlook functionality that would have been created had the software originally been designed for the new market. To compensate for the missing functionality, providers adopt the terminology of the new market, incorporating a slightly different meaning to fit

1 The FBI Trilogy project is a recent example of large project failure. After consuming $170 million dollars the FBI chose to cancel this project in 2005 outright (Sharma, 2005, Charette, 2005).

their software. When dealing with software that has been redirected, it is important to ensure terminology has not been redirected as well.

The final criterion is a technical evaluation of the software. The technical evaluation should answer two questions: "Will the software work well in my organization?" and "Is the software designed to hold up overtime?" To answer the first question, the acquiring organization should create a list of the pre-fab components and software languages it currently uses and supports and compare it to the list of components and languages utilized/required by the software. The software will not fit well in the organization's environment if there is a mismatch. For example, if an organization already uses Oracle as its database, the addition of new software that requires Microsoft SQL Server will increase the operational costs to the organization of supporting two database management systems, and require additional technical skills. To answer the second question, the organization must understand the design of the software, which we discuss next.

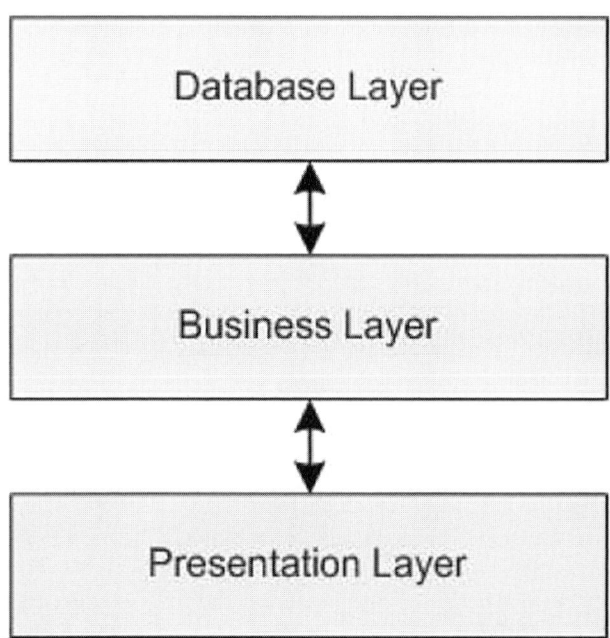

FIGURE 35.2 Three-layered design of software systems. Modern software design involves layering. This figure shows a three-tier design involving a database layer, a business layer for distributing and processing data, and a presentation layer for displaying data and results of analyses.

3.2. Software Design

In the construction of a biosurveillance system, we are concerned with individual pieces of software but also software systems. A *software system* is a group of software components that operate together for a common purpose. These components may operate on separate computers and even in different geographic regions, but they work according to a coherent design. Improperly designed software systems are frequently abandoned because of high maintenance costs.

A typical modern software system uses three layers (also called *tiers*): a database layer, a business layer, and a presentation layer (Figure 35.2). A layer is a conceptual way of grouping similar functions together so that each layer can be built and modified relatively independently of the others. The database layer stores data and responds to requests from the business layer. Requests to the database layer may be to create records, read records, update records, and delete records. The business layer is responsible for distributing data, processing data, and making requests to the database layer. Algorithmic analysis of biosurveillance data is an example of a function that would reside in the business layer. The presentation layer is responsible for displaying the data obtained from the business layer to the user. A map-generating program would reside in the presentation layer.

System designers make many decisions about which pre-fab components to include in these layers and how to arrange these components. The wide-scale adoption of this three-layer design, however, has resulted in standard approaches to the layers. The PHIN standard for the database layer, for example, recommends commercial off-the-shelf software (e.g., Oracle, SQLServer, or Sybase).

System designers often design the business and presentation layers together. This design practice does not violate the layering principle because the resulting design still maintains the separation of the layers. There are two standard approaches for the construction of the business and presentation layers: the Web-application approach and the desktop-application approach. Web-based applications use an Internet browser as the presentation layer. The browser accesses a Web server and Web application server in the business layer. Desktop applications locate the presentation layer on a user's local computer. In the desktop application approach, the business layer may reside on the user's local computer or on a central server. An advantage of a desktop application approach is the ability for the program to interact with items on the user's computer, including other applications such as Excel, Access, and Word. However, the desktop application approach entails maintenance issues on the local desktop, which may involve significant expense if there are many users of the system. The software must be installed and maintained on each machine. Web-based applications have the advantage that the technical staff can install software updates on the central servers.

3.3. Toolsets

A toolset is basically a programming language. Modern toolsets (e.g., the *Microsoft .NET* toolset) also include a set of pre-fab components that are commonly used to build software. These components make the difference between building a house with the old-fashioned hammer and nails approach and framing a house with a high-performance pneumatic nailer or preassembled walls. The components in the toolset vary in size and complexity. Some are small and simple, such as a tool for formatting data that are to be presented to a user on a screen. Others, such as Axis, which creates Web services, are quite large and complex.

The choice of toolset affects the cost and effort to build software initially and to add functionality or customize it later. The most important factors in choosing a toolset (or in evaluating the toolset being used by a provider of software that you are considering) are toolset popularity, experience, and productivity.

Toolset popularity is the number of software developers using the toolset. Popular toolsets often have articles, code examples, and complete solutions to common problems freely available on the Internet. This free support improves productivity. Toolsets used in the 1990s for constructing software systems are unpopular enough today that it is difficult to build or maintain a product based on them. These toolsets include FoxPro, Delphi, Progress, and PowerBuilder. The most popular toolsets at present are Java and Microsoft. NET.

Toolset experience is the measure of the years of experience that a particular software developer has using the toolset. Software developers with many years experience using a particular toolset are significantly more efficient as system developers. The Microsoft .NET toolset faced this problem when it was initially released.

Toolset productivity is the final factor to consider. Toolset productivity is the measure of how efficiently programmers can accomplish a programming task by using the toolset. Research on software languages has demonstrated that programmers are much less productive using languages such as C and C++ than Java or Microsoft .NET in the construction of software systems.

Current toolsets for constructing Web-based applications are LAMP, Java, and Microsoft .NET. Desktop applications are being constructed using Java or Microsoft .NET.

3.4. Open Source Versus Proprietary Software

Open source is a movement in the software community that has been growing for the past 20 years. The fundamental idea of open source is that software—and its source code—should be freely available.

The obvious benefit of open source to the biosurveillance community is free software. A less obvious—but perhaps more important—benefit derives from the availability of source code, which allows organizations to more easily tailor the software to meet their needs.

If a biosurveillance organization has access to the source code, it can modify the software to meet its needs. An organization with a need not anticipated by the software designer, say incorporating a new algorithm such as BARD, can modify the code so that the system can interact with BARD. Or it can ask or contract with another organization (such as BARD's developers) to do the work. In contrast, if a biosurveillance organization does not have access to the source code, it is at the mercy of the organization that owns the source code, whose development priorities and cost structures may not be compatible with the organization's schedule or budget. Even worse, the organization that developed the software may no longer exist.

The benefits of open source are so compelling that it is an emerging criteria for the selection of software in any industry.

4. SUPPORTING THE BIOSURVEILLANCE PROCESSES

The technical topics that we have discussed in Part VI to this point—standards, architecture, server facility, and software—are foundational to the correct design of a biosurveillance system. They are necessary, but not sufficient, for long-term success. In this section, we discuss additional decisions that designers face. In particular, we discuss the different options available for supporting several core processes of biosurveillance (Figure 35.3): collection of surveillance data, persistent storage of data, the collection of additional data, and data linkage. These processes correspond to the key components of most biosurveillance systems illustrated by Figure 35.4.

4.1. Data Collection System

Data collection refers to the processes discussed in the previous chapter—extraction, transmission, transformation, and loading—with which a biosurveillance system obtains data for analysis. A data collection system should be flexible (accommodate the needs of different data providers), reliable, and have minimal data transmission latencies. In this section, we discuss the two data collection processes that occur at the data provider facility—data extraction and transmission. For each of the processes (extraction and transmission), we discuss the technologies commonly employed in biosurveillance.

4.1.1. Data Extraction Methods

Data extraction refers to the initial process of obtaining data from the data provider's systems before the data provider transmits the data to the recipient. There are two methods for that have been used in biosurveillance to extract data from a data provider's systems—query-based methods and message filtering (Lober et al., 2004).

The simplest data extraction method is a query to a database (e.g., an SQL query that retrieves all emergency department (ED) registration for the period midnight to midnight). Most organizations are capable of extracting data using queries, however, the requirements are as follows: (1) that the data provider has access to its database, (2) that the database supports database queries (most modern databases do, but some organization may operate older systems), and (3) that the data provider has technical staff that can implement a script or other program that periodically queries the database. To ensure reliability (i.e., humans are fallible), this program should run automatically at a predefined time interval. An example of an query-based extraction is the extraction of over-the-counter medication sales information from a retailer

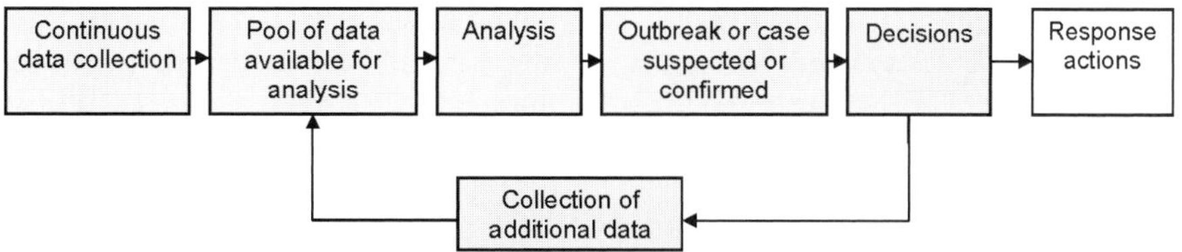

FIGURE 35.3 The biosurveillance process (Figure 1.1 from Chapter 1, reproduced here for convenient reference).

that contributes data to the National Retail Data Monitor (NRDM). Most of the retailers run a data extraction program on their databases every 24 hours that retrieves the number of medications sold on the previous day at each of their stores and stores the data in a file that is transmitted later in the data collection process. A limitation of query-based extraction is that it may put a load on the data provider's database systems, depending on which data are being queried for and how the data system is organized. For this reason, most organizations run query extraction procedures during off peak hours (e.g., midnight).

An alternative method is message filtering. A message is a discrete object of communication presented in a standard format. For example, a message may contain a information about a patient registering for care, a request back for more information, or a request to perform an action or an acknowledgement that an action was performed. The term *message filtering* refers to any process that selects individual messages from a stream of messages. For example, a message filter might be configured by a hospital to select only admission messages from a stream containing all admission-discharge-transfer messages.

Prerequisites for using filtering as a data extraction method include (1) a messaging system at the data provider, and (2) the ability to filter the system for the desired messages. The filtered messages can be transmitted immediately, queued for later transmission (e.g., when the receiving system is not functioning), or saved to a file for later transmission.

4.1.2. Transmission Methods

A transmission method is a way of moving data between computers over a network. The two most important characteristics of a transmission method are security and guaranteed delivery. A method is secure if only the intended recipient can view the data. Secure transmission is important because most biosurveillance data are confidential. Guaranteed delivery is the assurance that the receiver receives the transmission in its entirety. Guaranteed delivery is important because incomplete data are more difficult to analyze. Transmission methods can be either file or message based.

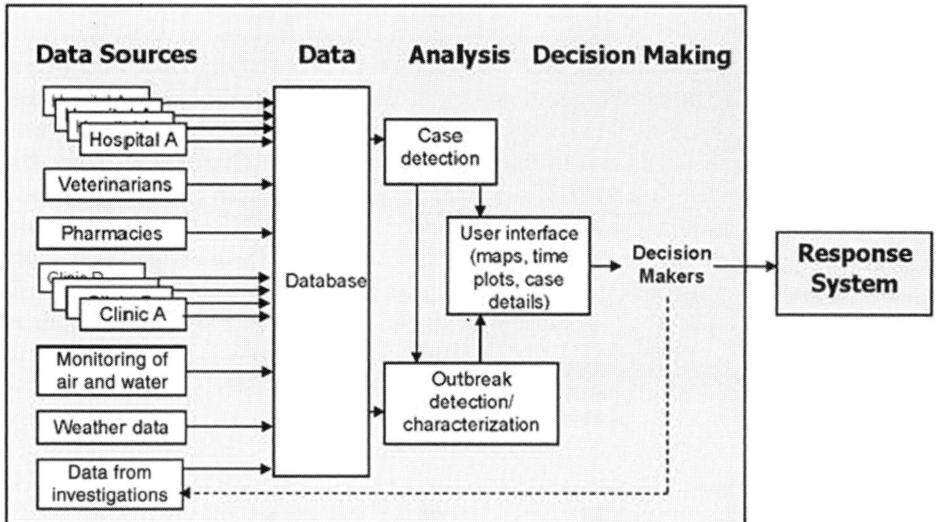

FIGURE 35.4 Basic components of a biosurveillance system (reproduced from Chapter 1 for convenient reference). The dashed line represents look-back for additional information to any data source.

File-Based Transmission. File-based transmission methods are ideal for the results of query-based data extraction but can also be used when a data provider filters messages and saves them to a file. Examples of file-based transmission methods include FTP, SFTP, HTTP, HTTPS, and WebDAV.

FTP (File Transmission Protocol) is a commonly used method for transmitting files. FTP guarantees delivery but is not secure. FTP does not encrypt data being transmitted. Moreover, FTP represents a security risk to the overall system because it sends passwords in clear text over the network. Anyone with access to the network can run a sniffer program to intercept passwords.

Secure FTP (SFTP) addresses the security problems of FTP. FTP and SFTP are currently the most popular file-based transmission methods used by healthcare organizations.

Hyper Text Transfer Protocol (HTTP) is familiar to most people in the context of a Web-browser downloading files from the Internet, but it also supports uploading of files to a Web server. Similar to FTP and SFTP, HTTP has a secure version, HTTPS. Both HTTP and HTTPS provide guaranteed delivery, but HTTP is not secure. Healthcare organizations rarely use HTTP or HTTPS for data transmission because FTP and SFTP are older protocols and the technologists in healthcare organizations are comfortable with their use.

WebDAV (Web-based Distributed Authoring and Versioning) is an extension to HTTP. WebDAV is secure and guarantees delivery. Although healthcare organizations do not currently use WebDAV, other organizations use it for data transmission.

HTTP and WebDAV are alternatives to FTP/SFTP for file-based data transmission. However, HTTP and WebDAV are not widely used by healthcare organizations or over-the-counter data providers. FTP/SFTP is the most widely used file-based data transmission method and will continue to be for the near future.

Message-Based Transmission. In general, message-based transmission is preferable when it is feasible. It does not put load on the data provider's database, and it has more built-in fault tolerance (it can store messages in a queue if the receiving system is temporarily unavailable).

There are two types of message transmission: point-to-point messaging and message busses.

Point-to-point messaging refers to any transmission method for sending messages from one sender to one recipient. Sending a message from a hospital's ED registration system directly to a biosurveillance system would be an example of point-to-point messaging.

A *message bus* is a transmission method that employs specialized software called a message router. Unlike point-to-point messaging, a message bus can accept messages from multiple senders and send each message to one or more receivers over a network. For example, the healthcare industry makes extensive use of HL7 message busses, which are managed by HL7-message routers (a.k.a. integration engine). The HL7-message router accepts messages from the ancillary systems of the hospital and then routes the messages to the appropriate receivers (other information systems in a hospital or an external biosurveillance organization). Message buses provide guaranteed delivery and security. An example of a biosurveillance system that uses message-based transmission is the RODS system, which takes advantage of the fact that healthcare systems often use HL7 message buses to route ED registration information internally (to their billing, laboratory, and other information systems). When a data provider has an existing message bus, it is an obvious choice for data transmission. All that is required is modification or configuration of the message router to direct messages to the biosurveillance system.

The capabilities of the organizations sending and receiving the data will dictate the set of transmission methods that you must include in your biosurveillance system. For example, if a biosurveillance system is interacting with a data provider that already employs messaging systems, you should employ message filtering and message-based transmission methods. If a data provider offers you a Web service (discussed in the previous chapter) because they have a service-oriented architecture (SOA), you should connect to their Web service for data.

4.2. Data Storage System

The most important function of the data storage system is to protect the data from loss. However, the data storage system must also store data in such a way that they are logically organized and quickly accessible to other services.

Computer scientists are a bit religious about the correct use of the terms *database* and *database management system* (DBMS). A *database* is simply a defined structure that houses a collection of data. A *DBMS*, such as Oracle, MySQL, or Microsoft SQL Server, is a computer program (usually very large) that not only allows a user to define a structure (a database) but provides utilities for managing the structure (e.g., adding a record, deleting a record, backing up the data, optimizing retrieval of records from the database, checking integrity).

DBMSs are sufficiently mature that the only decision you face about the DBMS is whether to use Oracle, Microsoft SQL server, or possibly an open source DBMS to economize. The selection is usually dictated by either the requirement of other biosurveillance software (some systems only can interact with one DBMS) or whether your information department or the ASP that you plan to use has a preference.

The decision that you should focus on is sizing of the data storage system, including disk space and hardware. Although the individual data messages being collected by a biosurveillance system are usually small (approximately 4 kb), the overall storage required may be quite massive depending on how many facilities are sending data to the system, how many types of data they are sending (e.g., ED registrations and laboratory results), and whether you are going to rely on data warehousing to provide fast access to the data (see caching discussion below). The amount of data collected and stored will increase over time because the analysts or analytic programs typically require access to historical data. Improper planning will limit the overall capacity of the database. A DBMS on a single server for the collection of ED data from 200 hospitals is an example of a system that will have problems. A better approach is to connect the server to a storage area network (SAN) to allow for expected growth of the data storage system.

4.3. Look Back/Investigation Support

A key property of biosurveillance is its cyclic (iterative) nature in which analyses of currently available information lead to additional, more directed, data collection (e.g., as during an outbreak investigation).

The topic of how to provide information-system support to an outbreak investigation is an open area of research. Here we focus on one particular topic: how to obtain additional data from a hospital about a patient of interest. This functionality is not widely available, and unless some thought is put into how it will be accomplished, you may have to redesign your data collection approach at a later time.

As a concrete example of the functionality, let us assume that an epidemiologist receives an alert from the planned system about a high level of respiratory illness in the community and wishes to obtain additional information about one or more patients. We and others refer to this functionality as *look-back* when the additional data can be obtained by request to another computer system or individual (i.e., the information is available somewhere, and it is simply a matter of getting it).

The previous sections focused on options for transferring data unidirectionally from one organization to another—from a hospital to a health department, from a laboratory to a centralized data storage facility, or from a biosurveillance system to an external party (health department or other). Most of these data transfers view the world as involving only one-way communication.

Look-back requires two-way communication, which is best implemented by using message-based transmission. Message-based data transmission has several properties that make it uniquely suited for this type of communication. Messages support asynchronous communication, which is required when human intervention may be necessary to fulfill a request (which may nearly always be the case for the near term when requesting additional data from healthcare organizations). Most message-based systems also associate response messages with request messages. In the event of a long delay in response, or multiple responses, this association provides a means of deciphering the history of communication between organizations.

Look-back requires not only a two-way messaging capability (asking for more information and receiving it) but also the capability to include in the request message the identity of the individual for whom additional data are requested. This seemingly simple requirement is surprisingly difficult to satisfy owing to the current ethical balance between protecting patient's right to confidentiality and the benefit to the public's health. At a minimum, a biosurveillance organization must send sufficient information back to a data provider to allow the data provider to uniquely identify the patient. We refer to information that is sufficient for the receiving organization to identify the individual as an *identifier.* A basic requirement for look-back is that the biosurveillance organization receives this identifier as part of its routine data collection process.

Table 35.1 lists a variety of identifiers that you could specify for your routine data collection and our impression of the administrative feasibility of obtaining them routinely for different types of biosurveillance data. Our opinion reflects the current balance between the need to protect confidentiality and the benefit to biosurveillance. Although many state health statutes allow a health department to collect any data needed for public health surveillance, in practice fully identifiable data are only collected for notifiable diseases (first row in Table 35.1). In current practice, all other data would require encryption of the identifier before transmission to a biosurveillance organization (Row 4), which would require that the data provider not only encrypt the identifier but also be capable

TABLE 35.1 Technical and Administrative Feasibility of Different Look-Back Identifiers for Different Types of Biosurveillance Data

Identifier Sent Routinely (To Enable Later Look-Back)	Additional Work for Data Provider	Privacy Protection	Data Endorsed by Current Health Statutes
Unencrypted identifier (e.g., social security number)*	None	No	Usually only for reporting notifiable diseases to local and state health departments
Unencrypted identifier; biosurveillance organization encrypts the identifier on receipt	None	Yes, but depends on trust	Usually only for reporting notifiable diseases to local and state health departments; for other data, it is a gray area
Encrypted identifier, but biosurveillance organization holds the decryption key	Yes	Yes, but depends on trust	Usually only for reporting notifiable diseases to local and state health departments; for other data, it is a gray area
Encrypted identifier	Yes†	Yes	All

*This identifier might be encrypted during data transmission (i.e., the data transmission protocol would encrypt it before transmitting it over-the-wire and then decrypt it on the receiving end), but it would appear in the biosurveillance organization's data systems in unencrypted form.
†A possible exception is when the data provider already has an internal identifier for an individual that is not publicly known.

of decrypting the identifier when it was provided back to the data provider in a request for additional data. This functionality is unlikely to exist in most data providers, and a biosurveillance organization would have to negotiate with the data provider to develop it. Row 2 is an approach that requires no additional work by a data provider but is a gray area in terms of health statutes. Row 3 requires much less technical effort on the part of a data provider than does Row 4, and is more protective of privacy, so is perhaps less of a gray area than is Row 2.

4.4. Data Linkage

Data linkage is the joining of two or more datums that are characteristics of the same entity (e.g., because they are observation about the same patient, same physical location, or same blood sample). Data linkage is trivially easy if datum A and datum B share an identifier that uniquely identifies the entity (e.g., the patient's social security number, street address, or specimen identification number). Data linkage is devilishly hard or impossible if the observations do not satisfy this requirement.

The United States does not have a unique identification system for people. As a result, some hospitals use a patient's social security number to identify the person. Other hospitals use internally generated unique identification numbers. As a result, linking clinical data about the same individual across multiple healthcare contacts (e.g., information obtained from a call center, a subsequent visit to an doctor's office, and a subsequent hospital admission) cannot be done without some errors.

There are statistical algorithms that take as input whatever information the organization about the two datums (e.g., the name, street address, phone number, and zip codes) and produce as output a probability that the observations are for the same entity. The operator of the biosurveillance system must set a threshold probability above which the linkage is established and below which the data will not be linked. This technology is at the heart of all master-person index projects.

5. MAKING USE OF DATA UTILITIES

Developers of biosurveillance systems should be aware of organizations that are undertaking the task of obtaining biosurveillance data on their behalf. Some of these efforts are mature, and their technical requirements for transmitting data should be taken into account in designing a system. Other efforts are more nascent, and whether you should take their technical requirements into account when designing your system will have to be assessed individually.

5.1. National Retail Data Monitor

The NRDM operated by the RODS Laboratory, University of Pittsburgh, is a data utility that it provides over-the-counter sales data for use by health departments for biosurveillance. As discussed in Chapter 33, the NRDM provides a Web service interface to other biosurveillance systems.

5.2. The National Weather Service

The National Weather Service operates a data utility that makes weather data freely available for download (see Chapter 19). The data are hourly observations made by automated weather stations at airports.

5.3. Surveillance Data, Inc.

Surveillance Data, Inc., sells healthcare data as well as market research data to organizations involved in healthcare and pharmaceuticals. They have an area of service in provisioning disease surveillance data *(http://www.surveillancedata.com/index.php=page_id=sur)*.

5.4. National Oceanic and Atmospheric Administration

The National Oceanic and Atmospheric Administration (NOAA) of the U.S. Department of Commerce operates a data utility primarily for use by aviation. Weather data from the United States, as well as around the world, are collected in real time, in standard formats, and integrated in a central location that are publicly available without technical or administrative barriers *(http://weather.noaa.gov/)*. The NOAA services is also discussed in Chapters 19 and 32.

5.5. Centers for Disease Control and Prevention

The BioSense program has plans to offer various types of biosurveillance data as a data utility (GAO, 2005). As of this writing, it receives International Classification of Diseases (ICD)-coded data from the Veterans Administration (VA) and Department of Defense Healthcare systems and laboratory test orders (not results) from one national laboratory company. Data access methods that will allow other organizations to receive the data are still under development.

5.6. National Bioterrorism Syndromic Surveillance Demonstration Project

Although the National Bioterrorism Syndromic Surveillance Demonstration Project is not functioning as a data utility, it collects data from large health maintenance organizations (HMOs) and health plans, some of which span many public health jurisdictions. It may be possible to negotiate data use agreements with this organization *(https://btsurveillance.org/btpublic/)*. We are not aware of the technical transmission methods that might be employed.

5.7. Department of Defense and VA

At present, the Department of Defense and VA are functioning as a data utility for BioSense, the BioNet project in San Diego, and for the Electronic Surveillance System for the Early Notification of Community-based Epidemics (ESSENCE) system in several jurisdictions. It may be possible to obtain data from them directly. We are not aware of the technical transmission methods used.

6. SUMMARY

This chapter explored key decisions that designers and developers of a biosurveillance system should make before system acquisition and/or development. This set of considerations are in addition to those discussed in Chapters 32 (Standards) and 33 (Architecture).

Perhaps the most critical decision from the perspective of project success and cost is whether an organization should attempt to develop and operate a biosurveillance system itself, use a professional ASP, or join forces with other organizations (e.g., health departments) to form a regional system. Because most biosurveillance systems must be available 24 hours a day seven days a week, a physical plant that is capable of supporting reliable operation is necessary. If uninterrupted operation of the system in the face of complete loss of the facility is a requirement (the Wall Street Standard), then a second site is mandatory. Because of cost considerations, these requirements should bias decision makers toward regional joint projects or use of ASP service providers.

The choice of software is also important to project success. *De novo* development of software is to be avoided if possible, owing to the cost and risk of failure. When selecting software, it is important to review the track record of the provider as well as the functionality of the actual software, not a prototype. Decisions should be biased toward open source because of expected lower cost and the expectation that you will modify the software considerably over time. Although ideally software products are so well designed that they can be configured by end-users to meet their needs without the need for custom programming, software for biosurveillance is still so immature that no product will be that configurable. If you elect to use a product for which source code is not available, pay attention to whether the vendor will modify the software to your specifications in a timely and affordable way, and consider your options should the provider stop supporting the product.

Additional key decisions relate to the design of specific elements in a biosurveillance system such as methods of data acquisition and exchange and persistent storage of data. Many data transmission methods exist for collecting data, and you may have to use more than one to be able to work with different data providers. Because of a strong direction that industry is taking, we expect data providers to move toward Web services as the prime method for data transmission. For data storage, the sizing of a DBMS is a commonly overlooked area, as is the need for caching to optimize the speed at which data can be retrieved by end-users and programs. An almost universally overlooked requirement is the need to obtain identifiers with routine surveillance data to enable look-back. This seemingly small detail involves considerably technical and administrative attention owing to the complex interplay of individual and societal interests and the technical work that data providers may need to perform to enable look-back. Finally, data utilities exist, and you should take into account their data transmission methods when designing or acquiring your system.

REFERENCES

Charette, R.N. (2005). Analyst Corner: Eyes Wide Open. Framingham, MA: CIO.com. *http://www2.cio.com/analyst/report3837.html.*

Liebert Corporation. (2003). Five Questions to Ask before Selecting Power Protection for Critical Systems: A Guide for IT and Data Center Managers. Columbus, OH: Liebert Corporation. *http://www.liebert.com/support/whitepapers/documents/5_quest.pdf.*

Lober, W.B., Trigg, L., and Karras, B. (2004). Information System Architectures for Syndromic Surveillance. *Morbidity and Mortality Weekly Report* vol. 53:203–208 *http://www.ncbi.nlm.nih.gov/entrez/query.fcgi?cmd=Retrieve&db=PubMed&dopt=Citation&list_uids=15717393.*

Sharma, S. (2005). Should You Kill That IT Project? Fairfax, VA: Gantthead.com. *http://www.gantthead.com/article.cfm?ID=223079.*

Standish Group. (2004). Standish Group 2004 Chaos Report. *http://www.standishgroup.com/quarterly_reports/index.php.*

Telecommunications Industry Association [TIA]. (2005). IP/VoIP Standards Development. Arlington, VA: TIA. *http://www.tiaonline.org/standards/ip/.*

U.S. Government Accountability Office [GAO]. (2005). Information Technology: Federal Agencies Face Challenges in Implementing Initiatives to Improve Public Health Infrastructure, GAO-05-308. Washington, DC: GAO. *http://www.gao.gov/new.items/d05308.pdf.*

36
CHAPTER

Project Management

Neil Jacobson
BBH Solutions, Inc., New York, New York

Sherry Daswani
RODS Laboratory, Center for Biomedical Informatics, University of Pittsburgh, Pittsburgh, Pennsylvania

Per H. Gesteland
Inpatient Medicine Division, Department of Pediatrics, University of Utah School of Medicine, Salt Lake City, Utah

1. CBBS AND CBBS PROJECTS

A computer-based biosurveillance system (CBBS) collects and analyzes surveillance data. A CBBS project manager must understand not only generic project management but also the information technology (IT) underlying a project and how such a project differs from a typical IT project.

Common IT elements found in a CBBS include the following:

- A data warehouse as a data repository for the often very large databases (VLDBs).
- Data mining and statistical tools to analyze the data in the data warehouse.
- Extract, transform, and load (ETL) tools to prepare data to be added to the data warehouse. This preparation could include field mapping, data standardization, data parsing, and data interpretation.
- Messaging components to "listen" for data newly available.
- Reporting and visualization tools.
- Geographical information systems (GIS) to support spatial analysis of the data.
- Notification systems sending alerts based on CBBS analysis.

Figure 36.1 shows the elements of the RODS (Real-Time Outbreak Detection System) developed by the University of Pittsburgh. That system includes nearly all the elements described above.

Defining characteristics of CBBS projects include the following:

- Data providers are typically entities separate from the CBBS owner. The U.S. Department of Health and Human Services recommends that CBBS use "some of the extensive information that is already collected in automated form in the process of medical care delivery or administration of medical care benefits" (Lazarus et al., 2001). Existing CBBSs use data collected by government entities, care providers, pharmacies, and clearinghouses (Lazarus et al., 2001; Hoffman et al., 2003; Lombardo et al., 2003; Mostashari et al., 2003). Existing CBBSs also use nonclinical data, including school and work absenteeism reports (Lombardo et al., 2003).
- The quality of the data others provide is not usually known from the outset. Recipients of data from outside the CBBS must evaluate the quality of that data, because validity of the CBBS data depends on source data quality (Buehler et al., 2004).
- Data often include health information, which is subject to Health Insurance Portability and Accountability Act (HIPAA) and other privacy regulations. Organizations providing the data may require legal agreements describing the intended use of the data, as well as the users that will have access to the data. To address data security and confidentiality, the CBBS may need to implement a variety of functions, including user authentication, user-specific filters restricting data access to appropriate subsets of the data, and de-identification schemes.
- De-identified data to protect data privacy complicates processing to protect against the same incident being reported by more than one data source. Without the identifying information checking for and removing duplicates, de-duplicating presents significant challenges (Lombardo et al., 2003).
- Data are provided by multiple sources. CBBS projects must address the key issues described above for each source of data. Project teams must re-evaluate the effectiveness of any de-identification schemes with the addition of new data sources, because additional data changes the outcome of de-identification statistical analysis.
- Systems are designed for incremental addition of data from new sources and analysis tools. CBBS often includes flexible frameworks designed to allow rapid addition of new data, and new analysis tools. Building effective frameworks for the future addition of unknown data and tools requires significant software engineering expertise.
- CBBS is a quickly evolving field, and for example, no single analysis methodology has proven to meet all the requirements (Moore et al., 2002). The implication of this evolution

Handbook of Biosurveillance
ISBN 0-12-369378-0

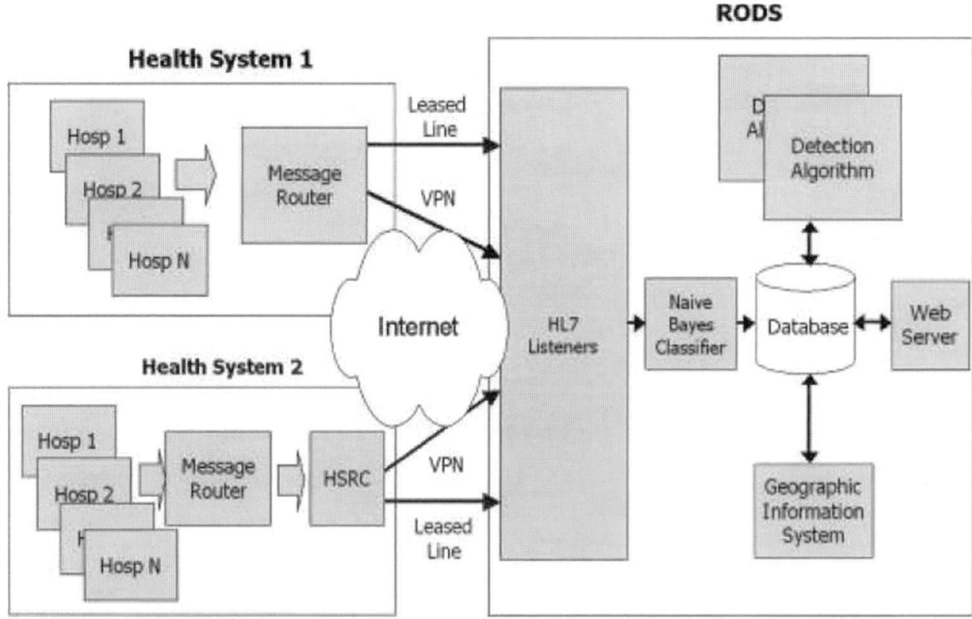

FIGURE 36.1 University of Pittsburgh RODS. (From Espino et al., 2004.)

is that CBBS project management will likely need to address not only changing requirements but also the rapid evolution of potential solutions.

- Few commercially available off-the-shelf CBBSs are available, with no clear market leader at this time.

The balance of this chapter describes the project management implications of the technology elements included within CBBS and the defining characteristics of CBBS projects.

2. PROJECT MANAGEMENT

Within the discipline of project management, a project is defined as "a temporary endeavor to create a unique product, service, or result," and project management as the "application of knowledge, skills, tools, and techniques to project activities to meet project requirements" (PMI, 2004). Project management success is completion of the project scope to the satisfaction of the project customer, on time and within budget.

To achieve success, project management integrates the project resources to produce the product and manages scope, time, costs, quality, human resources, communication, risks, and procurement.

2.1. Project Management Processes

As temporary undertakings, all projects have a beginning and an ending, and project management has corresponding "initiating" and "closing" processes. The middle, or work, of the project is managed by "planning," "executing," and "monitoring and controlling" processes. Effective project management involves repeated performance of these processes, as illustrated in Figure 36.2. Iterative application of the project management processes delivers the benefits of the "plan/do/check/act" model of quality management practices (PMI, 2004).

2.1.1. Initiating Processes

The initiating processes define and authorize a project. The initiating processes may occur outside the formal boundary of the project and may be performed by people other than the project team, such as the project *sponsor* (the person or organization funding the project), as illustrated in Figure 36.2. Key initiating activities include the following: establishing the organizational case for the project and its relationship to the organization's strategic plan, developing the project requirements, evaluating alternatives and describing why the specific project is best suited to meet the objectives, determining the initial scope description, determining what resources the organization is willing to dedicate to the project (people and funds), and assigning the project manager.

Involving customers and other stakeholders generally improves the probability of buy-in and customer/stakeholder acceptance. Project stakeholders include anyone a project is likely to affect. Customers and users are stakeholders, as are data providers, data privacy staff, and IT infrastructure providers. The product of the initiating processes is the project charter, which is a document that formally authorizes the project, defines the project's purpose and boundaries, and assigns resources. Its role is to empower the project team to undertake the project.

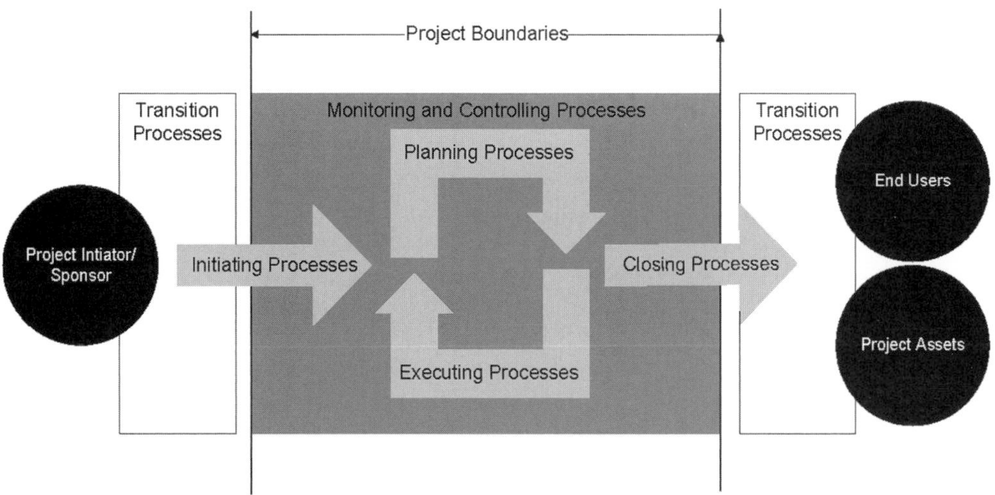

FIGURE 36.2 Project life cycle. (Adapted from Project Management Institute, *A Guide to the Project Management Body of Knowledge (PMBOK Guide),* 3rd ed., Project Management Institute, Inc., 2004. Copyright and all rights reserved. Material from this publication has been reproduced with the permission of PMI.)

2.1.2. Planning Processes

The planning processes define objectives and determine the course of action required to attain the objectives as well as the scope of the project (PMI, 2004). These processes comprise tasks commonly associated with project management, including the tasks shown in Table 36.1.

The planning processes begin early in the project, and the product of the processes is the project management plan.

We can think of the project management plan as the roadmap describing the path to successful project completion. The project management plan should include a project schedule as well as the various plans addressing each of the relevant project management areas: integration, scope, time, costs, quality, human resources, communication, risks, and procurement. The next section describes each of these areas in more detail. The project team performs the planning processes throughout the project to evaluate the need to adjust the plan in response to project realities.

TABLE 36.1 Common Project Management Tasks

Common Project Management Tasks
Refining the project scope
Breaking down the work of the project into pieces
Defining the activities that need to be performed to complete the project
Determining activity sequence and interrelationships
Determining necessary resources
Cost budgeting
Identifying and securing staffing
Scheduling
Identifying risks and developing plans to address them
Determining quality requirements and developing quality assurance plans
Determining project communications requirements
Developing project procurement plans
Developing project change control processes

Through the planning processes, the project team will define project-specific phases of work appropriate for the project life cycle. We discuss IT project life cycles and considerations in developing life cycles appropriate for CBBS projects shortly.

2.1.3. Executing Processes

The execution processes integrate resources to carry out the project management plan (PMI, 2004).

2.1.4. Monitoring and Controlling Processes

The monitoring and controlling processes regularly measure and monitor progress to identify deviations (known as variances) from the project management plan so corrective action can be taken to meet project objectives (PMI, 2004).

2.1.5. Closing Processes

The closing processes formalize acceptance of the project result by the project customer. Closing processes bring the project or phase to an orderly end, and transition the product of the project into the organization (PMI, 2004).

2.2. Project Management Areas

Projects are typically part of an organization. The relationship between the project and organization varies, and that relationship affects the scope of project management responsibilities. Although the nature of project management responsibility varies by organization and by project within an organization, project management encompasses many areas. The Project Management Institute's (PMI) list of project management areas, summarized in Table 36.2, provides a sense of the breadth of the project management task (PMI, 2004).

The *Project Management Book of Knowledge* provides comprehensive description of activities associated with each

TABLE 36.2 Project Management Institute Project Management Areas

Project Management Area	Description
Integration management	Unifying the activities and resources of the project to complete
Scope management	Ensuring that the project includes all the work required, but only the work required required to meet the project requirements
Time management	Ensuring timely completion
Cost management	Ensuring that the project can be completed within the project budget
Quality management	Ensuring that the project product will satisfy the project requirements
Human resource management	Organizing and managing the project team
Communications management	Ensuring timely and appropriate project communication
Procurement management	Managing the purchasing and acquisition of outside goods and services required to complete the project
Risk management	Identifying and managing project risks

Adapted from Project Management Institute, *A Guide to the Project Management Body of Knowledge (PMBOK Guide)*, 3rd ed., Project Management Institute, Inc., 2004. Copyright and all rights reserved. Material from this publication has been reproduced with the permission of PMI.

area, the interaction of the areas with the key project management processes described earlier, available management tools and techniques for each area, and inputs and outputs of each area (PMI, 2004).

3. PROJECT LIFE CYCLES

A project life cycle is as follows:

> The sequence of phases through which a project will evolve. Project life cycles are absolutely fundamental to the management of projects...They significantly affect how the project is structured. The basic life cycle follows a common generic sequence ... the exact wording varies between industries and organizations. There should be evaluation and approval points between phases often termed "gates." (Patel and Morris, 1999)

Each phase typically specifies what type of work is to be accomplished within the phase, what stakeholder should be involved, and what products (deliverables) result from the work of the phase. The deliverables support hand-off to subsequent phases. The IT project life cycle usually includes requirements, specification, development, and testing phases (Berkun, 2005). Each phase represents a significant shift in focus as the project progresses through succeeding levels of maturity and level of detail (Wideman, 2004).

Project management "manages" the work of a project phase by performing the project management processes concurrent with the work of the phase as illustrated in Figure 36.3. The initiation processes at the beginning of each phase can help keep the project aligned with organizational objectives. The closing processes for each phase often include phase-end reviews (known as *gates*) with formalized authorization to close the preceding stage and continue to the next phase.

Although most IT project life cycles include at least requirements, specify, develop, and test phases, there are many variations on this theme. This section describes two common IT project life cycles: the waterfall project life cycle, and the incremental life cycle.

3.1. Waterfall Life Cycle

The waterfall life cycle is arguably the first widely accepted IT life cycle (Bechtold, 1999). The waterfall life cycle is a sequence of phases in which the output of each phase becomes the input for the next phase. The system development life cycle (SDLC) is a widely known waterfall life cycle. The SDLC includes the usual *requirements, specify, develop, and test* phases, as well as three preparatory phases. The SDLC phases and a brief description of each are shown in Table 36.3.

Many organizations have adopted this life cycle as their organizational standard for IT projects. Two of the examples included at the end of this chapter applied this life cycle with successful results.

3.2. Incremental Life Cycle

The Incremental life cycle attempts to shorten the elapsed time between the requirements phase and delivery of tested

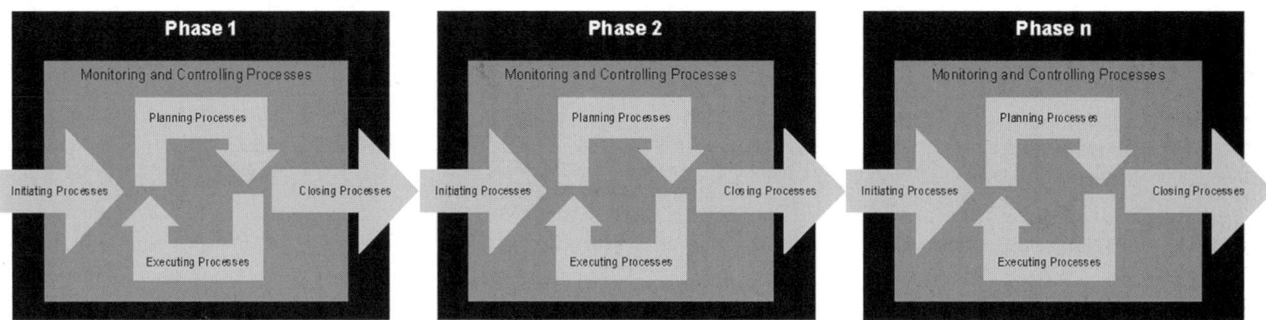

FIGURE 36.3 Project management processes and phases within project life cycles. (Adapted from Project Management Institute, *A Guide to the Project Management Body of Knowledge (PMBOK Guide)*, 3rd ed., Project Management Institute, Inc., 2004. Copyright and all rights reserved. Material from this publication has been reproduced with the permission of PMI.)

TABLE 36.3 System Development Life Cycle Phases

SDLC Phase	Description
Initiation	Sponsor identifies need
Concept	Defines scope and boundaries
Planning	Develops project management plan
Requirements	Analyzes user needs and develops user requirements
Design and specification	Transforms requirements into detailed system specifications
Development and implementation	Builds the system
Integration and testing	Demonstrates that the system conforms to requirements

functionality to increase the probability of project success (Beck and Andres, 2005). The incremental life cycle achieves this shortening by performing the requirements, specify, develop, and test phases to deliver fully functional *pieces* of the overall solution, and by iteratively applying the phases until the full system is completed. The iterations also promote overall project success by driving the integration of system components to demonstrate working functionality within each phase (Matta and Ashkenas, 2003). Figure 36.4 compares the waterfall life cycle with the incremental life cycle.

The unified software development process (UP) is a well-known incremental life cycle. The key characteristics of UP are that it is use-case driven, architecture centric, iterative, and incremental (Jacobson, 1999).

- *Use cases* describe a piece of functionality provided by a system that delivers a result of value to a user. Use cases mature during the project cycle as the stakeholders and project team gain greater understanding of the required functions.
- *System architecture* defines the elements (IT components) of the solution. If a use case is function, system architecture is form. The architecture describes the organization and interaction of the solution elements, participating system(s), and platforms (e.g., operating systems, database management systems, Java runtime environment).
- The *iterative and incremental approach* divides the system implementation undertaking into mini projects called *iterations*. Each iteration produces a product, called an *increment*. Controlled iterations reduce the risk on a single increment (Jacobson, 1999).

The UP life cycle includes the following phases, each with different focus:

- Inception—The inception phase is focused on establishing and documenting the business case and baseline vision for the solution.
- Elaboration—During this phase the team translates business requirements into solution needs. In the elaboration phase, the focus of the project shifts to discovery of the system architecture to provide a stable basis for the implementation effort in the development phase. The architecture evolves out of a consideration of the most significant requirements, those that have a great impact on the architecture of the system.

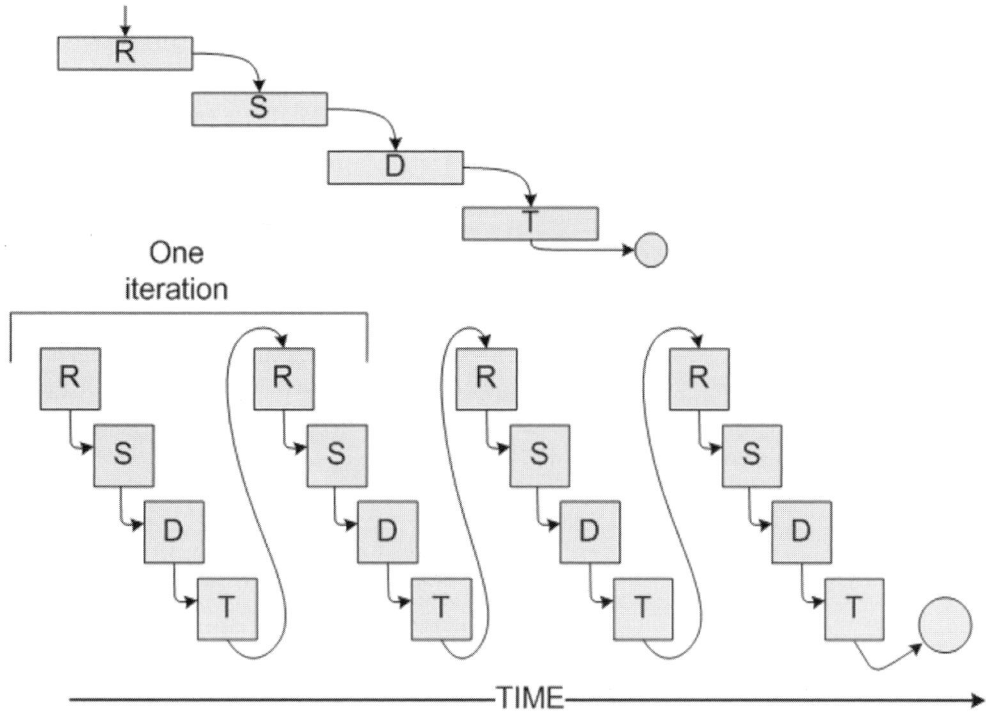

FIGURE 36.4 Waterfall compared with incremental life cycle.

- Construction—During the construction phase, the focus is on developing detailed specification/configuration documents tied to the requirements identified in the previous phase, implementing the functions/configurations, and performing unit testing of the components. In the construction phase, the architecture and requirements have an established baseline, the specification/configuration documents, and a strict change-management process controls changes.
- Transition—During the transition phase, the project team transfers the product to the user community. Once users inspect the product, issues often arise that require additional development to adjust the product, correct undetected problems, or finish some of the features that may have been postponed. This phase typically starts with a "beta release" of the product.

Although the focus evolves with each succeeding phase, the phases all include the iterative performance of "work flows," to produce increments. These work flows correspond to the basic phases of the waterfall: requirements, specify, develop, and test.

Incremental life cycles are often used for large systems developed over many years, such as the RODS system described earlier.

3.3. Other IT Project Life Cycles

Table 36.4 summarizes a number of other IT project life cycles. The "Methodologies" column describes well-known methodologies combining life cycles with methodology-specific processes project teams must accomplish within the life cycle phases.

3.4. Evaluating Project Life Cycles

By our definition, life cycles are fundamental to project management and significantly affect how the project is structured. This section provides a method to choose a life cycle that is well suited for a project. Dr. Richard Bechtold developed a method for considering project-specific characteristics ("key strengths" in Bechtold's terminology) to determine what development life cycle(s) are most appropriate (Bechtold, 1999). Table 36.5 is a summary of project-specific characteristics.

The following table matches project life cycles with project characteristics. For example, the waterfall project life cycle could be appropriate for a project with high ratings for "appropriate processes," "project brevity," and "requirements stability."

The two project examples presented later in this chapter were both brief projects. In Table 36.6, the project characteristic "Project Brevity" indicates the applicability of the waterfall life cycle. The examples describe how the project teams successfully applied the waterfall life cycle by adapting the project phases to address project-specific requirements.

Note that the IT industry is rife with "methodology wars," and IT practitioners may identify deeply with their life-cycle methodology of choice. The authors of this chapter provide the assignment of a sample methodology to life cycle category or the lumping of two competing methodologies in a single category to help project managers compare their project characteristics to major advantages of a life cycle. Clearly, a good team can deliver successful projects by using a life cycle not listed as most appropriate. Likewise, an appropriate life cycle does not guarantee project success.

Note that the IT life cycles are actually based on *software development* life cycles, which may not cover all aspects of a CBBS implementation. Therefore, applying them without recognizing this limitation may hide both risks and opportunities:

Hidden risk—CBBSs are data driven, and many CBBS projects obtain their data from people and organizations outside the project. External data creates unique and formidable project risks, including uncertain data acquisition tasks, unknown source data, misunderstood source data, and uncertain data cleansing tasks. Software development project life cycles do not directly address data.

TABLE 36.4 Implementation Life Cycles

Lifecycle	Description	Appropriate Use	Methodologies
Waterfall	Linear ordering of major implementation activies.	Requirements well understood and stable. Project of short duration.	SDLC
Throw-away prototype	Builds something in order to to understand the requirements (the reverse of the waterfall)	Vague, uncertain or volitale requirements.	RAD, Extreme Programming
Incremental build	Systems built and tested one piece at a time	Large projects	Unified Process
Multiple build	Similar to Incremental Build except that functional pieces of the system are released to the customer	Large projects; incremental benefits to customer	
Spiral	Multiple cycles through a plan,design, build, evaluate	When customer does not know what the final product should look like, and prototype will not answer the question	Microsoft Solutions Framework
Legacy maintenance	Mini-projects related to an existing (legacy) system	Legacy system exists, requirements are stable, fix/enhancement tasks short in duration	each fix a mini-SDLC

Adapted from Bechtold, R. (1999). *Essentials of Software Project Management.* Vienna, VA: Management Concepts. Reproduced by permission of Management Concepts.

TABLE 36.5 Project Characteristics

Project Characteristic	Description	Strength	Relevance Note
Appropriate processes	Is the project team using a consistent and well understood project methodology?	"Yes" answer(s)	
Capitalization	Is the project backed by sufficient funding?	"Yes" answer(s)	
Customer focus	Are you familiar with the customer? Is there evidence that the organization is prepared to support customer priorities?	"Yes" answer(s)	
Data availability	Is the data available from within the organization? Is it well understood? Will the data provider participate in the team?	"Yes" answer(s)	Especially relevant for data warehouse projects where source data will be provided from other systems
Historically based, relevant predictive data	Is there historical data available to use in planning the project?	"Yes" answer(s)	
Modern support environment and tools	Will the project be using recent and robust tools?	"Yes" answer(s)	Relevant only for development projects
Project brevity	Less than 2 months? 2-6 months? 6-12 months? 12-18 months?	Relatively short project, "Less than 2 months," "2-6 months"	
Requirements stability	How stable will the requirements be? Does the customer fully understand what they are asking for?	"Yes" answer(s)	
Strategic teaming	Is your organization set up for teaming? Do you have prequalified teaming partners?	"Yes" answer(s)	Relevant only if you need capabilities not available in house
Team cohesiveness	Has the project team worked together long? Do they have a history of welcoming and acclimating new members?	"Yes" answer(s)	
Team expertise	Does your team have both domain and technical area expertise? Has it delivered similar projects for other customers?	"Yes" answer(s)	
Training support	Is there appropriate training available? Can you arrange for training easily? Does your organization have a history of being receptive to providing training for employees?	"Yes" answer(s)	May not be relevant if "team expertise" is very high. However, consider that long projects may have staff turnover.
Expertise of the project manage	Has the designated project manager managed similar projects? With this team? With other team?	"Yes" answer(s)	

Adapted from Bechtold, R. (1999). *Essentials of Software Project Management.* Vienna, VA: Management Concepts. Reproduced by permission of Management Concepts.

TABLE 36.6 Project Characteristics and Project Life Cycles

	Waterfall	Throw Away Prototype	Incremental Build	Multiple Build	Spiral	Legacy
Appropriate Processes	•					
Capitalization		•				
Customer Focus						•
Data Availability	•					
Historical Data						•
Modern Support Environment		•	•	•		
Project Brevity	•					
Requirements Stability	•					
Strategic Teaming				If applicable		If applicable
Team Cohesiveness			•			
Team Expertise					•	
Training Support						
Expertise as Manager						

Adapted from Bechtold, R. (1999). *Essentials of Software Project Management.* Vienna, VA: Management Concepts. Reproduced by permission of Management Concepts.

There is a programming adage, applicable here: "programming is easy; data are hard."

Hidden risk—Data warehouse technology, a core IT element of most CBBSs, offers unique challenges not fully addressed by the life cycles. The success of a data warehouse project requires project management attention to unique risks facing these projects, including dependence on externally created data (see above), and the architecture risks associated with VLDBs (Adelman and Moss, 2000).[1] Moving from simple programs developed as pilots to production environments capable of receiving, processing, and reporting on millions of data points requires careful attention to scale.

Hidden risk—CBBS rely on IT infrastructure, including local area networks, wide-area networks, telecommunications links, the Internet, network operating systems, computer hardware, and storage devices. Software development methodologies often assume infrastructure is in place. Project teams must consider designing and building infrastructure as a separate but highly related, activity within a CBBS project. Given the amounts of data passing into CBBSs at a regular basis, the infrastructure must be robust.

Hidden opportunity—Rigid application of project life cycles designed to guide software development projects will likely lead to development of software. The best solution, however, may not be to develop software at all, but to purchase existing commercial off-the-shelf software (COTS) packages, or to use the software as a service or application service provider (ASP) solution alternative to building or buying.

4. UNIQUE CBBS PROJECT MANAGEMENT CONSIDERATIONS

This section describes specific approaches to address unique characteristics of CBBS and their associated risks.

4.1. Project Charter

Project and IT governance are the responsibility of the sponsoring organization. Good governance processes make the following decisions at the organizational level: how much money should we spend on projects, and on which projects should we spend the money; which projects should we adopt organization-wide; how good does the IT product resulting from the project need to be; what security and privacy risks can we accept; and who we should blame if the project fails (Ross and Weill, 2002; Ross et al., 2002). Uniquely relevant to CBBS are the answers to questions of "how comprehensive does the product need to be," and "what security and privacy risk is acceptable." Answers to these questions will guide project scope, data cleansing efforts, and the design of security. "Who to blame if the project fails" is also interesting, because most CBBS cannot succeed without data from outside the organization. The organization therefore must be willing to commit senior resources to establish relationships with data providers to secure their commitment, which is pivotal to project success.

In not-for-profit organizations, questions of "how much to spend" and "which project to spend the money on" introduce the difficult problem of determining public value. Measuring the performance and value of an IT project is "more of an art than a science," because financial measures, such as return on investment, do not completely apply (Weill and Ross, 2004). Weill and Ross (2004) and Moore (1995) provide frameworks for addressing public value and project and IT governance for not-for-profit organizations.

Project managers should push organizational sponsors to answer these key questions and document the answers in a formal "project charter."

4.2. Project Stakeholders

Stakeholders in a CBBS include more than just the system users and developers:

- Organization data privacy and compliance officers
- Data acquisition team (including the attorneys negotiating the usage agreements)
- Data providers
- IT infrastructure providers

Project teams need to consider the needs of these critical stakeholders during the project initiation phase, because overall project success may depend on their participation and acceptance of the solution.

4.3. IT Infrastructure

Changes to IT infrastructure to support CBBS can be costly and time-consuming. In addition to including the providers of IT infrastructure among the project's stakeholders, project teams should perform a high level assessment of the current IT environment early in the project to determine what infrastructure is in place to support the CBBS. If infrastructure upgrades are required to support the CBBS, determine funding responsibility and include the upgrades in the project schedule and project management plan.

4.4. Consider Solution Alternatives

CBBS can be custom developed, built from open source or COTS software, or provided as a service from an ASP. A single CBBS may integrate components of all these types. Defining requirements in implementation-neutral terms supports

1 Adelman and Moss comprehensively cover the project management issues of data warehouses in the referenced work, Database Warehouse Project Management.

consideration of all three alternatives. Requirements should address "what" the solution must do. Evaluate the capability of alternative solution approaches (COTS, open source, ASP, combination) to meet the requirements.

Project teams should add detail only after selecting a solution approach, because the required level of detail varies by solution approach:

- Custom development requires detailed definitions on exactly "how" the system will meet the requirements. We call this detail is called "system specifications."
- Customers have less control over how a COTS/open source and ASP solution meets their requirements. The details required for COTS/open source and ASP solutions is "configuration specifications"—how the existing systems will be configured to meet the requirements

Preparing system specifications describing how a system implements a requirement is a wasted activity if COTS or ASP solutions are to provide the functionality. In general, there is limited control over how COTS and ASP solutions implement a functional requirement.

4.5. Data

Teams must design their projects to perform "data discovery" activities for each potential data source as early in the CBBS project as possible to determine data quality. It is necessary to analyze a sample of the real data to estimate quantities of data expected, data structure, and data quality and completeness. The project team will also want to create sample data sets from the retrospective data to be used during system development and testing to simulate prospective data feeds.

4.6. Solution-Specific Development Phases

Prepare detailed development and implementation plans considering the selected solution type: custom software, integrated open source/COTS systems, ASP, or combinations of these. Design development and implementation phases by considering project risks and addressing the highest risk elements first. Consider including tests demonstrating all the data processing steps on process sample data sets as soon as possible in the plan (end-to-end tests). Then add additional functionality to the proven end-to-end framework.

For solutions integrating open source or COTS, development will include procurement activities covering request for proposal (RFP) preparation, proposal evaluation, vendor demonstrations, negotiations, and contracting. The durations of the negotiation and contracting phases are difficult to estimate and normally outside the direct control of the project team.

Consider basing the development phases for these solution types on the vendor's standard implementation approach.

The project team should determine detailed development activities based on the selected vendor's standard implementation processes. Why? Because, the vendor develops these processes to best suit its products/services, and these processes are well understood by the vendor's team. Even if using the vendor's process, do not forget the early demonstration of capabilities by using the data sets developed in the data discovery phase.

The development activities for an ASP solution, such as the RODS implementation for the 2002 Winter Olympic Games described later in this chapter, should be similar to those for open source/COTS solutions. The procurement activities for ASP solutions need to consider the required level of service, and the final contract must include a service level agreement (SLA) defining system availability and recourse if the SLA is not met.

4.7. System Support

CBBS-specific considerations when planning the closing phase include addressing methods and responsibility for notifications by data providers of changes to the data provided.

5. APPLYING PROJECT MANAGEMENT PRINCIPLES IN THE REAL WORLD

Project managers should take heed of Dwight Eisenhower's conclusion that "plans are useless, but planning is indispensable." The project management processes, project charters, project management plans, and project life cycles described in this chapter are nothing more than tools and techniques that can help project teams deliver successful projects—accomplishing the project objectives on time and within budget. The goal of project management is to help deliver successful projects not to demonstrate rigid adherence to a particular process or to produce a particular project management deliverable. Project management should focus first on planning for a successful project and then on adjusting the plans, as necessary, to deal with the project reality. Knowing what processes to follow, which tools to apply, and, most importantly, when to apply the processes and tools is the "art" of project management.

This section includes two examples of CBBS implementations managed by using the concepts in this chapter: rapid deployment of a CBBS for the 2002 Winter Olympic Games and the implementation of RODS as part of an Advanced Practice Center for BioSurveillance in the in Dallas–Fort Worth metroplex. The project examples illustrate how project teams tackled real world problems and applied project management "art" to help deliver successful CBBS projects. Some things to notice in the examples include the following:

- Project teams need to start slowly to progress really quickly. In the face of an extremely short implementation window, the CBBS for the 2002 Winter Olympic Games team did not dispense with the planning phase and rush straight into development. The natural tendency when faced with schedule pressure is to rush to get things started; resisting that tendency is critical. The example shows that time spent up front planning is key to project success, because through the careful planning process the team designed an implementation phase that allowed many activities to run parallel and successfully integrate into a successful product.

- The schedule may determine a "best" CBBS solution approach. Selecting the ASP approach for the CBBS for 2002 Winter Olympic Games was perhaps the most important decision leading to project success. With only seven weeks to deliver a working solution, custom development was essentially impossible and implementing a COTS solution unlikely. Developing a CBBS by using the ASP approach was the approach most likely to succeed.
- Not every project needs to include every life cycle phase. The Advanced Practice Center for BioSurveillance example included all the SDLC phases, as described below. In the CBBS for 2002 Winter Olympic Games example, requirements very closely matched the capabilities of the selected ASP solution, and the design and specification phases were not necessary at all.
- Project phases can overlap if carefully managed. The CBBS for the 2002 Winter Olympic Games project schedule example illustrates how concurrent activities can speed implementation, and the example explains the steps taken to make sure that the activities would integrate into the final solution.

5.1. Rapid Deployment of a CBBS for the 2002 Olympic Winter Games

In preparation for the 2002 Olympic Winter Games in Salt Lake City, the Utah Department's of Health, Environmental Quality, and Food and Agriculture, along with six local health districts and several federal agencies, formed an Environmental and Public Health Alliance to provide enhanced biosurveillance in the region. The events of September 11, 2001, fomented a spirit of collaboration and unity, while the rapidly approaching Winter Olympic Games provided the motivation to further augment this alliance's existing surveillance strategies. In addition, the Games provided an opportunity to demonstrate how a prototypic, real-time CBBS system could be rapidly deployed. Within months of the opening ceremonies, a group comprised of the University of Pittsburgh RODS Laboratory, the Utah Department of Health, and two healthcare systems deployed an automated RODS using chief complaint data from patients registering at emergency departments and urgent cares (an urgent care is a clinic that sees patients without appointments) throughout the region (Gesteland et al., 2002, 2003). The project was completed in seven weeks, and RODS was operational in time for the opening ceremonies of the Games. Although considered at the time a prototype, the system is still in use today.

The project team faced an extremely short implementation window, and this characteristic drove the project management approach. As the example shows, the team focused on project schedule risks in the planning phase. With an understanding of the unique project characteristics, the team designed a project life cycle with phases specific to the selected solution, and aligned with the intense timeline of the project.

5.1.1. Initiation

Utah had two large health systems, Intermountain Health Care (IHC) and University Health Care. The emergency departments and urgent cares from just these two health systems provided an estimated 70% of acute care in a seven-county region that encompassed all 10 Winter Olympic venues, including the Olympic Village. The project team implemented an existing CBBS and took advantage of the timely data already routinely collected in a standardized electronic format by these two institutions.

5.1.2. Concept

The team built the CBBS system by using the University of Pittsburgh RODS system, operating over a redundant network infrastructure comprising the Internet and leased T1 lines donated by Siemens Medical Systems. The team's objective was to deploy a syndromic surveillance system by using chief complaint data for disease outbreak detection.

5.1.3. Planning

The participating healthcare organizations had significant concerns that the team had to address in the planning and execution phases of the project. First, both organizations were in the midst of either replacing or upgrading their existing hospital information systems and could commit only limited resources to support this project. Second, the regional healthcare systems were already responding to mandates from the Utah Department of Health to share clinical data via other mechanisms to facilitate health surveillance efforts for the Games. Third, these requests for data sharing came at a time when the healthcare organizations were struggling to prepare for and adapt to new federal regulations promulgated under HIPAA. The team addressed these concerns, in part, by (1) defining explicit, scope-limited resource requirements, (2) defining the role the CBBS would play in the surveillance efforts and how it would offset other reporting requirements, and (3) defining the terms of data sharing in compliance with the HIPAA regulations.

The project management plan broke the project into four main task groups: public health, network, data, and application. The project team managed the project by using a Gantt chart (Figure 36.5) to provide an understanding of the task breakdown.

The *public health task grouping* included two levels. The first was establishing the legal foundation for the data sharing necessary for the project, which included executing the data sharing agreements, establishing the administrative structure to oversee the data, and obtaining institutional review board (IRB) and other necessary approvals. The second was

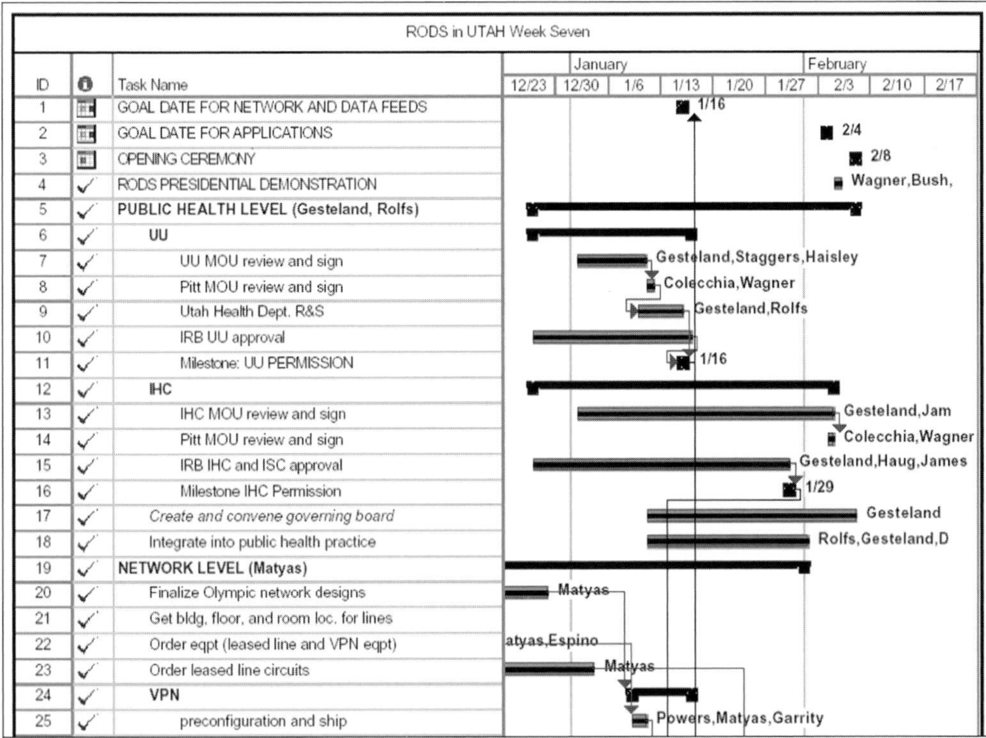

FIGURE 36.5 Sample of Gantt chart for Utah-RODS project. (Created with Microsoft Project 2000.)

integration and coordination of the CBBS with the existing Environmental and Public Health Surveillance strategy.

The *network task group* included implementing and testing the virtual private networking (VPN) and leased line networks employed, including design, equipment, procurement, configuration, installation, and testing.

Data tasks included verifying data availability, performing modifications to the data as needed, and building and testing the HL-7 interfaces.

The application task grouping entailed adapting RODS to display and analyze the Utah data, including adapting the GIS, user interface, and analytical components.

5.1.4. Implementation

The Utah RODS project manager (Gesteland) and the Director of the RODS Laboratory (Wagner) drafted data sharing agreements to describe the technical methods for maintaining privacy for the individual and the health system, providing the legal basis for the surveillance system. The complex and time-consuming negotiations over data sharing involved IRB, information security committees, privacy officers, senior administrators, and attorneys from both health systems: the University of Pittsburgh RODS Laboratory and the Utah Department of Health. The implementation process began on December 17, 2001, and the completed solution was operational within seven weeks. The decision to run the development task groups in parallel enabled the team to complete all aspects of the project within the required deadline. It took five weeks to obtain IRB approval, and data sharing agreements were signed during weeks six and seven. If the team had scheduled the technical work to begin only after obtaining legal and administrative approvals, the project would have failed.

Within three weeks, the team confirmed data availability and constructed and tested the HL-7 interfaces. Within five weeks, the team had established and tested the secure network connections. Modification of the existing RODS application was surprisingly easy because no historical or live data was available until February. All parties donating time, energy, and expertise with a commitment to success ensured the project's viability.

Each task group completed testing as soon as it could, but the team as a whole could ensure full completion across the system and with users only during the last few days of the project.

5.2. RODS as Part of an Advanced Practice Center for BioSurveillance in the Dallas–Fort Worth Metroplex

5.2.1. Initiation

The Dallas–Fort Worth metroplex encompasses a large metropolitan area in Texas covering four counties and including a population of 4.3 million. The local health department, Tarrant County Public Health (TCPH), desired to develop biosurveillance services not only for the metroplex but also for the entire North Texas region—a population of more than eight million in 16 counties.

TCPH undertook research and internal communications to review existing biosurveillance systems. It reviewed systems that varied in terms of capability, required infrastructure, and cost.

The TCPH had funds available through a grant from National Association of City County Health Officials (NACCHO) to build an Advance Practice Center for Biosurveillance for the community. This funding was available at the same time that the Dallas–Fort Worth metroplex was eligible to participate in a U.S. Department of Homeland Security funded project called BioWatch Support Implementation and Integration Program (BWSIIP), focusing specifically on enhancing syndromic surveillance capabilities but with infrastructure built for more complete biosurveillance purposes.

5.2.2. Planning

The initial goals for the CBBS were to set up the CBBS and obtain clinical surveillance data. These goals translated into four groups of tasks: (1) planning/training/reporting, (2) application installation, (3) hospital connections, and (4) verification activities. In each group of tasks, the project managers defined tasks and assigned responsibilities. We detail further the groups of tasks in the "Development" section below.

Negotiation for adequate resources and time—TCPH first evaluated its available resources, finding and using both local and state-wide resources to develop their CBBS solution.

Contracting and Recruitment of Personnel—TCPH was fortunate to have excellent and participative IT staff and systems, cooperation of the local stakeholders, and influential hospital council and state-based resources in the form of a fiberoptic Health Alert Network. However, it had need for appropriate personnel to manage the project. TCPH chose to use some of the supplemental grant funds to hire a business-oriented project director and manager to oversee the development of the system and interact with those inside and outside the organization and, once needed, a technical project manager to coordinate acquisition of data. Within the multicounty project area, TCPH was fortunate to have several epidemiologists that were both interested and knowledgeable in using additional nontraditional data sources to provide alerts for potential disease outbreaks and utilize technologically based methods for assessment, alerting, and reporting.

The project team used a project plan, project schedule, and weekly conference calls to communicate tasks and track progress.

5.2.3. Requirements

Key requirements, given the scope of ultimate coverage, included hosting the CBBS locally at the TCPH and building it to support the entire region. The project leadership determined they wanted to share data across counties with other health departments, provide access to multiple surveillance data streams 24 hours a day seven days a week through a secure Web site, obtain local emergency department data real-time where possible, access national data, and quickly conduct sophisticated analyses. In addition, they were interested in building hot backup or disaster recovery features for their system.

5.2.4. Design and Specification

The project leadership team reviewed these functional requirements against the resources available for the project. Although they had available adequate funding and excellent technical resources, CBBS development expertise was lacking in the project team. Because of this barrier, they realized that they were not equipped to build a system from scratch. Given this realization, they did not translate the functional requirements into detailed technical specifications. Instead, they chose to evaluate existing systems. Through that process they chose a system, RODS, that most closely matched the functional requirements identified, in particular local implementation, technical support, assistance with data collection from local hospitals, an available national data source (the NRDM), analysis tools, a Web-based interface, security, alerting, and reporting mechanisms. In addition, the system chosen was also open source, and the preferred data collection methodology was a real-time method. The TCPH project manager worked with the RODS implementation team to execute the project goals.

5.2.5. Development

System development began once the decision makers on the project had agreed to the system components and an overall project management plan with defined milestones and deadlines. TCPH and the RODS Laboratory worked together to execute development according to the tasks identified in the project plan. The following describes major project activity groups during the development phase.

Planning/Reporting. The planning/reporting group of tasks included preparing the health department for participation in the project, assessing the available human and technical resources, identifying solutions to resource gaps, making local modifications to a sample data use agreement, and prioritizing the hospitals to be connected. Communications and updates to external entities served as milestones and kept the project on track. The project managers determined the most important and most time-consuming data set to obtain would be the emergency department chief complaints from local hospitals. To keep the process moving, they set target dates for the first batch of hospitals to be connected, the next group, as well as the succeeding groups.

Application Installation. Application installation tasks included local installation of the RODS software, which would allow the county's system to receive the data from the hospitals, addition of visualization and analytical tools to the local application once data was being received, and supplemental upgrades to the base system as new features became available. Sample tasks included discussions with the state provider of the Health Alert Network and establishment of remote management of the system application on the local server.

Hospital Connections. The hospital connections group of tasks included working with the hospitals/health systems to be connected and defining milestones for each health system, including dates of initial contact, data use agreement signing, and transmission of data. For each healthcare data provider, the team defined additional technical tasks such as assignment of the project to the IT department within the hospital, verifying data availability, obtaining a historical data set, performing any necessary modifications to and testing of the health information system interfaces, and establishing a real-time connection. This outreach activity and education process to the hospitals strengthened the relationship between the local health department and the key hospitals in the area.

Verification Activities. The validating activities group of tasks included supplemental tasks to the core goal of creating the clinically based surveillance system. Specific tasks included inclusion of the over-the-counter (OTC) data from the NRDM service, and the public health activities related to creating an advisory group for evaluating the collected biosurveillance data.

Training and Maintenance. The RODS group provided training to the technical staff in the core counties and to the larger group of system users. The goal was to transfer knowledge and create a center of excellence within the TCPH department who could then provide advice, information, resources, and support to other communities as an Advanced Practice Center supported by NACCHO. RODS provided training to the system staff to facilitate knowledge transfer; training sessions for the local technical staff, the core set of epidemiologists; and an executive-level overview session for decision makers. The technical staff would ultimately be responsible for system operation and for subsequent upgrades. They also learned to work with hospitals to create data connections and automate data feeds from other data sources. RODS personnel provided the system users with an executive overview of Biosurveillance, the functions of the system, use of the algorithms for detecting anomalies in the data, and a user manual. Training continued through collaboration efforts and further systems development.

RODS personnel trained TCPH technical staff to include additional data sources and upgrade the system. Because it would maintain the system locally and hire trained staff specifically for the project, TCPH managed system maintenance internally.

Key factors for success included appropriate staffing of the program, coordination of local influencers for the benefit of the program, and constant collaboration between the system developers and the local health department to overcome unexpected hurdles.

6. SUMMARY

Successfully implementing CBBSs requires that project sponsors and project managers address unique characteristics of CBBS projects, as well as the IT elements commonly included in CBBS. Standard project management and development life cycles form a useful starting point for project planning and project management, and project teams can extend these standard methods to address CBBS-specific issues and project specific requirements.

As the two examples show, project management works. Project managers for time-sensitive projects succeed by combining careful up-front planning with diligent monitoring and controlling during execution. Considering project-specific characteristics, the project managers identified appropriate project life cycles, and they adapted these generic life cycles to meet project-specific needs. The project schedule was certainly one key to success. The examples show critical attention to other project management areas: the requirement for data sharing agreements, recognition of competition for limited resources to complete the project, coordination with IT infrastructure providers outside the project team, coordination with data providers outside the project team, and outreach to smooth transition of the project product into customers' operations.

ACKNOWLEDGMENTS

We thank the Project Management Institute for permission to reproduce material from the *Project Management Book of Knowledge*. We'd also like to thank Management Concepts for allowing us to reproduce and adapt work from Bechtold's Software Project Management.

REFERENCES

Adelman, S. and Moss, L. (2000). *Database Warehouse Project Management.*, Boston, MA: Addison-Wesle.

Bechtold, R. (1999). *Essentials of Software Project Management.* Vienna, VA: Management Concepts.

Beck, K. and Andres, C. (2005). *Extreme Programming Explained: Embrace Change (2nd Edition).* Upper Saddle River, NJ: Addison-Wesley Professional.

Berkun, S. (2005). *The Art of Project Management.* Sebastopol, CA: O'Reilly.

Buehler, J.W., Hopkins, R.S., Overhage, J.M., Sosin, D.M., and Tong, V. (2004). Framework for Evaluating Public Health Surveillance Systems for Early Detection Of Outbreaks: Recommendations from the CDC Working Group. *Morbidity and Mortality Weekly Report. Recommendations and Reports* vol. 53:1–11. *http://www.ncbi.nlm.nih.gov/entrez/query.fcgi?cmd=Retrieve&db=PubMed&dopt=Citation&list_uids=15129191.*

Gesteland, P.H., et al. (2002). Rapid Deployment of an Electronic Disease Surveillance System in the State of Utah for the 2002 Olympic Winter games. *Proceedings of the American Medical Informatics Association* 285–289. *http://www.ncbi.nlm.nih.gov/entrez/query.fcgi?cmd=Retrieve&db=PubMed&dopt=Citation&list_uids=12463832.*

Gesteland, P.H., et al. (2003). Automated Syndromic Surveillance for the 2002 Winter Olympics. *Journal of the American Medical Informatics Association* vol. 10:547–554. *http://www.ncbi.nlm.nih.gov/entrez/query.fcgi?cmd=Retrieve&db=PubMed&dopt=Citation&list_uids=12925547.*

Hoffman, M., et al. (2003). Multijurisdictional Approach to Biosurveillance, Kansas City. *Emerging Infectious Diseases* vol. 9:1281–1286. *http://www.cdc.gov/ncidod/EID/vol9no10/03-0060.htm.*

Jacobson, I., Booch, G., and Rumbaugh, J. (1999). *The Unified Software Development Process.* Boston, MA: Addison-Wesley Professional.

Lazarus, R., Kleinman, K.P., Dashevsky, I., DeMaria, A., and Platt, R. (2001). Using Automated Medical Records for Rapid Identification of Illness Syndromes (Syndromic Surveillance): The Example of Lower Respiratory Infection. *BMC Public Health* vol. 1:9. *http://www.ncbi.nlm.nih.gov/entrez/query.fcgi?cmd=Retrieve&db=PubMed&dopt=Citation&list_uids=11722798.*

Lombardo, J., et al. (2003). A Systems Overview of the Electronic Surveillance System for the Early Notification of Community-Based Epidemics (ESSENCE II). *Journal of Urban Health* vol. 80(Suppl 1):i32–i42. *http://www.ncbi.nlm.nih.gov/entrez/query.fcgi?cmd=Retrieve&db=PubMed&dopt=Citation&list_uids=12791777.*

Matta, N.F. and Ashkenas, R.N. (2003). Why Good Projects Fail Anyway. *Harvard Business Review* vol. 81:109–114, 134. *http://www.ncbi.nlm.nih.gov/entrez/query.fcgi?cmd=Retrieve&db=PubMed&dopt=Citation&list_uids=12964398.*

Moore, A., Cooper, G., and Wagner, M. (2002). Summary of Biosurveillance-Relevant Statistical and Data Mining Technology, Technical Report. Pittsburgh, PA: Auton (Autonomous Systems) Lab, Carnegie Mellon University http://www.autonlab.org/autonweb/showPaper.jsp?ID=moore-biosurv.

Moore, M. (1995). *Creating Public Value.* Boston, MA: Harvard University Press.

Mostashari, F., Fine, A., Das, D., Adams, J., and Layton, M. (2003). Use of Ambulance Dispatch Data as an Early Warning System for Communitywide Influenzalike Illness, New York City. *Journal of Urban Health* vol. 80(Suppl 1):i43–i49. *http://www.ncbi.nlm.nih.gov/entrez/query.fcgi?cmd=Retrieve&db=PubMed&dopt=Citation&list_uids=12791778.*

Patel, M. and Morris, O. (1999). *Guide to the Project Management Body of Knowledge.* Manchester, UK: Centre for Research in the Management of Projects, University of Manchester.

Project Management Institute [PMI]. (2004). *Project Management Body of Knowledge 3.0.* Newtown Square, PA: PMI.

Ross, J.W. and Weill, P. (2002). Six IT Decisions Your It People Shouldn't Make. *Harvard Business Review* vol. 80:84–91, 133. *http://www.ncbi.nlm.nih.gov/entrez/query.fcgi?cmd=Retrieve&db=PubMed&dopt=Citation&list_uids=12422792.*

Weill, P. and Ross, J. (2004). *Governance: How Top Performers Manage IT Decisions for Superior Results.* Cambridge, MA: Harvard Business School Press.

Wideman, M. (2004). The Role of the Project Life Cycle (Life Span) in Project Management. Vancouver, Canada: Max Wideman. *http://www.maxwideman.com/.*

37

CHAPTER

Methods for Field Testing of Biosurveillance Systems

Michael M. Wagner

RODS Laboratory, Center for Biomedical Informatics, University of Pittsburgh, Pittsburgh, Pennsylvania

1. INTRODUCTION

This chapter completes our discussion of evaluation methods, which began in Chapter 20 ("Methods for Evaluating Algorithms") and continued in Chapter 21 ("Methods for Evaluating Surveillance Data"). Although this chapter is conceptually self-contained, readers may benefit from reading the other chapters, either before or alongside this chapter.

In the earlier chapters, we discussed experimental methods that can provide insight into algorithms and surveillance data—both fundamental elements of any biosurveillance system. Analogous to the testing conducted by an engineer, the methods described in those chapters are best understood as laboratory or "bench" tests of components that we intend to incorporate into more complex systems, once they are thoroughly debugged and optimized.

In this chapter, we discuss methods for evaluating other elements of a biosurveillance system, such as computers that collect, store, display, or transfer data, as well as methods for the overall evaluation of a completed system. The methods apply to biosurveillance systems that are completely manually or a mixture of manual and automatic elements. The human element in a biosurveillance system always includes epidemiologists, who may or may not find a system useful or who may be misled by information that is not presented clearly. It commonly includes information technology staff, who must maintain the system, and data entry personnel, who can make errors or find the system difficult to use. We refer, collectively, to evaluations of operational systems as "field testing." Many important characteristics of biosurveillance systems can only be determined once they are deployed, at least partially, in the field.

Similar to a consumer report about refrigerators, a field evaluation may measure multiple characteristics of a system. Ultimately, the results can be summarized in a comparison table of price, reliability, sensitivity, false-alarm rate, timeliness and other attributes of the system. These results, if published, can then guide future consumers, who know which attributes are most important to them. Unlike a consumer product, however, biosurveillance is a societal function. Therefore, evaluators should estimate the benefit and cost of a system from the perspective of societal utility; ultimately, the aim is to understand the value of a biosurveillance system based on its cost and expected benefit to a population.

This chapter begins with examples of questions that can only be addressed through field testing. These questions clarify the role of field testing. After this initial discussion, we examine specific attributes of systems and components that can be measured and discuss briefly the methods that can be used. The list of attributes is lengthy, numbering 21 in total. But, as you will see from the example we provide of a field evaluation at the end of this chapter, the set of attributes relevant to any specific study is a small subset drawn from this list, and the goals of the evaluation will readily lead you to the appropriate attributes.

2. QUESTIONS NOT ADDRESSED BY BENCH TESTING OF DATA AND ALGORITHMS

The following questions can only be answered by field testing a biosurveillance system:

- Do epidemiologists use the system?
- How often do the components of the system fail and result in degradation of the system or complete failure of the system?
- What is the quality of the data under field conditions?
- What are the time latencies inherent in the various processing steps of a system?
- What is the actual detection and characterization performance in the field?
- What are the benefits of a system?
- How long does it take to deploy or build a system?
- What are the costs and level of effort required to build and maintain a system?
- Are the correct decisions and actions taken in response to surveillance data?
- How well does the system interoperate with other systems?

A typical laboratory evaluation focuses on a single use, a single type of data, or the information needs of a specific user. A fielded biosurveillance system may have multiple uses and multiple users. The range of uses, data, and users of a fielded biosurveillance system are, typically, quite large. The uses may include, in addition to prevention and control of disease, policy making, needs assessments, and accountability. These observations imply that many studies may be required to fully understand the properties of a biosurveillance system.

3. GOALS OF FIELD TESTING

As with any scientific experiment, it is essential that the individual responsible for the evaluation establish a clear objective for a study. Likewise, it is important for an evaluator to ensure that the evaluation's goal is appropriate for the stage

ISBN 0-12-369378-0

of development or maturity of a system or component being evaluated. If a system is complete and its component parts well studied, the appropriate goal might be measuring its detection performance and an appropriate method might be a prospective trial. This goal and method, however, would be extremely premature for a recently installed surveillance system that was receiving data from one hospital or a small fraction of hospitals in the surveillance region.

In general, a system (or any of its components) has a development life cycle. Evaluators may subject the system to many studies during that life cycle, perhaps with each study having a different goal. For systems that reach maturity and have been subjected to many evaluations along the way, the evaluation goal ultimately becomes the following: "What is the value of the system, and what is its cost?" This is the highest (and most expensive) level of evaluation.

Professional evaluators use the concepts of *formative study* and *summative study* to help them achieve clarity about their goals in designing and conducting a study. A formative study is one with the goal to improve a system or a system component. A formative study produces insight about whether a component or system is working as expected and whether we can improve it. A formative study might simply determine whether a biosurveillance system receives all data from hospitals that are transmitted and, if not, what is causing data loss.

Formative studies are most useful to the organization that is using or developing the system. Formative studies produce information that the organization can use to improve the system, and typically include an analysis of errors or failures. A formative study may or may not be suitable for publication, depending on the extent to which the results would be useful to other organizations.

The goal of a summative study is to establish the value of a system. An evaluator most often measures value relative to some other system or approach, but he can also measure it in more absolute terms (e.g., cost and/or benefit). For example, a study may focus on how well a system detects cases of disease under field conditions compared with an alternative approach. Such a study would use measurement techniques identical to the techniques described in Chapter 20 ("Methods for Evaluating Algorithms").

The results of summative studies are generally useful not only to the organization conducting the study but also to other organizations who are grappling with questions such as what type of biosurveillance system to acquire, or how to further develop their existing systems. For this reason, evaluators generally publish the results of summative studies. Summative evaluations require considerable effort and should not be undertaken prematurely on a system that, for example, does not have a mature data collection subsystem or has not been subject to extensive formative testing. One would like the system to be as good as it can be before going to the effort to benchmark its performance.

These distinctions between formative and summative are not absolute. A study may have both formative and summative features. A summative study may have a strong formative influence on a system. The intense scrutiny that a system receives during the evaluation may reveal areas of improvement. In some sense, the concept "formative" is simply a device to remind and encourage evaluators to conduct the simplest study that meets the need.

Because the experiments described in previous chapters on evaluation were obviously *scientific* experiments, we did not point out that evaluations, in general, are scientific experiments. It is easy to lose sight of this fact when field testing biosurveillance systems because of the complexity of the systems and the logistics of an evaluation. But, please keep in mind that any evaluation is a scientific experiment—nothing more and nothing less—and should be conducted like one, using appropriate methods, clarity, and scientific rigor.

4. ATTRIBUTES OF SYSTEMS AND COMPONENTS

We use the term *attribute* to refer to any property of a biosurveillance system (or system component) that we can measure. We can measure something as simple as the time lag between when a datum is collected about a patient to its receipt by a health department, or something as complex as the overall cost of a biosurveillance system.

There are several well-known guidelines for evaluation of biosurveillance systems (CDC, 2001, Buehler et al., 2004, Bravata et al., 2004). These guidelines include many examples of what we would term *attributes,* but they include other characteristics that are not easily measured and therefore do not qualify as attributes in the sense we are using the term.[1]

Table 37.1 summarizes the attributes we will discuss and their relationships to characteristics discussed in the published guidelines. The list in Table 37.1 is ordered roughly by formative to summative.

We included attributes with a "looking-to-buy-a-refrigerator" mind set, based on our experience building, deploying, and operating surveillance systems. We discuss our rationale for including each attribute.

4.1. Data Quality

Data quality refers to the completeness and accuracy of the data recorded in an information system (Hogan and Wagner, 1997).

1 The guidelines also provide advice on how to plan an evaluation and suggest information to include in a publication such as report such as a system description, a list of stakeholders; a chart describing the flow of data and the lines of response in a surveillance system; and a timeline for surveillance data. These aspects are outside the scope of this chapter and the interested reader should consult the references.

TABLE 37.1 Measurable Attributes of Biosurveillance Systems

Attribute	CDC working groups	Examples of What Can Be Measured
Data quality	Same	Completeness and accuracy of data
Sampling bias	Representativeness	Reporting by zip code, sociodemographic group
Disease coverage	—	Number of diseases in system database
Reliability	Stability	Number of errors by people or computers
Sensitivity, specificity, timeliness of case/outbreak detection	Sensitivity, PPV, and timeliness	Sensitivity, specificity, timeliness of case and outbreak detection
Diagnostic precision for case and outbreak detection	Alluded to	Sensitivity specificity and timeliness at different levels of diagnostic precision
Support for outbreak characterization	—	Relevant data collected by the system, collects usage during outbreak investigations
Time latencies	Timeliness	Delays between collection and receipt of surveillance data
Meets functional requirements	—	The difference between prespecified requirements and actual functions
Acceptability of system or components	Same	Subject or agency participation rates, interview completion rates and question refusal rates; log-ins.
Compliance with standards	Same	Results of conformance testing of user interface, data format, data coding
Portability	—	Cost to install in a different location
Privacy and Confidentiality	Same	Compliance with local and state regulations; ability to reidentify individuals
Security	Same	System vulnerabilities or security failures as identified by security audits
Benefits (morbidity and mortality reduction)		Expected reductions in mortality and morbidity through earlier detection
Benefits (other)	Usefulness, simplicity, representativeness	Expected reduction in operational costs owing to policy improvements or workflow efficiency.
Cost to build or acquire	Flexibility, comparable hardware and software, simplicity	Actual cost to build or purchase and install
Cost to operate		Salaries and overhead; hardware and licenses
Cost to add functionality		Actual cost
Cost to integrate with other systems		Actual cost
Cost of false alarms		Staff time, costs of treatments or other measures taken in response to a false alarm

CDC indicates Centers for Disease Control and Prevention.

Completeness is the proportion of data that we expect to find in the system that are actually in the system. An evaluator can measure completeness (or speaking more precisely, incompleteness) by counting the number of missing records or the number of missing fields within records. *Accuracy* refers to how closely the received data reflect the truth, as defined by some gold standard. An evaluator measures data accuracy with the same methods used to measure a classifier's accuracy (described in Chapter 20). The evaluator would compare data values recorded in the surveillance system to "true" values established by, for example, expert review of patient charts. Alternatively, the evaluator might use examination of other records or even direct observation of the actual patient encounter, as was done by Payne et al. (1993) in an evaluation of an immunization registry. From these comparisons, evaluators compute the sensitivity and specificity of the data received by the system, relative to the truth established by the gold standard determination.

In an operational biosurveillance system, surveillance data may arrive late. A network problem or human error may delay a transmission for hours, days, or weeks. Data may be subject to revisions and corrections by the sender after the initial submission. As a result, the completeness and accuracy of data in the databases of biosurveillance systems may improve as a function of time. An evaluator must be careful to measure the quality of data by using a snapshot of the data that reflects the state of the database at the time that an epidemiologist or other user would be conducting analyses and making decisions based on the data. The evaluator must be careful that studies of data accuracy do not include subsequent corrections and additions. This concern is especially important if the evaluator is conducting a retrospective study in which data in a database will clearly have been subject to such post hoc improvements. Unless the surveillance system timestamps all incoming data and an evaluator has used these elements to recreate an image of the data at the time of real-world use, a study may overestimate data quality.

For biosurveillance systems that receive data in batches, the unit of analysis can be the file rather than the individual patient record. Data completeness can be measured as the number of files received of the total sent. We discuss an evaluation of data completeness of batch and real-time data feeds in the final section of this chapter.

A study of data quality is an appropriate early study of a biosurveillance system. It can exert a strong formative influence on a system. The evaluation should include an error analysis of those records that are missing and those that are inaccurate to establish the causes of these errors. The evaluation should attempt to classify the causes as "potentially correctable" or "not correctable." Potentially correctable causes may include poorly crafted surveillance forms or computer-user interfaces, insufficient training and supervision of persons who complete these surveillance forms, or careless management of data.

An evaluator can reanalyze the study data set after manually correcting the potentially correctable errors to estimate an upper bound of data quality that can be expected from the system if the correctable causes of errors can be corrected (for a study of data quality in an electronic medical record system, see Wagner and Hogan, 1996).

4.2. Sampling Biases in Surveillance Data

Unless a biosurveillance system collects data about every individual in a population, it is subject to possible biases in its pattern of sampling that can affect the utility of the data. A system may sample poorer neighborhoods more or less often than affluent ones, children more or less often than adults, and institutionalized people more or less often than noninstitutionalized people. The differences can be great, as would be the case early in deployment of a biosurveillance system when hospital participation is spotty. Sampling bias is referred to by the Centers for Disease Control and Prevention (CDC) working group as "representativeness."

An evaluator can crudely measure sampling bias as the percentage of potential reporting entities from different zip codes or other geographic regions that participate in a system. As a more fine-grained measure, the evaluator can measure the fraction of received reports of a health-event with a known prevalence. For example, to estimate the sampling bias in a biosurveillance system that receives data from hospitals, published emergency department utilization data or hospital discharge data sets (described in Chapter 21) may be used to define the total number of individuals seeking care in a region, and those numbers may be compared with the number of reports received by the biosurveillance system. Hospital discharge data sets include sex, age, and zip code, enabling detailed analysis of sampling biases in those sociodemographic and geographic strata.

The decision between a using a crude or fine-grained measure of bias depends on how the surveillance data are used by epidemiologists and other end-users. If an epidemiologist or a detection algorithm uses the surveillance data to estimate incidence of disease in a strata of the population, it is essential to understand the sampling bias at that level. If, on the other hand, the analysis is not affected by biases (e.g., as in the National Retail Data Monitor) (Wagner et al., 2003), then it may not even be necessary to measure sampling bias.

4.3. Disease Coverage

The utility of a biosurveillance system is a function of the number of diseases that it covers. A special purpose system for a single disease is less valuable than is a system that covers 100 diseases if all other factors, such as cost, are equal. For a manual biosurveillance system, an evaluator can determine disease coverage by inspection of reporting forms and procedures. For an automatic system, the evaluator can inspect the structure and content of the database of the system.

4.4. Reliability of System or Components

A biosurveillance system, if you do not forget to include the human element, may involve thousands, if not hundreds of thousands of "components," all of which can potentially malfunction or fail.

Reliability of the computer elements of a biosurveillance system can be measured by how often the components are in a usable state when users or other computer systems attempt to access them. An evaluation can record system downtime and response delay characteristics experienced by users and other systems. Specific methods include automatic logging of unavailability of systems and functions accompanied by error analyses designed to identify correctable causes. Results can be reported as percentage of time that a system or feature was available.

Computer system reliability is relatively easy to measure, and the results of a study can be used to improve component reliability or to contribute to an estimation of overall system reliability. Information technology professionals can also audit the design and implementation of a system to determine its fault tolerance.

"Component" reliability of humans is more time-consuming to measure. In general, an evaluator would first identify the role that the person plays in the overall system and then measure the person's error rate. For example, if the person was the on-call epidemiologist and one of her roles was to field phone calls from emergency departments, a study could measure the number of phone calls missed and the reasons why.

Although an engineer can estimate overall system reliability from data about failure rates for individual components, it is also possible to test or measure overall system reliability directly. An evaluator can manipulate the inputs to a component or the overall system by injecting test data into the system at some point in the processing chain and following its appearance at various downstream points to measure time latencies or data loss and even whether and when a human notices the data or takes action. A test might involve buying, for example, 50 thermometers from a retail store and tracking their appearance in the biosurveillance system and effects on the human components. We refer to this test as the Moore test in honor of the first individual to challenge a biosurveillance system in this manner (Andrew Moore, personal communication).

4.5. Sensitivity, Specificity, and Timeliness of Case and Outbreak Detection

We discussed methods for measuring sensitivity, specificity (or false-alarm rate), and timeliness in detail in Chapter 20. We note that published guidelines recommend that evaluators measure predictive value positive and sensitivity (CDC, 2001; Buehler et al., 2004; Bravata et al., 2004). We recommend instead that, whenever possible, evaluators measure the sensitivity and specificity of a biosurveillance system. Appendix B provides a rationale for this recommendation.

Evaluators measure sensitivity, specificity, and timeliness in the field because these key properties of a biosurveillance system may differ from those measured in laboratory evaluations. In the field, a biosurveillance system is subject to the possibility of failure of many system components, whether they be human, network, or computer. If a component or the entire system fails often enough, receiver operating characteristic (ROC) and activity monitoring operating characteristic (AMOC) curves derived under more ideal laboratory conditions will overestimate the expected detection performance of a system. We note that laboratory evaluations may use carefully prepared data from which obvious errors in data files, such as duplicate records, have been removed.

Measuring field performance of *case*-detection algorithms is not difficult for common diseases because there will be sufficient numbers of cases. For this reason, an evaluator can conduct a field evaluation of a case-detection system almost immediately. Field evaluation of an outbreak-detection system, in contrast, is quite challenging because of the need for a large sample of outbreaks from which to measure sensitivity and timeliness.

Field evaluations of outbreak-detection algorithms are appropriate for large biosurveillance systems. Any single biosurveillance system, unless it is regional or nationwide, will not encounter a sufficient number of outbreaks unless it is deployed in several states. Field evaluations of outbreak-detection systems are appropriate for mature systems. It is hard to imagine freezing the development of a rapidly evolving biosurveillance system for a period of a year or more to accumulate an adequate sample.

One strategy for field testing of outbreak-detection algorithms is a multicenter trial in which essentially identical surveillance strategies for a common type of outbreak would be implemented in biosurveillance systems in multiple cities (or hospital in the case of nosocomial infections). A second approach would use a standard case-study format, such as the one developed by the RODS Laboratory *(http://rods.health.pitt.edu/)* (Rizzo et al., 2005), which encourages a uniform method for measuring time of detection, false-alarm rates, and cost of false alarms and for reporting of sufficient detail about any differences in surveillance data or systems to allow meta analysis across studies of individual outbreaks or small numbers of outbreaks.

Sensitive, specific, and timely detection of outbreaks is, of course, a raison d'être for a biosurveillance system; if its performance is not satisfactory along these dimensions, the other attributes we discuss are somewhat moot. Given that field evaluations will likely not be widely feasible until the current generation of biosurveillance systems matures, evaluators should conduct laboratory evaluations of the outbreak-detection algorithms planned for use in a system even before system development to provide insights into the expected field performance of the system.

4.6. Diagnostic Precision of Case and Outbreak Detection

As we discussed in Chapter 3, evaluators can measure sensitivity, specificity, and timeliness for health-events that vary in their level of diagnostic precision (ranging from "cow is sick" to "cow died from foot-and-mouth disease"). Thus, evaluations, and any publications resulting from the evaluations, should clearly state the level of diagnostic precision and discuss its implications for the value of the system to users.

An evaluation can study the increase in diagnostic precision that a system achieves as a function of time. Biosurveillance systems typically produce increasing levels of diagnostic precision over time because data with higher diagnostic value accumulate as a result of testing of individuals. An evaluator can study how the uncertainty about these measurements changes over time by using methods identical to those described in Chapter 20. Although this type of study has not yet been done, the analysis would involve the use of ROC curve analysis generalized to four dimensions (sensitivity, specificity, time, and diagnostic precision).

4.7. Support for Outbreak Characterization

The purpose of a biosurveillance system is not limited to detection of cases and outbreaks. Outbreak characterization is an equally important function and involves identifying the set of affected individuals, the geographic scope, biologic agent, and other characteristics of an outbreak.

As suggested by the riveting descriptions of outbreak investigations in Chapter 2, biosurveillance systems have been collecting and analyzing data to elucidate all of the aforementioned characteristics for many decades, albeit in a highly labor-intensive manner involving paper forms and "shoe-leather" epidemiology. Despite the informality of many current systems, evaluators can measure the time during the outbreak when each of these characteristics was elucidated (Dato et al., 2001, 2004; Ashford et al., 2003).

As automation results in biosurveillance systems that are more formal and, therefore, more amenable to modeling, evaluators will be able to apply the methods described in Chapter 20 to measure the sensitivity, specificity, and timeliness at which these characteristics are elucidated.

4.8. Time Latencies

We use the term *time latencies* to refer to the time delays between individual steps in the processing of information by a biosurveillance system. For computerized functions, an evaluator can use the timestamps associated with data (when they were recorded, received or transmitted) to measure time latencies between steps. For example, the timestamp of an emergency department registration and the timestamp of receipt by the biosurveillance system of that registration can be used to measure the delay from when information first became available about a patient in electronic form to its receipt by a biosurveillance organization. For manual functions of the system, time-motion studies may be required, unless the processing of information involved routine time or date stamping.

Note that the earlier CDC workgroup defined *timeliness* as "the speed between steps in a public health surveillance system" (CDC, 2001), whereas the other workgroup defined *timeliness of outbreak detection* as "the time lapse from the exposure to disease agent to the initiation of a public health intervention" (Buehler et al., 2004).

4.9. Meeting Functional Requirements

The design of modern information systems involves a process of functional requirements definition that precedes implementation (discussed in Chapter 4). During this process, a system analyst works with future users of the system to define their needs and then develops system specifications. An evaluator can compare the actual functions of a system against the prespecified requirements.

4.10. Acceptability of System or Components

Acceptability refers to the willingness of people and organizations to participate in a surveillance system or to use it (CDC, 2001). In particular, it refers to users' willingness to interact with a system and the willingness of hospitals, physicians, clinicians, industry, and government agencies to provide data as input.

Evaluators can use surveys or rates of participation to measure how willing these various entities are to provide data, modify their information systems to provide data (if necessary), answer questions about the data, modify format of data (when requested), and continue providing data. They can measure acceptability by using subject or agency participation rates, interview completion rates and question refusal rates (if the system involves interviews), and completeness of report forms; physician, laboratory, or hospital/facility reporting rate, and timeliness of data reporting (CDC, 2001).

A survey of users could ascertain whether usage was influenced by the health or economic importance of the health-related event, acknowledgment by the system of the person's contribution, dissemination of aggregate data back to reporting sources and interested parties, responsiveness of the system to suggestions or comments, burden on time relative to available time, ease and cost of data reporting, federal and state statutory assurance of privacy and confidentiality, the ability of the system to protect privacy and confidentiality, federal and state statute requirements for data collection and case reporting, and participation from the community in which the system operates (CDC, 2001).

4.11. Compliance with Standards

Compliance with standards means the extent to which a system makes use of standards for data representation, message format, code sets, and case definitions (CDC, 2001). A rigorous evaluation would involve conformance testing. A less rigorous evaluation might involve self-reporting by the system's developers.

The standards themselves could also be evaluated, although this type of study is outside of the scope of this chapter. As biosurveillance systems begin to incorporate automatic data feeds from other computer systems, it is becoming possible to study the availability and adequacy of standards for various types of surveillance data. Methods for evaluating standards could include surveys to determine market penetration of standards, measures of costs and effort required for implementation, and measures of the degree to which the standards facilitate data exchange.

4.12. Portability

Biosurveillance systems are expensive to develop and debug. From a societal utility perspective, the value of a system (often supported by tax dollars) includes its potential to be used in another location.

Portability refers to the effort required to install a system in a second location. There may be many differences between locations in the health events of interest (e.g., malaria), the prior probabilities of disease, and the types of surveillance data available. The organization of the human and animal healthcare systems (e.g., high or low penetration of managed care, existence of call centers) may vary. There may be differences in workflow in the participating organizations, the legal environment, and even the native language. Portability can perhaps best be measured by the cost and effort to install a system in another location.

4.13. Privacy and Confidentiality

Confidentiality is the expectation that information shared with another individual or organization will not be divulged. *Privacy,* although sometimes confused with confidentiality, is the right of an individual or organization not to divulge information. In the United States, these rights are relative, not absolute, in the area of biosurveillance because governmental public health is granted broad power to collect data for purposes of disease control (see Chapter 5). However, many consider it a violation of an individual's right to confidentiality if governmental public health collects data that are not essential for its function. The delineation between what governmental public health should and should not collect is dictated by the ethical principle of "need-to-know," which asserts that the intrusion into a person's privacy should be the minimum required to accomplish a function. For example, a health department does not need to know the name and addresses of every patient that visits an emergency department, although some health departments in the United States currently receive such information.

Privacy and confidentiality in electronic environments are addressed and protected in three ways: public policy, including state and federal law; technology itself, including adequate security, audit trails, and various forms of access restriction; and education and training (Alpert, 1998).

An evaluation of confidentiality could measure the rate of confidentiality violations by using surveys, anthropological observational methods, or incident reports. An evaluator could also examine the data access privileges granted to various users of the system against the data required to do their jobs. This relationship is an example of the need-to-know principle.

An evaluation of privacy and confidentiality might involve convening a panel comprising ethicists, legal experts, community representatives, and outside experts on biosurveillance to examine the data collection and management practices of a biosurveillance organization (Yasnoff et al., 2001).

4.14. Security

Security is the degree to which a biosurveillance system is vulnerable to corruption through physical damage, corruption or data theft by hackers, theft of passwords, or denial-of-service attacks. Security is typically measured by a comprehensive audit of physical security of the server site, firewalls, monitoring procedures, examination of documentation, and account authorization conducted by experts on computer and system security. Results of security evaluations are rarely if ever published because they identify vulnerabilities that could be exploited.

4.15. Benefits (Mortality and Morbidity Reduction)

The purpose of biosurveillance is to improve health. Therefore, a focus of evaluation should be on assessing a system's contribution to this goal.

The improvement in mortality and morbidity owing to a system would be the most direct measure of benefit, but it is difficult to quantify. To measure directly, evaluators need models of the expected effect with and without the system. For a further discussion of this topic, see Chapter 31, "Economic Studies in Biosurveillance."

Evaluators often use less direct measures of benefit that correspond to different points in the chain linking information to benefit depicted in Figure 1.3 (see Chapter 1). An evaluator can measure the improvement that a system achieves in the quality of information or decision making at many points in this chain and, assuming that all of the downstream steps in the chain function optimally, project the expected benefit in terms of improvement in health or reduction in mortality and morbidity.

Instances in which use of the system led to early detection of cases or outbreaks also provide evidence of system value. Examples of case reports of this type have been compiled and can be accessed by authorized public health officials at *https://www.rods.pitt.edu/cases/*.

The CDC guidelines discuss benefit mainly under "usefulness," which includes both benefit to policy making and disease-control programs. (CDC, 2001)

4.16. Benefits (Other)

Biosurveillance systems can improve the efficiency of an organization, thus lowering its operational costs. The systems can also provide information that facilitates assessment of the effect of prevention and control programs; leads to improved clinical, behavioral, social, policy, or environmental practices; or stimulates research intended to lead to prevention or control (CDC, 2001). Evaluators can measure improvements in efficiency by examining staffing before and after system deployment or modification and, at a minimum, can enumerate the additional uses of a system.

4.17. Cost To Build or Acquire

We discuss the costs to build, operate, expand, and integrate biosurveillance systems in separate sections because they correspond to different phases of biosurveillance-system development, and the costs of the earlier phases will be known and available for publication before the costs of later phases.

However, some general principles apply to estimating costs in all phases. When estimating costs, a first step is to identify all the cost items, which can be derived from (1) lists of personnel/consultants, their specific roles, skills, and costs; and (2) descriptions of hardware and software required and their costs (CDC, 2001). More detailed cost-related factors are listed under "simplicity" in the CDC guidelines: for example, amount and type of data necessary to establish that a health-related event has occurred; time spent on collecting data; amount of follow-up that is necessary to update data on a case; time spent on transferring, entering, editing, storing, and backing up data; time spent on analyzing and preparing the data for dissemination; staff training requirements; and time spent on maintaining the system.

The costs of the build-or-acquire phase include the costs of planning; acquisition of software, hardware, and licenses; the fees charged by contractors; and the labor and facility costs of the organization. The total is usually known by the organization through its normal accounting procedures.

4.18. Cost To Operate

The cost to operate a system is also usually known by the organization. It comprises costs of personnel, license fees, hardware upgrades, and costs involved with any repair of the system's computers, including parts and service.

4.19. Cost To Add Functionality

If a system has been in field operation for a period of time in excess of five days, resources have already likely been expended to plan to increase the functionality of the system (at least person-hours have been expended complaining about the system around the water cooler). If the system has been in field operation for several years, the developers have no doubt made substantial extensions, and there is experience with the cost and effort required to modify the system. The CDC guidelines use the term *flexibility* to refer to the cost and effort to modify a system in response to changing information needs or operating conditions, such as new health-related events, changes in case definitions or technology, and variations in funding or reporting sources. The change could be as simple as adding a new question to a form or as complex as the cost of fundamentally changing some aspect of the system.

4.20. Cost To Integrate with Other Systems

Because disease outbreaks do not respect the jurisdictional boundaries of hospitals or of health departments, biosurveillance systems operated by these entities must exchange data and, more generally, interoperate with each other. Integration consumes personnel time and financial resources of both organizations, and evaluators can study the costs and effort to develop this functionality and operate it through examination of project records and budgets of the two organizations.

4.21. Cost of False Alarms

Because of the inherent uncertainty of case and outbreak detection and characterization, users of biosurveillance systems may take actions that, in retrospect, were unnecessary or incorrect. The wasted effort, costs, or potentially more dire consequences of such actions are of great interest to users of these systems and their managers. These effects influence decisions to use a system, and how to configure its alarm thresholds and other parameters.

There have been no published studies that comprehensively investigate this effect.

5. AN EXAMPLE EVALUATION: FIELD TESTING OF THE DATA LAYER

An evaluator can study an entire system, a subsystem, or a component of a biosurveillance system. The following example is a pilot study conducted by Tsui et al. (2005) of the "data-layer" (subsystem) of a biosurveillance system. This subsystem encompassed computers located in hospitals, the Internet, and computers in a biosurveillance facility that received and stored the data.

This study compared the time latency, reliability, data loss, implementation cost, and effort for two subsystems that collected surveillance data from hospitals: one using direct connections to hospital HL7 message routers and the other batch transfer of files. The evaluators studied direct (54) and batch (2) data feeds from 56 hospitals in Pennsylvania and Utah to the RODS Public Health Data Center (PHDC). The data being transmitted were registration records containing patient chief complaints.

They measured time latency as the delay between patient registration and receipt of registration information by the PHDC; reliability, as the number of investigations of malfunctioning data feeds per connection-month; and data loss

TABLE 37-2 Characteristics of Real-Time and Batch Feeds

Measure	Real Time	Batch
Median time latency (hours)	0.03	23.95
Connection reliability (investigations/connection/month)	1.558 (206/13/8)	1.625 (26/2/8)
Data loss rate (data day/hospital/month)	0.002 (1/54/8)	1.625 (26/2/8)
Implementation time (hours)	8	8
Installation cost	$4,700	$4,700

(a measure of data quality), as days of data lost per hospital-month. They estimated implementation cost and effort by using the fee that the RODS PHDC charges for creating connections—which is set based on a prior analysis of total costs amortized over 200 connections—and the person-hours required to implement each data connection.

Table 37.2 is a *Consumer Reports*–type summary of the results taken from the technical report describing this study. It compares five key attributes that determine the benefits and costs of the two alternatives. In this particular experiment, the real-time system was superior to or equal to the batch system on all five attributes, making the choice for the consumer relatively simple, assuming the consumer was willing to use the PHDC facility for managing surveillance data or could create similar functionality at cost or effort that was acceptable.

6. SUMMARY

Evaluators who are planning field tests of biosurveillance systems or their components should achieve clarity about the purpose of the evaluation, in particular, whether the goals are formative or summative and whether they are appropriate for the maturity of the system or system component that is to be studied. The ultimate goal of evaluation—applicable only to systems that have been extensively studied by formative evaluations—is to understand their costs and benefits. Many attributes of systems and components contribute to their costs and benefits and form the subject of the many studies that may be conducted during the evolution of a system. The most important attributes are those that relate to the quality and diagnostic precision of surveillance data, performance (as measured by sensitivity, specificity, and timeliness for outbreak and case detection), and contribution to outbreak characterization, acceptability, and cost.

ADDITIONAL RESOURCES

Friedman C.P. and Wyatt, J.C. (1997). *Evaluation Methods in Medical Informatics.* New York: Springer-Verlag. A concise textbook and reference covering all aspects of evaluation of information systems, including study design and data interpretation.

ACKNOWLEDGMENTS

We thank Kenneth W. Goodman, MD, for reviewing a draft of this chapter and for discussions about privacy, confidentiality, and ethics and Fu Chiang Tsui, PhD, for permission to use results from his evaluation.

REFERENCES

Alpert, S.A. (1998). Health Care Information: Access, Confidentiality, and Good Practice. In *Ethics, Computing and Medicine: Informatics and the Transformation of Health Care* (K.W. Goodman, ed.). Cambridge: Cambridge University Press.

Ashford, D.A., et al. (2003). Planning against Biological Terrorism: Lessons from Outbreak Investigations. *Emerging Infectious Diseases* vol. 9:515–519.

Bravata, D.M., McDonals, K.M., Szeto, H., Smith, W.M., Rydzak, C., and Owens, D.K. (2004). A Conceptual Framework for Evaluating Information Technologies and DSSs for Bioterrorism Preparedness and Response. *Medical Decision Making* vol. 24:192–06.

Buehler, J., Hopkins, R.S., Overhage, J.M., Sosin, D., and Tong, V. (2004). Framework for Evaluating Public Health Surveillance Systems for Early Detection of Outbreaks. *Morbidity and Mortality Weekly Report* vol. 53:1–11.

Centers for Disease Control and Prevention [CDC]. (2001). Updated Guidelines for Evaluating Public Health Surveillance Systems: Recommendations from the Guide-Lines Working Group. *Morbidity and Mortality Weekly Report. Recommendations and Reports* 50(RR-13):1–35.

Dato, V., Wagner, M., Allswede, M., Aryel, R., and Fapohunda, A. (2001). The Nation's Current Capacity for the Early Detection of Public Health Threats including Bioterrorism. Washington, DC: Agency for Healthcare Research and Quality.

Dato, V., Wagner, M.M., and Fapohunda, A. (2004). How Outbreaks of Infectious Disease Are Detected: A Review of Surveillance Systems and Outbreaks. *Public Health Report* vol. 119:464–471.

Hogan, W.R. and Wagner, M.M. (1997). Accuracy of Data in Computer-based Patient Records. *Journal of the American Medical Informatics Association* vol. 4:342–355.

Payne, T., et al. (1993). Development and Validation of an Immunization Tracking System in a Large Health Maintenance Organization. *American Journal of Preventive Medicine* vol. 9:96–100.

Rizzo, S.L., Grigoryan, V.V., Wallstrom, G.L., Hogan, W.R., and Wagner, M.M. (2005). Using Case Studies to Evaluate Syndromic Surveillance, RODS Technical Report. Pittsburgh, PA: RODS Laboratory, University of Pittsburgh.

Tsui, F.-C., Espino, J., and Wagner, M.M. (2005). The Timeliness, Reliability, and Cost of Real-Time and Batch Chief Complaint Data. Pittsburgh, PA: RODS Laboratory, University of Pittsburgh.

Wagner, M., Robinson, J., Tsui, F., Espino, J.U., and Hogan, W. (2003). Design of a National Retail Data Monitor for Public Health Surveillance. *Journal of the American Medical Informatics Association* vol. 10:409–418.

Wagner, M.M. and Hogan, W.R. (1996). The Accuracy of Medication Data in an Outpatient Electronic Medical Record. *Journal of the American Medical Informatics Association* vol. 3:234–44.

Yasnoff, W.A., et al. (2001). A National Agenda for Public Health Informatics. *Journal of Public Health Management Practices* vol. 7:1–1.

Epilogue: The Future of Biosurveillance

1. INTRODUCTION

An author's decision whether to speculate about the future is an example of a "high-stakes" decision taken under uncertainty. There is potential cost to the author, who risks being a source of amusement to future generations. The potential benefit is to the reader, who may be better equipped to allocate her organization's resources, prioritize research, develop a curriculum, or manage her career. So with these caveats (and the prospects of author humiliation) in mind, let us proceed.

2. MEGATRENDS

Population growth, globalization, urbanization and bioterrorism have shaped the recent history of biosurveillance. These megatrends and others will also shape its future. The following is a discussion of some of the more important trends. As in the past, advances in science and technology will influence whether the burden of infectious disease in the world increases or decreases, as will the world's ability to incorporate them into biosurveillance practice.

2.1. Population Growth

The U.N. predicts that the world's population will grow by 40% over the next 45 years, from 6.5 billion to 9.1 billion (United Nations Population Division, 2004). This growth will be accompanied by ecological changes such as deforestation that facilitate the emergence of new diseases. Global warming may cause additional ecological changes and natural disasters, which are often accompanied by disease outbreaks. Although improved biosurveillance is not the primary remedy for these problems, absent direct solutions, better biosurveillance will be required if the world is not willing to accept an increase in levels of disease.

2.2. Population Density

The U.N. projects that the 40% growth in the world's population over the next 45 years will occur entirely in urban areas. In 1975, there were only five urban areas in the world with more than 10 million inhabitants, known as "megacities." (For an annual series of distinguished lectures on megacities that began in 1997, see *http://www.megacities.nl/*.) By 2005, there were 25 (Brinkhoff, 2005). The five largest megacities are the greater Tokyo area with a population of 34.1 million people, Mexico City with 22.7 million, Seoul with 22.3 million, New York-Newark with 21.9 million, and São Paulo with 20.2 million. A total of 437 urban areas have populations greater than or equal to one million (Brinkhoff, 2005). See Figure 1.

The significance of this trend toward megacities with increased population density is that the rate at which outbreaks of contagious diseases grow is a function of the size and density of a population. The number of individuals who are exposed to accidental or intentional contaminations of food, water, or air is also a function of the size and density of a population. Better biosurveillance will be required if cities wish to grow without accepting an increase in disease.

2.3. Globalization of Travel, Trade, and the Food Supply

The volume of travel and trade among cities and countries is increasing steadily. The globalization megatrend will increase the number of outbreaks caused by imported disease. It will also amplify the economic impact of such outbreaks. For example, the economic impact of SARS on East Asia was approximately US$18 billion—equivalent to 0.6% of the gross domestic product of the region (Fan, 2003). This impact was disproportionate to its effect on health alone, which comprised approximately 8000 people infected worldwide with 800 deaths. The economic impact of a pandemic of avian influenza could be as high as $US800 billion (World Bank East Asia and Pacific Region, 2005).

At present, the world reacts to diseases such as SARS by curtailing travel and trade. If globalization is to continue at its current pace, the international community must develop biosurveillance systems that not only contribute to the containment of outbreaks before they become pandemics, but also provide real-time, accurate risk information to countries, businesses, and individuals so that their reactions to such events match the actual risk.

2.4. Bioterrorism

The threat of bioterrorism will continue to stimulate research and development of biosurveillance systems. Designers of biosurveillance systems assume that intentional attacks will attempt to avoid detection by biosurveillance systems, thereby creating a blue-team red-team situation in which each side tries to understand and exploit the weakness of the other.

Interestingly, data compiled in the Monterey WMD Terrorism Database—the largest unclassified catalog of worldwide incidents involving the attempted acquisition, possession, threat and use of unconventional weapons by non-state actors—suggests that the frequency of events has been low, that is, there has been a low level of use of biological agents by terrorists and in discovered plots (Figure 2). Note that these figures are for publicly known events and do

Handbook of Biosurveillance
ISBN 0-12-369378-0

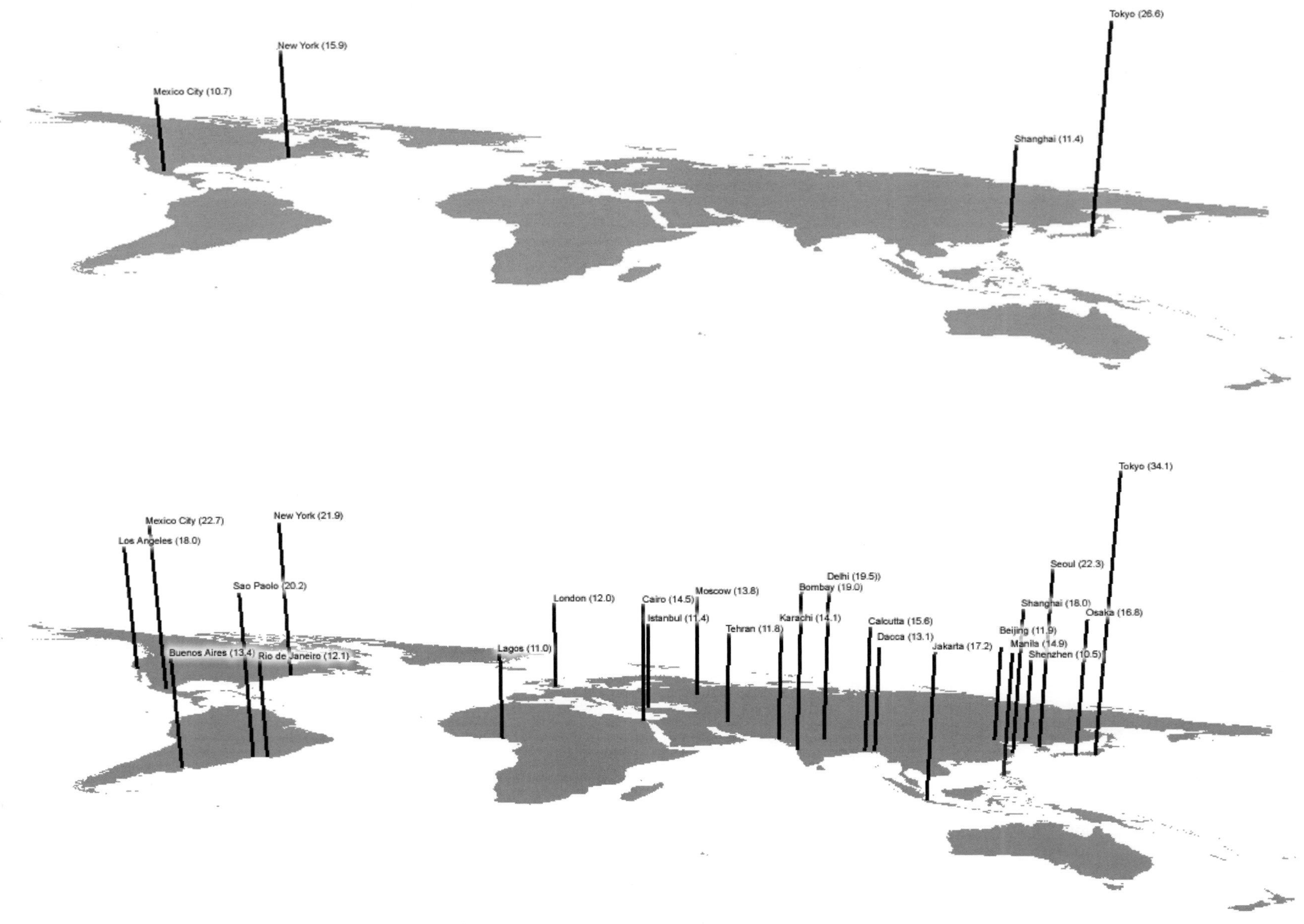

FIGURE 1. Cities with population of 10M or more in 1970 (top) and 2005 (bottom). Sources: United Nations Department of Economic and Social Affairs Population Division. United Nations. World Urbanization Prospects: The 2003 Revision. ST/ESA/SER.A/237 and Brinkhoff, T. The Principal Agglomerations of the World 2005-10-01. *http://www.citypopulation.de/World.html.*

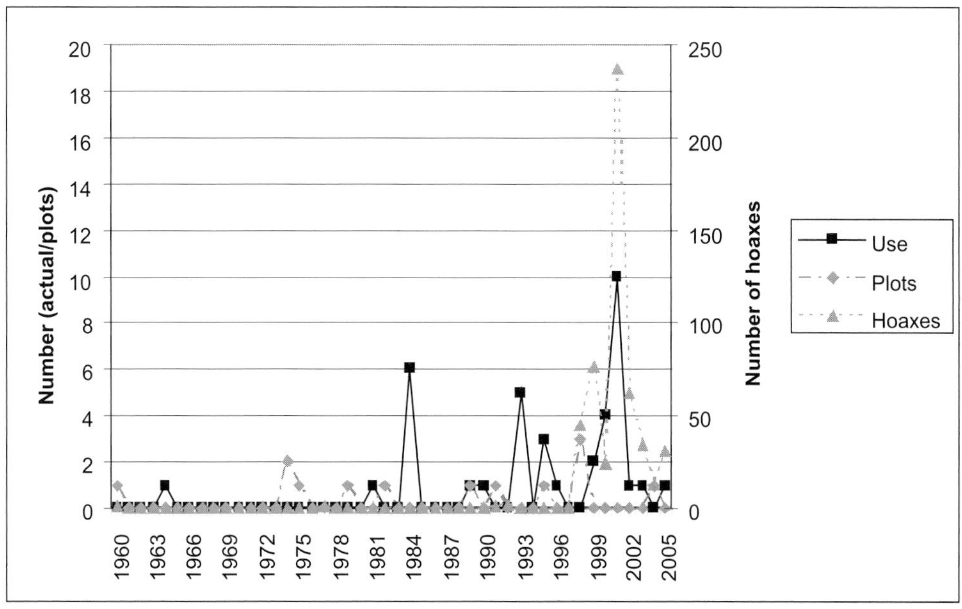

FIGURE 2. Actual bioterrorism uses (**solid line, squares**), plots (**dashed line, diamonds**), and hoaxes (**dotted line, triangles**), 1960–2005. The **left axis** (0–20) is for actual uses of biological agents and plots involving their use, the **right axis** (0–250) is for hoaxes. There were three additional uses of biological agents in 1933, 1947, and 1952 (not shown). The ten uses in 2001 include multiple anthrax letters. (Data courtesy of Monterey Weapons of Mass Destruction Terrorism Database.)

not include events whose existence may be secret. (Public information about unreported plots is important to researchers and to designers of biosurveillance systems because a plot represents a potential outbreak scenario that biosurveillance systems ideally must be capable of detecting.) Readers interested in the details of specific events can request access by writing to *wmdt@miis.edu*, but note that a *.gov* or *.mil* email account or special permission is required for access. There is a published analysis of events in the database through 1998 (Tucker, 2003). It is fortunate that such events have not occurred with frequency as biosurveillance methods are not adequate for these threats at present.

2.5. Information Technology

Advances in information technology may partly offset the negative effects of the above trends by accelerating and improving virtually every step of the biosurveillance process. Computers and networks have increased in speed and storage capacity in an exponential manner for the past 40 years. The approximate rate—a doubling every 12 to 24 months—was predicted by Gordon Moore in 1965 and is referred to as Moore's Law. (Moore's paper is available at *ftp://download.intel.com/museum/research/arc_collect/history_docs/pdf/icfuture.pdf.*)

Figure 3 is Moore's hand-drawn chart. It shows that in 1962 the minimum manufacturing cost per transistor was obtained by fabricating "chips" (integrated circuits) that contained 10 transistors. Three years later, the minimum manufacturing cost per transistor had fallen more than 10-fold and was achieved by fabricating chips containing 100 transistors. By 2004, the Intel Itanium ® 2 processor contained 592 million transistors.

This hardware megatrend is making biosurveillance more cost effective. If Moore's Law holds for another decade—as experts believe it will—it may make the hardware needed for biosurveillance dirt (or at least silicon) cheap.

Although there is no equivalent law that describes trends in the cost of software development, we note that biosurveillance systems are developed by knitting together existing software components such as web browsers and database management systems. These components are developed for other purposes and are available either for free or under license at a miniscule fraction of the actual development cost. There is little risk of future author remorse in predicting that the final cost of software development for biosurveillance systems will be small relative to the economic impact of SARS or a pandemic of avian influenza.

It is worth noting that advances in information technology are rarely, if ever, driven by biosurveillance, but rather occur for other reasons and must be harnessed by biosurveillance. For this reason, new developments in information technology (and other scientific fields) should be tracked for potential application to biosurveillance. Some developments to track include the following:

- The continuous improvement and proliferation of the Internet, which will further increase its value as an

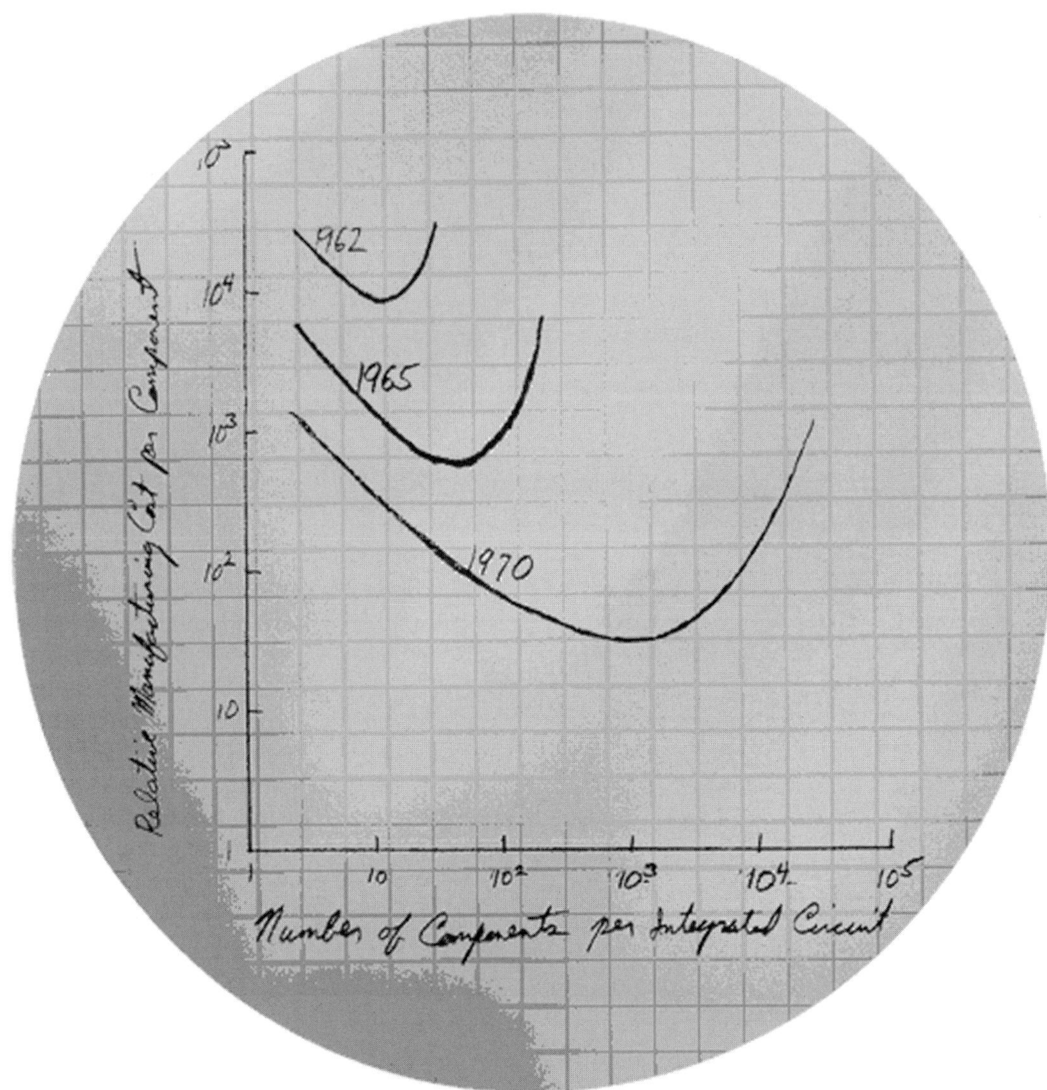

FIGURE 3. Moore's graph. In 1965, Gordon Moore sketched out his prediction of the pace of silicon technology. Decades later, Moore's Law remains true, driven largely by Intel's unparalleled silicon expertise. Copyright © 2005 Intel Corporation.

electronic network that connects the myriad organizations and individuals that participate in biosurveillance and enable new types of biosurveillance systems.

- Trends toward the creation and adoption of paperless medical records, which will result in more and more data that are useful in biosurveillance being captured electronically.
- Regional health information organizations (RHIOs), which will integrate healthcare data regionally for the purpose of improving the quality of care. RHIOs will represent a single point of access to healthcare data needed for biosurveillance.
- Ubiquitous personal devices connected to networks, which will support real-time physiological (e.g., skin temperature, heart rate, and coarse motion) and environmental (e.g., biological agent) monitoring. Consumer products such as cell phones and PDAs can now maintain a trace of all outdoor locations (through GPS and cell-tower communication logs) that their owner visits and when. Newer Bluetooth devices can even record a complete list of when, where, and how long they come in proximity of other devices. These capabilities suggest a possible future of massive parallel real-time contact-tracing. (Examples of these devices and methods can be found at *http://www.bodymedia.com/main.jsp*, *http://flat-earth.ece.cmu.edu/~eWatch*, and *http://www.cs.helsinki.fi/group/context/*.)
- The development and deployment of new product-tracking technologies such as radio frequency identification (RFID). This technology will enable biosurveillance systems to trace the movement and history of items associated with disease outbreaks.

- Advances in bioinformatics and computer science that will improve algorithms and techniques used to process biological data. These techniques include advances in privacy-protected data mining, which allow organizations to pool data into aggregates needed for large-scale event detection without any participant (including the organization performing the data analysis) being able to infer the individual data contributions.

2.6. Artificial Intelligence

The encoding of human knowledge into computer-interpretable format will play an increasing role in biosurveillance over the next decades. As we discussed in this book, the professionals working in biosurveillance master large bodies of knowledge during their professional lives. The very best of these individuals apply this knowledge to the interpretation of seemingly sketchy clues to achieve startling leaps of insight about the nature and causes of outbreaks. The methods developed by the field of artificial intelligence—including diagnostic expert systems, Bayesian networks, speech recognition, natural language processing, and computational decision theory—are only beginning to encode this knowledge so that it can be used to automate and support the biosurveillance process. To date, only a tiny fraction of the knowledge of physicians, epidemiologists, microbiologists, sanitarians, and others has been represented in computer-interpretable format. This area of research and development has immense potential to increase the speed and efficacy of biosurveillance.

2.7. Diagnostic and Sensor Technology

We are in the midst of a diagnostic revolution in medicine in which rapid testing methods based on DNA and protein identification increasingly allow clinicians (and laypeople, such as in home pregnancy testing) to establish precise diagnoses early in the course of a condition. These technologies will enable the earlier detection of cases of disease at high levels of diagnostic precision. When coupled with electronic data transmission to central locations, they will make possible the detection of outbreaks that are too small or diffuse (spatially or temporally) to be detected at present. Laboratories already employ rapid genotyping of biological agents to determine with high confidence whether a set of individuals were infected by each other or by a common source. Many of these tests are microchip-based and therefore subject to Moore's Law. As the costs of these tests drops, it will become economical to use them for screening and routine monitoring of the microbial ecosystem for emergence of microbial variants that may represent a threat to human or animal populations.

2.8. Molecular Genetics

As the scientific community further elucidates the relationship between genotype and phenotype, there will be an ability to predict the emergence of new diseases and even their characteristics. With time, there will be predictive mathematical models that can answer critical questions such as the probability that the current H5N1 avian influenza will recombine with various human influenza strains and the expected transmissibility and virulence of the resulting strains.

3. THE FUTURE OF BIOSURVEILLANCE RESEARCH

Although biosurveillance will benefit immensely from cross-pollination by the fields just discussed, there will remain areas of both applied and basic research that represent the core science of biosurveillance. Applied research in biosurveillance will translate techniques from other fields and address engineering and organizational issues related to the construction of biosurveillance systems. Basic research in biosurveillance will continue to address questions such as: (1) which biosurveillance data to collect, (2) how to analyze the data, and (3) when and how to act in response to the data.

Over the past five years, basic research has developed experimental designs to address each of these questions. It has developed a large number of algorithms and begun to obtain empirical answers to these questions for selected types of biosurveillance data and algorithms. However, a tremendous amount of research remains to be done.

On the question of which biosurveillance data to collect, there is now fairly good understanding of the types of data that are needed to detect outbreaks and cases and the technical and administrative challenges involved in obtaining them in real time. However, only a few dozen experiments have measured the value of selected data for selected outbreaks or diseases. The matrix of data versus outbreaks is sparsely populated at present. Designers (and users) of biosurveillance systems will benefit most when a more complete matrix is available. On the question of which data are needed to characterize outbreaks, there has been almost no work to elucidate the data needed or to assess the feasibility of obtaining these data automatically.

On the question of how to analyze biosurveillance data, there is now good understanding of the types of analyses that algorithms must perform. In particular, algorithms must detect cases; notice anomalous clusters of disease activity (not only in space and time, but also in sociodemographic strata); compute posterior probabilities of disease/outbreak/outbreak characteristic, given available surveillance data; and utilize all available data and knowledge (e.g., wind patterns, observations of astute individuals, and incubation period of disease).

In the area of automatic detection of cases, researchers are adapting algorithms such as diagnostic expert systems developed by the field of medical informatics, but these approaches have not been evaluated in biosurveillance applications. The process of integrating automated case detection into clinical information systems and the infrastructure being built by national health information network efforts in multiple countries has barely begun, if at all.

Researchers have developed many algorithms that detect possible clusters of disease (univariate, multivariate, spatial scanning, search-based methods); however, algorithms that use information about social contacts (social networks) do not yet exist. There is a need for increased emphasis on algorithm evaluation. The literature on laboratory or field testing to measure algorithm performance, however, is still small, and most experiments have used synthetic or semisynthetic data. As a result, the sensitivity and timeliness of these algorithms for detecting most outbreaks are poorly understood. Such understanding is critical if such algorithms are to be used to guide decision making.

Algorithms capable of computing the posterior probability of an outbreak, given surveillance data, are at an earlier stage of development. This capability is also essential if the algorithms are to guide decision making. Multivariate algorithms capable of analyzing all of the surveillance data that are available to a human analyst will become increasingly important because analysts may not have confidence in an algorithm that is ignoring key data. Similarly, analysts may require algorithms that have "explanation" capabilities (a research area in artificial intelligence). Development of algorithms with all these properties should be a research priority because the ability to collect surveillance data is far outstripping the ability of biosurveillance organizations to analyze them manually—a trend that will continue and result in a data-analytic bottleneck. Algorithms will increasingly incorporate the knowledge of experts about disease and outbreak characteristics (e.g., incubation periods). We note that the effort to encode such knowledge for all diseases and types of outbreaks (encoding all that medical, animal, diagnostic knowledge, and epidemiological knowledge) is a formidable research and development task that nevertheless must be tackled.

There has been almost no research on methods to animal and human biosurveillance data in an integrated manner to identify potential "hot spots" for crossover of diseases between the populations. Eventually, biosurveillance systems will track the distribution and intensification of animal farming, location of live animal markets, environmental changes, the distribution of human populations, and the food supply chain.

Finally, there has been little work on algorithms designed to support outbreak characterization. There is a need for methods that can accelerate the identification of the source and route of transmission of an outbreak. There is also a need for methods that can estimate the true size and spatial distribution of an outbreak from observable surveillance data and project its future size and distribution.

On the question of when and how to act in response to biosurveillance data, there have been few quantitative studies. Formal methods to address these questions exist (e.g., decision analysis). However, only a handful of studies have applied such techniques to address problems in biosurveillance. Many studies of the cost of illness, the cost of investigation, and the tradeoffs involved in waiting versus acting to provide guidance to decision makers and designers of biosurveillance systems.

4. EDUCATION, ORGANIZATION, AND LEADERSHIP

The future of biosurveillance will depend not only on scientific advances but also on how the world addresses educational, leadership, data privacy, and organizational issues. Developments in these areas can have as much impact on the level of disease in the world as the emergence of a new scourge or the development of a new technology. Examples of specific issues follow:

- *Education:* Due to the introduction of new techniques, many individuals working in biosurveillance face a steep learning curve. The public health workforce does not yet have expertise with the newer mathematical formalisms or with information technology. Conversely, mathematicians and information technologists who wish to contribute in this area do not yet have sufficient familiarity with epidemiology and medicine. Without explicit attention to education, even massive expenditures by governments may not produce rapid improvements in biosurveillance capability. Several organizations have recognized the need for extensive retraining of the work force and for curricular changes in the programs that educate the future workforce. The American Medical Informatics Association (AMIA) devoted a four-day workshop in spring 2001 to developing recommendations for training in public health informatics (Yasnoff, 2001, Yasnoff et al., 2001). The Robert Woods Johnson Foundation and the National Library of Medicine recently funded training programs in public health informatics (National Library of Medicine, 2005). Graduate schools of public health—where many individuals destined for careers in biosurveillance are trained—face significant challenges in evolving their curricula.
- *Leadership:* Perhaps of greater importance, government leaders and funding agencies also are not familiar with many of the techniques discussed in this book, their implications, or how to guide their development and adoption. These individuals exert a profound influence on biosurveillance legislation, assignment of responsibility and authority, and funding. For example, making high-level patterns of telephone calls available for biosurveillance would be very useful and would not seem to threaten individual privacy. However, there has to be societal and governmental insight and will to establish laws and policies that make such data available. The telephone companies are unlikely to make such data available on their own.
- *Legal and ethical:* There has been insufficient attention to the question of how best to balance the need to collect biosurveillance data about human health with the rights of individuals to privacy and confidentiality. It remains an

unfortunate possibility that in the next few years, the world will experience a devastating pandemic, an intentional release of a disease such as smallpox back into nature, or a mass-casualty bioagent release. In the aftermath of any of these scenarios, the nature of biosurveillance might change dramatically with new societal expectations of the level of reporting and individual surveillance that are necessary. Entirely speculative possibilities might include Internet-based systems for all members of a population to provide daily information of their state of health and symptoms, "medical forensics" to detect possible bioterrorists from medical records and lab tests, medical portals for screening international or even domestic travelers, or more intrusive surveillance of personal health-related information than current society would tolerate. The authors of the book sincerely hope that the field of biosurveillance never needs to enter such a "war footing," but it is possible that some advance debate is needed about these kinds of scenarios.

- *Organizational:* Governments are investing in biosurveillance, but the net result has been multiple uncoordinated efforts within countries and coordination across countries is even less. There have been several unheeded calls for recognizing that the problem at hand may require a "space-race" or "Manhattan-project" level of organization (Wagner, 2001, Wagner, 2002, Frist, 2005). What is needed may be the creation of government organizations with the ability to fund and focus the development of biosurveillance.
- *Improving biosurveillance in developing countries:* Because of globalization, it is simply no longer possible to erect a defensive perimeter against disease. There really is now only one population on the earth. The problem of infectious disease is a problem for humanity that must be solved for all people. The world's "biosurveillance system" will only be as good as its weakest link, which at present is biosurveillance in developing countries. Fortunately, Moore's Law is delivering cellular technology and the Internet to these countries; however, it will require significant investment to layer advanced methods of biosurveillance on top of this infrastructure.

Although the accomplishments of the past five years are significant, and there are positive trends that suggest that the world is rising to meet the challenge of biosurveillance, there remains a great deal to do and the threat is clear and present. We hope that this book has in some small way invigorated those who are devoting their careers to achieving this humanitarian goal by recognizing their work, and—by bringing the information from diverse fields together under one cover—will facilitate their work and encourage cooperation and understanding.

REFERENCES

Brinkhoff, T. (2005). The Principal Agglomerations of the World 2005-10-01. *http://www.citypopulation.de/World.html.*

Fan, E. (2003). *SARS: Economic Impact and Implications.* ERD Policy Brief No. 15. Manila: Economics and Research Department, Asian Development Bank.

Frist, W. (2005). The Manhattan Project for the 21st Century. *http://frist.senate.gov/index.cfm?FuseAction=Speeches.Detail&Speech_id=261&Month=8&Year=2005.*

National Library of Medicine. (2005). $3.68 Million Grant to Boost Public Health "Informatics." *http://www.nlm.nih.gov/news/press_releases/pubhlth_inform_grants05.html.*

Tucker, J. B. (2003). Historical trends related to bioterrorism: an empirical analysis. *Emerg Infect Dis* 5:498–504.

United Nations Population Division. (2004). World Population Prospects: The 2004 Revision Population Database. *http://esa.un.org/unpp/.*

Wagner, M. (2001). A Review of Federal Bioterrorism Preparedness Programs: Building an Early Warning Public Health Surveillance System. Testimony given to the Hearing of the Oversight and Investigations Subcommittee of the House Committee on Energy and Commerce. *http://energycommerce.house.gov/107/hearings/11012001Hearing406/Wagner684.htm.*

Wagner, M. M. (2002). The space race and biodefense: lessons from NASA about big science and the role of medical informatics. *J Am Med Inform Assoc* 9:120–2.

World Bank East Asia and Pacific Region. (2005). *Spread of Avian Flu Could Affect Next Year's Economic Outlook.* Washington, DC: World Bank.

Yasnoff, W. A. (2001). The promise of public health informatics. *J Public Health Manag Pract* 7:iv–v.

Yasnoff, W. A., Overhage, J. M., Humphreys, B. L., et al. (2001). A national agenda for public health informatics: summarized recommendations from the 2001 AMIA Spring Congress. *J Am Med Inform Assoc* 8:535–45.

Appendices

APPENDIX

CDC™ Severe Acute Respiratory Syndrome

Public Health Guidance for Community-Level Preparedness and Response to Severe Acute Respiratory Syndrome (SARS) Version 2

SUPPLEMENT B: SARS SURVEILLANCE

Appendix B1: Revised CSTE SARS Surveillance Case Definition

December 2003

Clinical Criteria

Early illness

- Presence of two or more of the following features: fever (might be subjective), chills, rigors, myalgia, headache, diarrhea, sore throat, rhinorrhea

Mild-to-moderate respiratory illness

- Temperature of >100.4° F (>38° C)[1] *and*
- One or more clinical findings of lower respiratory illness (e.g., cough, shortness of breath, difficulty breathing)

Severe respiratory illness

- Meets clinical criteria of mild-to-moderate respiratory illness, *and*
- One or more of the following findings:
  - Radiographic evidence of pneumonia, *or*
  - Acute respiratory distress syndrome, *or*
  - Autopsy findings consistent with pneumonia or acute respiratory distress syndrome without an identifiable cause

Epidemiologic Criteria

Possible exposure to SARS-associated coronavirus (SARS-CoV)

One or more of the following exposures in the 10 days before onset of symptoms:

- Travel to a foreign or domestic location with documented or suspected recent transmission of SARS-CoV[2] *or*
- Close contact[3] with a person with mild-to-moderate or severe respiratory illness and with history of travel in the 10 days before onset of symptoms to a foreign or domestic location with documented or suspected recent transmission of SARS-CoV[2]

Likely exposure to SARS-CoV

One or more of the following exposures in the 10 days before onset of symptoms:

- Close contact[3] with a confirmed case of SARS-CoV disease *or*
- Close contact[3] with a person with mild-moderate or severe respiratory illness for whom a chain of transmission can be linked to a confirmed case of SARS-CoV disease in the 10 days before onset of symptoms

Laboratory Criteria

Tests to detect SARS-CoV are being refined, and their performance characteristics assessed; therefore, criteria for

1 A measured documented temperature of >100.4° F (>38° C) is expected. However, clinical judgment may allow a small proportion of patients without a documented fever to meet this criterion. Factors that might be considered include patient's self-report of fever, use of antipyretics, presence of immunocompromising conditions or therapies, lack of access to health care, or inability to obtain a measured temperature. Initial case classification based on reported information might change, and reclassification might be required.

2 Types of locations specified will vary (e.g., country, airport, city, building, floor of building). The last date a location may be a criterion for exposure for illness onset is 10 days (one incubation period) after removal of that location from CDC travel alert status. The patient's travel should have occurred on or before the last date the travel alert was in place. Transit through a foreign airport meets the epidemiologic criteria for possible exposure in a location for which a CDC travel advisory is in effect. Information regarding CDC travel alerts and advisories and assistance in determining appropriate dates are available at *http://www.cdc.gov/ncidod/sars/travel.htm.*

3 Close contact is defined as having cared for or lived with a person with SARS or having a high likelihood of direct contact with respiratory secretions and/or body fluids of a person with SARS (during encounters with the patient or through contact with materials contaminated by the patient) either during the period the person was clinically ill or within 10 days of resolution of symptoms. Examples of close contact include kissing or embracing, sharing eating or drinking utensils, close conversation (<3 feet), physical examination, and any other direct physical contact between persons. Close contact does not include activities such as walking by a person or sitting across a waiting room or office for a brief time.

ISBN 0-12-369378-0

laboratory diagnosis of SARS-CoV are changing[4]. The following are the general criteria for laboratory confirmation of SARS-CoV:

- Detection of serum antibody to SARS-CoV by a test validated by CDC (e.g., enzyme immunoassay [EIA]), *or*
- Isolation in cell culture of SARS-CoV from a clinical specimen, *or*
- Detection of SARS-CoV RNA by a reverse-transcription-polymerase chain reaction (RT-PCR) test validated by CDC and with subsequent confirmation in a reference laboratory (e.g., CDC)

Information regarding the current criteria for laboratory diagnosis of SARS-CoV is available at *http://www.cdc.gov/ncidod/sars/labdiagnosis.htm.*

Exclusion Criteria

A person may be excluded as a SARS report under investigation (SARS RUI), including as a CDC-defined probable SARS-CoV case, if any of the following applies:

- An alternative diagnosis can explain the illness fully[5]
- Antibody to SARS-CoV is undetectable in a serum specimen obtained >28 days after onset of illness[6]
- The case was reported on the basis of contact with a person who was excluded subsequently as a case of SARS-CoV disease; then the reported case also is excluded, provided other epidemiologic or laboratory criteria are not present

Case Classification

SARS RUI

Reports in persons from areas where SARS is not known to be active:

- SARS RUI-1: Patients with severe illness compatible with SARS in groups likely to be first affected by SARS-CoV[7] if SARS-CoV is introduced from a person without clear epidemiologic links to known cases of SARS-CoV disease or places with known ongoing transmission of SARS-CoV

Reports in persons from areas where SARS activity is occurring:

- SARS RUI-2: Patients who meet the current clinical criteria for mild-to-moderate illness and the epidemiologic criteria

---

4 The identification of the etiologic agent of SARS (SARS-CoV) led to the rapid development of EIAs and immunofluorescence assays (IFAs) for serologic diagnosis and RT-PCR assays for detection of SARS-CoV RNA in clinical samples. These assays can be very sensitive and specific for detecting antibody and RNA, respectively, in the later stages of SARS-CoV disease. However, both are less sensitive for detecting infection early in illness. The majority of patients in the early stages of SARS-CoV disease have a low titer of virus in respiratory and other secretions and require time to mount an antibody response. SARS-CoV antibody tests might be positive as early as 8–10 days after onset of illness and often by 14 days after onset of illness, but sometimes not until 28 days after onset of illness. Information regarding the current criteria for laboratory diagnosis of SARS-CoV is available at *http://www.cdc.gov/ncidod/sars/labdiagnosis.htm.*

5 Factors that may be considered in assigning alternate diagnoses include the strength of the epidemiologic exposure criteria for SARS-CoV disease, the specificity of the alternate diagnostic test, and the compatibility of the clinical presentation and course of illness for the alternative diagnosis.

6 Current data indicate that >95% of patients with SARS-CoV disease mount an antibody response to SARS-CoV. However, health officials may choose not to exclude a case based on lack of a serologic response if reasonable concern exists that an antibody response could not be mounted.

7 Consensus guidance between CDC and CSTE on which groups are most likely to be first affected by SARS-CoV if it re-emerges is in development. In principle, SARS-CoV disease should be considered at a minimum in the differential diagnosis for persons requiring hospitalization for radiographically confirmed pneumonia or acute respiratory distress syndrome without identifiable etiology and who have one of the following risk factors in the 10 days before the onset of illness:

- Travel to mainland China, Hong Kong, or Taiwan, or close contact with an ill person with a history of recent travel to one of these areas, *or*
- Employment in an occupation associated with a risk for SARS-CoV exposure (e.g., healthcare worker with direct patient contact or worker in a laboratory that contains live SARS-CoV), *or*
- Part of a cluster of cases of atypical pneumonia without an alternative diagnosis

Guidelines for the identification, evaluation, and management of these persons are available at *http://www.cdc/gov/ncidod/sars/absenceofsars.htm.*

for possible exposure (spring 2003 CDC definition for suspect cases[8])

- SARS RUI-3: Patients who meet the current clinical criteria for severe illness and the epidemiologic criteria for possible exposure (spring 2003 CDC definition for probable cases[8])
- SARS RUI-4: Patients who meet the clinical criteria for early or mild-moderate illness and the epidemiologic criteria for likely exposure to SARS-CoV

SARS-CoV disease classification

- Probable case of SARS-CoV disease: in a person who meets the clinical criteria for severe respiratory illness and the epidemiologic criteria for likely exposure to SARS-CoV
- Confirmed case of SARS-CoV disease: in a person who has a clinically compatible illness (i.e., early, mild-to-moderate, or severe) that is laboratory confirmed

8 During the 2003 SARS epidemic, CDC case definitions were the following:

Suspect case

- Meets the clinical criteria for mild-to-moderate respiratory illness and the epidemiologic criteria for possible exposure to SARS-CoV but does not meet any of the laboratory criteria and exclusion criteria *or*
- Unexplained acute respiratory illness resulting in death in a person on whom an autopsy was not performed and who meets the epidemiologic criteria for possible exposure to SARS-CoV but does not meet any of the laboratory criteria and exclusion criteria

Probable case

- Meets the clinical criteria for severe respiratory illness and the epidemiologic criteria for possible exposure to SARS-CoV but does not meet any of the laboratory criteria and exclusion criteria

For more information, visit www.cdc.gov/ncidod/sars or call the CDC public response hotline at (888) 246-2675 (English), (888) 246-2857 (Español), or (866) 874-2646 (TTY)

APPENDIX

Sample Questionnaire/Survey

Board of Health Letterhead here.

July 25, 1996

The Massachusetts Department of Public Health in conjunction with the XXXXX Board of Health is investigating an outbreak of gastrointestinal illness which occurred among the attendees of a business conference held at Establishment A on July 19, 1996. Please complete these questions **ONLY IF YOU ATTENDED** the metting. Your completion of the following questions, **EVEN IF YOU DID NOT HAVE ANY STMPTOMS**, will greatly assist us in our efforts to identify the source of this illness. **All information which you provide will be kept strictly confidential and used solely for the purposes of this investigation.** Please return the completed questionaire to the XXXXX Board of Health at the address or fax below. Thank you for your assistance.

RETURN TO:

Board of Health
(123) 456-7890
Fax: (123) 456-7781

PLEASE DO NOT LEAVE ANY QUESTIONS BLANK

Date completed: ____ /____ / ____

1. LAST NAME ________________________ FIRST NAME ____________________

ADDRESS ________________________ TOWN ________________________

STATE ____________ ZIP CODE ____________ SEX: ☐ M ☐ F AGE __________

PHONE: (__________) - __________ - __________

2. Please mark any of the following **symptoms** that you have had **SINCE ATTENDING** the meeting on July 19, 1996.

☐ NONE ***(IF NONE, PLEASE MARK NONE AND GO TO QUESTION #10)**

☐ Chills ☐ Nausea ☐ Vomiting ☐ Abdominal Cramps

☐ Fever ☐ Muscle Aches ☐ Headaches ☐ Loss of Appetite

☐ Fatigue ☐ Dizziness ☐ Highest Temperature

☐ Diarrhea: # Episodes of diarrhea per day: ______________

Type of Diarrhea:

☐ Soft Stool
☐ Watery Diarrhea
☐ Bloody Diarrhea

Other Symptoms: ________________________

ISBN 0-12-369378-0

3. **What day** did your symptoms begin?
☐ Friday, 7/19/96
☐ Saturday, 7/20/96
☐ Sunday, 7/20/96
☐ Monday, 7/22/96
☐ Other, _________

4. **What time** did your symptoms begin? Time: ________:________ ☐ AM ☐ PM

5. **How long** did your symptoms last? ☐ Hours: # ____________ (Choose either hours or days)
☐ Days: # ____________

6. Did you seek **medical care** for your symptoms? ☐ YES ☐ NO

If YES, name of doctor? __________________ Phone # / Town

7. Were you **hospitalized** for your symptoms? ☐ YES ☐ NO

8. Did you provide a **stool sample** for testing? ☐ YES ☐ NO
Results of the test? ____

9. Did any family/household members **who did not attend the meeting** experience similar symptoms following your illness?

☐ YES ☐ NO If YES, #ILL ____________ DATE(S): ___________________________

10. Have **you** recently had any diarrhea, vomiting, or other symptoms **before the meeting?**

☐ YES ☐ NO if YES, DATE(S): ___________________________

11. Did you eat any food(s) or drink any beverage(s) at the meeting held on July 19, 1996 at Establishment A?

☐ YES ☐ NO If YES, what time did you eat? TIME: _____ : _____ ☐ AM ☐ PM

12. Please mark YES **OR** NO to indicate whether you consumed the following items:

Turkey	☐ YES	☐ NO			
Ham	☐ YES	☐ NO			
Salami	☐ YES	☐ NO			
Cheese	☐ YES	☐ NO			
Bread Rolls	☐ YES	☐ NO			
Sandwich Condiments:					
Lettuce	☐ YES	☐ NO	Pickle	☐ YES	☐ NO
Onion	☐ YES	☐ NO	Mayonnaise	☐ YES	☐ NO
Tomato	☐ YES	☐ NO	Mustard	☐ YES	☐ NO
Other	____________________				
Potato Salad	☐ YES	☐ NO			
Tuna Salad	☐ YES	☐ NO			
Coleslaw	☐ YES	☐ NO			
Tossed Green Salad	☐ YES	☐ NO			
Type of Dressing:	☐ French	☐ Italian	☐ No Dressing		
Broccoli Soup	☐ YES	☐ NO			
Chocolate Cake	☐ YES	☐ NO			

Carrot Cake	☐ YES	☐ NO			
Danish	☐ YES	☐ NO			
Sliced Fruit	☐ YES	☐ NO			
Type of Fruit:	☐ Pineapple	☐ Canteloupe	☐ Honeydew Melon		
Water	☐ YES	☐ NO	☐ Ice	☐ YES	☐ NO
Coffee	☐ YES	☐ NO	☐ Tea	☐ YES	☐ NO
Cream	☐ YES	☐ NO	☐ Milk	☐ YES	☐ NO
Soda	☐ YES	☐ NO	☐ Juice	☐ YES	☐ NO

Other Beverages of Food: ______________________________

THANK YOU FOR RESPONDING TO THESE QUESTIONS

APPENDIX

Surveillance Data Tables

TABLE C.1 Data and Source Data Systems for Biosurveillance

Category	Data for Detection and/or Characterization	Information Systems that Collect Data
Data about conditions that predispose to outbreaks	*Environmental conditions favorable for outbreaks*—Vegetation, climate, sea surface temperature, cloudiness, rainfall *Information about susceptibility of population*—Immunization records *Information about outbreaks in other regions* • Outbreak reports from World Health Organization • Data about emergence of new infections in other regions *Information about pathogens and their occurrence in the environment* • Antimicrobial resistance patterns • Results of routine testing of food and water supplies • Data about temporal and geographic patterns of specific viruses (e.g., for serotypes of influenza likely to spread during the next year) *Information about animal hosts and vectors* • Avian morbidity and mortality • Data from monitoring diseases of animals that can be transmitted to humans, including those of animals that will eventually be distributed to consumers in the form of pets or food	Satellites Meteorological data systems Immunization registries Public health information systems Clinical information systems Electronic veterinary records Food and water monitoring Information retrieval systems
Data that support detection of a release	Emergency or police calls (activities that might be indicative of a release) Emergency (911) call data (e.g., report envelop with white power) Poison control center calls Data from routine testing of water, food, and air for pathogens	Public health information systems Food, water, engineering systems Call systems Environmental sensors Police call systems
Data that support identification of presymptomatic individuals	*Detection of pathogen in animals or crops*—Dog bite reports (or other reports exposures to organisms in host animals) *Detection of pathogen in humans before symptoms* • Tuberculin skin test results • Test results from individuals identified through contact tracing or screening *Detection of physiologic changes in humans*—Biometric data (temperature, increase heart rate)	Public health information systems Laboratory information systems Electronic veterinary records Physiologic monitoring Clinical information systems
Data that support early case detection but are of low diagnostic precision	*Behaviors following recognition of symptoms* • Sales of over-the-counter products • Web-related health sites (queries related to words such as "cough") • Poison control system usage (reports of illness) • Phone calls to physicians • Absenteeism • Emergency medical services dispatching • Increased health maintenance organization (HMO) usage • Increased outpatient volumes, emergency department volume, appointments *Initial clinical care and testing*—Chief complaints, medical history, physical examinations, imaging studies, some clinical laboratory results, diagnosis (e.g., lymphadenopathy, adult respiratory distress syndrome), medications, unexplained critical illness *Increase in deaths of nonspecific cause*—Unexplained deaths, sudden deaths, newspaper reports (obituaries, articles of unusual deaths of animals and humans, clustering of victims)	Pharmaceutical sales/marketing systems Clinical information systems Attendance recording systems Call center/emergency service Public health information systems Internet-based systems Attendance monitoring systems
Data that support case detection at intermediate levels of diagnostic precision	*Clinical care* • Specific, but not definitive, test results (e.g., CD4 counts) • Combinations of test results (e.g., findings consistent with anthrax and Gram-positive rods in blood) • Clinical diagnoses from emergency departments or outpatient facilities • Data about treatments administered (e.g., tuberculosis medications) • Cause of death on death certificate *Increase in deaths of nonspecific cause*—Unexplained deaths, sudden deaths, newspaper reports (obituaries, articles about unusual deaths of animals and humans, clustering of victims)	Clinical systems Public health systems Pharmaceutical sales/marketing systems County coroner's office/medical examiner information

Continued

ISBN 0-12-369378-0

TABLE C.1 Data and Source Data Systems for Biosurveillance—Cont'd

Category	Data for Detection and/or Characterization	Information Systems that Collect Data
Data that support case detection at high diagnostic precision	*Culture and sensitivity results* • Bacterial, viral, and other cultures (sputum, stool, blood) • Antibiotic resistance data *Other definitive tests*—Direct antigen, enzyme immunoassay, fluorescent antibodies, direct fluorescent antibody test, rapid influenza A antigen-detection tests, serologic titers *Post-mortem examination* • Transbronchial biopsy, microscopic examination, lung biopsy specimens, skin biopsy • Postmortem COHB levels, toxicologic results, and autopsy	Laboratory Information management systems Public health information systems County coroner's office/medical examiner information
Data that support investigation/ characterization	*Demographic information* • Age, sex, race, ethnicity • Descriptive epidemiologic information such as onset of illness, incubation period, duration of illness, symptoms of illness, number of ill, deaths, hospitalizations, submitting physician, outcome (died, alive), hospital transfer status, county, travel history *Risk factors/information about possible common exposures*—Water sources, food sources, attendance at large indoor and outdoor gatherings, geographic location of home, work and or entertainment/leisure, food preferences, food histories, food handler interviews, travel, work histories, pet histories exposures to animals, vaccination status, illness in family members, recent illness, religion, employer, geographic dispersion of victims *Results of contact tracing*—Sexual partners, household or other contacts, contact information *Laboratory information*—Type and method of laboratory test, specimen collection date, submitting laboratory, source of specimen, specimen identification number *Food distribution data*—Source, marketing, processing, preparation, sale, packaging, refrigeration, storage, cooking, culture, poor cooking practices, food cultures, food handling methods, sanitary inspections, food delivery schedule, check of water temperature at time of seafood harvest, farm inspection, shell fish tags, transportation, etc. *Water distribution system data*—Wells, surface type, chlorination, contaminated, inadequate pool or hot tub filtration system *Air data*—Heating, ventilation, and air conditioning system records; lack of personal protective equipment; poor plant ventilation; air-conditioner system and exhaust fan	Public health information systems Vital statistics Clinical information systems Call center/emergency service Veterinary electronic records Food, water, engineering systems

Adapted from Wagner, M.M., et al. (2001). Availability and Comparative Value of Data Elements Required for an Effective Bioterrorism Detection. Washington, DC: Agency for Healthcare Research and Quality.

TABLE C.2 Types of Data/Data Sources Identified By Different Methods

Category	First Principle Analysis	Collected by Public Health	Used in Outbreak Investigations	Referenced in Centers for Disease Control and Prevention (CDC) Case Definitions
Preoutbreak	911-call system (suspicious activities, hazardous conditions) Police activity Veterinary data systems, especially public veterinary practices (diseases that animals transmit to humans, including those of animals that will eventually be distributed to consumers in the form of pets or food) Contact tracing to find infectious individuals before additional outbreaks occur.	Environmental condition as detected by satellite. (e.g., vegetation, climate, sea surface temperature, cloudiness, rainfall)[1] Data from immunization registries[2] (e.g., child's name, date and place of birth, identifiers, contact information, dates, types and lots of vaccinations) Outbreaks occurring in other regions that have the potential for spread[3,4] Emergence of new infections in other regions[4,5] Keywords in the Internet and electronic reports related to or indicative of outbreaks in other areas[6] Antimicrobial resistance patterns[7–11] Routine testing of animal food and water supplies Monitoring temporal and geographic patterns associated with the detection diseases (respiratory syncytial virus [RSV], human parainfluenza viruses [HPIV], respiratory and enteric adenoviruses, and rotavirus)[12] Avian morbidity and mortality, captive or free ranging sentinel animals[13] Port inspections for foreign pests and disease[14]		

First-principle analysis refers to data that we identified by review of the care-seeking literature and current understanding of the literature on biosurveillance. Collected by public health refers to types of data being collected routinely for biosurveillance by health departments. Used in outbreak investigations refer to types of data that have been pivotal in detecting outbreaks.

Adapted from Wagner, M.M. et al. (2001). Availability and Comparative Value of Data Elements Required for an Effective Bioterrorism Detection. Washington, DC: Agency for Healthcare Research and Quality.

[1]GEIS Rift Valley Fever Monitor. *http://www.geis.ha.osd.mil/riftvalleyfever/index.htm.*

[2]Georgia Registry of Immunization Transactions & Services (GRITS). *http://www.ph.dhr.state.ga.us/inpho/statewide/index.shtml.*

[3]Global Outbreak Alert and Response Network. *http://www.who.int/emc.*

[4]ProMed. *http://www.cdc.gov/ncidod/eid/vol4no3/kay.htm, http://www.promedmail.org/pls///promed/promed.homepromed.*

[5]Global Emerging Infections Sentinel Network (GeoSentinel). *http://www.istm.org/.*

[6]GPHIN Global Public Health Information Network. *http://www.carleton.ca/Capital_News/12031999/f4.htm;http://gphin-rmisp.hc-sc.gc.ca/index.html.*

[7]Anonymous. (1995). WHONET: An Information System for Monitoring Antimicrobial Resistance. *Emerging Infectious Diseases* vol. 1:66. *http://www.who.int/emc/WHONET/WHONET.html.*

[8]The Surveillance Network (TSN)TMDatabase. *http://www.mrlpharmaceutical.com/,http://www8.techmall.com/techdocs/TS990913-8.html".*

[9]Surveillance for Emerging Antimicrobial Resistance Connected to Healthcare (SEARCH). *http://www.cdc.gov/ncidod/hip/ARESIST/SEARCH.htm.*

[10]National Antimicrobial Resistance Monitoring System: Enteric Bacteria (NARMS:EB). *http://www.cdc.gov/ncidod/dbmd/narms/.*

[11]INSPEAR- International Network for the Study and Prevention of Emerging Antimicrobial Resistance. *http://www.cdc.gov/ncidod/hip/SURVEILL/inspear.HTM.*

[12]"National Respiratory and Enteric Virus Surveillance System (NREVSS). *http://www.cdc.gov/ncidod/dvrd/nrevss/.*

[13]National West Nile Virus (WNV) Surveillance System. *http://www.cdc.gov/ncidod/dvbid/westnile/surv&control.htm.*

[14]U.S. Department of Agriculture National Veterinary Services Laboratories (NVSL). http://www.aphis.usda.gov/vs/nvsl/.

Continued

TABLE C.2 Types of Data/Data Sources Identified By Different Methods—Cont'd

Category	First Principle Analysis	Collected by Public Health	Used in Outbreak Investigations	Referenced in Centers for Disease Control and Prevention (CDC) Case Definitions
Attack/release/exposure	911-call data (e.g., report envelop with white power) Poison center data (e.g., report of exposure to infectious or toxic agent) Vector surveillance data Environmental monitoring: special event, permanent	Monitoring of mosquito pools or other vectors for agents[13] Routine testing of water supplies[15] Animal exposures (rabies postexposure prophylaxis practices)[16] Injury-related exposures to infectious disease (reports of percutaneous injuries in hospitals)[17] Horse or other terminal host surveillance to observe for the presence in the environment of organisms that also infect humans[13]	Carbon monoxide monitor[18–21]	
Presymptomatic	Biometric data (temperature, increased heart rate) If animals are more susceptible than are humans, veterinary data systems Routine testing	Tuberculin skin testing	Contact tracing[22–25] programs[17]	
Nonspecific syndrome	Pharmaceutical sales/marketing systems (over-the-counter drug sales) Clinical (medical records, physical examination, nonspecific laboratory tests, nonspecific diagnosis, health maintenance organization (HMO) usage, outpatient clinical volumes, emergency department volume and chief complaints, unusual numbers of unexplained deaths) Absenteeism reporting systems (nonspecific increase in absence at work or school) Vital statistics (fact of death, nonspecific diagnosis on death certificate, acute increase in low birth rate or birth defects) Web-related health sites (queries related as "cough") Newspaper reports (obituaries, articles of unusual deaths of animals and humans, clustering of victims) Sentinel population monitoring	Chief complaints Symptoms (febrile rash,[26] fever, watery diarrhea, altered mental status,[26] bloody diarrhea,[16,26] dehydration, vomiting, jaundice, seizure,[16] paralysis, flu like illness,[27] cramps) Adult respiratory distress syndrome[26] Unexplained deaths and critical illness[28] Deaths from pneumonia and influenza[29] Number of patients reporting influenza-like illness,[30] fevers, headaches, diarrhea, vomiting, stuffy noses, coughs, or rashes[31] Information on sudden deaths[31] School absentee rates and predominant reason for school absentee[32] Nursing home respiratory illness rates[32] Emergency medical services dispatching[33] Widened mediastinum Gram-positive rods Number of medical visits with gastrointestinal illness	Chest abnormalities Chest X-ray,[34] deaths[35,36] Signs and symptoms of illness (fatigue, fever,[35] arthralgia, myalgia nausea, headache, abdominal cramps diarrhea, vomiting, fever, pneumonia, ataxia, lymphadenopathy,[37] conjunctivitis, hepatic abnormalities, prolonged shock, unexplained fever, gastroenteritis)[34,38–44] Clinical laboratory results (abnormal blood test results such as white blood count, pathological tests) Emergency department medications Absentee records	Diarrhea, abdominal cramps, loss of appetite, low-grade fever, nausea, and vomiting (cryptosporidiosis) Diarrhea and/or vomiting (cholera) Acute onset of fever, headache, myalgia, and/or malaise; nausea, vomiting, or rash may be present in some cases (ehrlichiosis) Fever, headache, stiff neck, and pleocytosis; altered mental status, confusion, coma, paresis or paralysis, cranial nerve palsies, sensory deficits, abnormal reflexes, generalized convulsions, and abnormal movements (encephalitis or meningitis, arboviral) Diarrhea (often bloody) and abdominal cramps (enterohemorrhagic *Escherichia coli*) Meningitis, bacteremia, epiglottitis, or pneumonia. (haemophilus influenzae, invasive disease) Watery diarrhea, loss of appetite, weight loss, abdominal bloating and cramping, increased flatus, nausea, fatigue, and low-grade fever; vomiting (cyclosporiasis) Meningitis or bacteremia, fetal loss through miscarriage or stillbirth, or neonatal meningitis or bacteremia (listeriosis) Fatigue, fever, headache, mildly stiff neck, arthralgia, or myalgia (listeriosis) Fever; headache, back pain, chills, sweats, myalgia, nausea, vomiting, diarrhea, and cough; coma, renal failure, pulmonary edema, and death (malaria) Fever, chills, headache, photophobia, cough, and myalgia (psittacosis) Encephalomyelitis, coma, death (Rabies, human) Diarrhea, abdominal pain, nausea, vomiting (salmonellosis) Diarrhea, fever, nausea, cramps, and tenesmus (shigellosis) Cutaneous infection (e.g., cellulitis, erysipelas, or infection of a surgical or nonsurgical wound),

Social behavior patterns and changes	
Poison control system usage (reports of illness)	
	deep soft-tissue infection (e.g., myositis or necrotizing fasciitis), meningitis, peritonitis, osteomyelitis, septic arthritis, postpartum sepsis (i.e., puerperal fever), neonatal sepsis, and nonfocal bacteremia (streptococcal disease, invasive, group A)
	Acute otitis media, pneumonia, bacteremia, or meningitis ("Streptococcus pneumoniae, drug-resistant invasive").
	Acute onset maculopapular rash Temperature greater than 99.0°F (greater than 37.2°C), arthralgia/arthritis, lymphadenopathy, or conjunctivitis (rubella)
	Sustained fever, headache, malaise, anorexia, relative bradycardia, constipation or diarrhea, and nonproductive cough (typhoid fever)

[15]Tillett, H.E., de Louvois, J., and Wall, P.G. (1998). Surveillance of Outbreaks of Waterborne Infectious Disease: Categorizing Levels of Evidence. *Epidemiology and Infection* vol. 120:pp. 37–42.

[16]Talan, D.A., et al. (1998). EMERGEncy ID NET: An Emergency Department-based Emerging Infections Sentinel Network. *Annals of Emergency Medicine* vol. 32:703–711.

[17]National Surveillance System for Health Care Workers. *http://www.cdc.gov/ncidod/hip/SURVEILL/nash.htm.*

[18]Anonymous. (1999). Carbon Monoxide Poisoning Associated with Use of LPG-Powered (Propane) Forklifts in Industrial Settings—Iowa, 1998. *Morbidity and Mortality Weekly Report* vol. 48:1121–1124.

[19]Anonymous. (1999). From the Centers for Disease Control and Prevention: Carbon Monoxide Poisoning Deaths Associated with camping—Georgia, March 1999. *Journal of the American Medical Association* vol. 282:1326.

[20]Anonymous. (2000). Houseboat-associated Carbon Monoxide Poisonings on Lake Powell—Arizona and Utah, 2000. *Morbidity and Mortality Weekly Report* vol. 49:1105–1108.

[21]From the Centers for Disease Control and Prevention. (2001). Houseboat-associated Carbon Monoxide Poisonings on Lake Powell—Arizona and Utah, 2000. *Journal of the American Medical Association* vol. 285:530–532.

[22]Anonymous. (1999). Cluster of HIV-Positive Young Women—New York, 1997–1998. *Morbidity and Mortality Weekly Report* vol. 48:413–416.

[23]Anonymous. (2000). Drug-Susceptible Tuberculosis Outbreak in a State Correctional Facility Housing HIV-Infected Inmates—South Carolina, 1999–2000. *Morbidity and Mortality Weekly Report* vol. 49:1041–1044.

[24]Anonymous. (2000). HIV-Related Tuberculosis in a Transgender Network–Baltimore, Maryland, and New York City area, 1998–2000. *Morbidity and Mortality Weekly Report* vol. 49:317–320.

[25]Anonymous. (2000). Fluoroquinolone-Resistance in *Neisseria gonorrhoeae*, Hawaii, 1999, and Decreased Susceptibility to Azithromycin in *N. gonorrhoeae*, Missouri, 1999. *Morbidity and Mortality Weekly Report* vol. 49:833–837.

[26]New Mexico Department of Health. (2001). RSVP. *http://epi.health.state.nm.us/rsvpdesc/default.shtml.*

[27]Rapid Syndrome Validation Project (RSVP). *http://epi.health.state.nm.us/rsvpdesc/default.shtml.*

[28]CDC Unexplained Deaths and Critical Illness Surveillance Project. *http://www.cdc.gov/ncidod/osr/EIP.htm http://www.cdc.gov/ncidod/eid/vol2no1/perkins.htmhttp://www.slackinc.com/child/idc/200012/unknown.asp.*

[29]121 Cities Mortality Reporting System. *http://www.cdc.gov/drugresistance/technical/surveillance.htm, http://wonder.cdc.gov/mmwr/about.htm.*

[30]United States Influenza Sentinel Physicians Surveillance Network. *http://www.cdc.gov/ncidod/diseases/flu/flusurv.htm,http://www.cdc.gov/drugresistance/technical/surveillance.htm, http://www.cdc.gov/ncidod/diseases/flu/weekly.htm.*

[31]Electronic Surveillance System for Early Notification of Community-based Epidemics (ESSENCE). *http://www.geis.ha.osd.mil/, http://www.geis.ha.osd.mil/getpage.asp?page?SyndromicSurveillance.htm&action?7&click?KeyPrograms.*

[32]New Jersey's Active Influenza-like Illness Surveillance System. *http://www.state.nj.us/health/fluinfo/.*

[33]New York City Emergency Management System. *http://www.ci.nyc.ny.us/html/oem/html/oem_creation.html.*

[34]Anonymous. (1982). Epidemiologic Notes and Reports Pneumocystis carinii Pneumonia among Persons with Hemophilia A. *Morbidity and Mortality Weekly Report* vol. 31:365–367.

[35]Anonymous. (1981). Pneumocystis Pneumonia—Los Angeles. vol. 30.

[36]Anonymous. (1999). Outbreak of Hendra-like virus—Malaysia and Singapore, 1998–1999. *Morbidity and Mortality Weekly Report* vol. 48:265–269.

[37]Antibodies to a Retrovirus Etiologically Associated with Acquired Immunodeficiency Syndrome (AIDS) in Populations with Increased Incidences of the Syndrome. *Morbidity and Mortality Weekly Report* vol. 33:337–339.

[38]Anonymous. (1999). Outbreak of *Escherichia coli* O157:H7 and *Campylobacter* among attendees of the Washington County Fair, New York, 1999. *Morbidity and Mortality Weekly Report* vol. 48:803–805.

[39]Anonymous. (1999). Outbreak of *Salmonella* Serotype Muenchen Infections Associated with Unpasteurized Orange Juice—United States and Canada, June 1999. *Morbidity and Mortality Weekly Report* vol. 48:582–585.

[40]Anonymous. (1999). Outbreaks of Gastrointestinal Illness of Unknown Etiology Associated with Eating Burritos—United States, October 1997-October 1998. *Morbidity and Mortality Weekly Report* vol. 48:210–213.

[41]Anonymous. (2000). Outbreak of Rift Valley fever—Saudi Arabia, August–October, 2000. *Morbidity and Mortality Weekly Report* vol. 49:905–908.

[42]Anonymous. (2000). *Escherichia coli* O111:H8 Outbreak Among Teenage Campers—Texas, 1999. *Morbidity and Mortality Weekly Report* vol. 49:321–324.

[43]Anonymous. (2000). Outbreak of *Shigella sonnei* Infections Associated with Eating a Nationally Distributed Dip—California, Oregon, and Washington, January 2000. *Morbidity and Mortality Weekly Report* vol. 49:60–61.

[44]Anonymous. (2000). Fatal Yellow Fever in a Traveler Returning from Venezuela, 1999. *Morbidity and Mortality Weekly Report* vol. 49:303–305.

Continued

TABLE C.2 Types of Data/Data Sources Identified By Different Methods—Cont'd

Category	First Principle Analysis	Collected by Public Health	Used in Outbreak Investigations	Referenced in Centers for Disease Control and Prevention (CDC) Case Definitions
Specific syndrome	Vital statistics (e.g., specific diagnosis related to bioterrorist agents) Physician database searches (Web-based medical queries) County coroner's office/medical examiner information Electronic medical records (specific diagnosis in outpatient, hospital or emergency department) Public health reporting (reportable diseases) Medication usage (usage of medications reserved for unusual diseases)	Cause of death on death certificate (if syndromic) Medical record (e.g., dialysis related infection,[45] opportunistic infection[46]) Disease reports that match a case definition (botulism)[47] Influenza-like activity as assessed by state epidemiologists[48] Discharge diagnoses (AIDS indicator diseases,[46] necrotizing fasciitis, streptococcal toxic shock syndrome [STSS],[49] neurocysteicercosis[16]) CD4 counts[46] Frequency of prescriptions for antiretroviral therapy and prophylaxis for *Pneumocystis carinii* pneumonia and *Mycobacterium avium* complex disease[46]	Medical diagnosis (PCP diagnosis, toxoplasmosis, carbon monocide poisoning[18–21]) Reportable disease (e.g., folliculitis/ dermatitis)[50]	Prodrome resembling a viral respiratory illness, hypoxia and dyspnea, radiographic evidence of mediastinal widening, severe abdominal distress, fever and signs of septicemia (anthrax) Diplopia, blurred vision, and bulbar weakness; symmetric paralysis (botulism, foodborne) Acute or insidious onset of fever, night sweats, undue fatigue, anorexia, weight loss, headache, and arthralgia (brucellosis) Painful genital ulceration and inflammatory inguinal adenopathy ulcer(s) not typical of disease caused by (chancroid) Urethritis, epididymitis, cervicitis, acute salpingitis, lymphogranuloma venereum and trachoma (Chlamydia trachomatis, genital infections) Fever, chest pain, cough, myalgia, arthralgia, and headache); pneumonia or other pulmonary lesion, diagnosed by chest radiograph, erythema nodosum or erythema multiforme rash; involvement of bones, joints, or skin by dissemination, meningitis; involvement of viscera and lymph nodes (coccidioidomycosis) Acute onset of fever, headache, myalgia, and/or malaise; nausea; vomiting; or rash (ehrlichiosis) Thrombocytopenia, leukopenia, and/or elevated liver enzymes; intracytoplasmic bacterial aggregates (morulae) may be visible in the leukocytes of some patients (ehrlichiosis) Hemolytic uremic syndrome (HUS) or thrombotic thrombocytopenic purpura (TTP; enterohemorrhagic *Escherichia coli*) Tuberculoid: one or a few well-demarcated, hypopigmented, and anesthetic skin lesions, frequently with active, spreading edges and a clearing center; peripheral nerve swelling or thickening also may occur (Hansen disease [leprosy]) Lepromatous: a number of erythematous papules and nodules or an infiltration of the face, hands, and feet with lesions in a bilateral and symmetrical distribution that progress to thickening of the skin (Hansen disease [Leprosy]) Borderline (dimorphous): skin lesions characteristic of both the tuberculoid and lepromatous forms; indeterminate: early lesions, usually hypopigmented macules, without developed tuberculoid or lepromatous features (Hansen disease [leprosy]) Febrile illness (i.e., temperature greater than 101.0°F [greater than 38.3°C]) characterized by bilateral diffuse interstitial edema that may radiographically resemble acute respiratory distress syndrome (ARDS), with respiratory compromise requiring

supplemental oxygen, developing within 72 hours of hospitalization, and occurring in a previously healthy person; an unexplained respiratory illness resulting in death, with an autopsy examination demonstrating noncardiogenic pulmonary edema without an identifiable cause (hantavirus pulmonary syndrome)

Acute onset of microangiopathic hemolytic anemia, renal injury, and low platelet count (hemolytic uremic syndrome postdiarrheal)

Anemia (acute onset) with microangiopathic changes (i.e., schistocytes, burr cells, or helmet cells) on peripheral blood smear and renal injury (acute onset) evidenced by either hematuria, proteinuria, or elevated creatinine level (hemolytic uremic syndrome postdiarrheal)

Discrete onset of symptoms and jaundice or elevated serum aminotransferase levels (hepatitis, viral, acute)

Fever, myalgia, cough, pneumonia (legionellosis)

Erythema migrans, recurrent, brief attacks (weeks or months) of objective joint swelling in one or a few joints, sometimes followed by chronic arthritis in one or a few joints, lymphocytic meningitis; cranial neuritis, particularly facial palsy (may be bilateral); radiculoneuropathy; or, rarely, encephalomyelitis, (second- or third-degree) atrioventricular conduction defects that resolve in days to weeks and are sometimes associated with myocarditis (listeriosis)

Generalized rash, length of time of rash, height of temperature, cough, coryza, or conjunctivitis (measles)

Meningitis, meningococcemia, purpura fulminans, shock, and death (meningococcal disease)

Unilateral or bilateral tender, self-limited swelling of the parotid or other salivary gland, lasting greater than or equal to two days and without other apparent cause (mumps)

Cough illness, length of illness, paroxysms of coughing, inspiratory whoop, or posttussive vomiting, without other apparent cause (pertussis)

Flaccid paralysis of one or more limbs, decreased or absent tendon reflexes, affected limbs, other apparent cause, sensory or cognitive loss, neurologic deficit, time after initial symptoms, death, follow-up status (polio, paralytic)

Fever, rigors, myalgia, malaise, and retrobulbar headache; acute hepatitis; pneumonia; meningoencephalitis; liver enzyme levels; chest film findings; endocarditis; chronic fatigue-like syndrome (Q fever)

---

[45]Dialysis Surveillance Network. *http://www.cdc.gov/ncidod/hip/DIALYSIS/dsn.htm.*

[46]Adult/Adolescent Spectrum of HIV Disease (ASD) Sentinel Surveillance Project. *Morbidity and Mortality Weekly Report Supplement* 1–22.

[47]CDC's National Botulism Surveillance and Reference Laboratory. vol. 49:778–780.

[48]Influenza Activity as Assessed by State and Territorial Epidemiologists. *http://www.cdc.gov/ncidod/diseases/flu/flusurv.htm.*

[49]IDSA.EIN or IDSA.NEThe Infectious Diseases Society of America (IDSA) Emerging Infections Network (EIN). *http://www.idsociety.org/EIN/TOC.htm.*

[50]Anonymous. (2000). Pseudomonas Dermatitis/Folliculitis Associated with Pools and Hot Tubs—Colorado and Maine, 1999–2000. *Morbidity and Mortality Weekly Report* vol. 49:1087–1091.

Continued

TABLE C.2 Types of Data/Data Sources Identified By Different Methods—Cont'd

Category	First Principle Analysis	Collected by Public Health	Used in Outbreak Investigations	Referenced in Centers for Disease Control and Prevention (CDC) Case Definitions
Specific syndrome (Continued)				Acute onset, myalgia, headache, and petechial rash (on the palms and soles in two-thirds of the cases); time from hospitalization (Rocky Mountain spotted fever)
				Hypotension renal impairment: creatinine, history of renal disease, platelet count, disseminated intravascular coagulation; alanine aminotransferase, aspartate aminotransferase, total bilirubin, preexisting liver disease, acute onset of diffuse pulmonary infiltrates and hypoxemia, cardiac failure, diffuse capillary leak, acute onset of generalized edema, or pleural or peritoneal effusions; hypoalbuminemia; generalized erythematous macular rash, desquamation, soft-tissue necrosis, necrotizing fasciitis or myositis; or gangrene (streptococcal toxic-shock syndrome)
				One or more ulcers (chancres) localized or diffuse mucocutaneous lesions, often with generalized lymphadenopathy; central nervous system infection inflammatory lesions of the cardiovascular system, skin, and bone (syphilis)
				Diffuse (generalized) maculopapulovesicular rash (varicella [chickenpox]–1999 non-notifiable)
				Acute onset of hypertonia and/or painful muscular contractions (usually of the muscles of the jaw and neck) and generalized muscle spasms without other apparent medical cause (tetanus)
				Fever: temperature greater than or equal to 102.0°F (greater than or equal to 38.9°C) rash: diffuse macular erythroderma; desquamation: one to two weeks after onset of illness, particularly on the palms and soles; hypotension: orthostatic syncope, or orthostatic dizziness; gastrointestinal: vomiting, diarrhea; muscular: severe myalgia or creatine phosphokinase; mucous membrane: vaginal, oropharyngeal, or conjunctival hyperemia, blood urea nitrogen or creatinine, urinary sediment with pyuria, total bilirubin, alanine amino-transferase enzyme, or aspartate aminotrans- ferase enzyme platelets; central nervous system: disorienta-tion or alterations in consciousness, focal neurologic signs, fever, hypotension (toxic-shock syndrome)
				Eosinophilia, fever, myalgia, and periorbital edema (trichinosis)
				A positive tuberculin skin test, chest radiographs, or clinical evidence of current disease (tuberculosis)
				Cutaneous ulcer, regional lymphadenopathy; regional lymphadenopathy with no ulcer: conjunctivitis with preauricular lymphadenopathy, stomatitis or pharyngitis or tonsillitis and cervical lymphadenopathy, intestinal pain, vomiting, diarrhea, primary pleuropulmonary disease (tularemia)
				Acute onset, constitutional symptoms, brief remission, recurrence of fever, hepatitis, albuminuria, renal failure, shock, and generalized hemorrhages (yellow fever)

Definitive	Microbial culture results Microbial resistance patterns CoHB levels	Bacterial culture and sensitivity (group A streptococcus,[51] group B streptococcus,[51] *Haemophilius influenzae*,[51] *Neisseria meningitidis*,[51] cholera,[52] MSRA,[53] *N. gonorrhoeae*,[54] *Staphylococcus aureus* with reduced susceptibility to vancomycin, *Salmonella* (serogroup/serotype),[10,55] *Vibrio* (species),[55] *Shigella* (serogroup/species),[10,55] *Listeria monocytogenes*,[55] *Campylobacter*,[55] (species),[10] *Yersinia enterocolitical*,[55] *E. coli* O157,[10,55] other H antigen–positive nonmotile, shiga-like toxin-producing)[16] *Cryptosporidum, Cyclospora* Serologic titers (HIV)[56] Detection of botulinum toxin in serum, stool, or patient's food[47] Rapid Diagnosis tests (Directigen, enzyme immunoassay [EIA], fluorescent antibodies) Number of isolates of *Salmonella* (current compared to historical)[57] DNA fingerprints[58], or RNA[13], (West Nile virus)	Serology[37,59,60] Pathology (transbronchial biopsy,[34] microscopic examination, lung biopsy specimens[35,36]) Postmortem examination,[35] (postmortem COHB levels, toxicologic) Cultures (sputum, stool, blood, joint for bacterial or viral pathogens)[61] Assays (cell culture, biologic toxicity, polymerase chain reaction [PCR],[41] immunologic) Chemical analyses for toxins DNA fingerprint,[23,24,62] pulsed-field gel electrophoresis,[63–65] reverse transcriptase[66,67] enzyme records, plasmid analysis, Viral cultures[68–70] Serologic testing[37] and assays,[71] rubella specific IGM, skin biopsy, Direct fluorescent antibody (DFA) test,[72] rapid influenza A antigen-detection tests[70] Pathological examination (intercellular parasites) PCR testing,[41] DNA sequencing, phylogenetic analysis, antimicrobial susceptibility pattern comparisons	Screening test for HIV antibody (EIA), followed by a positive result on a confirmatory (sensitive and more specific) test for HIV antibody (e.g., Western blot or immunofluorescence antibody [IFA] test) HIV virologic (nonantibody) tests: HIV nucleic acid (DNA or RNA) detection (e.g., DNA PCR or plasma HIV-1 RNA), HIV p24 antigen test, including neutralization assay, HIV isolation (viral culture; HIV infection, adult) Botulinum toxin in serum, stool, or patient's food, or isolation of *Clostridium* botulinum from stool (botulism, foodborne) Isolation of *Brucella* sp. from a clinical specimen, or fourfold or greater rise in *Brucella* agglutination titer between acute- and convalescent-phase serum specimens obtained greater than or equal to two weeks apart and studied at the same laboratory, or demonstration by immunofluorescence of *Brucella* sp. in a clinical specimen (brucellosis) Isolation of *H. ducreyi* from a clinical specimen (chancroid), *Treponema pallidum* infection by darkfield microscopic examination of ulcer exudates, by a serologic test for syphilis performed greater than or equal to seven days after onset of ulcers herpes simplex virus (HSV) culture negative for HSV (chancroid) Isolation of *C. trachomatis* by culture, or demonstration of *C. trachomatis* in a clinical specimen by detection of antigen or nucleic acid (chlamydia trachomatis, genital infections)

[51]Active Bacterial Core Surveillance (ABCs). *http://www.cdc.gov/drugresistance/technical/surveillance.htm, http://aspe.os.dhhs.gov/datacncl/datadir/cdc4.htm.*

[52]Cholera. *http://aspe.os.dhhs.gov/datacncl/datadir/cdc4.htm.*

[53]Community-Acquired MRSA Infection in Patients without Established Risk Factors. *Morbidity and Mortality Weekly Report* vol. 48:707–710.

[54]Gonococcal Isolate Surveillance Project (GISP). *http://www.cdc.gov/mmwr/preview/mmwrhtml/mm4937a1.htm.*

[55]Foodborne Diseases Active Surveillance Network (FoodNet). *http://www.cdc.gov/ncidod/dbmd/foodnet/.*

[56]HIV/AIDS Reporting System. *http://www.os.dhhs.gov/progorg/aspe/minority/mincdc4.htm.*

[57]Salmonella Outbreak Detection Algorithm (SODA). *http://www.cdc.gov/ncidod/dbmd/phlisdata/default.htm.*

[58]PulseNet, The National Molecular Subtyping Network for Foodborne Disease. *http://www.cdc.gov/ncidod/dbmd/pulsenet/pulsenet.htm#What%20is.*

[59]Anonymous. (2000). Serogroup W-135 Meningococcal Disease among Travelers Returning from Saudi Arabia–United States, 2000. *Morbidity and Mortality Weekly Report* vol. 49:345–346.

[60]Anonymous. (2000). Outbreak of Aseptic Meningitis Associated with Multiple Enterovirus Serotypes—Romania, 1999. *Morbidity and Mortality Weekly Report* vol. 49:669–671.

[61]Anonymous. (1999). Trichinellosis Outbreaks—Northrhine-Westfalia, Germany, 1998–1999. *Morbidity and Mortality Weekly Report*vol. 48:488–492.

[62]Anonymous. (1999). Tuberculosis Outbreaks in Prison Housing Units for HIV-Infected Inmates—California, 1995–1996. *Morbidity and Mortality Weekly Report* vol. 48:79–82.

[63]Anonymous. (1999). Outbreaks of *Shigella sonnei* Infection Associated with Eating Fresh Parsley–United States and Canada, July-August 1998. *Morbidity and Mortality Weekly Report* vol. 48:285–289.

[64]Anonymous. (1999). Use of Pulsed-Field Gel Electrophoresis for Investigation of a Cluster of Invasive Group A Streptococcal Illness—Spokane, Washington, 1999. *Morbidity and Mortality Weekly Report* vol. 48:681–683.

[65]Anonymous. (2000). Multistate Outbreak of Listeriosis—United States, 2000. *Morbidity and Mortality Weekly Report* vol. 49:1129–1130.

[66]From the Centers for Disease Control and Prevention. (2000). Fatal illnesses associated with a New World arenavirus—California, 1999–2000. *Journal of the American Medical Association* vol. 284:1237–1238.

[67]Arness, M.K., et al. (2000). Norwalk-like Viral Gastroenteritis Outbreak in U.S. Army Trainees. *Emerging Infectious Diseases* vol. 6:204–207.

[68]Anonymous. (1999). Outbreak of Poliomyelitis—Iraq, 1999. *Morbidity and Mortality Weekly Report* vol. 48:858–859.

[69]Anonymous. (1999). Outbreak of Poliomyelitis—Kunduz, Afghanistan, 1999. *Morbidity and Mortality Weekly Report* vol. 48:761–762.

[70]Anonymous. (1999). Outbreak of Influenza A Infection among Travelers—Alaska and the Yukon Territory, May-June 1999. *Morbidity and Mortality Weekly Report* vol. 48:545–456, 555.

[71]Anonymous. (1999). Outbreak of West Nile-like Viral Encephalitis—New York, 1999. *Morbidity and Mortality Weekly Report* vol. 48: 845–849.

Continued

TABLE C.2 Types of Data/Data Sources Identified By Different Methods—Cont'd

Category	First Principle Analysis	Collected by Public Health	Used in Outbreak Investigations	Referenced in Centers for Disease Control and Prevention (CDC) Case Definitions
Definitive (Continued)			Autopsy[35]	Isolation of toxigenic (i.e., cholera toxin-producing) *Vibrio* cholerae O1 or O139 from stool or vomitus, or serologic evidence of recent infection (cholera) Cryptosporidiosis oocysts in stool by microscopic examination or in intestinal fluid or small-bowel biopsy specimens, or oocyst or sporozoite antigens by immunodiagnostic methods (e.g., enzyme-linked immunosorbent assay [ELISA]), or PCR techniques when routinely available, or demonstration of reproductive stages in tissue preparations (cryptosporidiosis) Demonstration of a fourfold change in antibody titer to *E. chaffeensis* antigen by indirect IFA in paired serum samples, or PCR assay and confirmation of *E. chaffeensis* DNA, or Identification of morulae in Leukocytes, and a positive IFA titer to *E. chaffeensis* antigen (based on cutoff titers established by the laboratory performing the assay), or immunostaining of *E. chaffeensis* antigen in a biopsy or autopsy sample, or culture of *E. chaffeensis* from a clinical specimen; HGE: demonstration of a fourfold change in antibody titer to *E. phagocytophila* antigen by IFA in paired serum samples, or positive PCR assay and confirmation of *E. Phagocytophila* DNA, or identification of morulae in leukocytes, and a positive IFA titer to *E. Phagocytophila* antigen (based on cutoff titers established by the laboratory performing the assay), or immunostaining of *E. Phagocytophila* antigen in a biopsy or autopsy sample, or culture of *E. Phagocytophila* from a clinical specimen; ehrlichiosis (other or unspecified agent): demonstration of a fourfold change in antibody titer to more than one *Ehrlichia* species by IFA in paired serum samples, in which a dominant reactivity cannot be established, or identification of an *Ehrlichia* species other than *E. chaffeensis* or *E. phagocytophila* by PCR, immunostaining, or culture (ehrlichiosis) Fourfold or greater change in virus-specific serum antibody titer, or isolation of virus from or demonstration of specific viral antigen or genomic sequences in tissue, blood, cerebrospinal fluid (CSF), or other body fluid, or virus-specific immunoglobulin M (IgM) antibodies demonstrated in CSF by antibody-capture EIA, or virus-specific IgM antibodies demonstrated in serum by antibody-capture EIA and confirmed by demonstration of virus-specific serum immunoglobulin G (IgG) antibodies in the same or a later specimen by another serologic assay (e.g., neutralization or hemagglutination inhibition; encephalitis or meningitis, arboviral)

Demonstration of *G. lamblia* cysts in stool or demonstration of *G. Lamblia* trophozoites in stool, duodenal fluid, or small bowel biopsy, or demonstration of *G. lamblia* antigen in stool by a specific immunodiagnostic such as ELISA (giardiasis)

Isolation of typical Gram-negative, oxidase-positive diplococci (presumptive *Neisseria gonorrhoeae*); demonstration of *N. Gonorrhoeae* in a clinical specimen by detection of antigen or nucleic acid; or observation of Gram-negative intracellular diplococci in a urethral smear obtained from a male (gonorrhea)

Isolation of *H. influenzae* from a normally sterile site (e.g., blood or CSF or, less commonly, joint, pleural, or pericardial fluid) (haemophilus influenzae, invasive disease); *Cyclospora* oocysts in stool by microscopic examination, or in intestinal fluid or small-bowel biopsy specimens, or demonstration of sporulation, or DNA (by PCR) in stool, duodenal/jejunal aspirates, or small-bowel biopsy specimens (cyclosporiasis)

Demonstration of acid-fast bacilli in skin or dermal nerve, obtained from the full-thickness skin biopsy of a lepromatous lesion (Hansen disease [leprosy])

Detection of hantavirus-specific IgM or rising titers of hantavirus-specific IgG, or detection of hantavirus-specific ribonucleic acid sequence by PCR in clinical specimens, or detection of hantavirus antigen by immunohistochemistry (hantavirus pulmonary syndrome)

Hepatitis A: IgM antibody to hepatitis A virus (anti-HAV) positive hepatitis B: IgM antibody to hepatitis B core antigen (anti-HBc) positive or hepatitis B surface antigen (HBsAg) positive IgM anti-HAV negative (if done); hepatitis C: revised 2000 Serum aminotransferase levels greater than seven times the upper limit of normal, and IgM anti-HAV negative, and IgM anti-HBc negative (if done) or HBsAg negative, and antibody to hepatitis C virus (anti-HCV) positive, verified by an additional more specific assay non-A, non-B hepatitis: serum aminotransferase levels greater than 2.5 times the upper limit of normal, and IgM anti-HAV negative, and IgM anti-HBc negative (if done) or HBsAg negative, and Anti-HCV negative (if done) delta hepatitis*: HBsAg or IgM anti-HBc positive and antibody to hepatitis delta virus positive (hepatitis, viral, acute)

Continued

TABLE C.2 Types of Data/Data Sources Identified By Different Methods—Cont'd

Category	First Principle Analysis	Collected by Public Health	Used in Outbreak Investigations	Referenced in Centers for Disease Control and Prevention (CDC) Case Definitions
Definitive (Continued)				Isolation of *Legionella* from respiratory secretions, lung tissue, pleural fluid, or other normally sterile fluids, or demonstration of a fourfold or greater rise in the reciprocal IFA titer to greater than or equal to 128 against *Legionella pneumophila* serogroup 1 between paired acute- and convalescent-phase serum specimens; or detection of *L. pneumophila* serogroup 1 in respiratory secretions, lung tissue, or pleural fluid by DFA testing; or demonstration of *L. pneumophila* serogroup 1 antigens in urine by radioimmunoassay or enzyme-linked immunosorbent assay (legionellosis)
				Isolation of *L. monocytogenes* from a normally sterile site (e.g., blood or CSF or, less commonly, joint, pleural, or pericardial fluid) isolation of *L. Monocytogenes* from placental or fetal tissue (listeriosis)
				Isolation of *Borrelia burgdorferi* from a clinical specimen or demonstration of diagnostic IgM or IgG antibodies to *B. Burgdorferi* in serum or CSF. A two-test approach using a sensitive EIA or IFA followed by Western blot is recommended (listeriosis)
				Demonstration of malaria parasites in blood films (malaria)
				Positive serologic test for measles, IgM antibody; or significant rise in measles antibody level by any standard serologic assay; or isolation of measles virus from a clinical specimen (measles)
				Isolation of *Neisseria meningitidis* from a normally sterile site (e.g., blood or CSF or, less commonly, joint, pleural, or pericardial fluid; meningococcal disease)
				Isolation of mumps virus from clinical specimen, or significant rise between acute- and convalescent-phase titers in serum mumps IgG antibody level by any standard serologic assay, or positive serologic test for mumps IgM antibody (mumps)
				Isolation of *Bordetella pertussis* from clinical Specimen positive PCR for *B. pertussis* (pertussis)
				Isolation of *Chlamydia psittaci* from respiratory secretions, or fourfold or greater increase in antibody against *C. psittaci* by complement fixation (CF) or microimmunofluorescence (MIF) to a reciprocal titer of greater than or equal to 32 between paired acute- and convalescent-phase serum specimens, or presence of IgM antibody against *C. psittaci* by MIF to a reciprocal titer of greater than or equal to 16 (psittacosis)

Fourfold or greater change in antibody titer to *C. burnetii* phase II or phase I antigen in paired serum specimens ideally taken three to six weeks apart, or isolation of *C. burnetii* from a clinical specimen by culture, or demonstration of *C. burnetii* in a clinical specimen by detection of antigen or nucleic acid (Q fever)

Rabies, human encephalomyelitis, coma, death (rabies, human); detection by DFA of viral antigens in a clinical specimen (preferably the brain or the nerves surrounding hair follicles in the nape of the neck); or isolation (in cell culture or in a laboratory animal) of rabies virus from saliva, CSF, or central nervous system tissue; or identification of a rabies-neutralizing antibody titer greater than or equal to five (complete neutralization) in the serum or CSF of an unvaccinated person. (rabies, human)

Fourfold or greater rise in antibody titer to *Rickettsia rickettsii* antigen by IFA, CF, latex agglutination (LA), microagglutination,[73] or indirect hemagglutination antibody (IHA) test in acute- and convalescent-phase specimens ideally taken greater than or equal to three weeks apart; or positive PCR to *R. rickettsii*; or demonstration of positive immunofluorescence of skin lesion (biopsy) or organ tissue (autopsy); or isolation of *R. rickettsii* from clinical specimen (Rocky Mountain spotted fever)

Isolation of rubella virus, or significant rise between acute- and convalescent-phase titers in serum rubella IgG antibody level by any standard serologic assay, or positive serologic test for rubella IgM antibody (rubella)

Isolation of *Salmonella* from a clinical specimen (salmonellosis); isolation of *Shigella* from a clinical specimen (shigellosis)

Streptococcus (streptococcus pyogenes) by culture from a normally sterile site (e.g., blood or CSF, or, less commonly, joint, pleural, or pericardial fluid) (streptococcal disease, invasive, group A)

Isolation of group A streptococcus (streptococcal toxic-shock syndrome); isolation of *S. Pneumoniae* from a normally sterile site (e.g., blood, CSF, or, less commonly, joint, pleural, or pericardial fluid) and nonsusceptible isolate (i.e., intermediate- or high-level resistance of the *S. pneumoniae* isolate to at least one antimicrobial agent currently approved for use in treating pneumococcal infection (streptococcal disease, invasive, group A)* Isolation of *S. pneumoniae* from a normally sterile site (e.g., blood, CSF, or, less commonly, joint, pleural, or pericardial fluid) (streptococcus pneumoniae, invasive, [children less than five years])

Continued

TABLE C.2 Types of Data/Data Sources Identified By Different Methods—Cont'd

Category	First Principle Analysis	Collected by Public Health	Used in Outbreak Investigations	Referenced in Centers for Disease Control and Prevention (CDC) Case Definitions
Definitive Continued)				Demonstration of *T. pallidum* in clinical specimens by darkfield microscopy, DFA, or equivalent methods (syphilis, primary); *Trichinella* larvae in tissue obtained by muscle biopsy, or positive serologic test for *Trichinella* (trichinosis) Isolation of *M. tuberculosis* from a clinical specimen or demonstration of *M. tuberculosis* from a clinical specimen by nucleic acid amplification test, or demonstration of acid-fast bacilli in a clinical specimen (tuberculosis) Presumptive: elevated serum antibody titer(s) to *F. tularensis* antigen (without documented fourfold or greater change) in a patient with no history of tularemia vaccination or detection of *F. tularensis* in a clinical specimen by fluorescent assay; confirmatory: isolation of *F. tularensis* in a clinical specimen or fourfold or greater change in serum antibody titer to *F. tularensis* antigen (tularemia) Isolation of *S. typhi* from blood, stool, or other clinical specimen (typhoid fever) Fourfold or greater rise in yellow fever antibody titer in a patient who has no history of recent yellow fever vaccination and cross-reactions to other flaviviruses have been excluded; or demonstration of yellow fever virus, antigen, or genome in tissue, blood, or other body fluid (yellow fever); isolation of varicella virus from a clinical specimen; or DFA or PCR; or significant rise in serum varicella IgG antibody level by any standard serologic assay (varicella [chickenpox]–1999 Non-Notifiable) Reactive serologic test (nontreponemal: Venereal Disease Research Laboratory [VDRL] or rapid plasma reagin [RPR]; treponemal: fluorescent treponemal antibody absorbed [FTA-ABS] or microhemagglutination assay for antibody to *T. pallidum* [MHA-TP]) (syphilis, [1] primary) Elevated CSF protein or leukocyte count (syphilis, [7] neuro) Isolation of *E. coli* O157:H7 from a specimen, or isolation of *Shiga* toxin-producing *E. coli* from a clinical specimen (enterohemorrhagic *E. coli*)
Epidemiologic Data	Medical records systems (age, sex, race, employer, religion, past history) Worksite logs, guest lists, itinerary Food history, food handler interviews, food preference questionnaire,	Mortality information[55] (date of death, location of death) Descriptive information (age,[26] sex, [26] race, zip code, [26] ethnicity, occupation[26]) Outbreak information[55,74] (number ill, exposed, hospitalized: type of outbreak date of outbreak, location of outbreak, food specific attack rates,[75] location of food consumption) Illness related information (incubation period, duration of illness, symptoms	Duration of hospitalization, age, sex, illness in family members, recent illness, standardized, medical history questionnaire Deaths,[36,78,79] travel/work history, age, sex Hospital admissions,[38,44] potential risk factors, pregnancy,[80] age, sex, deaths, miscarriage,[80,81] Discharge record, histories of	Epidemiologic link (e.g., ingestion of a home-canned food within the previous 48 hours), persons who ate the same food as persons who have laboratory-confirmed botulism (botulism) Epidemiologic link to a confirmed case (brucellosis) Exposure is defined as having been (less than or equal to 30 days before onset) in wooded, brushy, or grassy areas (i.e., potential tick habitats) in a county in which Lyme disease is endemic; d history

Social aggregations: families, coworkers, recreational groups Location of recent travel or gatherings Food and water (specific intake of ill individuals, population distribution patterns)	of illness, submitting physician, patient status, hospital date of admission and discharge, outcome [died, alive], hospital transfer status, case-control study, treatment history, county, travel history, onset of illness, use of chemoprophylaxis) Identifying and contact information (phone number, etc.) Disease specific information (hemophila: inhibitor/immune tolerance usage,[76] severity, functional health status, service expenditure, service utilization (hospital bed usage),[16] infection control policies, volume depletion, device-associated infection rates,[45] surgical site infection rates, incidence rates, frequency of specific tuberculosis strains geographically,[77] spread of related strains in communities,[77] geographic mobility of related strains,[77] relatedness of strains in persons at high risk for tuberculosis,[77]	exposures,[36] health center visits, food handler interviews,[63] food preference questionnaire, food history[82–84] Hospitalization,[42] birth place, age race, vaccination status,[85–88] travel history,[44,59,70,89,90,91], guest list, worksite, hospital and provider referrals Age, hospitalization diagnoses, death certificate, location of residence, travel history, Sexual partners,[22,25,92] sexual orientation,[24] onset of illness, hospitalization Pet history,[72] location of residence, travel history,[70] visitors sign-in-log, at zoo, direct and indirect animal contact[72,80,93–101]	of tick bite is not required; disease endemic to county; a county in which Lyme disease is endemic is one in which at least two confirmed cases have been previously acquired or n which established populations of a known tick vector are infected with *B. burgdorferi* (listeriosis) Location infection acquired, number of episodes, acquired through artificial means (e.g., blood transfusion, common syringes, or malariotherapy), relapses, presence of other cases (malaria) Rash onset occurs within 18 days after entering the jurisdiction, and illness cannot be linked to local transmission. Imported cases should be classified as: -International. A case that is imported from another country (out-of-state). A case that is imported from another state in the United States. The possibility that a patient was exposed within his or her state of residence should be excluded; therefore, the patient either must have been out of state continuously for the entire period of possible

[72]Anonymous. (1999). Public Health Response to a Potentially Rabid Bear Cub—Iowa, 1999. *Morbidity and Mortality Weekly Report* vol. 48:971–973.

[73]Ma, J.Z., Peterson, D.R., and Ackerman, E. (1993). Parameter Sensitivity of a Model of Viral Epidemics Simulated with Monte Carlo techniques, IV: Parametric Ranges and Optimization. *International Journal of Bio-Medical Computing* vol. 33:297–311.

[74]Water-borne Disease Outbreak Surveillance System. *http://www.cdc.gov/mmwr/preview/mmwrhtml/00001596.htm.*

[75]Foodborne- Disease Outbreak Surveillance System (FBDO). *http://www.cdc.gov/mmwr/preview/mmwrhtml/ss4901a1.htm.*

[76]Hemophilia Surveillance System (HSS). *http://www.cdc.gov/ncidod/osr/data-reports/dr-bysurvsys.htm.*

[77]National TB Genotyping and Surveillance Network. *http://www.cdc.gov/ncidod/dastlr/tb/tb_tgsn.htm.*

[78]Anonymous. (1999). Outbreak of poliomyelitis—Angola, 1999. *Morbidity and Mortality Weekly Report* vol. 48:327–329.

[79]From the Centers for Disease Control and Prevention. (1999). Four Pediatric Deaths from Community-Acquired Methicillin-Resistant *Staphylococcus aureus*—Minnesota and North Dakota, 1997–1999. *Journal of the American Medical Association* vol. 282:1123–1125.

[80]Anonymous. (1999). Update: Multistate Outbreak of Listeriosis—United States, 1998–1999. *Morbidity and Mortality Weekly Report* vol. 47:1117–1118.

[81]Anonymous. (1999). Rubella Outbreak—Westchester County, New York, 1997–1998. *Morbidity and Mortality Weekly Report* vol. 48: 560–563.

[82]Anonymous. (2000). Scombroid Fish Poisoning—Pennsylvania, 1998. *Morbidity and Mortality Weekly Report* vol. 49:398–400.

[83]Anonymous. (2000). Human Ingestion of *Bacillus anthracis*-Contaminated Meat—Minnesota, August 2000. *Morbidity and Mortality Weekly Report* vol. 49:813–816.

[84]Anonymous. (2000). Foodborne Botulism from Eating Home-Pickled Eggs—Illinois, 1997. *Morbidity and Mortality Weekly Report* vol. 49:778–980.

[85]Anonymous. (2000). Outbreak of Poliomyelitis—Dominican Republic and Haiti, 2000. *Morbidity and Mortality Weekly Report* vol. 49:1094, 1103.

[86]Anonymous. (2000). Outbreak of Poliomyelitis—Cape Verde, 2000. *Morbidity and Mortality Weekly Report* vol. 49:1070.

[87]Anonymous. (2000). Measles Outbreak—Netherlands, April 1999–January 2000. *Morbidity and Mortality Weekly Report* vol. 49:299–303.

[88]Anonymous. (2000). Rubella among Hispanic Adults—Kansas, 1998, and Nebraska, 1999. *Morbidity and Mortality Weekly Report* vol. 49:225–228.

[89]From the Centers for Disease Control and Prevention. (2000). Serogroup W-135 Meningococcal Disease among Travelers Returning from Saudi Arabia—United States, 2000. *Journal of the American Medical Association* vol. 283:2647.

[90]Anonymous. (2000). Probable Locally Acquired Mosquito-Transmitted *Plasmodium vivax* Infection—Suffolk County, New York, 1999. *Morbidity and Mortality Weekly Report* vol. 49:495–498.

[91]Anonymous. (2000). Coccidioidomycosis in Travelers Returning from Mexico—Pennsylvania, 2000. *Morbidity and Mortality Weekly Report* vol. 49:1004–1006.

[92]Anonymous. (2000). Cluster of HIV-Infected Adolescents and Young Adults—Mississippi, 1999. *Morbidity and Mortality Weekly Report* vol. 49:861–864.

[93]Anonymous. (2001). Outbreaks of *Escherichia coli* O157:H7 Infections among Children Associated with Farm Visits—Pennsylvania and Washington, 2000. *Morbidity and Mortality Weekly Report* vol. 50:293–297.

[94]Anonymous. (2000). Salmonellosis Associated with Chicks and Ducklings—Michigan and Missouri, Spring 1999. *Morbidity and Mortality Weekly Report* vol. 49:297–299.

[95]Anonymous. (1999). Reptile-Associated Salmonellosis—Selected States, 1996–1998. *Morbidity and Mortality Weekly Report* vol. 48:1009–1013.

[96]Anonymous. (1999). Human rabies—Virginia, 1998. *Morbidity and Mortality Weekly Report* vol. 48:95–97.

[97]Anonymous. (1999). Blastomycosis Acquired Occupationally during Prairie Dog Relocation—Colorado, 1998. *Morbidity and Mortality Weekly Report* vol. 48:98–100.

[98]Anonymous. (2000). Outbreak of Rift Valley Fever—Yemen, August–October 2000. *Morbidity and Mortality Weekly Report* vol. 49:1065–1066.

[99]Anonymous. (2000). Update: Outbreak of Rift Valley Fever—Saudi Arabia, August–November 2000. *Morbidity and Mortality Weekly Report* vol. 49:982–985.

[100]Anonymous. (2000). Varicella Outbreaks among Mexican Adults—Alabama, 2000. *Morbidity and Mortality Weekly Report* vol. 49, no. 32, pp. 735–736.

[101]Anonymous. (2000). Reptile-Associated Salmonellosis—Selected States, 1996–1998. Canada Communicable Disease Report, vol. 26, no. 3, pp. 19–23.

Continued

TABLE C.2 Types of Data/Data Sources Identified By Different Methods—Cont'd

Category	First Principle Analysis	Collected by Public Health	Used in Outbreak Investigations	Referenced in Centers for Disease Control and Prevention (CDC) Case Definitions
Epidemiologic Data Continued		Laboratory information (type and method of laboratory test, specimen collection date, submitting laboratory, source of specimen, specimen identification number)	Food source[39,40,43,61,63,65,67,80,82–84,102–105] Party attendance, illnesses in other health care workers Onset of illness	exposure (at least seven to 18 days before onset of rash) or have had one of the following types of exposure while out of state: (1) face-to-face contact with a person who had either a probable or confirmed case or (2) attendance in the same institution as a person who had a case of measles (e.g., in a school, classroom, or day care center). An indigenous case is defined as a case of measles that is not imported. Cases that are linked to imported cases should be classified as indigenous if the exposure to the imported case occurred in the reporting state. Any case that cannot be proved to be imported should be classified as indigenous (measles) Age (streptococcus pneumoniae, invasive, [children less than five years]"; syphilis, [3] latent) date of initial infection (syphilis, [3] latent) death (toxic-shock syndrome; varicella [deaths only]) Treatment with two or more antituberculosis medications, lost to supervision or discharged from treatment greater than 12 months (tuberculosis), evidence or history of a tick or deerfly bite, exposure to tissues of a mammalian host of *Francisella tularensis*, or exposure to potentially contaminated water (tularemia)
Zoonotic Data	(See syndromic, incubation and preoutbreak sections)	Reportable zoonotic diseases, see also data in the preoutbreak, exposure and incubation data sections	Specimens from reptiles[101]	Rabies, animal A positive DFA test (preferably performed on central nervous system tissue) Isolation of rabies virus (in cell culture or in a laboratory animal) Confirmed: a case that is laboratory confirmed (rabies, animal)
Environmental Data	Civil engineering data (heating, ventilation and air conditioning system records) Food cultures, food handling methods, sanitary inspections, food delivery schedule Water supply: water quality and distribution system	Food (source, marketing, processing, preparation, sale, packaging, refrigeration, storage, cooking, culture)[75] Water (type, chlorination, contaminated, circumstances)[74] Work stations	Sources of environmental contamination (in door gasoline-powered forklifts,[18] propane stove, charcoal grill, furnace with poor venting, contaminated water[38,107]) Conditions conducive to development of infections (use of gasoline-powered auger, hand trowels, dog tunnels, and burrows; lack of personal protective equipment; poor plant ventilation[18]; air-conditioner system and exhaust fan; contact with manure; inadequate pool or hot tub filtration system[50,107]; poor cooking practices[40,61,105])	

Trace-back investigation (including check of water temperature at time of seafood harvest,[108,109] farm inspection,[63,110] shell fish tags,
Contamination via transportation and worksites[36]
Environmental sampling (mosquito trapping)[71,90,111,112]

[102]From the Centers for Disease Control and Prevention. (2000). Human ingestion of *Bacillus anthracis*-Contaminated meat—Minnesota, August 2000. *Journal of the American Medical Association* vol. 284:1644–1646.

[103]Anonymous. (2000). Outbreaks of *Salmonella* Serotype Enteritidis Infection Associated with Eating Raw or Undercooked Shell Eggs—United States, 1996–1998. *Morbidity and Mortality Weekly Report* vol. 49:73–79.

[104]Anonymous. (2000). Outbreak of *Escherichia coli* O157:H7 Infection Associated with Eating Fresh Cheese Curds—Wisconsin, June 1998. *Morbidity and Mortality Weekly Report* vol. 49:911–913.

[105]Anonymous. (2000). Foodborne Outbreak of Group A Rotavirus Gastroenteritis among College Students—District of Columbia, March-April 2000. *Morbidity and Mortality Weekly Report* vol. 49:1131–1133.

[106]Anonymous. (2000). Outbreaks of Norwalk-like Viral Gastroenteritis—Alaska and Wisconsin, 1999. *Morbidity and Mortality Weekly Report* vol. 49:207–211.

[107]Anonymous. (2000). Outbreak of Gastroenteritis Associated with an Interactive Water Fountain at a Beachside Park—Florida, 1999. *Morbidity and Mortality Weekly Report* vol. 49:565–568.

[108]Anonymous. (1999). Outbreak of *Vibrio parahaemolyticus* Infection Associated with Eating Raw Oysters and Clams Harvested from Long Island Sound—Connecticut, New Jersey, and New York, 1998. *Morbidity and Mortality Weekly Report* vol. 48:48–51.

[109]From the Centers for Disease Control and Prevention. (2000). Scombroid Fish Poisoning—Pennsylvania, 1998. *Journal of the American Medical Association* vol. 283:2927–2928.

[110]Anonymous. (1999). Update: Outbreak of Nipah Virus—Malaysia and Singapore, 1999. *Morbidity and Mortality Weekly Report* vol. 48, no. 16, pp. 335–337.

[111]Anonymous. (2000). Probable Locally Acquired Mosquito-Transmitted *Plasmodium vivax* Infection—Suffolk County, New York, 1999. *Canada Communicable Disease Report* vol. 26:37–39.

[112]From the Centers for Disease Control and Prevention. (2000). Probable Locally Acquired Mosquito-Transmitted *Plasmodium vivax* Infection—Suffolk County, New York, 1999. *Journal of the American Medical Association* vol. 284:431–432.

D

APPENDIX

Derivation of Bayes' Rule

This appendix provides a derivation of Bayes' Rule and illustrates its use by computing the posterior probability that a cow has Foot and Mouth Disease, given the findings of drooling saliva and another cow in the herd as similarly affected (the mini-BOSSS example discussed in Chapter 13).

1. BAYES' RULE

Suppose that there are several diseases, $D_1, D_2, \ldots, D_n$ (including a 'no disease' state) exactly one of which must be true. Bayes' Rule is a formula for computing the conditional probability of each disease given a set of patient findings F.

$$\textbf{Bayes' Rule:}\ P(D_i \mid F) = \frac{P(F \mid D_i) \times P(D_i)}{\sum_{j=1}^{n} P(F \mid D_j) \times P(D_j)}, \qquad (1)$$

given that the denominator is greater than zero. The derivation of Bayes' Rule follows from the definition of conditional probability and the axioms of probability theory.

Given two events of interest, say a disease D and a set of findings F, the *conditional probability* of F given D (denoted by $P(F|D)$) is defined as follows:

$$P(F \mid D) = \frac{P(\text{F AND D})}{P(D)}. \qquad (2)$$

It clearly follows that $P(F|D)P(D) = P(F \text{ AND } D)$, which is a result we will use in the derivation of Bayes' Rule below.

The conditional probability $P(F|D)$ is not defined when $P(D) = 0$.

$$\textbf{Law of Total Probability:}\ P(F) = \sum_{j=1}^{n} P\left(D_j \text{ AND } F\right).$$

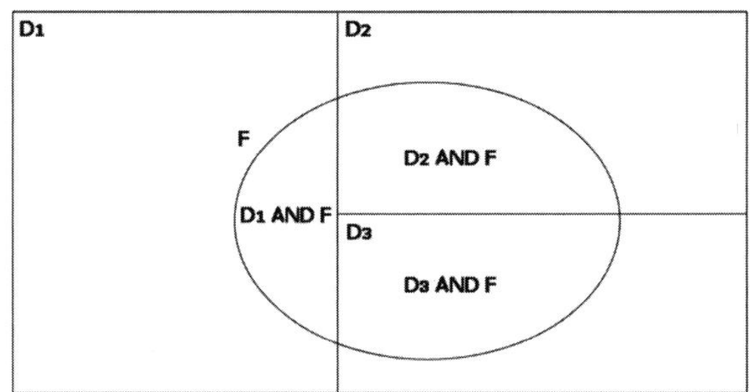

FIGURE D.1 A Venn diagram that illustrates the law of total probability.

Figure D.1 is a Venn diagram that illustrates the law of total probability. The area of oval F equals the sum of the areas of regions (D_1 AND F), (D_2 AND F), and (D_3 AND F).

Derivation of Bayes' Rule:

$$\begin{aligned}
P(D_i \mid F) &= \frac{P\left(D_i \text{ AND } F\right)}{P(F)}, \text{ by definition of conditional probability} \\
&= \frac{P\left(F \mid D_i\right)P(D_i)}{P(F)}, \text{ by definition of conditional probability} \\
&= \frac{P\left(F \mid D_i\right)P(D_i)}{\sum_{j=1}^{n} P\left(D_j \text{ AND } F\right)}, \text{ by law of total probability} \\
&= \frac{P\left(F \mid D_i\right)P(D_i)}{\sum_{j=1}^{n} P\left(F \mid D_j\right)P(D_j)}, \text{ by definition of conditional probability}
\end{aligned}$$

2. MINI-BOSSS EXAMPLE, REVISITED

In Chapter 13, we used the Odds Likelihood form of Bayes' Rule to compute the posterior probability that a cow had Foot and Mouth Disease, given the findings of drooling saliva and another cow in the herd also affected. The posterior probability was 0.083 (see Table 13.3). In this section, we show how the same posterior probability is computed using Equation 1 after introducing an assumption of conditional independence of findings, given a disease (the same assumption on which the Odds Likelihood form of Bayes' Rule rests).

The assumption of conditional independence of findings $f_1, f_2, \ldots, f_m$ (which we collectively call F), given a disease D_i, can be expressed as follows:

$$P(f_1, f_2, \ldots, f_m \mid D_i) = \prod_{j=1}^{m} P(f_j \mid D_i), \qquad (3)$$

where $f_1, f_2, \ldots, f_m$ are the individuals findings. For example, in the mini-BOSSS example we might define f_1 to be the variable "drooling" and f_2 to be the variable "more than one animal affected." In words, Equation 3 says that if we condition on the state of disease D_i then the probability distribution of any finding f_j is independent of the state of all the other findings.

Handbook of Biosurveillance
ISBN 0-12-369378-0

Substituting Equation 3 into Bayes' Rule (Equation 1), we obtain:

$$P(D_i \mid f_1, f_2, \ldots, f_m) = \frac{P(f_1, f_2, \ldots, f_m \mid D_i)P(D_i)}{\sum_{j=1}^{n} P(f_1, f_2, \ldots, f_m \mid D_j) \times P(D_j)}$$
$$= \frac{\prod_{k=1}^{m} P(f_k \mid D_i)P(D_i)}{\sum_{j=1}^{n}\prod_{k=1}^{m} P(f_k \mid D_j) \times P(D_j)}. \quad (4)$$

Thus, for our example we obtain the following equation, using the notation from Chapter 13:

$$P(FMD \text{ present} \mid DS, MA)$$
$$= \frac{P(DS \mid FMD \text{ present})P(MA \mid FMD \text{ present})P(FMD \text{ present})}{\left[\begin{array}{l} P(DS \mid FMD \text{ present})P(MA \mid FMD \text{ present})P(FMD \text{ present}) + \\ P(DS \mid FMD \text{ absent})P(MA \mid FMD \text{ absent})P(FMD \text{ absent}) \end{array}\right]}$$
$$= \frac{0.95 \times 0.95 \times 0.001}{0.95 \times 0.95 \times 0.001 + 0.05 \times 0.2 \times 0.999}$$
$$= 0.083$$

The fact that both forms of Bayes Rule yield the same posterior probability is not coincidental. The Odds Likelihood form of Bayes Rule can be derived mathematically from Equation 4.

APPENDIX

Predictive Value Positive and Negative

Students are often confused by *predictive value positive* (PVP) and *predictive value negative* (PVN)—two ratios that an evaluator can compute from the data in Table E.1. An evaluator may compute these ratios from the numbers in the horizontal rows in Table E.1. PVP is the ratio of the number of cases that a classifier or test identifies correctly (90 true cases of SARS in this example) to the total number of cases that it identifies either correctly or incorrectly (95 cases labeled as SARS by the algorithm in this example), or 0.947. PVN is the ratio of the number of cases the case detection algorithm classified correctly as not having SARS (95) over the total number of cases that it classified as not having SARS (105), or 0.905. The word "predictive" in PVP and PVN suggests that these probabilities indicate how well the classifier would *predict* the actual diagnosis of a patient from biosurveillance data about patients if it was used on a new set of patients, which seems to be exactly what we want to know if our goal is to detect a patient with SARS or to reassure a patient that he does not have SARS.

The problem (and a point about which there is sometimes confusion) is that the ratios PVP and PVN are extraordinarily dependent on the prevalence of the disease in the study set, and the prevalence in the study set is completely distorted by the artificial conditions of this type of study where often half of the individuals studied (100 out of 200) have the disease of interest. The level of misrepresentation is quite absurd, as suggested by Table E.2, where we consider a more realistic prevalence of SARS of 1 per 100,000 in patients presenting to emergency departments (even this estimate is high, unless we are considering a city that is in the midst of a SARS outbreak). In this more realistic example, the PVP is 90/500,090, or 0.00018 (in contrast to 0.947), and the PVN is 9,500,000/9,500,010, or nearly 1.0 (in contrast to 0.905).

As discussed in Chapter 20, the preferred method (and standard practice) is for the evaluator to measure and report only sensitivity and specificity, and for consumers of this information (system developers or biosurveillance personnel) to use the Bayes theorem to compute predictive probabilities from the reported sensitivity and specificity of the classifier and the current estimates of prevalence (also known as prior probability) of disease.

TABLE E.1 Results of Hypothetical Evaluation of SARS Case Detection Algorithm

	100 Individuals with SARS (cases)	100 Individuals without SARS (controls)	Total Classifications
Algorithm positive for SARS	90	5	95
Algorithm negative for SARS	10	95	105
Total patients	100	100	200

Sensitivity = 0.9, specificity = 0.95, PVP = 0.947, and PVN = 0.905.

TABLE E.2 Results of a Hypothetical Evaluation of SARS Case Detection Algorithm in Which Numbers of Control and SARS Cases Match Prevalence of SARS in Population (One case per 100,000 persons)

	Individuals with SARS	Individuals without SARS	Total Classifications
Algorithm positive for SARS	90	500,000	500,090
Algorithm negative for SARS	10	9,500,000	9,500,010
Total patients	100	10,000,000	10,000,100

Sensitivity = 0.9, specificity = 0.95, PVP = 0.00018, and PVN = ~1.0

ISBN 0-12-369378-0

APPENDIX

Data Communication to RODS: Technical Specifications

RODS Laboratory
University of Pittsburgh, Pittsburgh, Pennsylvania

1. HL-7 ADMISSION-DISCHARGE-TRANSFER (ADT) DATA TYPE

RODS collects a limited subset of patient admission information from emergency departments (EDs), typically via the A04 message type. The collected patient information includes patient age, gender, admitted date and time, free-text chief complaint, visit number, home zip code, and work zip code. The following describes the fields in a HL-7 message corresponding to the limited subset of patient registration data.

1.1. Basic Data Segments and Fields

- MSH
  - a. Field separator [Seq#1] (e.g., "|")
  - b. Encoding characters [Seq#2] (e.g., "^~\&")
  - c. Sending facility [Seq#4] (e.g., "PA_XMC")
  - d. HL-7 message date and time [Seq#7]
  - e. Trigger event code (only **A04**) [Seq#9]
  - f. Message control ID [Seq#10] (for ack purpose)
- PID
  - a. Patient encrypted number [Seq#3] *(optional;* used to reidentify the patient for the hospital if necessary)
  - b Patient DOB or age [Seq#7] (if use age, put age in the subfield of the DOB using caret "^" or other subfield delimiter defined in the MSH-2)
  - c. Patient gender [Seq#8]
  - d. Patient home zip code [Seq#11]
- PV1
  - a. Patient class [Seq# 2] (only **E** for ED visit)
  - b. Admitted date and time [Seq#44]
    Visit number [Seq#19]

Chief complaint [Seq#3-2] (free text) (Note some hospitals use DG1-4 for chief compliant.)

- DG1
  - a. ICD-9 code [Seq#3] (if available at the time of patient registration)
  - b. Diagnosis description [Seq#4]

IN1: Employer's zip code [Seq#44] (Note iIf IN1 is not available, then use NK1 segment with employer as the contact role.)

- NK1
  - a. Employer's zip code [Seq#4]
  - b. Contact role [Seq#7] as employer (The contact role is used to identify if the address in Seq#4 is work address.)

1.2. Lower Layer Protocol (Wrapping) Characters

1.2.1. Start-of-Message Character(s)

The start-of-message character(s) is(are) the character(s) at the beginning of a HL-7 message. The commonly used start-of-message character is <VT> (**\x0B** in hexadecimal value). Please provide the information of start-of-message character(s) used in your message router to RODS interface engineer so our RODS HL-7 listener can capture your HL-7 messages correctly.

1.2.2. End-of-message Character(s)

After receiving an HL-7 message, RODS listener will send an HL-7 acknowledgement (ACK) message back to your message router. The end-of-message character(s) in the ACK message will be the same end-of-message character(s) used in the HL-7 message transmitted from your message router. **Error! Reference source not found.** shows a sample HL-7 ADT message that uses <ETX> character (**\x03**) as the end-of-message character. Please let RODS interface engineer know the end-of-message character(s) used in your message router. The commonly used end-of-message characters are <FS><CR> (**\x1C** and **\x0D**).

The next HL-7 message should be transmitted AFTER your message router receives our ACK message as demonstrated in FIGURE F.1 **Sample HL-7 message to RODS. Red color represents necessary data field; green color, explanation of the field; <CR>, CR (\x0D) in the ASCII character set; and <ETX>, ETX (\x03) in the ASCII character set.**

Thus your message router can assure that the message transmitted to the RODS laboratory has been received.

1.2.3. End-of-Segment Character

Each HL-7 data segment must be ended with character <CR> (**\x0D**).

2. NON-HL7-FORMATTED ADT DATA

For the hospitals that do not have HL7 infrastructure or are not familiar with HL7 message protocol, the RODS software can also process your data in non-HL7 format. For example, RODS can process data files that use delimiters like bar or comma (e.g., csv files) or use fixed column position. The basic elements that RODS collects are *hospital ID, patient class, patient admitted date and time, free-text chief complaint, home*

ISBN 0-12-369378-0

```
MSH[1]|^~\&||UPMC(hosp)||RODS|200003171458||ADT^A04|20000317145841029270(m
essage control ID)[2]|P|2.3|<CR>[3]
PID|||||||^017(age)|M|||^^^^12345^^^^|<CR>
PV1||E||||||||||||||||||||||||||||||||||||||||||200109130045(admit
time)||<CR>
PV2|||^ABDOMINAL PAIN(chief complaint)|<CR>
DG1||I9|789.09|ABDOMINAL PAIN-SITE NEC(diagnosis desc.)|<CR>
NK1||||^^^^12345^^|||EMP|<CR>
IN1||||||||||||||||||||||||||||||||||||||||||||^^^^12345<CR>
<ETX>[4]
```

FIGURE F.1 Sample HL-7 message to RODS. Red color represents necessary data field; green color, explanation of the field; <CR>, CR (\x0D) in the ASCII character set; and <ETX>, ETX (\x03) in the ASCII character set.

zip code, work zip code (if available), *age or date of birth, gender, visit number* (if available).

Non-HL7-formatted data is usually transmitted in batch mode through ftp or secure ftp (sftp) channel. Please contact the RODS group for further ftp server information.

File name convention for batch-mode data files will follow the format: <state_id>_<health system ID>_<date and time the file was created>.txt. For example, the file name for the records in Figure F.3 is *pa_hosp1_200401311100.txt.*

3. NETWORK CONNECTION

Basic information of the RODS server that hosts the HL-7 listeners:

Connection type: *TCP socket connection*
IP: *192.168.1.10*
Host name: *hl7.rods.pitt.edu*
Production port number: *Designated by interface engineer*
Test port number: *Designated by interface engineer*

Two methods of transmitting data in real-time: SSL, VPN
Information required from the data provider to complete a SSL setup:

1. Contact information for the interface and network engineers
2. Firewall exception for the above listed IP address and associated SSL port
3. Host IP address

Information required from the data provider to complete a VPN setup:

1. Contact information for the interface and network engineers
2. Hardware make/model of VPN device
3. Tunnel Endpoint IP address
4. Global IP address of the interface engine (message router)
5. Test host IP address (if using one other than that of the message router)
6. Message control characters
7. IKE and IPSec settings
8. Preshared key (will be exchanged over the phone)

SSL is the preferred transfer methodology if the data provider's system supports it (most newer system versions do).

4. PRODUCTION SYSTEM MAINTENANCE AND CONTACT INFORMATION

The following is the RODS contact information for production system maintenance.

```
MSH|^~\&|COMMON||MEDIPAC||200003171458||ACK|20000317145841029270||2.3||
<CR>
MSA|AA|20000317145841029270<CR>
<ETX>
```

FIGURE F.2 Sample Ack HL-7 message to the message router. Red color represents necessary data field; green color, explanation of the field; <CR>, CR (\x0D) in the ASCII character set; and <ETX>, ETX (\x03) in the ASCII character set.

[1]Red color represents necessary data field.
[2]Green color represents explanation of the field.
[3]<CR> represents CR(\x0D) in the ASCII character set.
[4]<ETX> represents ETX (\x03) in the ASCII character set.

```
HOSP-1|E|200401310849|ABNORMAL LABS|15217||19230515|F|8456212|
HOSP-1|E|200401310949|SOB|15241|15213|19600114|M|8456216|
```

FIGURE F.3 Sample bar-delimited records.

During off hours, please page our alpha numeric pager at 412-XXX-XXXX or *412XXXXXXX@mobilecomm.net* if the downtime is more than 4 hours.

During office hours (9 a.m. to 5 p.m.) Monday through Friday, the following is the list of contact persons.

rods-system@cbmi.pitt.edu for all interface work
rods-network@cbmi.pitt.edu for all network connections
Steve DeFrancesco (IT Director): 412-383-8125, *sdefrancesco@cbmi.pitt.edu*

5. FAQ

Q1: How to add <ETX> ASCII character at the end of a HL-7 message when using *SMS OPENLink* v.21 message router?

A1: In the object protocol definition, there is a Start of Text and End of Text character field we set those into. In the field we do CTRL + V, and it asks us what character to enter, and in this case it's the actual number 3. It puts it into the field in reverse color. Save the def and go from there.

Q2: What VPN device should I use, or is my current one compatible with RODS' VPN devices?

A2: The lab uses a Lucent Brick 80 as the VPN device. If your VPN device is in the list of the ICSA Labs Certified IPSec Products *(https:www.icsalab.com/icsa/product.php?tid=428c$4e933c47-c58cef17$5298-2f493ab2),* it will be compatible with the Lucent device. We've tested the Cisco VPN Concentrator 3030 and found that it can work with the Lucent Brick 80.

Q3: Will RODS only work with specific integration/interface engines?

A3: No, RODS is very flexible, and it can work with any vendors that use the standard HL-7 message protocol.

Q4: Does RODS only take a real-time HL-7 data feed?

A4: No, RODS can also take data files submitted in batch mode through secure ftp. To discuss this option further, please contact Steve DeFrancesco.

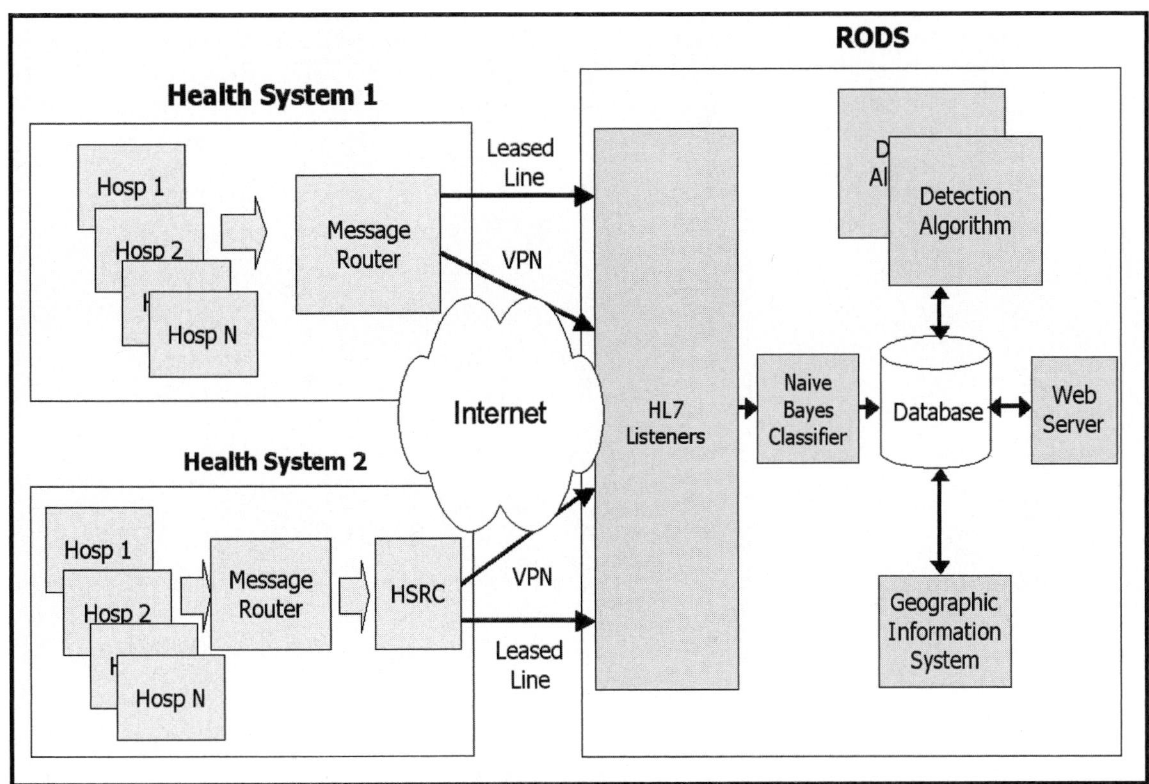

FIGURE F.4 RODS data communication diagram.

APPENDIX

Data Use Agreement

___(state)____ DEPARTMENT OF HEALTH PROTOCOL AGREEMENT FOR DATA USE AND CONFIDENTIALITY FOR PARTICIPATION IN THE REAL-TIME OUTBREAK & DISEASE SURVEILLANCE SYSTEM <FULL NAME OF HEALTH SYSTEM>

This Data Use and Confidentiality Protocol ("Protocol") is made as of the <DAY> day of <MONTH YEAR> by the ___(state)___ Department of Health ("_DH") and accepted <FULL NAME OF HEALTH SYSTEM> ("<HEALTH SYSTEM NAME>") as of the date of signature. The <HEALTH SYSTEM NAME> and the _DH shall be referred to individually as a "Party" or collectively as the "Parties."

WHEREAS, _DH is a public health authority as established by [*Chapter* __; *state code*] of the ___(state)___ Revised Code ("ORC") and as defined by HIPAA Privacy Regulation at 45 C.F.R. §164.501, it is permitted to receive protected health information from covered entities, without authorization, for the purpose of preventing or controlling disease, injury, or disability, including conducting public health surveillance, investigations, or interventions as stated in 42 USC 1320d-7 and 45 C.F.R. §164.512(b);

WHEREAS, the University of Pittsburgh Center for Biomedical Informatics is acting, in part, under a grant of authority from _DH to assist _DH in conducting public health surveillance, investigations, or interventions as stated in 45 C.F.R. §164.512(b); and

WHEREAS, the University of Pittsburgh Center for Biomedical Informatics is also engaged in research on the use of computer based systems to aid public health authorities in the monitoring of public health, utilizing "limited data set" information, as that term is defined by the HIPAA Privacy Regulation; and

WHEREAS, the Real-time Outbreak and Disease Surveillance ("RODS") system is a public health surveillance system developed by the Center for Biomedical Informatics of the University of Pittsburgh. The RODS system is capable of collecting data relevant to public health automatically and in real time. RODS is intended to be used by public health officers by providing tools for the display and analysis of the data; and

WHEREAS, the <HEALTH SYSTEM NAME>, _DH and the University of Pittsburgh Center for Biomedical Informatics are implementing a public health surveillance system (the "Project," as further defined below) utilizing RODS; and

WHEREAS, the RODS user interfaces will be available to Authorized Users designated by _DH for public health surveillance purposes, Authorized Users, as defined in section 3a below, as designated by _DH will be able to review aggregate time-trended and spatially displayed data;

WHEREAS, the parties shall abide by all applicable laws, rules and regulations, including and without limitation all patient confidentiality and medical records requirements.

NOW, THEREFORE, intending to be legally bound hereby, the parties establish the following protocol:

1. Definitions.
   a. "<HEALTH SYSTEM NAME> Data" are the data elements in which _DH has a public health surveillance interest and also constitutes limited, de-identified data, as that term is defined in the HIPAA Privacy Regulations at 45 C.F.R. 164.514(e), which <HEALTH SYSTEM NAME> maintains and disseminates to the Project.
   b. "Receiving Party" shall mean any Party member who is the recipient of <HEALTH SYSTEM NAME> Data or any derivative of <HEALTH SYSTEM NAME> Data.
   c. "Confidential Information" as used in this Protocol, shall consist of all <HEALTH SYSTEM NAME> Data released as a part of the records, except any information that:
      i. Was in the recipient Party's lawful possession prior to disclosure by the owner Party; or
      ii. Is lawfully received by a Party without restriction regarding use or confidentiality; or
      iii. Is now or hereafter becomes generally available to the public through no action, inaction, or fault of any Party hereto receiving the Confidential Information of the other Party.
   d. "HIPAA" or "Privacy Rule" shall mean the Standards for Privacy of Individually Identifiable Health Information set forth at 45 C.F.R. §160.101 et seq. and §164.102 et seq.
2. Provision of <HEALTH SYSTEM NAME> Data.
   a. The <HEALTH SYSTEM NAME> will use best efforts to provide <HEALTH SYSTEM NAME> Data to the _DH and The University of Pittsburgh, for use in connection with the Project.

ISBN 0-12-369378-0

b. The <HEALTH SYSTEM NAME> will use best efforts to include the following limited data set elements and abstract of each Emergency Department patient visit:
   i. The name of the facility or hospital ID;
   ii. Trigger event code;
   iii. Patient date of birth or age;
   iv. Patient gender;
   v. Patient home zip code;
   vi. Patient class;
   vii. Admitted date;
   viii. Admitted time;
   ix. Free text chief complaint;
   x. Patient temperature
   xi. Work zip code;
   xii. Visit identifier.
c. The <HEALTH SYSTEM NAME> may provide, now or in future enhancements to the Project, the following additional data, to the extent possible, and on an optional basis:
   i. The name of the facility or hospital ID, orders for cultures of stool, throat, blood that may include time of order, patient age, gender, home zip code, work zip code, and visit identifier;
   ii. The name of the facility or hospital ID, microbiology culture reports that may include time of culture, organism identified, patient age, gender, home zip code, work zip code, and visit identifier;
   iii. The name of the facility or hospital ID, results of natural language processing, if available, of chest radiograph reports that may include time of radiograph, radiographic findings identified by natural language processing, patient age, gender, home zip code, work zip code, and visit identifier; and
   iv. The name of the facility or hospital ID, results of natural language processing, if available, of emergency department dictations that may include time of visit, findings identified by natural language processing, patient age, gender, home zip code, work zip code, and visit identifier.
   v. Such other non-identifiable patient data as the <HEALTH SYSTEM NAME> may in its sole discretion elect to release to the Project.
d. To the extent historical data is available, the <HEALTH SYSTEM NAME> shall use best efforts to provide at least one (1) year of the data back-load from the <HEALTH SYSTEM NAME> that will be incorporated into RODS for historical comparison purposes.
e. The <HEALTH SYSTEM NAME> shall use best efforts to transmit to the University of Pittsburgh the <HEALTH SYSTEM NAME> Data electronically and in real time to the extent practical using <HEALTH SYSTEM NAME>'s current and available data transmission capabilities.

3. In order to facilitate this Protocol, the recipients of <HEALTH SYSTEM NAME> Data establish the following:
   a. Authorized Users: The following parties are permitted to access and use <HEALTH SYSTEM NAME> Data:
      i. The University of Pittsburgh RODS surveillance project staff.
      ii. _DH personnel and designees who agree to abide by the terms of this Protocol.
      iii. Subsidiary and affiliated entities of the Parties who agree to abide by the terms of this Protocol. All terms of this Protocol shall be binding upon the subsidiary and affiliated entities of the Parties. Each Party shall take such necessary actions to ensure compliance with the terms of this Protocol by any such subsidiary and affiliated entities.
   b. Permitted Uses:
      i. The <HEALTH SYSTEM NAME> Data as defined in this Protocol that is provided pursuant to this Protocol to _DH and The University of Pittsburgh shall only be used in connection with the Project for public health surveillance/investigation purposes and practice, as well as for public health research as permitted by and pursuant to ___(state)___ law as well as HIPAA, and for no other purpose.
      ii. Except as described in section 3.b.iii, below, the University of Pittsburgh may only provide access to <HEALTH SYSTEM NAME> Data to the Authorized Users designated by _DH in the form of summaries by rates and counts, or time trended or spatially aggregated summaries, and only when combined with data of other health systems participating in the Project.
      iii. When an alert is generated by the automatic detection algorithms of RODS, or when manual inspection of aggregate data viewed through the RODS interface identifies a cluster of patients with a higher than expected rate of symptoms, _DH and the local health department with jurisdictional authority may have access through the RODS interface to the name of facility or hospital, gender, age, home zip code, work zip code, and chief complaint for each patient in the suspicious cluster for public health investigation purposes as permitted by ___(state)___ law and HIPAA.
      iv. Each Party shall use the Confidential Information of the other Parties only for the purpose of performing the public health surveillance and investigation functions and research aspects of the Project as permitted by ___(state)___ law and HIPAA and for no other purposes.
      v. Each Party shall strictly limit access to the relevant portions of the Confidential Information of the other Parties to such employees as delineated herein who have a need to know such portions of the Confidential Information regarding the Project.

vi. Notwithstanding any other provision contained in this Protocol, the University of Pittsburgh shall have the right to publish the results of its research involving RODS, providing that any such publication shall not identify <HEALTH SYSTEM NAME> without the <HEALTH SYSTEM NAME'S> prior permission.

c. Assurances:

i. The Receiving Party shall not use or further disclose to any person or entity <HEALTH SYSTEM NAME> Data or Confidential Information other than as permitted by this Protocol or as otherwise required by law.

ii. The Receiving Party shall use appropriate safeguards to prevent unauthorized use or disclosure of <HEALTH SYSTEM NAME> Data or Confidential Information other than as provided for by this Protocol or as otherwise required by law and shall take appropriate action to address the cause and mitigate the effects of the unauthorized use or disclosure. The Reporting Party may, at its discretion, assist the Receiving Party in addressing or mitigating the effects of the unauthorized use or disclosure.

iii. The Receiving Party shall report to The <HEALTH SYSTEM NAME> any unauthorized use or disclosure of <HEALTH SYSTEM NAME> Data of which it becomes aware. Furthermore, the Reporting Party, if responsible for the unauthorized use or disclosure, shall cease such activity and to the best of its abilities, take appropriate steps to address the cause and mitigate the effects of the unauthorized use or disclosure. If the Reporting Party is responsible for the unauthorized use or disclosure, the Receiving Party is released of further responsibility for the unauthorized used or disclosed <HEALTH SYSTEM NAME> Data. The Receiving Party may, at its discretion, assist the Reporting Party in addressing or mitigating the effects of the unauthorized use or disclosure.

iv. The Receiving Party shall ensure and require that such Receiving Party's agents, including subcontractors, to whom it provides <HEALTH SYSTEM NAME> Data under this Protocol, agree to the same restrictions and conditions contained herein that apply to the Receiving Party.

v. The Receiving Party will not attempt to identify the individuals whose information is contained in the <HEALTH SYSTEM NAME> Data or attempt to contact the individual other than as permitted by this Protocol or as otherwise required by law.

4. Management of Data.

a. <HEALTH SYSTEM NAME> Data will be received and stored in a computer located in a machine room maintained by the Center for Biomedical Informatics at the University of Pittsburgh. The University of Pittsburgh shall use reasonable efforts (based on industry best practices) to secure, protect and manage the <HEALTH SYSTEM NAME> Data in compliance with the terms of this Protocol.

b. All <HEALTH SYSTEM NAME> Data shall, to the extent permitted by law, at all times remain the property of the <HEALTH SYSTEM NAME> and shall be returned to the <HEALTH SYSTEM NAME>, in its original form, immediately upon written request to the Receiving Party. If requested by the Reporting Party, the Receiving Party shall, to the extent permitted by law, destroy all original and copies of <HEALTH SYSTEM NAME> Data immediately upon written request by the <HEALTH SYSTEM NAME>, and provide a certificate of destruction upon written request.

c. No Party shall make copies of any <HEALTH SYSTEM NAME> Data or Confidential Information of the other Parties except for its internal use regarding or as related to this Project, or as permitted or required by law.

5. <HEALTH SYSTEM NAME> Requested Reports. The University of Pittsburgh shall, on a mutually agreeable schedule, and on a reasonable best efforts basis, provide an electronic file containing the <HEALTH SYSTEM NAME> Data and derivative analysis of the <HEALTH SYSTEM NAME> Data over the secure network to a server designated by <HEALTH SYSTEM NAME>.

6. IRB Approvals. As relevant and appropriate, the Parties shall abide by all applicable IRB requirements and/or requirements. Where IRB oversight is required, the Parties shall provide such documentation of compliance to each other.

7. <HEALTH SYSTEM NAME>'s Disclaimer of Warranties. Except as set forth in this Protocol, <HEALTH SYSTEM NAME> disclaims all warranties whether written, oral, expressed or implied including, without limiting the generality of the foregoing, any warranty of data accuracy, completeness, merchantability or fitness for a particular purpose.

8. University of Pittsburgh's Disclaimer of Warranties. University of Pittsburgh disclaims all warranties with respect to RODS, whether written, oral, expressed or implied including, without limiting the generality of the foregoing, any warranty of data accuracy, completeness, merchantability or fitness for a particular purpose.

9. Regulatory Compliance. The University of Pittsburgh shall take such actions as are necessary, including making modifications of <HEALTH SYSTEM NAME> Data, to comply with all applicable <HEALTH SYSTEM NAME> polices and procedures, and all applicable federal, state or local statutes or regulations (whether promulgated before or after the date the <HEALTH SYSTEM NAME> Data was received), including without limitation, all applicable HIPAA regulations and requirements. The University of Pittsburgh shall perform such work at its own expense. Such actions will be completed within the times specified for compliance within the statute or regulation.

<HEALTH SYSTEM NAME> shall have the right at all times to review and inspect the steps taken and procedures implemented by the University of Pittsburgh to assure the anonymity of patient data and the compliance with all such requirements and the terms of this Protocol.

10. Audits & Compliance.
    a. <HEALTH SYSTEM NAME>, at <HEALTH SYSTEM NAME>'s expense, shall be permitted to perform reasonable audits to ensure that appropriate controls have been established so that data management and use is consistent with the terms of this Protocol. Based on such audits, <HEALTH SYSTEM NAME> may request the University of Pittsburgh modify the <HEALTH SYSTEM NAME> Data and controls to comply with any of <HEALTH SYSTEM NAME>'s requested changes. Any requested changes are subject to review and approval by _DH prior to implementation.
    b. The Receiving Party agrees to make its internal practices, books and records relating to the use and disclosure of <HEALTH SYSTEM NAME> Data under this Protocol available to the United States Department of Health and Human Services or its agents and/or _DH or its agents for the purpose of determining compliance with and/or enforcing the provisions of this Protocol, _______(state)______ law, and HIPAA

11. Term and Termination. This Protocol constitutes public health surveillance as authorized by ___(state)___ law; Reporting Parties may not terminate participation in this Protocol without cause and approval by _DH. In such event, Paragraphs 3, 4(b), 4(c), 7, 8, and 9 shall survive the termination of participation in this Protocol.

IN WITNESS WHEREOF, the Parties have acknowledged participation in this Protocol as evidenced by their duly authorized representative as of the date first above written.

FULL NAME OF HEALTH SYSTEM

Signature ______________________________

Name ______________________________

Title ______________________________

Date ______________________________

THE ___(STATE)___ DEPARTMENT OF HEALTH

Signature ______________________________

Name ______________________________

Title ______________________________

Date ______________________________

Department of Health Authorized Use Agreement for Clinical Data

______(*state*)_____ Department of Health
Request to Become an ____ Department of Health (_DH) Designated Hospital Registration
Real-time Outbreak Detection System (___(*state*)___-RODS) User
And Data Use and Confidentiality Agreement

Step 1: Fill out the demographic information below for the person who is requesting access to the ____(*state*)____ RODS system. All of the fields ***must*** be completed to be granted access to the system.

To be filled out by the person requesting access to the system (Please Print):			
Requester's Name:	First	Middle	Last
Health Department:			Region:
Position/Title:			
Work Address:			
City:		State:	Zip:
Telephone:	e-mail address:		

Step 2: The person requesting access must read, agree to, and sign the Data Use and Confidentiality agreement below.

Data Use & Confidentiality Agreement for ___(*state*)___ Hospital Registration RODS Data

READ CAREFULLY– The ______(*state*)______ Department of Health is committed to protecting the privacy and security of individual identifiable health information and other information of a confidential nature for the hospital organization. As an ______(*state*)______ Department of Health designated ______(*state*)______ RODS user, authorized to access ______(*state*)______ RODS, the requester holds a position of trust relative to this information and must recognize the responsibilities entrusted to him/her in preserving the security and confidentiality of this information. Information pertaining to patients and other sensitive information must be held in strict confidence.

The ______(*state*)______ RODS information is collected for public health disease control and disease surveillance. Information in ______(*state*)______ RODS is protected health information. Confidentiality requirements that apply to this data include, but are not limited to applicable state, county, city and federal regulations.

All ______(*state*)______ RODS system account holders and authorized users are required to read the following data use and confidentiality agreement and acknowledge acceptance of the terms herein by signing where indicated.

An authorized user agrees to abide by the terms and conditions as written in the individual Data Use and Confidentiality Agreements signed between the ______(*state*)______ Department of Health and applicable hospital/hospital system. These terms and conditions include the following taken from the agreements:

a) Permitted Uses of the data:
   i) Hospital data as defined in the Agreements with the ______(*state*)______ Department of Health, is for use in connection with public health surveillance & investigation purposes and practice as permitted by and pursuant to Health Insurers Privacy and Accountability Act of 1996 (HIPAA) and for no other purpose.

ISBN 0-12-369378-0

ii) Except as described in section a.iii, below, the University of Pittsburgh may only provide access to Hospital Data to the Authorized Users designated by the ______(*state*)______ Department of Health in the form of summaries by rates and counts, or time trended or spatially aggregated summaries, and only when combined with data of other health systems participating in the Project.

iii) When an alert is generated by the automatic detection algorithms of RODS, or when manual inspection of aggregate data viewed through the RODS interface identifies a cluster of patients with a higher than expected rate of symptoms, the ______(*state*)______ Health Department designated authorized users may have access through the RODS interface to the name of facility or hospital, gender, age, home zip code, work zip code, and chief complaint for each patient in the suspicious cluster for public health investigation purposes as permitted by the Health Information Privacy and Accountability Act of 1996 (HIPAA).

iv) Each Party shall use the Confidential Information of the other Parties only for the purpose of performing the public health surveillance and investigation functions of the Project as permitted by HIPAA and for no other purposes whatsoever.

v) Each Party shall strictly limit access to the relevant portions of the Confidential Information of the other Parties to such of its employees as delineated herein who have a need to know such portions of the Confidential Information regarding the Project.

b) Assurances:

i) The Receiving Party shall not use or further disclose to any person or entity Hospital Data or Confidential Information other than as permitted by the Agreement or as otherwise required by law.

ii) The Receiving Party shall use appropriate safeguards to prevent use or disclosure of Hospital Data or Confidential Information other than as provided for by the Agreement.

iii) The Receiving Party shall report to the Hospital any use or disclosure of Hospital Data that is not provided for in, or is in violation of, this Agreement of which it becomes aware. Furthermore, the reporting Party, if responsible for the use or disclosure in violation of this Agreement, shall cease such activity and to the best of their abilities, take appropriate steps to address and mitigate the cause and effects of the use or disclosure in violation of this Agreement.

iv) The Receiving Party will not attempt to identify the individuals whose information is contained in the Hospital Data or attempt to contact the individual except as provided in a.iii above.

An authorized user's conduct may threaten the security and confidentiality of this information. It is the responsibility of every user to know and understand the following:

1. Users must not make or permit unauthorized use of any information in ______(*state*)______ RODS.
2. Users must not divulge or share login ID or password.
3. ______(*state*)______ RODS is for use in connection with public health surveillance & investigative purposes and practice as permitted by and pursuant to HIPAA and for no other purpose. Users must not access, request others to access, or allow others to access ______(*state*)______ RODS for non-public health surveillance purposes.
4. Users must not seek to benefit personally or permit others to benefit personally by any information contained in ______(*state*)______ RODS.
5. Users must not aid, abet, or act in conspiracy with another to violate any part of this code.
6. Authorization for access to ______(*state*)______RODS may be terminated at any time and specifically when a user's employment is terminated, when access to the data is not required for work related responsibilities, or when the user has been found in violation of this agreement.
7. Both the individual user and the public health agency by which they are employed have an obligation to protect the confidentiality and security of the information in ______(*state*)______ RODS.
8. Users must report any violations of this ______(*state*)______ RODS confidentiality and security code to the _DH Privacy Officer (Socrates Tuch, 614-466-4882) immediately.

I have read and understand the ______(*state*)______RODS Code of Responsibility for Security and Confidentiality of Data. I will abide by this code and will protect all ______(*state*)______ RODS records and data as confidential.

Requester's Signature Date

Step 3: Requester's Authorizing Agent (e.g., Agency Health Commissioner or Agency Bureau Chief) read and sign below.

I have reviewed the information on this form, and find it to be correct to the best of my knowledge. The user requesting access to ______*(state)*______RODS is either employed by, contracted by, or otherwise performing public health surveillance work in collaboration with this agency, and has need for access to the system. I understand that the user will have access to hospital data that may include personal identifiable health information, and agree to be bound by all appropriate data use and confidentiality agreements. I understand that it is my responsibility to assure that the requester/user named above abides by this agreement. In the event that the requester, named above, is no longer employed at this agency or changes job duties not requiring access to ______*(state)*______ RODS, the ______*(state)*______Department of Health Bureau of Health Surveillance Chief will be contacted by me or a designee to update/terminate user status.		
Authorized Signature	Title (Please Print)	Date

Step 4: This request should be mailed or faxed (614-644-1909) to the ______*(state)*______ Department of Health at the address below. _DH must retain the originals of this form. If faxed, please mail the originals as soon as possible. Requesters will be notified by e-mail when access is granted.

_________(name of official giving access to clinical data)_________

______*(state)*______ Department of Health
address
city, state, and zip code
telephone and email address

Step 5:

This request has been reviewed at _DH and access to RODS limited to the above named region is recommended.	
________________________________*(state)*______ RODS Coordinator, ______*(state)*______ Dept. of Health	
Name of _DH Official	Date____________

APPENDIX

National Retail Data Monitor/RODS Account Access Agreement

Account Access Agreement
for public health surveillance data through the
National Retail Data Monitor and/or RODS System

1. ______________________________(name; "Account Holder" – type or print clearly) will receive upon the execution of this agreement a password-protected account for access to the National Retail Data Monitor (NRDM) to view aggregate sales data of over-the-counter ("OTC") medications for the state of __________________. Account Holder has represented that he/she is authorized by the appropriate state/regional public health authority to review public health information for this state/region.
2. The Account Holder agrees that this password will not be shared with anyone.
3. Access to the RODS system for views of clinical data (if available for the region) will be provided to Account Holder upon submission of a form of authorization signed by Account Holder's public health authority. This form is region specific and must be obtained by the Account Holder from their regional health authority. The form may be attached to this agreement or submitted at a later time.
4. Access to the data is provided to Account Holder for the sole purpose of public health surveillance.
5. The data are provided to the University of Pittsburgh under Confidentiality Agreements with the data providers. The Account Holder understands and agrees that the data may not be disclosed to any third party, or used for any purpose other than public health surveillance, without the prior written permission of the University of Pittsburgh.
6. Any breach of the terms of this Agreement by Account Holder will result in the immediate termination of Account Holder's account. The University retains the right to take other legal action, including seeking an injunction to prevent recipient from breach of this agreement.
7. The Account Holder may not use the name or marks of the University of Pittsburgh or participating retailers in any press release or publicity without the prior written permission of the University and/or participating retailers. The RODS or NRDM systems, including interface screens, may not be used in any publicity without prior written approval of the University of Pittsburgh.
8. NO WARRANTIES OF ANY KIND ARE MADE BY UNIVERSITY OR THE ENTITIES SUPPLYING THE OTC OR CLINICAL DATA WITH RESPECT TO THE OTC OR CLINICAL DATA OR ANY USE THEREOF, AND UNIVERSITY AND THE DATA PROVIDERS HEREBY DISCLAIM THE IMPLIED WARRANTIES OF MERCHANTABILITY, FITNESS FOR A PARTICULAR PURPOSE AND NONINFRINGEMENT. NEITHER UNIVERSITY NOR THE DATA PROVIDERS shall be liable for any claims, losses or damages of any kind arising from RECIPIENT'S use of the OTC Data. It is understood that no patent, copyright, trademark or other proprietary right or license is granted by this Agreement, and ownership of all rights, title and interests in and to the OTC and clinical data remain vested in the entities that supplied such data.

PLEASE TYPE OR PRINT LEGIBLY BELOW

____

Account Holder Signature | Title | Date

____

Health Department Name | County

____

Address | City/State

____

Phone Number | Fax Number | E-mail Address

Fax signed agreement to Cleat Szczepaniak, Program Manager, NRDM. **Fax (412) 383-8126**

Handbook of Biosurveillance
ISBN 0-12-369378-0

APPENDIX

Data Security Agreement—Personnel

Acknowledgment of Data Use and Confidentiality Agreements
-2005-

I hereby acknowledge that I have been advised of the content of data use and confidentiality agreements involving the University of Pittsburgh and the data providers listed below. These agreements govern the provision of data to the Real-time Outbreak and Disease Surveillance (RODS) Laboratory of the University of Pittsburgh. I understand that I am bound by the provisions of these agreements, and I will maintain confidentiality of the data provided by these data suppliers, as set forth in the agreements.

I further understand that I may only access data from the providers listed below for purposes of research and development of the RODS system. I agree not to use this data for any other purpose, or to further disseminate this information to any other person outside of the RODS Laboratory. I understand that any publication based on this data must be submitted to the RODS Laboratory in advance of publication to ensure that confidential data is not inadvertently disclosed.

Data Providers:

Name (Print) ____________________ Signature ____________________ Date __________

Handbook of Biosurveillance
ISBN 0-12-369378-0

APPENDIX

Data Use Agreement with Commercial Data Provider

CONFIDENTIAL DISCLOSURE AND USE AGREEMENT

THIS AGREEMENT is made this _____________ day of _____________
2005, by and between _____________(retailer/drug store) _____________ and,
the _____________ (institution offering the biosurveillance system) ("UNIVERSITY").

WHEREAS, the Real-time Outbreak Disease Surveillance ("RODS") System is a public health surveillance system developed by the Center for Biomedical Informatics at the UNIVERSITY;

WHEREAS, UNIVERSITY is working with various state and local Departments of Health ("DOH") (collectively, the "RECIPIENTS") to monitor sales of consumer packaged goods with RODS in an effort to gain early detection of disease outbreaks ("PROJECT"); and

WHEREAS, COMPANY will provide such data as the parties shall mutually agree (collectively, the "DATA"), and UNIVERSITY shall perform analyses using the DATA ("ANALYSES") for the PROJECT;

WHEREAS, the DATA and RODS ANALYSES information will be available to RECIPIENTS for their use in the PROJECT.

NOW, THEREFORE, in consideration of such disclosure and in further consideration of the promises and obligations set forth below, the parties agree as follows:

1. COMPANY acknowledges and agrees that, in connection with the PROJECT, RECIPIENTS shall have access to certain confidential information of COMPANY, as defined more fully below as COMPANY PROPRIETARY INFORMATION.
2. COMPANY PROPRIETARY INFORMATION that is technical and business information and data describing or relating to the business of COMPANY and its affiliates, which information is held in confidence by COMPANY and its affiliates and is not generally known outside of COMPANY or its affiliates, as more specifically described on Exhibit A.
3. UNIVERSITY shall maintain the COMPANY PROPRIETARY INFORMATION in confidence, using that same degree of care that UNIVERSITY would use with its own confidential information, but in no case less than reasonable care. UNIVERSITY agrees not to disclose, display or otherwise make available the COMPANY PROPRIETARY INFORMATION to any third party, in any form, except to those UNIVERSITY employees with a need to know in connection with the PROJECT and only after such employees are made aware of and agree to be bound by these non-disclosure provisions and the use restrictions set forth in this Agreement. Notwithstanding the foregoing, UNIVERSITY shall have the right to publish its research involving the COMPANY PROPRIETARY INFORMATION, provided that UNIVERSITY complies with the following: UNIVERSITY affords COMPANY the opportunity to review and reasonably approve any materials to be published which include or reflect COMPANY PROPRIETARY INFORMATION so that COMPANY can ensure that the publication does not reveal brand-specific or retailer-specific identifiable information and such information cannot be reasonably determined, based on a statistical analysis; and UNIVERSITY identifies the applicable data parameters in such publication: Category/Product; Outlet/Geography; Measure; and Time Period.
4. COMPANY expressly agrees that UNIVERSITY shall be permitted to redistribute the COMPANY PROPRIETARY INFORMATION to those DOHs that are participating in the PROJECT under the following conditions: UNIVERSITY shall distribute the COMPANY PROPRIETARY INFORMATION in a form aggregated with similar data from other companies participating in the PROJECT, and shall aggregate such data by zip code. Each participating DOH shall be permitted to access only those portions of the COMPANY PROPRIETARY INFORMATION that are relevant to that DOH's jurisdiction. The participating DOHs may also review aggregated data, including the COMPANY PROPRIETARY INFORMATION using RODS interfaces and may conduct additional research using the COMPANY PROPRIETARY INFORMATION.
5. RECIPIENTS further agree not to use the COMPANY PROPRIETARY INFORMATION for any purposes other than the PROJECT. No warranties of any kind are made by COMPANY with respect to the COMPANY PROPRIETARY INFORMATION or any use thereof. COMPANY shall not be liable for any claims, losses or damages of any kind arising from the use or disclosure of COMPANY PROPRIETARY INFORMATION by RECIPIENTS or any third party. It is understood that no patent, copyright, trademark or other proprietary right or license is granted by this Agreement.

Handbook of Biosurveillance
ISBN 0-12-369378-0

6. The restrictions contained in this Agreement shall not apply to any COMPANY PROPRIETARY INFORMATION to the extent such information:
   (a) was otherwise known to RECIPIENTS at the time of disclosure by COMPANY, as evidenced by a written document; or
   (b) subsequently becomes known or available to RECIPIENTS from a third party acting lawfully, as evidenced by a written document; or
   (c) is or becomes part of the public domain without violation of this Agreement.
7. This Agreement shall be governed by and construed in accordance with the internal laws of the Commonwealth of Pennsylvania. RECIPIENTS' obligations hereunder with respect to COMPANY PROPRIETARY INFORMATION shall survive the termination of this Agreement.

IN WITNESS WHEREOF, the parties hereto have caused this Agreement to be executed by their respective duly authorized officers effective as of the date first written above.

______name of retailer/drug store______.	name of organization offering the system
By: ______________________	By: ______________________
Name: ______________________	Name: ______________________
Title: ______________________	Title: ______________________
Date: ______________________	Date: ______________________

Glossary

Abattoir: A slaughterhouse.

Accredited Standards Committee X12: A standards developing organization formed in 1979 by private industry to create standards to facilitate electronic commerce.

Aerosol: A suspension of tiny solid or liquid particles in a gas such as the air.

Agribusiness: Corporations that raise or process plants or animals for commercial purposes.

Alert history: The past alerts from a biosurveillance system in a defined period of time.

Algorithm: A well-defined procedure or method for accomplishing a task.

Allergen: A material or substance that has the potential to induce an allergic reaction in a person.

American Medical Association: The largest physicians' advocacy group in the United States. It creates and maintains the Current Procedural Terminology vocabulary standard.

AMOC curve: Activity monitoring operator characteristic curve. A graph characterizing the timeliness of a detector. The graph plots the expected time to detection as a function of the false-positive rate.

Annualized weekly death rate: The number of deaths in one week times 52, divided by the population.

ANOVA: Analysis of variance. A statistical method for analyzing differences between groups of subjects or observations.

Anthrax: A disease of livestock caused by *Bacillus anthracis* that can also infect humans. Human infections may be cutaneous, pulmonary, or gastrointestinal.

Architect: In information technology, a designer of an information system.

Architectural style: A set of general principles (rules) for designing an information system. These rules usually serve as constraints on (1) how we design the functions of each component (e.g., functions are stateless, functions operate independently); (2) how we arrange the components (e.g., in layers, in a star pattern or in a chain); and (3) how the components communicate with each other.

Architecture: (Information system.) A description of an information system's components (e.g., communications network(s), servers, applications and databases) and how they interact.

ArcImport: A standard file format for geospatial data.

Area release: A release of material into the atmosphere from an area such as a burning field.

ARMA model: Auto-regressive moving average model.

Artificial intelligence: A branch of computer science that develops algorithms that enable computers to perform tasks that involve knowledge, learning, or adaptation.

Atmospheric dispersion model: An algorithm that predicts the downwind concentrations of a substance that result when a given quantity of the substance is released into the air under specified meteorological conditions.

Atmospheric turbulence: The random motion of molecules of gas and water vapor in the air. Increased thermal energy and wind speed increases turbulence.

At-risk population: The set of people or animals who are biologically capable of contracting a specified disease.

Attack rate: The proportion of individuals exposed to an infectious agent who become clinically ill in a defined period of time.

AUC: The area under an ROC curve. Guessing randomly will achieve an expected AUC score of 0.5. A perfect detector would have an AUC of 1.0.

Auto regression model: A time series method that predicts the current value as a weighted linear sum of recent values. The coefficients of the linear sum are usually estimated from historical data.

Automated Surface Observation System: A joint effort among the National Weather Service, the Federal Aviation Administration, and the Department of Defense to provide frequent, accurate weather observations for the purposes of aviation.

Bacillus anthracis: The bacterium that causes the disease anthrax.

Bacteria: A group of living microscopic or unicellular organisms, with a relatively simple cell structure lacking a nucleus. Bacteria are the most abundant of all organisms, many of which are pathogens. They are commonly found in soil, water, and air, and are often closely associated with other organisms.

Bar code: A group of alternating black and white vertical lines that encode the Global Trade Item Number of a product. Manufacturers print it on the packaging of products.

ISBN 0-12-369378-0

Bayes rule/theorem: A formula that converts the conditional probability of an effect given a cause into the conditional probability of a cause given an effect. Principal citation: Bayes, Thomas (1763). An essay towards solving a problem in the doctrine of chances. *Philosophical Transactions of the Royal Society of London* 53:370–418.

Bayesian detection algorithm: An outbreak detection algorithm that computes the posterior probability of an outbreak.

Bayesian network: A popular representation for expressing the probabilistic interrelationships of many variable. An online tutorial, with further discussion, is available at *www.cs.cmu.edu/~awm/bayesnet.html.*

Bayesian reasoning: The process of applying Bayes rule to combine sources of evidence to deduce the probability of some unobserved event. Bayesian reasoning requires the availability of prior probabilities of events and a specification of the probabilistic relationships of all variables under consideration.

Bayesian wrapper: A method for computing posterior probabilities from non-Bayesian detection algorithms.

Behavioral psychology: A field that studies how humans behave (e.g., make decisions).

Benefit: The value (e.g., money saved, lives saved, adverse events avoided) provided by a service, item, procedure, or event.

Bioaerosol: Suspension of biological agents in a gas.

Bio-ALIRT: A DARPA-sponsored three-year program (2001–2003) with a goal of developing technology to reduce the time of detection of outbreaks.

Biodefense: A set of activities that together function to provide security against disease due to biological agents. Biosurveillance is one of these activities—along with sanitation, vaccination, quarantine, intelligence, interdiction (of terrorists and materiel), forensic science, and control of technologies used to create biological weapons.

Biohazard Detection System (BDS): An air monitoring system based on a DNA polymerase chain reaction test for *B. anthracis* used by the U.S. Postal Service.

Biological agent: Bacterium, fungus, parasite, or virus; or a toxin produced by a bacteria or fungus.

BioSense: A biosurveillance system being developed by the U.S. Centers for Disease Control and Prevention.

Biosurveillance: A systematic process that monitors the environment for bacteria, viruses, and other biological agents that cause disease; detects disease in people, plants, or animals caused by those agents; and detects and characterizes outbreaks of such disease.

Biosurveillance coordinator: A health department employee, often an epidemiologist, who is charged to develop a local or state health department's biosurveillance capabilities.

Biosurveillance system: Any systematic way to collect and analyze biosurveillance data.

BioWatch: A program managed by the Department of Homeland Security to place outdoor sensors in cities that can detect the presence of biological agents in the air, and thus provide the theoretically earliest possible detection of a windborne bioterrorism attack.

Blueprint: A set of specifications in sufficient detail that a diverse group of entities (such as network engineers, database administrators, and programmers) can work collaboratively to construct a system.

Boolean: The term Boolean has several connotations in computer science and mathematics. 1. A two-valued variable (true and false). 2. Refers to *Boolean algebra*, a mathematical formalism that manipulates sets using operators such as conjunction, disjunction, and negation.

Botulinum toxin: A zinc proteinase that cleaves one or more of the fusion proteins by which neuronal vesicles release acetylcholine into the neuromuscular junction. It is the most poisonous substance known.

Bovine Syndromic Surveillance System (BOSSS): A biosurveillance system developed in Australia for cattle diseases, which includes a web-based disease-reporting tool based on a diagnostic expert system for diseases of cattle.

Bushel: Unit of dry measure for produce equal to 32 quarts.

Cache: A memory area (typically random access memory) used for temporary storage of frequently accessed data.

Case: A person in the population or study identified as having the particular disease, health disorder, or condition under investigation.

Case definition: A necessary and sufficient set of criteria for identifying an individual as a case, usually based on clinical and laboratory data.

Case detection (system or algorithm): A process that detects individuals with disease.

Case-control study: An epidemiologic study of people with disease and a control group of people without the disease, comparing past history of exposure to a risk factor between "cases" and "controls."

Case-fatality rate: The observed death rate in diagnosed cases.

Cat-scratch disease: A self-limiting infectious disease of humans characterized by edema and pain in the lymph nodes (i.e., regional lymphadenopathy, lymphadenitis) and caused by the bacterium *Bartonella henselae.*

Causal representation: A knowledge representation centered around cause-and-effect relationships.

CDC: Centers for Disease Control and Prevention.

CDC/DOD code sets: A set of mappings from ICD-9 codes to syndrome groups developed by a workgroup including representatives from the Centers for Disease Control and Prevention, the Department of Defense, and other stakeholders.

Cell culture: The growth of microorganisms or cells in a special nutrient medium, such the growth or colonization of bacteria.

Centers for Medicare and Medicaid Services: The U.S. agency responsible for the management of the Medicare and Medicaid programs.

Certification Commission for Healthcare Information Technology: A voluntary, private-sector initiative to certify that healthcare information technology products comply with standards.

Charges: The amount billed for good or services. Charges usually exceeds the actual cost of the goods or services.

Charting template: A form (paper-based or electronic), often based on a diagnostic/problem management algorithm, which healthcare providers use to document a patient's history.

Chest rub: An over-the-counter healthcare product intended for application to the skin of the chest for the relief of the symptoms of runny nose and nasal congestion.

Chief complaint: A concise statement in English or other natural language of the symptoms that caused a patient to seek medical care.

Cholera toxin: A protein molecule produced by the bacteria *Vibrio cholerae*, whose action on the mucosal epithelium of the intestines is responsible for the characteristic diarrhea of the disease cholera.

Clarity test: A technique for checking whether the elements of a decision model have been defined in an unambiguous manner.

Classical swine fever: A multisymptom disease of domestic and feral pigs that is not known to be transmitted to humans.

Clean Water Act (CWA): A U.S. law enacted in 1972 (subsequently amended) that established the basic outline for pollution regulation of U.S. waters and the authority to perform pollution control.

Clinical call center: A facility that receives telephone calls from sick individuals or individuals seeking advice or appointments. Also known as *nurse triage* and *nurse hot lines*.

Cluster: An anomalous number of cases of disease.

Code of Federal Regulations (CFR): Rules written by federal agencies to interpret and apply laws.

Code set: A mapping from a set of codes (e.g., ICD-9, GTIN) to a syndrome.

Cohort study: A type of epidemiologic study used to compare the rate of illness over time of those exposed to specific items to the rate of illness among those not exposed.

Coliform: Designating or of the aerobic bacteria present in the colon (large intestine).

Combined source water: Drinking water supply drawing water from more than one source, such as springs, reservoirs, and rivers.

Commission on Systemic Interoperability: A body of individuals appointed by the leadership of the executive and legislative branches of the U.S. government charged by the Medicare Modernization Act with developing a strategy for adoption and implementation of healthcare information technology standards.

Common source: A single definable source for an outbreak.

Communicable disease: A disease caused by a biological agent (e.g., virus, bacterium, or parasite). Synonymous with *infectious disease*.

Conditional probability: The chance of one event occurring given the occurrence of another event.

Confidentiality: The expectation that information shared with another individual or organization will not be divulged.

Consolidated Health Informatics Initiative: A program of the executive branch of the U.S. government that seeks to establish healthcare information technology standards that all government healthcare information technology systems will use.

Control chart: A time series technique used in manufacturing systems to signal an alert when a time series departs from typical behavior.

Correlation analysis: In biosurveillance research, it is a method for determining the informational value of biosurveillance data based on the correlation function. The correlation function finds the time lag at which the correlation between two time series is maximized as well as the strength of the correlation.

Cost: The cash value of money, time, and labor spent for goods or services.

Cost-benefit analysis: An economic study that compares multiple alternatives when the costs and benefits all easily translate to monetary terms (e.g., dollars, yen, pounds).

Cost-effectiveness analysis: An economic study that compares multiple alternatives when the costs translate easily to monetary terms but the potential benefits are expressed in clinical units such as life years saved, deaths avoided, or operations avoided.

Cost-minimization analysis: An economic study that compares only the costs of multiple alternatives when each alternative yields identical benefits.

Cost-of-illness study: An economic study that quantifies the total monetary effect of a medical or public health problem.

Cost-of-treatment study: An economic study the calculates and profiles all the monetary costs associated with executing a treatment or solution.

Cost-to-charge ratio: The ratio of the amount billed to the actual cost of goods or services.

Cost-utility analysis: An economic study that compares multiple alternatives when the costs translate easily to monetary terms, but the potential benefits are expressed in health status measures such as quality-adjusted life years saved or utilities.

Covered entity (HIPAA): HIPAA's regulations directly cover three basic groups of individual or corporate entities—health plans, health care providers, and health care clearinghouses. CMS provides a series of flow charts to assist organizations in determining whether they are covered entities at *http://www.cms.hhs.gov/hipaa/hipaa2/support/tools/decisionsupport/CoveredEntityFlowcharts.pdf.*

Crude death rate: The number of deaths from any cause in a given population during a given time period (such as January 1–December 31) divided by the total population and multiplied by 1000.

Creutzfeldt-Jakob disease: A rare degenerative invariably fatal disorder that typically affects individuals over the age of 50.

Cryptosporidia: A genus of parasites some of which can cause the disease cryptosporidiosis.

Cryptosporidiosis: A gastrointestinal infection in humans caused by some species in the genus *Cryptosporidia.*

Current Procedural Terminology: A standard vocabulary for surgical procedures, minor procedures that physicians perform in the office, radiology tests, and a small number of laboratory tests (approximately 1000).

CUSUM: A time series method that signals an alert when the recent cumulative sum of positive deviations from the mean exceeds a threshold.

Cytomegalovirus (CMV): A common virus which usually causes severe disease in fetuses and in individuals with immunodeficiency.

Data accuracy: The degree to which data match an independent (gold standard) determination.

Data completeness: The proportion of data that we expect to find in an information system that is actually in the system.

Data model: A specification of how data are represented in an information system or a database management system.

Data provider: An organization that is being asked to provide, or is providing, data.

Data quality: The completeness and accuracy of the data recorded in an information system.

Data requester: An organization that is asking another organization to provide data.

Data transmission method: A way of moving data between computers over a network.

Data use agreement (DUA): A document that legally governs the use, care, and confidentiality of data.

Data utility: An intermediary organization positioned between organizations that provide data and organizations that are customers of the data. A data utility acts on behalf of many customers of data and streamlines the interaction with data providers.

Database: A defined structure that houses a collection of data.

Database management system (DBMS): A computer program such as Oracle, MySQL, or Microsoft SQL Server that not only allows a user to define a structure in which to store data (a database), but provides utilities for managing the structure (e.g., adding a record, deleting a record, backing up the data, optimization of retrieval of records from the database, integrity checking, and others).

Decision analysis: 1. A mathematical model of a decision problem and its interpretation, usually written. 2. An applied field of practice and research concerned with developing and using methodologies and software to help people make better decisions.

Decision support system: A computer-based system that aids the process of decision making.

Decision theory: An extension of probability theory that adds axioms and theorems that enable mathematicians to represent decision alternatives, costs, and benefits.

Decision tree: A graphical formalism for representing decisions and their possible consequences.

De-identified data: Data stripped of individual identifiers, such as social security number, personal address, and personal telephone number, to render it difficult or impossible to reidentify an individual from the data.

Detection algorithm method: In biosurveillance, an evaluation method that uses the date that a statistical detection algorithm first detects an anomaly in surveillance data as a measure of the value of a detection algorithm, biosurveillance data, or both.

Diagnostic expert system: A type of computer program that automates the task of diagnosis.

Diagnostic knowledge: The knowledge that diagnosticians bring to bear when they solve diagnostic problems. In clinical medicine, the relationship between findings and diseases.

Diagnostic precision: Nosological specificity of a diagnosis.

Differential diagnosis: 1. A dynamic process used by clinicians to determine the diseases suggested by the symptoms of a patient, listing the most likely causes, and using further testing, observation, and treatment to include (rule in) or exclude (rule out) possible causes. 2. The list of diagnoses currently being entertained by a clinician for a patient.

Discount rate: A rate, denoted typically by r, used to convert past or future costs or benefits to their present value.

Disease coverage (of a biosurveillance system): The set of diseases that a biosurveillance system is capable of detecting.

Disease incidence: The number of new cases in a population during a defined period such as a week. If we plot disease incidence for every day or week during an outbreak, the result is an epidemic curve.

Disease outbreak: See *Outbreak.*

Disease reporting: See *Notifiable disease reporting.*

Disease surveillance: The systematic collection and analysis of data for the purpose of detecting cases and outbreaks of disease. Often used to refer to *public health surveillance*, the disease surveillance conducted by governmental public health.

DOD-GEIS ICD code sets: A set of mappings from ICD-9 codes to syndrome groups developed by the Department of Defense for use in biosurveillance as part of its Global Emerging Infections System (DOD-GEIS).

Drop-in surveillance: The practice of asking physicians in emergency rooms to complete a form for each patient seen during the two- to six-week period surrounding a special event.

Dysrhythmia: Abnormal heart rhythm.

E-commerce service (ECS): A service in Amazon.com's service-oriented architecture that provides detailed product information and an electronic shopping cart to outside computer systems.

Economic study: A formal scientific analysis of different choices that individuals, organizations, or societies have to make when resources are scarce.

Effectiveness: The degree to which an intervention improves health (e.g., disease cases prevented, years of life saved, or quality-adjusted life years saved).

Electronic laboratory reporting: The automatic transmission of the results of laboratory tests that indicate the presence of a case of notifiable disease to governmental public health via a computer network.

Electronic veterinary record (EVR): Software that allows the recording of veterinary history, physical examination notes, lab results, surgical notes, diagnoses, and treatment.

Electrophoresis: The movement of colloidal (nondiffusable, insoluble, in suspension) particles caused by application of an electric field. Used to isolate and identify unknown substances.

EM algorithm: Expectation-maximization algorithm. A statistical method for maximizing the likelihood function. It is often used in the presence of missing data.

Email list: A computer-based method for distributing an email message to a set of people who have indicated an interest in receiving emails from a source. An organization sets up an email list by installing software (e.g., LISTSERV® or Majordomo) on its own hardware or by contracting with another organization to install and maintain the software and hardware.

Emerging pathogen: An infectious agent that has exhibited an initial occurrence or a recent increased incidence of reoccurrence.

Encephalitides: Infections which cause inflammation of the brain.

Encryption: A procedure used in cryptography that renders the contents of a message or file unintelligible to anyone not authorized to read it.

Enterprise information system: An information system that is intended to support the function of a single organization.

Environmental Protection Agency (EPA): A U.S. government organization that oversees human health and the environment. The EPA develops and enforces regulations regarding environmental science, research, education, and assessment activities.

Epidemic: The occurrence of a disease or condition at higher than normal levels in a population. In this book, synonymous with *outbreak*.

Epidemic curve: A graph (histogram) that describes an outbreak of disease by plotting the number of cases of a disease by date of onset.

Epidemiology: The study of the distribution and determinants of health-related states in specified populations, and the application of this study to control health problems.

Epidemiological differential diagnosis: Analogous to differential diagnosis. 1. A dynamic process used by outbreak investigators to determine key characteristics of an outbreak including biological agent, source, and route of transmission. 2. The list of possible biological agents, sources, and routes of transmission currently being entertained by outbreak investigators.

ESSENCE: A biosurveillance system developed initially by DOD GEIS and further developed by the Applied Physics Laboratory of Johns Hopkins University.

Ethics: The field of ethics, also called moral philosophy, involves systematizing, defending, and recommending concepts of right and wrong behavior.

European Article Number: The precursor to the Global Trade Item Number in Europe, a coding system for products.

Expected utility: The sum of the utilities of each of the possible consequences of a decision, multiplied by their probabilities of occurrence.

Explanation module: A module in diagnostic expert systems and decision support systems that can provide explanations for the systems conclusions or recommendations.

Exponentially weighted moving average (EWMA): A type of moving average (see *moving average*) in which recent counts are weighted more than less recent counts using an exponential function.

False alarm: An alarm or an alert from a biosurveillance system or algorithm in the absence of an outbreak or case. Synonym: False positive.

False-negative rate: A numerical measure of the performance of a detector. The fraction of alarm-free events for which an alarm should have been raised. A perfect detector would score zero.

False-positive rate: A numerical measure of the performance of a detector. The fraction of alarms raised that turn out to be false alarms. A perfect detector would score zero.

Federal Information Processing Standards: Information technology standards that the National Institute of Standards and Technology (NIST) creates for the U.S. government for use in its systems. NIST only creates a standard when no pre-existing industry standard meets a U.S. government need.

Federal Meteorological Handbook Number 1: A document prepared by the Office of the Federal Coordinator for Meteorological Services and Supporting Research that describes the use of the METAR standard in the United States.

Finding: A clinical observation, including symptoms, sociodemographics (age, gender), travel history, physical findings, and results of testing.

Flocculation: Appearance of small, fluffy, wool-like masses of soils or other precipitates in water; in water treatment, a way to remove solid waste matter from water.

Flu remedy: An over-the-counter healthcare product intended for the treatment of symptoms of influenza, including runny nose, headache, muscle aches, and fever.

Fluorescence: The production of light by a compound in the presence of radiant energy.

Fluorogenic: Any substance that causes fluorescence (glowing) after chemical reaction or uptake.

Food and Drug Administration (FDA): A U.S. government agency focused on the protection of public health by assessment and verification of the safety, efficacy, and security of the food supply, medical devices, pharmaceuticals, and cosmetics, among other items. The FDA ensures that these items are safe, effective, and accurately represented.

Foot and mouth disease: A highly contagious, virus-caused infectious disease of cloven-hoofed animals such as cattle, pigs, and sheep.

Forensics: The use of science and technology to investigate and establish facts in criminal or civil courts of law.

Formative study: An evaluative study whose goal is to improve a system or a system component.

Free text: Data in the form of human-readable phrases and sentences.

FTP/SFTP: File transfer protocol/secure. A protocol used for transferring files between computers (SFTP enables this transfer in a secure fashion).

Fully synthetic data: Data for testing a biosurveillance algorithm derived entirely from simulation.

Functional requirements: A set of specifications for a system at the level of functionality, such as the required timeliness of detection for an outbreak of anthrax caused by a surreptitious aerosol release.

Gaussian plume model: A simple atmospheric dispersion model that assumes that dispersion of materials in the atmosphere occurs in a Gaussian pattern.

Geographic information system (GIS): A computer system designed to allow users to collect, manage and analyze large volumes of spatially referenced information and associated attribute data.

Geolocation: The process of determining the real-world geographic location of a website visitor or a website itself (or any internet connected computer) by tracking his/its Internet protocol address and other factors.

Giardia: A genus of parasites that can cause giardiasis.

Giardiasis: A parasitic intestinal infection that can cause diarrhea.

Global Public Health Intelligence Network (GPHIN): A biosurveillance system developed by Health Canada and the World Health Organization that continuously monitors global media sources such as news wires and websites for articles about disease outbreaks and other events of international public health concern.

Global Trade Item Number (GTIN): A standard vocabulary for products sold in retail stores. Manufacturers print a bar code that encodes the GTIN on product packaging.

Gold standard: Data and methods that an evaluator can use to establish "ground truth" about the existence and timing of an outbreak or of an individual case.

Governmental public health: As used in this book, departments of health at the county, city, or state level, and the Centers for Disease Control and Prevention.

Grammar standard: See *Message format standard.*

GS1: A standards-developing organization that develops standards for information technology systems used in global commerce.

Hantavirus: A genus of RNA viruses carried by rodents and cause a variety of illness including pneumonia, hemorrhagic fever, and renal disease in humans.

Hazard analysis critical control point (HACCP): A formal process of analyzing and monitoring specific steps of food production and handling in order to identify hazards and prevent food contamination.

Hazardous materials response team (Hazmat): Emergency response and materials transportation team specialized in incidents involving hazardous materials.

Health codes: Health laws and regulations enacted by the governments of states and cities.

Health Level Seven: A message-format standard for transmitting healthcare data.

Health Level Seven, Inc.: The organization that creates and maintains Health Level Seven standards.

Healthcare-associated infection (HCAI): The technical definition of an HCAI in the United States is as follows: A localized or systemic condition resulting from an adverse reaction to the presence of an infectious agent(s) or its toxin(s) that (1) occurs in a patient in a health care setting (e.g., a hospital or outpatient clinic), (2) was not found to be present or incubating at the time of admission unless the infection was related to a previous admission to the same setting, and (3) if the setting is a hospital, meets the criteria for a specific infection site as defined by CDC.

Hemagglutinin: A glycoprotein (carbohydrate bound to a protein) on the surface of the influenza virus that allows the virus to attach to the cells of a host (chicken, human, etc.).

Hepatitis A: A viral illness cause by the hepatitis A virus whose symptoms include jaundice, fatigue, loss of appetite, darkening of urine, and fever.

Hepatitis C: A viral illness cause by the hepatitis C virus that is predominantly spread by blood (transfusion, drug injection).

Heuristic: In computer science, an algorithm that is not provably correct. In psychology, rules of thumb that people seem to employ when making decisions.

Hidden Markov model: A time series method which assumes the source of the time series observations is a state variable which cannot be observed, but can be estimated (using Bayesian reasoning) from the observations. For more information, see *www.cs.cmu.edu/~awm/hmm.html*.

HiFIDE: High-fidelity injection detectability experiments. A method for analyzing the expected detectability characteristics of a biosurveillance system.

HIPAA: Health Insurance Portability and Accountability Act (of 1996).

HIPAA Privacy Rule: National health information privacy regulations issued in 2003 by the U.S. Department of Health and Human Services (DHHS), pursuant to HIPAA of 1996. These standards provide protection for the confidentiality of individually identifiable health data.

HL7 reference information model: A standard data model for health care created by Health Level Seven, Inc.

HL7 tables: A standard vocabulary for some data elements used in the Health Level Seven message-format standard. For example, there are tables of codes for gender, marital status, race/ethnicity, and religion.

Hospital-acquired infection (HAI): The currently preferred term for nosocomial infection.

Host: A person or animal that harbors a parasite, virus, or bacteria. A host can also refer to a cell infected by a virus.

HTTP/HTTPS: Hypertext transfer protocol/secure. A protocol enabling the transmission of web pages (enables transfer in a secure fashion).

Hundred-weight: Unit of weight equivalent to 100 lb in the US and 112 lb in Canada.

Hypothesis test: A statistical procedure in which we compute the probability that we could have observed a given deviation from the null hypothesis by chance.

Incremental cost: Change in cost when moving from one alternative to another.

Incremental cost effectiveness: Change in cost per change in effectiveness when shifting from one alternative to another.

Incremental life cycle: An IT project life cycle in which the specify, develop, and test phases are applied to deliver fully functional pieces of the overall solution, and iteratively applied until the full system is completed.

Incubation period: The time from infection to onset of clinical illness.

Index case: The *index case* or *patient zero* is the initial patient in the population sample of an epidemiological investigation.

Infection control practitioner (ICP): An individual who specializes in preventing and controlling infectious diseases in hospitals and institutions.

Infection surveillance control program (ICSP): A set of activities within a healthcare organization, often managed by a staff dedicated to this function, to prevent healthcare-associated infections.

Infectious disease: A disease caused by a biological agent (e.g., virus, bacterium, or parasite). Synonymous with *communicable disease*.

Infectious period: The period during which an ill individual can infect another host.

Influence diagram: A graphical and mathematical representation of probabilistic and decision problems.

Information retrieval algorithm: An algorithm that selects documents from a collection based on a user's query.

Institutional review board (IRB): An appropriately constituted group that has been formally designated to review and monitor biomedical and behavioral research involving human subjects. In accordance with FDA and HHS regulations, an IRB has the authority to approve, require modifications in (to secure approval), or disapprove research. This group review serves an important role in the protection of the rights and welfare of human research subjects.

International Aquatic Animal Health Code: Standards and techniques promulgated by OIE that countries can apply to help protect themselves from animal diseases by establishing valid import barriers for the trade of certain animals or animal products.

International Classification of Diseases: A standard vocabulary for diseases, health status, types of patient visits to doctors and other health providers, and external causes of injuries. In the United States, ICD-9-CM also includes a standard vocabulary for procedures.

International Terrestrial Animal Health Code: Standards and techniques promulgated by OIE that countries can apply to help protect themselves from animal diseases by establishing valid import barriers for the trade of certain animals or animal products.

Internet: The publicly accessible worldwide system of interconnected computer networks that transmit data by packet switching using a standardized Internet protocol (IP) and other protocols.

Internet protocol (IP): A standard form for packets of data transmitted across the Internet.

Investigational drug: Experimental drug, undergoing trials, not yet approved for sale by the U.S. Food and Drug Administration.

ISO 3166: A vocabulary standard for geopolitical entities such as countries and their subdivisions (e.g., U.S. states and Canadian provinces).

IT project life cycle: The sequence of phases through which an information technology project will evolve.

Japanese encephalitis: An acute encephalitis caused by infection with the Japanese encephalitis virus, contracted through the bite of a mosquito.

Java database connectivity: A standard for submitting queries to, and getting results from, a database management system using the Java programming language.

Joint Commission on Accreditation of Healthcare Organizations (JCAHO): A U.S.-based nonprofit organization formed in 1951 with a mission to maintain and elevate the standards of healthcare delivery through evaluation and accreditation of healthcare organizations.

Kaposi's sarcoma: A type of cancer that was very rare before the start of the AIDS epidemic.

Knowledge base: A special kind of database for knowledge management. It is a collection of facts organized in such a way that algorithms can use them in computations.

Knowledge engineering: The process of knowledge elicitation from domain experts for the purpose of building artificial intelligence programs.

Language standard: A shared convention about how two or more entities communicate. With respect to information technology standards, it encompasses language, message format, and semantic standards.

Layered client-server (LCS) architecture: An architectural style that follows the following principles (rules): (1) some components (the servers) provide functions or data at the request of other components (the clients), and (2) we must group components into layers with components of one layer accessing the functions of components in the layer below it (i.e., the upper components are clients of the lower layer components). Virtually all modern enterprise systems utilize a LCS architecture.

LD_{50}: The dose of a substance or organism that kills 50% of the population exposed to it.

Likelihood ratio negative: The false-negative rate of a detector (1-sensitivity) divided by its true-negative rate (specificity). A measure of the discriminatory power of a test, when the result is negative.

Likelihood ratio positive: The true-positive rate of a detector (sensitivity) divided by its false-positive rate (1-specificity). A measure of the discriminatory power of a test, when the result is positive.

Limited data set: A technical term in the HIPAA Privacy Rule that refers to personal health information from which 18 identifiers have been removed (e.g., social security number, personal address, personal telephone number), and the covered entity releasing the data does not have actual knowledge that the remaining information can be used alone or in combination with other data to identify the subject.

Line release: A release of material into the atmosphere from a vehicle or airplane moving in a line.

Linear regression: A method for predicting the value of one field in a database record based on the remaining fields, in which the predicted value is assumed to be a weighted linear sum of the remaining fields.

Linked record analysis: In biosurveillance evaluation, a type of study that obtains the ground truth about the presence of a condition in patient via linkages to data sets that contain, for example, final diagnoses.

Listeriosis: An infection caused by the bacterium *Listeria monocytogenes*. Listeriosis is a rare disease and occurs primarily in newborns, elderly patients, and patients who are immunocompromised.

Logical Observation Identifiers, Names, and Codes (LOINC): A standard vocabulary for clinical laboratory tests and for clinical observations, such as EKG measurements and vital signs.

Look-back capability: In biosurveillance, the ability of a biosurveillance system to query a data provider for additional information about an individual.

Loyalty card: A card that a company provides to a customer. At checkout, the store computer automatically applies applicable discounts and data about the purchase is tagged with the customer's information.

Macrocosting: A "top-down" method to determine the total cost of an event, problem, procedure, or item. One finds a budget or total cost that includes the event, problem, procedure, or item, and determines what proportion of the budget or total cost should be assigned to that event, problem, procedure or item (e.g., to determine the cost of CT scans, taking a hospital budget and multiplying it by the percentage of the hospital budget allocated to CT scanning).

Malaria: An infection caused by a protozoa. Usually contracted through the bite of a mosquito.

Marginal cost: The change in cost when one changes a certain parameter (e.g., number of medications given, dollars invested, items used, people employed) by a single unit.

Marginal cost effectiveness: The change in cost and effectiveness when one changes a certain parameter (e.g., number of medications given, dollars invested, items used, people employed) by a single unit.

Market basket: The set of all products purchased in a single transaction.

Market share: The percentage or proportion of the total available market or market segment that is being serviced by a company.

Markov decision process: A discrete time stochastic control process characterized by a set of states, actions, and transition probability matrices that depend on the actions chosen within a given state.

Mass retailer: A company that sells a wide variety of products, including healthcare products, and also often includes a pharmacy. Examples include Wal-Mart and Target.

Maximum expected utility: A principle that states that a rational person or group should make a decision that maximizes his/its expected benefit.

Medical informatics: A field of study concerned with the broad range of issues in the management and use of biomedical information, including medical computing and the study of the nature of medical information itself.

Medicare Modernization Act of 2003: An Act of the U.S. Congress establishing a prescription drug benefit in the Medicare program. It also created the Commission for Systemic Interoperability and charged the National Committee for Vital and Health Statistics with establishing standards for electronic transmission of prescriptions from physician offices to pharmacies.

MEDLINE: A database of academic publications relating to all aspects of medicine and medical research, run by the National Library of Medicine. See *http://www.nlm.nih.gov/databases/databases_medline.html.*

MeSH categories: An industry standard vocabulary used for indexing articles for MEDLINE.

Message: A discrete object of communication presented in a standard format.

Message bus: A data transmission method that employs specialized software called a message router. Unlike point-to-point messaging, a message bus can accept messages from multiple senders and send each message to one or more receivers over a network.

Message format standard: An agreed upon specification for the position of data elements in a larger package called a *message* that facilitates the exchange of data among computer systems.

METAR: An international vocabulary, message format, and semantic standard for the encoding and transmission of weather observations.

Meteorology: The scientific study of the atmosphere, especially weather.

Microcosting: A "bottom-up" method to determine the total cost of an event, problem, procedure, or item. One determines the exact number and types of resources consumed by the event, problem, procedure, or item, and assigns costs to each resource type to calculate the entire cost (e.g., to determine the cost of CT scans, counting all the items and personnel used for the CT scans and multiplying by the cost of each item and personnel).

Model parameter: A numerical parameter used within a statistical model.

Monte Carlo integration: A stochastic computational method for numerically evaluating definite integrals.

Moore's Law (Andrew): A conjecture that developing software will never be easy or cheap.

Moore's Law (Gordon): A rule of thumb in the computer industry about the growth of computing power over time.

Moving average (MA): A time series method that signals an alert when the most recent recorded value is significantly larger than the average of recent values. Moving average is also used to smooth times series prior to analysis by other detection algorithms.

Multiple hypothesis testing: The phenomenon in which many statistical tests are performed, and in which we expect it is likely that at least one test will be positive due to the sheer volume of tests even if there is no departure from the null hypothesis.

Multispectral: Using more than one segment of electromagnetic energy.

National Atmospheric Release Advisory Center: A set of services and tools of the Lawrence Livermore National Laboratory allows biosurveillance, hazard management, and first responders to model atmospheric releases during or prior to an event to support management and planning of response.

National Bioterrorism Syndromic Surveillance Demonstration Program: A biosurveillance system that collects and monitors data from healthcare plans.

National Center for Health Statistics: One of the Centers for Disease Control and Prevention in the United States. It generates statistics about birth, death, and causes of death and morbidity in the United States.

National Council on Prescription Drug Programs: A standards-developing organization that creates information technology standards for use in the pharmacy services industry, including the Batch Transaction, Telecommunications, and SCRIPT standards.

National Drug Code Directory: A standard vocabulary of prescription drug products sold in the United States.

National Electronic Disease Surveillance System (NEDSS): NEDSS (prior to the introduction of PHIN) was the CDC's technical architecture for public health information systems. The NEDSS architecture focuses on notifiable disease reporting and workflow management.

National Institute for Standards and Technology: A nonregulatory federal agency within the U.S. Department of Commerce.

National Retail Data Monitor: A biosurveillance data utility in the United States that collects and provisions data about sales of over-the-counter healthcare products.

Natural language processing: Automated methods for converting free-text data into computer-understandable format.

NCPDP Batch Transaction Standard: A message-format standard that facilitates submitting claims electronically to prescription drug benefit programs.

NCPDP Script Standard: A message format standard for transmitting prescriptions electronically to a pharmacy.

NCPDP Telecommunications Standard: A message-format standard for submitting claims electronically to prescription drug benefit programs.

Necropsy: A postmortem examination of an animal to determine cause of death; analogous to autopsy in humans.

Neuraminidase: A glycoprotein (carbohydrate bound to a protein) on the surface of the influenza virus that is important in the release of progeny viruses from infected cells.

NHS Direct Syndromic Surveillance System: A biosurveillance system in use in the United Kingdom that monitors calls to nurse call centers.

Normal distribution: A probability distribution in which samples are real numbers. Two parameters, m and s, specify the mean and spread of the distribution, respectively. The average value of samples drawn from the distribution is m, which is also the most likely value. The distribution is symmetric, and there is a 95% chance that a sample will differ from the mean by less than $1.96s$. For more information, see *www.cs.cmu.edu/~awm/tutorials/gaussian.html*.

Nosocomial infection: An infection that is associated with a stay in a hospital. An infection is considered nosocomial if it occurs 48 hours or more after a hospital admission.

Nosology: A branch of medicine that deals with classification of diseases.

Notifiable condition mapping tables: A computer-interpretable set of facts (knowledge base) about which tests and results indicate that a patient has a notifiable disease.

Notifiable disease: A disease that by law must be reported to some jurisdiction, typically a local public health authority or department of agriculture, by hospitals, laboratories, and individual clinicians.

Null hypothesis: A mathematical model (usually a probability distribution) of what we expect to observe under normal circumstances.

Odds ratio: The incidence of a disease or other condition in exposed individuals divided by that in unexposed individuals.

OIE (World Organization for Animal Health): Equivalent to the World Health Organization of the U.N. for human health.

Ontology: A specification of concepts and the relationships between concepts (e.g., the phylogenetic tree).

Oocyst: The spore phase of *Cryptosporidia*. In this state it can survive for lengthy periods outside a host.

Open Database Connectivity: An information technology standard for submitting queries to, and getting results from, a database management system.

Open source software: Software released under an open source license or to the public domain.

Opportunity cost: The value of the next best choice that one gives up when making a decision.

Outbreak: Synonymous with epidemic. Sometimes used to refer to an epidemic that is limited to a localized increase in the incidence of a disease, such as, in a village, town, or closed institution, but in this book the terms are synonymous.

Outbreak characterization: The process or set of processes that elucidate the causative biological agent, source, route of transmission and other characteristics of an outbreak.

Outbreak detection: A process or set of processes that detect the existence of an outbreak.

Outbreak investigation: The process used by a health department to characterize an outbreak.

Over-the-counter healthcare product: Any item that an individual purchases without a prescription for the management of an illness.

PANDA: Population-wide ANomaly Detection and Assessment, described in Chapter 18.

Pandemic: A disease outbreak that affects people in many countries.

Pan-enterprise information system: An information system that is designed to support the interoperation of multiple organizations.

Paradigm shift: The process and result of a change in basic assumptions within the prevailing theory of a scientific discipline. Also called a *scientific revolution*.

Paramyxovirus: A family of viruses that have envelopes and are comprised of a single strand of RNA.

Pathogenicity: The degree to which a biological agent is capable of causing disease.

Pathognomonic findings: One or more findings that allow a clinician to establish a diagnosis with certainty.

Pediatric electrolyte product: A solution of salts and water that are indicated for the treatment of dehydration in children ages five years and under.

Pedigree: A documented history or genealogy.

Permissive environment: Settings that may allow collection of data about people that cannot otherwise be collected. Examples of permissive environments include some types of worksites, university campuses, or military bases. They are typically characterized by a geographically constrained location (a "site"), some centralized control of IT and telecommunications infrastructure, as well as a shared sense of community and trust.

Perspective: The point of view of a study (e.g., a study from an individual's perspective will only take into account the costs and benefits that affect that individual, whereas a study from society's perspective will take into account everything that affects society).

Pharmacy: A retail outlet that specializes in the sales of drug products.

PHIN Logical Data Model: A standard data model for public health that is based on the HL7 reference information model.

Plague: An infectious disease caused by the bacterium *Pasteurella pestis* causing pneumonia and swelling of lymph nodes in humans and animals.

Plume model of atmospheric dispersion: Atmospheric dispersion model that predicts steady-state downwind concentrations from continuous releases of material into the atmosphere.

Pneumocystis carinii: A fungus that causes pneumonia in immunocompromised patients.

Point release: A release of material into the atmosphere from a single fixed location.

Point-of-care system: A clinical information system that clinicians use to record or review clinical data, partly or completely replacing paper-based records.

Point-of-sale data: Data that retailers collect from cash registers and store computers about sales of products at the time the customer checks out.

Point-to-point messaging: Any transmission method for sending messages from one sender to one recipient.

Poisson distribution: A probability distribution in which all samples are integers, modeling the number of events occurring within a given time interval. Characterized by one parameter.

Polysemy: A problem in natural language processing in which one word, one phrase, or one acronym can have multiple meanings.

Portability: Refers to the effort required to install a system in a second location.

Posterior probability: A probability of an event when some particular evidence is available.

Postmortem: After death.

Preclinical data: Data that are the result of illness behavior that precedes clinical care.

Predictive economic study: An economic study of a hypothetical future event.

Presymptomatic data: Data that might be obtained about a person during the incubation period of a disease (after infection, but before the onset of symptoms).

Prior probability: The probability of an event, in the absence of any evidence.

Privacy: The right of an individual or organization not to divulge information.

Probabilistic independence: Two signals, or observations, are independent if observing the value of one of them gives you no information about the other.

Probability: The degree of belief in an event, expressed as a number between 0 (event is certainly false) up to 1 (event is certainly true). An online tutorial, with further discussion, is available at *http://www.cs.cmu.edu/~awm/prob.html*.

Probability theory: The mathematical study of probability.

Product category: A group of similar products. For example, *pediatric electrolytes* is a category of over-the-counter healthcare products used for the rehydration of children age five years and under.

Project: A temporary endeavor to create a unique product, service, or result.

Project management: The application of knowledge, skills, tools, and techniques to project activities to reach project goals.

Project SHARE (School Health and Absenteeism Reporting Exchange): A research project that links schools and public health for early event detection for bioterrorism events, as well as identification and management of both large- and small-scale outbreaks.

ProMED-mail: An Internet-based reporting system dedicated to rapid global dissemination of information on outbreaks of infectious diseases and acute exposures to toxins that affect human health, including those in animals and in plants grown for food or animal feed.

Promotional effect: A change in sales (relative to baseline sales) of a product caused by a discount, coupon, special display, or other activity intended to increase sales of the product.

Prospective economic study: An economic study of an event while it is occurring.

Protected health information (PHI): In HIPAA, information that healthcare organizations and insurers hold about individuals.

Protozoa: Unicellular organisms present in the environment, including parasites pathogenic to humans and domestic animals.

PSAP: Primary service answering point. A center receiving calls and coordinating the dispatch of emergency personnel.

Public Health Information Network (PHIN): CDC's vision for advancing fully capable and interoperable information systems in the many organizations that participate in public health (*http://www.cdc.gov/phin/*).

Public health surveillance: The systematic collection and analysis of data by governmental public health for the purpose of monitoring the health status of populations, including chronic diseases and injuries.

Puff model of atmospheric dispersion: An atmospheric dispersion model that predicts time-varying downwind concentrations from an episodic release of material into the atmosphere.

PulseNet: Internet-based network linking food safety investigators at the CDC, FDA, U.S. Department of Agriculture, and state public health laboratories.

p-value: During a statistical test, the p-value is the probability that we would have seen such a large deviation (or larger) under the null hypothesis. If the p-value of a hypothesis test is less than 0.05, we say that the test is significant at the 5% level.

Quality-adjusted life year (QALY): A year of life adjusted by the quality of that year of life. For example, a year of life in perfect health would be 1.0 QALY. A year of life with impaired health would be less than 1.0 QALY.

Rat bite fever: An acute, febrile illness of humans caused by bacteria transmitted by rodents.

Relational database: A computerized store of information whose logical arrangement can be considered independently of the physical location of information on computer media.

Relevance score: A score assigned to a document that represents how well the document matches a search query. A document's relevance score determines whether and how (e.g., the order if more than one document is found) the document is presented to the user.

Reservoir: A natural or man-made body of water used for the supply and control of water.

Retail chain: An organization that owns two or more retail stores or outlets.

Retail industry: The sector of the economy that obtains products from manufacturers and sells them to consumers in stores, through product catalogs, or via the Internet.

Retrospective economic study: An economic study of an event that has already transpired.

Ricin: A toxin produced by the castor oil plant and extractable from its beans. It is a metabolic poison.

ROC curve: Receiver operator characteristic curve. A graph characterizing a detector. The graph plots the true positive rate as a function of the false-positive rate.

RODS: A biosurveillance system developed by the University of Pittsburgh and the Auton Laboratory at Carnegie Mellon University.

Route of transmission: 1. The path from a source of biological agent to a host. 2. Also refers to generic paths such as person-to-person and vector.

RxNorm: A standard vocabulary and ontology of prescription drug products sold in the United States.

Safe Drinking Water Act (SDWA): U.S. law passed in 1974 that serves as the basis for drinking water health standards, including federal enforcement and public notification.

Sampling bias (in surveillance data): The degree to which surveillance data are not representative of a population.

Sanitarian: A professional who inspects and advises food service facilities, and recreational and potable water facilities. Also known as *environmental health specialist.*

Screening: Interviewing and testing people to distinguish apparently well individuals who probably have a disease from those who probably do not.

Sea-breeze effect: The phenomenon of the wind blowing from the ocean to shore during the day (and from the shore to the ocean at night) because of a temperature differential between the air over land and over water.

Search engine: A computer system that (1) locates and indexes web pages and (2) processes queries from users who are searching for information on the web.

Security: The degree to which a biosurveillance system is vulnerable to corruption through physical damage, corruption or data theft by hackers, theft of passwords, or denial-of-service attacks.

Self-reporting: A biosurveillance strategy in which individuals report the fact that they are ill to a biosurveillance organization.

Semantic standard: See *Standard data model.*

Semisynthetic data: Data for testing a biosurveillance algorithm derived from real data, with the effects of a simulated attack injected into the data.

Sensitivity: A numerical measure of the performance of a detector of some kind of event. The fraction of occurrences of the event in which an alarm is raised. A perfect detector would score 1.

Sensitivity analysis: Analysis that involves changing variables along a range of values and measuring the consequent effects on the results.

Sentinel clinician surveillance: A biosurveillance strategy in which a subset of clinicians in a region agree to report the number of individuals with certain conditions that they see.

Serfling method: An algorithmic method for detecting outbreaks.

Service agreement: An agreement among two or more organizations to provide services. In biosurveillance, services could include provision of data, diagnosis, case finding, and investigation.

Service-oriented architecture (SOA): An architectural style based on the client-server architecture. Servers in the SOA adhere to the principles of loose coupling, autonomy, abstraction, reusability, composability, statelessness, and discovery. Their functions are defined by service contracts.

Shape file: A standard file format for geospatial data.

Signal-to-noise ratio (SNR): An engineering term for the power ratio between a signal (meaningful information) and the background noise.

SNOMED International: The organization that creates and maintains the SNOMED-CT standard vocabulary.

SNOMED-CT: A standard vocabulary for diseases, symptoms, signs, specimen types, living organisms, procedures (includes diagnostic, surgical, and nursing procedures), chemicals (includes biological chemicals like enzymes and proteins and compounds used in drug preparations), drugs, anatomy, physiological processes and functions, occupations, and social contexts (e.g., religion and ethnic group).

Source: The starting point in the path via which a biological agent is eventually conveyed into the body of a victim.

Spatial cluster detection: Finding spatial areas where some monitored quantity is significantly higher than expected.

Spatial scan statistic: A method for spatial cluster detection.

Spatiotemporal data: Data in which individual records include information about both the location and time of recorded events.

Specificity: A numerical measure of the performance of a detector of some kind of event. Of those occasions in which the event does not occur, what fraction do not result in an alarm? A perfect detector would score 1.

Spirochete: A phylum of bacteria with long, helically coiled cells. The spirochete *Borrelia burgdorferi* causes Lyme disease and *Treponema pallidum* causes syphilis.

Spore: A resting state in the life cycle of some bacteria and fungi.

Stability class: A measurement of atmospheric turbulence along an ordinal scale. Two examples of stability classification schemes are those created by Pasquill and Turner.

Standard: A specification devised to facilitate construction of complex systems from component parts.

Statistical model: A representation of an uncertain process, usually with free parameters that must be supplied by an expert, or inferred from historical data. Linear regression, Bayesian networks, and hidden Markov models are all examples of statistical models.

Storage area network (SAN): A network designed to attach computer storage devices such as disk array controllers and tape libraries to servers.

Strain: An identifiable subgroup of a species of organism.

Strategic National Stockpile: A U.S. government reserve of millions of doses of vaccines, antidotes, antibiotics, respirators and other supplies that may be required to respond to a biological disaster.

Structured data: Data in a format (e.g., relational database tables) that can be interpreted by a computer.

Structured Query Language (SQL): A computer language used to manage and retrieve information from relational databases.

Subjective probability: The concept of subjective probability expands the frequentist interpretation of probability to include the degree of a person's belief in some proposition. Also called *Bayesian probabilities*.

Summative study: A study whose goal is to establish the value of a system or a system component relative to an alternative or in absolute terms (e.g., cost and benefit).

Swine: Domesticated pig or hog.

Syndrome: A defined set of clinical findings (e.g., symptoms, risk factors, demographic characteristics); for example, SARS (severe acute respiratory syndrome) and AIDS (acquired immune deficiency syndrome).

Syndromic surveillance: A biosurveillance method that detects outbreaks through analysis of prediagnostic case data. Sometimes used to refer to methods that use any type of data that is of low diagnostic precision.

Synthetic aperture: Method of scanning geographic features with an airborne or space-borne radar antenna using a duration of flight to simulate the presence of a much larger antenna.

System specification: A set of technical specifications, equivalent to a blueprint in architecture, that describes the components of an information system and how they interact.

Systematic error: Bias in measurement that leads to measured values being systematically high or low.

Table-top exercise: A group exercise in which individuals play the roles of decision makers in a simulated scenario.

Temperature inversion: A situation where a warmer layer of air exists above a cooler layer of air. The temperature difference serves as a relatively strong barrier to dispersion of material from the cooler layer of air to the warmer layer and vice versa.

Time horizon: The length of time that an economic study considers and includes (e.g., a study with a one-year time horizon will only consider costs and benefits within that year).

Time latency: In biosurveillance, the time between individual steps in the processing of information by a biosurveillance system.

Time series: A sequence of observations, recorded at regular intervals.

Timeliness: In biosurveillance, the period defined by the occurrence of an event (e.g., outbreak or case) and its detection. Timeliness is also known as earliness.

Toolset: An environment for developing software. It typically includes a programming language, debugging and testing tools, documentation support, and a set of pre-fab components that are commonly used to build software.

Trace-back/forward study: A systematic method to identify the source of contamination in food or other product, either tracing back from a product to its origin or tracing forward through the distribution system to the consumer.

Transmission control protocol (TCP): TCP is a set of standards for how communications between computers are broken up into many packets at one end and reassembled at the other end, with built-in redundancy to handle lost packets, duplicate packets, and automatic request for resending of packets.

Triangulation: The process of determining one's position by measuring the relative distances to two other fixed points.

Tularemia: A bacterial infection carried by rabbits and other rodents, caused by the *Pasturella tularensis*. It is transmitted to humans via contact with infected animals or insect bites. Victims suffer fever, pain, pneumonia, and lymph node inflammation.

Turbidity: A measure of the cloudiness of water, which is correlated with the amount of sediment present.

Unified Medical Language System: An amalgamation of pre-existing standard vocabularies. It includes as "source" vocabularies SNOMED-CT, RxNorm, ICD-9-CM, CPT, and LOINC.

Uniform resource locator (URL): URLs are the address system used by the Internet to locate resources such as websites.

Univariate time series: A time series in which each observation is a single number. One example is a daily count of the number of patients with respiratory problems admitted into an emergency room.

Universal Product Code: The precursor to the Global Trade Item Number in North America.

Upper respiratory tract infection: A viral, bacterial, or fungal illness of the nose, sinuses, throat, and/or larynx. Includes the common cold, sinusitis, influenza, and sore throat.

Use case: In software engineering, a technique for capturing the requirements of a new system or software change. Each use case describes one or more scenarios that convey how the system should interact with the end user or another system to achieve a specific goal.

Utility: A real number in the range of 0 to 1 that expresses the relative preference of a decision maker for an outcome.

Value of information: A calculation of the extent to which knowing the value of a currently unobserved piece of evidence would reduce our uncertainty about an event of interest.

Vector: An animal, such as a mosquito or tick that carries disease-causing microorganisms from one host to another.

Virtual private networking: A standard method for establishing secure communications among applications over the Internet.

Virus: Submicroscopic parasite of plants, animals, and bacteria that often cause disease and that consists of a core of RNA or DNA enclosed by a protein coat. Viruses are unable to replicate without a host, and are not typically considered living organisms.

Vocabulary standard: An agreed upon set of codes for things, actions, and states of being that facilitate the exchange of data among computer systems.

VoIP: Voice-over Internet protocol. Method of placing a voice telephone call over the Internet.

Volume release: A release of material into the atmosphere such as an aerosol of material created by an explosion.

Volumetric telephone usage: A potential biosurveillance strategy in which telephone calling patterns are analyzed to provide an early indication of change in health status of a population.

Wagnerian office: An office large enough to stage "Parsifal."

Waterfall life cycle: An IT project life cycle in which the output of each phase becomes the input for the next phase. The system development life cycle (SDLC) is a widely known waterfall life cycle.

Watershed: The surrounding region that drains into a reservoir, river system, or other body of water.

Web application server: A computer that delivers the interface to an application via web pages.

Website: A related collection of web pages and files that can be accessed from a common root URL, the home page, and usually resides on the same physical server.

Web spider: A program that browses the World Wide Web in a methodical, automated manner to find web pages for use by other programs such as search engines. It starts with a list of URLs to visit. As it visits these URLs, it identifies all the hyperlinks in the page and adds them to the list of URLs to visit, recursively browsing the web according to a set of policies.

WebDAV: Web-based Distributed Authoring and Versioning. An extension to HTTP that allows for files on remote web servers to be edited and managed by users.

Windborne: A mode of transmission of biological agent where the wind carries the agent over long distances.

Windborne outbreak: An outbreak caused by windborne transmission of a biological agent.

World Health Organization: The U.N. agency concerned with promoting the health of the world's people.

World-Wide Web (WWW): A hypertext-based, distributed information system originally created by researchers at CERN, the European Laboratory for Particle Physics, to facilitate sharing research information.

WSARE: What's Strange About Recent Events. A multivariate time series algorithm described in Chapter 15.

XML: Extensible Markup Language. A standard for annotating and structuring data.

Zoonotic disease: Any infectious disease that may be transmitted from animals, both wild and domestic, to humans.

Index